AN INTRODUCTION TO ENTOMOLOGY

AN INTRODUCTION TO ENTOMOLOGY

BY

JOHN HENRY COMSTOCK

LATE PROFESSOR OF ENTOMOLOGY AND GENERAL INVERTEBRATE
ZOOLOGY IN CORNELL UNIVERSITY

NINTH EDITION
REVISED

ITHACA   NEW YORK
COMSTOCK PUBLISHING COMPANY, INC.
1947

COPYRIGHT 1920 BY
COMSTOCK PUBLISHING COMPANY

COPYRIGHT 1924 BY
J. H. COMSTOCK

COPYRIGHT 1933, 1936, 1940 BY
COMSTOCK PUBLISHING COMPANY, INC.

PRINTED IN THE UNITED STATES OF AMERICA BY
THE VAIL-BALLOU PRESS, INC., BINGHAMTON, N. Y.

TO
MY OLD STUDENTS
WHOSE YOUTHFUL ENTHUSIASM WAS A CONSTANT INSPIRATION
DURING THE LONG PERIOD OF MY SERVICE AS A TEACHER
THIS EFFORT TO CONTINUE TO AID THEM IS
AFFECTIONATELY INSCRIBED

PREFACE TO THE 1940 EDITION

INCREASING interest in the biological control of insects has created an insistent demand for a wider knowledge of the parasitic forms active in destroying those insects which are injurious to agriculture. The greater number of these parasitic species is found in the order Hymenoptera. Opportunity has, therefore, been taken in this edition (1940) to revise and extend the discussion of the superfamilies Ichneumonoidea, Proctotrupoidea and Chalcidoidea, the groups which contain the most important parasites. In addition, a shorter key to the commoner families of the suborder Clistogastra together with keys to the subfamilies of the Ichneumonidæ and Chalcididæ have been included.

The text for the keys and for the new matter on the parasitic Hymenoptera has been contributed by Dr. Henry K. Townes, who has given much study to these groups. It is hoped that the keys will prove helpful to those interested in the parasitic forms.

GLENN W. HERRICK

Ithaca, June 1940.

PREFACE TO THE 1936 EDITION

IT SEEMED opportune with this reprinting (1936) of the Introduction to make some slight revision, more specifically of the orders of the wingless insects.

The class Myrientomata of Berlese has been advanced to its more logical position as the order Protura of Silvestri, under the Hexapoda. This has been done with some reservation, but in accord with the trend of opinion among morphologists and probably most systematists.

Probably the data for treating the suborder Entognatha of the Thysanura as a definite order are more numerous and more reliable than for the foregoing change in position of the Protura. Out of respect, however, for the careful conservatism of Professor Comstock, the writer has retained the Entognatha as a subordinate group in the Thysanura. This is, no doubt, illogical because Professor Comstock always kept pace with legitimate progress in his field of science.

PREFACE

The Hemimeridae has been advanced to its place in the Dermaptera and the discussion briefly extended on the basis of the paper by Rehn and Rehn.

At the suggestion of Dr. J. C. Bradley the suborder Idiogastra has been abandoned and the family Oryssidae included among the Chalastogastra.

GLENN W. HERRICK

Ithaca, July 1936.

THE ORIGINAL PREFACE

IT IS now nearly thirty years since "A Manual for the Study of Insects," in the preparation of which I was aided by Mrs. Comstock, was published. The great advances in the science of entomology during this period have made a revision of that work desirable. In the revision of the "Manual" so many changes and additions have been found necessary that the result is a book differing greatly from the original work; for this reason, it is published under a different title. The title selected is that of an earlier work, an "Introduction to Entomology" published in 1888 and long out of print.

Part I of the present volume was published separately in 1919, in order that it might be available for the use of classes in insect morphology and also that an opportunity might be offered for the suggestion of desirable changes to be made before the incorporation of it in the completed work. Such suggestions have been received, with the result that some very important changes have been made in the text.

In the preparation of this work I have received much help from my colleagues in the entomological department of Cornell University, for which I wish to make grateful acknowledgment, and especially to Dr. J. G. Needham for aid in the study of wing-venation, to Dr. O. A. Johannsen for help in the preparation of the chapter on the Diptera, to Dr. W. T. M. Forbes for help in the preparation of the chapter on the Lepidoptera, to Dr. J. C. Bradley for help in the preparation of the chapter on the Hymenoptera, and to Dr. J. T. Lloyd for the use of his figures of the cases of caddice worms.

From the published works of Professors Herrick, Crosby and Slingerland, Crosby and Leonard, Sanderson, and Matheson I have gleaned much information; references to these and to the more important of the other sources from which material has been drawn are indicated in the text and in the bibliography at the end of the volume.

PREFACE

References to the bibliography are made in the text by citing the name of the author and the year in which the paper quoted was published.

The wood cuts used in the text were engraved from nature by Mrs. Anna B. Comstock for our joint work, "A Manual for the Study of Insects." The other original figures and the copies of published figures were drawn by Miss Anna C. Stryke, Miss Ellen Edmonson, Miss Mary Mekeel, Mr. Albert Force, Mrs. Louise Nash, and Miss E. L. Keyes. I am deeply indebted to each of these artists for the painstaking care shown in their work.

As an aid to the pronunciation of the technical terms used and of the Latin names of insects, the accented syllable is marked with a sign indicating the quality of the vowel according to the English system of pronouncing Latin.

Two objects have been kept constantly in mind in the preparation of this book: first, to aid the student in laying a firm foundation for his entomological studies; and second, to make available, so far as possible in the limited space of a handbook, a knowledge of the varied phenomena of the insect world. It is hoped that those who use this book will find delight in acquiring a more intimate acquaintance with these phenomena.

JOHN HENRY COMSTOCK

Entomological Department
　Cornell University
　　August 1924.

TABLE OF CONTENTS

PART I. THE STRUCTURE AND METAMORPHOSIS OF INSECTS

CHAPTER I

	PAGES
THE CHARACTERISTICS OF INSECTS AND THEIR NEAR RELATIVES	1
Phylum Arthropoda	1
List of the classes of the Arthropoda	2
Table of the classes of the Arthropoda	3
Class Onychophora	4
Class Crustacea	6
Class Palæostracha	8
Class Arachnida	9
Class Pycnogonida	10
Class Tardigrada	12
Class Pentastomida	14
Class Diplopoda	15
Class Pauropoda	14
Class Chilopoda	28
Class Symphyla	23
Class Myrientomata	24
Class Hexapoda	26

CHAPTER II

THE EXTERNAL ANATOMY OF INSECTS	29

I. THE STRUCTURE OF THE BODY-WALL

a. The three layers of the body-wall	29
The hypodermis	29
The trichogens	30
The cuticula	30
Chitin	30
Chitinized and non-chitinized cuticula	30
The epidermis and the dermis	31
The basement membrane	31
b. The external apophyses of the cuticula	31
The cuticular nodules	31
The fixed hairs	31
The spines	32
c. The appendages of the cuticula	32
The spurs	32

The setæ.. 32
    The taxonomic value of setæ................................. 33
    A classification of setæ...................................... 33
        (1) The clothing hairs................................... 33
        (2) The glandular hairs.................................. 33
        (3) The sense-hairs..................................... 33
d. The segmentation of the body..................................... 34
    The body-segments, somites or metameres..................... 34
    The transverse conjunctivæ................................... 34
e. The segmentation of the appendages.............................. 34
 . The divisions of a body-segment................................. 34
    The tergum, the pleura, and the sternum..................... 34
    The lateral conjunctivæ...................................... 35
    The sclerites.. 35
    The sutures... 35
        The median sutures..................................... 35
    The piliferous tubercles of larvæ.............................. 35
    The homologizing of sclerites................................ 35
g. The regions of the body... 36

2. THE HEAD

a. The corneas of the eyes... 36
    The corneas of the compound eyes........................... 36
    The corneas of the ocelli..................................... 37
b. The areas of the surface of the head............................. 37
    The front... 37
    The clypeus... 38
    The labrum... 38
    The epicranium... 38
    The vertex.. 39
    The occiput... 39
    The genæ... 39
    The postgenæ... 39
    The gula.. 39
    The ocular sclerites.. 39
    The antennal sclerites....................................... 39
    The trochantin of the mandible.............................. 40
    The maxillary pleurites...................................... 40
    The cervical sclerites.. 40
c. The appendages of the head..................................... 40
    The antennæ.. 40
    The mouth-parts.. 42
        The labrum.. 42
        The mandibles... 42
        The maxillulæ... 42
        The maxillæ... 42
        The labium or second maxillæ.......................... 45

	The epipharynx..	46
	The hypopharynx...	47
d.	The segments of the head.......................................	47

3. THE THORAX

c.	The segments of the thorax.....................................	48
	The prothorax, mesothorax, and metathorax.....................	48
	The alitrunk..	49
	The propodeum or the median segment...........................	49
b.	The sclerites of a thoracic segment............................	49
	The sclerites of a tergum......................................	49
	The notum..	49
	The postnotum or the postscutellum..........................	50
	The divisions of the notum..................................	50
	The patagia..	50
	The parapsides..	51
	The sclerites of the pleura....................................	51
	The episternum..	51
	The epimerum...	51
	The preepisternum...	51
	The paraptera...	51
	The spiracles...	52
	The peritremes..	52
	The acetabula...	52
	The sclerites of a sternum.....................................	52
c.	The articular sclerites of the appendages......................	53
	The articular sclerites of the legs............................	53
	The trochantin..	53
	The antecoxal piece...	54
	The second antecoxal piece..................................	54
	The articular sclerites of the wings...........................	54
	The tegula..	54
	The axillaries..	54
d.	The appendages of the thorax...................................	55
	The legs..	56
	The coxa..	56
	The styli...	56
	The trochanter..	57
	The femur...	57
	The tibia...	57
	The tarsus..	57
	The wings...	58
	The different types of wings................................	59
	The margins of wings..	60
	The angles of wings...	60
	The axillary cord...	60
	The axillary membrane.......................................	60
	The alula...	60
	The axillary excision.......................................	61

The posterior lobe..................................... 61
The methods of uniting the two wings of each side........... 61
The hamuli.. 61
The frenulum and the frenulum hook..................... 61
The jugum.. 61
The fibula... 62
The hypothetical type of the primitive wing-venation......... 62
Longitudinal veins and cross-veins....................... 64
The principal wing-veins............................... 64
The chief branches of the wing-veins..................... 64
The veins of the anal area.............................. 65
The reduction of the number of the wing-veins............. 65
Serial veins... 67
The increase of the number of the wing-veins.............. 68
The accessory veins.................................... 68
The intercalary veins................................... 69
The adventitious veins.................................. 70
The anastomosis of veins................................ 70
The named cross-veins.................................. 71
The arculus... 72
The terminology of the cells of the wing................... 72
The corrugations of the wings........................... 73
Convex and concave veins............................... 73
The furrows of the wing................................ 73
The bullæ... 74
The ambient vein...................................... 74
The humeral veins..................................... 74
The pterostigma or stigma.............................. 74
The epipluræ.. 74
The discal cell and the discal vein....................... 74
The anal area and the preanal area of the wing............. 75

    4. THE ABDOMEN........................ 75

a. The segments of the abdomen............................ 75
b. The appendages of the abdomen.......................... 76
  The styli or vestigial legs of certain Thysanura............. 76
  The collophore of the Collembola........................ 76
  The spring of the Collembola........................... 76
  The genitalia... 76
  The cerci.. 77
  The median caudal filament............................. 78
  The prolegs of larvæ................................... 78

    5. THE MUSIC AND THE MUSICAL ORGANS OF INSECTS....... 78

a. Sounds produced by striking objects outside of the body......... 79
b. The music of flight..................................... 80

TABLE OF CONTENTS xiii

- c. Stridulating organs of the rasping type................... 81
  - The stridulating organs of the Locustidæ................. 82
  - The stridulating organs of the Gryllidæ and the Tettigoniidæ..... 83
  - Rasping organs of other than orthopterous insects............. 87
- d. The musical organs of a cicada........................... 89
- e. The spiracular musical organs........................... 91
- f. The acute buzzing of flies and bees....................... 91
- g. Musical notation of the songs of insects................... 92
- h. Insect choruses... 93

CHAPTER III

The Internal Anatomy of Insects............................ 94

1. THE HYPODERMAL STRUCTURES................. 95

- a. The internal skeleton.................................... 95
  - Sources of the internal skeleton........................... 95
    - Chitinized tendons.................................... 95
    - Invaginations of the body-wall or apodemes............. 95
  - The tentorium... 96
    - The posterior arms of the tentorium................... 96
    - The anterior arms of the tentorium.................... 97
    - The dorsal arms of the tentorium...................... 97
    - The frontal plate of the tentorium..................... 97
  - The endothorax.. 97
    - The pragmas.. 97
    - The lateral apodemes.................................. 98
    - The furcæ.. 98
- b. The hypodermal glands................................... 98
  - The molting-fluid glands................................. 99
  - Glands connected with setæ............................... 99
    - Venomous setæ and spines............................ 100
    - Androconia.. 100
    - The specific scent-glands of females................... 100
    - Tenent hairs... 100
  - The osmeteria... 101
  - Glands opening on the surface of the body................ 102
    - Wax-glands.. 102
    - Froth-glands of spittle insects........................ 102
    - Stink-glands... 102
    - The cephalic silk-glands.............................. 103
  - The salivary glands...................................... 104

2. THE MUSCLES............................. 104

3. THE ALIMENTARY CANAL AND ITS APPENDAGES....... 107

- a. The more general features............................... 107
  - The principal divisions................................... 108
  - Imperforate intestines in the larvæ of certain insects........ 108

b.	The fore-intestine	109
	The layers of the fore-intestine	109
	The intima	109
	The epithelium	109
	The basement membrane	109
	The longitudinal muscles	109
	The circular muscles	109
	The peritoneal membrane	109
	The regions of the fore-intestine	109
	The pharynx	109
	The œsophagus	110
	The crop	110
	The proventriculus	110
	The œsophageal valve	111
c.	The mid-intestine	111
	The layers of the mid-intestine	111
	The epithelium	112
	The peritrophic membrane	112
d.	The hind-intestine	112
	The layers of the hind-intestine	112
	The regions of the hind-intestine	113
	The Malpighian vessels	113
	The Malpighian vessels as silk-glands	113
	The cæcum	113
	The anus	113

4. THE RESPIRATORY SYSTEM ... 113

a. The open or holopneustic type of respiratory organs ... 114

1. *The spiracles* ... 114

The position of the spiracles ... 114
The number of spiracles ... 114
Terms indicating the distribution of the spiracles ... 115
The structure of spiracles ... 116
The closing apparatus of the tracheæ ... 116

2. *The tracheæ* ... 116

The structure of the tracheæ ... 117

3. *The tracheoles* ... 118

4. *The air-sacs* ... 118

5. *Modifications of the open type of respiratory organs* ... 119

b. The closed or apneustic type of respiratory organs ... 119

    1. *The Tracheal gills* ... 119
    2. *Respiration of parasites* ... 120
    3. *The blood-gills* ... 120

TABLE OF CONTENTS

	5. THE CIRCULATORY SYSTEM	121

The general features of the circulatory system 121
The heart 121
    The pulsations of the heart 122
The aorta 122
The circulation of the blood 122
Accessory circulatory organs 122

        6. THE BLOOD 122

        7. THE ADIPOSE TISSUE 123

        8. THE NERVOUS SYSTEM 123

a. The central nervous system 123
b. The œsophageal sympathetic nervous system 125
c. The ventral sympathetic nervous system 127
d. The peripheral sensory nervous system 128

    9. GENERALIZATIONS REGARDING THE SENSE-ORGANS OF INSECTS ... 129

A classification of the sense-organs 129
The cuticular part of the sense-organs 130

       10. THE ORGANS OF TOUCH 131

       11. THE ORGANS OF TASTE AND SMELL 132

       12. THE ORGANS OF SIGHT 134

a. The general features 134
    The two types of eyes 134
    The distinction between ocelli and compound eyes 134
    The absence of compound eyes in most of the Apterygota 135
    The absence of compound eyes in larvæ 135
b. The ocelli 135
    The primary ocelli 135
    The adaptive ocelli 136
    The structure of a visual cell 137
    The structure of a primary ocellus 137
    Ocelli of Ephemerida 139
c. The compound eyes 139
    The physiology of compound eyes 141
    The theory of mosaic vision 141
    Day-eyes 142
    Night-eyes 143
    Eyes with double function 143
    Divided eyes 144
    The tapetum 144

13. THE ORGANS OF HEARING.................................. 145

 a. The general features... 145
  The tympana... 145
  The chordotonal organs... 145
  The scolopale and the scolopophore.............................. 146
  The integumental and the subintegumental scolopophores.......... 146
  The structure of a scolopophore................................. 146
  The structure of a scolopale.................................... 147
  The simpler forms of chordotonal organs......................... 147
  The chordotonal ligament.. 147
 b. The chordotonal organs of larvæ................................. 148
 c. The chordotonal organs of the Locustidæ......................... 148
 d. The chordotonal organs of the Tettigoniidæ and of the Gryllidæ... 149
  The trachea of the leg.. 150
  The spaces of the leg... 151
  The supra-tympanal or subgenual organ........................... 151
  The intermediate organ.. 152
  Siebold's organ or the crista acustica.......................... 152
 e. Johnston's organ.. 152

14. SENSE-ORGANS OF UNKNOWN FUNCTIONS

 The sense-domes or the olfactory pores............................... 154

15. THE REPRODUCTIVE ORGANS

 a. The general features... 156
  Secondary sexual characters..................................... 157
 b. The reproductive organs of the female............................ 157
  The general features of the ovary............................... 157
  The wall of an ovarian tube..................................... 158
  The zones of an ovarian tube.................................... 158
  The contents of an ovarian tube................................. 158
  The egg-follicles... 158
  The functions of the follicular epithelium...................... 159
  The ligament of the ovary....................................... 159
  The oviduct.. 159
  The egg-calyx.. 159
  The vagina... 159
  The spermatheca.. 159
  The bursa copulatrix... 159
  The colleterial glands... 160
 c. The reproductive organs of the male.............................. 160
  The general features of the testes.............................. 160
  The structure of a testicular follicle.......................... 161
  The spermatophores.. 162
  Other structures.. 162

16. THE SUSPENSORIA OF THE VISCERA

The dorsal diaphragm 162
The ventral diaphragm 163
The thread-like suspensoria of the viscera 163

17. SUPPLEMENTARY DEFINITIONS

The œnocytes 163
The pericardial cells 164
The phagocytic organs 164
The light-organs 165

CHAPTER IV

THE METAMORPHOSIS OF INSECTS 166
I. THE EXTERNAL CHARACTERISTICS OF THE METAMORPHOSIS OF INSECTS
  a. The egg 166
    The shape of the egg 167
    The sculpture of the shell 167
    The microphyle 167
    The number of eggs produced by insects 168
    Modes of laying eggs 168
    Duration of the egg-state 170
  b. The hatching of young insects 171
    The hatching spines 171
  c. The molting of insects 171
    General features of the molting of insects 171
    The molting fluid 172
    The number of postembryonic molts 172
    Stadia 172
    Instars 172
    Head measurements of larvæ 173
    The reproduction of lost limbs 173
  d. Development without metamorphosis 174
    The Ametabola 174
  e. Gradual metamorphosis 175
    The Paurometabola 176
    The term nymph 176
    Deviations from the usual type 176
      The Saltitorial Orthoptera 177
      The Cicadas 177
      The Coccidæ 177
      The Aleyrodidæ 177
      The Aphididæ 177
      The Thysanoptera 177
  f. Incomplete metamorphosis 178
    The Hemimetabola 179
    The term naiad 179
    Deviations from the usual type 180
      The Odonata 180
      The Ephemerida 180

g.	Complete metamorphosis................................	180
	The Holometabola...	180
	The term larva..	180
	The adaptive characteristics of larvæ......................	181
	The different types of larvæ..............................	183
	The prepupa..	185
	The pupa...	186
	The chrysalis...	186
	Active pupæ..	187
	The cremaster..	187
	The cocoon...	188
	Modes of escape from the cocoon.......................	188
	The puparium..	190
	Modes of escape from the puparium.....................	190
	The different types of pupæ...............................	190
	The imago..	191
h.	Hypermetamorphosis.......................................	191
i.	Viviparous insects...	191
	Viviparity with parthenogenetic reproduction................	192
	Viviparity with sexual reproduction........................	193
j.	Neoteinia...	194

    2. THE DEVELOPMENT OF APPENDAGES............. 194

a.	The development of wings.................................	195
	The development of the wings of nymphs and naiads...........	195
	The development of the wings in insects with a complete metamorphosis...	195
b.	The development of legs...................................	197
	The development of the legs of nymphs and naiads............	198
	The development of the legs in insects with a complete metamorphosis...	198
c.	The development of antennæ..............................	199
d.	The development of mouth-parts...........................	200
e.	The development of the genital appendages..................	201

    3. THE DEVELOPMENT OF THE HEAD IN THE MUSCIDÆ........ 202

    4. THE TRANSFORMATION OF THE INTERNAL ORGANS....... 204

PART II. THE CLASSIFICATION AND THE LIFE-HISTORIES OF INSECTS

Chapter	V.	—The sub-classes and the orders of the class Hexapoda..	206
Chapter	VI.	—Class Hexapoda..................................	217
Chapter	VII.	—The Apterygota...................................	218
Chapter	VII.	—Order Protura............................	218
Chapter	VII.	—Order Thysanura.........................	219
Chapter	VII.	—Order Collembola.........................	225
Chapter	VIII.	—The Pterygota....................................	
Chapter	VIII.	—Order Orthoptera.........................	230
Chapter	IX.	—Order Zoraptera..........................	270
Chapter	X.	—Order Isoptera...........................	273
Chapter	XI.	—Order Neuroptera.........................	281
Chapter	XII.	—Order Ephemerida.........................	308
Chapter	XIII.	—Order Odonata............................	314
Chapter	XIV.	—Order Plecoptera.........................	325
Chapter	XV.	—Order Corrodentia........................	331
Chapter	XVI.	—Order Mallophaga.........................	335
Chapter	XVII.	—Order Embiidina..........................	338
Chapter	XVIII.	—Order Thysanoptera.......................	341
Chapter	XIX.	—Order Anoplura...........................	347
Chapter	XX.	—Order Hemiptera..........................	350
Chapter	XXI.	—Order Homoptera..........................	394
Chapter	XXII.	—Order Dermaptera.........................	460
Chapter	XXIII.	—Order Coleoptera.........................	464
Chapter	XXIV.	—Order Strepsiptera.......................	546
Chapter	XXV.	—Order Mecoptera..........................	550
Chapter	XXVI.	—Order Trichoptera........................	555
Chapter	XXVII.	—Order Lepidoptera........................	571
Chapter	XXVIII.	—Order Diptera............................	773
Chapter	XXIX.	—Order Siphonaptera.......................	877
Chapter	XXX.	—Order Hymenoptera........................	884
Bibliography...			1008
Index...			1029

PART I
THE STRUCTURE AND METAMORPHOSIS OF INSECTS

CHAPTER I

THE CHARACTERISTICS OF INSECTS AND OF THEIR NEAR RELATIVES

Phylum ARTHROPODA

The Arthropods

If an insect, a scorpion, a centipede, or a lobster be examined, the body will be found to be composed of a series of more or less similar rings or segments joined together; and some of these segments will be found to bear jointed legs (Fig. 1). All animals possessing these characteristics are classed together as the *Arthrŏpoda*, one of the chief divisions or phyla of the animal kingdom.

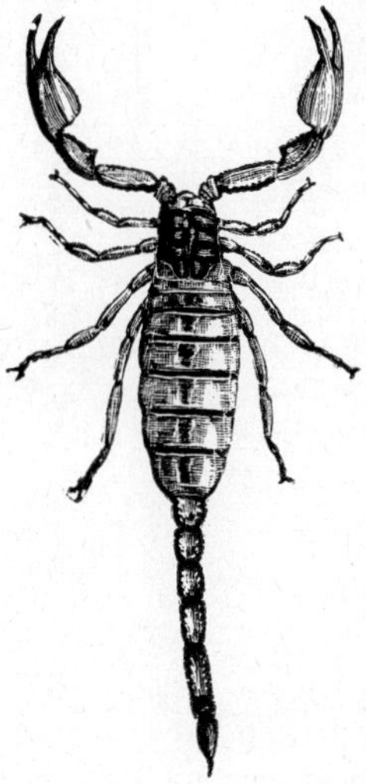

A similar segmented form of body is found among worms; but these are distinguished from the Arthropoda by the absence of legs. It should be remembered that many animals commonly called worms, as the tomato-worm, the cabbage-worm, and others, are not true worms, but are the larvæ of insects (Fig. 2). The angle-worm is the most familiar example of a true worm.

In the case of certain arthropods the distinctive characteristics of the phylum are not evident from a cursory examination. This may be due to a very generalized condition, as perhaps is true of *Peripatus;* but in most instances it is due to a secondary modification of form, the result of adaptation to special modes of life. Thus the segmentation of the body may be

Fig. 1.—An arthropod.

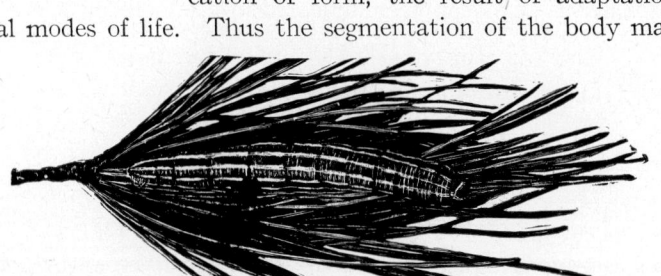

Fig. 2.—A larva of an insect.

(1)

obscured, as in spiders and in mites (Fig. 3); or the jointed appendages may be absent, as in the larvæ of flies (Fig. 4), of bees, and of many other insects. In all of these cases, however, a careful study of the structure of the animal, or of its complete life-history, or of other animals that are evidently closely allied to it removes any doubt regarding its being an arthropod.

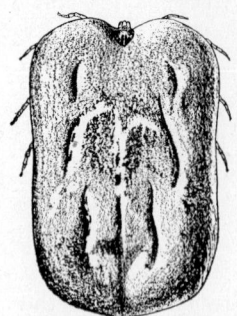

Fig. 3.—A mite, an arthropod in which the segmentation of the body is obscured. The southern cattle-tick, *Boophilus annulatus*.

The phylum Arthropoda is the largest of the phyla of the animal kingdom, including many more known species than all the other phyla taken together. This vast assemblage of animals includes forms differing widely in structure, all agreeing, however, in the possession of the essential characteristics of the Arthropoda. Several distinct types of arthropods are recognized; and those of each type are grouped together as a class.

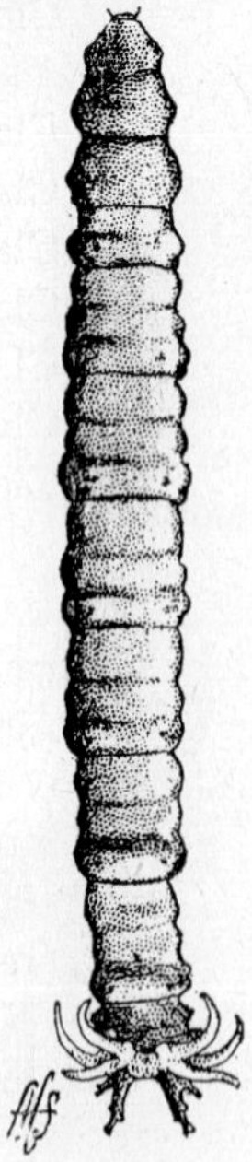

The number of distinct classes that should be recognized, and the relation of these classes to each other are matters regarding which there are still differences of opinion; we must have much more knowledge than we now possess before we can speak with any degree of certainty regarding them.

Fig. 4.—Larva of a fly, *Tipula abdominalis*; an arthropod in which the development of the legs is retarded.

Each of the classes enumerated below is regarded by all as a distinct group of animals; but in some cases there may be a question whether the group should be given the rank of a distinct class or not. The order in which the classes are discussed in this chapter is indicated in the following list.

LIST OF THE CLASSES OF THE ARTHROPODA

I. THE MOST PRIMITIVE ARTHROPODS
    Class Onychophora, page 4

II. THE AQUATIC SERIES
    Class Crustacea, page 6
    Class Palæostracha, page 8

III. AN OFFSHOOT OF THE AQUATIC SERIES, SECONDARILY AERIAL
    Class Arachnida, page 9

IV. DEGENERATE ARTHROPODS OF DOUBTFUL POSITION
    Class Pycnogonida, page 10
    Class Tardigrada, page 12
    Class Pentastomida, page 14

V. THE PRIMARILY AERIAL SERIES
    Class Onychophora (See above)
    Class Diplopoda, page 15
    Class Pauropoda, page 18
    Class Chilopoda, page 20
    Class Symphyla, page 23
    Class Myrientomata, page 24
    Class Hexapoda, page 26

TABLE OF CLASSES OF THE ARTHROPODA

A.   Worm-like animals, with an unsegmented body, but with many unjointed legs............................ONYCHOPHORA

AA. Body more or less distinctly segmented except in a few degenerate forms.

  B. With two pairs of antennæ and at least five pairs of legs; respiration aquatic........................CRUSTACEA

  BB. Without or apparently without antennæ.

    C. With well-developed aquatic respiratory organs.
                                                PALÆOSTRACHA

    CC. With well-developed aerial respiratory organs or without distinct respiratory organs.

      D. With well-developed aerial respiratory organs.

        E. Body not resembling that of the Thysanura in form.
                                                        ARACHNIDA

        EE. Body resembling that of the Thysanura in form (Family Eosentomidæ)........MYRIENTOMATA

      DD. Without distinct respiratory organs.

        E. With distinctly segmented legs.

          F. Body resembling that of the Thysanura in form, but without antennæ, and with three pairs of thoracic legs and three pairs of vestigial abdominal legs (Family Acerentomidæ)........MYRIENTOMATA

          FF. With four or five pairs of ambulatory legs; abdomen vestigial............PYCNOGONIDA

        EE. Legs not distinctly segmented.

          F. With four pairs of legs in the adult instar.
                                                         TARDIGRADA

FF. Larva with two pairs of legs, adult without legs..................PENTASTOMIDA
BBB. With one pair, and only one, of feeler-like antennæ. Respiration aerial.
  C. With more than three pairs of legs, and without wings.
    D. With two pairs of legs on some of the body-segments.
....DIPLOPODA
    DD. With only one pair of legs on each segment of the body.
      E. Antennæ branched..................PAUROPODA
      EE. Antennæ not branched.
        F. Head without a Y-shaped epicranial suture. Tarsi of legs with a single claw each. Opening of the reproductive organs near the caudal end of the body..........................CHILOPODA
        FF. Head with a Y-shaped epicranial suture, as in insects. Tarsi of legs with two claws each. Opening of the reproductive organs near the head.
....................................SYMPHYLA
  CC. With only three pairs of legs, and usually with wings in the adult state........................HEXAPODA

CLASS ONYCHOPHORA

The genus Peripatus of authors

The members of this class are air-breathing animals, with a nearly cylindrical, unsegmented body, which is furnished with many pairs of unjointed legs. The reproductive organs open near the hind end of the body.

The class Onychŏphora occupies the position of a connecting link between the Arthropoda and the phylum Annulata or worms; and is therefore of the highest interest to students of systematic zoology. All known members of this class have been included until recently in a single genus *Perĭpatus;* but now the fifty or more known species are distributed among nearly a dozen genera.

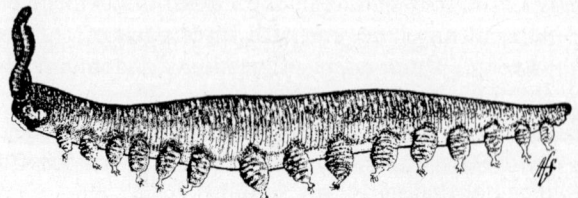

Fig. 5.—*Peripatoides novæ-zealandicæ.*

The body (Fig. 5) is nearly cylindrical, caterpillar-like in form, but is unsegmented externally. It is furnished with many pairs of legs, the number of which varies in different species. The legs have a ringed appearance, but are not distinctly jointed.

The head bears a pair of ringed antennæ (Fig. 6); behind these on the sides of the head, there is a pair of short appendages termed oral papillæ. The mouth opening is surrounded by a row of lobes which constitute the lips, and between these in the anterior part of the mouth-cavity there is an obtuse projection, which bears a row of chitinous points. Within the mouth cavity there are two pairs of hooked plates, which have been termed the mandibles, the two plates of each side being regarded as a single mandible.

Although the body is unsegmented externally, internally there are evidences of a metameric arrangement of parts. The ventral nerve cords, which at first sight appear to be without ganglia, are enlarged opposite each pair of legs, and these enlargments are regarded as rudimentary ganglia. We can, therefore speak of each section of a body corresponding to a pair of appendages as a segment. The

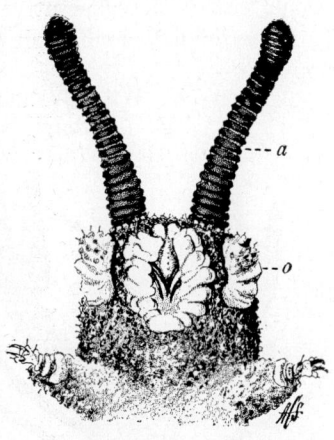

Fig. 6.—Ventral view of the head and first pair of legs of *Peripatoides*; *a*, antenna; *o*, oral papilla.

metameric condition is farther indicated by the fact that most of these segments contain each a pair of nephridia; each nephridium opening at the base of a leg.

The respiratory organs are short tracheæ, which are rarely branched, and in which the tænidia appear to be rudimentary.* In some species, the spiracles are distributed irregularly; in others, they are in longitudinal rows.

The sexes are distinct. The reproductive organs open near the hind end of the body, either between the last or the next to the last pair of legs.

The various species are found in damp situations, under the bark of rotten stumps, under stones or other objects on the ground. They have been found in Africa, in Australia, in South America, and in the West Indies.

Their relationship to the Arthropoda is shown by the presence of paired appendages, one, or perhaps two, pairs of which are modified as jaws; the presence of tracheæ which are found nowhere else except

---

*It is quite possible that the "short tracheæ" described by writers on the structure of these animals are tracheoles. See the account of the distinguishing features of tracheæ and tracheoles in Chapter III.

in the Arthropoda; the presence of paired ostia in the wall of the heart; and the presence of a vascular body cavity and pericardium.

They resemble the Annulata in having a pair of nephridia in most of the segments of the body corresponding to the pairs of legs, and in having cilia in the generative tracts.

An extended monograph of the Onychophora was published by Bouvier ('05–'07).

Class CRUSTACEA
The Crustaceans

Fig. 7.—A cray-fish.

The members of this class are aquatic arthropods, which breathe by true gills. They have two pairs of antennæ and at least five pairs of legs. The position of the openings of the reproductive organs varies greatly; but as a rule they are situated far forward.

The most familiar examples of the Crustācea are the crayfishes, the lobsters, the shrimps, and the crabs. Cray-fishes (Fig. 7) abound in our brooks, and are often improperly called crabs. The lobsters, the shrimps, and the true crabs live in salt water.

Excepting *Lĭmulus*, the sole living representative of the class described next, the Crustacea are distinguished from all other arthropods by their mode of respiration, being the only ones that breathe by true gills. Many insects live in water and are furnished with gill-like organs; but these are either tracheal gills or blood-gills, organs which differ essentially in structure from true gills, as

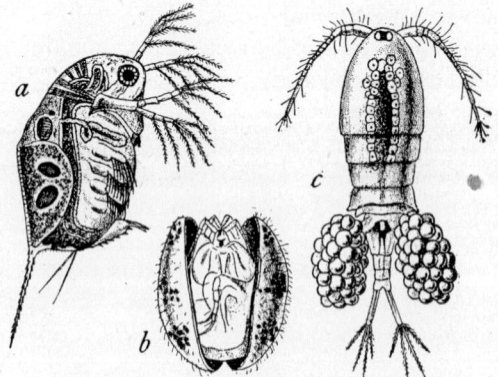

Fig. 8.—Minute crustaceans: *a*, *Daphnia*; *b*, *Cypridopsis*; *c*, *Cyclops*.

CHARACTERISTICS OF INSECTS AND THEIR RELATIVES

described later. The Crustacea also differ from other Arthropoda in having two pairs of antennæ. Rudiments of two pairs of antennæ have been observed in the embryos of many other arthropods; but in these cases one or the other of the two pairs of antennæ fail to develop.

The examples of crustaceans named above are the more conspicuous members of the class; but many other smaller forms abound both in the sea and in fresh water. Some of the more minute freshwater forms are almost sure to occur in any fresh-water aquarium. In Figure 8 are represented three of these greatly enlarged. The minute crustaceans form an important element in the food of fishes.

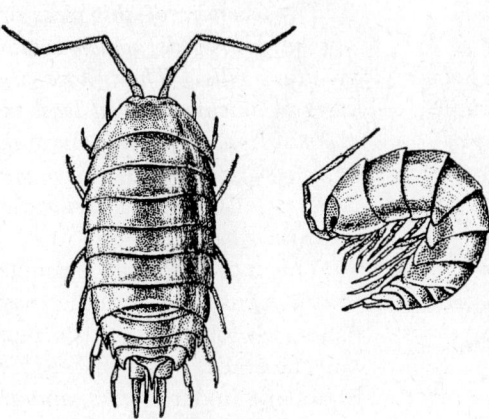

Fig. 9.—A sow-bug, *Cylisticus convexus* (From Richardson after Sars).

Some crustaceans live in damp places on land, and are often found by collectors of insects; those most often observed are the sow-bugs (Oniscoida), which frequently occur about water-soaked wood. Figure 9 represents one of these.

As there are several, most excellent text books devoted to the Crustacea, it is unnecessary to discuss farther this class in this place.

Class PALÆOSTRACHA

The King-crabs or Horseshoe-crabs

The members of this class are aquatic arthropods, which resemble the Crustacea in that they breathe by true gills, but in other respects are closely allied to the Arachnida. They are apparently without antennæ, the appendages homologous to antennæ being not feeler-like. The reproductive organs open near the base of the abdomen.

The class Palæŏstracha is composed almost entirely of extinct forms, there being living representatives of only a single order, the Xiphosūra, and this order is nearly extinct; for of it there remains only the genus *Lĭmulus*, represented by only five known species.

The members of this genus are known as king-crabs or horseshoe-crabs; the former name is suggested by the great size of some of the species; the latter, by the shape of the cephalothorax (Fig. 10).

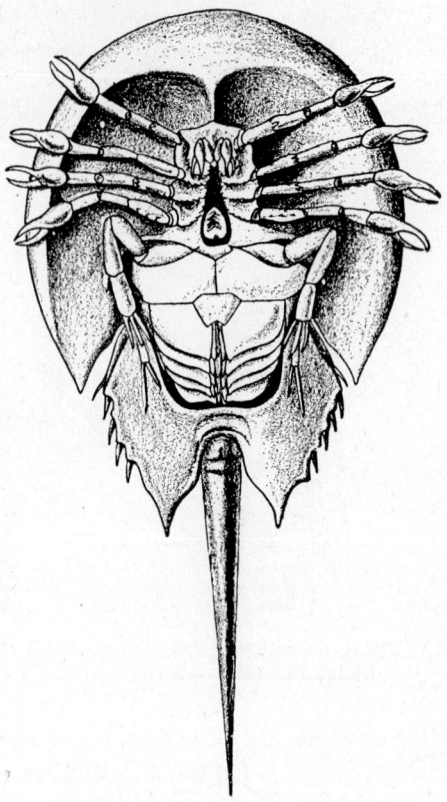

Fig. 10.—A horseshoe crab, *Limulus* (After Packard).

The king-crabs are marine; they are found on our Atlantic Coast from Maine to Florida, in the West Indies, and on the eastern shores of Asia. They are found in from two to six fathoms of water on sandy and muddy shores; they burrow a short distance in the sand or mud and feed chiefly on worms. The single species of our coast is *Lĭmulus polyphēmus*.

Class ARACHNIDA

Scorpions, Harvestmen, Spiders, Mites, and others

The members of this class are air-breathing arthropods, in which the head and thorax are usually grown together, forming a cephalothorax, which have four pairs of legs, and which apparently have no antennæ. The reproductive organs open near the base of the abdomen.

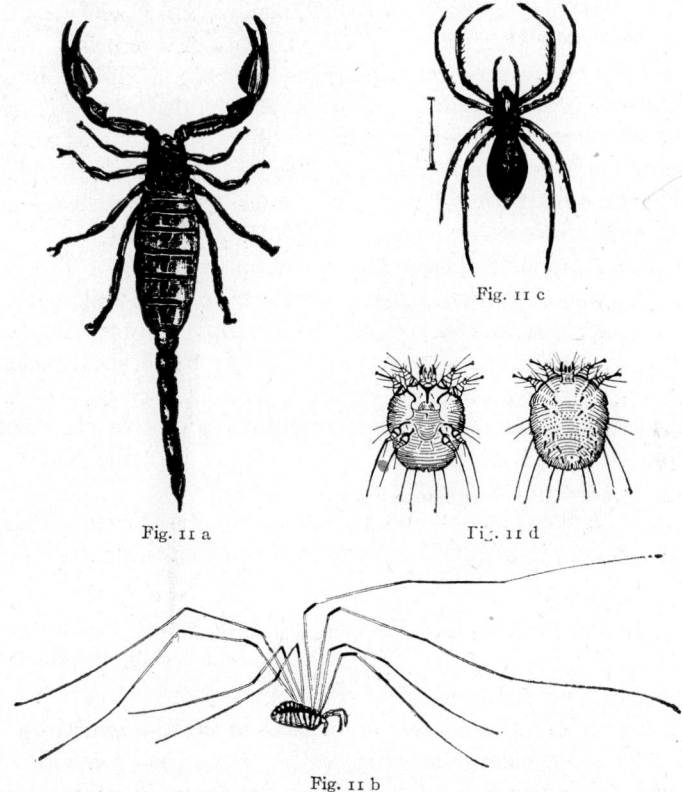

Fig. 11.—Arachnids: *a*, a scorpion; *b*, a harvestman. *c*, a spider; *d*, an itch-mite, from below and from above.

The Arăchnida abound wherever insects occur, and are often mistaken for insects. But they can be easily distinguished by the characters given above, even in those cases where an exception occurs to some one of them. The more important of the exceptions are the following: in one order, the Solpugida, the head is distinct from the thorax; as a rule the young of mites have only six legs, but a fourth pair is added during growth; and in the gall-mites there are only four legs.

The Arachnida are air-breathing; but it is believed that they have been evolved from aquatic progenitors. Two forms of respiratory organs exist in this class: first, book-lungs; and second, tubular tracheæ. Some members of it possess only one of these types; but the greater number of spiders possess both.

A striking characteristic of the Arachnida, which, however, is also possessed by the Palæostracha, is the absence of true jaws. In other arthropods one or more pairs of appendages are jaw-like in form and are used exclusively as jaws; but in the Arachnida the prey is crushed either by the modified antennæ alone or by these organs and other more or less leg-like appendages. The arachnids suck the blood of their victims by means of a sucking stomach; they crush their prey, but do not masticate it so as to swallow the solid parts.

In the Arachnida there exist only simple eyes.

The reproductive organs open near the base of the abdomen on the ventral side. In this respect the Arachnida resemble *Limulus*, the millipedes, and the Crustacea, and differ from the centipedes and insects.

Among the more familiar representatives of this class are the scorpions (Fig. 11, *a*), the harvestmen (Fig. 11, *b*), the spiders (Fig. 11, *c*), and the mites (Fig. 11, *d*).

As the writer has devoted a separate volume (Comstock, '12) to the Arachnida, it will not be discussed farther in this place.

Class PYCNOGONIDA

The Pycnogonids

The members of this class are marine arachnid-like arthropods, in which the cephalothorax bears typically seven pairs of jointed appendages, but in a few forms there are eight pairs, and in some the anterior two or three pairs are absent; and in which the abdomen is reduced to a legless, unsegmented condition. They possess a circulatory system, but no evident respiratory organs. The reproductive organs open through the second segment of the legs; the number of legs bearing these opening varies from one to five pairs.

The Pycnogŏnida or pycnŏgonids are marine animals, which bear a superficial resemblance to spiders (Fig. 12). Some of them are found under stones, near the low water line, on sea shores; but they are more abundant in deep water. Some are found attached to sea-anemones, upon which they probably prey; others are found climbing

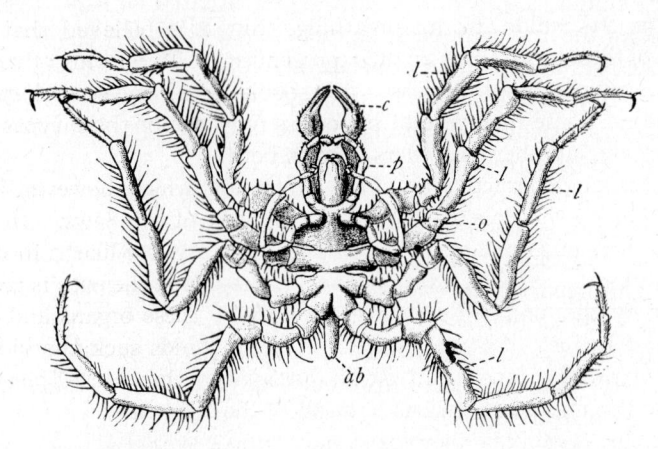

Fig. 12.—A pycnogonid, *Nymphon hispidum: c*, chelophore; *p*, palpus; *o*, ovigerous legs; *l, l, l, l*, ambulatory legs; *ab*, abdomen (After Hoeck).

over sea-weeds and Hydroids; and sometimes they are dredged in great numbers from deep water.

They possess a suctorial proboscis. In none of the appendages are the basal segments modified into organs for crushing the prey.

The cephalothorax comprises almost the entire body; the abdomen being reduced to a mere vestige, without appendages, and with no external indication of segmentation. But the presence of two pairs of abdominal ganglia indicates that originally the abdomen consisted of more than one segment.

There are typically seven pairs of appendages; but a few forms possess eight pairs; and in some the first two or three pairs are absent. The appendages, when all are present, consist of a pair of *chelophores*, each of which when well-developed consists of one or two basal segments and a chelate "hand"; the *palpi*, which are supposed to be tactile, and which have from five to ten joints when well-developed; the *ovigerous legs*, which are so-called because in the males they are used for holding the mass of eggs beneath the body; and the *ambulatory legs*, of which there are usually four pairs, but a few forms possess a fifth pair. The ambulatory legs consist each of eight segments and a terminal claw.

The only organs of special sense that have been found in these animals are the eyes. These are absent or at least very poorly developed in some forms, especially those that are found in very deep water, *i.e.* below four or five hundred fathoms. When well-developed they are simple, and consist of two pairs, situated on a tubercle, on the head or the first compound segment of the body, the segment that bears the first four pairs of appendages.

The reproductive organs open in the second segment of the legs. In some these openings occur only in the last pair of legs; in others, in all of the ambulatory legs.

Very little is known regarding the habits of these animals. The most interesting features that have been observed are perhaps the facts that the males carry the eggs in a mass, held beneath the body by the third pair of appendages, the ovigerous legs, and also carry the young for a time.

As to the systematic position of the class Pycnogonida, very little can be said. These animals are doubtless arthropods, and they are commonly placed near the Arachnida.

Class TARDIGRADA

The Tardigrades or Bear Animalcules

The members of this class are very minute segmented animals, with four pairs of legs, but without antennæ or mouth-appendages, and without special circulatory or respiratory organs; the reproductive organs open into the intestine.

The Tardigrāda or tărdigrades are microscopic animals, measuring from one seventy-fifth to one twenty-fifth of an inch in length. They are somewhat mite-like in appearance; but are very different from mites in structure (Fig. 13 and 14).

The head bears neither antennæ nor mouth-appendages. The four pairs of legs are short, unjointed, and are distributed along the entire length of the body, the fourth pair being at the caudal end. Each leg is terminated by claws, which differ in number and form in different genera.

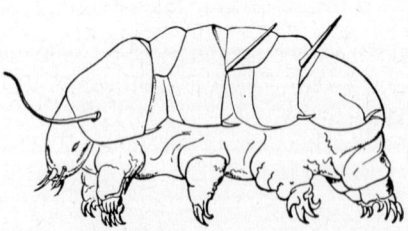

The more striking features of the internal structure of these animals is the absence of special circulatory and respiratory organs; the presence of a pair of chitinous teeth, either in the oral cavity or a short distance back of

Fig. 13.—A tardigrade (After Doyère).

it; the presence of Malpighian tubules; the unpaired condition of the reproductive organs of both sexes; and the fact that these organs open into the intestine. The central nervous system consists of a brain, a suboesophageal ganglion, and a ventral chain of four ganglia, connected by widely separated connectives.

The tardigrades are very abundant, and are very widely distributed. Some live in fresh water, a few are marine, but most of them live in damp places, and especially on the roots of moss, growing in gutters, on roofs or trees, or in ditches. But although they are common, their minute size and retiring habits result in their being rarely seen except by those who are seeking them.

Many of them have the power of withstanding desiccation for a long period. This has been demonstrated artificially by placing them on a microscopic slide and allowing the mositure to evaporate slowly. The body shrinks, its skin becomes wrinkled, and finally it assumes the appearance of a grain of sand in which no parts can be distinguished. In this state they can remain, it is said, for years; after which, if water be added, the body swells, assumes its normal form, and after a time, the creatures resume their activities.

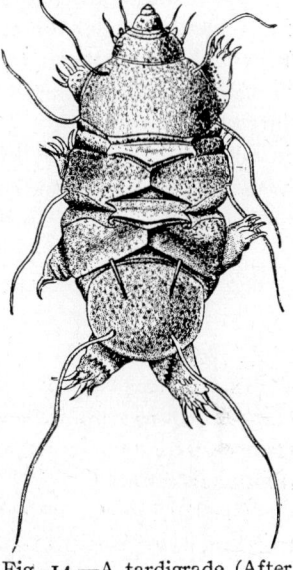

Fig. 14.—A tardigrade (After Doyère).

Regarding the systematic position of this class of animals nothing definite can be stated beyond the fact that they are doubtless arthropods. Their relationship to the other classes of arthropods has been masked by degenerative modifications. They are placed here near the end of the series of classes of arthropods, merely as a matter of convenience, in what may be termed an appendix to the arthropod series, which includes animals of doubtful relationships.

Class PENTASTOMIDA

The Pentastomids or Linguatulids

The members of this class are degenerate, worm-like, parasitic arthropods, which in the adult state have no appendages, except two pairs of hooks near the mouth; the larvæ have two pairs of short legs. These animals possess neither circulatory nor respiratory organs. The reproductive organs of the male open a short distance behind the mouth; those of the female near the caudal end of the body.

The Pentastŏmida or pentăstomids are worm-like creatures, whose form has been greatly modified by their parasitic life. The adults bear little resemblance to any other arthropods. Representatives of three genera are known. These are *Linguătula* in which the body is fluke-like in form(Fig. 15) and superficially annulated; *Porocĕphalus*, in which the body is cylindrical (Fig. 16) and ringed; and *Reighărdia*, which is devoid of annulations, and with poorly developed hooks and a mouth-armature.

The arthropodan nature of these animals is indicated by the form of the larvæ, which although greatly degenerate, are less so than the adults, having two pairs of legs (Fig. 17).

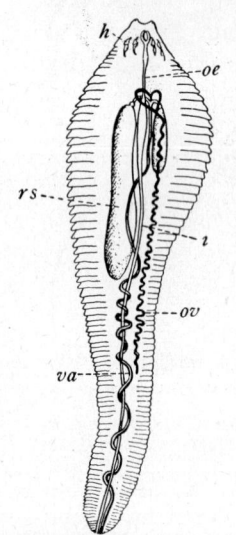

Fig. 15.—A pentastomid, *Linguatula tænioides*, female at the time of copulation: *h*, hooks; *oe*, œsophagus; *rs*, receptacula seminis, one of which is still empty; *i*, intestine; *ov*, ovary; *va*, vagina (From Lang after Leuckart).

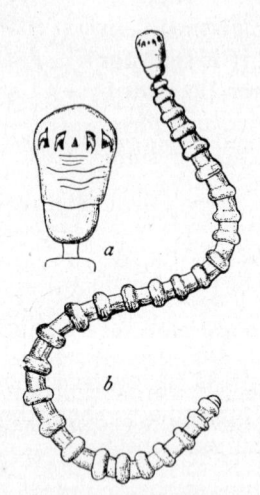

Fig. 16.—A pentastomid, *Porocephalus annulatus;* *a*, ventral view of head, greatly enlarged; *b*, ventral view of animal, slightly enlarged (After Shipley).

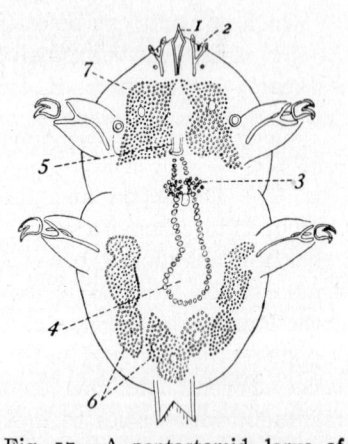

Fig. 17.—A pentastomid, larva of *Porocephalus proboscideus*, seen from below, highly magnified: 1, boring anterior end; 2, first pair of chitinous processes seen between the forks of the second pair; 3, ventral nerve ganglion; 4, alimentary canal; 5, mouth; 6 and 7, gland cells (From Shipley after Stiles).

CHARACTERISTICS OF INSECTS AND THEIR RELATIVES

Like many of the parasitic worms, these animals, in some cases at least, pass their larval life in one host, and complete their development in another of a different species; some larvæ being found in the bodies of herbivorous animals and the adults in predacious animals that feed on these herbivorous hosts.

The systematic position of the pentastomids is very uncertain. They have been considered by some writers to be allied to the mites. But it seems better to merely place them in this appendix to the arthropod series until more is known of their relationships.

Class DIPLOPODA

The Millipedes or Diplopods

The members of this class are air-breathing arthropods in which the head is distinct, and the remaining segments of the body form a continuous region. The greater number of the body-segments are so grouped that each apparent segment bears two pairs of legs. The antennæ are short and very similar to the legs. The openings of the reproductive organs are paired, and situated behind the second pair of legs.

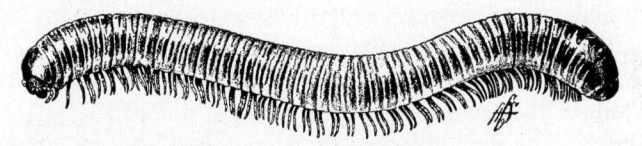

Fig. 18.—A millipede, *Spirobolus marginatus*.

The Diplŏpoda and the three following classes were formerly grouped together as a single class, the *Myriăpoda*. But this grouping has been abandoned, because it has been found that the Chilopoda are more closely allied to the insects than they are to the Diplopoda; and the Pauropoda and Symphyla are both very distinct from the Diplopoda on the one hand and the Chilopoda on the other. Owing to the very general and long continued use of the term Myriapoda, the student who wishes to look up the literature on these four classes should consult the references under this older name.

The most distinctive feature of the millipedes is that which suggested the name Diplopoda for the class, the fact that throughout the greater part of the length of the body there appears to be two pairs of legs borne by each segment (Fig. 18).

This apparent doubling of the appendages is due to a grouping of the segments in pairs and either a consolidation of the two terga of each pair or the non-development of one of them; which of these alternatives is the case has not been definitely determined.

It is clear, however, that there has been a grouping of the segments in pairs in the region where the appendages are doubled, for corresponding with each tergum there are two sterna and two pairs of spiracles.

A few of the anterior body segments, usually three or four in number, and sometimes one or two of the caudal segments remain single. Frequently one of the anterior single segments is legless, but the particular segment that lacks legs differs in the different families.

The head, which is as distinct as is the head of insects, bears the antennæ, the eyes, and the mouth-parts. The antennæ are short, and usually consist each of seven segments. The eyes are usually represented by a group of ocelli on each side of the head; but the ocelli vary greatly in number, and are sometimes absent. The mouth-parts consist of an upper lip or *labrum;* a pair of *mandibles;* and a pair of jaws, which are united at the base, forming a large plate, which is known as the *gnathŏchilārium*. In the genus *Polyxenus* there is a pair of lobes between the mandibles and the gnathochilarium, which have been named the *maxillulæ*. (paragnatha?).

The labrum is merely the anterior part of the upper wall of the head and, as in insects, is not an appendage. The mandibles, in the forms in which they are best developed, are fitted for biting, and consist of several parts (Fig. 19); but in some forms they are vestigial. The gnathochilarium (Fig. 20) is complicated in structure, the details of which vary greatly in different genera.

Fig. 19.—A mandible of *Julus; c,* cardo; *d, d,* teeth; *m,* muscle; *ma,* mala; *p,* pectinate plate; *s,* stipes (After Latzel).

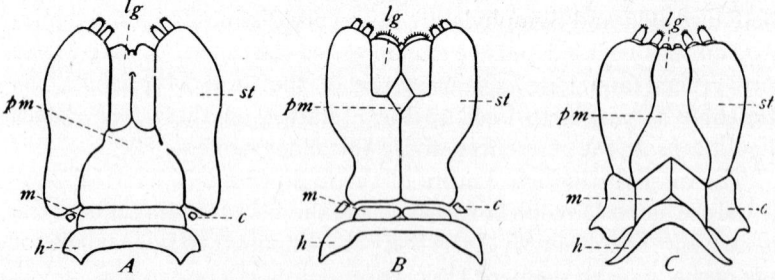

Fig. 20.—The gnathochilarium or second jaws of three diplopods; A, *Spirostreptus;* B, *Julus;* C, *Glomeris: c,* cardo; *h,* hypostoma; *lg,* linguæ; *m,* mentum; *pm,* promentum; *st,* stipes (After Silvestri).

CHARACTERISTICS OF INSECTS AND THEIR RELATIVES 17

In one subdivision of the class Diplopoda, which is represented by the genus *Polyxenus* and a few others, the mandibles are one-jointed; and between the mandibles and the gnathochilarium there is a pair of one-jointed lobes, which have not been found in other diplopods; these are the "maxillulæ" (Fig. 21). The correspondence of the parts of the gnathochilarium of *Polyxenus* and its allies with the parts of the gnathocillarium of other diplopods has not been satisfactorily determined.

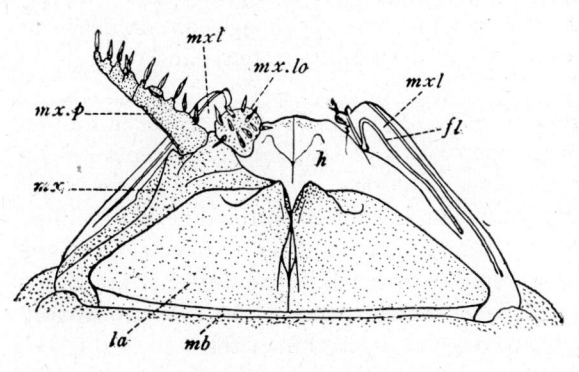

Fig. 21.—The second pair of jaws, maxillulæ, and the third pair of jaws, maxillæ or gnathochilarium, of *Polyxenus;* the parts of the maxillæ or gnathochilarium are stippled and some are omitted on the right side of the figure: *mb,* basal membrane of the labium; *la,* "labium" of Carpenter, perhaps the mentum and promentum of the gnathochilarium; *mx,* basal segment of the maxilla, perhaps the stipes of the gnathochilarium; *mx. lo,* lobe of the maxilla; *mx. p,* maxillary palpus; *h,* tongue or hypopharynx; *mxl,* maxillula; *fl.* flagellate process (After Carpenter).

Most of our more common millipedes possess stink-glands, which open by pores on a greater or less number of the body segments. These glands are the only means of defence possessed by millipedes, except the hard cuticula protecting the body.

The millipedes as a rule are harmless, living in damp places and feeding on decaying vegetable matter; but there are a few species that occasionally feed upon growing plants.

For a more detailed account of the Diplopoda see Pocock ('11).

Class PAUROPODA

The Pauropods

The members of this class are small arthropods in which the head is distinct, and the segments of the body form a single continuous region. Most of the body-segments bear each a single pair of legs. Although most of the terga of the body-segments are usually fused in couples, the legs are not grouped in double pairs as in the Diplopoda. The antennæ are branched. The reproductive organs open in the third segment back of the head.

The Pauropoda or pauropods are minute creatures, the described species measuring only about one twenty-fifth inch in length, more or less. They resemble centipedes in the elongated form of the body and in the fact that the legs are not grouped in double pairs as in the Diplopoda, although the terga of the body-region are usually fused in couples. These characteristics are well-shown by the dorsal and ventral views of *Pauropus* (Fig. 22 and 23).

Although the pauropods resemble the chilopods in the distribution of their legs, they differ widely in the position of the openings of the reproductive organs. These open in the third segment back of the head; that of the female is single, those of the male are double.

The head is distinct from the body-region. It bears one pair of antennæ and two pairs of jaws; the eyes are absent but there is an eye-like spot on each side of the head (Fig. 24). The first pair of jaws are large, one-jointed mandibles; the second pair are short pear-shaped organs. Between these two pairs

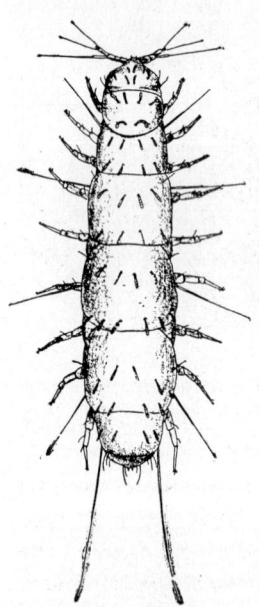

Fig. 22.—A pauropod, *Pauropus huxleyi*, dorsal aspect (After Kenyon).

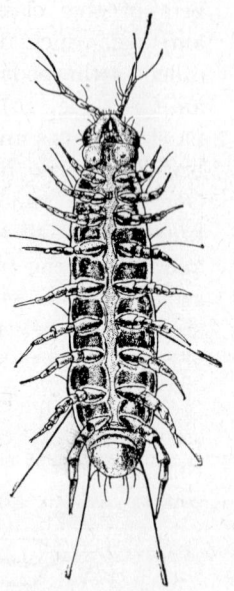

Fig. 23.—*Pauropus huxleyi*, ventral aspect (After Lubbock).

of jaws, there is a horny framework forming a kind of lower lip to the mouth (Fig. 25). The homologies of the mouth-parts with those of the allied classes of arthropods have not been determined.

The body-region consists of twelve segments. This is most clearly seen by an examination of the ventral aspect of the body. When the body is viewed from above the number of segments appears to be less, owing to the fact that the terga of the first ten segments are fused in couples. Nine of the body-segments bear well-developed legs. The appendages of the first segment are vestigial, and the last two segments bear no appendages.

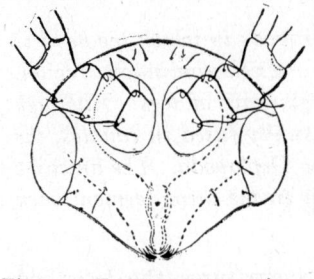

Fig. 24.—*Eurypauropus spinosus;* face showing the base of the antennæ, the mandibles, and the eye-like spots (After Kenyon).

The most distinctive feature of members of this class is the form of the antennæ, which differ from those of all other arthropods in structure. Each antenna (Fig. 26) consists of four short basal segments and a pair of one-jointed branches borne by the fourth segment. One of these branches bears a long, many-ringed filament with a rounded apical knob; and the other branch bears two such filaments with a globular or pear-shaped body between them. This is probably an organ of special sense.

Fig. 25.—Mouth-parts of *Eurypauropus ornatus;* *md*, mandible; *mx*, second jaws; *l*, lower lip (After Latzel).

The pauropods live under leaves and stones and in other damp situations. Representatives of two quite distinct families are found in this country and in various other parts of the world. In addition to these a third family, the *Brachypauropŏdidæ*, is found in Europe. In this family the pairs of terga consist each of two distinct plates. Our two families are the following:

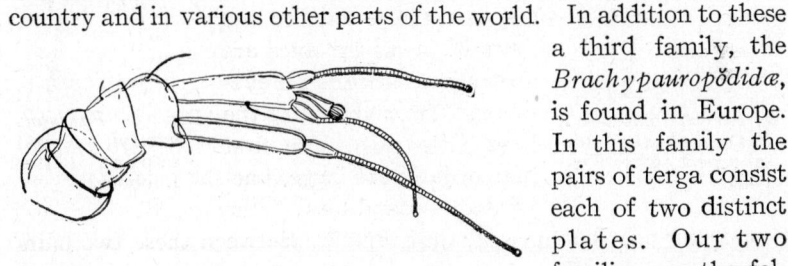

Fig. 26.—Antenna of *Eurypauropus spinosus* (After Kenyon).

Family Pauropŏdidæ.—In members of this family the head is not covered by the first tergal plate and the anal segment is not covered by the sixth tergal plate.

The best known representatives of this family belong to the genus *Pauropus* (Fig. 22). This genus is widely distributed, representatives having been found in Europe and in both North and South America. They are active, measure about one twenty-fifth inch in length, and are white.

Family Eurypaurŏpidæ.—The members of this family are characterized by the wide form of the body, which bears some resemblance to that of a sow-bug. The head is concealed by the first tergum of the body-region; and the anal segment, by the penultimate tergum. Our most familiar representative is *Eurypauropus spinosus* (Fig. 27). This, unlike *Pauropus*, is slow in its movements.

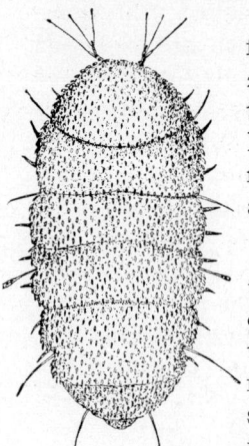

Fig. 27.—*Eurypauropus spinosus* (After Kenyon).

Class CHILOPODA

The Centipedes or Chilopods

The members of this class are air-breathing arthropods in which the head is distinct, and the remaining segments of the body form a continuous region. The numerous pairs of legs are not grouped in double pairs, as in the Diplopoda. The antennæ are long and many-jointed. The appendages of the first body-segment are jaw-like and function as organs of offense, the poison-jaws. The opening of the reproductive organs is in the next to the last segment of the body.

The animals constituting the class Chilŏpoda or chĭlopods are commonly known as centipedes. They vary to a considerable degree in the form of the body, but in all except perhaps the sub-class Notostigma the body-segments are distinct, not grouped in couples as in the diplopods (Fig. 28). They are sharply distinguished from the three preceding classes in the possession of poison-jaws and in having the opening of the reproductive organs at the caudal end of the body.

The antennæ are large, flexible, and consist of fourteen or more segments. There are four pairs of jaws including the jaw-like

CHARACTERISTICS OF INSECTS AND THEIR RELATIVES 21

appendages of the first body-segment. These are the *mandibles* (Fig. 29, A), which are stout and consist each of two segments; the *maxillæ* (Fig. 29, B, *a*), which are foliaceous, and usually regarded as biramous; the *second maxillæ* or *palpognaths*, which are leg-like in form, consisting of five or six segments, and usually have the coxæ united on the middle line of the body (Fig. 29, B, *b*), and the *poison-claws* or *toxicognaths*, which are the appendages of the first body-segment (Fig. 29, C).

The poison-claws consist each of six segments, of which the basal one, or coxa is usually fused with its fellow, the two forming a large coxal plate, and the distal one is a strong piercing fang in which there is the opening of the duct leading from a poison gland, which is in the appendage.

The legs consist typically of six segments, of which the last, the tarsus, is armed with a single terminal claw. The last pair of legs are directed backwards, and are often greatly modified in form.

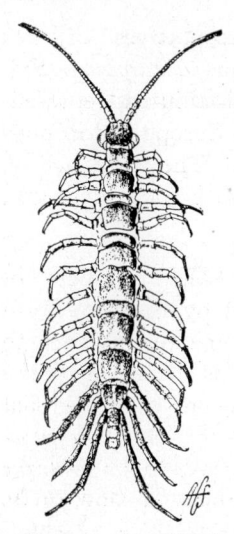

Fig. 28.—A centipede *Bothropolys multidentatus*.

The class Chilopoda includes two quite distinct groups of animals which are regarded by Pocock ('11) as sub-classes, the Pleurostigma and the Notostigma. The names of the sub-classes refer to the position of the spiracles.

Sub-Class
PLEUROSTIGMA

The typical Centipedes

In the typical centipedes, the sub-class Pleurostigma, the spiracles are paired and are situated in the sides of the segments that bear them. Each leg-bearing segment contains a distinct tergum and sternum, the number of sterna never exceeding that of the terga. The eyes

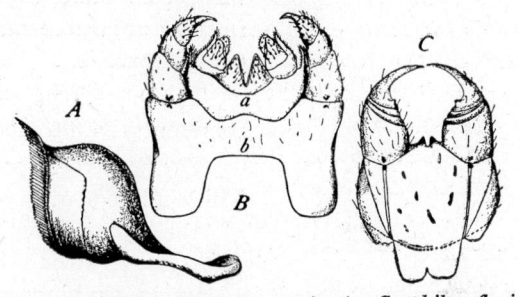

Fig. 29.—Mouth-parts of a centipede, *Geophilus flavidus*. A, right mandible, greatly enlarged. B, the two pairs of maxillæ, less enlarged; *a*, the united coxæ of the maxillæ; *b*, the united coxæ of the second maxillæ or palpognaths. C, the poison claws or toxicognaths (After Latzel)

when present are simple ocelli; but there may be a group of ocelli on each side of the head. Figure 28 represents a typical centipede.

Sub-Class NOTOSTIGMA

Scutigera and its Allies

In the genus *Scutigera* and its allies, which constitute the sub-class Notostigma, there is a very distinctive type of respiratory organs. There is a single spiracle in each of the spiracle-bearing segments, which are seven in number. These spiracles open in the middle line of the back, each in the hind margin of one of the seven prominent terga of the body-region. Each spiracle leads into a short sac from which the tracheal tubes extend into the pericardial blood-sinus.

There are fifteen leg-bearing segments in the body region; but the terga of these segments are reduced to seven by fusion and suppression.

The eyes differ from those of all other members of the old group Myriapoda in being compound, the ommatidia resembling in structure the ommatidia of the compound eyes of insects.

The following species is the most familiar representative of the Notostigma.

The house centipede, *Scutigera forceps*.—This centipede attracts attention on account of the great length of its appendages (Fig. 30), and the fact that it is often seen, in the regions where it is common, running on the walls of rooms in dwelling houses, where it hunts for flies and other insects. It prefers damp situations; in houses it is most frequently found in cellars, bathrooms, and closets. Sometimes it becomes very abundant in conservatories, living among the stored pots and about the heating pipes. It is much more common in the South than in the North.

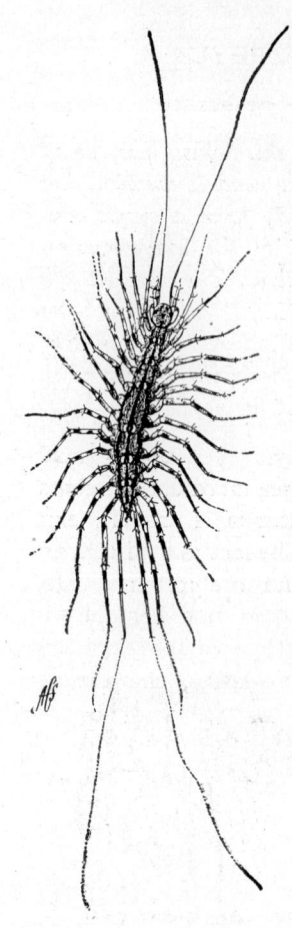

Fig. 30.—*Scutigera forceps*.

The body of the adult measures an inch or a little more in length. It is difficult to obtain perfect specimens, as they shed their legs when seized.

Class SYMPHYLA

The Symphylids

The members of this class are small arthropods in which the head is distinct, and the segments of the body form a single continuous region. Most of the body-segments bear a single pair of legs. The antennæ are very long and many-jointed. The head bears a Y-shaped epicranial suture, as in insects. The opening of the reproductive organs is in the third segment behind the head.

The class Symphyla includes a small number of many-legged arthropods which exhibit striking affinities with insects, and especially with the Thysanura. The body is centipede-like in form (Fig. 31). The head is distinct, and is not bent down as it is in the diplopods and pauropods; it is shaped as in Thysanura and bears a Y-shaped epicranial suture. The body-region bears fifteen terga, which are distinct, not grouped in couples as in the two preceding classes; and there are eleven or twelve pairs of legs.

The antennæ are long and vary greatly in the number of the segments. There are no eyes. The mandibles, the "maxillulæ" (paragnatha), the maxillæ, and the second maxillæ or labium are present.

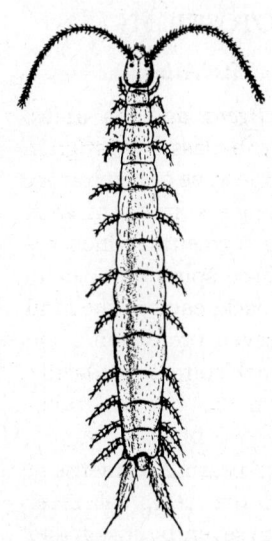

Fig. 31.—*Scolopendrella* (After Latzel).

Fig. 32.—Mouth-parts of *Scolopendrella* seen from below; *md*, mandible; *mx*, maxillæ; *s*, stipes; *p*, palpus; *l*, second maxillæ or labium. The mandible on the right side of the figure is omitted (After Hansen).

The *mandibles* (Fig. 32, *md*) are two-jointed; the *maxillulæ* (Fig. 33, *m*) are small, not segmented, and are attached to a median lobe or *hypopharynx* (Fig. 33, *h*); they are hidden when the mouth-parts are viewed from below as represented in Figure 32; the *maxillæ* (Fig.

32, *mx*) resemble in a striking degree the maxillæ of insects, consisting of a long stipes, (*s*), which bears a minute palpus, (*p*), and an outer and inner lobe; the *second maxillæ* or *labium* (Fig. 32, *l*) also resembles the corresponding part of the more generalized insects, being composed of a pair of united gnathites.

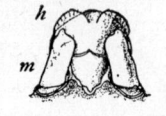

Fig. 33.—The hypopharynx (h) and maxillulæ (m) of *Scolopendrella* (After Hansen).

The legs of the first pair are reduced in size and in the number of their segments. The other legs consist each of five segments; the last segment bears a pair of claws. Excepting the first two pairs of legs, each leg bears on its proximal segment a slender cylindrical process, the *parapōdium* (Fig. 34, p). These parapodia appear to correspond with the styli of the Thysanura.

Fig. 34.—A leg of *Scolopendrella*; *p*, parapodium.

At the caudal end of the body there is a pair of appendages, which are believed to be homologous with the *cerci* of insects (Fig. 35, *c*).

A striking peculiarity of the symphylids is that they possess only a single pair of tracheal tubes, which open by a pair of spiracles, situated in the head beneath the insertion of the antennæ.

The members of this class are of small size, the larger ones measuring about one-fourth inch in length. They live in earth under stones and decaying wood, and in other damp situations. Immature individuals possess fewer body-segments and legs than do adults.

Less than thirty species have been described; but doubtless many more remain to be discovered.

The known species are classed in two genera: *Scolopendrĕlla* and *Scutigerĕlla*. In the former the posterior angles of the terga are produced and angular; while in the latter they are rounded.

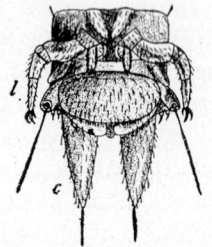

Fig. 35.—The caudal end of the body of *Scolopendrella*; *l*, leg; *c*, cercus (After Latzel).

A monograph of the Symphyla has been published by Hansen ('03).

Class MYRIENTOMATA

Professor Comstock, in the former editions of this book, gave this group of arthropods the rank of a class, coordinate with the other classes of the Arthropoda.

The position and rank of these animals were uncertain at the time the Introduction was written. Indeed, the affinities of the

Protura are not yet clearly understood, but the general opinion among morphologists and systematists is tending more and more to place these tiny creatures in an ordinal group, the *Protura*, among the insects in the class *Hexapoda*. It has seemed best to follow this general trend and we have, therefore, transferred this group to the class *Hexapoda*, order *Protura*, on p. 220.

In commenting on the position of the group in the original edition of the *Introduction*, Professor Comstock gave the following discussion and explanation of his conclusions at that time:

"The first discovered species was described in 1907 by Professor F. Silvestri of Portici, who regarded it as the type of a distinct order of insects, for which he proposed the name Protura. Later Professor Antonio Berlese of Florence described several additional species, and published an extended monograph of the order (Berlese '09 b).

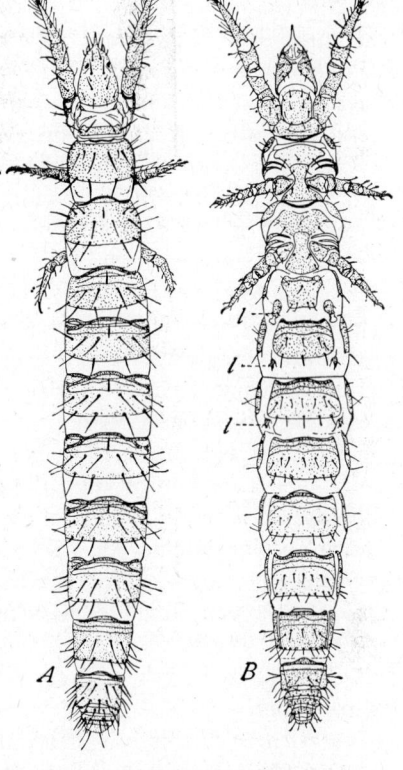

Fig. 36.—*Acerentomon doderoi*: A, dorsal aspect; B, ventral aspect; 1, 1, 1, vestigial abdominal legs (After Berlese).

"Professor Berlese concluded that these arthropods are more closely allied to the Myriapoda and especially to the Pauropoda than they are to the insects, and changed the name of the order, in an arbitrary manner, to Myrientomata.

"It seems clear to me that in either case whether the order is classed among the insects or assigned to some other position it should be known by the name first given to it, that is, the Protūra.

"In the present state of our knowledge of the affinities of the classes of arthropods, it seems best to regard the Protura as representing a separate class, of rank equal to that of the Pauropoda, Symphyla, etc.; and for this class I have adopted the name proposed for the group by Berlese, that is the Myrientomāta."

Class HEXAPODA

The Insects

The members of this class are air-breathing arthropods, with distinct head, thorax, and abdomen. They have one pair of antennae, three pairs of legs, and usually one or two pairs of wings in the adult state. The opening of the reproductive organs is near the caudal end of the body.

We have now reached in our hasty review of the classes of arthropods the class of animals to which this book is chiefly devoted, the Hexapoda,* or Insects, the study of which is termed entomology.

The number of species of insects now known is around 600,000, perhaps more rather than less. The number of species yet undescribed is purely problematical. Probably there are hundreds of thousands of unknown forms distributed over the tropical portions of the earth.

Insects vary greatly in size. Folsom says that some insects are smaller than the largest protozoans, while some are larger than the smallest vertebrates. A beetle, *Dynastes hercules* from Venezuela, which may be 155mm. long, and a Venezuelan grasshopper, *Tropidacris latreillei*, which may attain a length of 166mm., are among the largest insects. Some moths of the genus *Attacus* may have a wing expanse of from 240 to 255mm. while a Brazilian noctuid, *Erebus agrippina*, is said to have a wing expanse of 280mm. On the other hand, certain beetles of the family Trichopterygidae may be but .25mm. in length, and some hymenopterous egg parasites are even smaller.

Insects are essentially terrestrial; and in the struggle for existence they are the most successful of all terrestrial animals, outnumbering both in species and individuals all others together. On the land they abound under the greatest variety of conditions, special forms having been evolved fitted to live in each of the various situations where other animals and plants can live; but insects are not restricted to dry land, for many aquatic forms have been developed.

The aquatic insects are almost entirely restricted to small bodies of fresh water, as streams and ponds, where they exist in great numbers. Larger bodies of fresh water and the seas are nearly destitute of them except at the shores.

---

*Hexăpoda: *hex* (ἕξ), six; *pous* (πούς), a foot.

As might be inferred from a consideration of the immense number of insects, the part they play in the economy of nature is an exceedingly important one. Whether this part is to be considered a beneficial or an injurious one when judged from the human standpoint would be an exceedingly difficult question to determine. For if insects were to be removed from the earth the whole face of nature would be changed.

While the removal of insects from the earth would eliminate many pests that prey on vegetation, would relieve many animals of annoying parasites, and would remove some of the most terrible diseases to which our race is subject, it would result in the destruction of many groups of animals that depend, either directly or indirectly, upon insects for food, and the destruction of many flowering plants that depend upon insects for the fertilization of their blossoms. Truly this world would speedily become a very different one if insects were exterminated.

It may seem idle to consider what would be the result of the total destruction of insects; but it is not wholly so. A careful study of this question will do much to open our eyes to an appreciation of the wonderful "web of life" of which we are a part.

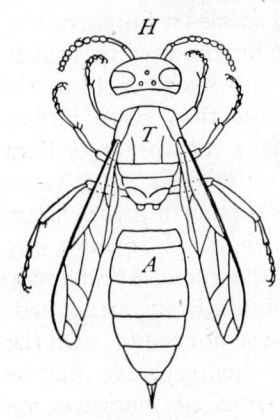

Fig. 37.—Wasp with head, thorax, and abdomen separated.

Most adult insects can be readily distinguished from other arthropods by the form of the body, the segments being grouped into three distinct regions, head, thorax, and abdomen (Fig. 37), by the possession of only three pairs of legs, and in most cases by the presence of wings.

The head bears a single pair of antennæ, the organs of sight, and the mouth-parts. To the thorax, are articulated the organs of locomotion, the legs and the wings when they are present. The abdomen is usually without organs of locomotion but frequently bears other appendages at the caudal end.

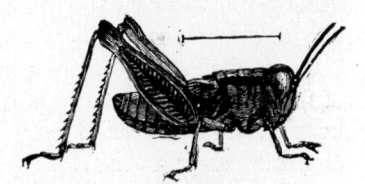

Fig. 38.—Nymph of the red-legged locust.

These characteristics are also possessed by the immature forms of several of the orders of insects; although with these the wings are rudimentary (Fig. 38). But in other orders of insects the immature forms have been greatly modified to adapt them to special modes of life, with the result that they depart widely from the insect type. For example, the larvæ of bees, wasps, flies, and many beetles are legless and more or less worm-like in form (Fig. 4); while the larvæ of butter-flies and moths possess abdominal as well as thoracic legs (Fig. 39).

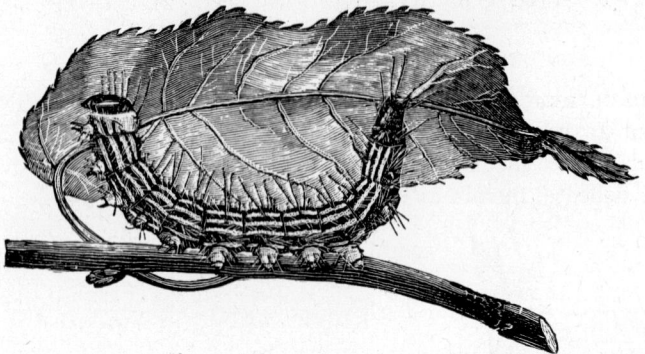

Fig. 39.—A larva of a handmaid moth, *Datana*.

Although the presence of wings in the adult state is characteristic of most insects, there are two orders of insects, the Thysanura and the Collembola, in which wings are absent. These orders represent a branch of the insect series that separated from the main stem before the evolution of wings took place; their wingless condition is, therefore, a primitive one. There are also certain other insects, as the lice and bird-lice, that are wingless. But it is believed that these have descended from winged insects, and have been degraded by their parasitic life; in these cases the wingless condition is an acquired one. Beside these there are many species belonging to orders in which most of the species are winged that have acquired a wingless condition in one or both sexes. Familiar examples of these are the females of the Coccidæ (Fig. 40), and the females of the canker-worm moths. In fact, wingless forms occur in most of the orders of winged insects.

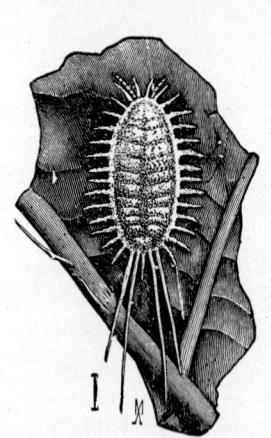

Fig. 40.—A mealy-bug, *Dactylopius*.

As the structure and transformations of insects are described in detail in the following chapters, it is unnecessary to dwell farther on the characteristics of the Hexapoda in this place.

CHAPTER II.

THE EXTERNAL ANATOMY OF INSECTS

I. THE STRUCTURE OF THE BODY-WALL

a. THE THREE LAYERS OF THE BODY-WALL

THREE, more or less distinct, layers can be recognized in the body-wall of an insect: first, the outer, protecting layer, the *cuticula;* second, an intermediate, cellular layer, the *hypodermis;* and third, an inner, delicate, membranous layer, the *basement membrane*. These layers can be distinguished only by a study of carefully prepared, microscopic sections of the body-wall. Figure 41 represents the appearance of such a section. As the outer and inner layers are derived from the hypodermis, this layer will be described first.

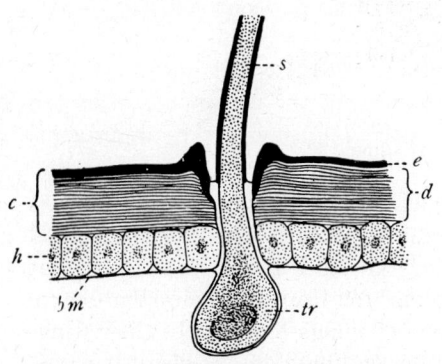

Fig. 41.—A section of the body-wall of an insect: *c*, cuticula; *h*, hypodermis; *bm*, basement membrane; *e*, epidermis, *d*, dermis; *tr*, trichogen; *s*, seta.

The hypodermis.—The active living part of the body-wall consists of a layer of cells, which is termed the *hypŏdermis* (Fig. 41, *h*).

The hypodermis is a portion of one of the germ-layers, the ectoderm. In other words, that portion of the ectoderm which in the course of the development of the insect comes to form a part of the body-wall is termed the hypodermis; while to invaginated portions of the ectoderm other terms are applied, as the epithelial layer of the tracheæ, the epithelial layer of the fore-intestine, and the epithelial layer of the hind-intestine.

The cells of which the hypodermis is composed vary in shape; but they are usually columnar in form, constituting what is known to histologists as a columnar epithelium. Sometimes the cells are so flattened that they form a simple pavement epithelium. I know of no case in which the hypodermis consists of more than a single layer of cells; although in wing-buds and buds of other appendages, where the cells are fusiform, and are much crowded, it appears to be irregularly stratified. This is due to the fact that the nuclei of the different cells are at different levels.

The Trichogens.—Certain of the hypodermal cells become highly specialized and produce hollow, hair-like organs, the setae, with which they remain connected through pores in the cuticula. Such a hair-forming cell is termed a *trichogen* (Fig. 41, *tr*); and the pore in the cuticula is termed a *trichopore*.

The cuticula.—Outside of the hypodermis there is a firm layer which protects the body and serves as a support for the internal organs; this is the *cuticula* (Fig. 41, *c*). The cuticula is produced by the hypodermis; the method of its production is discussed on p. 171, where the molting of insects is treated. The cuticula is not destroyed by caustic potash; it is easy, therefore, to separate it from the tissues of the body by boiling or soaking it in an aqueous solution of this substance.

Chitin.—This word was introduced into entomology by Odier in 1823 for the colorless, flexible covering of the arthropods after the integument had been boiled in caustic potash and the albuminous, oily, coloring and mineral substances had been removed thereby. By a not unusual turn in the use of words, chitin has come to mean, as stated by Newport (1836-1839): "The peculiar substance that constitutes the hard portion of the dermo-skeleton [in insects]." From 1870 and onward the words *chitinize* and *chitinization* have come to mean the hardening of the cuticula by the incorporation of chitin; and they are used with that meaning throughout this work. (For references to chitin, see p. 1010).

Rigid and flexible cuticula.—When freshly formed by the hypodermis, the cuticula is flexible and elastic, and certain portions of it, as at the nodes of the body and of the appendages, remain so. But the greater part of the cuticula, especially in adult insects, usually becomes firm and inelastic; this is due to a change in which the hardening substance is developed within or upon the original soft cuticula. What the exact nature of this change is or how it is produced is not known. This change is usually spoken of as chitinization; and the hard parts of the cuticula are then said to be chitinized, and the soft parts, as at the nodes, non-chitinized. The hardened or chitinized cuticula is rigid and inelastic while the soft or non-chitinized cuticula is flexible and elastic. The elasticity of the soft cuticula is well shown by the stretching of the body wall after a molt. It is also strikingly shown by the expansion of the soft, intersegmental cuticula to accommodate the growing eggs, as in the queens of **Termites**.

The formation of chitin is not restricted to the hypodermis, but is a property of the invaginated portions of the ectoderm; the fore-intestine, the hind-intestine, and the tracheæ are all lined with a cuticular layer, which is continuous with the cuticula of the body-wall and is chitinized. The most marked case of internal formation of chitin is the development of large and powerful teeth in the proventriculus of many insects.

The epidermis and the dermis.—Two quite distinct parts of the cuticula are recognized by recent writers; these are distinguished as the *epidermis* and the *dermis* respectively.

The epidermis is the external portion; in it are located all of the cuticular pigments; and from it are formed all scales, hairs, and other surface structures. It is designated by some writers as the *primary cuticula* (Fig. 41, *e*).

The dermis is situated beneath the epidermis. It is formed in layers, which give sections of the cuticula the well-known laminate appearance. It is sometimes termed the *secondary cuticula* (Fig. 41, *d*).

The basement membrane.—The inner ends of the hypodermal cells are bounded by a more or less distinct membrane; this is termed the *basement membrane* (Fig. 41, *bm*). The basement membrane is most easily seen in those places where the inner ends of the hypodermal cells are much smaller than the outer ends; here it is a continuous sheet connecting the tips of the hypodermal cells.

b. THE EXTERNAL APOPHYSES OF THE CUTICULA

The outer surface of the cuticula bears a wonderful variety of projections. These, however, can be grouped under two heads: first, those that form an integral part of the cuticula; and second, those that are connected with the cuticula by a joint. Those that form an integral part of the cuticula are termed *apophyses;* those that are connected by a joint are termed *appendages of the cuticula*.

The cuticular nodules.—The most frequently occurring outgrowths of the cuticula are small, more or less conical nodules. These vary greatly in size, form, and distribution over the surface of the body in different species of insects, and are frequently of taxonomic value.

The fixed hairs.—On the wings of some insects, as the Trichoptera and certain of the Lepidoptera, there are in addition to the more obvious setæ and scales many very small, hair-like structures, which differ from setæ in being directly continuous with the cuticula, and not connected with it by a joint; these are termed the *fixed hairs*, or *aculeæ*. The mode of origin and development of the fixed hairs has not been studied.

The spines.—The term spine has been used loosely by writers on entomology. Frequently large setæ are termed spines. In this work such setæ are called spine-like setæ; and the term spine is applied only to outgrowths of the cuticula that are not separated from it by a joint. Spines differ also from spine-like setæ in being produced by undifferentiated hypodermal cells and are usually if not always of multicellular origin, while each seta is produced by a single trichogen cell. The accompanying diagram (Fig. 42) illustrates this difference.

c. THE APPENDAGES OF THE CUTICULA

Under this head are included those outgrowths of the cuticula that are connected with it by a joint. Of these there are two quite distinct types represented by the spurs and the setæ respectively.

The spurs.—There exist upon the legs of many insects appendages which on account of their form and position have been termed spurs. Spurs resemble the true spines described above and differ from setæ in being of multicellular origin; they differ from spines in being appendages, that is, in being connected with the body-wall by a joint.

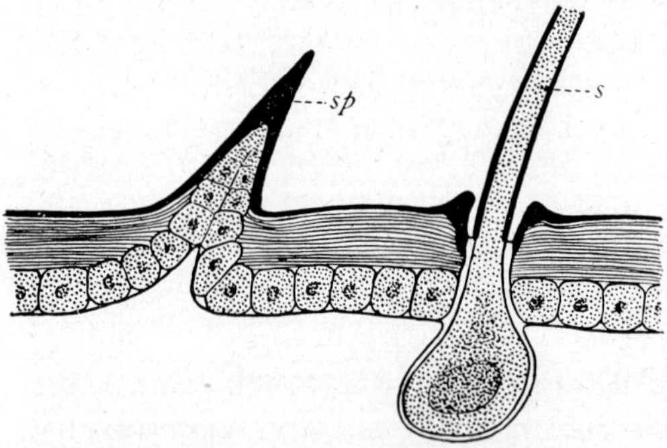

The setæ.—The setæ are what are commonly called the hairs of insects. Each seta (Fig. 42, *s*) is an appendage of the body-wall, which arises from a cup-like cavity in the cuticula, the *alvēolus*, situated at the outer end of a perforation of the cuticula, the *trĭchopore;* and each seta is united at its base with the wall of the trichopore by a ring of thin membrane, the *articular membrane* of the seta.

Fig. 42.—Diagram illustrating the difference between a spine (sp) and a seta (s).

The setæ are hollow; each is the product of a single hypodermal cell, a trichogen (Fig. 42), and is an extension of the epidermal layer of the cuticula.

In addition to the trichogen there may be a gland-cell opening into the seta, thus forming a glandular hair, or a nerve may extend to the seta, forming a sense-hair; each of these types is discussed later.

The most common type of seta is bristle-like in form; familiar examples of this type are the hairs of many larvæ. But numerous modifications of this form exist. Frequently the setæ are stout and firm, such are the *spine-like setæ;* others are furnished with lateral prolongations, these are the *plumose hairs;* and still others are flat, wide, and comparatively short, examples of this form are the *scales* of the Lepidoptera and of many other insects.

The taxonomic value of setæ.—In many cases the form of the setæ and in others their arrangement on the cuticula afford useful characteristics for the classification of insects. Thus the scale-like form of the setæ on the wing-veins of mosquitoes serves to distinguish these insects from closely allied midges; and the clothing of scales is one of the most striking of the characteristics of the Lepidoptera.

The arrangement of the setæ upon the cuticula, in some cases at least, is a very definite one. Thus Dyar ('94) was able to work out a classification of lepidopterous larvæ by a study of the setæ with which the body is clothed.

A classification of setæ.—If only their function be considered the hairs or setæ of insects can be grouped in the three following classes:

(1) *The clothing hairs.*—Under this head are grouped those hairs and scales whose primary function appears to be merely the protection of the body or of its appendages. So far as is known, such hairs contain only a prolongation of the trichogen cell that produced them. It should be stated, however, that this group is merely a provisional one; for as yet comparatively little is known regarding the relation of these hairs to the activities of the insects possessing them.

In some cases the clothing hairs have a secondary function. Thus the highly specialized overlapping scales of the wings of Lepidoptera, which are modified setæ, may serve to strengthen the wings; and the markings of insects are due almost entirely to hairs and scales. The fringes on the wings of many insects doubtless aid in flight, and the fringes on the legs of certain aquatic insects also aid in locomotion.

(2) *The glandular hairs.*—Under this head are grouped those hairs that serve as the outlets of gland cells. They are discussed in the next chapter, under the head of hypodermal glands.

(3) *The sense-hairs*—In many case a seta, more or less modified in form, constitutes a part of a sense-organ, either of touch, taste, or smell: examples of these are discussed in the next chapter.

d. THE SEGMENTATION OF THE BODY

The cuticular layer of the body-wall, being more or less rigid, forms an external skeleton; but this skeleton is flexible along certain transverse lines, thus admitting of the movements of the body, and producing the jointed appearance characteristic of insects and of other arthropods.

An examination of a longitudinal section of the body-wall shows that it is a continuous layer and that the apparent segmentation is due to infoldings of it (Fig. 43).

The body-segments, somites, or metameres.—Each section of the body between two of the infoldings described above is termed a *body-segment*, or *sōmite*, or *mĕtamere*.

Fig. 43.—Diagram of a longitudinal section of the body-wall of an insect.

The transverse conjunctivæ.—The infolded portion of the body-wall connecting two segments is termed a *conjunctiva*. These conjunctivæ may be distinguished from others described later as the *transverse conjunctivæ*.

The conjunctivæ are less densely chitinized than the other portions of the cuticula; their flexibility is due to this fact, rather than to a comparative thinness as has been commonly described.

e. THE SEGMENTATION OF THE APPENDAGES

The segmentation of the legs and of certain other appendages is produced in the same way as that of the body. At each node of an appendage there is an infolded, flexible portion of the wall of the appendage, a conjunctiva, which renders possible the movements of the appendage.

f. THE DIVISIONS OF A BODY-SEGMENT

In many larvæ, the cuticula of a large part of the body-wall is of the non-chitinized type; in this case the wall of a segment may form a ring which is not divided into parts. But in most nymphs, naiads, and adult insects, there are several densely chitinized parts in the wall of each segment; this enables us to separate it into well-defined portions.

The tergum, the pleura, and the sternum.—The larger divisions of a segment that are commonly recognized are a dorsal division, the *tergum;* two lateral divisions, one on each side of the body, the *pleura;* and a ventral division, the *sternum.*

Each of these divisions may include several definite areas of chitinization. In this case the sclerites of the tergum are referred to collectively as the *tergites*, those of each pleurum, as the *pleurites*, and those constituting the sternum, as the *sternites.*

The division of a segment into a tergum, two pleura, and a sternum are most easily seen in the wing-bearing segments, but it can be recognized also in the prothorax of certain generalized insects. This is especially the case in many Orthoptera, as cockroaches and walking-sticks, where the pleura of the prothorax are distinct from the tergum and the sternum. In the abdomen it is evident that correlated with the loss of the abdominal appendages a reduction of the pleura has taken place.

The lateral conjunctivæ.—On each side of each abdominal segment of adults the tergum and the sternum are united by a strip of non-chitinized cuticula; these are the lateral conjunctivæ. Like the transverse conjunctivæ, the lateral ones are more or less infolded.

The sclerites.—Each definite area of chitinization of the cuticula is termed a *sclērite.*

The sutures.—The lines of separation between the sclerites are termed *sutures.* Sutures vary greatly in form; they may be infolded conjunctivæ; or they may be mere lines indicating the place of union between two sclerites. Frequently adjacent sclerites grow together so completely that there is no indication of the suture; in such cases the suture is said to be *obsolete.*

The median sutures.—On the middle line of the tergites and also of the sternites there frequently exist longitudinal sutures. These are termed the *median sutures.* They represent the lines of the closure of the embryo, and are not taken into account in determining the number of the sclerites.

The dorsal median suture has been well-preserved in the head and thorax, as it is the chief line of rupture of the cuticula at the time of molting.

The piliferous tubercles of larvæ.—The setæ of larvæ are usually borne on slightly elevated annular sclerites; these are termed *piliferous tubercles.*

The homologizing of the sclerites.—While it is probable that the more important sclerites of the body in winged insects have been derived from a common winged ancestor and, therefore, can be homologized, many secondary sclerites occur which can not be thus homologized.

g. THE REGIONS OF THE BODY

The segments of the body in an adult insect are grouped into three, more or less well-marked regions: the *head*, the *thorax*, and the *abdomen*. Each of these regions consists of several segments more or less closely united.

The head is the first of these regions; it bears the mouth-parts, the eyes, and the antennæ. The thorax is the second region; it bears the legs and the wings if they are present. The abdomen is the third region; it may bear appendages connected with the organs of reproduction.

II. THE HEAD

The external skeleton of the head of an insect is composed of several sclerites more or less closely united, forming a capsule, which includes a portion of the viscera, and to which are articulated certain appendages.

a. THE CORNEAS OF THE EYES

The external layer of the organs of vision, the corneas of the eyes, is, in each case, a translucent portion of the cuticula. It is a portion of the skeleton of the head, which serves not merely for the admission of light but also to support the more delicate parts of the visual apparatus.

The corneas of the compound eyes.—The compound eyes are the more commonly observed eyes of insects. They are situated one on each side of the head, and are usually conspicuous. Sometimes, as in dragon-flies, they occupy the larger part of the surface of the head.

The compound eyes are easily recognized as eyes; but when one of them is examined with a microscope it is found to present an appearance very different from that of the eyes of higher animals, its surface being divided into a large number of six-sided divisions (Fig. 44); hence the term compound eyes applied to them.

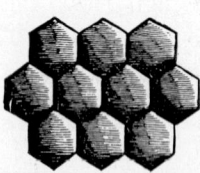

Fig. 44.—Part of a cornea of a compound eye.

A study of the internal structure of this organ has shown that each of these hexagonal divisions is the outer end of a distinct element of the eye. Each of these elements is termed an *ommatĭdium*. The number of ommatidia of which a compound eye is composed varies greatly; there may be not more than fifty, as in certain ants, or there may be many thousand, as in a butterfly or a dragon-fly.

As a rule, the immature stages of insects with a gradual metamorphosis and also those of insects with an incomplete metamorphosis, that is to say nymphs and naiads possess compound eyes. But the larvæ of insects with a complete metamorphosis, do not possess well-developed compound eyes; although there are frequently a few separate ommatidia on each side of the head. These are usually termed ocelli; but the ocelli of larvæ should not be confused with the ocelli of nymphs, naiads, and adults.

The corneas of the ocelli.—In addition to the compound eyes most nymphs, naiads, and adult insects possess other eyes, which are termed *ocelli*. The cornea of each ocellus is usually a more or less nearly circular, convex area, which is not divided into facets. The typical number of ocelli is four; but this number is rarely found. The usual number is three, a median ocellus, which has been derived from a pair of ocelli united, and a distinct pair of ocelli. Frequently the median ocellus is lacking, and less frequently, all of the ocelli have been lost. The position of the ocelli is discussed later.

b. THE AREAS OF THE SURFACE OF THE HEAD

In descriptions of insects it is frequently necessary to refer to the different regions of the surface of the head. Most of these regions were named by the early insect anatomists; and others have been described by more recent writers.

This terminology is really of comparatively little morphological value; for in some cases a named area includes several sclerites, while in others only a portion of a sclerite is included. This is due to the fact that but few of the primitive sclerites of the head have remained distinct, and some of them greatly overshadow others in their development. The terms used, however, are sufficiently accurate to meet the needs of describers of species, and will doubtless continue in use. It is necessary, therefore, that students of entomology become familiar with them.

The best landmark from which to start in a study of the areas of the surface of the head is the *epicranial suture*, the inverted Y-shaped suture on the dorsal part of the head, in the more generalized insects (Fig. 45, *e. su*). Behind the arms of this suture there is a series of *paired* sclerites, which meet on the dorsal wall of the head, the line of union being the stem of the Y, a median suture; and between the arms of the Y and the mouth there are typically three *single* sclerites (Fig. 45, F, C, L). It is with these unpaired sclerites that we will begin our definitions of the areas of the head.

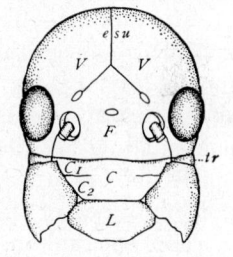

Fig. 45.—Head of a cricket.

The front.—The front is the unpaired sclerite between the arms of the epicranial suture (Fig. 45, F).

In the more generalized insects at least, if not in all, the front bears the median ocellus; and in the Plecoptera, the paired ocelli also. Frequently the suture between the front and the following sclerite, the clypeus, is obsolete; but as it ends on each side in the invagination which forms an anterior arm of the tentorium or endo-skeleton (Fig. 46, *at*), its former position can be inferred, at least in the more generalized insects, even when no other trace of it remains. In Figure 46 this is indicated by a dotted line.

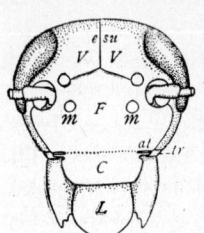

Fig. 46.—Head of a cockroach; *m*, muscle impressions.

The clypeus.—The *clypeus* is the intermediate of the three unpaired sclerites between the epicranial suture and the mouth (fig. 46, *c*). To this part one condyle of the mandible articulates.

Although the clypeus almost always appears to be a single sclerite, except when divided transversely as indicated below, it really consists of a transverse row of three sclerites, one on the median line, and one on each side articulating with the mandible. The median sclerite may be designated the *clypeus proper*, and each lateral sclerite, the *antecoxal piece of the mandible*. Usually there are no indications of the sutures separating the clypeus proper from the antecoxal pieces; but in some insects they are distinct. In the larva of *Corydalus*, the antecoxal pieces are not only distinct but are quite large (Fig. 47, *ac, ac*).

In some insects the clypeus is completely or partly divided by a transverse suture into two parts (Fig. 45). These may be designated as the *first clypeus* and the *second clypeus*, respectively; the first clypeus being the part next the front (Fig. 45, C_1) and the second clypeus being that next the labrum (Fig. 45, C_2).

The suture between the clypeus and the epicranium is termed the *clypeal suture*.

The labrum.—The *lābrum* is the movable flap which constitutes the upper lip of the mouth (Fig. 45, *L*). The labrum is the last of the series of unpaired sclerites between the epicranial suture and the mouth. It has the appearance of an appendage but is really a portion of one of the head segments.

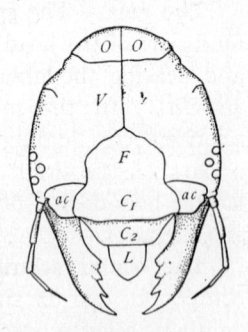

Fig. 47.—Head of a larva of *Corydalus*, dorsal aspect.

The epicranium.—Under the term *epicrānium* are included all of the paired sclerites of the skull, and sometimes also the front. The paired sclerites constitute the sides of the head and that portion of the dorsal surface that is behind the arms of the epicranial suture. The sclerites constituting this region are so closely united that they were regarded as a single piece by Straus-Durckheim (1828), who also included the front in this region, the epicranial suture being obsolete in the May beetle, which he used as a type.

The vertex.—The dorsal portion of the epicranium; or, more specifically, that portion which is next the front and between the compound eyes is known as the *vertex* (Fig. 45, *V, V*). In many insects the vertex bears the paired ocelli. It is not a definite sclerite; but the term vertex is a very useful one and will doubtless be retained.

The occiput.—The hind part of the dorsal surface of the head is the *occiput*. When a distinct sclerite, it is formed from the tergal portion of the united postgenæ described below (Fig. 47, *O, O*).

The genæ.—The *genæ* are the lateral portions of the epicranium. Each gena, in the sense in which the word was used by the older writers, includes a portion of several sclerites. Like vertex, however, the term is a useful one.

The postgenæ.—In many insects each gena is divided by a well-marked suture. This led the writer, in an earlier work ('95), to restrict the term *gena* to the part in front of the suture (Fig. 48, *G*), and to propose the term *postgena* for the part behind the suture (Fig. 48, *Pg*).

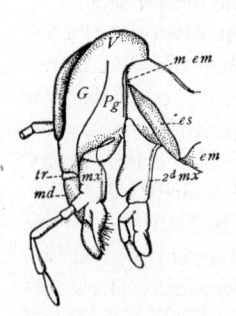

Fig. 48.—Head and neck of a cockroach.

The gula.—The *gula* is a sclerite forming the ventral wall of the hind part of the head in certain orders of insects, and bearing the labium or second maxillæ (Fig. 49, *Gu*). In the more generalized orders, the sclerite corresponding to the gula does not form a part of the skull. The sutures forming the lateral boundaries of the gula are termed the *gular sutures*.

The ocular sclerites.—In many insects each compound eye is situated in the axis of an annular sclerite; these sclerites bearing the compound eyes are the *ocular sclerites* (Fig. 50, *os*).

The antennal sclerites.—In some insects there is at the base of each antenna an annular sclerite; these are the *antennal sclerites* (Fig. 50, *as*). The antennal sclerites are most distinct in the Plecoptera.

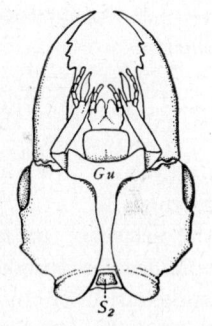

Fig. 49.—Head of *Corydalus*, adult, ventral aspect.

The trochantin of the mandible.—In some insects, as Orthoptera there is a distinct sclerite between each mandible and the gena: this is the *trochantin of the mandible* (Fig. 45, *tr*).

The maxillary pleurites.—In some of the more generalized insects, as certain cockroaches and crickets, it can be seen that each maxilla is articulated at the ventral end of a pair of sclerites, between which is the invagination that forms the posterior arm of the tentorium; these are the *maxillary pleurites;* the posterior member of this pair of sclerites can be seen in the lateral view of the head of a cockroach (Fig. 48, *m. em*).

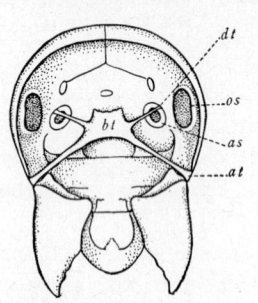

Fig. 50.—Head of a cricket, ental surface of the dorsal wall.

The cervical sclerites.—The *cervical sclerites* are the small sclerites found in the neck of many insects. Of these there are dorsal, lateral, and ventral sclerites. The cervical sclerites were so named by Huxley ('78); recently they have been termed the *intersegmental plates* by Crampton ('17), who considers them to be homologous with sclerites found in the intersegmental regions of the thorax of some generalized insects.

The lateral cervical sclerites have long been known as the *jugular sclerites* (*pièces jugulaires*, Straus Durckheim, 1828).

c. THE APPENDAGES OF THE HEAD

Under this category are classed a pair of jointed appendages termed the *antennæ*, and the organs known collectively as the *mouth-parts*.

The antennæ.—The *antennæ* are a pair of jointed appendages articulated with the head in front of the eyes or between them. The antennæ vary greatly in form; in some insects they are thread-like, consisting of a series of similar segments; in others certain segments are greatly modified. The thread-like form is the more generalized.

In descriptive works names have been given to particular parts of the antennæ, as follows (Fig. 51):

The Scape.—The first or proximal segment of an antenna is called the scape (a). The proximal end of this segment is often subglobose, appearing like a distinct segment; in such cases it is called the bulb (a^1).

The Pĕdicel.—The pedicel is the second segment of an antenna (*b*). In some insects it differs greatly in form from the other segments.

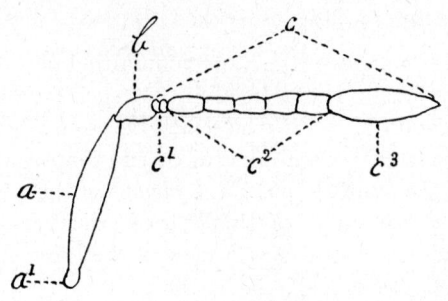

Fig. 51.—Antenna of a chalcis-fly.

The Clăvola.—The term clavola is applied to that part of the antenna distad of the pedicel (*c*); in other words, to all of the antenna except the first and second segments. In some insects certain parts of the clavola are specialized and have received particular names. These are the ring-joints, the funicle, and the club.

The Ring-joints.—In certain insects (*e.g.*, Chalcididæ) the proximal segment or segments of the clavola are much shorter than the succeeding segments; in such cases they have received the name of ring-joints (c^1).

The Club.—In many insects the distal segments of the antennæ are more or less enlarged. In such cases they are termed the club (c^3).

The Fūnicle.—The funicle (c^2) is that part of the clavola between the club and the ring-joints; or, when the latter are not specialized, between the club and the pedicel.

The various forms of antennæ are designated by special terms. The more common of these forms are represented in Fig. 52. They are as follows:

1. *Setāceous* or bristle-like, in which the segments are successively smaller and smaller, the whole organ tapering to a point.

2. *Filiform* or thread-like, in which the segments are of nearly uniform thickness.

3. *Monĭliform* or necklace-form, in which the segments are more or less globose, suggesting a string of beads.

4. *Sĕrrate* or saw-like, in which the segments are triangular and project like the teeth of a saw.

5. *Pĕctinate* or comb-like, in which the segments have long processes on one side, like the teeth of a comb.

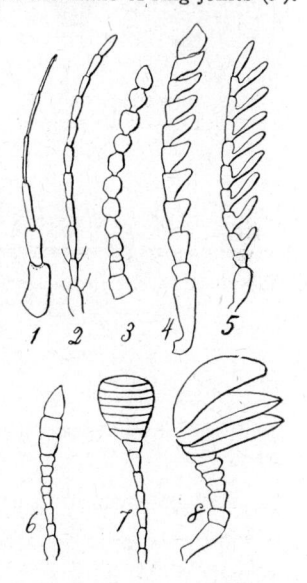

Fig. 52.—Various forms of antennæ.

6. *Clāvate* or club-shaped, in which the segments become gradually broader, so that the whole organ assumes the form of a club.

7. *Căpitate* or with a head, in which the terminal segment or segments form a large knob.

8. *Lămellate* in which the segments that compose the knob are extended on one side into broad plates.

When an antenna is bent abruptly at an angle like a bent knee (Fig. 51) it is said to be *genĭculate*.

The mouth-parts.—The mouth-parts consist typically of an upper lip, *labrum*, an under lip, *labium*, and two pairs of jaws acting horizontally between them. The upper jaws are called the *mandibles;* the lower pair, the *maxillæ*. The maxillæ and labium are each furnished with a pair of feelers, called respectively the *maxillary palpi*, and the *labial palpi*. There may be also within the mouth one or two tongue-like organs, the *epipharynx* and the *hypopharynx*. The mouth-parts of a locust will serve as an example of the typical form of the mouth-parts (Fig. 53).

The mouth-parts enumerated in the preceding paragraph are those commonly recognized in insects; but in certain insects there exist vestiges of a pair of lobes between the mandibles and the maxillæ, these are the paragnatha.

No set of organs in the body of an insect vary in form to a greater degree than do the mouth-parts. Thus with some the mouth is formed for chewing, while with others it is formed for sucking. Among the chewing insects some are predaceous, and have jaws fitted for seizing and tearing their prey; others feed upon vegetable matter, and have jaws for chewing this kind of food. Among the sucking insects the butterfly merely sips the nectar from flowers, while the mosquito needs a powerful instrument for piercing its victim. In this chapter the typical form of the mouth-parts as illustrated by the biting insects is described. The various modifications of it presented by the sucking insects are described later, in the discussions of the characters of those insects.

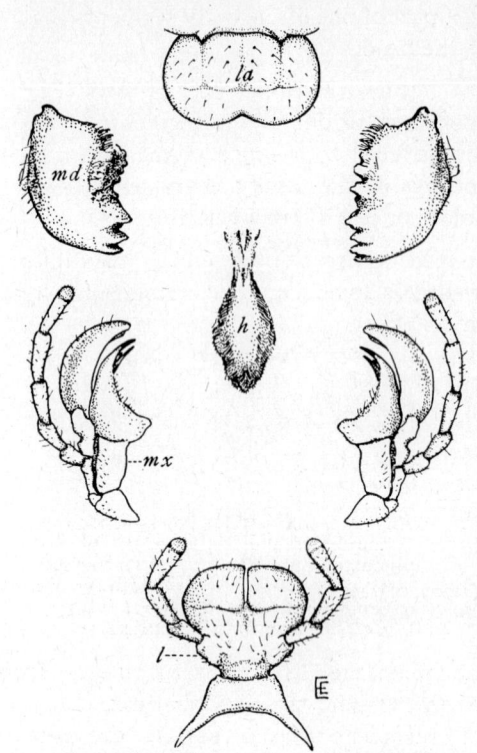

Fig. 53.—Mouth-parts of a locust: *la*, labrum; *md*, mandible; *mx*, maxilla; *h*, hypopharynx; *l*, labium.

The labrum.—The lăbrum or upper lip (Fig. 53), is a more or less flap-like organ above the opening of the mouth. As it is often freely movable, it has the appearance of an appendage of the body; but it is not a true appendage, being a part of one of the body segments that enter into the composition of the head.

The mandibles.—The măndibles are the upper pair of jaws (Fig. 53). They represent the appendages of one of the segments of the head. In most cases they are reduced to a single segment; but in some insects, as in certain beetles of the family Scarabæidæ, each mandible consists of several more or less distinct sclerites.

The parăgnatha.—In some insects there is between the mandibles and the maxillæ a pair of more or less appendage-like organs borne by the hypopharynx. These are the "paraglossæ" of writers on the Thysanura and Collembola and the "superlinguæ" of Folsom ('oo). They were termed the maxillulæ, a diminutive of maxillæ by Hansen ('93), who regards them as homologous with the first maxillæ of the Crustacea. But it has been shown by Crampton ('21) that they are homologous with the paragnatha of Crustacea. In Figure 54, A. represents a ventral view of the hypopharynx, paragnatha, and mandibles of the crustacean *Ligyda;* and B. the same parts of a naiad of a May-fly, *Heptagenia.* Paragnatha have been found in the Thysanura, Dermoptera, Orthoptera, Corrodentia, the naiads of Ephemerida, and the larvæ of Coleoptera.

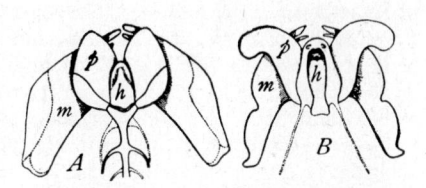

Fig. 54.—A. Posterior (ventral) view of mandibles and hypopharynx of the crustacean *Ligyda; h,* hypopharynx; *p,* paragnatha; *m,* mandibles; B. Same of a nymph of the Mayfly *Heptagenia* (From Crampton).

The Maxillæ.—The *maxillæ* are the second pair of jaws of insects. Like the mandibles they are the appendages of one of the segments of the head.

The maxillæ are much more complicated than the mandibles, each maxilla consisting, when all of the parts are present, of five primary parts and three appendages. The primary parts are the *cardo* or hinge, the *stipes* for footstalk, the *palpifer* or palpus-bearer, the *subgalea* or helmet-bearer, and the *lacinia* or blade. The appendages are the *maxillary palpus* or feeler, the *galea* or superior lobe, and the *digitus* or finger. The maxilla may also bear claw-like or tooth-like projections, spines, bristles, and hairs.

In the following description of the parts of the maxillæ, only very general statements can be made. Not only is there an infinite variation in the form of these parts, but the same part may have a very different outline on the dorsal aspect of the maxilla from what it has on the ventral. Compare Fi .55 and Fig. 56, which represent the two aspects of the maxilla of *Hydrophilus*. Excepting Fig. 56, the figures of maxillæ represent the ventral aspect of this organ.

The *cardo* or hinge (*a*) is the first or proximal part of the maxilla. It is usually more or less triangular in outline, and is the part upon which nearly all of the motions of this organ depend. In many cases, however, it is not the only part directly joined to the body; for frequently muscles extend direct to the subgalea, without passing through the cardo.

The *stipes* or footstalk (*b*) is the part next in order proceeding distad. It is usually triangular, and articulates with the cardo by its base, with the palpifer by its lateral margin, and with the subgalea by its mesal side. In many insects the stipes is united with the subgalea, and the two form the larger portion of the body of the maxilla (Fig. 53). The stipes has no appendages; but the palpifer on the one side, and the subgalea on the other, may become united to the stipes without any trace of suture remaining, and their appendages will then appear to be borne by the stipes. Thus in Fig. 53 it appears to be the stipes that bears the galea, and that receives muscles from the body.

The *pălpifer* or palpus-bearer (*c*) is situated upon the lateral (outer) side of the stipes; it does not, however, extend to the base of this organ, and frequently projects distad beyond it. It is often much more developed on the dorsal side of the maxilla than on the ventral (Figs. 55 and 56). It can be readily distinguished when it is distinct by the insertion upon it of the appendage which gives to it its name.

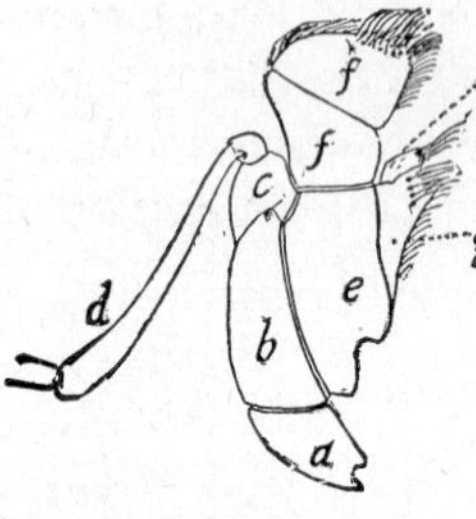

Fig. 55.—Ventral aspect of a maxilla of *Hydrophilus*.

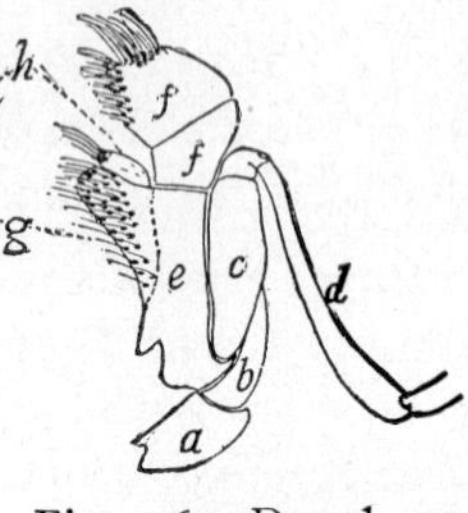

Fig. 56.—Dorsal aspect of a maxilla of *Hydrophilus*.

The *maxillary palpus* or feeler (*d*) is the most conspicuous of the appendages of the maxilla. It is an organ composed of from one to six freely movable segments, and is articulated to the palpifer on the latero-distal angle of the body of the maxilla.

The *subgālea* or helmet-bearer (*e*) when developed as a distinct sclerite is most easily distinguished as the one that bears the galea. It bounds the stipes more or less completely on its mesal (inner) side, and is often directly connected with the body by muscles. In many Coleoptera it is closely united to the lacinia; this gives the lacinia the appearance of bearing the galea, and of being connected with the body (Fig. 56). In several orders the subgalea is united to the stipes; consequently in these orders the stipes appears to bear the galea, and to be joined directly to the body if any part besides the cardo is so connected.

The *gālea* or helmet (*f*) is the second in prominence of the appendages of the maxilla. It consists of one or two segments, and is joined to the maxilla mesad of the palpus. The galea varies greatly in form: it is often more or less flattened, with the distal segment concave, and overlapping the lacinia like a hood. It was this form that suggested the name galea or helmet. In other cases the galea resembles a palpus in form (Fig. 57). The galea is also known as the *outer lobe*, the *upper lobe*, or the *superior lobe*.

The *lacinia* or blade (*g*) is borne on the mesal (inner) margin of the subgalea. It is the cutting or chewing part of the maxilla, and is often furnished with teeth and spines. The lacinia is also known as the *inner lobe*, or the *inferior lobe*.

The *dĭgitus* or finger (*h*) is a small appendage sometimes borne by the lacinia at its distal end. In the Cicindelidæ it is in the form of an articulated claw (Fig. 57); but in certain other beetles it is more obviously one of the segments of the maxilla (Figs. 55 and 56).

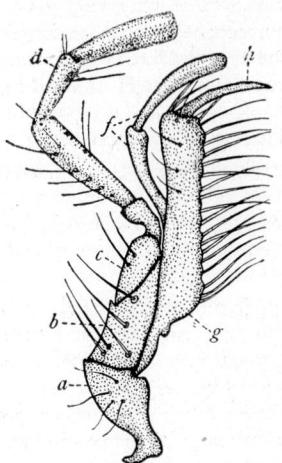

Fig. 57.—Maxilla of *Cicindela*.

The labium or second maxillæ.—The *lābium* or under lip (Fig. 53), is attached to the cephalic border of the gula, and is the most ventral of the mouth-parts. It appears to be a single organ, although sometimes cleft at its distal extremity; it is, however, composed of a pair of appendages grown together on the middle line of the body. In the Crustacea the parts corresponding to the labium of insects consists of two distinct organs, resembling the maxillæ; and in the embryos of insects the labium arises as a pair of appendages.

In naming the parts of the labium, entomologists have usually taken some form of it in which the two parts are completely grown together, that is, one which is not cleft on the middle line (Fig. 58). I will first describe such a labium, and later one in which the division into two parts is carried as far as we find it in insects.

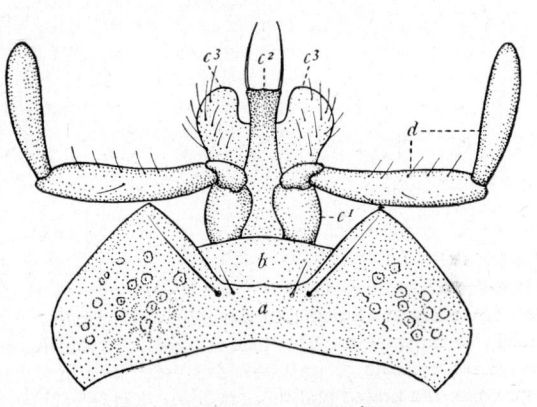

Fig. 58.—Labium of *Harpalus*.

The labium is usually described as consisting of three principal parts and a pair of appendages. The principal parts are the *submentum*, the *mentum*, and the *ligula;* the appendages are the *labial palpi*.

The *submentum*. The basal part of the labium consists of two transverse sclerites; the proximal one, which is attached to the cephalic border of the gula, is the *submentum* (*a*). This is often the most prominent part of the body of the labium.

The *mentum* is the more distal of the two primary parts of the labium (*b*). It is articulated to the cephalic border of the submentum, and is often so slightly developed that it is concealed by the submentum.

The *ligula* includes the remaining parts of the labium except the labial palpi. It is a compound organ; but in the higher insects the sutures between the different sclerites of which it is composed are usually obsolete. Three parts, however, are commonly distinguished (Fig. 58), a central part, often greatly prolonged, the *glossa* (c^2) and two parts, usually small membranous projections, one on each side of the base of the glossa, the *paraglossæ* (c^3). Sometimes, however, the paraglossæ are large, exceeding the glossa in size.

The *labial palpi*. From the base of the ligula arise a pair of appendages, the *labial palpi*. Each labial palpus consists of from one to four freely movable segments.

In the forms of the labium just described, the correspondence of its parts to the parts of the maxillæ is not easily seen; but this is much more evident in the labium of some of the lower insects, as for example a cockroach (Fig. 59). Here the organ is very deeply cleft; only the submentum and mentum remain united on the median line; while the ligula consists of two distinct maxilla-like parts. It is easy in this case to trace the correspondence referred to above. Each lateral half of the submentum corresponds to the cardo of a maxilla; each half of the mentum, to the stipes; while the remaining parts of a maxilla are represented by each half of the ligula, as follows: near the base of the ligula there is a part (c^1) which bears the labial palpus; this appears in the figure like a basal segment of the palpus; but in many insects it is easily seen that it is undoubtedly one of the primary parts of the organ; it has been named the *palpiger*, and is the homologue of the palpifer of a maxilla. The trunk of each half of the ligula is formed by a large sclerite (c^4); this evidently corresponds to the subgalea. At the distal extremity of this subgalea of the labium there are two appendages. The lateral one of these (c^3) is the *paraglossa*, and obviously corresponds to the galea. The mesal one (c^2) corresponds to the lacinia or inner lobe. This part is probably wanting in those insects in which the glossa consists of an undivided part; and in this case the glossa probably represents the united and more or less elongated subgaleæ.

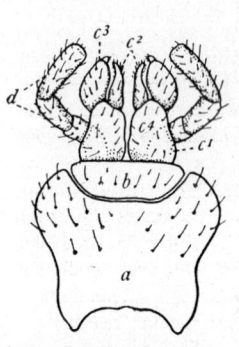

Fig. 59.—Labium of a cockroach.

The epipharynx.—In some insects there is borne on the ental surface of the labrum, within the cavity of the mouth, an unpaired fold, which is membranous and more or less chitinized; this is the *epipharynx*.

The hypopharynx.—The *hypophărynx* is usually a tongue-like organ borne on the floor of the mouth cavity. This more simple form of it is well-shown in the Orthoptera (Fig. 53). To the hypopharnyx are articulated the paragnatha when they are present. The hypopharynx is termed the *lingua* by some writers.

d. THE SEGMENTS OF THE HEAD

The determination of the number of segments in the head of an insect is a problem that has been much discussed since the early days of entomology. The first important step towards its solution was made by Savigny (1816), who suggested that the movable appendages of the head were homodyanmous with legs. This conclusion has been accepted by all; and as each segment in the body of an insect bears only a single pair of appendages, there are at least four segments in the head; *i. e.*, the antennal, the mandibular, the maxillary, and the second maxillary or labial.

In more recent times workers on the embryology of insects have demonstrated the presence of two additional segments. First, there has been found in the embryos of many insects a pair of evanescent appendages situated between the antennæ and the mandibles. These evidently correspond to the second antennæ of Crustacea, and indicate the presence of a second antennal segment in the head of an insect. This conclusion is confirmed by a study of the development of the nervous system. And in the Thysanura and Collembola vestiges of the second antennæ persist in the adults of certain members of these orders.

Second, as the compound eyes are borne on movable stalks in certain Crustacea, it was held by Milne-Edwards that they represent another pair of appendages; but this view has not been generally accepted. It is not necessary, however, to discuss whether the eyes represent appendages or not; the existence of an ocular segment has been demonstrated by a study of the development of the nervous system.

It has been shown that the brain of an insect is formed from three pairs of primary ganglia, which correspond to the three principal divisions of the brain, the *protecerebrum*, the *deutocerebrum*, and the *tritocerebrum*. And it has also been shown that the protocerebrum innervates the compound eyes and ocelli; the deutocerebrum, the antennæ; and the tritocerebrum, the labrum. This demonstrates the existence of three premandibular segments: an ocular segment or protocerebral segment, without appendages, unless the compound eyes represent them; an antennal or deutocerebral segment, bearing antennæ; and a second antennal or tritocerebral segment, of which the labrum is a part, and to which the evanescent appendages between the antennæ and the mandibles doubtless belong. As Viallanes has shown that the tritocerebrum of Crustacea innervates the second antennæ, we are warranted in considering the tritocerebral segment of insects to be the second antennal segment.

Folsom ('00) in his work on the development of the mouth-parts of *Anurida* described a pair of primary ganglia which he believed indicated the presence of a segment between the mandibular and maxillary segments. He named the appendages of this segment the *superlinguæ;* they are the paragnatha described above.

The existence of the supposed ganglia indicating the presence of a superlingual segment has not been confirmed by other investigators and is no longer maintained by Folsom.

The subœsophageal ganglion is formed by the union of three pairs of primitive ganglia, pertaining respectively to the mandibular, the maxillary, and the labial segments of the embryo.

LIST OF THE SEGMENTS OF THE HEAD

First, ocular, or protocerebral.
Second, antennal, or deutocerebral.
Third, second antennal, or tritocerebral.
Fourth, mandibular.
Fifth, maxillary.
Sixth, labial, or second maxillary.

III. THE THORAX

a. THE SEGMENTS OF THE THORAX

The prothorax, the mesothorax, and the metathorax.—The thorax is the second or intermediate region of the body; it is the region that in nymphs, naiads, and adults bears the organs of locomotion, the legs, and the wings when they are present. This region is composed of three of the body-segments more or less firmly joined together; the segments are most readily distinguished by the fact that each bears a pair of legs. In winged insects, the wings are borne by the second and third segments. The first segment of the thorax, the one next the head, is named the *prothorax;* the second thoracic segment is the *mesothorax;* and the third, the *metathorax.*

The simplest form of the thorax in adult insects occurs in the Apterygota (the Thysanura and the Collembola) where although the segments differ in size and proportions, they are distinct and quite similar (Fig. 60).

In the Pterygota, or winged insects, the prothorax is either free or closely united to the mesothorax; in many cases it is greatly reduced in size; it bears the first pair of legs. The mesothorax and the metathorax are more or less closely united, forming a box, which bears the wings and the second and third pairs of legs. This union of these two segments is often so close that it is very difficult to distinguish their limits. Sometimes the matter is farther complicated by a union with the thorax of a part or of the whole of the first

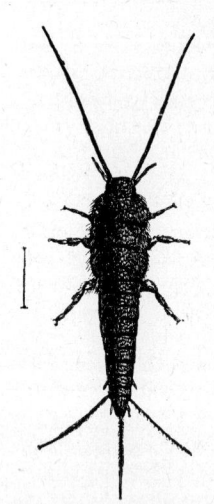

Fig. 60.—*Lepisma saccharina* (After Lubbock).

abdominal segment. In the Acridiidæ, for example, the sternum of the first abdominal segment forms a part of the intermediate region of the body, and in the Hymenoptera the entire first abdominal segment pertains to this region.

The alitrunk.—When, as in the Hymenoptera, the intermediate region of the body includes more than the three true thoracic segments it is designated the *ălitrunk*.

The propodeum or the median segment.—When the alitrunk consists of four segments the abdominal segment that forms a part of it is termed the *propōdeum* or the *median segment*. In such cases the true second abdominal segment is termed the first.

b. THE SCLERITES OF A THORACIC SEGMENT

The parts of the thorax most generally recognized by entomologists were described nearly a century ago by Audouin (1824); some additional parts not observed by Audouin have been described in recent times, by the writer ('02), Verhoeff ('03), Crampton ('09), and Snodgrass ('09, '10 a, and '10 b). The following account is based on all of these works.

In designating the parts of the thorax the prefixes *pro*, *meso*, and *meta* are used for designating the three thoracic segments or corresponding parts of them; and the prefixes *pre* and *post* are used to designate parts of any one of the segments. Thus the scutum of the prothorax is designated the proscutum; while the term prescutum is applied to the sclerite immediately in front of the scutum in each of the thoracic segments. This system leads to the use of a number of hybrid combinations of Latin and Greek terms, but it is so firmly established that it would not be wise to attempt to change it on this account.

Reference has already been made to the division of a body-segment into a tergum, two pleura, and a sternum; each of these divisions will be considered separately; and as the maximum number of parts are found in the wing-bearing segments, one of these will be taken as an illustration.

The sclerites of a tergum.—In this discussion of the external anatomy of the thorax reference is made only to those parts that form the external covering of this region of the body. The infoldings of the body-wall that constitute the internal skeleton are discussed in the next chapter.

The notum.—In nymphs and in the adults of certain generalized insects the tergum of each wing-bearing segment contains a single chitinized plate; this sclerite is designated the *notum*. The term notum is also applied to the tergal plate of the prothorax and to that of each abdominal segment. The three thoracic nota are designated as the *pronotum*, the *mesonotum*, and the *metanotum* respectively.

The notum of a wing-bearing segment is the part that bears the wings of that segment, even when the tergum contains more than one sclerite. Each wing is attached to two processes of the notum, the *anterior notal process* (Fig. 61, *a n p*) and the *posterior notal process* (Fig. 61, *p n p*); and the posterior angles of the notum are produced into the *axillary cords*, which form the posterior margins of the basal membranes of the wings.

The postnotum or postscutellum.—In the wing-bearing segments of most adult insects the tergum consists of two principal sclerites; the notum already described, and behind this a narrower, transverse sclerite which is commonly known as the *postscutĕllum*, and to which Snodgrass has applied the term *postnotum* (Fig. 61, *P N*).

The divisions of the notum.—In most specialized insects the notum of each wing-bearing segment is more or less distinctly divided by transverse lines or sutures into three parts; these are known as the *prescūtum* (Fig. 61, *Psc*), the *scutum* (Fig. 61, *Sct*), and the scutĕllum (Fig. 61, *Scl*).

It has been commonly held, since the days of Audouin, that the tergum of each thoracic segment is composed typically of four sclerites, the prescutum, scutum, scutellum, and postscutellum. But the investigations of Snodgrass indicate that in its more generalized form the tergum contains a single sclerite, the notum; that the postscutellum or postnotum is a secondary tergal chitinization in the dorsal membrane behind the notum, in more specialized insects; and that the separation of the notum into three parts, the prescutum, scutum, and scutellum, is a still later specialization that has arisen independently in different orders, and does not indicate a division into homologous parts in all orders where it exists.

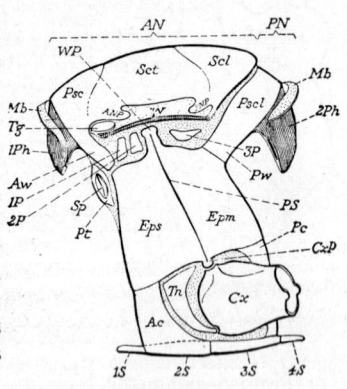

The patagia.—In many of the more specialized Lepidoptera the pronotum is produced on each side into a flat lobe, which in some cases is even constricted at the base so as to become a stalked plate, these lobes are the *patagiā*.

Fig. 61.—Diagram of a generalized thoracic segment (From Snodgrass).

The parapsides.—In some Hymenoptera the scutum of the mesothorax is divided into two parts by the prescutum; these separated halves of the scutum are called the parăpsides (see Fig. 1130A).

The sclerites of the pleura.—In the accompanying figure (Fig. 61) the sclerites of the left pleurum of a wing-bearing segment are represented diagrammatically; these sclerites are the following:

The episternum.—Each pleurum is composed chiefly of two sclerites, which typically occupy a nearly vertical position, but usually are more or less oblique. In most insects the dorsal end of these sclerites extends farther forward than the ventral end, but in the Odonata the reverse may be true. The more anterior in position of these two sclerites is the *episternum* (Fig. 61, *Eps*).

In several of the orders of insects one or more of the episterna are divided by a distinct suture into an upper and a lower part. These two parts have been designated by Crampton ('09) as the *anĕpistĕrnum* and the *katĕpistĕrnum* respectively (Fig. 62).

The epimerum.—The *epimērum* is the more posterior of the two principal sclerites of a pleurum (Fig. 61). It is separated from the episternum by the *pleural suture* (Fig. 61, *PS*) which extends from the *pleural wing process* above (Fig. 61, *Wp*) to the *pleural coxal process* below (Fig. 61, *CxP*).

In some of the orders of insects one or more of the epimera are divided by a distinct suture into an upper and a lower part. These two parts have been designated by Crampton ('09) as the *anepimērum* and the *katepimērum* respectively (Fig. 62).

The preepisternum.—In some of the more generalized insects there is a sclerite situated in front of the episternum; this is the *preepisternum*.

The paraptera.—In many insects there is on each side a small sclerite between the upper end of the episternum and the base of the wing; these have long been known as the *parăptera*. Snodgrass (10 *a*) has shown that there are in some insects two sclerites in this region, which he designates the *episternal paraptera* or *preparaptera* (Fig. 61, *1P* and *2P*); and that one or occasionally two are similarly situated between the epimerum and the base of the wing, the *epimeral paraptera* or *postparaptera* (Fig. 61, *3P*).

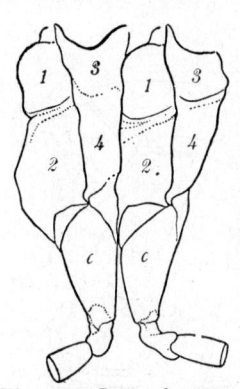

Fig. 62.—Lateral aspect of the meso- and metathorax of *Mantispa rugicollis*; *1, 1*, anepisternum; *2,2*,katepisternum; *3, 3*, anepimerum; *4,4*,katepimerum; *c, c*, coxa.

The spiracles.—The external openings of the respiratory system are termed *spiracles*. Of these there are two pairs in the thorax. The first pair of thoracic spiracles open, typically, one on each side in the transverse conjunctiva between the prothorax and the mesothorax; the second pair open in similar positions between the mesothorax and the metathorax. In some cases the spiracles have migrated either forward or backward upon the adjacent segment. For a discussion of the number and distribution of the spiracles, see the next chapter.

The peritremes.—In many cases a spiracle is surrounded by a circular sclerite; such a sclerite is termed a *pĕritreme*.

The acetabula or coxal cavities.—In some of the more specialized insects, as many beetles for example, the basal segment of the legs is inserted in a distinct cavity; such a cavity is termed an *acetăbulum* or *coxal cavity*. When the epimera of the prothorax extend behind the coxæ and reach the prosternum, the coxal cavities are said to be *closed* (Fig. 63); when the epimera do not extend behind the coxæ to the prosterum, the coxal cavities are described as *open* (Fig. 64).

The sclerites of a sternum.—In the more generalized insects the sternum of a wing-bearing segment may consist of three or four sclerites. These have been designated, beginning with the anterior one, the *presternum* (Fig. 61, *Ps*), the *sternum or eusternum* (Fig. 61, *S*), the *sternellum* (Fig. 61, *Sl*), and the *poststernellum* (Fig. 61, *Psl*).

In the more specialized insects only one of these, the sternum, remains distinctly visible. It is an interesting fact that while in the specialization of the tergum there is an increase in the number of the sclerites in this division of a segment, in the specialization of the sternum there is a reduction.

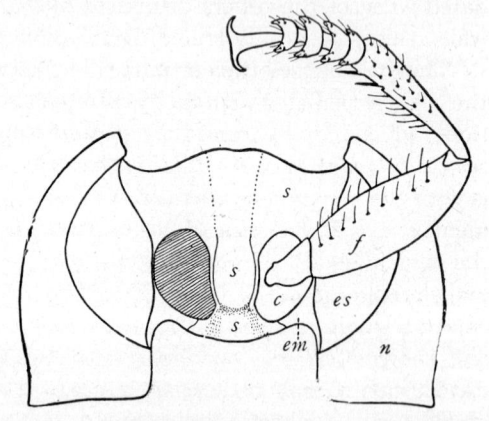

Fig. 63.—Prothorax of *Harpa...is*, ventral aspect; *c*, coxa; *em*, epimerum; *es*, episternum; *f*, femur; *n*, pronotum; *s, s, s,* prosternum.

It is a somewhat unfortunate fact that the term sternum has been used in two senses: first, it is applied to the entire ventral division of a segment; and second, it is applied to one of the sclerites entering into the composition of this division when it consists of more than a single sclerite. To meet this difficulty Snodgrass has proposed that the term *eusternum* be applied to the sclerite that has been known as the sternum; and that the word sternum be used only to designate the entire ventral division of a segment.

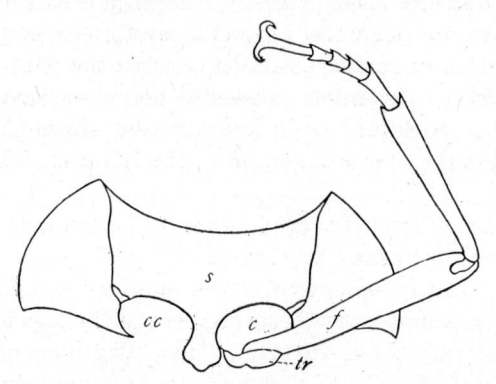

Fig. 64.—Prothorax of *Penthe*; *c*, coxa; *cc*, coxal cavity; *f*, femur; *s*, prosternum; *tr*, trochanter.

c. THE ARTICULAR SCLERITES OF THE APPENDAGES

At the base of each leg and of each wing there are typically several sclerites between the appendage proper and the sclerites of the trunk of the segment; these sclerites, which occupy an intermediate position between the body and its appendage, are termed the *articular sclerites*.

Frequently one or more of the articular sclerites become consolidated with sclerites of the trunk so as to appear to form a part of its wall; this is especially true of those at the base of the legs.

The articular sclerites of the legs.—The proximal segment of the leg, the coxa, articulates with the body by means of two distinct articulations, which may be termed the *pleural articulation of the coxa* and the *ventral articulation of the coxa* respectively. The pleural articulation is with the ventral end of the foot of the lateral apodeme of the segment, *i. e.* with the *pleural coxal process*, which is at the ventral end of the suture between the episternum and the **epimerum** (Fig. 61, *CxP*). The ventral articulation is with a sclerite situated between the coxa and the episternum; this sclerite and others associated with it may be termed the *articular sclerites of the legs*. The articular sclerites of the legs to which distinctive names have been applied are the following:

The trochantin.—The maximum number of articular sclerites of the legs are found in the more generalized insects; in the more specialized insects the number is reduced by a consolidation of some of them with

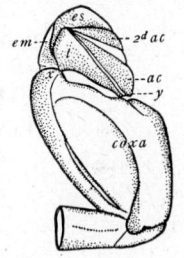

Fig. 65.—The base of a leg of a cockroach.

adjacent parts. The condition found in a cockroach may be taken as typical. In this insect the *trochantin* (Fig. 65, *t*) is a triangular sclerite, the apex of which points towards the middle line of the body, and is near the ventral articulation of the coxa (Fig. 65, *y*). In most specialized insects the trochantin is consolidated with the antecoxal piece, and the combined sclerites, which appear as one, are termed the trochantin.

The antecoxal piece.—Between the trochantin and the episternum there are, in the cockroach studied, two sclerites; the one next the trochantin is the *antecoxal piece*. This is the articular sclerite that articulates directly with the coxa (Fig. 65, *ac*). As stated above, the antecoxal piece is usually consolidated with the trochantin, and the term trochantin is applied to the combined sclerites. Using the term trochantin in this sense, the statement commonly made that the ventral articulation of the coxa is with the trochantin is true.

The second antecoxal piece.—The sclerite situated between the antecoxal piece and the episternum is the *second antecoxal piece* (Fig. 65, $2^d ac$). This is quite distinct in certain generalized insects; but it is usually lacking as a distinct sclerite.

The articular sclerites of the wings.—In the Ephemerida and Odonata the chitinous wing-base is directly continuous with the walls of the thorax. In all other orders there are at the base of each wing several sclerites which enter into the composition of the joint by which the wing is articulated to the thorax; these may be termed collectively the *articular sclerites of the wings*. Beginning with the front edge of this joint and passing backward these sclerites are as follows:

The tegula.—In several orders of insects there is at the base of the costal vein a small, hairy, slightly chitinized pad; this is the *tegula* (Fig. 66, *Tg*). In the more highly specialized orders, the Lepidoptera, the Hymenoptera, and the Diptera, the tegula is largely developed so as to form a scale-like plate overlapping the base of the wing.

The tegulæ of the front wings of Lepidoptera are specially large and are carried by special *tegular plates* of the notum. These, in turn, are supported by special internal *tegular arms* from the bases of the pleural wing-processes (Snodgrass, '09)

The axillaries.—Excepting the tegula, which is at the front edge of the wing-joint, the articular sclerites of the wings have been termed collectively the *axillaries*. Much has been written about these sclerites, and many names have been applied to them. The simplest terminology is that of Snodgrass ('09 and '10 *a*) which I here adopt.

The first axillary.—This sclerite (Fig. 66, 1 Ax) articulates with the anterior notal wing-process and is specially connected with the base of the subcostal vein of the wing. In rare cases it is divided into two.

The second axillary.—The second axillary (Fig. 66, 2 Ax) articulates with the first axillary proximally and usually with the base of the radius distally; it also articulates below with the wing-process of the pleurum, constituting thus a sort of pivotal element.

The third axillary.—The third axillary (Fig. 66, 3 Ax) is interposed between the bases of the anal veins and the fourth axillary when this sclerite is present. When the fourth axillary is absent, as it is in

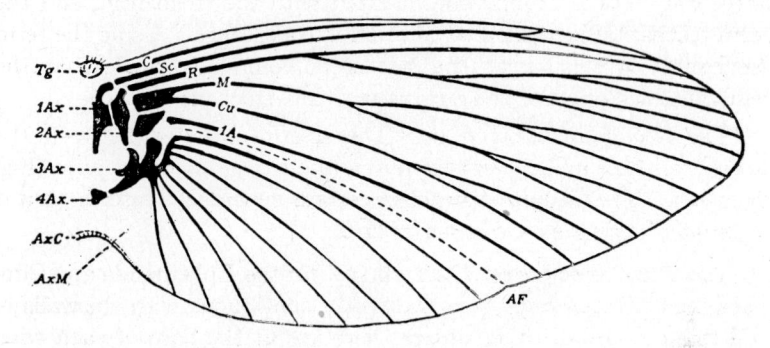

Fig. 66.—Diagram of a generalized wing and its articular sclerites (From Snodgrass).

nearly all insects except Orthoptera and Hymenoptera, the third axillary articulates directly with the posterior notal wing-process.

The fourth axillary.—When this sclerite is present it articulates with the posterior notal wing-process proximally and with the third axillary distally (Fig. 66, 4 Ax). Usually this sclerite is absent; it occurs principally in Orthoptera and Hymenoptera.

The median plates.—The median plates of the wing-joint are not of constant shape and occurrence; when present, these plates are associated with the bases of the media, the cubitus, and the first anal vein when the latter is separated from the other anals. Often one of them is fused with the third axillary and sometimes none of them are present.

d. THE APPENDAGES OF THE THORAX

The appendages of the thorax are the organs of locomotion. They consist of the *legs* and the *wings*. Of the former there are three pairs, a pair borne by each of the three thoracic segments; of the latter there are never more than two pairs, a pair borne by the mesothorax and a pair borne by the metathorax. One or both pairs of wings may be wanting.

The legs.—Each leg consists of the following named parts and their appendages: *coxa, trochanter, femur, tibia,* and *tarsus.*

The coxa.—The coxa is the proximal segment of the leg; it is the one by which the leg is articulated to the body (Fig. 67). The coxa varies much in form, but it is usually a truncated cone or nearly globular. In some insects the coxæ of the third pair of legs are more or less flattened and immovably attached to the metasternum; this is the case in beetles of the family Carabidæ for example. In such cases the coxæ really form a part of the body-wall, and are liable to be mistaken for primary parts of the metathorax instead of the proximal segments of appendages.

In several of the orders of insects the coxa is apparently composed of two, more or less distinct, parallel parts; this is the case, for example, in insects of the trichopterous genus *Neuronia* (Fig. 68, *Cx* and *epm*). But it has been shown by Snodgrass ('09) that the posterior part of the supposed double coxa, the "meron" (Fig. 68, *epm*) is a detached portion of the epimerum.

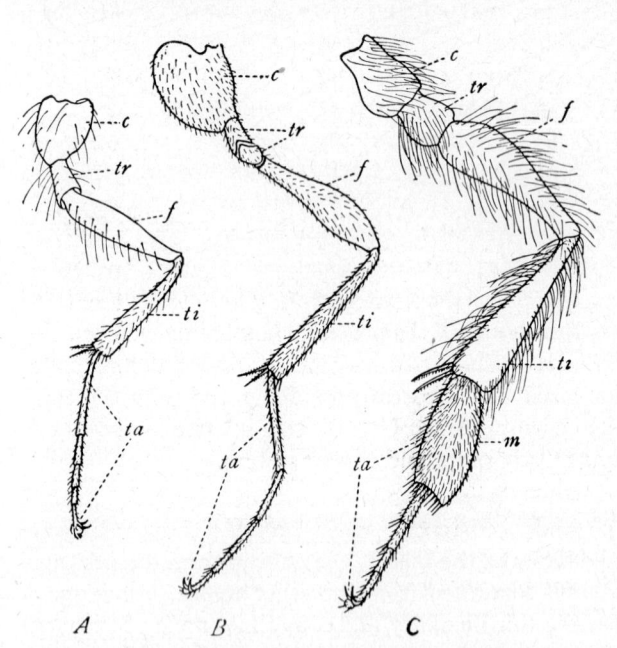

Fig. 67.—Legs of insects: A, wasp; B, ichneumon-fly; C, bee; *c*, coxa; *tr*, trochanter; *f*, femur; *ti*, tibia; *ta*, tarsus; *m*, metatarsus.

The styli.—In certain generalized insects, as *Machilis* of the order

Thysanura, the coxa of each middle and hind leg bears a small appendage, the *stylus* (Fig. 69). The styli are of great interest as they are believed to correspond to one of the two branches of the legs of Crustacea; thus indicating that insects have descended from forms in which the legs were biramous.

In several genera of the Thysanura one or more of the abdominal segments bear each a pair of styli; in *Machilis* they are found on the second to the ninth abdominal segments. These styli are regarded as vestiges of abdominal legs.

The trochanter.—The trochanter is the second part of the leg. It consists usually of a very short, triangular or quadrangular segment, between the coxa and the femur. Sometimes the femur appears to articulate directly with the coxa; and the trochanter to be merely an appendage of the proximal end of the femur (*e. g.* Carabidæ). But the fact is that in these insects, although the femur may touch the coxa, it does not articulate with it; and the organs that pass from the cavity of the coxa to that of the femur must pass through the trochanter. In some Hymenoptera the trochanter consists of two segments (67, *B*).

The femur.—The femur is the third part of the leg; and is usually the largest part. It consists of a single segment.

The tibia.—The tibia is the fourth part of the leg. It consists of a single segment; and is usually a little more slender than the femur, although it often equals or exceeds it in length. In such species as burrow in the ground, the distal extremity is greatly broadened and shaped more or less like a hand. Near the distal end of the tibia there are in most insects one or more spurs, which are much larger than the hairs and spines which arm the leg; these are called the *tibial spurs*, and are much used in classification.

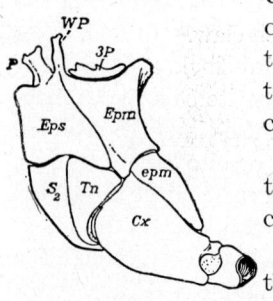

Fig. 68.—Lateral aspect of the mesothorax of *Neuronia* (From Snodgrass).

The tarsus.—The tarsus is the fifth and most distal part of the leg, that which is popularly called the foot. It consists of a series of segments, varying in number from one to six. The most common number of segments in the tarsus is five.

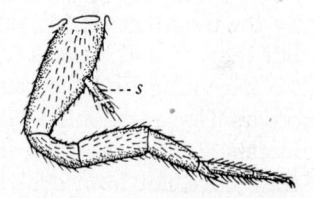

Fig. 69.—A leg of *Machilis*; *s*, stylus.

In many insects, the first segment of the tarsus is much longer, and sometimes much broader, than the other segments. In such cases this segment is frequently designated as the *metatarsus* (Fig. 67, *C*, *m*).

In some insects the claws borne by the distal end of the tarsus are outgrowths of a small terminal portion of the leg, the sixth segment of the tarsus of some authors. This terminal part with its appendages has received the name *prætarsus* (De Meijere '01). As a rule the prætarsus is withdrawn into the fifth segment of the tarsus or is not present as a distinct segment.

On the ventral surface of the segments of the tarsus in many insects are cushion-like structures; these are called *pulvilli*. The cuticula of the pulvilli is traversed by numerous pores which open either at the surface of the cuticula or through hollow hairs, the *tenent-hairs*, and from which exudes an adhesive fluid that enables the insect to walk on the lower surface of objects.

With many insects (*e. g.* most Diptera) the distal segment of the tarsus bears a pair of pulvilli, one beneath each claw. In such cases there is frequently between these pulvilli a third single appendage of similar structure; this is called the *empōdium;* writers on the Orthoptera commonly called the appendage between the claws the *arolium*. In other insects the empodium is bristle-like or altogether wanting.

In many insects the pulvillus of the distal segment of the tarsus is a circular pad projecting between the tarsal claws. In many descriptive works this is referred to as *the* pulvillus, even though the other pulvilli are well-developed. The pulvilli are called the *onychii* by some writers.

The claws borne at the tip of the tarsus are termed the *tarsal claws* or *ungues;* they vary much in form; they are usually two in number, but sometimes there is only one on each tarsus.

The wings.—The wings of insects are typically two pairs of membranous appendages, one pair borne by the mesothorax and one pair by the metathorax; prothoracic wings are unknown in living insects but they existed in certain paleozoic forms.

Excepting in the subclass Apterygota which includes the orders Thysanura and Collembola, wings are usually present in adult insects. Their absence in the Apterygota is due to the fact that they have not been evolved in this division of the class Hexapoda; but when they are absent in adult members of the subclass Pterygota, which includes the other orders of insects, their absence is due to a degradation, which has resulted in their loss.

The loss of wings is often confined to one sex of a species; thus with the canker-worm moths, for example, the females are wingless, while the males have well-developed wings; on the other hand, with the fig-insects, *Blastŏphaga*, the female is winged and the male wingless.

Studies of the development of wings have shown that each wing is a saclike fold of the body-wall; but in the fully developed wing, its saclike nature is not obvious; the upper and lower walls become closely applied throughout the greater part of their extent; and since they become very thin, they present the appearance of a single delicate membrane. Along certain lines, however, the walls remain separate, and are thickened, forming the firmer framework of the wing. These thickened and hollow lines are termed the *veins* of the wing; and their arrangement is described as the *venation* of the wing.

The thin spaces of the wings which are bounded by veins are called *cells*. When a cell is completely surrounded by veins it is said to be *closed;* and when it extends to the margin of the wing it is said to be *open*.

The different types of insect wings.—What may be regarded as the typical form of insect wing is a nearly flat, delicate, membranous appendage of the body, which is stiffened by the so-called wing-veins; but striking modifications of this form exist; and to certain of them distinctive names have been applied, as follows:

In the Coleoptera and in the Dermaptera, the front wings are thickened and serve chiefly to protect the dorsal wall of the body and the membranous hind wings, which are folded beneath them when not in use. Front wings of this type are termed wing-covers or *ĕlytra*.

The front wings of the Hemiptera, which are thickened at the base like elytra, are often designated the *hemĕlytra*.

The thickened fore wings of Orthoptera are termed *tĕgmina* by many writers.

The hind wings of Diptera, which are knobbed, thread-like organs, are termed *haltēres*. The hind wings of the males of the family Coccidæ are also thread-like.

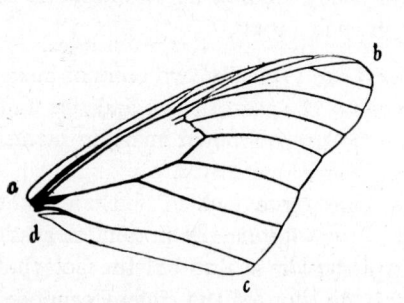

Fig. 70.—Diagram of a wing showing margins and angles.

The reduced front wings of the Strepsiptera are known as the *pseudo-haltēres*.

The margins of wings.—Most insect wings are more or less triangular in outline; they, therefore, present three margins: the *costal margin* or *costa* (Fig. 70, a-b); the *outer margin* (Fig. 70, b-c); and the *inner margin* (Fig. 70, c-d).

The angles of wings.—The angle at the base of the costal margin of a wing is the *humeral angle* (Fig. 70, a); that between the costal margin and the outer margin is the *apex* of the wing (Fig. 70, b);

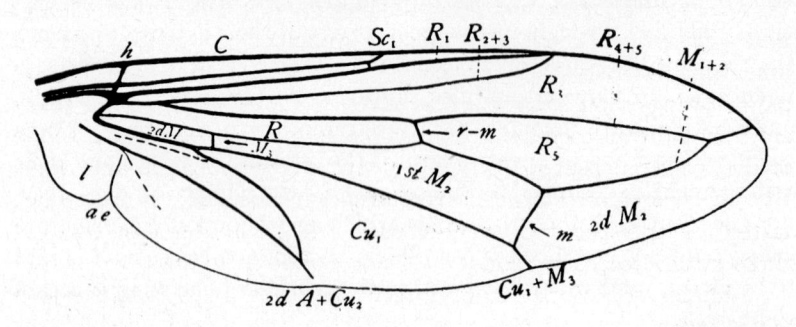

Fig. 71.—Wing of *Conops*; *ae*, axillary excision; *l*, posterior lobe.

and that between the outer margin and the inner margin is the *anal angle* (Fig. 70, c).

The axillary cord.—The posterior margin of the membrane at the base of the wing is usually thickened and corrugated; this cord-like structure is termed the *axillary cord*. The axillary cord normally arises, on each side, from the posterior lateral angle of the notum, and thus serves as a mark for determining the posterior limits of the notum.

The axillary membrane.—The membrane of the wing base is termed the axillary membrane; it extends from the tegula at the base of the costal margin to the axillary cord; in it are found the axillary sclerites.

The alula.—In certain families of the Diptera and of the Coleoptera the axillary membrane is expanded so as to form a lobe or lobes which fold beneath the base of the wing when the wings are closed; this part of the wing is the *alula* or *alulet*. The alulæ are termed the *squamæ* by some writers, and the *calypteres* by others.

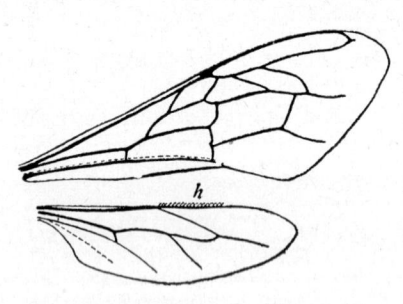

Fig. 72.—Wings of the honeybee; *h*, hamuli.

The axillary excision.—In the wings of most Diptera and in the wings of many other insects there is a notch in the inner margin of the wing near its base (Fig. 71, *ae*), this is the *axillary excision*.

The posterior lobe of the wing.—That part of the wing lying between the axillary excision when it exists, and the axillary membrane is the *posterior lobe* of the wing. The posterior lobe of the wing and an alula are easily differentiated as the alula is margined by the axillary cord.

The methods of uniting the two wings of each side.—It is obvious that a provision for ensuring the synchronous action of the fore and hind wings adds to their efficiency; it is as important that the two pairs of wings should act as a unit as it is that the members of a boat's crew should pull together. In many insects the synchronous action of the wings is ensured by the fore wing overlapping the hind wing. But in other insects special structures have been developed which fasten together the two wings of each side. The different types of these structures have received special names as follows:

The hamuli.—With certain insects the costal margin of the hind wings bears a row of hooks, which fasten into a fold on the inner margin of the fore wings (Fig. 72); these hooks are named the *hămuli*.

The frenulum and the frenulum hook.—In most moths there is a strong spine-like organ or a bunch of bristles borne by the hind wing at the humeral angle (Fig. 73, *f*); this is the *frenulum* or little bridle. As a rule the frenulum of the female consists of several bristles; that of the male, of a single, strong, spine-like organ. In the males of certain moths, where the frenulum is highly developed, there is a membranous fold on the fore wing for receiving the end of the frenulum, this is the *frenulum hook* (Fig. 73, *fh*).

The jugum.—In one family of moths, the Hepialidæ, the posterior lobe of the fore wing is a slender, finger-like organ which is stiffened by a branch

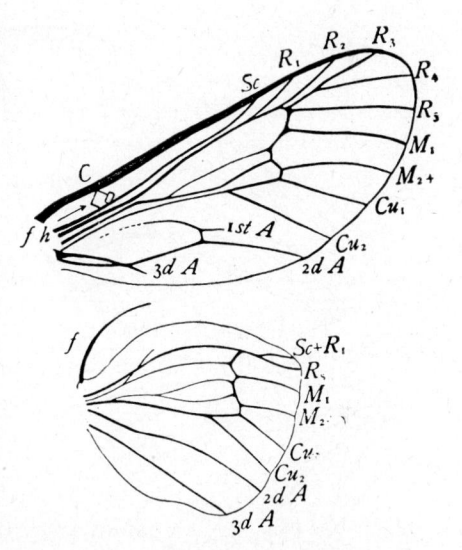

Fig. 73.—Wings of *Thyridopteryx ephemeræ-formis*; *f*, frenulum; *fh*, frenulum hook.

of the third anal vein, and which projects beneath the costal margin of the hind wing. As the greater part of the inner margin of the fore wing overlaps the hind wing, the hind wing is held between the two (Fig. 74). This type of the posterior lobe of the fore wing is termed the *jugum* or yoke. The structure of the jugum is shown in Figure 75.

The fibula.—In several groups of insects an organ has been developed that serves to unite the fore and hind wings, but which functions in a way quite different from that of the jugum. Like the jugum it is found at the base of the fore wing; but unlike the jugum it extends back above the base of the hind wing and is clasped over an elevated part of the hind wing; this organ is the *fibula* or clasp.

In some insects, as in the Trichoptera, the fibula consists only of a specialized posterior lobe of the fore wing; in others, as in the genus *Corydalus* of the order Neuroptera, the proximal part of the fibula is margined by the axillary cord, showing that the axillary membrane enters into the composition of this organ (Fig. 76).

The hypothetical type of the primitive wing-venation.—A careful study of the wings of many insects has shown that the fundamental type of venation is the same in all of the orders of winged insects. But this fact is evident only when the more primitive or generalized members of different orders are compared with each other. In most of the orders of insects the greater number of species have become so modified or specialized as regards the structure of their wings that it is difficult at first to trace out the primitive type.

This agreement in the important features of the venation of the wings of the generalized members of the different orders of insects is still more evident when the wings of nymphs, naiads, and pupæ are studied. It has been demonstrated that in the development of wings of generalized insects the longitudinal wing-veins are formed about preexisting tracheæ. In the development of the wing, these tracheæ grow out into the wing-bud, and later the wing-veins are formed about them.

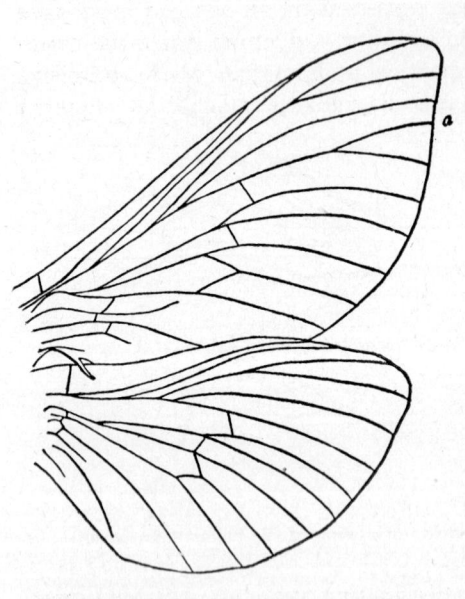

Fig. 74.—Wings of a hepialid, seen from below; *a*, accessory vein.

THE EXTERNAL ANATOMY OF INSECTS 63

The wings of nymphs, naiads, and pupæ are broad at the base, and consequently the tracheæ that precede the wing-veins are not crowded together as are the wing-veins at the base of the wings of

Fig. 75.—Jugum of a hepialid.　　Fig. 76.—Fibula of *Corydalus*.

adults. For this reason the identity of the wing-veins can be determined more surely in the wings of immature insects than they can be in the wings of adults. This is especially true where two or more veins coalesce in the adult wing while the tracheæ that precede these veins are distinctly separate in the immature wing.

A study was made of the tracheation of the wings of immature insects of representatives of most of the orders of insects, and, assuming that those features that are possessed by all of them must have been inherited from a common ancestor, a diagram was made representing the hypothetical tracheation of a nymph of the primitive winged insect (Fig. 77). In this diagram the tracheæ are lettered

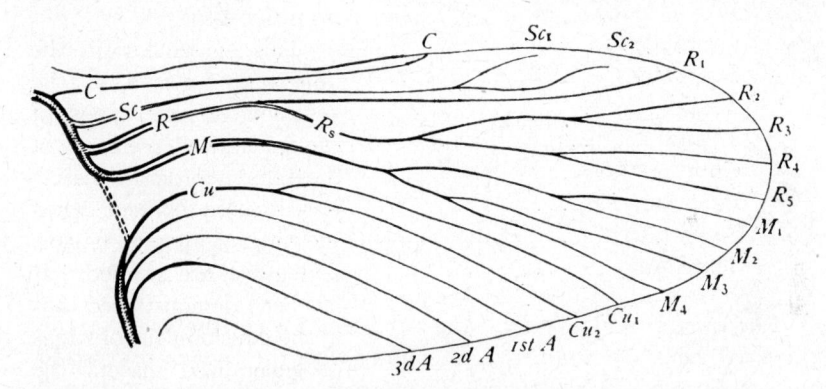

Fig. 77.—Hypothetical tracheation of a wing of the primitive nymph.

with the abbreviations used in designating the veins that are formed about them in the course of the development of the wing. The diagram will serve, therefore, to indicate the typical venation of an insect wing, except that the tracheæ are not crowded together at the base of the wing as are the veins in the wings of adults.*

Longitudinal veins and cross-veins.—The veins of the wing can be grouped under two heads: first, *longitudinal veins*, those that normally extend lengthwise the wing; and second, *cross-veins*, those that normally extend in a transverse direction.

The insertion of the word *normally* in the above definitions is important; for it is only in comparatively generalized wings that the direction of a vein can be depended upon for determining to which of these two classes it belongs.

The principal wing-veins.—The longitudinal wing-veins constitute the principal framework of the wings. In the diagram representing the typical venation of an insect wing (Fig. 77), only longitudinal veins are indicated; this is due to the fact that the diagram was based on a study of the tracheation of wings, and in the more generalized wings the cross-veins are not preceded by tracheæ; moreover in the wings of more generalized paleozoic insects there were no definite cross-veins, but merely an irregular network of thickened lines between the longitudinal veins.

There are eight principal veins; and of these the second, third, fourth, and fifth are branched. The names of these veins and the abbreviations by which they are known are as follows, beginning with the one nearest the costal margin of the wing:

Names of veins	Abbreviations
Costa	C
Subcosta	Sc
Radius	R
Media	M
Cubitus	Cu
First Anal	1st A
Second Anal	2d A
Third Anal	3d A

The chief branches of the wing-veins.—The chief branches of the principal veins are numbered, beginning with the branch nearest to the costal margin of the wing. The term used to designate a branch of a vein is formed by compounding the name of the vein with a

---

*For many details regarding the development of the wings of insects, their structure, and the terminology of the wing-veins, that can not be included in this work, see a volume by the writer entitled *The Wings of Insects*. This is published by The Comstock Publishing Company, Ithaca, N. Y.

numeral indicating the number of the branch; thus, for example, the first branch of the radius is radius-one or vein R_1.

In the case of radius and media, each of which has more than two branches, each division of the vein that bears two or more branches has received a special name. Thus after the separation of radius-one from the main stem of radius there remains a division which is typically four-branched; this division is termed the radial sector, or vein R_s; the first division of the radial sector, which later separates into radius-two and radius-three, is designated as radius-two-plus-three or vein R_{2+3}; and the second division is termed radius-four-plus-five or vein R_{4+5}. Media is typically separated into two divisions, each of which is two-branched; the first division is media-one-plus two or vein M_{1+2}, the second is media-three-plus-four or vein M_{3+4}.

The veins of the anal area.—The three anal veins exhibit a wide range of variation both as to their persistence and to their form when

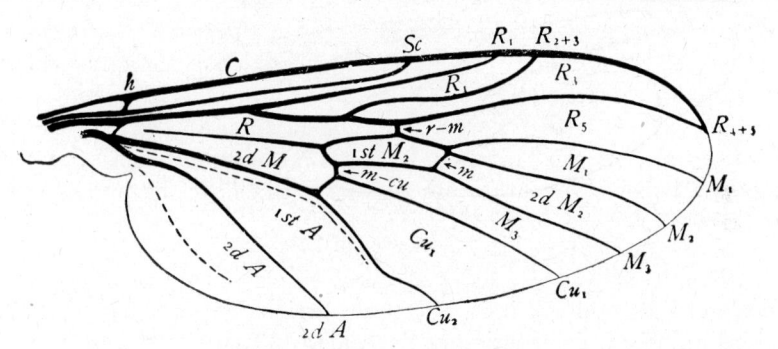

Fig. 78.—A wing of *Rhyphus*.

present. In those cases where the anal veins are branched there is no indication that the branching has been derived from a uniform primitive type of branching. For this reason in describing a branched anal vein merely the number of branches is indicated.

In some cases, as in the Odonata, there is a single anal vein the identity of which can not be determined. In such cases this vein is designated merely as the *anal vein* or *vein A*, and its branches as A_1, A_2, A_3, etc.

The reduction of the number of wing-veins.—In many wings the number of the veins is less than it is in the hypothetical type. In some cases this is due to the fact that one or more veins have faded out in the course of the evolution of the insects showing this deficiency; frequently in such wings vestiges of the lacking veins remain, either as faint lines in the positions formerly occupied by the veins or as short fragments of the veins. A much more common way in which the number of veins has been reduced is by the coalescence of adjacent veins. In many wings the basal parts of two or more principal veins are united so as to appear as a single vein; and the number of the branches of a vein has been reduced in very many cases by two or more branches becoming united throughout their entire length.

When a vein consists of two or more of the primitive veins united, the name applied to the compound vein should indicate this fact. In the wing of *Rhyphus* (Fig. 78), for example, radius is only three-branched; but it would be misleading to designate these branches as R_1, R_2, and R_3, for this would indicate that veins R_4 and R_5 are lacking. The first branch is evidently R_1; the second branch is composed of the

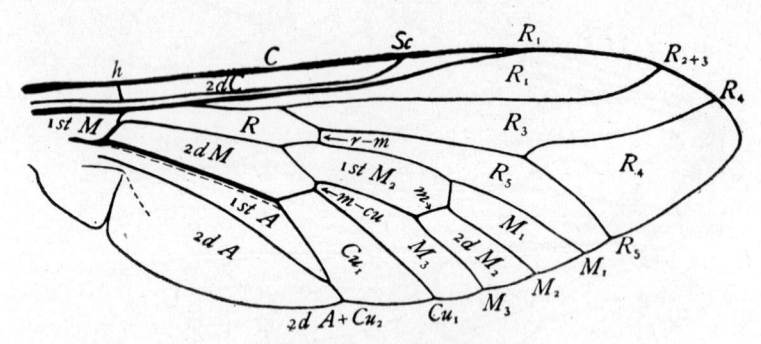

Fig. 79.—A wing of *Tabanus*.

coalesced R_2 and R_3, it is, therefore, designated as R_{2+3}; and the third branch, which consists of the coalesced R_4 and R_5, is designated as R_{4+5}.

A second method of coalescence of veins is illustrated by a wing of *Tabanus* (Fig. 79). In this wing the tips of cubitus-two and the second anal vein are united; here the coalescence began at the margin of the wing and is progressing towards the base. The united portions of the two veins are designated as $2d\ A_+Cu_2$.

When it is desired to indicate the composition of a compound vein it can be readily done by combining the terms indicating its elements. But in descriptions of hymenopterous wings where a compound vein may be formed by the coalescence of several veins the logical carrying out of this plan would result in a very cumbersome terminology, one that it is impracticable to use in ordinary descriptions. In such cases the compound vein is designated by the term indicating its most obvious element. Thus, for example, in the fore wing of *Pamphilius*, where veins M_4, Cu_1, and Cu_2 coalesce with the first anal vein, the united tips of these veins is designated as vein 1st A, the first anal vein being its most obvious element (Fig. 80), although it is really vein $M_4+Cu_1+Cu_2+$1st A.

Serial veins.—In the wings of some insects, where the wing-venation has been greatly modified, as in certain Hymenoptera, there exist what appears to be simple veins that in reality are compound veins composed of sections of two or more veins joined end to end with no indication of the point of union. Compound veins formed in this

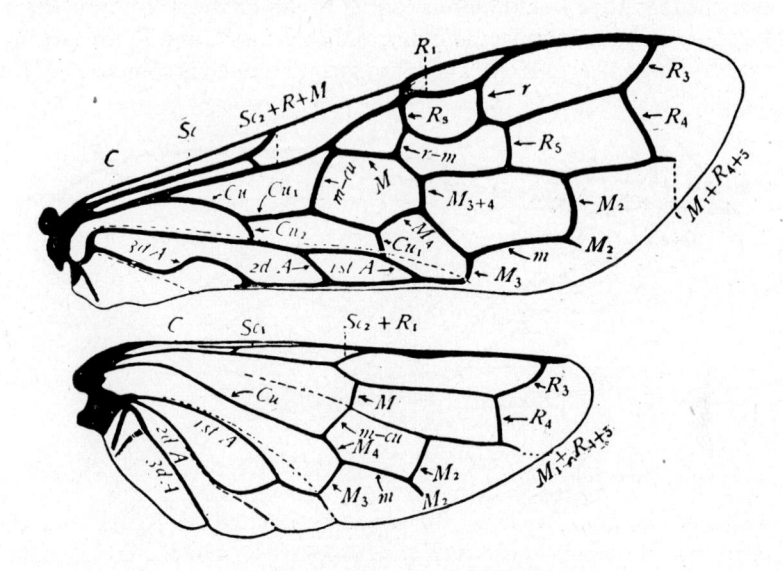

Fig. 80.—Wings of *Pamphilius*.

manner are termed *serial veins*. Examples of wings in which there are serial veins are figured in the chapter treating of the Hymenoptera.

In designating serial veins either the sign & or a dash is used between the terms indicating the elements of the vein, instead of the sign + as the latter is used in designating compound veins formed by the coalescence of veins side by side. If the serial vein consists of only two elements the sign & is used; thus the serial vein in the wings of braconids, which consists of the medial cross-vein and vein M_2, is designated as m & M_2.

In those cases where sections of several veins enter into the composition of a serial vein, the serial vein is designated by the abbreviation of the name of the basal element connected by a dash with the abbreviation of the name of the terminal element. Thus a serial vein, the basal element of which is the cubitus and the terminal element vein M_1, is designated as vein $Cu—M_1$. A serial vein thus formed exists in the hind wings of certain ichneumon flies.

The increase of the number of wing-veins. In the wings of many insects the number of veins is greater than it is in the hypothetical type. This multiplication of veins is due either to an increase in the

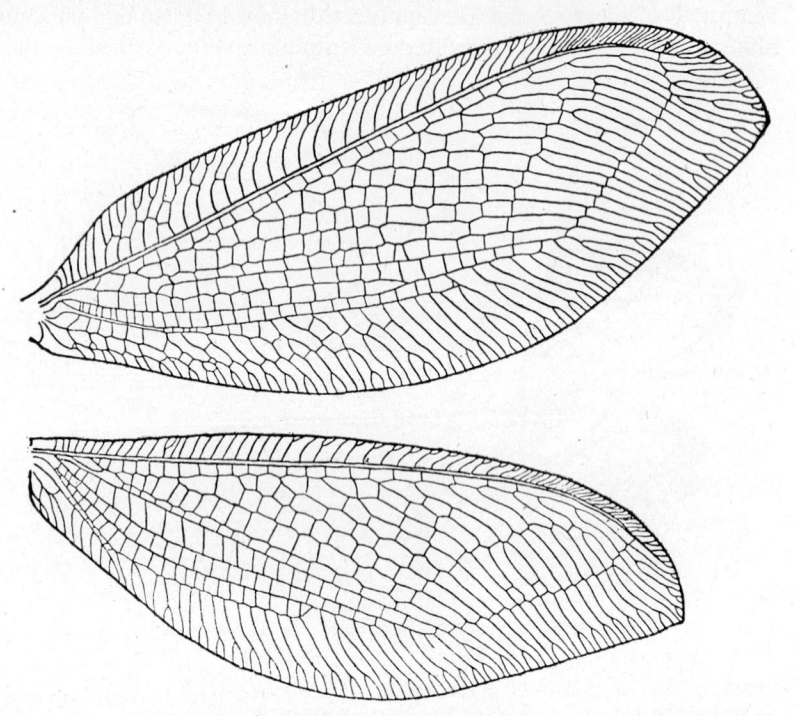

Fig. 81.—Wings of *Osmylus hyalinatus*.

number of the branches of the principal veins by the addition of secondary branches, termed *accessory veins*, or to the development of secondary longitudinal veins between these branches, termed *intercalary veins*. In no case is there an increase in the number of principal veins.

The accessory veins.—The wings of *Osmylus* (Fig. 81) are an example of wings in which accessory veins have been developed; here the radial sector bears many more branches than the typical number; those branches that are regarded as the primitive branches are lettered R_1, R_2, R_3, R_4, and R_5 respectively (Fig. 82); the other branches are the secondarily developed accessory veins. Two types of accessory veins are recognized the *marginal accessory veins* and the *definitive accessory veins*.

The *marginal accessory veins* are twig-like branches that are the result of bifurcations of veins that have not extended far back from the margin of the wing; many such short branches of veins exist in the wings of *Osmylus* (Fig. 81). The number and position of the marginal accessory veins are not constant, differing in the wings of the two sides of the same individual.

The *definitive accessory veins* differ from the marginal accessory

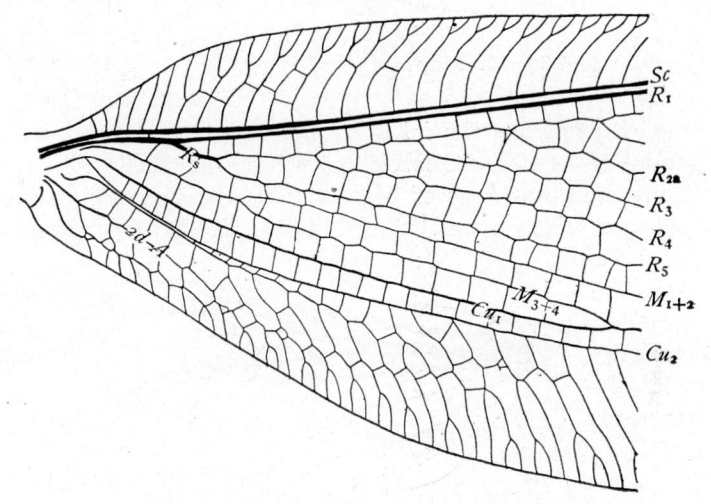

Fig. 82.—Base of fore wing shown in Figure 81.

veins in having attained a position that is comparable in stability to that of the primitive branches of the principal veins.

In those cases where the accessory veins are believed to have been developed in regular order they are designated by the addition of a letter to the abbreviation of the name of the vein that bears them; thus if vein R_2 bears three accessory veins they are designated as veins R_{2a}, R_{2b}, and R_{2c}, respectively.

The intercalary veins.—The intercalary veins are secondarily developed longitudinal veins that did not arise as branches of the primitive veins, but were developed in each case as a thickened fold in a corrugated wing, more or less nearly midway between two pre-existing veins, with which primarily it was connected only by cross-veins. Excellent examples of unmodified intercalary veins are common in the Ephemerida, where most of the intercalary veins remain distinct from the veins between which they were developed, being connected with them only by cross-veins, the proximal end of the intercalary vein being free (Fig. 83).

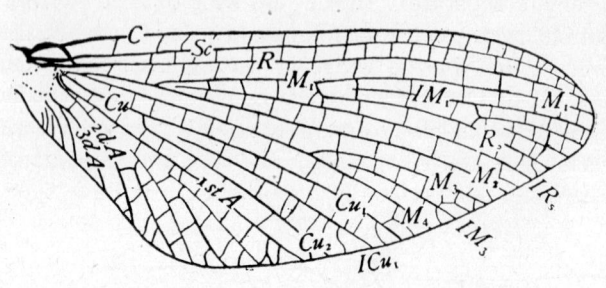

Fig. 83.—Wing of a May-fly (After Morgan).

When it is desirable to refer to a particular intercalary vein it can be done by combining the initial I, indicating intercalary, with the designation of the area of the wing in which the intercalary vein occurs. For example, in the wings of most May-flies there is an intercalary vein between veins Cu_1 and Cu_2, *i.e.* in the area Cu_1; this intercalary vein is designated as ICu_1.

The adventitious veins.—In certain insects there are secondary veins that are neither accessory veins nor intercalary veins as defined above; these are termed *adventitious veins*. Examples of these are the *supplements* of the wings of certain Odonata and the *spurious vein* of the Syrphidæ.

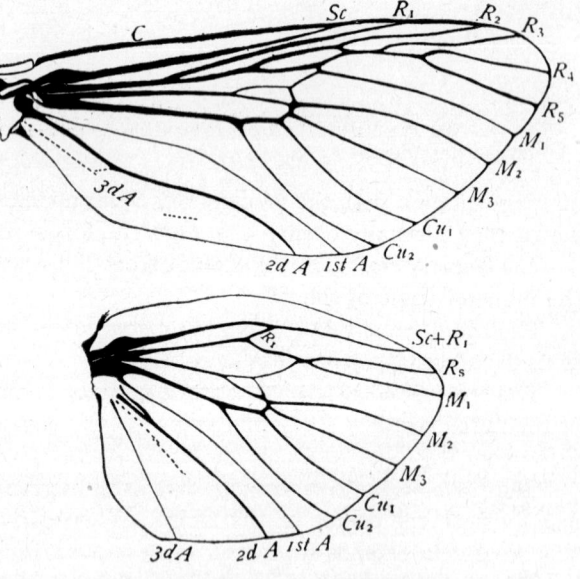

Fig. 84.—Wings of *Prionoxystus*.

The anastomosis of veins.—The typical arrangement of wing-veins is often modified by an anastomosis of adjacent veins; that is, two veins will come together at some point more or less remote from their extremities and merge into one for a greater or less distance, while their extremities remain separate. In the fore wing of *Prionoxystus* (Fig. 84), for example, there is an anastomosis of veins R_3 and R_{4+5}.

The named cross-veins.—In the wings of certain insects, as the dragon-flies, May-flies, and others, there are many cross-veins; it is impracticable in cases of this kind to name them. But in several of the orders of insects there are only a few cross-veins, and these have been named. Figure 85 represents the hypothetical primitive type

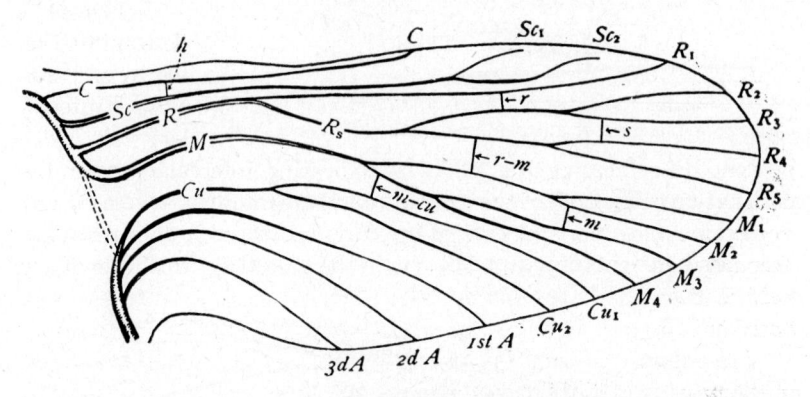

Fig. 85.—The hypothetical primitive type of wing-venation with the named cross-veins added.

of wing-venation with the named cross-veins added in the positions in which they normally occur; these are the following:

The humeral cross-vein (h) extends from the subcosta to costa near the humeral angle of the wing.

The *radial cross-vein* (r) extends between the two principal divisions of radius, *i. e.* from vein R_1 to vein R_s.

The *sectorial cross-vein* (s) extends between the principal divisions of the radial sector— *i. e.*, from vein R_{2+3} to vein R_{4+5} or from vein R_3 to vein R_4.

The *radio-medial cross-vein* (r—m) extends from radius to media, usually near the center of the wing. When in its typical position this cross-vein extends from vein R_{4+5} to vein M_{1+2}.

The *medial cross-vein* (m) extends from vein M_2 to vein M_3. This cross-vein divides cell M_2 into cells, 1st M_2 and 2d M_2; see Figure 87 where the cells are lettered.

The *medio-cubital cross-vein* (m—cu) extends from media to cubitus.

The arculus.—In many insects there is what appears to be a cross-vein extending from the radius to the cubitus near the base of the wing; this is the *arculus*. The arculus is designated in figures of wings by the abbreviation *ar*. Usually when the arculus is present the media appears to arise from it; the fact is, the arculus is compound, being composed of a section of media and a cross-vein. Figure 86 is a diagram representing the typical structure of the arculus. That part of the arculus which is a section of media is designated as the *anterior arculus* (*aa*) and that part formed by a cross-vein, the *posterior arculus* (*pa*).

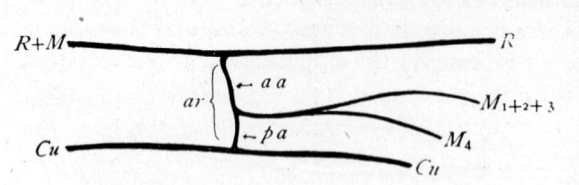

Fig. 86.—Diagram of an arculus of a dragon-fly.

The terminology of the cells of the wing.—Each cell of the wing is designated by the name of the vein that normally forms its front margin when the wings are spread. See Figure 87 where both the veins and the cells of the wing are lettered.

The cells of the wing fall naturally into two groups: first, those on the basal part of the wing; and second, those nearer the distal end of the wing. The former are bounded by the stems of the principal veins, the latter, by the branches of these veins; a corresponding distinction is made in designating the cells. Thus a cell lying behind the main stem of radius and in the basal part of the wing is designated as cell R; while a cell lying behind radius-one is designated as cell R_1.

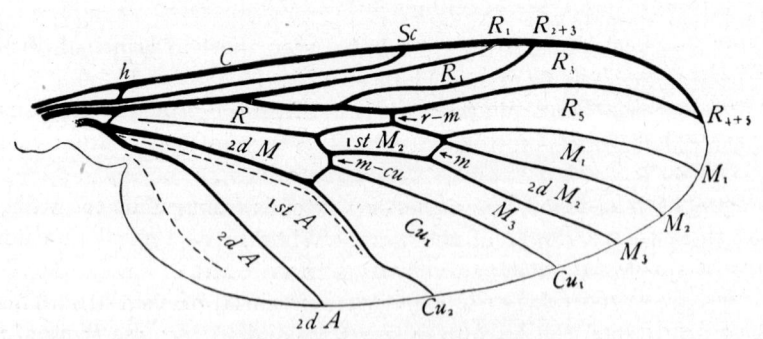

Fig. 87.—A wing of *Rhyphus*.

It should be remembered that the coalescence of two veins results in the obliteration of the cell that was between them. Thus when veins R_2 and R_3 coalesce, as in the wings of *Rhyphus* (Fig. 87), the cell lying behind vein R_{2+3} is cell R_3, and not cell R_{2+3}, cell R_2 having been obliterated.

When one of the principal cells is divided into two or more parts by one or more cross-veins, the parts may be numbered, beginning with the proximal one. Thus in *Rhyphus* (Fig. 87), cell M_2 is divided by the medial cross-vein into cell *1stM*$_2$ and cell *2dM*$_2$.

When two or more cells are united by the atrophy of the vein or veins separating them, the compound cell thus formed is designated by a combination of the terms applied to the elements of the compound cell. When, for example, the stem of media is atrophied, the cell resulting from the combination of cells R and M is designated as cell $R+M$.

The application of this system of naming the cells of the wing is an easy matter in those orders where there are but few cross-veins; but in those orders where there are many cross-veins it is not practicable to apply it. In the latter case we have to do with *areas* of the wing rather than with separate cells. These areas are designated as are the cells of the few-veined wings with which they correspond; thus the area immediately behind vein R_2 is area R_2.

The corrugations of the wings.—The wings of comparatively few insects present a flat surface; in most cases the membrane is thrown into a series of folds or corrugations. This corrugating of the wing in some cases adds greatly to its strength, as in the wings of dragon-flies; in other cases the corrugations are the result of a folding of the wing when not in use, as in the anal area when this part is broadly expanded.

It rarely happens that there is occasion to refer to individual members of either of these classes of folds, except perhaps the one between the costa and the radius, which is the *subcostal fold* and that which is normally between the cubitus and the first anal vein, the *cubito-anal fold*.

Convex and concave veins.—When the wings are corrugated, the wing-veins that follow the crests of ridges are termed *convex veins;* and those that follow the furrows, *concave veins*.

The furrows of the wing.—There are found in the wings of many insects one or more suture-like grooves in the membrane of the wing; these are termed the furrows of the wing. The more important of these furrows are the four following:

The *anal furrow* when present is usually developed in the cubito anal fold; but in the Heteroptera it is found in front of the cubitus.

The *median furrow* is usually between radius and media.

The *nodal furrow* is a transverse suture beginning at a point in the costal margin of the wing corresponding to the nodus of the Odonata and extending towards the inner margin of the wing across a varying number of veins in the different orders of insects.

The *axillary furrow* is a line that serves as a hinge which facilitates the folding of the posterior lobe of the wing of many insects under that part of the wing in front of it.

The bullæ.—The *bullæ* are weakened places in veins of the wing where they are crossed by furrows. The bullæ are usually paler in color than the other portions of the wing; they are common in the wings of the Hymenoptera (Fig. 88), and of some other insects.

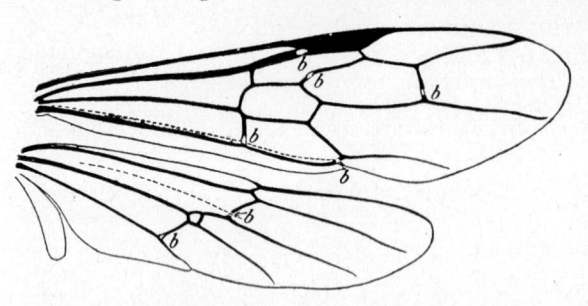

Fig. 88.—Wings of *Myrmecia*; b, b, b, bullæ.

The ambient vein.—Sometimes the entire margin of the wing is stiffened by a vein-like structure; this is known as the *ambient vein*.

The humeral veins.—In certain Lepidoptera and especially in the Lasiocampidæ, the humeral area of the hind wings is greatly expanded and in many cases is strengthened by the development of secondary veins. These are termed the *humeral veins*.

The pterostigma or stigma.—A thickened, opaque spot which exists near the costal margin of the outer part of the wing in many insects is known as the *pterostigma* or *stigma*.

The epipleuræ.—A part of the outer margin of the elytra of beetles when turned down on the side of the thorax is termed the *epipleura*.

The discal cell and the discal vein.—The term *discal cell* is applied to a large cell which is situated near the center of the wing; and the term *discal vein*, to the vein or series of veins that limits the outer end of the discal cell. These terms are not a part of the uniform terminology used in this book, and can not be made so, being applied to different parts of the wing by writers on different orders of insects. They are included here as they are frequently used, as a matter of convenience, by those who have adopted the uniform terminology. The discal cell of the Lepidoptera is cell $R+M+1stM_2$; that of the Diptera is cell 1st M_2; and that of the Trichoptera is cell R_{2+3}.

The anal area and the preanal area of the wing.—In descriptions of wings it is frequently necessary to refer to that part of the wing supported by the anal veins; this is designated as the *anal area* of the wing; and that part lying in front of the anal area, including all of the wing except the anal area, is termed the *preanal area*.

IV. THE ABDOMEN

a. THE SEGMENTS OF THE ABDOMEN

The third and terminal region of the body, the abdomen, consists of a series of approximately similar segments, which as a rule are without appendages excepting certain segments near the caudal end of the body.

The body-wall of an abdominal segment is usually comparatively simple, consisting in adults of a tergum and a sternum, united by lateral conjunctivæ. Sometimes there are one or two small sclerites on each lateral aspect of a segment; these are probably reduced pleura.

The number of segments of which the abdomen appears to be composed varies greatly in different insects. In the cuckoo-flies (Chrysididæ) there are usually only three or four visible; while in many insects ten or eleven can be distinguished. All intergrades between these extremes occur.

The apparent variation in the number of abdominal segments is due to two causes: in some cases, some of the segments are telescoped; and in others, adjacent segments coalesce, so that two or more segments appear as one.

A study of embryos of insects has shown that the abdomen consists typically of eleven segments; although this number may be reduced during the development of the insect by the coalescence of adjacent segments.

In some insects there is what appears to be a segment caudad of the eleventh segment; this is termed the *telson*. The telson differs from the segments preceding it in that it never bears appendages.

Special terms have been applied, especially by writers on the Coleoptera, to the caudal segments of the abdomen. Thus the terminal segment of a beetle's abdomen when exposed beyond the elytra is termed the *pygidium;* the tergite cephalad of the pygidium, especially in beetles with short elytra, the *propygidium;* and the last abdominal sternite, the *hypopygium*. The term hypopygium is also applied to the genitalia of male Diptera by writers on that order of insects.

b. THE APPENDAGES OF THE ABDOMEN

In the early embryonic stages of insects, each segment of the abdomen, except the telson, bears a pair of appendages (Fig. 89). This indicates that the primitive ancestor of insects possessed many legs, like a centipede. But the appendages of the first seven abdominal segments are usually lost during embryonic life, these segments being without appendages in postembryonic stages, except in certain Thysanura and Collembola, and in some larvæ.

Reference is made here merely to the primary appendages of the segments, those that are homodynamous with the thoracic legs; secondarily developed appendages, as for example, the tracheal gills, are present in the immature instars of many insects.

The styli or vestigial legs of certain Thysanura.—In certain Thysanura the coxa of each middle and hind thoracic leg bears a small appendage, the *stylus* (Fig. 90); and on from one to nine abdominal segments there is a pair of similar styli. These abdominal styli are believed to be homodynamous with those of the thoracic legs, and must, therefore, be regarded as vestiges of abdominal legs.

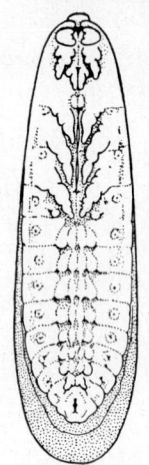

Fig. 89.-Embryo of *Hydrophilus* showing abdominal appendages.

The collophore of the Collembola.—Although in the postembryonic stages of Collembola the collophore is an unpaired organ on the middle line of the ventral aspect of the first abdominal segment, the fact that it arises in the embryo as a pair of appendages comparable in position to the thoracic legs, has led to the belief that it represents the legs of this segment. The structure of the collophore is described more fully later in the chapter treating of the Collembola.

The spring of the Collembola.—The spring of the Collembola, like the collophore, is believed to represent a pair of primary appendages. This organ is discussed in the chapter treating of the Collembola.

The genitalia.—In most insects there are more or less prominent appendages connected with the reproductive organs. These appendages constitute in males the *genital claspers* and in females the *ovipositor;* to them have been applied the general term *genitālia,* they are also known as the *gonapŏphyses.*

The genitalia, when all are developed consist of three pairs of appendages. Writers vary greatly in their views regarding the segments of the abdomen to which these appendages belong. One cause of difference is that some writers regard the last segment of the abdomen as the tenth abdominal segment while others believe it to be the eleventh, which is the view adopted in this work, this segment bears the cerci when they are present. The three pairs of appendages that constitute the genitalia are borne by the eighth and ninth segments, two pairs being borne by the ninth segment. The outer pair of the ninth segment constitute the sheath of the ovipositor. See account of the genitalia of the Orthoptera in Chapter eight.

The genitalia of many insects have been carefully figured and described and special terms have been applied to each of the parts. But as most of these descriptions have been based upon studies of representatives of a single order of insects or even of some smaller group, there is a great lack of uniformity in the terms applied to homologous parts in the different orders of insects; such of these terms as are commonly used are defined later in the characterizations of the several orders of insects.

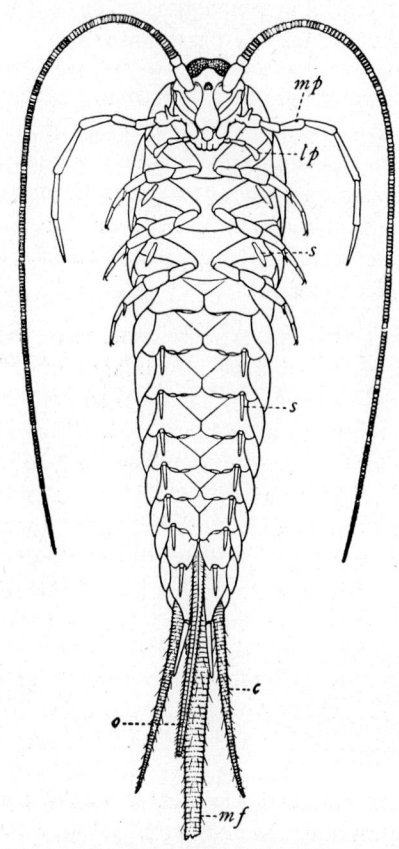

Fig. 90.—Ventral aspect of Machilis; *c*, cercus; *lp*, labial palpus; *mf*, median caudal filament; *mp*, maxillary palpus; *o*, ovipositor; *s, s*, styli. That part of the figure representing the abdomen is after Oudemans.

The cerci.—In many insects there is a pair of caudal appendages which are known as the *cerci;* these are the appendages of the eleventh abdominal segment, the last segment of the body except in the few cases where a telson is presemt.

The cerci vary greatly in form; in some insects, as in most Thysanura, in the Plecoptera, and in the Ephermerida, they are long and many jointed; while in others they are short and not segmented. The function of the cerci is different in different insects; they are believed to be tactile in some, olfactory in others, and in some males they aid in holding the female during copulation.

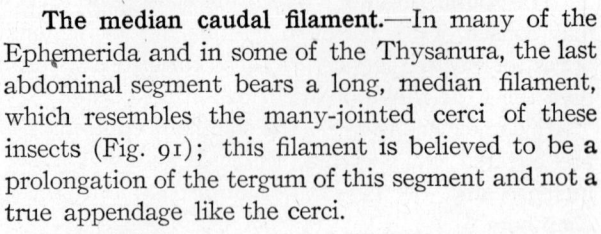

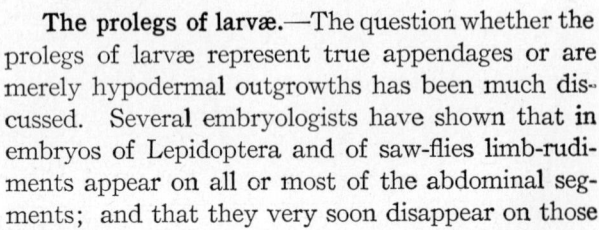

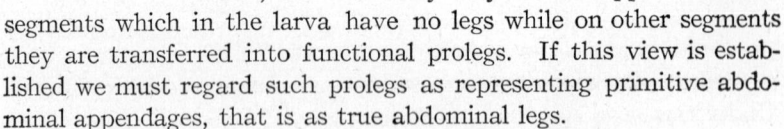

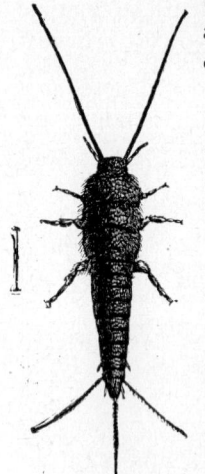

The median caudal filament.—In many of the Ephemerida and in some of the Thysanura, the last abdominal segment bears a long, median filament, which resembles the many-jointed cerci of these insects (Fig. 91); this filament is believed to be a prolongation of the tergum of this segment and not a true appendage like the cerci.

The prolegs of larvæ.—The question whether the prolegs of larvæ represent true appendages or are merely hypodermal outgrowths has been much discussed. Several embryologists have shown that in embryos of Lepidoptera and of saw-flies limb-rudiments appear on all or most of the abdominal segments; and that they very soon disappear on those segments which in the larva have no legs while on other segments they are transferred into functional prolegs. If this view is established we must regard such prolegs as representing primitive abdominal appendages, that is as true abdominal legs.

Fig. 91.—*Lepisma saccharina*.

V. THE MUSIC AND THE MUSICAL ORGANS OF INSECTS

Much has been written about music; but the greater part of this literature refers to music made by man for human ears. Man, however, is only one of many musical animals; and, although he excels all others in musical accomplishments, a study of what is done by our humbler relatives is not without interest.

The songs of birds command the attention of all observers. But there is a great orchestra which is performing constantly through the warmer portions of the year, which is almost unnoticed by man. Occasionally there is a performer that cannot be ignored, as:—

> "The shy Cicada, whose noon-voice rings
> So piercing shrill that it almost stings
> The sense of hearing." (ELIZABETH AKERS.)

But the great majority fiddle or drum away unnoticed by human ears.

Musical sounds are produced by many different insects, and in various ways. These sounds are commonly referred to as the songs of insects; but properly speaking few if any insects sing; for, with some possible exceptions, the note of an insect is always at one pitch, lacking musical modulations like those of the songs of man and of birds.

The sound produced by an insect may be a prolonged note, or it may consist of a series of short notes of varying length, with intervals of rest of varying lengths. These variations with differences in pitch give the wide range of insect calls that exists.

In some cicadas where the chambers containing the musical organs are covered by opercula, the insect can give its call a rhythmic increase and decrease of loudness, by opening and closing these chambers.

As most insect calls are strident, organs specialized for the production of these calls are commonly known as stridulating organs. But many sounds of insects are produced without the aid of organs specialized for the production of sound. The various ways in which insects produce sounds can be grouped under the following heads:

First.—By striking blows with some part of the body upon surrounding objects.

Second.—By rapid movements of the wings. In this way is produced what may be termed the music of flight.

Third.—By rasping one hard part of the body upon another. Under this head fall the greater number of stridulating organs.

Fourth.—By the rapid vibration of a membrane moved by a muscle attached to it. This is the type found in the cicadas.

Fifth.—By the vibration of membranes set in motion by the rush of air through spiracles. The reality of this method has been questioned.

Sixth.—By rapid changes of the outline of the thorax due to the action of the wing muscles.

a. SOUNDS PRODUCED BY STRIKING OBJECTS OUTSIDE THE BODY

Although the sounds produced by insects by striking blows with some part of the body upon surrounding objects are not rapid enough to give a musical note, they are referred to here for the sake of completeness.

The most familiar sounds of this kind are those produced by the insects known as the death-watch. These are small beetles of the family Ptinidæ, and especially those of the genus *Anobium*. These are wood-boring insects, frequently found in the woodwork of old houses and in furniture, where they make a ticking sound by striking their heads against the walls of their burrows. The sound consists of several, sharp, distinct ticks, followed by an interval of silence, and is believed to be a sexual call.

The name death-watch was applied to these insects by superstitious people who believed that it presaged the death of some person in the house where it is heard. This belief probably arose from the fact that the sound is most likely to be heard in the quiet of the night, and would consequently be observed by watchers by sick-beds.

The name death-watch has also been applied to some species of the Psocidæ, *Clothilla pulsatoria* and *Atropos divinatoria*, which have been believed to make a ticking sound. This, however, is doubted by some writers, who urge that it is difficult to believe that such minute and soft insects can produce sounds audible to human ears.

The death-watches produce their sounds individually; but an interesting example of an insect chorus is cited by Sharp ('99, p. 156), who, quoting a Mr. Peal, states that an ant, presumably an Assamese species, "makes a concerted noise loud enough to be heard by a human being at twenty or thirty feet distance, the sound being produced by each ant scraping the horny apex of the abdomen three times in rapid succession on the dry, crisp leaves of which the nest is usually composed."

b. THE MUSIC OF FLIGHT

The most obvious method by which insects produce sounds is by beating the air with their wings during flight. It can be readily seen that if the wing-strokes are sufficiently rapid and are uniform, they will produce, like the flapping reeds of a mouth organ, a musical note.

When, however, we take into account the fact that to produce the lowest note regularly employed in music, the C of the lowest octave, requires 32 vibrations a second, *i. e.*, nearly 2,000 vibrations per minute, it will seem marvellous that muscular action can be rapid enough to produce musical notes. Nevertheless, it is a fact that many insects sing in this way; and too their notes are not confined to the lower octaves. For example, the common house fly hums F of the middle octave, to produce which, it must vibrate its wings 345 times per second or 20,700 times per minute.

As a rule, the note produced by the wings is constant in each species of insect. Still with insects, as with us, the physical condition of the singer has its influence. The vigorous honey-bee makes the A of 435 vibrations, while the tired one hums on the E of 326 vibrations.

While it is only necessary to determine the note produced by vibrating wings to ascertain the rate of vibration, a graphical demonstration of the rate is more convincing. Such a demonstration has been made by Marey ('69) who fixed a fly so that the tip of the wing just touched the smoked surface of a revolving cylinder, and thus obtained a wavy line, showing that there were actually 320 strokes in a second. This agrees almost exactly with the number inferred from the note produced.

The music of flight may be, in many cases, a mere accidental result of the rapid movement, and in no sense the object of that movement, like the hum of a trolley car; but there are cases where the song seems to be the object of the movement. The honeybee produces different sounds, which can be understood by man, and probably by bees, as indicating different conditions. The contented hum of the worker collecting nectar may be a song, like the well-known song of a hen wandering about on a pleasant day, or may be an accidental sound. But the honeybee produces other sounds that communicate ideas. The swarming sound, the hum of the queenless colony, and the note of anger of a belligerent bee can be easily distinguished by the experienced beekeeper, and doubtless also by the bee colony. It seems probable, therefore, that in each of these cases the rate of vibration of the wings is adjusted so as to produce a desired note. This is also probably true of the song of the female mosquito, which is pitched so as to set the antennal hairs of the male in vibration.

While the music of flight is a common phenomenon, many insects have a silent flight on account of the slowness of the wing-movement.

c. STRIDULATING ORGANS OF THE RASPING TYPE

The greater number of the insect sounds that attract our attention are produced by the friction of hard parts of the cuticula by which a vibrating surface is set in motion. In some cases, as in many of the Orthoptera, the vibrating surface is a part of the wings that is specialized for this purpose; but in other cases, a specialized vibrating surface has not been observed.

Stridulating organs of the rasping type are possessed by representatives of several of the orders of insects; but they are most common in the order Orthoptera, and especially in the families Acridiidæ, Locustidæ, and Gryllidæ, where the males of very many species possess them. Very few other Orthoptera stridulate; and with few exceptions it is only the males that sing.

In each of these families the vibrating element of the stridulating organ is a portion of one or of both of the fore wings; but this is set in motion in several different ways. In some exotic Acridiidæ abdominal stridulating organs exist.

The stridulating organs of the Locustidae.—With many species of the Locustidæ we find the males furnished with stridulating organs; but these are comparatively simple, and are used only in the day time.

Two methods of stridulation are used by members of this family. The simpler of these two methods is employed by several common species belonging to the Œdipodinæ; one of which is the Carolina locust, *Dissosteira carolina*, whose crackling flight is a common feature of country roadsides. These locusts, as they fly, rub the upper surface of the costal margin of the hind wings upon the lower surface of the thickened veins of the fore wings, and thus produce a loud but not musical sound.

The second method of stridulation practiced by locusts consists in rubbing the inner surface of the hind femora, upon each of which there is a series of bead-like prominences (Fig. 92), against the outer surface of the fore wings. With these insects, there is a thickening of the radius in the basal third of each fore wing, and a widening of the two areas between this vein and the costal margin of the wing, which serves as a sounding board (Fig. 93). The two wings and femora constitute a pair of violin-like organs; the thickened radius in each case corresponding to the strings; the membrane of the wing, to the body of the instrument; and the file of the femur, to the bow. These two organs are used simultaneously. When about to stridulate, the insect

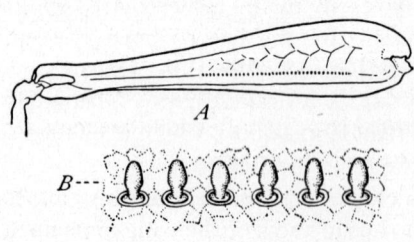

Fig. 92.—A, hind femora of *Stenobothrus;* B, file greatly enlarged.

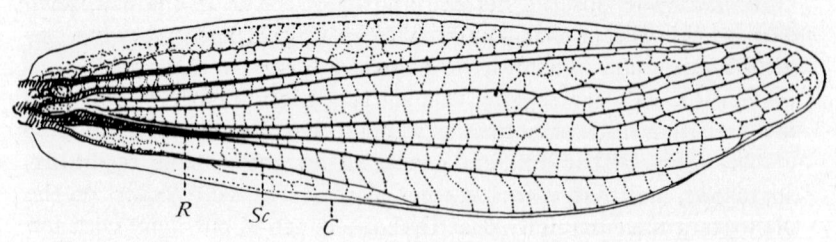

Fig. 93.—Fore wing of a male of *Stenobothrus*. R, radius; Sc, subcosta; C, costa.

places itself in a nearly horizontal position, and raising both hind legs at once rasps the femora against the outer surface of the wings. The most common representatives of insects that stridulate in this way belong to the genus *Stenobothrus*.

The stridulating organs of the Gryllidae and the Tettigoniidae.—The stridulating organs of the Gryllidæ and the Tettigoniidæ are of the same type, and are the most highly specialized found in the Orthoptera. They consist of modified portions of the fore wings; both the vibrating and the rasping elements of the organs pertaining to the wings.

It is by rubbing the two fore wings together that sound is produced.

In what is probably the more generalized condition of the organs, as seen in *Gryllus*, each fore wing bears a rasping organ, the file (Fig. 94, *f*) a hardened area, the *scraper* (Fig. 94, *s*), against which the file of the other wing acts, and vibrating areas, the *tympana* (Fig. 94, *t*, *t*). As the file of either wing can be used to set the tympana of the wings in vibration, we may say that *Gryllus* is ambidextrous.

When the cricket wishes to make his call, he elevates his fore wings so that they make an angle of about forty-five degrees with the body; then holding them in such a position that the scraper of one rests on the file of the other, he moves the wings back and forth laterally, so that the file and scraper rasp upon each other. This throws the wings into vibration and produces the call.

Fig. 94.—Fore wing of *Gryllus*; A, as seen from above, that part of the wing which is bent down on the side of the abdomen is not shown; *s*, scraper; *t*, *t*, tympana. B, base of wing seen from below; *s*, scraper; *f*, file. C, file greatly enlarged.

It is easy to observe the chirping of crickets. If one will move slowly towards a cricket that is making his call, and stop when the cricket stops chirping until he gains confidence and begins again, one can get sufficiently near to see the operation clearly. This can be done either in the day time or at night with the aid of a light.

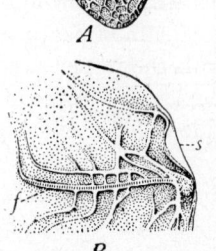

The songs of the different genera of crickets can be easily distinguished, and that of each species, with more care. Writers on the Orthoptera have carefully described the songs of our more common crickets, and especially those of the tree crickets. The rate of chirping is often influenced by temperature, being slower in cool nights than in warm ones; and becoming slower towards morning if the temperature falls.

In certain genera of crickets as *Nemobius* and *Œcanthus*, while each fore wing is furnished with a file and tympana, the scraper of the right wing is poorly formed and evidently not functional. As these insects use only the file of the right wing to set the tympana of the wings in vibration, they may be said to be right-handed.

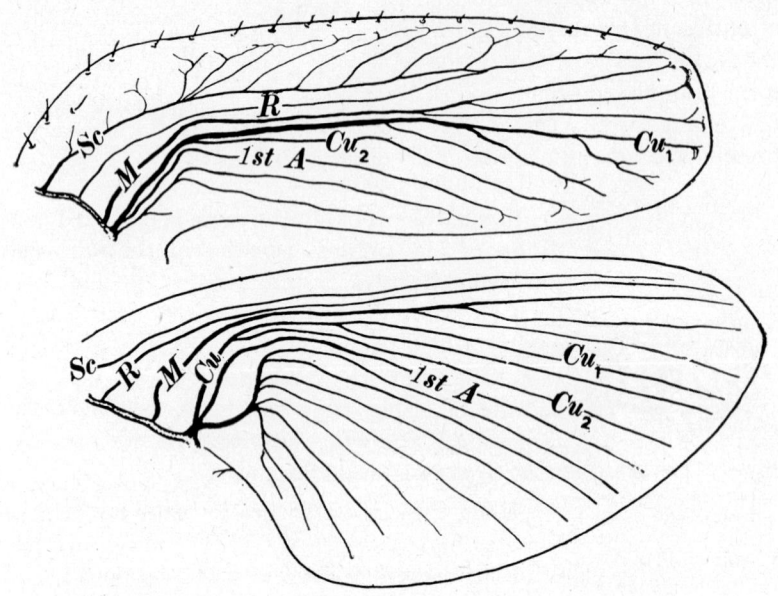

Fig. 95.—Wings of a female nymph of *Œcanthus* (From Comstock and Needham).

In the Locustidæ a similar modification of the function of the stridulating organs has taken place. In all of our common representatives of the family, at least, only one of the files is used. But in these cases it is the file of the left wing that is functional; we may say, therefore, that so far as observed the Locustidæ are left-handed. Different genera exhibit great differences as to the extent of the reduction of the unused parts of the stridulating organs. The file is present in both wings of all of the forms that I have studied; but the unused file is sometimes in a vestigial condition. The scraper is less persistent, being frequently entirely lacking in one of the wings. In some cases, the tympana of one wing have been lost; but in others the tympana of both wings are well preserved, although only one file is used. In these cases it is probable that the tympana of both wings are set in vibration by the action of the single functional file.

The determination of the homologies of the parts of the wing that enter into the composition of the stridulating organs was accomplished by a study of the tracheation of the wings of nymphs (Comstock and Needham, '98–'99). The results obtained by a study of the wings of Œcanthus will serve as an illustration.

Figure 95 represents the wings of a female nymph of this genus, with the tracheæ lettered. The only parts to which we need to give attention in this discussion are the cubital and anal areas of the fore wing; for it is this part of the wing that is modified in the male to form the musical organ. Both branches of cubitus are present, and Cu_1 bears three accessory branches. The three anal tracheæ are present and are quite simple.

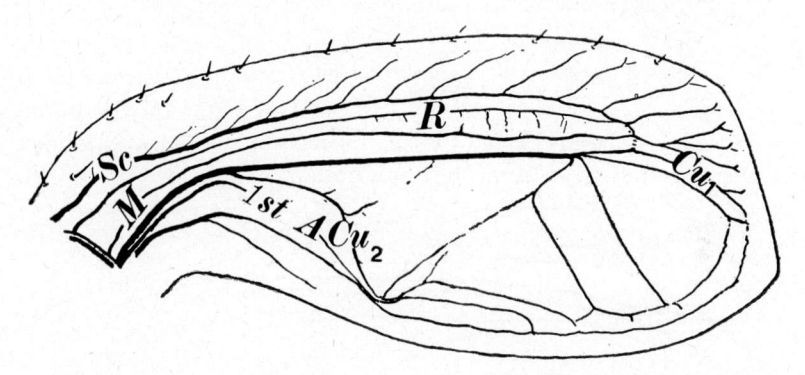

Fig. 96.—Fore wing of a male nymph of Œcanthus (From Comstock and Needham).

The homologies of the tracheæ of the fore wing of a male nymph, Figure 96, were easily determined by a comparison with the tracheæ of the female. The most striking difference between the two sexes is a great expanding of the area between the two branches of cubitus in the male, brought about by the bending back of the basal part of Cu_2.

The next step in this study was to compare the wing of an adult male, Figure 97, with that of the nymph of the same sex; and the solution of the problem was soon reached. It can be easily seen that the file is on that part of Cu_2 that is bent back toward the inner margin of the wing (Fig. 97, *f*); the tympana are formed between the branches of cubitus (Fig. 97, *t, t*); and the scraper is formed at the outer end of the anal area (Fig. 97, *s*).

A similar study was made of the wings of *Conocephalus*, as an example of the Tettigoniidæ, Figure 98 represents the wings of a male nymph; and Figure 99 the fore wing of an adult. The most striking feature and one characteristic of the family, is that the musical organ occupies an area near the base of the wing which is small com-

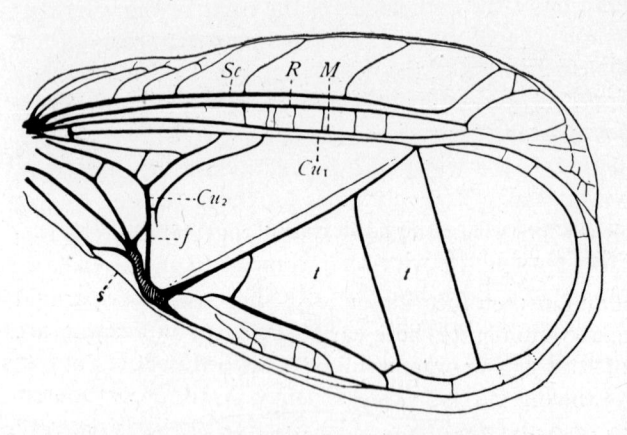

Fig. 97.—Fore wing of an adult male of *Œcanthus*; f, vein bearing the file; s, scraper; t, t, tympana.

pared with the area occupied by the musical organs of the Gryllidæ. But here, as in the Gryllidæ, the file is borne by the basal part of Cu_2,

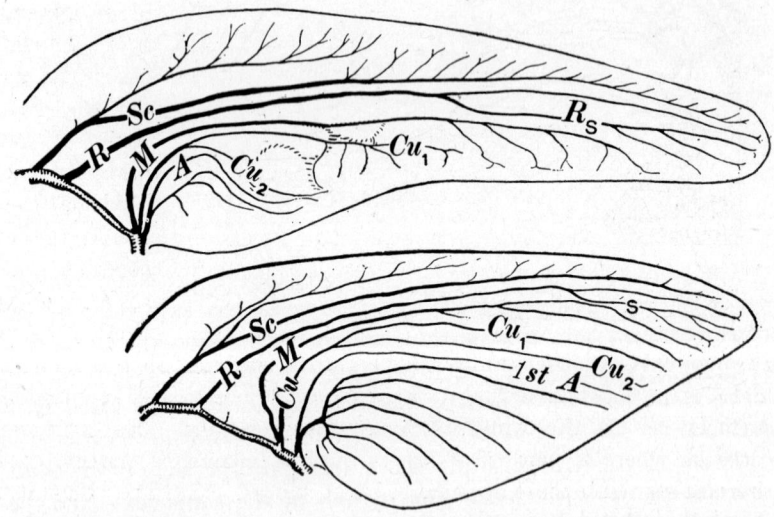

Fig. 98.—Wings of a male nymph of *Conocephalus*, (From Comstock and Needham).

the tympana are formed between the branches of cubitus, and the scraper is formed at the outer end of the anal area.

Rasping organs of other than orthopterous insects.—Rasping organs are found in many other than orthopterous insects and vary

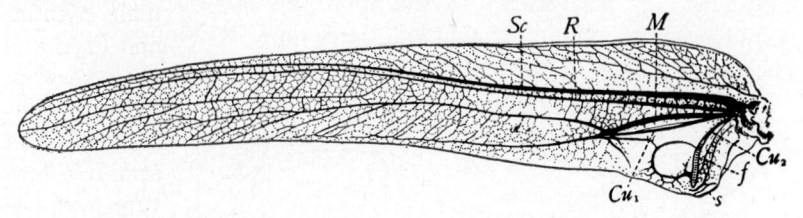

Fig. 99.—Right fore wing of an adult male of *Conocephalus*, seen from below; *f*, file; *s*, scraper.

greatly in form and in their location on the body. Lack of space forbids any attempt to enumerate these variations here; but examples of various types of stridulating organs will be described in later chapters when treating of the insects that possess them. As in the Orthoptera, they consist of a rasp and a scraper. The rasp is a file-like area of the surface of a segment of the body or of an appendage; and the scraper is a hard ridge or point so situated that it can be drawn across the rasp by movements of the body or of an appendage. In some cases the apparatus consists of two rasps so situated that they can be rubbed together.

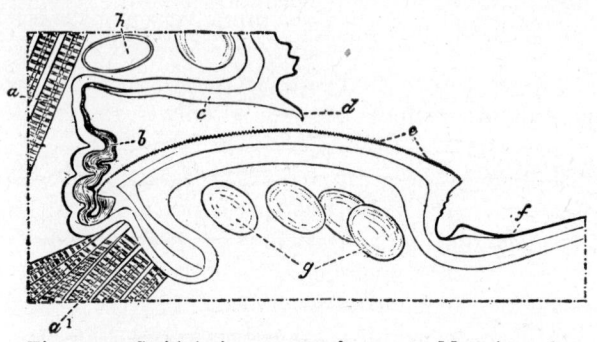

Fig. 100.—Stridulating organ of an ant, *Myrmica rubra* (From Sharp after Janet); *d*, scraper; *e*, file.

With many beetles one of the two parts of the stridulating organ is situated upon the elytra; and it is quite probable that in these cases the elytra acts as vibrating surfaces, as do the wings of locusts and crickets. But in many cases as where a part of a leg is rubbed against a portion of a thoracic segment, there appears to be no vibrating surface unless it is the wall of the body or of the appendage that acts as a sounding board. In the stridulating organ of *Myrmica rubra*, var. *lævinodis*, figured by Janet (Fig. 100), the scraper is the posterior border of one abdominal segment, and the file is situated on the dorsum of the following segment. It is quite conceivable that in this case the dorsal wall of the segment bearing the file is made to vibrate by the successive impacts of the scraper upon the ridges of the file. In fact this seems to me more probable than that the sound produced is merely that of the scraper striking against the successive ridges of the file. There is at least one recorded case where the body wall is specialized to act as a sounding board. According to Sharp ('95, p. 200), in the males of the Pneumorides, a tribe of South African Acridiidæ, where the phonetic organ is situated on the abdomen, this part is inflated and tense, no doubt with the result of increasing the volume and quality of the sound.

Ordinarily the stridulating organs of insects are fitted to produce notes of a single degree of pitch; but Gahan ('00) figures those of some beetles that are evidently fitted to produce sounds of more than one degree of pitch; the file of *Hispopria foveicollis*, consists of three parts, one very finely striated, followed by one in which the striæ are much coarser, and this in turn followed by one in which the striation is intermediate in character between the other two.

While the stridulating organs of the Orthoptera are possessed almost exclusively by the males, in the Coleoptera, very many species of which stridulate, the phonetic organs are very commonly possessed by both sexes, and serve as a mutual call. In one genus of beetles, *Phonapate*, stridulating organs have been found only in the females (Gahan, '00).

It seems evident that in the great majority of cases the sounds produced by insects are sexual calls; but this is not always so. It was pointed out long ago by Charles Darwin that "beetles stridulate under various emotions, in the same manner as birds use their voices for many purposes besides singing to their mates. The great *Chiasognathus* stridulates in anger or defiance; many species do the same from distress or fear, if held so that they cannot escape; by striking the hollow stems of trees in the Canary Islands, Messrs. Wollaston and Crotch were able to discover the presence of beetles belonging to the genus *Acalles* by their stridulation. Lastly the male *Ateuchus* stridulates to encourage the female in her work and from distress when she is removed" (*The Descent of Man*).

The most remarkable case where stridulating organs have been developed for other than sexual purposes is that of the larvæ of certain Lucanidæ and Scarabæidæ described by Schiodte ('74). In these larvæ there is a file on the coxa of each middle leg, and the hind legs are shortened and modified so as to act as scrapers. The most highly specialized example of this type of stridulating organ is possessed by the larvæ of *Passalus*, in which the legs of the third pair are so much shortened that the larvæ appear to have only four legs; each hind leg is a paw-like structure fitted for rasping the file (Fig. 101).

These insects are social, a pair of beetles and their progeny living together in decaying wood. The adults prepare food for the larvæ; and the colony is able to keep together by stridulatory signals.

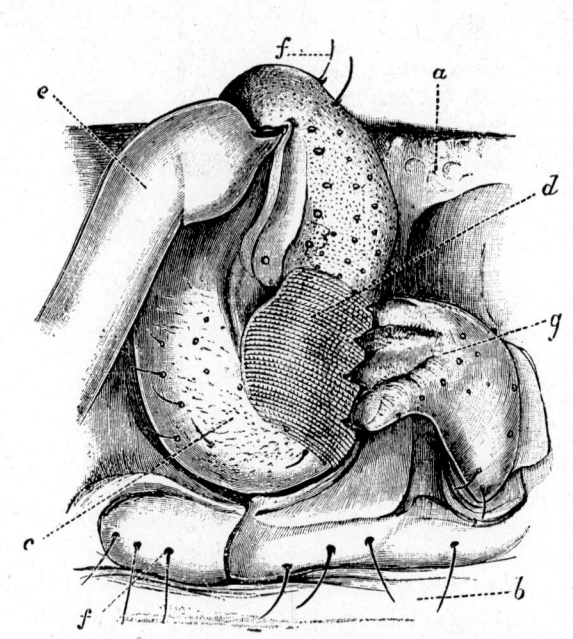

Fig. 101.—Stridulating organ of a larva of *Passalus*; *a*, *b*, portions of the metathorax; *c*, coxa of the second leg; *d*, file; *e*, basal part of femur of middle leg; *f*, hairs with chitinous process at base of each; *g*, the diminutive third leg modified for scratching the file (From Sharp).

d. THE MUSICAL ORGANS OF A CICADA

With the cicadas there exists a type of stridulating organ peculiar to them, and one that is the most complicated organ of sound found in the animal kingdom. Yet, while the cicadas are the most noisy of the insect world, the results obtained by their complicated musical apparatus are not comparable with those produced by the comparatively simple vocal organs of birds and of man.

It is said that in some species of *Cicada* both sexes stridulate; but as a rule the females are mute, possessing only vestiges of the musical apparatus.

The structure of the stridulating organs varies somewhat in details in different species of Cicada; but those of *Cicada plebeia*, which were described and figured by Carlet ('77), may be taken as an example of the more perfect form. In the male of this species there is a pair of large plates, on the ventral side of the body, that extend back from the hind border of the thorax and overlap the basal part of the abdomen; these are the *opercula* (Fig. 102, *o*). The opercula are expansions of the sternellum of the metathorax, and each serves as a lid covering a pair of cavities, containing the external parts of the musical apparatus of one side of the body.

The two cavities covered by a single operculum may be designated as the *ventral cavity* (Fig. 102, *v. c.*) and the *lateral cavity* (Fig. 102, *l. c.*) respectively. Each cavity is formed by an infolding of the body-wall.

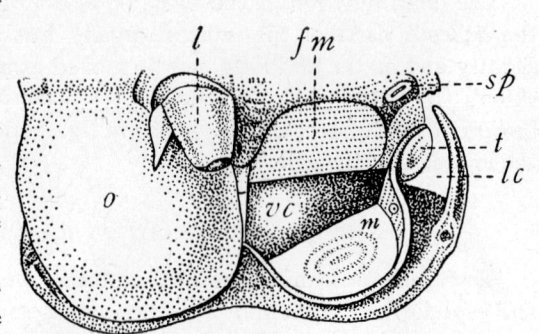

Fig. 102.—The musical apparatus of a cicada; *fm*, folded membrane; *l*, base of leg; *lc*, lateral cavity; *m*, mirror; *o*, operculum, that of the opposite side removed; *sp*, spiracle; *t*, timbal; *vc*, ventral cavity (After Carlet).

In the walls of these cavities are three membranous areas; these are known as the *timbal*, the *folded membrane*, and the *mirror*.

The timbal is in the lateral cavity on the lateral wall of the partition separating the two cavities (Fig. 102, *t*); the other two membranes are in the ventral cavity. The folded membrane is in the anterior wall of the ventral cavity (Fig. 102, *f. m.*); and the mirror is in the posterior wall of the same cavity (Fig. 102, *m*). Within the body, there is in the region of the musical apparatus a large thoraco-abdominal air chamber, which communicates with the exterior through a pair of spiracles (Fig. 102 *sp*); and a large muscle, which extends from the furca of the second abdominal segment to the inner face of the timbal.

By the contraction of this muscle the timbal is pulled towards the center of the body; and when the muscle is relaxed, the elasticity of the chitinous ring supporting the timbal causes it to regain its former position. By a very rapid repetition of these movements of the timbal the sound is produced.

It is probable that the vibrations of the timbal are transmitted to the folded membrane and to the mirror by the air contained in the large air chamber mentioned above; as the strings of a piano are made to vibrate by the notes of a near-by violin. The sound, however, is produced primarily by the timbal, the destruction of which renders the insect a mute; while the destruction of the other membranes, the timbal remaining intact, simply reduces the sound.

The chief function of the opercula is doubtless the protecting of the delicate parts of the musical organ; but as they can be lifted slightly and as the abdomen can be moved away from them to some extent, the chambers containing the vibrating parts of the organ can be opened and closed, thus giving a rhythmic increase and decrease of the loudness of the call.

e. THE SPIRACULAR MUSICAL ORGANS

There has been much discussion of the question whether insects, and especially Diptera and Hymenoptera, possess a sound-producing organ connected with the spiracles or not. Landois ('67) believed that he found such an organ and figures and describes it in several insects. It varies greatly in form in different insects. In the Diptera it consists of a series of leaf-like folds of the intima of the trachea; these are held against each other by a special humming ring, which lies close under the opening of the spiracle; and is found within two or all four of the thoracic spiracles. These membranous folds of the intima are set in vibration by the rush of air through the spiracles.

In the May-beetle, according to Landois, a buzzing organ is found near each of the fourteen abdominal spiracles. It is a tongue-like fold projecting into the lumen of the trachea under the base of the closing apparatus. On its upper surface it is marked with very fine arched furrows. He concludes that this tongue is put in vibration by the breathing of the insect, and hence the buzzing of the flying beetle.

If insects produce sounds in the way described by Landois they have a voice quite analogous to our own. But the validity of the conclusions of Landois has been seriously questioned; the subject, therefore, demands further investigation. See also Duncan ('24).

f. THE ACUTE BUZZING OF FLIES AND BEES

Many observers have found that when the wings of a fly or of a bee are removed or held so that they can not vibrate the insect can still produce a sound. The sound produced under these circumstances is higher, usually an octave higher, than that produced by the wings. It is evident, therefore, that these insects can produce sounds in two ways; and an extended search has been made for the organ or organs producing the higher note.

Landois believed that the spiracular organs referred to above were the source of the acute sound. But more recently Pérez ('78) and Bellesme ('78) have shown that when the spiracles are closed artificially the insect can still produce the high tone. Pérez attributes the sound to the vibrations of the stumps of the wings against the solid parts which surround them or of the sclerites of the base of the wing against each other. But Bellesme maintains that the sound is produced by changes in the form of the thorax due to the action of the wing-muscles.* When the wing-muscles are at rest the section of this region, according to this writer, represent an ellipse elongated vertically; the contraction of the muscles transforms it to an ellipse elongated laterally; the thorax, therefore, constitutes a vibrating body which moves the air like a tine of a tuning fork. Bellesme states that by fastening a style to the dorsal wall of the thorax he obtained a record of the rate of its vibrations, the number of which corresponded exactly to that required to produce the acute sound which the ear perceives.

The fact that the note produced when the wings are removed is higher than that produced by the wings is supposed by Bellesme to be due to the absence of the resistance of air against the wings, which admits of the maximum rate of contraction of the wing-muscles.

g. MUSICAL NOTATION OF THE SONGS OF INSECTS

Mr. S. H. Scudder ('93) devised a musical notation by which the songs of stridulating insects can be recorded. As the notes are always at one pitch the staff in this notation consists of a single horizontal line, the pitch being indicated by a separate statement. Each bar represents a second of time, and is occupied by the equivalent of a semibreve; consequently a quarter note ♩, or a quarter rest ♪, represents a quarter of a second; a sixteenth note ♬, or a sixteenth rest ♪ a sixteenth of a second and so on. For convenience's sake he introduced a new form of rest, shown in the second example given below, which indicates silence through the remainder of a measure; this differs from the whole rest commonly employed in musical notation by being cut off obliquely at one end.

---

*This view was maintained by Siebold at a much earlier date in his *Anatomy of the Invertebrates*.

The following examples taken from his paper on "The Songs of our Grasshoppers and Crickets" will serve to illustrate this method of notation.

The chirp of *Gryllotalpa borealis* (Fig. 103) "is a guttural sort of sound, like grü or grēēu, repeated in a trill indefinitely, but seldom

Fig. 103.—The chirp of *Gryllotalpa borealis* (From Scudder).

for more than two or three minutes, and often for less time. It is pitched at two octaves above middle C."

Fig. 104.—The chirp of the katydid (From Scudder).

The note of the true katydid, *Cyrtophyllus concavus*, (Fig. 104) "which sounds like *xr*, has a shocking lack of melody; the poets who have sung its praises must have heard it at a distance that lends enchantment." "They ordinarily call 'Katy' or say 'She did' rather than 'Katy did'; that is they rasp their fore wings twice more frequently than thrice." Mr. Scudder in his account of this song fails to indicate its pitch.

h. INSECT CHORUSES

Most insect singers are soloists, singing without reference to other singers or in rivalry with them. But there are a few species the members of which sing in unison with others of their kind that are near them. The most familiar sound of autumn evenings in rural places in this country is a chorus of the snowy tree cricket, *Œcanthus niveus*. Very many individuals of this species, in fact all that are chirping in any locality, chirp in unison. Early in the evening, when the chirping first begins, there may be a lack of unanimity in keeping time; but this lasts only for a short period, soon all chirp in unison, and the monotonous beat of their call is kept up uninterrupted throughout the night. Individual singers will stop to rest, but when they start again they keep time with those that have continued the chorus.

Other instances of insect choruses have been recorded. Sharp ('99, 156) quotes accounts of two produced by ants; one of these is given on an earlier page (p. 80).

CHAPTER III

THE INTERNAL ANATOMY OF INSECTS

BEFORE making a more detailed study of the internal anatomy of insects, it is well to take a glance at the relative positions of the different systems of organs within the body of insects and other arthropods.

One of the most striking features in the structure of these animals is that the body-wall serves as a skeleton, being hard, and giving support to the other organs of the body. This skeleton may be represented, therefore, as a hollow cylinder. We have now to consider the arrangement and the general form of the organs contained in this cylinder.

The accompanying diagram (Fig. 105), which represents a vertical, longitudinal section of the body, will enable the student to gain an

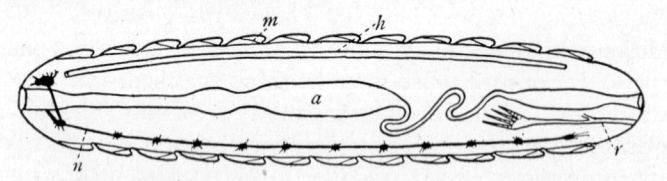

Fig. 105.—Diagram showing the relations of the internal organs; a, alimentary canal; h, heart; m, muscle; n, nervous system; r, reproductive organs.

idea of the relative positions of some of the more important organs. The parts shown in the diagram are as follows: The body-wall, or skeleton; this is made up of a series of overlapping segments; that part of it between the segments is not hardened with **chitin, thus** remaining flexible and allowing for the movements of the body. Just within the body-wall, and attached to it, are represented a few of the muscles (m); it will be seen that these muscles are so arranged that the contraction of those on the lower side of the body would bend it down, while the contraction of those on the opposite side would act in the opposite direction, other muscles not shown in the figure provide for movements in other directions. The alimentary canal (a) occupies the centre of the body, and extends from one end to the other. The heart (h) is a tube open at both ends, and lying between the alimentary canal and the muscles of the back. The central part of the nervous system (n) is a series of small masses of nervous matter connected by two longitudinal cords: one of these masses, the brain, lies in the head above the alimentary canal; the others are situated, one in each segment, between the alimentary canal and the layer of muscles of the ventral side of the body; the two cords connecting these masses, or ganglia, pass one on each side of the oesophagus to the brain. The reproductive organs (*r*) lie in the cavity of the abdomen and open near the caudal end of the body. The respiratory organs are omitted from this diagram for the sake of simplicity. We will now pass to a more detailed study of the different systems of organs.

I. THE HYPODERMAL STRUCTURES

The active living part of the body-wall is the hypodermis, already described in the discussion of the external anatomy of insects. In addition to the external skeleton, there are derived from the hypodermis an internal skeleton and several types of glands.

a, THE INTERNAL SKELETON

Although the skeleton of an insect is chiefly an external one, there are prolongations of it extending into the body-cavity. These inwardly directed processes, which serve for the attachment of muscles and for the support of other viscera are termed collectively the *internal skeleton* or *endo-skeleton*. The internal skeleton is much more highly developed in adult insects than it is in the immature instars.

Sources of the internal skeleton.—The parts of the internal skeleton are formed in two ways: first by the chitinization of tendons of muscles; and second, by invaginations of the body-wall.

Chitinized tendons.—Chitinized tendons of the muscles that move the mouth-parts, of muscles that move the legs, and of other muscles are of frequent occurrence. As these chitinized tendons help support the internal organs they are considered as a part of the internal skeleton.

Invaginations of the body-wall or apodemes.—The second and more important source of the parts of the internal skeleton consists of invaginations of the body-wall. Such an invagination is termed an *ăpodeme*. The more important apodemes, if not all, arise as invaginations of the body-wall between sclerites or at the edge of a sclerite on the margin of a body-segment; although by the fusion of sclerites about an apodeme, it may appear to arise from the disc of a sclerite.

Frequently, in the more generalized insects, the mouth of an apodeme remains open in the adult insects. In Figure 106 are represented two apodemes that exist in the thorax of a locust, *Melanoplus*. Each of these (*ap* and *ap*) is an invagination of the body-wall, between the episternum and the epimeron of a segment, immediately above the base of a leg. These are known as the lateral apodemes of the thorax and serve as points of attachment of muscles.

Fig. 106.—Ental surface of the pleurites of the meso- and metathorax of *Melanoplus*, showing the lateral apodemes, *ap, ap*.

The number of apodemes may be very large, and it varies greatly in different insects. Among the more important apodemes are the following:—

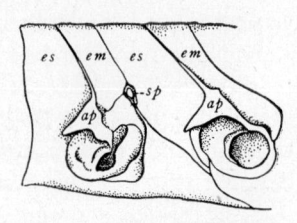

The tentorium.—The chief part of the internal skeleton of the head is termed the *tentorium*. This was studied by Comstock and Kochi ('02). We found that in the generalized insects studied by us it is composed of two or three pairs of apodemes that, extending far into the head, meet and coalesce. The three pairs of apodemes that may enter into the formation of the tentorium were termed the *anterior*, the *posterior*, and the *dorsal arms of the tentorium* respectively. The coalesced and more or less expanded tips of these apodemes constitute the *body of the tentorium*. From the body of the tentorium there extend a variable number of processes or chitinized tendons.

The posterior arms of the tentorium.—The posterior arms of the tentorium (Fig. 107, 109, 110, *pt*) are the lateral apodemes of the

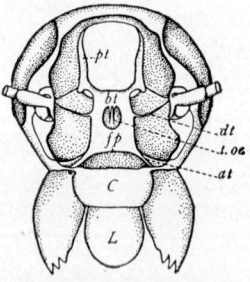

Fig. 107.— Tentorium of a cockroach, dorsal aspect.

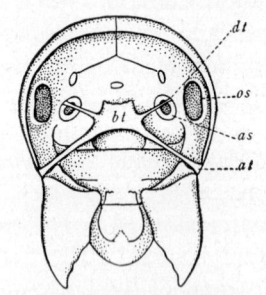

Fig. 108.—Part of the tentorium of a cricket, ventral aspect.

maxillary segment. In many Orthoptera the open mouth of the apodeme can be seen on the lateral aspect of the head, just above the articulation of the maxilla (Fig. 48). In the Acridiidæ (Fig. 109) these apodemes bear a striking resemblance to the lateral apodemes of the thorax (Fig. 106), except that the ventral process of the maxillary apodeme is much more prominent, and the two from the opposite sides of the head meet and coalesce, thus forming the caudal part of the body of the tentorium.

Fig. 109.—Head of *Melanoplus*, caudal aspect.

The anterior arms of the tentorium.—Each anterior arm of the tentorium (Fig. 107, 108, 110, *at*) is an invagination of the body-wall which opens on the margin of the antecoxal piece of the mandible when it is distinct; if this part is not distinct the apodeme opens between the clypeus and the front (Fig. 46, *at*).

The dorsal arms of the tentorium.—Each dorsal arm of the tentorium arises from the side of the body of the tentorium between the anterior and posterior arms and extends either to the front or to the margin of the antennal sclerite (Fig. 107, 108, 110, *dt*).

The frontal plate of the tentorium.—In the cockroaches the anterior arms of the tentorium meet and fuse, forming a broad plate situated between the crura cerebri and the mouth; this plate was termed by us the *frontal plate of the tentorium* (Fig. 107, *fp*). On each side, an extension of this plate connects it with the body of the tentorium; these enclose a circular opening through which pass the crura cerebri.

Other cervical apodemes and some chitinized tendons are described in the paper cited above.

The endothorax.—The internal skeleton of the thorax is commonly termed the *endothorax;* under this head are not included the internal processes of the appendages.

The endothorax is composed of invaginations of each of the sections of a thoracic ring. Those portions that are derived from tergites are termed *phragmas;* those derived from the pleurites, *lateral apodemes;* and those, from the sternites, *furcæ.*

Fig. 110.—Tentorium of *Melanoplus*, cephalic aspect. The distal end of the dorsal arms detached.

The phragmas.—A phragma is a transverse partition extending entad from the front or the hind margin of a tergite; three of them are commonly recognized; these were designated by Kirby and Spence (1826) the *prophragma,* the *mesophragma,* and the *metaphragma;* but, as they do not arise one from each segment of the thorax, and arise differently in different insects, these terms are misleading. No phragma is borne by the prothorax; the mesothorax may bear two and the metathorax one, or the mesothorax one and the metathorax two. A more definite terminology is that used by Snodgrass ('09) by which the anterior phragma of any segment is termed the *prephragma* of that segment, and the posterior phragma of any segment is termed the *postphragma* of that segment.

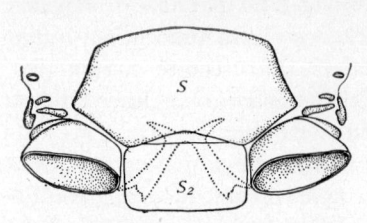

Fig. 111.—Ventral aspect of the metathorax of *Stenopelmatus*. The position of the furca within the body is represented by a dotted line.

The lateral apodemes.—Each lateral apodeme is an invagination of the body-wall between the episternum and the epimeron. The lateral apodemes are referred to above (Fig. 106).

The furcæ.—Each furca is an invagination of the body-wall arising between the sternum and the sternellum (Fig. 111); when the sternellum is obsolete, as it is in most insects, the furca arises at the caudal margin of the segment (Fig. 112).

b. THE HYPODERMAL GLANDS

A gland is an organ that possesses the function of either transforming nutritive substances, which it derives from the blood, into some useful substance, as mucus, wax, or venom, or of assimilating and removing from the body waste material.

The different glands vary greatly in structure; many are *unicellular*, the gland consisting of a single cell, which differs from the other cells of the epithelium of which it is a part in being larger and in possessing the secreting and excreting functions; others are *multicellular*, consisting of more than one cell, usually of many cells. In these cases the glandular area usually becomes invaginated, and provided with an efferent duct; and often the invagination is much branched.

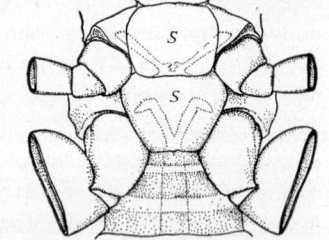

Fig. 112.—Ventral aspect of the meso—and metathorax of *Gryllus;* the positions of the furcæ within the body are indicated by dotted lines.

The glands found in the body of an insect can be grouped under three heads; the hypodermal glands, the glands of the alimentary canal, and the glands of the reproductive organs. In this place reference is made only to the hypodermal glands, those developed from the hypodermis.

The Molting-fluid glands.—Under this head are classed those unicellular, hypodermal glands that secrete a fluid that facilitates the process of molting, as described in the next chapter (Fig. 113).

While molting-fluid glands are very numerous and conspicuous in certain insects, those living freely exposed where there exists the greatest liability to rapid desiccation, Tower ('06) states that he has never found these glands in larvæ that live in burrows, or in the soil, or in cells; in these cases the molting fluid is apparently secreted by the entire hypodermal layer.

Fig. 113.—Molting-fluid glands of the last larval instar of *Leptinotarsa decimlineata*, just before pupation; *le*, larval epidermis; *ld*, larval dermis; *mf*, molting fluid; *pe*, forming pupal epidermis; *h*, hypodermis; *g*, molting fluid gland (After Tower).

Glands connected with setæ.—There are in insects several kinds of glands in which the outlet of the gland is through the lumen of a seta. The function of the excretions of these glands is various as indicated below. There are also differences in the manner of issuance of the excretion from the seta. In some cases, as in the tenent hairs on the feet of certain insects, the excretion can be seen to issue through a pore at the tip of the seta. In some kinds of venomous setæ the tip of the seta breaks off in the wound made by it and thus sets free the venom. But in most cases the manner of issuance has not been determined, although it is commonly believed to be by means of a minute pore or pores in the seta, the thickness of the wall of the seta making it improbable that the excretion passes from the seta by osmosis.

The structure of a glandular seta is illustrated by Figure 114; the essential difference between such a seta and an ordinary one, that is a

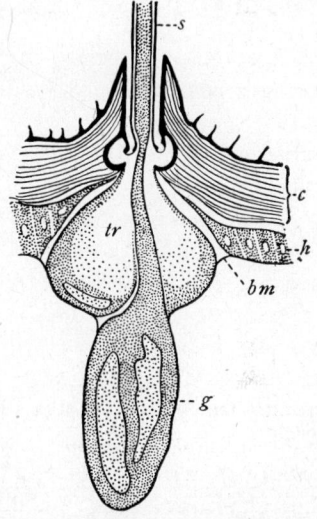

Fig. 114.—Glandular seta; *s*, seta; *c*, cuticula; *h*, hypodermis; *bm*, basement membrane; *tr*, trichogen; *g*, gland (After Holmgren).

clothing hair, is that there is connected with it, in addition to the trichogen cell which produced it, the gland cell which opens through it.

In most of the published figures of glandular setæ there is no indication that these organs are supplied with nerves; but in some cases a nerve extending to the gland cell is clearly shown. This condition may be found to be general when more extended investigations of glandular cells have been made. The best known kinds of glandular setæ are the following:

Venomous setæ and spines.—These are best known in larvæ of Lepidoptera, several common species of which possess stinging hairs; among these are *Lagoa crispata, Sibine stimulea, Automeris io*, and the brown-tail moth, *Euproctis chrysorrhœa*.

Androconia.—The term androcōnia* is applied to some peculiarly modified scales on the wings of certain male butterflies. These are the outlets of glands, which secrete a fluid with an agreeable odor; the supposed function of which is to attract the opposite sex, like the beautiful plumage and songs of male birds. The androconia differ marvelously from ordinary scales in the variety of their forms (Fig. 115). They usually occur in patches on the upper surface of the fore wings; and are usually concealed by other scales; but they are scattered in some butterflies. The most familiar examples of grouped androconia are those that occur in the discal stigma of the hair-streaks, in the brand of certain skippers and in the costal fold of others, and in the scent-pouch of the male of the monarch butterfly

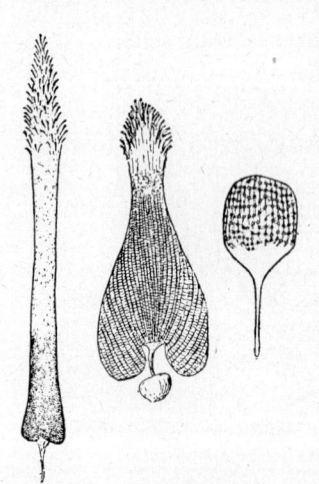

Fig. 115.—Androconia from the wings of male butterflies (After Kellogg).

The specific scent-glands of females.—The well-known fact that if an unfertilized female moth be confined in a cage or otherwise in the open many males of the same species as the female will be attracted to it, and sometimes evidently from a great distance, leads to the conclusion that there must emanate from the female a specific odor. The special glands producing this odor have not been recognized.

Tenent hairs.—In many insects the pulvilli or the empodia are clothed with numerous hairs that are the outlets of glands which

---

*Androcōnia: *andro*- (ἀνήρ, ἀνδρός), male; *conia* (κονία), dust.

secrete an adhesive fluid; this enables the insect to walk on the lower surface of objects (Fig. 116).

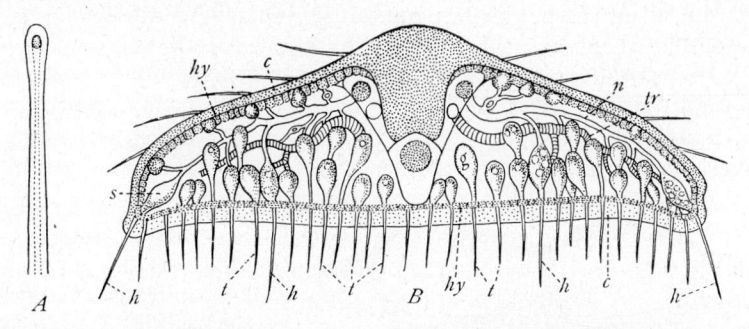

Fig. 116.—A, terminal part of a tenent hair from *Eupolus*, showing canal in the hair and opening near the tip; B, cross-section through a tarsal segment of *Telephorus;* c, cuticula; g, gland of tenent hair; h, h, tactile hairs; hy, hypodermis; n, nerve; s, sense-cell of tactile hair; t, t, tenent hairs (After Dewitz).

The osmeteria.—In many insects there are hypodermal glands that open into sac-like invaginations of the body-wall which can be evaginated when the insect wishes to make use of the secretion produced by these glands; such an organ is termed an *osmeterium*. The invagination of the osmeterium admits of an accumulation of the products of the gland within the cavity of the sac thus formed; when the osmeterium is evaginated the secretion becomes exposed to the air, being then on the outside of the osmeterium, and rapid diffusion of the secretion results.

The most familiar examples of osmeteria are those of the larvæ of the swallow-tailed butterflies, which are forked, and are thrust out from the upper part of the prothorax when the caterpillar is disturbed, and which diffuse a disagreeable odor (Fig. 117). They are obviously organs of defense.

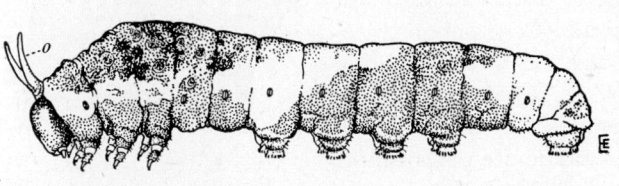

Fig. 117.—Larva of *Papilio thoas;* o, osmeterium expanded.

Osmeteria are present in the larvæ of certain blue butterflies, Lycænidæ. These are in the seventh and eighth abdominal segments, and secrete a honey-dew, which attracts ants that attend and probably protect the larvæ. The osmeteria of many other caterpillars have been described.

Glands opening on the surface of the body.—There are several kinds of hypodermal glands, differing widely in function, that open on the surface of the body; among the best known of these are the following:

Wax-glands.—The worker honeybee has four pairs of wax-glands; these are situated on the ventral wall of the second, third, fourth, and fifth abdominal segments, and on that part of the segment which is overlapped by the preceding segment; each gland is simply a disc-like area of the hypodermis (Fig. 118). The cuticle covering each gland is smooth and delicate, and is known as a wax plate. The wax exudes through these plates and accumulates, forming little scales, which are used in making the honey-comb.

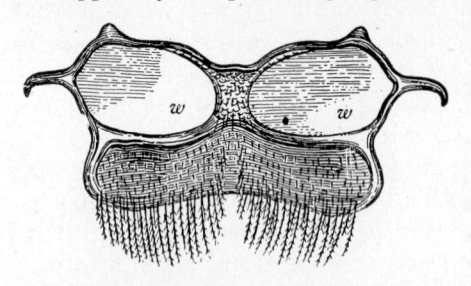

Fig. 118.—Wax-plates of the honeybee (After Cheshire).

Wax-glands exist in many of the Homoptera. In some of these the unicellular wax-glands are distributed nearly all over the body; and the product of these glands forms, in some, a **powdery covering; in others**, a clothing of threads; and in still others a series of plates (Fig. 119). Certain coccids excrete wax in considerable quantities. China wax, which was formerly an article of commerce, is the excretion of a coccid known as Pe-la (*Ericerus Pe-la*).

Froth-glands of spittle-insects.—In the spittle-insects (Cercopidæ) there are large hypodermal glands in the pleural regions of the seventh and eighth abdominal segments, which open through numerous minute pores in the cuticula. These glands secrete a muci-laginous substance, which is mixed with a fluid excreted from the anus, and thus fits it for the retention of bubbles of air included in it by means of abdominal appendages (Guilbeau '08).

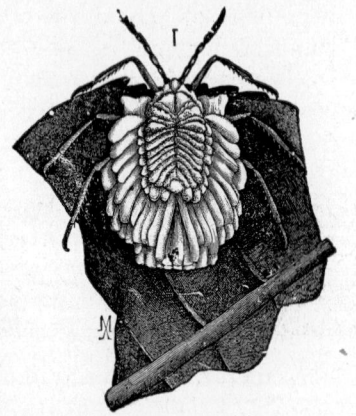

Fig. 119.—*Orthesia*, greatly enlarged.

Stink-glands.—Glands that secrete a liquid having a fetid odor and that are doubtless defensive exist in many insects. In the stink-bugs

(Pentatomidæ) the fluid is excreted through two openings, one on each side of the lower side of the body near the middle coxæ; in the bed-bug (*Cimex*), the stink-glands open in the dorsal wall of the first three abdominal segments; in *Dytiscus,* the glands open on the prothorax; and in certain Coleoptera they open near the caudal end of the body. These are merely a few examples of the many glands of this type that are known.

The cephalic silk-glands.—In the Lepidoptera, Trichoptera, and Hymenoptera, there is a pair of glands that secrete silk, and which open through the lower lip. These glands are designated as the cephalic silk-glands to distinguish them from the silk-glands of certain Neuroptera and Coleoptera in which the silk is produced by modified Malpighian vessels and is spun from the anus.

The cephalic silk-glands are elongate and coiled; they often extend nearly the whole length of the body; the two ducts unite and the single terminal duct opens through the lower lip, and is not connected with the mouth cavity. These glands are a pair of salivary glands which have been transformed into silk organs. According to Carrière

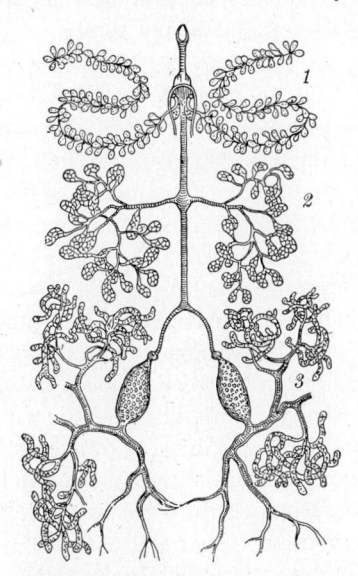

Fig. 120.—The salivary glands of the honeybee (After Cheshire).

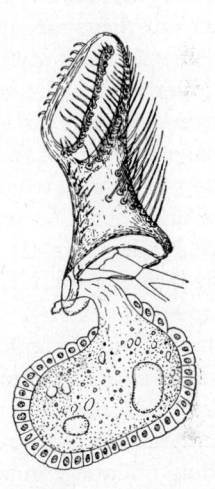

Fig. 121.—The mandibular gland of a honeybee.

and Burger ('97), who studied their development in the embryo of a bee, they are developed from the rudiments of the spiracles of the first thoracic segment. In the later development they move cephalad and the paired openings become a single one. This is the reason that in the adult there are no spiracles in the prothorax.

The Salivary glands.—The term salivary glands is a general one, applied to various glands opening in the vicinity of the mouth. The number of these varies greatly in different insects; the maximum number is found in the Hymenoptera. In the adult worker honey-bee, for example, there are four pairs of glands opening into the mouth; three of these are represented in Figure 120 and the fourth in Figure 121. These are designated as the supracerebral glands (Fig. 120, *1*), the postcerebral glands (Fig. 120, *2*), the thoracic glands (Fig. 120, *3*), and the mandibulary glands (Fig. 121), respectively.

II. THE MUSCLES

There exist in insects a wonderfully large number of muscles; some of these move the segments of the body, others move the appendages of the body, and still others are found in the viscera. Those of the viscera are described later in the accounts of the organs in which they occur.

The muscles that move the segments of the body form several layers just within the body-wall, to which they are attached. The inner layer of these is well shown in Figure 122, which is a copy of one of the plates in the great work by Lyonet (1762) on the anatomy of a caterpillar, *Cossus ligniperda*. The two figures on this plate represent two larvæ which have been split open lengthwise, one on the middle line of the back (Fig. 5), and one on the middle line of the ventral surface (Fig. 4); in each case the alimentary canal has been removed, so that only those organs that are attached quite closely to the body-wall are left. The bands of parallel fibers are the muscles that move the segments. It should be borne in mind, however, that only a single layer of muscles is represented in these figures, the layer that would be seen if a caterpillar were opened in the way indicated. When these muscles are cut away many other muscles are found extending obliquely in various directions between these muscles and the body-wall.

In the head and thorax of adult insects the arrangement of the muscles is even more complicated; for here the muscles that move the appendages add to the complexity of the muscular system.

As a rule, the muscles of insects are composed of many distinct fibers, which are not enclosed in tendinous sheaths as with Verte-

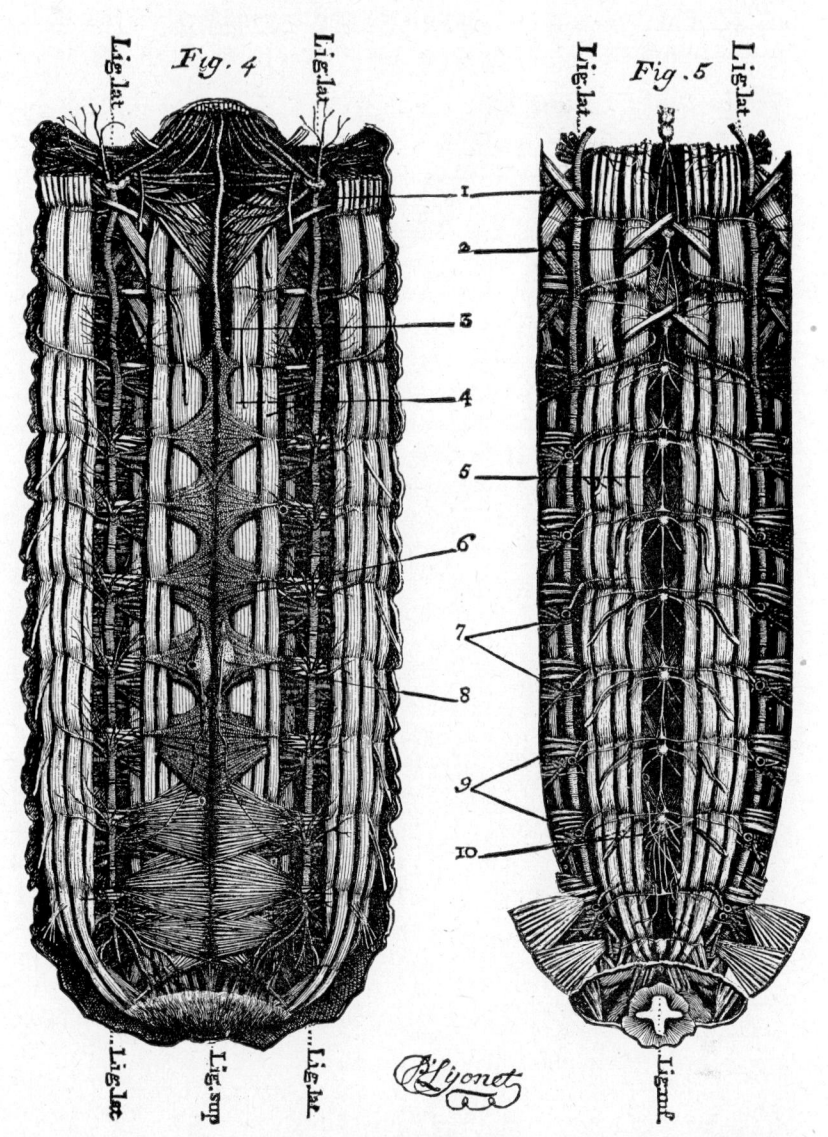

Fig. 122.—Internal anatomy of a caterpillar, *Cossus ligniperda;* 1, principal longitudinal trachæ; 2, central nervous system; 3, aorta; 4, longitudinal dorsal muscles; 5, longtiudinal ventral muscles; 6, wings of the heart; 7, tracheal trunks arising near the spiracles; 8, reproductive organs; 9, vertical muscles; 10, last abdominal ganglion (From Lyonet).

brates. But the muscles that move the appendages of the body are furnished with a tendon at the end farthest from the body (Fig. 123).

The muscles of insects appear very differently from those of Vertebrates. In insects, the muscles are either colorless and transparent, or yellowish white; and they are soft, almost of a gelatinous consistency; notwithstanding this they are very efficient. The fibers of insect muscles are usually, if not always, of the striated type.

Fig. 123—A leg of a May-beetle (After Straus-Durckheim).

Much has been written regarding the muscular power of insects, which has been supposed to be extraordinarily great; the power of leaping possessed by many and the great loads, compared to the weight of the body of the insect, that insects have drawn when harnessed to them by experimenters, have been cited as illustrating this. But it has been pointed out that these conclusions are not warranted; that the comparative contractile force of muscles of the same kind depends on the number and thickness of the fibers, that is, on the comparative areas of the cross-sections of the muscles compared; that this sectional area increases as the square of any linear dimension, while the weight of similar bodies increases as the cube of any linear dimension; and consequently, that the muscles of the legs of an insect one fourth inch long and supporting a load 399 times its own weight, would be subjected to the same stress, per square inch of cross-section, as they would be in an insect 100 inches long of precisely similar shape, that carried only its own weight. We thus see that it is the small size of insects rather than an unusual strength of their muscles, that makes possible the apparently marvelous exhibitions of muscular power.

Detailed accounts of the arrangement of the muscles in particular insects have been published by various writers: among the more important of these monographs are the following: Lyonet (1762), on the larva of a cossid moth; Straus-Durckheim (1828), on a May-beetle; Newport (1839), on the larva of a Sphinx moth; Lubbock (1858), on the larva *Pygaera bucephala;* Berlese ('09a), on several insects; and Forbes ('14) on caterpillars.

THE INTERNAL ANATOMY OF INSECTS

III. THE ALIMENTARY CANAL AND ITS APPENDAGES

a. THE MORE GENERAL FEATURES

The alimentary canal is a tube extending from one end of the body to the other. In some larvæ, its length is about the same as that of the body; in this case it extends in a nearly straight line, occupying

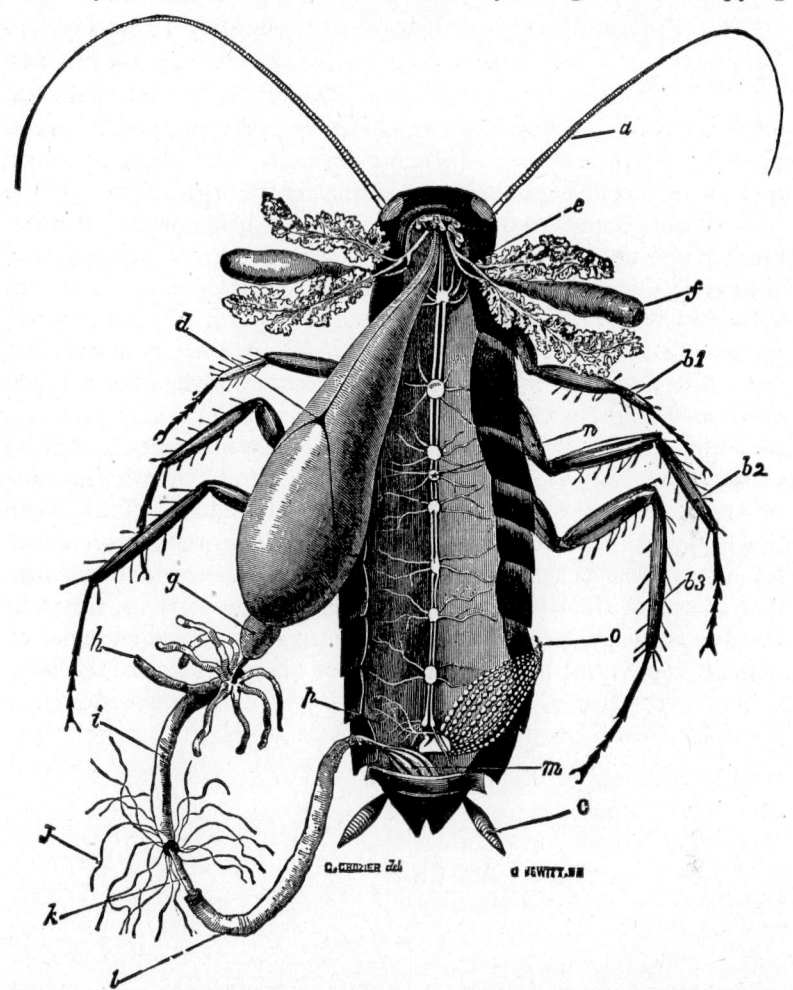

Fig. 124.—Internal anatomy of a cockroach, *Periplaneta orientalis;* a, antennæ; b_1, b_2, b_3, first, second, and third legs; c, cerci; d, ventricular ganglion; e, salivary duct; f, salivary bladder; g, gizzard or proventriculus; h, hepatic cœca; i, mid-intestine; j, Malpighian vessels; k, small intestine; l, large intestine; m, rectum; n, first abdominal ganglion; o, ovary; p, sebaceous glands (From Rolleston).

the longitudinal axis of the body, as is represented in the diagram given above (Fig. 105). In most insects, however, it is longer than the body, and is consequently more or less convoluted (Fig. 124); great variations exist in the length of the alimentary canal as compared to the length of the body; it is longer in herbivorous insects than it is in those that are carnivorous.

The principal divisions.—Three chief divisions of the alimentary canal are recognized; these are termed the *fore-intestine*, the *mid-intestine*, and the *hind-intestine*, respectively. In the embryological development of the alimentary canal, the fore-intestine and the hind-intestine each arises as an invagination of the ectoderm, the germ layer from which the hypodermis of the body-wall is derived (p. 29). The invagination at the anterior end of the body, which develops into the fore-intestine, is termed the *stomodæum;* that at the posterior end, which develops into the hind-intestine, the *proctodæum*. Between these two deep invaginations of the outer germ layer of the embryo, the stomodæum and the proctodæum, and ultimately connecting them, there is developed an entodermal tube, the *mesĕnteron*, which becomes the mid-intestine.

These embryological facts are briefly stated here merely to elucidate two important features of the alimentary canal: first, the fore-intestine and the hind-intestine are invaginations of the body wall and consequently resemble it in structure, the chitinous lining of these two parts of the alimentary canal is directly continuous with the cuticula of the body wall, and the epithelium of these two parts and the hypodermis are also directly continuous; and second, the striking differences, pointed out later, in the structure of the mid-intestine from that of the fore- and hind-intestines are not surprising when the differences in origin are considered.

Imperforate intestines in the larvæ of certain insects.—In the larvæ of certain insects the lumen of the alimentary canal is not a continuous passage; in these larvæ, while food passes freely from the fore-intestine to the mid-intestine, there is no passage of the waste from the mid-intestine to the hind-intestine; there being a constriction at the point where the mid-intestine and hind-intestine join, which closes the passage during a part or the whole of the larval life. This condition has been observed in the following families:—

(a) *Hymenoptera.*—Proctotrypidæ (in the first larval instar), Ichneumonidæ, Formicidæ, Vespidæ, and Apidæ.

(b) *Diptera.*—Hippoboscidæ.

(c) *Neuroptera.*—Myrmeleonidæ, Osmylidæ, Sisyridæ, and Chrysopidæ. In these families the larvæ spin silk from the anus.

(d) *Coleoptera.*—In the Campodeiform larvæ of Stylopidæ and Meloidæ.

b. THE FORE-INTESTINE

The layers of the fore-intestine.—The following layers have been recognized in the fore-intestine:

The intima.—This is a chitinous layer which lines the cavity of the fore-intestine; it is directly continuous with the cuticula of the body-wall; and is molted with the cuticula when this is molted.

The epithelium.—This is a cell layer which is continuous with the hypodermis; it is sometimes quite delicate so that it is difficult to demonstrate it.

The basement membrane.—Like the hypodermis the epithelium is bounded on one side by a chitinous layer and on the other by a basement membrane.

The longitudinal muscles.—Next to the basement membrane there is a layer of longitudinal muscles.

The circular muscles.—Outside of the longitudinal muscles there is a layer of circular muscles.

The peritoneal membrane.—Surrounding the alimentary canal there is a coat of connective tissue, which is termed the peritoneal membrane. This is one of a few places in which connective tissue, so abundant in Vertebrates, is found in insects.

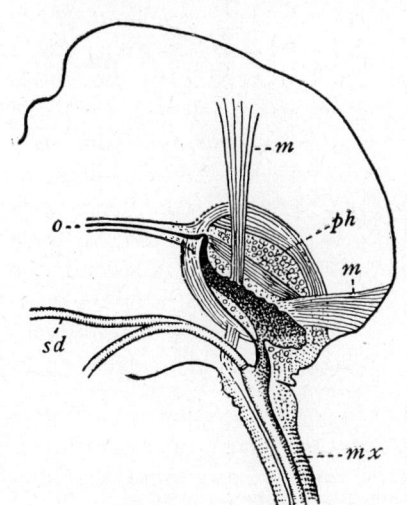

Fig. 125.—Longitudinal section through the head of *Anosa plexippus*, showing the interior of the left half; *mx*, left maxilla, the canal of which leads into the pharynx; *ph*, pharynx; *o*, oesophagus; *m*, *m*, muscles of the pharynx; *sd*, salivary duct (After Burges).

The regions of the fore-intestine.—Several distinct regions of the fore-intestine are recognized; but the extent of these regions differ greatly in different insects.

The pharynx.—The pharynx is not a well-defined region of the intestine; the term pharynx is commonly applied to a region between the mouth and the œsophagus; in mandibulate insects the pharynx is not distinct from the mouth-cavity; but in sucking insects the pharynx is a highly specialized organ, being greatly enlarged, muscular, and attached to the wall of the head by muscles. It is the pumping organ by which the liquid food is drawn into the alimentary canal. The pharynx of the milkweed butterfly (Fig. 125) is a good example of this type of pharynx.

The œsophagus.—The œsophagus is a simple tube which traverses the caudal part of the head and the cephalic part of the thorax. There are variations in the application of the term œsophagus depending on the presence or absence of a crop and of a proventriculus, which are modified portions of the œsophagus; when either or both of these are present, the term œsophagus is commonly restricted to the unmodified part of the fore-intestine.

The crop.—In many insects a portion of the œsophagus is dilated and serves as a reservoir of food; this expanded part, when present, is termed the *crop*. In the cockroach (Fig. 124) it is very large, comprising the greater part of the fore-intestine; in the ground-beetle *Carabus* (Fig. 126, *c*), it is much more restricted; this is the case also in the honeybee, where it is a nearly spherical sac in which the nectar is stored as it is collected from flowers and carried to the hive. In some insects the crop is a lateral dilatation of the œsophagus, and in some of these it is stalked.

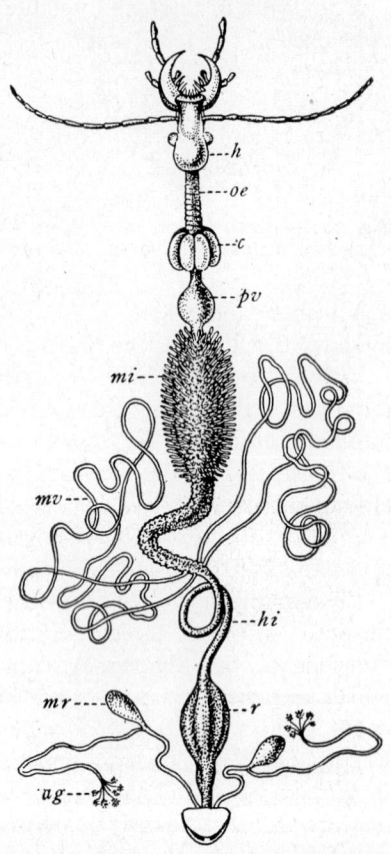

Fig. 126.—Alimentary canal of *Carabus auratus;* *h,* head; *oe,* œsophagus; *c,* crop; *pv,* proventriculus; *mi,* mid-intestine covered with villiform gastric cœca; *mv,* Malpighian vessels; *hi,* part of hind-intestine; *r,* rectum; *ag,* anal glands; *mr,* muscular reservoir (After Dufour).

The proventriculus.—In certain insects that feed on hard substances, the terminal portion of the fore-intestine, that part immediately in front of the mid-intestine or ventriculus, is a highly specialized organ in which the food is prepared for entrance into the more delicate ventriculus; such an organ is termed the *proventriculus* (Fig. 126, *pv*). The characteristic features of a proventriculus are a remarkable development of the chitinous intima into folds and teeth and a great increase in the size of the muscles of this region. The details of the structure of this organ vary greatly in different insects; a cross-section of the proventriculus of the larva of *Corydalus* (Fig. 127) will serve to illustrate its form. In the proventriculus, the food is both masticated and more thoroughly mixed with the digestive fluids.

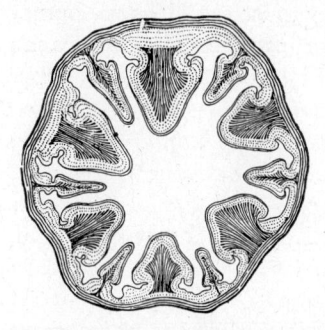

Fig. 127.—Cross-section of the proventriculus of a larva of *Corydalus*.

The œsophageal valve.—When the fore-intestine projects into the mid-intestine, as shown in Figure 128, the folded end of the fore-intestine is termed the *œsophageal valve*.

c. THE MID-INTESTINE

The mid-intestine is the intermediate of the three principal divisions of the alimentary canal, which are distinguished by differences in their embryological origins, as stated above. The mid-intestine is termed by different writers the *mesĕnteron*, the *stomach*, the *chylific ventricle*, the *chylestomach*, and the *ventrĭculus*.

The layers of the mid-intestine.—The structure of the mid-intestine differs markedly from that of the fore-intestine. In the mid-intestine there is no chitinous intima, and the relative positions of the circular and longitudinal muscles are reversed.

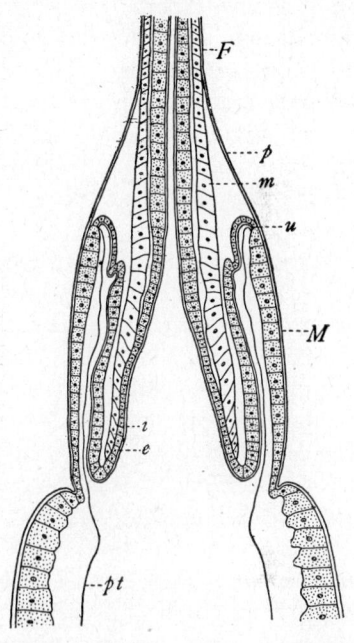

Fig. 128.—The œsophageal valve of a larva of *Simulium;* F, fore-intestine: M, mid-intestine; *u*, point of union of fore-intestine and mid-intestine; *p*, peritoneal membrane; *i*, intima of fore-intestine; *e*, epithelium of fore-intestine; *pt*, peritrophic membrane; *m*, muscles.

The sequence of the different layers is as follows: a lining *epithelium*, which is supported by a *basement membrane*, a layer of *circular muscles*, a layer of *longitudinal muscles*, and a *peritoneal membrane*.

The epithēlium.—The epithelium of the mid-intestine is very conspicuous, being composed of large cells, which secrete a digestive fluid. These cells break when they discharge their secretion and are replaced by new cells, which are developed in centers termed *nidi* (Fig. 129, *n*).

The extent of the digestive epithelium is increased in many insects by the development of pouch-like diverticula of the mid-intestine, these are the *gastric cæca* (Fig. 124, *h*). These differ greatly in number in different insects and are wanting in some. In some predaceous beetles they are villiform and very numerous (Fig. 126, *mi*).

The peritrophic membrane.—In many insects there is a membranous tube which is formed at or near the point of union of the fore-intestine and the mid-intestine and which incloses the food so that it does not come in contact with the delicate epithelium of the mid-intestine; this is known as the *peritrophic membrane* (Fig. 128, *pt*). As a rule this membrane is found in insects that eat solid food and is lacking in those that eat liquid food. It is obvious that the digestive fluid and the products of digestion pass through this membrane. It is continuously formed at its point of origin and passes from the body inclosing the excrement.

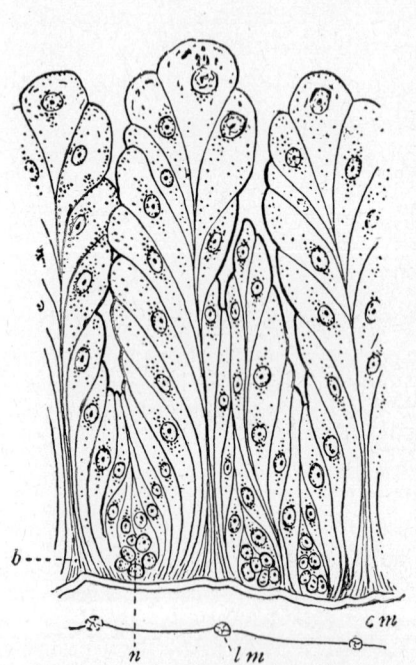

Fig. 129.—Resting epithelium of mid-intestine of a dragon-fly naiad; *b*, bases of large cells filled with digestive fluid; *cm*, space filled by circular muscles; *lm*, longitudinal muscles; *n*, nidus in which new cells are developing (From Needham).

d. THE HIND-INTESTINE

The layers of the hind-intestine.—The layers of the hind-intestine are the same as those of the fore-intestine described above, except that a greater or less number of circular muscles exist between the basement membrane of the epithelial layer and the layer of longitudinal muscles. The sequence of the layers of the hind-intestine is, therefore, as follows: the *intima*, the *epithelium*, the *basement membrane*, the *ental circular muscles*, the *longitudinal muscles*, the *ectal circular muscles*, and the *peritoneal membrane*.

The regions of the hind-intestine.—Three distinct regions are commonly recognized in the hind-intestine, these are the *small intestine* (Fig. 124, *k*), the *large intestine* (Fig. 124, *l*), and the *rectum* (Fig. 124, *m*).

The Malpighian vessels.—There open into the beginning of the hind-intestine two or more simple or branched tubes (Fig. 124, *j*), these are the *Malpighian vessels*. The number of these vessels varies in different insects but is very constant within groups; there are either two, four, or six of them; but, as a result of branching, there may appear to be one hundred or more. The function of the Malpighian vessels has been much discussed; it was formerly believed to be hepatic, but now it is known that normally it is urinary.

The Malpighian vessels as silk-glands.—There are certain larvæ that in making their cocoons spin the silk used from the anus. These larvæ are chiefly found among those in which the passage from the mid-intestine to the hind-intestine is closed. The silk spun from the anus is secreted by the Malpighian vessels.

Among the larvæ in which the Malpighian vessels are known to secrete silk are those of the Myrmeleonidæ, *Osmylus* (Hagen 1852), *Sisyra* (Anthony '02), *Lebia scapularis* (Silvestri '05), and the Coccidæ (Berlese '96). Berlese states that the Malpighian vessels secrete the woof of the scale of the Coccidæ.

The cæcum.—In some insects there is a pouch-like diverticulum of the rectum, this is the *cæcum*.

The anus.—The posterior opening of the alimentary canal, the *anus*, is situated at the caudal end of the abdomen.

IV. THE RESPIRATORY SYSTEM

Insects breathe by means of a system of air-tubes, which ramify in all parts of the body and its appendages; these air-tubes are of two kinds, which are termed *tracheæ* and *tracheoles*, respectively. In adult **insects** and in most nymphs and larvæ, the air is received through openings in the sides of the segments of the body, which are known as *spiracles* or *stigmāta*.

Many insects that live in water are furnished with special devices for obtaining air from above the water; but with naiads and a few aquatic larvæ the spiracles are closed; in these insects the air is purified by means of gill-like organs, termed *tracheal gills*. A few insects have *blood-gills*.

Two types of respiratory systems, therefore, can be recognized: first, the *open type*, in which the air is received through spiracles; and second, the *closed type*, in which the spiracles are not functional.

a. THE OPEN OR HOLOPNEUSTIC TYPE OR RESPIRATORY ORGANS

That form of respiratory organs in which the trachæ communicate freely with the air outside the body through open spiracles is termed the open or holopneustic type.*

As the open type of respiratory organs is the most common one, those features that are common to both types will be discussed under this head as well as those that are peculiar to this type. Under the head of closed respiratory organs will be discussed only those features distinctly characteristic of that type.

I. *The Spiracles*

The position of the spiracles.—The spiracles are situated one on each side of the segments that bear them or are situated on the lateral aspects of the body in the transverse conjunctivæ.

The question of the position of the spiracles has not been thoroughly investigated; but I believe that normally the tracheæ, of which

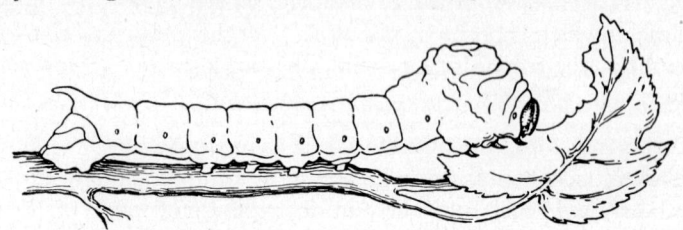

Fig. 130.—Lateral view of a silk worm showing the spiracles (After Verson).

the spiracles are the mouths, are invaginations of the transverse conjunctivæ between segments. From this normal position a spiracle may migrate either forward or backward upon an adjacent segment (Fig. 130).

The number of spiracles.—The normal number of spiracles is ten pairs; when in their normal position, there is a pair in front of the

---

*Hŏlopneustic: *holo* (ὅλος), *whole*; *pneuma* (πνεῦμα), *breath*.

second and third thoracic segments and the first to the eighth abdominal segments, respectively. There are none in the corresponding position in front of the first thoracic segment. See account of cephalic silk-glands p. 103.

The two pairs of thoracic spiracles are commonly distinguished as the mesothoracic and the metathoracic spiracles; that is each pair of spiracles is attributed to the segment in front of which it is normally situated. Following this terminology there are no prothoracic spiracles; although sometimes the first pair of spiracles is situated in the hind margin of the prothorax, having migrated forward from its normal position. It would be better to designate the thoracic spiracles as the first and second pairs of thoracic spiracles, respectively; in this way the same term would be applied to a pair of spiracles whatever its position. There are many references in entomological works to "prothoracic spiracles," but these refer to the pair of spiracles that are more commonly designated the mesothoracic spiracles.

In many cases the abdominal spiracles have migrated back upon the segment in front of which they are normally situated, being frequently situated upon the middle of the segment.

The statements made above refer to the normal number and distribution of spiracles; but a very wide range of variations from this type exists. Perhaps the most abnormal condition is that found in the genus *Sminthurus* of the Collembola, where there is a single pair of spiracles which is borne by the neck. In the Poduridæ, also of the Collembola, the respiratory system has been lost, there being neither tracheæ nor spiracles.

Terms indicating the distribution of the spiracles.—The following terms are used for indicating the distribution of the spiracles; they have been used most frequently in descriptions of larvæ of Diptera. These terms were formed by combining with *pneustic* (from *pneo*, to breathe) the following prefixes: *peri-*, around, about; *pro-*, before; *meta-* after; and *amphi*, both.

Peripneustic.—Having spiracles in a row on each side of the body, the normal type.

Propneustic.—With only the first pair of spiracles.

Metapneustic.—With only the last pair of spiracles.

Amphipneustic.—With a pair of spiracles at each end of the body.

The structure of spiracles.—In their simplest form the spiracles or stigmata are small round or oval openings in the body-wall. In many cases they are provided with hairs to exclude dust; in some, as in the larva of *Corydalus*, each spiracle is furnished with a lid (Fig. 131, *a*); in fact, very many forms of spiracles exist. Usually each spiracle opens by a single aperture; but in some larvæ and pupæ of Diptera they have several openings (Fig. 131, *b*).

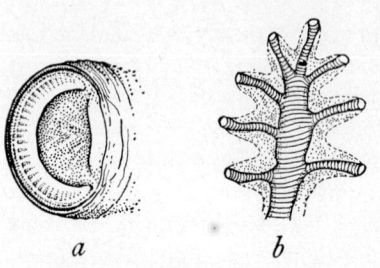

Fig. 131.—Spiracles; *a*, of the larva of *Corydalus*; *b*, of the larva of *Drosophila amœna*.

The closing apparatus of the tracheæ.—Within the body, a short distance back of the spiracle, there is an apparatus consisting of several chitinous parts, surrounding the trachea, and moved by a muscle, by which the trachea can be closed by compression (Fig. 132). This is the *closing apparatus of the trachea*. The closing of this apparatus and the contraction of the body by the respiratory muscles is supposed to force the air into the tracheoles, which are the essential respiratory organs.

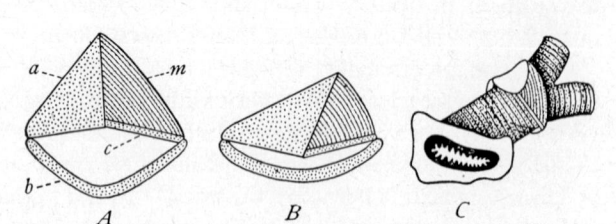

Fig. 132.—Diagrams representing the closing apparatus of the tracheæ; *a, b, c*, chitinous parts of the apparatus; *m*, muscle; A, apparatus open; B, apparatus closed; C, spiracle and trunk of trachea showing the position of the apparatus. (From Judeich and Nitsche).

2. THE TRACHEÆ

Each spiracle is the opening of an air-tube or trachea. The main tracheal trunk which arises from the spiracle soon divides into several branches, these in turn divide, and by repeated divisions an immense number of branches are formed. Every part of the body is supplied with tracheæ.

In a few insects the group of tracheæ arising from a spiracle is not connected with the groups arising from other spiracles; this is the case in *Machilis* (Fig. 133). In most insects, however, each group of tracheæ is connected with the corresponding groups in adjacent segments by one or more longitudinal tracheæ, and is also connected

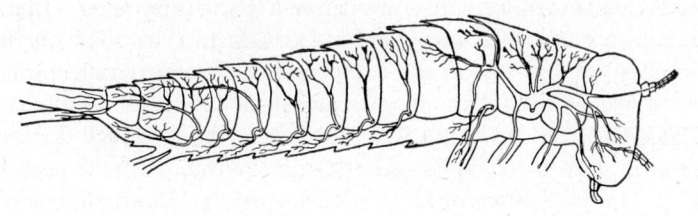

Fig. 133.—The tracheæ of *Machilis* (From Oudemans).

with the group on the opposite side of the same segment by one or more transverse tracheæ (Fig. 134).

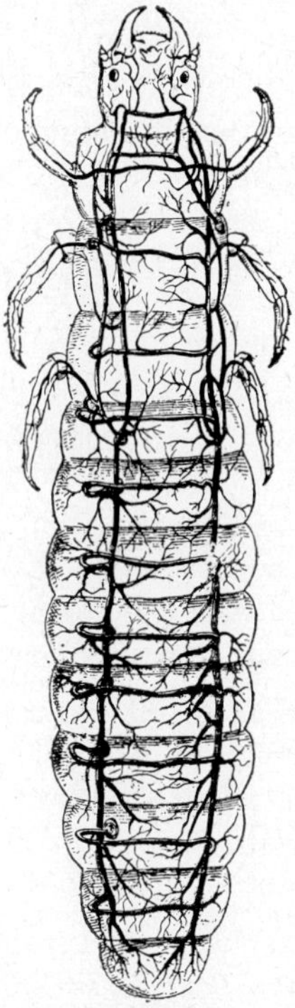

The structure of the tracheæ.—The fact that in their embryological development the tracheæ arise as invaginations of the body-wall, makes it easy to understand the structure of the tracheæ. The three layers of the body-wall are directly continuous with corresponding layers in the wall of a trachea (Fig. 135). These layers of a trachea are designated as the *intima*, the *epithelium*, and the *basement membrane*.

The *intima* is the chitinous inner layer of the tracheæ. It is directly continuous with the cuticula of the body-wall, and like the cuticula is molted at each ecdysis.

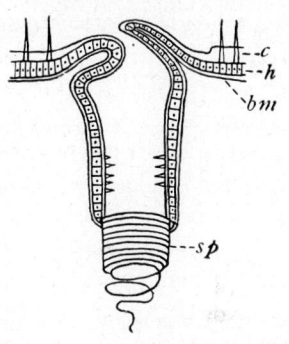

A peculiar feature of the intima of tracheæ is the fact that it is furnished with thickenings which extend spirally. These give the tracheæ their characteristic transversely striated appearance. If a piece of one of the larger tracheæ be

Fig. 134.—Larva of *Cantharis vesicatoria*, showing the distribution of tracheæ (From Henneguy after Beauregard).

Fig. 135.—Section of a trachea and the body-wall; *c*, cuticula; *h*, hypodermis; *bm*, basement membrane; *sp*, spiral thickening of the intima, the tænidium.

pulled apart the intima will tear between the folds of the spiral thickening, and the latter will uncoil from within the trachea like a thread (Fig. 135). The spiral thickening of the intima of a trachea is termed the *tænĭdium*. In some insects there are several parallel tænidia; so that when an attempt is made to uncoil the thread a ribbon-like band is produced, composed of several parallel threads. This condition exists in the larger tracheæ of the larva *Corydalus*.

The *epithēlium* of the trachea is a cellular layer, which is directly continuous with the hypodermis of the body-wall.

The *basement membrane* is a delicate layer, which supports the epithelium, as the basement membrane of the body-wall supports the hypodermis.

3. The Tracheoles

The tracheoles are minute tubes that are connected with the tips of tracheæ or arise from their sides, but which differ from tracheæ in their appearance, structure, and mode of origin; they are not small tracheæ, but structures that differ both histologically and in their origin from tracheæ.

The tracheoles are exceedingly slender, measuring less than one micron in diameter; ordinarily they do not taper as do tracheæ; they contain no tænidia; and they rarely branch, but often anastomose which gives them a branched appearance (Fig. 136, *t* and 138 B, *t*).

Each tracheole is of unicellular origin, and is, at first, intracellular in position, being developed coiled within a single cell of the epithelium of a trachea. In this stage of its development it has no connection with the lumen of the trachea in the wall of which it is developing, being separated from it by the intima of the trachea. A subsequent molting of the intima of the trachea opens a connection between the lumen of the tracheole and the trachea. At the same time or a little later the tracheole breaks forth from its mother cell, uncoils, and extends far beyond the cell in which it was developed.

The tracheoles are probably the essential organs of respiration, the tracheæ acting merely as conduits of air to the tracheoles.

4. The Air-Sacs

In many winged insects there are expansions of the tracheæ, which are termed *air-sacs*. These vary greatly in number and size. In the honeybee there are two large air-sacs which occupy a considerable part of the abdominal cavity; while in a May-beetle there are hundreds of small air-sacs. The air-sacs differ from tracheæ in lacking tænidia.

As the air-sacs lessen the specific gravity of the insect they probably aid in flight; as filling the lungs with air makes it easier for a man to float in water; in each case there is a greater volume for the same weight.

5. Modifications of the open type of respiratory organs in aquatic insects

There are many insects in which the spiracles are open that live in water; these insects breathe air obtained from above the surface of the water. Some of these insects breathe at the surface of the water,

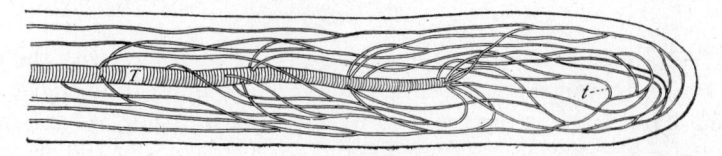

Fig. 136.—Part of a tracheal gill of the larva of *Corydalus;* T, trachea; *t,* tracheoles.

as the larvæ and pupæ of mosquitoes, the larvæ of *Eristalis*, and the Nepidæ; others get a supply of air and carry it about with them beneath the surface of the water, as the Dytiscidæ, the Notonectidæ and the Corisidæ. The methods of respiration of these and of other aquatic insects with open spiracles are described in the accounts of these insects given later.

b. THE CLOSED OR APNEUSTIC TYPE OF RESPIRATORY ORGANS

That type of respiratory organs in which the spiracles do not function is termed the closed or apneustic* type; it exists in naiads and in a few aquatic larvæ.

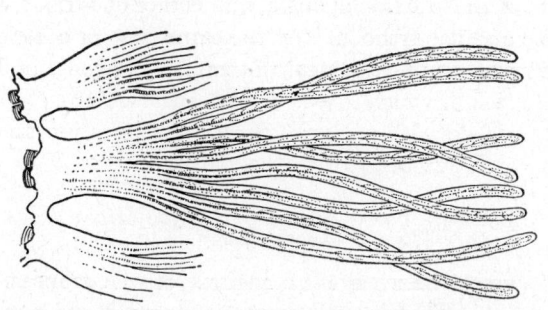

1. The Tracheal Gills

Fig. 137.—Part of a tuft of tracheal gills of a larva of *Corydalus.*

In the immature insects mentioned above, the air in the body is purified by means of organs known as *tracheal gills*.

*Apneūstic: *apneustos* (ἄπνευστος), without breath

These are hair-like or more or less plate-like expansions of the body-wall, abundantly supplied with tracheæ and tracheoles. Figures 136 **and 137 represent** a part of a tuft of hair-like tracheal gills of a larva of *Corydalus* and figure 138 a plate-like tracheal gill of a naiad of a damsel-fly. In these tracheal gills the tracheoles are separated from the air in the water only by the delicate wall of the tracheal gill which admits of the transfer of gases between the air in the tracheoles and the air in the water.

Tracheal gills are usually borne by the abdomen, sometimes by the thorax, and in case of one genus of stone-flies by the head. They pertain almost exclusively to the immature stages of insects; but stone-flies of the genus *Pteronarcys* retain them throughout their existence. In the naiads of the Odonata the rectum is supplied with many tracheæ and functions as a tracheal gill.

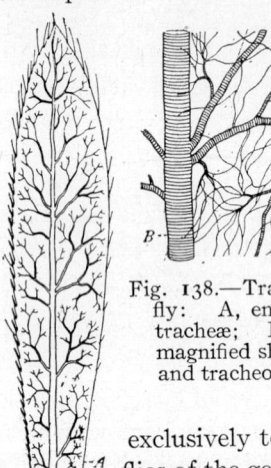

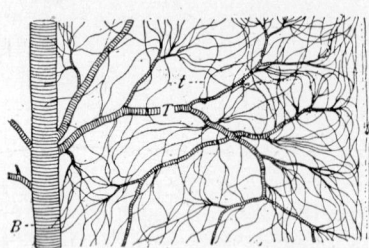

Fig. 138.—Tracheal gill of a damsel-fly: A, entire gill showing the tracheæ; B, part of gill more magnified showing both tracheæ (T) and tracheoles (t).

2. *Respiration of Parasites*

It is believed that internal parasitic larvæ derive their air from air that is contained in the blood of their hosts, and that this is done by osmosis through the cuticula of the larva, the skin of the larva being furnished with a network of fine tracheæ (Seurat '99).

3. *The blood-gills*

Certain aquatic larvæ possess thin transparent extensions of the body wall, which are filled with blood, and serve as respiratory organs. These are termed *blood-gills*.

Blood-gills have been observed in comparatively few insects; among them are certain trichopterous larvæ; the larva of an exotic beetle, *Pelobius;* and a few aquatic dipterous larvæ, *Chironomus* and *Simulium*. It is probable that the ventral sacs of the Thysanura, described in the account of that order, are also blood-gills.

V. THE CIRCULATORY SYSTEM

The general features of the circulatory system.—In insects the circulatory system is not a closed one, the blood flowing in vessels during only a part of its course. The greater part of the circulation of this fluid takes place in the cavities of the body and of its appendages, where it fills the space not occupied by the internal organs.

Almost the only blood-vessel that exists in insects lies just beneath the body-wall, above the alimentary canal (Fig. 105, *h*). It extends from near the caudal end of the abdomen through the thorax into the head. That part of it that lies in the abdomen is the *heart;* the more slender portion, which traverses the thorax and extends into the head is the *aorta*.

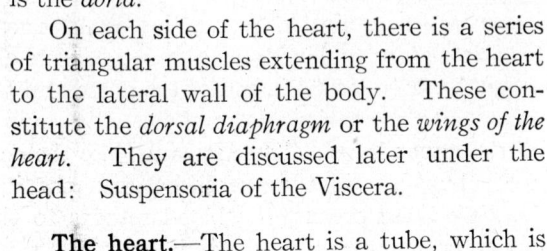

On each side of the heart, there is a series of triangular muscles extending from the heart to the lateral wall of the body. These constitute the *dorsal diaphragm* or the *wings of the heart*. They are discussed later under the head: Suspensoria of the Viscera.

The heart.—The heart is a tube, which is usually closed at its posterior end; at its anterior end it is continuous with the aorta. The heart is divided by constrictions into chambers which are separated by valves (Fig. 139). The number of these chambers varies greatly in different insects; in some, as in *Phasma* and in the larva of *Corethra*, there is only one, in others, as in the cockroach, there are as many as thirteen, but usually there are not more than eight. The blood is admitted to the heart through slit-like openings, the *ostia of the heart;* usually there is a pair of ostia in the lateral walls of each chamber. Each ostium is furnished with a valve which closes it when the chamber contracts.

Fig. 139.—Heart of a May-beetle; *a*, lateral aspect of the aorta; *b*, interior of the heart showing valves; *c*, ventral aspect of the heart and wing-muscles, the muscles are represented as cut away from the caudal part of the heart; *d*, dorsal aspect of the heart (After Straus-Durckheim).

The wall of the heart is composed of two distinct layers: an inner muscular layer; and an outer, connective tissue or peritoneal layer. The muscular layer consists chiefly of annular muscles; but longitudinal fibers have also been observed.

The pulsations of the heart.—When a heart consists of several chambers, they contract one after another, the wave of contraction passing from the caudal end of the heart forwards. As the valves between the chambers permit the blood to move forward but not in the opposite direction, the successive contraction of the chambers causes the blood received through the ostia to flow toward the head, into the aorta.

The aorta.—The cephalic prolongation of the heart, the *aorta* (Fig. 139, *a*), is a simple tube, which extends through the thorax into the head, where it opens in the vicinity of the brain. In some cases, at least, there are valves in the aorta.

The circulation of the blood.—The circulation of the blood can be observed in certain transparent insects, as in young naiads, in larvæ of Trichoptera, and in insects that have just molted. The blood flows from the open, cephalic end of the aorta and passes in quite definite streams to the various parts of the body-cavity and into the cavities of the appendages. These streams, like the ocean currents, have no walls but flow in the spaces between the internal organs. After bathing these organs, the blood returns to the sides of the heart, which it enters through the ostia.

Accessory circulatory organs.—Accessory pulsating circulatory organs have been described in several insects. These are sac-like structures which contract independently of the contractions of the heart. They have been found in the head in several Orthoptera; in the legs of Hemiptera, and in the caudal filaments of Ephemerida.

VI. THE BLOOD

The blood of insects is a fluid, which fills the perivisceral cavity, bathing all of the internal organs of the body, and flowing out into the cavities of the appendages of the body. In only a comparatively small portion of its course, is the blood enclosed in definite blood-vessels; these, the heart and the aorta are described above. The blood consists of two elements, a fluid *plasma* and cells similar to the white corpuscles of the blood of vertebrates, the *leucocytes*.

The blood of insects differs greatly in appearance from the blood of vertebrates, on account of the absence of red blood-corpuscles. In most insects the blood is colorless; but in many species it has a yellowish, greenish, or reddish color. In the latter case, however, the color is not due to corpuscles of the type which gives the characteristic color to the blood of vertebrates.

The leucocytes are nucleated, colorless, amœboid cells similar to the white corpuscles of vertebrates, in appearance and function; they take up and destroy foreign bodies and feed upon disintegrating tissue. It is believed that the products of digestion of disintegrating tissue by the leucocytes pass into the blood and serve to nourish new tissue.

The blood receives the products of digestion of food, which pass in a liquid form, by osmosis, through the walls of the alimentary canal. On the other hand it gives up to the tissues which it bathes the materials needed for their growth. In insects oxygen is supplied to the tissues and gaseous wastes are removed chiefly by the respiratory system and not by means of the blood as in vertebrates.

VII. THE ADIPOSE TISSUE

On opening the body of an insect, especially of a larva, one of the most conspicuous things to be seen is fatty tissue, in large masses. These often completely surround the alimentary canal, and are held in place by numerous branches of the tracheæ with which they are supplied. Other and smaller masses of this tissue adhere to the inner surface of the abdominal wall, in the vicinity of the nervous system, and at the sides of the body. In adult insects it usually exists in much less quantity than in larvæ.

The chief function of the adipose tissue is the storage of nutriment; but it is believed that it also has a urinary function, as concretions of uric acid accumulate in it during the life of the insect.

VIII. THE NERVOUS SYSTEM

a. THE CENTRAL NERVOUS SYSTEM

The more obvious parts of the central nervous system are the following: a ganglion in the head above the œsophagus, the *brain;* a ganglion in the head below the œsophagus, the *subœsophageal ganglion;* a series of ganglia, lying on the floor of the body cavity in the thorax and in the abdomen, the *thoracic* and the *abdominal ganglia;* two longitudinal cords, the *connectives,* uniting all of these ganglia in a series; and many *nerves* radiating from the ganglia to the various parts of the body.

The connectives between the brain and the subœsophageal ganglion pass one on each side of the œsophagus; these are termed the *crura cerebri,* or the legs of the brain; in the remainder of their course, the two connectives are quite closely parallel (Fig. 124).

The series of ganglia is really a double one, there being typically a pair of ganglia in each segment of the body; but each pair of ganglia **is more or** less closely united on the middle line of the body, and often appear to be a single ganglion.

In some cases the ganglia of adjacent segments coalesce, thus reducing the number of distinct ganglia in the series. It has been demonstrated that the brain is composed of the coalesced ganglia of three of the head segments, and the subœsophageal ganglion of the coalesced ganglia of the remaining four segments.

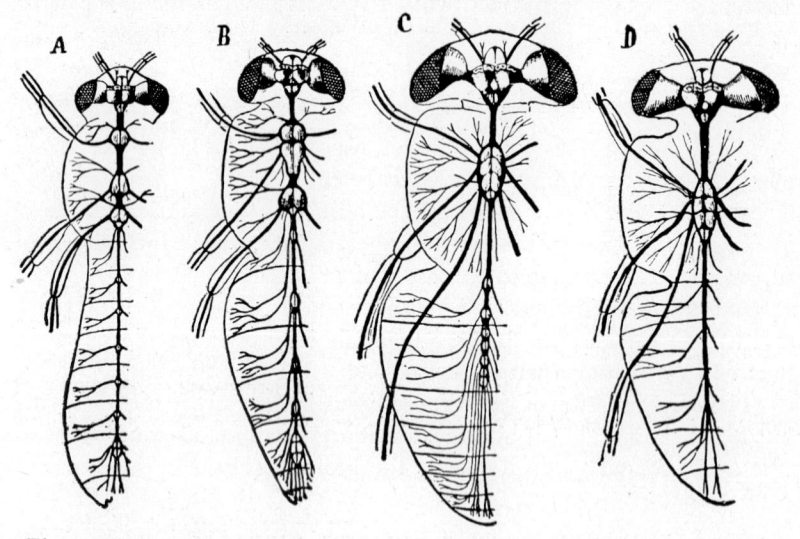

Fig. 140.—Successive stages in the coalescence of thoracic and of abdominal ganglia in Diptera; A, *Chironomus;* B, *Empis;* C, *Tabanus;* D, *Sarcophaga* (From Henneguy after Brandt).

The three parts of the brain, each of which is composed of the pair of ganglia of a head segment, are designated as the *protocerebrum*, the *deutocerebrum*, and the *tritocerebrum*, respectively. The protocerebrum innervates the compound eyes; the deutocerebrum, the antennæ; and the tritocerebrum, the labrum.

The subœsophageal ganglion is composed of four pairs of primary ganglia; these are the ganglia of the segments of which the mandibles, the maxillulæ, the maxillæ, and the labium, respectively, are the appendages.

The three pairs of thoracic ganglia often coalesce so as to form a single ganglionic mass; and usually in adult insects the number of abdominal ganglia is reduced in a similar way.

Successive stages in the coalescence of the thoracic and abdominal ganglia can be seen by a study of the nervous system of the larva, pupa, and adult of the same species, a distinct cephalization of the central nervous system taking place during the development of the insect. Varying degrees of coalescence of the thoracic and of the abdominal ganglia can be seen by a comparative study of the nervous systems of different adult insects (Fig. 140).

The transverse band of fibers that unite the two members of a pair of ganglia is termed a *commissure*. In addition to the commissures that pass directly from one member of a pair of ganglia to the other, there is in the head a commissure that encircles the œsophagus in its passage from one side of the brain to the other, this is the *subœsophageal commissure* (Fig. 141).

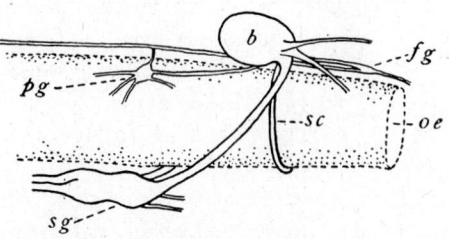

Fig. 141.—Lateral view of the œsophagus of a caterpillar, showing the subœsophageal commissure; *b*, brain; *oe*, œsophagus; *sc*, subœsophageal commissure; *sg*, subœsophageal ganglion; *pg*, paired ganglion (After Liénard).

The nerves that extend from the central chain of ganglia to the different parts of the body are a part of the central nervous system; the core of each nerve fiber being merely a process of a ganglionic cell, however long it may be.

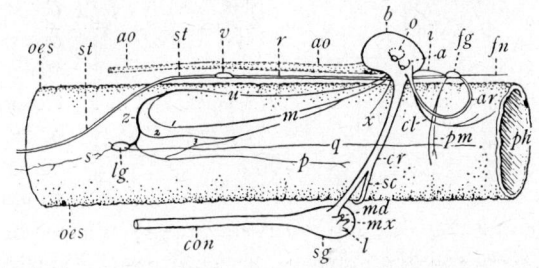

b. THE ŒSOPHAGEAL SYMPATHETIC NERVOUS SYSTEM

In addition to the central nervous system as defined above there are three other nervous complexes which are commonly described as separate systems although they are connected to the central nervous system by nerves. These are the œsophageal sympathetic nervous system, the ventral sympathetic nervous

Fig. 142.—Lateral view of the nerves of the head in the larva of *Corydalus*; *a*, antennal nerve; *ao*, aorta; *ar* paired nerves connecting the frontal ganglion with the brain; *b*, brain; *cl*, clypeo-labral nerve; *con*, connective; *cr*, crura cerebri; *fg*, frontal ganglion; *fn*, frontal nerve; *i*, unpaired nerve connecting the frontal ganglion with the brain; *l*, labial nerve; *lg*, the paired ganglia; *md*, mandibular nerve; *m, p, q, s, u, z*, nerves of the œsophageal sympathetic system; *mx*, maxillary nerve; *o*, optic nerves; *oes*, œsophagus; *ph*, pharynx; *pn*, pharyngeal nerve; *r*, recurrent nerve; *sc*, subœsophageal commissure; *sg*, subœsophageal ganglion; *st*, stomagastric nerve; *v*, ventricular ganglion (From Hammar).

system, and the peripheral sensory nervous system. The first of these is connected with the brain; the other two, with the thoracic and abdominal ganglia of the central nervous system.

The œsophageal sympathetic nervous system is intimately associated with the œsophagus and, as just stated, is connected with the brain. It is described by different writers under various names; among these are *visceral*, *vagus*, and *stomatogastric*. It consists of two, more or less distinct, divisions, an unpaired median division and a paired lateral division.

The unpaired division of the œsophageal sympathetic nervous system is composed of the following parts, which are represented in Figures 141, 142, 143, and 144: the *frontal ganglion* (*fg*), this is a minute ganglion situated above the œsophagus a short distance in front of the brain; the *unpaired nerve connecting the frontal ganglion with the brain* (*i*), this is a small nerve extending from the brain to the frontal ganglion; the *paired nerves connecting the frontal ganglion with the brain* (*ar*), these are arching nerves, one on each side, extending from the upper ends of the crura cerebri to the frontal ganglion; the *frontal nerve* (*fn*), this nerve arises from the anterior border of the frontal ganglion and extends cephalad into the clypeus, where it bifurcates; the *pharyngeal nerves* (*pn*), these extend, one on each side, from the frontal ganglion to the lower portions of the pharynx; the *recurrent nerve* (*r*), this is a single median nerve, which arises from the caudal border of the frontal ganglion, and extends back, passing under the brain and between the

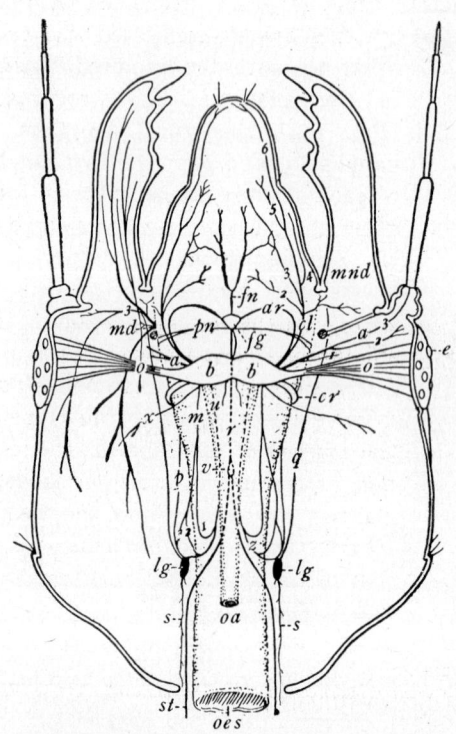

Fig. 143.—Dorsal view of the nerves of the head in the larva of *Corydalus*; *e*, ocelli; *mnd*. mandible; other lettering as in Figure 142 (From Hammar).

aorta and the œsophagus, to terminate in the ventricular ganglion; the *ventricular ganglion* (*v*), this is a minute ganglion on the middle line, a short distance cauda of the brain, and between the aorta and the œsophagus; and the *stomogastric nerves* (*st*), these are two nerves which extend back from the caudal border of the ventricular ganglion, they are parallel for a short distance, then they separate and pass, one on each side, to the sides of the alimentary canal which they follow to the proventriculus.

The paired division of the œsophageal sympathetic nervous system varies greatly in form in different insects. In the larva of *Corydalus*, there is a single pair of ganglia (Fig. 142 and 143, *lg*), one on each side of the œsophagus; each of these ganglia is connected with the brain by two nerves (*m* and *u*) but they are not connected with each other nor with the unpaired division of this system. In a cockroach (Fig. 144), there are two pairs of ganglia (*ag* and *pg*); the two ganglia of each side are connected with each other and with the recurrent nerve of the unpaired division.

As yet comparatively little is known regarding the function of the œsophageal sympathetic nervous system of insects; nerves extending from it have been traced to the clypeus, the muscles of the pharynx, the œsophagus, the mid-intestine, the salivary glands, the aorta, and the heart. Its function is probably analogous to that of the sympathetic nervous system of Vertebrates.

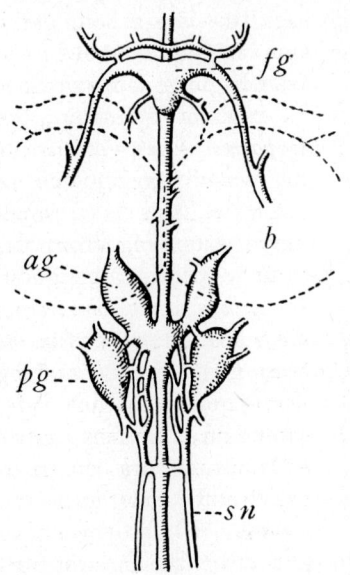

Fig. 144.—The œsophageal sympathetic nervous system of *Periplaneta orientalis;* the outlines of the brain (*b*) and the roots of the antennal nerve which cover a portion of the sympathetic nervous system are given in dotted lines; *ag*, anterior ganglion; *pg*, posterior ganglion; *fg*, frontal ganglion; *sn*, nerves of the salivary glands; *r*, recurrent nerve (After Hofer).

c. THE VENTRAL SYMPATHETIC NERVOUS SYSTEM

The ventral sympathetic nervous system consists of a series of more or less similar elements, each connected with a ganglion of the ventral chain of the central nervous system. Typically there is an element of this system arising in each thoracic and abdominal ganglion; and each element consists of a median nerve extending from the ganglion of its origin caudad between the two connectives, a pair of lateral branches of this median nerve, and one or more ganglionic enlargements of each lateral branch. Frequently the median nerve extends to the ganglion of the following segment.

A simple form of this system exists in the larva of *Cossus ligniperda* (Fig. 122); and a more complicated one, in *Locusta viridissima* (Fig. 145).

From each lateral branch of the median nerve a slender twig extends to the closing apparatus of the tracheæ.

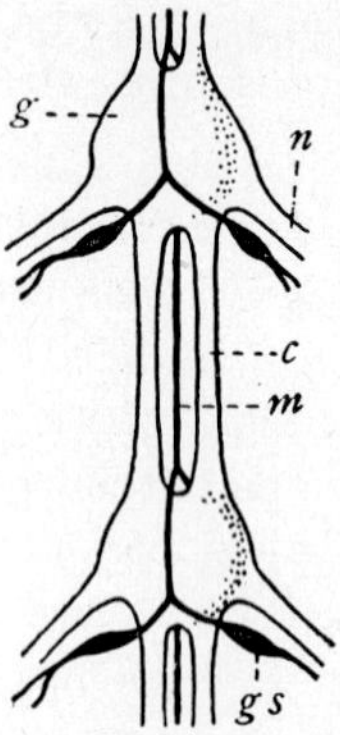

d. THE PERIPHERAL SENSORY NERVOUS SYSTEM

Immediately beneath the hypodermal layer of the body-wall, there are many bipolar and multipolar nerve-cells whose prolongations form a network of nerves; these constitute the *peripheral sensory nervous system* or the *subhypodermal nerve plexus*.

Fig. 145.—Part of the ventral chain of ganglia of *Locusta viridissima* and of the ventral sympathetic nervous system; *g*, ganglion of the central nervous system; *n*, nerve; *c*, connective; *m*, median nerve of the sympathetic system; *gs*, ganglion of the sympathetic system (From Berlese).

The fine nerves of this system are branches of larger nerves which arise in the central nervous system; and the terminal prolongations of the bipolar nerve-cells innervate the sense-hairs of the body-wall.

Figure 146 represents a surface view of a small part of the peripheral sensory nervous system of the silkworm, *Bombyx mori*, as figured by Hilton ('02); the bases of several sense hairs are also shown. The details of this figure are as follows: *h, h, h*, the bases of sense-hairs; *s, s, s*, bipolar nerve cells; *m, m, m*, multipolar cells; *n, n, n*, nerves. All of these structures are united, forming a net work. Of especial interest is the fact that the terminal prolongation of each bipolar nerve-cell enters the cavity of a sense-hair and that the other prolongation is a branch of a larger nerve which comes from the central nervous system.

The peripheral sensory nervous system is so delicate that it can not be seen except when it is stained by some dye that differentiates nervous matter from other tissues. For this purpose the intravitam methylen blue method of staining is commonly used.

IX. GENERALIZATIONS REGARDING THE SENSE-ORGANS OF INSECTS

The sense-organs of insects present a great variety of forms, some of which are still incompletely understood, in spite of the fact that they have been investigated by many careful observers. In the limited space that can be devoted to these organs here only the more general features of them can be described and some of the disputed questions regarding them briefly indicated.

A classification of the sense-organs.—The different kinds of sense-organs are distinguished by the nature of the stimulus that acts on

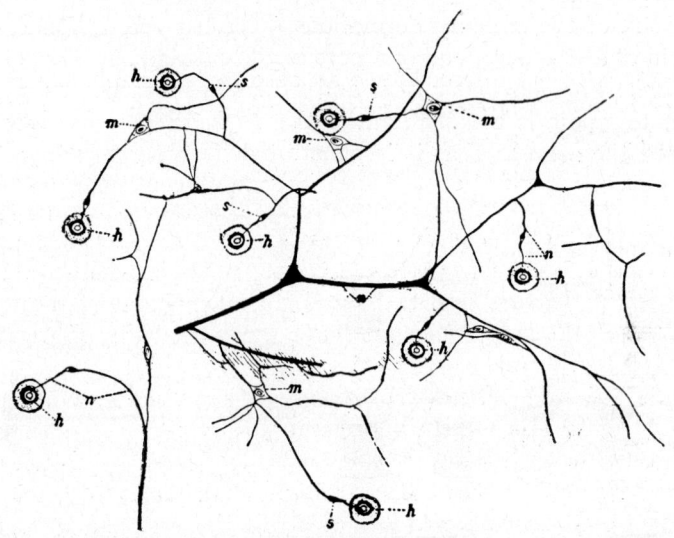

Fig. 146.—Surface view of subhypodermal nerves and nerve-cells from the silkworm (From Hilton)

each. This stimulus may be either a mechanical stimulus, a chemical one, or light. The organs of touch and of hearing respond to mechanical stimuli; the former, to simple contact with other objects; the latter, to vibratory motion caused by waves of sound. The organs of taste and of smell are influenced only by soluble substances and it seems probable that chemical changes are set up in the sense-cells by these substances; hence these organs are commonly referred to as the chemical sense-organs; no criterion has been discovered by which the organs of taste and of smell in insects can be distinguished. The organs of sight are acted upon by light; it is possible that the action of light in this case is a chemical one, as it is on a photographic plate, but the eyes have not been classed among the chemical sense-organs. For these reasons the following groups of sense-organs are recognized:

The mechanical sense-organs.—The organs of touch and of hearing.
The chemical sense-organs.—The organs of taste and of smell.
The organs of sight.—The compound eyes and the ocelli.

The cuticular part of the sense-organs.—In most if not all of the sense-organs of insects there exists one or more parts that are of cuticular formation. The cuticular parts of the organs of sight and of hearing are described later, in the accounts of these organs; in this place, a few of the modifications of the cuticula found in other sense-organs are described.

Each of the cuticular formations described here is found either within or at the outer end of a pore in the cuticula; as some of these formations are obviously setæ and others are regarded as modified setæ, this pore is usually termed the *trichopore*; it has also been termed the *neuropore*, as it is penetrated by a nerve-ending.

As the cuticular part of this group of sense-organs, those other than the organs of hearing and of sight, is regarded as a seta, more or less modified, these organs are often referred to as the *setiferous sense-organs;* they are termed the *Hautsinnesorgane* by German writers.

Special terms have been applied to the different types of setiferous sense-organs, based on the form of the cuticular part of each; but these types cannot be sharply differentiated as intergrades exist between them. In Figure 147 are represented the cuticular parts of several of these different types; these are designated as follows:

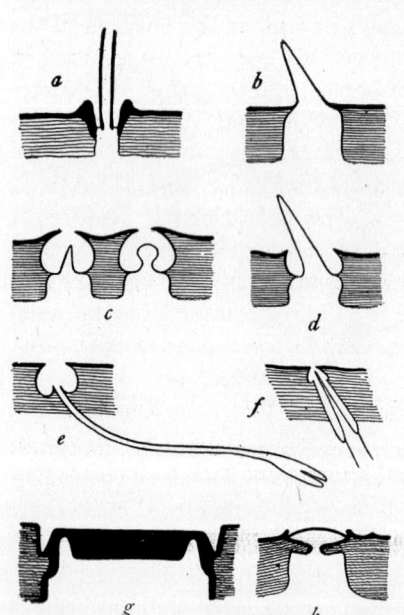

Fig. 147.—Various forms of the cuticular portion of the setiferous sense-organs. The lettering is explained in the text.

The thick-walled sense-hair, sensillum trichodeum.—In this type the cuticular part is a seta, the base of which is in an alveolus at the end of a trichopore and is connected with the wall of the trichopore by a thin articular membrane (Fig. 147, *a.*)

If the sense-hair is short and stout, it is termed by some writers a *sense-bristle, sensillum chæticum;* but there is little use for this distinction.

In the thick-walled sense-hairs, the wall of the seta is fitted to receive only mechanical stimuli, being relatively thick, and as these organs lack the characteristic features of the organs of hearing, they are believed to be organs of touch.

The sense-cones.—The sense-cones vary greatly in form and in their relation to the cuticula of the body-wall; their distinctive feature is that they are thin-walled. For this reason, they are believed to be chemical sense-organs, the thinness of the wall of the cone permitting osmosis to take place through it. In the sense-cones, too, there is no joint at the base, as in the sense-hairs, the articular membrane being of the same thickness as the wall of the cone; there is, therefore, no provision for movement in response to mechanical stimuli.

In one type of sense-cone, the *sensillum basiconicum*, the base of the cone is at the surface of the body-wall (Fig. 147, *b*). In another type, *sensillum cœloconicum*, the cone is in a pit in the cuticula of the body-wall (Fig. 147, *c*). Two forms of this type are represented in the figure; in one, the sense-cone is conical; in the other, it is fungiform. Intergrades between the basiconicum and the cœloconicum types exist (Fig. 147, *d*).

The flask-like sense-organ, sensillum ampullaceum.—This is a modification of the sense-cone type, the characteristic feature of which is that the cone is at the bottom of an invagination of the articular membrane; in some cases the invagination is very deep so that the cone is far within the body-wall (Fig. 147, *e*); intergrades between this form and the more common *sensillum cœloconicum* exist (Fig. 147, *f*).

The pore-plate, sensillum placodeum.—In this type the cuticular part of the organ is a plate closing the opening of the trichopore; in some cases, this plate is of considerable thickness with a thin articular membrane (Fig. 147, *g*); in others it is thin throughout (Fig. 147, *h*).

The olfactory pores.—This type of sense-organ is described later.

X. THE ORGANS OF TOUCH

The organs of touch are the simplest of the organs of special sense of insects. They are widely distributed over the surface of the body and of its appendages. Each consists of a seta, with all the characteristics of setæ already described, a trichogen cell, which excreted the seta, and a bipolar nerve-cell. These organs are of the type known as *sensillum trichodeum* referred to in the preceding section of this chapter.

According to the observations of Hilton ('02) the terminal prolongation of the nerve-cell enters the hair and ends on one side of it at some distance from its base (Fig. 148). The proximal part of this nerve-cell is connected with the peripheral sensory nervous system, as already described (page 128).

The presence of this nervous connection is believed to distinguish tactile hairs from those termed clothing hairs, and from the scales that are modified setæ. If this distinction is a good one, it is quite probable that many hairs and scales that are now regarded as merely clothing will be found to be sense-organs, when studied by improved histological methods. In fact Guenther ('01) and others have shown that some of the scales on the wings of Lepidoptera, especially those on the veins of the wings, are supplied with nerves; but the function of these scales is unknown.

Hilton states that he "found no evidence to indicate nerves ending in gland cells or trichogen cells by such branches as have been described and figured by Blanc ('90), but in every case the very fine nerve termination could be traced up past the hypodermal cell layer with no branches." Many figures of unbranched nerve fibers ending in sense-hairs are also given by O. vom Rath ('96).

A very different form of nerve-endings in sense-hairs is given by Berlese ('09, *a*). This author represents the nerve extending to a sense-hair as dividing into many bipolar nerve-endings.

XI. THE ORGANS OF TASTE AND OF SMELL

(*The chemical sense-organs*)

It is necessary to discuss together the organs of taste and of smell, as no morphological distinction between them has been discovered. If a chemical sense-organ is so located that it comes in contact with the food of the insect, it is commonly regarded as an organ of taste, if not so situated, it is thought to be an organ of smell. In the present state of our knowledge, this is the only distinction that can be made between these two kinds of organs.

Many experiments have been made to determine the function of the various chemical sense-organs but the results are, as yet, far from conclusive. The problem is made difficult by the fact that these organs **are** widely distributed over the body and its appendages, and in some parts, as on the antennæ of many insects, several different types of sense-organs are closely associated.

Those organs that are characterized by the presence of a thin-walled sense-cone (Fig. 147, *b-f*) or by a pore-plate (Fig. 147, *g, h*) are believed to be chemical sense-organs. It is maintained by Berlese ('09, *a*) that an essential feature of these chemical sense-organs is the presence of a gland-cell, the excretion of which, passing through the thin wall of the cuticular part, keeps the outer surface of this part, the sense-cone or pore-plate, moist and thus fitted for the reception of chemical stimuli. According to this view a chemical sense-organ consists of a cuticular part, a trichogen cell or cells which produced

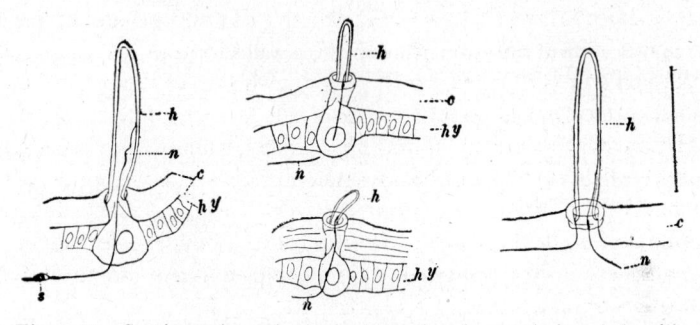

Fig. 148.—Sections through the body-wall and sense-hairs of the silkworm; *c*, cuticula; *h*, hair; *hy*, hypodermis; *n*, nerve; *s*, bipolar nerve-cell (From Hilton). The line at the right of the figure indicates one tenth millimeter.

this part, a gland-cell which excretes a fluid which keeps the part moist, and a nerve-ending.

It is interesting to note that tactile hairs may be regarded as specialized clothing hairs, specialized by the addition of a nervous connection, and that sense-cones and pore-plates may be regarded as specialized glandular hairs with a nervous connection; in the latter case, the specialization involves a thinning of the wall of the hair so as to permit of osmosis through it.

In the different accounts of chemical sense-organs there are marked differences as regards the form of the nerve-endings. In many of the descriptions and figures of these organs the nerve-ending is represented as extending unbranched to the chitinous part of the organ, resembling in this respect those represented in Figure 148. In other accounts the gland-cell is surrounded by an involucre of nerve-cells (Fig. 149).

In the types of chemical sense-organs described above the action of the chemical stimuli is supposed to be dependent upon osmosis through a delicate cuticular membrane. It should be noted, however, that several writers have described sense-cones in which there is a pore; but the accuracy of these observations is doubted by other writers.

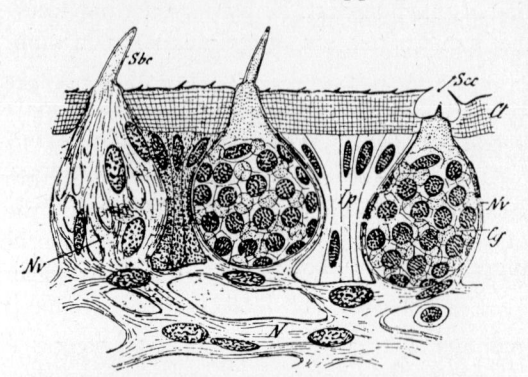

Fig. 149.—Section of the external layers of the wall of an antenna of *Acrida turrita;* *Ct*, cuticula; *Ip*, hypodermis; *N*, nerve; *Nv*, involucre of nerve-cells surrounding the glandular part of a sense-organ; *Sbc*, sensillum basiconicum; *Scc*, sensillum cœloconicum. Three sense-organs are figured; a surface view of the first is represented, the other two are shown in section. (From Berlese).

A very different type of sense-organs which has been termed *olfactory pores* is described in the concluding section of this chapter.

XII. THE ORGANS OF SIGHT

a. THE GENERAL FEATURES

The two types of eyes of insects.—It is shown in the preceding chapter that insects possess two types of eyes, the ocelli or simple eyes and the compound or facetted eyes.

Typically both types of eyes are present in the same insect, but either may be wanting. Thus many adult insects lack ocelli, while the larvæ of insects with a complete metamorphosis lack compound eyes.

When all are present there are two compound eyes and, typically two pairs of ocelli; but almost invariably the members of one pair of ocelli are united and form a single median ocellus. The median ocellus is wanting in many insects that possess the other two ocelli.

The distinction between ocelli and compound eyes.—The most obvious distinction between ocelli and compound eyes is the fact that in an ocellus there is a single cornea while in a compound eye there are many. Other features of compound eyes have been regarded as distinctively characteristic of them; but in the case of each of these features it is found that they exist in some ocelli.

Each ommatidium of a compound eye has been considered as a separate eye because its nerve-endings constituting the retinula are isolated from the retinulæ of other ommatidia by surrounding accessory pigment cells; but a similar isolation of retinulæ exist in some ocelli.

It has also been held that in compound eyes there is a layer of cells between the corneal hypodermis and the retina, the crystalline-cone-cells, which is absent in ocelli; but in the ocelli of adult Ephemerida there is a layer of cells between the lens and the retina, which, at least, is in a position analogous to that of the crystalline-cone-cells; the two may have had a different origin, but regarding this, we have, as yet, no conclusive data.

The absence of compound eyes in most of the Apterygota.—Typically insects possess both ocelli and compound eyes; when either kind of eyes is wanting it is evidently due to a loss of these organs and not to a generalized condition. Although compound eyes are almost universally absent in the Apterygota in the few cases where they are present in this group they are of a highly developed type and not rudimentary; the compound eyes of *Machilis*, for example, are as perfect as those of winged insects.

The absence of compound eyes in larvæ.—The absence of compound eyes in larvæ is evidently a secondary adaptation to their particular mode of life, like the internal development of wings in the same forms. In the case of the compound eyes of larvæ, the development of the organs is retarded, taking place in the pupal stage instead of in an embryonic stage, as is the case with nymphs and naiads.

While the development of the compound eyes as a whole is retarded in larvæ, a few ommatidia may be developed and function as ocelli during larval life.

b. THE OCELLI

There are two classes of ocelli found in insects: first, the ocelli of adult insects and of nymphs and naiads, which may be termed the *primary ocelli;* and second, the ocelli of most larvæ possessing ocelli, which may be termed *adaptive ocelli*.

The primary ocelli.—The ocelli of adult insects and of nymphs and naiads having been originally developed as ocelli are termed the primary ocelli. Of these there are typically two pairs; but usually when they are present there are only three of them, and in many cases only a single pair.

When there are three ocelli, the double nature of the median ocellus is shown by the fact that the root of the nerve is double, while that of each of the other two is single.

In certain generalized insects, as some Plecoptera, (Fig. 150) all of the ocelli are situated in the front; but in most insects, the paired ocelli have either migrated into the suture between the front and the vertex (Fig. 151), or have proceeded farther and are situated in the vertex.

The structure of primary ocelli is described later.

The adaptive ocelli.—Some larvæ, as those of the Tenthredinidæ, possess a single pair of ocelli, which in their position and in their structure agree with the ocelli of the adult insects; these are doubtless primary ocelli. But most larvæ have lost the primary ocelli; and if they possess ocelli the position of them and their structure differ greatly from the positions and structure of primary ocelli.

Except in the few cases where primary ocelli have been retained by larvæ, the ocelli of larvæ are situated in a position corresponding to the position of the compound eyes of the adult (Fig. 152); and there are frequently several of these ocelli on each side of the head. This has led to the belief that they represent a few degenerate ommatidia, which have been retained by the larva, while the development of the greater number of ommatidia has been retarded. For this reason they are termed *adaptive ocelli*.

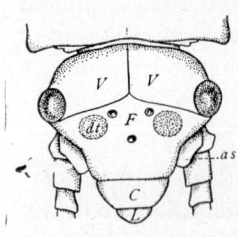

Fig. 150.—Head of a naiad of *Pteronacys; dt*, spots in the cuticula beneath which the dorsal arms of the tentorium are attached; the three ocelli are on the front (F), between these two spots.

The number of adaptive ocelli varies greatly, and sometimes is not constant in a species; thus in the larva of *Corydalus*, there may be either six or seven ocelli on each side of the head.

There are also great variations in the structure of adaptive ocelli. These variations probably represent different degrees of degeneration or of retardation of development. The extreme of simplicity is found in certain dipterous larvæ; according to Hesse ('01) an ocellus of *Ceratopogon* consists of only two sense-cells. As examples of complicated adaptive ocelli, those of lepidopterous larvæ can be cited.

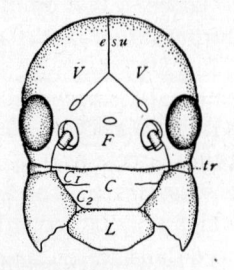

Fig. 151.—Head of a cricket.

The ocellus of *Gastropacha rubi*, which is described and figured by Pankrath ('90), resembles in structure, to a remarkable degree, an ommatidium, and the same is true of the ocellus of the larva of *Arctia caja* figured by Hesse ('01).

The structure of a visual cell.—The distinctively characteristic feature of eyes is the presence of what is termed *visual cells*. In insects, and in other arthropods, a visual cell is a nerve-end-cell, which contains a nucleus and a greater or less amount of pigment, and bears a characteristic border, termed the *rhabdomere;* this is so called because it forms a part of a rhabdom.

Fig. 152.—Head of a larva of *Corydalus*, dorsal aspect.

The visual cells are grouped in such a way that the rhabdomeres of two or more of them are united to form what is known as a *rhabdom* or optic rod. A group of two visual cells with the rhabdom formed by their united rhabdomeres is shown in Figure 153, A and B.

The form of the rhabdomere varies greatly in the visual cells of different insect eyes; and the number of rhabdomeres that enter into the composition of a rhabdom also varies.

Figure 153, C represents in a diagrammatic manner the structure of rhabdomere as described by Hesse ('01). The rhabdomere (r) consists of many minute rodlets each with a minute knob at its base and connected with a nerve fibril.

Fig. 153.—Two visual cells from an ocellus of a pupa of *Apis mellifica*. A, longitudinal section ; B, transverse section; *n, n,* nerves; *nu,* nucleus; *r,* rhabdom; *p,* pigment (After Redikorzew), C, diagram illustrating the structure of a rhabdomere; *r,* rhabdomere; *c,* cell-body (From Berlese after Hesse).

The structure of a primary ocellus.—The primary ocelli vary greatly in the details of the form of their parts, but the essential features of their structure are illustrated by the accompanying diagram (Fig. 154).

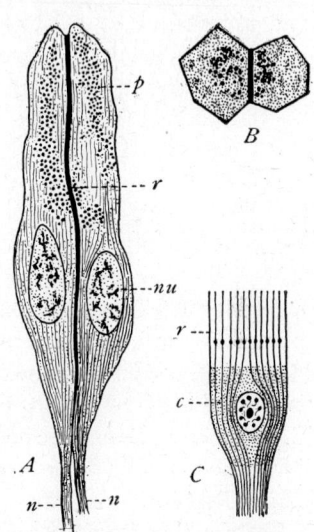

In some ocelli, as for example the lateral ocelli of scorpions, the visual cells are interpolated among ordinary hypodermal cells, the two kinds forming a single layer of cells beneath the cornea; but in the ocelli of insects, the sense-cells form a distinct layer beneath the hypodermal cells. In this type of ocellus the following parts can be distinguished:

The cornea.—The cornea (Fig. 154, *c*) is a transparent portion of the cuticula of the body-wall; this may be lenticular in form or not.

The corneal hypodermis.—The hypodermis of the body-wall is continued beneath the cornea (Fig. 154, *c. hy.*); this part of the hypodermis is termed by many writers the *vitreous layer* of the ocellus; but the term *corneal hypodermis*, being a self-explanatory term, is preferable. Other terms have been applied to it, as the *lentigen layer* and the *corneagen*, both referring to the fact that this part of the hypodermis produces the cornea.

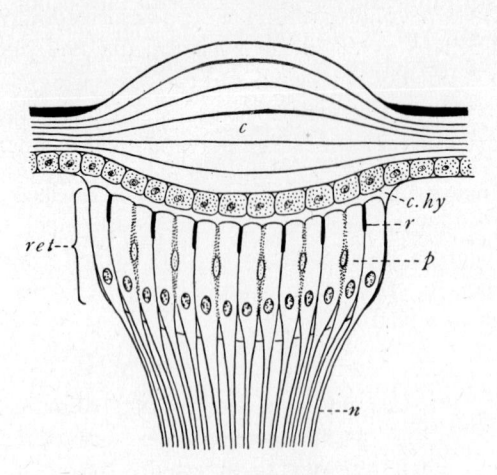

Fig. 154.—A diagram illustrating the structure of a primary ocellus; *c*, cornea; *c. hy*, corneal hypodermis; *ret*, retina; *n*, ocellar nerve; *p*, accessory pigment cell; *r*, rhabdom.

The retina.—Beneath the corneal hypodermis is a second cellular layer, which is termed the *retina*, being composed chiefly or entirely of visual cells (Fig. 154, ret).

The visual cells of the retina are grouped, as described above (Fig. 153), so that the rhabdomeres of several of them, two, three or four, unite to form a rhabdom; such a group of retinal cells is termed a *retinula*.

The visual cells are nerve-end-cells, each constituting the termination of a fiber of the ocellar nerve, and are thus connected with the central nervous system.

Accessory pigment cells.—In some ocelli there are densely pigmented cells between the retinulæ, which serve to isolate them in a similar way to that in which the retinula of an ommatidium of a compound eye is isolated (Fig. 154, *p*). Even in cases where accessory pigment cells are wanting a degree of isolation of the rhabdoms of the retinulæ of an ocellus is secured by pigment within the visual cells (Fig. 153, *p*).

THE INTERNAL ANATOMY OF INSECTS 139

Ocelli of Ephemerida.—It has been found that the ocelli of certain adult Ephemerida differ remarkably from the more common type of ocelli described above. These peculiar ocelli have been described and figured by Hesse ('01) and Seiler ('05). In them the cuticula over the ocellus, the cornea, is arched but not thickened and the corneal hypodermis is a thin layer of cells immediately beneath it. Under the hypodermis there is a lens-shaped mass of large polygonal cells; and between this lens and the retina there is a layer of closely crowded columnar cells.

The development of these ocelli has not been studied; hence the origin of the lens-shaped mass of cells and of the layer of cells between it and the retina is not known.

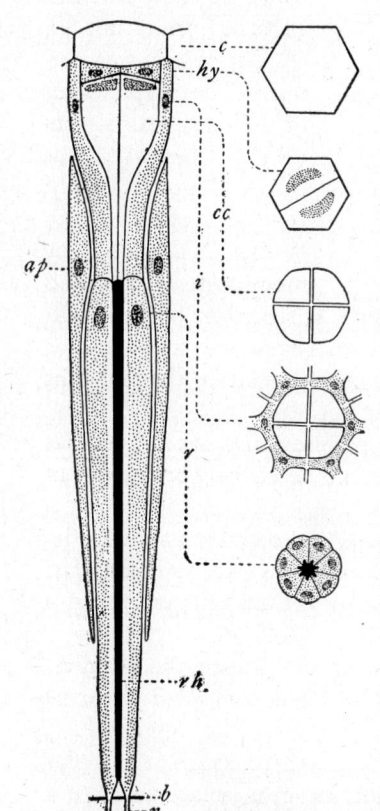

Fig. 155.—An ommatidium of *Machilis*. The lettering is explained in the text.

c. THE COMPOUND EYES

A compound eye consists of many quite distinct elements, the ommatidia, each represented externally by one of the many facets of which the cuticular layer of the eye is composed. As the ommatidia of a given eye are similar, a description of the structure of one will serve to illustrate the structure of the eye as a whole.

The structure of an ommatidium.—The compound eyes of different insects vary in the details of their structure; but these variations are merely modifications of a common plan; this plan is well-illustrated by the compound eyes of *Machilis*, the structure of which was worked out by Seaton ('03). Figure 155 represents a longitudinal section and a series of transverse sections of an ommatidium in an eye of this insect, which consists of the following parts.

The cornea.—The cornea is a hexagonal portion of the cuticular layer of the eye and is biconvex in form (Fig. 155, *c*).

The corneal hypodermis.—Beneath each facet of the cuticular layer of the eye are two hypodermal cells which constitute the corneal hypodermis of the ommatidium. These cells are quite distinct in *Machilis* and their nuclei are prominent (Fig. 155, *hy*); but in many insects they are greatly reduced, and consequently are not represented in many of the published figures of compound eyes.

The crystalline-cone-cells.—Next to the corneal hypodermis there are four cells, which in one type of compound eyes, the eucone eyes, form a body known as the crystalline-cone, for this reason these cells are termed the *crystalline-cone-cells* (Fig. 155, *cc*). Two of these cells are represented in the figure of a longitudinal section and all four, in that of a transverse section. In each cell there is a prominent nucleus at its distal end.

The iris-pigment-cells.—Surrounding the crystalline-cone-cells and the corneal hypodermis, there is a curtain of densely pigmented cells, which serves to exclude from the cone light entering other ommatidia; for this reason these cells are termed the *iris-pigment* (Fig. 155, *i*). They are also known as the *distal retinula cells;* but as they are not a part of the retina this term is misleading.

There are six iris-pigment-cells surrounding each crystalline-cone; but as each of these cells forms a part of the iris of three adjacent ommatidia, there are only twice as many of these cells as there are ommatidia. This is indicated in the diagram of a transverse section (Fig. 155, *i*).

The retinula.—At the base of each ommatidium, there is a group of visual cells forming a retinula (Fig. 155, *r*); of these there are seven in *Machilis;* but they vary in number in the eyes of different insects. The visual cells are so grouped that their united rhabdomeres form a rhabdom, which extends along the longitudinal axis of the ommatidium (Fig. 155, *rh*). The distal end of the rhabdom abuts against the proximal end of the crystalline-cone; and the nerve-fibers of which the visual cells are the endings pass through the basement membrane (Fig. 155, *b*) to the optic nerve.

The visual cells are pigmented and thus aid in the isolation of the ommatidium.

The accessory pigment-cells.—In addition to the two kinds of pigment-cells described above there is a variable number of accessory pigment-cells (Fig. 155, *ap*), which lie outside of and overlap them.

From the above it will be seen that each ommatidium of a eucone eye is composed of five kinds of cells, three of which, the corneal hypodermis, the crystalline-cone-cells, and the retinular cells produce solid structures; and three of them are pigmented.

Three types of compound eyes are recognized: first, the *eucone eyes*, in these each ommatidium contains a true crystalline-cone, as described above, and the nuclei of the cone-cells are in front of the cone; second, the *pseudocone eyes*, in these the four cone-cells are filled with a transparent fluid medium, and the nuclei of these cells are behind the refracting body; and third, the *acone eyes*, in which although the four cone-cells are present they do not form a cone, either solid or liquid.

d. THE PHYSIOLOGY OF COMPOUND EYES

The compound eyes of insects and of **Crustacea** are the most complicated organs of vision known to us. It is not strange therefore, that the manner in which they function has been the subject of much discussion. It is now, however, comparatively well-understood; although much remains to be determined.

In studying the physiology of compound eyes, three sets of structures, found in each ommatidium, are to be considered: first, the dioptric apparatus, consisting of the cornea and the crystalline-cone; second, the percipient portion, the retinula, and especially the rhabdom; and third, the envelope of pigment, which is found in three sets of cells, the iris pigment-cells, the retinular cells, and the accessory or secondary pigment-cells.

The dioptrics of compound eyes is an exceedingly complicated subject; a discussion of it would require too much space to be introduced here. It has been quite fully treated by Exner ('91). to whose work those especially interested in this subject are referred. The important point for our present discussion is that by means of the cornea and the crystalline-cone, light entering the cornea from within the limits of a certain angle passes through the cornea and the crystalline-cone to the rhabdom, which is formed of the combined rhabdomeres of the nerve-end-cells, constituting the retinula, the precipient portion of the ommatidium.

The theory of mosaic vision.—The first two questions suggested by a study of physiology of compound eyes have reference to the nature of the vision of such an eye. What kind of an image is thrown upon the retinula of each ommatidium? And how are these images combined to form the image perceived by the insect? Does an insect with a thousand ommatidia perceive a thousand images of the object viewed or only one?

The theory of mosaic vision gives the answers to these questions. This theory was proposed by J. Müller in 1826; and the most recent investigations confirm it. The essential features of it are the following: only the rays of light that pass through the cornea and the crystalline-cones reach the precipient portion of the eye, the others fall on the pigment of the eye and are absorbed by it; in each ommatidium the cornea transmits to the crystalline-cone light from a very limited field of vision, and when this light reaches the apex of the crystalline-cone it forms a point of light, not an image; hence the image formed upon the combined retinulæ is a mosaic of points of light, which combined make a single image, and this image is an erect one.

Figure 156 will serve to illustrate the mosaic theory of vision. In this figure are represented the corneas (c), the crystalline-cones (cc), and the rhabdoms (r.) of several adjacent ommatidia. It can be seen, from this diagram, that each rhabdom receives a point of light which comes from a limited portion of the object viewed (O); and that the image (I) received by the percipient portion of the eye is a single erect image, formed by points of light, each of which corresponds in density and color to the corresponding part of the object viewed.

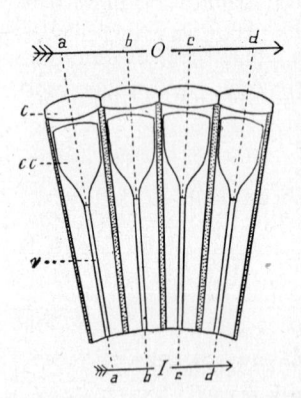

Fig. 156.—Diagram illustrating the theory of mosaic vision.

The distinctness of vision of a compound eye depends in part upon the number and size of the ommatidia. It can be readily seen that the image formed by many small ommatidia will represent the details of the object better than one formed by a smaller number of larger ommatidia; the smaller the portion of the object viewed by each ommatidium the more detailed will be the image.

The distinctness of the vision of a compound eye depends also on the degree of isolation of the light received by each ommatidium, which is determined by the amount and distribution of the pigment. Two types of compound eyes, differing in the degree of isolation of the light received by each ommatidium, are recognized; to one type has been applied the term day-eyes, and to the other, night-eyes.

Day-eyes.—The type of eyes known as day-eyes are so-called because they are fitted for use in the day-time, when there is an abundance of light. In these eyes the envelope of pigment surrounding the transparent parts of each ommatidium is so complete that only the light that has traversed the cornea and crystalline-cone of that ommatidium reaches its rhabdom. The image formed in such an eye is termed by Exner an *apposed image;* because it is formed by apposed points of light, falling side by side and not overlapping. Such an image is a distinct one.

Night-eyes.—In the night-eyes the envelope of pigment surrounding the transparent parts of each ommatidium is incomplete; so that rays of light entering several adjacent corneas can reach the same retinula. In such an eye there will be an overlapping of the points of light; the image thus formed is termed by Exner a *superimposed image.* It is obvious that such an image is not as distinct as an apposed image. It is also obvious that a limited amount of light will produce a greater impression in this type of eye than in one where a considerable part of the light is absorbed by pigment. Night-eyes are fitted to perceive objects and the movement of objects in a dim light, but only the more general features of the object can be perceived by them.

Eyes with double function.—It is a remarkable fact that with many insects and **Crustacea** the compound eyes function in a bright light as day-eyes and in a dim light as night-eyes. This is brought about by movements in the pigment. If an insect having eyes of this kind be kept in a light place for a time and then killed while still in the light, its eyes will be found to be day-eyes, that is eyes fitted to form apposed images. But if another insect of the same species be kept in a dark place for a time and then killed while still in the dark, its eyes will be found to be night-eyes, that is eyes fitted to form superimposed images.

Figure 157 represents two preparations showing the structure of the compound eyes of a diving-beetle, studied by Exner. In one (Fig. 157, *A*), each rhabdom is surrounded by an envelope of pigment, so that it can receive only the light passing through the crystalline-cone of the ommatidium of which this rhabdom is a part. This is the condition found in the individual killed in the light, and illustrates well the structure of a day-eye. In the other preparation (Fig. 157, *B*), which is from an individual killed in the dark, it can be seen that the pigment has moved up between the crystalline-cones so that

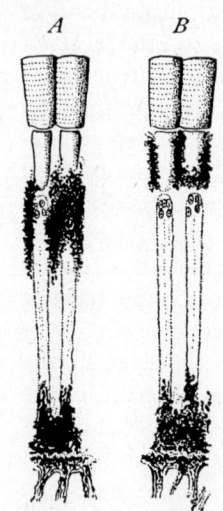

Fig. 157.—Ommatidia from eyes of *Colymbetes;* A, day-eye condition; B, night-eye condition (From Exner).

the light passing from the tip of a cone may reach several rhabdoms, making the eye a night-eye. These changes in the position of the pigment are probably due to amœboid movements of the cells.

Divided Eyes.—In many insects each compound eye is divided into two parts; one of which is a day-eye, and the other a night-eye. The two parts of such an eye can be readily distinguished by a difference in the size of the facets; the portion of the eye that functions as a day-eye being composed of much smaller facets than that which functions as a night-eye.

A study of the internal structure of a divided eye shows that the distribution of the pigment in the part composed of smaller facets is that characteristic of day-eyes; while the part of the eye composed of larger facets is fitted to produce a superimposed image, which is the distinctive characteristic of night-eyes.

Great differences exist in the extent to which the two parts of a divided eye are separated. In many dragon-flies the facets of a part of each compound eye are small, while those of the remainder of the eye are much larger; but the two fields are not sharply separated. In some *Blepharocera* the two fields are separated by a narrow band in which there are no facets, and the difference in the size of the facets of the two areas is very marked. The extreme condition is reached in certain May-flies, where the two parts of the eye are so widely separated that the insect appears to have two pairs of compound eyes (Fig 158).

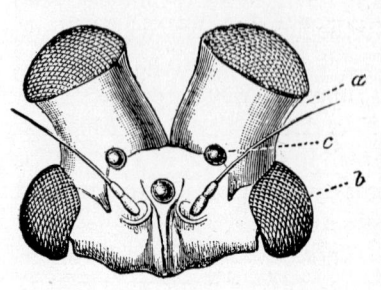

Fig. 158.—Front of head of Cloëon, showing divided eyes; *a*, night-eye; *b*, day-eye; *c*, ocellus (From Sharp).

The tapetum.—In the eyes of many animals there is a structure that reflects back the light that has entered the eye, causing the well-known shining of the eyes in the dark. This is often observed in the eyes of cats and in the eyes of moths that are attracted to our light at night. The part of the eye that causes this reflection is termed a *tapetum*. The supposed function of a tapetum is to increase the effect of a faint light, the light being caused to pass through the retina a second time, when it is reflected from the tapetum.

The structure of the tapetum varies greatly in different animals; in the cat and other carnivores it is a thick layer of wavy fibrous tissue; in spiders it consists of a layer of cells behind the retina containing

THE INTERNAL ANATOMY OF INSECTS 145

small crystals that reflect the light; and in insects it is a mass of fine tracheæ surrounding the retinula of each ommatidium.

XIII. THE ORGANS OF HEARING

a. THE GENERAL FEATURES

The fact that in many insects there are highly specialized organs for the production of sounds indicates that insects possess also organs of hearing; but in only a few cases are these organs of such form that they have been generally recognized as ears.

The tympana.—In most of the jumping Orthoptera there are thinned portions of the cuticula, which are of a structure fitted to be put in vibration by waves of sound. For this reason these have been commonly regarded as organs of hearing, and have been termed *tympana*. In the Acridiidæ, there is a tympanum on each side of the first abdominal segment (Fig. 159); and in the Locustidæ and in the Gryllidæ, there is a pair of tympana near the proximal end of each tibia of the first pair of legs (Fig. 160).

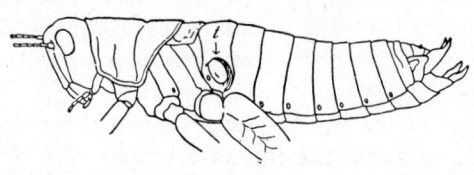

Fig. 159.—Side view of a locust with the wings removed; *t*, tympanum.

The chordotonal organs.—An ear to be effective must consist of something more than a membrane that will be put in vibration by means of sound; the vibrations of such a tympanum must be transferred in some way to a nervous structure that will be influenced by them if the sound is to be perceived. Such structures, closely associated with the tympana of Orthoptera, were discovered more than a half century ago by Von Siebold (1844) and have been studied since by many investigators. The morphological unit of these essential auditory

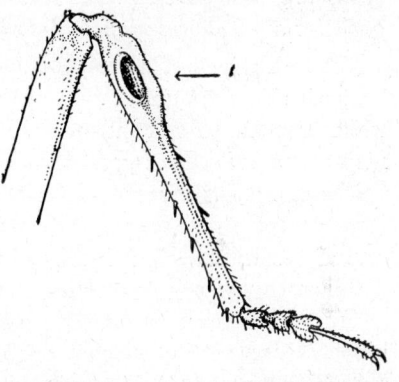

Fig. 160.—Fore leg of a katydid; *t*, tympanum.

structures of insects is a more or less peg-like rod contained in a tubular nerve-ending (Fig. 161, A and B); this nerve-ending may or may not be associated with a specialized tympanum. To all sense-organs characterized by the presence of these auditory pegs, Graber ('82) applied the term *chordotonal organs* or fiddle-string-like organs.

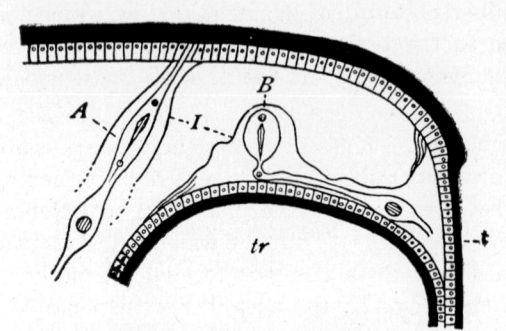

Fig. 161.—Diagrammatic representation of the auditory organs of a locustid (After Graber).

The scolopale and the scolopophore.—The peg-like rod characteristic of a chordotonal organ of an insect was named by Graber the *scolopale;* and to the tubular nerve-ending containing the scolopale, he applied the term *scolopophore.*

The integumental and the subintegumental scolopophores.—With respect to their position there are two types of scolopophores; in one, the nerve-ending is attached to the body-wall (Fig. 161, A); in the other, it ends free in the body-cavity (Fig. 161, B). These two types are designated respectively as *integumental scolopophores* and *subintegumental scolopophores.*

The structure of a scolopophore.—In a scolopophore there can be distinguished an outer sheath (Fig. 161, I), which appears to be continuous either with the basement membrane of the hypodermis or with that of the epithelium of a trachea, and within this sheath the complicated nerve-ending; this nerve-ending is represented diagrammatically in Figure 161 from Graber and in detail in Figure 162 from Hess ('17).

In Figure 162 the following parts are represented: a bipolar sense-cell (*sc*) with its nucleus (scn); the proximal pole of this sense-cell is connected with the central nervous system by a nerve; and its distal pole is connected with the scolopale (*s*) by an axis-fiber (*af*); surrounding the distal prolongation of the sense-cell and the scolopale there is an enveloping or accessory cell (*ec*), in which there is a prominent nucleus (*ecn*); distad of the enveloping cell is

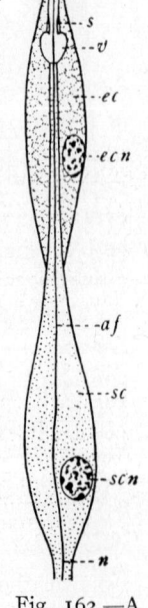

Fig. 162.—A scolopophore of the integumental type (From Hess).

the cap-cell (*cc*), in which there is a nucleus (*ccn*); extending from the end-knob (*ek*) of the scolopale and surrounded by the cap-cell there is an attachment fiber or terminal ligament (*tl*), by which the scolopophore is attached to the body-wall, the scolopophore represented being of the integumental type; at the base of the scolopale and partly surrounding it, there is a vacuole (*v*).

The structure of a scolopale.—The scolopalæ or auditory pegs are exceedingly minute and are quite uniform in size, regardless of the size of the insect in which they are; but they vary in form in different insects. They are hollow (Fig. 162, *s*); but the wall of the scolopale is almost always thickened at its distal end, this forming an end-knob (Fig. 162, *ek*). They are traversed by the axis-fiber of the sense-cell. The vacuole at the base of the scolopale connects with the lumen of the scolopale; this vacuole is filled with watery fluid.

In Figure 163 is shown a part of the scolopophore represented in Figure 162, more enlarged (A), and three cross-sections (B, C, D) of the scolopale. The wall of the scolopale is composed at either end of seven ribs (*r*), each of which is divided in the central portion, making fourteen ribs in this part. The entire scolopale, except possibly the terminal ligament, is bathed in the watery liquid, and is free to vibrate (Hess '17).

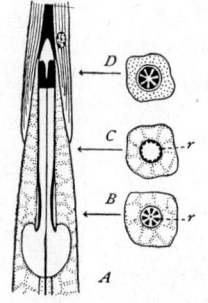

Fig. 163.—Part of the scolopophore shown in Figure 162 more enlarged (From Hess).

It should be remembered that the scolopalæ of different insects vary greatly in form; the one figured here is merely given as an example of one type.

The simpler forms of chordotonal organs.—In the simplest form of a chordotonal organ there is a single scolopophore; usually, however, there are two or more closely parallel scolopophores. In figure 164, which represents a chordotonal organ found in the next to the last segment of the body of a larva of *Chironomus*, these two types are represented, one part of the organ being composed of a single scolopophore, the other of several.

The chordotonal ligament.—In Figure 164 the nerve connecting the chordotonal organ with the central nervous system is represented at *n;* and at *li* is shown a structure not yet mentioned, the *chordotonal ligament*, which is found in many chordotonal organs. Figure 165 is a diagrammatic representation of the relations of the chordotonal organs of a larva of *Chironomus* to the central nervous system and to the body-wall. Here each chordotonal organ is approximately T-shaped; the proximal nerve forming the body of the T; the scolopophore, one arm; and the chordotonal ligament, the other arm.

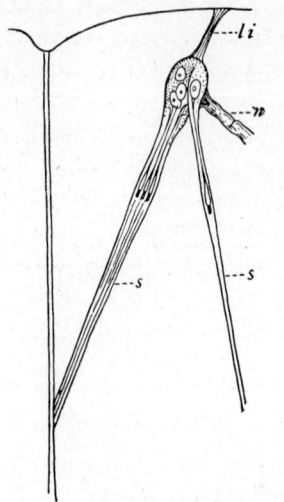

It will be observed that in this type of chordotonal organ the scolopophore and the ligament form a fiddle-string-like structure between two points in the wall of a single segment. It is believed that in cases of this kind the integument acts as a tympanum or sounding board.

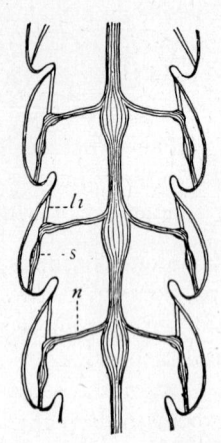

Fig. 164.—Chordotonal organ of a larva of *Chironomus*. (From Graber).

Fig. 165. — Diagram representing the chordotonal organs of a larva of *Chironomus* (After Graber).

b. THE CHORDOTONAL ORGANS OF LARVÆ

Chordotonal organs have been observed in so many larvæ that we may infer that they are commonly present in larvæ. These organs are very simple compared with those of certain adult insects, described later. Those figured in the preceding paragraphs will serve to illustrate the typical form of larval chordotonal organs. Even in the more complicated ones, there are comparatively few scolopophores; and, as a rule, they are not connected with specialized tympana, but extend between distant parts of the body-wall, which probably acts as a sounding board.

In certain larvæ, however, the scolopophores are attached to specialized areas of the body-wall. Hess ('17) has shown that the pleural discs of cerambycid larvæ, which are situated one on each side of several of the abdominal segments, serve as points of attachment of scolopophores.

c. THE CHORDOTONAL ORGANS OF THE LOCUSTIDÆ

In the Locustidæ there are highly specialized ears situated one on each side of the first abdominal segment. The external vibrating part of these organs, the tympanum, is conspicuous, being a thinned portion of the body-wall (Fig. 166).

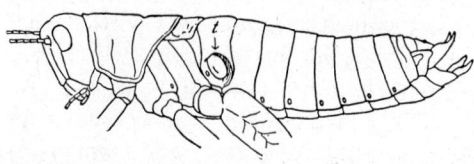

Fig. 166.—Side view of a locust with the wings removed; *t*, tympanum.

Closely applied to the inner surface of each tympanum (Fig. 167, *T*), there is a ganglion known as Müller's organ (*ga*), first described by Müller (1826). This ganglion contains many ganglion-cells and scolopalæ and is the termination of a nerve extending from the central nervous system, the auditory nerve (*n*). Figure 168 represents a section of Müller's organ, showing the ganglion-cells and scolopalæ.

Intimately associated with the Müller's organ are two horny processes (Fig. 167, *o* and *u*) and a pear-shaped vesicle (Fig. 167, *bi*); and near the margin of the tympanum, there is a spiracle (Fig. 167, *st*), which admits air to a space inside of the tympanum, the *tympanal air-chamber*.

As the nerve-endings in Müller's organ are attached to the tympanum, it is a chordotonal organ of the integumental type; it is attached to a vibratile membrane, between two air-spaces.

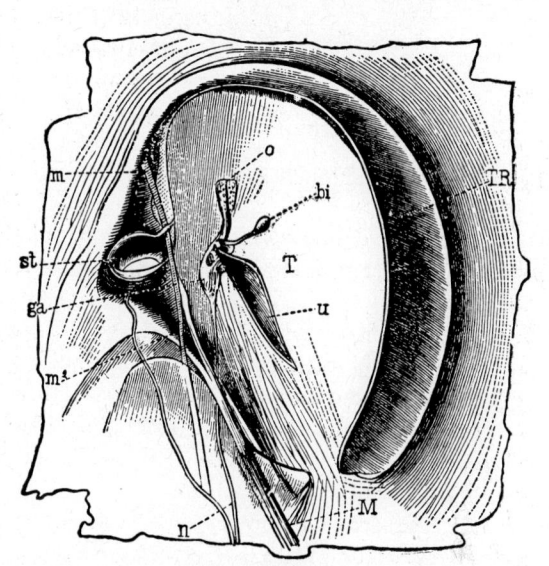

Fig. 167.—Ear of a locust, *Caloptenus italicus*, seen from inner side; T, tympanum; TR, its border; *o*, *u*, two horn-like processes; *bi*, pear-shaped vesicle; *n*, auditory nerve; *ga*, terminal ganglion or Müller's organ; *st*, spiracle; M, tensor muscle of the tympanum (From Packard after Graber).

d. THE CHORDOTONAL ORGANS OF THE TETTIGONIIDÆ AND OF THE GRYLLIDÆ

In the long-horned grasshoppers and in the crickets, there is a pair of tympana near the proximal end of the tibia of each fore leg. In many genera, these tympana are exposed and easily observed (Fig. 169); but in some genera each is covered by a fold of the body-wall and is consequently within a cavity, which communicates with the outside air by an elongated opening (Fig. 170, *a* and *b*).

Within the legs bearing these tympana, there are complicated chordotonal organs. Very detailed accounts of these organs have been published by Graber ('76), Adelung ('92) and Schwabe ('06); in this place, for lack of space, only their more general features can be described.

Figure 171 represents a longitudinal section of that part of a fore tibia of *Decticus verrucivorus* in which the chordotonal organs are situated, and Figure 172 represents a cross-section of the same tibia, passing through the tympana and the air-chambers formed by the folds of the body-wall. In the following account the references, in most cases, are to both of these figures.

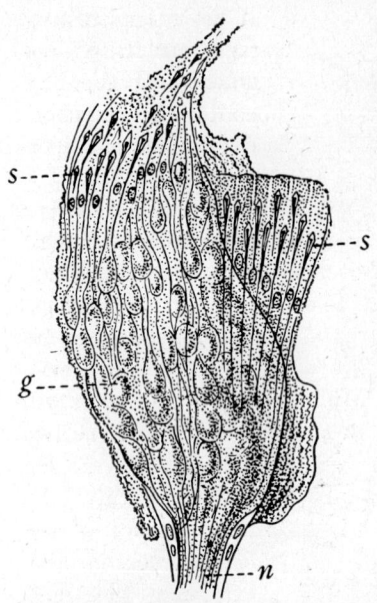

Fig. 168.—Section of Muller's organ; *g*, ganglion-cells; *n*, nerve; *s*, *s*, scolopalæ (After Graber).

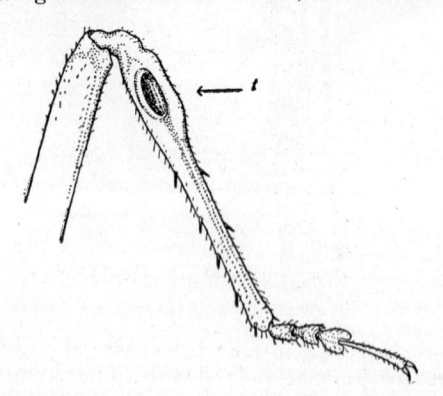

Fig. 169.—Fore leg of a katydid; *t*, tympanum.

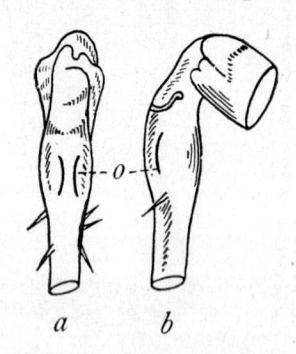

Fig. 170.—Tibia of a locustid with covered tympana; *a*, front view; *b*, side view; *o*, opening (After Schwabe).

The trachea of the leg.—The trachea of the leg figured in part here is remarkable for its great size and for its division into two branches, the front trachea (Ti) and the hind trachea (Te); these two branches reunite a short distance beyond the end of the chordotonal organs.

It is an interesting fact that these large tracheæ of the legs containing the chordotonal organs open through a pair of supernumery spiracles, differing in this respect from the tracheæ of the other legs.

The spaces of the leg. — By reference to Figure 172, it will be seen that the two branches of the leg trachea occupy the middle space of the leg between the two tympana (Tie and Tii) and separate an outer space, the upper one in the figure, from an inner space. The outer space (E) contains a chordotonal organ, of which the scolopale is represented at S; and the inner space contains small tracheæ (t), muscles (m), the tibial nerve (Ntb), and a tendon (Tn). The interstices of the outer and inner spaces are filled with blood.

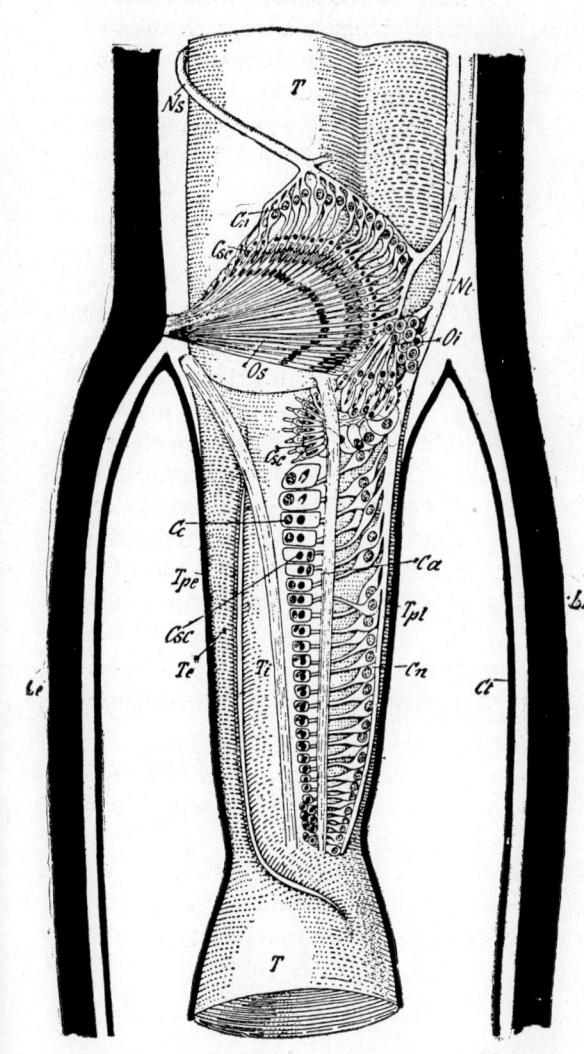

Fig. 171.—Longitudinal section of a fore tibia of *Decticus verrucivorus* (From Berlese after Schwabe).

In the outer space some leucocytes and fat-cells (Gr) are represented.

The supra-tympanal or subgenual organ.—In the outer space of the tibia, a short distance above the tympana, there is a ganglion (Fig.

171, Os) composed of nerve-endings, which are scolopophores of the integumental type. Two nerves extend to this ganglion, one from each side of the leg, and each divides into many scolopophores. The attachment fibers of the scolopophores converge and are attached to the wall of the leg. Two terms have been applied to this organ, both indicating its position in the leg; one refers to the fact that it is above the tympana, the other, that it is below the knee.

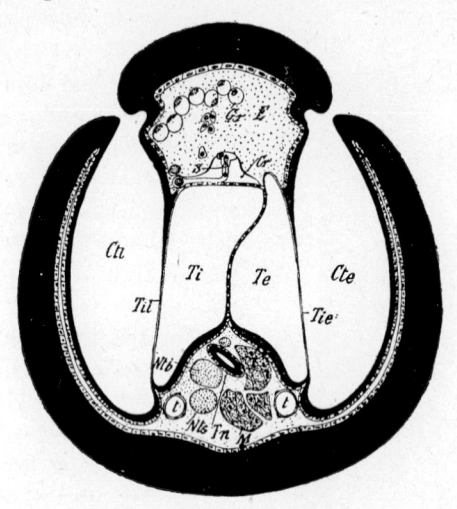

Fig. 172.—Transverse section of the fore tibia of *Decticus verrucivorus* (From Berlese after Schwabe). In comparing this figure with the preceding, note that in that one the external parts are at the left, in this one, at the right.

The intermediate organ.—Immediately below the supra-tympanal organ, and between it and the organ described in the next paragraph, is a ganglion composed of scolopophores of the subintegumental type; this is termed the *intermediate organ* (Fig. 171, Oi).

Siebold's organ or the crista acustica.—On the outer face of the front branch of the large trachea of the leg there is a third chordotonal organ, the *Siebold's organ* or the *crista acusitca*. A surface view of the organ is given in Figure 171 and a cross-section is represented in Figure 172. It consists of a series of scolopophores of the subintegumental type, which diminish in length toward the distal end of the organ (Fig. 171). The relation of Siebold's organ to the trachea is shown in Figure 172. It forms a ridge or crest on the trachea, shown in setion at *cr* in Figure 172; this suggested the name *crista acustica*, used by some writers.

e. THE JOHNSTON'S ORGAN

There has been found in the pedicel of the antenna of many insects, representing several of the orders, an organ of hearing, which is known as the *Johnston's organ*, having been pointed out by Christopher Johnston (1855). This organ varies somewhat in form in different insects and in the two sexes of the same species; but that of a male mosquito will serve as an example illustrating its essential features.

The following account is based on an investigation by Professor Ch. M. Child ('94).

In an antenna of a mosquito (Fig. 173) the scape or first segment, which contains the muscles of the antenna, is much smaller than the pedicel or second segment, and is usually overlooked, being concealed by the large, globular pedicel; the clavola consists of thirteen slender segments. Excepting one or two terminal segments, each segment of the clavola bears a whorl of long, slender setæ; these are more prominent in the male than in the female.

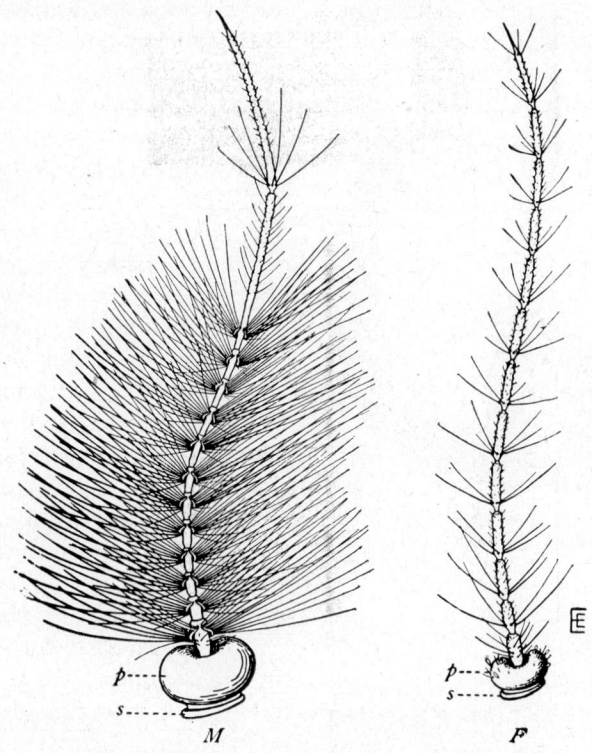

Fig. 173.—Antennæ of mosquitoes, *Culex;* M, male; F, female; *s*, scape; *p*, pedicel.

Figure 174 represents a longitudinal section of the base of an antenna; in this the following parts are shown: S, scape; P, pedicel, C, base of the first segment of the clavola; *cp*, conjunctival plate connecting the pedicel with the first segment of the clavola; *pr*, chitinous processes of the conjunctival plate; *m*, muscles of the antenna; N, principal antennal nerve; *n*, nerve of the clavola; immediately within the wall of the segments there is a thin layer of hypodermis; the lumen of the pedicel is largely occupied by a ganglion composed of scolopophores, the attachment fibers of which are attached to the chitinous processes of the conjunctival plate.

As to the action of the auditory apparatus as a whole, it was shown experimentally by Mayer ('74) that the different whorls of setæ borne by the segments of the clavola, and which gradually decrease in length on successive segments, are caused to vibrate by different notes; and it is believed that the vibrations of the setæ are transferred to the conjunctival plate by the clavola, and thence to the nerve-endings.

It was formerly believed that the great specialization of the Johnston's organ in male mosquitoes enabled the males to hear the songs of the females and thus more readily to find their mates. But it has been found that in some species, at least, of mosquitoes and of midges in which the males have this organ highly specialized the females seek the males. This has led some writers to doubt that the Johnston's organ is auditory in function. But the fact remains that its distinctive feature is the presence of scolopalæ, which is the distinctive characteristic of the auditory organs of other insects.

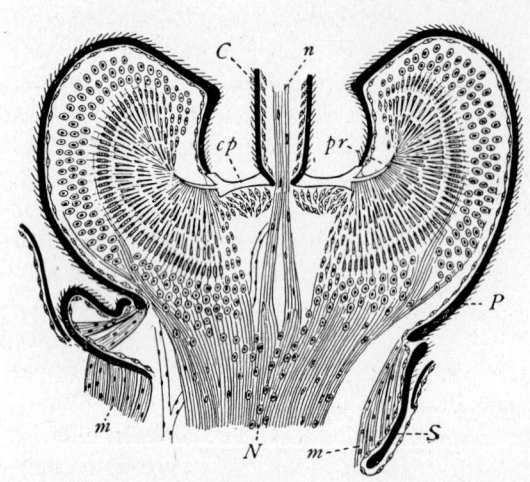

Fig. 174.—Longitudinal section of the base of an antenna of a male mosquito, *Corethra culiciformis* (After Child).

XIV. SENSE-ORGANS OF UNKNOWN FUNCTIONS

In addition to the sense-organs discussed in the foregoing account there have been described several types of supposed sense-organs which are as yet very imperfectly understood. Among these there is one that merits a brief discussion here on account of the frequent references to it in entomological literature. Many different names have been applied to the organs of this type; of these that of *sense domes* is as appropriate as any, unless the conclusions of McIndoo, referred to on the following page, are confirmed, in which case his term *olfactory pores* will be more descriptive.

The sense-domes are found in various situations, but they occur chiefly on the bases of the wings and on the legs. Each sense-dome consists of a thin, hemispherical or more nearly spherical membrane, which either projects from the outer end of a pore in the cuticula (Fig. 175, *a*) or is more or less deeply enclosed in such a pore (Fig. 175, *b*); intergrades between the two types represented in the accompanying figures occur.

When a sense-dome is viewed in section a nerve-ending is seen to be connected with the dome-shaped or bell-like membrane. A striking feature of these organs is the absence of any gland-cells connected with them, such as are found in the chemical sense-organs described on an earlier page.

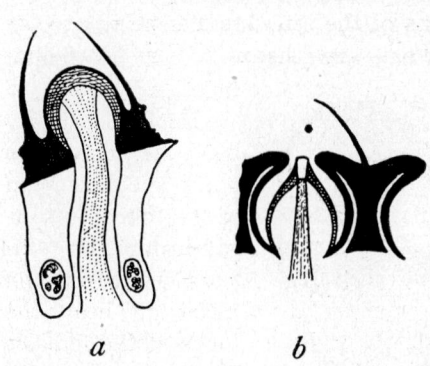

Fig. 175.—Sense-domes (From Berlese).

In one very important respect there is a marked difference in the accounts of these organs that have been published. The organs were first discovered long ago by Hicks ('57); but they have been more carefully studied in recent years by several writers, who have been able to make use of a greatly improved histological technic; among these writers are Berlese ('09 *a*), Vogel ('11), Hochreuter (12'), Lehr ('14), and McIndoo ('14).

All of the writers mentioned above except the last named maintain that the sense-cell ends in a structure, in the middle of the sense-dome, which differs in appearance from both the membrane of the sense-dome and the body of the sense-cell. This structure varies in form in different sense-domes; in some it is cylindrical, and is consequently described as a peg; in others, it is greatly flattened so that it is semilunar in form when seen in section. In Figure 175, *b*, which represents a section made transversely to the long axis of this part it appears peglike; but in Figure 175, *a*, which represents a longitudinal view of it, it is semilunar in form.

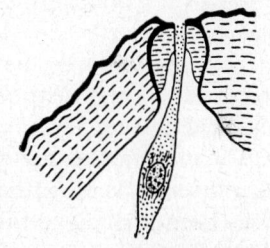

Fig. 176—Olfactory pore of McIndoo (From McIndoo)

According to McIndoo (Fig. 176) no structure of this kind is present, but the sense-fiber of the sense-cell pierces the bottom of the cone and enters the round, oblong, or slit-like pore-aperture. "It is thus seen that the cytoplasm in the peripheral end of the sense-fiber comes in direct contact with the air containing odorous particles and that odors do not have to pass through a hard membrane in order to stimulate the sense-cell as is claimed for the antennal organs."

XV. THE REPRODUCTIVE ORGANS

a. THE GENERAL FEATURES

In insects the sexes are distinct. Formerly *Termitoxenia*, a genus of wingless, very aberrant Diptera, the members of which live in nests of Termites, was believed to be hermaphroditic, but this is now doubted.

Individuals in which one side has the external characters of the male and the other those of the female are not rare; such an individual is termed a *gynăndromŏrph;* in some gynandromorphs, both testes and ovaries are present but in no case are both functional; these therefore are not true hermaphrodites.

In females the essential reproductive organs consist of a pair of *ovaries*, the organs in which the ova or eggs are developed, and a tube leading from each ovary to an external opening, the *oviduct*. In the male, the essential reproductive organs are a pair of *testes*, in which the spermatozoa are developed and a tube leading from each testis to an external opening, the *vas deferens*. In addition to these essential organs, there are in most insects accessory organs, these consist of glands and of reservoirs for the reproductive elements.

The form of the essential reproductive organs and the number and form of the accessory organs vary greatly in different insects. It is impossible to indicate the extent of these variations in the limited space that can be devoted to this subject in this work. Instead of attempting this it seems more profitable to indicate by diagrams, one for each sex, the relations of the accessory organs that may exist to the essential organs.

In adult insects the external opening of the reproductive organs is on the ventral side of the abdomen near the caudal end of the body. The position of the opening appears to differ in different insects and in some cases in the two sexes of the same species. The lack of uniformity in the published accounts bearing on this point is partly due to differences in numbering the abdominal segments; some authors describing the last segment of the abdomen as the tenth while others

THE INTERNAL ANATOMY OF INSECTS   157

believe it to be the eleventh; embryological evidence supports the latter view.

In most insects there is a single external opening of the reproductive organs; but in the Ephemerida and in a few other insects the two efferent ducts open separately.

Secondary sexual characters.—In addition to differences in the essential reproductive organs and in the genital appendages of the two sexes, many insects exhibit what are termed *secondary sexual characters*. Among the more striking of these are differences in size, coloring, and in the form of certain organs. Female insects are usually larger than the males of the same species; this is due to the fact that the females carry the eggs; but in those cases where the males fight for their mates, as stag-beetles, the males are the larger. Striking differences in the coloring of the two sexes are common, especially in the Lepidoptera. In many insects the antennæ of the male are more highly specialized than those of the female; and this is true also of the eyes of certain insects. These are merely a few of the many known secondary sexual characters found in insects.

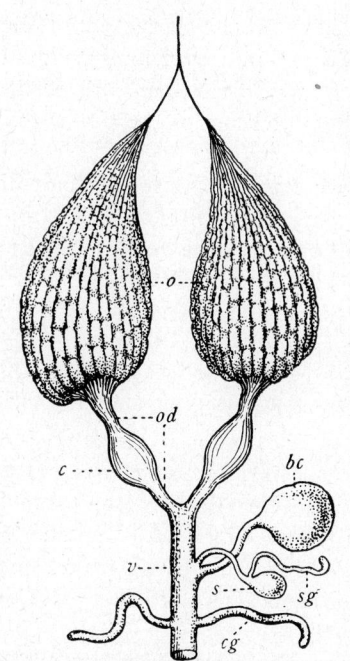

Fig. 177.—Diagram of the reproductive organs of a female insect; *o*, ovary; *od*, oviduct; *c*, egg-calyx; *v*, vagina; *s*, spermatheca; *bc*, bursa copulatrix; *sg*, spermathecal gland; *cg*, colleterial glands.

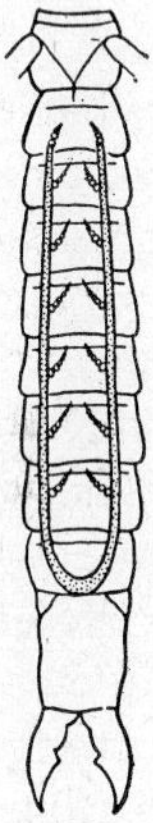

Fig. 178.—Reproductive organs of *Japyx*, female (After Grassi).

b. THE REPRODUCTIVE ORGANS OF THE FEMALE

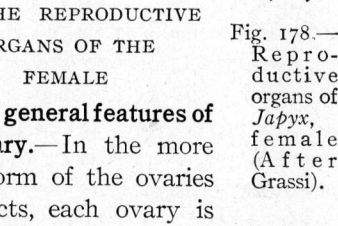

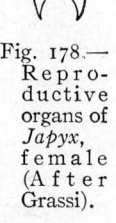

The general features of the ovary.—In the more usual form of the ovaries of insects, each ovary is a compact, more or less spindle-shaped body composed of many parallel *ovarian tubes* (Fig. 177, *o*), which open into a common efferent tube, the oviduct. In *Campodea*, however, there is a single ovarian tube; and in certain other Thysanura the ovarian tubes have a metameric arrangement (Fig. 178). The number of ovarian tubes differs greatly in different insects; in many Lepidoptera there are only four in each ovary; in the honeybee, about 150; and in some Termites, 3000 or more.

The wall of an ovarian tube..—The ovarian tubes are lined with an *epithelial layer*, which is supported by a *basement membrane;* outside of this there is a *peritoneal envelope*, composed of connective tissue; and sometimes there are muscles in the peritoneal envelope.

The zones of an ovarian tube.—Three different sections or zones are recognized in an ovarian tube; first, the *terminal filament*, which is the slender portion which is farthest from the oviduct (Fig. 179, *t*); second, the *germarium*, this is a comparatively short chamber, between the other two zones (Fig. 179, *g*); and third, the *vitellarium*, which constitutes the greater portion of the ovarian tube.

The contents of an ovarian tube.—In the germarium are found the *primordial germ-cells* from which the eggs are developed; and in the vitellarium are found the developing eggs. In addition to the cells that develop into eggs there are found, in the ovarian tubes of many insects, cells whose function is to furnish nutriment to the developing eggs; these are termed *nurse-cells*.

Depending upon the presence or absence of nurse-cells and on the location of the nurse-cells when present, three types of ovarian tubes are recognized: first, those without distinct nurse-cells (Fig. 179, A); second, those in which the eggs and masses of nurse-cells alternate in the ovarian tube (Fig. 179, B); and third, those in which the nurse-cells are restricted to the germarium (Fig. 179, C), which thus becomes a nutritive chamber. In the latter type the developing eggs are each connected by a thread with the nutritive chamber.

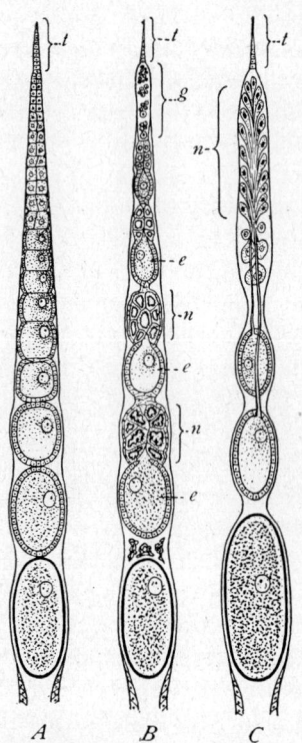

Fig. 179.—Three types of ovarian tubes; *e, e, e,* eggs; *n, n, n,* nurse-cells (After Berlese).

The egg-follicles.—The epithelium lining of the ovarian tube becomes invaginated between the eggs in such a way that each egg is enclosed in an epithelial sac or *egg-follicle*, which passes down the tube with the egg (Fig. 179). There is thus a tendency to strip the tube of its epithelium, but a new one is constantly formed.

The functions of the follicular epithelium.—It is believed that in some cases, and especially where the nurse-cells are wanting, the follicular epithelium serves a nutritive function. But the most obvious function of this epithelium is the formation of the chorion or egg-shell, which is secreted on its inner surface. The pit-like markings so common on the shells of insect eggs indicate the outlines of the cells of the follicular epithelium.

The ligament of the ovary.—In many insects, the terminal filaments of the several ovarian tubes of an ovary unite and form a slender cord, the ligament of the ovary, which is attached to the dorsal diaphragm; but in other insects this ligament is wanting, the terminal filaments ending free in the body cavity.

The oviduct.—The common outlet of the ovarian tubes is the *oviduct* (Fig. 177, *od*). In most insects the oviducts of the two ovaries unite and join a common outlet, the vagina; but in the Ephemerida and in some Dermaptera each oviduct has a separate opening.

The egg-calyx.—In some insects each oviduct is enlarged so as to form a pouch for storing the eggs, these pouches are termed the *egg-calyces* (Fig. 177, c.)

The vagina.—The tube into which the oviducts open is the vagina (Fig. 177, *v*). The vagina differs in structure from the oviducts, due to the fact that it is an invagination of the body-wall, and, like other invaginations of the body-wall, is lined with a cuticular layer

The spermatheca.—The spermatheca is a sac for the storage of the seminal fluid (Fig. 177, *s*). As the pairing of the sexes takes place only once in most insects and as the egg-laying period may extend over a long time, it is essential that provision be made for the fertilization of the eggs developed after the union of the sexes. The eggs become full-grown and each is provided with a shell before leaving the ovarian tubes. At the time an egg is laid a spermatozoan may pass from the spermatheca, where thousands of them are stored, into the egg through an opening in the shell, the micropyle, which is described in the next chapter (Figs. 184 and 185).

In some social insects, eggs that are developed years after the pairing took place are fertilized by spermatozoa that have been stored in the spermatheca.

The bursa copulatrix.—In many insects there is a pouch for the reception of the seminal fluid before it passes to the spermatheca.

This pouch is known as the *bursa copulatrix* or copulatory pouch. In some insects this pouch is a diverticulum of the vagina (Fig. 177, *bc*); in others it has a distinct external opening, there being two external openings of the reproductive organs, the opening of the vagina and the opening of the bursa copulatrix.

When the bursa copulatrix has a distinct external opening there may or may not be a passage from it to the vagina. In at least some Orthoptera (*Melanoplus*) there is no connection between the two; when the eggs are laid they are pushed past the opening of the bursa copulatrix where they are fertilized.

In the Lepidoptera (Fig. 180), there is a passage from the bursa copulatrix to the vagina. In this case the seminal fluid is received by the bursa copulatrix at the time of pairing, later it passes to the spermatheca, and from here it passes to the vagina.

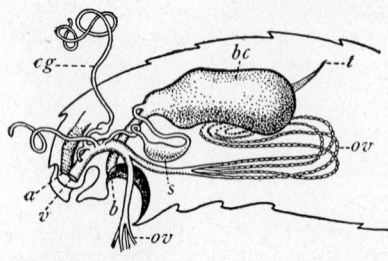

Fig. 180.—Reproductive organs of the female of the milkweed butterfly; *a*, anus; *b*, opening of the bursa copulatrix; *ov*, ovarian tubes; *t*, terminal filaments of the ovary; *v*, opening of the vagina (After Burgess).

A bursa copulatrix is said to be wanting in Hymenoptera, Diptera, Heteroptera and Homoptera except the Cicadas.

The colleterial glands.—There are one or two pairs of glands that open into the vagina near its outlet (Fig. 177, *cg*); to these has been applied the general term *colleterial glands*. Their function differs in different insects; in some insects they secrete a cement for gluing the eggs together, in others they produce a capsule or other covering which protects the eggs.

The spermathecal gland.—In many insects there is a gland that opens either into the spermatheca or near the opening of the spermatheca, this is the *spermathecal gland* (Fig. 177, *sg*).

c. THE REPRODUCTIVE ORGANS OF THE MALE

The reproductive organs of the male are quite similar in their more general features to those of the female; but there are striking differences in details of form.

The general features of the testes.—As the reproductive elements developed in the testes, the spermatazoa, always remain small, the testes of a male are usually much smaller than the ovaries of the female of the same species.

In the more common form, each testis is a compact body (Fig. 181, *t*) composed of a variable number of tubes corresponding with the ovarian tubes, these are commonly called the *testicular follicles;* but it would have been better to have termed them the testicular tubes, reserving the term follicle for their divisions.

The testicular follicles vary in number, form, and in their arrangement. In many insects as the Neuroptera, the Hemiptera, the Diptera, and in *Campodea* and *Japyx*, each testis is composed of a single follicle. In some beetles, Carabidæ and Elateridæ, the follicle is long and rolled into a ball. In some Thysanura the testicular follicles have a metameric arrangement.

In some Coleoptera, each testis is separated into several masses, each having its own outlet leading to the vas deferens; while in some other insects the two testes approach each other during the pupal stage and constitute in the adult a single mass.

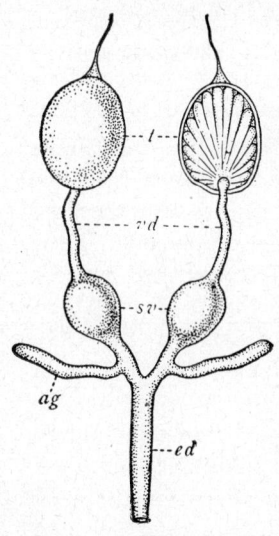

Fig. 181.—Diagram of the reproductive organs of a male insect; the right testis is shown in section; *ag*, accessory glands; *ed*, ejaculatory duct; *sv*, seminal vesicles; *t*, testes; *vd*, vasa deferentia.

The structure of a testicular follicle.—Like the ovarian tubes, the testicular follicles are lined with an epithelial layer, which is supported by a basement membrane, outside of which there is a peritoneal envelope composed of connective tissue. And in these follicles a series of zones are distinguished in which the genital cells are found in different stages of development, corresponding to the successive generations of these cells. In addition to the terminal filament four zones are recognized as follows:

The germarium.—This includes the primordial germ-cells and the spermatogonia.

The zone of growth.—Here are produced the spermatocytes of the first order and the spermatocytes of the second order.

The zone of division and reduction.—In this zone are produced the spermatids or immature spermatozoa.

The zone of transformation.—Here the spermatids become spermatozoa.

A discussion of the details of the development of the successive generations of the genital cells of the male, or spermatogenesis, does not fall within the scope of this volume.

The spermatophores.—In some insects the spermatozoa become enveloped in a sac in which they are transferred to the female; this sac is the *spermatophore*. Spermatophores have been observed in Gryllidæ, Locustidæ, and certain Lepidoptera.

Other structures.—*A ligament of the testis*, corresponding to the ligament of the ovary, is often present; the common outlet of the testicular follicles, corresponding to the oviduct is termed the *vas deferens* (Fig. 181, *vd*); an enlarged portion of the vas deferens serving as a reservoir for the products of the testis is known as a *seminal vesicle* (Fig. 181, *sv*); the invaginated portion of the body-wall, corresponding with the vagina of the female, is the *ejaculatory duct* (Fig. 181, *ed*); *accessory glands*, corresponding to the colleterial glands of the female, are present (Fig. 181, *ag*); the function of these glands has not been determined, they may secrete the fluid part of the semen, and they probably secrete the spermatophore when one is formed; *the penis*, this is merely the chitinized terminal portion of the ejaculatory duct, which can be evaginated with a part of the invaginated portion of the body-wall. It is furnished with powerful muscles for its protrusion and retraction.

XVI. THE SUSPENSORIA OF THE VISCERA

The organs discussed here do not constitute a well-defined system, but are isolated structures connected with different viscera. As in most cases they appear to serve a suspensory function, they are grouped together provisionally as the suspensoria of the viscera.

The dorsal diaphragm.—This is a membranous structure which extends across the abdominal cavity immediately below the heart, to which it is attached along its median line. The lateral margins of this diaphragm are attached to the sides of the body by a series of triangular prolongations, which have been commonly known as the *wings of the heart* (Fig. 139, *c*). The dorsal diaphragm is composed largely of very delicate muscles. Its relation to the heart is illustrated by the accompanying diagram (Fig. 182, *d*).

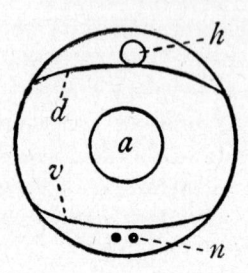

Fig. 182.—Diagram showing the relation of the dorsal diaphragm and the ventral diaphragm to other viscera; *a*, alimentary canal; *d*, dorsal diaphragm; *h*, heart; *n*, ventral nervous system; *v*, ventral diaphragm.

There are differences of opinion as to the function of the dorsal diaphragm. An important function is probably to protect the heart from the peristaltic movements of the alimentary canal. It also supports the heart; and it may play a part in its expansion.

The dorsal diaphragm is also known as the *pericardial diaphragm*.

The ventral diaphragm.—The ventral diaphragm is a very delicate membrane which extends across the abdominal cavity immediately above the ganglia of the central nervous system. It is quite similar in form to the dorsal diaphragm; it is attached along each side of the body, just laterad of the great ventral muscles, by a series of prolongations resembling in form the wings of the heart. The position of the ventral diaphragm is illustrated in Figure 182, *v*.

This diaphragm has been described as a ventral heart; but I believe that its function is to protect the abdominal ganglia of the central nervous system from the peristaltic movements of the alimentary canal.

The thread-like suspensoria of the viscera.—Under this head may be classed the ligament of the ovary and the ligament of the testis, already described. In addition to these, there is, in some insects at least, a thread-like ligament that is attached to the intestine.

XVII. SUPPLEMENTARY DEFINITIONS

There are found in the bodies of insects certain organs not referred to in the foregoing general account of the internal anatomy of insects. These organs, though doubtless very important to the insects in which they occur, are not likely to be studied in an elementary course in entomology and, therefore, a detailed account of them may well be omitted from an introductory text-book. This is especially true as our knowledge of the structure and functions of these organs is so incomplete that an adequate discussion of the conflicting views now held would require more space than can be devoted to it here. The organs in question are the following:

The œnocytes.—The term *œnocytes* is applied to certain very large cells, that are found in clusters, often metamerically arranged, and connected with the tracheæ and the fat body of insects. The name was suggested by the light yellow color which often characterizes these cells, the color of certain wines; but the name is not a good one, as œnocytes vary greatly in color. Several other names have been applied to them but they are generally known by the name used here. Two types of œnocytes are recognized: first, the larval œnocytes; and second, the imaginal œnocytes.

The larval œnocytes are believed by Verson and Bisson ('91) to be ductless glands which take up, elaborate, and return to the blood definite substances, which may then be taken up by other cells of the body. Other views are held by other writers, but the view given above seems, as this time to be the one best supported by the evidence at hand.

As to the function of the imaginal œnocytes, there are some observations that seem to show that they are excretory organs without ducts, cells that serve as storehouses for excretory products, becoming more filled with these products with the advancing age of the insect.

The pericardial cells.—The term *pericardial cells* is applied to a distinct type of cells that are found on either side of the heart in the pericardial sinus or crowded between the fibers of the pericardial diaphragm.

These cells can be rendered very conspicuous by injecting ammonia carmine into the living insect some time before killing and dissecting it; by this method the pericardial cells are stained deeply while the other cells of the body remain uncolored.

It is believed that the pericardial cells absorb albuminoids originating from the food and transform them into assimilable substances.

The phagocytic organs.—The term *phagocyte* is commonly applied to any leucocyte or white blood corpuscle that shows special activity in ingesting and digesting waste and harmful materials, as disintegrating tissue, bacteria, etc. The action of phagocytes is termed *phagocytosis;* an excellent example of phagocytosis is the part played by the leucocytes in the breaking down and rebuilding of tissues in the course of the metamorphosis of insects; this is discussed in the next chapter.

Phagocytosis may take place in any part of the body bathed by the blood and thus reached by leucocytes. In addition to this widely distributed phagocytosis, it is believed that in certain insects there are localized masses of cells which perform a similar function; these masses of cells are known as the *phagocytic organs*.

Phagocytic organs have been found in many Orthoptera and in earwigs; they are situated in the pericardial region; and can be made conspicuous by injecting a mixture of ammonia carmine and India ink into the body cavity; by this method the pericardial cells are stained red and the phagocytic organs black.

The light-organs.—The presence of organs for producing light is widely distributed among living forms both animal and vegetable.

The most commonly observed examples of light-producing insects are certain members of the Lampyridæ, the fireflies and the glow-worms, and a member of the Elateridæ, the "cucujo" of the tropics. With these insects the production of light is a normal function of highly specialized organs, the light-organs.

Examples of insects in which the production of light is occasionally observed are larvæ of mosquitoes, and certain lepidopterous larvæ. In these cases the production of light is abnormal, being due either to the presence in the body of light-producing bacteria or to the ingestion of luminescent food.

The position of the specialized light-organs of insects varies greatly; in the fireflies, they are situated on the ventral side of the abdomen; in the glow-worms, along the sides of the abdomen; and in the cucujo, the principal organs are in a pair of tubercles on the dorsal side of the prothorax and in a patch in the ventral region of the metathorax.

The structure of the light-organs of insects varies in different insects, as is shown by the investigations of several authors. A good example of highly specialized light-organs are those of *Photinus marginellus*, one of our common fireflies. An excellent account of these is that of Miss Townsend ('04), to which the reader is referred. A detailed account of the origin and development of the light-organs of *Photurus pennsylvanica* is given by Hess ('22).

CHAPTER IV.

THE METAMORPHOSIS OF INSECTS

Many insects in the course of their lives undergo remarkable changes in form; a butterfly was once a caterpillar, a bee lived first the life of a clumsy footless grub, and flies, which are so graceful and active, are developed from maggots.

In the following chapters considerable attention is given to descriptions of the changes through which various insects pass; the object of this chapter is merely to discuss the more general features of the metamorphosis of insects, and to define the terms commonly used in descriptions of insect transformations.

I. THE EXTERNAL CHARACTERISTICS OF THE METAMORPHOSIS OF INSECTS

The more obvious characteristics of the metamorphosis of insects are those changes in the external form of the body that occur during postembryonic development. In some cases there appears to be but little in common between the successive forms presented by the same insect, as the caterpillar, chrysalis, and adult stages of a butterfly. On the other hand, in certain insects, the change in the form of the body during the postembryonic life is comparatively little. Based on these differences, several distinct types of metamorphosis have been recognized; and in those cases where the insect in its successive stages assumes different forms, distinctive terms are applied to the different stages.

a. THE EGG

Strictly speaking, all insects are developed from eggs, which are formed from the primordial germ-cells in the ovary of the female. As a rule, each egg is surrounded by a shell, formed by the follicular epithelium of the ovarian tube in which the egg is developed; and this egg, enclosed in its shell, is deposited by the female insect, usually on or near the food upon which the young insect is to feed. In some cases, however, the egg is retained by the female until it is hatched; thus flesh-flies frequently deposit active larvæ upon meat, especially when they have had difficulty in finding it; and there are other viviparous insects, which are discussed later. In this place is discussed the more common type of insect eggs, those that are laid while still enclosed in their shell.

The shape of the egg.—The terms ovoid and ovate have a definite meaning which has been derived from the shape of the eggs of birds; but while many eggs of insects are ovate in form, many others are not.

The more common form of insect eggs is an elongate oval, somewhat curved; this type is illustrated by the eggs of crickets (Fig. 183, 1); many eggs; are approximately spherical, as those of some butterflies (Fig. 183, 2); while some are of remarkable shape, two of these are represented in Figure 183, 3, 4.

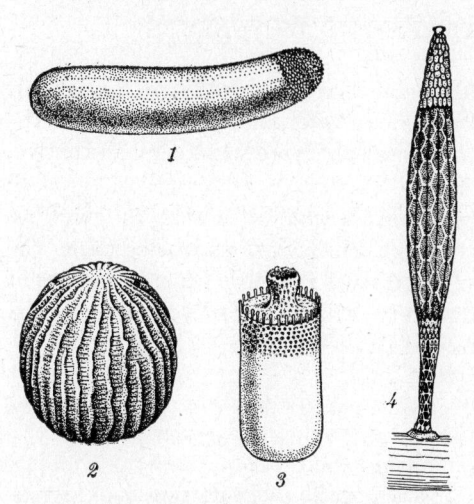

Fig. 183.—Eggs of insects; 1, *Œcanthus nigricornis*; 2, *Œnis semidea*; 3, *Piezosterum subulatum*; 4, *Hydrometra martini*.

The sculpture of the shell.—Almost always the external surface of the shell of an insect egg is marked with small, hexagonal areas; these are the imprints of the cells of the follicular epithelium, which formed the shell. In many cases the ornamentation of the shell is very conspicuous, consisting of prominent ridges or series of tubercles; this is well-shown in the eggs of many Lepidoptera (Fig. 184).

The micropyle.—It has been shown, in the course of the discussion of the reproductive organs of the female, that the egg becomes full-grown, and the protecting chorion or egg-shell is formed about it before it is fertilized. This renders necessary some provision for the entrance of the male germ-cell into the egg; this provision consists of one or more openings in the shell through which a spermatozoan may enter. This opening or group of openings is termed the *micropyle*.

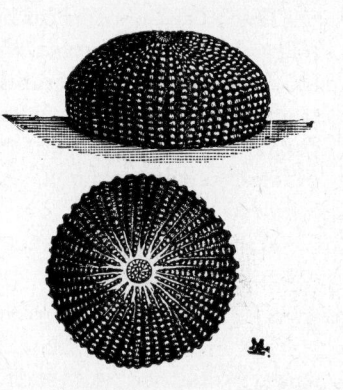

Fig. 184.—Egg of the cottonworm moth; the micropyle is shown in the center of the lower figure.

The number and position of the micropylar openings varies greatly in the eggs of different insects. Frequently they present an elaborate pattern at one pole of the egg (Fig. 184); and sometimes they open through more or less elongated papillæ (Fig. 185).

While in most cases it is necessary that an egg be fertilized in order that development may continue, there are many instances of parthenogenesis among insects.

The number of eggs produced by insects.—A very wide variation exists in the number of eggs produced by insects. In the sheep-tick, for example, a single large egg is produced at a time, and but few are produced during the life of the insect; on the other hand, in social insects, as ants, bees, and termites, a single queen may produce hundreds of thousands of eggs during her lifetime.

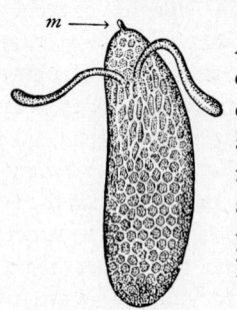

Fig. 185.—Egg of *Drosophila melanogaster; m*, micropyle.

These, however, are extreme examples; the peculiar mode of development of the larva of the sheep-tick within the body of the female makes possible the production of but few eggs; while the division of labor in the colonies of social insects, by which the function of the queen is merely the production of eggs, makes it possible for her to produce an immense number; this is especially true where the egg-laying period of the queen extends over several years.

The following may be taken as less extreme examples. In the solitary nest-building insects, as the fossores, the solitary wasps, and the solitary bees, the great labor involved in making and provisioning the nest results in the reduction of the number of eggs produced to a comparatively small number; while many insects that make no provision for their young, as moths, for example, may lay several hundred eggs.

With certain chalcis-flies the number of young produced is not dependent upon the number of eggs laid; for with these insects many embryos are developed from a single egg. This type of development is termed *polyembryony*.

Modes of laying eggs.—Perhaps in no respect are the wonderful instincts of insects exhibited in a more remarkable way than in the manner of laying their eggs. If insects were reasoning beings, and if each female knew the needs of her young to be, she could not more accurately make provision for them than is now done by the great majority of insects.

This is especially striking where the life of the young is entirely different from that of the adult. The butterfly or moth may sip nectar from any flower; but when the female lays her eggs, she selects with unerring accuracy the particular kind of plant upon which her larvæ feed. The dragonfly which hunts its prey over the field, returns to water and lays her eggs in such a position that the young when it leaves the egg is either in or can readily find the element in which alone it is fitted to live.

The ichneumon-flies frequent flowers; but when the time comes for a female to lay her eggs, she seeks the particular kind of larva upon which the species is parasitic, and will lay her eggs in no other. It is a remarkable fact that no larva leads so secluded a life that it cannot be found by its parasites. Thus the larvæ of *Tremex columba* bore in solid wood, where they are out of sight and protected by a layer of wood and the bark of the tree in which they are boring; nevertheless the ichneumon-fly *Megarhÿssa lunator*, which is parasitic upon it, places her eggs in the burrows of the *Tremex* by means of her long drill-like ovipositor (Fig. 186).

In contrast with the examples just cited, some insects exhibit no remarkable instinct in their egg-laying. Our common northern walking-stick, *Diapheromera*, drops its eggs on the ground under the shrubs and trees upon which it feeds. This, however, is sufficient provision, for the eggs are protected throughout the winter by the fallen leaves, and the young when hatched, readily find their food.

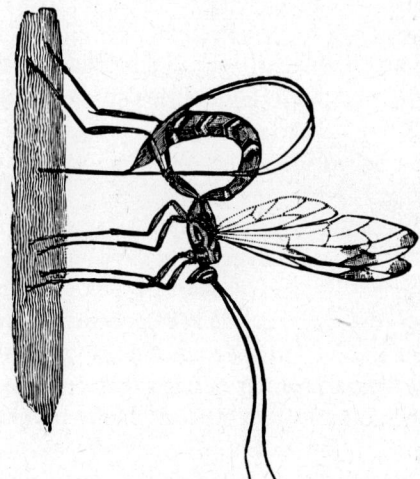

Fig. 186.—*Megarhyssa lunator.*

Many species, the young of which feed upon foliage, lay their eggs singly upon leaves; but many others, and this is especially true of those, the young of which are gregarious, lay their eggs in clusters. In some cases, as in the squash bug, the mass of eggs is not protected (Fig. 187); in others, where the duration of the egg-state is long, the eggs are protected by some covering. The females of our tent-caterpillars cover their eggs with a water-proof coating; and the tussock moths of the genus *Hemerocampa* covertheir egg-clusters with a frothy mass.

170　　*AN INTRODUCTION TO ENTOMOLOGY*

The laying of eggs in compact masses, however, is not correlated, in most cases, with gregarious habits of the larvæ. The water-scavenger beetles, Hydrophilidæ, make egg-sacks out of a hardened silk-like secretion (Fig. 188); the locusts, Acridiidæ, lay their eggs in oval masses and cover them with a tough substance; the scale-insects of the genus *Pulvinaria* excrete a large cottony egg-sac (Fig. 189);

Fig. 187—Egg-mass of the squash-bug.

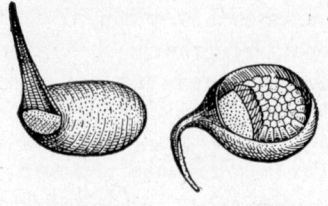

Fig. 188.—Egg-sac of *Hydrophilus* (After Miall).

the eggs of the praying mantis are laid in masses and overlaid with a hard covering of silk (Fig. 190); and cockroaches produce pod-like egg-cases, termed ootheca, each containing many eggs (Fig. 191).

Among the more remarkable of the methods of caring for eggs is that of the lace-winged flies, *Chrysopa*. These insects place each of their eggs on the summit of a stiff stalk of hard silk (Fig. 192).

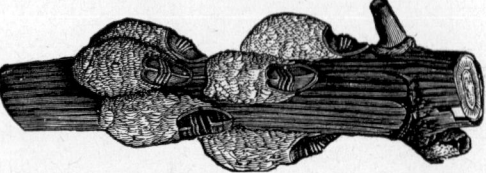

Fig. 189.—*Pulvinaria innumerabilis*, females on grape with egg-sacs

Fig. 190.—Egg-mass of a praying mantis.

Duration of the egg-state.—In the life-cycle of most insects, a few days, and only a few, intervene between the laying of an egg and the emergence of the nymph, naiad, or larva from it. In some the duration of the egg-state is even shorter, the hatching of the egg taking place very soon after it is laid, or even, as sometimes in flesh-flies, before it is laid. On the other hand, in certain species, the greater part of the life of an individual is passed within the egg-shell. The common apple-tree tent-caterpillars, *Clisiocampa americana*, lays its eggs in early summer; but these eggs do not hatch till the following spring; while the remainder of the life-cycle occupies only a few weeks. The eggs of *Bittacus* are said to remain unhatched for two years; and a similar statement is made regarding the eggs of our common walking-stick.

b. THE HATCHING OF YOUNG INSECTS

Only a few accounts have been published regarding the manner in which a young insect frees itself from the embryonic envelopes. In some cases it is evident that the larva cuts its way out from the egg-shell by means of its mandibles; but in others, a specialized organ has been developed for this purpose.

Fig. 191.—Ootheca of a cockroach.

The hatching spines.—An organ for rupturing the embryonic envelopes is probably commonly present. It has been described under several names. It was termed an *egg-burster* by Hagen, the *ruptor ovi* by C. V. Riley an *egg-tooth* by Heymons, and the *hatching spines* by Wheeler.

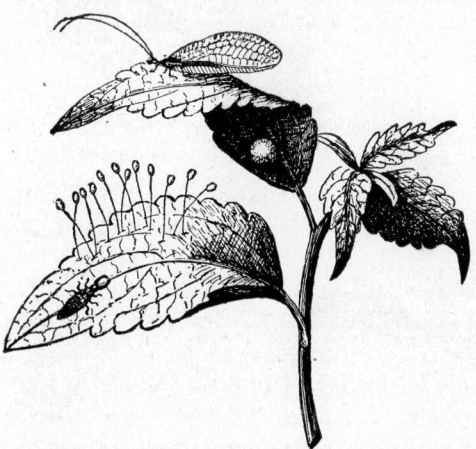

Fig. 192.—Eggs, larva, cocoon, and adult of *Chrysopa*.

c. THE MOLTING OF INSECTS

The young of insects cast periodically the outer parts of the cuticula; this process is termed *molting* or *ecdysis*.

General features of the molting of insects.—The chitinization of the epidermis or primary cuticula adds to its efficiency as an armor, but it prevents the expansion of the body-wall rendered necessary by the growth of the insect; consequently as the body grows, its cuticula becomes too small for it. When this occurs a second epidermis is formed by the hypodermis; after which the old epidermis splits open, usually along the back of the head and thorax, and the insect works itself out from it. The new epidermis being elastic, accommodates itself to the increased size of the body; but in a short time it becomes chitinized; and as the insect grows it in turn is cast off. The cast skin of an insect is termed the *exuviæ*, the plural noun being used as in English is the word clothes.

Coincident with the formation of the new epidermis, new setæ are formed beneath the old epidermis; these lie closely appressed to the outer surface of the new epidermis until released by the molting of the old epidermis.

In the above account only the more general features of the process of molting are indicated, the details, according to the observations of Tower ('06) are as follows. (See Figure 113, p. 99). In the formation of the new epidermis it appears as a thin, delicate lamella, spread evenly over the entire outer surface of the hypodermis; it grows rapidly in thickness until finally, just before ecdysis takes place, it reaches its final thickness. After ecdysis the epidermis hardens rapidly and its coloration is developed. As soon as ecdysis is over the deposition of the dermis or secondary cuticula begins. This layer is a carbohydrate related to cellulose, and is deposited in layers of alternating composition, through the period of reconstruction and growth, during which it reaches its maximum thickness. Preliminary to ecdysis a thin layer of molting fluid is formed, and through its action the old dermis is corroded and often almost entirely destroyed, thus facilitating ecdysis. This dissolving of the dermis, is, according to Tower, a most constant phenomenon in ecdysis and has been found in all insects examined by him in varying degrees.

It is said that the Collembola molt after reaching sexual maturity, in this respect agreeing with the Crustacea and the "Myriapoda," and differing from the Arachnida and from all other insects (Brindley '98).

The molting fluid.—As indicated above, the process of molting is facilitated by the excretion of a fluid known as the molting fluid. This is produced by unicellular glands (Fig. 113, p. 99) which are modified hypodermal cells. These glands are found all through the life of the insect and upon all parts of the body; but are most abundant upon the pronotum, and are more abundant at pupation than at any other period.

The number of postembryonic molts.—A very wide range of variation exists as to number of molts undergone by insects after they leave the egg-shell. According to Grassi ('98, p. 292), there is only a single partial molt with *Campodea* and *Japyx*, while the May-fly *Chloeon* molts twenty times. Between these extremes every condition exists. Probably the majority of insects molt from four to six times; but there are many records of insects that molt many more times than this.

Stadia.—The intervals between the ecdyses are called *stadia*. In numbering the stadia, the first stadium is the period between hatching and the first postembryonic ecdysis.

Instars.—The term *instar* is applied to the form of an insect during a stadium; in numbering the instars, the form assumed by the insect between hatching and the first postembryonic molt is termed the first instar.

Head measurements of larvæ.—It was demonstrated by Dyar ('90) that the widths of the head of a larva in its successive instars follow a regular geometric progression in their increase. The head was selected as a part not subject to growth during a stadium; and the width as the most convenient measurement to take. By means of this criterion, it is possible to determine, when studying the transformations of an insect, whether an ecdysis has been overlooked or not. Experience has shown that slight variations between the computed and the actual widths may occur; but these differences are so slight that the overlooking of an ecdysis can be readily discovered. The following example will serve to illustrate the method employed.

A larva of *Papilio thoas* was reared from the egg; and the widths of the head in the successive instars was found to be, expressed in millimeters, as follows: .6; 1.1; 1.6; 2.2; 3.4.

By dividing 2.2 by 3.4 (two successive members of this series), the ratio of increase was found to be .676+; the number, .68 was taken, therefore, as sufficiently near the ratio for practical purposes. By using this ratio as a factor the following results were obtained:

Width found in fifth instar = ... 3.4
Calculated width in fourth instar (3.4 × .68) = ... 2.312
"   "   " third   "   (2.312 × .68) = ... 1.57
"   "   " second   "   (1.57 × .68) = ... 1.067
"   "   " first   (1.067 × .68) = ... .725

By comparing the two series, as is done below, so close a correspondence is found that it is evident that no ecdysis was overlooked.

Widths found:—.6; 1.1; 1.6; 2.2; 3.4
"   calculated:—.7; 1.1–; 1.6–; 2.3.

Fig. 93.—A spider in which lost legs were being reproduced.

The reproduction of lost limbs.—The reproduction of lost limbs has been observed in many insects; but such reproduction occurs here much less frequently than in the other classes of the Arthropoda. The reproduction takes place during the period of ecdysis, the reproduced part becoming larger and larger with each molt; hence with insects, and with Arachnida as well, the power of reproducing lost limbs ceases with the attainment of sexual maturity; but not so with the Crustacea and the "Myriapoda" which molt after becoming sexually mature. In none of the observed examples of the reproduction of appendages has an entire leg been reproduced.

It appears to be necessary that the original coxa be not removed in order that the reproduction may take place. Figure 193 represents a spider in our collection in which two legs, the left fore leg and the right hind leg, were being reproduced when the specimen was captured.

d. DEVELOPMENT WITHOUT METAMORPHOSIS
(Ametabolous* Development)

While most insects undergo remarkable changes in form during their postembryonic development, there are some in which this is not the case. In these the young insect just hatched from the egg is of practically the same form as the adult insect. These insects grow larger and may undergo slight changes in form of the body and its appendages; but these changes are not sufficiently marked to merit being termed a metamorphosis. This type of development is known technically as *ametăbolous development*.

Development without metamorphosis is characteristic of the two orders Thysanura and Collembola, which in other respects, also, are the most generalized of insects.

The nature of the changes in form undergone by an insect with an ametabolous development is illustrated by the development of *Machilis alternata*, one of the Thysanura. The first instar of this insect, according to Heymons ('07), lacks the clothing of scales, the styli on the thoracic legs, and the lateral rows of eversible sacs on the abdominal segments; and the antennæ and cerci are relatively shorter and consist of a much smaller number of segments than those of the adult. These changes, however, are comparable with those undergone by many animals in the course of their development that are not regarded as having a metamorphosis. In common usage in works on Entomology the term metamorphosis is used to indicate those marked changes that take place in the appearance of an insect that are correlated with the development of wings.

In addition to the Thysanura and the Collembola there are certain insects that develop without metamorphosis, as the Mallophaga and the Pediculidæ. But their ametabolous condition is believed to be an acquired one. In other words, it is believed that the bird-lice and the true lice are descendants of winged insects whose form of body and mode of development have been modified as a result of parasitic life.

The Ametabola.—Those insects that develop without metamorphosis are sometimes referred to as the *Ametabola*. This term was first proposed by Leach (1815), who included under it the lice as well as the Thysanura and Collembola. But with our present knowledge, if it is used it should be restricted to the Thysanura and Collembola, those insects in which a development without metamorphosis is a primitive not an acquired condition.

---

*Ametăbolous: Greek *a*, without; *metabole* (μεταβολή), change.

e. GRADUAL METAMORPHOSIS

(Paurometabolous* Development)

In several orders of insects there exists a type of development that is characterized by the fact that the young resemble the adult in the general form of the body and in manner of life. There is a gradual growth of the body and of the wing rudiments and genital appendages.

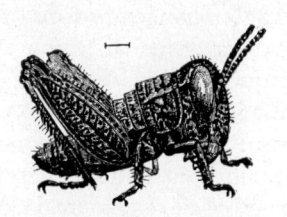

Fig. 194.—Nymph of *Melanoplus*, first instar (After Emerton).

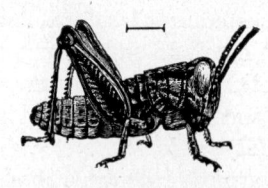

Fig. 195.—Nymph of *Melanoplus*, second instar (After Emerton).

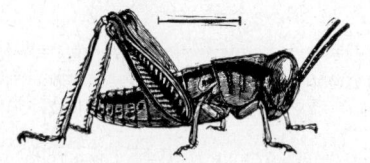

Fig. 196.—Nymph of *Melanoplus*, third instar (After Emerton)

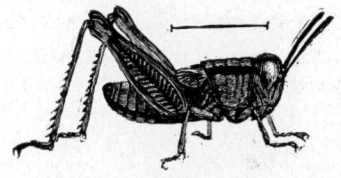

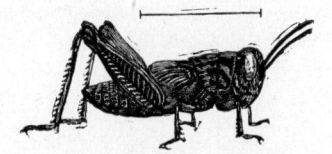

Fig. 197.—Nymph of *Melanoplus*, fourth instar (After Emerton).

Fig. 198.—Nymph of *Melanoplus*, fifth instar (After Emerton).

Fig. 199.—*Melanoplus*, adult.

But the changes in form take place gradually and are not very great between any two successive instars except that at the last ecdysis there takes place a greater change, especially in the wings, than at any of the preceding ecdyses. This type of metamorphosis is designated as *gradual metamorphosis* or *paurometabolous development*.

The characteristic features of paurometabolous development are correlated with the fact that the mode of life of the young and of the

---

*Paurometăbolous: *pauros* (παῦρος), little; *metabole* (μεταβολή), change.

adult are essentially the same; the two living in the same situation, and feeding on the same food. The adult has increased power of locomotion, due to the completion of the development of the wings; this enables it to more readily perform the functions of the adult, the spread of the species, and the making of provision for its continuance; but otherwise the life of the adult is very similar to that of the young.

The development of a locust or short-horned grasshopper will serve as an example of gradual metamorphosis. Each of the instars of our common red-legged locust, *Melanoplus femur-rubrum*, is represented in the accompanying series of figures. The adult (Fig. 199) is represented natural size; each of the other instars somewhat enlarged; the hair line above the figure in each case indicates the length of the insect.

The young locust just out from the egg-shell can be easily recognized as a locust (Fig. 194). It is of course much smaller than the adult; the proportion of the different regions of the body are somewhat different; and it is not furnished with wings; still the form of the body is essentially the same as that of the adult. In the second and third instars (Fig. 195 and 196) there are slight indications of the development of wing-rudiments; and these rudimentary wings are quite conspicuous in the fourth and fifth instars (Fig. 197 and 198). The change at the last ecdysis, that from the fifth instar to the adult, is more striking than that at any preceding ecdysis; this is due to the complete expansion of the wings, which takes place at this time.

The Paurometabola.—Those orders of insects that are characterized by a gradual metamorphosis are grouped together as the *Paurometabola*. This is not a natural division of the class Hexapoda but merely indicates a similarity in the nature of the metamorphosis in the orders included. This group includes the Isoptera, Dermaptera, Orthoptera, Corrodentia, Thysanoptera, Homoptera, and Hemiptera.

The term nymph.—An immature instar of an insect that undergoes a gradual metamorphosis is termed a *nymph*.

In old entomological works, and especially in those written in the early part of the last century, the term nymph was used as a synonym of pupa; but in more recent works it is applied to the immature instar of insects that undergo either a gradual or incomplete metamorphosis. In this book I restrict the use of this term to designate an immature instar of an insect that undergoes a gradual metamorphosis.

Deviation from the usual type.—It is to be expected that within so large a group of organisms as the Paurometabola there should have been evolved forms that exhibit deviations from the usual type of development. The more familiar examples of these are the following:

The Saltatorial Orthoptera.—In the crickets, locusts, and long-horned grasshoppers, the wings of the nymphs are developed in an inverted position; that surface of the wing which is on the outside in the adult is next to the body in the nymphal instars; and the rudimentary hind wings are outside of the fore wings, instead of beneath them, as in the adult. At the last ecdysis the wings assume the normal position.

The Cicadas.—In the Cicadas there exists a greater difference between the nymphal instars and the adult than is usual with insects in which the metamorphosis is gradual. The nymphs live below the surface of the ground, feeding upon the roots of plants; the adults live in the open air, chiefly among the branches of trees. The forelegs of the nymphs are fossorial (Fig. 200); this is an adaptation for subterranean life, which is not needed and not possessed by the adults. And it is said that the last nymphal instar is quiescent for a period.

The Coccidæ.—In the Coccidæ the mode of development of the two sexes differ greatly. The female never acquires wings, and in so far as external form is concerned the adult is degenerate. The male, on the other hand, exhibits a striking approach to complete metamorphosis, the last nymphal instar being enclosed in a cocoon, and the legs of the adult are not those of the nymph, being developed from imaginal disks. But the wings are developed externally.

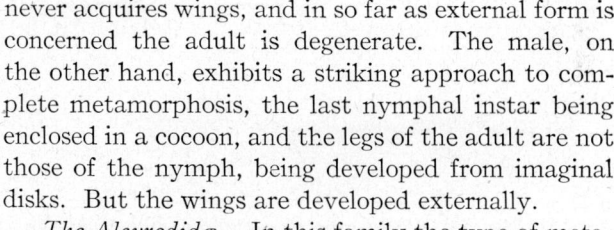

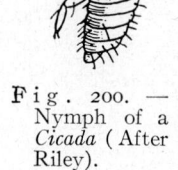

Fig. 200. — Nymph of a *Cicada* (After Riley).

The Aleyrodidæ.—In this family the type of metamorphosis corresponds quite closely with that described later as complete metamorphosis; consequently the term larva is applied to the immature instars except the last, which is designated the pupa.

The wings arise as histoblasts in the late embryo, and the growth of the wing-buds during the larval stadia takes place inside the body-wall. The change to the pupal instar, in which the wing-buds are external, takes place beneath the last larval skin, which is known as the pupa case or puparium. The adult emerges through a T-shaped opening on the dorsum of the puparium. Both sexes are winged.

The Aphididæ.—In the Aphididæ there exists a remarkable type of development known as *hĕtĕrŏgamy* or cyclic reproduction. This is characterized by an alternation of several parthenogenetic generations with a sexual generation. And within the series of parthenogenetic forms there may be an alternation of winged and wingless forms. In some cases the reproductive cycle is an exceedingly complicated one; and different parts of it occur on different food plants.

The Thysanoptera.—In the Thysanoptera, as in most other insects with a gradual metamorphosis, the nymphs resemble the adults in the form of the body, and the wings are developed externally; but the last nymphal instar is quiescent or nearly so and takes no nourishment. This instar is commonly described as the pupa.

f. INCOMPLETE METAMORPHOSIS
(*Hemimetabolous* Development*)

In three of the orders of insects, the Plecoptera, Ephemerida, and Odonata, there exists a type of metamorphosis in which the changes

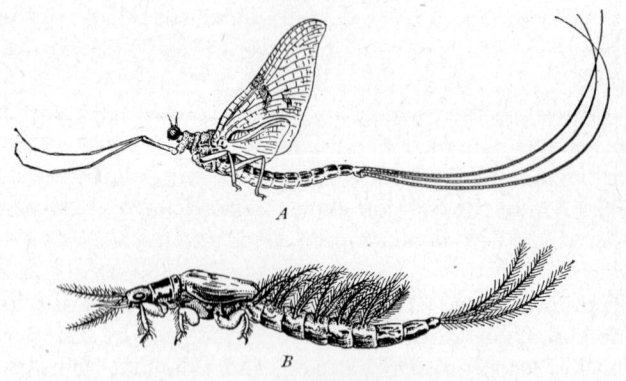

Fig. 201.—Transformation of a May-fly, *Ephemera varia*; A, adult; B, naiad (After Needham).

that take place in the form of the body are greater than in gradual metamorphosis but much less marked than in complete metamorphosis. For this reason the terms *incomplete metamorphosis* and *hemimetabolous development* have been applied to it.

Both incomplete metamorphosis and complete metamorphosis are characterized by the fact that the immature instars exhibit adaptive modifications of form and structure, fitting them for a very different mode of life than that followed by the adult. This is often expressed by the statement that the immature instars are "sidewise developed"; for it is believed that in these cases the development of the individual does not repeat the history of the race to which the individual belongs.

*Hemimetăbolous: *hemi* (ἡμί), half; *metabole* (μεταβολή), change.

This mode of development is termed *cenogenesis*.* It contrasts strongly with gradual metamorphosis, where there is a direct development from the egg to the adult.

In each of the orders that are characterized by an incomplete metamorphosis, the adaptive characteristics of the young insects fit them for aquatic life; while the adults lead an aerial existence. The transformations of a May-fly (Fig. 201) will serve to illustrate this type of metamorphosis.

The primitive insects were doubtless terrestrial; this is shown by the nature of the respiratory system, which is aerial in all insects. In the course of the evolution of the different orders of insects, the immature forms of some of them invaded the water in search of food. This resulted in a sidewise development of these immature forms to better fit them to live in this medium; while the adult continued their development in, what may be termed by contrast, a direct line. In some of the Plecoptera, as *Capnia* and others, the results of the cenogenetic development are not marked except that the immature forms are aquatic.

In the three orders in which the metamorphosis is incomplete, the cenogenetic development of the immature instars involved neither a change in the manner of development of the wings nor a retarding of the development of the compound eyes; consequently these immature forms, although sidewise developed, constitute a class quite distinct from larvæ.

The Hemimetabola.—The three orders in which the development is a hemimetabolous one are grouped together as the *Hemimetăbola;* these are the Plecoptera, Ephemerida, and Odonata. This grouping together of these three orders is merely for convenience in discussions of types of metamorphosis and does not indicate a natural division of the class Hexapoda. The radical differences in the three types of aquatic respiratory organs characteristic of the three orders indicate that they were evolved independently.

The term naiad.—The immature instars of insects with an incomplete metamorphosis have been termed nymphs; but as a result of their sidewise development they do not properly belong in the same class as the immature instars of insects with a gradual metamorphosis. I, therefore, proposed to designate them as *naiads* (Comstock '18, *b*).

The adoption of the term naiad in this sense affords a distinctive term for each of the three classes of immature insects corresponding to the three types of metamorphosis, *i. e.*, nymphs, naiads, and larvæ.

---

*Cēnogĕnesis: *kainos* (καινος), new; *genesis*.

Deviation from the usual type of incomplete metamorphosis.—The more striking deviations from the usual type of hemimetabolous development are the following:

The Odonata.—In the Odonata the wings of the naiads are inverted; these insects resembling in this respect the Saltitorial Orthoptera. What is the upper surface of the wings with naiads becomes the lower surface in the adults, the change taking place at the last ecdysis.

The Ephemerida.—In the Ephemerida, there exists the remarkable phenomenon of an ecdysis taking place after the insect has left the water and acquired functional wings. The winged instar that is interpolated between the last aquatic one and the adult is termed the *sub-imago.*

g. COMPLETE METAMORPHOSIS

(Holometabolus* Development)

The representatives of several orders of insects leave the egg-shell in an entirely different form from that they assume when they reach maturity; familiar examples of these are caterpillars which develop into butterflies, maggots which develop into flies, and grubs which develop into beetles. These insects and others that when they emerge from the egg-shell bear almost no resemblance in form to the adult are said to undergo a *complete metamorphosis* or a *holometăbolous development.*

The Holometabola.—Those orders that are characterized by a holometabolous development are grouped together as the *Holometăbola.* This group includes the Neuroptera, Mecoptera, Trichoptera, Lepidoptera, Diptera, Siphonaptera, Coleoptera, and Hymenoptera.

This grouping together of these orders, while convenient for discussions of metamorphosis, is doubtless artificial. It is not at all probable that the Holometabola is a monophylitic group. In other words complete metamorphosis doubtless arose several times independently in the evolution of insects.

The term larva.—The form in which a holometabolous insect leaves the egg is called *larva.* The term was suggested by a belief of the ancients that the form of the perfect insect was masked, the Latin word *larva* meaning a mask.

Formerly the term larva was applied to the immature stages of all insects; but more recent writers restrict its use to the immature in-

---
*Holometăbolous: *holos* (ὅλος), complete; *metabole* (μεταβολή), change.

stars of insects with a complete metamorphosis; and in this sense only is it used in this book.

The adaptive characteristics of larvæ.—The larvæ of insects with complete metamorphosis, like the naiads of those with incomplete metamorphosis, exhibit an acquired form of body adapting them to special modes of life; and in this case the cenogenetic or "sidewise development" is much more marked than it is in insects with an incomplete metamorphosis. Here the form of the body bears but little relation to the form to be assumed by the adult, the nature of the larval life being the controlling factor.

The differences in form between larvæ and adults are augmented by the fact that not only have larvæ been modified for special modes of life, but in most cases the adults have been highly specialized for a different mode of life; and so great are these differences that a quiescent pupa stage, during which certain parts of the body can be made over, is necessary.

Here, as in the case of insects with an incomplete metamorphosis, we have an illustration of the fact that natural selection can act on any stage in the development of animal to better adapt that particular stage to the conditions under which it exists. Darwin pointed out in his "Origin of Species" that at whatever age a variation first appears in the parent it tends to reappear at a corresponding age in the offspring. This tendency is termed *homochronous heredity**.

It is obvious that the greater the adaptive characteristics of the immature forms, the less does the ontogeny of a species represent the phylogeny of the race to which it belongs. This fact led Fritz Muller, in his "Facts for Darwin", to make the aphorism "There were perfect insects before larvæ and pupæ." The overlooking of this principle frequently results in the drawing of unwarranted conclusions, by those writers on insects who cite adaptive larval characteristics as being more generalized than the corresponding features of the adult.

The more obvious of the adaptive characteristics of larvæ are the following:

The form of the body.—As indicated above the form of the body of a larva bears but little relation to the form to be assumed by the adult, the nature of the larval life being the controlling factor in determining the form of the body. As different larvæ live under widely differing situations, various types of larvæ have been developed; the more important of these types are described later.

The greater or less reduction of the thoracic legs.—In the evolution of most larvæ there has taken place a greater or less reduction of the thoracic legs; but the extent of this reduction varies greatly. The larvæ of certain Neuroptera, as *Corydalus* for example, have as perfect

---

*Homŏchronous: *homos* (ὁμο's), one and the same; *chronos* (χρόνος), time.

legs as do naiads of insects with an incomplete metamorphosis. The larvæ of Lepidoptera have short legs which correspond to only a part of the legs of the adult. While the larvæ of Diptera have no external indications of legs.

The development of prolegs in some larvæ.—A striking feature of many larvæ is the presence of abdominal organs of locomotion; these have been termed *prolegs;* the prolegs of caterpillars are the most familiar examples of these organs.

The prolegs were so named because they were believed to be merely adaptive cuticular formations and not true legs; this belief arose from the fact that they are shed with the last larval skin. Some recent writers, however, regard the prolegs as true legs. It is now known that abdominal appendages are common in the embryos of insects; and these writers believe that the prolegs are developed from these embryonic appendages, and that, therefore, they must be regarded as true legs.

If this is true, there has taken place a remarkable reversal in the course of development. The abdominal legs, except those that were modified into appendages of the reproductive organs, the gonapophyses, were lost early in the phylogeny of the Hexapoda. The origin of complete metamorphosis must have taken place at a much later period; when, according to this belief, the abdominal appendages, which had been latent for a long time, were redeveloped into functional organs.

The development of tracheal gills.—A striking feature of many larvæ is the possession of tracheal gills. This is obviously an adaptive characteristic the development of which was correlated with the assumption of aquatic life by forms that were primarily aerial; and it is also obvious that the development of tracheal gills has arisen independently many times; for they exist in widely separated families belonging to different orders of insects that are chiefly aerial. They are possessed by a few lepidopterous larvæ, and by the representatives of several families of Neuroptera, Coleoptera and Diptera. On the other hand, in the Trichoptera the possession of tracheal gills by the larvæ is characteristic of nearly all members of the order.

The internal development of wings.—This is perhaps the most remarkable of the sidewise developments of larvæ. Although larvæ exhibit no external indications of wings, it has been found that the rudiments of these organs arise at as early a period in insects with a complete metamorphosis as they do in those with an incomplete metamorphosis; and that during larval life the wing rudiments attain an advanced stage in their development. But as these rudiments are invaginated there are no external indications of their presence during larval life. The details of the internal development of wings are discussed later.

Occasionally atavistic individual larvæ are found which have external wing-buds.

As to the causes that brought about the internal development of wings we can only make conjectures. It has occurred to the writer that this type of wing-development may have arisen as a result of boring habits, or habits of an analogous nature, of the stem forms from which the orders of the Holometabola sprang. Projecting wing-buds would interfere with the progress of a boring insect; and, therefore, an embedding of them in the body, thus leaving a smooth contour, would be advantageous.

In support of this theory attention may be called to the fact that the larvæ of the most generalized Lepidoptera, the Hepialidæ, are borers; the larvæ of the Siricidæ, which are among the more generalized of the Hymenoptera are borers; so too are many Coleoptera; most larvæ of Diptera are burrowers; and the larvæ of Trichoptera live in cases.

The retarding of the development of the compound eyes.—One of the most distinctively characteristic features of larvæ is the absence of compound eyes. The life of most larvæ is such that only limited vision is necessary for them; and correlated with this fact is a retarding of the development of the greater portion of the compound eyes; only a few separate ommatidia being functional during larval life.

In striking contrast with this condition are the well-developed eyes of nymphs and naiads.

The larvæ of *Corethra* and *Panorpa* are the only larvæ known to me that possess compound eyes.

The invaginated conditions of the head in the larvæ of the more specialized Diptera.—The extreme of sidewise development is exhibited by the larvæ of the more specialized Diptera. Here not only are the legs and wings developed internally but also the head. This phenomenon is discussed later.

The different types of larvæ.—As a rule, the larvæ of the insects of any order resemble each other in their more general characteristics, although they bear but little resemblance to the adult forms. Thus the grubs of Coleoptera, the caterpillars of Lepidoptera, or the maggots of Diptera, in most cases, can be recognized as such. Still in each of these orders there are larvæ that bear almost no resemblance to the usual type. As examples of these may be cited the water-pennies (Parnidæ, Coleoptera), the slug-caterpillars (Cochlidiidæ, Lepidoptera), and the larvæ of *Microdon* (Diptera).

To understand the variations in form of larvæ it should be borne in mind that the form of the body in all larvæ is the result of secondary adaptations to peculiar modes of life; and that this modification of form has proceeded in different directions and in varying degrees in different insects.

Among the many types of larvæ, there are a few that are of such common occurrence as to merit distinctive names; the more important of these are the following:

Campodeiform.—In many paurometabolous insects and in some holometabolous ones, the early instars resemble *Campodea* more or less in the form of the body (Fig. 202); such naiads and larvæ are described as *campodeiform*.

In this type, the body is long, more or less flattened, and with or without caudal setæ; the mandibles are well developed; and the legs are not greatly reduced. Among the examples of this type are the larvæ of most Neuroptera, and the active larvæ of many Coleoptera (Carabidæ, Dysticidæ, and the first instar of Meloidæ).

Eruciform.—The *eruciform* type of larvæ is well-illustrated by most larvæ of Lepidoptera and of Mecoptera; it is the caterpillar form (Fig. 203). In this type the body is cylindrical; the thoracic legs are short, having only the terminal portions of them developed; and the abdomen is furnished with prolegs or with proleg-like cuticular folds. Although these larvæ move freely, their powers of locomotion are much less than in the campodeiform type.

Fig. 202.—*Campodea staphylinus* (After Lubbock).

Scarabeiform.—The common white grub, the larva of the May-beetle (Fig. 204) is the most familiar example of a *scarabeiform* larva.

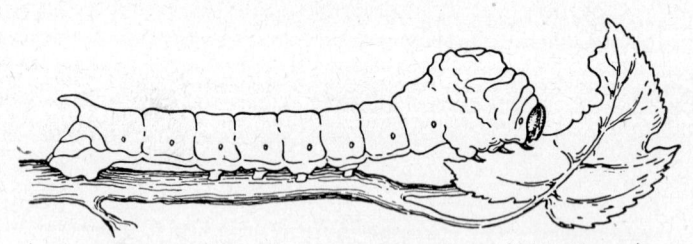

Fig. 203.—The silk-worm, an eruciform larva (After Verson).

In this type the body is nearly cylindrical, but usually, especially when at rest, its longitudinal axis is curved; the legs are short; and prolegs are wanting. This type is quite characteristic of the larvæ of the Scarabæidæ, hence the name; but it occurs in other groups of insects.

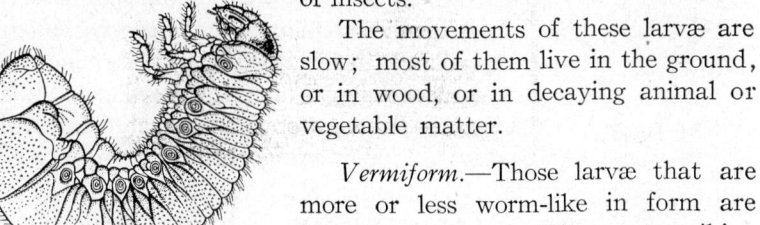

The movements of these larvæ are slow; most of them live in the ground, or in wood, or in decaying animal or vegetable matter.

Vermiform.—Those larvæ that are more or less worm-like in form are termed *vermiform*. The most striking features of this type are the elongated form of the body and an absence of locomotive appendages (Fig. 205).

Fig. 204.—Larva of *Melolontha vulgaris* (After Schiodte).

Naupliiform.—The term *naupliiform* is applied to the first instar of the larva of *Platygaster* (Fig. 206), on account of its resemblance to the nauplius of certain Crustacea.

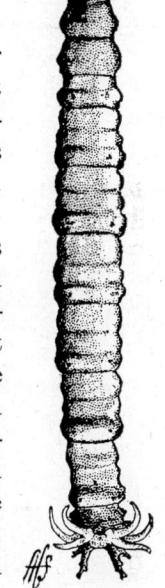

The prepupa.—Usually the existence of an instar between the last larval one and the pupal instar is not recognized. But such a form exists; and the recognition of it becomes important when a careful study is made of the development of holometabolous insects. As is shown later, during larval life the development of the wings is going on within the body. As the larva approaches maturity, the wings reach an advanced stage of development within sac-like invaginations of the body-wall. Near the close of the last larval stadium the insect makes preparation for the change to the pupa state. Some form a cell within which the pupa state is passed, the larvæ of butterflies suspend themselves, and most larvæ of moths spin a cocoon. Then follows a period of apparent rest before the last larval skin is shed and the pupal state assumed. But this period is far from being a quiet one; within the apparently motionless body important changes take place. The most easily observed of these changes is a change in the position of the wings. Each of these passes out through the mouth of the sac in which it has been developed, and lies outside of the newly developed pupal cuticula, but beneath the last larval cuticula. Then follows a period of variable duration in different insects, in which the wings are really

Fig. 205.—Larva of a crane-fly.

outside of the body although still covered by the last larval cuticula, this period is the *prepupal* stadium. The prepupal instar differs markedly from both the last larval one and from the pupa; for after the shedding of the last larval cuticula important changes in the form of the body take place before the pupal instar is assumed.

The pupa.—The most obvious characteristics of the pupa state are, except in a few cases, inactivity and helplessness. The organs of locomotion are functionless, and may even be soldered to the body throughout their entire length, as is usual with the pupæ of Lepidoptera (Fig. 207). In other cases, as in the Coleoptera (Fig. 208) and in the Hymenoptera, the wings and legs are free, but enclosed in more or less sac-like cuticular sheaths, which put them in the condition of the proverbial cat in gloves. More than this, in most cases, the legs of the adult are not fully formed till near the end of the pupal stadium.

Fig. 206.—Larva of Platygaster (After Ganin.)

The term *pupa*, meaning girl, was applied to this instar by Linnæus on account of its resemblance to a baby that has been swathed or bound up, as is the custom with many peoples.

Although the insect during the pupal stadium is apparently at rest, this, from a physiological point of view, is the most active period of its postembryonic existence; for wonderful changes in the structure of the body take place at this time.

Fig. 207.—Pupa of a moth.

In the development of a larva the primitive form of the body has been greatly modified to adapt it to its peculiar mode of life; this sidewise development results in the production of a type of body that is not at all fitted for the duties of adult life. In the case of an insect with incomplete metamorphosis, the full grown naiad needs to be modified comparatively little to fit it for adult life; but the change from a maggot to a fly, or from a caterpiller to a butterfly, involves not merely a change in external form but a greater or less remodeling of its entire structure. These changes take place during the period of apparent rest, the prepupal and pupal stadia.

Fig. 208 Pupa of a beetle.

The chrysalis.—The term *chrysalis* is often applied to the pupæ of butterflies. It was suggested by the golden spots with which the pupæ of certain butterflies are ornamented.

Two forms of this word are in use: first, chrysalis, the plural of which is chrysalides; and second, chrysalid, the plural of which is chrysalids. The singular of the first form and the plural of the second are those most often used.

Active pupæ.—The pupæ of mosquitoes and of certain midges are remarkable for being active. Although the wings and legs are functionless, as with other pupæ, these creatures are able to swim by means of movements of the caudal end of the body.

In several genera of the Neuroptera (*Chrysopa, Hemerobius,* and *Raphidia*) the pupa becomes active and crawls about just before transforming to the adult state.

Movements of a less striking character are made by many pupæ, which work their way out of the ground, or from burrows in wood, before transforming. In some cases, as in the pupæ of the carpenter-moths (Cossidæ) the pupa is armed with rows of backward projecting teeth on the abdominal segments, which facilitate the movements within the burrow.

The cremaster.—Many pupæ, and especially those of most Lepidoptera, are provided with a variously shaped process of the posterior end of the body, to which the term *cremaster* is applied. This process is often provided with hooks which serve to suspend the pupa, as in butterflies, or to hold it in place, after it has partly emerged from the cocoon, and while the adult is emerging from the pupal skin, as in cocoon-making moths. In its more simple form, where hooks are lacking, it aids the pupa in working its way out of the earth, or from other closed situations.

The method of fixing the cremaster in the disk of silk from which the pupa of a butterfly is suspended was well-illustrated by C. V. Riley ('79). The full grown larva spins this disk and hangs from it during the prepupal stadium by means of its anal prolegs (Fig. 209, *a*). When the last larval skin is shed, it is worked back to the caudal end of the body (Fig. 209, *b*); and is then grasped between two of the abdominal segments (Fig. 209, *c*,) while the caudal end of the body is removed from it; and thus the cremaster is freed, and is in a position from which it can be inserted in the disk of silk.

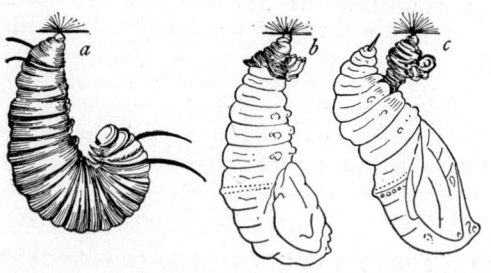

Fig. 209.—Transformations of the milkweed butterfly (From Riley).

The cocoon.—The pupal instar is an especially vulnerable one. During the pupal life the insect has no means of offence, and having exceedingly limited powers of motion, it has almost no means of defense unless an armor has been provided.

Many larvæ merely retreat to some secluded place in which the pupal stadium is passed; others bury themselves in the ground; and still others make provision for this helpless period by spinning a silken armor about their bodies. Such an armor is termed a *cocoon*.

The cocoon is made by the full-grown larva; and this usually takes place only a short time before the beginning of the pupal stadium. But in some cases several months elapse between the spinning of the cocoon and the change to pupa, the cocoon being made in the autumn and the change to pupa taking place in the spring. Of course a greater or less portion of this period is occupied by the prepupal stadium.

Cocoons are usually made of silk, which is spun from glands already described. In some cases, as in the cocoons of *Bombyx*, the silk can be unwound and utilized by man.

While silk is the chief material used in the making of cocoons, it is by no means the only material. Many wood-boring larvæ make cocoons largely of chips. Many insects that undergo their transformation in the ground incorporate earth in the walls of their cocoons. And hairy caterpillars use silk merely as a warp to hold together a woof of hair, the hairs of the larva being the most conspicuous element in the cocoon.

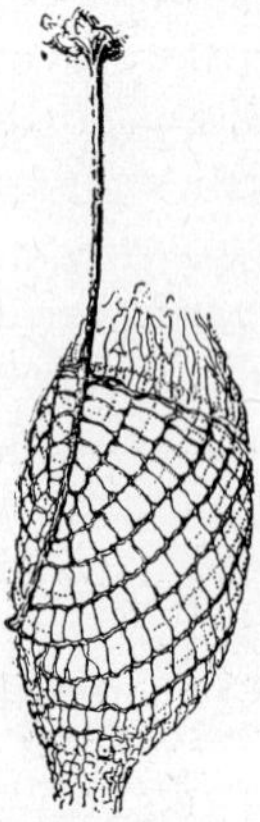

In those cases in which silk alone is used there is a great variation in the nature of the silk, and in the density of the cocoon. The well-known cocoons of the saturniids illustrate one extreme in density, the cocoons of certain Hymenoptera, the other.

The fiberous nature of the cocoon is usually obvious; but the cocoons of saw-flies appear parchment-like, and the cocoons of the sphecids appear like a delicate foil. While in the more common type of cocoons the wall is a closely woven sheet, there are cocoons that are lace-like in texture (Fig. 210).

Fig. 210.— Lace-like cocoon of *Trichostibas parvula*, from which the adult has emerged.

Modes of escape from the cocoon.—The insect, having walled itself in with a firm layer of silk, is forced to meet the problem of a means of escape from this inclosure; a problem which is solved in greatly varied ways.

In many insects in which the adult has biting mouth parts, the adult merely gnaws its way out by means of its mandibles In some cases, as the Cynipidæ, it is said that this is the only use made of its mandibles by the adult.

In some cases the mandibles with which the cocoon is pierced pertain to the pupal instar, this is true of *Chrysopa* and *Hemerobius;* and the Trichoptera break out from their cases, by means of their mandibles, while yet in the pupal state.

For those insects in which the adult has sucking mouth parts, the problem is even more difficult. Here it has been met in several quite distinct ways. The pupæ of many Lepidoptera possess a specialized organ for breaking through the cocoon; in some the anterior end of the pupa is furnished with a toothed crest (*Lithocolletes hamadryella*); in certain saturniids there is a pair of large, stout, black spines, one on each side of the thorax, at the base of the fore wings with which the adult cuts a slit in the cocoon through which the **moth** emerges, this was observed by Packard in *Tropæa luna;* but as these spines are present in other saturniids, where the cocoon is too dense to be cut by them, and where an opening is made in some other way, it is probable that, as a rule, their function is locomotive, aiding the **moth** to work its way out from the cocoon, by a wriggling motion.

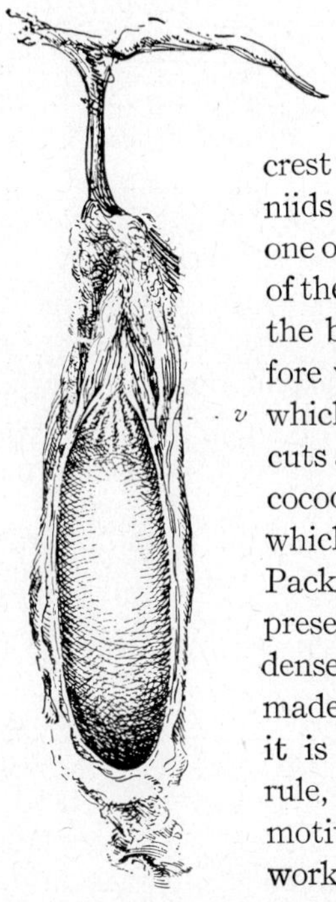

One of the ways in which saturniids pierce their cocoons is that practiced by *Bombyx* and *Telea*. These insects soften one end of the cocoon by a liquid, which issues from the mouth; and then, by forcing the threads apart or by breaking them, make an opening.

Fig. 211.—Longitudinal section of a cocoon of *Callosamia promethea*; *v*, valve-like arrangement for the escape of the adult.

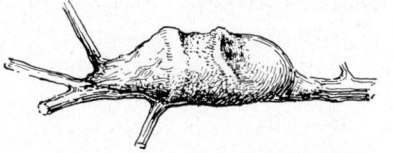

Fig. 212.—Cocoon of *Megalopyge opercularis*.

Fig. 213.—Old cocoon of *Megalopyge opercularis*.

Far more wonderful than any of the methods of emergence from the cocoon described above are those in which the larva makes provision for the escape of the adult. The most familiar of these is that practiced by the larvæ of *Samia cecropia* and *Callosamia promethea*. These larvæ when they spin their cocoons construct at one end a conical valve-like arrangement, which allows the adult to emerge without the necessity of making a hole through the cocoon (Fig. 211, *v*). A less familiar example, but one that is fully as wonderful, is that of a *Megalopyge*. The larva of this species makes a cocoon of the form shown in Figure 212. After an outer layer of the cocoon has been made, the larva constructs, near one end of it, a hinged partition; this serves as a trap door, through which the moth emerges. That part of the cocoon that is outside of the partition is quite delicate and is easily destroyed. Hence most specimens of the cocoons in collections present the appearance represented in Figure 213.

The puparium.—The pupal stadium of most Diptera is passed within the last larval skin, which is not broken till the adult fly is ready to emerge. In this case the larval skin, which becomes hard and brown, and which serves as a cocoon, is termed a *puparium*. In some families the puparium retains the form of the larva; in others the body of the larva shortens, assuming a more or less barrel-shaped form, before the change to a pupa takes place (Fig. 214).

Modes of escape from the puparium.—The pupæ of the more generalized Diptera escape from the puparium through a T-shaped opening, which is formed by a lengthwise split on the back near the head end and a crosswise split at the front end of this (Fig. 215), or rarely, through a cross-wise split between the seventh and eighth abdominal segments. In the more specialized Diptera there is developed a large bladder-like organ, which is pushed out from the front of the head, through what is known as the frontal suture, and by which the head end of the puparium is forced off. This organ is known as the *ptilinum*. After the adult escapes, the ptilinum is withdrawn into the head.

Fig. 214.—Pupa-rium of *Trypeta*.

The Different types of pupæ.—Three types of pupæ are commonly recognized; these are the following:

Fig. 215.—Puparium of a stratiomyiid.

Exarate pupæ.—Pupæ which, like those of the Coleoptera and Hymenoptera, have the legs and wings free, are termed exarate pupæ.

Obtected pupæ.—Pupæ which like the pupæ of Lepidoptera, have the limbs glued to the surface of the body, are termed obtected pupæ.

Coarctate Pupæ.—Pupæ that are enclosed within the hardened larval skin, as is the case with the pupæ of most of the Diptera, are termed coarctate pupæ.

The imago—The fully developed or adult insect is termed the *imago*.

The life of the imago is devoted to making provision for the perpetuation of the species. It is during the imaginal stadium that the sexes pair, and the females lay their eggs. With many species this is done very soon after the last ecdysis; but with others the egg-laying is continued over a long period; this is especially true with females of the social Hymenoptera.

h. HYPERMETAMORPHOSIS

There are certain insects, representatives of several different orders that exhibit the remarkable peculiarity in their development that the successive larval instars represent different types of larvæ. Such insects are said to undergo a hypermetamorphosis.

The transformations of several of these insects will be described later in the accounts of the families to which they belong; and for this reason, in order to avoid repetition, are not discussed here. The more striking examples are *Mantispa, Meloe, Stylops,* and *Platygaster.*

i. VIVIPAROUS INSECTS

There are many insects that produce either nymphs or larvæ instead of laying eggs. Such insects are termed *viviparous*. This term is opposed to *oviparous*, which is applied to those insects that lay eggs that hatch after exclusion from the body.

It has been pointed out in the discussion of the reproductive organs that, from the primordial germ-cells, there are developed in one sex spermatoza and in the other eggs; and it should be borne in mind that the germ-cells produced in the ovary of a female from the primordial germ-cells are eggs. These eggs grow and mature; in some cases they become covered with a shell, in others they are not so covered; in some cases they are fertilized by the union of a spermatozoan with them, and in others they are never fertilized; but in all these cases they are eggs. We may say, therefore, that all insects are developed from eggs.

A failure to recognize this fact has introduced confusion into entomological literature. Some writers have termed the germ-cells produced by agamic aphids *pseudova* or false eggs. But these germ-cells are as truly eggs as are those from which the males of the honeybee develop; they are merely unfertilized eggs. The term pseudovum conveys a false impression; while the phrase, an unfertilized egg, clearly states a fact.

Some writers make use of the term ovoviviparous indicating the production of eggs that have a well-developed shell or covering, but which hatch within the body of the parent; but the distinction is not fundamental, since viviparous animals also produce eggs as indicated above.

Among viviparous insects there are found every gradation from those in which the larvæ are born when very young to those in which the entire larval life is passed within the body of the parent. There also exist examples of viviparous larvæ, viviparous pupæ, and viviparous adults. And still another distinction can be made; in some viviparous insects the reproduction is parthenogenetic; in others it is sexual.

Viviparity with parthenogenetic reproduction.—In certain viviparous insects the reproduction is parthenogenetic; that is, the young are produced from eggs that are not fertilized. This type of reproduction occurs in larvæ, pupæ, and apparently in adults.

Pædogenetic Larvæ.—In 1862 Nicholas Wagner made the remarkable discovery that certain larvæ belonging to the Cecidomyiidæ give birth to living young. This discovery has been confirmed by other observers, and for this type of reproduction the term *pædogenesis*, proposed by Von Baer, has come into general use. This term is also spelled *pedogenesis;* the word is from *pædo* or *pedo*, a child, and *genesis*.

The phenomenon of pædogenesis is discussed later in the accounts of the Cecidomyiidæ and of the Micromalthidæ.

Pædogenetic pupæ.—The most frequently observed examples of pædogenetic reproduction are by larvæ; but that pupæ also are sometimes capable of reproduction is shown by the fact that Grimm ('70) found that eggs laid by a pupa of *Chironomus grimmii*, and of course not fertilized, hatched.

Anton Schneider ('85) found that the adults of this same species of *Chironomus* reproduced parthenogenetically. This species, therefore, exhibits a transition from pædogenesis to normal parthenogenesis.

Viviparous adult agamic females.—There may be classed under this class provisionally, the agamic females of the Aphididæ; as these are commonly regarded as adults. It has been suggested, however, that the agamic reproduction of the Aphids may be a kind of pædogenesis; the agamic females being looked upon as nymphs. This however, is not so evident in the case of the winged agamic generation. On the other hand, the reproductive organs of the agamic aphids are incompletely developed, as compared with those of the sexual forms, lacking a spermatheca and colleterial glands.

This discussion illustrates the difficulty of attempting to make sharp distinctions, whereas in nature all gradations exist between different types of structure and of development. Thus Leydig ('67) found a certain aphid to be both oviparous and viviparous; the eggs and the individuals born as nymphs being produced from neighboring tubes of the same ovary.

Viviparity with sexual reproduction.—Although most insects that reproduce sexually are oviparous, there are a considerable number in which sexual reproduction is associated with viviparity.

Among these sexual viviparous insects there exist great differences in method of reproduction; with some the young are born in a very immature stage of development, a stage corresponding to that in which the young of oviparous insects emerge from the egg; while with others the young attain an advanced stage of development within the body of the mother.

Sexual viviparous insects giving birth to nymphs or larvæ.—That type of viviparity in which sexual females give birth to very immature nymphs or larvæ exists in more or less isolated members of widely separated groups of insects. As the assumption of this type of reproduction involves no change in the structure of the parent, but merely a precocious hatching of the egg, it is not strange that it has arisen sporadically and many times. In some cases, however, the change is not so slight as the foregoing statement would imply; as, for example, in the case of the viviparous cockroach, which does not secrete oothecæ as do other cockroaches.

Among the recorded examples of this type of viviparity are representatives of the Ephemerida, Orthoptera, Hemiptera, Lepidoptera, Coleoptera, Strepsiptera, and Diptera.

Sexual viviparous insects giving birth to old larvæ.—The mode of reproduction exhibited by these insects is doubtless the most exceptional that occurs in the Hexapoda, involving, as it does, very important changes in the structure of the reproductive organs of the females.

With these insects the larvæ reach maturity within the body of the parent, undergoing what is analogous to an intra-uterine development, and are born as full-grown larvæ. This involves the secretion of a "milk" for the nourishment of the young.

This mode of reproduction is characteristic of a group of flies, including several families, and known as the Pupipara. This name was suggested for this group by the old belief that the young are born as pupæ; but it has been found that the change to pupa does not take place till after the birth of the larva.

The reproduction of the sheep-tick, *Melophagus ovinus*, may be taken as an illustration of this type of development; this is described in the discussion of the Hippoboscidæ, the family to which this insect belongs.

The giving birth to old larvæ is not restricted to the Pupipara. Surgeon Bruce (quoted by Sharp, '99) has shown that the Tsetse fly, *Glossina morsitans*, reproduces in this way, the young changing to pupæ immediately after birth.

An intermediate type of development is illustrated by *Hylemyia strigosa*, a dung-frequenting fly belonging to the Anthomyiidæ. This insect, according to Sharp ('99), produces living larvæ, one at a time. "These larvæ are so large that it would be supposed they are full-grown, but this is not the case, they are really only in the first stage, an unusual amount of growth being accomplished in this stadium."

j. NEOTEINIA

The persistence with adult animals of larval characteristics has been termed *neoteinia** or *neotenia*. When this term first came into use it was applied to certain amphibians, as the axolotle, which retains its gills after becoming sexually mature; but it is now used also in entomology.

The most familiar examples of neoteinic insects are the glow-worms, which are the adult females of certain beetles, the complemental females of Termites, and the females of the Strepsiptera.

II. THE DEVELOPMENT OF APPENDAGES

In the preceding pages the more obvious of the changes in the external form of the body during the metamorphosis of insects and some deviations from the more common types of development have been discussed. The changes in the form of the trunk that have been described are those that can be seen without dissection; but it is impracticable to limit a discussion of the development of the appendages of the body in this way, for in the more specialized types of metamorphosis a considerable part of the development of the appendages takes place within the body-wall.

---

*Nōoteinia: neos (νέος), youthful; teinein (τείνειν), to stretch.

a. THE DEVELOPMENT OF WINGS

Two quite distinct methods of development of wings exist in insects; by one method, the wings are developed as outward projecting appendages of the body; by the other, they reach an advanced stage of development within the body. The former method of development takes place with nymphs and naiads, the latter with larvæ.*

1. The Development of the Wings of Nymphs and Naiads

In insects with a gradual or with an incomplete metamorphosis the development of the appendages proceeds in a direct manner. The wings of nymphs and naiads are sac-like outgrowths of the body-wall, which appear comparatively early in life and become larger and larger with successive molts, the expanding of the wing-buds taking place immediately after the molt; an illustration of this has been given in the discussion of gradual metamorphosis, page 175.

2. Development of the Wings in Insects with a Complete Metamorphosis

Although there are differences in details in the development of the wings in the different insects undergoing a complete metamorphosis, the essential features are the same in all. The most striking feature is that the rudiments of the wings, the wing-buds, arise within the body and become exposed for the first time when the last larval skin is shed. The development of the wings of the cabbage butterfly (*Pontia rapæ*) will serve as an example of this type of development of wings. The tracing of that part of this development which takes place during the larval life can be observed by making sections of the body-wall of the wing-bearing segments of the successive instars of this insect.

The first indication of a wing-bud is a thickening of the hypodermis; this thickening, known as a *histoblast* or an *imaginal disc*, has been observed in the embryos of certain insects, in the first larval instar of the cabbage butterfly it is quite prominent (Fig. 216, *a*). During the second stadium, it becomes more prominent and is invaginated, forming a pocket-like structure (Fig. 216, *b*). During the third stadium a part of this invagination becomes thickened and evaginated into the pocket formed by the thinner

---

*Only the more general features of the development of wings are discussed here. For a fuller account see "The Wings of Insects" (Comstock '18, *a*).

portions of tne invagination (Fig. 216, c). During the fourth stadium, the evaginated part of the histoblast becomes greatly extended (Fig. 216, d). It is this evaginated portion of the histoblast that later becomes the wing. During the fifth stadium the wing-bud attains the form shown in Figure 216, e, which represents it dissected out of the wing-pocket At the close of the last larval stadium, the fifth, the wing is pushed out from the wing-pocket, and lies under the old larval cuticula during the prepupal stadium. It is then of the form shown in Figure 216, f. The molt that marks the beginning of the pupal stadium, exposes the wing-buds, which in the Lepidoptera become closely soldered to the sides and breast of the pupa. Immediately after the last molt when the adult emerges, the wings expand greatly and assume their definitive form.

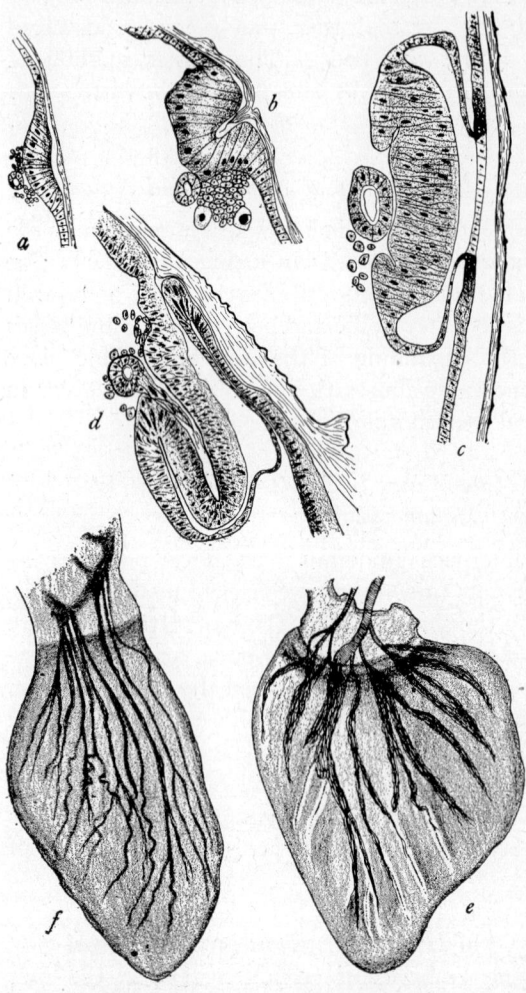

Fig. 216.—Several stages in the development of the wings of a cabbage butterfly (After Mercer).

While this increase in size and changes in form of the developing wing are taking place, there occur other remarkable developments in its structure. A connection is made with a large trachea near which the histoblast is developed, shown in cross-section in the first four parts (*a*, *b*, *c*, and *d*) of Figure 216; temporary respiratory organs, consisting of bundles of tracheoles, are developed (*e* and *f*); and later, near the close of the larval period, the tracheæ of the wing are developed, and the bundles of tracheoles disappear. During the later stages in the development of the wing the basement membranes of the hypodermis of the upper and lower sides of the wing come together, except along the lines where the veins are to be developed later, and become united. In this way the wing is transformed from a bag-like organ to a sheet-like one. The lines along which the two sides of the wing remain separate are the vein cavities; in these the trunks of the wing-tracheæ extend. During the final stages of the development of the wing, the walls of the vein-cavities are thickened, thus the wing-veins are formed; and the spaces between the wing-veins become thin.

By reference to Figure 216, *c* and *d*, it will be seen that the histoblast consists of two quite distinct parts, a greatly thickened portion which is the wing-bud and a thinner portion which connects the wing-bud with the hypodermis of the body-wall, and which constitutes the neck of the sac-like histoblast, this is termed the *peripodal membrane*, a term suggested by the similar part of the histoblast of a leg; and the enclosed cavity is known as the *peripodal cavity*.

In the more specialized Diptera, the peripodal membranes are very long and both the wing-buds and the leg-buds are far removed from the body-wall. A condition intermediate between that which exists in the Lepidoptera, as shown in Figure 216, and that of the more specialized Diptera was found by Kellogg ('07) in the larva of *Holorusia rubiginosa*, one of the crane-flies (Fig. 217).

b. THE DEVELOPMENT OF LEGS

The development of the legs proceeds in widely different ways in different insects. In the more generalized forms, the legs of the embryo reach an advanced stage of development before the nymph or naiad leaves the egg-shell, and are functional when the insect is born; on the other hand, in those specialized insects that have vermiform larvæ, the development of the legs is retarded, and these organs do not become functional until the adult stage is reached. Almost every conceivable intergrade between these two extremes exist.

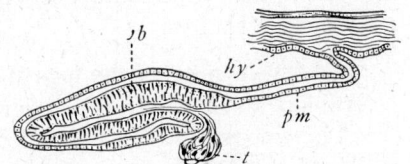

Fig. 217.—Wing-bud in the larva of the giant crane-fly, *Holorusia rubiginosa; hy*, hypodermis; *pm*, peripodal membrane; *t*, trachea; *wb*, wing-bud (After Kellogg).

1. *The Development of the Legs of Nymphs and of Naiads*

In insects with a gradual metamorphosis and also in those with an incomplete metamorphosis the nymph or naiad when it emerges from the eggshell has well-developed legs, which resemble quite closely those of the adult. The changes that take place in the form of the legs during the postembyronic development are comparatively slight; there may be changes in the relative sizes of the different parts; and in some cases there is an increase in the number of the segments of the tarsus; but the changes are not sufficiently great to require a description of them here.

2. *The Development of the Legs in Insects with a Complete Metamorphosis*

It is a characteristic of most larvæ that the development of their legs is retarded to a greater or less extent. This retardation is least in campodeiform larvæ, more marked in eruciform larvæ, and reaches its extreme in vermiform larvæ.

The development of the legs of insects with campodeiform larvæ.—Among the larvæ classed as campodeiform the legs are more or less like those of the adults of the same species; there may be differences in the proportions of the different segments of the leg, in the number of the tarsal segments, and in the number and form of the tarsal claws; but these differences are not of a nature to warrant a discussion of them here. These larvæ lead an active life, like that of nymphs, and consequently the form of legs has not been greatly modified from the paurometabolous type.

The development of the legs of insects with eruciform larvæ.—In caterpillars and other eruciform larvæ the thoracic legs are short and fitted for creeping; this mode of locomotion being best suited to their mode of life, either in burrows or clinging to foliage. This form of leg is evidently an acquired one being, like the internal development of wings, the result of those adaptive changes that fit these larvæ to lead a very different life from that of the adults.

In the case of caterpillars the thoracic legs are short, they taper greatly, and each consists of only three segments. It has been commonly believed and often stated that the three segments of the larval leg correspond to the terminal portion of the adult leg; but studies of the development of the legs of adults have shown that the divisions of the larval leg have no relation to the five divisions of the adult leg.

It has been shown by Gonin ('92), Kellogg ('01 and '04), and Verson ('04) that histoblasts which are the rudiments of the legs of the adult exist within the body-wall of the caterpillar at the base of the larval legs. Late in the larval life the extremity of the legs of the adult are contained in the legs of the caterpillar. It has been shown that the cutting off of a leg of a caterpillar at this time results in a mutilation of the terminal part of the leg of the adult.

The development of the legs of the adult within the body of caterpillars has not been studied as thoroughly as has been the development of the wings; but enough is known to show that in some respects the two are quite similar; this is especially true of the development of the tracheoles and of the tracheæ.

The development of the legs in insects with vermiform larvæ.—In vermiform larvæ the development of the entire leg is retarded. The leg arises as a histoblast, which is within the body and bears, in its more general features, a resemblance to the wing-buds of the same insect. The development of the legs of vermiform larvæ has been studied most carefully in the larvæ of Diptera. During the larval life the leg becomes quite fully developed within the peripodal cavity; in *Corethra*, they are spirally coiled; in *Musca*, the different segments telescope into each other. At the close of the larval period, the evagination of the legs takes place.

c. THE DEVELOPMENT OF ANTENNÆ

1. *The Transformation of the Antennæ of Nymphs and of Naiads*

In the case of nymphs and of naiads the insect when it emerges from the eggshell has well-developed antennæ. The changes that take place during the postembryonic development are, as a rule, comparatively slight; in most insects, an increase in the number of the segments of the antennæ takes place; but in the Ephemerida, a reduction in number of the antennal segments occurs.

2. *The Development of the Antennæ in Insects with a Complete Metamorphosis*

One of the marked characteristics of larvæ is the reduced condition of the antennæ; even in the campodeiform larvæ of the Neuroptera, where the legs are comparatively well-developed, the antennæ are greatly reduced.

In eruciform larvæ the development of the antennæ follows a course quite similar to that of the legs. The larval antennæ are small;

the antennæ of the adult are developed from histoblasts within the head and during the latter part of the larval life are folded like the bellows of a closed accordian; at the close of this period they become evaginated, but the definitive form is not assumed until the emergence of the adult. A similar course of development of the antennæ takes place in vermiform larvæ (Fig. 218).

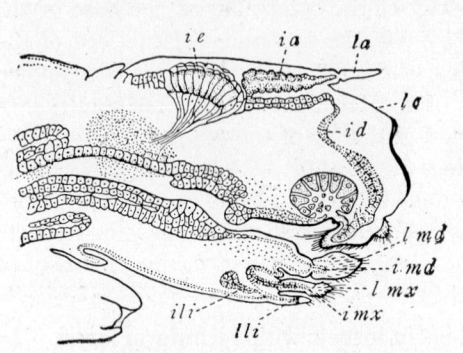

Fig. 218.—Sagittal section through head of old larva of *Simulium*, showing forming imaginal head parts within. *lc*, larval cuticula; *id*, imaginal head-wall; *la*, larval antenna; *ia*, imaginal antenna; *ie*, imaginal eye; *lmd*, larval mandible; *imd*, imaginal mandible; *lmx*, larval maxilla; *imx*, imaginal maxilla; *lli*, larval labium; *ili*, imaginal labium (From Kellogg).

d. THE DEVELOPMENT OF THE MOUTH-PARTS

Great differences exist among insects with reference to the comparative structure of their mouth-parts in their immature and adult instars. In some insects the immature instars have essentially the same type of mouth-parts as the adults; in most of these cases, the mouth-parts are of the biting types, but in the Homoptera and Heteroptera both nymphs and adults have them fitted for sucking; in many other insects, the mouth-parts of the larvæ are fitted for biting while those of adults are fitted for sucking; and in still others, as certain maggots, the development of the mouth-parts is so retarded that they are first functional in the adult insect. Correlated with these differences are differences in the method of development of these organs.

In those insects that have a gradual or incomplete metamorphosis and in the Neuroptera, the Coleoptera, and the Hymenoptera in part, the mouth-parts of the immature and adult instars are essentially of the same type. In these insects the mouth-parts of each instar are developed within the corresponding mouth-parts of the preceding instar. At each ecdysis there is a molting of the old cuticula, a stretching of the new one before it is hardened, a result of the growth in size of the appendages, and sometimes an increase in the number of the segments of the appendage. In a word, the mouth-parts of the adult are developed from those of the immature instar in a comparatively direct manner. In some cases, however, where the mouth-parts of the larva are small and those of the adult are large, only the tips of the developing adult organs are within those of the larva at the close of the larval period, a considerable part of the adult organs being embedded in the head of the old larva.

In a few Coleoptera and Neuroptera (the Dytiscidæ, Myrmeleonidæ, and Hemerobiidæ) the larvæ, although mandibulate, have the mouth-parts fitted for sucking. In these cases the form of the mouth-parts have been modified to fit them for a peculiar method of taking nourishment during the larval life. The mouth-parts of the adults are of the form characteristic of the orders to which these insects belong.

In those insects in which the larvæ have biting mouth-parts and the adults those fitted for sucking, the development is less direct. In the Lepidoptera, for example, to take an extreme case, there are great differences in the development of the different organs; within the mandibles of the old larvæ there are no developing mandibles, these organs being atrophied in the adult; but at the base of each larval maxilla, there is a very large, invaginated histoblast, the developing maxilla of the adult; these histoblasts become evaginated at the close of the larval period, but the maxillæ do not assume their definitive form till after the last ecdysis.

The extreme modification of the more usual course of development of the mouth-parts is found in the footless and headless larvæ of the more specialized Diptera. Here the mouth-parts do not appear externally until during the pupal stadium and become functional only when the adult condition is reached. See the figures illustrating the development of the head in the Muscidæ (Fig. 220).

It should be noted that the *oral hooks* possessed by the larvæ of the more specialized Diptera are secondarily developed organs and not mouth-parts in the sense in which this term is commonly used. These oral hooks serve as organs of fixation in the larvæ of the Œstridæ and as rasping organs in other larvæ.

e. THE DEVELOPMENT OF THE GENITAL APPENDAGES

The development of the genital appendages of insects has been studied comparatively little and the results obtained by the different investigators are not entirely in accord; it is too early therefore to do more than to make a few general statements.

In the nymphs of insects with a gradual metamorphosis rudimentary genital appendages are more or less prominent and their development follows a course quite similar to that of the other appendages of the body.

In insects with a complete metamorphosis the genital appendages are represented in the larvæ by invaginated histoblasts; the developing appendages become evaginated in the transformation to the pupa state and assume their definitive form after the last ecdysis.

III. THE DEVELOPMENT OF THE HEAD IN THE MUSCIDÆ

In the more generalized Diptera the head of the larva becomes, with more or less change, the head of the adult; the more important of these changes pertain to the perfecting of the organs of sight and the development of the appendages, the antennæ and mouth-parts.

But in the more specialized Diptera there is an anomalous retarding of the development of the head, which is so great that the larvæ of these insects are commonly referred to as being acephalous. This retarded development of the head has been carefully studied by Weisman ('64), Van Rees ('88) and Kowalevsky ('87). The accompanying diagrams (Fig. 220) based on those given by the last two authors illustrate the development of the head in *Musca*, which will serve as an illustration of this type of development of the head.

The larvæ of *Musca* are conical (Fig. 219); and the head-region is represented externally only by the minute apical segment of the conical body. It will be shown later that this segment is the neck of the insect, the developing head being invaginated within this and the following segments. This invagination of the head takes place during the later embryonic stages.

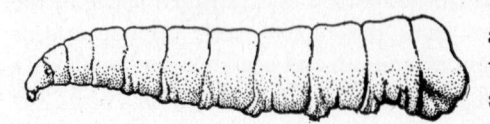

Fig. 219.—Larva of the house-fly, *Musca domestica* (After Hewitt).

In Figure 220 are given diagrams, adapted from Kowalevsky and Van Rees, representing three stages in the development of the head of *Musca*. Diagram A represents the cephalic end of the body of a larva; and diagram B and C, the corresponding region in a young and in an old pupa respectively; the parts are lettered uniformly in the three diagrams.

The three thoracic segments (1, 2, and 3) can be identified by the rudiments of the legs (l^1, l^2, and l^3). In the larva (A) the leg-buds are far within the body, the peripodal membrane being connected with the hypodermis of the body-wall by a slender stalk-like portion. In the young pupa (B) the peripodal membranes of the histoblasts of the legs are greatly shortened and the leg-buds are near the surface of the body; and in the old pupa (C) the leg-buds are evaginated. The wing-buds are omitted in all of the diagrams.

In the first two segments of the body of the larva (A) there is a cavity (*ph*) which has been termed the "pharynx"; this is the part in which the oral hooks characteristic of the larvæ of the Muscidæ develop. The name pharynx is unfortunate as this is not a part of the alimentary canal; it is an invaginated section of the head, into the base of which the œsophagus (*œ*) now opens.

In the figure of the larva (A) note the following parts: the œsophagus (*œ*); the ventral chain of ganglia (*vg*), the brain (*b*), and a

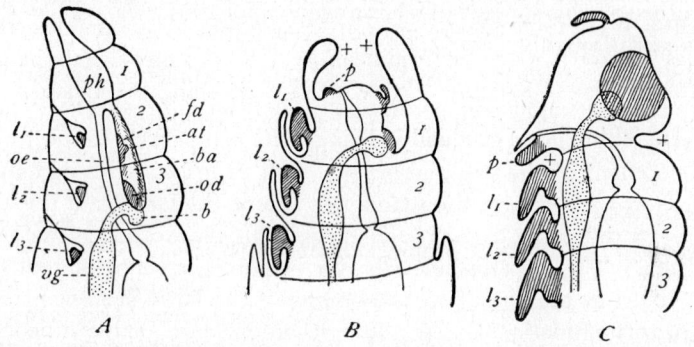

Fig. 220.—Development of the head in the Muscidæ. A, larva; B, young pupa; C, old pupa (From Korschelt and Heider after Kowalevsky and Van Rees).

sac (*ba*) extending from the so-called pharynx to the brain. There are two of these sacs, one applied to each half of the brain, but only one of these would appear in such a section as is represented by the diagram. These sacs were termed the *brain-appendages* by Weismann. In each of the "brain-appendages" there is a disc-like thickening near the brain, the *optic disc* (*od*); this is a histoblast which develops into a compound eye; in front of the optic disc there is another prominent histoblast; the *frontal disc* (*fd*), upon which the rudiment of an antenna (*at*) is developed.

In the larva the brain and a considerable part of the "brain-appendages" lie within the third thoracic segment. In the young pupa (B) these parts have moved forward a considerable distance; and in the old pupa (C) the head has become completely evaginated. The part marked *p* in the two diagrams of the pupa is the rudiment of the proboscis.

By comparing diagrams B and C it will be seen that what was the tip of the first segment of the larva and of the young pupa (++) becomes the neck of the insect after the head is evaginated.

IV. THE TRANSFORMATIONS OF THE INTERNAL ORGANS

Great as are the changes in the external form of the body during the life of insects with a complete metamorphosis, even greater changes take place in the internal organs of some of them.

In the space that can be devoted to this subject in this work, only the more general features of the transformation of the internal organs can be discussed; there is an extensive and constantly increasing literature on this subject which is available for those who wish to study it more thoroughly.

In insects with a gradual or with an incomplete metamorphosis there is a continuous transformation of the internal organs, the changes **in form taking place gradually:** being quite comparable to the gradual development of the external organs; but in insects with a complete metamorphosis, where the manner of life of the larva and the adult are very different, extensive changes take place during the pupal stadium. The life of a butterfly, for example, is very different from that it led as a caterpillar; the organs of the larva are not fitted to perform the functions of the adult; there is consequently a necessity for the reconstruction of certain of them; hence the need of a pupal stadium. Pupæ are often referred to as being quiet; but physiologically the pupal period is the most active one in the post-embryonic life of the insect.

In those cases where a very marked change takes place in the structure of internal organs, there is a degeneration and dissolution of tissue, this breaking down of tissues is termed *histŏlysis*.

In the course of histolysis some cells, which are frequently leucocytes or white blood corpuscles, feed upon the debris of the disintegrating tissue; such a cell is termed a **phagocyte,** and the process is termed *phăgocytōsis*. It is believed that the products of the digestion of disintegrating tissue by the phagocytes pass by diffusion into the surrounding blood and serve to nourish new tissue.

After an organ has been more or less broken down by histolysis, the extent of the disintegration differing greatly in different organs and in different insects, there follows a growth of new tissue; this process is termed *hĭstogĕnesis*.

The histogenetic reproduction of a tissue begins in the differentiation and multiplication of small groups of cells, which were not affected by the histolysis of the old tissue; such a group of cells is termed an *imaginal disc* or a *histoblast*. They were termed imaginal discs on account of the disc-like form of those that were first described and because they are rudiments of organs that do not become functional till the imago stage; but the term histoblast is of more general application and is to be preferred.

The extent of the transformation of the internal organs differs greatly in different insects. In the Coleoptera, the Lepidoptera, the Hymenoptera, and the Diptera Nemocera, the mid-intestine and some other larval organs are greatly modified, but there is no general histolysis. On the other hand, in the Diptera Brachycera, there is a general histolysis. In *Musca* all organs break down and are reformed except the central nervous system, the heart, the reproductive organs, and three pairs of thoracic muscles. Regarding the extent of the transformations in the other orders where the metamorphosis is complete we have, as yet, but little data.

For a more detailed and exhaustive discussion of the morphology of insects the special student should consult the authoritative book by R. E. Snodgrass, "Principles of Insect Morphology" (1935).

PART II
THE CLASSIFICATION AND THE LIFE-HISTORIES OF INSECTS

Class HEXAPODA
The Insects

The members of this class are air-breathing arthropods, with distinct head, thorax, and abdomen. They have one pair of antennæ, three pairs of legs, and usually one or two pairs of wings in the adult state. In most adult insects the head bears a pair of compound eyes. The opening of the reproductive organs is near the caudal end of the body.

The more general character of insects, together with their structure and morphology, has been discussed at some length in Part I of this book. It is now appropriate to consider the classification, habits and life histories of insects. Part II will therefore be devoted to a discussion of the subclasses, orders, and families of this great group of animals.

The class Hexapoda is divided into two subclasses, the small wingless insects, *Apterygota*, and the winged insects, *Pterygota*. The subclass Apterygota is a small one, containing but three orders (four by some authors) and mostly unfamiliar forms except to those especially interested in these tiny creatures. The subclass Pterygota is a very large group including all of the remaining twenty-three orders discussed in this book. This subclass contains all of the more familiar forms, such as grasshoppers, beetles, butterflies, moths, flies, wasps, and bees.

CHAPTER V

THE SUBCLASSES AND THE ORDERS OF THE CLASS HEXAPODA

INSECTS constitute one of the classes of the Arthropoda, that division of the animal kingdom in which the body is composed of a series of more or less similar segments and in which some of these segments bear jointed legs. This class is known as the Hexapoda.

The distinctive characteristics of the Class Hexapoda and its relation to the other classes of the Arthropoda are discussed in the first chapter of this work; we have now to consider the division of this class into subclasses and orders.

The orders that constitute the Hexapoda represent two well-marked groups; this class is divided, therefore, into two subclasses. This division was first proposed by Brauer ('85), who recognized the fact that while the wingless condition of certain insects, the fleas, lice, bird-lice, and the wingless members of orders in which the wings are usually present, is an acquired one, the wingless condition of the Thysanura and Collembola is a primitive one. In other words, from the primitive insects, which were wingless, there were evolved on the one hand the orders Thysanura and Collembola, which remained wingless, and on the other hand, a winged form from which have descended all other orders of insects.

An extended study of the wings of insects has shown that the wings of all of the orders of winged insects are modifications of a single type; it is believed, therefore, that all of the orders of winged insects have descended from a common winged ancestor. As to the lice, bird-lice, and fleas, the relation of each of these groups to certain winged insects, as shown by their structure, has led to the belief that their wingless condition is an acquired one, being the result of parasitic habits. The lice or Anoplura are commonly regarded as closely allied to the Homoptera and Heteroptera; the bird-lice or Mallophaga to the Corrodentia; and the fleas or Siphonaptera to the Diptera. Hence these wingless insects are placed with the winged insects in a single subclass.

The two subclasses thus recognized were named by Brauer the Apterygogenea and the Pterygogenea respectively. The cumbersomeness of these names led to the substitution for them of the shorter names Apterygota and Pterygota. The Apterygota includes the orders Thysanura and Collembola; and the Pterygota, all other orders of insects. Some writers regard the Thysanura and Collembola as suborders of a single order, which they term the Aptera.

The distribution of insects into orders is based on the classification of Linnæus, as set forth in his "Systema Naturæ" (1735–1768). Linnæus, who has been called the Adam of zoological science, divided his class Insecta into seven orders; these he named Coleoptera, Hemiptera, Lepidoptera, Neuroptera, Hymenoptera, Diptera, and Aptera, respectively.

Since the time of Linnæus many modifications of his classification of insects have been proposed; and new ones are constantly appearing. The result is that now there is a great lack of uniformity in the classification used by different writers.

The modifications of the Linnæan distribution of insects into orders are based on the belief that in certain cases Linnæus grouped into a single order forms that really represent two or more distinct orders. The result has been a great increase in the number of orders recognized.

Linnæus included in his class Insecta, under the order Aptera, not only wingless insects but also arachnids, crustaceans, centipedes, and millipedes. The animals thus grouped by Linnæus are now distributed into several classes; and to the class composed of the animals now commonly known as insects, those characterized by the possession of only six legs, the term Hexapoda is commonly applied. Some writers, however, apply the term Insecta to the class of insects as now limited.

Some of the more recently recognized orders of insects are represented among living insects by comparatively few species; but in each case the structure of the insects included in the group is so different from that of all other insects that we are led to believe that they represent a division of the class Hexapoda that is of ordinal value.

There are given below the names of the orders of insects recognized in this work. The sequence in which these orders are discussed is of necessity a more or less arbitrary one. In general the plan adopted here is to make the series an ascending one; that is, the more generalized or primitive insects are placed first and the more highly specialized ones later in the series; but as the different orders of insects have been specialized in very different ways, the relative degrees of their specialization cannot be shown by arranging them in a single linear series, as must be done in a book. To indicate the different ways in which the different members of a group have been specialized and the relative rank of those specialized in a similar way, use must be made of a diagram representing a genealogical tree. Many such diagrams have been made, but no one of them has received general acceptance; much remains to be learned before such a diagram can be made that will inspire confidence in its accuracy.

In the course of the preparation of a special treatise on the wings of insects (Comstock '18 a), I wrote a table indicating the more striking of the methods of specialization of the wings characteristic of each of the orders of winged insects; and in the discussion of the different orders, I followed the sequence indicated by this table. In doing this I did not advocate the basing of a classification of insects upon the characters presented by the wings alone, but merely made use of these characters for the purposes of that work.

A renewed study of the relationships of the different orders to each other, in which an effort has been made to correlate other characters with those presented by the wings, has not indicated the desirability of changes in the sequence of the orders indicated in that table, except in the allocation of those orders in which wings are wanting.

The importance of the wings of insects for taxonomic purposes was early recognized by entomologists, as is well shown by the fact that the names of the Linnæan orders are all drawn from the nature of the wings, except one, Aptera, and that from the absence of wings.

The different methods of specialization of the wings arose very early in that part of the geological history of insects that is known to us. And as most of the fossil remains of the older insects consist of wings, we are forced to depend very largely on the characters presented by these organs for data regarding the separation of the primitive insects into the groups from which the orders of recent insects have been developed. But in characterizing the orders as they now exist all the results of the study of the structure of insects and of their transformations are available.

Aside from the structure of the wings, the characters most used in characterizing the orders of insects are those presented by the structure of the mouth-parts and the nature of the post-embryonic development. While these characters are of value in defining the orders, but little use has been made of them, as yet, in working out the lines of descent of the various orders from the primitive insects.

The primitive insects had chewing mouth-parts and this type has been retained in the greater number of the orders. But although many detailed accounts of the structure of the mouth-parts of chewing insects have been published, no one has worked out the various ways in which they have been specialized in such a manner as to indicate the phylogeny of the orders.

Several different types of sucking mouth-parts exist among living insects; but these are apparently of comparatively late origin, and while they are of great value in defining the orders in which they occur, they do not afford characters for determining the primitive divisions of the Pterygota.

The nature of the post-embryonic development of insects, like the structure of the mouth-parts, affords characters for defining the orders of recent insects, but is of little value in determining the phylogeny of the orders.

The primitive insects doubtless developed without any marked metamorphosis as do the Thysanura and Collembola of today. With the development of wings, there arose that type of development known as gradual metamorphosis, and this type is retained by eight of the orders recognized in this work. Incomplete metamorphosis is the result of a sidewise development of the immature instars of the insects exhibiting it, in order to fit them for life in the water, and it doubtless arose independently in each of the three orders in which it occurs; it is therefore an ordinal characteristic in each case and not one indicating a natural group of orders. This is also true of complete metamorphosis, which also doubtless arose independently in different divisions of the insect series, as, for example, in the Neuroptera, which it is believed is a very ancient order, the origin of which was much earlier than the attainment of complete metamorphosis.

TABLE OF THE METHODS OF SPECIALIZATION OF THE WINGS CHARACTERISTIC OF THE ORDERS OF WINGED INSECTS*

This table is merely the result of an effort to indicate the more striking of the methods of specialization of the wings characteristic of each of the orders of insects. It is not a key for determining the orders of insects. It is not available for this purpose; because, in many cases, the wings of an insect do not show the type of specialization characteristic of the order to which the insect belongs. Thus, for example, while the most characteristic modification of the courses of the wing-veins in the Diptera and Hymenoptera is due to the coalescence of veins proceeding from the margin of the wing towards the base of the wing, there is no indication of this type of coalescence of veins in some of the nemocerous Diptera.

A. Wings specialized by the development of supernumerary veins in the preanal area.
  B. Supernumerary veins of the accessory type.
    C. Wings developed externally.
      D. Wings retained throughout life. Wings without a striking contrast in the thickness of the veins of the anterior part of the wing and those of the middle portion................................ORTHOPTERA
      DD. Wings deciduous, there being near the base of each wing a transverse suture along which the wing is broken off after the swarming flight. Wings with the veins of the anterior part of the wing greatly thickened and those of the middle portion reduced to narrow lines ....ISOPTERA
    CC. Wings developed internally ....NEUROPTERA
  BB. Supernumerary veins of the intercalary type.
    C. Flight-function cephalized; the hind wings being greatly reduced in size..EPHEMERIDA
    CC. Flight-function not cephalized; the hind wings as large as or larger than the fore wings..ODONATA
AA. Wings specialized by a reduction in the number of veins in the preanal area.
  B. Wings developed externally.
    C. The two pairs of wings similar in texture.
      D. With the tendency to develop accessory veins retained..PLECOPTERA
      DD. With the tendency to develop accessory veins in the preanal area lost.
        E. With the courses of some of the longitudinal veins modified so that they function as cross-veins...................CORRODENTIA
        EE. The transverse bracing of the wing attained in the usual way.
          F. The veins of the wing bordered with dark bands...EMBIIDINA
          FF. The veins of the wing not bordered with dark bands.
            G. Wings long and narrow, supplemented by a wide fringe of hairs.....................................THYSANOPTERA
            GG. Wings not greatly narrowed and not supplemented by a wide fringe of hairs...........................HOMOPTERA
    CC. The front wings more or less thickened.
      D. The front wings not greatly reduced in length as compared with the hind wings.
        E. The front wings thickened throughout............HOMOPTERA
        EE. The front wings thickened at the base, the terminal portion membranous....................................HETEROPTERA
      DD. The front wings greatly reduced in length.........DERMAPTERA

---

*From "The Wings of Insects," pp. 120-122.

 BB. Wings developed internally.
  C. Fore wings greatly thickened.
   D. Fore wings modified so as to serve as covers of the posterior wings
    ....COLEOPTERA
   DD. Fore wings reduced to slender, leathery, club-shaped appendages
    ....STREPSIPTERA
  CC. The two pairs of wings similar in texture.
   D. With the tendency to develop accessory veins retained..MECOPTERA
   DD. With the tendency to develop accessory veins lost.
    E. The most characteristic method of reduction of the wing-veins of the preanal area being by coalescence outward.
     F. Anal veins of the fore wings tending to coalesce at the tip. Wings usually clothed with hairs........................TRICHOPTERA
     FF. Anal veins of the fore wings not tending to coalesce at the tip. Wings clothed with scales...................LEPIDOPTERA
    EE. The most characteristic method of reduction of the wing-veins of the preanal area being by coalescence from the margin of the wing inward.
     F. With only one pair of wings......................DIPTERA
     FF. With two pairs of wings...................HYMENOPTERA

  The sequence in which the orders of insects are discussed in the following chapters has been determined by the above table. This sequence, like all linear arrangements of groups of organisms, is more or less arbitrary. Thus while there is an effort to place first the more generalized orders and later those that are more specialized, the putting together of orders exhibiting the same type of specialization results in some cases in the placing of comparatively generalized forms after those that are obviously more highly specialized. The position of the Plecoptera is an illustration of this. The insects of this order are evidently more generalized than, for example, the Neuroptera or the Odonata, which are placed earlier in the linear series.

  The comparatively high position assigned to the Plecoptera is, however, only apparent. A reference to the table will show that the orders of insects are grouped in two series, "A" and "AA". Under "A" are placed those orders in which the wings are specialized by addition in the preanal area, and under "AA" those orders in which the wings are specialized by reduction in the preanal area. Each of these series includes some quite generalized insects and others that are highly specialized. The completion of the discussion of the first series before taking up the second series results in the generalized members of the second series following the highly specialized members of the first series.

  The more generalized members of these two series, the Orthoptera of the first series and the Plecoptera of the second series, are probably more closely allied to each other than is either of these orders to the more specialized orders of the series in which it is placed; the two series arose from a common starting point, the Palæodictyoptera, but have widely diverged in the course of their development.

  An even more striking illustration of the difficulty of indicating the relative ranks of orders by the use of a single linear series is the position of the Isoptera in the above table. This order is a very ancient one; it separated from the Palæodictyoptera before definite cross-veins in the wings had been developed and has not attained them. It is placed in the table next to the Orthoptera because the wings are specialized by the development of supernumerary veins of the accessory type and are developed externally; but the peculiar specialization of the wings is very different from that of the Orthoptera as is indicated in the table. And in other respects the termites have reached a stage of development far in advance of that shown by any of the Orthoptera. They have attained a social mode of life, with the correlated separation of the species into several castes and the development of remarkable instincts. In this respect they rival the social Hymenoptera.

In fact the living members of each of the orders of insects must be regarded as a group of organisms representing the results of specialization in a direction different from that of any other order; and to attempt to decide which order is the "highest" seems as futile as the discussion by children of the question: "Which is better, sugar or salt?" The list below indicates the sequence in which the orders are discussed in the following chapters.

THE SUBCLASSES AND ORDERS OF THE HEXAPODA

SUBCLASS APTERYGOTA.—Wingless insects in which the wingless condition is believed to be a primitive one, there being no indication that they descended from winged ancestors.
    ORDER PROTURA.—The Telson-tails. p. 218.
    ORDER THYSANURA.—The Bristle-tails. p. 219.
    ORDER COLLEMBOLA.—The Spring-tails. p. 225.
SUBCLASS PTERYGOTA.—Winged insects and wingless insects in which the wingless condition is believed to be an acquired one.
    ORDER ORTHOPTERA.—The Cockroaches, Crickets, Grasshoppers, and others. p. 230.
    ORDER ZORAPTERA.—The genus *Zorotypus*. p. 270.
    ORDER ISOPTERA.—The Termites or White Ants. p. 273.
    ORDER NEUROPTERA.—The Dobson, Aphis-lions, Ant-lions, and others. p. 281.
    ORDER EPHEMERIDA.—The May-flies. p. 308.
    ORDER ODONATA.—The Dragon-flies and the Damsel-flies, p. 314.
    ORDER PLECOPTERA.—The Stone-flies. p. 325.
    ORDER CORRODENTIA.—The Psocids. p. 331.
    ORDER MALLOPHAGA.—The Bird-lice. p. 335.
    ORDER EMBIIDINA.—The Embiids. p. 338.
    ORDER THYSANOPTERA.—The Thrips. p. 341.
    ORDER ANOPLURA.—The Lice. p. 347.
    ORDER HOMOPTERA.—The Cicadas, Leaf-hoppers, Aphids, Scale-bugs, and others. p. 394.
    ORDER HEMIPTERA.—The True Bugs. p. 350.
    ORDER DERMAPTERA.—The Earwigs. p. 460.
    ORDER COLEOPTERA.—The Beetles. p. 464.
    ORDER STREPSIPTERA.—The Twisted Winged Insects. p. 546.
    ORDER MECOPTERA.—The Scorpion-flies. p. 550.
    ORDER TRICHOPTERA.—The Caddice-flies. p. 555.
    ORDER LEPIDOPTERA.—The Moths, the Skippers, and the Butterflies. p. 571.
    ORDER DIPTERA.—The Flies. p. 773.
    ORDER SIPHONAPTERA.—The Fleas. p. 877.
    ORDER HYMENOPTERA.—The Bees, Wasps, Ants, and others. p. 884.

TABLE FOR DETERMINING THE ORDERS OF THE HEXAPODA

This table is merely intended to aid the students in determining to which of the orders a specimen that he is examining belongs. No effort has been made to indicate in the table the relation of the orders to one another.

A. Winged. (The wing-covers, *Elytra*, of beetles and of earwigs are wings.)
  B. With two wings.
    C. Wings horny, leathery, or parchment-like.
      D. Mouth-parts formed for sucking. Wings leathery, shortened, or membranous at the tip. p. 350........................HEMIPTERA
      DD. Mouth-parts formed for biting. Jaws distinct.
        E. Wings horny, without veins. Hind legs not fitted for jumping. p. 464..COLEOPTERA
        EE. Wings parchment-like with a network of veins. Hind legs fitted for jumping. p. 230............................ORTHOPTERA
    CC. Wings membranous.
      D. Abdomen with caudal filaments. Mouth-parts vestigial.
        E. Halteres wanting. p. 308........................EPHEMERIDA
        EE. Halteres present (males of Coccidæ). p. 394......HOMOPTERA
      DD. Abdomen without caudal filaments. Halteres in place of second wings. Mouth-parts formed for sucking. p. 773.....DIPTERA
  BB. With four wings.
    C. The two pairs of wings unlike in structure.
      D. Fore wings reduced to slender club-shaped appendages; hind wings fan-shaped with radiating veins. Minute insects. p. 546..STREPSIPTERA
      DD. Front wings leathery at base, and membranous at tip, often overlapping. Mouth-parts formed for sucking. p. 350...HEMIPTERA
      DDD. Front wings of same texture throughout.
        E. Front wings horny or leathery, being veinless wing-covers. (*Elytra*).
          F. Abdomen with caudal appendages in form of movable forceps. p. 460...DERMAPTERA
          FF. Abdomen without forceps-like appendages. p. 464. COLEOPTERA
        EE. Front wings leathery or parchment-like with a network of veins.
          F. Under wings not folded; mouth-parts formed for sucking.
            G. Beak arising from the front part of the head. p. 350.HEMIPTERA
            GG. Beak arising from the hind part of the lower side of the head. p. 394..HOMOPTERA
          FF. Under wings folded lengthwise. Mouth-parts formed for chewing. p. 230.................................ORTHOPTERA
    CC. The two pairs of wings similar, membranous.
      D. Last joint of tarsi bladder-like or hoof-like in form and without claws. p. 341................................... THYSANOPTERA
      DD. Last joint of tarsi not bladder-like.
        E. Wings entirely or for the greater part clothed with scales. Mouth-parts formed for sucking. p. 571....................LEPIDOPTERA
        EE. Wings naked, transparent, or thinly clothed with hairs.
          F. Mouth-parts arising from the hinder part of the lower surface of the head, and consisting of bristle-like organs inclosed in a jointed sheath. p. 394.................................HOMOPTERA
          FF. Mouth-parts in normal position. Mandibles not bristle-like.
            G. Wings net-veined, with many veins and cross-veins.
              H. Tarsi consisting of less than five segments.
                I. Antennæ inconspicuous, awl-shaped, short and slender.
                  J. First and second pairs of wings of nearly the same length; tarsi three-jointed. p. 314......................ODONATA
                  JJ. Second pair of wings either small or wanting; tarsi four-jointed. p. 308......................EPHEMERIDA
                II. Antennæ usually conspicuous, setiform, filiform clavate, capitate, or pectinate.
                  J. Tarsi two- or three-jointed.
                    K. Second pair of wings the smaller. p. 331.CORRODENTIA

       KK. Second pair of wings broader, or at least the same size as the first pair. p. 325....PLECOPTERA
   JJ. Tarsi four-jointed; wings equal. p. 273..ISOPTERA
 HH. Tarsi consisting of five segments.
  I. Abdomen with setiform, many-jointed anal filaments. (Certain May-flies). p. 308..................EPHEMERIDA
  II. Abdomen without many-jointed anal filaments.
   J. Head prolonged into a trunk-like beak. p. 550.MECOPTERA
   JJ. Head not prolonged into a beak. p. 281...NEUROPTERA
GG. Wings with branching veins and comparatively few cross-veins, or veinless.
 H. Each of the veins of the wing extending along the middle of a brown line. p. 338.........................EMBIIDINA
 HH. Wings not marked with brown lines.
  I. Tarsi two-or three-jointed.
   J. Hind wings smaller than the fore wings.
    K. Cerci present; body less than three millimeters in length. p. 270..........................ZORAPTERA
    KK. Cerci absent; larger insects. p. 331..CORRODENTIA
   JJ. Posterior wings as large as or larger than the anterior ones. (Certain Stone-flies). p. 325........PLECOPTERA
  II. Tarsi four- or five-jointed.
   J. Abdomen with setiform, many-jointed anal filaments. (Certain May-flies). p. 308.............EPHEMERIDA
   JJ. Abdomen without many-jointed anal filaments.
    K. Prothorax horny. First wings larger than the second, naked or imperceptibly hairy. Second wings without, or with few, unusually simple, veins. Jaws (mandibles) well developed. Palpi small. p. 884....HYMENOPTERA
    KK. Prothorax membranous or, at the most, parchment-like. Second wings as large as or larger than the first, folded lengthwise, with many branching veins. First wings naked or thinly clothed with hair. Jaws (mandibles) inconspicuous. Palpi long. Moth-like insects. p. 555....................TRICHOPTERA
AA. Wingless or with vestigial or rudimentary wings.
 B. Insects with a distinct head and jointed legs, and capable of locomotion.
  C. Aquatic insects.
   D. Mouth-parts fitted for piercing and sucking.
    E. Free-swimming nymphs. p. 350..................HEMIPTERA
    EE. Larvæ parasitic in sponges (Sisyridæ). p. 281.....NEUROPTERA
   DD. Mouth-parts fitted for chewing.
    E. Either somewhat caterpillar-like larvæ that live in portable cases or campodeiform larvæ that spin nets for catching their food. (Caddice-worms). p. 555....................................TRICHOPTERA
    EE. Neither case-bearing nor net-spinning larvæ.
     F. Naiads, that is, immature insects that resemble adults in having the thorax sharply differentiated from the abdomen, and, except in very young individuals, with rudimentary wings.
      G. Lower lip greatly elongated, jointed, capable of being thrust forward, and armed at its extremity with sharp hooks. p. 314.ODONATA
      GG. Lower lip not capable of being thrust forward.
       H. Usually with filamentous tracheal gills on the ventral side of the thorax. p. 325........................PLECOPTERA
       HH. Tracheal gills borne by the first seven abdominal segments. p. 308..........................EPHEMERIDA
     FF. Larvæ, that is, immature forms that do not resemble adults in the form of the body, and in which the developing wings are not visible externally.
      G. Several segments of the abdomen furnished with prolegs. p. 571...LEPIDOPTERA
      GG. With only anal prolegs or with none.

       H. With paired lateral filaments on most or on all of the abdominal segments. (Sialidæ). p. 281..........NEUROPTERA
       See also Haliplidæ and Gyrinidæ. p. 464......COLEOPTERA
      HH. Without paired lateral filaments on the abdomen. p. 464.
....COLEOPTERA
CC. Terrestrial insects.
  D. External parasites.
    E. Infesting the honey-bee. (*Braula*). p. 773............DIPTERA
    EE. Infesting birds or mammals.
      F. Body strongly compressed. (Fleas). p. 877.....SIPHONAPTERA
      FF. Body not strongly compressed.
        G. Mouth-parts formed for chewing. (Bird-lice). p. 335.
....MALLOPHAGA
        GG. Mouth-parts formed for piercing and sucking.
          H. Antennæ inserted in pits, not visible from above. (Pupipara). p. 773................................DIPTERA
          HH. Antennæ exserted, visible from above.
           G. Tarsi with a single claw which is opposed by a toothed projection of the tibia. (Lice). p. 347...............ANOPLURA
           GG. Tarsi two-clawed. p. 350........·..........HEMIPTERA
  DD. Terrestrial insects not parasites.
    E. Mouth-parts apparently retracted within the cavity of the head so that only their apices are visible, being overgrown by folds of the genæ.
      F. Abdomen consisting of ten or eleven segments. (Campodeidæ and Japygidæ). p. 220..............................THYSANURA
      FF. Abdomen consisting of not more than six segments. p. 225.
....COLLEMBOLA
    EE. Mouth-parts mandibulate, either fitted for chewing or with sickle-shaped mandibles formed for seizing prey. (See also EEE.)
      F. Larvæ with abdominal prolegs.
        G. Prolegs armed at the extremity with numerous minute hooks. (Caterpillars). p. 571..........................LEPIDOPTERA
        GG. Prolegs not armed with minute hooks.
          H. With a pair of ocelli, one on each side. (Larvæ of saw-flies). p. 884...HYMENOPTERA
          HH. With many ocelli on each side of the head. p. 550.....
....MECOPTERA
      FF. Without abdominal prolegs.
        G. Body clothed with scales. (Machilidæ and Lepismatidæ). p. 220..................................·..THYSANURA
        GG. Body not clothed with scales.
          H. Antennæ long and distinct.
           I. Abdomen terminated by strong movable forceps. p. 460.
....DERMAPTERA
           II. Abdomen not terminated by forceps.
            J. Abdomen strongly constricted at base. (Ants. etc.). p. 884..............................HYMENOPTERA
            JJ. Abdomen not strongly constricted at base.
             K. Head with a long trunk-like beak. (*Boreus*). p. 550.
....MECOPTERA
             KK. Head not prolonged into a trunk.
              L. Insects of small size, more or less louse-like in form, with a very small prothorax, and without cerci. (Book-lice and Psocids). p. 331.....CORRODENTIA
              LL. Insects of various forms, but not louse-like, prothorax not extremely small; cerci present.
               M. Hind legs fitted for jumping, hind femora enlarged. (Wingless locusts, grasshoppers, and crickets). p. 230...................ORTHOPTERA
               MM. Hind femora not greatly enlarged, not fitted for jumping.

  N. Prothorax much longer than the mesothorax; front legs fitted for grasping prey. (Mantidæ). p. 230........................ORTHOPTERA
  NN. Prothorax not greatly lengthened.
   O. Cerci present; antennæ usually with more than fifteen joints, often many-jointed.
    P. Cerci with more than three joints.
     Q. Body flattened and oval. (Blattidæ). p. 230...................ORTHOPTERA
     QQ. Body elongate.
      R. Head very large. (*Termopsis*). p. 273. ....ISOPTERA
      RR. Head of moderate size. p. 268. ...................GRYLLOBLATTIDÆ
    PP. Cerci short, with one to three joints.
     Q. Body linear with very long linear legs. (Walking-sticks). p. 230...ORTHOPTERA
     QQ. Body elongate or not, if elongate the legs are not linear.
      R. Body elongate; front tarsi with first joint swollen. p. 338...EMBIIDINA
      RR. Front tarsi not enlarged.
       S. Minute insects, less than 3 mm. in length; antennæ nine-jointed. p. 270...............ZORAPTERA
       SS. Larger insects; antennæ usually more than nine-jointed. (White-ants). p. 273...................ISOPTERA
   OO. Cerci absent; antennæ usually with eleven joints. p. 464.................COLEOPTERA
  HH. Antennæ short, not pronounced; larval forms.
   I. Body cylindrical, caterpillar-like. p. 550.MECOPTERA
   II. Body not caterpillar-like.
    J. Mandibles sickle-shaped; each mandible with a furrow over which the maxilla of that side fits, the two forming an organ for piercing and sucking. (Ant-lions, aphis-lions, hemerobiids). p. 281..............NEUROPTERA
    JJ. Mouth-parts not of the ant-lion type.
     K. Larva of *Raphidia*. p. 281...........NEUROPTERA
     KK. Larvæ of beetles. p. 464...........COLEOPTERA
EEE. Mouth-parts haustellate, fitted for sucking; mandibles not sickle-shaped.
  F. Body covered with a waxy powder or with tufts or plates of wax. (Mealy-bugs, *Orthezia*). p. 350....................HEMIPTERA
  FF. Body more or less covered with minute scales, or with thick long hairs; proboscis if present coiled beneath the head. (Moths). p. 571...LEPIDOPTERA
  FFF. Body naked, or with isolated or bristle-like hairs.
   G. Prothorax not well developed, inconspicuous or invisible from above. p. 773..............................DIPTERA
   GG. Prothorax well developed.
    H. Last joint of tarsi bladder-like or hoof-like in form and usually without claws; mouth-parts forming a triangular unjointed beak. p. 550....................THYSANOPTERA
    HH. Last joint of tarsi not bladder-like, and furnished with one or two claws; mouth-parts forming a slender; usually jointed beak.
     I. Beak arising from the front part of the head. p. 350. ....HEMIPTERA
     II. Beak arising from the back part of the head. p. 394... ....HOMOPTERA

BB. Either without a distinct head, or without jointed legs, or incapable of locomotion.
C. Forms that are legless but capable of locomotion; in some the head is distinct, in others not. Here belong many larvæ representing several of the orders, and the active pupæ of mosquitoes and certain midges. It is impracticable to separate them in this key.
CC. Sedentary forms, incapable of locomotion.
D. Small abnormal insects in which the body is either scale-like or gall-like in form, or grub-like clothed with wax. The waxy covering may be in the form of powder, or large tufts or plates, or a continuous layer, or of a thin scale, beneath which the insect lives. (Coccidæ). p. 350.HEMIPTERA
DD. Pupæ, the inactive stage of insects with a complete metamorphosis; capable only of a wriggling motion, and incapable of feeding.
E. Obtected pupæ, pupæ in which the legs and wings are glued to the surface of the body; either in a cocoon or naked. p. 571. LEPIDOPTERA
EE. Coarctate pupæ, pupæ enclosed in the hardened larval skin. p. 773..DIPTERA
EEE. Exarate pupæ, pupæ that have the legs and wings free; either in a cocoon or naked. This type of pupa is characteristic of all of the orders in which the metamorphosis is complete except the Lepidoptera and Diptera.

The order Protura does not occur in the foregoing table because it has only now, in this edition (1936), been placed among the Hexapoda. Since the members of the order are uncommon insects rarely met with it did not seem advisable to incur the added expense of rearranging and reprinting a large part of the table.

CHAPTER VI

Subclass I. APTERYGOTA

Wingless Insects

The members of this subclass are small wingless insects in which the wingless condition is believed to be a primitive one, there being no indication that they have descended from winged ancestors. The mouthparts vary. In some they are sucking, in others chewing. The metamorphosis is always slight and in some cases absent.

This subclass contains but the three orders, Protura, Thysanura, and Collembola. These insects are all primitive and usually generalized. They are all small and wingless and because of their usually concealed habits are not as generally known as the winged forms. They are widely distributed and about 1200 species are now known. Probably many more remain to be discovered. A characteristic feature of these primitive insects is the abdominal appendages, especially the abdominal styli present in the Machilidae and in others of the Thysanura.

CHAPTER VII

ORDER PROTURA

The Telson-Tails

The members of this order are small arthropods in which the body is elongate, as in the Thysanura, fusiform, pointed behind, and depressed; it may be greatly extended and retracted. The antennæ, cerci and compound eyes are absent. The oral apparatus is suctorial, and consists of three pairs of gnathites. There are three pairs of thoracic legs, and three pairs of vestigial abdominal legs. The abdomen is composed of eleven segments and a telson. The opening of the reproductive organs is unpaired, and near the hind end of the body. The head bears a pair of organs, termed pseudoculi, the nature of which has not been definitely determined. The metamorphosis is slight, consisting of an increase in the number of abdominal segments.

The known members of this order are very small arthropods, the body measuring from one-fiftieth to three-fiftieths of an inch in length. The form of the body is shown by Figure 36, p. 25.

These exceedingly interesting creatures are found in damp situations, as in the humus of gardens. They are widely distributed: they are now known to occur in India, England, Italy, and other European countries. A number of species have been described from the southwestern United States.

The mouthparts are withdrawn into the head and the mandibles are stylet-like and fitted for piercing. In the newly hatched insect the abdomen is 9-segmented, but during later growth three more segments are added between the last two segments. This mode of change in form is known as anamorphosis.

The systematic position of the Protura is still unsettled. The differentiated thorax with three pairs of legs and the form of the mouthparts are characteristic of insects. The lack of antennae and the intercalary addition of body segments during growth (anamorphosis) are very unlike insects. The name Protura refers to the last telson-like segment of the abdomen.

The order contains two families as follows:

FAMILY 1. ACERENTŎMIDÆ, in which the tracheæ and spiracles are absent and the second and third abdominal appendages are 1-jointed. This family includes the two genera *Acerentomon* and *Acerentulus*.

FAMILY 2. EOSENTŎMIDAE, in which tracheæ are present with two pairs of thoracic spiracles and the second and third abdominal appendages are 2-jointed. This family includes the genus *Eosentomon* and probably *Protapteron*.

ORDER THYSANURA*

The Bristle-Tails

The members of this order are wingless insects in which the wingless condition is believed to be a primitive one, there being no indication that they have descended from winged ancestors; the mouth-parts are formed for chewing; and the adult insects resemble the young in form. In these three respects, these insects resemble the next order, the Collembola; but they differ from the Collembola in that the abdominal segments are not reduced in number and the cerci are usually filiform and many-jointed; some members of the order have also a caudal filament.

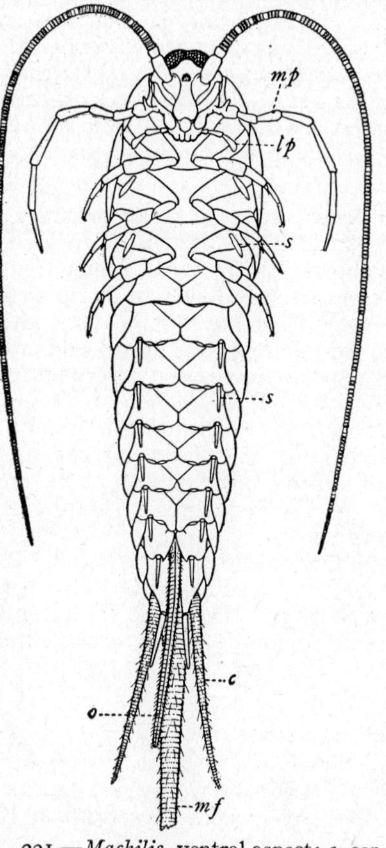

The members of this order are known as bristle-tails, a name suggested by the presence, in most of them, of either two or three many-jointed filiform appendages at the caudal end of the body (Fig. 221, c, and mf). The paired caudal appendages are the cerci; the median one, when three are present, is the median caudal filament, a prolongation of the eleventh abdominal segment. In *Jāpyx* (Fig. 222), the cerci are not jointed but are strong, curved appendages, resembling the forceps of earwigs.

The bristle-tails are most often found under stones and other objects lying on the ground; but some species live in houses. While most species prefer cool situations, there is one, the fire-brat, that frequents warm ones, about fireplaces and in bakehouses. The antennæ are long and many-jointed. In the Machilidæ

Fig. 221.—*Machilis*, ventral aspect: c, cercus; lp, labial palpus; mf, median caudal filament: mp, maxillary palpus; o, ovipositor; s, s, styli.

*Thysanūra: *thysanos* (θυσάνος), a tassel; *oura* (οὐρά), the tail.

(*Machilis*), the eyes are very perfect; for this reason, they are used in Chapter III to illustrate the structure of the compound eyes of insects. In all other Apterygota they are more or less degenerate or are lost entirely. In the Lepismatidæ (*Lepisma*), the degeneration of the eyes has progressed far, they being reduced to a group of a dozen ommatidia, on each side of the head. In the Campodeidæ and the Japygidæ, the eyes have disappeared. The mouth-parts are formed for chewing; those of *Machilis* will serve to illustrate their form. The mandibles are elongate with a toothed apex and a sub-apical projection terminated by a grinding surface (Fig. 223, A); the paragnatha are comparatively well developed (Fig. 224); on the outer edge of each there is a small lobe, which Carpenter ('03), who regarded the organs as true appendages, believed to be a vestigial palpus, and at the tip there are two distinct lobes, which this author homologized with the galea and the lacinia of a typical maxilla; the maxillæ (Fig. 223, B) bear prominent palpi.

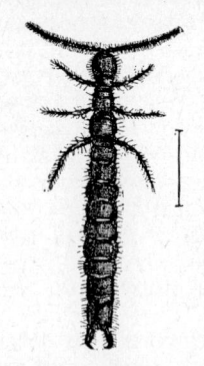

Fig. 222.—*Japyx solifugus*. (After Lubbock.)

In the Campodeidæ and the Japygidæ, the jaws are apparently sunk in the head. This condition is due to their being overgrown by folds of the genæ. In the Machilidæ and the Lepismatidæ the jaws are not overgrown; these two families are known, on this account, as the Ectotrophi or Ectotrophous Thysanura; while the Campodeidæ and the Japygidæ are grouped together as the Entotrophi or Entotrophous Thysanura. The overgrowing of the mouth-parts by folds of the genæ is characteristic of the Collembola also and is discussed more fully in the next chapter.

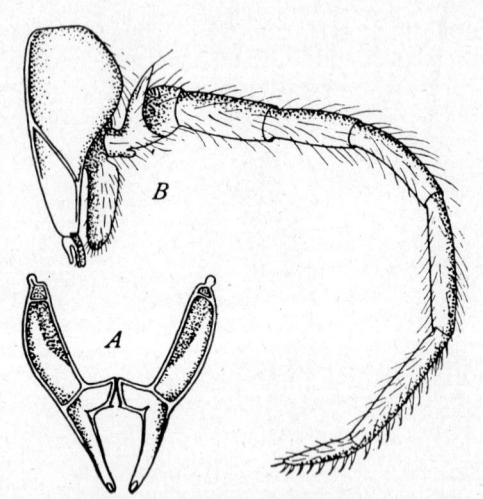

The three thoracic segments are distinctly separate. There is nothing in the structure of the thorax to indicate that these insects have descended from winged ancestors. The three pairs of legs are well developed. In the genus *Machilis* the coxæ of the second and third pairs of legs each bears a stylus (Fig. 221, *s*).

Fig. 223.—A, mandibles of *Machilis;* B, maxilla of *Machilis*. (After Oudemans.)

The abdomen consists of eleven segments. The eleventh segment bears the cerci, which are filiform and many-jointed except in the Japygidæ, where they are forceps-like. In the Machilidæ and the Lepismatidæ the eleventh abdominal segment bears a long, many-jointed median caudal filament; styli and eversible ventral sacs are also usually present; these vary in number in different genera.

The styli are slender appendages (Fig. 221, s). Each stylus consists of two segments, a very short basal one and a much longer terminal one. The maximum number of styli is found in *Machilis* (Fig. 221), where they are borne by the second and third thoracic legs and the second to the ninth abdominal segments. In *Lepisma* there are only three pairs; these are borne by the seventh, eighth, and ninth abdominal segments.

The abdominal styli are borne by large plates, one on each side of the ventral aspect of each abdominal segment. These plates are termed *coxites*, as they are believed to be flattened coxæ of abdominal legs which have otherwise disappeared.

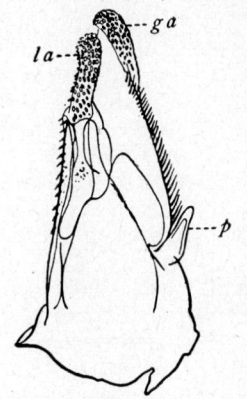

Fig. 224.—One of the paragnatha of *Machilis*. (After Carpenter.)

A result of the large size and position of the coxites is a reduction in the size of the sternum in the abdominal segments. This is well shown in *Machilis* (Fig. 221); in the first seven abdominal segments, there is in each a median triangular sclerite; this is the sternum; in the eighth and ninth segments no sternum is visible.

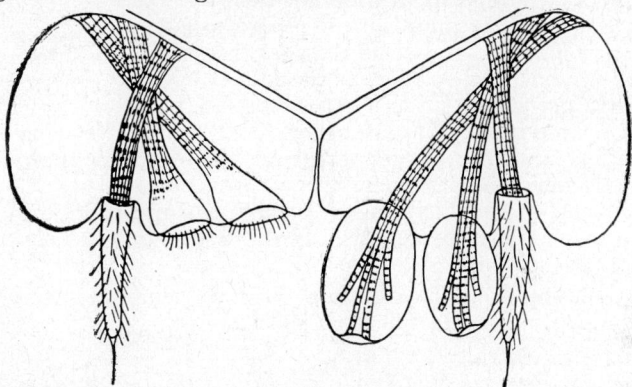

Fig. 225.—Cross-section of an abdominal segment of *Machilis* showing the styli and the ventral sacs. The ventral sacs of the left side are retracted; those of the right side, expanded. (After Oudemans.)

In the families Machilidæ and Lepismatidæ the females have an ovipositor, which consists of two pairs of filiform gonapophyses arising from between the coxites of the eighth and ninth abdominal segments respectively.

The ventral sacs are sac-like expansions of the wall of the coxites which can be everted, probably by blood-pressure, and are withdrawn into the cavity of the coxite by muscles (Fig. 225). In Figure 221, the openings into the retracted ventral sacs are represented; there is one pair in the first abdominal segment; two pairs in each of the four following segments; and a single pair each in the seventh and eighth abdominal segments. In *Lepisma* the ventral sacs are wanting. The function of the ventral sacs has not been definitely determined; but it seems probable that they are blood-gills.

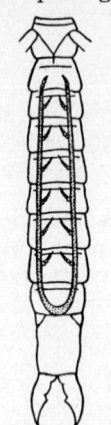

The presence in the Thysanura of styli and of ventral sacs, which are evidently homologous with those of the Symphyla, is an indication of the primitive condition of these insects. The generalized form of the reproductive organs of the Thysanura is another indication of this. In *Japyx* the ovarian tubes have a metameric arrangement (Fig. 226); and in *Machilis* (Fig. 227) we find an intermediate form between a metameric arrangement of the ovarian tubes and a compact ovary. These facts, and especially the presence of styli and ventral sacs, are opposed to the view held by some writers that the Thysanura are degenerate instead of primitive insects. It is true that degenerate features are present in the order, as the loss of eyes in *Japyx* and *Campodea;* but this loss is correlated with the life of these insects in dark places, like the loss of eyes in certain cave-beetles, and is not important in the determination of the zoological position of the order.

Fig. 226.—Ovary of *Japyx*. (After Grassi.)

Fig. 227.—Ovary of *Machilis: c*, coxite of the eighth abdominal segment; *s*, stylus; *o*, ovipositor. (After Oudemans.)

The young of the Thysanura resemble the adults in form, there being no marked metamorphosis. In *Campodea* and *Japyx* the molt is partial (Grassi '89).

This is a small order; less than twenty American species have been described. The classification is as follows:

Suborder I. ECTOGNATHA

Body usually clothed with scales; mouthparts outside of the head, not overgrown with folds of the genae; caudal end of abdomen with three long, filiform appendages; compound eyes present.

FAMILY I, MACHILIDAE. The abdominal tergites reflexed to the under surface so as to form an imbrication covering the sides of the coxites (Fig. 221). Compound eyes large and contiguous. Prothorax smaller than the mesothorax. Middle and hind legs with styli. Saltatorial insects.

This family is represented by the genus *Măchilis*, of which several species occur in North America. These insects are found in heaps of stones and in other concealed places; they are very active and leap with agility when disturbed. They are about 12 mm. in length.

FAMILY 2, LEPISMATIDAE. Abdominal tergites not covering the sides of the coxites. Eyes small and distant. Prothorax as large as or larger than the mesothorax. Middle and hind legs without styli. Not saltatorial insects.

The best-known representative of this family is the silverfish or fish-moth *Lepisma saccharina* (Fig. 228). It is silvery white with a yellowish tinge about the antennæ and legs and measures about

Fig. 228.—*Lepisma saccharina*. (After Lubbock.)

8 mm. in length. It is often a troublesome pest in laundries, libraries, and museums, as it injures starched clothes, the bindings of books, labels, and other things on which paste or glue is used. The popular names were suggested by the clothing of scales with which the body is covered.

Another common representative of this family is the fire-brat, *Thermōbia domĕstica*. This species resembles the fish-moth in general appearance except that it has dusky markings on its upper surface. It is remarkable for frequenting warm and even hot places about ovens, ranges, and fireplaces.

Suborder II. ENTOGNATHA*

Body not clothed with scales; mouthparts within the head, being overgrown by folds of the genæ; median caudal filament wanting; compound eyes absent.

Family 3, Campodeidæ. Cerci filiform, long and many-jointed; first abdominal segment without styli.

The best-known member of this family is *Campōdea staphylīnus* (Fig. 229). It lives in damp places under stones, fallen trees, or in rotten wood and leaves. It is a very delicate, small, white insect, about 6 mm. in length. It has on the first abdominal segment a pair of appendages which occupy a position corresponding to that of the thoracic legs and each consists of two or three segments.

Family 4, Projapygidae. Cerci short, rather stout, few-jointed; first abdominal segment with styli.

This family is represented by the genera *Projapyx* and *Anajapyx*.

The species *A. vesiculosus* described by Silvestri may be considered representative.

Family 5, Japygidae. Cerci forceps-like; styli present on first abdominal segment.

This family is represented by the genus *Jāpyx*, of which two species have been found in this country. These insects can be recognized by the forceps-like form of the cerci (Fig. 222). They are small, delicate, uncommon insects found under stones.

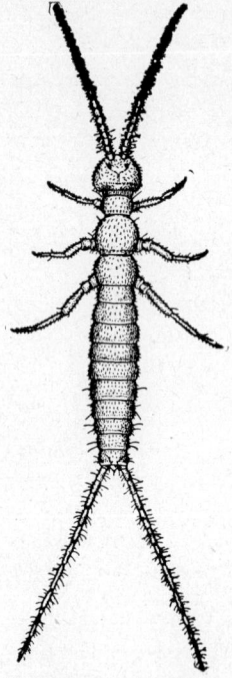

Fig. 229. — *Campodea staphylinus*. (After Lubbock.)

---

*This suborder is raised to the rank of an order, Diplura, by Silvestri. Other systematists believe it should rank as a definite order.

ORDER COLLEMBOLA*

The Spring-Tails

The members of this order resemble the Thysanura in being wingless insects in which the wingless condition is believed to be a primitive one, there being no indication that they have descended from winged ancestors, and in that the adult insects resemble the young in form. They differ

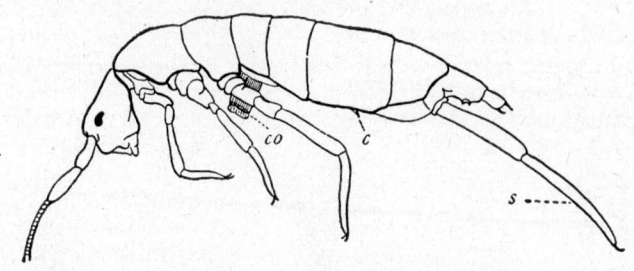

Fig. 230.—Side view of *Tomocerus plumbens*: *co*, collophore; *c*, catch; *s* spring. (After Willem.)

from the Thysanura as follows: the abdominal segments are reduced in number, there being only six of them; the first abdominal segment bears a ventral tube, the collophore, furnished with a pair of eversible sacs which assist the insects in walking on smooth surfaces; the fourth abdominal segment usually bears a pair of appendages, which constitute a springing organ; and the third abdominal segment usually bears a short pair of appendages, the catch, which hold the spring when it is folded under the abdomen.

The common name *spring-tails* has been applied to these insects on account of the caudal springing organ that is possessed by most members of the order. The spring-tails are minute insects, often of microscopic size and rarely as large as 5 mm. in length. Most of the species live on decaying matter. These insects are common under stones and decayed leaves and wood, in the chinks and crevices of bark, among moss, and on herbage in damp places. Sometimes they occur abundantly in winter on the surface

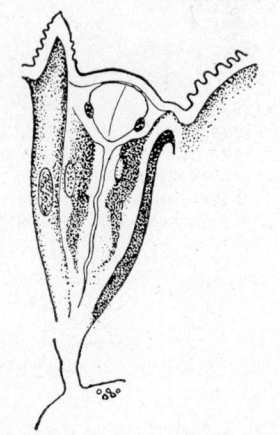

Fig. 231.—An ommatidium of *Podura aquatica*. (After Willem.)

*Collĕmbola: *colla* (κόλλα), glue; *embolon* (ἔμβολον), a bolt, bar;—from their collophores.

of snow, where they appear as minute black specks, which spring away on either side from our feet as we walk; and some species collect in great numbers on the surface of standing water. Several species are known to be photogenic.

The body consists of the head, three thoracic segments, and six abdominal segments (Fig. 230). The prothorax is usually small and in several genera is overlapped by the tergum of the mesothorax; in the Sminthuridæ the body-segments are more or less fused together. The structure of the abdomen is remarkable, as it consists of only six segments; there is no indication of the manner in which the reduction of the number of segments has taken place. The anus is at the caudal end of the body; the genital opening is on a small papilla on the fifth abdominal segment.

The antennæ consist of from four to six segments, usually of four. They vary greatly in their comparative length; in some genera the last segment or the last two segments are divided into many rings or subsegments (Fig. 230).

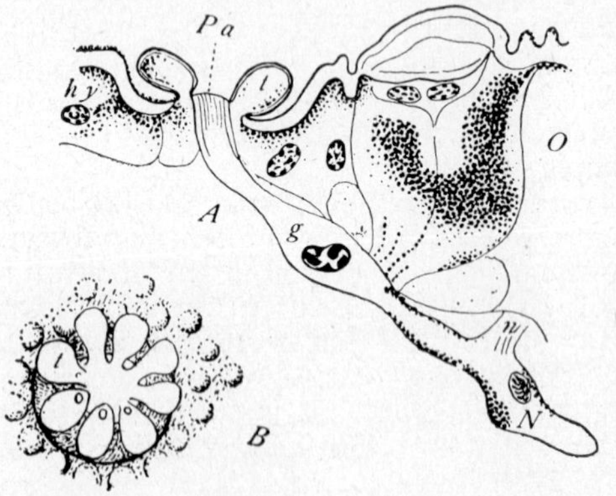

The eyes of the Collembola are commonly described as a group of eight, or fewer, distinct simple eyes on each side of the head. But these so-called simple eyes are not ocelli; they are more or less degenerate ommatidia, each group being the vestige of a compound eye. In *Podura aquatica*, these eyes, as figured by Willem ('00), are clearly ommatidia of the eucone type (Fig. 231). In some other Collembola, as in *Anurida maritima* (Fig. 232, O), the reduction of the ommatidia has progressed so far that they present the appearance of ocelli; and in still others the eyes are lost entirely. Primary ocelli have not been found in the Collembola.

Fig. 232.—A, longitudinal section of an ommatidium and of the postantennal organ of *Anurida maritima*; B, a surface view of the postantennal organ. (After Willem.) O, ommatidium; Pa, postantennal organ; hy, hypodermal cells; N, optic nerve; n, branch of the optic nerve; t, t, tubercles surrounding the postantennal organ; g, nerve-end-cell of the postantennal organ. (After Willem.)

The mouth-parts are typically mandibulate; the jaws consisting of a pair each of mandibles, paragnatha, and maxillæ. The paragnatha of *Orchesella cincta* were described by Folsom ('99); and those of *Anurida maritima* by the same writer ('00). These organs were termed the *superlinguæ* by Folsom.

One of the most striking characteristics of the Collembola is that the jaws are apparently retracted into the cavity of the head so that only their tips are visible. But it has been shown by Folsom ('00), who studied the development of the mouth-parts of *Anurida maritima*, that, strictly speaking, the jaws are not "retracted," as is usually stated, but are overgrown by the genæ. In an early embryonic stage, a downward projection of the gena appears on each side of the head, and these "mouth-folds" become larger and larger in successive stages until the condition seen in the fully developed insect is reached.

The development of mouth-folds is not restricted to the Collembola, but occurs also in the Entotrophous Thysanura, and to a less marked extent in many of the Pterygota, especially in some Orthoptera, where the gena of each side is prolonged into a small, but distinct, flat fold over the base of the mandible.

In some of the Poduridæ the mouth-parts are fitted for piercing and sucking, the mandibles and maxillæ being styliform and projecting in a conspicuous cone.

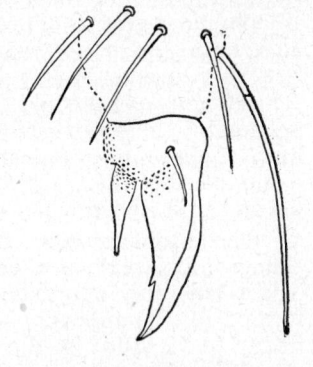

Fig. 233.—Hind foot of *Achorutes maturus*. (After Folsom.)

In some of the Collembola there is a sense organ situated between the base of the antenna and the ocular field; this is known as the *postantennal organ;* its presence or absence and its form when present afford characters used in the description of these insects. In its simplest form it is a claviform hyaline tubercle (*Sminthurus*). A more complicated type is that of *Anurida maritima*, which has been figured by Willem ('oo). In Figure 232, *Pa* represents a longitudinal section of this organ. It is a nerve-end-cell, branching from the optic nerve and extending to the surface of the body, where it is covered by a very thin cuticular layer. It is protected by a ring of tubercles (*t, t*), two of which are shown in the sectional view (A) and eight in the surface view (B). The function of this organ has not been determined; it has been suggested that it is an organ of smell.

The legs of the Collembola consist each of five segments, which correspond to the five principal divisions of the legs of the higher insects. Willem ('oo) considers the two antecoxal pieces as segments of the legs and consequently states that the legs are composed of seven segments. The tarsi in most genera bear two claws, an outer, larger one, the *unguis*, and an inner, smaller one, the *unguiculus;* these claws are apposable (Fig. 233); in some genera the inner claw is wanting.

One of the most characteristic features of the Collembola is the collophore, or ventral tube, which is situated on the ventral aspect of the first abdominal segment (Fig. 230, *co*). This organ varies greatly in form in the different genera; in some it is a simple tubercle, divided into two halves by a central slit; in others it is enlarged and becomes a jointed tube divided at its free end into two lobes. The collophore bears at its extremity a pair of eversible sacs through the walls of which exude a viscid fluid. By means of this organ these insects are enabled to cling to the lower surface of smooth objects. The collophore is developed from a pair of appendages, which in the course of their development become fused together at their base.

The third abdominal segment usually bears a pair of short appendages, whose basal segments are fused; this is the *tenaculum*, or catch (Fig. 230, c), which holds the spring when it is folded under the abdomen.

The spring or *furcula* (Fig. 230, s) is formed by the appendages of the fourth abdominal segment which are united at the base but separate distally. These appendages are three-jointed. The united basal segment is termed the *manubrium* (Fig. 234, *ma*); the intermediate segments, the *dentes* (Fig. 234, *d*); and the terminal segments, the *mucrones* (Fig. 234, *mu*).

In the Entomobryidæ the furcula appears to be formed by the appendages of the fifth abdominal segment; but a study of the muscles that move it shows that it really pertains to the fourth segment. In some genera of the Poduridæ the furcula is wanting.

The order Collembola includes two quite distinct types of insects; in one of these types the body is elongate with distinct segmentation; in the other the body is shortened, the abdomen globose and its segments in part fused. Based on this distinction the order is divided into two suborders as follows:

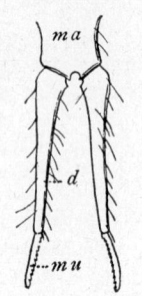

Fig. 234.—The furcula of *Papirius: ma*, manubrium; *d*, left dens; *mu*, left mucro. (After Lubbock.)

A. Body elongate. ....Suborder Arthropleona.
AA. Body globose..........................Suborder Symphypleona.

Suborder I. ARTHROPLEONA*

Body elongate with distinct segmentation, rarely with the last two or three segments of abdomen partially fused; tracheae absent.

Family 1, Poduridae. Furcula, when present, clearly appended to the fourth abdominal segment; prothorax well developed; cuticula usually granulated.

Among the better-known members of this family are the following: The "Snow-flea," *Achorūtes nivĭcola*, which occurs abundantly in winter on the surface of snow (Fig. 235); this species is also known as *Achorūtes sociālis*. *Achorūtes armātus* is often found on fungi.

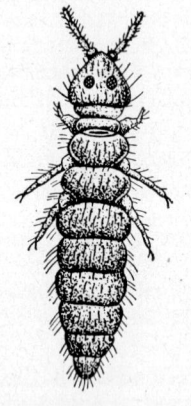

Fig. 235.—The snow-flea, *Achorutes nivicola*. (After Folsom.)

*Arthropleona: *arthron* (ἄρθον), a joint; *pleon*, a crustacean's abdomen.

Anūrida marĭtima occurs abundantly on the seashore, chiefly between tide marks; several important embryological and anatomical monographs have been published regarding this species. *Podūra aquătica* is one of the most abundant members of the Collembola; it occurs on the surface of standing water on the margins of ponds and streams.

FAMILY 2, ENTOMOBRYIDAE. Furcula present and apparently appended to the fifth abdominal segment; prothorax reduced and cuticula not granulated.

This is the largest family of the Collembola, containing many genera and species. In some genera the body is clothed with scales. To this family belongs the genus *Orchesĕlla*, the only genus in the Collembola in which the antennæ consist of six segments.

SUBORDER II. SYMPHYPLEONA*

Body shortened, subglobular in shape with segments of body, except the last two, fused closely together and segmentation mostly obliterated; tracheæ present in some genera.

FAMILY 3, NEELIDAE. Antennae short and stout; thorax large and longer than abdomen. The principal genera are *Neelus* and *Neelides*.

FAMILY 4, SMINTHURIDAE. Antennæ long and slender; thorax shorter than abdomen. The principal genera *Sminthurus* and *Papirius*.

In *Sminthurus*, tracheæ are present; in the other genera they are absent or extremely vestigial. The presence of tracheæ in *Sminthurus* enables these insects to live in drier situations than can other Collembola. The "garden-flea" *Sminthūrus hortĕnsis* is found upon the leaves of young cabbage, turnip, cucumber, and various other plants.

Fig. 236.—*Papirius fuscus.* (After Lubbock.)

---

*Symphypleōna; *symphyo*, to grow together; *pleon*, a crustacean's abdomen.

Subclass II. PTERYGOTA

Winged Insects

The members of this subclass are winged; or, if without wings, they have had winged ancestors and this is an acquired condition. The mouthparts vary; in some they are sucking, in others chewing. The metamorphosis varies from slight to gradual to complete.

Much the greater number of species of insects belong to this subclass and much the greater number of them possess wings, for example, the beetles, flies, wasps, bees, and many others. Many aphids, all fleas, lice, worker ants, female scale insects, and some others have no wings. The wingless condition of these forms, however, is an acquired one, for the evidence is clear that they have descended from winged ancestors. The subclass Pterygota contains all of the remaining twenty-three orders discussed in this book.

CHAPTER VIII

ORDER ORTHOPTERA*

Grasshoppers, Crickets, Cockroaches, and others

The winged members of this order have two pairs of wings; the fore wings are more or less thickened, but have a distinct venation; the hind wings are folded in plaits like a fan when at rest; there are many forms in which the wings are vestigial or even wanting. The mouth-parts are formed for chewing. The metamorphosis is gradual (*paurometabolous*); the nymphs are terrestrial.

The order Orthoptera includes some of the very common and best-known insects. The most familiar representatives are the long-horned grasshoppers, locusts, crickets, katydids, and cockroaches.

With the exception of a single family, the Mantidæ, the members of this order are as a rule injurious to vegetation; and many species are quite apt to multiply to such an extent that their destruction of plant life becomes of great economic importance.

The two pairs of wings of the Orthoptera differ in structure. The front wings are leathery or parchment-like, forming covers for the more delicate hind wings. These wing-covers have received the special name *tegmina*. The tegmina usually overlap, at least at the tips, when at rest. The hind wings are thinner than the tegmina and usually have a broadly expanded anal area, which is folded in plaits like a fan when at rest. Many Orthoptera have vestigial wings, and many are wingless. In the males of the Saltatorial Orthoptera, the Locustidæ, the Tettigoniidæ, and the Gryllidæ, musical organs have been formed by modifications of certain parts of the wings; these have been described in Chapter II.

The mouth-parts are of the mandibulate type, that is, they are formed for chewing. The mouth-parts of a locust are figured on page 42.

In the Orthoptera the metamorphosis is gradual, paurometabolous. In the case of those species in which the wings of the adult are either vestigial or wanting, the adults resemble very greatly immature insects. It is often important to determine whether a short-winged specimen is an adult or not. Fortunately this determination can usually be made with ease with the Saltatorial Orthoptera, the Locustidæ, the Tettigoniidæ, and the Gryllidæ. In these three families the wing-pads of the nymphs are inverted, as shown by the curving down of the extremities of the wing-veins, instead of up as with the adult; and the rudimentary hind-wings are outside of the tegmina, instead of beneath them. The development of the wings of a locust is described in Chapter IV, p. 175.

---

*Orthŏptera: *orthos* (ὀρθός), straight; *pteron* (πτερόν), a wing.

The segmentation of the abdomen and the development and structure of the genitalia or gonapophyses in the jumping Orthoptera are of especial interest; as, on account of the generalized condition of these parts in these insects, they can serve as a type with which the corresponding parts in more specialized insects can be compared. In some members of this group of families all of the abdominal segments are preserved more or less distinct, and in nearly all of them the genitalia are well-developed.*

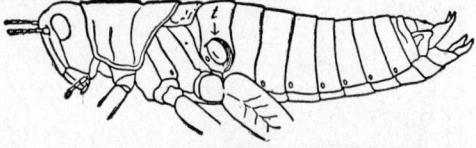

The segmentation of the abdomen can be seen best on the dorsal aspect of this region; for in some cases the tergum of a segment is well-preserved while the sternum is vestigial. Figure 237 represents a side view of a female locust with the wings removed in order to show the segmentation of the abdomen. The first eight segments of the abdomen of this insect are very distinct; but the caudal segments are much less so. Figure 238 represents the caudal part of the abdomen of the same insect more enlarged, in order to facilitate the lettering of the parts.

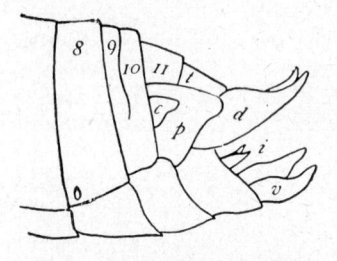

Fig. 237.—Side view of a locust with the wings removed: t, tympanum.

In this insect the eighth abdominal tergum resembles the preceding ones. The ninth and tenth abdominal terga are shorter and are joined together on each side; but in many other jumping Orthoptera these terga are not thus united. Caudad of the tenth abdominal tergum there is a shield-shaped part, which is commonly known as the *supra-anal plate;* this plate is divided into two sclerites by a transverse suture; the first of these sclerites is believed to be the tergum of the eleventh abdominal segment, and the other the telson (Fig. 238, t). Thus all of the abdominal segments are preserved, in part at least, in this insect.

Fig. 238.—Side view of the caudal end of the abdomen of a female locust: *8, 9, 10, 11,* the tergites of the eighth, ninth, tenth, and eleventh abdominal segments; *t,* telson; *p,* podical plate; *c,* cercus; *d, i, v,* dorsal, inner, and ventral valves of the ovipositor.

The last two abdominal segments, the eleventh and the telson, are even more distinctly preserved in the early instars of some orthopterous insects than they are in the adult (Fig. 239). In many adult Orthoptera there is no suture between the eleventh tergum and the telson.

On each side of the body, in the angle between the supra-anal plate and the lateral part of the tenth tergum, there is a triangular sclerite (Fig. 238, p); this pair of sclerites has long been

---

*The genitalia are vestigial in *Tridactylus* and are entirely wanting in *Gryllotalpa*. In these genera the reduction or loss of the genitalia is probably correlated with the subterranean life of these insects, they having no need for an ovipositor.

known as the *podical plates;* but they have recently been named the *paraprocts* because they are situated one on each side of the anus. They are the sternum of the eleventh abdominal segment, which is divided on the midventral line, to admit of the expansion of the posterior end of the alimentary canal during defecation.

In this insect the cerci (Fig. 238, *c*) project from beneath the caudal border of the tenth tergum; they appear, therefore, to be appendages of the tenth abdominal segment; but it is believed that in all insects where cerci are present they are appendages of the eleventh abdominal segment. This, for example, is obviously the case in the Plecoptera (Fig. 240). The homology of the paraprocts is also well shown in this figure.

Fig. 239.—Caudal segments of a nymph of a female locust, dorsal aspect: *11*, eleventh abdominal segment; *t*, telson; *c*, cercus.

The ovipositor consists of three pairs of processes or gonapophyses; these are termed the valves or *valvulæ* of the ovipositor; they are distinguished as the dorsal, ventral, and inner valvulæ, respectively. In the locust the dorsal valvulæ (Fig. 238, *d*) and the ventral valvulæ (Fig. 238, *v*) are strong, curved, and pointed pieces; the inner valvulæ (Fig. 238, *i*) are much smaller.

The relation of the gonapophyses to the segments of the abdomen can be seen more clearly in the female of *Ceuthophilus* (Fig. 241). The ventral valvulæ arise from the posterior margin of the eighth sternum and the dorsal and inner valvulæ arise from the ninth sternum. These relations can be seen even more clearly in very young nymphs where the rudiments of the gonapophyses are mere tubercles, one pair on the hind margin of the eighth abdominal sternum and two pairs on the ninth sternum (Fig. 242).

In the male, as in the female, the form of the caudal end of the abdomen and its appendages differs greatly in different members of this order. Space can be taken here to illustrate these parts in only a single species. For detailed accounts of these parts in other members of this order, special papers on this subject should be consulted. Among the more recent and generally available of these are those of

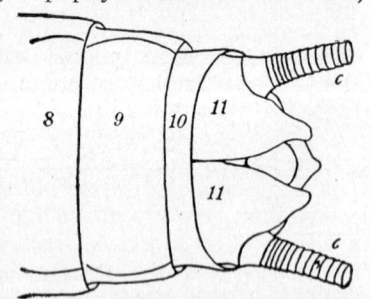

Fig. 240.—End of abdomen of *Pteronarcys dorsata*, female, ventral view: *11, 11*, the divided sternum of the eleventh abdominal segment, the podical plates; *c, c*, basal parts of the cerci.

Crampton ('18) and Walker ('19 and '22 b). These papers include references to the very extended literature on this subject.

Figures 243 and 244 represent the caudal end of the abdomen of the male of the Carolina locust, *Dissosteira carolina*. In this insect the ninth and tenth terga are joined together on each side (Fig. 244) and the eleventh tergum is separated from the apical part of the supra-anal plate (Fig. 243, *s*) by a distinct suture. The ninth sternum is large, is turned upward behind, and bears a large conical part (Fig. 244, *cx*) termed the *coxale*, which is believed to be united coxites of the ninth segment.

There are two genera of rare and remarkable insects, each of which has been placed in the Orthoptera by some writers and each of which

Fig. 241.—Side view of end of abdomen of *Ceuthophilus lapidicola*: *7, 8, 9, 10*, above, tergites of the seventh to the tenth abdominal segments; *7, 8*, below, sternites of the seventh and eighth abdominal segments; *b*, basal segment of the ventral valve of the ovipositor; *c*, cercus; *p*, podical plate; *d, i, v*, dorsal, inner, and ventral valves of the ovipositor. (After Walker.)

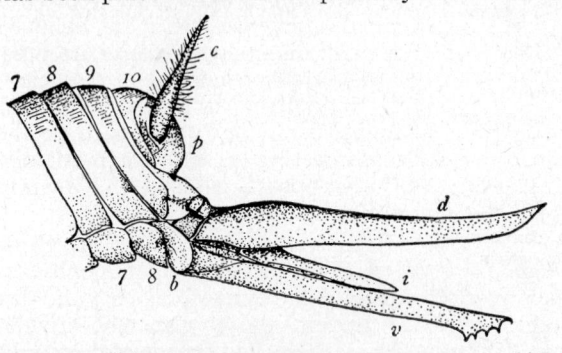

Fig. 242.—Ventral view of end of abdomen of young nymph of *Conocephalus fasciatus*. (After Walker.)

is regarded by others as constituting a separate order; these are *Grylloblatta* and *Hemimerus*. These genera are briefly discussed at the close of this chapter.

Leaving out of account the two genera named above, the order Orthoptera includes only six families, all of which are represented in the United States. These families can be separated by the following table.*

TABLE OF FAMILIES OF ORTHOPTERA

A. Hind femora fitted for jumping, i. e., very much stouter or very much longer, or both stouter and longer, than the middle femora; organs of flight of immature forms inverted; stridulating insects. (The SALTATORIAL ORTHOPTERA.)

---

*The limits assigned to the order Orthoptera in this work are those that have been commonly recognized for a long period and are those adopted in recently published manuals treating of this order, except that in some of them the Dermaptera is included in the Orthoptera. But Handlirsch ('08) in his great work on fossil insects proposed a new classification of insects, which differs greatly from the classification adopted here. In this classification the families Blattidæ, Mantidæ, and Phasmidæ are removed from the Orthoptera and **each is** made to constitute a distinct order.

B. Antennæ long and setaceous, except in the mole-crickets and sand-crickets; tarsi three- or four-jointed; organs of hearing situated in the fore tibiæ; ovipositor elongate, except in the mole-crickets and sand-crickets, with its parts compact.
  C. Tarsi four-jointed; ovipositor, when exserted, forming a strongly compressed, sword-shaped blade. p. 234..... TETTIGONIIDÆ
  CC. Tarsi usually three-jointed, except in the pigmy mole-crickets where they are reduced; ovipositor, when exserted, forming a nearly cylindrical, straight, or occasionally upcurved needle, except in the Trigonidiinæ. p. 242.. GRYLLIDÆ.
 BB. Antennæ short; tarsi three-jointed; organs of hearing situated in the first abdominal segment; ovipositor short, with its parts separate. p. 252... LOCUSTIDÆ
AA. Hind femora closely resembling those of the other legs, and scarcely if at all stouter or longer than the other femora, *i. e.*, not fitted for jumping; organs of flight in a normal position when immature; stridulating organs not developed.
 B. Body elongate; head free; pronotum elongate; legs slender, rounded; cerci jointed or without joints; walking insects.
  C. Front legs simple; cerci without joints. p. 260......... PHASMIDÆ
  CC. Front legs fitted for grasping; cerci jointed. p. 262... MANTIDÆ
 BB. Body oval, depressed; head wholly or almost wholly withdrawn beneath the pronotum; pronotum shield-like, transverse; legs compressed; cerci jointed; rapidly running insects. p. 263.................. BLATTIDÆ

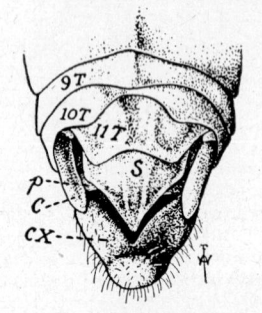

Fig. 243.—Dorsal view of end of abdomen of *Dissosteira carolina*, male: *9T, 10T, 11T*, ninth, tenth, and eleventh terga; *s*, supra-anal plate; *p*, podical plate, *c*, cercus; *cx*, coxale.

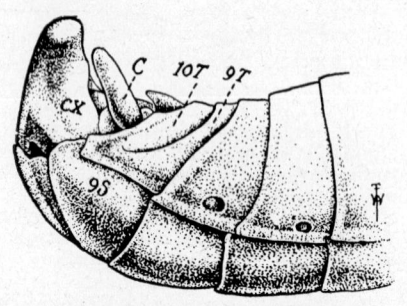

Fig. 244.—Side view of end of abdomen of *Dissosteira carolina*, male; lettering as in Figure 243.

FAMILY TETTIGONIIDÆ

THE LOCUSTIDAE OF AUTHORS*

The Long-horned Grasshoppers

To this family belong the most attractive in appearance of our common Orthoptera. In many of them the wings are graceful in

---

*The name Locustidæ has been commonly applied to this family. This usage is the result of an erroneous application of the generic name *Locusta* to certain members of this family. The insects of the genus *Locusta*, established by Linnæus, and other insects commonly known as locusts, are members of the family to which the common name *short-horned grasshoppers* is applied and which is properly termed the *Locustidæ*.

The Tettigoniidæ is the Phasgonuridæ of Kirby's catalogue.

form and delicate in color, and the antennæ are exceedingly long and slender, looking more like ornaments than like organs of practical use.

These beautiful creatures are much less frequently seen than are the crickets and locusts because of their protective green color, which renders them inconspicuous in their haunts among foliage or on the blades of grass. Their presence is most often indicated by the chirping of the males.

The long-horned grasshoppers are those jumping Orthoptera with long, slender antennæ, longer than the body, in which the tarsi are four-jointed and the ovipositor is sword-shaped.

The tegmina of the males are furnished, in nearly all winged species, with stridulating organs; but these occupy a much smaller part of the tegmina than with the crickets. The six plates of which the ovipositor is composed are closely united so that this organ has the appearance of a single sword-shaped blade.

The different members of the Tettigoniidæ exhibit a great variety of methods of oviposition; some lay their eggs in the ground; some in the pith of twigs; some singly in the edges of leaves; some in rows on leaves and stems; and others between the root-leaves and stems of various plants.

The Tettigoniidæ found in America north of Mexico represent eight subfamilies; these can be separated by the following table, which is based on one by Scudder ('97).

A. Body generally winged; tarsi more or less depressed.
  B. Fore tibiæ furnished with auditory tympana; fore wings of male, when present, furnished with stridulating organs.
    C. First two segments of the tarsi without a lateral groove; the two series of spines on the hind side of the posterior tibiæ continued to the apex. p. 236.
                                                             PHANEROPTERINÆ
    CC. First two segments of the tarsi with a lateral groove; one or both of the two series of spines on the hind side of the posterior tibiæ not continued to the apex.
      D. Fore tibiæ without apical spines above.
        E. The apex of the vertex short, crowded by the prominent antennary fossæ; pronotum crossed by two distinct sutures. p. 238......
                                                             PSEUDOPHYLLINÆ
        EE. The apex of the vertex extended and free from the not prominent antennary fossæ; pronotum without transverse sutures, or with only one.
          F. Fore and middle femora unarmed beneath; the vertex terminating in a rounded tubercle, which is hollowed out on the sides. p. 238.
                                                               CONOCEPHALINÆ
          FF. Fore and middle femora spined beneath, the vertex produced forward into a long sharp cone. p. 239.......COPIPHORINÆ
      DD. Fore tibiæ with an apical spine above on the outer side; usually wingless or with vestigial wings. p. 239............DECTICINÆ
  BB. Fore tibiæ without auditory tympana; fore wings of male, when present, without stridulating organs. p. 240....................GRYLLACRINÆ
AA. Body usually wingless; tarsi distinctly compressed.
  B. Tarsi without pulvilli; inserting angle of the hind femora situated on the inner side. p. 241.................................RHAPHIDOPHORINÆ
  BB. Tarsi provided with pulvilli; inserting angle of the hind femora situated on the outer side. p. 242............................STENOPELMATINÆ

Some of the more common and better-known representatives of these families are referred to below. To save space the distinguishing characteristics of the subfamilies are not repeated; these are indicated in the table above.

Subfamily PHANEROPTERINÆ
The False Katydids

Fig. 245.—*Microcentrum rhombifolium* and its eggs.

To this subfamily belong certain long-horned grasshoppers that have broad leaf-like wings and arboreal habits. In these respects they resemble the well-known katydid whose strident call suggested the popular name. Several of these species have received popular names in which the word *katydid* enters, as indicated below. These species may be termed collectively the false katydids; the true katydids constitute the next subfamily.

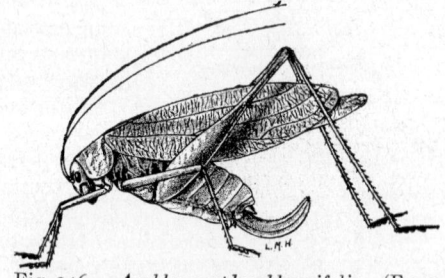

Fig. 246.—*Amblycorypha oblongifolia*. (From Lugger.)

Blatchley ('20) describes twenty species and varieties of the false katydids that are found in northeastern America; these represent eight genera. Among our common species there are representatives of three genera; these can be separated as follows.

A. Tegmina broadened in the middle; the extreme point of the vertex much broader than the first segment of the antennæ.

B. Hind femora much shorter than the tegmina; ovipositor short and turned abruptly upward (Fig. 245). p. 237 .... MICROCENTRUM
BB. Hind femora but little if any shorter than the tegmina; ovipositor well developed, and curved gradually upward. p. 237 ...... AMBLYCORYPHA
AA. Tegmina of nearly equal breadth throughout; the extreme point of the vertex but little if any broader than the first segment of the antennæ. p. 237.
.... SCUDDERIA

Microcĕntrum.—Two species of this genus are found in the United States east of the Rocky Mountains; these are known as the angular-winged katydids. Figure 245 represents the female of the larger angular-winged katydid, *Microcĕntrum rhombifōlium*, and the remarkable way in which it deposits its eggs on leaves and twigs. In this species the slightly hollowed front of the pronotum has a very small central tooth, which is lacking in smaller species. The smaller angular-winged katydid, *Microcĕntrum retinĕrve*, is only slightly smaller than the larger one.

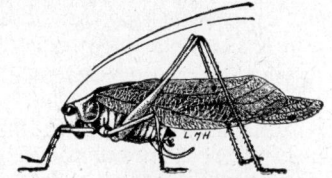

Fig. 247.—***Scudderia septentrionalis.*** (From Lugger.)

Amblycŏrypha.—The three most common species of the genus are the following: The oblong-winged katydid, *Amblycŏrypha oblongifōlia* (Fig. 246), is the largest of the three most common species. The tegmina measure from 34 to 37 mm. in length; the ovipositor is less serrate and less curved than in the next species. The round-winged katydid, *Amblycŏrypha rotundifōlia*, is a smaller species; the tegmina are not more than 30 mm. in length and are wide for their length, as indicated by the specific name; the ovipositor is quite broad, much curved, and roughly serrated. Uhler's katydid, *Amblycŏrypha ūhleri*, is our smallest species; the body measures from 14 to 16 mm. in length; the tegmina from 24 to 26 mm.; and the ovipositor about 8 mm.

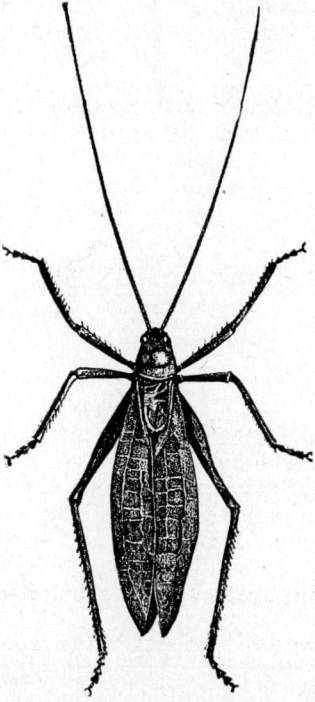

Scuddĕria.—Species of this genus are found throughout the United States and in Canada; but the greater number of our species are found east of the Great Plains. One species, *Scuddĕria mexicāna*, is found in California and Oregon. A common eastern species which may serve as an example of the insects of this genus, is the northern bush-katydid, *Scuddĕria septentrionālis*. Figure 247 represents the male of this species, natural size.

Fig. 248.—*Pterophylla camellifolia*. (After Harris.)

Subfamily PSEUDOPHYLLINÆ
The True Katydids

The best-known representative of this subfamily in the United States is the northern true katydid, *Pterophylla camellifōlia*. (Fig. 248). This insect is found throughout the United States east of the Rocky Mountains; but in the North it lives in colonies which occupy quite limited areas. This is the insect whose song suggested the popular name *katydid*. It differs from members of the preceding subfamily in having the hind wings shorter than the tegmina, and in having the tegmina very convex, so that it has an inflated appearance.

Subfamily CONOCEPHALINÆ
The Meadow Grasshoppers

From the middle of the summer to the autumn there can be found upon the grass in our meadows and moist pastures many light-green long-horned grasshoppers of various sizes; these, on account of the situations in which they are usually found, are termed *the meadow-grasshoppers*. Our common species represent only two genera; but each of these includes many species.

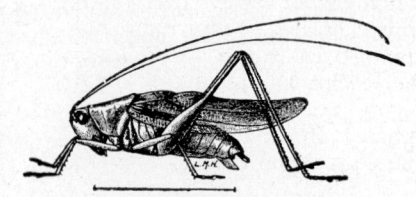

Fig. 249.—*Orchelimum vulgare*, male. (From Lugger.)

Orchĕlimum.—This genus includes the larger and stouter species of meadow grasshoppers; but they are of medium size compared with other Tettigoniidæ. In these the ovipositor is usually

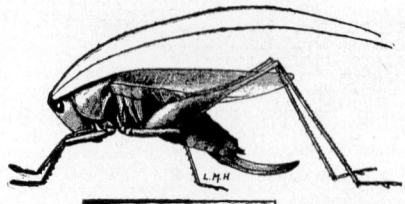

Fig. 250.—*Orchelimum vulgare*, female. (From Lugger.)

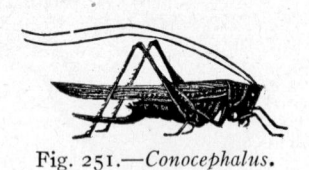

Fig. 251.—*Conocephalus*.

up-curved. Our most abundant species is the common meadow grasshopper, *Orchĕlimum vulgāre*. This is found from the Rocky Mountains to the Atlantic Coast. Figure 249 represents the male, natural size; and Figure 250, the female.

Conocĕphalus.—This genus comprises the smaller and slenderer species of this subfamily. In these the ovipositor is slender, and straight or slightly curved (Fig. 251). Until recently this genus has been generally known as *Xiphidium*.*

---

*It is unfortunate that according to the rules of nomenclature the name *Conocephalus* must be applied to this genus instead of to the typical genus of the next subfamily, now known as *Neoconocephalus*, with the result that the subfamily name Conocephalinæ is applied to the meadow grasshoppers instead of to the cone-headed grasshoppers.

Subfamily COPIPHORINÆ

The Cone-headed Grasshoppers

The cone-headed grasshoppers are so called because the vertex is prolonged forward and upward into a cone. These are much larger insects than the meadow grasshoppers and are found in trees as well as upon grass. This subfamily is represented in our fauna by four genera; but three of these are found only in the South. All of the northern species belong to the genus *Neoconocĕphalus*, of which eleven species occur in the United States. The most common species in the north, east of the Rocky Mountains, is the sword-bearer, *Neoconocephalus ĕnsiger*. Figure 252 represents the male of this species, and Figure 253 the female. Both sexes have very long wings, and the ovipositor of the female is remarkable for its length.

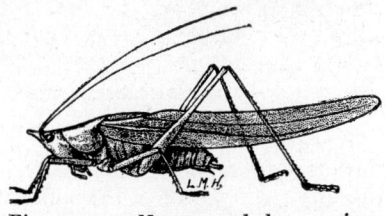

Fig. 252.—*Neoconocephalus ensiger*, male. (From Lugger.)

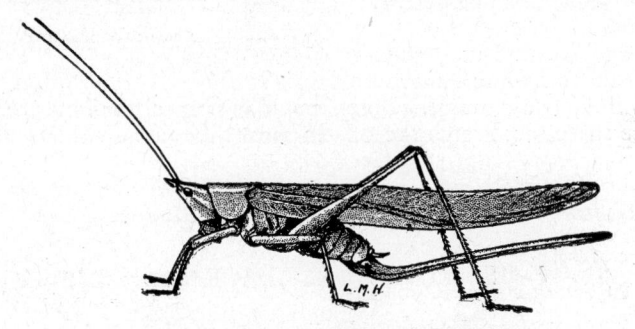

Fig. 253.—*Neoconocephalus ensiger*, female. (From Lugger.)

In most of the species of *Neoconocephalus* there are two distinct forms: one pea-green in color and the other of a brownish straw-color.

Subfamily DECTICINÆ

The Shield-backed Grasshoppers

A few members of this subfamily have well-developed wings; but in most species the wings are small, especially in the female, where they are sometimes even absent. Most of the species bear some resemblance to crickets. They present, however, a strange appearance, due to the pronotum extending backward over the rest of the thorax, like a sun-bonnet worn over the shoulders with the back side forward. It was the large size of the pronotum that suggested for the group the popular name *the shield-backed grasshoppers*.

These insects live in grassy fields or in open woods, where they hop about in exposed positions. Even in some of the short-winged forms the stridulating organs of the tegmina of the males are well developed.

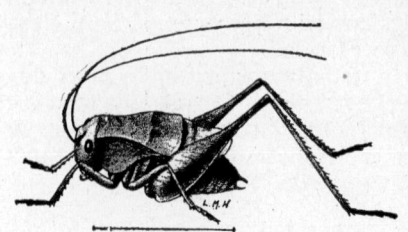

Fig. 254.—*Atlanticus testaceus*, male. (From Lugger.)

The North American species represent twenty genera; most of these are found west of the Mississippi River; a few species occur in the east; nearly all of these belong to the genus *Atlănticus*. Figure 254 represents the male of *Atlănticus testāceus*, and Figure 255 the female of *Atlanticus davisi*.

Most of the species of this subfamily are local or very rare and not of economic importance; but species of the genus *Anabrus* and of *Peranabrus* at times invade cultivated areas in the western United States and do immense damage. Many popular names have been applied to these insects; perhaps the one in most general use is *the western cricket*.

Fig. 255.—*Atlanticus davisi*, female.

A very complete monograph of the North American species of this subfamily has been published by Caudell ('07).

Subfamily GRYLLACRINÆ

The Leaf-rolling Grasshoppers

The members of this subfamily agree with the preceding subfamilies and differ from the two following in having the tarsi more or less depressed. They agree with the following subfamilies and differ from the preceding in the absence of auditory tympana in the fore tibiæ and in the absence of stridulating organs even when the tegmina are present.

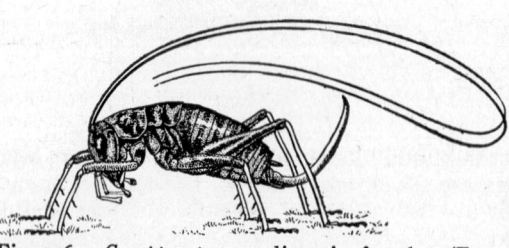

Fig. 256.—*Camptonotus carolinensis*, female. (From Blatchley.)

Only a single species, the Carolina leaf-roller, *Camptonōtus carolinĕnsis* (Fig. 256), occurs in our fauna. This species is wingless; it measures from 13 mm. to 15 mm. in length. Its known range extends from New Jersey west to Indiana and south to Florida.

This insect is very remarkable in its habits, which have been described by Caudell ('04) and McAtee ('08). It makes a nest by rolling a leaf and fastening the roll with silken threads which it spins from its mouth. It remains in its nest during the day and emerges at night to capture aphids upon which it feeds.

Subfamily RHAPHIDOPHORINÆ

The Cave-Crickets or Camel-Crickets

Many common names have been applied to members of this subfamily; among these are *cave-crickets*, because they abound in caves and are found in other dark places; *camel-crickets*, because of the high, arched back of some species (Fig. 257); and *stone-crickets*, from their habit of hiding beneath stones. This last name is not at all distinctive.

These are wingless long-horned grasshoppers that bear some resemblance to the true crickets (Fig. 258). They have a short, thick body and remarkably stout hind femora, like a cricket, but are entirely destitute of tegmina and wings, and the females, like other Tettigoniidæ, have a sword-shaped ovipositor. The more common species are either of a pale brown or a dirty white color and more or less mottled with either lighter or darker shades.

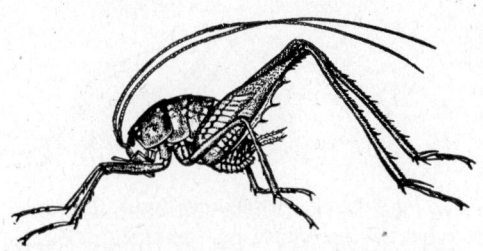

Fig. 257.—*Ceuthophilus uhleri*, male. (From Blatchley.)

Fig. 258.—*Ceuthophilus*, female.

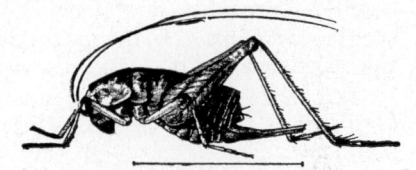

Fig. 259.—*Ceuthophilus maculatus*, female. (From Lugger.)

These insects live in dark and moist places, under stones and rubbish, especially in woods, in cellars, in the walls of wells, and in caves. On one occasion I saw many thousands of them on the roof of a cave in Texas.

Caudell ('16) in his monograph of this subfamily lists twelve genera including many species that occur in the United States. Most of our common species in the East belong to the genus *Ceuthŏphilus*. Figure 257 represents the male of *Ceuthŏphilus uhleri*, and Figure 259 the female of *Ceuthŏphilus maculātus*.

Subfamily STENOPELMATINÆ

The Sand-Crickets

These are large, clumsy creatures with big heads (Fig. 260). They live under stones and in loose soil. They are represented in our fauna by a single genus, *Stenopelmātus*, several species of which are found in the Far West and especially on the Pacific Coast.

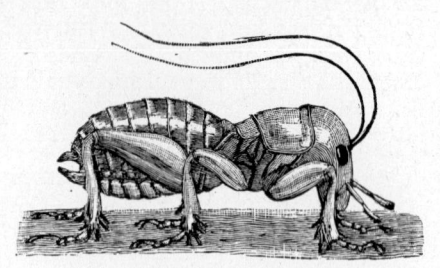

Fig. 260.—*Stenopelmatus*.

Family GRYLLIDAE*

The Crickets

Although the word *cricket* forms a part of some popular compound names of members of the Tettigoniidæ, as "western crickets" and "sand-crickets," when the word is used alone it is correctly applied only to members of this family.

In the more typical crickets, the hind legs are fitted for leaping; the antennæ are long and slender; the tegmina lie flat on the back and are bent down abruptly at the sides of the body; the ovipositor is spear-shaped; and the tarsi are three-jointed. Wingless forms are common.

The more striking departures from these characteristics are the following: in the Tridactylinæ the antennæ are short; in the Trigonidiinæ the ovipositor is sword-shaped; in the Gryllotalpinæ and the Tridactylinæ the ovipositor is wanting in our species; and in the Tridactylinæ the tarsi are reduced.

It is evident that one step in the reduction of the number of tarsal segments is the growing together of the metatarsus and the second segment. This is shown in the hind tarsi of *Anaxīpha*, *Œcănthus*, *Nemōbius*, and doubtless others, where the suture between these two segments can be seen although the segments are anchylosed.

Tympana are usually present in the fore tibiæ, one on each side of each tibia, as in the Tettigoniidæ. In some genera one tympanum of each pair is wanting; this is sometimes the outer and sometimes the inner one; in the wingless, and therefore mute, species, the tympana are wanting; and in the Tridactylinæ there are none.

---

*This family is termed the Achetidæ by some writers.

With most species of crickets the two sexes differ greatly in appearance; the female has a long ovipositor and the venation of the wings is simple, while the male has the horizontal part of the fore wings modified to form musical organs. The structure of these has been described in Chapter II.

The Gryllidæ includes eight subfamilies, all of which are represented in the United States. These subfamilies can be separated by the following table.

A. The next to the last segment of the tarsi distinct, depressed, and heart-shaped.
  B. Hind tibiæ armed with two series of spines without teeth between them. p. 243..TRIGONIDIINÆ
  BB. Hind tibiæ with teeth between the spines. p. 244.........ENEOPTERINÆ
AA. Tarsi compressed, the next to the last segment minute, compressed.
  B. Fore legs fitted for walking.
    C. Hind tibiæ without spines except the apical spurs.
      D. With well-developed wings; hind tibiæ with only two very small apical spurs. (*Neoxabea*.) p. 245......................ŒCANTHINÆ
      DD. Wingless or subapterous; hind tibiæ with three pairs of apical spurs. p. 250.....................................MOGOPLISTINÆ
    CC. Hind tibiæ armed with two series of spines.
      D. Body subspherical; wingless; hind femora ovate, very strongly swollen. p. 249..............................MYRMECOPHILINÆ
      DD. Body more elongate, usually winged; hind femora more elongate, not exceptionally swollen.
        E. Hind tibiæ with minute teeth between the spines. p. 245 ŒCANTHINÆ
        EE. Hind tibiæ without teeth between the spines. p. 247......GRYLLINÆ
  BB. Fore legs fitted for digging.
    C. Antennæ many-jointed; all of the tarsi three-jointed. p. 250.......
    ....GRYLLOTALPINÆ
    CC. Antennæ eleven-jointed; fore and middle tarsi two-jointed, hind tarsi one-jointed or wanting. p. 251......................TRIDACTYLINÆ

Subfamily TRIGONIDIINÆ

The Sword-bearing Crickets

These are small crickets, our species measuring from 4 mm. to 8.5 mm. in length of body. They live chiefly on shrubs and tall grasses and weeds growing in or near water. Their distinguishing features are the following: The next to the last segment of the tarsi is distinct, depressed, and heart-shaped, the hind tibiæ are slender with three pairs of mobile spines besides the terminal spurs, and with no teeth between these spines; and the ovipositor of the female is compressed and curved upwards. In the sword-shaped form of the ovipositor these crickets present a striking exception to the characteristics of the Gryllidæ.

The following are our best-known representatives of this subfamily.

Anaxipha exigua.—This cricket resembles somewhat in general appearance the common small field-crickets (*Nemobius*), but unlike them it does not live on the ground. The antennæ are very long (Fig. 261); the ovipositor is one half as long as the hind femora; the hind femora of the male are longer than the tegmina; and the stridulating area of the tegmina is large. The length of the body is 5-8 mm.

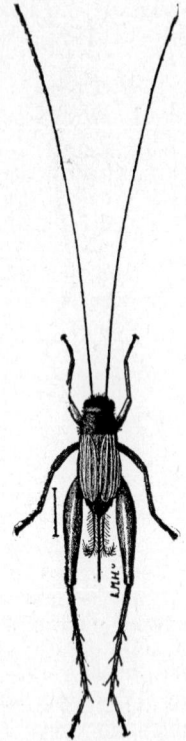

There are two forms of this species: in one, the hind wings are wanting and only the tympana on the outer face of the fore tibiæ are present; in the other, long hind wings are present and there is a tympanum on each face of the fore tibiæ.

This species is found from southern New England west to Minnesota and Nebraska and south to Florida and Texas.

Falcícula hebārdi.—This is a smaller species than the preceding, the body measuring only 4-5 mm. in length. It is uniform pale yellowish brown in color. The hind wings are wanting. The stridulating area is small, confined to the basal fourth of the tegmina. The fore tibiæ are without visible tympana. Its range extends from New Jersey south and southwest to Florida and Texas.

Cyrtŏxipha columbiāna.—This is a small, pale green fading to brownish yellow, cricket; it is found on shrubs and small trees, usually near water. The wings are always present and prolonged in the form of a tail or queue. Tympana are present on both faces of the fore tibiæ. The tegmina extend 2-3 mm. beyond the end of the abdomen. The length of the body to apices of tegmina is 8.5 mm. Its range extends from Washington, D. C., to Florida and Texas.

Fig. 261.—*Anaxipha exigua.* (From Lugger.)

Phylloscўrtus pulchĕllus.—This cricket differs from the three preceding species in having the last segment of the maxillary palpi spoon-shaped. The head and the thorax are bright crimson-red; the margin of the thorax is pale yellow; the abdomen is black; and the tegmina are chestnut-brown. The length of the body is 6-7 mm. This species is found throughout the United States east of the Mississippi River, except in the northern portions.

Subfamily ENEOPTERINÆ

The Larger Brown Bush-Crickets

These crickets resemble those of the preceding subfamily in the heart-shaped form of the next to the last segment of the tarsi; but differ in having teeth between the spines of the tibiæ, and in the ovipositor being spear-shaped.

Only a few species are found in our fauna. These represent three genera: *Orocharis*, in which both tympana of the fore tibiæ are present; *Hăpithus*, with a tympanum on the inner face only of the fore tibiæ; and *Tafalĭsca*, with no tympana and no stridulating organs.

The most common species is *Orŏcharis saltātor* (Fig. 262). This is usually pale reddish brown, but some individuals are grayish. The length of the body is 14–16 mm. It is found from New Jersey west to Nebraska and south to Florida and Texas.

The only common species of *Hapithus* is *H. agitātor*, which is found from Long Island west to Nebraska and south to Florida and Texas.

Our only species of *Tafalisca* is *T. lūrida*, which is found in southern Florida.

Subfamily ŒCANTHINÆ
The Tree-Crickets

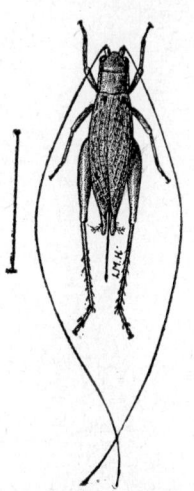

Fig. 262.—*Orocharis saltator.* (From Lugger.)

These are delicate crickets, many of which are of a light green color, with the body and legs sometimes dusky. Figure 263 represents a male; in the females the front wings are more closely wrapped about the body, giving the insect a narrower appearance. They live in more or less elevated positions, varying, according to the species, from among herbaceous plants to the higher parts of fruit and forest trees, hence the name *tree-crickets* commonly applied to them. Their frequent occurrence among flowers suggested the name of the principal genus, *Œcănthus*, implying *I dwell in flowers*. Two genera of tree-crickets are represented in our fauna, *Neoxabea* and *Œcanthus;* these can be distinguished by differences in the armature of the hind tibiæ.

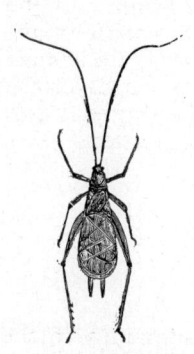

Fig. 263.—*Œcanthus niveus,* male.

Neoxābea.—In this genus the hind tibiæ bear neither teeth nor spines except the apical spurs, and the first segment of the antennæ is armed in front with a stout, blunt tooth (Fig. 264, *h*). *Neoxābea bipunctāta* is the only species known. In this species the hind wings are almost twice as long as the fore wings; the fore wings of the female are each marked with two rather large blackish spots; the wings of the male are unmarked. The general color is pale pinkish brown. The length of the body is about 16 mm.

Œcănthus.—In this genus the hind tibiæ bear both spines and teeth. Several species occur in the United States and Canada; these differ in the color of the body, in the markings on the first two segments of the antennæ, in their song, and in the elevation above the surface of the ground in which they are usually found. Most of our species are found east of the Great Plains; one, *Œcănthus califŏrnicus*, occurs in California; and one, *Œcanthus*

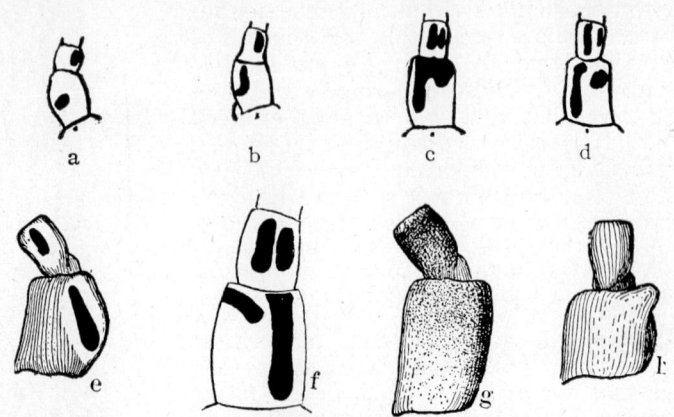

Fig. 264.—Basal segments of antennæ of *Œcanthus* and *Neoxabea*. (The lettering is explained in the text. (After Lugger and Fulton.)

argentīnus, in Texas. The species of eastern North America can be distinguished by the following table, which is copied from a detailed account of these insects by B. B. Fulton ('15).

A. Basal segment of antennæ with a swelling on the front and inner side. First and second segments each with a single black mark.
    B. Basal antennal segment with a round black spot. (Fig. 264, a)..*Œ. nĭveus*
    BB. Basal antennal segment with a J-shaped black mark. (Fig. 264, b) ....*Œ. angustipĕnnis*
    BBB. Basal antennal segment with a straight club-shaped black mark. (Fig. 264, e)..*Œ. exclamatiŏnis*
AA. Basal antennal segment without a swelling on the front and inner side. First and second antennal segments each with two black marks or entirely black. Tegmina of males 5 mm. or less in width.
    B. Head and thorax pale yellowish green or black or marked with both colors.
        C. First antennal segment with a narrow black line along inner edge and a black spot near the distal end. Body entirely pale yellowish green. (Fig. 264, d)...*Œ. quadripunctatus*
        CC. First antennal segment with black markings similar to above, but broader and usually confluent, sometimes covering the whole segment. Head and thorax often with three longitudinal black stripes; ventral side of abdomen always solid black in life. (Fig. 264, c)..*Œ. nigricornis*
    BB. Head, thorax, and antennæ reddish brown. Wings in life with conspicuous green veins. Marks on basal antennal segment broad but seldom confluent. (Fig. 264, f)..*Œ. pini*
AAA. Basal antennal segment without a swelling on the front and inner side. Basal portion of antenna red unmarked with black. (Fig. 264, g). Tegmina of male about 8 mm. wide......................*Œ. latipennis*

The species of *Œcanthus* that most often attracts attention is the snowy tree-cricket, *Œcănthus nĭveus* (Fig. 263). The presence of this insect, though usually unseen, is made very evident in late summer and in the autumn by the song of the males. This song is begun early in the evening and is continued throughout the night; it consists of a monotonous series of high-pitched trills rhythmically repeated indefinitely. It is a remarkable fact that all of these crickets that are chirping in any locality chirp in unison. Individual singers will stop to rest, but when they start again they keep time with those that have continued the chorus. Except where the true katydid is heard, this is the most conspicuous insect song heard in the night in the regions where this species occurs. This cricket inhabits chiefly high shrubs and trees; it deposits its eggs singly in the bark or cambium of trees and bushes.

While the presence of the snowy tree-cricket is made evident by its song, there is another species that has attracted much attention by its manner of oviposition; this is *Œcănthus nigricŏrnis*. The female lays her eggs in a longitudinal series in the twigs or canes of various plants (Fig. 265). She selects the raspberry more often than any other plant; and as that portion of the cane beyond the incisions made for the eggs usually dies, it often happens that these crickets materially injure the plants. In such cases the dead canes should be cut out and burned early in the spring before the eggs hatch.

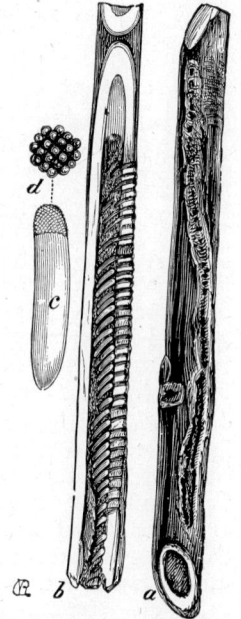

Fig. 265.—Stem of black raspberry with the eggs of *Œcanthus nigricornis: c, d,* egg enlarged. (From Riley.)

Subfamily GRYLLINÆ

The Field-Crickets

The field-crickets abound everywhere, in pastures, meadows, and gardens; and certain species enter our dwellings. They lurk under stones or other objects on the ground or burrow into the earth. They are chiefly solitary, nocturnal insects; yet many can be seen in the fields in the daytime. They usually feed upon plants but are sometimes predacious. With most species the eggs are laid in the autumn, usually in the ground, and are hatched in the following summer. The greater number of the old crickets die on the approach of winter; but a few survive the cold season. In many of the species there are both short-winged and long-winged forms.

This subfamily is represented in our fauna by several genera; but nearly all of our common species are included in the two genera *Gryllus* and *Nemobius*.

The larger field-crickets, *Gryllus*.—The members of this genus are dark-colored, thick-bodied insects of medium or large size. In these the hind tibiæ are armed with strong *fixed* spines and the first segment of the hind tarsi is armed with two rows of teeth above. There are two auditory tympana in each fore tibia. The length of the body is rarely less than 14 mm.

Fig. 266.—*Gryllus assimilis luctuosus.*

Many supposedly distinct species of *Gryllus* have been described as occurring in our fauna; but now all of our *native* forms are believed to be merely varieties of one species, *Gryllus assimilis*, and the different varieties are distinguished by subspecific names. Six of these varieties that occur in the East are described by Blatchley ('20). Two of these will serve to illustrate our native forms.

Gryllus assimilis luctuōsus.—This is one of our more common forms of the genus. It is distinguished by the great length of the ovipositor of the female, which is nearly or fully half as long again as the hind femora (Fig. 266); and by the fact that the head of the male is distinctly wider than the front of the pronotum.

Gryllus assimilis pennsylvănicus.—In this variety the ovipositor is less than half as long again as the hind femora, and the head of the male is but little if any wider than the front of the pronotum (Fig. 267). In fresh specimens the color is not shining black, but with a very fine grayish pubescence.

In addition to our native forms of *Gryllus*, there is an Old World species that has been introduced into this country; this is the house-cricket, *Gryllus domĕsticus*. References to the "cricket of the hearth" are common in English literature and refer to this species, which is now widely distributed in this country, though it is rarely abundant. It is pale yellowish brown or straw-colored, and slender in form (Fig. 268). The length of the body is 15–17 mm.

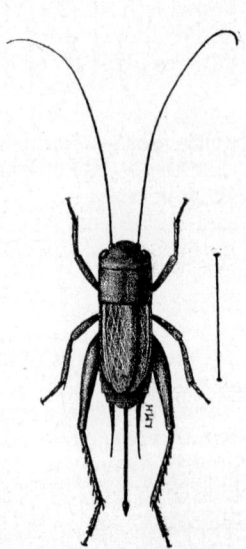

Fig. 267.—*Gryllus assimilis pennsylvanicus.* (From Lugger.)

Our native field-crickets sometimes enter our dwellings in the autumn; but the house-cricket can be easily distinguished from these.

The smaller field-crickets, *Nemōbius.*—To this genus belong the little field-crickets, which are the most abundant of all of our crickets. In these the hind tibiæ are furnished with long, *mobile*, hairy spines, and the first segment of the hind tarsi is unarmed above or with only one row of teeth. There is only one tympanum in each fore tibia. The length of the body is less than 12 mm.

There are many species and varieties of this genus in our fauna. The following enlarged figures of two of our species will serve to illustrate the form of these insects. (Fig. 269 and 270.)

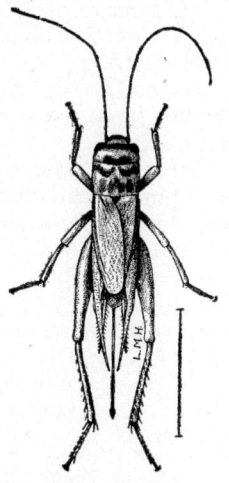

Fig. 268.—*Gryllus domesticus*. (From Lugger.)

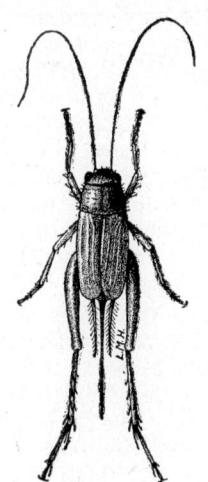

Fig. 269.—*Nemobius fasciatus*. (From Lugger.)

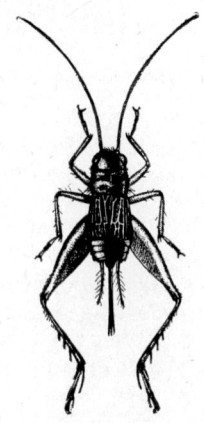

Fig. 270.—*Nemobius palustris*. (Fom Blatchley.)

Subfamily MYRMECOPHILINÆ

The Ant-loving Crickets

The members of this subfamily are very small crickets, which live as guests in the nests of ants. The form of these crickets is very remarkable. The body is ovate, greatly convex above, and wingless (Fig. 271); the hind femora are ovate and greatly enlarged, the cerci are long; and the ovipositor is short and stout.

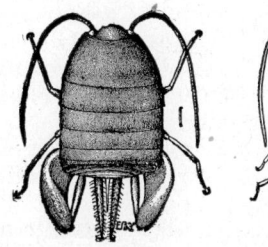

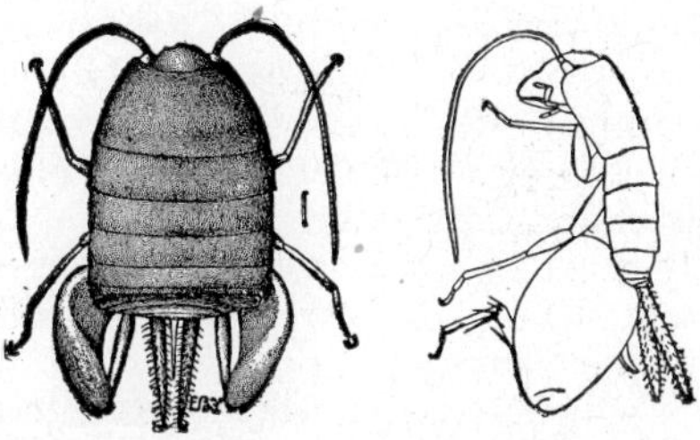

Wheeler ('oo) states that these crickets feed on an oily secretion covering the surface of the body of the ants; they also obtain this substance from the greasy walls of the ant-burrows. Apparently the ants derive no benefit from the presence of these

Fig. 271.—*Myrmecophila pergandei*. (From Lugger.)

guests, and destroy them when they can; but the crickets are very agile. These are the smallest of the true Orthoptera.

This subfamily includes a single genus, *Myrmecŏphila*, of which five species have been described from the United States. Only one species has been found in the East; this is *Myrmecŏphila pergăndei*. In this species the length of the body is 3–5 mm.

Subfamily MOGOPLISTINÆ

The Wingless Bush-Crickets

These crickets are found chiefly on bushes or among rubbish under bushes; some are found beneath debris in sandy places. They are small; those found in the United States measure from 5 mm. to 13 mm. in length of body. They are either wingless or furnished in the male sex with short tegmina, in which the stridulating organs are well developed. The body is covered with translucent, easily abraded scales.

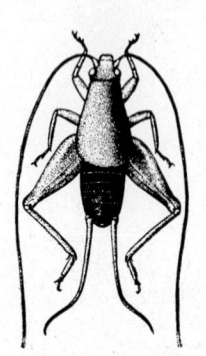

Most of the species are tropical or subtropical in distribution; our species are found chiefly in the South and Southwest; but the range of one of them extends north to Long Island. Only four species have been described from the East and one of these is restricted to Florida. A few others are known from the western part of our country. A monograph of the North American species was published by Rehn and Hebard ('12).

Fig. 272.—*Cryptoptilum trigonipalpum.* (From Rehn and Hebard.)

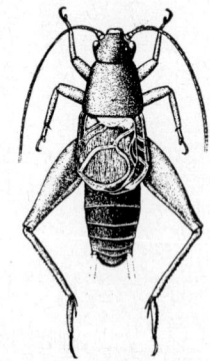

Fig. 273.—*Holosphyrum boreale.* (From Rehn and Hebard.)

Figure 272 represents the male of *Cryptŏptilum trigonipălpum*, a wingless species found from Virginia southward; and Figure 273, the male of *Holosphȳrum boreāle*, found in the Southwest.

Subfamily GRYLLOTALPINÆ

The Mole-Crickets

The mole-crickets differ greatly in appearance from the more typical crickets, the form of the body and of the fore legs being adapted to burrowing in the ground. The front tibiæ, especially, are fitted for digging; they are greatly broadened and shaped somewhat like a hand or a foot of a mole; they are terminated by strong blade-like teeth, termed the dactyls (Fig. 274).

Two of the tarsal segments are blade-like and so situated that they can be moved across the dactyls like the cutting blades of a mowing machine (Fig. 275). Sharpe ('95) states that this organ enables the mole-cricket to cut the small roots it meets in digging its burrows; but this is doubted by Morse ('20), who believes that the roots are cut by the powerful mandibles.

The antennæ of mole-crickets are much shorter than the body; the hind femora are but little enlarged, not well fitted for jumping; and the ovipositor is not visible externally. The name of the type genus, *Gryllotălpa*, is from *Gryllus*, a cricket, and *talpa*, a mole.

Two genera of mole-crickets are found in the United States: *Gryllotalpa*, in which the front tibiæ are furnished with four dactyls; and *Scapteriscus*, in which each fore tibia bears only two dactyls. Each of these genera is represented in our fauna by several species.

Our best-known and most widely distributed species is *Gryllotălpa hexadăctyla* (Fig. 274). This species has been generally known in this country as *Gryllotalpa boreālis;* but this name is now believed to be a synonym. The range of this species extends from British America to the southern part of South America. The length of the body is 20–30 mm.

Fig. 274.—*Gryllotalpa hexadactyla.*

The mole-crickets are not common insects in this country; but occasionally they are found in great numbers in a limited locality. They make burrows in moist places from six to eight inches below the surface of the ground, and feed upon the tender roots of various plants, and also on other insects. The eggs are deposited in a neatly constructed subterranean chamber, about the size of a hen's egg.

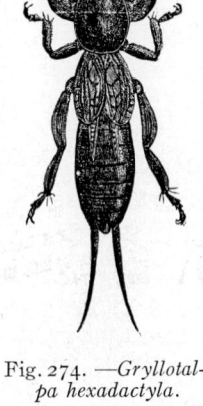

Fig. 275.—Front leg of a mole-cricket; A, inner aspect; B, outer aspect; e, ear-slit. (From Sharp.)

Subfamily TRIDACTYLINÆ

The Pigmy Mole-Crickets

The members of this subfamily resemble the mole-crickets in the form of the body and in their burrowing habits; but they are much smaller, the larger species measuring only 10 mm. in length; and the hind femora are greatly enlarged, being strongly saltatorial (Fig. 276). The antennæ are short and composed of only eleven segments. The fore wings are usually short and never extend to the end of the abdomen; they are horny, are almost veinless, and are not furnished with stridulating organs in the male. The hind wings are much longer, usually extending beyond the end of the abdomen. The fore tibiæ lack auditory tympana. The first four tarsi, in our genera, are two-jointed; the hind tarsi are one-jointed or wanting. The hind tibiæ are furnished with movable plates, "natatory lamellæ," near the distal end; these are ordinarily closely appressed to the tibia but can be spread out like a fan. It is probable that these plates are used to aid the insect in leaping from the surface of water upon which they have jumped; they may also serve a similar purpose on land, making a firm planting of the end of the leg upon the ground.

Fig. 276.— *Tridactylus apicalis.* (From Lugger.)

The ovipositor is vestigial in our species; but Walker ('19) states that in the exotic genus *Ripipteryx* there is a well-developed ovipositor, which is remarkably similar to that of the short-horned grasshoppers. These insects apparently have two pairs of cerci; this is due to the fact that in addition to the true cerci each of the two podical plates is greatly elongated and bears a terminal segment, which appears like a stylus or cercus.

These insects burrow rapidly in sand and possess great powers of leaping. They live on and in the damp sand on the shores of ponds and streams. Their burrows extend only a short distance below the surface of the ground.

Only two genera, each represented by a single species, have been found in America north of Mexico.

Tridăctylus.—In this genus the hind tibiæ are furnished with four pairs of long, slender plates, the "natatory lamellæ;" and the hind tarsi are one-jointed. Our species is *Tridăctylus apicālis* (Fig. 276). The length of the body is 6-9.5 mm.

Ĕllipes.—In this genus the hind tibiæ are furnished with a single pair of "natatory lamellæ"; and the hind tarsi are wanting. Our species is *Ellipes minūta.* The length of the body is 4-5 mm.

Walker ('19) as a result of his studies of the genitalia of *Ripipteryx* believes that the pigmy mole-crickets are more closely allied to the Locustidæ than they are to the Gryllidæ, and ranks them as constituting a distinct family, the Tridactylidæ.

Family LOCUSTIDÆ*

The Locusts or Short-horned Grasshoppers

The family Locustidæ includes the locusts or short-horned grass-

---

*This family is termed the Acrididæ by some writers, this name being based on the generic name *Acrida* of Linnæus; other writers use the family name Acry-

hoppers. These are common and well-known insects. They differ from most of the members of the two preceding families in having the antennæ much shorter than the body, and consisting of not more than twenty-five segments. The ovipositor of the female is short and composed of separate plates; and the basal segment of the abdomen is furnished on each side with a tympanum, the external parts of the organs of hearing (Fig. 277, *t*).

It is to these insects that the term *locust* is properly applied; for the locusts of which we read in the Bible, and in other books published in the older countries, are members of this family. Unfortunately, in the United States the term *locust* has been applied to the Periodical Cicada, a member of the order Homoptera, described later. And, what is more unfortunate, the scientific name Locustidæ has been applied by many writers to the long-horned grasshoppers.

Locusts lay their eggs in oval masses and cover them with a tough substance. Some species lay their eggs in the ground. The female makes a hole in the ground with her ovipositor, which is a good digging tool. Some species even make holes in fence-rails, logs, and stumps; then, after the eggs are laid the hole is covered up with a plug of gummy material. There is but one generation a year, and in most cases the winter is passed in the egg-state. This family is of great economic importance, as the members of it usually appear in great numbers in nearly every region where plants grow, and often do much damage.

With many species of the Locustidæ the males are furnished with stridulating organs. These have been described in Chapter II, page 82.

There are very many species of locusts in the United States and Canada; these represent four of the subfamilies of the family Locustidæ, which can be separated by the following table.

A. Claws of the tarsi with a small pad (arolium) between them; pronotum extending at most over the extreme base of the abdomen.
  B. Prosternum armed anteriorly with a distinct conical or cylindrical tubercle. p. 254.. LOCUSTINÆ.
  BB. Prosternum without a distinct tubercle; arolium usually small or rather small.
    C. Head rounded at the union of the vertex and front; front perpendicular or nearly so. p. 257................................... ŒDIPODINÆ.
    CC. Vertex and front of head meeting at an acute angle; vertex extending horizontally; front strongly receding. p. 259.......... TRUXALINÆ.
AA. Claws of tarsi without an arolium between them; pronotum extending over the abdomen. p. 259...................................... ACRYDIINÆ.

---

diidæ, based on the generic name *Acrydium* of Fabricius; and still others use the family name Acridiidæ, based on *Acridium*, an emended spelling of *Acrydium*. The oldest name given to this family is Acrydiana, applied to it by Latreille in 1802; but the group of insects that Latreille used as the type of the family is the Locusta of Linnæus (1758); for this reason the name given to the family by Latreille has been changed to Locustidæ. See also the footnote on page 234.

Subfamily LOCUSTINÆ
The Spur-throated Locusts

The members of this subfamily are distinguished from other North American locusts by the presence of a tubercle on the prosternum. Here belong many of our more common species; and among them are found the most injurious insects of the order Orthoptera. Among our best-known species are the following.

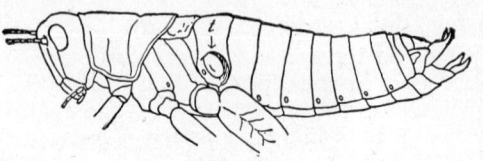

Fig. 277.—Side view of a female locust with the wings removed.

The Rocky Mountain locust or western grasshopper, Melănoplus sprētus.—The most terrible of insect scourges that this country has known have been the invasions

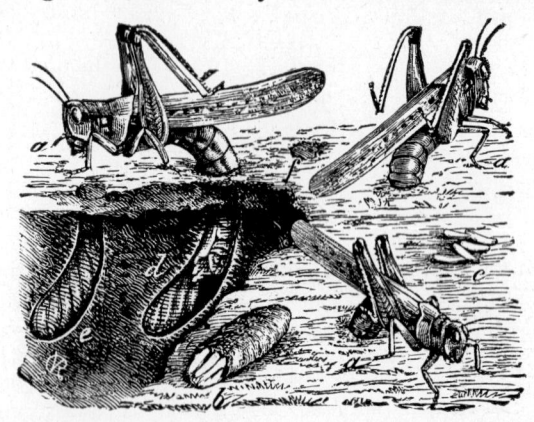

Fig. 278.—Egg-laying of the Rocky Mountain Locust: *a, a, a,* female in different positions, ovipositing; *b,* egg-pod extracted from the ground with the end broken open; *c,* a few eggs lying loose on the ground; *d, e,* show the earth partially removed, to illustrate an egg-mass already in place, and one being placed; *f* shows where such a mass has been covered up. (From Riley.)

of this species. Large areas of country have been devastated, and the inhabitants reduced to a state of starvation. The cause of all this suffering is not a large insect. It is represented in natural size by Figure 278. It measures to the tip of its wing-covers 20–35 mm., and resembles very closely our common red-legged locust, the most abundant of all our species. It can easily be distinguished from this species by the greater length of the wings, which extend about one-third of their length beyond the tip of the abdomen, and by the fact that the apex of the last abdominal segment in the males is distinctly notched.

Fig. 279.—*Melanoplus femur-rubrum.*

The permanent home or breeding grounds of this species is in the high, dry lands on the eastern slope of the Rocky Mountains, extending from the southern limit of the true forests in British America south through Montana, Wyoming, the western part of the Dakotas, and the Parks of Colorado. There are also regions in which the species exists permanently west of the Rocky Mountains in Idaho and Utah.

When the food of this insect becomes scarce in its mountain home, it migrates to lower and more fertile regions. Its long wings enable it to travel great distances; and thus the larger part of the region west of the Mississippi River is liable to be invaded by it. Fortunately, the species cannot long survive in the low, moist regions of the valleys. Although the hordes of locusts which reach these sections retain their vigor, and frequently consume every bit of green vegetation, the young, which hatch from the eggs that they lay, perish before reaching maturity. In this way the invaded region is freed from the pest until it is stocked again by another incursion. There is, however, a large section of country lying immediately east of the great area indicated above as the permanent home of this species, which it frequently invades and in which it can perpetuate itself for several years, but from which it in time disappears. This sub-permanent region, as it has been termed, extends east in British America so as to include nearly one-third of Manitoba; and, in the United States, it embraces nearly the whole of the Dakotas, the western half of Nebraska, and the northeast fourth of Colorado.

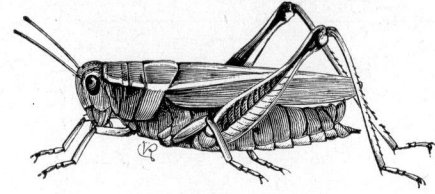

Fig. 280.—*Melanoplus bivittatus*. (From Riley.)

Fig. 281.—*Melanoplus bivittatus* killed by a fungus. (From Lugger.)

The temporary region, or that only periodically visited and from which the species generally disappears within a year, extends east and south so as to include more than half of Minnesota and Iowa, the western tier of counties of Missouri, the whole of Kansas and Oklahoma, and the greater part of Texas. The country lying east of the section thus indicated has never been invaded by this locust, and there is no probability that it will ever be reached by it.

256   AN INTRODUCTION TO ENTOMOLOGY

Detailed directions for the control of this pest have been published in many State and Federal Government reports. Among these methods of control are the plowing of land in which its eggs have been deposited, the use of poisoned bran-mash as a bait, and catching of the insects by machines commonly known as "hopper dozers."

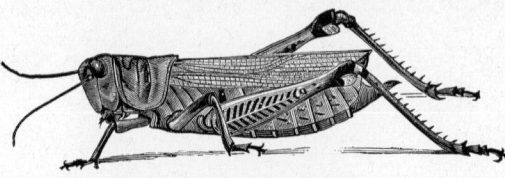

Fig. 282.—*Melanoplus differentialis.* (From Riley.)

The red-legged locust, *Melănoplus femur-rūbrum.*—This is the most common short-horned grasshopper throughout the United States, except where *Melanoplus spretus* occurs. It ravages our meadows and pastures more than all other species combined. It is found in most parts of North America. The female is represented, natural size, by Figure 279.

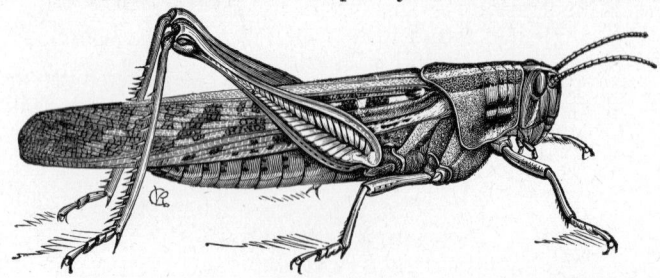

Fig. 283.—*Schistocerca americana.* (From Riley.)

Melănoplus bivittātus.—This species is also found from the Atlantic to the Pacific. It is marked with a yellowish stripe, extending along each side from the upper angle of the eye to the tip of the front wing (Fig. 280). The length of the body varies from 23 mm. to 40 mm.

This locust is often killed by a parasitic fungus. Dead fungus-infected individuals are frequently found clinging to weeds, up which they have climbed to die (Fig. 281).

Melănoplus differentiālis.—This species is slightly larger than the preceding; and it lacks the prominent yellow stripe (Fig. 282).

Schistŏcerca americāna.—This magnificent species occurs in the Southern States and has been found as far north as Con-

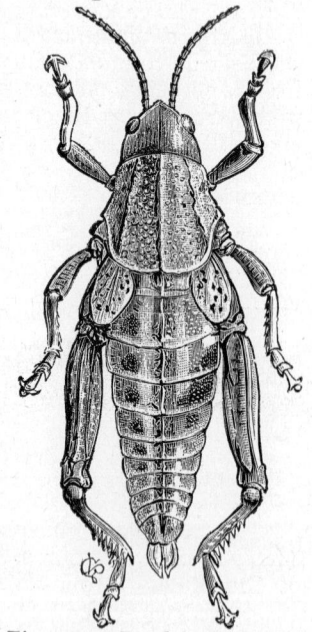

Fig. 284.—*Brachystola magna.* (From Riley.)

necticut and Iowa. It can be recognized by Figure 283, which represents it natural size. This locust sometimes assumes the migratory habit, and is sometimes injurious to agriculture.

The lubber grasshopper, *Brachȳstola mǎgna*.—This is a large, clumsy species in which the wings are vestigial (Fig. 284); it is confined to the central portion of North America.

Leptȳsma marginicǒllis.— In most of the spur-throated locusts the face is nearly vertical; but in a few species it is very oblique. This species is a good illustration of this type (Fig. 285); it is found in the Southern States east of the Mississippi River.

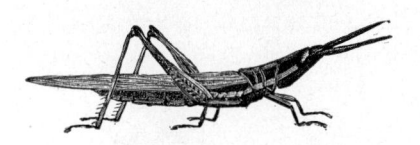

Fig. 285.—*Leptysma marginicollis.*

Subfamily ŒDIPODINÆ

The Band-winged Locusts

In this subfamily the prosternum is without a distinct tubercle; the head is rounded at the union of the vertex and the front; and the front is perpendicular or nearly so. In most of our species the hind wings are in part black, and a portion of them yellow or red; this gives them a banded appearance. There are many representatives of this subfamily in our fauna; the following are some of the more common ones.

The clouded locust, *Encoptolǒphus sǒrdidus*.—This species (Fig. 286) is very common in the eastern United States during the autumn. It abounds in meadows and pastures, and attracts attention by the crackling sound made by the males during flight. It is of a dirty brown color, mottled with spots of a darker shade. The length of the body of the male is 19-22 mm.; of the female, 24-32 mm.

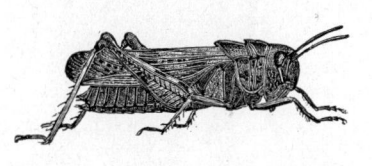

Fig. 286.—*Encoptolophus sordidus.*

The northern green-striped locust, *Chortǒphaga viridifasciāta*.— This is a very common species in the United States and Canada east of the Rocky Mountains. There are two well-marked varieties. In one, the typical form, the head, thorax, and femora are green, and there is a broad green stripe on each fore wing, extending from the base to beyond the middle; this often includes two dusky spots on the edge. In the other variety, the ground color is dusky brown. Intergrades occur, in which the head and thorax are of a reddish velvety brown. The length of the body is 17–32 mm.

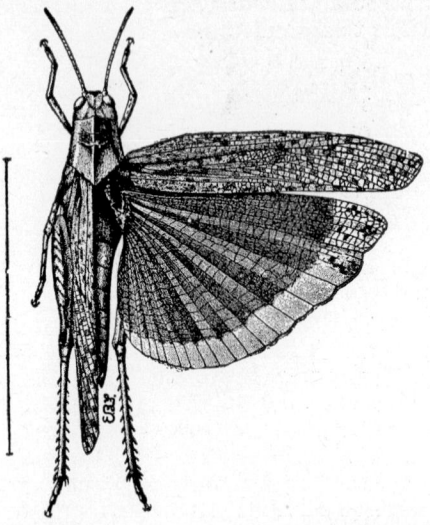

The Carolina locust, *Dissostēīra carolīna*.—Notwithstanding its specific name, this species is common throughout the United States and Canada. It is a large species; the length of the body of the males is 24–33 mm., of the females 33–40 mm. It abounds in highways and in barren places. It takes flight readily, and the males stridulate while in the air. The color of this insect varies greatly, simulating that of the soil upon which it is found. It is usually of a pale yellowish or reddish brown, with small dusky spots. The hind wings are black, with a broad yellow margin which is covered with dusky spots at the tip (Fig. 287).

Fig. 287.—*Dissosteira carolina*. (From Lugger.)

Boll's locust, *Sphărăgemon bŏlli*.—This species is widely distributed in the United States and southern Ontario east of the Rocky Mountains. The length of the body of the male is 20–28 mm., of the female 27–36 mm. The hind wings are pale greenish yellow at the base and are crossed by a dark band; the apical third is transparent smoky in color (Fig. 288).

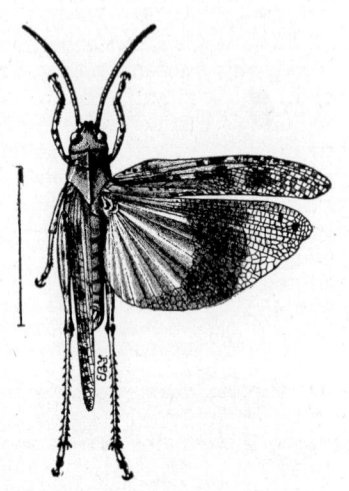

The coral-winged locust, *Hippĭscus apiculātus*.—This is one of the larger of our band-winged locusts (Fig. 289). The length of the body of the male is 25–30 mm., of the female 36–44 mm. The general color is ash-brown. The basal portion of the hind wings is bright coral-red, rarely yellow; this part is bordered without by a dark band. This species is widely distributed east of the Rocky Mountains.

Fig. 288.—*Spharagemon bolli*. (From Lugger.)

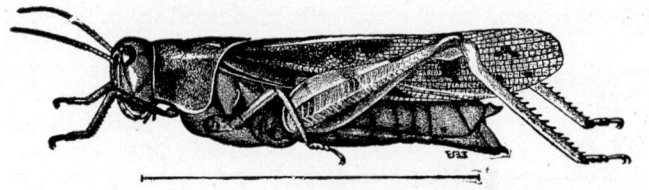

Fig. 289.—*Hippiscus apiculatus.* (From Lugger.)

Subfamily TRUXALINÆ

The Slant-faced Locusts

In this subfamily, as in the preceding one, the prosternum is unarmed but the head is of a different form. In the Truxalinæ, the vertex and the front meet on an acute angle. In some species this angle is a sharp one, the shape of the head being similar to that of *Leptysma* (Fig. 285). In other species, however, the front is less receding; this is the case in the following species.

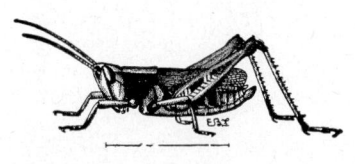

Fig. 290.—*Chloealtis conspersa*, male. (From Lugger.)

The sprinkled locust, *Chloëăltis conspĕrsa*.—This is a very abundant species in the northern United States and Canada east of the Great Plains. It is brown, with the sides of the pronotum and the first two or three abdominal segments shining black in the male; and with the body and tegmina of the female sprinkled or mottled with darker brown. The tegmina and hind wings are a little shorter than the abdomen in the male (Fig. 290), and much shorter in the female (Fig. 291). The males measure 15–20 mm. in length; the females, 20–28 mm.

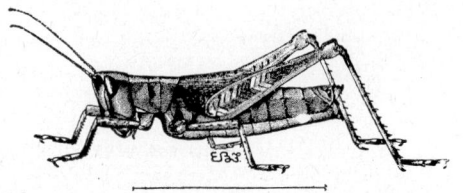

Fig. 291.—*Chloealtis conspersa*, female. (From Lugger.)

Subfamily ACRYDIINÆ

The Pigmy Locusts

The Acrydiinæ includes small locusts of very unusual form. They differ so much from other locustids that some students of the

Orthoptera believe they constitute a separate family. The most striking character of the subfamily is the shape of the pronotum. This is prolonged backwards over the abdomen to or beyond its extremity (Fig. 292). The head is deeply set in the pronotum; and the prosternum is expanded into a broad border, which partly envelops the mouth-parts like a muffler. The antennæ are very slender and short. The tegmina are vestigial, being in the form of small, rough scales; while the wings are usually well-developed.

Fig. 292.—A pigmy locust.

These locusts differ, also, from all others in having no arolium between the claws of the tarsi.

The pigmy locusts are commonly found in low, wet places, and on the borders of streams. Their colors are usually dark, and are often protective, closely resembling the soil upon which the insects occur. They are very active and possess great leaping powers.

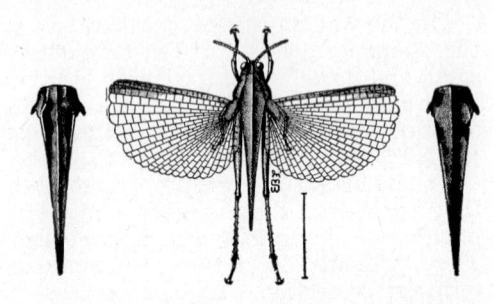

Fig. 293.—*Acrydium granulatum.* (From Blatchley, after Kirby.)

Some of the species vary greatly in coloring; this has resulted often in a single species being described under two or more names. This is an exceedingly difficult group in which to determine the species.

Figure 293 represents *Acrȳdium granulātum* with its wings spread, and the pronota of two color varieties.

Figure 294 represents *Acrȳdium arenōsum obscūrum*, greatly enlarged, with its wings closed.

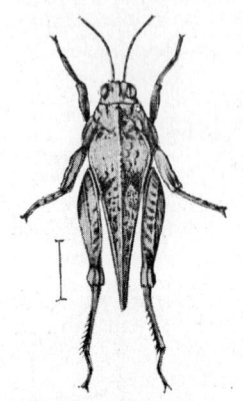

Fig. 294.—*Acrydium arenosum obscurum.* (From Hancock.)

FAMILY PHASMIDÆ*

The Walking-Sticks and the Leaf-Insects

The Phasmidæ is of especial interest on account of the remarkable mimetic forms of the insects comprising it. In those species that are found in the United States, except one in Florida, the body is linear (Fig. 295), wingless, and furnished with long legs and antennæ. This peculiar form has suggested the name *walking-sticks* which is commonly applied

---

*This family is separated from the Orthoptera by Handlirsch ('06–'08) and made to constitute a distinct order, the Phasmoidea.

to these insects; they are also known as stick-insects. In some exotic species the body has the appearance of being covered with moss or with lichens, which increases the resemblance to a stick or a piece of bark.

While our species are all wingless, except *Aplopus mayeri*, found in southern Florida, many exotic species are furnished with wings; and with some of these the wings resemble leaves. Among the more remarkable of the leaf-insects, as they are known, are those of the genus *Phyllium* (Fig. 296), the members of which occur in the tropical regions of the Old World.

In the walking-sticks, the body is elongate and subcylindrical. the abdomen consists of ten segments, but the basal segment is small and usually coalesced with the metathorax and sometimes it is entirely invisible; the legs are all fitted for walking, the tarsi are five-jointed except in the genus *Timema*, where they are three-jointed; the cerci are without joints.

These insects are strictly herbivorous; they are slow in their motions, and often remain quiet for a long time in one place. They evidently depend on their mimetic form for protection. In addition to this some species have the power of ejecting a stinking fluid, which is said to be very acrid; this fluid comes from glands placed in the thorax.

The eggs are scattered on the ground beneath the plants upon which the insects feed, the female, unlike most Orthoptera, making no provision for their safety. In our common northern species the eggs are dropped late in the summer and do not hatch till the following spring, and they often remain till the second spring before they hatch.

About 600 species of phasmids have been described; but they are largely restricted to the tropical and subtropical regions. Caudell ('03) in his monograph of the species of the United States enumerates sixteen species that occur in our fauna; but these are found chiefly in the southern part of the country.

Our common northern walking-stick is *Diapherŏmera femorāta* (Fig. 295). The range of this species extends into Canada. It is a quite common insect, and on several occasions has appeared in such great numbers as to be seriously destructive to the foliage of forest trees; but these outbreaks have been temporary.

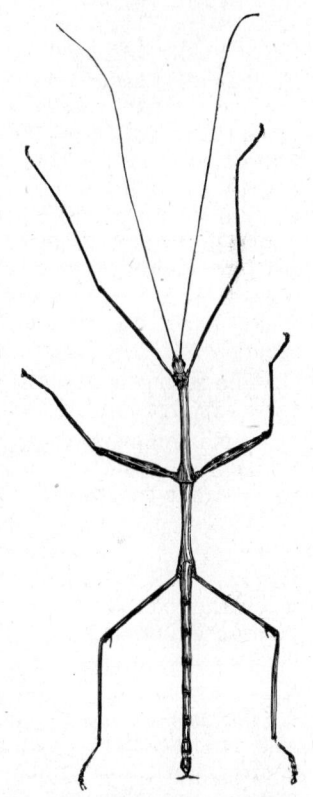

Fig. 295.—*Diapheromera femorata.*

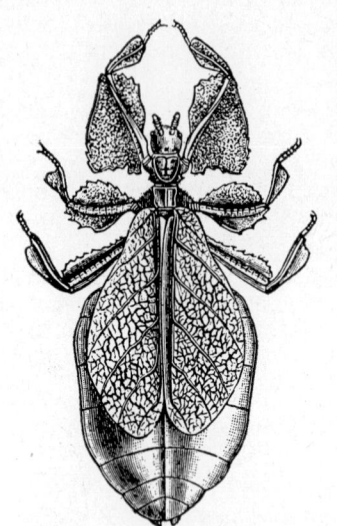

Among the more striking in appearance of the walking-sticks found in the South are *Megaphăsma dĕntricus*, our largest species, measuring from 125 to 150 mm. in length, and *Anisomŏrpha buprestoides*, a yellowish brown species, about half as long as the preceding, with conspicuous, broad, black stripes extending from the front of the head to the tip of the abdomen.

The reproduction of lost legs occurs frequently in this family.

Fig. 296.—*Phyllium scythe*. (From Sharp, after Westwood.)

Family MANTIDÆ*

The Praying Mantes or Soothsayers

The praying mantes are easily recognized by the unusual form of the prothorax and of the first pair of legs (Fig. 297). The prothorax is elongate, sometimes nearly as long as the remainder of the body; and the front legs are large and fitted for seizing prey. The coxæ of the front legs are very long, presenting the appearance of femora; and the femora and tibiæ of these legs are armed with spines; the tibia of each leg can be folded back against the femur so that the spines of the two will securely hold any insect seized by the praying mantis.

The second and third pairs of legs are simple and similar; the tarsi are five-jointed; and the cerci are jointed.

With some species the wings resemble leaves of plants in form and coloring. This resemblance is protective, causing the insects to resemble twigs of the plants upon which they are.

All of the species are carnivorous, feeding on other insects. They do not pursue their prey but wait patiently with the front legs raised like uplifted hands in prayer, until it comes within reach, when they seize it. This position, which they assume while waiting, gives them most of their popular names, of which there are many.

The eggs of the Mantidæ are encased in chambered oöthecæ, which are usually fastened to the stems or twigs of plants (Fig. 298). In the case of the species that occur in the North, there is only one generation in a year and the winter is passed in the egg-state.

Most of the members of this family are tropical insects; a few species, probably less than twenty, live in the southern half of

---

*This family is separated from the Orthoptera by Handlirsch ('06–'08) and made to constitute a distinct order, the Mantoidea.

the United States; and one of our native species, *Stagmomăntis*

Fig. 297.—*Stagmomantis carolina*.

carolīna (Fig. 297), is found as far north as Maryland and southern Indiana.

Recently two exotic species have been introduced into the Northern States, probably by the importation of oöthecæ on nursery stock, and have become established here. These are the *Măntis religiōsa* of Europe, which was first observed in this country near Rochester, N. Y., in 1899, and *Paratenodēra sinĕnsis* of China and Japan, which was first observed here at Philadelphia about 1895.

Family BLATTIDÆ*

The Cockroaches

The cockroaches are such well-known insects that there is but little need for a detailed account of their characteristics. As already indicated in the table of families, the body is oval and depressed; the head is nearly horizontal, and wholly or almost wholly withdrawn beneath the pronotum; the head is bent so that the mouth-parts project caudad between the bases of the first pair of legs; the antennæ are long and bristle-like; and the pronotum is shield-like. This family includes only the cockroaches; but these insects are known in some localities as "black beetles," and our most common species in the northern cities bears the name of Croton-bug.

Fig. 298.—Egg-cases of *Stagmomantis carolina*. (From Riley.)

---

*This family is separated from the Orthoptera by Handlirsch ('06–'08) and made to constitute a distinct order, the Blattoidea.

In the Northern States our native species are usually found in the fields or forests under sticks, stones, or other rubbish. But certain imported species become pests in dwellings. In the warmer parts of the country, however, native and foreign species alike swarm in buildings of all kinds, and are very common out of doors.

Fig. 299.—Oötheca of a cockroach.

Cockroaches are very general feeders; they destroy nearly all forms of provisions, and injure many other kinds of merchandise. They often deface the covers of cloth-bound books, eating blotches upon them for the sake of the sizing used in their manufacture; and I have had them eat even the gum from postage stamps. They thrive best in warm, damp situations; in dwellings they prefer the kitchens and laundries, and the neighborhood of steam and water pipes. They are chiefly nocturnal insects. They conceal themselves during the day beneath furniture or the floors, or within the spaces in the walls of a house; and at night they emerge in search of food. The depressed form of their bodies enables them to enter small cracks in the floors or walls.

Not only are these insects very destructive to our possessions, but owing to their fetid odor merely the sight of them awakens disgust; but it is due them to state that they are said to devour greedily bed-bugs. This will better enable us to abide their presence in our staterooms on ocean voyages, or in our chambers when we are forced to stop at poor hotels.

The eggs of cockroaches are enclosed in purse-like capsules (Fig. 299). These capsules, or oöthecæ, vary in form in different genera, but are more or less bean-shaped. Within, the oötheca is divided into two parallel spaces, in each of which there is a row of separate chambers, each chamber enclosing an egg. The female often carries an oötheca protruding from the end of the abdomen for several days. It has been found that a single female may produce several oöthecæ.

The nymphs resemble the adults except in size, and, in the case of winged species, in the degree of development of the wings. In adults also of some species the wings are reduced, atrophied, or absent; this condition exists more frequently in females than in males (Fig. 300).

As in most other insects, the homologies of the wing-veins can be most easily determined by a study of the tracheation of the wings of nymphs; Figure 301 will serve to illustrate this.

Experiments conducted by the Bureau of Entomology at Washington have shown that one of the most effective means of ridding premises of cockroaches is dusting the places they frequent with commercial sodium fluorid. Several other substances are used for this purpose;

Fig. 300.—A wingless cockroach.

among these are borax, pyrethrum, sulphur, and phosphorus paste.

Cockroaches are chiefly inhabitants of warm countries; although nearly one thousand species have been described, few are found in the

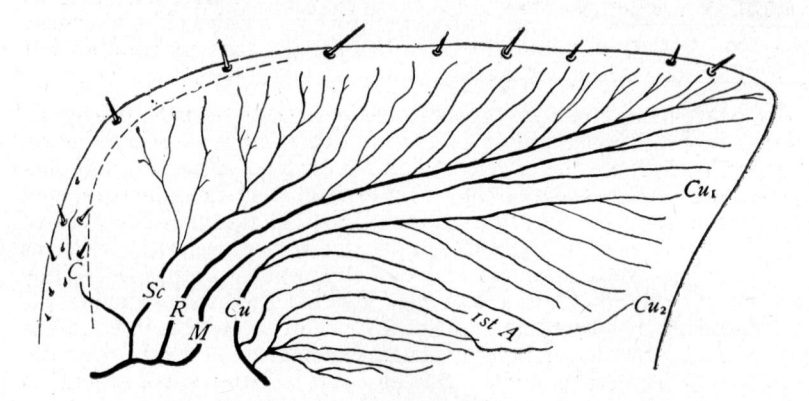

Fig. 301.—Fore wing of a nymph of a cockroach.

temperate regions. Only forty-three species have been found in North America north of the Mexican boundary, and ten of these are probably introduced species (Hebard '17). The cockroaches that are most often found in buildings are two introduced species, the Croton-bug and the Oriental cockroach, and two native species, the American cockroach and the common wood-cockroach. The adults

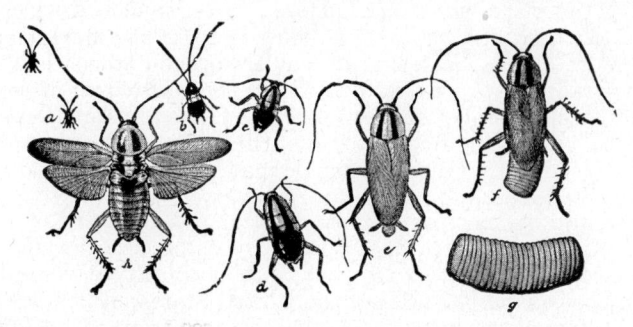

Fig. 302.—The Croton-bug: *a*, first instar; *b*, second instar; *c*, third instar; *d*, fourth instar; *e*, adult; *f*, adult female with egg-case; *g*, egg-case, enlarged; *h*, adult with the wings spread. All natural size except *g*. (From Howard and Marlatt.)

of these four species can be separated by the following table. For tables separating all North American species see Hebard ('17).

  A. With well-developed tegmina.
    B. Tegmina extending to or beyond the tip of the abdomen.
      C. Body about 12 mm. in length.................... *The Croton-bug*
      CC. Body 16 mm. or more in length.

D. Margin of the pronotum light in color while the disk is dark. .. .. *The common wood-cockroach*, male
DD. Pronotum reddish-brown with two blotches of a lighter color. ....*The American cockroach*
BB. Wings not extending to the tip of the abdomen.
C. With a light band on each lateral border of the pronotum......... ....*The common wood-cockroach*, female
CC. With no bands on the pronotum.........*The Oriental cockroach*, male
AA. Tegmina represented by small ovate pads....... *The Oriental cockroach, female*

The Croton-bug, *Blattĕlla germănica* (Fig. 302), is the best-known of all of the cockroaches in our northern cities. It is easily recognized by its small size, about 12 mm. in length, and by its pale color with two dark, parallel bands on the pronotum. Its popular name originated in New York City, and was suggested by the fact that this pest is very abundant, in houses, about water pipes connected with the Croton Aqueduct. This is a species introduced from Europe; it has spread to nearly all parts of the world, living upon ships, and spreading from them.

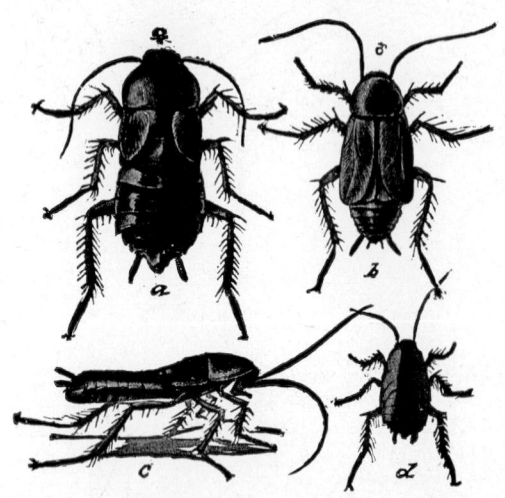

Fig. 303.—The oriental cockroach: *a*, female; *b*, male; *c*, side view of female; *d*, half-grown specimen. All natural size. (From Howard and Marlatt.)

The oriental cockroach, *Blătta orientālis* (Fig. 303), is also a cosmopolitan species; its original habitat is supposed to have been in Asia; but it has been distributed by commerce throughout the world except in the colder regions. In this country it is most abundant in the central latitudes of the United States; it has been found in only a few places in Canada. It measures from 18 to 25 mm. in length. It is blackish brown in color. In the male the wings cover about two-thirds of the abdomen; while in the female they are small, ovate-lanceolate, lateral pads.

The American cockroach, *Periplanēta americāna* (Fig. 304), is a native of tropical or subtropical America that has become distributed both in tropical and mild climates over the entire world. This is a large species measuring from 25 to 33 mm. in length.

The common wood-cockroach, *Parcoblătta pennsylvănica*, is a common species throughout the eastern half of the United States,

ORTHOPTERA

and its range extends into southern Canada. It is a na-

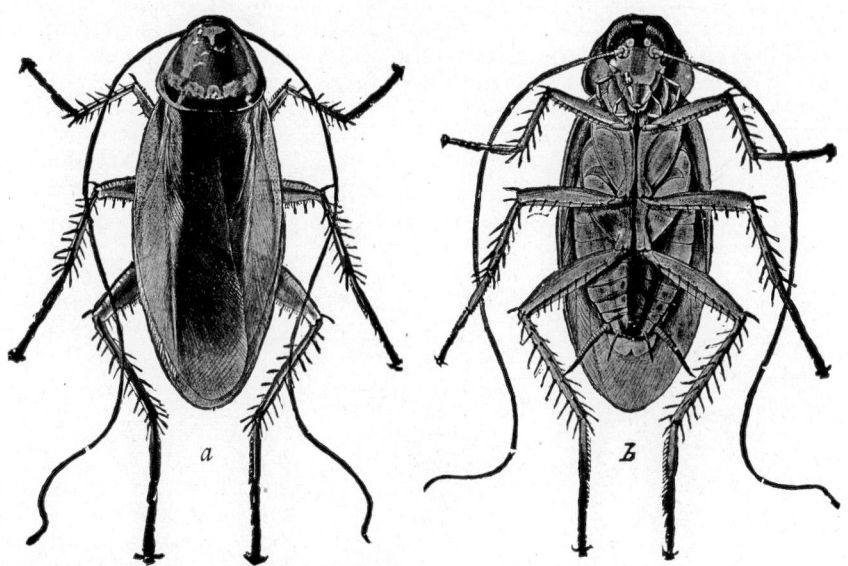

Fig. 304.—The American cockroach. (From Howard and Marlatt.)

tive of our woods but is frequently attracted to lights in our houses. The two sexes differ so greatly in appearance that they were long believed to be distinct species. In both sexes the lateral margins of the pronotum are light in color while the disk is dark. In the male the body measures from 15 to 25 mm. in length and the wings extend beyond the tip of the abdomen (Fig. 305). The female is smaller and the wings are much shorter than in the male (Fig. 306)

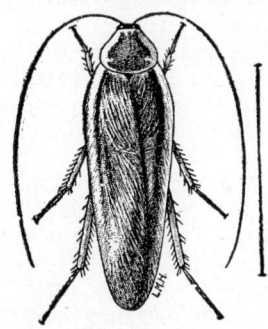

Fig. 305.—The common wood-cockroach, male. (From Lugger.)

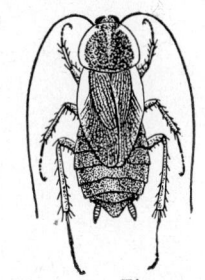

Fig. 306.—The common wood-cockroach, female. (From Blatchley.)

ORTHOPTEROID INSECTS OF UNCERTAIN KINSHIP

Under this head is placed a family of insects the zoological position of which has not been definitely determined.

Family GRYLLOBLATTIDÆ

This family was recently established by Dr. E. M. Walker ('14) for the reception of the species described below, which, while showing striking affinities to the Orthoptera, differs remarkably from all other known members of this order. Some writers who favor the breaking up of the order Orthoptera into several orders, regard this species as the type of a distinct order of insects, the Notoptera.

Grylloblătta campodeifŏrmis.—In this the only species of the family known, the body is elongate, slender, depressed, and thysanuriform

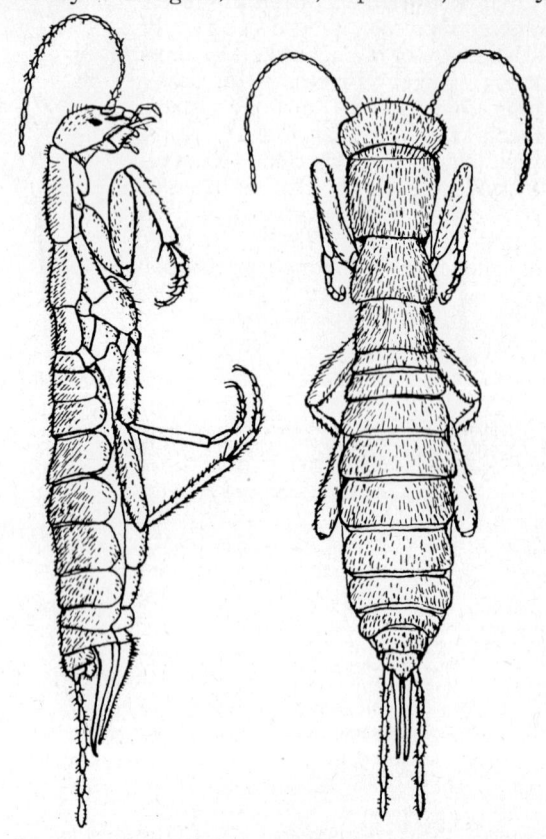

Fig. 307.—*Grylloblatta campodeiformis.* (After Walker.)

(Fig. 307). The legs are fitted for running, the tarsi are five-jointed and lack pulvilli. The cerci are long, about as long as the hind tibiæ, slender, and eight-jointed. The ovipositor is exserted and resembles that of the Tettigoniidæ. The eyes are small and the ocelli are absent. The adult male measures 16.5 mm. in length; the female, 30 mm.

As yet, this species has been found only in the vicinity of Banff, Alberta, and in Plumas County, California. It is found under stones, at high altitudes, and runs like a centipede.

Family HEMIMERIDÆ

Further studies of this family during the years since the publication of former editions of the *Introduction* seem clearly to have established the affinity of the *Hemimeridae* with the order Dermaptera. Recent explorations and collections from rats of Africa have brought to light several new species in the family and much additional knowledge concerning their habits. We have, therefore, placed the family with a brief discussion of it under the Dermaptera, on page 463. In 1925 Professor Comstock said concerning this family, "although these are exotic insects, they are mentioned here on account of their exceptional manner of development and mode of life." The additional knowledge concerning them emphasizes this observation and justifies retaining a brief discussion of the family among its proper relatives.

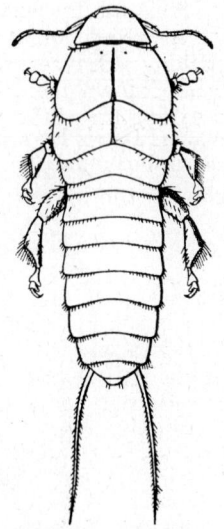

Fig. 308.—*Hemimerus hanseni*. (From Hansen.)

CHAPTER IX

ORDER ZORAPTERA*

So little is known regarding the insects of this order, only a single genus having been found, that it would be premature at this time to define definitely the characters of the order. This is well shown by the fact that recent discoveries have greatly modified our views regarding the ordinal characters of these insects.

This order was established by Silvestri in 1913. At that time only wingless individuals were known; and it was supposed by this author that the wingless condition was a distinctive ordinal character; he, therefore, proposed the name Zoraptera for the order. But recently Caudell ('20) has described winged individuals of each of the two species found in this country. The name Zoraptera, however, must be retained even though it is inappropriate.

Family ZOROTYPIDÆ

The single known genus, *Zorŏtypus*, is the type of this family and until other genera are found the characters of this genus must be taken as those of this family and of the order Zoraptera as well.

At the time this is written, only six species of *Zorotypus* have been described. These have been found in widely separated parts of the world, one each in Africa, Ceylon, Java, and Costa Rica, and two in Florida. One of the species from Florida has been found also in Texas.

The known species are all minute, the largest measuring only 2.5 mm. in length. In our two species both wingless and winged adults have been found; and it is probable that these two forms exist in the other species. The winged adults that have been observed are all females; but it would not be wise to conclude that only this sex is winged. Of the wingless form both male and female have been found. As these are social insects, living in colonies of various sizes, it may be that the wingless and the winged adults represent distinct castes, analogous to the castes of termites. Another similarity to termites is that the winged individuals shed their wings as do the winged termites.

The wingless adults (Figure 309, 4) resemble in general appearance small worker termites; but they have longer legs and are more active. The legs are formed for running; the tarsi are two-jointed and each bears two claws. The mandibles are strong. The antennæ are moniliform and nine-jointed. Compound eyes and ocelli are wanting. The cerci are short, fleshy, and unsegmented.

The winged adult female (Fig. 309, 1) has large compound eyes, three ocelli, nine-jointed antennæ, and two pairs of wings. The vena-

---

*Zorăptera: *zoros* (swpòs), pure; *apterous* (ἄπτερος), without wings.

ZORAPTERA 271

tion of the wings is represented in the figure. As the tracheation of the wings of nymphs has not been studied, I will not venture to make any suggestions regarding the homologies of the wing-veins.

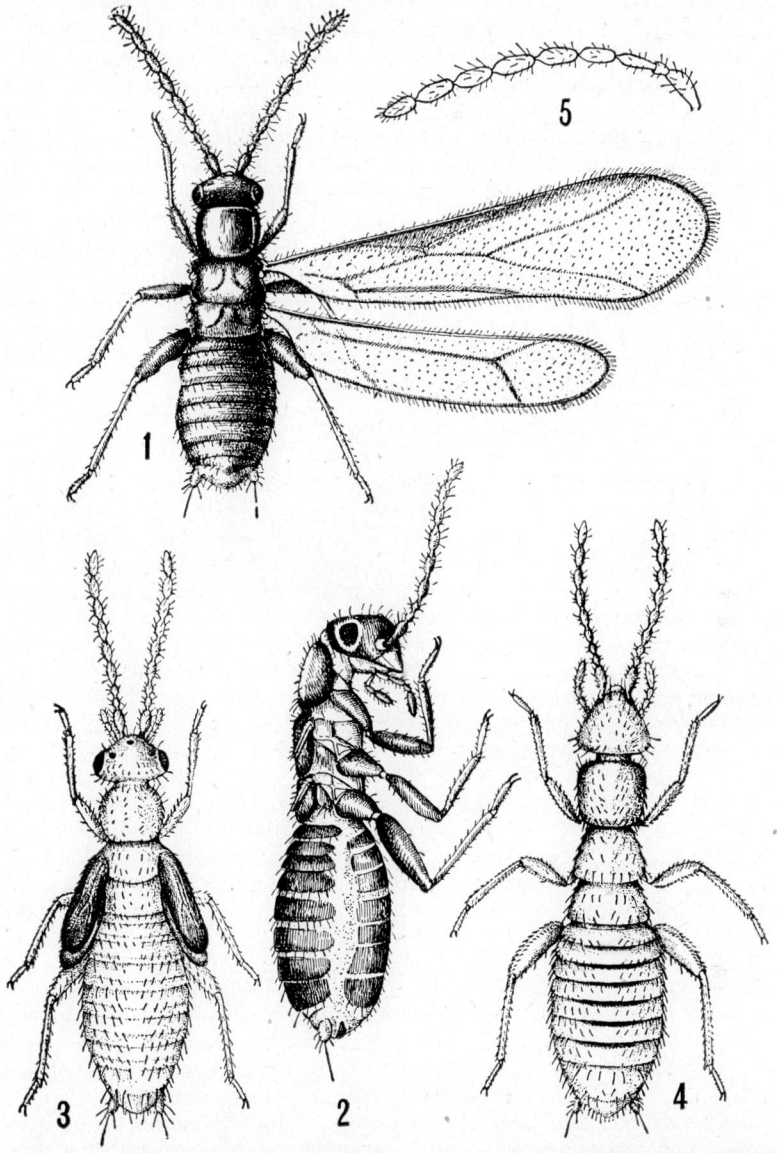

Fig. 309.—*Zorotypus hubbardi:* 1, winged adult female; 2, adult female that had shed her wings; 3, nymph of winged form; 4, wingless adult female. 5, Antenna of adult wingless *Zorotypus snyderi*. (From Caudell, in Proc. Ent. Soc. Wash., Vol. 22.)

Figure 309, 2, represents an adult female that had shed her wings; and Figure 309, 3, a nymph with well-developed wing-pads.

The two known American species are *Zorotypus hubbardi* and *Zorotypus snyderi*. Detailed descriptions of each of the forms of each of these species are given by Caudell ('20), and the external anatomy of *Zorotypus hubbardi* is described by Crampton ('20 a), who also discusses the relationships of the order Zoraptera to the other orders of insects.

The colonies of *Zorotypus* are found under the bark of logs and stumps and frequently near the galleries of termites. For this reason they were formerly believed to live as inquilines in the nests of termites; but recent observations do not support this view.

CHAPTER X

ORDER ISOPTERA*

The Termites or White-Ants

The members of this order are social insects, living in colonies like ants. Each species consists of several distinct castes, the number of which differs in different species. Each caste includes both male and female individuals. In most species there are four castes as follows: first, the first reproductive caste, in which the wings become fully developed and are used for a swarming flight and then shed; second, the second reproductive caste, in which the wing-buds remain short; the members of this caste are neoteinic, becoming sexually mature while retaining the nymphal form of the body; third, the worker caste; and fourth, the soldier caste. Except in a single Australian genus, the two pairs of wings are similar in form and in the more general features of their venation; they are long and narrow, and are laid flat on the back when not in use. The abdomen is broadly joined to the thorax; the mouth-parts are formed for chewing; the metamorphosis is gradual.

The termites or white-ants are chiefly tropical insects; but some species live in the temperate zones. These insects can be easily recognized by the fact that they live in ant-like colonies, by the pale color of the greater number of individuals of which a colony is composed, and by the form of the abdomen, which is broadly joined to the thorax instead of being pedunculate as in ants.

The termites are commonly called white-ants on account of their color and of a resemblance in form and habits to the true ants. These resemblances, however, are only very general. In structure the termites and ants are widely separated. In habits there is little more in common than that both are social, and the fact that in each the function of reproduction is restricted to a few individuals, while the greater number differ in form from the sexually perfect males and females, and are especially adapted to the performance of the labors and defense of the colony.

The cuticula of termites is delicate even in adults; the mature winged forms can withstand exposure to dry air for a limited period, as is necessary during their swarming flight; but other members of a colony quickly become shriveled and die if exposed. It is for this reason that they build tubes constructed of earth and excrement for passage-ways, and only rarely appear in the open, and then merely for a brief period.

The mouth-parts, which are fitted for chewing, are quite generalized, resembling somewhat those of the Orthoptera; but in the case of the soldier caste the mandibles are very large and vary greatly in form in the different species.

---

*Isŏptera: *isos* (ἴσος), equal; *pteron* (πτερόν), a wing.

The antennæ are moniliform; in the winged adults the number of segments varies in different species from twelve to twenty-five or more. In newly hatched nymphs the number of antennal segments is less than in the later instars.

The members of the winged sexual caste have pigmented compound eyes and a pair of ocelli. It is commonly stated that both the workers and soldiers of termites are blind; but in some species the soldiers have compound eyes; these, however, are not pigmented. There is an African species, the marching termite, in which both workers and soldiers possess eyes (Fuller '12).

The median ocellus is wanting in termites; but in many forms there is in its place a more or less distinct opening of a gland, the frontal gland, whose secretion is used for defense; this opening of the frontal gland is termed the *fontanĕl* or *fontanelle* or *fenĕstra*.

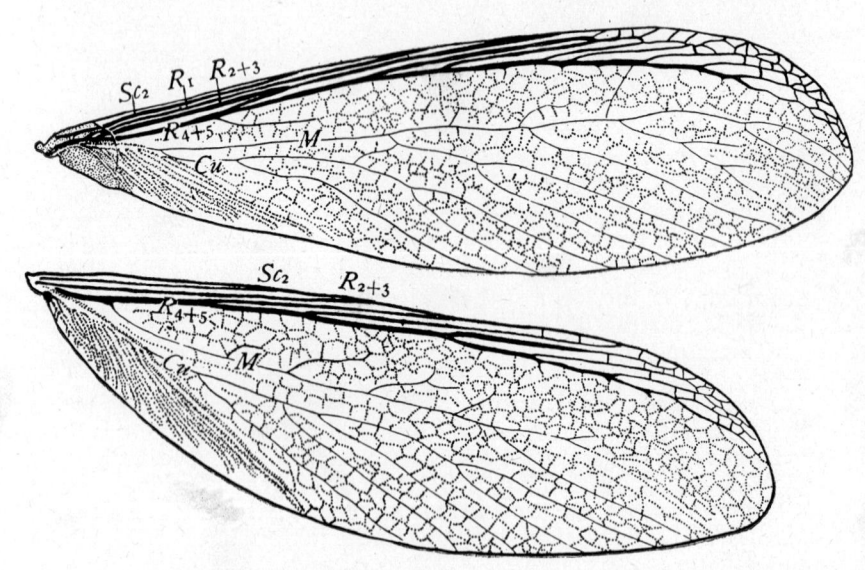

Fig. 310.—Wings of *Termopsis angusticollis*.

The wings are long and narrow, and, when folded on the back of the insect, extend far beyond the end of the abdomen. In the Australian genus *Mastotermes*, the anal area of the hind wings is broadly expanded; in other termites the fore and hind wings are similar in form (Fig. 310). In each case, the veins of the anterior part of the wing are greatly thickened and those of the middle portion reduced to indistinct bands or to narrow lines. Regular cross-veins are lacking, the membrane of the wings being strengthened by an irregular network of slightly chitinized wrinkles. The wings are deciduous, being shed after the swarming flight. The shedding of the wings is facilitated by the presence in each wing, except in the hind wings of certain genera, of a curved transverse suture, the *humeral suture*.

The homologies of the wing-veins are discussed by the writer in his "The Wings of Insects," Chapter VIII.

The abdomen consists of ten visible segments and bears a pair of two- to six-jointed cerci. The genitalia are vestigial and are concealed by a backward prolongation of the sternum of the seventh abdominal segment.

If a colony of termites be examined, many kinds of individuals will be found. This multiplicity of forms is due partly to the fact that these insects undergo a gradual metamorphosis, and nymphs of various sizes and degrees of development will be found running about among the mature individuals. But even if only adults be considered, it will be found that each species consists of several distinct castes.

With the termites the number of castes is greater than with the social bees, social wasps, and ants; and each caste includes both male and female individuals. The termites differ also from other social insects in that there are at least two and sometimes three castes whose function is reproduction. The following castes have been found among these insects.

The first reproductive caste.—At a certain season of the year, late spring or early summer for our most common species in the eastern United States, there can be found in the nests individuals with fully developed wings. These are sexually perfect males and females and constitute what is known as the *first reproductive caste*. In these the cuticula is black or dark chestnut in color, the eyes are functional, and the wings project more than half their length beyond the end of the body. A little later, these winged individuals leave the nest in a body; sometimes clouds of them appear. After flying a greater or less distance they alight on the ground, and then shed their wings.

At this time the males seek the females and they become associated in pairs; but the fertilization of the females does not take place till later. It seems probable that in some cases swarms issue from different nests at the same time, as we know to be the case with the true ants, and that in this way males and females from different nests may pair, and thus the danger of inbreeding be lessened; but Holmgren and others doubt that this occurs. The greater number of individuals comprising one of these swarms soon perish; they fall victims to birds and other insectivorous animals.

Each of the more fortunate couples that have escaped their enemies, find a suitable place for the beginning of a nest and become the founders of a new colony. Such a pair are commonly known as the *king* and the *queen* of the colony; they are also known as the *primary royal pair* to distinguish them from the second reproductive caste. The primary royal pair can be recognized by the presence on the thorax of the stumps of the wings that they have shed.

After the nest has been begun, the abdomen of the female becomes greatly enlarged, as a result of the growth of the reproductive organs and their products; this is greater in certain exotic species than it is in those found in this country. Figure 311 represents in natural size the queen of a species found in India. The dark spots along the middle of the dorsal wall of the abdomen are the chitinized parts of that region; the lighter portions are made up of the very much stretched membrane uniting the segments. This queen is a comparatively small one; in some species the queens become from 150 to 200 mm. in length; of course such a queen is incapable of locomotion, but lives with its mate inclosed in a royal chamber; their food is brought to them and the eggs are carried away by workers. In our native species the queens do not become so greatly enlarged and do not lose the power of movement.

A remarkable peculiarity in the habits of termites is that the association of the male and the female is a permanent one; the king and the queen live together in the nest, and there is repeated coition.

The second reproductive caste.—There are frequently found in the nests of termites neoteinic sexual forms; that is, individuals which are sexually mature but which retain the nymphal form of the body, having short wing-buds which do not develop further. These individuals constitute *the second reproductive caste*, which is represented by both males and females. The members of this caste are pale in color; their compound eyes are only slightly pigmented; and they never leave the nest unless by subterranean tunnels. If a primary king or queen dies, its place is taken by individuals of the second reproductive caste. For this reason, the members of this caste are commonly known as *substitute kings* and *queens* or as *complemental kings* and *queens*. The substitute queens produce comparatively few eggs, and consequently it requires several of them to replace a primary queen. Many pairs of substitute kings and queens are commonly found in orphaned nests.

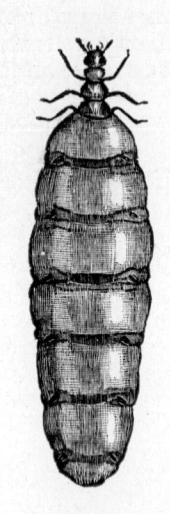

Fig. 311.—Queen termite, *Termes gilvus*.

The third reproductive caste.—In some cases there have been found adult neoteinic sexual forms which resemble workers in lacking wing-buds. These are known as *ergatoid kings* and *queens*.

The workers.—If a termite nest be opened at any season of the year there will be found a large number of wingless individuals of a dirty white color, usually blind, and of the form represented by Figure 312. These are named the *workers*, for upon them devolve nearly all of the labors of the colony. A study of the internal anatomy of workers has shown that both sexes are represented in this caste; the reproductive organs are, however, only little developed as a rule; but occasionally workers capable of laying eggs are found. The worker caste is not always present; it is absent in the genera *Kalo-*

Fig. 312.—A worker.

termes and *Cryptotermes*, where the nymphs of the reproductive forms apparently attend to the duties of workers; and in the genera *Termopsis* and *Neotermes* ordinary sterile workers are not found, but the third reproductive caste, large, worker-like, grayish brown, fertile forms, with no wing-buds, is present. The nymphs of this caste often perform the duties of the workers (Banks and Snyder '20). In some tropical species there are two types of workers, which differ in size.

The soldiers.—Associated with the workers, and resembling them in color and in being wingless, there occur numerous representatives of another caste, which can be recognized by the enormous size of their heads and mandibles (Fig. 313); these are the soldiers. They are so named because it is believed that their chief function is the protection of the colony; but they do not seem to be very effective in this. Among the soldiers, as among the workers, both sexes are represented; but as a rule the reproductive organs are not functional. Sometimes, however, soldiers capable of laying eggs are found. In the genus *Kalotermes* soldiers with small wing-buds are often found. The soldier caste has not been found in the genus *Anoplotermes*, and in the genera *Constrictotermes* and *Nasutitermes* the soldier caste is wanting unless the nasuti, described in the next paragraph, are regarded as soldiers. In certain tropical species there are two types of soldiers which differ in size.

Fig. 313.—A soldier.

The nasūti.—In certain species of termites there are found individuals in which the head is elongated into a nose-like process, from the tip of which a fluid exudes, which is used as a means of defense and also, it is said, as a cement in constructing the nest and the earth-like tubes through which the insects travel (Fig. 314). Such individuals are known as *nasuti*. In this caste the mandibles are small, differing greatly from those of soldiers. The nasuti are usually smaller than the workers and are pigmented. They have been commonly described as a special type of soldiers; but it seems better, in order to avoid confusion, to regard them as constituting a distinct caste. Among the North American termites, nasuti are found in the genera *Constrictotermes* and *Nasutitermes*, which lack the true soldier caste. In some tropical termites two types of nasuti, large and small, have been found.

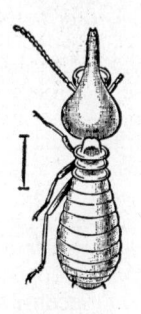

Fig. 314.—A nasutus. (After Sharp.)

As to the origin of the different castes of termites there has been much discussion and two radically different views are held. The first view was probably suggested by the well-known fact that, in the case of the honey-bee, queens can be developed from eggs or young larvæ that ordinarily would produce workers. According to this view the newly hatched termites are not differentiated into castes; but this differentiation takes place later as the result of extrinsic factors, such as food, the presence or absence of parasitic protozoa in the alimentary tract, and the care received from the older workers. According to the second view the young of the different castes are different and the castes are therefore "predetermined in the egg or embryo by intrinsic factors."

Some comparatively recent investigations support the second view. It was found by Thompson ('17 and '19) that although the newly hatched nymphs are externally all alike, they are differentiated by internal structural characters into two clearly defined types: first, the *reproductive* or *fertile forms*, with large brain and large sex organs, and usually a dense opaque body; and second, the *worker-soldier* or *sterile forms*, with small brain and small sex organs, and usually a clear transparent body.

It was also found by Thompson that later, when the nymphs had become from 2 mm. to 3 mm. in length, they were differentiated into "small-headed" but large-brained reproductive forms, and "large-headed" but small-brained worker-soldier forms. In the case of worker-soldier nymphs of *Eutermes pilifrous*, a Jamaican species, which were 2 mm. long and externally all alike, they were distinguishable, after staining, into worker nymphs with a small vestigial frontal gland, and soldier nymphs with a large frontal gland.

In a study of *Reticulitermes (Leucotermes) flavipes*, Thompson ('17) found that the nymphs of the reproductive forms that are only 1.3 to 1.4 mm. in length are differentiated into two groups by differences in the size of the brain and sex organs. These are early instars of the first reproductive caste and the second reproductive caste, respectively. The early instars of the third reproductive caste have not been distinguished from the nymphs of workers.

There is space here for but little regarding the nest-building habits of these wonderful insects. In the tropics certain species build nests of great size. Some of these are mounds ten or twelve feet or more in height. Other species build large globular masses upon the trunks or branches of trees or upon other objects. Figure 315 represents such a nest which I observed on a fence in Cuba. Owing to the delicacy of their cuticula and the consequent danger of becoming shriveled if exposed, the termites build covered ways from their nests to such places as they wish to visit, if they are in exposed situations like that shown in the figure. These exposed nests are composed chiefly of the excreted undigested wood upon which the insects have fed. This is molded into the desired form and on drying it becomes solid.

The termites that live in the United States do not build exposed nests; and, as the queens do not lose the power of movement, there is no permanent royal cell, centrally located, in which the king and queen are imprisoned, as is the case with many tropical species. Some of our species mine in the earth, their nests being made under stones or other objects lying on the ground; some burrow only in wood; and others that burrow in the ground extend their nests into wood. To the last category belong the species of the genus *Reticulitermes*, which includes all of the termites found north of Georgia and east of Nevada. These often infest the foundation timbers of buildings, flooring in basements, and other woodwork of buildings and furniture. These pests will feed upon almost any organic matter; books are sometimes completely ruined by them. In infesting anything composed of wood, they eat out the interior, leaving a thin film on the outside. Thus a table may appear to be sound, but crumble to pieces beneath a slight weight, entrance having been made through the floor of the house and the legs of the table.

While termites infest chiefly dead wood, there are many records of their infesting living plants. I found them common throughout Florida, infesting orange-trees, guava-bushes, pampas-grass, and sugar-cane. When termites infest living plants, they attack that part which is at or just below the surface of the ground. In the case of pampas-grass the base of the stalk is hollowed; with woody plants, as orange-trees and guava bushes, the bark at the base of the tree is eaten and frequently the tree is completely girdled; with sugar-cane the most serious injury is the destruction of the seed-cane.

Fig. 315.—Nest of a Cuban termite.

Certain African termites have been found to cultivate fungus-gardens in their nests, similar to those of the well-known leaf-cutting ants.

The care of the young and of the queens by the workers in colonies of social insects has attracted the attention and admiration of observers in all times. This care has been quite generally attributed to something resembling the parental feelings of our own species. But the observations of several naturalists in recent years have shown that with the social insects the devotion of the workers to the brood and to the queen is far from being purely altruistic; that it is largely or entirely due to a desire to feed upon certain exudates produced by the individuals that are fed; the feeding of the young and the queens being accompanied by a licking of their bodies by the nurses.

Among the more important papers on this subject are one on termites by Holmgren ('09) and one on ants by Wheeler ('18). Holmgren shows that all of the castes of termites, but especially the queens, have extensive exudate tissues, consisting of the peripheral layers of the abdominal fat-body, the products of which pass through pores in the cuticula, where they are licked up by other members of the colony; and that the intensity of the licking and feeding of the individuals of a termite colony is directly proportional to the amount of their exudate tissue. Wheeler in his paper on ants shows that the larvæ of certain species of ants possess remarkable exudate organs, and proposes for this exchange of nourishment the term *trophallaxis*.*

It is believed by the writers quoted above and by others that this exchange of nourishment between those individuals that feed and those that are fed was the source of the colonial habit in social insects. Roubaud ('16) in a paper on the wasps of Africa points out the probable steps by which the social habit was developed in wasps. Beginning with certain solitary eumenids that feed their larvæ from day to day and while doing this feed upon saliva exuded by the larvæ, he suggests that there naturally follows a tendency to increase the number of larvæ to be reared simultaneously in order at the same time to satisfy the urgency of oviposition and to profit by the greater abundance of the secretion of the larvæ.

Now that this explanation of the origin of the social habit has been suggested, it, doubtless, will be much discussed. The student is urged, therefore, to consult the current literature for opinions regarding it.

The most extended account of the termites of this country is the recently published paper by Nathan Banks and Thomas E. Snyder ('20). In the first part of this paper, Mr. Banks gives a revision of the nearctic termites, in which all of our known species are described, seventeen of them for the first time. This brings the total number of our known species up to thirty-six, representing ten genera.

In the second part of this paper, Mr. Snyder brings together the known data regarding the habits and distribution of the termites of the United States; much of which data is based on his personal observations.

Many species of insects live in the nests of termites. The relations of the termitophiles, of which several hundred species have been described, to their hosts vary greatly; some are predatory, some are parasites, and others are guests. Among the guests some are indifferently tolerated, while others are true guests which produce exudates that are eagerly devoured by their hosts and in return either receive regurgitated food or manage to prey on the defenseless brood. Among the termitophiles are some that are very remarkable in form, having the abdomen excessively enlarged and being furnished with large exudate organs.

---

*Trophallaxis: *trophe* (τροφή), nourishment; *allattein* (ἀλλάττειν), to exchange.

CHAPTER XI

Order NEUROPTERA*

The Horned Corydalus, the Lacewing-Flies, the Ant-Lions, and others

The members of this order have four wings; these are membranous and are usually furnished with many veins and cross-veins. In most members of the order, the wings have been specialized by the addition in the preanal area of many supernumerary veins of the accessory type. The mouth-parts are formed for chewing. The tarsi are five-jointed. The cerci are absent. The metamorphosis is complete.

The order Neuroptera as now restricted differs greatly in extent from the Neuroptera of the early entomologists. Formerly there were included in this order many insects that are no longer believed to be closely related. This has resulted in the establishment of several distinct orders for the insects that have been removed from the old order Neuroptera. This fact should be kept in mind when consulting the older text-books.

The wings of the Neuroptera are membranous and usually furnished with many wing-veins. The two pairs of wings are similar in texture and usually in outline; in some the fore wings are slightly larger than the hind wings, in others the two pairs of wings are of the same size. The anal area is small in both fore and hind wings; it is rarely folded (Sialidæ), and then only slightly so. A distinct anal furrow is rarely developed. Definitive accessory veins are usually present, and, as a rule, there are many marginal accessory veins. Intercalary veins are never developed. When at rest, with few exceptions, the wings are folded roof-like over the abdomen. In some cases organs for uniting the fore and hind wings are present.

Correlated with the extensive development of accessory veins in the Neuroptera, there has resulted in nearly all of the families of this order the production of a pectinately branched radial sector; that is, this vein is so modified that it consists of a supporting stem upon which are borne a greater or less number of parallel branches. This is shown in most of the figures of wings illustrating this chapter, and is a distinctive characteristic of this order; in no one of the other orders of living insects in which accessory veins occur is a well-developed pectinately branched radial sector found. Such a radial sector existed, however, in many of the Paleozoic insects. In certain genera of the Neuroptera a dichotomously branched radial sector has been retained.

In many Neuroptera one or more series of cross-veins extend across the wing and form with sections of the longitudinal veins that

---

*Neurŏptera: *neuron* (νεῦρον), a nerve; pteron (πτερόν), a wing.

they connect a very regular zigzag line; such cross-veins are termed *gradate veins*. Examples of series of gradate veins are well shown in the wings of the Hemerobiidæ and in those of the allied families.

The mouth-parts are formed for chewing. In several families the larvæ suck the blood of their prey by means of their peculiarly modified mandibles and maxillæ. These are very long and those of each side form an organ for piercing and sucking. The mouth-parts of the larva of an ant-lion will serve to illustrate this type of mouth-parts (Fig. 316).

In this insect the mandibles (*md*) are very long, curved at the distal end, fitted for grasping and piercing the body of the prey, and armed with strong spines and setæ. On the ventral aspect of each mandible there is a furrow extending the entire length of the mandible; and over this furrow the long and slender maxilla (*mx*) fits. On the dorsal aspect of the maxilla there is also a furrow. These two furrows form a tube which extends from the tip of the combined mandible and maxilla to the base of this organ where it communicates with the mouth cavity. Through this tube the blood of the prey is conveyed to the mouth. On the middle line of the body, between the mentum (*m*) and the front margin of the dorsal wall of the head (*l*), there is a tightly closed slit which is the mouth; this, however, is not functional, the food being received into lateral expansions of the mouth-cavity at the base of the mandibles and maxillæ.

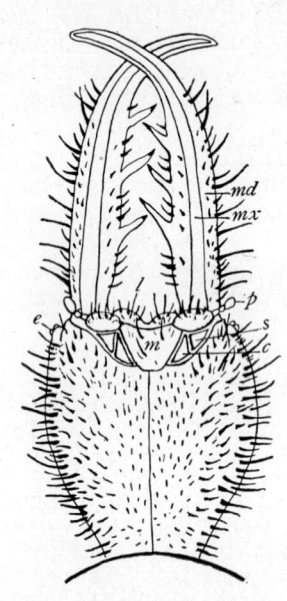

Fig. 316.—Head and mouth-parts of a larva of an ant-lion, ventral aspect: *c*, cardo of the maxilla; *e*, eye; *l*, front margin of the dorsal wall of the head, labrum (?); *m*, mentum; *md*, mandible; *mx*, maxilla; *p*, labial palpus; *s*, stipes of the maxilla.

For a more detailed account of the structure of the mouth-parts of an ant-lion, see Lozinski ('08).

The metamorphosis is complete. The larvæ that are known are predacious or parasitic and in most cases are campodeiform; a few of them are aquatic, Sialidæ, Sisyridæ, and certain exotic forms, but most of them are terrestrial; some when full-grown enter the ground and make earthen cells in which they transform, but most of them spin cocoons. The silk of which these cocoons are made, in the case of those in which the silk-organs have been described, is secreted by modified Malpighian vessels and is spun from the anus.

The silk-organs of *Sisyra* will serve as an example of neuropterous silk-organs; these were described by Miss Anthony ('02). Figure 317 is a diagram of a sagittal section of a larva through the median plane. In this larva the posterior fourth of the mid-intestine is merely a solid cord of atrophied cells; the passage from the mid-intestine to the hind-intestine is thus closed. The atrophied part of the mid-intestine ends in the walls of a dilation, the silk-receptacle (*sr*). Into this receptacle empty the five Malpighian tubes, three of which are attached by both ends and two of which extend posteriorly and end free in the body cavity; all are modified in their middle portions for the secretion of silk; here the cells are much larger and more irregular in shape than the ordinary Malpighian tubule cells, and show singular, branched nuclei like those characteristic of silk-gland-

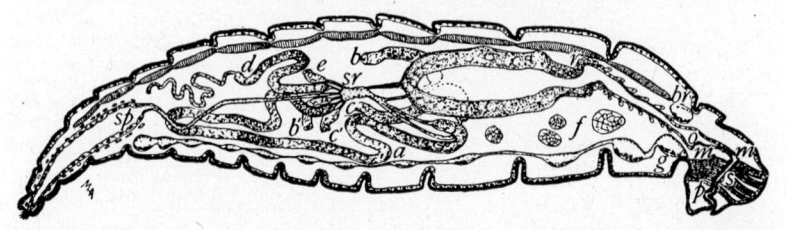

Fig. 317.—Sagittal section of a larva of *Sisyra*: *a*, *b–b'*, *c–c'*, three silk-glands attached at both ends; *d*, *e*, two silk-glands attached at one end; *sr*, silk-receptacle; *sp*, spinneret; *f*, fat-bodies; *br*, brain; *g*, subœsophageal ganglion; *r*, band of regenerative cells of the stomach; *p*, point of junction of sucking tubes; *s*, sucking pharynx; *m*, muscle attachment of pharynx; *o*, œsophagus. (From Anthony.)

cells of caterpillars and other insects. That part of the hind-intestine extending back from the silk-receptacle is a slender tube for the greater part of its length; but in the last three abdominal segments it is enlarged, forming a reservoir for accumulated silk (*sp*), which is spun from the anus when needed for making the cocoon.

The pupæ of Neuroptera are exarate, that is, their legs and wings are free. In some cases (*Chrysopa*, *Hemerobius*, and *Mantispa*) the pupa crawls about for a time after leaving its cocoon and before changing to the adult.

The known Neuroptera of the world represent twenty families; the wings of one or more members of each of these families have been figured by the writer in his "The Wings of Insects." Thirteen of these families are represented in North America; these can be separated by the following table.

TABLE OF THE FAMILIES OF NORTH AMERICAN NEUROPTERA

A.   Prothorax as long as or longer than the mesothorax and metathorax combined.
    B.   Fore legs greatly enlarged and fitted for grasping. p. 289. MANTISPIDÆ
    BB.   Fore legs not enlarged and not fitted for grasping. p. 289. RAPHIDIIDÆ
AA.   Prothorax not as long as the mesothorax and metathorax combined.
    B.   Hind wings broad at base and with the anal area folded like a fan when not in use. p. 284.. SIALIDÆ
    BB.   Hind wings narrow at base and not folded like a fan when closed.
        C.   Wings with very few veins and covered with **whitish powder. p. 307.**
.... **CONIOPTERYGIDAE**
        CC.   Wings with numerous veins and not covered with powder.
            D.   Antennæ gradually enlarged towards the **end or filiform with a** terminal knob.

  E. Antennæ short; wings with an elongate cell behind the point
    of fusion of veins Sc and R_1. p. 303......Myrmeleonidæ
  EE. Antennæ long; wings without an elongate cell behind the
    point of fusion of veins Sc and R_1. p. 305.....Ascalaphidæ
 DD. Antennæ not enlarged towards the end.
  E. Male with pectinate antennæ; female with an exserted
    ovipositor. p. 297.........................Dilaridæ
  EE. Antennæ not pectinate in either sex; female without exserted ovipositor.
   F. Radius of the fore wings with apparently two or more sectors.
    G. Radius of the fore wings with apparently two sectors, one of which is vein R_{2+3} and the other vein R_{4+5}. p. 293......................Sympherobiidæ
    GG. Radius of the fore wings with three or more sectors. Veins R_4 and R_5 arise separately from vein R_1; one or more definitive accessory branches of the radius of the fore wings present. p. 294........
     ....Hemerobiidæ
   FF. Radius of the fore wings with a single sector.
    G. Radial sector of the fore wings without definitive accessory veins although marginal accessory veins are present. p. 291..........Sisyridæ
    GG. Radial sector of fore wings with definitive accessory veins.
     H. Transverse veins between the costa and subcosta simple. p. 299..........Chrysopidæ
     HH. Many of the transverse veins between the costa and subcosta forked.
      I. Humeral cross-vein recurved and branched; first radio-medial cross-vein of the hind wings longitudinal and sigmoid. p. 298.............Polystœchotidæ
      II. Humeral cross-vein not recurved; first radio-medial cross-vein of the hind wings transverse. p. 298..........Berothidæ

Family SIALIDÆ*

The Sialids

The members of the Siălidæ differ greatly in size and appearance; but they agree in having the hind wings broad at the base with the anal area folded like a fan when not in use. In this respect they differ from all other Neuroptera.

The type of the wing-venation of the sialids differs greatly in the two subfamilies into which the family is divided, as described below.

The larvæ are aquatic, predacious, campodeiform, and possess paired, lateral filaments on most or on all of the abdominal segments. They leave the water when full-grown and transform in earthen cells on the banks of the streams or lakes in which they lived as larvæ. The eggs are deposited in clusters on any convenient support near the water, in such situations that the young larvæ can easily find

---

*This family is separated from the Neuroptera by Handlirsch ('06–'08) as a distinct order, the Megaloptera. H. W. Van der Weele ('10) also separates it from the Neuroptera, but he associates with it the family Raphidiidæ in his order Megaloptera.

access to the water. The adults fly but little; they are most often found resting on some support near the water, with the wings folded over the abdomen.

The Sialidæ of the world have been monographed by H. W. Van der Weele ('10).

The family Sialidæ is divided into two subfamilies; these can be separated as follows:

A. Ocelli wanting; fourth tarsal segment prominently bilobed; radial sector not pectinately branched. Insects rather small, having an expanse of wings of about 25 mm. p. 285 .... SIALINÆ
AA. With three ocelli; fourth tarsal segment simple, not bilobed; radial sector pectinately branched. Insects large or moderately large. p. 286 .... CORYDALINÆ

Subfamily SIALINÆ

The Alder-Flies

The alder-flies are so-called because the adults are commonly found on alders on the banks of streams; this name was given to them by English anglers.

The subfamily Sialinæ includes only two genera, both of which are represented in this country, each by a single species. These genera are distinguished as follows:

A. Costal area of the fore wings greatly expanded before the middle (Fig. 318). .... SIALIS
AA. Costal area of the fore wings slightly expanded before the middle .... PROTOSIALIS

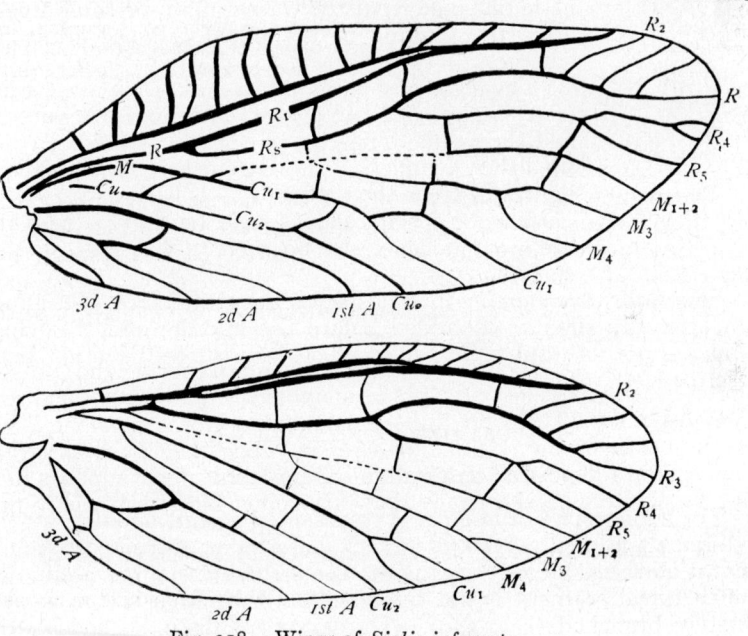

Fig. 318.—Wings of *Sialis infumata*.

The smoky alder-fly, *Sialis infumāta*.—This is a small insect having a wing-expanse of about 25 mm.; the males are sometimes smaller than this, and the females slightly larger. It is dusky brownish in color. It can be easily recognized by the form and venation of its wings (Fig. 318).

The costal area of the fore wings is greatly expanded before the middle, and most of the wing-veins are stout. A striking feature of these wings, one that is characteristic of the subfamily Sialinæ, is that the radial sector is nearly typical in form; the only modification being the development of one or more marginal accessory veins upon it. These accessory veins, however, are in a quite different position from that occupied by the accessory veins borne by the radial sector in the Corydalinæ, where a pectinately branched radial sector has been developed.

The larva (Fig. 319) is furnished with the paired lateral filaments characteristic of the larvæ of the Sialidæ on the first seven abdominal segments. These filaments are more or less distinctly five-segmented. The last abdominal segment is prolonged into a tapering lash-like filament.

The larvæ are found in swiftly flowing streams adhering to the lower side of stones in the bed of the streams and in trashy places filled with aquatic plants in the borders of streams and ponds; they are very active. The larvæ transform in earthen cells at some little distance from the water. Two or three weeks after the making of the pupal cell the adult fly emerges.

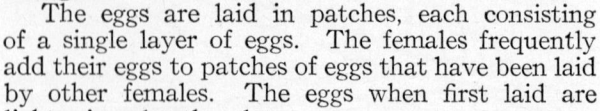

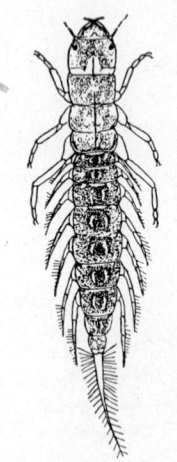

The eggs are laid in patches, each consisting of a single layer of eggs. The females frequently add their eggs to patches of eggs that have been laid by other females. The eggs when first laid are lighter in color than later.

Fig. 319.—Larva of *Sialis infumata*. (After Needham.)

Several specific names have been given to what are now believed to be merely varieties of this species.

Protosialis americāna.—In this species the costal area of the fore wings is only slightly expanded before the middle; and the wing-veins are not as stout as in *Sialis*. The early stages have not been described.

Subfamily CORYDALINÆ

Corydalus and the Fish-Flies

The subfamily Corydalinæ is represented in this country by the well-known horned corydalus and several smaller species, commonly known as the fish-flies. In these insects there are three ocelli; the fourth tarsal segment is not bilobed; and the radial sector is pectinately branched (Fig. 320). The larvæ are distinguished by the presence of a pair of anal prolegs, each of which bears a pair of hooks. Six species are found in the United States and Canada; these represent four genera, which can be separated as follows.

A. Latero-caudal angles of the head with a sharp tooth. Large insects. p. 287 .... CORYDALUS
AA. Latero-caudal angles of the head unarmed. Insects moderately large; the fish-flies.
    B. Wings somewhat ashy in color with more or less dusky markings.
        C. Veins of fore wings marked with dark and light, uniformly alternate. p. 288... CHAULIODES
        CC. Veins of fore wings uniform in color except where dusky markings cross them. p. 288.. NEOHERMES
    BB. Wings black or brown with white markings. p. 288........ NIGRONIA

Corydalus.—The only member of this genus in our fauna is *Corўdalus cornūtus*. This is a magnificent insect, which has a wing-expanse of from 100 to 130 mm. Figure 321 represents the male,

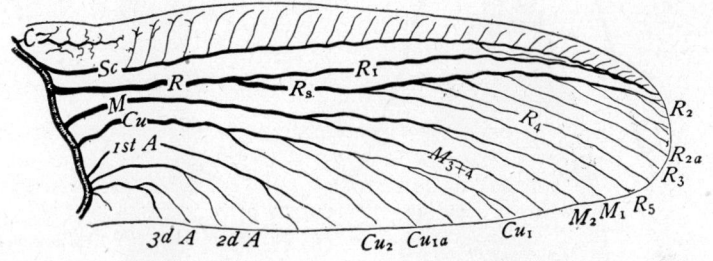

Fig. 320.—Fore wing of a pupa of *Corydalus*.

which has remarkably long mandibles. The female resembles the male, except that the mandibles are comparatively short.

The larvæ are called *dobsons* or *hellgrammites* by anglers and are used by them for bait, especially for bass. Figure 322 represents a full-grown dobson, natural size. These larvæ live under stones in the beds of streams. They are most abundant where the water flows swiftest. They feed upon the naiads of stone-flies and May-flies and on other insects.

The larvæ of *Corydalus* differ from those of the following genera in the possession of a tuft of hair-like tracheal gills at the base of each of the lateral appendages on the first seven abdominal segments.

When about two years and eleven months old, the larva leaves the water and makes a cell under a stone or some other object on or near the bank of the stream. This occurs during the early part of the summer; here the larva changes to a pupa. In about a month after the larva leaves the water, the adult insect appears. The eggs are then soon laid; these are attached to stones or other objects overhanging the water. They are laid in blotch-like masses, which are chalky white in color and measure from 12 mm. to nearly 25 mm. in diameter. A single mass contains from two thousand to three thousand eggs. When the larvæ hatch they at once find their way into the water, where they remain until full-grown.

Chauliodes.—See the table on page 287 for the distinguishing characteristics of this genus. There are two species in our fauna; these are distinguished as follows:

Chauliodes rastricŏrnis.—In this species the antennæ are serrate; the embossed markings on the head and prothorax are black on a paler ground; and the prothorax is longer than wide.

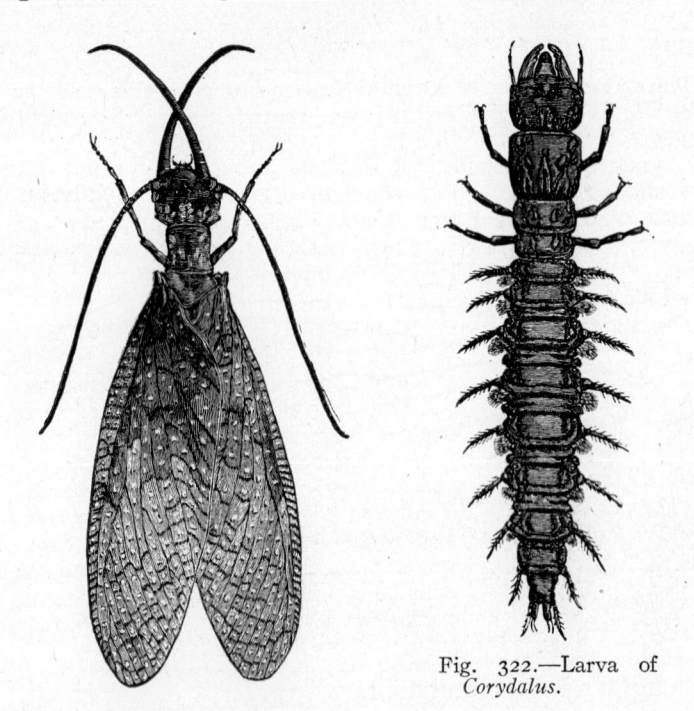

Fig. 322.—Larva of *Corydalus.*

Fig. 321.—*Corydalus cornutus*, male.

Chauliodes pectinicŏrnis.—In this species the antennæ are serrate; the embossed markings on the head and prothorax are yellow on a black ground; and the prothorax is not longer than wide.

Neohĕrmes.—This genus is represented by *Neohermes califŏrnicus*. In addition to the characters given in the table of genera above, this species is distinguished by the great length of its antennæ, which are about two-thirds as long as the fore wings.

Nigrōnia.—In this genus the wings are black or brown with white markings. Two species only are known; these can be distinguished by the form of the white markings on the wings.

Nigrōnia fasciātus.—In this species there is a continuous, broad, somewhat arcuate, white band extending across the middle of each wing and almost attaining the hind margin.

Nigrōnia serricŏrnis.—In this species there is an irregular band of white spots, generally broadest in front, extending across the middle of each front wing. On the hind wings, the white band is represented by only a few minute dots or is entirely wanting.

Family RAPHIDIIDÆ*

The Snake-Flies

The members of the Raphidiidæ are found in this country only in the Far West. They are strange-appearing insects, the prothorax being greatly elongate, like the neck of a camel (Fig. 323). The female bears a long, slender, sickle-shaped ovipositor at the end of the abdomen. The fore legs resemble the other pairs of legs, and are borne at the hind end of the prothorax. The wings are long and narrow and furnished with a pterostigma.

Fig. 323.—*Raphidia*, female.

The wing-venation of a representative of each of the two genera belonging to this family is figured by the writer in his "The Wings of Insects."

The larvæ are found under bark and are carnivorous. They are common in California under the loose bark of the eucalyptus. They also occur in orchards, and doubtless do good by destroying the larvæ and pupæ of the codlin-moth. The pupæ are not enclosed in silken cocoons but lie concealed in sheltered places. Figure 324 represents a larva and a pupa of *Raphidia* as figured by Professor Kellogg.

This family includes only two genera, *Raphidia* and *Inocellia*. In the former there are three ocelli on the top of the head between the compound eyes; in the latter these ocelli are wanting. Six species of *Raphidia* and three of *Inocellia* are found in America north of Mexico.

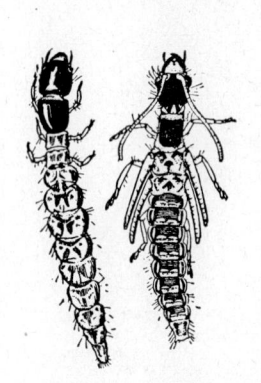

Fig. 324.—Larva and pupa of *Raphidia*. (From Kellogg.)

Family MANTISPIDÆ

The Mantis-like Neuroptera

The members of the Mantispidæ are even more strange in appearance than are those of the preceding family. Here, as in that family,

---

*This family is separated from the Neuroptera by Handlirsch ('06–'08) and made to constitute a separate order, the Raphidioidea.

the prothorax is greatly elongated; but the members of this family can be easily recognized by their remarkable fore legs, which are greatly enlarged and resemble those of the praying mantes in form (Fig. 325). These legs are fitted for seizing prey; and, in order that they may reach farther forward, they are joined to the front end of the long prothorax. In the adult stage, these insects are predacious; while the larvæ, so far as is known, are parasitic.

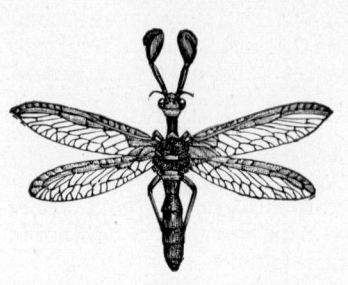

Fig. 325.—*Mantispa*. In the specimen figured the fore legs were twisted somewhat in order to show the form of the parts.

Brauer ('69) described the transformations of *Mantispa styriaca*, a European species. This insect undergoes a hypermetamorphosis. It was accidentally discovered that the larvæ were parasitic in the egg-sacs of spiders of the genus *Lycosa*. These are the large black

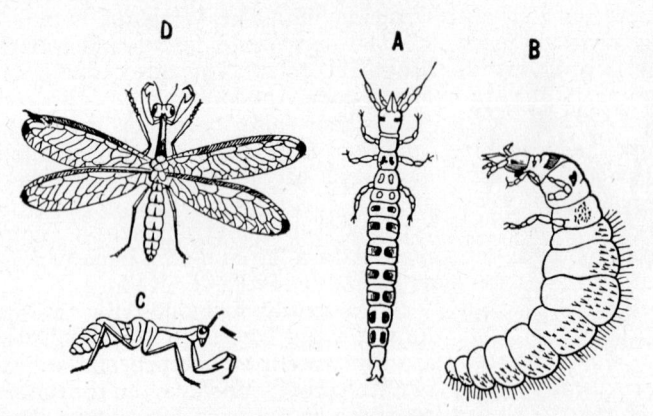

Fig. 326.—Hypermetamorphosis of *Mantispa*. (From Henneguy, after Brauer.)

spiders which are common under stones, and which carry their egg-sacs with them. Brauer obtained eggs from a female *Mantispa* kept in confinement. These eggs were rose-red in color, and fastened upon stalks, like the eggs of *Chrysopa*. The eggs were laid in July; and the larvæ emerged 21 days later. The young larvæ are campodeiform (Fig. 326, A); they are very agile creatures, with a long, slender body, well-developed legs, and long, slender antennæ. They pass the winter without food. In the spring they find their way into the egg-sacs of the above-named spiders. Here they feed upon the young spiders; and the body becomes proportionately thicker. Later the larva molts and undergoes a remarkable change in form, becoming what is known as the second larva; in this stage the larva is scarabeiform (Fig. 326, B); the legs are much reduced in size; the antennæ are short; and the head is very small. When fully grown this larva measures from 7 to 10 mm. in length. It then spins a cocoon, and changes to a pupa within the skin of the larva. Later the larval skin is cast; and, finally, after being in the cocoon about a month, the pupa becomes active, pierces the cocoon and the egg-sac, and crawls about for a time (Fig. 326, C); later it changes to the adult form (Fig. 326, D).

The life-history of *Sȳmphasis vāria*, a Brazilian species, is partly known. The larvæ of this species live parasitically in the nests of wasps; when full-grown each larva spins a cocoon in one of the cells of the nest.

Only a few representatives of this family occur in the United States, and all are rare insects.

Family SISYRIDÆ

The Spongilla-Flies

The Sisȳridæ include a very limited number of small, smoky brown insects, of the form shown in Figure 327. They are called *Spongilla-flies* because the larvæ live as parasites in fresh-water sponges, the typical genus of which is *Spongilla*. Two interesting features of these insects are the comparative simplicity of the wing-venation of the adults, and the anomalous habits of the larvæ.

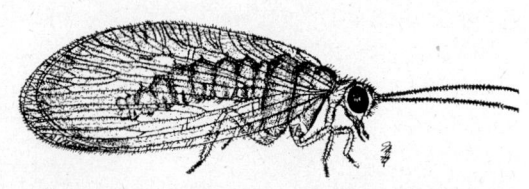

Fig. 327.—*Sisyra umbrata*, greatly enlarged. (From Anthony.)

The more striking characteristics of the wings (Fig. 328) are the following: The costal area of the fore wings is not greatly broadened; the humeral vein is not recurrent and is not branched. Veins Sc and R_1 coalesce near the apex of the wing. The radial sector is pectinately branched; but no definitive accessory veins have been developed; this is the simplest form of pectinately branched radial sector found in the fore wings in this order. Marginal accessory veins are present.

The larvæ are aquatic and live in fresh-water sponges, upon which they feed. The life-history of a representative of each of the two genera, *Sisyra* and *Climacia*, which constitute this family, was worked out by Professor Needham ('01); and the anatomy and transformations of a species of *Sisyra* were carefully studied by Miss Anthony ('02). The following notes are based on the accounts published by these authors.

Sĭsyra umbrāta.—The form of the adult is shown in Figure 327; its color is nearly uniform blackish brown. The legs and the apex of the abdomen are dirty yellowish. The length of the male to the tips of the wings is 6 mm; that of the female, 8 mm.

The larva (Fig. 329) is campodeiform. Its mouth-parts are formed for sucking as in the larvæ of ant-lions (see page 282); they

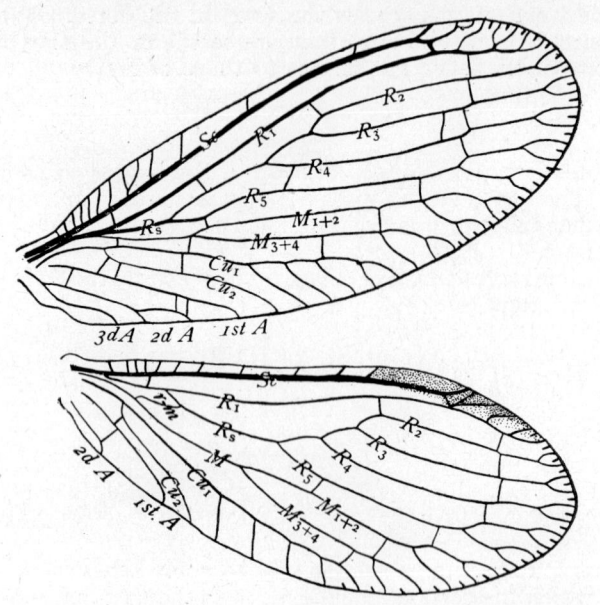

Fig. 328.—Wings of *Sisyra flavicornis*.

are very long; and two sucking organs, each formed of the mandible and maxilla of one side, are closely parallel for the greater part of their length. Each of the first seven abdominal segments bears on the ventral side a pair of jointed filaments which are believed to be tracheal gills. When full-grown, the larva leaves the water and spins over itself, on some object near the water, a hemispheric cover of closely woven silk, attached by its edges to the supporting surface, and a complete inner cocoon of considerably smaller size, likewise closely woven.

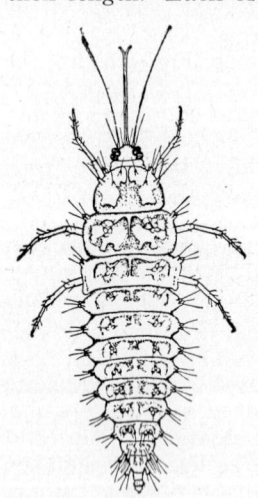

Fig. 329.—Larva of *Sisyra umbrata*. (After Anthony.)

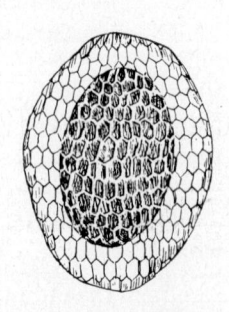

Fig. 330.—Cocoon and cocoon-cover of *Climacia*.

The silk-organs of the larva are described on pages 282-3.

Climācia dictyona.—This species resembles the preceding in size but is yellowish in coloration; the two can be distinguished by the form of the labia (Fig. 331).

In the larva of this species the setæ on the dorsum of the thorax are situated on tubercles; they are sessile in the larva of *Sisyra*. The habits of the larva are similar to those of *Sisyra*.

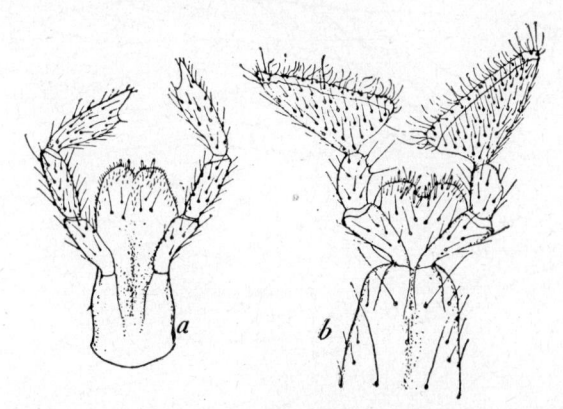

Fig. 331.—Labia of Spongilla-flies: *a, Climacia dictyona; b, Sisyra umbrata*. (After Needham.)

Before spinning its cocoon this larva spins a hemispheric cover beneath which the cocoon is made, as does the larva of *Sisyra*. But in the case of *Climacia* this cocoon-cover is lace-like; it is a beautiful object (Fig. 330).

Excepting the sialids, the larvæ of *Sisyra* and *Climacia* are the only known aquatic neuropterous larvæ found in this country.

Family SYMPHEROBIIDÆ

The Sympherobiids

This family includes certain insects which were formerly classed with the Hemerobiidæ but which exhibit a type of specialization of the wings that is quite different from that which is distinctively characteristic of that family.

The distinctive characteristic of the Sympherobiidæ is that vein R_{2+3} of the fore wings has become separated from the remainder of the radial sector and is attached separately to vein R_1. This results in the radius of the fore wing having two sectors, each of which is forked (Fig. 332).

In this family the number of the branches of the radial sector has not been increased, this vein being four-branched in both fore and hind wings; but the tips of all of the branches are forked. The costal area of the fore wing is broad towards the base of the wing; and the humeral vein is recurved and branched.

The North American species of this family represent two genera.

Sympherōbius.—In this genus there are two series of gradate veins in the fore wings; the outer series consists of four cross-veins (Fig. 332). Seven American species have been described. The wing-expanse of these insects ranges from 9 mm. to 12 mm.

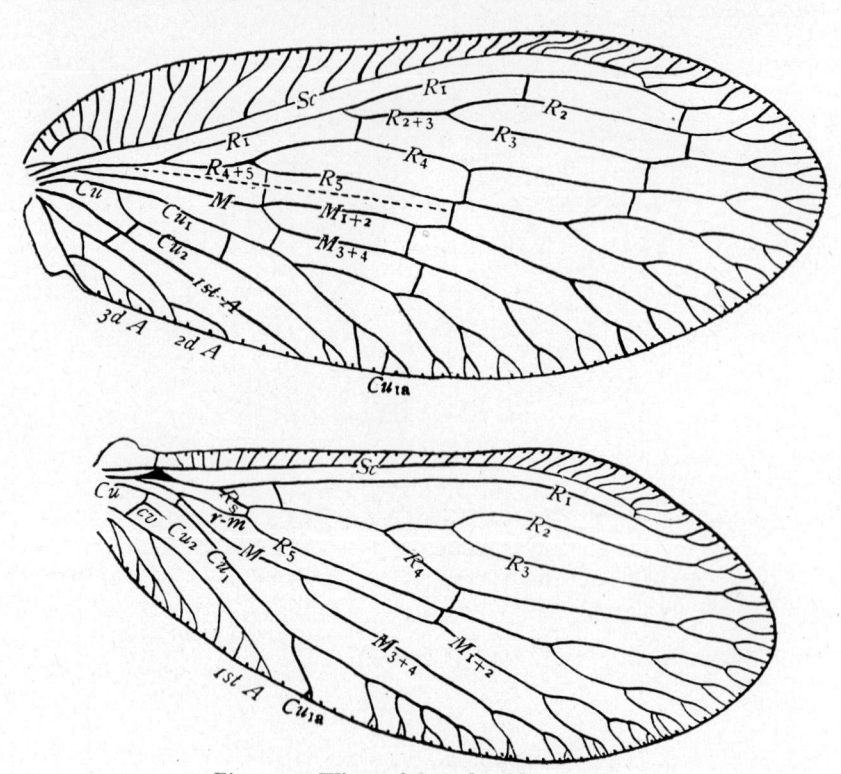

Fig. 332.—Wings of *Sympherobius amiculus.*

Psĕctra.—In this genus there is only one series of gradate veins in the fore wings. The only species is *Psectra dĭptera.* The specific name of this species was suggested by the fact that in the female the hind wings are atrophied. This is a widely distributed species both in this country and in Europe. Its wing expanse is from 5 mm. to 6 mm.

Family HEMEROBIIDÆ

The Hemerobiids

The Hemerobiidæ include insects of moderate size; in most of our species the wing-expanse is between 12 mm. and 22 mm.; in one species of *Megalomus* it is only 6 mm. In most of the species the body is brown or blackish and is often marked with yellow; in some the body is pale yellow. The wings are usually hyaline or pale yellowish.

This family has been greatly restricted in recent times; formerly there were included in it the members of the two preceding and the three following families.

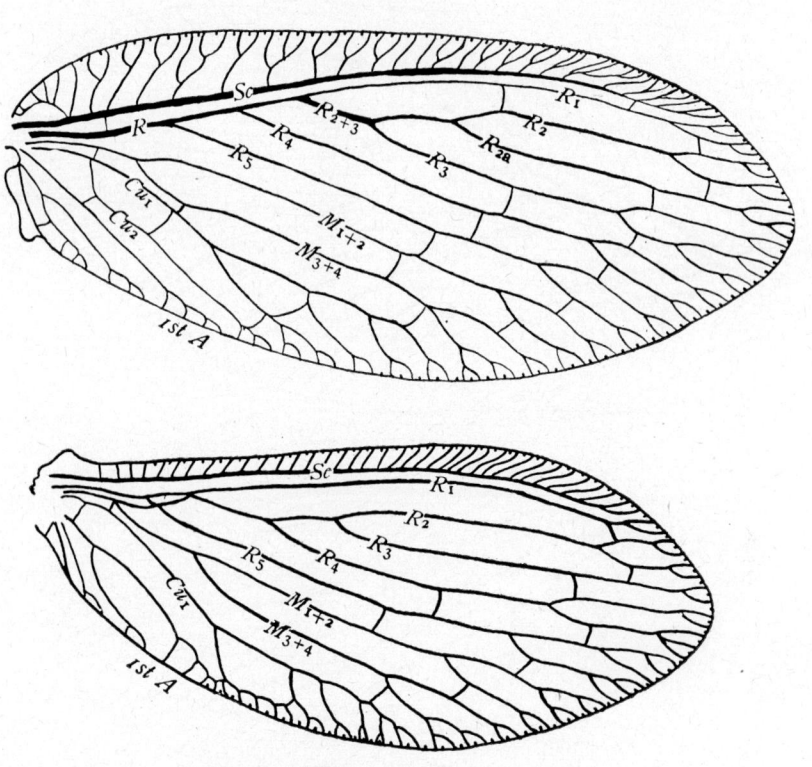

Fig. 333.—Wings of *Hemerobius humuli*.

As now restricted this family is composed of a group of genera that are characterized by a distinctive manner of specialization of the radius of the fore wings. This feature is a coalescence of vein R_1 and the stem of the pectinately branched radial sector, which results in what I have termed *the suppression of the stem of the radial sector*.

A comparatively simple example of this condition is exhibited by *Hemerobius humuli;* in the fore wings of this species (Fig. 333), veins R_5, R_4, and R_{2+3} arise separately from what appears to be the main stem of the radius but which is really vein R_1 and the basal part of the radial sector coalesced.

An early stage in the suppression of the stem of the radial sector is shown in the hind wing of *Hemerobius humuli* (Fig. 333). Here vein R_{2+3+4} is bent forward near its base and is joined to vein R_1. The extending of the union of veins R_1 and R_{2+3+4} from the point where they now anastomose towards the base of the wing, so as to obliterate the small cell between them, and also towards the apex of

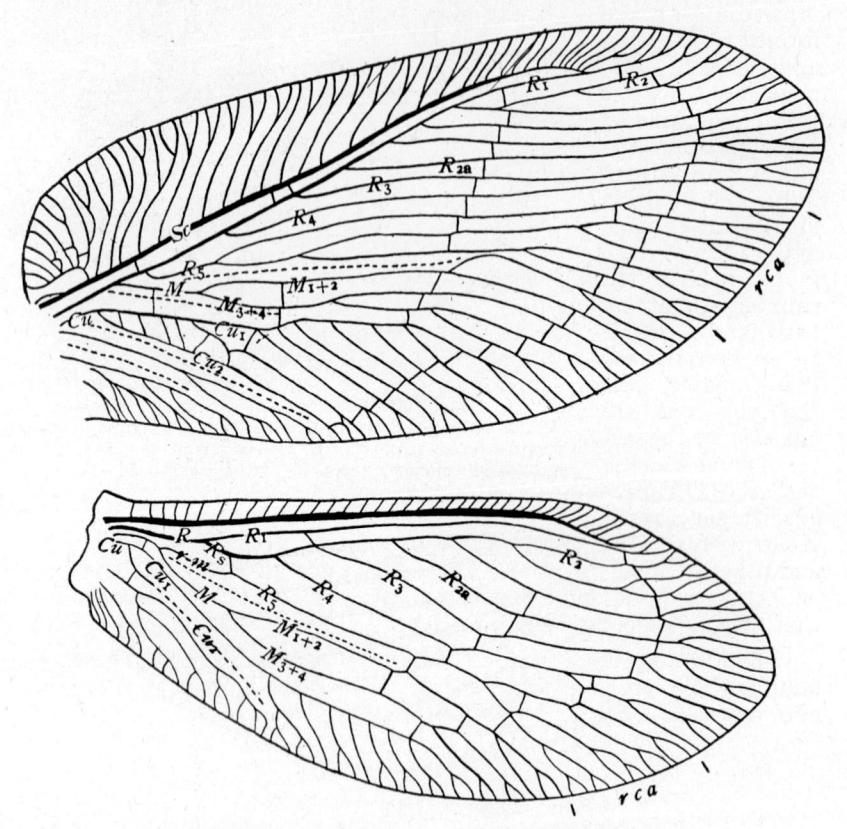

Fig. 334.—Wings of *Megalomus mæstus*.

the wing for a certain distance, would produce the condition that exists in the fore wing.

The wings of *Hemerobius* represent a comparatively simple type of hemerobiid wings; those of *Megalōmus mæstus* (Fig. 334), a more complicated one. Here there have been developed a larger number of definitive accessory veins and of marginal accessory veins.

Under the title "A Revision of the Nearctic Hemerobiidæ" Mr. N. Banks ('05) has published an account of this family, the two preceding families, and the three following families, in which all of our species known at that time are described.

The larvæ of the hemerobiids, as far as they are known, resemble in their general appearance aphis-lions (Chrysopidæ), and, like the aphis-lions, feed on plant-lice and other small insects. Their mouth-parts are formed for piercing and sucking (see page 282), and the posterior part of the alimentary canal is transformed into a silk-organ, as in *Sisyra* (see page 283). They are found most often running about on trees, and especially on coniferous trees. Some, like the aphis-lions, are naked; but the larvæ of some species, at least, of *Hemerobius* cover themselves with a cloak, composed of the empty skins of their victims and other debris (Fig. 335). These larvæ are furnished at the sides with projections which serve as pedicels to elongate, divergent hairs that help to keep the cloak in place.

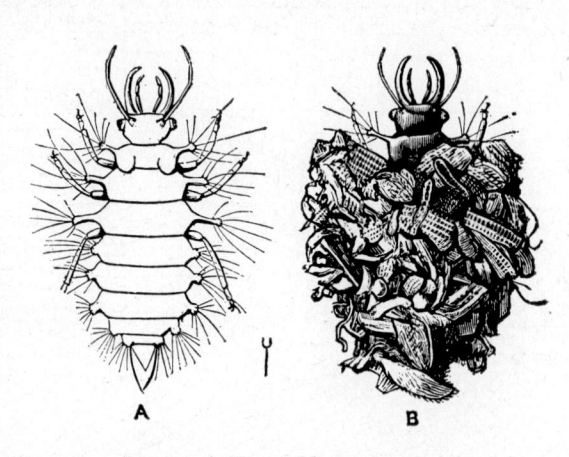

Fig. 335.—Larva of *Hemerobius:* A, the larva bare; B, the same partially concealed by the remains of its victims, etc.; a portion of the covering has been removed in order to show the head. (From Sharp.)

There are thirty described American species belonging to this family; these represent four genera, *Hemerobius, Boriomyia, Megalomus,* and *Micromus.*

Family DILARIDÆ

The Dilaridæ is a small family, representatives of which are found chiefly in the Old World. In this family the antennæ of the male are pectinate; and the female is furnished with an exerted ovipositor.

Only a single, exceedingly rare species, *Dīlar americānus*, has been found in North America; and of this only a single female individual is known. This is a small insect; the length of the body, not including the ovipositor, is about 3 mm.; the length of the ovipositor is a little greater than that of the body; the expanse of the wings is about 14 mm. There is a single five-branched radial sector in both fore and hind wings. In several exotic species the radius of the fore wings bears two or more sectors.

The type of our species was taken at Bee Spring, Kentucky, in June, 1874.

Family BEROTHIDÆ

The Berothidæ is a small family, which is represented in North America by a single genus, *Lomamyia*, of which only two species are

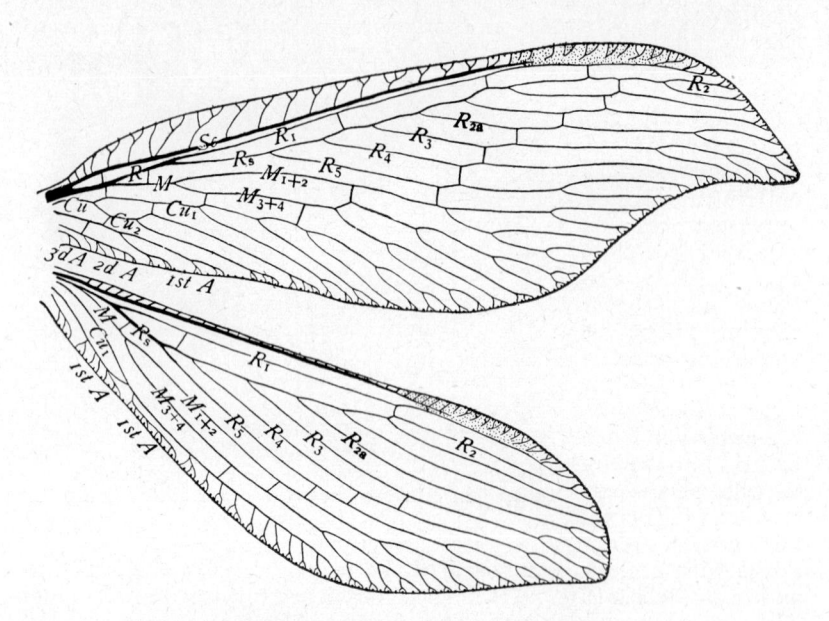

Fig. 336.—Wings of *Berotha insolita*.

known. Figure 336 represents the venation of the wings of the type species of the family, *Berōtha insŏlita*, which is found in India, and to which our species are closely allied.

The fore wings are falcate, which is not true of certain exotic genera; the humeral cross-vein is not recurved; many of the transverse veins between the costa and the subcosta are forked; the radial sector bears definitive accessory veins; and there is a single series of gradate veins in the radial area. In the hind wings the first radio-medial cross-vein is transverse; vein Cu_2 is wanting; and the area between the margin of the wing and veins 1st A and Cu_1 is narrow and largely occupied by the fanlike tips of the accessory veins.

Nothing is known regarding the early stages of these insects.

Fig. 337.—*Polystœchotes punctatus*.

Family POLYSTŒCHOTIDÆ

The family Polystœchotidæ was established to receive the genus *Polystœchotes*, of which only two species, both American, are known. These are larger insects than are the members of the allied families, measuring in wing-expanse from 40 mm. to 75 mm., varying greatly in size. They are nocturnal and are attracted to lights. The two species can be distinguished as follows:

Polystœchotes punctātus (Fig. 337). This is blackish, with three longitudinal lines on the prothorax, and with the lateral margins of this segment yellowish.

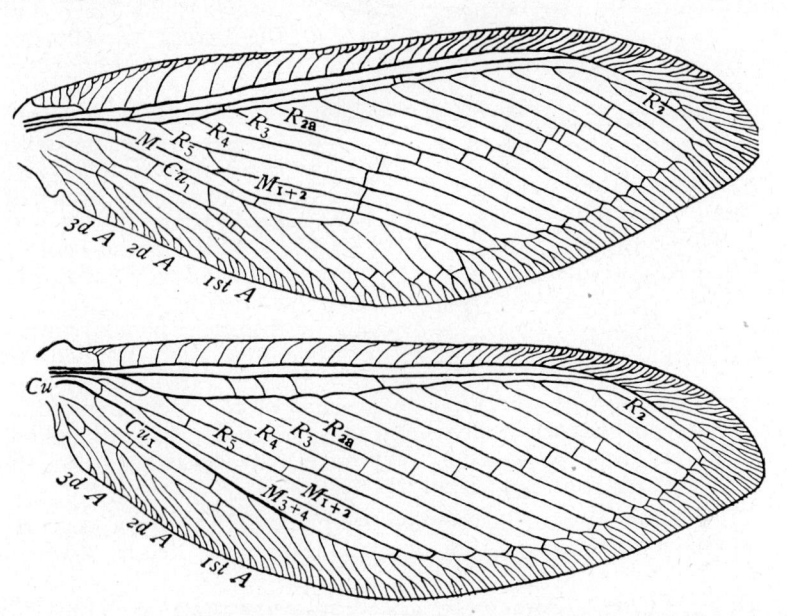

Fig. 338.—Wings of *Polystœchotes punctatus*.

Polystœchotes vittātus.—This is pale yellowish, with a black stripe on the sides of the thorax, and with the abdomen dark brown.

The larva of neither of these species is known. This is a strange fact considering the size and the abundance of these insects.

The wings of *Polystœchotes punctatus* (Fig. 338) represent the type of wing-venation characteristic of this family. In these wings the humeral cross-vein is recurved and branched; veins Sc and R_1 coalesce at the tip; the radial sector is pectinately branched; the number of cross-veins is greatly reduced; but there is in both fore and hind wings a very perfect series of gradate veins.

In these wings the development of definitive accessory veins on the radial sector and the regularity of the border of marginal accessory veins have reached a very high degree of perfection.

Family CHRYSOPIDÆ

The Lacewing-Flies or Aphis-Lions

The family Chrysopidæ includes the insects commonly known as lacewing-flies; these and their larvæ, the aphis-lions, are common and well-known insects; they are found upon herbage and the foliage of shrubs and trees throughout the summer months (Fig. 339).

The adults are easily recognized by their delicate lacelike wings and their green or yellowish green color. Members of several of the preceding families have delicate lacelike wings; but with those insects the wings are more or less brown or are hyaline.

While these insects are most commonly known as the lacewing-flies, other popular names have been applied to them; they are sometimes called *golden-eyed flies*, on account of the peculiar metallic color of their eyes while alive; and as some species, when handled, emit a very disagreeable odor, they have been called *stink-flies*, an undesirable name for such beautiful insects.

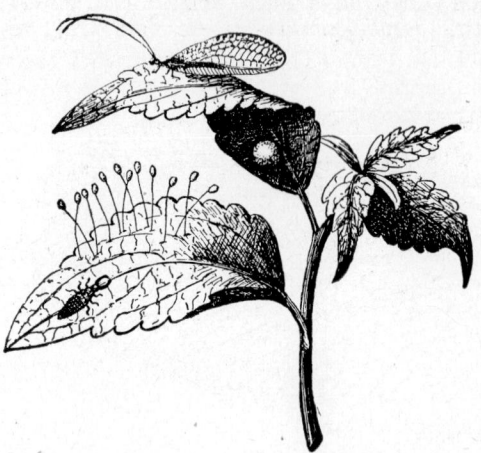

Fig. 339.—Eggs, larva, cocoon, and adult of *Chrysopa*.

The wings of the Chrysopidæ are characterized by a very remarkable and distinctive type of specialization, the details of which

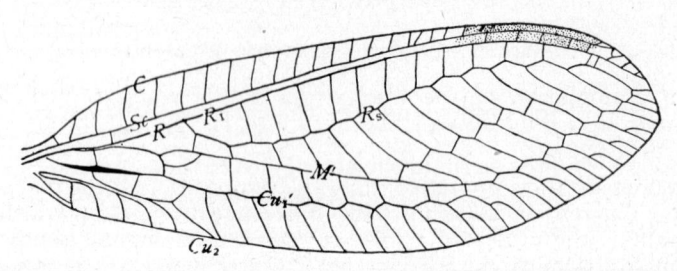

Fig. 340.—Fore wing of *Chrysopa nigricornis:* M', pseudo-media; Cu_1', pseudo-cubitus.

can be understood only by a study of the tracheation of the wings of the pupæ. Such a study has been made by McClendon ('06), Tillyard ('16), and R. C. Smith ('22).

A superficial examination of a wing of *Chrysōpa* (Fig. 340) reveals the presence of two longitudinal veins between the radial sector and the inner margin of the wing, one of which appears to be the media and the other vein Cu_1; but each of these, as is shown later, is a serial vein composed of sections of several veins.

As it would be impracticable to apply to these serial veins names indicating their composition, they have been termed the *pseudo-media* or vein M' and the *pseudo-cubitus-one* or vein Cu_1', respectively (Fig. 340, M' and Cu_1').

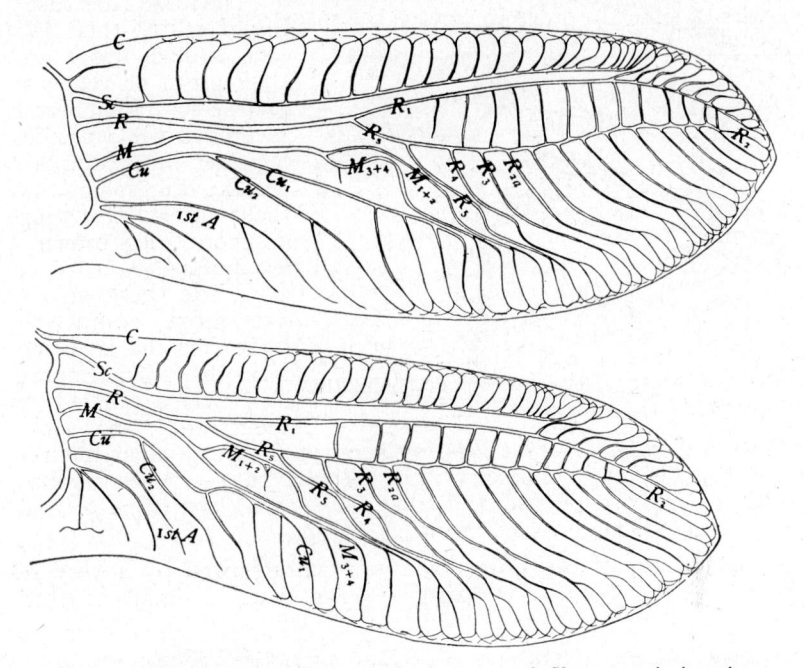

Fig. 341.—Tracheation of the wings of a pupa of *Chrysopa nigricornis*.

An examination of the tracheation of the wings of a pupa of *Chrysopa nigricŏrnis* reveals the nature of the two serial veins M' and Cu_1' (Fig. 341).

In order to show more definitely the composition of the two serial veins, a diagram of an adult wing is given (Fig. 342), in which the elements of the coalesced veins are represented slightly separated, and the cross-veins connecting the coalesced veins are represented by dotted lines. By comparing this diagram with Figure 340 the homologies of the different veins can be recognized.

The larvæ of the lacewing-flies are known as aphis-lions, because they feed upon aphids; they are found on the foliage of plants infested by these pests; they also feed upon other small insects and the eggs of insects; they are spindle-shaped (Fig. 339) and are furnished with piercing and sucking mouth-parts like those of ant-lions.

Nearly all aphis-lions are naked; but a few species cover themselves with the skins of their victims and other debris, as do the larvæ of *Hemerŏbius*. This has been observed by European writers (Sharp

'95); and recently Mr. R. C. Smith ('21) has found that the larvæ of several of our native species have a similar habit.

The cocoons are generally found on the lower sides of leaves or on the supports of plants; they are spherical and composed of dense layers of silk. In order to emerge the insect cuts a circular lid from

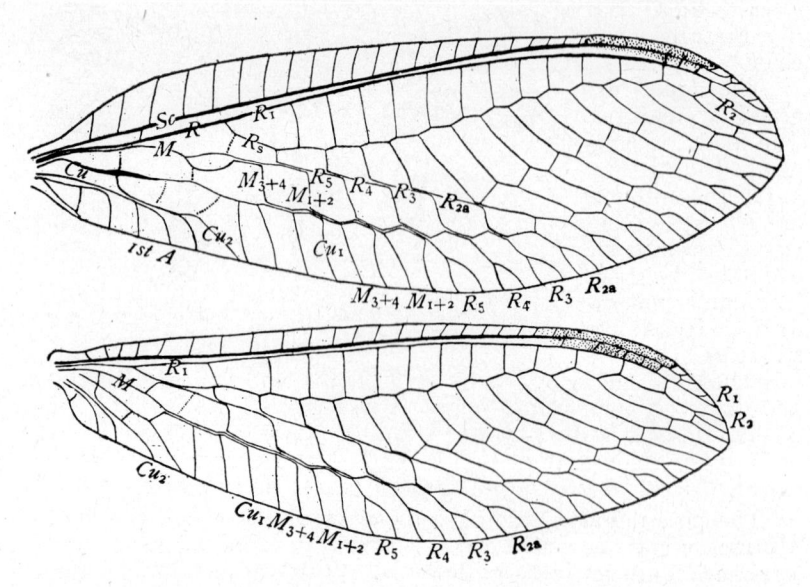

Fig. 342.—Diagram of the wings of *Chrysopa nigricornis*, showing the coalesced veins slightly separated.

one side of the cocoon; this is done by the pupa by means of its mandibles. After emerging from its cocoon, the pupa crawls about for a short time before changing to the adult state.

The adults are often attracted to lights at night. A remarkable fact in the life-history of these insects is the way in which the female cares for her eggs. When about to lay an egg she emits from the end of her body a minute drop of a tenacious substance, which is probably a product of the colleterial glands; this she applies to the object on which she is standing and then draws it out into a slender thread by lifting the abdomen; then an egg is placed on the summit of this thread. The thread dries at once and firmly holds the egg in mid-air. These threads are usually about 12 mm. in length, and occur singly or in groups; a group is represented attached to a leaf in Figure 339.

About fifty species belonging to this family have been found in the United States and Canada; the greater number of these belong to the genus *Chrysōpa*.

Family MYRMELEONIDÆ

The Ant-Lions

The members of the family Myrmeleonidæ are commonly known as ant-lions. This name was suggested by the fact that the larvæ of the best-known species, those that dig pitfalls, feed chiefly on ants.

The adults are graceful creatures. The body is long and slender (Fig. 343); the antennæ are short and enlarged towards the end; the wings are long and narrow and delicate in structure; they are

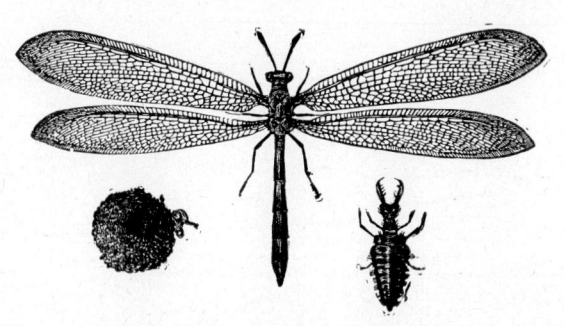

Fig. 343.—Larva, cocoon with pupa-skin projecting, and adult, of an ant-lion.

furnished with many accessory veins, both definitive and marginal, and with very many cross-veins. A distinctive feature of the wings of these insects is the presence of an elongated cell behind the point of fusion of veins Sc and R_1 (Fig. 344); this characteristic serves to distinguish this family from the closely allied Ascalaphidæ.

The determination of the homologies of the wing-veins of the Myrmeleonidæ was completed only recently. The results of this determination are set forth in detail by the writer in his "The Wings of Insects," where they are illustrated by many figures.

Our native species, as a rule, are not striking in appearance; the wings are hyaline and are often more or less spotted with black or brown marks; but certain exotic forms, as those of the genus *Palpares*, are large and have conspicuously marked wings.

The larvæ have broad and somewhat depressed bodies which taper towards each end (Fig. 343). The mouth-parts are large and powerful and are of the piercing and sucking type; they are described on page 282. The pupa state is passed in a spherical cocoon, made of sand fastened together with silk, and neatly lined with the same material (Fig. 343). The silk is spun from the posterior end of the alimentary canal and is secreted by modified Malpighian vessels, as in *Sisyra* (see page 283.)

This is a large family including several hundred described species. In his "Catalogue of the Neuropteroid Insects of the United States," Banks ('07) lists fifty-eight species of this family known at that time to occur in our fauna; these are distributed among eleven genera.

The life-histories of comparatively few of the species are known; but certain species, the larvæ of which dig pitfalls in sandy places, have attracted much attention since the earliest days of entomology.

304   AN INTRODUCTION TO ENTOMOLOGY

Ant-lions are much more common in the Southern and Southwestern States than they are in the North. The pitfalls of the larvæ are usually found in sandy places that are protected from rain, as beneath buildings or overhanging rocks. In making these pitfalls the sand is thrown out by an upward jerk of the head, this part of the body

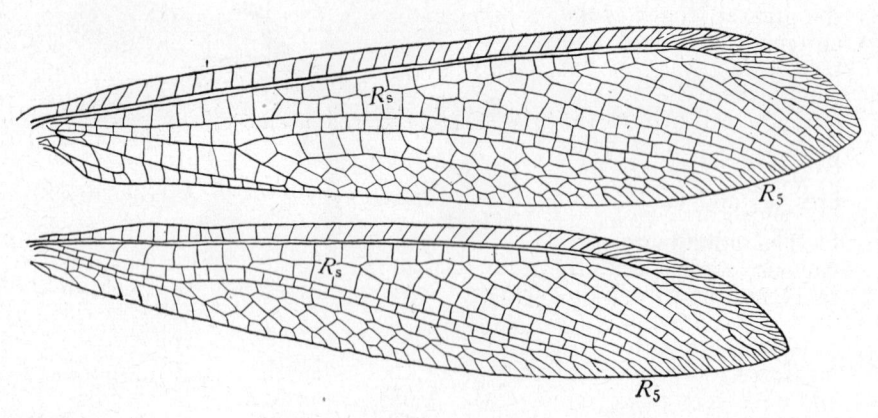

Fig. 344.—Wings of *Myrmeleon*.

serving as a shovel. The pits differ greatly in depth, according to the nature of the soil in which they are made. Their sides are as steep as the sand will lie. When an ant or other wingless insect steps upon the brink of one of these pits, the sand crumbles beneath its feet, and it is precipitated into the jaws of the ant-lion, which is buried in the sand, with its jaws at the bottom of the pit (Fig. 345). In case the ant does not fall to the bottom of the pit, the ant-lion undermines it by throwing out some sand beneath it. I have even seen an ant-lion throw the sand in such a way that in falling it would tend to hit the ant and knock it down the side of the pit. These larvæ can be easily kept in a dish of sand, and their habits watched.

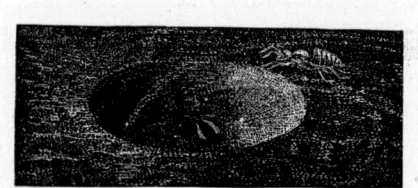

Fig. 345.—Pitfall of an ant-lion.

The most common ant-lion in the North is *Myrmēleon immaculātus;* the larva of this species makes a pitfall. The habits of the larvæ of *Glenūrus*, *Dendrōleon*, and *Acanthăclisis*, three genera that are represented in this country, have been described by European writers. These larvæ do not dig pitfalls, but partially bury themselves in the sand, from which position they throw themselves quickly upon their victims.

Family ASCALAPHIDÆ

The Ascalaphids

The family Ascalaphidæ is quite closely allied to the preceding family; but the members of this family can be easily distinguished from myrmeleonids by the greater length of the antennæ (Fig. 346) and by the fact that in the wings there is not an elongate cell behind the point of fusion of veins Sc and R_1; compare Figures 347 and 344.

Fig. 346.—*Ululodes hyalina*. (From Kellogg, after McClendon.)

The adults are predacious; some species fly in the daytime in bright sunshine, but it is said that others fly in the twilight. Some species resemble myrmeleon-

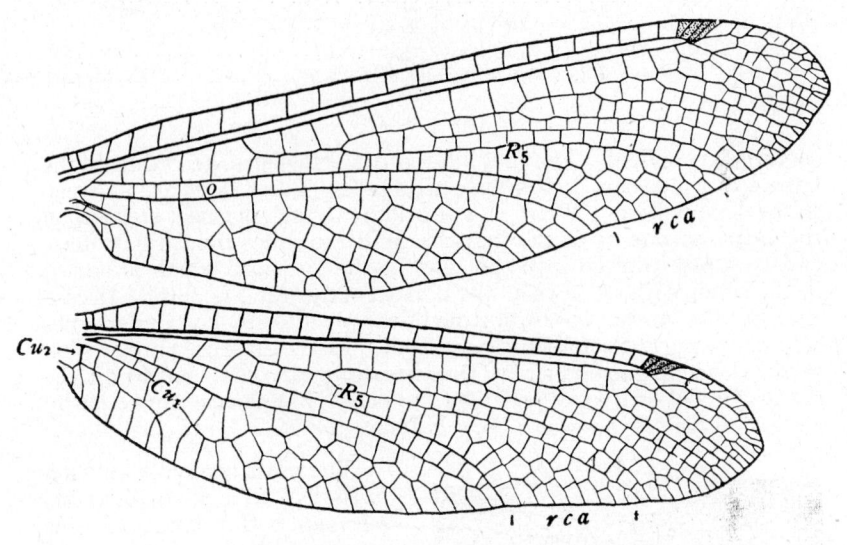

Fig. 347.—Wings of *Ululodes hyalina*.

ids in appearance, while others resemble dragon-flies. When at rest they remain motionless on some small branch or stalk, head down, with the wings and antennæ closely applied to the branch, and the abdomen erected and often bent so as to resemble a short brown twig or branch (Fig. 346).

The larvæ resemble ant-lions in the form of the body and possess the same type of mouth-parts (Fig. 348). They have on each segment of the body a pair of lateral finger-like appendages, which are clothed with hairs. They do not dig pitfalls, but lie in ambush on the surface of the ground, with the body more or less covered, and wait for small insects to come near them. When a larva is full-grown, it spins a spherical silken cocoon. An account of the life-history of one of our native species, *Ululōdes hyălina*, has been published by McClendon ('02).

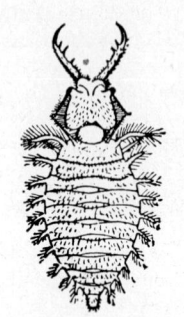

Fig. 348.—Larva of *Ululodes hyalina*. (After McClendon.)

The Ascalaphidæ of the world have been monographed by H. W. Van der Weele ('08). In this monograph more than two hundred species are described. The members of this family are chiefly tropical insects, but a few species occur in the United States; these represent three genera, which can be separated by the following table.

A. Eyes entire.. Neuroptynx
AA. Each eye divided into two parts by a groove.
    B. Hind margin of wings entire........................... Ululodes
    BB. Hind margin of wings excised....................... Colobopterus

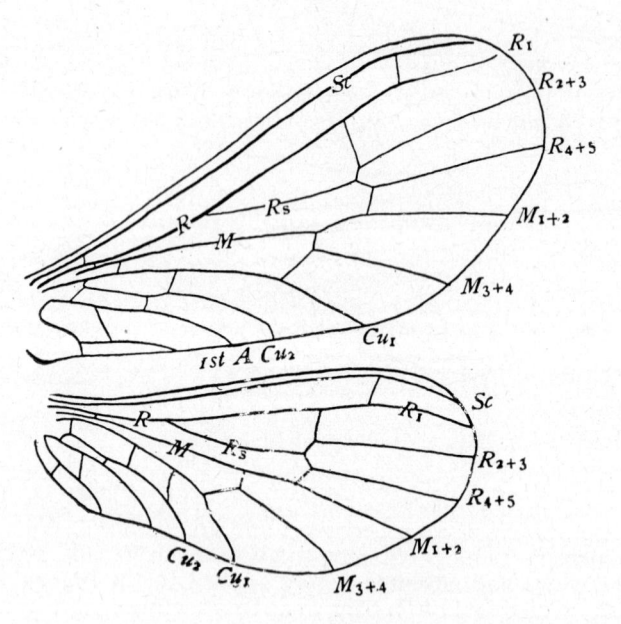

Fig. 349.—Wings of *Semidalis aleurodiformis*. (After Enderlein.)

Family CONIOPTERYGIDÆ

The Mealy-winged Neuroptera

The Coniopterygidæ is a family of limited extent; and it includes only small insects, the smallest of the Neuroptera; the described American species measure only 3 mm. or less in length. They are characterized by a reduced wing-venation (Fig. 349) and by having the body and wings covered by a whitish powder.

While the adults resemble very slightly other neuropterous insects, the larvæ resemble those of the Hemerobiidæ and allied families in form, in the structure of their mouth-parts, in their predacious habits, and in their metamorphosis.

The larvæ have been seen to feed upon coccids, aphids, and the eggs of the red-spider; they doubtless feed on other small insects. When full-grown they make a double cocoon consisting of an outer flat layer and an inner spherical case.

Mr. Nathan Banks ('07) has published a revision of the species that have been found in our fauna. This includes eight species, representing five genera.

CHAPTER XII

ORDER EPHEMERIDA*

The May-Flies

The members of this order have delicate membranous wings, which are triangular in outline and are usually furnished with a considerable number of intercalary veins and with many cross-veins; the hind wings are much smaller than the fore wings and are sometimes wanting. The mouth-parts of the adults are vestigial; those of the naiads are fitted for chewing. The metamorphosis is incomplete.

The May-flies or ephemerids are often very common insects in the vicinity of streams, ponds, and lakes; frequently the surface of such bodies of water is thickly strewn with them. They are attracted by lights; and it is not an uncommon occurrence in summertime to see hundreds of them flying about a single street-lamp.

The May-flies are easily distinguished from other net-winged insects by the shape of the wings and the relative sizes of the two pairs (Fig. 350).

Fig. 350.—A May-fly.

The mouth-parts of the adult are vestigial, as these insects eat nothing in this state. The antennæ are very small; they are composed of two short, stout segments succeeded by a slender, many-jointed bristle. The thorax is robust, with the mesothorax predominant; the great development of this segment is correlated with the large size of the fore wings. The abdomen is long, soft, and composed of ten

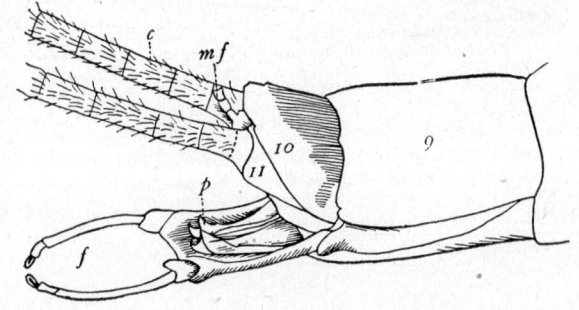

Fig. 351.—Caudal end of abdomen of *Siphlurus alternatus*, male: *9, 10, 11*, abdominal segments; *c*, cerci; *mf*, median caudal filament; *p*, penis; *f*, forceps-limbs. (After Morgan.)

*Ephemĕrida, Ephemera: *ephemeron* (ἐφήμερον), a May-fly.

visible segments; the eleventh segment, which bears the cerci, is overlapped by the tenth (Fig. 351). The cerci are long, slender, and many-jointed; and in some species there is a median caudal filament, which resembles the cerci in form; these three organs, the two cerci and the median caudal filament, are commonly referred to as the caudal setæ. In the male there is a pair of clasping organs placed ventrally at the extremity of the tenth segment; these are usually two-, three-, or four-jointed and are termed the forceps-limbs. Each vas deferens and each oviduct has a separate opening; in the male these openings are at the caudal end of the body; in the female, between the seventh and eighth sternites.

In some May-flies the compound eyes are divided; one part of each, in such cases, is a day-eye, and the other a night-eye (see page 144).

As the adult May-fly takes no food, its alimentary canal is not needed in this stage for purposes of digestion, and, instead of serving this function, acts as a balloon, being inflated with air, thus lessening the specific gravity of the body and aiding in flight.

In this order a marked cephalization of the flight function has taken place, which has resulted in a great reduction of the hind wings in all living forms. In some cases (*Cænis et al.*), this has gone so far that the hind wings are wanting (Fig. 352); but at least one pair of wings is present in all members of this order.

Fig. 352.—*Cænis*, a two-winged May-fly.

When at rest, the wings are held upright; they are never folded over the abdomen. No anal furrow has been developed. A striking feature of the wings of May-flies is their well-known corrugated or fan-like form, there being a remarkably

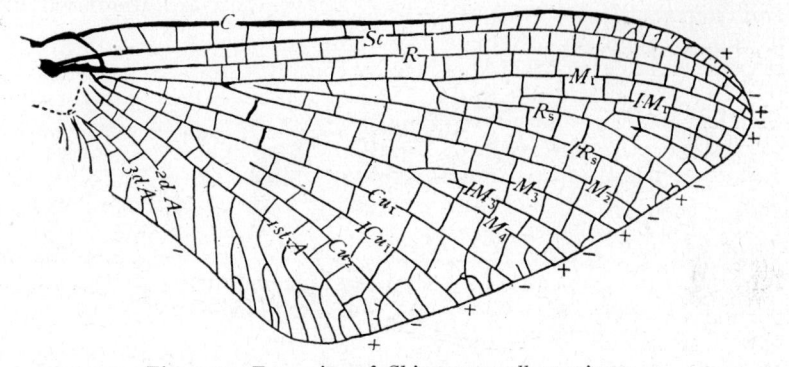

Fig. 353.—Fore wing of *Chirotonetes albomanicatus*.

perfect alternation of so-called convex and concave veins. Correlated with the development of the fan-like form of the wings has been the development of intercalary veins, that is, veins that did not arise as branches of the primitive veins, but were developed in each case as a thickened fold, more or less nearly midway between two preexisting veins, with which primarily it was connected only by cross-veins. The veins labeled IM_1, IM_3, and ICu_1 in Figures 353 and 354 are good illustrations of this type of veins. The initial I in these designations is an abbreviation of the word intercalary. Thus the intercalary vein between veins Cu_1 and Cu_2, *i. e.*, in the area Cu_1, is designated as vein ICu_1.

Figures 353 and 354 will aid in the determination of the homologies of the wing-veins of May-flies. In these figures convex veins are designated by plus signs and concave veins by minus signs. In attempting to determine the homology of a vein in a wing where the venation is reduced, it should first be determined whether the vein is convex or concave, as the corrugations of the wings of May-flies are the most persistent features of them. For a more detailed account of this subject, see Chapter X of "The Wings of Insects."

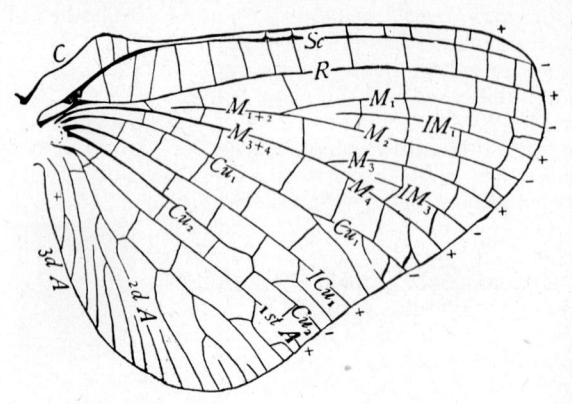

Fig. 354.—Hind wing of *Chirotonetes albomanicatus*.

The Greek name *Ephĕmeron* applied to these insects in the days of Aristotle was derived from *ephemeros*, signifying lasting but a day; and from that time to this, frequent references have been made to the insects that live only a single day. This brevity of the life of these insects is true only of their existence in the adult state. Strictly speaking, the May-flies are long-lived insects; some species pass through their life-cycle in a few weeks in midsummer; but as a rule one, two, or even three years are required for the development of a generation. The greater part of this time is passed, however, beneath the surface of water, and after the insect emerges into the air and assumes the adult form its existence is very brief. With many species the individuals leave the water, molt twice, mate, lay their eggs, and die in the course of an evening or early morning; and although the adults of many genera live several days, the existence of these insects is very short compared with that of the adults of other insects.

The females lay their eggs in water. Some short-lived species discharge the contents of each ovary in a mass. Individuals are often found in which there project from the caudal end of the body two parallel subcylindrical masses of eggs, one protruding from each of the openings of the oviducts. "The less perishable species extrude their eggs gradually, part at a time, and deposit them in one or the other of the following manners: either the mother alights upon the water at intervals to wash off the eggs that have issued from the mouths of the oviducts during her flight or else she creeps down into the water—enclosed within a film of air with her wings collapsed so as to overlie the abdomen in the form of an acute narrowly linear bundle, and with her setæ closed together—to lay her eggs upon the under side of the stones, disposing of them in rounded patches, in a single layer evenly spread, and in mutual contiguity." (Eaton '83).

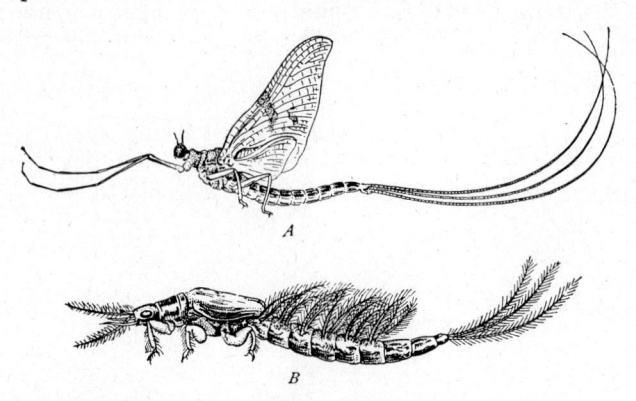

Fig. 355.—Metamorphosis of a May-fly, *Ephemera varia:* A, adult; B, naiad. (After Needham.)

The metamorphosis of May-flies is incomplete. The wings are developed externally, as in the Orthoptera; the development of the compound eyes is not retarded; but the immature forms, or naiads, are "sidewise developed" to fit them for aquatic life. In most species the form of the body of the naiads is elongate and furnished with two or three long "caudal setæ," that is, cerci and in some a median caudal filament; in these respects the naiads resemble, to a greater or less degree, the adults (Fig. 355); but except in the early instars the abdomen of a naiad is furnished with tracheal gills (Figs. 355 and 356.)

The tracheal gills are usually large and prominent; in most species there are seven pairs, borne by the first seven abdominal segments. They vary greatly in form in the different genera. In some each gill is divided into two long narrow branches, which lie in one plane (Fig. 355); in others the gills consist of a scoop-shaped covering piece beneath which is a more delicate part consisting of many thread-like branches. A detailed account of the various forms of tracheal gills of May-flies is given by Miss Morgan ('13).

The naiads of May-flies are all aquatic; they are very active; and are almost entirely herbivorous, feeding largely on the decaying stems and leaves of aquatic plants, the epidermis of moss and of roots,

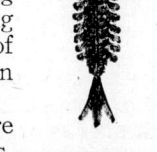

Fig. 356.—Naiad of a May-fly.

algæ, and diatoms. The variations in the details of their habits are described as follows by Dr. Needham ('18).

"A few, like *Hexagēnia*, *Ephĕmera*, and *Polymitărcys* are burrowers beneath the bottom silt. A few like *Cænis* and *Ephemerĕlla*, are of sedentary habits and live rather inactively on the bottom, and on silt-covered stems. Many are active climbers among green vegetation; such are *Callibætis* and *Blastūrus;* and some of these can swim and dart about by means of synchronous strokes of tail and gills with the swiftness of a minnow. The species of *Leptophlēbia* love the beds of slow-flowing streams, and all the flattened nymphs of the Heptageninæ live in swiftly moving water, and manifest various degrees of adaptation to withstanding the wash of strong currents. The form is depressed, and margins of the head and body are thin and flaring, and can be appressed closely to the stones to deflect the current."

There are two features of special interest in the structure of the naiads of May-flies: first, the hypopharynx bears a pair of lateral lobes, which are believed to be vestiges of paragnatha; and second, the presence of accessory circulatory organs in the cerci and median caudal filament (Fig. 357).

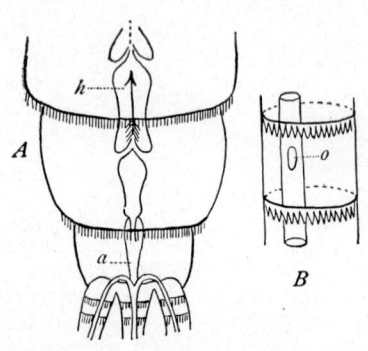

May-flies exhibit a remarkable peculiarity in their development. After the insect leaves the water and has apparently assumed the adult form, that is, after the wings have become fully expanded, it molts again. These are the only insects that molt after they have attained functional wings. The term *subimago* is applied to the instar between the naiad and the final form of the insect, the adult. With some species the duration of the subimago stage is only a few minutes; the insect molts on leaving the water; flies a short distance; and molts again. In others this stage lasts twenty-four hours or more.

Fig. 357.—A, caudal end of abdomen of *Cloëon dipterum:* h, heart; a, accessory circulatory organs. B, twenty-sixth segment of a cercus: o, orifice in blood vessel. (After Zimmerman.)

With many species of May-flies there is great uniformity in the date of maturing of the individuals. Thus immense swarms of them will leave the water at about the same time, and in the course of a few days pass away, this being the only appearance of the species until another generation has been developed. The great swarms of "lake-flies," *Ephĕmera sĭmulans*, which appear along our northern lakes about the third week of July, afford good illustration of this peculiarity.

Family EPHEMERIDÆ

The May-Flies

The order Ephemerida includes a single family, the Ephemeridæ; the characteristics of this family, therefore, are those of the order, which are given above.

Comparatively few writers have made extended studies of the classification of the ephemerids; this is doubtless partly due to the fact that pinned specimens usually become shriveled and are very fragile; consequently this order is poorly represented in most collections of insects. In spite of this, more than one hundred species have been described from the United States. An important paper on the classification of May-flies is that by Dr. Needham ('05) in Bulletin 86 of the New York State Museum. Here are given keys for separating the North American genera, one for the adult insects and one for the naiads.

CHAPTER XIII

ORDER ODONATA*

The Dragon-Flies and the Damsel-Flies

The members of this order have four membranous wings, which are finely netted with veins; the hind wings are as large as or larger than the fore wings; and each wing has near the middle of the costal margin a joint-like structure, the nodus. There are no wingless species. The mouth-parts are formed for chewing. The metamorphosis is incomplete.

Dragon-flies and damsel-flies are very common insects in the vicinity of streams, ponds, and lakes; they are well known to all who frequent such places. The dragon-flies, especially, attract attention on account of their large size (Fig. 358) and rapid flight, back and forth, over the water and the shores; the damsel-flies (Fig. 359) are less likely to be noticed, on account of their less vigorous flight.

The name of this order is evidently from the Greek word *odous*, a tooth; but the reason for applying it to these insects is obscure; it may refer to the tusk-like form of the abdomen.

Fig. 358.—A dragon-fly, *Plathemis lydia*. (From Sanborn.)

In these insects, the head is large; it differs in shape in the two suborders as described below. The compound eyes are large; they often occupy the greater part of the surface of the head; in many cases the upper facets of the eye are larger than the lower, and in a few forms the line of division between the two kinds is sharply marked. It is probable that the ommatidia with the larger facets are night-eyes, and those with the smaller facets, day-eyes. See pages 142 and 143. Three ocelli are present. The antennæ are short; they consist of from five to eight segments; of these the two basal ones are thick, the others form a bristle-like organ. The mouth-parts are well developed; the labrum is prominent; the mandibles and maxillæ are both strongly toothed; and the labium consists of

---

*Odonāta: *odous* (ὀδούς), a tooth.

(314)

three large lobes, which with the labrum nearly enclose the jaws when at rest. The thorax is large. The wings are, as a rule, of nearly similar size and structure; they are richly netted with veins; and the costal border of each is divided into basal and apical parts by what is termed the *nodus* (Fig. 364, *n*). The legs are rarely used for walking, but are used chiefly for perching, and are set far forward; the tarsi are three-jointed. The abdomen is long, slender, and more or less cylindrical; the caudal end is furnished with clasping organs in the males.

A remarkable peculiarity of the order is the fact that the copulatory organs of the male are distinct from the opening of the vasa deferentia; the former are situated on the second abdominal segment, the latter on the ninth. Before pairing, the male conveys the seminal fluid to a bladder-like cavity on the second abdominal segment; this is done by bending the tip of the abdomen forward. Except in the subfamily Gomphinæ, the pairing takes place during flight. The male seizes the prothorax or hind part of the head of the female with his anal clasping organs; the female then curves the end of the abdomen to the organs on the second abdominal segment of the male. Pairs of dragon-flies thus united and flying over water are a common sight.

Fig. 359.—A damsel-fly.

The Odonata are predacious, both in the immature instars and as adults. The adults feed on a great variety of insects, which they capture by flight; and the larger dragon-flies habitually eat the smaller ones, but a large part of their food consists of mosquitoes and other small Diptera.

The eggs are laid in or near water. All of the damsel-flies and many dragon-flies are provided with an ovipositor, by means of which punctures are made in the stems of aquatic plants, in logs, in wet mud, etc., for the reception of the eggs. The females of those dragon-flies that lack a well-developed ovipositor deposit their eggs in various ways. In some species the female flies back and forth over the surface of the water, sweeping down at intervals to touch it with the tip of her abdomen and thus wash off one or more eggs into it. In other species the eggs are laid in a mass on some object just below the surface of the water; some species do this by alighting upon a water-plant, and, pushing the end of the abdomen below the surface of the water, glue a bunch of eggs to the submerged stem or leaf; in other species the mass of eggs is built up gradually; the female will poise in the air a short distance above the point where the mass of eggs is being laid, and at frequent intervals descend with a swift curved motion and add to the egg-mass and then return to her former position to repeat the operation. Still other species hang their eggs in long gelatinous strings, on some plant stem at the surface of the water.

The metamorphosis is incomplete. The naiads are all aquatic except those of a few Hawaiian damsel-flies, which live on moist soil under the leaves of liliaceous plants. The wings are developed externally, and the development of the compound eyes is not retarded, as it is with larvæ. The adaptations for aquatic life differ in the two suborders and are described later.

All naiads of the Odonata are predacious. The mouth is furnished with well-developed mandibles and maxillæ, all of which are armed with strong teeth. But none of these is visible when the insect is at rest. The lower lip is greatly enlarged, and so formed that it closes over the jaws, concealing them. For this reason it has been termed the *mask*. But it is much more than a mask; it is a powerful weapon of offence. It is greatly elongated and is jointed in such a way that it can be thrust out forward in front of the head. It is armed at its extremity with sharp hooks, for seizing and retaining its prey (Fig. 360).

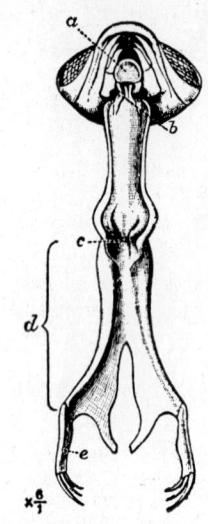

Fig. 360.—Under side of head of a naiad of a damsel-fly with labium unfolded. (After Sharp.)

The order Odonata is divided into three suborders. One of these suborders, the Anisozygŏptera, is composed almost entirely of fossil forms, being represented among living insects by a single genus, *Epiophlēbia*, which is found in Japan. The other two suborders are well represented in this country; one of them consists of the dragon-flies, the other of the damsel-flies.

Suborder ANISOPTERA*
The Dragon-Flies

The dragon-flies are easily recognized by the relative size of the two pairs of wings, and by the attitude of the wings when at rest (Fig. 361). The hind wings are larger than the fore wings and are of a somewhat different shape; but the most striking characteristic is the fact that the wings are extended horizontally when at rest.

Fig. 361.—A dragon-fly, *Libellula luctuosa*.

---

*Anisŏptera: *anisos* (ἄνισος), unequal; *pteron* (πτερόν), a wing.

The head is large, broad, often semi-globose, and concave behind. The wings are very strong. An important factor in the strengthening

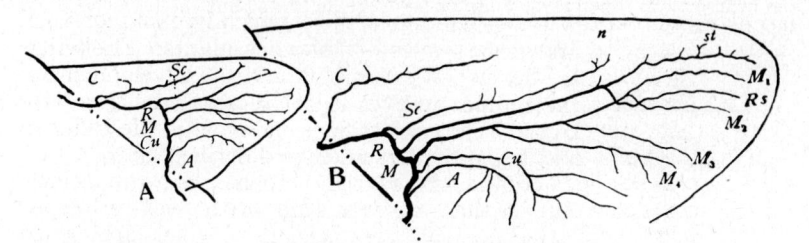

Fig. 362.—Wings of naiads of *Gomphus descriptus*, early stages. (From Comstock and Needham.)

of the wings of these insects is the development of a series of corrugations, which has resulted in certain veins becoming convex and others concave; this has progressed so far that there is a very perfect alternation of convex and concave veins.

The habits of dragon-flies have been carefully studied by Professor Needham ('18), who writes as follows:

"Among the dragon-flies are many superb flyers. The speed on the wing of *Trāmea* and *Anax* equals, and their agility exceeds, that of swallows. They all capture their prey in flight; and are dependent on their wings for getting a living. But the habit of flight is very different in different groups. Only a few of the

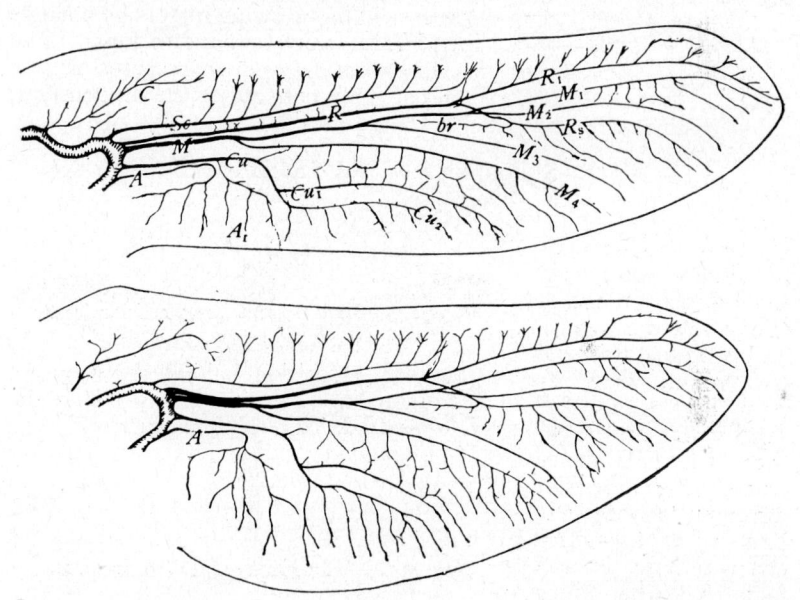

Fig. 363.—Tracheation of the wings of a grown naiad of *Gomphus descriptus*. (After Needham.)

strongest forms roam the upper air at will. There is a host of beautiful species, the skimmers or *Libellulidæ*, that hover over ponds in horizontal flight, the larger species on tireless wings, keeping to the higher levels. The stronger flying Æschnidæ course along streams on more or less regular beats; but the Gomphines are less constantly on the wing, flying usually in short sallies, from one resting place to another, and alighting oftener on stones or other flat surfaces than on vertical stems."

The characters presented by the venation of the wings of the Odonata are much used in the classification of these insects. In general the veins and areas of the wings are designated as in the accounts of the wings of other orders of insects; but there are certain features in the wings of these insects that are peculiar to them.

The most distinctive feature of the wings of the Odonata is the fact that in the course of their development one or more branches, usually two, of the medial trachea invade the area of the radial sector.

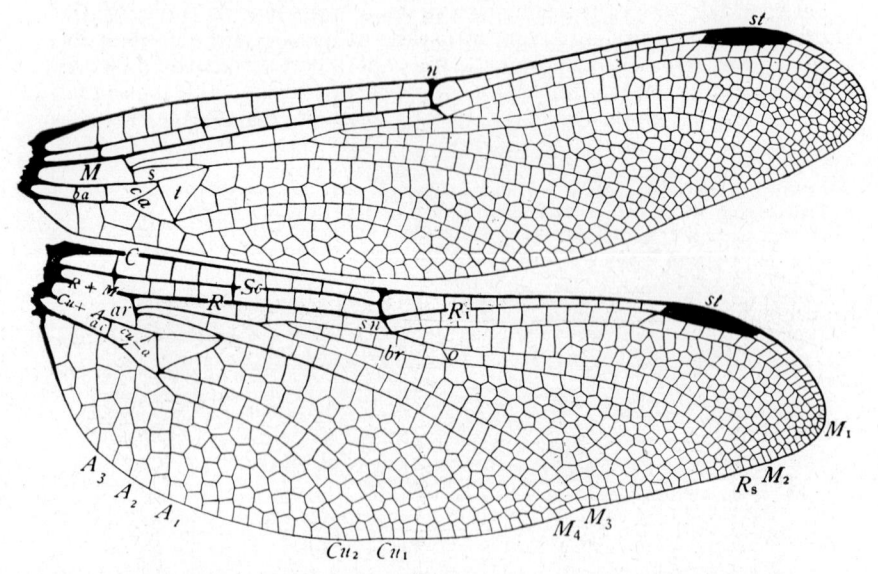

Fig. 364.—Wings of *Gomphus descriptus*. In the front wing, cells or areas are labeled; in the hind wing, veins.

This results in vein R_s occupying a position behind one or more, usually two, of the branches of media. Figure 362 represents the tracheation of the wings of two naiads of *Gŏmphus descrĭptus;* the wing shown at A is of a very young naiad; that at B is of a somewhat older one. In the wing shown at A, the branches of trachea M are in their typical position; in the wing shown at B, trachea M_1 is in front of trachea R_s. Figure 363 represents the tracheation of a full-grown naiad of the same species. In this stage of the development of the wings, both tracheæ M_1 and M_2 are in front of trachea R_s; and it is in this position that the veins of the adult wing are developed (Fig. 364).

ODONATA

By comparing the figure of the wing of an adult (Fig. 364) with that of the full-grown naiad (Fig. 363), it will be seen that the *oblique vein* marked *o* is not a cross-vein but a section of vein R_s; so too, what appears to be another cross-vein, labeled *s n*, is also a section of vein R_s; this section of vein R_s is known as the *subnodus*. It will also be seen that what appears to be the base of the radial sector, labeled *b r*, is a secondarily developed vein which connects the radial sector with a branch of media; this secondary vein is known as the *bridge*. The beginning of the formation of the bridge is shown in Figure 363.*

The more important of the other special terms used in descriptions of the wings of dragon-flies are the following: Much use is made in taxonomic work of the two series of cross-veins that are nearest the costal margin of the wing; those of these cross-veins that are situated between the base of the wing and the nodus are termed the *antenodal cross-veins;* the first of these two series of antenodal cross-veins extend from the costa to the subcosta; the second from the subcosta to the radius; the antenodal cross-veins are termed the *antecubital cross-veins* by some writers. The two series of cross-veins nearest to the costal margin of the wing and between the nodus and the apex of the wing are termed the *postnodal cross-veins;* the first of the two series of postnodal cross-veins extend from the costa to vein R_1; the second, from vein R_1 to vein M_1; the postnodal cross-veins are termed the *postcubital* cross-veins by some writers. Near the base of the wing there is in dragon-flies a well-marked area of the wing, which is usually triangular in outline (Fig. 364, *t*); this is the *triangle;* frequently the triangle is divided by one or more cross-veins into two or more cells. The area lying immediately in front of the triangle (Fig. 364, *s*) is termed the *supertriangle;* like the triangle this area may consist of a single cell or may be divided by one or more

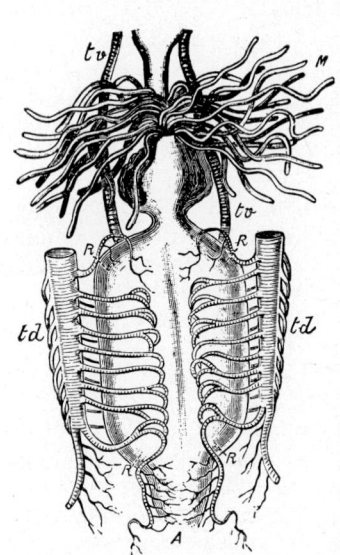

Fig. 365.—Hind-intestine and part of the tracheal system of a naiad of *Æschna cyanea: R, R, R, R,* rectum; *A*, anus; *td*, dorsal tracheal tubes; *tv*, ventral tracheal tubes; *M*, Malpighian tubes. (From Sharp, after Oustalet.)

Fig. 366.— Exuviæ of a naiad of a dragon-fly, *Tetragoneuria.*

*The conclusions regarding the homologies of the wing-veins given here are based on investigations by Dr. Needham the results of which were published by

cross-veins. Other named areas are the *basal anal area* (Fig. 364, *ba*) and the *cubital area* (Fig. 364, *ca*).

The writer has given in his "The Wings of Insects" an extended discussion of the wings of Odonata, illustrated by many figures, including a plate in which adjacent veins are represented in different colors, so that the course of each can be easily followed.

With the naiads of dragon-flies there is a remarkable modification of the organs of respiration, which fits these insects for aquatic life. The caudal part of the alimentary canal, the rectum, is modified so as to constitute a tracheal gill. It is somewhat enlarged; and its walls are abundantly supplied with tracheæ and tracheoles (Fig. 365). Water is alternately taken in and forced out through the anal opening; by this process the air in the tracheæ, with which the walls of the rectum are supplied, is purified in the same manner as in an ordinary tracheal gill.

The rectal tracheal gill of the naiads of dragon-flies is an organ of locomotion, as well as of respiration. By drawing water into the rectum gradually, and expelling it forcibly, the insect is able to dart through the water with considerable rapidity. This can be easily observed when naiads are kept in an aquarium.

When the naiad of a dragon-fly is fully grown it leaves the water to transform. The skin of the naiad splits open on the back of the thorax and head, and the adult emerges, leaving the empty skin of the naiad clinging to the object upon which the transformation took place. Figure 366 represents such a skin clinging to the stem of a water plant.

The suborder Anisoptera includes two families, the Æschnidæ and the Libellulidæ; each of these families is represented in our fauna by many genera and species. These are enumerated in the "Catalogue of the Odonata of North America" by Muttkowski ('10). The two families can be separated by the characters given below.

Family ÆSCHNIDÆ

The Æschnids

In this family the triangle (Fig. 364, *t*) is about equally distant from the arculus (Fig. 364, *ar*) in the fore and hind wings; and, except in the genus *Cordulegăster*, there is an oblique brace-vein extending back from the inner end of the stigma (Fig. 364).

The æschnids are mostly large species; among them are the largest, fleetest, and most voracious of our dragon-flies. Some of them roam far from water and are commonly seen coursing over lawns in the evening twilight; but most of them fly over clear water.

Comstock and Needham ('98–'99) and by Needham ('03). These conclusions have been questioned by Tillyard ('22) and by Schmieder ('22); but I do not feel that it would be wise to modify them before a much more extended investigation of the subject has been made.

ODONATA

Family LIBELLULIDÆ

The Libellulids or Skimmers

In this family the triangle in the hind wing is much nearer the arculus than is the triangle of the fore wing; and there is no oblique brace-vein extending back from the inner end of the stigma, as in the æschnids.

This is a large family including many of our commonest and best-known species of dragon-flies; many of them are familiar figures flying over ponds and ditches and by roadsides. Most of them are of well-sustained flight, and are seen continually hovering over the surface of still water; this suggested the common name *skimmers* which has been applied to them.

Suborder ZYGOPTERA*

The Damsel-Flies

The damsel-flies differ from the dragon-flies in that the two pairs of wings are similar in form and are either folded parallel with the abdomen when at rest (Fig. 367) or uptilted (*Lestes*). The head is transverse, each eye being borne by a lateral prolongation of the head. The females possess an ovipositor by means of which the eggs are placed in the stems of aquatic plants, sometimes beneath the surface of the water.

The name of the suborder probably refers to the fact that the wings are brought together when at rest.

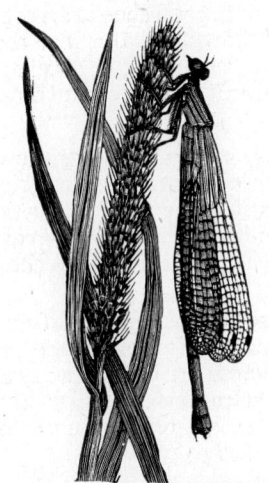

Fig. 367.—A damsel-fly.

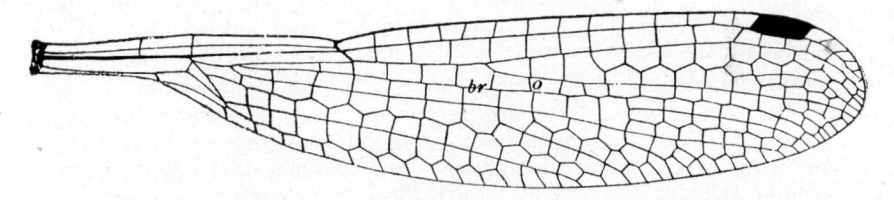

Fig. 368.—Wing of *Lestes rectangularis: o*, oblique vein; *br*, the bridge.

*Zygŏptera: *zygon* (ζυγόν), yoke; *pteron* (πτερόν), a wing.

Unlike the dragon-flies, the damsel-flies are comparatively feeble in their flight. They are found near the margins of streams and ponds, in which the immature stages are passed.

Most of the features in the venation of the wings of dragon-flies described on earlier pages are also characteristic of the wings of damsel-flies. Figure 368 represents an entire wing of *Lestes rectangularis*;

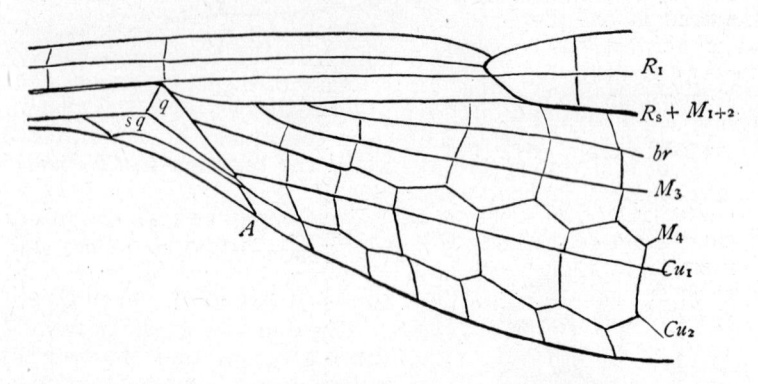

Fig. 369.—Base of fore wing of *Lestes rectangularis*: *br*, the bridge; *q*, quadrangle; *sq*, subquadrangle.

in this figure *o* indicates the oblique vein, and *br* the bridge. In Figure 369 the base of this wing is represented more enlarged, and the principal veins are lettered.

In the suborder Zygoptera the cubitus and the first branch, vein Cu_1, extend in a comparatively direct course from the base of the wing outward (Fig. 369); the abrupt bends in these veins in the region of the triangle, which are so characteristic of the Anisoptera, are only slightly developed here. This results in the areas corresponding to the triangle and the supertriangle of the Anisoptera being in direct line and forming an area which is often quadrangular; this area is termed the *quadrangle* (Fig. 369, *q*). In a large part of this order the cross-vein separating the parts of the quadrangle corresponding to the triangle and the supertriangle of the Anisoptera is lacking, in which case the quadrangle consists of a single cell (Fig. 369, *q*). In some members of this sub-

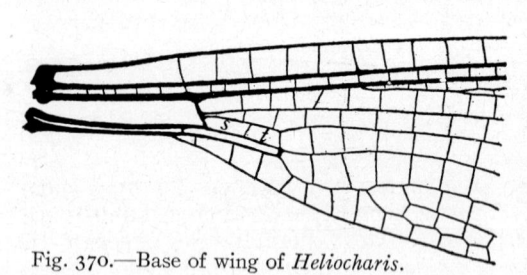

Fig. 370.—Base of wing of *Heliocharis*.

order it is present; in Figure 370, representing the base of a wing of *Heliocharis*, the two cells of the quadrangle are labeled *t* and *s* to facilitate comparison with figures of wings of Anisoptera. In certain other members of this suborder the quadrangle is divided into several cells by cross-veins (Fig. 371).

The cubital area of the wing is usually quadrangular in outline in the Zygoptera, and is termed the *subquadrangle* (Fig. 369, sq). Like the quadrangle, it may consist of a single cell or it may be divided by cross-veins (Fig. 371).

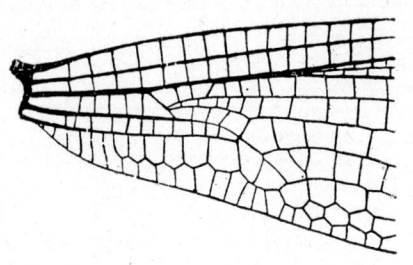

Fig. 371.—Base of wing of *Hetærina*.

The naiads of damsel-flies have three plate-like tracheal gills at the caudal end of the body (Fig. 372). The structure of these gills is illustrated by Figure 373; at A is represented an entire gill showing the tracheæ; and at B, part of a gill more magnified, showing both tracheæ (T) and tracheoles (t).

Fig. 372.—Naiad of a damsel-fly, *Argia*.

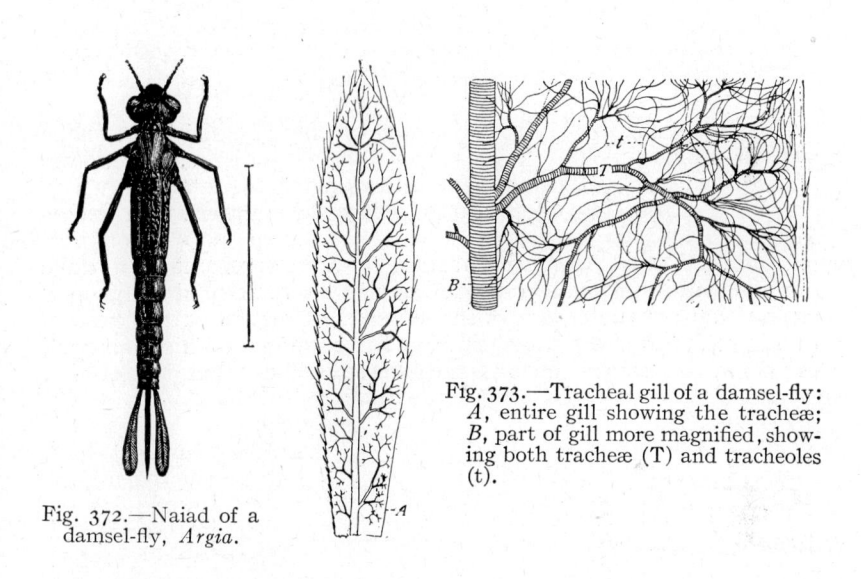

Fig. 373.—Tracheal gill of a damsel-fly: *A*, entire gill showing the tracheæ; *B*, part of gill more magnified, showing both tracheæ (T) and tracheoles (t).

The suborder Zygoptera includes two families, the Agrionidæ and the Cœnagrionidæ. The genera and species of these families are enumerated by Muttkowski ('10). The two families can be separated as follows.

A. Wings with many, at least five, antenodal cross-veins........AGRIONIDÆ
AA. Wings usually with only two antenodal cross-veins, rarely with three or four...CŒNAGRIONIDÆ

Family AGRIONIDÆ

The True Agrionids

In the Agrionidæ the wings are furnished with many antenodal cross-veins; and, although the wings are narrow at the base, they are not so distinctly petiolate as in the next family. These insects may be termed the true agrionids, as owing to a misapplication of the generic name *Agrion* the members of the next family have been incorrectly known as the agrionids.

Here belong the most beautiful of our damsel-flies, whose metallic blue or green colors are sure to attract attention. They are feeble in flight and do not go far from the banks of the pond or stream in which they were developed.

There are only two genera of this family in our fauna. These are *Ăgrion*, which has been commonly known as *Calŏpteryx*, and *Hetærīna*. In *Ăgrion* the wings are broad and spoon-shaped. In *Hetærīna* the wings are rather narrow, and in the males the base of one or both pairs is red.

Family CŒNAGRIONIDÆ

The Stalked-winged Damsel-Flies

The members of this famiily are easily recognized by the shape of their wings, which are long, narrow, and very distinctly petiolate (Fig. 368); and by the fact that in each wing there are only twc antenodal cross-veins, except in a few cases where there are three or four.

To this family belong the smallest of our damsel-flies; but while our species are of small or moderate size, there exist in the tropics species that are the largest of the Odonata. Some of our species are dull in color; but many are brilliant, being colored with green, blue, or yellow. This family includes the greater number of our damsel-flies.

CHAPTER XIV
ORDER PLECOPTERA*
The Stone-Flies

The members of this order have four membranous wings. In some genera the branches of the principal veins are reduced in number and there are comparatively few cross-veins; in others, accessory veins are developed and there are many cross-veins; in most genera the hind wings are much larger than the fore wings, and are folded in plaits and lie upon the abdomen when at rest. The mouth-parts are of the chewing type of structure, but are frequently vestigial in the adult. The cerci are usually long and many-jointed. The metamorphosis is incomplete.

Members of this order are common insects in the vicinity of rapid streams and on wave-washed rocky shores of lakes; but they attract little attention on account of their inconspicuous colors and secretive habits. They are called stone-flies because the immature forms are very abundant under stones in the beds of streams.

In the adults the body is depressed, elongate, and with the sides nearly parallel (Fig. 374). The prothorax is large. The antennæ are long, tapering, and many-jointed. The mouth-parts are usually greatly reduced. In some genera the mandibles are almost membranous, but in others they are firm and toothed, being well fitted for biting. The maxillæ exhibit variations in the degree of their reduction similar to those shown by the mandibles. The maxillary palpi are five-jointed. The labial palpi are three-jointed. The legs are widely separated, except the fore legs in the Pteronarcidæ; the tarsi are three-jointed.

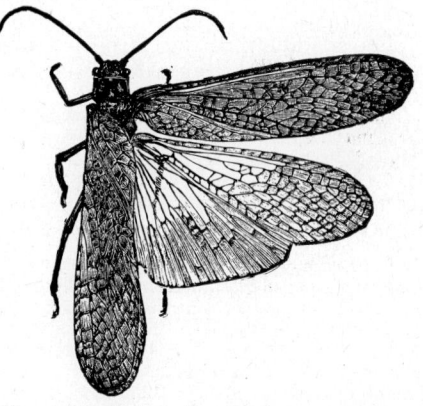

Fig. 374.—A stone-fly, *Pteronarcys dorsata*.

The hind wings are a little shorter than the fore wings, but usually, owing to the expansion of the anal area, they are considerably larger than the fore wings; in a few genera the hind wings are smaller than the fore wings; in some species the wings of the male are greatly reduced in size, and in others the males are wingless. When at rest, the wings are folded in plaits and lie upon the

---

*Plecoptera: *plecos* (πλέκος), plaited; *pteron* (πτερόν), a wing.

abdomen, as shown on the left side of Figure 374. The cerci are usually long and many-jointed; but they are rudimentary in the Nemouridæ.

The stone-flies are unattractive in appearance; in most of them the colors are obscure, being predominantly black, brown, or gray; but some of them that are active in the daytime and inhabit foliage are green. Their powers of flight are quite limited; they are usually found crawling about on stones or on plants near streams. Several of the smaller species appear in the adult state upon snow on warm days in the latter half of winter. They become more numerous in early spring and often find their way into our houses. The most common one of these in central New York is the small snow-fly, *Căpnia pygmæa*.

It is probable that most adult stone-flies eat nothing; this can be inferred from the reduced condition of their mouth-parts. But it has been shown by Newcomer ('18) that several species of *Tæniŏpteryx*, which are equipped with well-developed mouth-parts, feed upon the buds and leaves of plants. One species in particular, *T. pacĭfica*, is a serious pest in the Wenatchee Valley, Wash., where it bites into the buds of fruit trees.

One of the more striking features of the venation of the wings of the Plecoptera is a lack of uniformity in the number and courses of the subordinate veins. Not only are striking differences in wing-venation to be observed between different individuals of the same species, but frequently the wings of the two sides of an individual will vary greatly in venation. This is especially true as to the number of cross-veins and the branching of the veins in the distal parts of the wings. On the other hand, the characters presented by the trunks of the principal veins are quite constant.

There is one characteristic of the wings of the Plecoptera that is so constant that it may be considered an ordinal character. This is the fact that in the wings of the adult the radial sector of the hind wings is attached to media instead of to radius (Fig. 376*b*). This switching of the radial sector of the hind wings is true only of the venation of the adult. In the wings of naiads the trachea R_s is a branch of trachea R.

There are certain features of the wings of Plecoptera, which, although not always constant, occur in so large a portion of the members of the order that they may be considered characteristic; these are the following, all of which are represented in Figure 376*b*: The presence of the radial cross-vein (*r*). The absence of cross-veins in cell R and in the basal part of area R_1. (Cross-veins are found in cell R in *Pteronarcys*.) The strengthening in the fore wings of the area between media and vein Cu_1 and of that between veins Cu_1 and Cu_2 by the development of many cross-veins. The reduction of media to a two-branched condition. The reduction of the radial sector to a two-branched condition. (This reduction of the radial sector is apparent only after an extended study of the wings of stone-flies. In many cases, of which the form represented by Figure 376*b* is one, accessory veins have been developed on vein R_2+_3 which appear to be the primitive branches of the radial sector; but these accessory veins are very inconstant in number and position.) And the unbranched condition of the first anal vein.

In concluding this brief summary of the special features of the wings of the Plecoptera it seems desirable to define some terms frequently used by writers on this order.

The transverse cord.—In many genera of this order there is a nearly continuous series of cross-veins extending across each wing just beyond the middle of its length; this series of cross-veins is termed the *anastomosis* by many writers on the Plecoptera. As it is not formed by an anastomosing of veins, the use of the term *transverse cord* is preferable.

The pterostigma.—In most members of this order a specialized pterostigma has not been developed; but the term *pterostigma* is commonly applied to the cell beyond the end of the subcosta and between the costa and vein R_1, even though it is of the same color and texture as the remainder of the wing.

The basal anal cell.—A very constant feature of the anal area of the wings of Plecoptera is the presence of a cross-vein near the base of the wing, which extends from the first anal vein to the second. The cell that is closed by this cross-vein is termed the *basal anal cell* (Fig. 376b, ba).

The females drop their eggs in a mass in water. I have taken females of *Perla* and of *Pteronarcys* at lights, each with a mass of eggs hanging from the abdomen.

The metamorphosis is incomplete. The immature forms are all aquatic. These naiads are common on the lower surface of stones in rapids. They can be found easily by lifting stones from such situations and turning them over quickly, when the naiads will be found clinging to the stones with their flat bodies closely appressed to them and their legs, antennæ, and cerci radiating on the surface of the stone, but they are apt to run away quickly.

The naiads of stone-flies live only in well-aerated water; they are not found in stagnant water or in foul streams. They are said to feed on other aquatic insects, including smaller individuals of their own species; but according to the observations of Dr. P. W. Claassen they are largely vegetable feeders.

The body is depressed (Fig. 375); the antennæ are long, so too are the cerci. Most species possess tracheal gills, situated usually on the ventral side of the thorax just behind the base of each leg; but tracheal gills are found in some species either on the under side of the head, on the basal abdominal segments, or at the tip of the abdomen. A large number of the smaller species are destitute of tracheal gills; in these the air supply is absorbed through the thin cuticula of the ventral surface. The colors of naiads are often brighter than those of adults.

Fig. 375. — Naiad of a stone-fly, *Acroneura*.

When full-grown the naiads leave the water and transform on some near-by object. The empty exuviæ are often found clinging to stones or logs projecting from water or on the banks of streams.

According to a recent classification of this order, that of Tillyard ('21), it includes seven families; but only four of these families are represented in our fauna. A monograph of the North American species of the order is in preparation by Professor J. G. Needham and Professor P. W. Claassen; this is nearly completed and probably will be published soon. The four families of our fauna can be **separated** by the following table.

A. Anal area of the fore wings with two or more series of cross-veins (Fig. 376a). p. 328..................................PTERONARCIDÆ
AA. Anal area of the fore wings with not more than a single series of cross-veins, usually with no cross-veins beyond the basal anal cell.
B. Media of the fore wings separating from radius gradually, the two forming a sharp angle (Fig. 376b). p. 328........................PERLIDÆ
BB. Media of the fore wings separating from radius abruptly, the two forming a blunt angle (Fig. 376c).
C. Anal area of the fore wings with a forked vein arising from the basal anal cell (Fig. 376a). Cerci vestigial. p. 330.   NEMOURIDÆ
CC. Anal area of the fore wings with only simple veins arising from the basal anal cell (Fig. 376d). Cerci well developed. p. 330......
....CAPNIIDÆ

Family PTERONARCIDÆ

This is a small family which is represented in North America by only two genera and by but few species.

Pteronărcys.—This genus includes the largest of our stone-flies. Figure 374 represents a common species. The venation of the wings

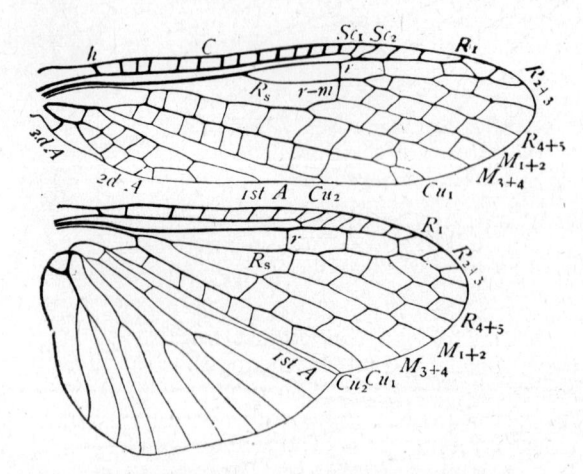

Fig. 376a.—Wings of *Pteronarcella badia.*

is reticulate; the reticulation is irregular and extends in the fore **wings** from the costa through the anal area.

A remarkable feature of members of this genus is that vestiges of tracheal gills are retained by the adults.

Pteronarcĕlla.—This genus includes smaller species than the preceding one, and the venation of the wings is more regular than in *Pteronarcys* (Fig. 376a).

Family PERLIDÆ

The members of this family differ from the Pteronarcidæ in the smaller number of cross-veins in the anal area of the fore wings, there being usually no cross-veins beyond the basal anal cells (Fig. 376b); and they differ from the following families in that media of

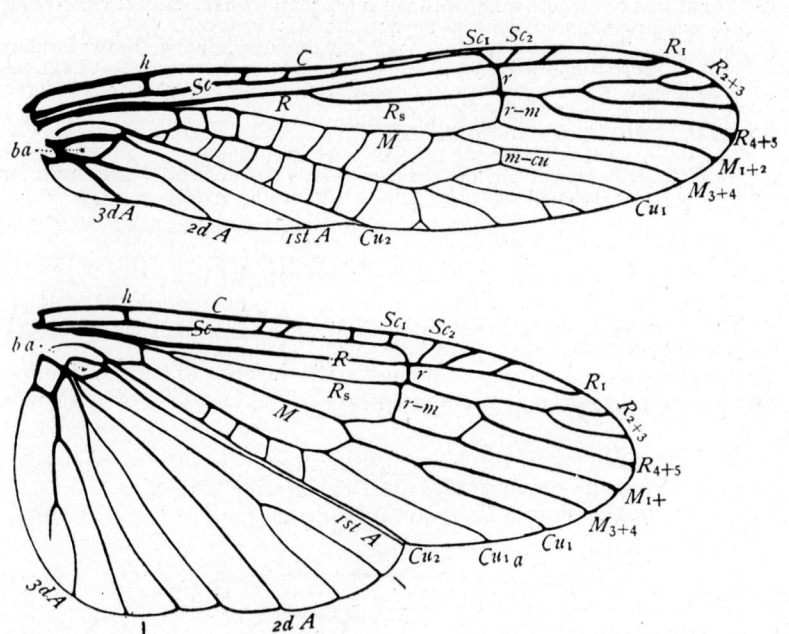

Fig. 376b.—Wings of *Isogenus sp.*

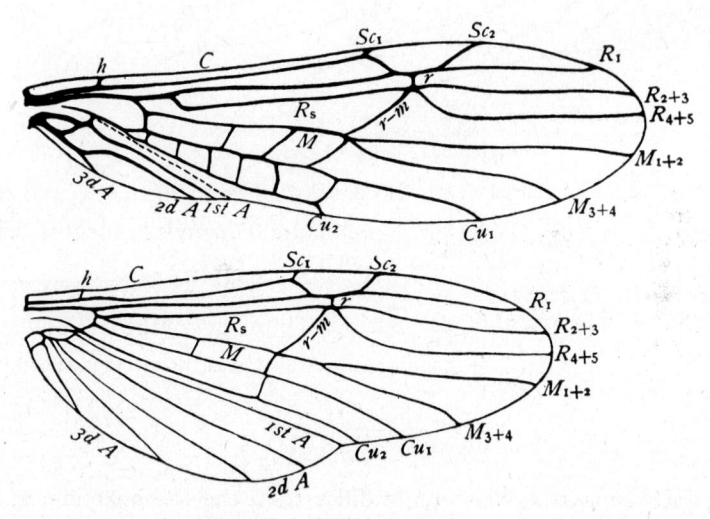

Fig. 376c.—Wings of *Nemoura sp.*

the fore wings separates from radius gradually, the two forming a sharp angle (Fig. 376b).

This is the largest of the families, including a large portion of the genera and species found in our fauna; fourteen genera have been described from this region.

Family NEMOURIDÆ

In this and the following family the media of the fore wings separates from radius abruptly, the two forming a blunt angle (Fig. 376c). In this family the second and third anal veins of the fore wings coalesce for some distance beyond the basal anal cell, forming a forked vein (Fig. 376c), and the cerci are vestigial.

The family is represented in our fauna by nine genera. Our more common representatives are small, dusky, and grayish species that are found emerging throughout the spring of the year.

Family CAPNIIDÆ

In this family, as in the Nemouridæ, the media of the fore wings separates from radius abruptly, the two forming a blunt angle (Fig. 376d); but in this family there are in the anal area of the fore wings

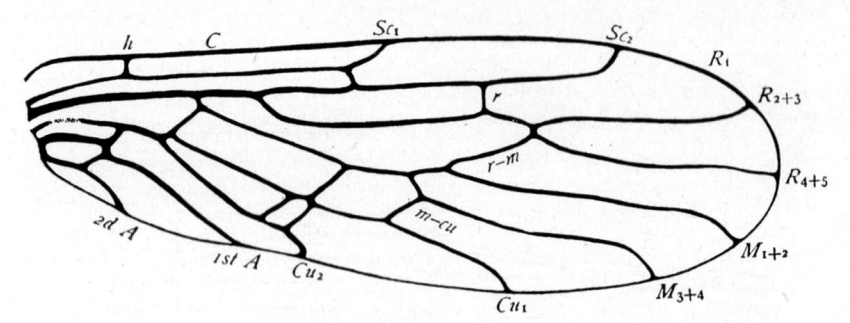

Fig. 376d.—Wing of *Capnia sp.*

only simple veins arising from the basal anal cell (Fig. 376d), and the cerci are well developed. This is a small family which is represented in our fauna by only three genera.

The members of this family that are most often seen are the little black species of *Căpnia* that appear on snow on warm days in the latter half of winter and in early spring. The naiads of these live chiefly in small brooks.

CHAPTER XV

ORDER CORRODENTIA*

The Psocids and the Book-Lice

The winged members of this order have four membranous wings, with the veins prominent, but with comparatively few cross-veins; the fore wings are larger than the hind wings; and both pairs when not in use are placed roof-like over the body, being almost vertical, and not folded in plaits. The mouth-parts are formed for chewing. The metamorphosis is gradual.

The best-known representatives of this order are the minute, soft-bodied insects which are common in old papers, books, and neglected collections and which have received the popular name *book-lice*. These low, wingless creatures form, however, but a small part of the order. The more typical winged forms (Fig. 377) bear a strong resemblance to plant-lice or aphids. The body is oval, the head free, and the prothorax small. The fore wings are larger than the hind wings; and both pairs when not in use are placed roof-like over the body, being almost vertical, and not folded in plaits. The wing-veins are prominent, but the venation of the wings is reduced. The tarsi are two- or three-jointed. Cerci are wanting.

Fig. 377.—A winged psocid, *Cerastipsocus venosus*.

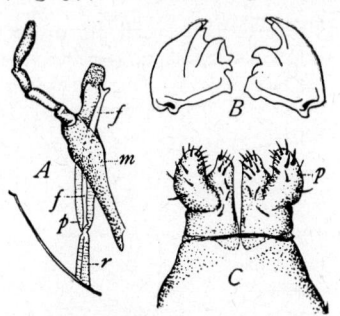

Fig. 378.—Mouth-parts of a book-louse, *Troctes divinatorius*: *A*, maxilla and paragnathus of right side, ventral view; *m*, maxilla; *f, f*, paragnathus; *p*, protractor muscle; *r*, retractor muscle. *B*, mandibles. *C*, labium, ventral view; *p*, palpus. (After Snodgrass.)

The mouth-parts are of especial interest on account of the presence of well-preserved paragnatha. Figure 378 represents the mouth-parts of the common book-louse, *Troctes divinatōrius*, as figured by Snodgrass ('05). The mandibles (*B*) are of the ordinary, strong, heavy, biting type. The maxillæ (*m*) consist each of a body piece, a weakly chitinized terminal lobe, and a four-jointed palpus. The paragnathus (*f, f*) is represented in the figure at *A*, with the maxilla; it lies above the maxilla and is, therefore, in its typical position between the maxilla and the mandible of the same

---

*Corrodĕntia: Latin *corrodens*, gnawing.

332   AN INTRODUCTION TO ENTOMOLOGY

side. Note that the figure is a ventral view, hence the paragnathus is represented as passing beneath the maxilla. The paragnatha have

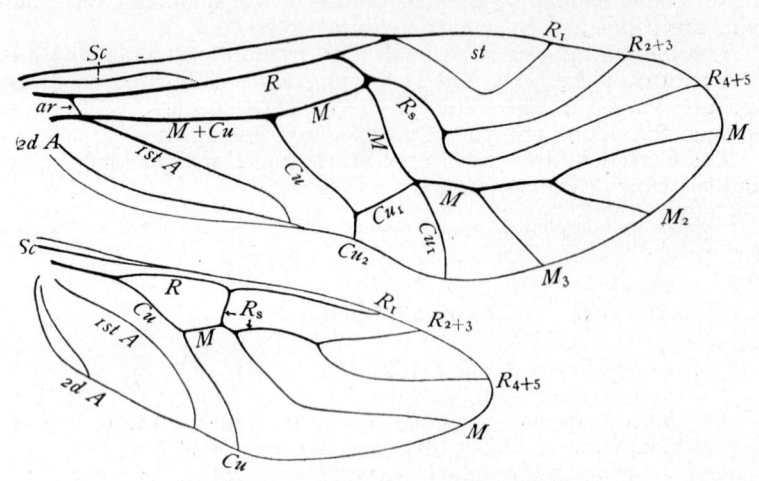

Fig. 379.—The wings of a psocid.

been known as the *furcæ maxillares*. The labium (C) bears a pair of one-jointed palpi.

The venation of the wings is distinctively characteristic in this order. The venation is more or less reduced; but its most characteristic feature is the bracing of the wing by anastomoses of the principal

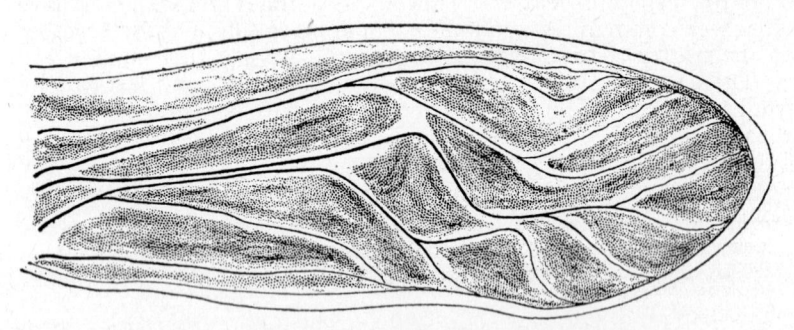

Fig. 380.—Fore wing of a full-grown nymph of a psocid.

veins instead of by cross-veins. This is well shown by the wings of *Psocus* (Fig. 379). The determination of the homologies of the wing-veins in this insect was accomplished by a study of the tracheation of the wings of nymphs. Figure 380 represents the tracheation of a fore wing of a full-grown nymph of *Psocus*.

There are no cross-veins in the wings of *Psocus*; the arculus (ar) in the fore wing is merely the base of media, and what appear as cross-veins in the central portion of the wing are sections of media and cubitus. In some genera, however, the radial cross-vein is present, and in some, instead of an anastomosis of veins M and Cu$_1$, these veins are connected by a medio-cubital cross-vein.

The metamorphosis is gradual. The nymphs resemble the adults in the form of the body, but lack wings and ocelli in those species that are winged in the adult; in the wingless species the differences between the young and the adult are even less marked.

The Corrodentia of the United States and Canada represent two families, which can be separated as follows.

A.   Wings well developed; ocelli present........................Psocidæ
AA. Wings absent or vestigial; ocelli absent......................Atropidæ

Family PSOCIDÆ

The Psocids

The family Psocidæ includes the more typical members of the Corrodentia, those in which the wings are well developed (Fig. 377). Usually the wings extend much beyond the end of the abdomen; but short-winged forms occur in species which ordinarily are long-winged. Of course the young of all are wingless, and there is a gradual development as the insect matures. The antennæ consist of only thirteen segments; this will enable one to separate the immature forms from the Atropidæ, in which the antennæ have a greater number of segments.

The psocids occur upon the trunks and leaves of trees, and on stones, walls, and fences. They feed upon lichens, fungi, and probably other dry vegetable matter. They are sometimes gregarious. I have often seen communities of a hundred or more closely huddled together on the trunks of trees, feeding on lichens.

The eggs are laid in heaps on leaves, branches, and the bark of trunks of trees. The female covers them with a tissue of threads. It is believed that both sexes have the power of spinning threads. The silk is spun from the labium.

More than seventy species, representing eleven genera, have been described from our fauna.

Family ATROPIDÆ

The Book-Lice and Their Allies

The family Atropidæ includes small Corrodentia, which are wingless or possess only vestigial wings. The most commonly observed species are those known as book-lice, which are the minute soft-bodied insects often found in old books (Fig. 381). Of these the two following species are the best known.

Trŏctes divinatōrius.—This is a wingless species which measures about 1 mm. in length; it is grayish white, with black eyes.

Ătropos pulsatōria.—In this species the fore wings are represented by small convex scales; it is of a pale yellowish white color and is a little more than 1 mm. in length.

Each of these species has been known as the *death-watch*, as they have been believed by superstitious people to make a ticking sound that presaged the death of some person in the house where it is heard. It is not probable that such minute and soft insects can produce sounds audible to human ears. The sounds heard were probably made by some wood-boring beetles, *Anobiidæ*, which are also known as the *death-watch*.

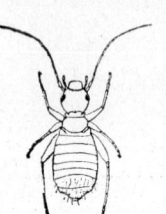

Fig. 381.—A book-louse.

Book-lice are found chiefly in damp, well-shaded rooms, not in general use. They do not attack man, but feed upon dead vegetable and animal matter, as the paste in book-bindings, wall-paper, and photographs. They rarely occur in sufficient numbers to do serious injury. They can be destroyed by fumigating the infected room with hydrocyanic acid gas. This, however, should be used only by experienced persons. Ordinarily a prolonged heating and drying of the room will be sufficient to destroy them.

CHAPTER XVI

ORDER MALLOPHAGA*

The Bird-Lice

The members of this order are wingless parasitic insects with chewing mouth-parts. Their development is without metamorphosis.

The bird-lice resemble the true lice in form, being wingless, and having the body more or less flattened; certain species that infest domestic fowls are well-known examples. These insects differ from the true lice in having chewing mouth-parts. They feed upon feathers, hair, and dermal scales, while the true lice, which constitute the order Anoplura, have sucking mouth-parts, feed upon blood, and infest only mammals.

The Mallophaga infest chiefly birds, and on this account the term *bird-lice* is applied to the entire group; a few genera, however, are parasitic upon mammals. Some writers term the Mallophaga *the biting lice*, which is a more accurate designation; but the name *bird-lice* is more generally used.

The bird-lice are small insects. The more common species range from 1 mm. to 5 mm. in length. The mouth-parts are on the under side of the head, the most anterior part of the head being a greatly enlarged clypeus; they are of the mandibulate type; and paragnatha ("furcæ maxillares") have been found in several species (Snodgrass '05). There is a pair of "simple eyes" located in the lateral margins of the head. The structure of these eyes has not been described; but judging from their position they are probably degenerate ommatidia and not ocelli. The front legs are shorter than the others and are used to convey food to the mouth.

There is an interesting correlation between the habits of these insects and the structure of their feet. The tarsi of those species that feed on mammals are one-clawed and fitted for folding against the tibiæ; they are organs well adapted for clinging to hairs. Those species that feed on birds have two-clawed tarsi and are better fitted for running. The above distinction is not quite accurate, as a few two-clawed species feed on kangaroos, wallabies, and wombats.

---

*Mallŏphaga: *mallos* (μαλλός), wool; phagein (φαγεῖν), to eat.

The accompanying figures represent some of our common species.

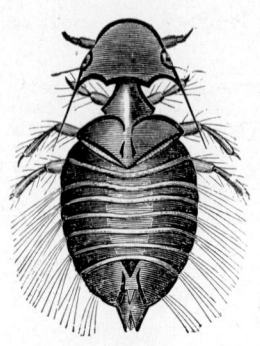

Fig. 382.—*Goniodes stylifer*. (From Law.)
Fig. 383.—*Trichodectes latus*. (From Law.) 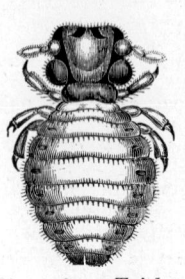
Fig. 384.—*Trichodectes spherocephalus*. (From Law.) 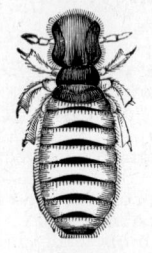
Fig. 385.—*Trichodectes scalaris*. (From Law.) 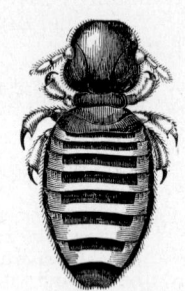

Goniōdes stȳlifer (Fig. 382) infests turkeys; *Trichodĕctes lātus* (Fig. 383), dogs; *Trichodĕctes spherocĕphalus* (Fig. 384), sheep; *Trichodĕctes scalāris* (Fig. 385), domestic cattle; and *Trichodĕctes ēqui* (Fig. 386), horses and asses.

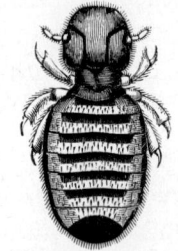

Fig. 386.—*Trichodectes equi*. (From Law.)

The eggs of the Mallophaga are glued to the hairs or feathers of their hosts. The development takes place on the body of the host and is without metamorphosis. The young are not so dark in color as the adults and the cuticula is less densely chitinized. The ametabolous condition of these insects is believed to be an acquired one, a result of their parasitic habits.

The bird-lice are well known to most people who have pet birds or who keep poultry. It is to free themselves from these pests that birds wallow in dust. When poultry are kept in closed houses they should be provided with a dust-bath. All poultry houses should be cleaned at least twice a year, and the old straw burned. Sprinkling powdered sulphur in the nests and oiling the perches with kerosene will tend to keep the pests in check. If a poultry house becomes badly infected, it should be cleaned thoroughly, every part whitewashed, and the poultry dusted with either insect-powder or sodium fluoride.

The Mallophaga is a small order. Professor V. L. Kellogg in his "Mallophaga" (Kellogg '08 b) estimates the number of known species to be 1250; these represent twenty-seven genera. But there are doubtless many species not yet discovered, as comparatively few birds and mammals have been thoroughly searched for these pests.

The work just quoted is the latest and most complete systematic treatise on this order. It followed a long series of papers on these insects published by this author. A more generally accessible account of the species that have been found in North America is a chapter in Professor Herbert Osborn's "Insects Affecting Domestic Animals" (Osborn '96).

The chief divisions of the order adopted by Kellogg ('08 b) are as follows.

A. With filiform, 3- or 5-segmented, exposed antennæ; no maxillary palpi; mandibles vertical........................SUBORDER ISCHNOCERA
   B. With 3-segmented antennæ; tarsi with one claw; infesting mammals. ....FAMILY TRICHODECTIDÆ
   BB. With 5-segmented antennæ; tarsi with two claws; infesting birds..... ....FAMILY PHILOPTERIDÆ
AA. With clavate or capitate, 4-segmented, concealed antennæ; with 4-segmented maxillary palpi; mandibles horizontal..........SUBORDER AMBLYCERA
   B. Tarsi with one claw; infesting mammals........FAMILY GYROPIDÆ
   BB. Tarsi with two claws; infesting birds, excepting a few species that infest kangaroos, wallabies, and wombats...........FAMILY LIOTHEIDÆ

CHAPTER XVII

ORDER EMBIIDINA*

The Embiids

This order is composed of small and feeble insects in which the body is elongate and depressed. The winged members of the order have two pairs of wings, which are quite similar in form and structure; they are elongate, membranous, extremely delicate, and folded on the back when at rest; the venation of the wings is considerably reduced. The mouthparts are formed for chewing. Cerci are present and consist each of two segments. The metamorphosis is of a peculiar type.

This is a small order of insects; Enderlein ('12 a) in his monograph of the Embiidina of the world lists only sixty-one species. The body is elongate and depressed (Figs. 387 and 388). Only the males are winged; and in some genera this sex also is wingless. The venation of the wings is reduced; this reduction has been brought

Fig. 387.—*Embia sabulosa,* male. (After Enderlein.)

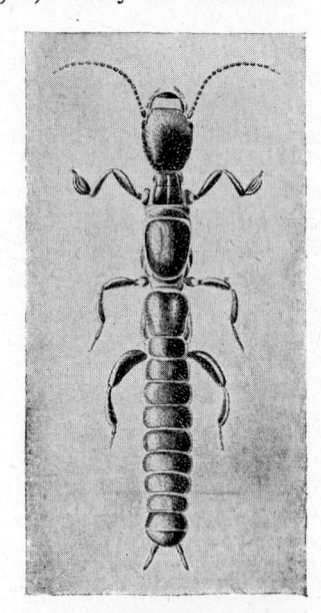

Fig. 388.—*Embia sabulosa,* female. (After Enderlein.)

about both by the coalescence of veins and by the atrophy of veins. Each of the veins of the wings extends along the middle of a brown band; between these bands the membrane of the wing is pale in color. The alternating brown and pale bands give the wing a very characteristic

---
*Embiidina: Embiidæ, *Embia, embios* (ἔμβιος), lively.

appearance (Fig. 389). In those forms where the venation of the wings has been reduced by the atrophy of veins, the brown bands persist after the veins have faded out; hence it is easy to determine by these bands the former position of veins that have been lost. A discussion of the venation of the wings of the Embiidina is given in my "The Wings of Insects."

The antennæ are filiform and are composed of from sixteen to thirty-two segments. The compound eyes consist of many ommatidia,

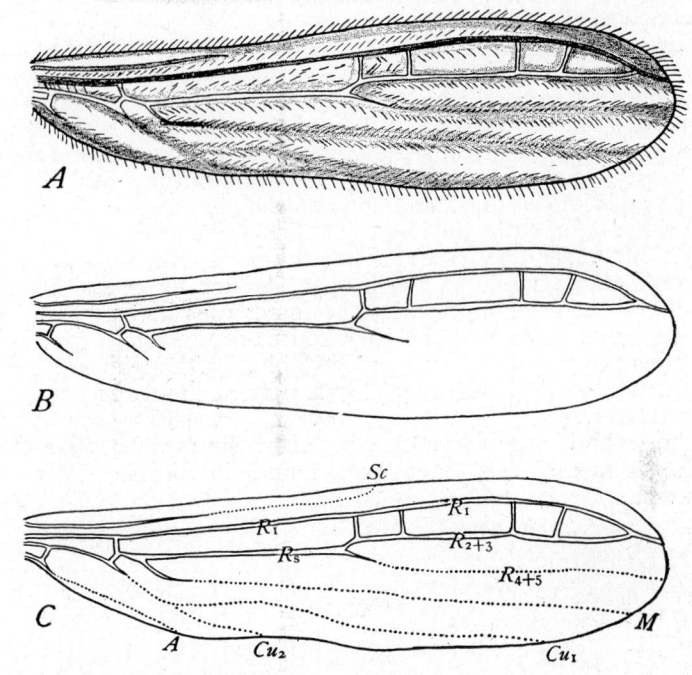

Fig. 389.—Fore wing of *Oligotoma saundersi:* A, the wing; B, outline of the wing showing the existing venation; C, outline of the wing showing the venation restored. (After Wood-Mason.)

which are of the eucone type. Ocelli are always wanting. The mouth-parts are mandibulate; the maxillary palpi are five-jointed and the labial palpi three-jointed. The abdomen is composed of ten distinct segments and bears at its tip a pair of two-jointed cerci.

Figure 387 represents the male of *Embia sabulōsa*, with the wing of one side removed; and Figure 388, the female of this species.

The metamorphosis is of a type intermediate between gradual and complete. This was shown by Melander ('02 b), who studied the development of *Embia texāna*. The young resemble the adults in the form of the body, except that the body is cylindrical instead of depressed; and the cuticula of the young is less densely chitinized and pigmented than is that of the adult. In the case of the females and of those males that are wingless in the adult instar, it might be said that these insects develop without metamorphosis. But in the case of the winged males the development resembles that of insects with a complete metamorphosis in one important respect; that is, the development of the wings is internal until the penultimate molt is reached. Melander states that he sectioned the fully grown larva and found the wings as large invaginated pockets completely beneath the hypodermis. In the penultimate instar of the winged females there are well-developed, external wing-pads. This instar may well be termed a pupa.

The embiids are very active insects both in running and in flight. They are often gregarious. They live in silken nests or galleries under stones or other objects lying on the ground, and burrow into the soil when the surface becomes too dry. Imms found in his studies of *Embia major* in the Himalayas that maternal care on behalf of the ova and larvæ is strongly exhibited by the females, in much the same manner as is known to occur among the Dermaptera.

Writers differ as to the source of the silk of which the nests are made. Melander ('02 a) and others have described glands in the metatarsi of the forelegs, which open through hairs, and have observed that in spinning its nest the insect uses its fore feet. But Enderlein maintains that the chief source of the silk is from glands that open through a spinneret on the labium, although the secretion of the metatarsal glands may play a part in the formation of the silken tissues.

The embiids are widely distributed in the warmer parts of the world. A few species have been found in Florida, Texas, and California.

CHAPTER XVIII

ORDER THYSANOPTERA*

The Thrips

The members of this order are minute insects with wings or wingless. The winged species have four wings; these are similar in form, long, narrow, membranous, not plaited, with but few or with no veins, and only rarely with cross-veins; they are fringed with long hairs, and in some species are armed with spines along the veins or along the lines from which veins have disappeared. The mouth-parts are formed for piercing and sucking. The tarsi are usually two-jointed and are bladder-like at the tip. The metamorphosis is gradual, but deviates from the usual type.

These insects are of minute size, rarely exceeding 2 mm. or 3 mm. in length. They can be obtained easily, however, from various flowers, especially those of the daisy and clover. Ordinarily it is only necessary to pull apart one of these flowers to find several thrips. They are in many cases very active insects, leaping or taking flight with great agility. In case they do not leap or take flight when alarmed, they are apt to run about and at the same time turn up the end of the abdomen in a threatening manner, as if to sting. In this respect they resemble the rove-beetles.

The body is long (Fig. 390). The head is narrower than the thorax, without any distinct neck. The antennæ are filiform or moniliform and consist of from six to nine segments; they are always much longer than the head and may be two or three times as long. The compound eyes are large, with conspicuous facets, which are circular, oval, or reniform in outline. Three ocelli are usually present in the winged forms, but sometimes there are only two ocelli; wingless species lack ocelli. The mouth-parts are fitted for piercing and sucking; they are in the form of a cone which encloses the piercing organs. The cone is composed of the clypeus, labrum, maxillary sclerites, and labium. The piercing organs consist of the left mandible (the right mandible is vestigial) and the two maxillæ. Each maxilla is composed of two parts: first, the palpus-bearing maxillary sclerite; and second, the maxillary seta. For detailed accounts of the mouth-parts see Hinds ('02) and Peterson ('15). The above statement regarding the mouth-parts is based on the paper by Peterson. The mouth-parts of the Thysanoptera bear a striking resemblance to those of the Hemiptera

Fig. 390.—A thrips.

---

*Thysanŏptera: *thysanos* (θύσανος), fringe; *pteron* (πτερόν), a wing.

342   *AN INTRODUCTION TO ENTOMOLOGY*

and the Homoptera, which are described in detail in later chapters.

The three thoracic segments are well developed. The wings are laid horizontally on the back when not in use; they are very narrow, but are fringed with long hairs (Fig. 391), which, diverging in flight, compensate for the smallness of the membrane. The fringing of the wings suggested the name Thysanoptera, by which the order is known. The two longitudinal veins that traverse the disk of the wing in

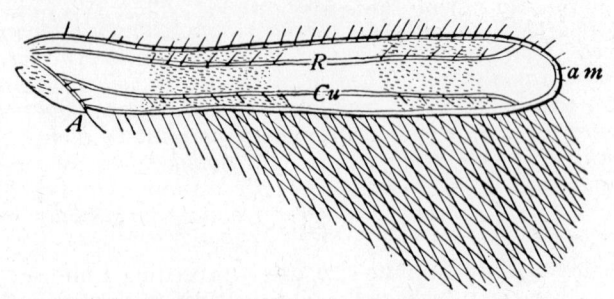

Fig. 391.—Fore wing of *Ælothrips nasturtii*. (After Jones.) The lettering is original.

the more generalized forms I believe to be the radius and the media respectively. The costal vein is continued by an ambient vein, which margins the entire preanal area of the wing (Fig. 391, *am*). The ambient vein is termed the "ring vein" by writers on this order, although the term *ambient vein* has been long in use for veins in this position. There is a short longitudinal vein separating the anal and preanal areas; this is doubtless the anal vein (Fig. 391, *A*). An organ for uniting the two wings of each side, and consisting of hooked spines situated near the base of the wings and a membranous fold on the under side of the anal area of the fore wing, is described by Hinds ('02).

In some species one or both sexes are wingless in the adult state; and in others, short-winged forms occur.

The legs are well developed, but are furnished with very peculiar tarsi. These are usually composed of two segments; the last segment terminates in a cup-shaped or hoof-like end and is usually without claws. Fitted into the cup-shaped end of the tarsus there is a very delicate, protrusile, membranous lobe or bladder, which is withdrawn into the cup when not in use but is protruded when the tarsus is brought into contact with an object. This is one of the most distinctive characteristic features of the members of this order. It was this feature that suggested the name Physopoda which is applied to this order by some writers.*

The abdomen consists of ten distinct segments. The form of the caudal segments differs in the two suborders as indicated below.

The manner of oviposition differs in the two suborders. In the Terebrantia the female cuts slits with her saw-like ovipositor and deposits her eggs singly in the tissue of the infested plant. In the Tubulifera it is evident that the eggs must be deposited on the surface.

---
*Physopoda: *physao* (φυσάω), to blow up; *pous* (πούς), a foot.

The metamorphosis of these insects is in some respects peculiar; but it conforms more closely to the paurometabolous type than to any other, the newly hatched young resembling the adult in the form of its body (Fig. 392, *A*) and in having similar mouth-parts and food habits. The first two or three instars have no external wings; these instars are commonly referred to as *larvæ*. The use of the term *larva* in this connection is not inappropriate if the wings are developing internally during these early stadia. That this may be the case is indicated by the large size of the wing-pads when they first

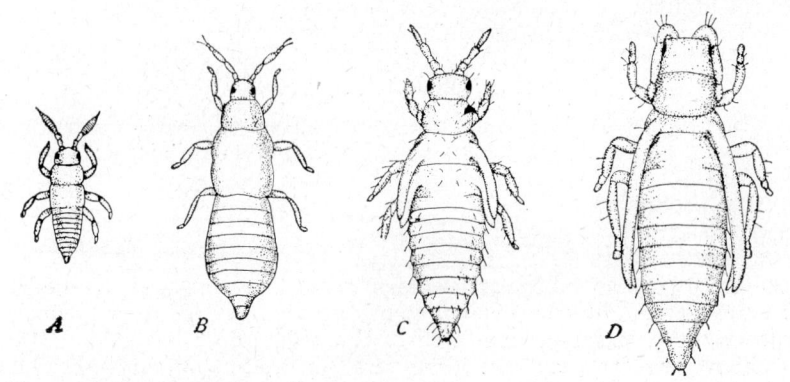

Fig. 392.—Immature forms of the citrus thrips: *A*, first larval instar; *B*, second larval instar; *C*, propupa; *D*, pupa. (After Horton.)

appear externally. After the last larval molt the insect assumes a form known as the *propupa* (Fig. 392, *C*). This resembles the larva in form; the antennæ are slender, and the insect is moderately active. Its most striking feature is the presence of large wing-pads, which extend at first to about the end of the second abdominal segment and increase in length somewhat during this stadium. With the next molt the insect becomes what is known as the *pupa*. In this stage the wing-pads are longer (Fig. 392, *D*), the antennæ extend back over the head and prothorax, and the insect is quiescent. With the next molt the adult form is assumed.

The different species of thrips vary greatly in habits, some being injurious to vegetation, while others are carnivorous, feeding on aphids and other small insects, the eggs of insects, and mites, especially the "red spider." Their most important economic role, however, is that of pests of cultivated plants. The thrips that infest plants puncture the tissue of the plant by their piercing mouth-parts and suck out the sap.

The order Thysanoptera is divided into two suborders, which can be separated as follows:

A. Female with a saw-like ovipositor; terminal abdominal segment of female conical; that of the male bluntly rounded.................TEREBRANTIA
AA. Female without a saw-like ovipositor; terminal abdominal segment tubular in both sexes. p. 345....................................TUBULIFERA

Suborder TEREBRANTIA*

In this suborder the female has a four-valved, saw-like ovipositor; the terminal abdominal segment of the female is conical; that of the male bluntly rounded. Wings are usually present; the front wings are stronger than the hind wings and usually have more or less well-developed veins; the membrane of the wings is clothed with microscopic hairs.

The members of this suborder are more agile than those of the other one. They run rapidly; and spring, by bending under the tip of the abdomen and suddenly straightening it out.

This suborder includes two families.

Family ÆOLOTHRIPIDÆ

In this family the wings are comparatively broad. Each fore wing has two longitudinal veins extending from its base to near the tip, where they unite with a prominent ambient vein on each side of the tip (Fig. 391); four or five cross-veins are present in each fore wing, in some species (Fig. 393); in others, cross-veins are wanting (Fig. 391). The ovipositor is upcurved.

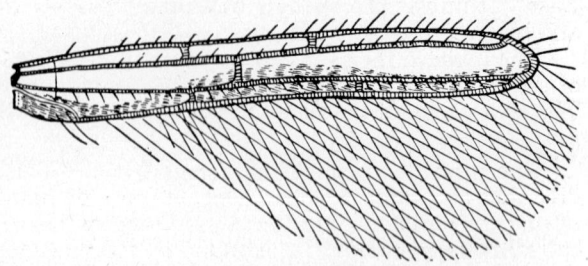

Fig. 393.—Fore wing of *Erythrothrips arizonæ*. (After Moulton.)

Comparatively few species belonging to this family have been found in our fauna; the best-known one is the following.

The banded thrips, *Æŏlothrips fasciātus*.—This species is widely distributed both in this country and in Europe. The adult is yellowish brown to dark brown in color, with three white bands on the wings, one at the base, one in the middle, and one at the tip. The larva is yellow with the abdomen deeper orange behind. This species infests many plants; it is common in the heads of red clover.

Family THRIPIDÆ

In this family the wings, when present, are usually narrow and pointed at the tip. The radius and cubitus of the front wings, when present, usually coalesce for about one third their length, so that cubitus appears to be a branch of radius. The ovipositor is downcurved.

To this family belong most of the species of thrips that have attracted attention on account of their economic importance. The better-known of these are the following.

---

*Terebrantia: *terebro*, to bore through.

The onion thrips, *Thrips tabāci*.—This is a serious pest of the onion. It is found on the bulbs in loose soil and at the axils of leaves, causing the disease known as white blast on account of the whitish appearance of the infested fields. Although called the *onion thrips*, it infests a great variety of plants.

The greenhouse thrips, *Heliothrips hæmorrhoidalis*.—This is a tropical insect, which is often a serious pest in greenhouses; it is also found out of doors in the milder California climate. Drops of a reddish fluid which turns black cover the infested leaves.

The bean thrips, *Heliothrips fasciātus*.—This is a serious pest on oranges, alfalfa, pear trees, and various garden crops in California.

The orange thrips, *Euthrips citri*.—This is a serious orange pest in California and Arizona; it deforms the new growth of foliage and causes scabbing and scarring of the fruits.

The pear thrips, *Euthrips pyri*.—This thrips infests pears, prunes, peaches, and other deciduous fruits, both in California and in the East. It infests the opening buds and blossoms, stunting the leaves and blasting the blossoms.

The tobacco thrips, *Euthrips fuscus*.—This is a destructive enemy of shade-grown tobacco causing the injury known as white vein. The white veins of the leaves show in the wrapper when manufactured into cigars.

The strawberry thrips, *Euthrips trĭtici*.—This species was first described as a pest of wheat, hence its specific name; but on account of its extensive injury to the flowers of strawberry it is now known as the strawberry thrips. It is found in the flowers of almost all wild and cultivated plants and is the commonest and most widely distributed of all American species of thrips.

The grass thrips, *Anăphothrips striātus*.—This species infests June grass, timothy, and other grasses by destroying the heads of the infested plants. The young insect pierces the stem just above the upper node, where it is tender, causing it to shrivel and all the parts above the injury to die. The dead and yellow heads of grasses thus destroyed can be seen in early summer everywhere in grass-growing regions. This disease is known as silver-top.

CONTROL.—Thrips are destroyed in those cases where it is practicable to spray the infested plants by the use of contact poisons, such as nicotine or kerosene emulsion, and soap solution. Detailed directions for making and applying these sprays are given in many published bulletins and in special text-books. The burning of old grass in early spring would probably destroy the hibernating grass thrips.

SUBORDER TUBULIFERA*

In this suborder the female is without a saw-like ovipositor and the terminal abdominal segment is tubular in both sexes. The wings are usually present; the fore pair only with a single vestigial, longi-

---

*Tubulifera: *tubulus*, a little tube; *fero*, to bear.

tudinal vein; the membrane of the wings is not clothed with microscopic hairs. This suborder includes a single family.

Family PHLŒOTHRIPIDÆ

The members of this family are, as a rule, considerably larger and more powerfully formed than the Terebrantia, some of them being the giants of the order. They live usually in secluded places, as between the parts of composite flowers, under the bark of trees, on the underside of foliage, in galls, moss, turf, fungi, etc. Their movements are very deliberate and they never run or spring (Hinds '02).

Nearly as many species and genera of this family have been found in this country as of the other suborder; but this family appears to be of much less economic importance than is the Thripidæ. One species, *Aleurŏdothrips fasciapĕnnis*, which is common in Florida, feeds on the eggs, larvæ, and pupæ of the citrus white fly, *Dialeurodes citri*.

CHAPTER XIX

ORDER ANOPLURA*

The True Lice

The members of this order are wingless parasitic insects with piercing and sucking mouth-parts. Their development is without metamorphosis.

The order Anoplura is composed of the true lice. These are small wingless insects, which live on the skin of mammals and suck their blood. They are sharply distinguished from the Mallophaga or bird-lice by the possession of piercing and sucking mouth-parts. The most familiar examples of the Anoplura are three species that infest man and several species that are found on domestic animals.

The name Siphunculata was proposed for this order by Meinert in 1891 and is now used by some authors; but the name Anoplura is much the older name, having been proposed by Leach in 1815, and is more generally used.

The body is more or less flattened (Fig. 394). The head is free and horizontal. The compound eyes are vestigial or are wanting. There are no ocelli. The antennæ are three-, four-, or five-jointed. The mouth is furnished with a fleshy, unjointed proboscis, which can be withdrawn into the head or extended to a considerable length. Within this proboscis are two knife-like stylets; and at its base, when extended, there is a wreath of recurved hooks. These hooks serve to anchor firmly the proboscis when inserted in the skin of the infested animal. Authors do not agree as to the homologies of the different mouth-parts of these insects.

The thoracic segments are fused. The legs are similar; the tarsi consist of a single segment, which is often greatly reduced. There is a single tarsal claw, which is opposed by a toothed projection of the tibia, forming an efficient organ for clinging to the hairs of the host. The abdomen consists of nine segments; there are no cerci.

The eggs of the true lice are commonly known as "nits." They are attached to the hairs of the host by a glue-like substance. The young lice resemble the adults except in size. As with the Mallophaga, the ametabolous condition of these insects is believed to be an acquired one, a result of their parasitic life.

This is a small order. Dalla Torre ('08) in his monograph of the Anoplura of the world lists only sixty-five species. These represent fifteen genera, which are grouped in four families. The two following families include all of the species that infest man and the common domestic animals.

---

*Anoplūra: *anoplos* (ἄνοπλος), unarmed; *oura* (οὐρά), tail.

348   *AN INTRODUCTION TO ENTOMOLOGY*

Family PEDICULIDÆ

In this family the eyes are comparatively large, convex, and distinctly pigmented; and the proboscis is short, hardly reaching the thorax. Here belong the three well-known species of lice that are parasites of man. These are the following.

The head-louse, *Pedĭculus căpitis.*—This is the most common species infesting man. It lives in the hair of the head, and is most common on the heads of neglected children. Under ordinary circumstances, cleanliness and the use of a fine-toothed comb are all that is necessary to insure freedom from this disgusting pest. But sometimes adults of most cleanly habits become infested by it. It can be destroyed by the use of tincture of larkspur or a larkspur lotion, which can be obtained from druggists.

The body-louse, *Pedĭculus cŏrporis.*—This insect lives upon the skin of most parts of the body, but especially on the chest and back. It is often troublesome on ships, in military camps, in prisons, and in the apartments of uncleanly people who neglect to change their clothes. It was a terrible scourge during the World War, when troops were obliged to live under most unsanitary conditions in trenches and camps. The female attaches her eggs to fibers in the seams of undergarments, from which the young hatch in about a week. This species is exceedingly prolific. It and the preceding species transmit several human diseases, including typhus fever, trench fever, and relapsing fever.

The method of destroying these vermin commonly employed in hospitals and poorhouses is to rub mercurial ointment in the seams

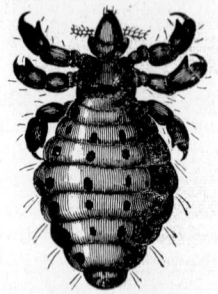

Fig. 394.—The short-nosed ox-louse. (From Law.)

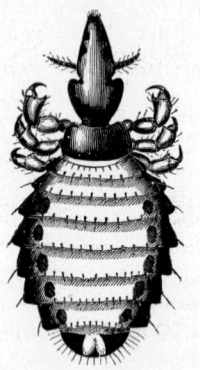

Fig. 395.—The horse-louse. (From Law.)

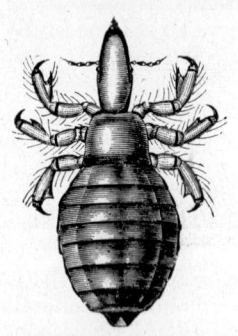

Fig. 396.—The hog-louse. (From Law.)

of undergarments. During the World War much attention was devoted to the problem of control of this pest and hundreds of papers were published on this subject. It has been found that both the lice and their eggs are destroyed by the ordinary laundering process used in washing clothes.

The crab-louse, *Phthirius pubis*.—The common name of this species is suggested by the form of the body, which is nearly as broad as long. When highly magnified, the resemblance of this insect to a crab is quite striking; but to the unaided eye it appears more like a large scale of dandruff. These offensive vermin affect the pubic region and armpits of man, stretching themselves out flat, holding tight to the cuticle, and inflicting most irritating punctures. They can be destroyed by mercurial ointment.

Family HÆMATOPINIDÆ

In this family the eyes are vestigial or wanting and the proboscis is very long. Here belong the true lice that infest our common domestic animals; the more important of these are the following.

The short-nosed ox-louse, *Hæmatopinus eurysternus* (Fig. 394).
The horse-louse, *Hæmatopinus asini* (Fig. 395).
The hog-louse, *Hæmatopinus suis* (Fig. 396).
The long-nosed ox-louse, *Linognathus vituli* (Fig. 397).
The dog-louse, *Linognathus piliferus* (Fig. 398).

For the destruction of these pests upon cattle, poisonous substances must not be used, as injury would result from the animals licking themselves. They may be safely treated by washing with a strong infusion of tobacco leaves, or by rubbing with an ointment made of one part sulphur and four parts lard, or by sprinkling with Scotch snuff or powdered wood-ashes. Stavesacre lotion and larkspur lotion are also used. The insecticide should be applied thoroughly, leaving no spot untouched where the lice can gather and remain and from which they can spread over the body again. The application should be repeated several times at intervals of three or four days, in order to destroy the young which may hatch after the first application. It is also necessary, in order to make sure of eradicating the pests, to dress with similar agents, or with strong lye or kerosene, all places where the cattle have been in the habit of rubbing, and the cracks in the stables where they have stood; or to whitewash the stables and rubbing-places.

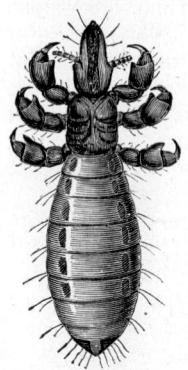

Fig. 397.—The long-nosed ox-louse. (From Law.)

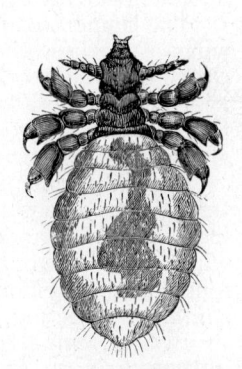

Fig. 398.—The dog-louse. (From Law.)

For a more extended account of the true lice found in North America, see Professor Herbert Osborn's "Insects Affecting Domestic Animals," pp. 164-188 (Osborn '96).

CHAPTER XX

ORDER HEMIPTERA*

The True Bugs

The winged members of this order have four wings; the first pair of wings are thickened at the base, with thinner extremities which overlap on the back. The mouth-parts are formed for piercing and sucking; the beak arises from the front part of the head. The metamorphosis is gradual.

People who know but little regarding entomology are apt to apply the term *bug* to any kind of insect; but strictly speaking, only members of the order Hemiptera are bugs.

The bugs are very common insects. Many species abound on grass and on the foliage of other plants; some species live on the surface of water; others live within water; and a few are parasitic on birds and mammals.

This order is a very important one; it includes many species injurious to vegetation; among these are some of our more important pests of cultivated plants. On the other hand, some of the species are ranked among beneficial insects on account of their predacious habits; for many of them feed upon noxious insects.

The name Hemiptera was suggested by the form of the front wings. In these the basal half is thickened so as to resemble the elytra of beetles, only the terminal half being wing-like. The hind wings are membranous, and are folded beneath the front wings. On this account the front wings are often termed *wing-covers;* they are also termed *hemelytra*, a word suggested by their structure.

Formerly, when the Homoptera was included in the order Hemiptera, the true bugs constituted the suborder Heteroptera; this name indicated the remarkable difference in the texture of the two pairs of wings of the true bugs and served to contrast this condition with that found in the Homoptera, where the two pairs of wings are usually similar in structure.

In the Hemiptera the front wings present characters much used in the classification of these insects; and consequently special names have been applied to the different parts of them. The thickened basal portion is composed of two pieces joined together at their sides; one of these is narrow and is the part next to the scutellum when the wings are closed; this is distinguished as the *clavus* (Fig. 399, *cl*); the other part is the *corium* (Fig. 399, *co*). The terminal portion of the front wing is termed the *membrane* (Fig. 399, *m*). In certain families, the Anthocoridæ for example, a narrow piece along the costal margin of the wing is separated by a suture from the remainder of the

---

*Hemĭptera: *hemi-* (ἡμι), half; *pteron* (πτερόν), a wing.

The order Hemiptera as now restricted includes only one of the suborders of the old order Hemiptera, the suborder Heteroptera. The following order, the Homoptera, was formerly regarded as a suborder of the Hemiptera.

corium; this is the *embolium* (Fig. 400, *e*). In certain other cases, as the Miridæ for example, a triangular portion of the terminal part of

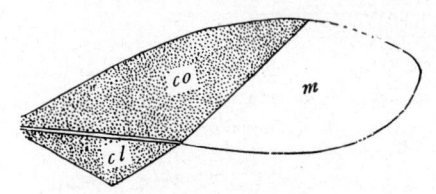

Fig. 399.—Diagram of a front wing of a bug: *cl*, clavus; *co*, corium; *m*, membrane.

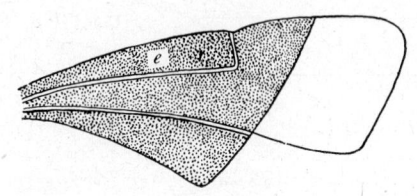

Fig. 400.—Diagram of a front wing of an anthocorid: *e*, embolium.

the corium is separated as a distinct piece; this is the *cuneus* (Fig. 401, *cu*).

The wings of the Hemiptera exhibit remarkable departures from the primitive type of wing-venation. So great are these that, at first, one sees very little in common between the wings of a bug and those of insects of any other order. But an examination of the tracheation of the wings of nymphs of bugs shows that these wings are merely modifications of the primitive type of insect wings. This is more obvious in some families than in others; it is well shown in the tracheation of a fore wing of a pentatomid nymph (Fig. 402).

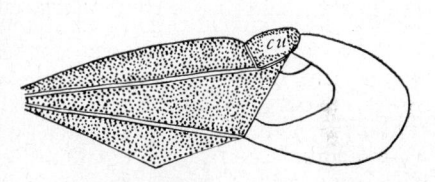

Fig. 401.—Diagram of a front wing of a mirid: *cu*, cuneus.

The head in the Hemiptera varies greatly in form in the different families; but the accompanying figures of the head of one of the Belostomatidæ, *Lethocerus* (Figs. 403 and 404), will serve to illustrate the position and form of the parts that are commonly referred to in descriptions of members of this order.

There are two factors which make difficult the determination of the areas of the surface of the head in these insects that have been recognized and defined in the more generalized insects (see pages 37 to 40): first, in some cases the sutures that limit these areas in the more generalized insects are here obsolete; second, the basal part of each mandible and of each maxilla enters into the composition of the wall of the head.

A similar modification of the head and mouth-parts exists in the Homoptera, and the students of the Hemiptera should study the relations of the mouth-parts to the head-capsule in that order, where they are more easily seen than in the Hemiptera.

352   AN INTRODUCTION TO ENTOMOLOGY

An important feature of the head in the Hemiptera is the extended development of the gular regions, which results in the beak being

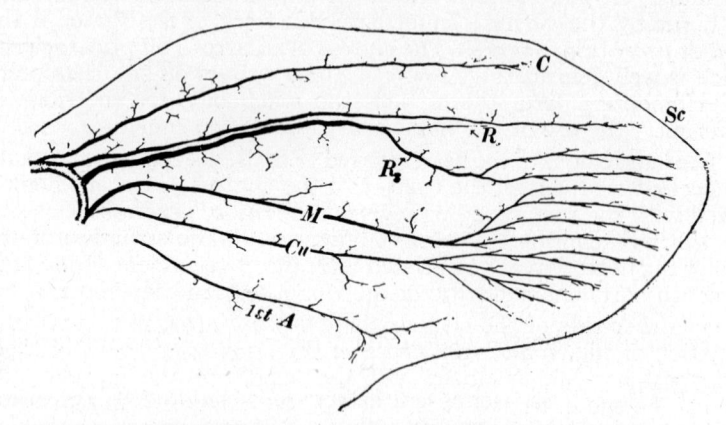

Fig. 402.—Tracheation of a fore wing of a pentatomid nymph.

borne by the front part of the head. This contrasts strongly with the condition found in the Homoptera, where the gula is so reduced that

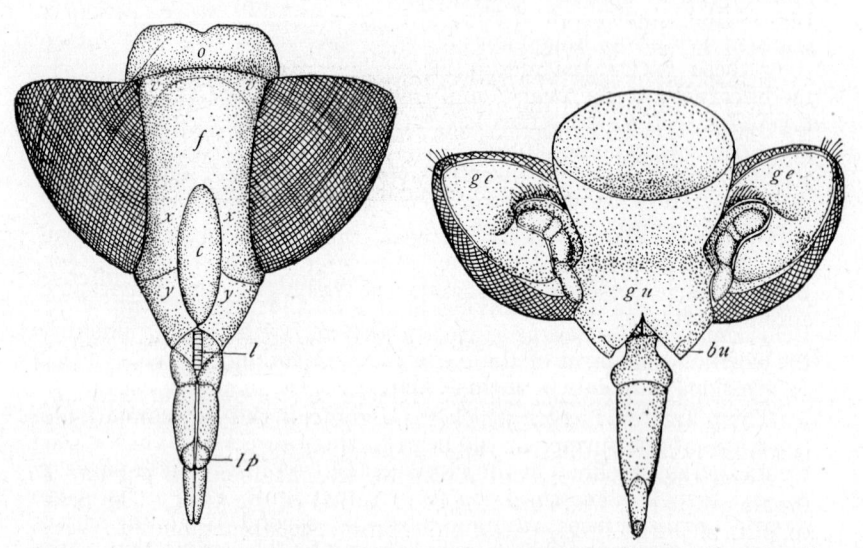

Fig. 403.—Head of *Leinocerus*, dorsal aspect.

Fig. 404.—Head of *Lethocerus*, ventral aspect.

the beak arises from the hind part of the lower side of the head.

In *Lethocerus* the *occiput* (Fig. 403, *o*) is separated from the vertex by a distinct transverse suture. The *vertex* (Fig. 403, *v, v*) is very short on the middle line of the body but is much longer on each side next to the compound eye; the epicranial suture is very indistinct in the adult. In those bugs in which the paired ocelli are present they are borne by the vertex. Immediately in front of the vertex is the *front* or *frons* (Fig. 403, *f*). The *clypeus* is a narrow, elliptical sclerite which is well defined (Fig. 403, *c*). Some writers on the Hemiptera and Homoptera term the clypeus the *tylus;* but, for the sake of uniformity, the use of this name should be discontinued.

The four regions of the head referred to in the preceding paragraph, the occiput, the vertex, the front, and the clypeus, are easily homologized with the corresponding regions in the more generalized insects. We will now consider certain modifications of the structure of the wall of the head that are correlated with the development of the type of mouth-parts characteristic of the Hemiptera and Homoptera.

On either side of the clypeus there is what appears to be a prolongation of the front. In *Lethocerus* (Fig. 403, *x, x*), each of these prolongations extends about half the length of the clypeus and bounds the eye in front. It is believed that each of them represents the basal part of a mandible; they are termed, therefore, the *mandibular sclerites*. In some Homoptera the mandibular sclerites are distinct; this condition exists in the head of a cicada figured in the next chapter (Fig. 463). The mandibular sclerites were so named by Smith ('92), who first recognized that they pertain to the mandibles. Before that time several different names were applied to them, which are still in use by some writers; these are *jugæ*, *loræ*, and *fulcra*.

In *Lethocerus* there is a pair of sclerites in front of the mandibular sclerites and bounding the distal end of the clypeus; each of these is the basal part of a maxilla; for this reason they are termed the *maxillary sclerites* (Fig. 403, *y, y*) In *Lethocerus* the tips of these sclerites meet on the dorsal wall of the head covering the tip of the clypeus.

On the ventral aspect of the head, the *gula* occupies the median area (Fig. 404, *gu*); and the *genæ*, the lateral areas (Fig. 404, *ge*). In each gena there is a deep groove in which the very remarkable antenna rests.

At the front end of the ventral wall of the head there is a pair of sclerites, each of which is articulated with a maxillary sclerite; these are known as the *bucculæ* and are believed to represent the maxillary palpi (Fig. 404, *bu*). In *Lethocerus* the caudal margin of each buccula is solidly joined to the front end of the gula.

From the above account it can be seen that only a portion of the mouth-parts enters into the constitution of the beak. The beak consists of the following parts: the labrum, the labium, and four very slender lancet-like organs enclosed in the labrum and labium, the mandibular setæ and the maxillary setæ.

The labrum is joined to the distal end of the clypeus; in *Lethocerus* the base of the labrum is covered by the maxillary sclerites, where they overlap the tip of the clypeus, and its distal end extends into the furrow of the labium, but the intermediate portion is exposed

354   AN INTRODUCTION TO ENTOMOLOGY

(Fig. 403, *l*). It is a slender, pointed, transversely striated organ.
The *labium* constitutes the most prominent part of the beak; in most Hemiptera it consists of four segments; but in several families it is reduced to three segments. At the distal end of the third segment in *Lethocerus* and some other aquatic Hemiptera there is a pair of small appendages, each of which consists of a single segment (Fig. 403, *lp*); these were described by Leon ('97) as vestiges of the labial palpi.* The dorsal surface of the labium is deeply grooved, forming a channel which encloses the mandibular and maxillary setæ. The labium is not a piercing organ; its function is to protect and direct the setæ and to determine, by means of tactile hairs at its tip, the place where the puncture should be made by the setæ (Fig. 405, *t*).

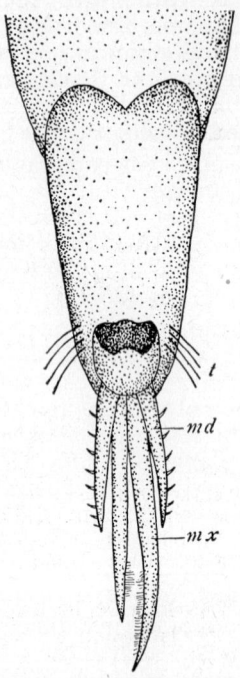

The *mandibular setæ* and the *maxillary setæ* are four slender, lance-like organs which arise within the head-capsule and pass out from the head through a furrow in the lower side of the labrum and extend in a furrow on the upper side of the labium to the tip of this organ, from which they are pushed out when not in use (Fig. 405). As the four setæ emerge from the head they lie side by side; the outer pair are the mandibular setæ, the inner pair the maxillary setæ. Farther from the head the maxillary setæ become twisted so that one of them lies above the other. Figure 406 represents a cross-section of the setæ of a squash-bug as figured by Tower ('14); the setæ are fastened together by interlocking grooves

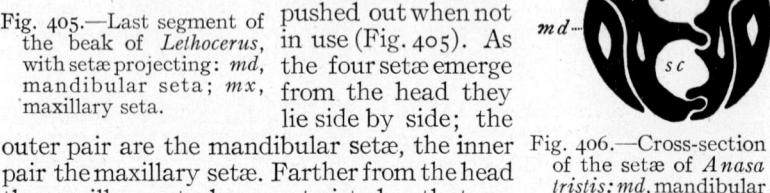

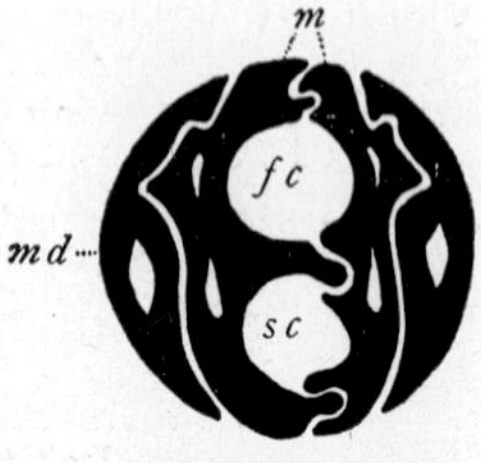

Fig. 405.—Last segment of the beak of *Lethocerus*, with setæ projecting: *md*, mandibular seta; *mx*, maxillary seta.

Fig. 406.—Cross-section of the setæ of *Anasa tristis*: *md*, mandibular setæ; *m*, maxillary setæ; *fc*, food canal; *sc*, salivary canal. (From Tower.)

---

*There has been much discussion regarding the homologies of the parts of the labium in the Hemiptera and the Homoptera. The early entomologists believed that the lower lip of bugs was composed of the labium and the grown-together labial palpi; but this view is no longer held. Leon, who published a series of papers on the labium of aquatic bugs, believes that the first two segments of the labium consists of the submentum and the mentum; the third segment, of the palpiger, which bears vestiges of the labial palpi; and the fourth segment, of the remainder of the ligula. Heymons ('99) argues at great length against the conclusions of Leon. He believes that the segmentation of the labium is merely the result of secondary divisions of this organ and that labial palpi do not exist in the Hemiptera and Homoptera.

and ridges; and between the maxillary setæ are two canals, the upper one (*fc*) for the passage in of food, the lower one (*sc*) for the passage out of saliva. The tip of the mandibular setæ are barbed (Fig. 405, *md*); their function is that of piercing the tissue fed upon and holding the setæ in place; while the tips of the maxillary setæ, which are acute and fluted, probe the tissue, take up the fluid food, and eject the saliva.

Within the head each seta is connected with a chitinous lever, or with a series of two levers which in turn articulate with the head-capsule; the in and-out movements of the setæ are produced by muscles extending from the head-capsule to them and to the levers connecting them with the wall of the head. Figure 407 represents the articulation of a mandibular seta of a squash-bug, as represented by Tower; and in the next chapter the relations of both the mandibular setæ and the maxillary setæ to the head-capsule in a cicada are represented (Fig. 465).

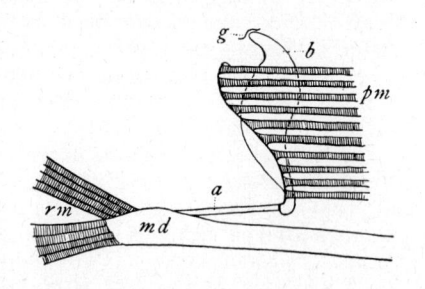

Fig. 407.—Articulation of a mandibular seta with the wall of the head: *md*, mandibular seta; *a* and *b*, chitinous levers; *g*, wall of the head; *rm*, retractor muscles; *pm*, protractor muscle. (From Tower.)

Correlated with the development of the hemipterous type of mouth-parts there is a remarkable specialization of the pharynx, which fits it as a sucking organ, and the development of an organ for forcing out the saliva, which is known as the salivary pump. A detailed account of these organs is given by Bugnion and Popoff ('11).

Most of the Hemiptera protect themselves by the emission of a disagreeable odor. In the adult stink-bugs (Pentatomidæ) this is caused by a fluid which is excreted through two openings, one on each side of the ventral aspect of the thorax, behind or near the middle coxa. These openings are termed the *osteoles*. Each of these is usually in some kind of an open channel styled the *osteolar canal*, and this is surrounded by a more or less rugged and granulated space, the evaporating surface. In the nymphs the stink-glands open on the dorsal aspect of the abdomen. In the bedbug (*Cimex*), the stink-glands open in the dorsal wall of the first three abdominal segments. The legs of the Hemiptera vary greatly in form, but the tarsi are rarely more than three-jointed. The lateral margin of the abdominal segments is much developed in several families, and forms a flat, reflexed or vertical border to the abdomen, which is called the *connexivum*.

In the Hemiptera the metamorphosis is gradual; the newly hatched young resembles the adult in the form of its body but lacks wings. After one or two molts the wing-buds appear and become larger and larger at successive molts. With the last molt there takes place a great expansion of the wings, the change at this time being much greater than at either of the previous molts. There are many forms in this order in which wings are not developed. In some species all individuals are wingless; in others there are two forms of adults, the winged and the wingless.

In this order we find variations in structure which correspond closely with variations in habits. There are certain families the members of which are truly aquatic, living within the water, through which they swim and to the surface of which they come occasionally for air. There are others which are truly terrestrial, living upon the surface of plants, or in other positions away from water. There are still other families the members of which hold an intermediate position between the aquatic and the terrestrial forms, living upon the surface of water or in marshy places.

In the systematic arrangement of the families of the Hemiptera adopted here the aquatic forms are placed first; the terrestrial forms, last; and the semiaquatic forms hold an intermediate position. The sequence of the families is more fully indicated in the following synopsis.

SYNOPSIS OF FAMILIES

THE SHORT-HORNED BUGS. Bugs with short antennæ, which are nearly or quite concealed beneath the head.
  Bugs that live within water.
    The Water-boatmen, Family CORIXIDÆ. p. 360.
    The Back-swimmers, Family NOTONECTIDÆ. p. 362.
    The Water-scorpions, Family NEPIDÆ. p. 364.
    The Giant Water-bugs, Family BELOSTOMATIDÆ. p. 365.
    The Creeping Water-bugs, Family NAUCORIDÆ. p. 367.
  Bugs that live near water.
    The Toad-shaped Bugs, Family GELASTOCORIDÆ. p. 368.
    The Ochterids, Family OCHTERIDÆ. p. 368.
THE LONG-HORNED BUGS. Bugs with antennæ at least as long as the head, and prominent except in the Phymatidæ, where they are concealed under the sides of the prothorax.
  The Semi-aquatic Bugs.
    The Shore-bugs, Family SALDIDÆ. p. 369.
    The Broad-shouldered Water-striders, Family VELIIDÆ. p. 369.
    The Water-striders, Family GERRIDÆ. p. 370.
    The Mesoveliids, Family MESOVELIIDÆ. p. 372.
    The Hebrids, Family HEBRIDÆ. p. 372.
    The Water-measurers, Family HYDROMETRIDÆ. p. 373.
  The Land-bugs.
    The Land-bugs with four-jointed antennæ.
      The Schizopterids, Family SCHIZOPTERIDÆ. p. 373.
      The Dipsocorids, Family DIPSOCORIDÆ. p. 374.
      The Isometopids, Family ISOMETOPIDÆ. p. 374.
      The Leaf-bugs, Family MIRIDÆ. p. 375.
      The Termatophylids, Family TERMATOPHYLIDÆ. p. 377.
      The Flower-bugs, Family ANTHOCORIDÆ. p. 377.
      The Bedbugs, Family CIMICIDÆ. p. 378.
      The Many-combed Bugs, Family POLYCTENIDÆ. p. 379.
      The Nabids, Family NABIDÆ. p. 380.
      The Assassin-bugs, Family REDUVIIDÆ. p. 380.
      The Ambush-bugs, Family PHYMATIDÆ. p. 382.
      The Unique-headed Bugs, Family ENICOCEPHALIDÆ. p. 383.

The Lace-bugs, Family TINGIDÆ. p. 384.
The Cotton-stainer Family, Family PYRRHOCORIDÆ. p. 385.
The Chinch-bug Family, Family LYGÆIDÆ. p. 386.
The Stilt-bugs, Family NEIDIDÆ. p. 388.
The Flat-bugs, Family ARADIDÆ. p. 388.
The Squash-bug Family, Family COREIDÆ. p. 389.
The Land-bugs with five-jointed antennæ.
The Stink-bug Family, Family PENTATOMIDÆ. p. 390.
The Burrower-bugs and the Negro-bugs, Family CYDNIDÆ. p. 391.
The Shield-backed-bugs, Family SCUTELLERIDÆ. p. 392.

TABLE FOR SEPARATING THE FAMILIES OF THE HEMIPTERA

A. Antennæ shorter than the head, and nearly or quite concealed in a cavity beneath the eyes.
 B. Hind tarsi with indistinct setiform claws (except in *Plea*, of the family Notonectidæ, which is less than 3 mm. in length).
  C. Fore tarsi consisting of one segment, which is flattened or shovel-shaped, and without claws; head overlapping the prothorax dorsally. p. 360.
  ....CORIXIDÆ
  CC. Fore tarsi of the usual form, and with two claws; head inserted in the prothorax. p. 362...............................NOTONECTIDÆ
 BB. Hind tarsi with distinct claws.
  C. Ocelli absent; bugs that live within water.
   D. Membrane of the hemelytra with distinct veins.
    E. Caudal appendages of the abdomen long and slender; tarsi one-segmented. p. 364................................ NEPIDÆ
    EE. Caudal appendages of the abdomen short, flat, and retractile; tarsi two-segmented. p. 365............. BELOSTOMATIDÆ
   DD. Membrane of the hemelytra without veins. p. 367..NAUCORIDÆ
  CC. Ocelli present; bugs that live on shores of streams and ponds.
   D. Fore legs stout, fitted for grasping; antennæ concealed. p. 368.
   ....GELASTOCORIDÆ
   DD. Fore legs slender, fitted for running; antennæ exposed. p. 368.
   ....OCHTERIDÆ
AA. Antennæ at least as long as the head, usually free, rarely (P hymatidæ) fitting in a groove under the lateral margin of the pronotum.
 B. Body linear; head as long as the three thoracic segments. p. 373.
 ....HYDROMETRIDÆ
 BB. Body of various forms, but, when linear, with the head shorter than the thorax.
  C. Last segment of the tarsi more or less split, and with the claws of at least the front tarsi inserted before the apex.
   D. Hind femora extending much beyond the apex of the abdomen; intermediate and hind pairs of legs approximated, very distant from the front pair; beak four-jointed. p. 370......GERRIDÆ
   DD. Hind femora not extending much beyond the apex of the abdomen; intermediate pair of legs about equidistant from front and hind pairs (except in *Rhagovelia*); beak three-jointed. p. 369... VELIIDÆ
  CC. Last segment of the tarsi entire, and with the claws inserted at the apex.
   D. Antennæ four-jointed.*
    E. Hemelytra resembling network, and very rarely with any distinction between the corium and the membrane. p. 384.
    ....TINGIDÆ
    EE. Hemelytra of various forms or absent, but not of the form presented by the Tingidæ.

---

*In certain families there are minute intermediate joints between the principal joints of the antennæ; for the purposes of this table, these intermediate joints are not counted.

F. Beak three-jointed.
  G. Hemelytra when well-developed with an embolium (Fig. 408); those forms in which the adult has vestigial hemelytra have no ocelli.
    H. Hemelytra vestigial; parasitic bugs preying on man, bats, and birds. p. 378..............CIMICIDÆ
    HH. Hemelytra usually well developed; not parasitic bugs. p. 377...................ANTHOCORIDÆ
  GG. Hemelytra when well developed without an embolium; those forms in which the adult has vestigial hemelytra have ocelli.
    H. Ocelli wanting.
      I. Body greatly flattened. p. 388........ARADIDÆ
      II. Body not greatly flattened. p. 380..REDUVIIDÆ
    HH. Ocelli present, though sometimes difficult to see.
      I. Antennæ whip-like, the first two segments short and thick, the third and fourth long and very slender and clothed with long hairs, the third segment thickened towards the base.
        J. Head when viewed from above wider than long, strongly deflexed; beak short. p. 373.
        ........................SCHIZOPTERIDÆ
        JJ. Head extended horizontally or slightly deflexed; beak long. p. 374...DIPSOCORIDÆ
      II. Antennæ not of the form described above.
        J. Beak long, reaching to or beyond the intermediate coxæ.
          K. Membrane of hemelytra with looped veins. p. 369.......................SALDIDÆ
          KK. Membrane of hemelytra without veins.
            L. Hemelytra with the clavus similar in texture to the membrane (Fig. 409). p. 372..................HEBRIDÆ
            LL. Clavus and membrane distinct. p. 372....................MESOVELIIDÆ
        JJ. Beak not reaching the intermediate coxæ.
          K. Front legs with greatly thickened femora. p. 382............PHYMATIDÆ
          KK. Front femora somewhat thickened, but much less than half as wide as long. p. 380...........REDUVIIDÆ
FF. Beak four-jointed.
  G. Front legs fitted for grasping prey.
    H. The fore tarsi, which are one-jointed, capable of being closed upon the end of the broad tibiæ. p. 383.
    ........................ENICOCEPHALIDÆ
    HH. The fore tibiæ armed with spines andcpable of being closed tightly upon the femora, wahich are stout. In the forms with long wings the membrane is usually furnished with four long veins bounding three discal cells which are often open. From these cells diverge veins which form several marginal cells (Fig. 410). p. 380..NABIDÆ
  GG. Front legs fitted for walking.
    H. Hemelytra with a cuneus; membrane with one or two closed cells at its base, otherwise without veins (Fig. 411).
      I. Ocelli wanting.
        J. Membrane of the hemelytra with two closed cells. p. 375....................MIRIDÆ
        JJ. Membrane with only one closed cell.

K. Tarsi furnished with an arolium. p. 375.
...................................MIRIDÆ
KK. Tarsi without an arolium. p. 377......
..........................TERMATOPHYLIDÆ
II. Ocelli present. p. 374..........ISOMETOPIDÆ

HH. Hemelytra without a cuneus; membrane with four or five simple or anastomosing veins arising from the base, or with a large number of veins arising from a cross-vein at the base.

I. Ocelli wanting.

J. Exceedingly flat bugs, p. 388......ARADIDÆ

JJ. Rather stout and heavily formed bugs. p. 385...................PYRRHOCORIDÆ

II. Ocelli usually present.

J. Head with a transverse incision in front of the ocelli, which are always present (Fig. 449). p. 388.................NEIDIDÆ

JJ. Head without transverse incision.

K. Membrane with four or five simple veins arising from the base of the membrane, the two inner ones sometimes joined to a cell near the base (Fig. 413). p. 386.
........................LYGÆIDÆ

KK. Membrane with many, usually forked veins, springing from a transverse basal vein (Fig. 414). p. 389....
........................COREIDÆ

HHH. Hemelytra vestigial; parasitic bugs preying on bats. p. 379............POLYCTENIDÆ

DD. Antennæ five-jointed.*

E. Hemelytra with the clavus similar in texture to the membrane, which is without veins (Fig. 409); small semiaquatic bugs, measuring less than 3 mm. in length (*Hebrus*). p. 372
....HEBRIDÆ

EE. Hemelytra with the clavus markedly thicker than the membrane.

F. Tibiæ armed with strong spines. p. 391.......CYDNIDÆ

FF. Tibiæ smooth or with small spines.

G. Scutellum narrowed behind, only rarely almost covering the abdomen. p. 390..........PENTATOMIDÆ

GG. Scutellum not narrowed as in the Pentatomidæ, very convex, nearly or quite covering the abdomen. p. 392..............SCUTELLERIDÆ

---

*In some cases there are minute intermediate joints between the principal joints of the antennæ; for the purposes of this table these intermediate joints are not counted.

360   *AN INTRODUCTION TO ENTOMOLOGY*

Fig. 408.—Anthocoridæ.   Fig. 409.—Hebridæ.

Fig. 410.—Nabidæ.   Fig. 411.—Miridæ.

Fig. 412.—Pyrrhocoridæ.   Fig. 413.—Lygæidæ.

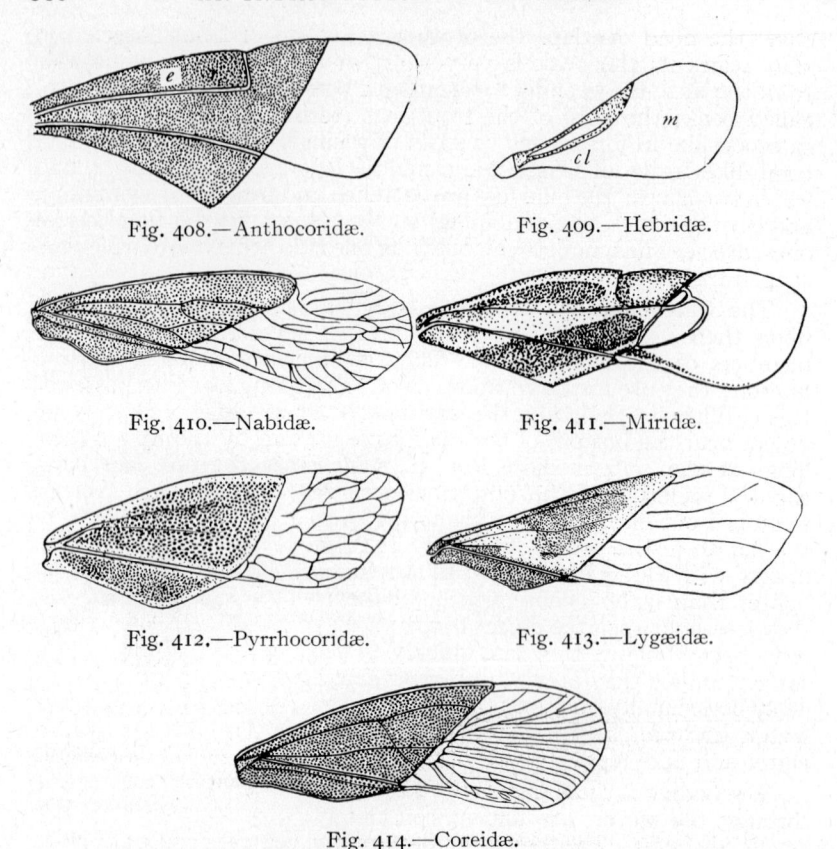

Fig. 414.—Coreidæ.

Figures 408 to 414.—Diagrams illustrating the types of hemelytra characteristic of several families of Hemiptera.

Family CORIXIDÆ*

The Water-Boatmen

The family Corixidæ includes oval, gray-and-black mottled bugs, usually less than half an inch in length, which live in lakes, ponds, and streams, in both stagnant and running water. The characteristic form and markings of these insects are shown in Figure 415.

The name of the typical genus of this family, *Corixa*, is evidently from the Greek word *coris*, meaning a bug. For this reason many writers have spelled the generic name *Corisa* and the family name Corisidæ. This name was probably given to these insects because they have an odor like that of the bedbug.

The water-boatmen exhibit some striking peculiarities in struc-

---
*Corixidæ, *Corixa*, a misspelling of *Corisa: coris* (κόρις), a bug.

ture: the head overlaps the prothorax instead of being inserted in that segment; the beak is very short and scarcely distinguishable from the face, the opening to the mouth being on the front of the so-called beak; the tarsi of the front legs (termed *palæ*) are flattened or scoop-like in form; each consists of a single segment and bears a comb-like fringe of bristles; the middle legs are long, slender, and end in two claws; the hind legs are flattened and fringed for swimming; and, in the males, the abdominal sterna, especially the four caudal ones, are very unsymmetrical, being on one side broken into irregular-shaped fragments.

The water-boatmen have the body flattened above, and swim upon their ventral surface; they differ in these respects from the members of the next family. They swim with a quick, darting motion; they use for this purpose chiefly their long, oar-like, posterior legs. When in their favorite attitude, they are anchored to some object near the bottom of the pond or aquarium by the tips of their long, slender, intermediate legs; at such times the fore legs hang slightly folded, and the posterior legs are stretched out horizontally at right angles to the length of the body. The body of these insects, with the air which clings to it, is much lighter than water; consequently whenever they lose hold upon the object to which they have been clinging, they rise quickly to the surface, unless they prevent it by swimming. They occasionally float on the surface of the water, and can leap into the air from the water and take flight.

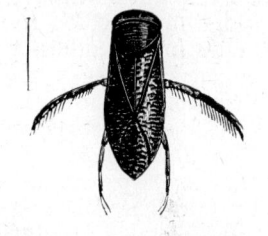

Fig. 415.—A water-boatman.

The bodies of these insects, as they swim through the water, are almost completely enveloped in air. The coating of air upon the ventral surface and sides can be easily seen, for it glistens like silver. By watching the insects carefully when they are bending their bodies, the air can be seen to fill the spaces between the head and the prothorax, and between the prothorax and the mesothorax. The space beneath the wings is also filled with air. When these insects are in impure water, they must come to the surface at intervals to change this supply of air. But I have demonstrated that in good water it is not necessary for them to do this. The air with which the body is clothed is purified by contact with the fine particles of air in the water; so that the insect can breathe its coat of air again and again indefinitely.

It has been commonly believed that the corixids are carnivorous; but Hungerford ('19) has shown, by an extended series of experiments, that these insects gather their food supply from the ooze at the bottom of pools in which they live. This flocculent material they sweep into their mouths by means of the flat rakes of their fore tarsi. This material is largely of plant origin; but the protozoa and other minute animals living on it are also consumed. This author also found that the corixids feed on the chlorophyll of *Spirogyra*.

In most cases the eggs of corixids are attached to the stems of aquatic plants; but *Ramphocorixa acuminata* usually attaches its eggs to the body of a crayfish.

The males of most of the Corixidæ are furnished with stridulating organs. These consist of one or two rows of chitinous "pegs" on the fore tarsi and a roughened area on the inner surface of the fore femora near the base. By rubbing the tarsal comb of one leg over the roughened area of the femur of the opposite leg, a chirping sound is produced. These stridulating organs differ in form in different species.

In addition to the stridulating organs of the fore legs there is in certain species a more or less curry-comb-like organ near the lateral margin of the dorsal wall of the sixth abdominal segment; this has been termed the "strigil." It is situated, when present, on the left side in *Corixa* and on the right side in several other genera. Its function has not been definitely determined.

Both the adults and the eggs of *Corixa* are used for food for man and for birds in Mexico and in Egypt. The eggs are gathered from water-plants. Glover states that in Mexico the natives cultivate a sedge upon which the insects will deposit their eggs; this sedge is made into bundles, which are floated in the water of a lake until covered with eggs; the bundles are then taken out, dried, and beaten over a cloth; the eggs, being thus disengaged, are cleaned and powdered into flour. Kirkaldy ('98) reports the importation into England of *Corixa mercenaria* and its eggs for food of insectivorous birds, game, fish, etc., *by the ton;* and computes "that each ton of the adults will contain little short of 250 million individuals!! As to the ova, they are beyond computation." The adults are captured at night with nets when they leave the water in swarms.

It is difficult to separate the different species of water-boatmen on account of their close resemblance to each other; this is especially true of the females. Fifty-five species are listed in the Van Duzee check-list; these represent six genera.

Family NOTONECTIDÆ

The Back-Swimmers

The Notonectidæ differ from all other aquatic Hemiptera in the fact that they always swim on their backs; and there is a corresponding difference in the form of these insects. The body is much deeper than in the allied families, and is more boat-shaped. The back, from the peculiar attitude of the insect when in the water, corresponds to the bottom of a boat, and is sloped so as to greatly resemble in form this part (Fig. 416).

Fig. 416.—*Notonecta undulata*.

The eyes are large, reniform, twice sinuated on the outer side, and project a little way over the front margin of the prothorax. Ocelli are absent. The prothorax has the lateral margins sharp and projecting. The legs are all long; the hind pair are much the longest and fitted for swimming. The tarsi consist each of three segments, but the basal segment is so small that it is often overlooked. There is a ridge along the middle line of the venter which is clothed with hairs, and along each side of this a furrow. Along the upper edge of the outside of this furrow and a short distance from the side of the body, there is a fringe of long hairs, and beneath this fringe the abdominal spiracles are situated.

The features presented by the ventral side of the abdomen just referred to can be seen on dead specimens; but it is well to examine them on living insects. This can be done by placing a back-swimmer in a glass of water, and, when it is resting at the surface of the water, studying it by means of a lens of low power. Under these conditions it can be seen that the furrow on either side of the venter is an air-chamber, which is enclosed by the two fringes of hairs, one borne by the ridge of the middle line on the body and the other by the outer margin of the furrow. It can also be seen that there is a hole near the tip of the abdomen through which the air passes into the chambers beneath the fringes of hairs. Sometimes when watching an individual under these conditions it will be seen to force the air out of the chambers beneath the fringes of hair, using the hind legs for this purpose, and sometimes an entire fringe will be lifted like a lid.

By examining the first ventral abdominal segment of a dead individual a little furrow can be seen on each side; these are air-passages extending between the chambers on the ventral side of the abdomen to that beneath the wings.

Air is also carried among the hairs on the lower side of the thorax, and in the spaces between the head and the prothorax and between the prothorax and the mesothorax; this is probably expired air.

In collecting back-swimmers, care must be taken or they will inflict painful stings with the stylets of their beak.

The manner of oviposition of these insects differs in different species. Some merely attach their eggs to the surface of aquatic plants by means of a colorless, water-proof glue; others have a long ovipositor by means of which they insert their eggs in the tissue of these plants.

The males of some back-swimmers possess stridulating areas; these are located on the femora and tibiæ of the fore legs and on the sides of the face at the base of the beak.

The notonectids of our fauna represent three genera; these can be separated by the following table:

A. Legs dissimilar; hind legs flattened and fringed for swimming.
  B. Last segment of the antennæ much shorter than the penultimate segment. ....NOTONECTA
  BB. Last segment of the antennæ longer than the penultimate segment. ....BUENOA
AA. Legs quite similar..PLEA

Notoněcta.—To this genus belong the greater number of our species, of which twelve have been described. These are the back-swimmers that are commonly seen floating at the surface of the water, with the caudal part projecting sufficiently to admit of the air being drawn into the air chambers. When in this position, their long, oar-like, hind legs are stretched outward and forward ready for action; when disturbed they dart away toward the bottom of the pond, carrying a supply of air with them.

Buĕnoa.—This genus, of which six species have been found in this country, is composed of much more slender forms than is the preceding. The habits of two of our species have been studied by Hungerford ('19). These insects do not rest at the surface of the water as do some species of *Notonecta*, but may be seen swimming slowly, or even poising in midwater some distance beneath the surface. They abound in water teeming with Entomostraca, upon which they largely feed.

Plēa.—The members of this genus are small insects, not exceeding 3 mm. in length. The shape of the body is quite different from that of other back-swimmers, being highly arched behind. They are found in tangles of aquatic vegetation, to the filaments of which they cling when at rest. They feed on small Crustacea. Only one species, *Plea striola*, has been described from our fauna.

Family NEPIDÆ

The Water-Scorpions

The members of this family can be distinguished from other aquatic Hemiptera by the presence of a long respiratory tube at the end of the abdomen. This tube consists of two long filaments, each with a groove on its mesal side. By applying these filaments together the grooves form a tube, which conducts the air to two spiracles situated at the caudal end of the abdomen. By means of this apparatus these insects are able to rest on the bottom of a shallow pond, or among rubbish or plants in water, and by projecting this tube to the surface obtain what air they need.

With regard to the form of the body, two very different types exist in this family. In one, represented by the genus *Nepa*, the body is a long oval, flat, and thin (Fig. 417); in the other, represented by the genus *Ranatra*, the body is almost linear and cylindrical (Fig. 418). An intermediate form, *Curicta*, represented by two species, is found in Louisiana, Texas, and Arizona.

Fig. 417.—*Nepa apiculata.*

The water-scorpions are carnivorous; and with them the first pair of legs is fitted for seizing prey. In these legs the coxæ are very long, especially in *Ranatra;* the femora are furnished with a groove into which the tibiæ and tarsi fit like the blade of a pocket-knife into its handle.

Although the Nepidæ are aquatic insects, the second and third pairs of legs are fitted for walking rather than for swimming.

Of the genus *Nepa* we have only a single species, *Nēpa apiculāta*. This insect is about 16 mm. in length, not including the respiratory tube, which measures a little more than 6 mm. It lives in shallow water concealed in the mud or among the dead leaves and twigs, lying in wait for its prey. The eggs are inserted in the tissues of decaying plants; they are an elongate oval and bear near one end a crown of eleven slender filaments.

Of the genus *Rănatra* eight American species have been described. These insects are found in the same situations as *Nepa;* where, owing to the linear form of the body and to the dirt with which it is usually covered, it is quite difficult to detect their presence. They have also been observed in deep water clinging to the stems of rushes and grasses, with the respiratory tube piercing the surface film (Bueno); and also upon floating dead leaves and stalks of cat-tail, where they were basking in the sun and entirely dry (Hungerford).

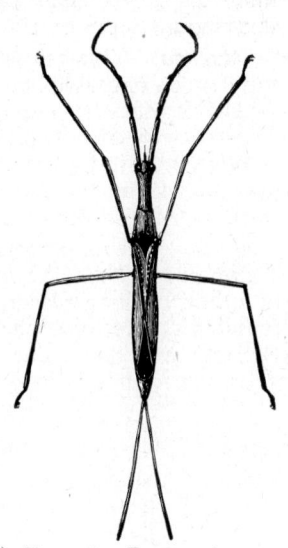

Fig. 418.—*Ranatra fusca.*

Ranatra has stridulating organs; these consist of a roughened patch on the outside of each fore coxa and a rasp on the inner margin of each shoulder of the prothorax; by means of these organs a squeaking sound is produced.

The eggs of *Ranatra* have been described by Pettit; they are elongate oval, about 3.5 mm. in length, and bear at one end a pair of slender appendages, about 4 mm. long; they are embedded in the rotting stems of aquatic plants, from which the appendages of the eggs project.

A monograph of the Nepidæ of North America was published by Hungerford ('22).

Family BELOSTOMATIDÆ

The Giant Water-Bugs

The common name "giant water-bugs" was applied to this family because to it belong the largest of the Hemiptera now living; a species that is found in Guiana and Brazil measures from 75 to 100 mm. in length; and the larger of our species exceed in size our other water-bugs.

The members of this family are all wide and flat-bodied aquatic insects, of more or less ovate outline. The fore legs are raptorial; the middle and hind legs are fitted for swimming, being flattened and ciliated; this is especially true of the hind legs. At the caudal end of the body there is, in the adult, a pair of narrow, strap-like respiratory appendages, which are retractile.

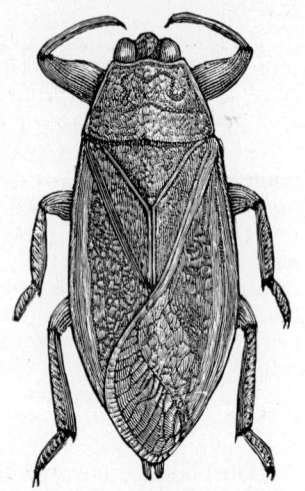

These insects are rapacious creatures, feeding on other insects, snails, and small fish. Like other water-bugs, they fly from pond to pond and are frequently attracted to lights. This is especially the case where electric lights are used, into which they sometimes fly and are killed by hundreds. On this account they are known in many parts of the country as "electric-light bugs."

The family Belostomatidæ is represented in this country by four genera. Recent studies of the nomenclature of the genera of this family have resulted in the making of changes in some of the generic names. This should be kept in mind when using the older text-books. Our genera are separated by Hungerford ('19) as follows:

Fig. 419.—*Lethocerus americanus.*

A. Mesothorax with a strong midventral keel; membrane of the hemelytra reduced.. ABEDUS
AA. Mesothorax without a midventral keel; membrane of the hemelytra not reduced.
   B. Basal segment of the beak longer than the second; base of the wing-membrane nearly or quite straight. Body about 25 mm. or less in length.. BELOSTOMA
   BB. Basal segment of the beak shorter than the second; base of the wing-membrane sinuous. Body more than 37 mm. in length.
      C. Anterior femora grooved for the reception of the tibia...LETHOCERUS
      CC. Anterior femora not grooved for the reception of the tibia.. BENACUS

Lethŏcerus.—To this genus and the following one belong our larger members of this family. The appearance of these insects is indicated by Figure 419, which represents *Lethocerus americānus*. In this genus the anterior femora are furnished with a groove for the reception of the tibia. Five species have been described from the United States and Canada. In most of the references to these insects in our literature the generic name *Belostoma* is used.

Fig. 420.—*Belostoma fluminea.*

Fig. 421.—Male of *Abedus*, with eggs.

Benacus.—Only a single species of this genus, *Benācus grĭseus*, is found in our fauna. This closely resembles *Lethocerŭs americānus* (Fig. 419), but can be distinguished from that species by the absence of the groove in the femora of the fore legs.

Belostoma.—To this genus as now recognized belong our more common representatives of the smaller members of this family. These have long been known incorrectly under the generic name *Zaitha*. Our most common species is *Belŏstoma flumĭnea* (Fig. 420).

In this genus and the following one the eggs are carried by the males on their backs, where they are placed by the females, sometimes in spite of vigorous opposition on the part of the male.

Abēdus.—Five species of this genus have been found in the southwestern parts of the United States. Figure 421 represents the male of one of these carrying his load of eggs.

Family NAUCORIDÆ

The Creeping Water-Bugs

The Naucoridæ includes flat-bodied, chiefly oval insects, of moderate size. The abdomen is without caudal appendages. The front legs are fitted for grasping, the femora being greatly enlarged; the middle and hind legs are suited for crawling rather than for swimming. There are no ocelli; the antennæ are very short, and well concealed beneath the eyes; the beak is three-jointed and covered at the base by the large labrum; and the hemelytra are furnished with a distinct embolium.

Although these are aquatic insects, they have been comparatively little modified for such a life. They carry air beneath their wings and obtain this air by pushing the tip of the abdomen above the surface of the water.

They are predacious and are fond of reedy and grassy, quiet waters, where they creep about like the dytiscid beetles, creeping and swimming around and between the leaves and sprays of the submerged plants, seeking their prey.

Fig. 422.—*Pelocoris femoratus*.

Only two genera of this family are represented in our fauna; these are *Pelŏcoris* and *Ambrȳsus*. In *Ambrysus* the front margin of the prothorax is deeply excavated for the reception of the head; in *Pelocoris* this is not the case.

Pelocoris.—Only three species of this genus are found in this country and these are restricted to the eastern half of the United States. The most common one is *Pelŏcoris femorātus* (Fig. 422). It measures about 9 mm. in length, and when alive is more or less greenish testaceous in color; but after death it is pale yellow or brownish in color, with black or dark brown markings.

Ambrȳsus.—Ten species of this genus have been found in this country; they are restricted to the Far West.

Family GELASTOCORIDÆ

The Toad-shaped Bugs

The Gelastocŏridæ was formerly known as the Galgulidæ; consequently most of the references to these insects will be found under the older family name, which has been dropped, as the generic name *Gălgulus*, on which it was based, is not tenable.

In these insects the body is broad and short, and the eyes are prominent and projecting; the form of the body and the protuberant eyes remind one of a toad (Fig. 423). Ocelli are present. The antennæ are short and nearly or quite concealed beneath the eyes. The beak is short, stout, and four-segmented. The fore legs are raptorial.

Fig. 423.—*Gelastocoris oculatus*.

The toad-shaped bugs live on the muddy margins of streams or other bodies of water. Some of them make holes for themselves, and live for a part of the time beneath the ground. They feed upon other insects, which they capture by leaping suddenly upon them. Their colors are protective and vary so as to agree with the color of the soil on which they live. Hungerford has found that the eggs are buried in the sand. Only five species are known to occur in this country.

The most common and most widely distributed representative of the family found in this country is *Gelastŏcoris oculātus* (Fig. 423). Two other species of *Gelastocoris* are found in the Southern and Western States. In this genus the hemelytra are not fused and the fore tarsi are two-clawed.

In the genus *Mononyx*, of which a single species, *Mŏnonyx fŭscipes*, is found in California, the hemelytra are free, but the fore tarsi are one-clawed.

The genus *Nerthra* is also represented in this country by a single species, *Nĕrthra stygīea*, which is found in Georgia and Florida. In this genus the hemelytra are fused together along a straight suture indicated by a groove.

Family OCHTERIDÆ

The Ochterids

These are shore-inhabiting bugs, which are closely allied to the preceding family, in which they were formerly classed. They differ from the toad-shaped bugs in having the fore legs slender and fitted for running, and in having the short antennæ exposed. They resemble the following family, the Saldidæ, in having the beak long, reaching the hind coxæ. The eyes are prominent, and two ocelli are present.

The family includes a single genus, *Ochtērus*, which, due to an error, has been commonly known as *Pelŏgonus*. Only three species occur in the United States; one of these was described from Virginia, one from Florida, and the third is widely distributed from the Atlantic Coast to Arizona.

The widely distributed species is *Ochterus americānus*. It measures 5 mm. in length, and is blackish in color sprinkled with golden yellow points. On each side of the prothorax, behind the front angles, there is a bright yellow spot.

The members of this family are predacious.

Family SALDIDÆ

The Shore-Bugs

With the Saldidæ we reach the beginning of the extensive series of families of Hemiptera in which the antennæ are prominent and are not concealed beneath the head. In this family the insects are of small size, and of dark colors with white or yellow markings. The head stands out free from the thorax on a cylindrical base. The antennæ are four-jointed; there are two ocelli; the rostrum is three-jointed and very long, reaching to or beyond the middle coxæ. The membrane of the wing-covers is furnished with looped veins, forming four or five long cells placed side by side. Occasionally there is little or no distinction between the corium and the membrane. Two forms sometimes occur in the same species, one with a distinct membrane, and another with the membrane thickened and almost as coriaceous as the corium proper. The shape of these shore-bugs is shown by Figure 424.

Fig. 424.—A shore-bug.

These insects abound in the vicinity of streams and other bodies of water, and upon damp soils, especially of marshes near our coasts. Some of the shore bugs dig burrows, and live for a part of the time beneath the ground. They take flight quickly when disturbed, but alight after flying a short distance, taking care also to slip quickly into the shade of some projecting tuft of grass or clod where the soil agrees with the color of their bodies.

Thirty-three species belonging to this family have been found in the United States and Canada; these represent eight genera.

Family VELIIDÆ

The Broad-shouldered Water-Striders

The Velliidæ includes insects which are very closely allied to the following family, the water-striders, both in structure and in habits. In both families the distal segment of the tarsi, at least of the fore tarsi, is more or less bifid, and the claws are inserted before the apex; these characters distinguish these two families from all other Hemiptera. In the Veliidæ the body is usually stout, oval, and broadest across the prothorax (Fig. 425). The beak is three-jointed; the legs are not extremely long, the hind femora not extending much beyond the end of the abdomen. In fact, the legs are fitted for running over the water, instead of for rowing, as with the Gerridæ. The intermediate legs are about equidistant from the front and hind pairs, except in *Rhagovelia*. These insects are dimorphic, both fully winged and short-winged or wingless adults occurring in the same species.

About twenty species of this family have been found in America north of Mexico; these represent four genera.

The broad-shouldered water-striders are found both on the banks of streams and ponds and on the surface of water. About one-half of our species belong to the genus *Microvēlia*. These are very small, plump-bodied bugs, which are usually black and silvery in color or mottled with brown. They are found at the water's edge but run out on the water when disturbed; and they are also often found upon rafts of floating vegetation.

To the genus *Rhagovēlia* belong somewhat larger forms, which are characterized by the long, deeply split, terminal segment of the tarsi of the middle legs. Our most common species of this genus is *Rhagovēlia obēsa* (Fig. 425). These bugs are found running over the surface of rapidly moving waters in streams. They can also dive and swim well under water. Four species of *Rhagovelia* are found in this country.

Fig. 425.—*Rhagovelia obesa.*

The genus *Vēlia* includes the larger members of the family. In these the tarsi of the middle legs are not cleft. Four species of this genus occur in our fauna. They are found on moderately rapid streams or little bogs and eddies connected therewith.

The fourth genus occurring in our fauna is represented by a single species, *Macrovēlia harrĭsii*, which is restricted to the Far West.

Family GERRIDÆ

The Water-Striders

This family includes elongated or oval insects which live upon the surface of water. Their legs are long and slender; the hind femora extend much beyond the apex of the abdomen; the middle and hind pairs of legs are approximated and distant from the fore legs; the terminal segment of the tarsi, at least of the fore tarsi, is more or less bifid, and the claws are inserted before the apex. The beak is four-jointed. The antennæ are long and four-jointed.

The water-striders prefer quiet waters, upon which they rest or over which they skim rapidly; they often congregate in great numbers. There are commonly two forms of adults belonging to the same species, the winged and the wingless; sometimes a third form occurs in which the adult has short wings.

These insects are predacious; they feed on insects that fall into the water, and I have seen them jump from the water to capture flies and other insects that were flying near them.

Twenty species of water-striders have been found in America north of Mexico; these represent seven genera. These genera are separated by Hungerford ('19) as follows:

A. Inner margin of the eyes sinuate behind the middle. Body comparatively long and narrow; abdomen long. (Subfamily Gerrinæ).
B. Pronotum sericeous, dull; antennæ comparatively short and stout.
C. First segment of the antennæ shorter than the second and third taken together.
D. Antennæ half as long as the body; sixth abdominal segment of the male roundly emarginate..LIMNOPORUS
DD. Antennæ not half as long as the body, not extending beyond the thorax; sixth abdominal segment of the male doubly emarginate. ....GERRIS
CC. First segment of the antennæ longer than the second and third taken together...GERRIS
BB. Pronotum glabrous, shining; antennæ long and slender...TENAGOGONUS
AA. Inner margin of the eyes convexly rounded. Body comparatively short and broad; abdomen so short as to appear almost nymphal in some forms. (Subfamily Halobatinæ).
B. First antennal segment much shorter than the other three taken together; not much longer than the second and third taken together, and sometimes shorter.
C. Fourth (apical) segment of the antennæ longer than the third.
D. Eyes fairly prominent; colors of body black and yellow..TREPOBATES
DD. Eyes smaller, widely separated; body lead-colored, sericeous. ocean dwellers..HALOBATES
CC. Fourth segment of antennæ never more than equal to the third; basal segment of anterior tarsi much shorter than the second; hind femur equal to or much shorter than the hind tibia and tarsus taken together...RHEUMATOBATES
BB. First antennal segment nearly equal to the remaining three taken together, much longer than the second and third; antennæ almost as long as the entire body; hind femur twice as long as hind tibia. ....METROBATES

Gĕrris.—Of the twenty species of water-striders found in this country, nine belong to this genus; a common species in the East is *Gerris confŏrmis* (Fig. 426).

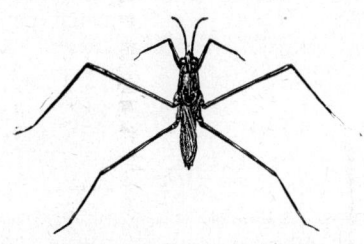

Fig. 426.—*Gerris conformis.*

Limnŏporus.—We have only a single species of this genus, *L. rufoscutillātus.*

Tenagogonus.—Three species are listed from our fauna, only one of which has been found in the North; this is *T. gillĕttei*, which is reported from Ohio. The others are found in Florida and California.

Metrŏbates.—Our only species, *M. hespērius*, is found in Ontario and the eastern part of the United States.

Trepŏbates.—This genus is represented only by *T. pĭctus*. This is a beautiful yellow and black species, which is quite widely distributed.

Rheumatŏbates.—Three species of this genus have been described. The males are remarkable for the strange form of the posterior femora, which are strongly bent, and the shape of the antennæ, which are fitted for clasping.

Halŏbates.—These are truly pelagic insects, living on the surface of the ocean, often hundreds of miles from land. They are most abundant in the region of calms near the equator; they feed on the juices of dead animals floating on the surface, and probably attach their eggs to floating sea-weed (*Sargassum*). *H. micans* is found off the coast of Florida and *H. serĭceus* off the coast of California.

Family MESOVELIIDÆ

The Mesoveliids

This is a small family of which only two species have been found in North America. These are the following.

Mesovēlia mulsănti.—This is a small bug, measuring only 4 or 5 mm. in length; it is of a pale yellow color marked with brown. The antennæ are long, filiform, and four-jointed; the beak is three-jointed; the legs are moderately long and slender; and the tarsi are three-jointed. This species is dimorphic, the adults being either winged or wingless. In the winged form, the membrane of the hemelytra is without veins.

This species lives on the surface of quiet waters and on rafts of floating vegetation and is predacious. It is furnished with an ovipositor and embeds its eggs in the stems of aquatic plants.

Mesovēlia douglasensis.—This is a smaller species than the preceding; the length of the female is 2.1 mm., of the male 1.8 mm. It is olive-brown in color. It was recently discovered and described by Professor Hungerford ('24). It was found near Douglas Lake, Michigan.

Family HEBRIDÆ

The Hebrids

This family includes very small plump-bodied bugs, measuring less than 3 mm. in length. The antennæ are either four-jointed or five-jointed; the beak is three-jointed; and the tarsi are two-jointed. Ocelli are present. The head and thorax are sulcate beneath. The clavus of the hemelytra is similar in texture to the membrane, which is without veins (Fig. 427). Two genera of this family are found in the United States.

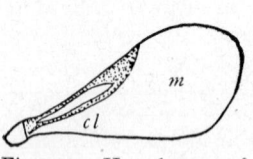

Fig. 427.—Hemelytron of *Hebrus*.

Hebrus.—In this genus the antennæ consist of five segments, not counting a minute segment at the base of the third. The adults are always winged. Four species occur in our fauna. These bugs are found on moist earth at the margins of pools and run out upon the water when disturbed; they are also found on floating vegetation.

Merragāta.—In this genus the antennæ are four-jointed not counting the small segment at the base of the third. The adults are dimorphic, short-winged and long-winged forms occcurring in the same species. These insects inhabit still and stagnant waters and often descend beneath the surface; at such times the body is surrounded by a film of air. Only two species have been found, as yet, in this country.

Family HYDROMETRIDÆ

The Water-Measurers

The members of this family are very slender insects, with linear legs and antennæ (Fig. 428). The head is as long as the entire thorax, although this region is long also. The eyes are round, projecting, and placed a little nearer the base than the tip of the head. Ocelli are absent. The antennæ are four-jointed; the beak is three-jointed; and the tarsi are three-jointed.

These insects creep slowly upon the surface of the water; they carry the body considerably elevated, and are found mostly where plants are growing in quiet waters. It was probably their deliberate gait when walking on water that suggested the generic name *Hydromētra*, or water-measurer. In this country these insects have been commonly known under the generic name *Limnŏbates*, or marsh-treaders; but *Hydrometra* is much the older name.

Only three species have been found in the United States. One of these, *Hydromētra martīni* (Fig. 428), is widely distributed. The other two, *Hydromētra australis* and *Hydromētra wilĕyi*, are found in the South. These insects are dimorphic both winged and wingless forms occurring in the same species. Descriptions of the three species are given by Hungerford ('23).

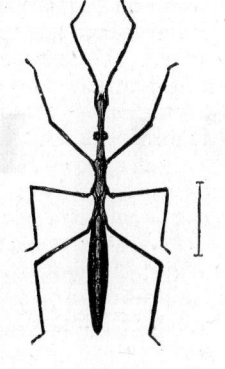

Fig. 428.—*Hydrometra martini.*

The egg of *Hydrometra martini* is remarkable in form; it is figured on page 167.

Family SCHIZOPTERIDÆ

The Schizopterids

This family and the following one, the Dipsocoridæ, constitute a quite distinct superfamily, the members of which are most easily recognized by the form of the antennæ (Fig. 429, *b*). These are four-jointed; the first two segments are short and thick; the third and fourth segments are long, slender, and clothed with long hairs; the third segment is thickened toward the base. In these two families ocelli are present; the beak is three-jointed; the legs are quite slender, and the tarsi are three-jointed. The species are small or very minute.

The Schizopteridæ is distinguished from the following family by the shape of the head and the form of the cavities in which the front legs are inserted. The head when viewed from above is wider than long and is strongly deflexed; the fore coxal cavities are very prominent and tumidly formed. The beak is short. The Schizopteridæ is represented in our fauna by a single species, *Glyptocŏmbus saltātor* (Fig. 429). This is a minute bug, measuring only 1.2 mm. in length and .6 mm. in width. The known specimens were taken on Plummers Island, Md. The describer of this species, Mr. O. Heidemann, states: "This species is most difficult to collect and is only to be found by sifting fallen leaves, rubbish and earth. The collector must watch patiently until the minute insect makes its presence known by jumping, and even then it takes a skillful hand to secure it in a vial."

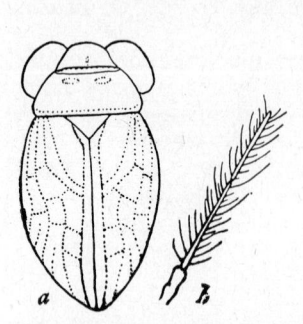

Fig. 429.—*Glyptocombus saltator:* *a*, dorsal aspect; *b*, antenna. (After Heidemann.)

Family DIPSOCORIDÆ

The Dipsocorids

This family is closely allied to the preceding family; the distinguishing features common to the two families are indicated in the account of that family.

In the Dipsocoridæ the head is extended horizontally or slightly deflexed, and the fore coxal cavities are not at all prominent. The beak is long.

This family is represented in our fauna by a single genus, *Ceratocŏmbus*, of which two or three species have been found in New Mexico; and one of these is doubtfully reported from Florida. These measure less than 2 mm. in length.

Family ISOMETOPIDÆ

The Isometopids

This is a family of limited extent, there being very few species known in the entire world. It includes very small bugs, those found in this country ranging from 2 mm. to 2.6 mm. in length.

The Isometopidæ is closely allied to the following family, the Miridæ; by some writers it has been classed as a subfamily of that family. In both families the antennæ are four-jointed; the beak is four-jointed; the hemelytra are composed of clavus, corium, cuneus, and membrane; at the base of the membrane there are one or two cells; otherwise the membrane is without veins. The Isometopidæ is distinguished from the following family by the presence of ocelli, two in number.

Only four species of this family have been found in our fauna; one in Texas, one in Arizona, and two in the East. The Eastern species are *Myiomma cixiiformis*, which is dull black in color with a narrow white band across the base of the cuneus; and *Isometopus pulchellus*, which is easily recognizable by its contrasting colors of dark brown and yellowish white (Fig. 430). Both are exceedingly rare insects.

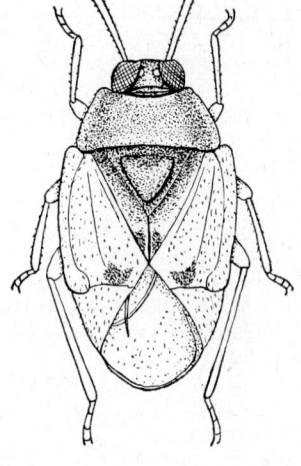

Fig. 430.—*Isometopus pulchellus*. (After Heidemann.)

Family MIRIDÆ

The Leaf-Bugs

This family, which has been known as the Capsidæ, is more largely represented in this country than any other family of the Hemiptera. Van Duzee in his "Catalogue of the Hemiptera North of Mexico" lists 398 species, which represent 129 genera. The species are usually of medium or small size. The form of the body varies greatly in the different genera, which makes it difficult to characterize the family.

The most available character for distinguishing these insects is the structure of the hemelytra. These are almost always complete, and composed of clavus, corium, cuneus, and membrane. At the base of the membrane there are one or two cells; otherwise the membrane is without veins (Fig. 431). Other characters of the family are as follows: the ocelli are wanting; the beak and the antennæ are each four-jointed; the coxæ are subelongate; and the tarsi are three-jointed.

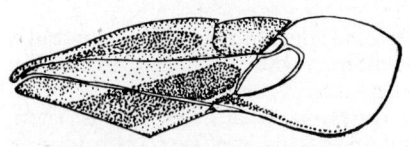

Fig. 431.—Hemelytron of *Pœcilocapsus lineatus*.

It is impracticable to discuss here the divisions of this family; reference can be made to only a few of the more common species.

The four-lined leaf-bug, *Pœcilocăpsus lineātus*.—This is a bright yellow bug, marked with black. It measures about 8 mm. in length. There are four longitudinal black lines which extend over the prothorax and the greater part of the hemelytra (Fig. 432). There is in many individuals a black dot on the cuneus of each hemelytron; and the membrane is also black.

This insect infests various plants, but abounds most on the leaves of currant, gooseberry, mint, parsnip, Weigela, Dahlia, and rose. It punctures the young and tender leaves, causing small brown spots; but these are sometimes so numerous and closely placed that the leaves become completely withered. It is a widely distributed species, its range extending from Canada to Georgia and westward to the Rocky Mountains.

There is only one generation a year. The eggs are laid in the terminal twigs of currant and other bushes in midsummer and hatch the following spring. They are laid in clusters, each containing six or eight eggs; these egg-clusters are forced out of the stem somewhat by the growth of the surrounding plant tissue; and as the projecting part of the egg is white, they can be easily found.

The methods of control are the pruning and burning of twigs containing egg-clusters, and, early in the season, the destruction of the nymphs by the use of kerosene emulsion or some one of the tobacco extracts.

Fig. 432.—*Pœcilocapsus lineatus.*

The tarnished plant-bug, *Lȳgus pratĕnsis.*—The tarnished plant-bug is a very common species which is found throughout the United States and in Canada. It is smaller than the preceding species, measuring 5 mm. in length and 2.5 mm. in its greatest width. It is exceedingly variable in color and markings; its color varies from a dull dark-brown to a greenish or dirty yellowish brown. In the more typical forms the prothorax has a yellowish margin and several longitudinal yellowish lines; there is a V-shaped yellowish mark on the scutellum; the distal end of the corium is dark; and the cuneus is pale, with a black point at the apex.

This pest is a very general feeder; it has been recorded as injuring about fifty species of plants of economic value; its injuries to the buds of Aster, Dahlia, and Chrysanthemum, and to the buds and blossoms of orchard-trees, and to nursery stock, are well-known. As yet no practical method of control of this pest has been found.

The apple-redbug, *Heterocŏrdylus mălinus.*—This species and the following one are sometimes a serious pest in apple orchards. They cause spotting of the leaves; but, what is far more serious, they puncture the young fruit, which results either in the dropping of the fruit or in its becoming badly deformed so as to be unmarketable. The eggs are inserted into the bark of the smaller branches late in June or early in July; they hatch in the following spring soon after the opening of the leaves of the fruit-buds. The nymphs are tomato-red in color. They first attack the tender leaves, but as soon as the fruit sets they attack it. The young nymphs can be killed by an application of "black leaf 40" tobacco-extract diluted at the rate of 1 pint in 100 gallons of water; the efficiency of this spray is increased by the addition of about 4 pounds of soap to each 100 gallons. Two applications of the spray should be made: the first, just before the blossoms open; the second, just after the petals fall. The spraying should be done on bright warm days, for in cool weather many of the nymphs hide away in the opening leaves.

The adult apple-redbug is about 6 mm. long. The general color varies from red to nearly black. Usually the thorax is black in front and red behind. The wings are red, usually black along the inner edge and with a pointed ovate black spot near the outer margin. The scutellum, legs, and antennæ are black. The entire dorsal surface is sparsely covered with conspicuous white, flattened, scale-like hairs.

The false apple-redbug, *Lygidea mendax*.—This species resembles the preceding one in general appearance and in habits. The nymphs can be distinguished by their brighter red color, by the absence of dusky markings on the thorax, and by having the body clothed with fine, short, black hairs. The adult of this species is lighter-colored and lacks the scale-like hairs on the dorsal surface.

The above account of these two species is an abstract of one published by Professor C. R. Crosby ('11).

The hop-redbug, *Paracalocoris hawleyi*.—The leaves of hop plants are sometimes perforated and the stems stunted and deformed by the nymphs of this species, which are red with white markings. The adult is 6 mm. long, black, with hemelytra hyaline or pale yellowish, and the cuneus reddish. For a detailed account see Hawley ('17).

Family TERMATOPHYLIDÆ

The Termatophylids

This family is closely allied to the following one, the Anthocoridæ, but differs in that the beak is four-jointed and ocelli are wanting. The hemelytra are well developed, furnished with an embolium, and usually with a single large cell in the membrane. The tarsi are three-jointed and are not furnished with an arolium.

The Termatophylidæ is a very small family, but it is world-wide in its distribution. A single very rare species has been found in this country. This is *Hesperŏphylum heidemănni*, which has been taken in New Hampshire and Arizona. Only the female of this species has been described. It is dark brown with the scutellum yellowish white; the cell in the membrane of the hemelytra is semicircular; the length of the body is 4 mm.

Family ANTHOCORIDÆ

The Flower-Bugs

This family is closely allied to the following one; but in the flower-bugs ocelli are present, though sometimes difficult to see, and the hemelytra are almost always fully developed and are furnished with an embolium (Fig. 433). As in the following family, the beak consists of three segments; the antennæ, of four; and the tarsi, of three.

Fig. 433.—Hemelytron of *Triphelps*.

The species are small. They are found in a great variety of situations, often upon trees and on flowers, sometimes under bark or rubbish. They are predacious. Thirty-six species have been catalogued in our fauna; these represent thirteen genera. The following species will serve as an example.

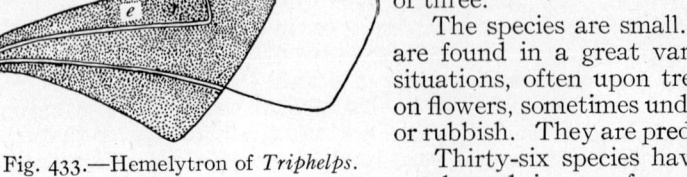

The insidious flower-bug, *Triphelps insidiōsus*.—This is perhaps the best-known of our species. It is a small black bug, measuring only 2 mm. in length; the hemelytra are yellowish white on the corium, at the tip of which is a large, triangular, blackish spot; the membrane is milky white. This species is widely distributed; it is common on flowers, and is often found preying upon the leaf-inhabiting form of the grape Phylloxera; it is also often found in company with the chinch-bug, upon which it preys and for which it is sometimes mistaken.

Family CIMICIDÆ

The Bedbug Family

The members of this family are parasitic bugs, which are either wingless or possess only vestigial hemelytra. In these insects the ocelli are absent, the antennæ are four-jointed, the beak is three-jointed, and the tarsi are three-jointed. Only four species belonging to this family have been found in America north of Mexico. These can be separated by the following table, which is based on a more detailed one given by Riley and Johannsen ('15).

A. Beak short, reaching to about the anterior coxæ.
  B. Pronotum with the anterior margin very deeply sinuate. The genus *Cīmex*.
    C. Body covered with very short hairs; second segment of the antennæ shorter than the third; hemelytra with the inner margin rounded and shorter than the scutellum. The common bedbug...*C. lectulārius*
    CC. Body covered with longer hairs; second and third segments of the antennæ of equal length; hemelytra with the inner margin straight and longer than the scutellum. Species found on bats...*C. pilosellus*
  BB. Anterior margin of the pronotum very slightly sinuate or nearly straight in the middle. Species found in swallows' nests...*Œcīacus vicārius*
AA. Beak long, reaching to the posterior coxæ. Infests poultry in southwest United States and in Mexico..................*Hæmatosīphon inōdorus*

The common bedbug, *Cīmex lectulārius*.—The body is ovate in outline and is very flat (Fig. 434); it is reddish brown in color, and is 4–5 mm. long by 3 mm. broad when full-grown. This pest is a nocturnal insect, hiding by day in cracks of furniture and beneath various objects. Ordinarily it is found only in the dwellings of man; but it has been known to infest chicken houses. The means commonly employed to destroy this pest is to wet the cracks of the bedstead and other places in which it hides with corrosive sublimate dissolved in alcohol. This is sold by druggists under the name of bedbug poison. As this substance is a virulent poison, it should be used with great care. In case a room is badly infested, it should be thoroughly cleaned; fumigated with sulphur or with hydrocyanic acid gas; the walls repapered, kalsomined, or whitewashed; and the woodwork repainted. Detailed directions for the use of gases against household insects are given by Herrick ('14). In traveling, where one is forced to lodge at places infested by this insect or by fleas, protection from them can be had by sprinkling a small quantity of pyrethrum powder between the sheets of the bed on retiring.

Fig. 434.—*Cimex lectularius*.

The other members of this family found in this country can be distinguished from the common bedbug by means of the table given above.

Family POLYCTENIDÆ

The Many-combed Bugs

The Polyctenidæ includes a small number of very rare species of bugs that are parasitic upon bats. Until recently it was not known to be represented in America north of Mexico; but Ferris ('19) records the finding of one species, *Hesperŏctenes lŏngiceps*, on the bat *Eumops californicus*, near San Bernardino, California. Figure 435 is a reduced copy of a figure of this insect by Ferris. The left half of the figure represents the dorsal aspect of this insect; the right half, the ventral aspect. This carefully made figure renders a detailed description unnecessary. The length of the body of the female is 4.5 mm.; of the male, 3.8 mm.

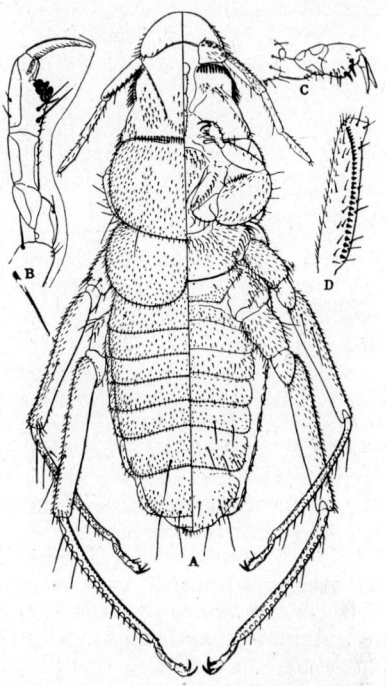

Fig. 435.—*Hesperoctenes longiceps:* A, female, left half dorsal, right half ventral; B, posterior tarsus; C, anterior tarsus; D, dorsal aspect of second antennal segment, distal end upward. (After Ferris.)

In this family the hemelytra are vestigial and the hind and middle tarsi are four-jointed. The name of the typical genus, *Polyctenes*, was probably suggested by the presence of several comb-like series of spines on the body.

Family NABIDÆ

The Nabids

In this family the body is oblong and somewhat oval behind. The beak is long, slender, and four-jointed. The hemelytra are longer than the abdomen, or are very short. Some species are dimorphic, being represented by both long-winged and short-winged forms. In the forms with long wings the membrane is usually furnished with four long veins bounding three discal cells, which are often open; from these discal cells diverge veins which form several marginal cells (Fig. 436). The fore tibiæ are armed with spines and are capable of being closed tightly upon the femora, which are stout; they are thus fitted for grasping prey.

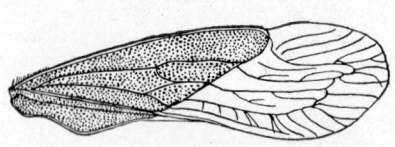

Fig. 436.—Hemelytron of *Nabis ferus*.

Nearly all of our common species belong to the genus *Nabis*; in fact this genus includes twenty-six of the thirty-one species found in this country. Due to an error made long ago, this genus has been commonly known as *Coriscus;* and most of the references to these insects are under this name.

Nābis fērus.—This is one of our most common species. It measures about 8 mm. in length. It is pale yellow with numerous minute brown dots; the veins of the membrane are also brownish. This species is distributed from the Atlantic Coast to the Pacific. It secretes itself in the flowers or among the foliage of various herbaceous plants, and captures small insects upon which it feeds.

Nābis subcoleoptrātus.—The short-winged form of this species is another very common insect (Fig. 437). This is of a shining jet-black color, with the edge of the abdomen and legs yellowish. The hemelytra barely extend to the second abdominal segment. The long-winged form of this species is not common; it is much narrower behind, and the hemelytra and the abdomen are rather dusky, or piceous, instead of jet-black.

Fig. 437.—*Nabis subcoleoptratus.*

Family REDUVIIDÆ

The Assassin-Bugs

The Reduviidæ is a large family, including numerous genera of diverse forms. Many of the members of it are insects of considerable size, and some are gayly colored. They are predacious, living on the blood of insects. In some cases they attack the higher animals; and, occasionally, even man suffers from them.

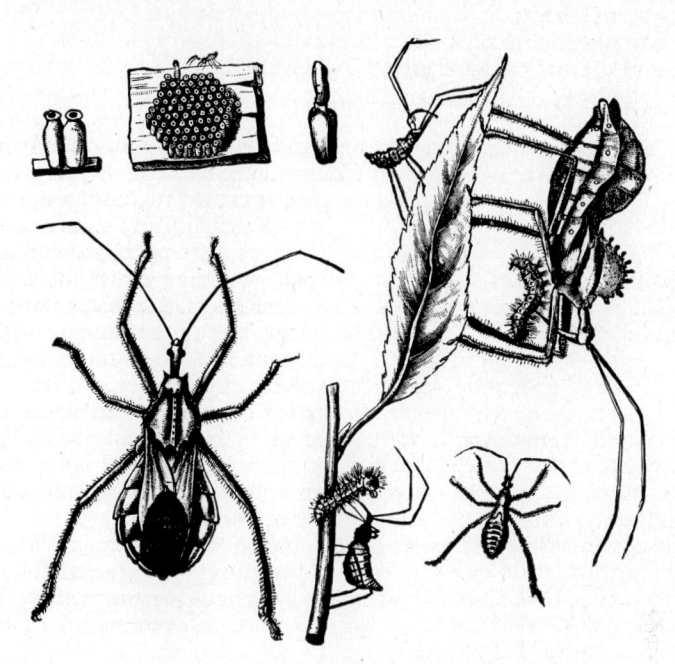

Fig. 438.—*Arilus cristatus.* (From Glover.)

In this family the beak is short, three-jointed, attached to the tip of the head, and with the distal end, when not in use, resting upon the prosternum, which is grooved to receive it. Except in a few species, ocelli are present in the winged forms. The antennæ are four-jointed.

More than one hundred species occur in our fauna; these represent forty-four genera. The following species will serve to illustrate the great diversity in form of members of this family.

The wheel-bug, *Arilus cristātus.*—The wheel-bug is so called on account of the cogwheel-like crest on the prothorax (Fig. 438). It is a common insect south of New York City, and is found as far west as Texas and New Mexico. The adult, a cluster of eggs, and several nymphs, are represented in the figure. The nymphs when young are blood-red, with black marks.

Fig. 439.—*Reduvius personatus.*

The masked bedbug-hunter, *Redūvius personātus.*—The adult of this species is represented by Figure 439; it measures from 15 to 20 mm. in length, and is black or very dark brown in color.

There are two marked peculiarities of this species that have caused it to attract much attention: first, in its immature instars the body is covered with a viscid substance which causes particles of dust and fibers to adhere to it; not only the body, but the legs and antennæ also, are masked in this way; in fact the nymph resembles a mass of lint, and attracts attention only when it moves; second, this species infests houses for the sake of preying upon the bedbug. It feeds also upon flies and other insects.

The big bedbug, *Triătoma sanguisūga.*—Closely allied to the masked bedbug-hunter is a large bug which insinuates itself into beds for a less commendable purpose than that of its ally, for it seeks human blood at first hand. This insect measures 25 mm. in length; it is black marked with red; there are six red spots on each side of the abdomen, both above and below. It inflicts a most painful wound.

This is one of several species of the Reduviidæ that received the name of "kissing-bug" as a result of sensational newspaper accounts which were widely published in the summer of 1899 and which stated that a new and deadly bug had made its appearance, which had the habit of choosing the lips or cheeks for its point of attack on man. It is found from New Jersey south to Florida and west to Illinois and Texas.

The genus *Triatoma* was renamed *Conŏrhinus* and most of the references to this species are under this generic name.

The thread-legged bug, *Ĕmesa brevipĕnnis.*—This is our most common representative of one of the subfamilies of the Reduviidæ in which the body is slender and the middle and hind legs are thread-like (Fig. 440). The front legs are less thread-like, and are fitted for grasping; they suggest by their form the front legs of the Mantidæ; the coxa is greatly elongated, more than four times as long as thick; the femur is spined; and the tibia shuts back upon the femur. In Figure 440 they are represented beneath the thread-like antennæ. *Ĕmesa brevipĕnnis* measures about 33 mm. in length; it is found upon trees, or sometimes swinging by its long legs from the roofs of sheds or barns.

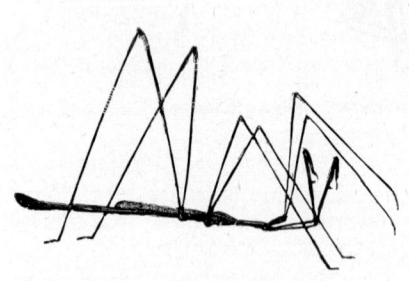

Fig. 440.—*Emesa brevipennis.*

A monograph of the Reduviidæ of North America has been published by Fracker ('12).

Family PHYMATIDÆ

The Ambush-Bugs

The Phymatidæ is poorly represented in this country but some of the species are very common. Here we find the body extended laterally into angular or rounded projections, suggesting the name of the typical genus. But the most striking character which distinguishes this group is the remarkable form of the front legs. These are fitted for seizing prey. The coxa is somewhat elongated; the femur is greatly thickened, so that it is half or two thirds as broad as long; the tibia is sickle-shaped, and fits closely upon the broadened and curved end of the femur; both tibia and femur are armed with a series of close-set teeth, so that the unlucky insect that is grasped by this organ is firmly held between two saws; the apparently useless tarsus is bent back into a groove in the tibia. Another striking character is presented by the antennæ, the terminal segment being more or less enlarged into a knob. Under the lateral margin of the pronotum in *Phymata* there is on each side a groove into which the antenna fits.

Only two genera are represented in our fauna, each by six species. These are *Phymata* and *Macrocĕphalus*. In *Phymata* the scutellum is of ordinary size; in *Macrocephalus* it is very large and extends to the tip of the abdomen.

Fig. 441.—*Phymata erosa.*

Our most common species is *Phȳmata erosa* (Fig. 441). It is a yellow insect, greenish when fresh, marked with a broad black band across the expanded part of the abdomen. It conceals itself in the flowers of various plants, and captures the insects which come to sip nectar. It is remarkable what large insects it can overcome and destroy; cabbage butterflies, honey-bees, and large wasps are overpowered by it.

Family ENICOCEPHALIDÆ

The Unique-headed Bugs

In this family the hemelytra are wholly membranous and provided with longitudinal veins and a few cross-veins (Fig. 442). The head is constricted at its base and behind the eyes, and is swollen between these two constrictions. This is a form of head not found in any other Hemiptera. Ocelli are present. The antennæ are four-jointed; the first, second, and third segments are each followed by a small ring-joint. The beak is four-jointed. The front tarsi are one-jointed, the middle and hind tarsi two-jointed. The front legs are fitted for grasping prey, the fore tarsi being capable of closing upon the end of the broad tibiæ.

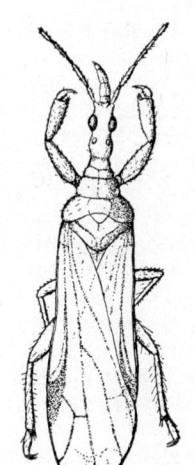

Fig. 442.—*Systelloderus biceps.* (After Johannsen.)

This is a small family; but few species are known from the entire world, and only two have been described from America north of Mexico. These are *Enicocĕphalus formicina*, found in California, and *Systelloderus bīceps*, which has been found from New York to Utah.

But little has been published regarding the habits of these insects. It is evident, from the structure of their fore legs, that they are predacious. Professor Johannsen ('09 b) found *Systelloderus biceps* (*Henicocephalus culicis*) flying in small swarms near Ithaca, N. Y. Their manner of flight resembled that of chironomids. They were observed repeatedly from July 5 to the last week in August, always in the latter part of the afternoon. This species measures 4 mm. in length.

The type genus of this family was first named *Enicocephalus;* this name was later emended to *Henicocephalus;* but the older form of the name, though incorrectly formed, is now used.

Family TINGIDÆ

The Lace-Bugs

The Tingidæ are doubtless the most easily recognized of all Hemiptera. The reticulated and gauze-like structure of the hemelytra, usually accompanied by expansions of the prothorax of a similar form, gives these insects a characteristic appearance which needs only to be once seen to be recognized in the future. Figure 443 represents one of these insects greatly enlarged, the hair-line at the side indicating the natural size of the insect. They are generally very small insects. But they occur in great numbers on the leaves of trees and shrubs, which they puncture in order to suck their nourishment from them.

In this family the ocelli are wanting; the beak and antennæ are four-jointed; the scutellum is usually wanting or vestigial, replaced by the angular hind portions of the pronotum; and the tarsi are two-jointed.

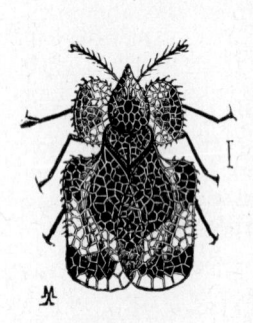

Fig. 443.—*Corythucha arcuata.*

About seventy-five species of lace-bugs, representing twenty-three genera, are now listed from this country. There are two well-marked subfamilies.

Subfamily TINGINÆ

This division includes nearly all of the known species. Here the scutellum is usually covered by an angular projection of the pronotum; and the hemelytra have no distinction between the clavus, corium, and membrane. The following species will serve as an illustration of this subfamily.

The hawthorn lace-bug, *Corythūcha arcuāta.*—This is a widely distributed species, which punctures the under surface of the leaves of different species of *Cratægus*. The infested leaves have a brown and sunburnt appearance. Eggs, nymphs, and adults are found together. The adult is represented, much enlarged, in Figure 443.

In Figure 444 the eggs and a nymph are shown. The eggs are covered by a brown substance, which hardens soon after oviposition.

Subfamily PIESMINÆ

In this subfamily the scutellum is not covered; the hemelytra have a distinct clavus, with a well-marked claval suture; the clavus is furnished with one, and the corium with three, longitudinal veins which are much stronger than the network of veins between them. In long-winged individuals the tip of the membrane lacks the network of veins and appears like the membrane in other families. As yet but a single American species has been described.

The ash-gray Piesma, *Piĕsma cinērea*.—This species measures about 3 mm. in length, and is of an ash-gray color. The prothorax is deeply pitted, so that it presents the same appearance as the base of the wing-covers. The head is deeply bifid at tip, and there is a short robust spine between the eye and the antenna on each side. This species sometimes infests vineyards to an injurious extent, destroying the flower-buds in early spring.

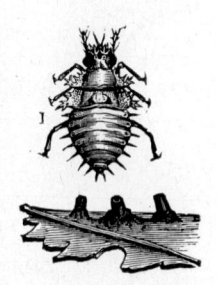

Fig. 444.—Eggs and nymph of *Corythucha arcuata*.

Family PYRRHOCORIDÆ

The Cotton-Stainer Family

In this family the antennæ are four-jointed; the beak is also four-jointed; ocelli are absent; and the hemelytra are not furnished with a cuneus. The members of the family are stout and heavily built insects, and are generally rather large and marked with strongly contrasting colors, in which red and black play a conspicuous part, in this respect resembling some of the larger species of the following family. The Pyrrhocoridæ can be distinguished from the Lygæidæ by the absence of ocelli, and by the venation of the membrane of the hemelytra (Fig. 445). At the base of the membrane there are two or three large cells, and from these arise branching veins.

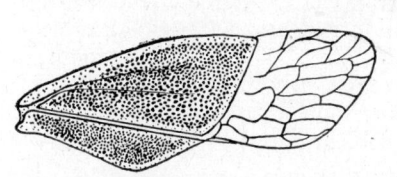

Fig. 445.—Hemelytron of *Euryopthalmus succinctus*.

Only twenty-two species, representing five genera, have been found in our fauna, and these are restricted to the Southern and Western States.

Our most important species, from an economic standpoint, is the red-bug or cotton-stainer, *Dysdĕrcus suturĕllus* (Fig. 446). It is oblong-oval in form, of a red color; the hemelytra and an arc on the base of the prothorax, and also the scutellum, are pale brown. The hemelytra have the costal margin, a narrow line bordering the base of the membrane and continuing diagonally along the outer margin of the clavus, and also a slender streak on the inner margin of the clavus, pale yellow. It varies much in size, ranging from 10 mm. to 16 mm. in length. The young bugs are bright red with black legs and antennæ. From time immemorial this has been one of the worst pests with which the cotton-planters of Florida and the West Indies have had to contend. It does much damage by piercing the stems and bolls with its beak and sucking the sap; but the principal injury to the crop is from staining the cotton in the opening boll by its excrement. It is also injurious to oranges; it punctures the rind of the fruit with its beak; and soon decay sets in, and the fruit drops. These insects can be trapped in cotton-fields by laying chips of sugar-cane upon the earth near the plants; in orange-groves small heaps of cotton-seed will be found useful, as well as pieces of sugar-cane. The insects that collect upon these traps can be destroyed with hot water.

Fig. 446.—*Dysdercus suturellus*.

The species whose range extends farthest north is *Euryophthălmus succinctus*. This is found from New Jersey south to Florida and west to Arizona. It is brownish black, with the lateral and hind margins of the prothorax, the costal margin of the hemelytra, and the edge of the abdomen, margined with orange or red. It measures about 15 mm. in length.

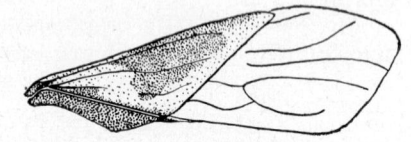

Fig. 447.—Hemelytron of *Lygæus kalmii*.

FAMILY LYGÆIDÆ

The Chinch-Bug Family

The Lygæidæ is one of the larger families of the Hemiptera. It includes certain forms which closely resemble members of the preceding family in size, form, and strongly contrasting colors. But the great majority of the species are of smaller size and less brightly colored; and all differ from that family in presenting distinct ocelli. The membrane of the hemelytra is furnished with four or five simple veins, which arise from the base of the membrane; sometimes the two inner veins are joined to a cell near the base (Fig. 447).

Nearly two hundred species belonging to this family have been found in our fauna; these represent fifty-five genera and seven subfamilies. Although these insects feed on vegetation, they have attracted but little attention as pests of cultivated plants excepting the following species.

The chinch-bug, *Blissus leucŏpterus.*—This well-known pest of grain-fields is a small bug, which when fully grown measures a little less than 4 mm. in length. It is blackish in color, with conspicuous, snowy white hemelytra. There is on the costal margin of each hemelytron near the middle of its length a black spot; from each of these spots there extends towards the head a somewhat Y-shaped dusky line. The body is clothed with numerous microscopic hairs. In Figure 448 this insect is represented natural size and enlarged. The species is dimorphic, there being a short-winged form.

Fig. 448.—*Blissus leucopterus.*

There are two generations of the chinch-bug each year. The insects winter in the adult state, hiding beneath rubbish of any kind; they even penetrate forests and creep under leaves, and into crevices in bark. In early spring they emerge from their winter quarters and pair; soon after, the females begin to lay eggs; this they do leisurely, the process being carried on for two or three weeks. The eggs are yellowish; about 500 are laid by a single insect; they are deposited in fields of grain, beneath the ground upon the roots, or on the stem near the surface. The eggs hatch in about two weeks after being laid. The newly hatched bugs are red; they feed at first on the roots of the plant which they infest, sucking the juices; afterwards they attack the stalks. The bugs become full-grown in from forty to fifty days. Before the females of this brood deposit their eggs, they leave their original quarters and migrate in search of a more abundant supply of food. About this time the wheat becomes dry and hard; and the migration appears to be a very general one. Although the insects sometimes go in different directions, as a general rule the masses take one direction, which is towards the nearest field of oats, corn, or some other cereal or grass that is still in a succulent state. At this time many of the bugs have not reached the adult state; and even in the case of the fully winged individuals the migration is usually on foot. In their new quarters the bugs lay the eggs for the second or fall brood.

The methods of control of this pest that are used are the following: the burning in autumn of all rubbish about fields, in fence corners, and in other places where the bugs have congregated to pass the winter; the stopping of the marching of the spring brood into new fields by means of a furrow or ditch with vertical sides, and with holes like post-holes at intervals of a few rods in the bottom of the furrow or ditch, in which the bugs are trapped; the use of a line of gas-tar on the ground to stop the marching of the spring brood; in some cases kerosene emulsion has been used to advantage; the sowing of decoy plots of attractive grains in early spring, and the later plowing under of the bugs and their food and harrowing and rolling the ground to keep the bugs from escaping; and the artificial dissemination of the fungus *Sporotrichum globuliferum,* which is the cause of a contagious disease of the chinch-bug.

Family NEIDIDÆ

The Stilt-Bugs

The family Neididæ consists of a small number of species, which on account of their attenuated forms are very striking in appearance (Fig. 449). The body is long and narrow; the legs and antennæ are also long and extremely slender. There is a transverse incision in the vertex in front of the ocelli. The antennæ are four-jointed, elbowed at the base of the second segment, and with the tip of the first segment enlarged. The beak is four-jointed; and the membrane of the hemelytra is furnished with a very few veins.

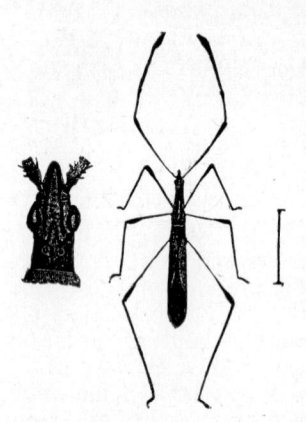

Only eight species of this family have been found in our fauna; but these represent six genera. Only two of the species are widely distributed in the United States and Canada. These are sluggish insects, found in the undergrowth of woods and in meadows and pastures.

Fig. 449.—*Jalysus spinosus.*

Jălysus spinōsus.—This is the best-known member of this family. It is distributed from the Atlantic to the Pacific in both the United States and Canada. It is as slender as a crane-fly (Fig. 449) and of a pale tawny color. The front of the head tapers off to an almost acute, upturned point. An erect spine projects from the base of the scutellum, and another from each side of the mesopleura, just in front of the posterior coxæ. The body is about 8 mm. in length.

Jălysus perclavātus.—This is one of the southern members of the family, but it has been found in New Jersey and the District of Columbia. It is smaller than the preceding species; the length of the male is 5 mm., of the female 6 mm. There is an erect spine between the bases of the antennæ; and the last segment of the antennæ is shorter and thicker than in *J. spinosus.*

Nēides mūticus.—Like *Jalysus spinosus*, this species is found from the Atlantic to the Pacific in both the United States and Canada. It lacks the spines of the scutellum and thorax; and the front of the head is bent down, in the form of a little horn.

The other representatives of this family in our fauna are found in Florida, Arizona, New Mexico, and California.

Family ARADIDÆ

The Flat-Bugs

The members of this family are very flat insects; in fact they are the flattest of all Hemiptera. They live in the cracks or beneath the bark of decaying trees; and the form of the body is especially adapted for gliding about in these cramped situations. They are usually dull brown or black; sometimes they are varied with reddish or pale markings. The hemelytra are usually well developed, with distinct corium, clavus, and membrane; but they are reduced in size, so that when folded they cover only the disk of the abdomen (Fig. 450). Ocelli are lacking; the antennæ are four-jointed; the tarsi are two-jointed; and the beak is four-jointed, but often apparently three-jointed.

Fig. 450.—*Aradus acutus.*

These insects are supposed to feed upon fungi or upon the juices of decaying wood and bark. The family is well represented in this country; fifty-nine species, representing nine genera, are now known, and doubtless many remain to be discovered.

Family COREIDÆ

The Squash-Bug Family

The members of this family vary greatly in form. Some of the species are among the most formidable in appearance of all of our Hemiptera; while others are comparatively weak and inconspicuous. The family is characterized as follows: the antennæ are inserted above an ideal line extending from the eye to the base of the rostrum, and are four-jointed; the vertex is not transversely impressed; the ocelli are present; the beak is four-jointed; the scutellum is small or of medium size; the hemelytra are usually **complete** and composed of clavus, corium, and membrane; the membrane is furnished with many veins, which spring from a transverse basal vein, and are usually forked (Fig. 451); the tarsi are three-jointed.

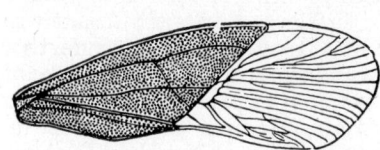

Fig. 451.—Hemelytron of *Leptocoris trivittatus.*

This is a large family; one hundred and twenty-four species, representing forty-eight genera, have been found in our fauna. It contains both vegetable feeders and carnivorous forms; in some cases the same species will feed upon both insects and plants. The most common and best-known species is the following.

The squash-bug, *Ănăsa trĭstis.*—The form of the body of the adult insect is represented in Figure 452. In this stage the insect appears blackish brown above and dirty yellow beneath. The ground color is really ochre-yellow, darkened by numerous minute black punctures. Upon the head are two longitudinal black stripes; the lateral margins of the prothorax are yellow, owing to the absence of the punctures along a narrow

Fig. 452.—*Anasa tristis.*

space; and the margin of the abdomen is spotted with yellow from a similar cause; the membrane of the hemelytra is black.

This species winters in the adult state. In early summer it lays its eggs in little patches on the young leaves of squash and allied plants. The young bugs are short and more rounded than the adult insects. There are several generations of this species each year.

This is one of the most annoying of the many pests of the kitchen-garden; and, unfortunately, no satisfactory method of control has been devised. The egg masses are conspicuous and can be collected and destroyed; the young nymphs can be killed by spraying with 10% kerosene emulsion; the adults can be trapped under bits of boards and stones; and many nymphs can be killed by destroying the vines as soon as the crop is harvested.

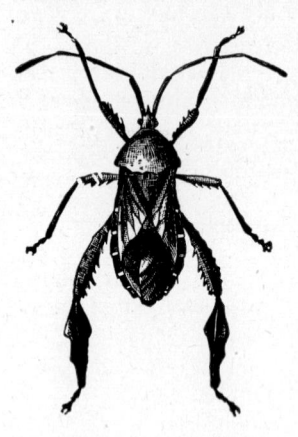

Fig. 453.—*Acanthocephala femorata*. (From Glover.)

Acanthocĕphala femorāta (Fig. 453) will serve as an example of one of the larger members of this family. This species is distributed from North Carolina to Florida and Texas. It has been known to destroy the cotton-worm, and is said to injure the fruit of the cherry by puncturing it with its beak and sucking the juices.

Family PENTATOMIDÆ

The Stink-Bug Family

With the Pentatomidæ we reach a series of families, three in number, in which the antennæ are usually five-jointed, differing in this respect from all of the preceding families. The form of the body presented by the great majority of the members of the Pentatomidæ is well shown by Figure 454. It is broad, short, and but slightly convex; the head and prothorax form a triangle. The scutellum is narrowed behind; it is large and in a few forms nearly covers the abdomen. The tibiæ are unarmed or are furnished with very fine short spines.

As with the Coreidæ, the members of this family vary greatly in their habits; some are injurious to vegetation; others are predacious; while some species feed indifferently upon animal or vegetable matter. Some species are often found on berries and have received the popular name of

Fig. 454.—A pentatomid.

stink-bugs on account of their fetid odor, which they are apt to impart to the berries over which they crawl. This nauseous odor is caused by a fluid which is excreted through two openings, one on each side of the lower side of the body near the middle coxæ.

The harlequin cabbage-bug, *Murgăntia histrionica.*—Among the species of the Pentatomidæ that feed upon cultivated plants, the harlequin cabbage-bug or "calico-back" is the most important pest. It is very destructive to cabbage and other cruciferous plants in the Southern States and on the Pacific Coast. It is black, with bands, stripes, and margins of red or orange or yellow. Its bizarre coloring has suggested the popular names given above. The full-grown bugs live through the winter, and in the early spring each female lays on the under surface of the young leaves of its food-plants about twelve eggs in two parallel rows. The eggs are barrel-shaped and are white banded with black. The young bugs are pale green with black spots. They mature rapidly; and it is said that there are several generations in one season.

This is an exceedingly difficult species to contend against. Much can be done by cleaning up the cabbage stalks and other remnants as soon as the crop is harvested, and, in the following spring, trapping the bugs that have hibernated by placing turnip or cabbage leaves in the infested gardens or fields, or by planting trap-crops of mustard or other cruciferous plants. The bugs that are not collected by these methods and their eggs should be collected by hand; this can be easily done as both the bugs and their eggs are conspicuous.

Fig. 455.- *Podisus maculiventris.* (From Glover.)

As if to atone for the destruction caused by their relative, the harlequin cabbage-bug, there are many members of this family that aid the agriculturist by destroying noxious insects. The species of the genus *Pŏdisus* have been reported often as destroying the Colorado potato-beetle, currant worms, and other well-known pests. Figure 455 represents a member of this genus, *Podisus maculivĕntris*

Family CYDNIDÆ

The Burrower-Bugs and the Negro-Bugs

The Cydnidæ is the second of the series of families in which the antennæ are five-jointed. In this family the outline of the body is more generally oval, rounded, or elliptical, and the form more convex, than in the Pentatomidæ. The scutellum is large but varies greatly in size and in outline. Each lateral margin of the scutellum is furnished with a furrow into which the margin of the hemelytron of that side fits. In this respect the Cydnidæ agrees with the preceding family and differs from the following one. The tibiæ are armed with strong spines.

The family includes two well-marked subfamilies.

Subfamily CYDNINÆ

The Burrower-Bugs

The subfamily Cydninæ includes the greater number of the members of the Cydnidæ found in the United States and Canada; of these there are twenty-nine species now listed, representing nine genera; most of these are restricted to the South and the Far West.

In this subfamily the scutellum is either broad and bluntly rounded, or triangular with the apex pressed down. The species are generally black or very dark brown. They are found burrowing in sandy places, or on the surface of the ground beneath sticks and stones, or at the roots of grass and other herbage. A European species is said to suck the sap from various plants near the ground. It is desirable that further observations be made upon the habits of this subfamily.

Fig. 456.—*Cyrtomenus mirabilis.*

Figure 456 represents *Cyrtŏmenus mirăbilis*, a species found in the South and the Southwest.

Subfamily THYREOCORINÆ

The Negro-Bugs

The subfamily Thyreocorinæ is represented in our fauna by a single genus, *Thyreŏcoris*, of which sixteen species have been found in this country. They are mostly black and beetle-like in appearance, some have a bluish or greenish tinge, and all are very convex. The body is short, broad, and very convex, in fact almost hemispherical. The scutellum is very convex and covers nearly the whole of the abdomen.

These insects infest various plants, and often injure raspberries and other fruits by imparting a disagreeable, bedbug-like odor to them. A common and widely distributed species is *Thyreocoris ater* (Fig. 457). Another species often found on berries is *T. pulicārius;* this species is sometimes a serious celery pest. It is shiny black and has a white stripe on each side of the body; it measures 3 mm. in length.

Fig. 457.—*Thyreocoris ater.*

Family SCUTELLERIDÆ

The Shield-backed Bugs

The members of this family are turtle-shaped bugs; that is, the body is short, broad, and very convex. The scutellum is very large, covering nearly the whole of the abdomen. The lateral margins of the scutellum are not furnished with grooves for receiving the edges of the hemelytra as is the case in the two preceding families. The tibiæ are smooth or furnished with small spines. Figure 458 represents *Eurygăster alternātus* somewhat enlarged, and serves to illustrate the typical form of members of this family.

The family is represented in this country by fourteen genera including twenty-six species. I have met no account of any of our species occurring in sufficient numbers to be of economic importance.

Fig. 458.—*Eurygaster alternatus.*

CHAPTER XXI

ORDER HOMOPTERA*

Cicadas, Leaf-Hoppers, Aphids, Scale-Bugs, and others

The winged members of this order have four wings, except in the family Coccidæ; the wings are of the same thickness throughout, and usually are held sloping at the sides of the body when at rest. The mouth-parts are formed for piercing and sucking; the beak arises from the hind part of the lower side of the head. The metamorphosis is gradual except in some highly specialized forms.

Although the Homoptera is a well-defined order, the families of which it is composed differ greatly in the appearance of their members. For this reason there is no popular name that is applied to the order as a whole.

The Homoptera was formerly regarded as a suborder of the Hemiptera, that order being divided into two suborders, the Heteroptera and the Homoptera. But these two groups of insects differ so markedly in structure that it seems best to regard them as distinct orders. The Hemiptera is, therefore, restricted to what was formerly known as the suborder Heteroptera, and the suborder Homoptera is raised to the rank of a separate order.

The wings of the Homoptera are usually membranous, but in some the front wings are subcoriaceous. In these cases, however, they are of quite uniform texture throughout, and not thickened at the base as in the Hemiptera.

Many wingless forms exist in this order; in the family Coccidæ the females are always wingless; and in the family Aphididæ the males may be either winged or wingless, while the sexually perfect females and certain generations of the agamic females are wingless. In the Coccidæ the males have only a single pair of wings, the hind wings being represented by a pair of club-like halteres. Each of these is furnished with a bristle, which is hooked and fits in a pocket on the hind margin of the fore wing of the same side.

In several of the families of the Homoptera the wing-venation is greatly reduced; and even in the case of the more generalized forms, if only the wings of adults be studied the venation of these wings appears to depart widely from the hypothetical primitive type; but by examining the tracheæ that precede the wing-veins in the wings of the nymphs, it is easy to determine the homologies of the wing-veins. This has now been done in the case of representatives of each of the families. The most generalized condition was found in the wings of a cicada, which will serve as the type of homopterous wing-venation.

---

*Homŏptera: *homos* (ὁμός), same, *pteron* (πτερόν), a wing.

Figure 459 represents the tracheation of the fore wing of a young nymph of a cicada. The dotted line a–b indicates approximately the line along which the hinge of the wing of the adult is formed. In this wing the only departures from the typical branching of the tracheæ are the following: trachea R_1 coalesces with the radial sector to a point beyond the separation of trachea R_{4+5} from the sector; the first anal trachea coalesces with trachea Cu for a short distance; and the second and third anal tracheæ are united at the base. These differences are remarkably slight compared with the great changes that have taken place in the specialization of the mouth-parts and other organs of the adult cicada.

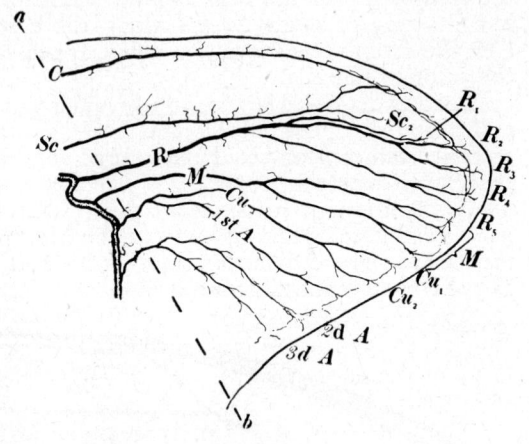

Fig. 459.—Tracheation of a fore wing of a young nymph of a cicada.

Figure 460 represents the fore wing of a mature nymph of a cicada. In this wing trachea R_1 is completely aborted. In fact one of the

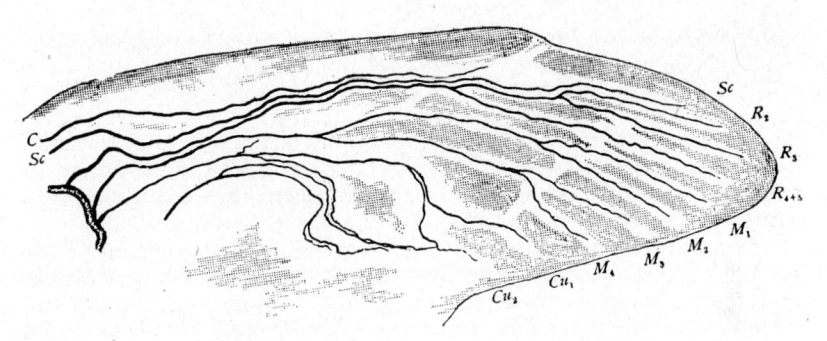

Fig. 460.—Tracheation of a fore wing of a mature nymph of a cicada.

most characteristic features in the venation of the Homoptera, and of the Hemiptera also, is the absence or very great reduction of vein R_1 in the adult wings of most members of these two orders. In the stage represented in this figure the developing cross-veins appear as pale bands.

Figure 461 represents the wings of an adult cicada. In this figure, where the veins are not numbered their homologies are indicated by the numbering of the cells behind them. In the adult wing there is a massing of several veins along the costal margin of the wing, and the cross-veins have the same appearance as the branches of the primary veins.

Further details regarding the development of the wings of a cicada, and accounts of the development of the wings of representatives of other families of the Homoptera, are given in "The Wings of Insects" (Comstock '18).

In the Homoptera the front part of the head is bent under and back so that the beak arises from the hind part of the lower side of the head. There is no distinct neck; and so closely is the head applied to the thorax that usually the front coxæ are in contact

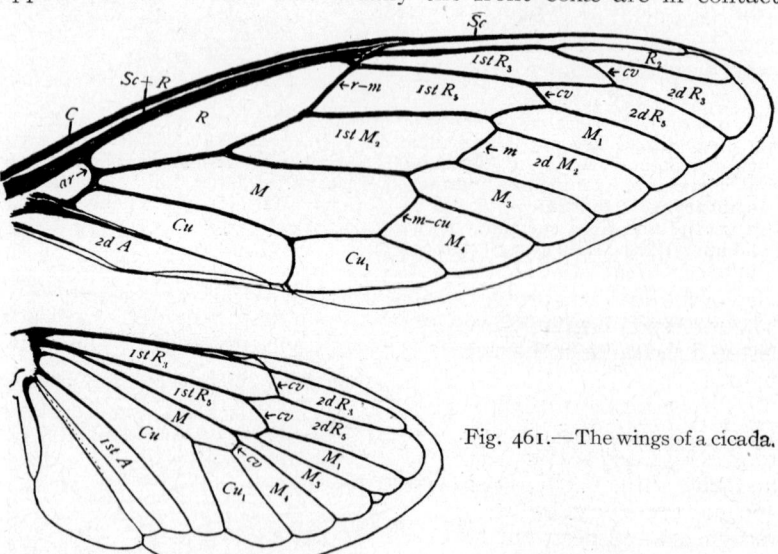

Fig. 461.—The wings of a cicada.

with the sides of the head, and in many forms the beak appears to arise from between the front legs.

The mouth-parts are formed for piercing and sucking. The piercing organs consist of four long, bristle-like setæ, the mandibular and maxillary setæ; these are enclosed in a long, jointed sheath, which is the labium. The labium and the enclosed setæ constitute what is commonly termed the beak.

The beak, however, corresponds to only a portion of the mouth-parts of a chewing insect, each mandibular and maxillary seta being only a part of a mandible or maxilla; in each case another part of the organ enters into the composition of the head-capsule.

As an example of the homopterous type of head and mouth-parts those of a cicada are probably the most available, on account of the large size of these insects and the comparative ease with which the parts of the head can be distinguished. Figure 462 represents a lateral

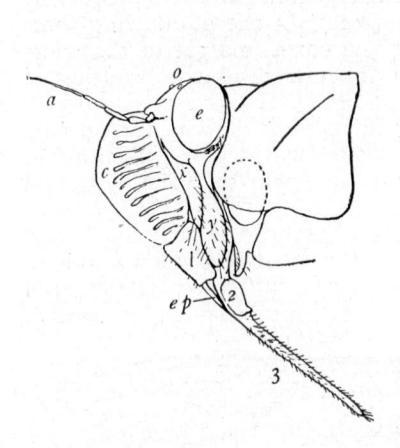

Fig. 462.—Head and prothorax of a cicada, lateral aspect: *a*, antenna; *c*, clypeus; *e*, compound eye; *ep*, epipharynx; *l*, labrum; *o*, ocelli; *2*, *3*, second and third segments of the labium. (After Marlatt, with changes in the lettering.)

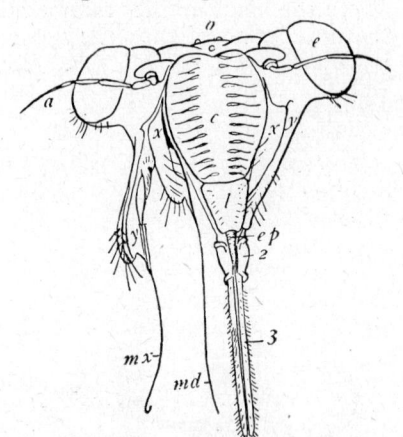

Fig. 463.—Head of a cicada, front view: *md*, mandibular seta; *mx*, maxillary seta; other letters as in Fig. 462. (After Marlatt, with changes in the lettering.)

view of the head and prothorax of a cicada, and Figure 463 a front view. The corresponding parts are lettered the same in the two figures.

The compound eyes (Figs. 462 and 463, *e*), the antennæ (Figs. 462 and 463, *a*), and the three ocelli (Figs. 466 and 467, *o*), can be easily recognized and need not be described in detail.

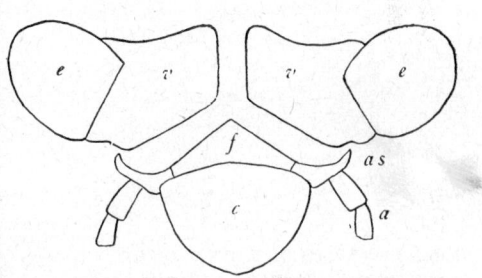

The *front* is a small sclerite near the summit of the head. It can be most easily recognized by the fact that it bears the median ocellus. In the adult insect the suture between it and the vertex is indistinct; but in the exuviæ of a nymph, where the epicranial suture has been opened by the emergence of the adult, the outline of this sclerite is evident (Fig. 464). In many homopterous insects the front is vestigial or wanting.

Fig. 464.—Part of the exuviæ of the head of a nymph of a cicada: *a*, antennæ; *as*, antennal sclerite; *c*, clypeus; *e*, *e*, compound eyes; *f*, front; *v*, *v*, vertex. (After Berlese.)

The *vertex* (Figs. 462 and 463, *v*) bears the paired ocelli.

The *clypeus* (Figs. 462 and 463, *c*) is very large, occupying the greater part of the anterior surface of the head. In several of the published accounts of the head of homopterous insects the clypeus has been incorrectly identified as the front.

The *labrum* (Figs. 462 and 463, *l*) is joined to the lower end of the clypeus; at its distal end it forms a sheath covering the base of the labium and the enclosed setæ. This part is described as the clypeus by those who have incorrectly identified the clypeus as the front.

The *epipharynx* (Figs. 462 and 463, *e*) arises at its normal position on the ental surface of the labrum; but it is greatly developed and projects beyond the end of the labrum. The projecting part has been mistaken for the labrum by some writers, those who have failed to recognize the front and have termed the clypeus the front and the labrum the clypeus.

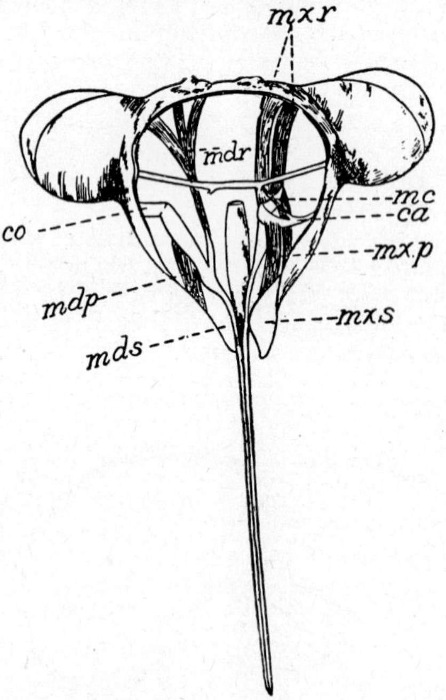

The *mandibular sclerites* are easily recognized in the cicada. On each lateral aspect of the head there are two quite distinct sclerites; the one that is next to the clypeus and the base of the labium is the mandibular sclerite (Figs. 462 and 463, *x*). This sclerite is termed the *lora* by some writers on the Homoptera.

The mandibular sclerites are believed to be in each case the basal part of a mandible. They were first recognized as such by Professor J. B. Smith ('92); and this conclusion has been adopted by Marlatt ('95), Heymons ('99), Meek ('03), Berlese ('09), and Bugnion and Popoff ('11). On the other hand, Muir and Kershaw ('12) regard the loræ as "lateral developments of the clypeal region" and not parts of mandibles.

Fig. 465.—Caudal view of the head of a cicada, with part of the head-capsule and muscles removed so as to show the left mandible and the right maxilla. (From Meek.)

The structure of the mandible as a whole has been worked out by Meek ('03) and is shown in the left half of Figure 465. Within the cavity of the head the maxillary seta is enlarged, and to it are attached a retractor muscle (*mdr*) and a protractor muscle (*mdp*). The seta is attached to the dorsal end of the mandibular sclerite (Fig. 465, *mds*) by a quadrangular sclerite (Fig. 465, *co*).

The *maxillary sclerites* (Figs. 462 and 463, *y*) are closely parallel with the mandibular sclerites, but extend farther down, joining the terminal part of the labrum. Each maxillary sclerite is a part of a maxilla. This is clearly shown by the fact that in the embryo each maxilla is at first a bilobed appendage; from one of these lobes the maxillary sclerite is developed, and from the other the maxillary seta (see Heymons '99). In the adult insect the maxillary sclerites are not separated from the epicranium by sutures as are the mandibular sclerites (Figs. 462 and 463).

The form and relations of the different parts of a maxilla, as worked out by Meek ('03), are shown in the right half of Figure 465. From the enlarged base of the maxillary seta a crescent-shaped sclerite (Fig. 465, *ca*) extends to the maxillary sclerite (Fig. 465, *mxs*). In this figure the maxillary retractor muscles (*mxr*), the maxillary protractor muscles (*mxp*), and a tendon (*mc*) connecting the crescent-shaped sclerite with the tentorium, are also represented.

It is interesting to note the similarity in the structure of the mandibles and the maxillæ. Each consists of a basal part which forms a portion of the wall of the head; a terminal piercing organ, the seta; and a sclerite connecting these two parts.

The *labium* forms the outer wall of the beak; it consists of three segments; the second and third are lettered in Figures 462 and 463. The proximal segment is probably homologous with the submentum of the chewing insect mouth; the second segment, with the mentum; and the third segment, with the ligula (see footnote, page 354). The dorsal surface of the labium, which is the lower surface, is deeply grooved, forming a channel which encloses the mandibular and maxillary setæ.

The labium, which is all that is commonly seen of the beak in either hemipterous or homopterous insects, is not a piercing organ; it is not pushed into the food substance of the insect, but serves merely as a sheath for the mandibular and maxillary setæ, which are the piercing organs and which are worked by the protractor and retractor muscles within the head (Fig. 465).

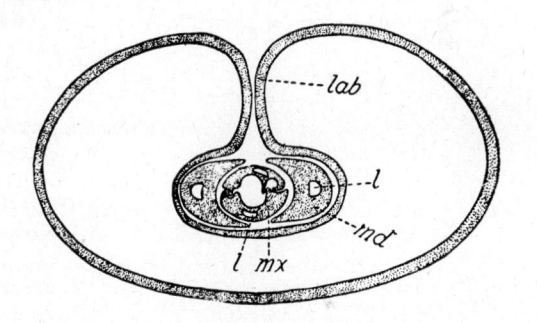

Fig. 466.—Cross-section of the third segment of the beak of a cicada: *lab*, labium; *md*, mandibular seta; *mx*, maxillary seta; *l, l,* lumina in the seta. (From Meek.)

Figure 466 represents a cross-section of the third segment of the beak of a cicada as figured by Meek ('03), and shows the relation of the labium to the mandibular and maxillary setæ. Each seta is crescent-shaped in cross-section; the mandibular setæ lie outside of the maxillary setæ; the maxillary setæ, which extend side by side at the base of the beak, are twisted so that at this point one lies above the other. The two are fastened together by interlocking grooves and ridges; and between them is a channel for the passage of the food. Within each of the four setæ, there is a lumen (Fig. 466, *l*, *l*).

The *hypopharynx* is a funnel-shaped, chitinized organ found near the base of the ental surface of the labium, at the end of the pharynx.

The nature of the metamorphosis differs to a considerable degree in the different families; in most cases it is gradual, but marked modifications of this type have been developed in the Aleyrodidæ and in the Coccidæ.

The members of this order feed on vegetation and to it belong some of our more important insect pests.

This order includes ten families, which are designated as follows:
The Cicadas, Family CICADIDÆ, p. 401.
The Spittle-insects, Family CERCOPIDÆ, p. 402.
The Tree-hoppers, Family MEMBRACIDÆ, p. 404.
The Leaf-hoppers, Family CICADELLIDÆ, p. 406.
The Lantern-fly Family, Family FULGORIDÆ, p. 408.
The Jumping Plant-lice, Family CHERMIDÆ, p. 410.
The Typical Aphids, Family APHIDIDÆ, p. 415.
The Adelgids and the Phylloxerids, Family PHYLLOXERIDÆ, p. 428.
The Aleyrodids, Family ALEYRODIDÆ, p. 437.
The Scale-bugs, Family COCCIDÆ, p. 440.

TABLE FOR DETERMINING THE FAMILIES OF THE HOMOPTERA

A. Beak evidently arising from the head; tarsi three-jointed; antennæ minute, bristle-like.
  B. With three ocelli, and the males with musical organs. Usually large insects, with all the wings entirely membranous. p. 401.....CICADIDÆ
  BB. Ocelli only two in number or wanting; males without musical organs.
    C. Antennæ inserted on the sides of the cheeks beneath the eyes. p. 408 ....FULGORIDÆ
    CC. Antennæ inserted in front of and between the eyes.
      D. Prothorax not prolonged above the abdomen.
        E. Hind tibiæ armed with one or two stout teeth, and the tip crowned with short, stout spines. p. 402......................CERCOPIDÆ
        EE. Hind tibiæ having a row of spines below. p. 406. CICADELLIDÆ
      DD. Prothorax prolonged into a horn or point above the abdomen. p. 404 ....MEMBRACIDÆ
AA. Beak apparently arising from between the front legs, or absent; tarsi one- or two-jointed; antennæ usually prominent and threadlike, sometimes wanting.
  B. Tarsi usually two-jointed; wings when present four in number.
    C. Wings transparent.
      D. Hind legs fitted for leaping; antennæ nine- or ten-jointed. p. 410.. ....CHERMIDÆ
      DD. Legs long an slender, not fitted for leaping; antennæ three- to seven-jointed. 412........................Superfamily APHIDOIDEA
    CC. Wings opaque, whitish; wings and body covered with a whitish powder. p. 437.......................................ALEYRODIDÆ

BB. Tarsi usually one-jointed; adult male without any beak, and with only two wings; female wingless, with the body either scale-like or gall-like in form, or grub-like and clothed with wax. The waxy covering may be in the form of powder, of large tufts or plates, of a continuous layer, or of a thin scale beneath which the insect lives. p. 440.............COCCIDÆ

Family CICADIDÆ

The Cicadas

The large size and well-known songs of the more common species of this family render them familiar objects. It is only necessary to refer to the periodical cicada and to the harvest-flies, one of which is represented by Figure 467, to give an idea of the more striking characteristics of this family. We have species of cicadas much smaller than either of these; but their characteristic form is sufficient to distinguish them from members of the other families of this order.

The species are generally of large size, with a subconical body. The head is wide and blunt, with prominent eyes on the outer angles, and three bead-like ocelli arranged in a triangle between the eyes. The structure of the mouth-parts is described on an earlier page and illustrated by several figures; and the form and venation of the wings are shown by Figure 461. But the most distinctive peculiarity is the form of the musical organs of the males; an example of these is described and figured on pages 89 to 91.

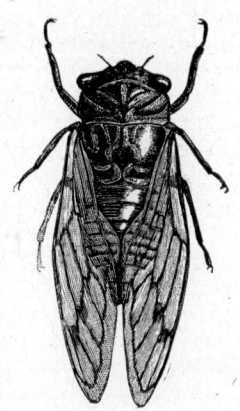

Fig. 467.—*Tibicen linnei.*

The family Cicadidæ is well represented in this country; seventy-four species, representing sixteen genera, are now listed from our fauna. The two following species will serve as illustrations.

There are several species of cicadas that are commonly known as dog-day cicadas or harvest-flies; the most abundant of these is the species that has received the popular name of the *lyreman;* this is *Tibīcen linnēi* (Fig. 467). The shrill cry of this species, which is the most prominent of the various insect sounds heard during the latter part of the summer, has brought its author into prominent notice. This insect varies both in size and colors. It commonly measures 50 mm. to the tip of the closed wings; it is black and green, and more or less powdered with white beneath. The transformations of this insect are similar to those of the following species, except that it probably completes its development in a much shorter period. It differs also in seldom, if ever, occurring in sufficient numbers to be of economic importance; but a brood of it appears each year.

The member of this family that has attracted most attention is the periodical cicada, *Tibicina septĕndecim*. This species is commonly known as the seventeen-year locust; but the term *locust* when applied to this insect is a misnomer, the true locusts being members of the order Orthoptera. The improper application of the term *locust* to this species was doubtless due to the fact that it appears in great swarms, which reminded the early observers in this country of the hordes of migratory locusts or grasshoppers of the Old World. This species is remarkable for the long time required for it to attain its maturity. The eggs are laid in the twigs of various trees; the female makes a series of slits in the twig, into which the eggs are placed. Sometimes this cicada occurs in such great numbers that they seriously injure small fruit trees, by ovipositing in the twigs and smaller branches. The nymphs hatch in about six weeks. They soon voluntarily drop to the ground, where they bury themselves. Here they obtain nourishment by sucking the juices from the roots of forest and fruit trees. And here they remain till the seventeenth year following. They emerge from the ground during the last half of May, at which time the empty pupa-skins may be found in great numbers, clinging to the bark of trees and other objects. It is at this period that the cicadas attract attention by the shrill cries of the males. The insects soon pair, the females oviposit, and all disappear in a few weeks.

More than twenty distinct broods of this species have been traced out; so that one or more broods appear somewhere in the United States nearly every year. In many localities, several broods co-exist; in some cases there are as many as seven distinct broods in the same place, each brood appearing in distinct years. There is a variety of the species in which the period of development is only thirteen years. This variety is chiefly a southern form, while the seventeen-year broods occur in the North.

Family CERCOPIDÆ

The Spittle-Insects or Frog-Hoppers

During the summer months one often finds upon various shrubs, grass, and other herbs, masses of white froth. In the midst of each of these masses there lives a young insect, a member of this family. In some cases as many as four or five insects inhabit the same mass of froth. It is asserted that these insects undergo all their transformations within this mass; that when one is about to molt for the last time, a clear space is formed about its body and the superficial part of the froth dries, so as to form a vaulted roof to a closed chamber within which the last molt is made.

The adult insects wander about on herbage, shrubs, and trees. They have the power of leaping well. The name frog-hoppers has doubtless grown out of the fact that formerly the froth was called "frog-spittle" and was supposed to have been voided by tree-frogs from their mouths. The name is not, however, inappropriate, for the broad and depressed form of our more common species is somewhat like that of a frog.

The origin and formation of the froth of spittle-insects has been discussed by many writers. Guilbeau ('08) found by many experiments that the froth is derived from two sources. The greater part of the fluid is voided from the anus; to this fluid is added a mucilaginous substance which renders it viscous and causes the retention of air bubbles, which are introduced into it by the insect by means of its caudal appendages. The mucilaginous substance is the excretion of large hypodermal glands, which are in the pleural region of the seventh and eighth abdominal segments. These are known as the glands of Batelli; they open through numerous minute pores in the cuticula.

It is evident that the covering of froth protects the spittle-insects from parasites and other enemies.

In this family the antennæ are inserted in front of and between the eyes; the prothorax is not prolonged back of the abdomen, as in the Membracidæ; the tibiæ are armed with one or two stout teeth, and the tip is crowned with short, stout spines, as shown in Figure 468.

The Cercopidæ is represented in our fauna by six genera, which include twenty-five species. The following species will serve as examples.

One of the more common and very widely distributed species is *Lepyrōnia quadrangulāris* (Fig. 468). The adult of this species is a brownish insect, densely covered with microscopic hairs, and black beneath; the hemelytra are marked with two oblique brown bands, which are confluent near the middle of the costal margin; the humeral region is dusky; and the tip of each hemelytron is marked with a small blackish curve; the ocelli are black, but indistinct. This species measures from 6 mm. to 8 mm. in length.

Fig. 468.—*Lepyronia quadrangularis*, natural size, and one tibia enlarged.

Somewhat resembling the preceding species, and also common and widely distributed, is *Aphrŏphora quadranotāta*. In this species the body is pale; the hemelytra are dusky, each with two large hyaline costal spots, margined with dark brown; the ocelli are blood-red; and the head and pronotum are furnished with a slightly elevated, median, longitudinal line.

To the genus *Clastŏptera* belong certain other common members of this family. In this genus the body is short and plump, sometimes nearly hemispherical; the species are small, our common forms ranging from 3 mm. to 6 mm. in length. *Clastōptera prōteus* is a conspicuous species on account of its bright yellow markings. It varies greatly in color and markings; but the most striking forms are black, with three transverse yellow bands, two on the head and one on the thorax, and with the scutellum and a large oblique band on each hemelytron yellow. Another common species is *Clastŏptera obtūsa*. This occurs on black alder in summer and autumn. It is of a claret-brown color above, marked with two pale bands on the vertex, two on the prothorax, and a wavy, broader band on the hemelytra. The membrane is often whitish, the waved band is extended exteriorly, and there is a pale V-shaped figure on the end of the scutellum.

Family MEMBRACIDÆ

The Tree-Hoppers

The most useful character for distinguishing members of this family is the prolongation of the prothorax backward above the abdomen; sometimes it extends back to the tip of the abdomen and completely covers the wings. This development of the prothorax resembles that which occurs in the pigmy locusts, the subfamily Acrydiinæ of the order Orthoptera. In many of the Membracidæ, however, the prothorax is not only prolonged backward but is extended sidewise or upwards, with the result that in some cases the insect presents a most bizarre appearance; this is especially true of certain tropical forms; Figure 469 represents two species found in Central America.

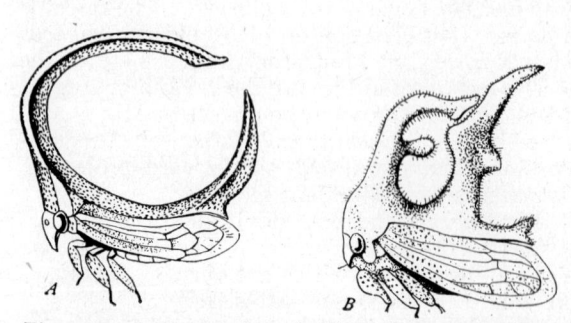

Fig. 469.—A, *Spongophorus ballista*; B, *Spongophorus querini*.

Many species of the Membracidæ live upon bushes or small trees; others inhabit grass and other herbaceous plants. Although these insects subsist upon the juices of plants, they rarely occur in sufficient numbers to be of economic importance. Sometimes the females injure young trees by laying their eggs in the bark of the smaller branches and in buds and stems. Many members of this family excrete honey-dew and are attended by ants, especially in the nymphal stages, as are the aphids. The adults are good leapers; hence the common name *tree-hoppers*.

Fig. 470.—*Ceresa bubalus*.

This family is well represented in this country; one hundred eighty-five species, representing forty-three genera, are now listed. Among our more common species are the following.

The Buffalo tree-hopper, *Cerēsa būbalus*.—The popular name of this species refers to the lateral prolongations of the prothorax, which suggest the horns of a buffalo (Fig. 470). The life-history of this insect has been worked out by Funkhouser ('17). The nymphs feed on succulent herbs, particularly sweet clover; the eggs are laid on young trees, particularly elm and apple, the stems of which are injured by the egg-punctures. Oviposition occurs most commonly in early September, at Ithaca, N. Y. The eggs hatch early in the following May. The young nymphs leave the trees on which the eggs were deposited and migrate to succulent weeds. The early life of the adult is spent on the weeds; but later the females migrate to trees for egg-laying.

The two-horned tree-hopper, *Cerēsa dĭceros.*—This species resembles the buffalo tree-hopper in size and form. It is a pale dirty yellow, spotted with brown; the lateral and caudal aspect of each horn is brown; the caudal tip of the prothorax, and a large spot midway between the tip and the horns, are also brown. The insect is densely clothed with hairs. It is common on black elder, *Sambucus canadensis*. Funkhouser followed the life-history from the egg to the adult on this plant. The eggs are laid about the middle of August in the second-year stems, and hatch about the middle of May.

The two-marked tree-hopper, *Enchenōpa binotāta.*—In this species the pronotum is prolonged in an upward- and forward- projecting horn (Fig. 471). This insect is very abundant on trees, shrubs, and vines. It is gregarious, and both adult and immature forms are found clustered together. The eggs are usually laid in frothy masses, which are very white and appear like wax. Funkhouser states that a variety of this species found on butternut lays its eggs in the buds and does not cover them with the heavy froth. The specific name of this species refers to the fact there are two yellow spots on the dorsal line of the pronotum.

Fig. 471.—*Enchenopa binotata.*

Another very common species, and one that is closely allied to the preceding, is *Campylĕnchia lătipes*. This is brownish, unspotted, and has a rather longer horn than that of the two-marked tree-hopper; but it varies much in color and in the length of the pronotal horn. This is a grass-inhabiting species and is common in pastures and especially on alfalfa. It is often taken by sweeping.

Fig. 472.—*Telamona.*

Telamōna.—To this genus belong our humpback

Fig. 473.—Portraits of some tree-hoppers.

forms (Fig. 472), of which about thirty species have been found in our fauna. They live chiefly on oaks, hickories, basswood, and other forest trees. The adults generally rest singly on the limbs and branches of the trees; they are strong flyers and are difficult to capture. The immature forms keep together in small groups.

Figure 473 represents a front view of several membracids in our collection.

Family CICADELLIDÆ*
The Leaf-Hoppers

This family is a very large one, and it is also of considerable economic importance; for it includes a number of species that are very injurious to cultivated plants. The members of it are of small or moderate size. The antennæ are inserted in front of and between the eyes; the pronotum is not prolonged above the abdomen; and the hind tibiæ are nearly or quite as long as the abdomen, curved, and armed with a row of spines on each margin. The form and armature of the hind tibiæ are the most salient characters of this family. The form of the body is commonly long and slender, often spindle-shaped; but some are plump.

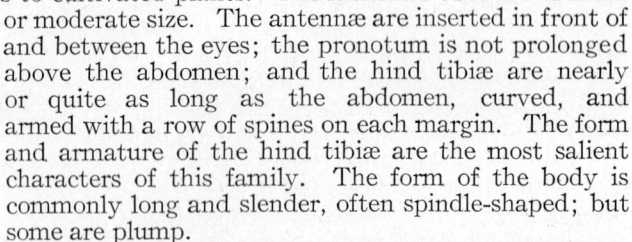

Fig. 474.—*Euscelis exitiosus.*

These insects are able to leap powerfully; and, as they are more often found on the leaves of herbage and on grass than elsewhere, they have been named leaf-hoppers. They infest a great variety of plants; some of them are important pests in gardens, orchards, and vineyards; but they are most destructive as pests of grains and grasses. Although this is true, much less attention has been paid to injuries caused by them to grains and grasses than to those inflicted upon vineyards and rose bushes.

More than seven hundred species, representing about seventy genera, have been found in the United States and Canada. Among the more important members of the family from an economic standpoint are the following.

The destructive leaf-hopper, *Euscelis exitiōsus*, which is represented, greatly enlarged, in Figure 474, sometimes infests winter wheat to a serious extent. It is a widely distributed species, its range including nearly the whole of the United States. It is a small, active, brownish insect, which measures with its wings folded about 5 mm. in length. It injures grass or grain by piercing the midrib of the leaf and sucking the juices from it.

The grape-vine leaf-hopper, *Erythronēura comes*, is a well-known pest which infests the leaves of grape, in all parts of this country where this vine is grown. It is a little more than 3 mm. in length, and has the back and wings marked in a peculiar manner with yellow and red. In the winter the darker markings are a dark orange-red, but after feeding has been resumed for a short time in the spring they change to a light lemon-yellow. The darker markings on the

---

*This family has been commonly known as the Jassidæ, but Cicadellidæ is the older name.

adults vary so much that eleven distinct varieties are now recognized; two of these are represented at *b* and *c* in Figure 475.

The rose leaf-hopper, *Ĕmpoa rosæ*, is a well-known pest of the rose. Swarms of these insects may be found, in various stages of growth, on the leaves of the rose-bush through the greater part of the summer, and their numerous cast skins may be seen adhering to the lower sides of the leaves; in fact attention is most frequently called to this pest by these white exuviæ. The adult measures less than 3 mm. in length. Its body is yellowish white, its wings are white and transparent, and its eyes, claws, and ovipositor are brown.

The apple leaf-hopper, *Empŏasca fabæ.*—Although this species is named *the apple leaf-hopper*, it infests to an injurious extent many different plants, both cultivated and wild. Slingerland and Crosby ('14) state that it infests apple, currant, gooseberry, raspberry, potato, sugar-beets, beans, celery, grains, grasses, shade trees, and weeds. The adult insect measures about 3 mm. in length, and is of a pale yellowish green color with six or eight distinguishing white spots on the front margin of the pronotum.

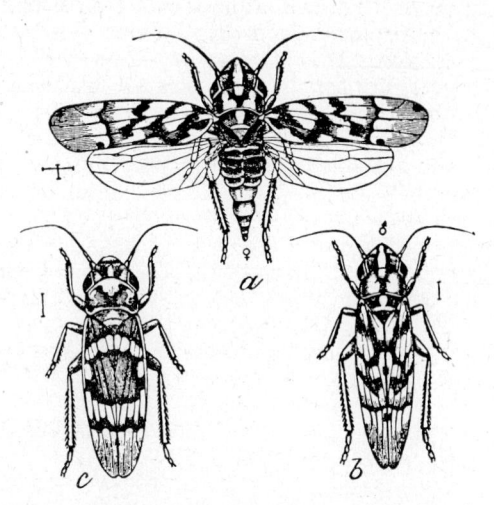

Fig. 475.—*Erythroneura comes: a* and *b*, female and male of the typical *comes* variety; *c*, the *vitis* variety. (From Slingerland.)

The genus *Dræculacephala* includes grass-green or pale green, spindle-shaped species, in which the head as seen from above is long and triangular. One of the species, *D. reticulāta*, sometimes greatly injures fields of grain in the South.

The genus *Oncometŏpia* includes species in which the head is more blunt than in the preceding genus and is wider across the eyes than the thorax. *O. undāta* (Fig. 476) is a common species. Its body, head, fore part of the thorax, scutellum, and legs are bright yellow, with circular lines of black on the head, thorax, and scutellum. The fore wings are bluish purple, when fresh, coated with whitish powder. It measures 12 mm. in length. It is said to lay its eggs in grape canes, and to puncture with its beak the stems of the bunches of grapes, causing the stems to wither and the bunches to drop off.

Fig. 476.—*Oncometopia undata*.

One division of this family, the subfamily Gyponinæ, includes forms which resemble certain genera belonging to the Cercopidæ by their plump proportions. Among these are *Pĕnthima americāna*, which is a plump, short-bodied insect, resembling a *Clastoptera;* and the genus *Gypona* includes a large number of species, some of which resemble very closely certain species of *Aphrophora*. A glance at the posterior tibiæ of these leaf-hoppers will enable one to distinguish them from the cercopids, which they so closely resemble.

Methods of combating leaf-hoppers.—Leaf-hoppers, being sucking insects, are fought with contact insecticides. But it is difficult to destroy the adults, for they are so well-protected by their wings that applications strong enough to kill them are liable to injure the foliage of the host-plant; and, too, they are very active and fly away when approached. The most effective remedial measures are those directed against the nymphs. These consist of the use of some spray, as a ten-per-cent. kerosene emulsion or a soap solution made by dissolving one pound of soap in six or eight gallons of water, or a solution made of one ounce of "black leaf 40" tobacco extract and six gallons of water in which has been dissolved a piece of soap the size of a hen's egg. The application should be so applied as to wet the lower surface of every leaf.

Family FULGORIDÆ

The Lantern-Fly Family

This family is remarkable for certain exotic forms which it includes. Chief among these is the great lantern-fly of Brazil, *Laternāria phosphōrea*. This is the largest species of the family and is one of the most striking in appearance of all insects (Fig. 477). It has immense wings, which expand nearly six inches; upon each hind wing there is

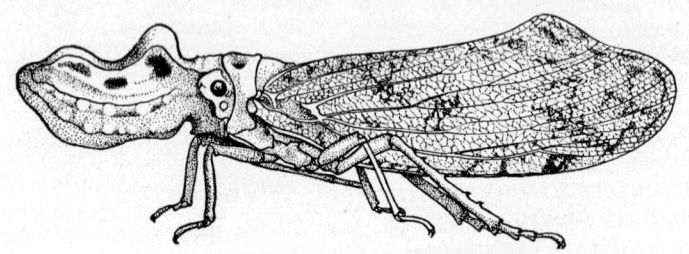

Fig. 477.—The lantern-fly, *Laternaria phosphorea*.

a large eye-like spot. But the character that makes this insect especially prominent is the form of the head. This has a great bladder-like prolongation extending forward, which has been aptly compared to the pod of a peanut. Maria Sibylla Merian, a careful observer, who wrote more than two hundred years ago (1705), stated that this prolongation of the head is luminescent. This statement was accepted by Linnæus without question; and he made use of names for this and some allied species, such as *laternaria, phosphorea, candelaria,* etc., to illustrate the supposed light-producing powers of these insects. The common name *lantern-fly* is based on the same belief.

The Brazilian lantern-fly has been studied by many more recent observers, and all have failed to find that it is luminescent. It may be that the individuals observed by Madame Merian were infested by luminescent bacteria, as has been observed to be the case occasionally in certain other insects. No member of this family is known to be luminescent.

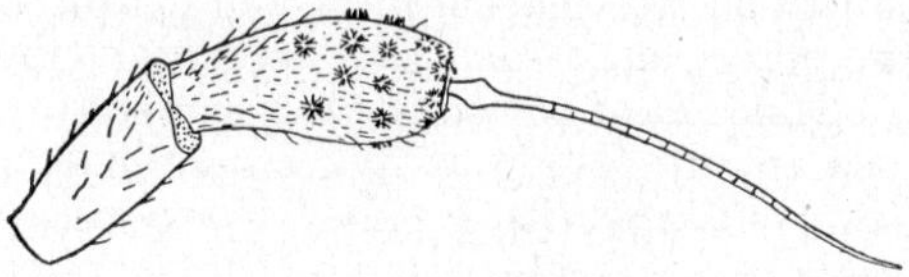

The Chinese candle-fly, *Fulgōria candelāria*, is another very prominent member of this family, which is commonly represented in collections of exotic insects and is often figured by the Chinese. This too has been reputed to give light.

Fig. 478.—Antenna of *Megamelus notula*. (After Hansen.)

Certain fulgorids found in China excrete large quantities of a white, flocculent wax, which is used by the Chinese for candles and other purposes.

There does not seem to be any typical form of the body characteristic of this family. The different genera differ so greatly that on superficial examination they appear to have very little in common. The most useful character for recognizing these insects is the form and position of the antennæ. These are situated on the side of the cheeks beneath the eyes; the two proximal segments, the scape and pedicel, are stout (Fig. 478); the clavola consists of a small, nearly pear-shaped basal segment and a slender, segmented or unsegmented, bristle-like terminal part. The pedicel is provided with numerous sense-organs.

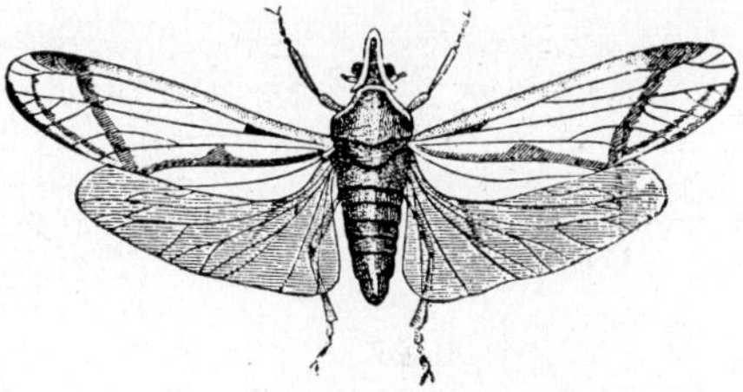

Fig. 479.—*Scolops*.

So far as numbers are concerned this family is well represented in our fauna, three hundred fifty-seven species and seventy-seven genera having been listed; but our species are all small compared with the exotics mentioned above. The following of our native genera will serve to illustrate some of the variations in form represented in this country. The species all feed on the juices of plants.

Fig. 480.—*Otiocerus coquebertii*. (From Uhler.)

Scolops.—In this genus the head is greatly prolonged (Fig. 479), as with the Chinese candle-fly. Our more common species, however, measure only about 8 mm. in length.

Otiŏcerus.—In this genus the body is oblong; the head is compressed, with a double edge both above and below. *Otiocerus coquebĕrtii* (Fig. 480) is a gay lemon-yellow or cream-colored species, with wavy red lines on the fore wings. It measures about 8 mm. to the tips of the wings, and lives upon the leaves of grape-vines, oaks, and hickory.

Ormenis.—In our common representatives of this genus the wing-covers are broad, and closely applied to each other in a vertical position; they are more or less truncate, and give the insects a wedge-shaped outline. *O. septentrionālis* (Fig. 481) is a beautiful, pale green species powdered with white, which feeds on wild grape-vines, drawing nourishment from the tender shoots and midribs of the leaves, during its young stages.

Fig. 481.—*Ormenis septentrionalis.*

Family CHERMIDÆ*

The Jumping Plant-Lice

The jumping plant-lice are small insects; many of them measure less than 2 mm. in length; and the larger of our species, less than 5 mm. They resemble somewhat the winged aphids; but they look more like miniature cicadas (Fig. 482). They differ from aphids in the firmer texture of the body, in the stouter legs, in having the hind legs fitted for jumping, and in the antennæ being ten-jointed or rarely nine- or eleven-jointed. The terminal segment of the antennæ bears two thick setæ of unequal length.

Both sexes are winged in the adult. The front wings are ample, and, while often transparent, are much thicker than the hind wings. The homologies of the wing-veins of the fore wings of *Psyllia floccōsa* are indicated in Figure 483.

Fig. 482.—*Psyllia.*

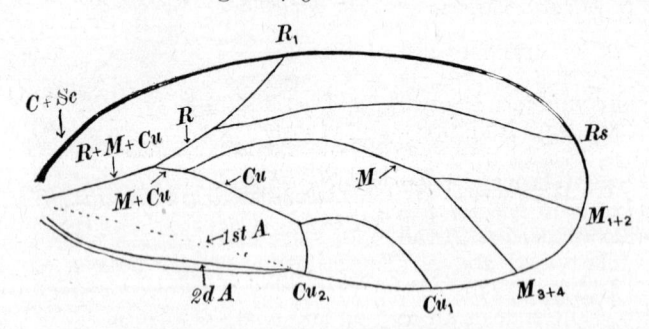

Fig. 483.—The venation of a fore wing of *Psyllia floccosa*. (After Patch.)

The beak is short and three-jointed. The basal segment of the beak is held rigidly between the fore coxæ.

---

*This family has been quite commonly known as the Psyllidæ, a result of an incorrect application of the name *Chermes* to a genus of the Phylloxeridæ.

The jumping plant-lice are very active little creatures, jumping and taking flight when disturbed; but their flight is not a prolonged one. They subsist entirely upon the juices of plants; some species form galls; but it is rare that any of the species appear on cultivated plants in sufficient numbers to attract attention, except in case of the pear-tree Psylla.

The family Chermidæ is of moderate size; in our latest list one-hundred thirty-seven species representing twenty-four genera, are enumerated from our fauna. The two following species will serve to illustrate variations in habits of these insects.

Pachypsẏlla cĕltidis-mămma.—This is a gall-making species which infests the leaves of hackberry (*Cĕltis occidentālis*).

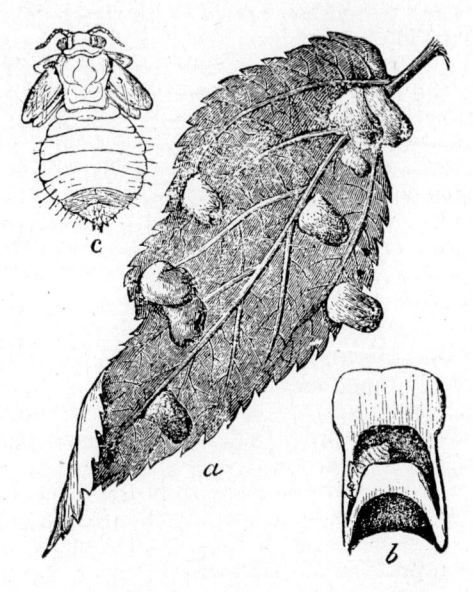

Fig. 484.—Gall of *Pachypsylla celtidis-mamma:* *a*, leaf with galls, from under-side; *b*, section of gall enlarged and insect in cavity; *c*, nymph, enlarged. (From Riley.)

Figure 484 represents an infested leaf with galls, and a single gall and a nymph enlarged. The adult insect (Fig. 485) has a wing expanse of about 6 mm.

The pear-tree psyllia, *Psẏllia pyrĭcola.*—This is our most impor-

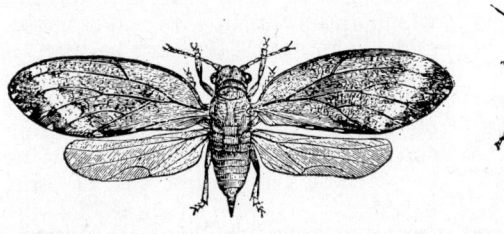

Fig. 485.—*Pachypsylla celtidis-mamma.* (From Packard.)

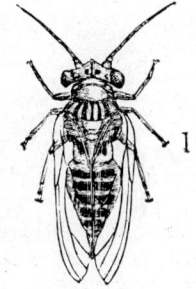

Fig. 486.—*Psyllia pyricola.*

tant species from an economic standpoint, being a serious enemy of the pear. It is a small species (Fig. 486); the summer generations measure to the tips of the folded wings from 2.1 mm. to 2.8 mm., the hibernating form 3.3 mm. to 4 mm. The general color is light orange to reddish brown, with darker markings. The eggs are laid early in the spring in the creases of the bark, in old leaf-scars, and about the base of the terminal buds. The young nymphs migrate to the axils of the leaf petioles and the stems of the forming fruit; later they spread to the under side of the leaves. They secrete large quantities of honey-dew, upon which a blackish fungus grows; this is often the first indication of the presence of the pest. There are at least four generations each year. Badly infested trees shed their leaves and young fruit in midsummer. In some cases orchards have been so badly injured by this pest that they have been cut down by their owners.

The methods of control that are recommended are the following: the scraping off of the rough bark from the trunks and larger branches of the trees and burning it, in order to destroy the hibernating adults; and thorough spraying of the trees with kerosene emulsion or "black leaf 40" tobacco extract when the petals have fallen from the blossoms, in order to destroy the newly hatched nymphs; this spraying should be repeated in three or four days; later sprayings are not so effective on account of the protection afforded the insects by the expanded leaves and by their covering of honey-dew.

A monograph of the North American species of this family has been published by Crawford ('14).

Superfamily APHIDOIDEA

The Plant-Lice or Aphids and their Allies

The plant-lice or aphids are well-known insects; they infest nearly all kinds of vegetation in all parts of the country. Our most common examples are minute, soft-bodied, green insects, with long legs and antennæ, which appear on various plants in the house and in the field. Usually, at least, in each species there are both winged and wingless forms (Fig. 487). There are many species of aphids, nearly all of which are of small size; some measure less than 1 mm. in length; and our largest species, only 5 or 6 mm.

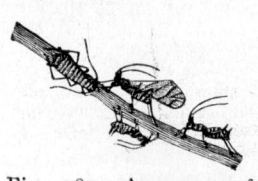

Fig. 487.—A group of aphids.

The body in most species is more or less pear-shaped. The winged forms have two pairs of delicate, transparent wings. These are furnished with a few simple or branched veins; but the venation is more extended than in either of the two following families. The fore wings are larger than the hind wings; and the two wings of each side are connected by a small group of hamuli. The wings are usually held roof-like when at rest (Fig. 488, *ab*), but are laid flat on the abdomen in some genera. The beak is four-jointed and varies greatly in length; in some species it is longer than the body. The antennæ consist of from three to six segments; the last segment is usually provided with a narrowed prolongation (Fig. 488, *aa*). The first two segments of the antennæ are always short, but the other segments show a great specific variation in length and are therefore very useful as systematic characters. Excepting the first two, the segments of the antennæ are usually provided with sense-organs, the *sensoria*, which vary in number and shape in different species and are

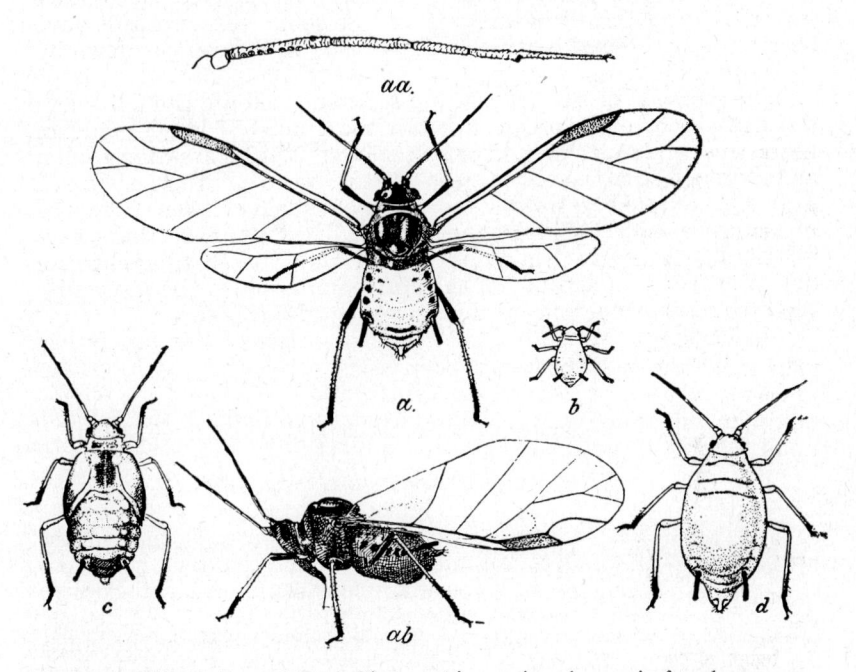

Fig. 488.—The melon aphis, *Aphis gossypi*: *a*, winged agamic female; *aa*, enlarged antenna of same; *ab*, winged agamic female, with wings closed, sucking juice from leaf; *b*, young nymph; *c*, last nymphal instar of winged form; *d*, wingless agamic female. (From Chittenden.)

much used in the classification of these insects. On the back of the sixth abdominal segment there is, in many species, a pair of tubes, the *cornicles*, through which a wax-like material is excreted. In some genera these organs are merely perforated tubercles, while in still other genera they are wanting. It was formerly believed that the honey-dew excreted by aphids came from the cornicles; for this reason they are termed the *honey-tubes* in many of the older books. The honey-dew of aphids is excreted from the posterior end of the alimentary canal. It is sometimes produced in such quantities that it forms a glistening coating on the leaves of the branches below the aphids, and stone walks beneath shade-trees are often densely spotted with it. This honey-dew is fed upon by bees, wasps, and ants. The bees and wasps take the food where they find it, paying little if any attention to its source; but the ants recognize in the plant-lice useful auxiliaries, and often care for them as men care for their herds. This curious relationship is discussed later, under the head of *Ants*.

In addition to honeydew, many aphids excrete a white waxy substance. This may be in the form of powder, scattered over the

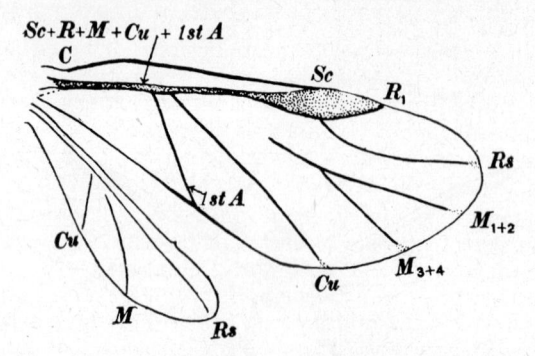

Fig. 489.—The wings of *Eriosoma americana*. (From Patch.)

surface of the body, or it may be in large flocculent or downy masses; every gradation between these forms exists.

The superfamily Aphidoidea includes two families, the Aphididæ and the Phylloxeridæ. These two families differ in the life-histories of their species and in the venation of the wings of the winged forms, as follows:

A. Only the sexually perfect females lay eggs; the parthenogenetic forms give birth to developed young, which, however, in some cases, are each enclosed in a pellicle. The radius of the fore wings is branched; and the outer part of the stigma is bounded behind by vein R_1 (Fig. 489)..............APHIDIDÆ
AA. Both the sexually perfect females and the parthenogenetic forms lay eggs. Vein R_1 of the fore wings is wanting; and the outer part of the stigma is bounded behind by the radial sector (Fig. 490)..............PHYLLOXERIDÆ

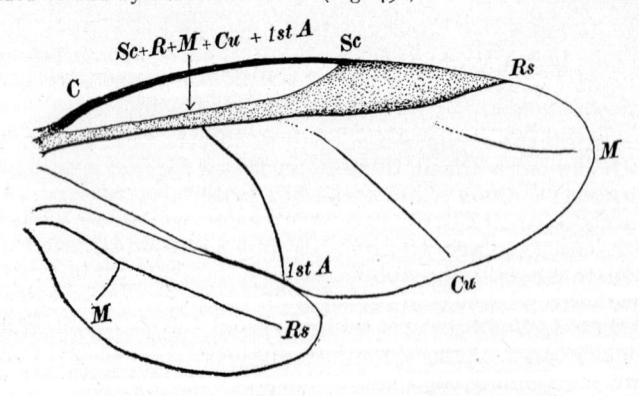

Fig. 490.—The wings of *Adelges*. (From Patch.)

Family APHIDIDÆ

The Typical Aphids

To this family belong the far greater number of the genera and species of the Aphidoidea. The distinctive characters of this family are given under A in the table above. For a detailed discussion of the wing-venation of these insects, see Patch ('09).

In the Aphididæ there exists a remarkable type of development known as *heterŏgamy* or cyclic reproduction. This is characterized by an alternation of parthenogenetic generations with a sexual generation. And within the series of parthenogenetic generations there may be an alternation of winged and wingless forms. In some cases the reproductive cycle is an exceedingly complicated one, and different parts of it occur on different species of food plants.

In those cases where different parts of the reproductive cycle occur on different food-plants, the plant on which the over-wintering fertilized egg is normally deposited and upon which the stem-mother and her immediate progeny develop is termed the *primary host;* and that plant to which the migrants fly and from which a later form in the series migrates to the primary host is known as the *secondary host*.

Different species of aphids differ greatly in the details of their development; it is difficult, therefore, to make generalizations regarding this matter. The following account will serve to indicate the sequence of the forms occurring in the reproductive cycle of a migrating aphid, one in which the different parts of the cycle occur on different food-plants. This account refers to what occurs in the North, where the winter interrupts the production of young, and eggs are developed which continue the life of the species through the inclement season. In hot climates also, where there is a wet and a dry season, eggs are produced to carry the species over the period when succulent food is lacking. And in some cases in the North, on exhausted vegetation the non-migratory species produce eggs during the summer months.

The stem-mother.—In the spring there hatches from an overwintering egg a parthenogenetic, viviparous female, which lives on the primary host. As this female is the stock from which the summer generations spring, she is known as the *stem-mother* or *fundatrix*. The stem-mother is winged in some species of one of the tribes (Callipterini); but usually she is wingless.

The wingless agamic form.—In most species the stem-mother gives birth to young which do not develop wings and which are all females. These reproduce parthenogetically and are known as *the wingless agamic form* or *spuriæ apteræ.** These reproduce their kind for a variable number of generations and then produce the next form. All of these generations live on the primary host. In a few species the wingless agamic form rarely appears if at all.

---

*Spuriæ (New Latin, fem. pl.); Lat. *spurius*, an illegitimate or spurious child.

The winged agamic form.—After a variable number of generations of the wingless agamic form have been developed and the food-plant has become overstocked by them, there appears a generation which becomes winged and which migrates to the secondary host. These are all parthenogenetic, viviparous females. They are known as *the winged agamic form* or *spuriæ alatæ* or *migrants* or *migrantes*. In some species, the second generation, the offspring of the stem-mother, are winged migrants.

When the migrating winged agamic form becomes **established** on the secondary host, it produces young which are all females of the wingless agamic form. After a variable number of **generations of** this form have been developed, there is produced a generation of winged agamic females which migrate from the secondary host to the primary host. The two forms developed on the secondary host, the wingless and the winged agamic forms, may closely resemble the corresponding forms previously developed on the primary host or may differ markedly from them.

The members of the last generation of the series of parthenogenetic forms, which produce the males and the oviparous females, are termed the *sexuparæ*. In some non-migrating species this generation is wingless.

The males and the oviparous females.—The winged agamic females that have migrated from the secondary host to the primary one, here give birth to true sexual forms, male and female. These pair, and each female produces one or more eggs. These are sometimes designated as *gamogenetic eggs* to distinguish them from the so-called *pseudova* developed in agamic females. See note on page 191.

The males and the oviparous females are termed collectively the *sexuales;* and some writers refer to the oviparous females as the *ovipara*. (Note that *ovipara* is a plural noun.)

The sexuales differ greatly in form and habits in the different tribes of aphids. In the more generalized aphids the ovipara of some species are winged, and the males are very commonly winged; both sexes have beaks and feed in the same way as do the other forms; and each female produces several eggs. In some of the more specialized aphids the sexuales are small, wingless, and beakless; consequently they can take no food. Each female produces a single egg, which in some cases is not deposited but remains throughout the winter within the shriveled body of the female.

In some cases the young produced by the agamic females are each enclosed in a pellicle when born; this is soon ruptured and the young aphid escapes from it. The young thus enclosed are termed *pseudova* by many writers.

The foregoing account, omitting exceptions and variations, can be summarized as follows:

A. DIFFERENT TYPES OF INDIVIDUALS IN THE APHIDIDÆ

First type.—The stem-mother or fundatrix, which is hatched from a fertilized egg, is usually wingless, and reproduces parthenogenetically.

Second type.—The parthenogenetically produced wingless agamic females.

Third type.—The parthenogenetically produced winged agamic females.
Fourth type.—The sexual forms, males and oviparous females.

B. SEQUENCE OF GENERATIONS IN A MIGRATING SPECIES

Only the first of a series of similar generations is counted.
First generation.—The stem-mother.
Second generation.—Wingless agamic females. There may be a series of generations of this form here.
Third generation.—Winged agamic females. These migrate to the secondary host.
Fourth generation.—Wingless agamic females. There may be a series of generations of this form here.
Fifth generation.—Winged agamic females. These migrate to the primary host and are the sexuparæ.
Sixth generation.—Males and oviparous females. The females produce the fertilized eggs from which the stem-mothers are hatched, thus completing the life-cycle.

A remarkable fact that has been demonstrated by several observers is that the number of generations of the wingless agamic form may be influenced by the conditions under which the aphids live. In an experiment conducted under my direction by Mr. Slingerland, in the insectary at Cornell University, we reared 98 generations of the wingless agamic form without the appearance of any other form. The experiment was carried on for four years and three months without any apparent change in the fecundity of the aphids, and was discontinued owing to the press of other duties. As the aphids were kept in a hothouse throughout the winters, seasonal influences were practically eliminated; and as members of each generation were placed singly on aphid-free plants and their young removed as soon as born, there was no crowding.

In order to determine the influence of crowding, members of the sixtieth generation were placed on separate plants and their young not removed. At the end of three weeks the winged agamic form appeared, evidently in response to need of migration to less densely populated plants; while in other cages where the young were removed promptly, no migrants appeared up to the end of the experiment.

The family Aphididæ includes a very large number of genera and species. The genera are grouped into tribes and these into subfamilies in various ways by different authors. Recent classifications by American authors are those of Oestlund ('18) and Baker ('20). Four subfamilies are recognized by Baker. The characters of these subfamilies given below are largely compiled from this author.

Subfamily APHIDINÆ

To this subfamily belong most of the species of aphids that are commonly seen living free (*i. e.*, not in galls) upon the foliage of plants. But while most of the species feed on foliage, some of them attack stems and roots. Their attacks on foliage in some cases merely cause a weakening of it; in other cases, the leaves become curled or otherwise distorted; such distortions are termed *pseudogalls*. True galls formed by aphids are described in the accounts of the last two subfamilies.

In the Aphidinæ the males and the oviparous females are comparatively generalized; they are furnished with functioning mouthparts and feed as do the other forms; the females lay several eggs; in a few species the oviparous females are winged; and winged males are common. Wax-glands are not abundant in members of this subfamily; and the antennal sensoria are oval or subcircular.

The following are a few of the more common representatives of the Aphidinæ. These are selected to illustrate some of the more striking differences in habits exhibited by the different species.

a. BARK-FEEDING APHIDINÆ

The following species will serve as an example of the bark-feeding species belonging to this subfamily, and also of the maximum size reached by any aphid.

The giant hickory-aphid, *Longistĭgma cāryæ*.—This is a very large species, one of the largest aphids known, measuring to the tip of the abdomen 6 mm., and more than 10 mm. to the tips of the wings (Fig. 491). It can be distinguished by the shape of the stigma of the fore wings, which is drawn out at the tip to an acute point extending almost to the tip of the wing. The top of the thorax and the veins of the wings are black and there are four rows of little transverse black spots on the back. The body is covered with a bluish white substance like the bloom of a plum. This is a bark-feeding species; it is found clustered on the under side of limbs in summer. It infests hickory, maple, and several other forest trees. The oviparous female is wingless; the male, winged.

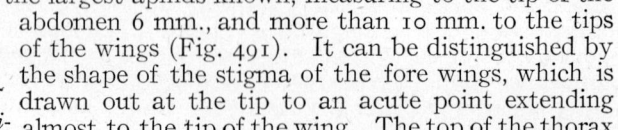

Fig. 491.—*Longistigma caryæ*.

b. LEAF-FEEDING APHIDINÆ

Examples of the leaf-feeding species belonging to this subfamily can be found on a great variety of plants. Among those most easily observed are the species infesting the leaves of fruit trees, and especially the following.

The apple-leaf aphis, *Aphis pomi*.—This is a bright green species, the entire life-cycle of which is passed on the apple. The migrants fly to other parts of the infested tree or to other apple-trees. As a result of the attacks of this species the leaves of the apple are often badly curled and sometimes drop off the tree.

The rosy apple-aphis, *Anūraphis rōseus*.—The common name of this species refers to the fact that the agamic females are usually of a pinkish color; but they may vary in color to a light brown, slaty gray, or greenish black, with the body covered with a whitish coating. This species is most common on apple; but it infests also pear, white thorn, and three species of *Sorbus*. It is a migrating species.

The apple-bud aphis, *Rhopalosīphum prunifōliæ*.—This is the species that most commonly infests the opening apple-buds, often nearly covering them. It also infests pear, plum, quince, and many other plants. It is a migrating species; various species of grain serve as its secondary host.

c. ROOT-FEEDING APHIDINÆ

The corn-root aphis, *Anuraphis maidi-radīcis*.—This is a serious pest of corn throughout the principal corn-growing States, sometimes totally ruining fields of corn. Broom-corn and sorghum are the only other cultivated crops injured by it; but it infests many species of weeds that grow in corn-fields. Our knowledge of this species is largely the result of investigations of Professor S. A. Forbes, who has published several detailed accounts of it in his reports as State Entomologist of Illinois. This author found that this aphid is largely dependent on a small brown ant, the corn-field ant (*Lāsius americānus*), the nests of which are common in corn-fields. The ants store the winter eggs of the aphids in their nests and care for them throughout the winter. In the spring, when the stem-mothers hatch, they are transferred by the ants to the roots of the weeds upon which they feed. As soon as corn-plants are available, the ants transfer the aphids to the roots of the corn, the ants digging burrows along the roots of the corn for this purpose. The ants in return for their labors derive honey-dew from the aphids.

One can understand how these ants that attend aphids that are excreting honey-dew should learn to drive away the enemies of the aphids, as is often done; but is it not wonderful that *Lasius americanus* should recognize the importance of preserving the eggs from which their herds are to develop!

The strawberry-root aphid, *Āphis fŏrbesi*.—The winter eggs of this species are found upon the stems and along the midribs of the green leaves of strawberry plants. The stem-mothers and one or more generations of the offspring feed upon the leaves in the early spring. But a little later in the season the corn-field ant appears and transfers the aphids to the roots of the strawberry, where it cares for them in the same way that in corn-fields it cares for the corn-root aphis. This ant is entirely responsible for the infesting of the roots by the aphids; and it is here that the greatest injury to the plants is done.

Subfamily MINDARINÆ

This subfamily was established by Baker ('20) for the reception of the genus *Mindarus*, which can be distinguished from all other living aphids by the venation of the wings. In this genus the radial sector of the fore wings separates from vein R_1 at the base of the long, narrow stigma (Fig. 492). In all other living aphids the origin of the radial sector is much nearer the tip of the wing; but in many of the fossil aphid wings it is as in *Mindarus*. The males and the oviparous females are small and wingless; but they retain the beak, at least in most individuals, and feed. The female lays several eggs.

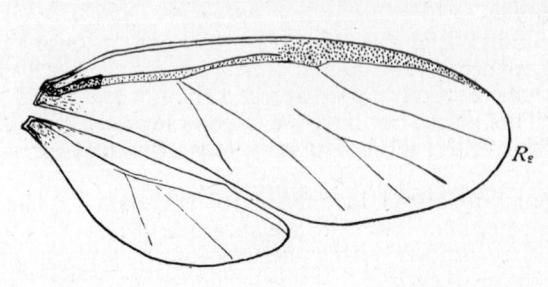

Fig. 492.—Wings of *Mindarus*. (After Patch.)

Only one species, *Mindārus abiĕtinus*, is known. This lives free upon the twigs of spruce and other conifers, which become somewhat distorted and are often killed by the attack of the insects. When disturbed this insect secretes large quantities of honeydew.

The life-cycle of this species usually includes only three generations, the stem-mother, the winged agamic females (*sexuparæ*), and the sexual forms. Sometimes there is a generation of wingless agamic females.

This species was redescribed by Thomas as *Schizoneura pinicola*.

Subfamily ERIOSOMATINÆ

This subfamily includes those genera of aphids in which the males and the oviparous females are greatly specialized by reduction. They do not have functioning mouth-parts; some have a beak when born but lose it at the first molting; in others the beak is vestigial at birth. As they cannot feed, they remain small. Both sexes are wingless. The oviparous females produce each a single egg, which in some species is not laid but remains throughout the winter in the shriveled body of the female.

In this subfamily, the cornicles are much reduced or are wanting; wax-glands are abundantly developed; and the antennal sensoria are prominent. These are often annular.

The members of this subfamily that are most likely to attract attention can be grouped under two heads: *a*, the woolly aphids; and *b*, the gall-making Erisomatinæ. These groups, however, do not represent natural divisions of the subfamily and do not include all members of it. They are merely used for convenience in the present discussion.

a. THE WOOLLY APHIDS

The woolly aphids are the most conspicuous members of the Aphididæ, on account of the abundant, white, waxy excretion that covers colonies of them. The three following species are widely distributed and are common.

The woolly apple aphis, *Eriosōma lanĭgera*.—This plant-louse, on account of its woolly covering and the fact that it is a serious pest of the apple, is known as the woolly apple aphis, although the apple is its secondary host. This insect not only has a complicated series of generations but the life-cycle is subject to variations; its usual course is as follows:

The winter-eggs are deposited in crevices of the bark of elm. From these eggs stem-mothers hatch in the spring and pass to the young leaves, where they produce either the well-known leaf-curl of the elm or, when a group of terminal leaves are affected, what has been termed a rosette, which is a cluster of deformed leaves. Within these pseudogalls the second generation is produced; this consists of wingless agamic females. The offspring of these, the third generation, become winged and migrate from the elm to the apple. Here they produce the fourth generation, the members of which live on the water-shoots or the tender bark of the apple, and are wingless. The fifth generation also consists of wingless agamic females. Some of these develop on the bark of the branches, which apparently ceases to grow at the point of attack but swells into a large ridge about the cluster of plant-lice, leaving them in a sheltered pit; the aphids also frequently congregate in the axils of the leaves and the forks of the branches. Other members of this generation pass to the roots of the tree, where they produce knotty swellings on the fibrous roots. The sixth generation consists, in part, of winged agamic females which migrate from the apple to the elm, where they produce the seventh generation. This generation, the last in the series, consists of the males and oviparous females, both of which are beakless and wingless. These pair and each female produces a single egg, which is found in a crevice of the bark with the remains of the body of the female.

The course of events outlined above may be modified in two ways: first, it is said that the sexual forms are sometimes produced on the apple; and second, some members of the sixth generation do not develop wings and migrate, but are wingless and produce young that hibernate on the apple. This species infests also mountain ash and hawthorn, as secondary hosts.

The elm-feeding generations of this species that cause the leaf-curls and rosettes have been known as *Schizoneura americāna*. And there are also found during the summer aphids on tender elm bark which are believed to belong to this species and which have been described under the name *Schizoneura rilēyi*. In the Pacific Coast States there is another species of aphid that produces leaf curl on elm. This is *Schizoneura ŭlmi*, a European species, which in Europe has been found to migrate to *Ribes*.

The alder-blight, *Procĭphilus tessellātus*.—A woolly aphid that is found in dense masses on the branches of several species of alder is known as the alder-blight. Colonies of this species are easily found in the regions where it occurs, as their covering of flocculent excretion renders them very conspicuous. These colonies are of especial interest, as within them is found the predacious larva of the wanderer butterfly, *Feniseca tarquinius*, which feeds on the aphids.

In the late summer or early autumn the last generation of wingless agamic females bring forth young, which winter among the fallen leaves at the base of the alder and return to the branches in the spring. From this there appears to be no need of an alternate host. But it was found by Dr. Patch that at the same time that the form that hibernates at the base of the alder is produced, winged migrants appear and fly to maple trees, where they give birth, in the crevices and rough places in the bark, to males and oviparous females. Each of these females produces a single egg. From these eggs there hatch in the spring aphids which pass to the lower side of the leaves of the maple, where they become conspicuous on account of their abundant and long woolly excretion. In this period of its existence this species is the well-known pest of the maple that has long been known as *Pemphigus acerfōlii*, which name must now be regarded as a synonym of *Prociphilus tessellātus*, the older name. In July winged migrants are developed on maple which fly to alder.

The alder-blight excretes honeydew abundantly; the result is that the branches infested by this insect, and those beneath the cluster of aphids, become blackened with fungi that grow upon this excretion. There is also a curious fungus which grows in large spongy masses immediately beneath the cluster of plant-lice; this is known to botanists as *Scorias spongiosa*. It is evidently fed by the honeydew that falls upon it.

The beech-tree blight, *Prociphilus imbricātor.*—This infests both twigs and leaves of beech. Like the preceding species it occurs in clusters of individuals, each of which is clothed with a conspicuous downy excretion. These clusters often attract attention by the curious habit which the insects have of waving their bodies up and down when disturbed. When an infested limb is jarred, the aphids emit a shower of honeydew. Owing to the abundance of this excretion, the branches and leaves of an infested tree become blackened by growths of fungi, as with the preceding species. The life-cycle of this species has not been determined.

b. THE GALL-MAKING ERIOSOMATINÆ

Certain members of this subfamily cause the growth of remarkable galls, resembling in this respect certain members of the following subfamily. Among the gall-making Eriosomatinæ that are most likely to attract attention are the following.

The cockscomb elm-gall colopha, *Cŏlopha ulmĭcola.*—There are two species of aphids that make similar galls on the leaves of elm. These galls are commonly known as cockscomb elm-galls on account of their shape. Those made by the two species of aphids are so similar that a description of one will apply to the other. In each case the gall is an excrescence resembling a cock's comb in form, which rises abruptly from the upper surface of the leaf (Fig. 493, a). It is compressed, and has its sides wrinkled perpendicularly and its summit irregularly gashed and toothed. It opens on the under side of the leaf by a long slit-like orifice.

The winter eggs can be found during the winter in the crevices of the bark of the elm; each egg is usually enclosed in the dry skin of the oviparous female (Fig. 493, b). In the spring the stem-mothers

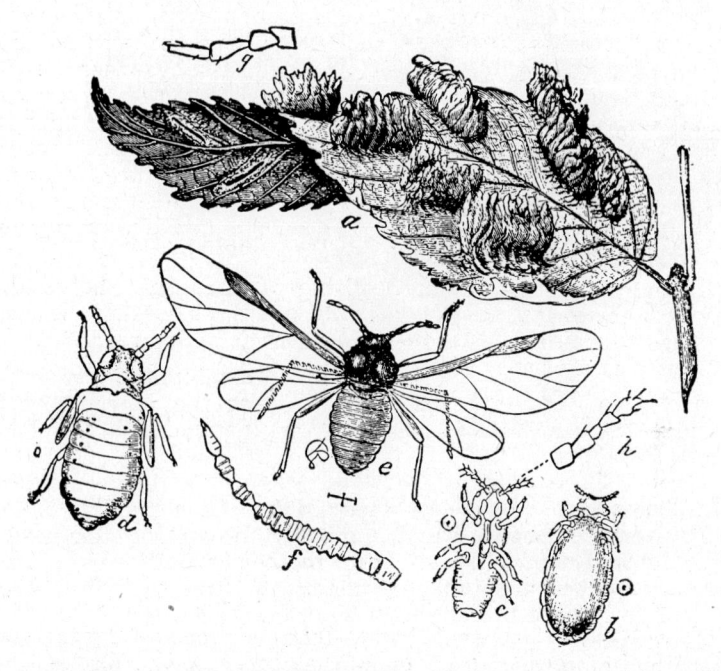

Fig. 493.—*Colopha ulmicola: a*, leaf showing galls from above and beneath; *b*, fertilized egg surrounded by the skin of the female; *c*, newly born young of the second generation; *h*, its antenna; *d*, full-grown nymph of the second generation; *e*, adult of second generation; *f*, antenna of migrant; *g*, antenna of stem-mother. (From Riley.)

pass to the leaves and each causes by its attack the growth of a gall. The second generation is produced within the gall; it consists of winged agamic females (Fig. 493, *e*). These migrants can be distinguished from those of the other cockscomb elm-gall aphid by the fact that in this species vein M of the fore wings is forked.

The migrants of this species pass from the elm to certain grasses, among them species of *Eragrostis* and *Panicum*. The forms found on these secondary hosts have been described under the name *Cŏlopha eragrŏstidis*, but this is a much later name than *Cŏlopha ulmĭcola*.

The cockscomb elm-gall tetraneura, *Tetraneura gramīnis.*—The life-cycle of this species is quite similar to that of the preceding one. The primary host is elm. The stem-mothers cause the growth of cockscomb-like galls; and the migrants produced in these galls pass to grasses. These migrants differ from those of the preceding species in that vein M of the fore wings is not forked. This species was first described from individuals found on the secondary hosts and was named *Tetraneura gramīnis.* Later, forms found on elms were named *Tetraneura colophoides.*

For a detailed account of the gall-aphids of the elm, see Patch ('10).

The poplar-leaf gall-aphid, *Thecābius populicaulis.*—This aphid is common on several species of poplar. It makes a swelling the size of a small marble on the leaf at the junction of the petiole with the blade. This gall is of a reddish tint, and has on one side

Fig. 494.—The vagabond poplar-gall. (From Walsh and Riley.)

a slit-like opening. In the early part of the season each gall is occupied by a single wingless female, probably the agamic stem-mother, which by midsummer becomes the mother of numerous progeny. These are winged and probably migrate to some other host-plant; but the life-cycle of this species has not been determined.

The vagabond gall-aphid, *Mordwilkōja vagabŭnda.*—This species infests the tips of the twigs of several species of poplar; here it causes the growth of large corrugated galls, which resemble somewhat the flower of the double cockscomb of our gardens. The galls are at first bright green, but later turn black, become woody, and remain on the trees during the winter (Fig. 494). Very little is known regarding the life-cycle of this species.

Subfamily HORMAPHIDINÆ

The members of this subfamily are usually gall-makers, resembling in this respect certain members of the Eriosomatinæ, and also resembling them in that the antennal sensoria are annular. But in this subfamily the sexual forms are not so specialized by reduction as in the preceding one. In the Hormaphidinæ, although the males and the oviparous females are small and wingless, they possess beaks, they feed, and the oviparous female lays more than one egg. In this subfamily great specialization of wax-producing organs occurs. In many species some of the agamic generations become greatly modified in form so that they do not resemble the more typical aphids. In some species these modified forms have the appearance of an *Aleyrodes;* in other species, that of a coccid.

Our best-known representatives of this subfamily are two species of gall-makers, each of which infests alternately witch-hazel and birch. The life-histories of these were very carefully worked out by

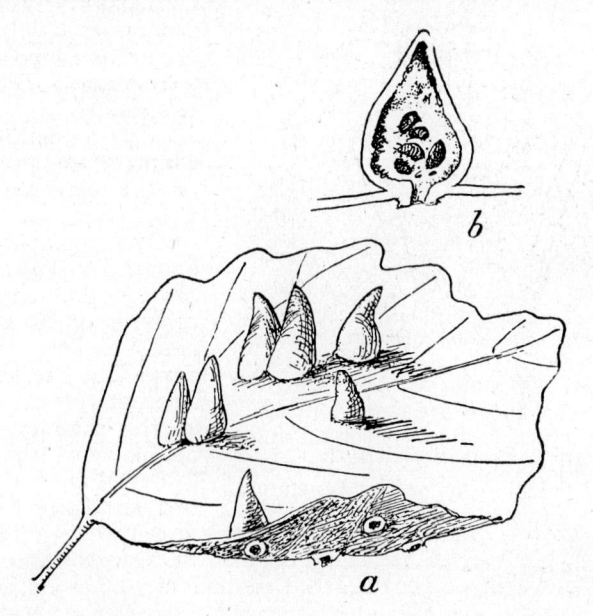

Fig. 495.—The witch-hazel cone-gall: *a*, natural size; *b*, section of gall, enlarged. (From Pergande.)

Pergande ('01); the following accounts are greatly condensed from that author.

The witch-hazel cone-gall aphid, *Hŏrmaphis hamamĕlidis*.—The winter-egg is deposited on the branches and twigs of witch-hazel and hatches early in the spring. The stem-mother, which hatches from this egg, attacks the lower surface of the leaf, causing the growth of a conical gall on the upper surface of the leaf with a mouth on the lower surface (Fig. 495). The second generation, the offspring of the stem-mother, consists of many individuals; these are produced within the gall, which becomes crowded with them. These are agamic females, which become winged, leave the gall, and migrate to birches, where they deposit their young on the lower side of the leaves. The first instar of the third generation, the offspring of the migrants, is broadly oval, with the entire margin of the body studded with short and stout excretory tubercles (Fig. 496); from each of these there issues a short, glassy, beautifully iridescent, waxy rod. The second and third instars of this generation are marked by a reduction of the antennæ, beak, and legs. The fourth instar, which is found about the middle of June, is aleyrodiform (Fig. 497). The fourth and fifth generations resemble the third, there being three aleyrodiform generations. The members of the sixth generation become winged and are the return migrants. These fly to witch-hazel, where they give birth to the seventh generation, which consists of males and oviparous females. These pair and the females lay the winter eggs; each female produces from five to ten eggs. The males and the oviparous females are both wingless. In this species the antennæ of the winged forms are three-jointed.

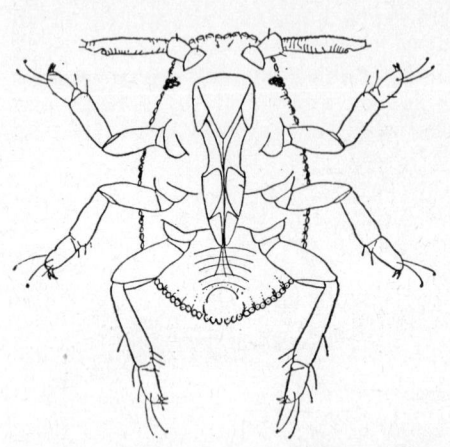

Fig. 496.—*Hormaphis hamamelidis*, first instar of the third generation. (From Pergande.)

Later experiments by Morgan and Shull ('10) indicate that this species can complete its life-cycle on the witch-hazel. According to these authors there are only three generations: first, the stem-mother, which causes the growth of the cone-gall; second, the winged forms, which are developed in the gall and which spread to the leaves; and third, the males and oviparous females. No aleyrodiform individuals were found on the witch-hazel.

The spiny witch-hazel-gall aphid, *Hamamelistes spinōsus*.—The winter eggs of this species are commonly deposited near the flower-buds of witch-hazel, late in June or early in July, but they do not hatch till May or June of the following year. The stem-mother attacks the flower-bud, which becomes transformed into a large gall of the form shown in Figure 498. Within this gall the stem-mother produces the second generation; these crowd the gall and develop into winged migrants, which leave the gall, from July to late fall, and fly to birches. The

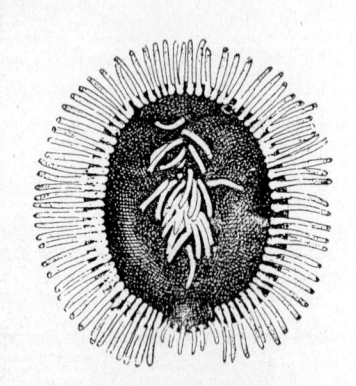

Fig. 497.—*Hormaphis hamamelidis*, fourth instar of the third generation. (After Pergande.)

young of the migrants, the third generation, feed a short time and then settle close to the leaf-buds, where they hibernate; the last in-

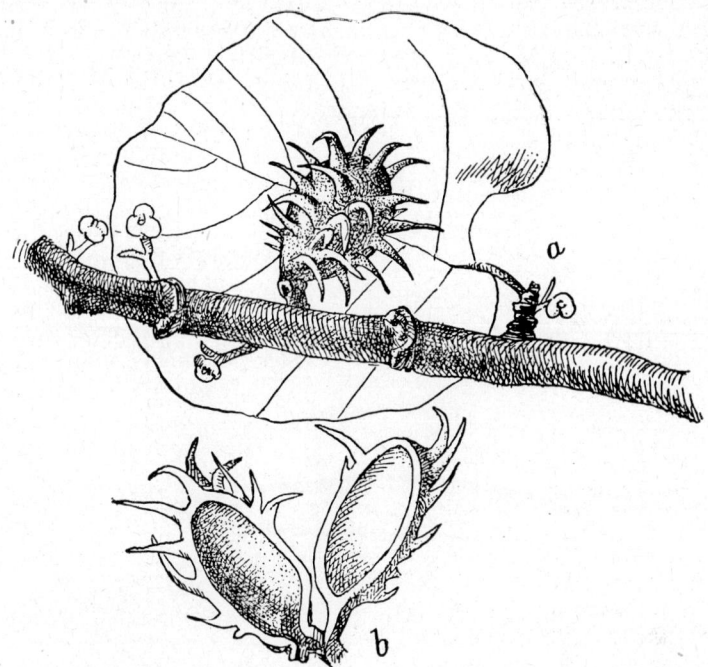

Fig. 498.—The spiny witch-hazel gall: *a*, mature gall; *b*, section of gall. (From Pergande.)

star of this generation resembles a coccid (Fig. 499). The fourth generation is produced early in the spring; the young of this generation move to the young and tender leaves of the birch, which, as a

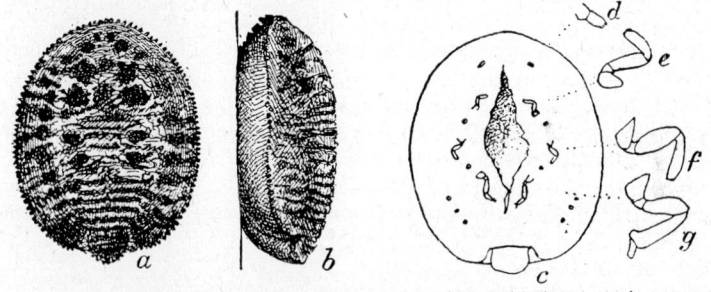

Fig. 499.—*Hamamelistes spinosus*, last instar of the third generation, much enlarged: *a*, dorsal view; *b*, lateral view; *c*, ventral view; *d*, antenna; *e*, *f*, *g*, legs. (From Pergande.)

result of the attack, become corrugated, the upper surface bulging out between the veins, and the folds closing up below. In these pseudogalls the fifth generation is produced; the members of this generation become winged and migrate to witch-hazel in early summer, where they produce the seventh and last generation of the series, the males and oviparous females. These pair and the females soon lay their eggs. Both sexes are wingless. The winged migrants of this species can be distinguished from those of the preceding species by their five-jointed antennæ

Family PHYLLOXERIDÆ

The Adelgids and the Phylloxerids

The members of this family differ from the typical aphids in that both the sexually perfect females and the parthenogenetic forms lay eggs, in lacking vein R_1 of the fore wings, and in that the outer part of the stigma is bounded behind by the radial sector (Fig. 500).

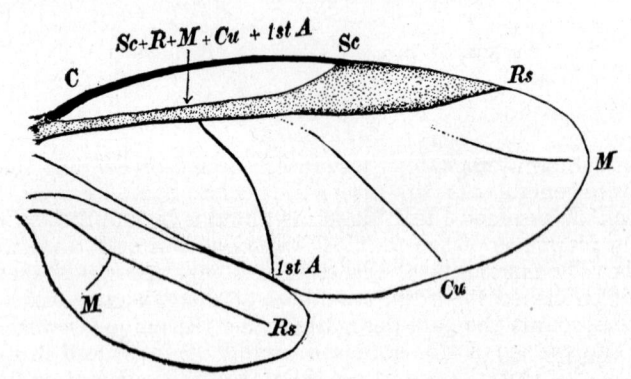

Fig. 500.—Wings of *Adelges*. (From Patch.)

In this family the cornicles are always wanting; and the males and sexually perfect females are dwarfed and wingless.

This family includes two subfamilies, which can be separated by the following table. These subfamilies are regarded as distinct families by some writers.

   A. The wingless agamic females excrete a waxy flocculence. The winged forms have five-jointed antennæ, the last three segments of which bear each a single sensorium. The wings are held roof-like when at rest. The free part of vein Cu of the fore wings is separate from vein 1st A (Fig. 500). The sexual forms have a beak. The alimentary canal is normal, producing a fluid excrement. The species infest only conifers..................................Adelginæ

   AA. The wingless agamic females do not secrete a waxy flocculence, but in the genus *Phylloxerina* they excrete a waxy powder. The winged forms have three-jointed antennæ, the second segment of which bears two sensoria. The wings when at rest are laid flat upon the abdomen. The free parts of veins Cu

and 1st A of the fore wings coalesce at base (Fig. 501). The sexual forms have no beak. The anus is closed. The species do not infest conifers........
....PHYLLOXERINÆ

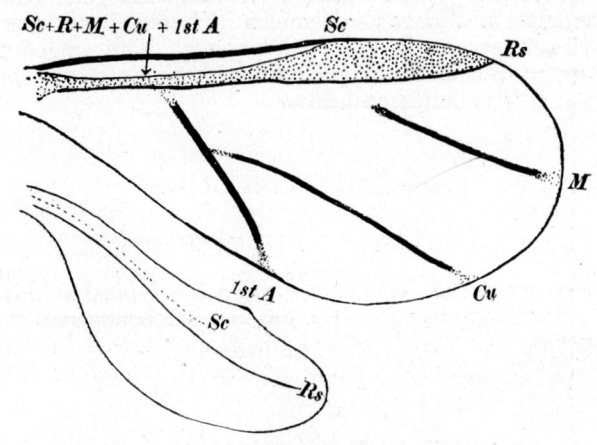

Fig. 501.—Wings of *Phylloxera*. (From Patch.)

Subfamily ADELGINÆ
The Adelgids

This subfamily includes those insects found on conifers that have been quite generally known under the generic name *Chermes*. But it has been determined that this name should be applied to certain jumping plant-lice of the family Chermidæ, formerly known as the Psyllidæ. The necessity of this change is very unfortunate, as much has been published regarding members of the Adelginæ and in most of these accounts they are described under the name *Chermes*.

All the species of this subfamily infest conifers; and in all cases in which the sexual generation is known, this generation lives on spruce. The secondary host may be either larch, pine, or fir.

Much has been written regarding the life-histories of these insects. It has been found that what may be regarded as the typical life-cycle of an *Adĕlges* or "*Chermes*" is a very complex one, including the developing of two parallel series of forms differing in habits; that in one of these series a single host-plant, spruce, is infested and the life-cycle is completed in one year; while in the other series the life-cycle extends over two years and is passed in part upon spruce and in part upon larch or some other host-plant.

In this typical life-cycle, beginning with the individual that hatches from a fertilized egg, there are developed five generations, the members of which differ in either form or habits or both from those of the other generations, before the cycle is completed by the production again of fertilized eggs. The actual number of generations may be greater than this, owing to the fact that in a part of the cycle there may be a series of similar generations only the first of which is counted in this enumeration.

This indicating of a typical life-cycle is an effort to outline as simply as possible the life-history of these insects. In some species it is much more complicated; thus, for example, Borner ('08) in his account of the life-history of *Cnaphalōdes strobilōbius* recognizes seven parallel series of forms.

The distinctive characters of the five differing generations in the typical life-cycle are indicated below.

A. GENERATIONS ON SPRUCE (*Picea*)

A one-year cycle or the first year of a two-year cycle.

First generation.—This consists of the true stem-mother (*fundatrix vera*), a wingless agamic female. In the case of those supposed parthenogenetic species which do not migrate to another host-plant and which complete their life-cycle in one year, this form is the offspring of the second generation, an agamic form; in the case of species that migrate to a secondary host-plant, and where there are two parallel series, the stem-mother is the offspring of either the second generation or the fifth generation, the sexual forms.

The stem-mothers hatch in the autumn; they hibernate immature in crevices at the bases of buds, complete their growth in the spring, and by their attack upon the buds cause the beginning of the growth of galls. Each stem-mother lays a large number of eggs.

Second generation.—The members of this generation hatch from the eggs laid by the stem-mothers, and by their attack upon the buds cause the completion of the growth of the galls. The galls are formed by the hypertrophy and coalescence of the spruce-needles. The members of this generation have been termed the *gallicolæ*, because they inhabit the galls. They reach the last nymphal instar within the galls. When this stage is reached, the galls open and the nymphs emerge and soon molt, becoming winged agamic females.

As to their habits, there are two types of gallicolæ: first, the non-migrants, which remain on the spruce and lay the eggs from which the stem-mothers of the one-year cycle are hatched; and second, the migrants, which fly to a secondary host-plant, which is not spruce, and where they lay many eggs, but not so many as are laid by the stem-mothers.

B. GENERATIONS ON A SECONDARY HOST

Part of the second year of a two-year cycle.

The secondary host may be a species of either larch (*Larix*), pine (*Pinus*), or fir (*Abies*); but no galls are produced on any of these.

Third generation.—The members of this generation hatch from eggs laid by migrants of the second generation that have flown from spruce to larch or other secondary host and laid their eggs there. The young that hatch from these eggs hibernate in crevices in the bark and complete their growth in the spring, becoming wingless agamic females. The members of this generation and of similar generations which follow immediately but which are not numbered here, are termed *colonici*, because they are settlers in a new region, or *exsules*, that is, exiles. Some writers term the first of this series of generations false stem-mothers (*fundatrices spuriæ*) to distinguish them from the true stem-mother, which is the beginning of the two-year cycle. The members of the third generation resemble those of the first generation, but usually lay fewer eggs and do not cause the growth of galls.

The offspring of the third generation are all wingless agamic females, which reproduce their kind. Of these there may be a series of generations, which are not numbered in this generalized statement; and there may be among these several parallel series of generations, differing in the life-cycle but all reproducing parthenogenetically on the secondary host. The secondary host may be thus infested throughout the year; while the primary host, if there is not an annual series, will be free during the interval between the migration of the second generation and the return migration of the fourth generation.

Among the offspring of the third generation two types are recognized by Marchal ('13): first, nymphs which remain undeveloped for a time, the *sistens* type; and second, nymphs which develop at once into wingless agamic females, the *progrediens* type.*

Fourth generation.—The members of this generation are produced by individuals of the progrediens type of the third generation. They develop into winged agamic females. The adults migrate to spruce and there lay a small number of eggs. Since their offspring are the sexual forms, this generation is known as the *sexuparæ*.

C. A GENERATION ON SPRUCE

The completion of the second year of a two-year cycle.

Fifth generation.—From eggs laid by the sexuparæ that have migrated from the secondary host to spruce, there are developed males and sexually perfect females, termed the *sexuales;* both of these forms are wingless. They pair and each female lays a single egg. These eggs hatch in the autumn; the young hibernate and become the true stem-mothers. Thus is completed the two-year life-cycle.

Omitting the annual series, the typical two-year life-cycle includes the following series of generations, which are described above.

First.—The wingless agamic stem-mother.
Second.—The winged agamic migrants.
Third.—The wingless agamic colonici or exsules.
   (a) The sistentes, several generations.
   (b) The progredientes, several generations.
Fourth.—The winged agamic sexuparæ.
Fifth.—The wingless sexuales, males and sexually perfect females. Each female produces a single fertilized egg, from which hatches a stem-mother, thus completing the life-cycle.

In the case of some species, which have been studied very carefully by different observers, only an annual series, consisting of the first and second generations described above, is known. It should be noted that in a life-cycle of this kind there are no sexual forms and that although a winged form appears it is not known to migrate. These facts indicate that either some members of the winged generation migrate to a secondary host-plant which has not been discovered, or that the species in question have become, by adaptation, purely parthenogenetic. Which of these alternatives is true has been much discussed.

The following species are some of the more common of our representatives of this subfamily.

The pine-leaf adelges, *Adĕlges pinifōliæ.*—Our knowledge of the life-history of this species is still fragmentary. In one part of its life-cycle it infests the leaves of white pine (*Pinus strobus*). The generations found here are winged agamic females. These attach themselves firmly to the pine-needles, each with its head directed towards the base of the needle. Within each there are developed about one hundred eggs, which are not extruded. After the death of the female, the mass of eggs remains adhering to the leaf, covered over and

---

*Sistens, Latin *sisto*, to stand; progrediens, Latin *pro*, forth, *gradior*, to go.

protected by the remains of the body and closed wings of the dead insect.

It has been determined that these plant-lice infesting the pine leaves are specifically identical with those that issue from a cone-like gall found on several species of spruce (Fig. 502). The spruce-inhabiting form has been known as *Chermes abieticolens;* but *pinifoliæ* is the older specific name and should be used for all forms of this species. It is probable that this species has a two-year life-cycle and that spruce is its primary host and pine its secondary host.

Fig. 502.—Gall of *Adelges pinifoliæ* on spruce.

The green-winged adelges, *Adĕlges abĭetis.*—This species causes the growth of pine-apple-shaped galls on several species of spruce (Fig. 503). It is a European species and its life-history has been the subject of much controversy. It is held by Borner ('08) that it has a typical life-cycle in which there are two parallel series: first, an annual series on spruce alone; and second, a two-year series in which larch is used as a secondary host. On the other hand, Cholodkovsky ('15) maintains that it is a parthenogenetic species; that its life-cycle includes only two generations, the agamic hibernating stem-mothers and the gallicolæ; and that the form with a typical life-cycle is a distinct species (*Chĕrmes vĭridis*). Dr. Patch ('09) has studied *Adelges abietis* in Maine and has found only the parthenogenetic forms, the hibernating stem-mothers and the gallicolæ; thus confirming the conclusion that it may have become a parthenogenetic species.

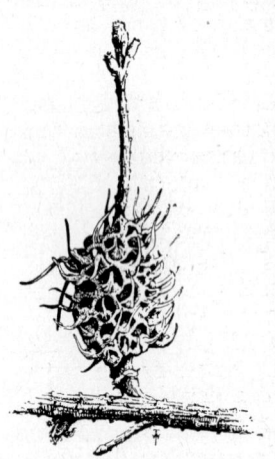

Fig. 503.—Gall of *Adelges abietis.*

The pine-bark adelges, *Adĕlges pinicŏrticis.*—This species infests several species of pine, but especially white pine. The trunks and larger limbs of the infested trees often appear as if whitewashed; this is due to the woolly excretion which covers the bodies of the insects. But little is known regarding the life-cycle of the species. Wingless females, which are doubtless agamic as they lay many eggs, hibernate on the pine and feed on the bark in the spring. They lay their eggs in April; these soon hatch and the young develop into winged agamic females in May. These soon disappear and the pine is said to be free from the pest during the summer. Return migrants to the pine have not been observed; but there must be a generation of these, the parents of the wingless hibernating generation, if, as stated, the pines are free from the pest during the summer.

Subfamily PHYLLOXERINÆ
The Phylloxerids

The distinguishing characters of this subfamily are given under AA in the table on page 428 and need not be repeated here. It includes two genera, *Phylloxera* and *Phylloxerina*.

The genus *Phylloxerīna* is distinguished by the fact that the wingless agamic females excrete a waxy powder, which gives them the appearance of mealy-bugs. Species of this genus have been found in this country on poplar, willow, and sour-gum.

The genus *Phylloxēra* is represented by the grape Phylloxera and thirty or more described species that infest forest-trees—hickory, oak, and chestnut. Most of these are found on hickory. Those on hickory cause the growth of galls either on the leaves or on the tender twigs and petioles. Other species produce either pseudogalls or white or yellowish circular spots on the infested leaves. The species infesting forest-trees were monographed by Pergande ('04).

Although in this subfamily there is a generation of winged migrants in the life-cycle of each species, few if any of them have a secondary host. The migrants fly to other parts of the infested plant or to other plants of the same species.

So far as is known, the life-cycle of the species infesting forest-trees is a comparatively simple one. The stem-mother hatches in the spring from an over-wintering, fertilized egg and causes the growth of a gall; she develops within the gall and produces unfertilized eggs. From these eggs hatch young that develop into winged agamic females. These produce eggs of two sizes; from the smaller eggs hatch males; and from the larger ones, females. The sexes pair and each female lays a single fertilized egg. In some species these eggs are laid in June and do not hatch till the following April.

The grape Phylloxera, *Phylloxēra vastātrix*.—From an economic standpoint this species is the most important member of the Phylloxerinæ; millions of acres of vineyards have been destroyed by it.* The most extensive ravages of this pest have occurred in France and in California. This species is a native of the eastern United States, where it infests various species of wild grapes. It does not injure these seriously; but when it was introduced into France it was found that the European grape, *Vitis vinifera*, is extremely susceptible to its attack. The great injury to the vineyards of California is due to the fact that it is the European grape that is chiefly grown there.

The presence of this insect is manifested by the infested vines in two ways: first, in the case of certain species of grapes, there

---

*"The Phylloxera when at its worst had destroyed in France some 2,500,000 acres of vineyards, representing an annual loss in wine products of the value of 150,000,000." (Marlatt '98.)

appear upon the lower surface of the leaves galls, which are more or less wrinkled and hairy (Fig. 504), which open upon the upper surface of the leaf, and each of which contains a wingless, agamic plant-louse and her eggs; second, when the fibrous roots of a sickly vine are examined, we find, if the disease is due to this insect, that the minute fibers have become swollen and knotty; or, if the disease is far advanced, they may be entirely decayed (Fig. 505, c). Upon these root-swellings there may be found wingless, agamic, egg-laying plant-lice, the authors of the mischief.

Fig. 504.—Leaf of grape with galls of *Phylloxera*. (From Riley.)

The life-history of this species is a complicated one, due to the fact that parallel series of generations with different life-cycles may be developed at the same time. While a fertilized winter egg may be considered a part of the typical life-cycle, some of the agamic females hibernate on the roots of the vine and form a part of a series of agamic generations that apparently may continue indefinitely year after year.

The typical life-cycle, that one in which males and sexually perfect females form a part, extends over two years and includes four forms as follows:

The gallicolæ.—From an over-wintering fertilized egg, there hatches in the spring a wingless agamic stem-mother, which passes to a leaf and by her attack causes the growth of a gall, within which she passes the remainder of her life. She reaches maturity in about fifteen days, fills the gall with eggs, and soon dies. The young that hatch from the eggs laid by the stem-mother resemble her in being wingless agamic females; they escape from the gall, spread over the leaves, and in turn cause the growth of galls. Six or seven generations of this form (Fig. 506) are developed during the summer. They are termed the *gallicolæ*.

The radicicolæ or colonici.—On the appearance of cold weather, young hatched from eggs laid by the gall-inhabiting form pass down the vines to the roots, where they hibernate. This completes the first year of the two-year cycle. In the following spring these colonici, that is, settlers in a new region, attack the fibrous roots, and cause the growth of knotty swellings on them (Fig. 505, b, c) and ultimately their destruction. This is the most serious injury to the vine caused by this species. There is a series of generations of the root-inhabiting form all of which are wingless agamic females. This form (Fig. 507) differs somewhat in appearance from the gallicolæ.

The migrants or sexuparæ.— During the late summer and fall there are hatched from eggs laid by some individuals of the root-inhabiting

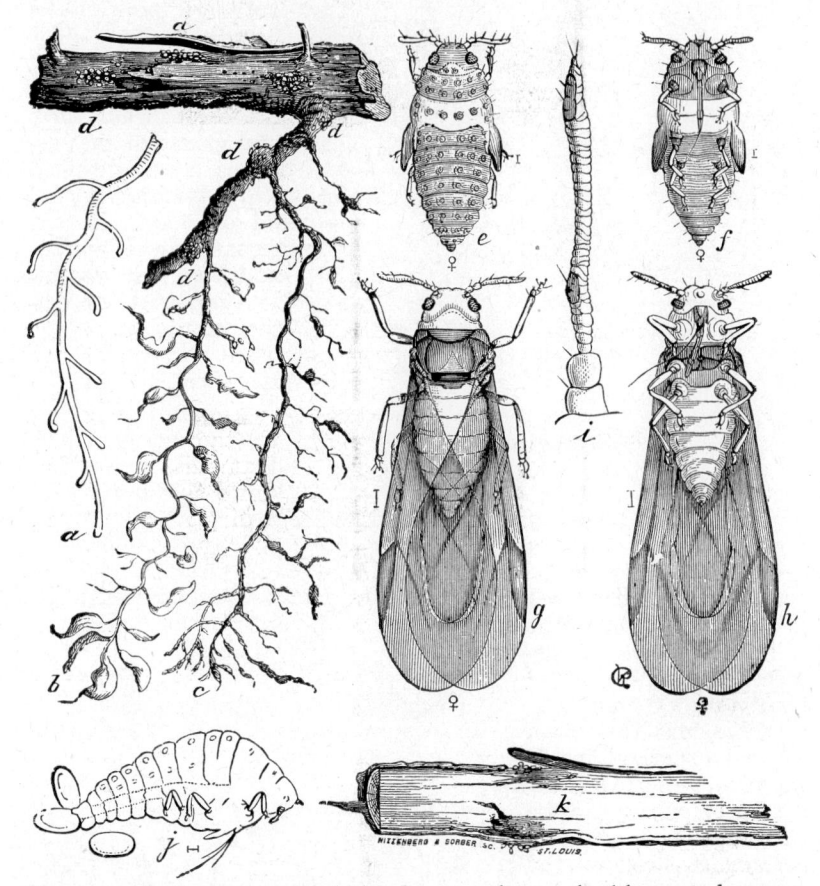

Fig. 505.—*Phylloxera*, root-inhabiting form: *a*, shows a healthy root; *b*, one on which the lice are working, representing the knots and swellings caused by their punctures; *c*, a root that has been deserted by them, and where the rootlets have commenced to decay; *d, d, d*, show how the lice are found on the larger roots; *e*, agamic female nymph, dorsal view; *f*, same, ventral view; *g*, winged agamic female, dorsal view; *h*, same, ventral view; *i*, magnified antenna of winged insect; *j*, side view of the wingless agamic female, laying eggs on roots; *k*, shows how the punctures of the lice cause the large roots to rot. (From Riley.)

form, young that develop into winged agamic females (Fig. 505, *g, h*). These come forth from the ground, fly to neighboring vines, and lay eggs in cracks in the bark or under loose bark. They lay only a few eggs, from three to eight.

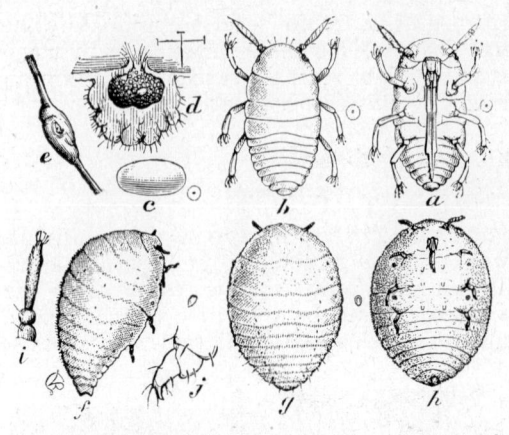

Fig. 506.—*Phylloxera*, gall-inhabiting form: *a*, *b*, newly hatched nymph, ventral and dorsal views; *c*, egg; *d*, section of gall; *e*, swelling of tendril; *f*, *g*, *h*, mother gall-louse, lateral, dorsal, and ventral views; *i*, her antenna; *j*, her two-jointed tarsus. Natural sizes indicated at sides. (From Riley.)

The sexuales.—The eggs laid by the winged migrants are of two sizes; from the smaller eggs there hatch males; and from the larger eggs, sexually perfect females. These pair and each female produces a single egg, which is laid in the fall on old wood. Here it remains over winter, and from it in the following spring a stem-mother is hatched. This completes the two-year life-cycle.

Control.—Owing to the great injury that this species has done to vineyards, hundreds of memoirs have been published regarding it; but, as yet, no satisfactory means of destroying it that can be generally used has been discovered. Where the soil conditions are favorable it can be destroyed by the use of carbon-bi-sulphide, but this is an expensive method; where the vineyards are so situated that they can be submerged with water at certain seasons of the year, the insect can be drowned; and it has been found that vines growing in very sandy soil are less liable to be seriously injured by this pest.

While it is usually impracticable to destroy this pest in an infested vineyard, there is a preventative measure that has given good results.

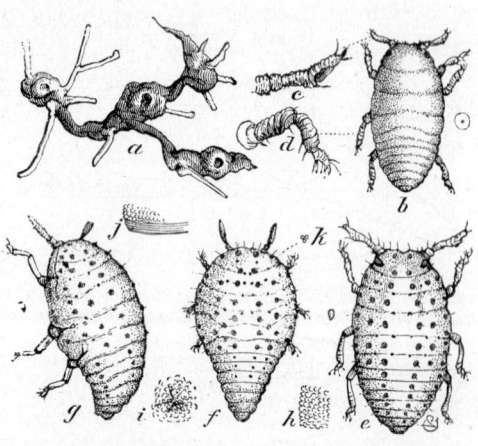

Fig. 507.—*Phylloxera*, root-inhabiting form: *a*, roots of Clinton vine showing the relation of swellings to leaf-galls, and power of resisting decomposition; *b*, nymph as it appears when hibernating; *c*, *d*, antenna and leg of same; *e*, *f*, *g*, forms of more mature lice; *h*, granulations of skin; *i*, tubercle; *j*, transverse folds at border of joints; *k*, simple eyes. (From Riley.)

Certain varieties of American grapes are not seriously injured by the root-form of the Phylloxera. By growing these varieties, or by using the roots of them as stocks on which to graft the susceptible European varieties, the danger of injury by this pest is greatly reduced.

Family ALEYRODIDÆ

The Aleyrodids or White Flies

The members of this family are small or minute insects; our more common species have a wing-expanse of about 3 mm. In the adult state both sexes have four wings, differing in this respect from the Coccidæ, with which they were classed by the early entomologists. The wings are transparent, white, clouded or mottled with spots or bands. The wings, and the body as well, are covered with a whitish powder. It is this character that suggested the name of the typical genus,* and the common name *white flies*.

In the immature stages, these insects are scale-like in form and often resemble somewhat certain species of the genus *Lecanium* of the family Coccidæ. Except during the first stadium, the larvæ remain quiescent upon the leaves of the infested plant and in most species are surrounded or covered by a waxy excretion. In Figure 508 there is represented one of the many forms of this excretion. Here it consists of parallel fibers, which radiate from the margin of the body, and its white color contrasts strongly with the dark color of the insect. In some species the fringe of excretion is wanting; and in others, the excretion from the margin of the body, instead of extending laterally and forming a fringe, is directed toward the leaf upon which the insect rests, and

Fig. 508.—An aleyrodid

Fig. 509.—*Aleurochiton forbesii*.

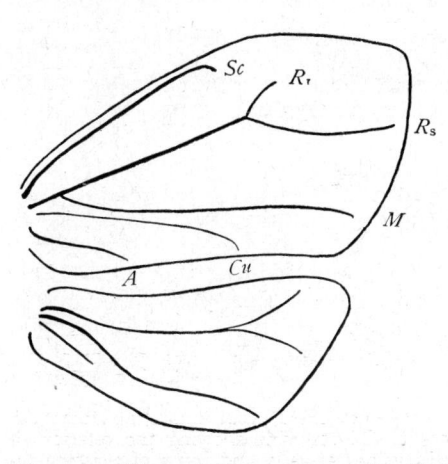

Fig. 510.—Wings of *Udamoselis*. (After Enderlein, with changed lettering.)

---

*Aleyrodes (αλευρώδης), like flour.

thus the body is lifted away from the leaf and is perched upon an exquisite palisade of white wax (Fig. 509).

The members of this family feed exclusively on the leaves of the host-plants. With few exceptions they are not of economic importance; and also with few exceptions, the injurious species are not widely distributed over the world as are many aphids and coccids. This is probably due to the fact that as they live exclusively on leaves they are not so liable to be transported on cuttings and nursery stock. They are most abundant in tropical and semi-tropical regions.

The adults present the following characters: The compound eyes are usually constricted in the middle and in some species each eye is completely divided. In some cases the facets of the two parts of a divided eye are different in size; it is probable that in such cases one part is a day-eye and the other part a night-eye (see page 144). The ocelli are two in number; each ocellus is situated near the anterior margin of a compound eye. The antennæ are usually seven-jointed. The labium is composed of three segments. The fore wings are larger than the hind wings; when at rest the wings are carried nearly horizontally. The venation of the wings is greatly reduced; the maximum number of wing-veins found in the family is in the fore wings of the genus *Udamŏselis* (Fig. 510). The three pairs of legs are similar in form; the tarsi are two-jointed; and each tarsus is furnished with a pair of claws and an empodium or paronychium. The anus opens on the dorsal wall of the abdomen at some distance from the caudal end of the body and within a tubular structure, which is termed the *vasiform orifice*. A tongue-like organ, the *lingula*, projects from the vasiform orifice; and at the base of the lingula there is a broad plate, the *operculum;* the anus opens beneath these two organs.

In this family the type of metamorphosis corresponds quite closely with that known as complete metamorphosis; consequently the term *larva* is applied to the immature instars except the last, which is designated the *pupa*.

The eggs are elongate-oval in shape and are stalked. The larvæ during the first stadium are active, after which they remain quiescent. There are four larval and one pupal instars. The wings arise as histoblasts in the late embryo, and the growth of the wing-buds during the larval stadia takes place inside the body-wall. The change to the pupal instar, in which the wing-buds are external, takes place beneath the last larval skin, which is known as the pupa-case or puparium. In many descriptions of these insects only three larval instars are recognized, the fourth being described as the pupa. As the change to a pupa takes place beneath the last larval skin, the puparium, and as the adult emerges through a T-shaped opening in the dorsal wall of the puparium, the pupa itself is rarely observed.

Parthenogenesis occurs in this family; but according to the observations of Morrill, unfertilized eggs produce only males.

As with the adults, the anus of the immature forms opens in a vasiform orifice on the dorsal aspect of the body at some distance from the caudal end of the body. The excrement is in the form of honey-dew, of which much is excreted.

Formerly all the members of this family were included in a single genus, *Aleyrōdes;* consequently, except in comparatively recent works, the various species are described under this generic name. In later days, very extended studies have been made of the family; and the

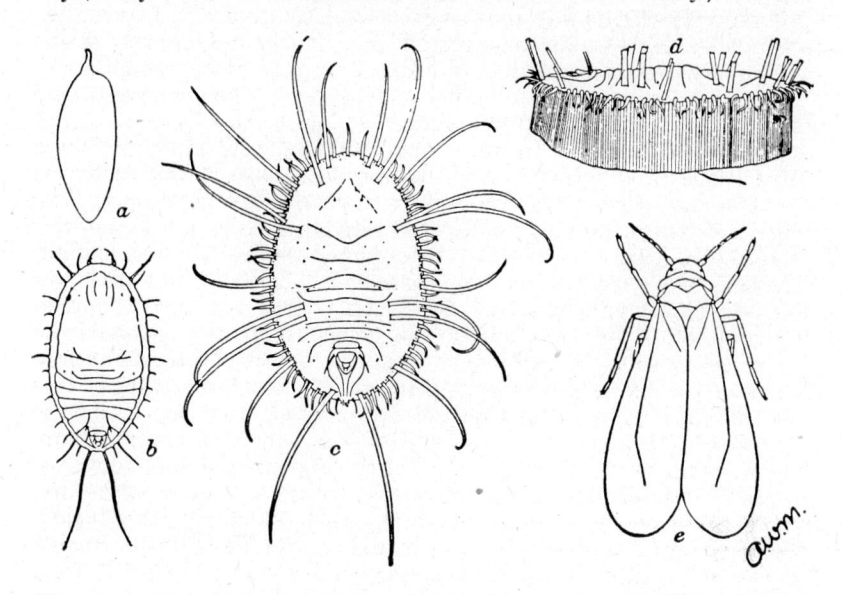

Fig. 511.—*Asterochiton vaporariorum: a,* egg; *b,* larva, first instar; *c,* puparium, dorsal view; *d,* puparium, lateral view; *e,* adult. (After Morrill.)

genus *Aleyrodes* has been divided into many genera, which are now grouped into three subfamilies. The most complete systematic works on the family are those of Quaintance and Baker ('13 and '17). The following species are among our more common representatives of the family.

The greenhouse white fly, *Asterŏchiton vaporariōrum.*—One of the most important of the greenhouse pests is this insect, which infests very many species of plants that are grown under glass; and sometimes it is a serious pest in the open on tomato and other plants that are set out after the weather is warm.

The adult measures about 1.5 mm. in length, and like other aleyrodids is covered with a white, waxy powder. The eggs are only .2 mm. in length, and are suspended from the leaf by a short stalk (Fig. 511, *a*). In the first instar the larva is flat, oval in outline, and with each margin of the body furnished with eighteen spines (Fig. 511, *b*), of which the last is much the longest. In the second and third instars there are only three pairs of marginal spines, a very small pair near the cephalic end of the body and two somewhat larger ones near the caudal end. The marginal fringe of wax is narrow. The puparium is box-like, the body of the insect being elevated on a palisade of vertical wax rods (Fig. 511, *d*). There are other rods of wax represented in the dorsal view of the puparium (Fig. 511, *c*).

The most successful means of destroying this pest is by fumigation of infested greenhouses with hydrocyanic acid gas.

The strawberry white fly, *Asterŏchiton packardi*.—This species is closely allied to the greenhouse white fly, but differs in minute characters presented by the spines and wax rods of the immature forms. It infests strawberry plants, and is a hardy species, passing the winter in the egg state out of doors.

The citrus white fly, *Dialeurōdes cĭtri*.—This is a well-known pest in the orange-growing sections of our country, and is also found in greenhouses in the North. It infests all citrus fruits grown in this country and is found on several other plants.

This insect injures its host in two ways: first directly, by sucking the sap from the leaves; and second indirectly, by furnishing nourishment, in the form of honeydew, to a fungus, the sooty mold (*Meliola camelliæ*), which forms a dark-brown or black membranous coating on the leaves and fruit, and thus interfering with the functioning of the leaves, retarding the ripening of the fruit, and decreasing the yield of the fruit. There are from two to six generations of this species in a year. An extended account of it is given by Morrill and Back ('11).

The maple white fly, *Aleurochiton forbesii*.—Figure 509 represents this species, which is fairly common on maple, but rarely in sufficient numbers to do serious injury.

Family COCCIDÆ

The Scale-Insects or Bark-Lice, Mealy-Bugs, and others

The family Coccidæ includes the scale-insects or bark-lice, the mealy-bugs, and certain other insects for which there are no popular names. To this family belong many of the most serious pests of horticulturists; scarcely any kind of fruit is free from their attacks; and certain species of scale-insects and of mealy-bugs are constant pests in greenhouses. Most of the species live on the leaves and stems of plants; but some species infest the roots of the host-plants. The great majority of the species remain fixed upon their host during a part of their life-cycle, and can thus be transported long distances while yet alive, on fruit or on nursery stock; this has resulted in many species becoming world-wide in distribution. Most of the species are minute or of moderate size; but some members of the family found in Australia measure 25 mm. or more in length.

While the economic importance of this family is due chiefly to the noxious species that belong to it, it contains several useful species. The most important useful species at this time is the lac-insect, *Tachărdia lăcca*. The stick-lac of commerce, from which shell-lac or shellac is prepared, is a resinous substance excreted by this species, which lives on the young branches of many tropical trees, most of which belong to the genus *Ficus*, the figs.

In the past, several coccids have been important as coloring agents. The bodies of the lac-insects, which are obtained from stick-lac in the manufacture of shellac, are the source of lac-dye. Another coccid, *Kĕrmes ĭlicis*, which lives on a species of oak in southern Europe, has been used as a dye from very early times. And the well-known

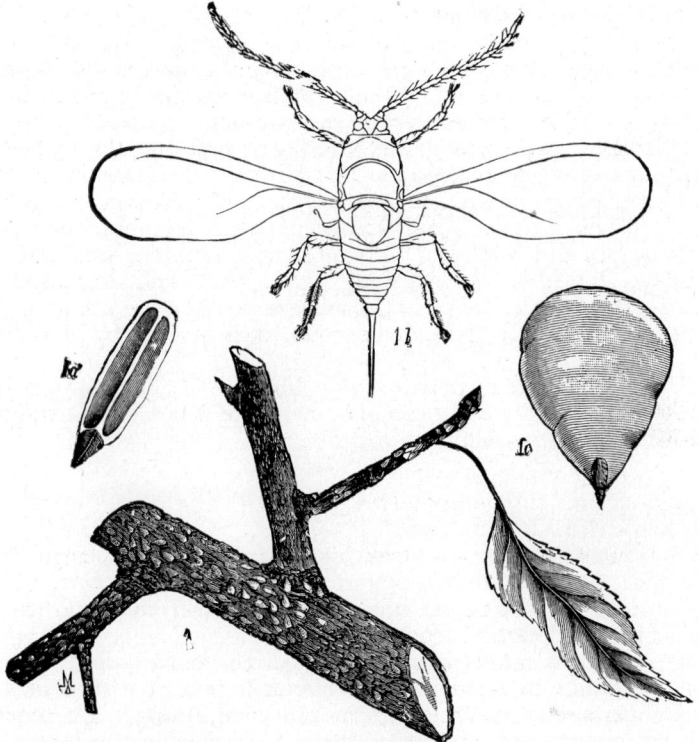

Fig. 512.—*Chionaspis furfura*: *1*, scales on pear, natural size; *1a*, scale of male, *1b*, adult male, *1c*, scale of female, enlarged.

cochineal is composed of the dried bodies of a coccid, *Cŏccus căcti*, which lives on various species of cactus. Recently these dyes have been largely supplanted by those obtained from coal-tar.

China-wax is also produced by a coccid. It is the excretion of an insect known as pe-la, *Ericērus pe-la*, and was formerly much used in China in the manufacture of candles, before the introduction of paraffin.

In the adult state, the two sexes of coccids differ greatly in form. The males are usually winged (Fig. 512); in a few species they are either wingless or have vestigial wings. The fore wings are usually large, compared with the size of the body; the hind wings are always greatly reduced in size; usually they are a pair of club-shaped halteres, but in a few forms they are more or less wing-like. Each hind wing is furnished with a bristle, which is hooked at the end and fits into a pocket or fold on the inner margin of the fore wing of the same side; in a few species there are two or three or more of these hamuli.

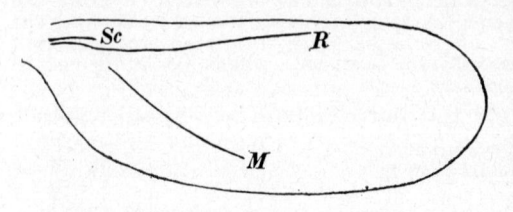

Fig. 513.—Wing of *Pseudococcus.* (From Patch.)

The venation of the fore wings is greatly reduced; a wing of *Pseudocŏccus* (Fig. 513) will serve to illustrate the usual type of wing-venation found in this family.

The legs are wanting in many adult females, having been lost during the metamorphosis. In adult males they are of ordinary form; except in a few species, the tarsi are one-jointed, and each is furnished with a single claw. Accompanying the tarsal claw there are often a few long, clubbed setæ, the *digitules* (Fig. 514); these are tenent hairs; some of the digitules arise from the tip of the tarsus, and some from the claw.

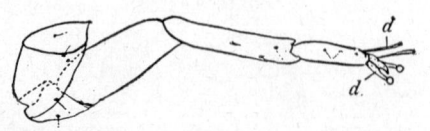

The caudal end of the abdomen of the male usually bears a slender tubular process, the *stylus*. In some species the stylus is as long as or even longer than the abdomen; in others it is short, and in some it is apparently wanting. The stylus serves as a support for the penis, which is protruded from it and in some species is very long.

Fig. 514.—Leg of a female *Lecanium*: d, d, digitules.

The female coccid is always wingless, and the body is either scale-like or gall-like in form, or grub-like and clothed with wax. The waxy covering may be in the form of powder, of large tufts or plates, of a continuous layer, or of a thin scale, beneath which the insect lives.

The eyes of coccids exhibit varying degrees of degeneration and retardation of development. The extreme of degeneration is found in the females, where there is only a single simple eye on each side of the head; this is probably a vestige of a compound eye. In the adult males of the more generalized forms, compound eyes are present; and in some of these forms, there are also ocelli, two in some and three in others. When compound eyes are present the facets are usually large, and not closely associated. In the more specialized forms, instead of compound eyes there are on each lateral half of the head from two to eight widely separated simple eyes, which may be scattered vestiges of compound eyes.

The structure and development of the eyes of the male of the common mealy-bug, *Pseudococcus (Dactylopius) destructor*, was studied by Krecker ('09). In this insect there is on each side of the head a very small eye; since these are the only eyes possessed by the young nymphs, they were termed by Krecker the primary eyes. In the adult, in addition to the primary eyes, there are two pairs of eyes, one pair on the dorsal aspect of the head, and a second pair on the ventral aspect; these he termed the accessory eyes.

The so-called primary eyes are very degenerate, in the adult at least. There is a lens below which there are a few retinal cells; but there is no corneal hypodermis, no rhabdoms, and no iris.

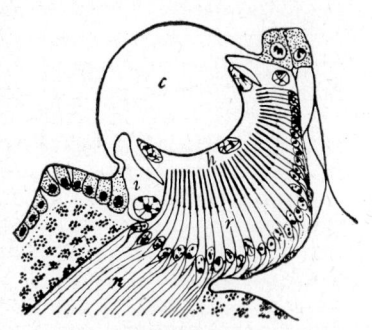

The development of the so-called accessory eyes is greatly retarded. The histoblasts from which they are developed appear in the latter part of the second nymphal stadium or in the beginning of the third; these are thickenings of the hypodermis. When fully developed as seen in the adult, the accessory eyes (Fig. 515) have a large circular cornea, followed by a comparatively thin layer of corneal hypodermis, encircling which is a single row of large iris cells. Below the corneal hypodermis there is a crescent-shaped area of polygonal rods (rhabdoms), which are terminally situated upon the retinal cells. From the proximal end of the retinal cells extend the nerve fibrils which join to form the optic nerve, which follows the contour of the head to enter the brain laterally. Reddish brown pigment fills the retina, the iris, and also a ridge surrounding the eyes. There are no cells which function as pigment cells alone.

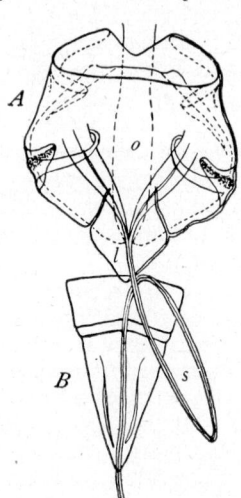

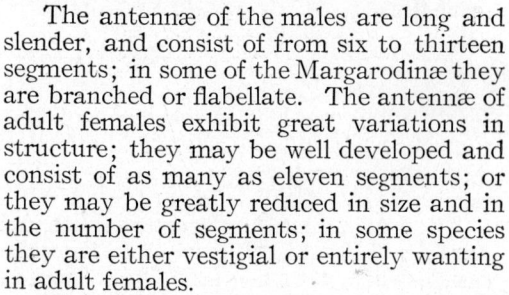

Fig. 515.—A depigmented "accessory eye" of *Pseudococcus destructor*: *c*, cornea; *h*, corneal hypodermis; *i*, iris cell; *r*, retinal cells; *n*, nerve.

The antennæ of the males are long and slender, and consist of from six to thirteen segments; in some of the Margarodinæ they are branched or flabellate. The antennæ of adult females exhibit great variations in structure; they may be well developed and consist of as many as eleven segments; or they may be greatly reduced in size and in the number of segments; in some species they are either vestigial or entirely wanting in adult females.

The mouth-parts are situated on the hind part of the ventral aspect of the head, and often extend caudad of the first pair of legs. In front of the beak there is a densely chitinized area, which includes the clypeus, the labrum, and the mandibular and maxillary sclerites. In cleared specimens there can be seen within this area a complicated endoskeleton (Fig. 516, A).

Fig. 516.—Mouth-parts of a coccid: A, the densely chitinized area in front of the beak; B, the beak; *l*, labrum; *o*, œsophagus; *s*, loop of mandibular and maxillary setæ in the crumena. (After Berlese.)

The labium (Fig. 516, B), which is commonly termed the beak or rostrum, consists of three segments in a few forms found in

New Zealand, but usually it is more or less reduced, consisting either of two segments or of only one; in a few subfamilies it is wanting in the adult. The mandibular and maxillary setæ are wanting in the later nymphal instars of some forms, in some adult females, and in all adult males. These setæ, when present, are usually long, frequently longer than the body, and in some species several times as long. When not exserted, they are coiled within a pouch, termed the *crumēna*, only their united tips extending to the labium. The crumena is a deep invagination of the body-wall, which extends far back into the body-cavity. Its walls are delicate, and not easily observed; but the coiled setæ within it can be easily seen in cleared specimens (Fig. 516, *s*).

In the classification of coccids, the characters most used are those presented by the female, although those of the male are used to some extent. The most available characters of the female are the following: first, the general form of the body; second, the form of the waxy excretions; third, the structure of the caudal end of the body; and fourth, the form and position of the pores through which the wax is excreted.

To study the third and fourth classes of characters listed above, it is necessary to remove the wax, to clarify the body, and, in some cases, to stain it. The method most commonly used for removing the wax and clarifying the body is to boil the specimen in a ten per cent. aqueous solution of caustic potash. For staining the body, Gage ('19) found that a solution of säurefuchsin was most satisfactory; his formula for the preparation of this solution is as follows:

```
Säurefuchsin..................................................0.5 gr.
Hydrochloric acid, 10%........................................25.0 c.c.
Distilled water...............................................300.0 c.c.
```

The cleaned and stained specimens are usually mounted in Canada balsam for microscopic examination.

Within the family Coccidæ there are to be found most remarkable variations in structure; this is especially true of the form of the caudal end of the body and of the form of the parts through which the wax and other excretions are exuded. These characters have been described by many authors; but, unfortunately, there is a great lack of uniformity in the terminology used by them. In this place, only sufficient space can be taken to define the more important struc-

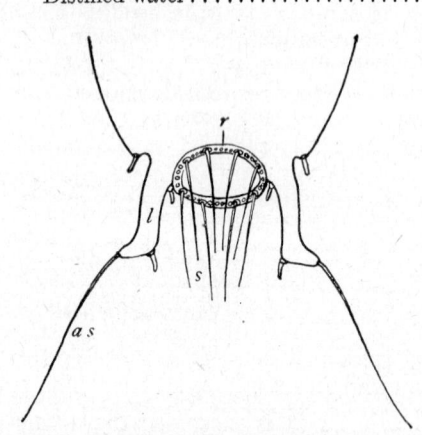

Fig. 517.—Caudal end of female of *Eriococcus araucariæ*: *r*, anal ring; *s*, anal-ring setæ; *l*, anal lobe; *as*, anal seta. Between the bases of the anal-ring setæ there are openings of wax-glands.

tures, using the terms that are more generally applied to them.

The anal ring.—In the mealy-bugs, the tortoise-scales, and the lac-insects, and in the nymphs of some others, the anus is surrounded by a well-defined ring, the *anal ring* (Fig. 517, *r*).

The anal-ring setæ.—The anal ring bears several, from two to thirty but usually six, long and stout setæ, the *anal-ring setæ* (Fig. 517, *s*).

The anal lobes.—In many coccids, the caudal end of the body is terminated by a pair of lobes, the *anal lobes* (Fig. 517, *l*).

The anal setæ.—Each anal lobe bears one or more prominent setæ, the *anal setæ* (Fig. 517, *as*).

The anal plates.—In the subfamily Lecaniinæ, the abdomen of the female is cleft at the caudal end, and, at the cephalic end of the cleft, there is a pair of triangular, or sometimes semi-circular plates, the *anal plates* (Fig. 518, *ap*).

The pygidium.—In the subfamily Diaspidinæ, the abdomen of the adult female is terminated by a strongly chitinized unsegmented region, which consists of four coalesced segments (Fig. 519); this region is termed the *pygidium* by writers on the Coccidæ. This application of the term pygidium is quite different from that used in descriptions of other insects, where it refers only to the tergite of the last abdominal segment. A more detailed account of the characters presented by the pygidium of the Diaspidinæ is given later.

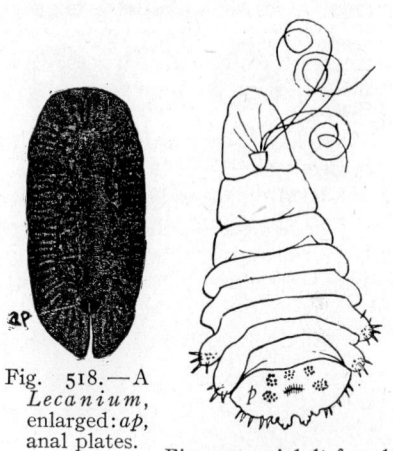

Fig. 518.—A *Lecanium*, enlarged: *ap*, anal plates.

Fig. 519.—Adult female *Lepidosaphes*: *p*, pygidium.

The spines and the setæ.—The position and number of spines and of setæ are often indicated in specific descriptions. Care should be taken to distinguish between these two kinds of structures. A seta can be recognized by the cup-like cavity in the cuticula, the alveolus, within which it is jointed to the body; while a spine is an outgrowth of the cuticula that is not separated from it by a joint. See figure 42, page 32. The writer in his early works on the Coccidæ ('81, '82, '83) termed certain spine-like setæ *spines*.

The outlets of wax-glands.—In the Coccidæ there are many minute openings in the cuticula through which wax is excreted; these vary greatly in form, in position on the body, and in the structure of the part of the cuticula through which they open. As the characters presented by these openings are much used in the classification of coccids, a very elaborate terminology referring to them has been developed. Unfortunately different authors use quite different terms, and, therefore, it is necessary to learn the terms used by an author in order to understand his descriptions. The most detailed and systematic terminology that has been proposed is that of MacGillivray ('21). Some of the many terms adopted by this author are defined below.

The ceratubæ.—In the Diaspidinæ and in some species of several other subfamilies, the terminal portion of the outlet of some of the wax-glands is an invaginated cuticular tube. The inner end of this tube is truncate, and, in the Diaspidinæ, bears a perforated knob. This invaginated cuticular tube is termed a *ceratuba*. The ceratubæ vary greatly in length and in shape; in some the greater part of the tube is reduced to a fine thread, with a bulb-like inner end. A few ceratubæ are represented in a diagram given later (Fig. 522). The openings of most ceratubæ are flush with the body-wall, but some of them open through plates in the marginal fringe. The different types of ceratubæ have received distinguishing names formed by combining a prefix with the word ceratubæ.

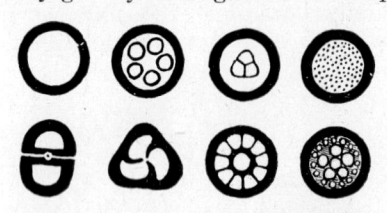

Fig. 520.—Several types of openings of cerores.

The cerores.—The various types of outlets of the wax-glands in which the cuticula is not invaginated so as to form a ceratuba are termed *cerores*. The openings of cerores through the cuticula vary

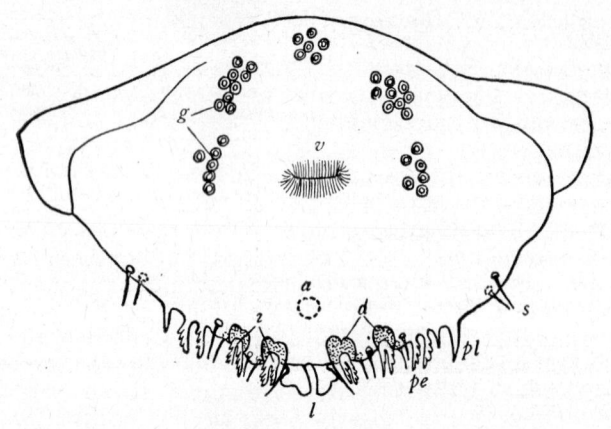

Fig. 521.—Diagram of a pygidium of a diaspid: *a*, anus showing through the body; *d*, densariæ; *g*, genacerores; *i*, incisions; *l*, first pair of lobes; *pe*, pectinæ; *pl*, plate; *s*, setæ; *v*, vagina.

greatly in form; several types of these openings are represented in Figure 520. While in most cases the openings of cerores are flush with the general surface of the cuticula, in some coccids (Ortheziinæ) the cerores open through spines. There are also variations in the grouping of the cerores. Each of the various types has received a technical name formed by combining a prefix with the word cerores.

Thus, for example, the cerores that occur in four or five groups about the genital opening in many of the Diaspidinæ (Fig. 521, g) are termed *genacerores*.

The features of the pygidium.—The pygidia of adult female diaspids present characters that are much used in distinguishing the species of this subfamily; among these are the following.

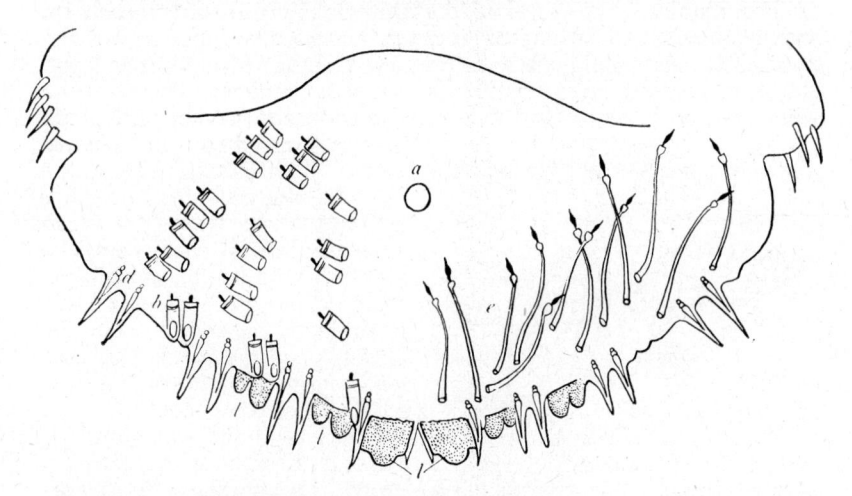

Fig. 522.—A composite diagram of a pygidium: *a*, anus; *b*, marginal ceratubæ, with elongated openings; *d*, ceratubæ opening through plates; *e*, linear ceratubæ; *l*, *l*, *l*, lobes; the lobes of the second and third pairs are divided.

The position of the anus, which opens on the dorsal aspect of the pygidium at varying distances from the end of the body (Fig. 522, *a*).

The opening of the vagina, on the ventral aspect of the pygidium (Fig. 521, *v*).

The presence or absence of groups of *genacerores* (Fig. 521, *g*), the number of these groups when present, and the number of cerores in each group. The different groups are distinguished as the median group (*mesogenacerores*), the cephalo-lateral groups, one on each side (*pregnacerores*), and the caudo-lateral groups, one on each side (*postgenacerores*), respectively. These all open on the ventral aspect of the pygidium. Each genaceroris has several openings.

The position and number of openings of ceratubæ, and the types of ceratubæ that are present (Fig. 522).

The number of pairs of *lobes* borne by the margin, the shape of the lobes, and whether they are divided or not (Fig. 522, *l*, *l*, *l*). The pairs of lobes are numbered, beginning with the pair at the end of the body; in some species this pair is represented by a single lobe.

The number of pairs of *incisions* (*incisuræ*) in the margin of the pygidium (Fig. 521, *i*).

The presence or absence of thickenings of the margins of the incisions (*densariæ*); these are thickenings of the ventral wall (Fig. 521, *d*).

The presence or absence of club-shaped thickenings of the dorsal wall (*paraphyses*) that extend forward from near the bases of the lobes (Fig. 523, *p*).

The presence or absence of a thickening of the lateral margin of the pygidium cephalad of the region in which the lobes are situated, and resembling the lobes in structure (Fig. 523, *m*).

The number and shape of the thin projections of the margin, known as *plates*. Two quite distinct types of plates can be distinguished: in one they are broad and fringed (Fig. 521, *pe*); the plates of this type have been termed *pectinæ*; in the other type they are spine-like in form (Fig. 521, *pl*); some writers restrict the term *plate* to this type, and use *pectinæ* for the first type. Each plate contains the outlet of a wax-gland.

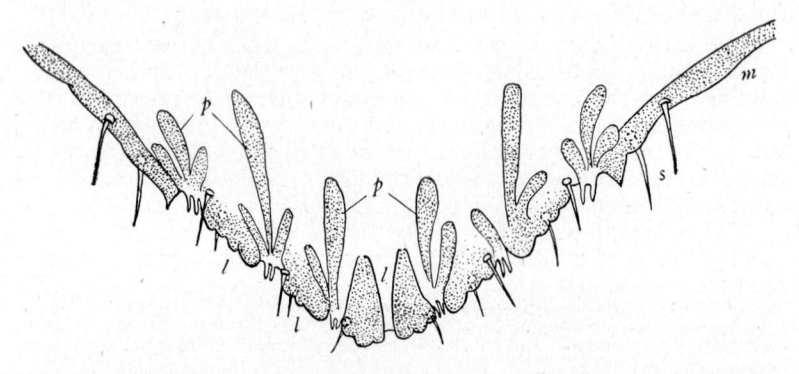

Fig. 523.—Part of the pygidium of *Chrysomphalus tenebricosus*, ventral aspect, with the paraphyses (*pp*) of the dorsal wall showing through: *l, l, l*, lobes; *m*, thickened margin; *s*, spine-like setæ.

The metamorphosis of coccids.—In this family the two sexes are indistinguishable during the first nymphal stadium. Both are furnished with legs, antennæ, and functional mouth-parts. It is during this period that the sedentary species spread over the plants that they infest. In their subsequent development the sexes differ greatly; hence the metamorphosis of each can be best discussed separately.

The females never become winged. Some, as the mealy-bugs and *Orthezia*, continue active throughout their entire or almost entire life; but most forms become sedentary early in life and remain fixed upon their host. Many species lose their legs and antennæ when they assume the quiescent form; and in some the mandibular and maxillary setæ are wanting in the adult. The number of nymphal instars in females varies from two to four; the smaller number occurs in the more specialized subfamilies.

In the males there are usually four nymphal instars. During the latter part of the nymphal life the male is quiescent, having formed a cocoon or a scale within or beneath which it remains till it emerges as an adult. The stage of development at which the quiescent period begins varies greatly. Thus, while in the mealy-bugs the cocoon is made during the second stadium, in *Icerya* it is not made till near the end of the third. In the Diaspidinæ the formation of the scale begins either at the close of the first stadium or immediately after the first molt. With the molt at the beginning of the quiescent period the male loses its legs, antennæ, and mandibular and maxillary setæ. The setæ are not replaced; and, consequently, the adult males can take no food. The legs and antennæ of the adult are developed from histoblasts, as in insects with a complete metamorphosis; the wing-buds appear in the last nymphal stadium; but they are developed externally, as in insects with a gradual metamorphosis. The type of metamorphosis of the male coccid is, therefore, neither strictly complete nor gradual. This illustrates the difficulty of attempting to make sharp distinctions; for in nature all gradations exist between the different types of structure and of development.

The classification of the Coccidæ.—The different writers on the Coccidæ have grouped the genera into a variable number of subfamilies. In the classification by MacGillivray ('21), this author recognizes seventeen subfamilies, and gives two tables for separating them, one based on the characters presented by the first nymphal instar, and one on those of adult females. Tables are also given for separating the genera and species of the different subfamilies.

The following are a few of the better-known representatives of this family found in this country. Several subfamilies not mentioned here are represented in our fauna.

Subfamily MONOPHLEBINÆ

The Giant Coccids

The common name of this family was suggested by the large size of many of the exotic species. The best-known species found in North America is of moderate size; this is the cottony-cushion scale, *Icērya pŭrchasi* (Fig. 524). The adult female measures from 4 to 8 mm. in length, is scale-like, dark orange-red, and has the dorsal surface more or less covered with a white or yellowish white powder. It secretes a large, longitudinally ribbed egg-sac, which is white tinged with yellow This beautiful insect was at one time the most dangerous insect pest in California, and did a great amount of injury. It is an introduced Australian species, and has been subdued

Fig. 524.—*Icerya purchasi:* females, adults, and young on orange.

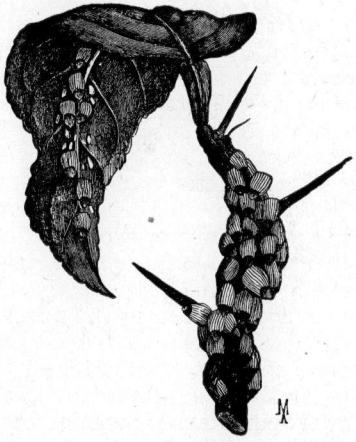

to a great extent by the introduction of an Australian lady-bug, *Rodōlia cardinālis*.

Subfamily COCCINÆ

The Cochineal Coccids

This subfamily is of especial interest because it includes the cochineal insect, *Cŏccus căcti*. This is a native of Mexico, but occurs in the southern United States. It feeds upon various species of the Cactaceæ. It has been extensively cultivated in India, Spain, and other countries. The adult female bears some resemblance to a mealy-bug, but differs in lacking anal lobes and an anal ring. It excretes a mass of white cottony threads, within which the eggs are laid. The dye-stuff consists of the female insects, which, when mature, are brushed off the plants, killed, and dried. The entire insect is used. Cochineal is now being superseded by aniline dyes, which are made from coal-tar.

Subfamily ORTHEZIINÆ

The Ensign Coccids

Members of this subfamily occur not uncommonly on various weeds. They are remarkable for the symmetrically arranged, glistening, white plates of excretion with which the body is clothed. Figure 525 represents a nymph; in the adult female, the excretion becomes more elongated posteriorly, and forms a sac containing the eggs mixed with fine down. Later, when the young are born, they excrete a sufficient amount of the lamellar excretion to cover them. In many species the egg-sac is held in a more or less elevated position; this fact suggested the common name *ensign-coccids* for these insects. Most of our species belong to the genus *Orthezia*,

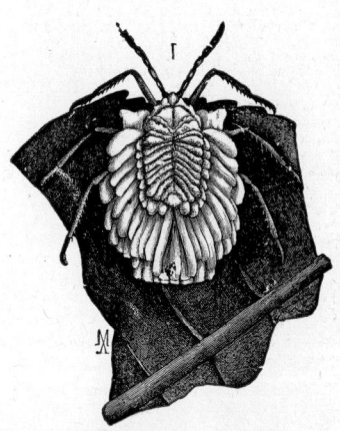

Fig. 525.—*Orthezia*, greatly enlarged.

Subfamily ERIOCOCCINÆ

The Mealy-Bugs

This subfamily includes many genera and species; the best-known members of it are certain mealy-bugs, which are the most common and noxious of greenhouse pests. These insects have received the

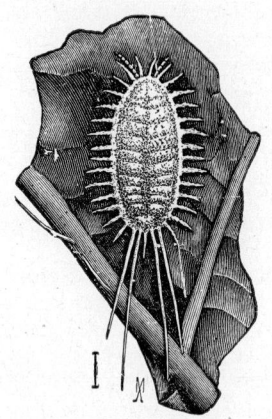

Fig. 526.—*Pseudococcus longispinosus.*

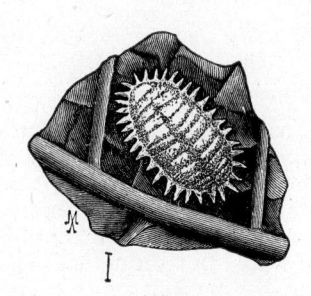

Fig. 527.—*Pseudococcus citri.*

name *mealy-bugs* because their bodies are covered with a fine granular excretion, appearing as if they had been dusted with flour. The females are active nearly throughout their entire life. The males make a cocoon early in their nymphal life in which they remain till they emerge as adults.

Figure 526 represents *Pseudocŏccus longispinōsus,* a common species in greenhouses; and Figure 527, *Pseudocŏccus cĭtri,* another species that is found in greenhouses in the North. The latter species is also a well-known pest of orange trees in the South.

Several species of mealy-bugs of the genus *Ripĕrsia* are found in the nests of ants of the genus *Lasius.*

Subfamily LECANIINÆ

The Tortoise-Scales

The tortoise-scales are so named on account of the form of the body in many species.

Fig. 528.—*Lecanium hesperidum,* adult females, natural size.

452   AN INTRODUCTION TO ENTOMOLOGY

The most striking characteristic of this subfamily is that the abdomen of the female is cleft at the caudal end, and at the cephalic end of this cleft there is a pair of triangular or semicircular plates, the anal plates (Fig. 518).

This is a large subfamily including many genera and species. While the various forms agree in the distinguishing characteristics given above, there are great differences in the appearance of the adult females. Many of them excrete very little wax, the body being practically naked, and the eggs, or the young in the viviparous species,

Fig. 529. *Saissetia oleæ:* *1*, adult females on olive, natural size; *1a*, female, enlarged.

are deposited beneath the body; in other species, although the body is nearly naked, the adult female excretes a large, cottony egg-sac; and in still others the body is deeply encased in wax.

The three following species will serve as examples of those in which the body is naked and which do not form an egg-sac.

The soft scale, *Lecānium hespĕridum.*—This is the commonest and most widely spread member of this subfamily; it infests a great variety of plants; in the North, it is very common in greenhouses; in the warmer parts of the country it lives out of doors. The adult female is nearly flat (Fig. 528), and is viviparous.

The black scale, *Saissētia ōleæ.*—This is a well-known pest, especially in California, where it infests various kinds of fruit-trees and other plants. The adult female is dark brown, nearly black, in color; nearly hemispherical in form (Fig. 529), often, however, quite a little longer than broad. There is a median longitudinal ridge on the back, and two transverse ridges, the three forming a raised surface of the form of a capital H.

The hemispherical scale, *Saissētia hemisphærica.*—The adult female is nearly hemispherical in form, with the edges of the body flattened (Fig. 530). This species is found in conservatories everywhere, and in the open air in warmer regions.

Pulvināria.—Those members of this subfamily in which the adult female is nearly naked but excretes a large cottony egg-sac beneath or behind the body, are represented in this country by the genus *Pulvinaria*, of which we have many species. Our best-known species are the two following.

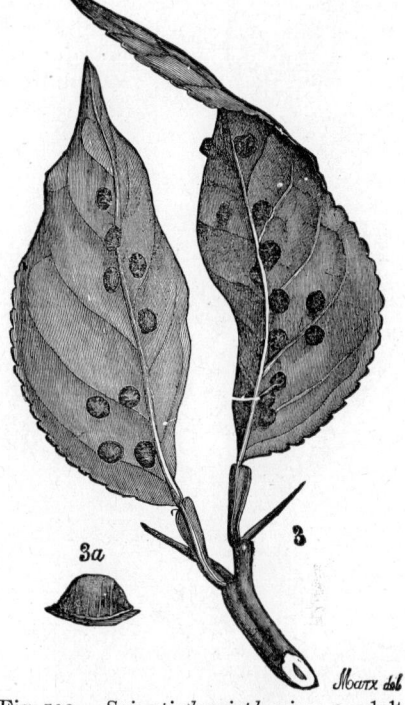

Fig. 530.—*Saissetia hemisphærica: 3*, adult females on orange, natural size; *3a*, adult female, enlarged.

The cottony maple-scale, *Pulvināria vītis.*—This species is common on maple, osage orange, grape, and other plants. Figure 531 represents several adult females with their egg-sacs on a cane of grape.

Fig. 531.—*Pulvinaria vitis.*

The maple-leaf pulvinaria, *Pulvināria acerĭcola.*—This species is also found on maple. It differs from the preceding species in that the egg-sac is much longer than the body of the female, and is formed on the leaves instead of on the stem of the host.

Ceroplastes.—In this genus the body of the female is covered with thick plates of wax. More than sixty species have been described, several of which are found in the southern United States; the following will serve as an example of these beautiful insects.

The barnacle scale, *Ceroplăstes cirripedifŏrmis.*—Several individuals of this species are represented natural size, and one enlarged, in Figure 532. It infests orange, quince, and many other plants.

SUBFAMILY KERMESIINÆ

The Pseudogall Coccids

This subfamily includes only one genus, *Kĕrmes*. Species of this genus are common on oaks wherever they grow. These insects are remarkable for the wonderful gall-like form of the adult females. So striking is this resemblance, that they have been mistaken for galls by many entomologists. Fig. 533 represents a species of this genus upon *Quercus agrifolia.* The gall-like bodies on the stem are adult females, the smaller scales on the leaves are immature males.

SUBFAMILY DIASPIDINÆ

The Armored Scales

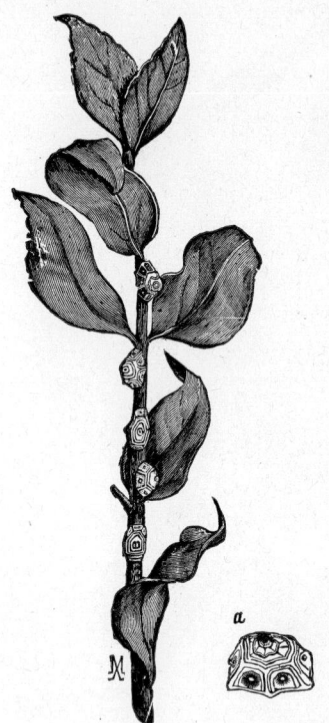

Fig. 532.—*Ceroplastes cirripediformis.*

The Diaspidinæ includes those coccids that form a scale, composed in part of molted skins and partly of an excretion of the insect, beneath which the insect lives. It is on account of this covering that these scale-insects are named the armored scales. The Diaspidinæ are also characterized by a coalescence of the last four abdominal segments so as to form what is known as the pygidium; this peculiar structure is described on an earlier page.

The formation of the scale begins immediately after the close of the active period of the first nymphal instar. At this time the young insect settles and begins to draw nourishment from its host. Soon after, there exude from the body fine threads of wax, the commencement of the formation of the scale. At the close of the first stadium, the molted skin is added to the scale and forms a part of it. This is also true, except as noted below, of the second molted skin of the female (Fig. 534, 2b and 2c). In the formation of the scale of the male only the first molted skin is added to the scale (Fig. 534, 2d). The scales of males can be distinguished by this fact, and, too, they are much smaller than the scales of females.

In a few genera the female does not molt the second exuviæ*; the body shrinks away from it, and transforms within it. In such cases is it termed a *puparium*. Figure 535 represents the scale of *Fiorinia fiorinæ;* here the puparium can be seen through the transparent scale.

Fig. 533.—*Kermes sp.* on *Quercus agrifolia*: adult females on the stem; immature males on the leaves.

The shape of the scale, and the position of the exuviæ on it, furnish characters that are very useful in the classification of the Diaspidinæ.

To this subfamily belong some of the most serious pests of shrubs and trees, as, for example, the San José scale and the oyster-shell scale. The following are a few of the many well-known species of this very important subfamily.

---

*The term *exuviæ* is a Latin word which had no singular form, the plural noun being used as is in English the word *clothes*. Some recent writers use the term *exuvia* for a single molted skin.

456　　AN INTRODUCTION TO ENTOMOLOGY

The purple scale, *Lepidŏsaphes pinnæfŏrmis*.—This scale-insect is well known in the orange-growing sections of this and of other

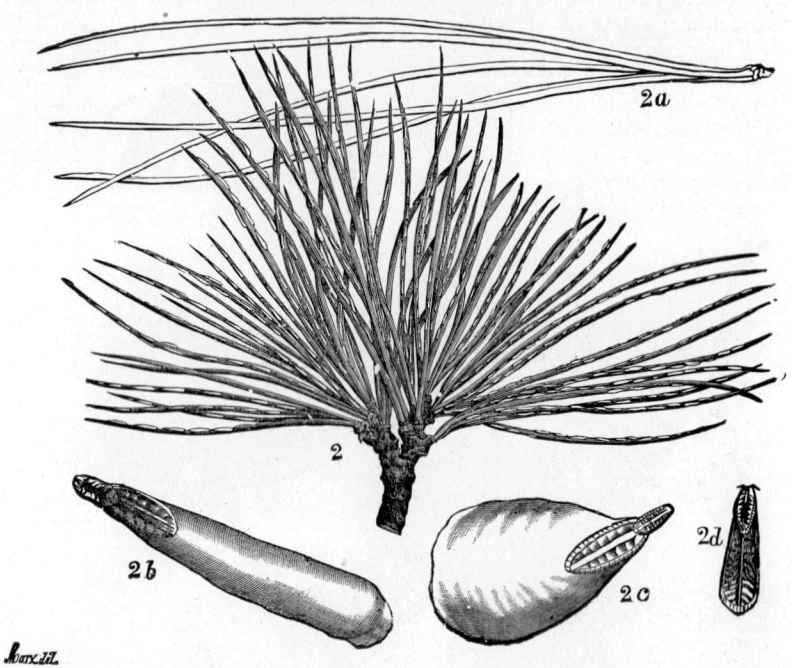

Fig. 534.—*Chionaspis pinifoliæ*: *2*, scales on *Pinus strobus*, natural size, leaves stunted; *2a*, leaves not stunted by coccids; *2b*, scale of female, usual form, enlarged; *2c*, scale of female, wide form, enlarged; *2d*, scale of male, enlarged.

countries. It is one of the two most common scale-insects found on citrus trees in Florida. The scales of this species are represented in Figure 536; they are represented natural size on the leaf, and greatly enlarged in the other figures. The scale of the female is long, more or less curved, and widened posteriorly (Fig. 536, *1a* and *1b*); the first of these two figures represents the dorsal scale, and the second the ventral scale, which is well developed in this species. Some eggs can be seen through a gap in the ventral scale. The scale of the male (Fig. 536, *1c*) is usually straight or nearly so. At about one-quarter of the length of the scale from the posterior extremity, the scale is thin, forming a hinge which allows the posterior part of it to be lifted by the male as he emerges. While this insect is chiefly known as a pest of citrus trees, it has been found on several other species of plants. It has been described under several different names; for a long time it was known as *Mytilăspis citrĭcola*.

Fig. 535.—*Fiorinia fioriniæ*.

Glover's scale, *Lepidŏsaphes glovērii*.—This is the second of the two most common species of scale-insects found on citrus trees in Florida. In this species the scale of the female is much narrower than that of the preceding species. This species is widely distributed over the warmer parts of the world.

The oyster-shell scale, *Lepidosaphes ulmi*.—This is a northern representative of the genus to which the two preceding species belong. It is closely allied to the purple scale; in fact Figure 536 would serve to illustrate this species except that it does not occur on orange and that it is found chiefly on the trunk of its host. The two species differ in the characters presented by the pygidium. The oyster-shell scale is

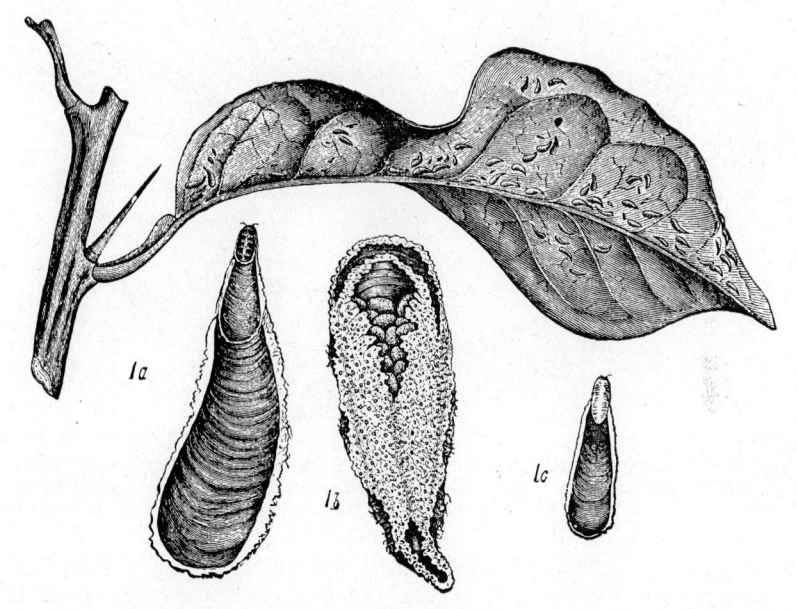

Fig. 536.—*Lepidosaphes pinnæformis: 1*, scales on orange, natural size; *1a*, scale of female, dorsal view, enlarged; *1b*, scale of female with ventral scale and eggs, enlarged; *1c*, scale of male, enlarged.

a cosmopolitan insect, and it infests very many species of shrubs and trees. In the North it is the commonest and best-known scale-insect infesting fruit-trees and various ornamental shrubs. It is discussed in all of our manuals of fruit-insects; in some of them it is described under the name *Mytilăspis pomōrum*.

The scurfy scale, *Chionăspis fŭrfura*.—This, like the preceding species, is a very common pest of the apple and various other trees and shrubs; but usually it is not very destructive. In this species the scale of the female is widened posteriorly, and bears the exuviæ at the anterior end (Fig. 512, *1c*). The scale of the male is very small, being only .75 mm. in length, narrow, and tricarinated (Fig. 512, *1a*).

The pine-leaf scale, *Chionăspis pinifōliæ.*—This is a very common pest of pine, spruce, and other coniferous trees, throughout the United States and Canada. It infests the leaves of its various hosts. The scale of the female is snowy white in color, with the exuviæ light yellow; it is usually long and narrow, as represented in Figure 534, 2b; but on the broader-leaved pines it is often of the form shown

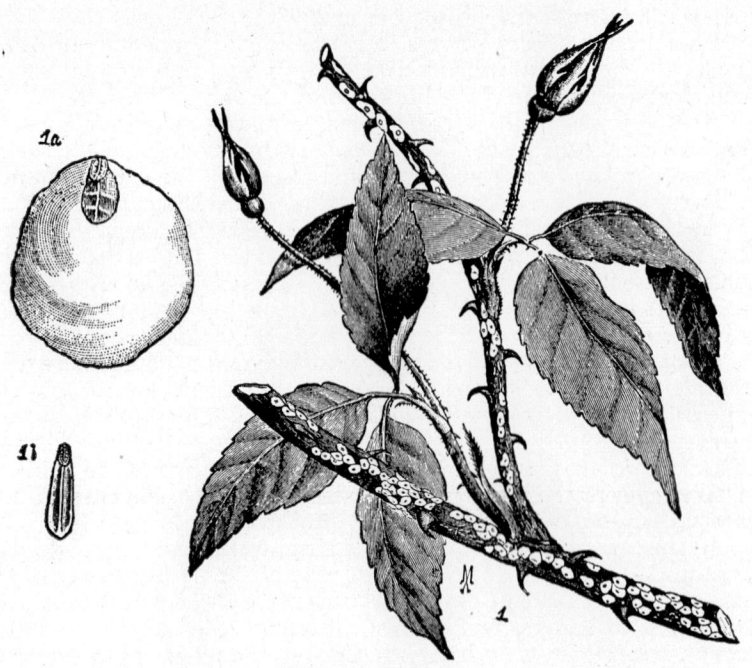

Fig. 537.—*Aulacaspis rosæ: 1*, scales on rose, natural size; *1a*, scale of female, enlarged; *1b*, scale of male, enlarged.

at 2c in the figure; this is the typical form of the scale of the female in the genus *Chionaspis*.

The rose-scale, *Aulacăspis rōsæ.*—This species infests the stems of roses, blackberry, raspberry, dewberry, and some other plants. The infested stems often become densely coated with the scales. The scale of the female is circular, snowy white, with the exuviæ light yellow and upon one side (Fig. 537, *1a*). The scale of the male is also white; it is long, tricarinated, and with the exuviæ at one end (Fig. 537, *1b*); it measures 1.25 mm. in length.

The San José scale, *Comstockăspis perniciōsa.*—The San José scale was first described by the writer in 1881, under the name *Aspidiōtus perniciōsus*. It has since been made the type of a new genus, *Comstockaspis*, by MacGillivray ('21). At the time it was described it was known only in Santa Clara County, California. But in describing it, notwithstanding its limited distribution, I stated: "From what I have seen of it, I think that it is the most pernicious scale-insect known in this country." Since that time it has become widely distributed. Slingerland and Crosby write of it as follows: "The San José scale has attained greater notoriety, has been the cause of more legislation, both foreign and interstate, and has demonstrated its capabilities of doing more injury to the fruit interests of the United States and Canada than any other insect."

This species infests various fruit-trees and ornamental shrubs; it infests the bark, leaves, and fruit of its hosts, and usually causes reddish discolorations of the bark and of the skin of the fruit.

This species can be distinguished from the other scale-insects that are important pests of our fruit-trees by the form of the scales. The scale of the female is circular and flat, with the exuviæ central, or nearly so. The scale is gray, excepting the central part, that which covers the exuviæ, which varies from a pale yellow to a reddish yellow. It measures 2 mm. in diameter. The scale of the male is black, and is somewhat elongate when fully formed. The exuviæ is covered with secretion; its position is marked by a nipple-like prominence which is between the center and the anterior margin of the scale.

Control of scale-insects.—The extensive damage that has been done by scale-insects to fruit-trees and to cultivated shrubs has led to many experiments in the destruction of these insects. The results have been quite satisfactory; with proper care, it is now possible to keep in check the ravages of these pests. Detailed accounts of the methods to be employed are given in many easily available publications, and especially in bulletins of experiment stations.

In the case of deciduous trees and shrubs, the best time to destroy scale-insects infesting them is during the winter, when the trees are bare and in a dormant state. At this time the entire tree can be reached with sprays, without the interference of leaves; and, too, certain sprays can be safely used that are liable to injure the trees during the growing season. This is especially true of the lime-sulphur mixture, which is very widely used for the destruction of scale-insects, and is very effective. Among the other insecticides used for this purpose are kerosene emulsion and dilute miscible oils. For summer spraying, whale-oil soap, one pound dissolved in four or five gallons of water, can be safely used.

In the case of trees that are constantly clothed with foliage, the effective use of sprays is more difficult. In the orange-growing sections of California the trees are fumigated with hydrocyanic acid gas, the tree to be treated being first covered with a large tent.

CHAPTER XXII

ORDER DERMAPTERA*

The Earwigs

The winged members of this order usually have four wings; but in some of them the wings are wanting. The fore wings are leathery, very small, without veins, and when at rest meet in a straight line on the back; the hind wings, when well developed, are large, with radiating veins, and when at rest are folded both lengthwise and crosswise. The mouth-parts are formed for chewing. The caudal end of the body is furnished with a pair of appendages, the cerci, which usually resemble forceps. The metamorphosis is gradual or wanting.

This order includes three families, Forficulidæ, Arixeniidae and Hemimeridæ. The order is made up largely of the earwigs of the family Forficulidae. These are long and narrow insects resembling rove-beetles in the form of the body and in having short and thickened fore wings (Fig. 538); but the earwigs are easily distinguished from rove-beetles by the presence of a pair of forceps-like appendages at the caudal end of the body.

The common name, earwig, was given to these insects in England, and has reference to a widely spread fancy that these insects creep into the ears of sleeping persons. Other similar names are applied to them in Europe, *Ohr-Wurm* in Germany and *perceoreille* in France.

The earwigs are rare in the northeastern United States and Canada, but are more often found in the South and on the Pacific Coast. In Europe they are common, and often troublesome pests, feeding upon the corollas of flowers, fruits, and other vegetable substances. Some species are carnivorous, feeding on other insects, and some are probably scavengers. They are nocturnal, hiding in the day-time among leaves and in all kinds of crevices, and coming out by night; sometimes they are attracted to lights.

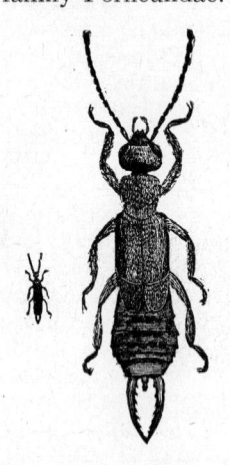

Fig. 538.—An earwig, *Labia minor*, male.

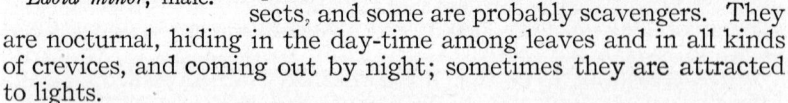

Earwigs are small or of moderate size; the living species measure from 2.5 to 37 mm. in length. The body is narrow and flat. The mouth-parts are fitted for chewing, and resemble in their more general features those of the Orthoptera; minute but distinct paragnatha are present; and the second maxillæ are incompletely fused.

---

*Dermăptera: *derma* (δέρμα), skin; *pteron* (πτερόν), a wing.

The compound eyes are rather large; but the ocelli are wanting. The antennæ are slender, and consist of from ten to thirty-five segments; the second segment is always small. The fore wings are leathery, very short, without veins, and when at rest meet in a straight line on the back. This pair of wings is commonly termed the *tegmina* or the *elytra*. The hind wings when at rest are folded both lengthwise and crosswise and project a short distance behind the fore wings (Fig. 538). The radiating veins of the hind wings extend from a point near the middle of the length of the wing (Fig. 539). When the wing is not in use, that part over which the radiating veins extend is folded in plaits like a fan. This part of the wing is the greatly expanded anal area. The preanal area is much reduced with but two longitudinal veins, and is quite densely chitinized. The tracheation of the hind wings has been described and figured by the writer

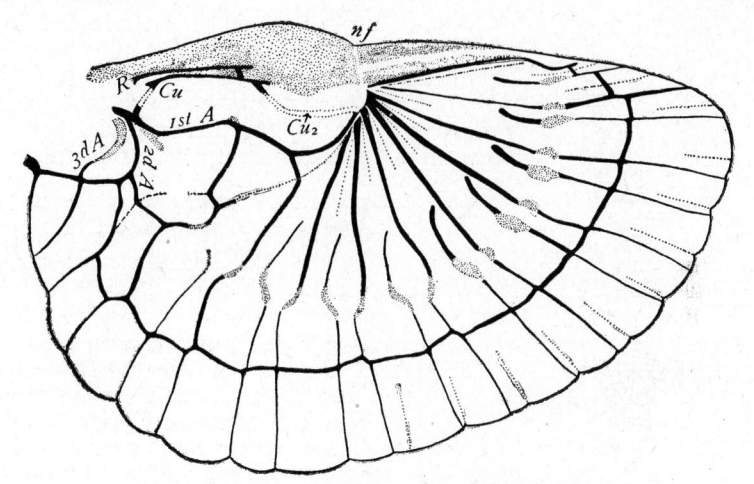

Fig. 539.—Hind wing of an earwig: *nf*, nodal furrow.

(Comstock '18). The wings vary much in size and development even in the same species; and there are many species that are wingless. The legs are similar in form, and the tarsi are three-jointed.

The most distinctive feature of earwigs is the form of the cerci, which are forceps-like, and usually very prominent. A similar form of cerci is found, however, in the genus *Japyx* of the order Thysanura. The size and shape of the forceps of earwigs differ in the different species and in the two sexes of the same species; they are usually more highly developed in the male than in the female; they are used as organs of defense and offense, in pairing, and are sometimes used as an aid in folding the wings.

Certain earwigs possess stink-glands, which open through tubercles situated one on each side near the hind margins of the second and third visible abdominal segments, from which, it is said, they can squirt a foul-smelling fluid to a distance of three or four inches.

The female earwig has smaller forceps and but six visible abdominal sterna while the male has eight. In some earwigs the two efferent ducts of the reproductive organs open separately. The metamorphosis is gradual, the wings developing externally. The female is said to brood over the eggs but to abandon them soon after hatching.

Fig. 541.— *Prolabia pulchella burgessi*, male. (From Rehn and Hebard.)

Earwigs are cosmopolitan insects, and are easily transported by commerce; consequently exotic species are liable to be found near seaports; and some such species have become established in this country. The species of the world have been monographed by Burr ('11).

Fig. 540.—*Labia minor*, female, and end of abdomen of the male. (From Lugger.)

The order is a comparatively small one; only about four hundred living species have been described, and these are mostly tropical or semitropical. The native and the exotic species that have become established in America north of Mexico number together only fifteen.

The seaside earwig, *Anisolabis maritima*.—In this species both pairs of wings are wanting, the antennæ are 24-jointed, and the length of the body is from 18 to 20 mm. This earwig is found along the coast from Maine to Texas.

The ring-legged earwig, *Anisolabis annulipes*.—This is also a wingless species. The antennæ are only 15- or 16-jointed, the body is about 10 mm. long and the legs are ringed with fuscous. Its range does not extend as far north as that of the seaside earwig, but it extends farther inland.

The little earwig, *Labia minor*.—In this species the body is thickly clothed with fine yellowish pubescence. The body measures only from 4 to 5 mm. in length. Figure 538 represents the male, and Figure 540, the female. This species is widely distributed in the United States and is established in Canada.

The handsome earwig, *Prolabia pulchella*.—This species is widely distributed over the southern United States; it is found under the bark of dead trees. The body is dark chestnut-brown, shining and glabrous. It measures from 6 to 6.5 mm. in length. This species is dimorphic; in one form, known as *burgessi* (Fig. 541), the hind wings are shorter than the tegmina.

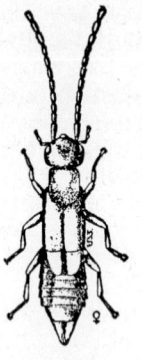

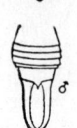

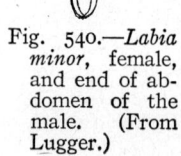

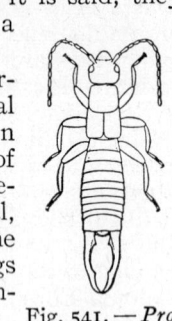

The spine-tailed earwig, *Dōru aculeātum*.—In the genus *Dōru* the pygidium of the male is armed with a distinct spine (Fig. 542). This species is dark chestnut-brown, with the palpi, legs, edges of pronotum, and the outer two-thirds of the tegmina yellow. The hind wings are usually aborted. The length of the body is 7.5 to 11 mm. It is distributed from New Jersey and southern Michigan to Nebraska, Georgia, and Louisiana.

The common European earwig, *Forficula auriculāria*.—In this species and in the preceding one as well, the second tarsal segment is lobed and prolonged beneath the third; but the two species can be distinguished by the shape of the forceps of the male (Fig. 543). The males of this species are dimorphic; in one form the forceps average about 4 mm. in length, in the other about 7 mm. This common European species appeared in great numbers at Newport, Rhode Island, about 1912.

FAMILY ARIXENIIDAE.—This family contains but a single genus, *Arixenia*, which, up to 1913 at least, contained but two species. One, *A. esau*, was taken from the breast-pouch of a bat, *Cheiromeles torquatus*, in Sarawak, Borneo. The other, *A. jacobsoni*, was taken on guano in a cave in Java which was frequented by bats. This species occurred in large numbers crawling over the ground and on the walls of the cave. The adults are very hairy, wingless, and viviparous. The antennæ have fourteen segments but the eyes are greatly reduced with from 70 to 80 facets. The mandibles are flattened and the cerci are rather long, unsegmented and curved, thus having the appearance of weak forceps. See Burr and Jordan, Trans. 2nd Entom. Cong., p. 398, 1913.

FAMILY HEMIMERIDAE.—This family includes but a single genus, *Hemimerus*, which now contains eight species, *hanseni* (see Figure 308, p. 269), *bouvieri*, *talpoides*, *vicinus*, *advectus*, *vosseleri*, *sessor*, and *deceptus*. These insects are small (from 8.5mm. to 15mm. in length), wingless, viviparous ectoparasites on rats of the genus *Cricetomys*. The known species are distributed throughout equatorial Africa as far south as the Transvaal. The eyes are absent and the cerci are elongated, unsegmented but not formed into forceps. The antennæ are well developed and have eleven segments while the legs, although short and stout, are fitted for rather rapid running among the hairs of the hosts. The body is broad and flattened, convex above and below, and uniformly orange-ochraceous in color, and the surface bears minute hairs. See Rehn and Rehn, Proc. Acad. Nat. SC. Phila., Vol. 87, pp. 457–508, 1936.

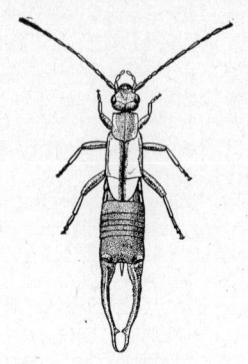

Fig. 542.—*Doru aculeatum*, male. (From Blatchley.)

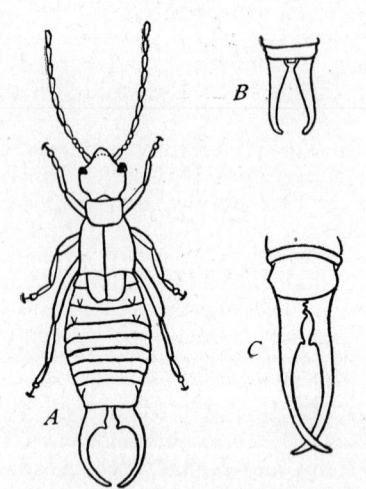

Fig. 543.—*Forficula auricularia: A*, male with short forceps; *B*, forceps of female; *C*, long type of forceps of male. (After Morse.)

CHAPTER XXIII

ORDER COLEOPTERA*

The Beetles

The winged members of this order have four wings; but the first pair of wings are greatly thickened, forming a pair of "wing-covers" or elytra, beneath which the membranous hind wings are folded when at rest. The elytra meet in a straight line along the middle of the back and serve as armor, protecting that part of the body which they cover. The mouthparts are formed for chewing. The metamorphosis is complete.

The order Coleoptera includes only the beetles. These insects can be readily distinguished from all others except the earwigs by the structure of the forewings, these being horny, veinless "wing-covers" or elytra, which meet in a straight line along the middle of the back (Fig. 544); and they differ from earwigs in lacking pincer-like appendages at the caudal end of the body. Beetles also differ from earwigs in having a complete metamorphosis.

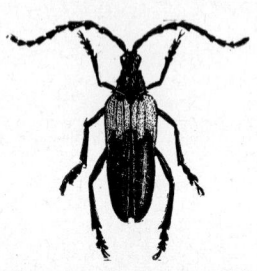

Fig. 544.—*Desmocerus palliatus.*

Only a few modifications of the typical characteristics exist in this order; among the more familiar of these are the following: in some of the Meloidæ the elytra do not meet in a straight line; in many of the Carabidæ, Curculionidæ, *et al.*, the hind wings are wanting, and in some of these the elytra are grown together; in a few females of the Lampyridæ and Phengodidæ both pairs of wings are wanting.

The different mouth-parts are very evenly developed; we do not find some of them greatly enlarged at the expense of others, as in several other orders of insects. The upper lip, or labrum, is usually distinct; the mandibles are powerful jaws fitted either for seizing prey or for gnawing; the maxillæ are also well developed and are quite complicated, consisting of several distinct pieces; the maxillary palpi are usually prominent; and the lower lip, or labium, is also well developed and complicated, consisting of several parts and bearing prominent labial palpi. Detailed figures of the maxillæ and labium of beetles are given in Chapter II.

In the classification of beetles much use is being made of the variations in form of the ventral and lateral sclerites of the thorax. Figure 545 will serve as an illustration of these sclerites. One feature merits special mention: the coxæ of the hind legs are flattened and immovably attached to the thorax so that they appear to be a part of the thorax instead of the basal segment of an appendage.

---

*Coleŏptera: *coleos* (κολεός), a sheath; *pteron* (πτερόν), a wing.

Almost the only use that has been made of the characteristics of the wings has been restricted to certain features of the elytra, those that can be seen without spreading the wings. These are the shape of the elytra, the presence or absence of striæ, the presence or absence of punctures and their distribution when present, and the presence or absence of **setæ,** pubescence, or scales on the surface of the elytra. A beginning has been made, however, to make use of the venation of the hind wings; this, as yet, is restricted to an indication of the type of wing-venation characteristic in each case of the superfamilies.

The venation of the wings of the Coleoptera has become greatly modified, and, consequently, the determination of the homologies of the wing-veins is a difficult matter. The transformation of the fore wings into elytra has resulted in a great reduction of their venation; and the foldings of the hind wings interrupt the veins and cause distortions in their courses.

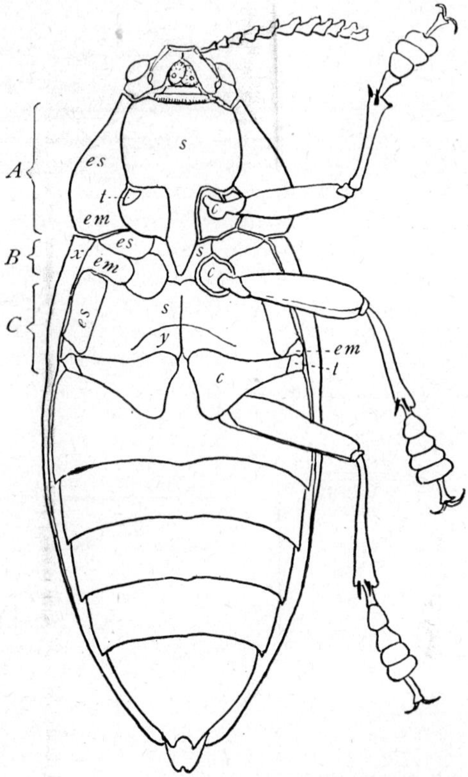

Fig. 545.—Ventral aspect of a beetle, *Enchroma gigantea*: *A*, prothorax; *B*, mesothorax; *C*, metathorax; *c, c, c,* coxæ; *em, em, em,* epimera; *es, es, es,* episterna; *s, s, s,* sterna; *t, t,* trochantins; *x,* elytrum; *y,* antecoxal piece of metasternum.

It is only recently that extended studies of the wing-venation of the Coleoptera have been made, and the conclusions reached by the different investigators are not fully in accord. But much progress has been made, and so much interest is being shown in the subject that we can confidently expect that conclusions will soon be reached that can be generally accepted.

Among the recent studies of the subject is an extended one by Dr. Wm. T. M. Forbes ('22 *b*), in which the venation of the hind wings of more than fifty species of beetles are figured. The accompanying figures (Figs. 546–547), copied from Dr. Forbes' paper, will serve to illustrate his conclusions regarding the homologies of the wing-veins of beetles. Another recent paper in which the venation of the wings of many beetles is figured is that of Graham ('22).

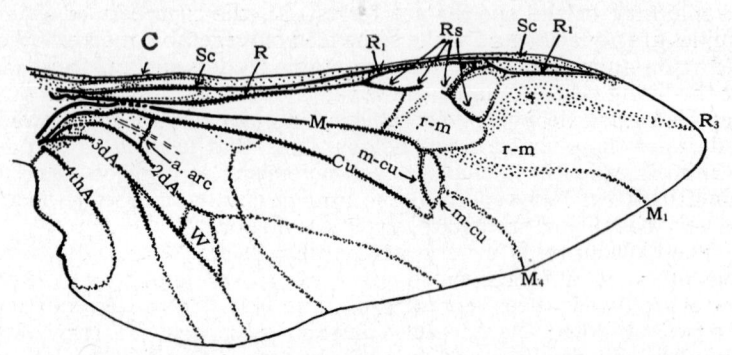

Fig. 546.—Tracheation of wing of imago of *Calosoma*. (From Forbes.)

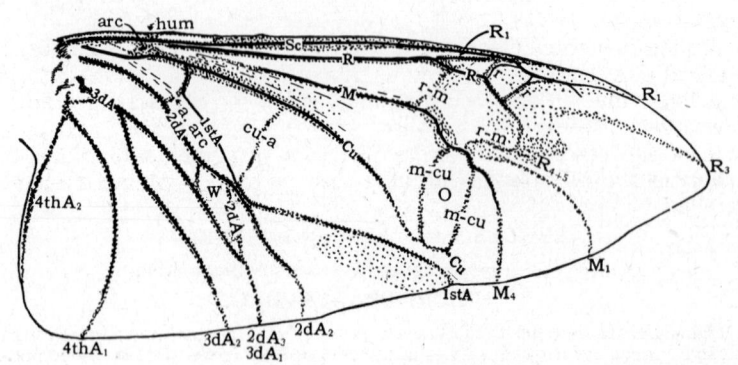

Fig. 547.—Tracheation of wing of imago of **Dytiscus verticalis**. (From Forbes.)

Beetles undergo a complete metamorphosis. The larvæ, which are commonly called grubs, vary greatly in form; some are campodeiform, others are scarabeiform, and some are vermiform. In some members of the order there is a hypermetamorphosis, the successive larval instars representing different types of larvæ; this is true of members of the Meloidæ and Micromalthidæ. Occasionally individuals of *Tenebrio molitor* are found in which the wings are developed externally. The pupæ are exarate, that is, the limbs are free (Fig. 548). These insects usually transform in rude cocoons made of earth or of bits of wood fastened together by a viscid substance excreted by the larvæ. Many wood-burrowing species transform in the tunnels made by the larvæ; and some of the Dermes-

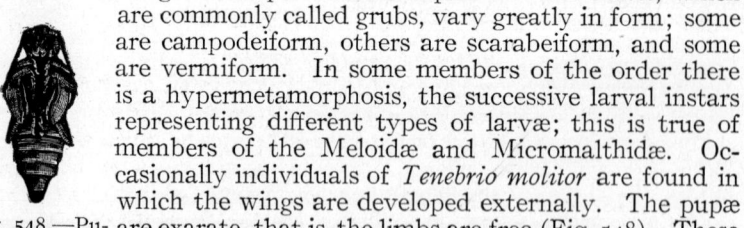

Fig. 548.—Pupa of a beetle.

tidæ as well as some of the Coccinellidæ transform in the last larval skin.

Both larvæ and adults present a very wide range of habits. While the majority of the species are terrestrial, the members of several families are aquatic; and while some feed on vegetable matter, others feed upon animal matter. The vegetable feeders include those that eat the living parts of plants, those that bore in dead wood, and those that feed upon decaying vegetable substances. Among the animal feeders are those that are predacious, those that feed on dried parts of animals, and those that act as scavengers, feeding on decaying animal matter. Viewed from the human standpoint, some species are very beneficial, others are extremely noxious.

The Coleoptera is a very large order; in the "Catalogue of the Coleoptera of America, North of Mexico" by Leng ('20), 18,547 species are listed; these represent 109 families. The order is divided into two suborders, the Adephaga and the Polyphaga. In each of the suborders the families are grouped into superfamilies, two in the Adephaga and twenty in the Polyphaga; and in the suborder Polyphaga the superfamilies are grouped into seven series of superfamilies.

Students of the Coleoptera are not fully agreed as to some of the details of this classification; but as this catalogue will doubtless serve, for a long time, in this country, as a guide for the arrangement of collections, it seems best to follow it in this introductory text-book. Some of the places where there is a lack of agreement among the authorities are indicated in the conspectus on page 38 of the Catalogue.

SYNOPSIS OF THE COLEOPTERA
(Tables for determining the families are given below)
I. SUBORDER ADEPHAGA

This suborder includes the first seven families, the Cicindelidæ to the Gyrinidæ inclusive, pages 476 to 484. The family Rhysodidæ (page 508) is also included in this suborder by some writers.

II. SUBORDER POLYPHAGA

This suborder includes all but the first seven families, or the first eight if the Rhysodidæ be included in the Adephaga.

The families of this suborder are grouped into seven series, as follows:—

SERIES I.—THE PALPICORNIA

This series includes a single family, the Hydrophilidæ; page 485.

SERIES II.—THE BRACHELYTRA

This series includes fifteen families, the Platypsyllidæ to the Histeridæ, inclusive, pages 486 to 490.

SERIES III.—THE POLYFORMIA

This series includes forty-three families, the Lycidæ to the Nosodendridæ, inclusive, pages 491 to 508.

SERIES IV.—THE CLAVICORNIA

This series includes thirty families, if the Rhysodidæ be placed here instead of in the suborder Adephaga; these are the families Rhysodidæ to Cisidæ, inclusive, pages 508 to 515.

468   AN INTRODUCTION TO ENTOMOLOGY

SERIES V.—THE LAMELLICORNIA

This series includes four families, the Scarabæidæ, the Trogidæ, the Lucanidæ, and the Passalidæ, pages 515 to 524.

SERIES VI.—THE PHYTOPHAGA

This series includes three families, the Cerambycidæ, the Chrysomelidæ, and the Mylabridæ, pages 524 to 535.

SERIES VII.—THE RHYNCHOPHORA

This series includes six families, the Brentidæ to the Scolytidæ, inclusive, pages 536 to 542.

TABLES FOR DETERMINING THE FAMILIES OF THE COLEOPTERA

TABLE I.—THE SUBORDERS AND THE SERIES OF SUPERFAMILIES

A. Ventral part of the first segment of the abdomen divided by the hind coxal cavities, so that the sides are separated from the very small medial part. Suborder ADEPHAGA; see Table II, below.
AA. Ventral part of the first segment of the abdomen visible for its entire breadth.   Suborder POLYPHAGA.

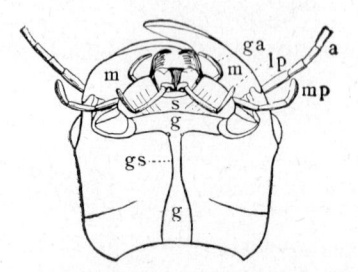

Fig. 549.—Head of *Harpalus*, ventral aspect: *a*, antenna; *g*, *g*, gula; *ga*, galea or outer lobe of the maxilla; *gs*, gular suture; *lp*, labial palpus; *m*, *m*, mandibles; *mp*, maxillary palpus; *s*, submentum.

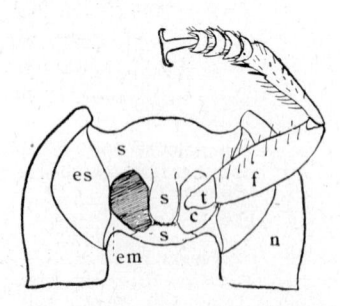

Fig. 550.—Prothorax of *Harpalus*, ventral aspect: *c*, coxa; *em*, epimerum; *es*, episternum; *f*, femur; *n*, pronotum; *s*, *s*, *s*, prosternum.

B. Head not prolonged into a narrow beak, palpi always flexible; two gula sutures at least before and behind (Fig. 549); sutures between the prosternum and the episterna and epimera distinct (Fig. 550); the epimera of the prothorax not meeting on the middle line behind the prosternum (Fig. 550).
   C. Abdomen with at least three corneous segments dorsally, and exposed more or less by the short elytra. Hind wings with simple, straight veins; antennæ variable, but never lamellate. Series BRACHELYTRA, See Table III, below.
   CC. Abdomen with at most two corneous segments dorsally, usually completely covered by the elytra; hind wings with veins in part connected by recurrent veins.
      D. Antennæ clubbed or not, but if clubbed not lamellate.
         E. Tarsi usually apparently four-jointed, the real fourth segment being reduced in size so as to form an indistinct segment at the base of the last segment, with which it is immovably united (Fig. 551, *A*); the first three segments of the tarsi dilated and brush-like beneath; the third segment bilobed. In two genera, *Parandra* and

Spondylis, the fourth segment of the tarsus, although much reduced and immovably united with the fifth, is distinctly visible, the first three

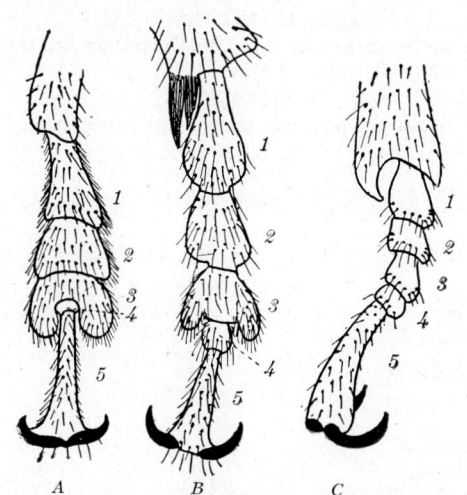

Fig. 551.—Tarsi of Phytophaga: *A*, typical; *B*, *Spondylis*; *C*, *Parandra*.

segments are but slightly dilated, and the third is either bilobed, *Spondylis* (Fig. 551, *B*), or not, *Parandra* (Fig. 551, *C*). Series PHYTOPHAGA. See Table VI, below.

EE. Tarsi varying in form and in the number of the segments, but when five-jointed not of the type described under E above, the joint between the fourth and fifth segments being flexible. Series PALPICORNIA, POLYFORMIA, and CLAVICORNIA. See Table IV, below.

DD. Antennæ with a lamellate club. Series LAMELLICORNIA. See Table V, below.

BB. Head either prolonged into a beak or not; palpi usually short and rigid; gular sutures confluent on the median line (Fig. 552, *gs*); prosternal sutures wanting; the epimera of the prothorax meeting on the middle line behind the prosternum (Fig. 552, *em*). Series RHYNCHOPHORA. See Table VII, below.

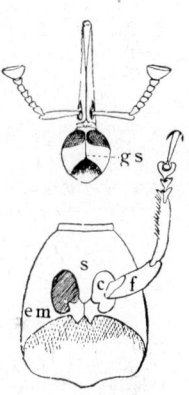

Fig. 552.—Head and prothorax of *Rhynchophorus*: *gs*, confluent gular sutures; *s*, prosternum; *em*, epimerum; *c*, coxa; *f*, femur.

TABLE II.—THE FAMILIES OF THE SUBORDER ADEPHAGA

A. Metasternum with an antecoxal piece, separated by a well-marked suture reaching from one side to the other and **extending** in a triangular process between the hind coxæ.

B. Antennæ eleven-jointed; hind coxæ mobile, and of the usual form; habits terrestrial.

C. Antennæ inserted on the front above the base of the mandibles. p. 476.
....CICINDELIDÆ

CC. Antennæ arising at the side of the head between the base of the mandibles and the eyes.
  D. Beetles of a round convex form in which the scutellum is entirely concealed. p. 481. .... Omophronidæ
  DD. Not such beetles. p. 478........................ Carabidæ
BB. Antennæ ten-jointed; hind coxæ fixed and greatly expanded so as to conceal the basal half of the hind femora and from three to six of the abdominal segments; habits aquatic. p. 481................. Haliplidæ
AA. Metasternum either with a very short antecoxal piece, which is separated by an indistinct suture and which is not prolonged posteriorly between the coxæ, or without an antecoxal piece.
 B. Metasternum with a very short antecoxal piece. p. 481.Amphizoidæ
 BB. Metasternum without an antecoxal piece.
   C. Legs fitted for swimming.
     D. With only two eyes. p. 482........................Dytiscidæ
     DD. With four eyes, two above and two below. p. 484. .Gyrinidæ
   CC. Legs fitted for walking. p. 508................... Rhysodidæ*

TABLE III.—THE FAMILIES OF THE BRACHELYTRA

A. Elytra short, leaving the greater part of the abdomen exposed; the suture between the elytra when closed straight; wings present, and when not in use folded beneath the short elytra; the dorsal part of the abdominal segments entirely horny.
 B. Abdomen flexible, and with seven or eight segments visible below. p. 488.
 ....Staphylinidæ
 BB. Abdomen not flexible, and with only five or six ventral segments visible.
   C. Antennæ with less than six joints. p. 490............Clavigeridæ
   CC. Antennæ eleven-jointed, rarely ten-jointed. p. 489..Pselaphidæ
AA. Elytra usually long, covering the greater part of the abdomen; when short the wings are wanting, or, if present, may or may not be folded under the short elytra when at rest; the dorsal part of the abdominal segments partly membranous.
 B. Hind tarsi five-jointed.
   C. Antennæ elbowed, and clavate. p. 490. ....Histeridæ
   CC. Antennæ rarely elbowed, and then not clavate.
     D. Abdomen with not more than five ventral segments.
       E. Antennæ capitate, the last three segments forming an abrupt club. p. 490 ....Sphæritidæ†
       EE. Antennæ but slightly clavate if at all. p. 490....Scaphidiidæ
     DD. Abdomen with six or more ventral segments.
       E. Anterior coxæ flat. p. 486......................Platypsyllidæ
       EE. Anterior coxæ either globular or conical.
         F. Anterior coxæ globular. p. 487................. Leptinidæ
         FF. Anterior coxæ conical.
           G. Posterior coxæ widely separated.
             H. Eyes wanting or inconspicuous. p. 487......Silphidæ
             HH. With well-developed eyes.
               I. Elytra covering the abdomen. p. 488...Scydmænidæ
               II. Elytra not covering the entire abdomen. p. 490......
               ....Scaphidiidæ
           GG. Posterior coxæ approximate.
             H. Posterior coxæ laminate. p. 488...........Clambidæ
             HH. Posterior coxæ not laminate.

---

*The Rhysodidæ is a very aberrant family, and its affinities have been much discussed. The form of the ventral part of the first abdominal segment is similar to that characteristic of the Adephaga; hence, according to Table I, this family should be placed in this suborder. But other characters led Leconte and Horn ('83) to place it in the Clavicornia, in which view they are followed by recent writers.

†See also p. 508, the Nitidulidæ of the series Clavicornia.

    I. Eyes with large facets. p. 486............BRATHINIDÆ
    II. Eyes with small facets. p. 487..............SILPHIDÆ
 BB. Hind tarsi either only three-jointed or four-jointed, but apparently three-jointed, the third segment being small and concealed in a notch at the end of the second segment. (See also BBB and BBBB.)
  C. Abdomen with six or seven ventral segments.
   D. Tarsi four-jointed, the third segment small and concealed in a notch at the end of the second segment. p. 488..........CORYLOPHIDÆ
   DD. Tarsi three-jointed. p. 490.................TRICHOPTERYGIDÆ
  CC. Abdomen with only three ventral segments. p. 490. SPHÆRIIDÆ
 BBB. All tarsi four-jointed. (See also BBBB.)
  C. Hind coxæ contiguous and with plates covering the femora entirely or in part. p. 487...SILPHIDÆ
  CC. Hind coxæ separate and not covering the femora. p. 488 CORYLOPHIDÆ
 BBBB. Hind tarsi with only four segments; the fore tarsi, and almost always the middle tarsi also, with five segments. p. 487.............SILPHIDÆ

TABLE IV.—THE FAMILIES OF THE PALPICORNIA, POLYFORMIA, AND CLAVICORNIA

 It is impracticable to separate these three series of families in these tables, owing to the fact that characters sharply separating them have not been found.
A. Hind tarsi five-jointed.
 B. Maxillary palpi as long as or longer than the antennæ. p.485 HYDROPHILIDÆ
 BB. Maxillary palpi much shorter than the antennæ.
  C. Tarsal claws very large; the first three abdominal segments grown together on the ventral side.
   D. Abdomen with more than five ventral segments; anterior coxæ with very large trochantin. p. 503...................PSEPHENIDÆ
   DD. Abdomen with five ventral segments.
    E. Anterior coxæ transverse, with distinct trochantin. p.504.DRYOPIDÆ
    EE. Anterior coxæ rounded, without trochantin. p. 504...ELMIDÆ
  CC. Tarsal claws of usual size; ventral abdominal segments usually free, sometimes (Buprestidæ) the first two grown together.
   D. Abdomen with not more than five ventral segments.
    E. Femur joined to the apex or very near the apex of the trochanter.
     F. Antennæ inserted upon the front. p. 514............PTINIDÆ
     FF. Antennæ inserted before the eyes.
      G. Tibiæ without spurs. p. 514..................ANOBIIDÆ
      GG. Tibiæ with distinct spurs.
       H. First ventral segment scarcely longer than the second. p. 515.......................................BOSTRICHIDÆ
       HH. First ventral segment elongated. p. 515......LYCTIDÆ
    EE. Femur joined to the side of the trochanter.
     F. Anterior coxæ globular or transverse, usually projecting but little from the coxal cavity.
      G. Anterior coxæ transverse, more or less cylindrical.
       H. Posterior coxæ grooved for the reception of the femora.
        I. Legs stout, retractile; tibiæ dilated, usually with a furrow near the outer end for the reception of the tarsi; tibial spurs distinct.
         J. Antennæ inserted at the side of the head.
          K. Head prominent, mentum large. p. 508.NOSODENDRIDÆ
          KK. Head retracted, mentum small. p. 508.BYRRHIDÆ
         JJ. Antennæ inserted on the front; head retracted. p. 506.
....CHELONARIIDÆ
        II. Tibiæ slender, with small and sometimes obsolete terminal spurs, or without spurs.
         J. Head constricted behind; eyes smooth. p. 494.CUPESIDÆ
         JJ. Head not constricted behind; eyes granulated.

K. Anterior coxæ with a distinct trochantin. p. 505.
.... DASCILLIDÆ
KK. Anterior coxæ without trochantin.
L. Lacinia of the maxillæ armed with a terminal hook. p. 505..................... EUCINETIDÆ
LL. Lacinia not armed with a terminal hook. p. 505.
.... HELODIDÆ
HH. Posterior coxæ flat; not grooved for the reception of the femora.
I. Tarsi more or less dilated, first segment not short.
J. Antennæ eleven-jointed, terminated by a three-jointed club. p. 508............................ NITIDULIDÆ
JJ. Antennæ ten-jointed, club two-jointed. p. 508.....
.... RHIZOPHAGIDÆ
II. Tarsi slender, first segment short. p. 508..OSTOMIDÆ
GG. Anterior coxæ globular.
H. Prosternum with a process which extends backward into a groove in the mesosternum.
I. The first two abdominal segments grown together on the ventral side. p. 502...................... BUPRESTIDÆ
II. Ventral segments free.
J. Prothorax loosely joined to the mesothorax; front coxal cavities entirely in the prosternum.
K. Posterior coxæ laminate; trochanters small.
L. Antennæ somewhat distant from the eyes, their insertion narrowing the front. p. 502..... EUCNEMIDÆ
LL. Antennæ inserted under the margin of the front.
M. Antennæ arising near the eyes. p. 499. ELATERIDÆ
MM. Antennæ arising at a distance in front of the eyes (*Perothops*). p. 502........... EUCNEMIDÆ
KK. Posterior coxæ not laminate; trochanters of middle and posterior legs very long. p. 499.... CEROPHYTIDÆ
JJ. Prothorax firmly joined to the mesothorax; front coxal cavities closed behind by the mesosternum. p. 502
.... THROSCIDÆ
HH. Prosternum without a process received by the mesosternum, although it may be prolonged so as to meet the mesosternum.
I. Posterior coxæ contiguous. p. 511.......... PHALACRIDÆ
II. Posterior coxæ separated.
J. Body depressed; middle coxal cavities not closed externally by a meeting of the mesosternum and metasternum. p. 509.......................... CUCUJIDÆ
JJ. Body more or less convex; middle coxal cavities entirely surrounded by the sterna.
K. Prosternum not prolonged behind. p. 510. ..
.... MYCETOPHAGIDÆ
KK. Prosternum prolonged, meeting the mesosternum.
L. Anterior coxal cavities open behind. p. 510....
.... CRYPTOPHAGIDÆ
LL. Anterior coxal cavities closed behind. p. 509
.... EROTYLIDÆ
FF. Anterior coxæ conical, and projecting prominently from the coxal cavity.
G. Posterior coxæ dilated into plates partly protecting the femora, at least at their bases.
H. Antennæ serrate or flabellate. p. 499....... RHIPICERIDÆ
HH. Antennæ with the last three segments forming a large club.
I. Tarsi with second and third segments lobed beneath. p. 510..................................... BYTURIDÆ
II. Tarsi simple. p. 506..................... DERMESTIDÆ

HHH. Antennæ with the last three segments somewhat larger than the preceding, but not suddenly enlarged. p. 510.
...................................Derodontidæ
GG. Posterior coxæ not dilated into plates partly protecting the femora.
H. Posterior coxæ flat, not prominent, covered by the femora in repose.
I. Tarsi with the fourth joint of normal size. p. 493. Cleridæ
II. Tarsi with the fourth joint very small. p. 493. Corynetidæ
HH. Posterior coxæ conical and prominent.
I. Anterior coxæ with distinct trochantins. p. 493. Melyridæ
II. Anterior coxæ without trochantins. p. 493. Lymexylidæ
DD. Abdomen with six or more ventral segments.
E. Anterior coxæ globular.
F. Tibial spurs well developed. p. 499...............Cebrionidæ
FF. Tibial spurs very delicate and short. p. 499....Plastoceridæ
EE. Anterior coxæ conical.
F. Posterior coxæ not prominent, flat, covered by the femora in repose.
G. Tarsi with the fourth joint of normal size. p. 493...Cleridæ
GG. Tarsi with the fourth joint very small. p. 493. Corynetidæ
FF. Posterior coxæ more or less conical and prominent at least internally, not covered by the femora in repose.
G. Anterior coxæ long, with distinct trochantins.
H. Abdomen with seven or eight ventral segments.
I. Middle coxæ contiguous; epipleuræ distinct.
J. Episterna of metathorax not sinuate on inner side, epipleuræ usually wide at the base.
K. Head more or less covered; antennæ approximate or moderately distant; metathoracic epimera long. p. 491.
...................................Lampyridæ
KK. Head exposed; antennæ distant; metathoracic epimera wide. p. 492.................Phengodidæ
JJ. Episterna of metathorax sinuate on the inner side; epipleuræ narrow at the base. p. 492......Cantharidæ
II. Middle coxæ distant; epipleuræ wanting. p. 491. Lycidæ
HH. Abdomen with only six ventral segments. p. 493.Melyridæ
GG. Anterior coxæ without trochantins.
H. Elytra entire; length of body 10 mm. or more. p. 493..
...................................Lymexylidæ
HH. Elytra shorter than the abdomen; length of body less than 3 mm. p. 494......................Micromalthidæ
AA. Hind tarsi either only three-jointed, or four-jointed but apparently only three-jointed, the third joint being small and concealed in a notch at the end of the second joint. (See also AAA and AAAA.)
B. Wings fringed with long hairs. A minute aquatic species from S. Cal. and Ariz. (*Hydroscapha*). p. 485.....................Hydrophilidæ
BB. Wings not fringed with hairs.
C. Tarsi with second segment dilated.
D. Tarsal claws appendiculate or toothed; first ventral abdominal segment with distinct curved coxal lines. p. 511..........Coccinellidæ
DD. Tarsal claws simple; first ventral abdominal segment without coxal lines. p. 511..............................Endomychidæ
CC. Tarsi with second segment not dilated.
D. Elytra entirely covering the abdomen; ventral abdominal segments nearly equal. p. 511..............................Lathridiidæ
DD. Elytra truncate, the first and fifth abdominal segments longer than the others.
E. Maxillæ with galea distinct; anterior coxæ small, rounded. p. 509.
...................................Monotomidæ
EE. Galea wanting, anterior coxæ subtransverse. p. 508. Nitidulidæ

AAA. All tarsi four-jointed. (See also AAAA.)
  B. The first four abdominal segments grown together on the ventral side.
    C. Tibiæ dilated, armed with rows of spines, and fitted for digging. p. 505.
....HETEROCERIDÆ
    CC. Tibiæ neither dilated nor fitted for digging.
      D. Antennæ inserted under a distinct frontal ridge; anterior coxæ distant from the metasternum. p. 510.........................COLYDIIDÆ
      DD. Antennæ inserted on the front; anterior coxæ inclosed behind by the metasternum. p. 511............................MURMIDIIDÆ
  BB. Ventral segments of abdomen not grown together.
    C. Anterior coxæ transverse. p. 511....................MYCETÆIDÆ
    CC. Anterior coxæ either globose or oval.
      D. Anterior coxæ globose.
        E. Tarsi slender. p. 511............................ENDOMYCHIDÆ
        EE. Tarsi more or less dilated and spongy beneath. p. 509. EROTYLIDÆ
      DD. Anterior coxæ oval.
        E. Anterior coxæ separated by the horny prosternum.
          F. Body depressed; head free. p. 510.........MYCETOPHAGIDÆ
          FF. Body cylindrical; thorax prolonged over the head. p. 515.
....CISIDÆ
        EE. Anterior coxæ contiguous; prosternum semimembranous. p. 505.
....GEORYSSIDÆ
AAAA. Hind tarsi with only four segments; the fore tarsi, and almost always the middle tarsi also, with five segments.
  B. Anterior coxal cavities closed behind.
    C. Tarsal claws simple.
      D. Abdomen with five ventral segments.
        E. Ventral abdominal segments in part grown together.
          F. Next to the last segment of the tarsi spongy beneath. p. 514.
....LAGRIIDÆ
          FF. Penultimate segment of tarsi not spongy. p. 513. TENEBRIONIDÆ
        EE. Ventral abdominal segments free.
          F. Anterior coxal cavities confluent. p. 498............OTHNIIDÆ
          FF. Anterior coxal cavities separated by the prosternum.
            G. Elytra truncate; tip of abdomen exposed. p. 508.RHIZOPHAGIDÆ
            GG. Elytra entire. p. 515.......................SPHINDIDÆ
      DD. Abdomen with six ventral segments. p. 498......EURYSTETHIDÆ
    CC. Tarsal claws pectinate. p. 512......................ALLECULIDÆ
  BB. Anterior coxal cavities open behind.
    C. Head not strongly and suddenly constricted at base.
      D. Middle coxæ not very prominent.
        E. Antennæ received in grooves. p. 514...............MONOMMIDÆ
        EE. Antennæ free.
          F. Prothorax margined at the sides.
            G. Middle coxal cavities entirely surrounded by the sterna. p. 510.
....CRYPTOPHAGIDÆ
            GG. Epimera of mesothorax reaching the coxæ.
              H. Metasternum long; epimera of metathorax visible. p. 514.
....MELANDRYIDÆ
              HH. Metasternum quadrate; epimera of metathorax covered. p. 509...CUCUJIDÆ
          FF. Prothorax not margined at the sides. p. 498..........PYTHIDÆ
      DD. Middle coxæ very prominent. p. 494...............ŒDEMERIDÆ
    CC. Head strongly constricted at base.
      D. Head prolonged behind and gradually narrowed. p. 494.CEPHALOIDÆ
      DD. Head suddenly narrowed behind.
        E. Prothorax with the side pieces not separated from the pronotum by a suture.
          F. Prothorax at base narrower than the elytra.
            G. Hind coxæ not prominent or but slightly so.
              H. Anterior coxæ globular, not prominent. p. 509.CUCUJIDÆ

HH. Anterior coxæ conical, prominent.
    I. Abdomen composed of five free segments; tarsi with the penultimate joint lobed beneath.
        J. Neck wide; eyes large, finely faceted, and generally emarginate. p. 498......................PEDILIDÆ
        JJ. Neck narrow, eyes not emarginate.
            K. Eyes large, oval, rather finely faceted. p. 498.PEDILIDÆ
            KK. Eyes small, rounded, generally coarsely faceted. p.498.
....ANTHICIDÆ
    II. Abdomen composed of four free segments, the first formed of two united, with the suture sometimes indicated; tarsi with the antepenultimate joint lobed beneath. p. 499..
....EUGLENIDÆ
    GG. Hind coxæ large, prominent.
        H. Tarsal claws simple; head horizontal. p. 498.PYROCHROIDÆ
        HH. Claws cleft or toothed, front vertical. p.495.MELOIDÆ
FF. Prothorax at base as wide as the elytra. p. 494..RHIPIPHORIDÆ
EE. Lateral suture of prothorax distinct; base of prothorax as wide as the elytra.
    F. Antennæ filiform.
        G. Hind coxæ plate-like. p. 494..................MORDELLIDÆ
        GG. Hind coxæ not plate-like. p. 514.......... .MELANDRYIDÆ
    FF. Antennæ flabellate in the male, subserrate in the female. p. 494.
....RHIPIPHORIDÆ

TABLE V.—THE FAMILIES OF THE LAMELLICORNIA

A. Plates composing the club of the antennæ flattened and capable of close apposition.
    B. Abdomen with six visible ventral segments. p. 515.......SCARABÆIDÆ
    BB. Abdomen with five visible ventral segments.
        C. Epimera of mesothorax attaining the oblique coxæ. p. 515.SCARABÆIDÆ
        CC. Epimera of mesothorax not attaining the coxæ. p. 522..TROGIDÆ
AA. Plates composing the club of the antennæ not capable of close apposition, and usually not flattened.
    B. Mentum deeply emarginate, ligula filling the emargination. p. 524..
....PASSALIDÆ
    BB. Mentum entire, ligula covered by the mentum or at its apex. p. 523.
.... LUCANIDÆ

TABLE VI.—FAMILIES OF THE PHYTOPHAGA

This series includes three families, which are so connected by intermediate forms that it is not easy to separate them. The following table will aid the student in separating the more typical forms.

A. Body elongate; antennæ almost always long, often as long as the body or longer. The larvæ are borers. p. 524...................... CERAMBYCIDÆ
AA. Body short and more or less oval; antennæ short.
    B. Front prolonged into a broad quadrate beak; elytra rather short so that the tip of the abdomen is always exposed. The larvæ live in seeds. p. 535
....MYLABRIDÆ.
    BB. Front not prolonged into a beak; usually the tip of the abdomen is covered by the elytra. Both larvæ and adults feed on the leaves of plants. p. 530...CHRYSOMELIDÆ

TABLE VII.—THE FAMILIES OF THE RHYNCHOPHORA
(Compiled from Blatchley and Leng)

A. Beak rarely absent, usually longer than broad; tibiæ never with a series of teeth externally.

B. Antennæ straight without a distinct club, though with the outer joints often more or less thickened; beak present at least in the female and pointing directly forward; form usually very slender and elongate. p. 536..Brentidæ

BB. Antennæ straight or elbowed, always with a distinct club.
   C. Palpi flexible; antennal club rarely compact; beak always short and broad; labrum present; thorax with a transverse raised line which is either ante-basal or basal. p. 536..................... Platystomidæ
   CC. Palpi rigid and labrum wanting except in the subfamily Rhinomacerinæ antennal club usually compact; beak variable in length, often long and curved downwards. p. 537........................ Curculionidæ*

AA. Beak absent or extremely short and broad; tibiæ with a series of teeth externally, or, if these are wanting, with a prominent curved spine at apex; antennæ short, but little longer than the head, always elbowed, and with a compact club except in *Phthorophœlus* where the club is lamellate; palpi rigid; body short, subcylindrical or rarely oval.
  B. Anterior tarsi with the first segment longer than the second, third and fourth together. p. 541—................................. Platypodidæ
  BB. Anterior tarsi with the first segment shorter than the second, third, and fourth together. p. 542................................. Scolytidæ

Suborder ADEPHAGA†

The name of this suborder, Adephaga, was suggested by the predacious habits of its members. These beetles are distinguished from other Coleoptera by the presence of a suture on each side of the prothorax separating the pleurum from the notum, and by the fact that the ventral part of the first segment of the abdomen is divided by the hind coxal cavities so that the sides are separated from the very small medial part (Fig. 553).

The Adephaga differ from other Coleoptera in that the nutritive cells of the ovaries alternate with the egg-chambers.

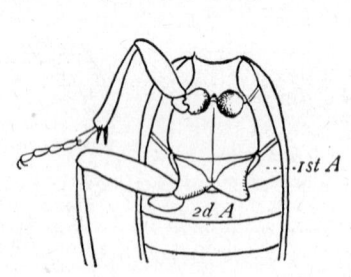

Fig. 553.—Ventral aspect of part of thorax and abdomen of *Galerita janus*: *1st A*, first abdominal segment; *2d A*, second abdominal segment.

The larvæ are campodeiform, and differ from all other beetle larvæ in that their legs are six-jointed except in a single exotic species; this is one more segment than is found in the legs of other beetle larvæ. The legs are usually furnished with two claws, whereas the legs of other coleopterous larvæ are one-clawed.

This suborder is represented in North America by seven families; these can be separated by Table II, page 469.

Family CICINDELIDÆ

The Tiger-Beetles

The graceful forms and beautiful colors of the greater number of the tiger-beetles, those of the genus *Cicindela*, have made the

---

*Since this table was published by Blatchley and Leng, the family Belidæ has been separated from the Curculionidæ. See page 537.
†Adĕphaga: *adephagous* (ἀδηφάγος), voracious.

family one of the favorites of students of Coleoptera. Their popular name is suggestive of their predacious habits, and of the stripes with which many are marked. They are usually a metallic green or bronze, banded or spotted with yellow. Some are black; and some that live on white sand are grayish white, being exactly like the sand in color. Figure 554 represents a common species of *Cicindela*.

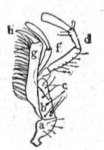

Fig. 554.

Fig. 555.

A useful character for distinguishing the members of this family is the fact that the terminal hook of the maxilla (the digitus) is united to this organ by a movable joint (Fig. 555, *h*).

The sexes of the tiger-beetles can be distinguished, except in *Amblycheila*, by the sixth abdominal segment of the males being notched so as to expose a small seventh segment; while in the females only six segments are visible. In the males, the first three segments of the anterior tarsi are usually dilated and densely clothed with hair beneath.

The tiger-beetle larvæ (Fig. 556) are as ugly and ungraceful as the adults are beautiful. The two have only one habit in common—their eagerness for prey. The larvæ live in vertical burrows in sandy places or in beaten paths. These burrows occur also in ploughed fields that have become dry and hard. They often extend a foot or more in depth. The larva takes a position of watchfulness at the mouth of its burrow. Its dirt-colored head is bent at right angles to its lighter-colored body and makes a neat plug to the opening of the hole. Its rapacious jaws extend upward, wide open, ready to seize the first unwary insect that walks over this living trap, or near it; for a larva will throw its body forward some distance in order to seize its prey. On the fifth segment of the abdomen there is a hump, and on this hump are two hooks curved forward. This is an arrangement by which the little rascal can hold back and keep from being jerked out of its hole when it gets some large insect by the leg, and by which it can drag its struggling prey down into its lair, where it may eat it at leisure. It is interesting to thrust a straw down into one of these burrows, and then dig it out with a trowel. The chances are that you will find the indignant inhabitant at the remote end of the burrow, chewing savagely at the end of the intruding straw.

Fig. 556.

One hundred and fourteen species of tiger-beetles are now listed in our fauna; these represent four genera, which can be separated as follows:

A. Posterior coxæ contiguous; eyes large, prominent.
  B. Third joint of the maxillary palpi shorter than the fourth...Cicindela
  BB. Third joint of the maxillary palpi longer than the fourth.....Tetracha
AA. Posterior coxæ separated; eyes small.

B. Sides of the elytra widely inflexed; thorax scarcely margined. AMBLYCHEILA
BB. Sides of the elytra narrowly inflexed; thorax distinctly margined...OMUS

Cicindēla.—To this genus belong the greater number of our tiger-beetles; seventy-six species and many varieties occur in our fauna; excepting the two species of *Tetracha*, all of the tiger-beetles found in the East belong to the genus *Cicindela*.

The members of this genus, unlike most other members of the family, are diurnal in habit. They are found on bright, hot days in dusty roads, in beaten paths, and on the shores of streams. They are the most agile of all beetles; and they are not merely swift of foot, but are also able to fly well. When approached, they remain still until we can see them well but are still out of reach; then like a flash they fly up and away, alighting several rods ahead of us. Before alighting they usually turn so that they face us, and can thus watch our movements. They hide by night and in cloudy or rainy weather in holes in the ground or beneath stones or rubbish. The beetles have been found hibernating, each in a separate burrow extending under a stone. I have seen them in September digging burrows in a hillside; these descended slightly and were about five inches deep. The beetles kicked the dirt out behind them as they dug, so that it lay in a heap at the opening of the hole.

Tĕtracha.—Two species of this genus are widely distributed in the United States. They are rather large, metallic-green beetles. Figure 557 represents *Tĕtracha carolīna*, which can be distinguished by the apical portion of the elytra being yellow. Our only other species is *Tĕtracha virgĭnica*. These beetles are nocturnal, hiding during the day and hunting by night.

Amblycheila.—The best-known representative of this genus is *Amblycheila cylindrifŏrmis*, which is found in Kansas, Colorado, Arizona, and New Mexico. It is a very large species, measuring 35 mm. in length. It is nocturnal, hiding in holes during the day and coming forth at night to capture its prey. Two other species of this genus have been described from Arizona and Utah.

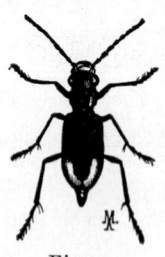

Fig. 557.

Omus.—Thirty-three species of this genus have been found on the Pacific Coast, nearly all of them in California. They are nocturnal insects, hiding under rubbish during the daytime.

Family CARABIDÆ
The Ground-Beetles

The ground-beetles are so called because they are very common on the surface of the ground, lurking under stones or rubbish, where they hide by day. At night they roam about in search of their prey. Our more common species are easily recognized by their shining black color and long legs. On the Pacific Coast, however, the darkling beetles (Family Tenebrionidæ), which are also black and have long legs, abound under stones and fragments of wood on the ground.

But the two families can be easily distinguished by the fact that in the ground-beetles all the tarsi are five-jointed, while in the darkling beetles the hind tarsi are only four-jointed; and the darkling beetles do not run rapidly as do the ground-beetles.

With the ground-beetles, the antennæ are thread-like, tapering gradually towards the tip, and each segment is of nearly uniform thickness throughout its length; the legs are fitted for running, and the antennæ are inserted between the base of the mandibles and the eyes. Although most of the species are black, there are those that are blue, green, or brown, and a few that are spotted. The wing-covers are almost always ornamented with longitudinal ridges and rows of punctures.

Most members of this family are predacious, feeding upon other insects, which they spring upon or capture by chase. A few species use vegetable food; but their depredations are rarely of economic importance. As there are more than two thousand described North American species, and as many of the species are very common, this family may be considered the most important family of the predacious insects.

The larvæ of ground-beetles are generally long, with the body of nearly equal breadth throughout (Fig. 558). They have sharp projecting mandibles; and the caudal end of the body is usually furnished with a pair of conical bristly appendages. They live in the same obscure situations as the adult insects, but are more shy, and are consequently less frequently seen. Like the adults, they are predacious.

Fig. 558.

Among the more common ground-beetles are the following.

The searcher, *Calosōma scrutātor*.—This is one of the larger and more beautiful of our ground-beetles; it has green or violet wing-covers margined with reddish, and the rest of the body is marked with violet-blue, gold, green, and copper (Fig. 559). This beetle and the two following have been known to climb trees in search of caterpillars.

Calosōma sycophănta, a common species in Europe, has been introduced and successfully colonized in New England, as a means of combating the gipsy-moth and the brown-tail moth. This species is somewhat smaller than the preceding, and lacks the reddish band on the margins of the elytra.

The fiery hunter, *Calosōma călidum*, is easily recognized by the rows of reddish or copper-colored pits on the wing-covers (Fig. 560).

Fig. 559.

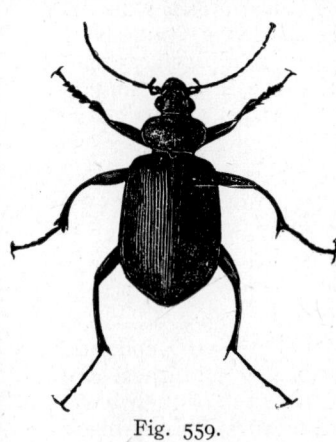

The bombardier-beetles, *Brachīnus*.—There are many species of beetles that have at the hind end of the body little sacs in which is secreted a bad-smelling fluid, which is used as a means of defence. These beetles spurt this fluid out onto their enemies when attacked. But in the case of the bombardier-beetles this fluid changes to a gas, which looks like smoke as soon as it comes in contact with the air, and is ejected with a sound like that of a tiny popgun. When some larger insect tries to capture one of these insect-soldiers, and gets very near it, the latter fires its little gun into the face of its enemy. The noise astonishes the pursuer, and the smoke blinds him. By the time he has recovered from his amazement, the little bombardier is at a safe distance. These beetles have quite a store of ammunition; for we have often had one pop at us four or five times in succession, while we were taking it prisoner. The bombardier-beetles belong to the genus *Brachinus*, of which we have in this country twenty-seven species. They are very similar in appearance; the head, prothorax, and legs are reddish yellow, and the wing-covers are dark blue, blackish, or greenish blue (Fig. 561).

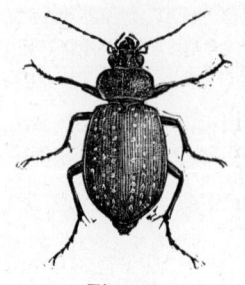

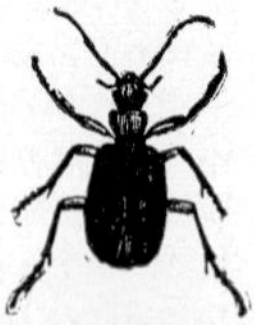

Fig. 560.

There is a common beetle which resembles the bombardier-beetles quite closely in size and color, but which may be distinguished by the comb-like form of the tarsal claws; this is *Lebia grandis* (Fig. 562). It has been reported more often than any other insect as destroying the Colorado potato-beetle.

Galerīta jānus is still another species that bears some resemblance to the bombardier-beetles. But it is much larger, measuring 16 mm. in length, and has only the prothorax and legs reddish yellow, the head being black; the prothorax is only about half as wide as the wing-covers.

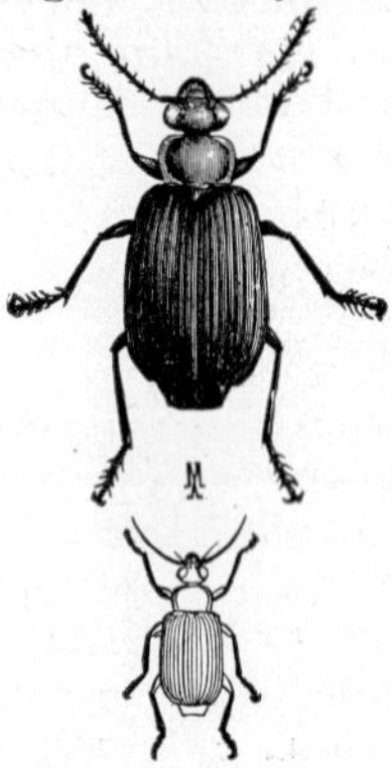

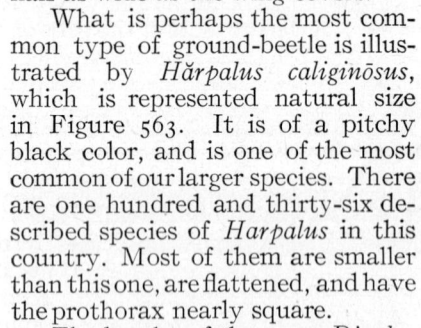

Fig. 561.

What is perhaps the most common type of ground-beetle is illustrated by *Hărpalus caliginōsus*, which is represented natural size in Figure 563. It is of a pitchy black color, and is one of the most common of our larger species. There are one hundred and thirty-six described species of *Harpalus* in this country. Most of them are smaller than this one, are flattened, and have the prothorax nearly square.

Fig. 562.—*Lebia grandis*, natural size and enlarged.

Fig. 563.

The beetles of the genus *Dicælus* are quite common; and some of the larger species resemble *Harpalus caliginosus* quite closely. They can be distinguished by a prominent keel-shaped ridge which extends back upon each wing-cover from near the corner of the prothorax.

Fig. 564.

The most common of all ground-beetles, in the Northeastern States at least, is *Pœcilus lucublăndus*. In this species (Fig. 564) the narrow, flat margin on each side of the prothorax is widened near the hind angle of this segment.

The family AMPHIZOIDÆ is represented in our fauna by two species of *Amphizōa*, which occur in California, Vancouver, and Alaska, clinging to logs or stones under the surface of streams. In these beetles the metasternum is truncate behind, not reaching the abdomen, and has a very short antecoxal piece.

The family OMOPHRONIDÆ consists of a single genus, *Ŏmophron*, the members of which are remarkable for their round form and the fact that the scutellum is entirely concealed. They measure about 6 mm. in length, and are found in holes in wet sand near the margins of streams and ponds. They are found from the Atlantic to the Pacific; fourteen species have been described.

Family HALIPLIDÆ

The Crawling Water-Beetles

This family includes a few species of small aquatic beetles, which are oval, more or less pointed at each end, and very convex; our larger species measure from 3.5 mm. to 5 mm. in length, but some are much smaller. The wing-covers have rows of punctures, and the hind coxæ are greatly expanded so as to conceal the basal half of the hind femora and from three to six of the abdominal segments. The anterior and middle tibiæ and the tarsi of all of the legs are furnished with long, swimming hairs.

These beetles are found in ponds and streams, but most frequently in spring-fed pools that do not dry up during the summer, and contain filamentous algæ and other aquatic plants. They swim poorly but crawl over the stems of aquatic plants. Little is known regarding the feeding habits of the adults. Matheson ('12) found that several species feed on the contents of the cells of *Nitella* and the softer portions of *Chara* and other filamentous algæ. He observed also that two species of *Peltodytes* attach their eggs to aquatic plants, mainly *Nitella* and *Chara*, while *Haliplus ruficollis* places its eggs within the dead cells of *Nitella*.

The larvæ are aquatic, living in the same pools as the adults. The body is slender; each segment except the head is furnished on the back with fleshy lobes with spiny tips (Fig. 565), which vary greatly in size in different species; in the larvæ of *Peltodytes* each of these spines bears a long, jointed filament, which is a tracheal gill. The larvæ of this genus have no spiracles; but the larvæ of *Haliplus* possess both thoracic and abdominal spiracles. The larvæ of the Haliplidæ feed on filamentous algæ; when mature, they leave the water and each makes a cell in the damp earth in which the pupa state is passed.

Fig. 565.

About forty species of the Haliplidæ have been found in our fauna; these represent three genera. In *Brӳchius*, which is represented by two species in California, the prothorax is quadrate; in the other genera it is narrowed in front. In *Hăliplus* the last segment of the palpi is small and awl-shaped; in *Peltŏdytes* it is longer than the third segment, and conical. The last two genera are widely distributed.

Family DYTISCIDÆ
The Predacious Diving-Beetles

If one will approach quietly a pool of standing water, there may be seen oval, flattened beetles hanging head downward, with the tip of the abdomen at the surface of the water. Such beetles belong to this family.

The predacious diving-beetles are usually brownish black and shining, but are often marked indefinitely with dull yellow They can be distinguished from the water scavenger-beetles, which they resemble in general appearance, by the thread-like form of the antennæ. The hind legs are the longest and are fitted for swimming, being flattened and fringed with hair. The middle and the hind pair of legs are widely separated. This is due to the very large hind coxæ which cover the greater part of the lower surface of the thorax.

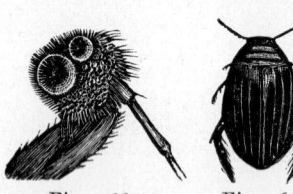

In the males of certain genera the first three segments of the fore tarsi are dilated and form a circular disk, upon the under side of which are little cup-like suckers (Fig. 566); these serve as clasping organs. In a few cases the middle tarsi are dilated also. The females of some species exhibit an interesting dimorphism in that some of the individuals have the elytra furnished with a number of deep furrows (Fig 567), while others of the same species have them smooth.

Fig. 566.   Fig. 567.

The diving-beetles abound in our streams and ponds, but they are more often found in standing water than in streams. When at rest they float in an inclined position, head downward, with the tip of the hind end of the body projecting from the water. The spiracles open on the dorsal side of the abdomen beneath the elytra. By lifting the elytra slightly a reservoir is formed for air, which the beetle can breathe as it swims through the water. When the air becomes impure the beetle rises to the surface, forces it out, and takes a fresh supply.

These beetles are very voracious. They destroy not only other insects, but some of them will attack larger animals, as small fish. When kept in aquaria they can be fed upon any kind of meat, raw or cooked. They fly from pond to pond, and are often attracted to light at night. Many of the species make sounds, both under the water and in the air. In some cases this is done by rubbing the abdominal segments upon the elytra; in others, by rubbing the hind legs upon a rough spot on the lower side of the abdomen.

The females deposit their eggs singly in punctures in the tissues of living plants. The larvæ are known as water-tigers, because of their blood-thirstiness. They are elongated, spindle-form grubs (Fig. 568). The

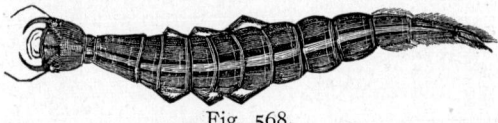

Fig. 568.

head is large, oval or rounded, and flattened; the mandibles are large and sickle-shaped; in each there is a slit-like opening near the tip; from this opening a canal leads along the inner surface to a basal opening on the upper surface, which communicates with the corner of the mouth when the mandible is closed. The central part of the mouth, between the mandibles, is closed, the upper and lower lips being locked together by a dovetail joint. The mandibles are admirably fitted for holding prey and at the same time sucking juices from its body. The thorax is furnished with six well-developed legs. The abdomen is terminated by a pair of processes; at the tip of the abdomen there is a pair of large spiracles, which the larva protrudes into the air at intervals, in order to breathe.

When a larva is fully grown it leaves the water, burrows into the ground, and makes a round cell, within which it undergoes its transformations. The pupa state lasts about three weeks in summer; but the larvæ that transform in autumn remain in the pupa state all winter.

This is the largest of the families of water-beetles; more than three hundred North American species are known.

The best way to obtain specimens is to sweep the vegetation growing on the bottom of a quiet pool with a dip-net.

The larger of our common species belong to *Cybĭster*, *Dytĭscus*, and allied genera. In *Cybister* the little cups on the under side of the tarsal disks of the male are similar, and arranged in four rows. In *Dytiscus* and its allies the cups of the tarsal disks vary in size. Figure 569 represents a common species of *Dytiscus*.

The most common of the diving-beetles which are of medium size belong to the genus *Acĭlius*. In this genus the elytra are densely punctured with very fine punctures, and the females usually have four furrows in each wing-cover (Fig. 567).

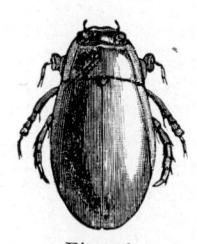

Fig. 569.

There are also common diving-beetles which are of about the same size as the preceding, but which have the wing-covers marked with numerous very fine transverse striæ; these belong to the genus *Colymbētes*.

Of the smaller diving-beetles, measuring less than 6 mm. in length, many species can be found in almost any pond. These represent many genera.

Family GYRINIDÆ
The Whirligig-Beetles

As familiar to the country rover as the gurgling of the brook or the flecks of foam on its "golden-braided centre," or the trailing ferns and the rustling rushes on its banks, are these whirligigs on its pools. Around and around each other they dart, tracing graceful curves on the water, which vanish almost as soon as made. They are social fellows, and are almost always found in large numbers, either swimming or resting motionless near together. They rarely dive, except when pursued; but are so agile that it is extremely difficult to catch them without a net. Many of them when caught exhale a milky fluid having a very disagreeable odor. They feed upon small flies, beetles, and other insects that fall into the water, and are furnished with well-developed wings, with which they fly from one body of water to another.

This is one of the most easily-recognized families of the whole order Coleoptera. The members of it are oval or elliptical in form (Fig. 570), more or less flattened, and usually of a very brilliant bluish black color above, with a bronze metallic lustre. The fore legs are very long and rather slender; the middle and hind legs are short, broad, and very much flattened. These insects are remarkable for having the eyes completely divided by the margin of the head, so that they appear to have four eyes—a pair upon the upper surface of the head with which to look into the air, and a pair upon the under side for looking into the water. The antennæ are very short and peculiar in form. The third segment is enlarged, so as to resemble an ear-like appendage, and the following ones form a short spindle-shaped mass. They are inserted in little cavities in front of the eyes.

Fig. 570.

The eggs of these insects are small, of cylindrical form, and are placed end to end in parallel rows upon the leaves of aquatic plants. The larvæ (Fig. 571) are long, narrow, and much flattened. Each abdominal segment is furnished with a pair of tracheal gills, and there is an additional pair at the caudal end of the body. The elongated form of the body and the conspicuous tracheal gills cause these larvæ to resemble small centipedes. When a larva is full-grown it leaves the water and spins a gray, paper-like cocoon attached to some object near the water. The pupa state of those species in which it has been observed lasts about a month.

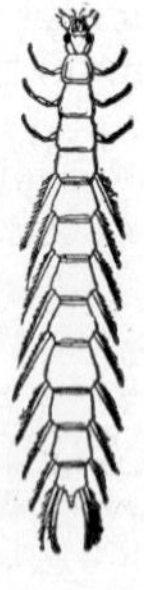

Fig. 571.

The family is a small one. At present only forty-one North American species are known. These represent three genera. The genus *Gyretes* is distinguished by having the last ventral segment of the abdomen elongated and conical. It is represented by two species. In the other two genera the last ventral segment is flattened and rounded at the tip. In *Dineūtus* the scutellum is invisible; there are thirteen species of this genus. In *Gyrīnus* the scutellum is visible; of this genus we have twenty-six species.

Suborder POLYPHAGA*

In the suborder Polyphaga the ventral part of the first segment of the abdomen is visible for its entire breadth (Fig. 574); the first three ventral segments are immovably united (except in the Cupesidæ), and the notum of the prothorax is not separated from the pleura by distinct sutures.

So far as known, the nutritive cells of the ovaries are massed together in the terminal chamber of each ovarian tube in all members of this suborder.

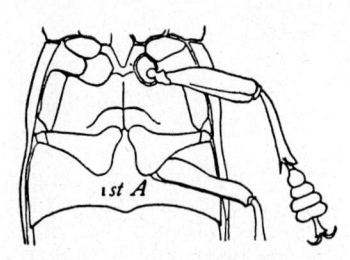

Fig. 574.—Ventral aspect of part of thorax and abdomen of *Enchroma gigantea: 1st A*, first abdominal segment.

The larvæ vary greatly in form; some are campodeiform, some are scarabeiform, and others are vermiform; in none are the legs more than five-jointed, and in none are the legs two-clawed.

This suborder includes all but the seven preceding families of the Coleoptera; the families included in it are grouped into seven series; see synopsis, page, 467.

Family HYDROPHILIDÆ
The Water-Scavenger Beetles

The water-scavenger beetles are common in quiet pools, where they may be found swimming through the water, or crawling among the plants growing on the bottom. They can be easily taken by sweeping such plants with a dip-net.

They are elongated, elliptical, black beetles, resembling the predacious diving beetles in appearance; but they are usually more convex, and differ also in having club-shaped antennæ and very long palpi. As the antennæ are usually concealed beneath the head, it often happens that the inexperienced student mistakes the long palpi for antennæ.

These beetles are supposed to live chiefly upon the decaying vegetation in the water; but a number of species have been known to catch and eat living insects. They breathe, by carrying a film of air on the lower surface of the body. This film gives them a silvery appearance when seen from below. They obtain the air by bringing the head to the surface of the water and projecting the antennæ, which they again fold back with a bubble of air when they descend. The female makes a case for her eggs out of a hardened silk-like secretion. Some species deposit as many as a hundred eggs in one of these water-proof packages (Fig. 572). The egg-cases in some instances are fastened beneath the

Fig. 572.

---
*Polyphaga: *polyphagus*, eating many kinds of food.

leaves of aquatic plants; in others they are provided with floats and let loose in the water; and in still other species the cases are carried by the mother underneath her body and steadied with her hind legs. Frequently some of the young larvæ devour their companions; in this way the size of the family is decreased before it escapes from the egg-case. Later they live upon insects that fall into the water, and upon snails. These larvæ resemble somewhat those of the Dytiscidæ; but the body is much more plump, and the mandibles are of moderate size.

The family Hydrophilidæ is represented in North America by one hundred and ninety species. The largest of our common species is *Hȳdrous triangulāris* (Fig. 573). In the genus *Hydrous* the metasternum is prolonged backward into a spine between the hind legs, and the sternum of the prothorax bears a deep furrow.

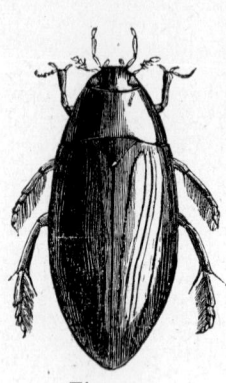

Fig. 573.

The beetles of the genus *Tropistĕrnus* agree with *Hydrous* in the form of the prosternum and metasternum, but differ in size, our species measuring less than 12 mm. in length. The most common species in the East is *Tropisternus glăbra*, and on the Pacific Coast *T. califŏrnicus*.

Next in size to *Hydrous* are several species of *Hydrŏphilus*. In this genus the metasternum is prolonged somewhat, but does not form a long, sharp spine as in *Hydrous* and *Tropisternus*, and the sternum of the prothorax bears a keel-shaped projection. Our most common species is *Hydrophilus obtusātus;* this measures about 15 mm. in length.

Some of the smaller species of this family are not aquatic, but live in moist earth and in the dung of cattle, where, it is said, they feed on dipterous larvæ.

FAMILY PLATYPSYLLIDÆ

The Beaver-Parasite

Only a single representative of this family is known; this is *Platypsȳlla căstoris*, which lives parasitically on the beaver. This beetle is about 2.5 mm. in length; the body is ovate, elongate, and much flattened; the wing-covers are short, about as long as the prothorax, and leave five abdominal segments exposed; the eyes and wings are wanting.

Specimens of this remarkable insect are most easily obtained by beating over a sheet of paper the dried skins of beavers, which can be found in fur-stores.

The family BRATHINIDÆ is composed of the genus *Brathīnus*, of which three species are described, two from the East and one from California. These beetles are somewhat elongate, with the outline of the prothorax and of the elytra elliptical; they measure from 3.6 mm. to 5 mm. in length. The Californian species was found living in wet moss, darkly overshadowed by bushes, at the margin of a mountain stream.

Family LEPTINIDÆ
The Mammal-Nest Beetles

This family is represented by only three species in North America. One of these, *Leptīnus testāceus*, is a European species, but is widely distributed in this country. It lives in the nests of mice and other small rodents and insectivora, and also in the nests of bumble-bees. Whether it is a parasite or merely a guest has not been definitely determined; but it seems probable that it feeds upon the eggs and young of mites and other small creatures found in these nests. This beetle is oblong-oval and much depressed in form, and pale yellow in color; it measures from 2 mm. to 2.5 mm. in length. Specimens can be obtained by shaking a nest of a mouse over a sheet of paper.

The other two species are *Leptinĭllus vălidus*, found in Hudson Bay territory, and *Leptinĭllus aplodŏntiæ*, found in California on a rodent (*Aplodontia*).

Family SILPHIDÆ
The Carrion-Beetles

The carrion-beetles are mostly of medium or large size, many species attaining the length of 35 mm. while the smaller species of the more typical genera are nearly 12 mm. in length; some members of the family, however, are minute. The segments near the tip of the antennæ are usually enlarged so as to form a compact club, which is neither comb-like nor composed of thin movable plates; sometimes the antennæ are nearly filiform.

These insects usually feed upon decaying animal matter; some, however, feed upon fungi; some on vegetables; and a few species have been known to be predacious when pressed by hunger, destroying living snails and insects, even members of their own species; while a few occur only in the nests of ants.

It is easy to obtain specimens of these insects by placing pieces of meat or small dead animals in the fields and examining them daily. There are several other families of beetles the members of which can be attracted in this way.

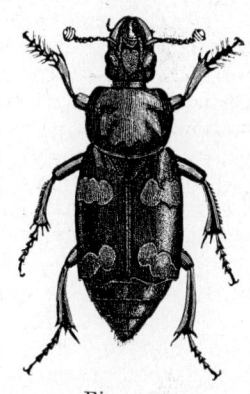

Fig. 575.

The larvæ also live upon decaying flesh and are found in the same situations as the adults.

We have in this country more than one hundred species of this family. Our larger and more familiar species represent two genera, *Necrŏphorus* and *Sĭlpha*.

The burying-beetles, *Necrŏphorus*.—To this genus belong the larger members of the family The body is very stout, almost cylindrical (Fig. 575). Our common species have a reddish spot on each end of each wing-cover; these spots are often so large that they appear as two transverse bands. In some species the prothorax and the head are also marked with red.

These insects are called burying-beetles because they bury carrion. When a pair of these beetles discover a dead bird, mouse, or other small animal, they dig beneath it, removing the earth so as to allow the carrion to settle into the ground. This they will continue until the object is below the surface of the ground. Then they cover it with earth, and finally the female digs down to it and lays her eggs upon it. The larvæ that hatch from these eggs feed upon the food thus provided for them. There are many accounts of exhibitions of remarkable strength and sagacity by burying-beetles. A pair of these insects have been known to roll a large dead rat several feet in order to get it upon a suitable spot for burying.

The members of the genus *Silpha* are very much flattened (Fig. 576). The prothorax is round in outline, with very thin edges which overlap the wing-covers somewhat. The body is not nearly as stout as that of a burying beetle, being fitted for creeping under dead animals instead of for performing deeds requiring great strength. *Silpha bituberōsa*, which is known as the spinach carrion-beetle, feeds on spinach, beets, and other plants, in the West.

Fig. 576.

In some of the minute members of this family the body is nearly hemispherical.

The family CLAMBIDÆ consists of very minute species, measuring about 1 mm. in length. They live in decomposing vegetable matter. The edge of the hind wings is fringed with long hairs. For other characters, see table, page 470. Only six species, representing three genera, occur in our fauna.

The family SCYDMÆNIDÆ includes very small insects found under bark or stones, in ants' nests, or near water. They are small, shining, usually ovate but sometimes slender insects, of a brown color, and more or less clothed with erect hairs. Other characters are given in the preceding table of families. About one hundred seventy-five North American species are known.

The family CORYLOPHIDÆ includes minute beetles found under damp bark and in decaying fungi and other vegetable matter. The body is oval or rounded, and in many species is clothed with a grayish pubescence. The wings are wide, and fringed with long hairs. Some of our common species measure less than 1 mm. in length. Fifty-seven North American species have been described.

FAMILY STAPHYLINIDÆ

The Rove-Beetles

The rove-beetles are very common about decaying animal matter, and are often found upon the ground, under stones or other objects.

They are mostly very small insects; a few species, however, are of larger size, measuring 12 mm. or more in length. Their appearance is very characteristic, the body being long and slender, and the wing-covers very short (Fig. 577). The wings, however, are fully developed, often exceeding the abdomen in length; when not in use the wings are folded beneath the short wing-covers. The abdominal segments are freely movable.

It is interesting to watch one of these insects fold its wings; frequently they find it necessary to make use of the tip of the abdomen or of one of the legs in order to get the wings folded beneath the wing-covers.

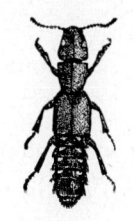

Fig. 577.

The rove-beetles can run quite swiftly; and they have the curious habit, when disturbed, of raising the tip of the abdomen in a threatening manner, as if they could sting. As some of the larger species resemble wasps somewhat in the form of the body, these threatening motions are often as effective as if the creatures really had a sting. William Beebe states (*Atlantic Monthly*, October 1919) that when some rove-beetles were attacked by ants they raised their tails and ejected a drop or two of a repellent fluid which drove the ants away. This observation indicates the probable explanation of the actions of these beetles when disturbed.

As these insects feed upon decaying animal and vegetable matter, they should be classed as beneficial. The larvæ resemble the adults in the form of the body and are found in similar situations, about decaying animal and vegetable matter, beneath bark and in fungi. Some species are guests in the nests of ants, and others in the nests of termites.

Nearly three thousand North American species of rove-beetles have been described. The great majority are small and exceedingly difficult to determine. Among the large species that are common are the following.

Creŏphilus maxillōsus.—This species varies from 12 mm. to nearly 25 mm. in length. It is of a shining black color, spotted with patches of fine gray hairs. There is a conspicuous band of these across the middle of the wing-covers, and another on the second and third abominal segments; this abdominal band is best marked on the lower side of the body.

Staphylīnus maculōsus is a larger species, which often measures fully 25 mm. in length. It is densely punctured, and of a dull brown color, with the scutellum black, and a row of obscure, square, blackish spots along the middle of the abdomen.

Staphylīnus vulpīnus resembles the preceding somewhat, but it has a pair of bright yellow spots at the base of each abdominal segment.

Ontholĕstes cingulātus is of about the same size as the preceding. It is brown, speckled with brownish black spots, and the tip of its abdomen is clothed with golden hairs.

The family PSELAPHIDÆ includes certain very small beetles, the larger ones not exceeding 3 mm. in length. They resemble rove-beetles in the shortness of the wing-covers and in having the dorsal part of the abdominal segments entirely horny; but they differ from them in that the abdomen is not flexible, and in having fewer abdominal segments, there being only five or six on the ventral side. The species are chestnut-brown, dull yellow, or piceous, and are usually slightly pubescent. The antennæ are usually eleven-jointed, rarely ten-jointed. The elytra and abdomen are convex and usually wider than the head and prothorax. These beetles are usually found under stones and bark, or flying in the twilight; a few species have been found in the nests of ants. There are three hundred and fifty-five described North American species.

The family CLAVIGERIDÆ, or the ant-loving beetles, includes a small number of beetles that resemble the Pselaphidæ in the characters given above except that the antennæ are only two-jointed. These beetles live in the nests of ants. They excrete from small tufts of hairs, on the three basal abdominal segments, a fluid of which the ants are very fond. The ants caress the tufts of hairs with their antennæ, causing the exudation of the fluid, which they greedily swallow. The ants are said to feed the beetles and to allow them to ride about on their backs, when the beetles wish to do so. Only seven North American species are described.

The family TRICHOPTERYGIDÆ, or the feather-wing beetles, includes the smallest beetles that are known; most of our species are less than 1 mm. in length. The most striking feature of the typical forms is the shape of the wings, which are long, narrow, and fringed with long hairs, being feather-like in appearance; but in some species the wings are wanting. Some species live in rotten wood, muck, manure, and other decaying organic matter; few have been found in ants' nests. There are about eighty described species in our fauna.

The family SCAPHIDIIDÆ, or shining fungus-beetles, includes small, oval, very shining beetles, found in fungi, rotten wood, dead leaves, and beneath the bark of logs. The elytra are broadly truncate behind, not covering the tip of the conical abdomen. But little is known regarding their life-history. There are fifty described North American species.

The family SPHÆRITIDÆ is represented in our fauna by a single species, *Sphærītes glabrātus*, which has been found in Alaska and California. This beetle is very similar in appearance to those of the genus *Hister*. For distinguishing characters, see table, page 470.

The family SPHÆRIIDÆ includes a single genus, *Sphærius*, which is represented in North America by only three known species. They are very minute beetles, measuring about .5 mm. in length; they are very convex, and may be found walking on mud or under stones near water.

The family HISTERIDÆ includes certain easily recognized beetles which are found about carrion and other decomposing substances. They are mostly small, short, rounded or somewhat square-shaped beetles, of a shining black color, with the wing covers marked by lines of fine punctures and truncate behind, leaving two segments of the abdomen exposed (Fig. 578). In some species the wing-covers are marked with red. There are nearly four hundred described North American species.

Fig. 578.

The family LYCIDÆ includes certain beetles that were formerly classed in the fire-fly family; but they differ from the Lampyridæ in having the middle coxæ distant, and in that the elytra lack epipleuræ. The elytra are usually furnished with several longitudinal ribs and a network of fine elevated lines. The members of this family are diurnal in habits; they are found on the leaves of plants, where they seek and feed upon other insects. A common species is *Calŏpteron reticulātum* (Fig. 579).

Fig. 579.

Family LAMPYRIDÆ

The Firefly Family

During some warm, moist evening early in our northern June we are startled to see here and there a tiny meteor shoot out of the darkness near at hand, and we suddenly realize that summer is close upon us, heralded by her mysterious messengers, the fireflies. A week or two later these little torch-bearers appear in full force, and the gloom that overhangs marshes and wet meadows, the dusk that shrouds the banks of streams and ponds, the darkness that haunts the borders of forests, are illumined with myriads of flashes as these silent, winged hosts move hither and thither under the cover of the night.

The fireflies are soft-bodied beetles of medium or small size, with slender, usually eleven-jointed, saw-like antennæ. The prothorax is expanded into a thin projecting margin, which in most cases completely covers the head (Fig. 580). The wing-covers are rather soft, and never strongly embrace the sides of the abdomen, as with most other beetles.

The fireflies are nocturnal insects and are sluggish by day. The property of emitting light is possessed by adults of both sexes and by larvæ. The latter and the wingless females of certain species are known as glow-worms. The light-organs of the winged adults are situated on the lower side of one or more of the abdominal segments; but they are lacking in some genera.

Fig. 580.

There have been many speculations as to the usefulness of the light-producing power of various organisms to the organisms themselves; and as regards many of these photogenic creatures no definite conclusions have been reached. But there is considerable evidence to show that in the case of adult lampyrids it serves to enable these insects to find their mates. It has been found that females enclosed in a perforated opaque box do not attract males, while those enclosed in a glass vial do; thus showing that it is the light emitted by the female, and not its odor, that attracts the male. It has also been shown that in some cases at least there are specific differences in the method of flashing which enables the insects to distinguish at a distance their proper mates.

More than fifty species of the Lampyridæ have been found in this country.

The family PHENGODIDÆ includes a small number of species that were formerly included in the firefly family. In this family the prothorax, though rounded in front, does not cover the head, which is exposed. The antennæ are usually plumose or flabellate in the males. The females of some species, at least, are glow-worms, resembling the larvæ in form, and are photogenic. Only twenty-three American species have been described; most of these are found in California, Texas, and Arizona, but some occur in the East.

Family CANTHARIDÆ

The Soldier-Beetles and others

The family Cantharidæ includes those genera that were formerly included in the family Lampyridæ as the subfamily Telephorinæ. For the distinctive characters separating this from the allied families, see the table, page 473.

The application of the name Cantharidæ to this family is the result of one of those unfortunate changes in generic names rendered necessary by our code of nomenclature. In this case the change is especially unfortunate, as the generic name *Cantharis* has been commonly applied to certain blister-beetles and is used in that sense in many medical works and in most text-books of entomology. The change is sure to result in much confusion.

The most common members of this family are the soldier-beetles, *Chauliŏgnathus*. These are very abundant in late summer and autumn on various flowers, but especially on those of goldenrod. The most common species in the East are the Pennsylvania soldier-beetle, *Chauliŏgnathus pennsylvănicus*, which is yellow, with a black spot in the middle of the prothorax and one near the tip of each wing-cover (Fig. 581); and the margined soldier-beetle, *C. marginātus*. This species (Fig. 582) can be distinguished from the former by the head and lower parts of the femora being orange. The beetles of this genus are remarkable for having an extensible, fleshy filament attached to each maxilla. These filaments are probably used in collecting pollen and nectar from flowers.

Fig. 581.

Fig. 582.

This family is represented in our fauna by nine genera which include more than one hundred and fifty species.

The family MELYRIDÆ is composed chiefly of small or very small beetles, some of which are found on flowers, and others on the ground in low, moist places. They are said to be carnivorous. They vary greatly in form, but bear a general resemblance in structure to the preceding four families, from which they can be distinguished by the presence of only six ventral abdominal segments. Some members of the family are furnished with soft, orange-colored vesicles, which they protrude from the sides of the body and which are supposed to be scent organs for defence. One of our most common representatives is *Cŏllops quadrimaculātus*, which is yellow-orange, with the top of the head and four spots on the elytra bluish black (Fig. 583). This species is found on grasses in damp localities. The family is represented in our fauna by more than three hundred species.

Fig. 583.

The family CLERIDÆ, or the checkered beetles, includes a considerable number of predacious species which are found on flowers and on the trunks of trees. Many of them are beautifully marked with strongly contrasting colors; this has suggested the common name *checkered beetles* for them. Frequently they are more or less ant-like in form, the prothorax being in these cases narrower than the wing-covers, and slightly narrower than the head. The abdomen has either five or six ventral segments; the anterior coxæ are conical, prominent, and contiguous, or very slightly separated; the hind coxæ are transverse, not prominent, and covered by the femora in repose; the legs are slender; and the tarsi are five-jointed.

In the larval state these insects are usually carnivorous, living under bark and in the burrows of wood-boring insects, upon which they prey; some are found in the nests of bees; and still others feed on dead animal matter.

Fig. 584.

The family is represented in our fauna by nearly two hundred species. Figure 584 represents one of our more common species, *Trichōdes nuttălli*.

The family CORYNETIDÆ has recently been separated from the Cleridæ, which they closely resemble. In this family the fourth joint of the tarsi is atrophied; this character distinguishes these beetles from the Cleridæ. About forty American species have been described.

To this family belongs the red-legged ham-beetle, *Necrōbia rŭfipes*. This is a small steel-blue beetle with reddish legs; it measures from 3.5 mm. to 6 mm. in length. It is found about dead animal matter in fields and in other situations. It sometimes invades storehouses and seriously infests hams.

The family LYMEXYLIDÆ includes elongated, narrow beetles, with short serrate antennæ. Only two species have been found in this country and these are rare. To this family belongs the ship-timber beetle, *Lymĕxylon navāle*, of northern Europe. The larva of this species was at one time a very serious pest in ship-yards, on account of its habit of drilling cylindrical holes in the timber. The method of control by immersing the timber during the time of oviposition of the beetle was suggested by Linnæus.

The family MICROMALTHIDÆ includes a single species, *Micromălthus dĕbilis*. This is a small beetle, measuring only 2.2 mm. in length. It is elongate, piceous, shining, with the antennæ and legs yellow. This species is of great interest on account of its remarkable life-history, only a part of which is yet known. Two papers on this subject have been published by Mr. H. S. Barber ('13a, '13b). Briefly, this author's observations indicate that eggs are produced by larvæ as well as by the adult females; that there are seven or eight forms of larvæ; that the two sexes of adults are developed through two distinct lines of larvæ; and that viviparous as well as oviparous pædogenesis occurs in the life-cycle. The larvæ are found in decaying oak, chestnut, and pine logs, where they make burrows in the decaying wood, on which they feed.

The family CUPESIDÆ includes only four American species. These are found under the bark of decaying trees, and sometimes in houses. The body is covered with small scales; other characteristics are given in the table of families, page 471.

The family CEPHALOIDÆ is a small family of which only eight American species have been described. See table, page 474, for distinctive characters.

The family ŒDEMERIDÆ is composed of beetles of moderate size, with elongate, narrow bodies. The head and prothorax are somewhat narrower than the wing-covers; the antennæ are long, nearly filiform, sometimes serrate; the anterior coxal cavities are open behind, and the middle coxæ are very prominent. Less than fifty North American species have been described. They are generally found on plants, but some live on the ground near water. The larvæ live for the most part in decaying wood.

Fig. 585.

The family MORDELLIDÆ includes a large number of small beetles which are easily recognized by their peculiar form (Fig. 585). The body is arched, the head being bent down; and the abdomen is usually prolonged into a slender point. Our most common species are black; but many are variegated, and all are pubescent. The adults are usually found on flowers; the larvæ live in rotten wood and in the pith of various plants, upon which they are supposed to feed. Nearly one hundred fifty American species have been described.

The family RHIPIPHORIDÆ includes a small number of beetles, which are very remarkable in structure and habits. The wing-covers are usually shorter than the abdomen, and narrowed behind (Fig. 586); sometimes they are very small, and in one exotic genus they are wanting in the female, which lacks the wings also and resembles a larva in form. The antennæ are pectinate or flabellate in the males, and frequently serrate in the females. The adult insects are found on flowers. The larvæ that are known are parasites, some in the nests of wasps, and some on cockroaches.

Fig. 586.

Family MELOIDÆ
The Blister-Beetles

The blister-beetles are of medium or large size. The body is comparatively soft; the head is broad, vertical, and abruptly narrowed into a neck; the prothorax is narrower than the wing-covers, which are soft and flexible; the legs are long and slender; the hind tarsi are four-jointed, and the fore and middle tarsi are five-jointed.

These beetles are found on foliage and on flowers, on which they feed in the adult state; some of the species are very common on goldenrod in the autumn; and several species feed on the leaves of potato.

The blister-beetles are so called because they are used for making blister-plasters. The beetles are killed, dried, and pulverized, and the powder thus obtained is made into a paste, which when applied to the skin produces a blister. The species most commonly used is a European one, commonly known as the Spanish-fly; but our American species possess the same blistering property.

The postembryonic development of those blister-beetles of which the complete life-history is known is a very remarkable one; for it has been found that in each of these cases there is a complicated hypermetamorphosis. The food of the larva consists, in some species, of the eggs of short-horned grasshoppers, in others of the egg and the food stored in the cell of some solitary bee. The female blister-beetle lays her eggs in the ground; a large number of eggs are laid by a single female; this fact is doubtless correlated with the difficulties to be overcome by the larvæ in their search for their proper food, in which comparatively few are successful. The newly hatched larva is campodeiform (Fig. 587, A), and is known as the *triungulin*, a term applied to the first instar of blister-beetle larvæ. This term was suggested by the fact that in this instar the tarsi appear to be three-clawed; but in reality each tarsus is armed with a single claw, on each side of which there is a claw-like seta.

The triungulins are very active. In the case of those that feed on the eggs of short-horned grasshoppers, they run over the ground seeking a place where one of these insects has deposited its egg-pod; if a triungulin is successful in this search it bores its way into the egg-pod; if more than one find the same egg-pod, battles occur till only one is left. In the case of those species that develop in the nests of bees, the triungulin, instead of hunting for a nest, merely climbs a plant, and remains near a flower till it has a chance to seize hold of a bee visiting the flower; it then clings to the bee until she goes to her nest, then, letting go of the bee, it remains in the cell and is shut up there with the egg of the bee and the store of food which she provides for her young. The triungulin first devours the egg; after which it molts and undergoes a change of form, becoming a clumsy creature, which feeds upon the food stored in the cell. Several other changes in form occur before the beetle reaches the adult stage; these changes are quite similar to those undergone by the larva of *Epicauta*, described below.

The wonderful instinct by which the triungulins of these blister-beetles find their way to the nests of solitary bees has not yet reached perfection; for many of them attach themselves to flies, wasps, honey-bees, and other flower-visiting insects, and merely gain useless transportation thereby.

The life-history of *Epicauta vittāta*, which was worked out by Mr. C. V. Riley ('77), will serve to illustrate the hypermetamorphosis of blister-beetles. The adult beetle is yellowish or reddish above, with the head and prothorax marked with black and with two black stripes on each wing-cover (Fig. 587, F). It feeds on the leaves of potato, and is sometimes a serious pest. The female lays her eggs in

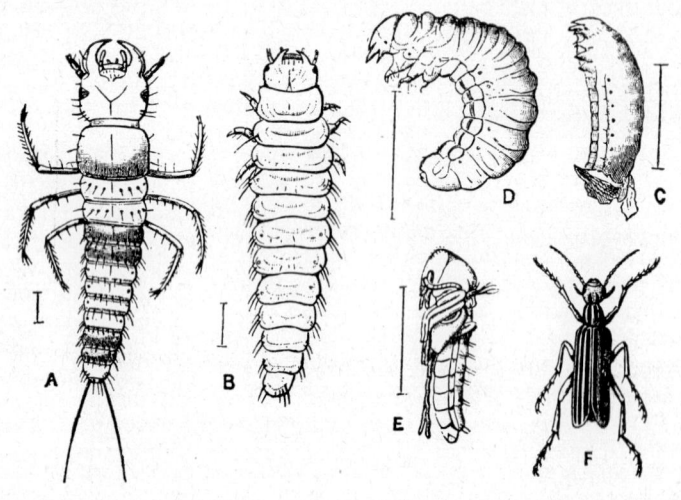

Fig. 587.—Hypermetamorphosis of *Epicauta vittata*. (From Sharp, after Riley.)

the ground in loose, irregular masses of about 130 each; several such masses are laid by a single female. She prefers for purposes of oviposition the very same warm, sunny locations chosen by the locusts for laying their eggs. The triungulins that hatch from the blister-beetle eggs (Fig. 587, A) are very active; when one of them finds an egg-pod of a locust it penetrates it, and in the course of several days devours two of the eggs; a period of rest follows during which it molts. The second instar (Fig. 587, B) differs greatly in form from the triungulin, and is known as the *caraboid larva*. A second molt takes place after about a week, but it is not accompanied by any very great change of form, though the larva is now curved, less active, and in form like the larva of a scarabæid beetle. About a week later, the third molt occurs; the change in form at this molt is not great, the fourth instar (Fig. 587, D) like the third being scarabæoid in form; these two instars can be distinguished as the *first scarabæoid larva* and the *second scarabæoid larva* respectively. The second scarabæoid larva grows apace, its head being constantly bathed in the rich juices of the locust eggs, which it rapidly sucks or more or less completely devours. In another week it forsakes the remnants of the locust egg-mass and forms a smooth cavity in the soil a short distance from it. The larva then molts; the skin is not shed entirely, but remains attached to the caudal end of the body (Fig. 587, C). The new skin of the larva becomes rigid and of a deeper yellow color, reminding one of a puparium of a dipterous insect; this instar, the fifth, is termed the *coarctate larva*. The insect has the power of remaining in this coarctate larval condition for a long time, and generally thus hibernates. At the fifth molt the larva becomes active again, and burrows about in the ground; it now resembles in form the second scarabæoid larva except that it is smaller and whiter; this, the sixth instar, was termed by Riley the *scolytoid larva*. In the cases observed by Riley, the scolytoid larvæ did not feed but transformed to pupæ (Fig. 587, E) in the course of a few days. The pupa state lasted five or six days.

More than two hundred species of blister-beetles have been found in this country; but by far the greater number of these are confined to the western half of this region. Our most common species in the East belong to the genus *Epicauta*. These insects feed in the adult state on the leaves of various plants, but especially those of potato, and upon the pollen of goldenrod; the larvæ, so far as is known, are parasitic in the egg-pods of locusts (*Melanoplus*). In addition to *Epicauta vittāta*, discussed above, our more common species are the Pennsylvania blister-beetle, *Epicauta pennsylvănica*, which is of a uniform black color (Fig. 588); and *Epicauta cinērea*, which is sometimes clothed throughout with an ash-colored pubescence, and sometimes the wing-covers are black, except a narrow gray margin; the two varieties were formerly considered distinct species; the first is commonly known as the gray blister-beetle, the last as the margined blister-beetle.

Fig. 588.

Closely allied to the beetles mentioned above are those of the genus *Macrŏbasis*. The most common species of this genus found in the East is *Macrŏbasis unīcolor*. This beetle measures from 8 mm. to 15 mm. in length; it is represented, enlarged, in Figure 589; it is black, but rather densely clothed with grayish hairs, which give an ashen hue to the upper surface; it is known as the ashy-gray blister-beetle.

Fig. 589.

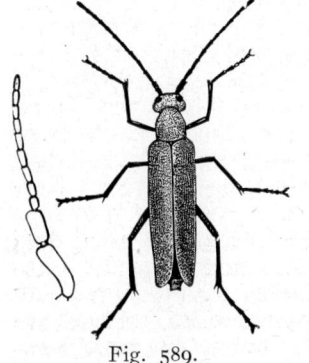

The beetles of the genus *Meloe* present an exception to the characters of the Coleoptera in that the wing-covers, instead of meeting in a straight line down the back, overlap at the base (Fig. 590). These wing-covers are short, and the hind wings are lacking. These beetles are called *oil-beetles* in England, on account of the yellowish liquid which oozes from their joints when they are handled. Our most common species is the buttercup oil-beetle, *Meloe angusticollis*. It is found in meadows and pastures feeding on the leaves of various species of buttercups.

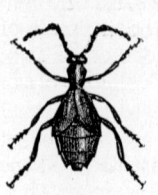

Fig. 590.

The species of the genus *Nemognatha* and some allied forms are remarkable for having the maxillæ developed into a long sucking-tube, which is sometimes as long as the body, and which resembles somewhat the sucking-tube of a butterfly.

The family EURYSTETHIDÆ includes only three American species, one found in Alaska and two found in California. One of the latter, *Eurystethus subopacus*, was found by Professor VanDyke on the seashore, in crevices of inter-tidal rocks.

The family OTHNIIDÆ is represented in our territory by five species of *Othnius*, one from the East and four from the Far West. They are small beetles, which are found running actively on the leaves of trees, and are probably predacious. In this family the anterior coxal cavities are closed behind, and none of the abdominal segments are grown together on the ventral side.

The family PYTHIDÆ includes less than a score of North American species. Some of these live under bark, and are said to prey on bark-beetles; others are found under stones. See table, p. 474, for distinctive characters.

The family PYROCHROIDÆ includes a small number of beetles, which are from 8 mm. to 18 mm. in length. The body is elongate; the head and prothorax are narrower than the wing-covers; the antennæ are serrate or subpectinate in the females and usually flabellate in the males (Fig. 591). The beetles are found about decaying trees, beneath the bark of which the larvæ live.

The family PEDILIDÆ.—In this and in the following family the abdomen is composed of five free segments, and the tarsi have the penultimate joint lobed beneath. In this family the eyes are large, finely faceted, and usually emarginate. These beetles are arboreal in habits. There are about fifty described species in our fauna.

Fig. 591.

The family ANTHICIDÆ.—In this family, as in the preceding one, the abdomen consists of five free segments, and the penultimate joint of the tarsi is bilobed. But in this family the eyes are small, rounded, usually coarsely faceted, and emarginate. These are active ground beetles of predacious habits. Among our more common species are those of the genus *Notoxus*, in which the prothorax is prolonged over the head into a horn. There are nearly two hundred described species of this family in our fauna.

The family EUGLENIDÆ is composed of small or minute beetles, found on leaves and flowers; many of them are less than 2 mm. in length. They resemble the members of the two preceding families; but differ in having the antepenultimate segment of the tarsi bilobed, instead of the penultimate, and in having the abdomen composed of only four free ventral segments, of which the first is formed of two, firmly united but with the suture sometimes evident. There are about forty described North American species.

The family CEROPHYTIDÆ includes only two rare species of *Ceröphytum*, one found in California and one in Pennsylvania. These were formerly included in the Elateridæ; but they differ from that family in that the posterior coxæ are not laminated, and the trochanters of the middle and posterior legs are very long.

The family CEBRIONIDÆ includes a few species found in the South. They were formerly included in the Elateridæ; but they differ from that family in that the abdomen consists of six or more ventral segments. This family differs from the following one in having the tibial spurs well developed.

The family PLASTOCERIDÆ includes about a score of species found in the South and in California. It is closely allied to the preceding family, but differs in having the tibial spurs short and very delicate.

The family RHIPICERIDÆ, or cedar-beetles, is represented in this country by a very small number of species, which are most commonly found on cedars. The antennæ are serrate in the females, frequently flabellate in the males. The anterior and middle coxæ are conical and prominent, the former with large trochantins; the posterior coxæ are transverse, and dilated into a small plate partly covering the femora.

Family ELATERIDÆ

The Click-Beetles or Elators

There is hardly a country child that has not been entertained by the acrobatic performances of the long, tidy-appearing beetles called snapping-bugs, click-beetles, or skip-jacks (Fig. 592). Touch one of them and it at once curls up its legs, and drops as if shot; it usually lands on its back, and lies there for a time as if dead. Suddenly there is a click, and the insect pops up into the air several inches. If it comes down on its back, it tries again and again until it succeeds in striking on its feet, and then it runs off.

Fig. 592.

Our common species of click-beetles are mostly small or of medium size, ranging from 2.5 mm. to 18 mm. in length. A few species are larger, some reaching the length of nearly 50 mm. The majority of the species are of a uniform brownish color; some are black or grayish, and some are conspicuously spotted

(Fig. 593). The body is elongated, somewhat flattened, and tapers more or less toward each end; the antennæ are moderately elongated and more or less serrate; the first and second abdominal segments are not grown together on the ventral side; and the hind coxæ are each furnished with a groove for the reception of the femur.

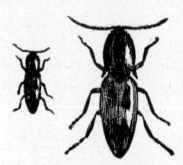

Fig. 593.—A click-beetle, *Aeolus dorsalis*, natural size and enlarged.

The ability to leap into the air when placed on their back, which is possessed by most members of this family and by a few members of the following family, is due to two facts: first, the prosternum is prolonged into a process which extends into a groove in the mesosternum; and second, the prothorax is loosely joined to the mesothorax, so that it can be freely moved up and down. When preparing to leap, the beetle bends its body so as to bring the prosternal process nearly out of the groove in the mesosternum; then it suddenly straightens its body, with the result that the prosternal process descends violently into the groove; the blow thus given to the mesothorax causes the base of the elytra to strike the supporting surface, and by their elasticity the whole body is propelled upward.

Adult elaters are found on leaves and flowers, and are exclusively phytophagous; the larvæ live in various situations; most of them are phytophagous, but some species are carnivorous.

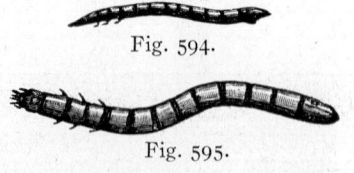

The larvæ are long, narrow, worm-like creatures, very even in width, with a very hard cuticula, and are brownish or yellowish in color (Figs. 594 and 595). They are commonly known as wire-worms, a name suggested by the form and hardness of the body.

Fig. 594.

Fig. 595.

Some wire-worms live under the bark of trees and in rotten wood; but many of them live in the ground, and feed on seeds and the roots of grass and grain. In fact there is hardly a cultivated plant that they do not infest, and, working as they do beneath the surface of the ground, it is extremely difficult to destroy them. Not only do they infest a great variety of plants, but they are very apt to attack them at the most susceptible period of their growth, before they have attained sufficient size and strength to withstand the attack; and often seed is destroyed before it is germinated. Thus fields of corn or other grain are ruined at the outset. The appearance of these insects when in the ground, as seen through the glass side of one of our root-cages, is shown in Fig. 596.

There is a vast number of species of click-beetles; more than five hundred have been described from North America alone. It is quite difficult to separate the closely allied species, as there is but little variation in shape and color.

The larvæ also show comparatively little variation in the general form; but in this stage the shape of the parts of the head and its appendages, and the structure of the caudal end of the body, afford useful characters. The value of these characters in indicating the principal divisions of the family is pointed out by Hyslop ('17).

An extended series of experiments were conducted by Comstock and Slingerland ('91) in an effort to discover a practicable method of preventing the ravages of wire-worms. In those species that we bred, it required several years for the larva to complete its growth. In these species the full-grown larva changes to a pupa in the latter part of the summer, in a little cell in the ground; the pupa soon afterwards changes to an adult; but the adult remains in the cell formed by the larva till the following spring.

Although we tried an extensive series of experiments, extending over several years, we were unable to find any satisfactory way of destroying the larvæ infesting field crops. But we found that if the cells containing pupæ or recently-transformed adults were broken, the insects perished. We conclude, therefore, that much can be done towards keeping these insects in check by fall-ploughing; for in this way many of the cells containing pupæ or young adults would be broken.

Fig. 596.—A corn-plant growing in a root-cage infested by wire-worms and click-beetles. (From a specimen in the Cornell Insectary.)

The eyed elater, *Alaus oculātus*.—Although most of our click-beetles are of moderate size, we have a few species that are large. The most common of these is the eyed elater. This is the great pepper-and-salt-colored fellow that has two large, black, velvety, eye-like spots on the prothorax (Fig. 597). These are not its eyes, however. The true eyes are situated one on each side of the head near the base of the antenna. This

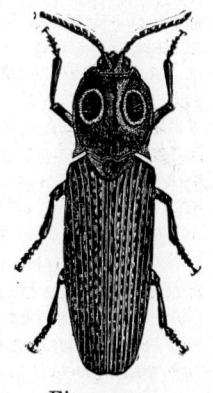

Fig. 597.

insect varies greatly in size, some individuals being not more than half as large as others.

The larvæ live in decaying wood, and are often found in the trunks of old apple trees. It was formerly believed that they fed on the decaying wood; but they have been found to be carnivorous. The larger larvæ are about 60 mm. in length.

There is an elater quite similar to the preceding that differs in having the eye-like spots less distinctly marked; this is *Alaus myops*. This species is not as common as the preceding one.

The family EUCNEMIDÆ was formerly regarded as a subfamily of the Elateridæ. It differs from the Elateridæ, as now restricted, in having the labrum concealed, and in that the antennæ are somewhat distant from the eyes, and their insertion narrowing the front. The adults are found under bark or on the leaves of plants; most of the species are rare. "The larvæ have a striking resemblance to those of the family Buprestidæ, both in form and habits, being abruptly enlarged in front, and usually occurring in wood which has just begun to decay." (Blatchley '10.)

The family THROSCIDÆ includes a few small species which resemble the elaters and buprestids in having the prosternum prolonged behind into a process, which is received in the mesosternum. They differ from the elaters in having the prothorax firmly joined to the mesothorax, and the front coxal cavities closed behind by the mesosternum instead of by the prosternum; and from the buprestids in having the ventral abdominal segments all free. The adult beetles are found on flowers.

Family BUPRESTIDÆ

The Metallic Wood-Borers or Buprestids

The buprestids resemble the click-beetles somewhat in form, being rather long and narrow; but they are easily recognized by their metallic coloring. Their bodies are hard and inflexible, and usually appear as if made of bronze; but some species exhibit the brightest of metallic colors. The antennæ are serrate; the first and second abdominal segments are grown together on the ventral side; and these beetles do not have the power of springing when placed on the back.

The adults are found upon flowers and upon the bark of trees, basking in the hot sunshine. Some of them fly very rapidly, with a loud buzzing noise; and some drop to the ground when disturbed, and feign death.

Most of the larvæ are borers, feeding beneath bark or within solid wood. In such species the body is of a very characteristic form, which is commonly designated as "flat-headed." The flattened portion, however, is composed largely of the segments immediately following the head. The first thoracic segment is very wide and flat; the next two or three segments are also flattened, but are successively smaller; while the rest of the body is quite narrow and cylindrical.

These "flat-headed" larvæ are legless, and have been compared to tadpoles on account of their form. Their burrows are flattened, corresponding with the shape of the larger part of the body. In some of the smaller species the larvæ are cylindrical, and are furnished with three pairs of legs. These are leaf-miners; and in the adult state the body is much shorter than in the more typical species.

This family is represented in our fauna by nearly three hundred species; among the more important of those that infest cultivated plants are the following.

The Virginian buprestid, *Chalcŏphora virgĭnica*.—This is the largest of our common buprestids (Fig. 598). It is copper-colored, often almost black, and has its upper surface roughened by irregular, lengthwise furrows. This beetle appears late in spring in the vicinity of pine-trees. The larvæ bore in the wood of pine, and are often very injurious.

Dicĕrca divaricāta is 18 mm. or more in length, copper-colored or brassy above, with the wing-covers marked with square, elevated, black spots. The wing-covers taper very much behind, and are separated at the tips (Fig. 599). The larva bores in peach, cherry, beech, and maple.

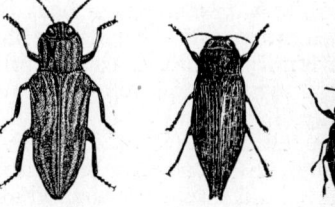

Fig. 598. Fig. 599. Fig. 600.

The flat-headed apple-tree borer, *Chrysobŏthris femorāta*.—This is one of the most injurious of all buprestids. The adult (Fig. 600) is about 12 mm. long, and is very dark green above, with bronze reflections, especially in the furrows of the wing-covers. It appears during June and July, and lays its eggs upon the trunk and limbs of apple, peach, oak, and other trees. The larvæ at first bore into the bark and sap-wood, and later into the solid wood. The transformations are completed in one year.

To prevent the ravages of this pest, the trees are rubbed with soap during June or July, or cakes of soap are placed in the forks of the trees, so that the rains will dissolve the soap and wash it down over the trunks. This is supposed to prevent the beetles from depositing their eggs on the trees. After a tree is once infested, the larvæ should be cut out with a gouge or a knife. Nursery stock that is infested should be promptly burned.

Fig. 601.

The red-necked agrilus, *Ăgrilus rufĭcŏllis*.—This beetle (Fig. 601) is about 7.5 mm. long. Its body is narrow and nearly cylindrical. The head is of a dark bronze color, the prothorax of a beautiful coppery bronze, and the wing-covers black. The larva bores in the stems of raspberry and blackberry, causing a large swelling, known as the raspberry gouty-gall. These galls should be collected and burned in early spring.

The family PSEPHENIDÆ includes only the genus *Psephēnus*, of which we have four species, one found in the East and three in

California. This genus was formerly included in the following family; but it differs from the Dryopidæ in having more than five ventral abdominal segments. Our eastern species is *Psephēnus lecŏntei*. These beetles are found in the vicinity of running water, and often, in the heat of the day, collect on stones that project from the water; they fly swiftly when disturbed. The body is oval, subdepressed, narrowed in front, and clothed with fine, silken hairs, which retain a film of air when the insect goes beneath water. The females deposit their eggs in a layer on the under side of submerged stones in shallow brooks. The beetles measure from 4.5 mm. to 6 mm. in length.

The larva is found clinging to the lower surface of stones in rapid streams, and I have found it in muck near a spring. It is very flat, circular in outline (Fig. 602), and measures about 7 mm. in length. It breathes by five pairs of branched tracheal gills on the ventral side of the abdomen. It is rarely recognized as an insect by the young collector; in fact it was originally described as a crustacean under the generic name *Fluvicola*. I have suggested the common name *water-penny* for the larva. When mature the larva leaves the water, and pupates under the last larval skin, beneath a stone or other object in a damp situation.

Fig. 602.

The family DRYOPIDÆ as now restricted includes only the subfamily Parninæ of the old family Parnidæ, in which were included the preceding family and the following family. The Dryopidæ differ from the Psephenidæ in that the members of it have only five ventral abdominal segments, and from the Elmidæ in that in the Dryopidæ the anterior coxæ are transverse, with a distinct trochantin.

This family includes small water-beetles in which the legs are not fitted for swimming. They are found most often in swift-running water, where they cling to stones, logs, or aquatic plants. The body is clothed with fine, silken hairs, which retain a film of air when the insect is beneath the water. They feed on decaying matter in the water.

The larvæ are also aquatic. The larva of *Hĕlichus lithŏphilus* (Fig. 603) resembles somewhat the water-penny (*Psephenus*), except that the body is more elongate and is deeply notched between the segments.

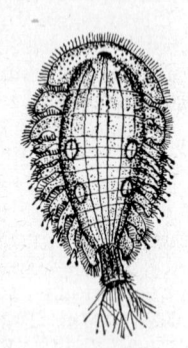

Fig. 603.

Seventeen species of this family have been found in our fauna.

The family ELMIDÆ includes beetles that are closely allied to the preceding family in structure and in habits; but in this family the anterior coxæ are rounded and without a trochantin, and the body is less densely pubescent than in the Dryopidæ.

The larvæ of some exotic species are said to live in damp earth; but the larva of *Stenĕlmis bicarinātus*, which was described by Professor Matheson ('14) is aquatic. This larva (Fig. 604) differs greatly in form from the representatives of the two preceding families figured above, being long and slender.

The family HETEROCERIDÆ, or the variegated mud-loving beetles, includes only the genus *Heterŏcerus*, of which eleven species have been found in our fauna. In this family all of the tarsi are four-jointed; the **first four abdominal segments are grown together on the ventral side**; and the tibiæ are dilated, armed with rows of spines, and fitted for digging. These beetles are oblong or sub-elongate, oval, densely clothed with short, silky pubescence, very finely punctate, and of a brown color, with the elytra usually variegated with undulated bands or spots of yellow color. They live in galleries which they excavate in sand or mud at the margins of bodies of water, and, when disturbed, run from their galleries and take flight.

The family GEORYSSIDÆ, or the minute mud-loving beetles, includes only the genus *Georўssus*, of which only two species have been found in the United States. They are very minute, rounded, convex, roughly sculptured, black insects, found at the margin of streams, on wet sand; they cover themselves with a coating of mud or fine sand, so that they can be detected only when they move.

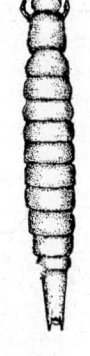

Fig. 604.

The family DASCILLIDÆ includes certain beetles that live on plants, usually near water. The legs are short, with slender tibiæ; The tarsi are five-jointed; the anterior coxæ bear a distinct trochantin; the posterior coxæ are transverse, and dilated into a plate partly covering the femora; and the abdomen has five free, ventral segments, the fifth rounded at the tip.

Sharp ('99) states that the larva of *Dascĭllus cervīnus* is subterranean, and is believed to live on roots; in form it is somewhat like a lamellicorn larva, but is straight, and has a large head.

Only twenty-one species of this family have been described from our fauna; but these represent fifteen genera.

The family EUCINETIDÆ has recently been separated from the Dascyllidæ. In the Eucinetidæ the anterior coxæ do not bear a trochantin; the posterior coxæ are dilated into immense oblique plates, concealing the hind legs in repose; and the internal lobe of the maxillæ is armed with a terminal hook. Only eight species of this family occur in our fauna; seven of these belong to the genus *Eucinetus*.

The larva of a European species of *Eucinētus* lives on fungoid matter on wood.

The family HELODIDÆ includes small beetles, less than 6 mm. in length, found on plants near water. As in the preceding family, the anterior coxæ are without a trochantin; but the lacinia of the maxillæ is not armed with a terminal hook; and the cuticula of the body is usually soft and thin. Sharp ('99) states that the larvæ are aquatic, and are remarkable for possessing antennæ consisting of a great many joints. Our fauna includes thirty-two described species of this family.

The family CHELONARIIDÆ is represented in our fauna by **a** single species, *Chelonārium lecŏntei*, found in Florida. See table, page 471, for distinguishing characters.

Family DERMESTIDÆ

The Dermestids

There are several families of small beetles that feed on decaying matter, or on skins, furs, and dried animal substances. The most important of these is the Dermestidæ, as several species belonging to this family destroy household stores or goods.

The dermestids can be distinguished from most of the other beetles with similar habits by the fact that the wing-covers completely cover the abdomen. They are chiefly small beetles, although one of the common species measures 8 mm. in length. They are usually oval, plump beetles, with pale gray or brown markings, which are formed of minute scales, which can be rubbed off. These beetles have the habit of pretending that they are dead when they are disturbed; they will roll over on their backs with their legs meekly folded and lie still for a long period.

The larvæ do much more damage than the adults. They are active, and are clothed with long hairs. These hairs are covered throughout their entire length with microscopic barbs.

This family is represented in our fauna by about one hundred thirty species; the following are some of the more important of these.

The larder-beetle, *Dermĕstes lardārius*.—This pest of the larder is the most common of the larger members of this family. It measures from 6 mm. to 7.5 mm. in length, and is black except the basal half of its wing-covers, which are pale buff or brownish yellow. This lighter portion is usually crossed by a band of black spots, three on each wing-cover (Fig. 605). The larva feeds on dead animal matter, as meat, skins, feathers, and cheese. It is often a serious pest where bacon or ham is stored. When full-grown it is about 12 mm. in length, dark brown above, whitish below, and rather thickly covered with long, brown hairs. It is said that these insects can be attracted by baits of old cheese, from which they may be gathered and destroyed.

Fig. 605.

The carpet-beetle, *Anthrēnus scrophulāriæ*.—This is a well-known household pest. It is an introduced European insect, which was first recognized as a serious pest in this country about 1874. It feeds in its larval state on carpets, woollens, furs, and feathers; and for a considerable period was exceedingly destructive. In recent years its ravages have been greatly reduced by the more general use of rugs instead of carpets. As rugs are taken up and cleaned frequently, the insect does not have a chance to breed as it does under carpets which are tacked to the floor and taken up only once or twice a year. The larva is well known to many housekeepers as the buffalo-moth. It is a short, fat grub, about 5 mm. in length when full-grown, and densely clothed with dark brown hairs. It lives in the cracks of floors, near the edges of rooms, and beneath furniture, where it eats holes in the carpet. It also enters wardrobes and destroys clothing. The adult is a pretty little beetle which may be found in infested houses, in the spring, on the ceilings and windows. It measures from 2.2 mm. to 3.5 mm. in length and is clothed with black, white, and brick-red scales. There is a whitish spot on each side of the prothorax, and three irregular, whitish spots on the outer margin of each wing-cover; along the suture where the two wing-covers meet, there is a band of brick-red scales, which is widened in several places. It is worth while to learn to know this beetle; for a lady-bug which often winters in our houses is frequently mistaken for it. The carpet-beetle in its adult state feeds on the pollen of flowers. Sometimes it abounds on the blossoms of currant, cherry, and other fruits. The best way to avoid the ravages of this pest is to use rugs instead of carpets, and to trap the larvæ by placing woollen cloths on the floors of closets. By shaking such cloths over a paper once a week, the larvæ can be captured.

The change from carpets to rugs is a very desirable one; for carpets that are tacked to the floor and taken up only once or twice a year are unwholesome. The change need not be a very expensive one. As carpets wear out they may be replaced with rugs; and good carpets can be made over into rugs. If the floors are not polished, as is usually the case where it was planned to cover them with carpets, they can be made presentable by filling the cracks with putty and painting the boards where they are to be exposed.

The museum pests, *Anthrēnus verbăsci* and *Anthrēnus museōrum.*— There are two minute species of this family that are a constant source of annoyance to those having collections of insects. The adult beetles measure from 2 mm. to 3 mm. in length, and are very convex. They deposit their eggs on specimens in our collections; and the larvæ feed upon the specimens, often destroying them. In order to preserve a collection of insects it is necessary that they should be kept in tight cases, so that these pests cannot gain access to them. Specimens should not be left exposed except when in use. And the entire collection should be carefully examined at least once a month. The injury is done by the larvæ, which are small, plump, hairy grubs. Their presence is indicated by a fine dust that falls on to the bottom of the case from the infested specimens. These larvæ can be destroyed by pouring a small quantity of carbon disulphide into the case, and keeping it tightly closed for a day or two. Benzine poured on a bit of cotton in the box will cause the pests to leave the specimens, when they may be taken from the box and destroyed. But we have found carbon bisulphide the better agent for the destruction of these pests.

The family BYRRHIDÆ, or the pill-beetles, are short, very convex beetles of small or moderate size; some, however, are 12 mm. in length. The body is clothed with hair or minute scales. The legs can be folded up very compactly, the tibia usually having a furrow for the reception of the tarsus. These beetles are found upon walks and at the roots of trees and grass; a few live under the bark of trees. Nearly one hundred species have been found in this country.

The family NOSODENDRIDÆ includes two species of *Nosodĕndron*, one found in the East and one in California. They were formerly included in the Byrrhidæ, but differ from that family in having the head prominent and the mentum large. These beetles live under the bark of trees.

The family RHYSODIDÆ includes only four species, two from each side of the continent. They are elongate, somewhat flattened beetles, with the head and prothorax deeply furrowed with longitudinal grooves; and the abdomen with six ventral segments, the first broadly triangular, widely separating the coxæ. They are found under bark. See footnote, page 470.

The family OSTOMIDÆ includes oblong, somewhat flattened beetles, of a black or reddish black color. Most of them live under bark; but some are found in granaries, and have been widely distributed by commerce. The larva of a species found under bark has been observed to feed on the larva of the codlin-moth.

One well-known species, *Tenebroides mauritănicus*, infests granaries. It is a shining brown beetle measuring about 8 mm. in length; it is commonly known as the *cadelle*. Both adult beetles and larvæ feed on grain, but are also predacious, feeding on other insects infesting grain. The larvæ when full-grown burrow into the sides of the bins, where they transform.

The family NITIDULIDÆ comprises small, somewhat flattened beetles. With many species the prothorax has wide, thin margins, and the wing-covers are more or less truncate, so as to leave the tip of the abdomen exposed; but sometimes the elytra are entire. The tarsi are usually five-jointed, with the fourth segment very small; they are more or less dilated; the posterior coxæ are flat, not sulcate; the anterior coxæ are transverse; and the abdomen has five free ventral segments.

Most species of this family feed on the juices of fruits and fermenting sap that exudes from trees; a few are found on flowers, and others on fungi or carrion. About one hundred thirty species are known from North America. One of the most common species is *Glischrochīus* (*Ips*) *fasciātus* (Fig. 606). This is a shining black species, with two conspicuous, interrupted, reddish bands across the wing-covers.

Fig. 606.

The family RHIZOPHAGIDÆ includes only the genus *Rhizŏphagus*, of which there are fourteen North American species. These are small, slender, elongate species, which live beneath bark. This genus was formerly included in the preceding family; it differs from that family in that the antennæ are only ten-jointed, and the club of the antennæ is two-jointed.

The family MONOTOMIDÆ is composed of small, depressed beetles, found mostly under the bark of trees, but some species live in the nests of ants. The wing-covers are truncate behind, leaving the last abdominal segment exposed. There are nearly forty described North American species.

Family CUCUJIDÆ
The Cucujids

The insects of this family are very flat and usually of an elongate form; most of the species are brown, but some are of a bright red color. As a rule they are found under bark and are believed to be carnivorous both in the larval and adult states; but some feed in grain. There are nearly one hundred species in our fauna.

The most conspicuous of our common species is *Cūcujus clăvipes* (Fig. 607). This insect is about 12 mm. in length and of a bright red color, with the eyes and antennæ black and the tibiæ and tarsi dark.

The most important member of this family is the corn silvanus, *Silvānus surinamĕnsis*, which is one of the small beetles that infest stored grain. This species is readily distinguished from other small beetles with similar habits by its flattened form and the saw-like edges of the prothorax. Besides grain it often infests dried fruits and other stores. It measures from 2.5 mm. to 3 mm. in length. The larva as well as the adult feed on grain. It differs from the larva of the granary-weevil (*Calendra*) in the more elongate form of its body and in the possession of three pairs of legs.

Fig. 607.

Family EROTYLIDÆ
The Erotylids

The members of this family are usually of moderate or small size; but some species are quite large, measuring 18 mm. or more in

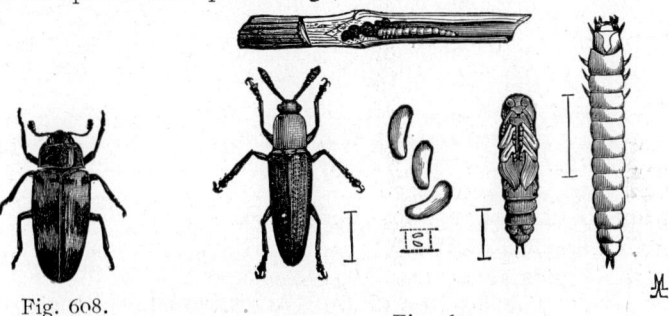

Fig. 608.

Fig. 609.

length. Some of our more common species are conspicuously marked with shining black and red.

To the genus *Megalodăchne* belong two common, large species, which are black, with two dull red bands extending across the wing-covers. *M. heros* (Fig. 608) is 16 mm. or more in length. *M. fasciāta* is about 12 mm. in length.

The genus *Languria* includes long, narrow species, which resemble click-beetles in form. Figure 609 represents *Langūria mozărdi*, greatly enlarged. This is a reddish species with dark blue wing-covers; the larva bores in the stalks of clover. It has not become a serious pest as the larvæ are destroyed whenever clover is cut at the proper time.

The larvæ of some species of this family feed on fungi.

The family DERODONTIDÆ includes only five American species, two found in the East and three in the Far West. They are small brown or dull brownish yellow beetles, having the head deeply impressed, with a small, smooth tubercle on each side inside the eye. These beetles are found on fungi.

The family CRYPTOPHAGIDÆ includes insects of small size, usually less than 2.5 mm. in length, and of variable form but never very flat. The thorax is nearly or quite as wide as the wing-covers, and the first ventral abdominal segment is somewhat longer than the others. They are generally of a light yellowish brown color, and live on fungi and decomposing vegetable matter.

The family BYTURIDÆ includes only the genus *Bytūrus*, of which there are five species in our fauna. This genus was formerly included in the Dermestidæ, but it differs from that family in having the second and third segments of the tarsi lobed beneath, the front coxal cavities closed behind, and the tarsal claws armed with a large basal tooth. The following is a well-known species.

The raspberry fruit-worm, *Bytūrus unīcolor*.—The fruit of the red raspberry is often infested by a small white worm, which clings to the inside of the berry after it is picked. This is the larva of an oval, pale, dull yellow beetle, which is densely clothed with short, fine, gray hairs. The beetle is represented enlarged in Figure 610; it measures from 3.7 to 4.5 mm. in length. This insect is also injurious in the adult state, as it feeds on the buds and tender leaves of the raspberry and later attacks the blossoms.

Fig. 610.

The family MYCETOPHAGIDÆ, or the hairy fungus-beetles, is composed of small, oval, rarely elongate, moderately convex beetles. They are densely punctured and hairy, and are usually prettily marked insects. They live on fungi and under bark. There are about thirty species in our fauna.

The family COLYDIIDÆ is composed of small insects which are usually of an elongate or cylindrical form, and are found under bark, in fungi, and in earth. Some of the species are known to be carnivorous, feeding on the larvæ of wood-boring beetles. The tarsi are four-jointed; the tibiæ are not fitted for digging, and the first four abdominal segments are grown together on the ventral side. More than eighty North American species are known.

The family MURMIDIIDÆ includes five introduced species representing five genera. They are very small, oval beetles, differing from the Colydiidæ in having the antennæ inserted on the front, and in having the anterior coxæ inclosed behind by the metasternum.

The family LATHRIDIIDÆ includes very small beetles which are found chiefly under bark and stones or in vegetable debris, especially decaying leaves. They are oblong; the wing-covers are usually wider than the prothorax and entirely cover the abdomen. There are about one hundred species in our fauna.

The family MYCETÆIDÆ includes only four American species, which have recently been separated from the following family; they differ from the Endomychidæ in having the tarsi distinctly four-jointed.

The family ENDOMYCHIDÆ includes a small number of species, whch are found chiefly in fungus, in decaying wood, or beneath logs and bark. They are small, oval or oblong beetles. The antennæ are about half as long as the body; the prothorax is nearly square, and usually has a wide, thin margin, which is slightly turned upwards at the sides.

The family PHALACRIDÆ, or the shining flower-beetles, includes very small, convex, shining black beetles; they are found on flowers and sometimes under bark. The larvæ live in the heads of flowers, especially in those of the Compositæ. More than one hundred North American species have been described.

Family COCCINELLIDÆ

The Lady-Bugs

These insects are well-known to nearly every child under the popular name given above. They are more or less nearly hemispherical, generally red or yellow, with black spots, or black, with white, red, or yellow spots.

The larvæ occur running about on foliage; they are often spotted with bright colors and clothed with warts or with spines (Fig. 611). When ready to change to a pupa the larva fastens itself by its tail to any convenient object, and the skin splits open on the back. Sometimes the pupa state is passed within this split skin, and sometimes the skin is forced back and remains in a little wad about the tail (Fig. 612).

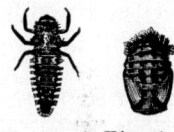

Fig. 611. Fig. 612.

With very few exceptions, the lady-bugs are predacious, both in the larval and adult states. They feed upon small insects and upon the eggs of larger species. The larvæ of certain species are known as "niggers" by hop-growers, and are greatly prized by them; for they are very destructive to the hop-louse. On the Pacific Coast lady-bugs are well known as the most beneficial of all insects to the fruit-growers. In addition to the native species found there, several species have been introduced as a means of combating scale-insects. One of these, *Rodōlia cardinālis,* which has been incorrectly known as *Vedālia cardinālis*, has proved of very great value in subduing the cottony-cushion scale (*Icerya purchasi*). This lady-bug was introduced from Australia.

The larva of **Brachyacantha** is found in the nests of ants. It is covered with dense tufts of delicate white wax; its food probably consists of the eggs of coccids living in the nests.

A very common lady-bug in the East is *Adālia bipunctāta*. This species is reddish yellow above, with the middle of the prothorax black, and with a black spot on each wing-cover. It frequently passes the winter in our dwellings, and is found on the walls and windows in early spring. Under such circumstances it is often mistaken for the carpet-beetle, and, unfortunately, destroyed.

The nine-spotted lady-bug, *Coccinĕlla novemnotāta*, has yellowish wing-covers, with four black spots on each, in addition to a common spot just back of the scutellum (Fig. 613).

Fig. 613.

Although almost all of the Coccinellidæ are predacious, there are some that are herbivorous. One of these is found in the East. This is the squash-ladybird, *Epilăchna boreālis*. This beetle and its larva (Fig. 614) feed on the foliage of various cucurbitaceous plants, but prefer that of the squash. The adult is yellowish, with large, black spots. The larva is yellow and is clothed with forked spines. A pupa is shown in the figure near the upper right-hand corner.

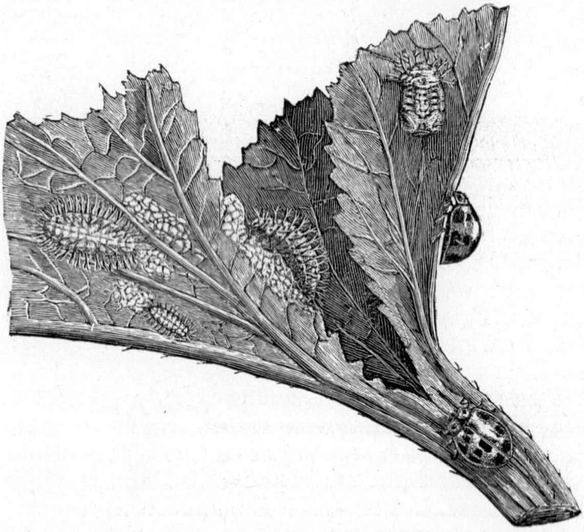

The bean-ladybug, *Epilăchna corrŭpta*, which is found in the South and Southwest, is another herbivorous species.

The family ALLECULI-DÆ, or the comb-clawed bark-beetles, includes brownish beetles, without spots, which are found on leaves and flowers and under bark. The body is usually elongate, elliptical, quite convex, and clothed above with minute hairs, which give a silken gloss to the surface. They are most easily distinguished from allied families by the tarsal claws being pectinate, and the anterior coxal cavities closed behind. The larvæ of some of our species at least live in rotten wood and resemble wire-

Fig. 614.

worms in appearance. There are more than one hundred described North American species.

Family TENEBRIONIDÆ

The Darkling Beetles

The darkling beetles are nearly all of a uniform black color, although some are gray, and a few are marked with bright colors. The different species vary greatly in size and in the form of the body. The hind tarsi are four-jointed, and the fore and middle tarsi are five-jointed. For other characters, see table, p. 474.

These insects occur chiefly in dry and warm regions. Thus while we have comparatively few species in the northeastern United States, there are many in the Southwest. Most of the species feed on dry vegetable matter, and often on that which is partially decomposed; some live in dung, some in dead animal matter, others in fungi, and a few prey upon larvæ. More than eleven hundred species occur in this country. The three following will serve to illustrate the variations in form and habits.

The meal-worm, *Tenĕbrio mŏlitor*.—This is a well-known pest in granaries and mills. The larva is a hard, waxy yellow, cylindrical worm, which measures when full-grown 25 mm. or more in length, and closely resembles a wire-worm; it feeds on flour and meal. The beetle is black and about 15 mm. in length, (Fig. 615). The larvæ and pupæ are used for bird-food and are grown in quantity by bird-supply houses.

The forked fungus-beetle, *Boletothērus cornūtus*, is common in the northeastern United States and in Canada about the large toadstools (*Polyporus*) which grow on the sides of trees. The surface of the body and wing-covers is very rough, and the prothorax bears two prominent horns (Fig. 616). The larva lives within the fungi referred to above.

Fig. 615. Fig. 616.

The pinacate-bugs.—Several species of *Eleōdes* are abundant on the Pacific Coast, where they are found under stones and pieces of wood lying on the ground. They are apt to congregate in large numbers under a single shelter, and are clumsy in their movements. They defend themselves when disturbed by elevating the hinder part of the body and discharging an oily fluid from it. They present an absurb appearance, walking off clumsily, and carrying the hind end of the body as high as possible. The most common species are large, smooth, club-shaped beetles (Fig. 617), and are commonly known as *pinacate-bugs*. These beetles and those belonging to several closely allied genera lack hind wings.

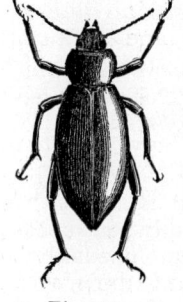

Fig. 617.

The family LAGRIIDÆ, or the lagriid bark-beetles, includes elongate beetles, with a narrow, subcylindrical prothorax, and a more or less brassy color. They are closely allied to the preceding family, but differ in having the next to the last segment of the tarsi spongy beneath. They are found under bark and on leaves. The larvæ feed freely on foliage and are much less retiring in habits than those of the darkling beetles. Seventeen species are listed from the United States; most of these are found in the South and Far West. Our most common species in the East is *Arthromācra ænea* (Fig. 618); this species measures from 9.5 mm. to 13.5 mm. in length.

Fig. 618.

The family MONOMMIDÆ is represented in this country by only six species, found chiefly in the Far West and in Florida. They are small, black, oval beetles, in which the anterior coxal cavities are open behind, the hind tarsi four-jointed, and the other tarsi five-jointed; and in which the antennæ are received in grooves on the under side of the prothorax. Except one species found in Florida, our species belong to the genus *Hyporhāgus*.

The family MELANDRYIDÆ includes about eighty North American species. These are found under bark and in fungi. They are usually of elongate form, although some, like the one figured here, are not so. The maxillary palpi are frequently very long and much dilated; and the first segment of the hind tarsi is always much elongated. Among our more common species are two belonging to the genus *Penthe*. These are rather large, oval, depressed beetles, upwards of 12 mm. in length, and of a deep black color. *Pĕnthe obliquāta* is distinguished by having the scutellum clothed with rust-red hairs (Fig. 619). *Pĕnthe pimĕlia* closely resembles this species, except that the scutellum is black.

Fig. 619.

The family PTINIDÆ has been restricted recently to one of the subfamilies of the old family Ptinidæ, which included, in addition to the insects now retained in it, those classed in the three following families. In the Ptinidæ, as now restricted, the antennæ are inserted upon the front of the head and rather close together, and the thorax is not margined at the sides. Only thirty-seven of our listed species are retained in this family. They are all small beetles and some of them are household pests, infesting stored provisions, clothing, and books. The best-known species is *Ptīnus fur*. This is a reddish brown beetle; in the female the elytra are marked with two patches of white hairs. It measures from 2.8 mm. to 3.5 mm.

The family ANOBIIDÆ, or the death-watch family, includes a large part of the old family Ptinidæ, there being more than two hundred species in our fauna. In this family the antennæ are inserted on the sides of the head in front of the eyes; the thorax is usually margined at the sides; and the tibiæ are without spurs. These beetles are small, and are generally of a cylindrical form, though some are broadly oval or nearly globular. They live chiefly on dry vegetable matter and are often found boring in the woodwork of buildings. Some are pests in drug-stores and groceries, where they infest a great variety of substances both vegetable and animal. Among the better-known species are the drug-store beetle, *Sitŏdrepa panīcea*, which not only infests many kinds of drugs but is also sometimes a pest in groceries where it infests cereals; and the cigarette-beetle, *Lasiodĕrma serricŏrne*, which infests dried tobacco and destroys cigarettes and cigars by boring holes through them.

To this family belongs the death-watch, *Xestŏbium rufovillōsum*, which bores in the timbers of buildings and makes a ticking sound by striking its head or jaws against the walls of its burrows. This sound heard in the night by superstitious watchers by sick-beds has been supposed to portend death.

The family BOSTRICHIDÆ, or the powder-post beetles, includes beetles which are elongate in form; the head is usually deflexed, and protected by the thorax, which is then hood-like in form; and the first ventral segment of the abdomen is scarcely longer than the second. These beetles live almost exclusively in dry wood either in cylindrical burrows or beneath the bark. Sometimes they infest timbers to such an extent that the wood is largely reduced to powder, hence the common name, powder-post beetles. The adult of one species, *Amphĭcerus bicaudātus*, bores into the living twigs of fruit-trees and grape-vines for food, but it breeds in dying wood, such as prunings and dying branches. This species is known as the apple-twig borer and also as the grape-cane borer.

The family LYCTIDÆ is composed of a small number of beetles which resemble the powder-post beetles in habits. In this family the head is prominent and not covered by the prothorax; and the first ventral segment of the abdomen is much longer than the second. Most of our species belong to the genus *Lyctus*.

The family SPHINDIDÆ is represented in North America only by six small species, which are found in dry fungi which grow on the trunks of trees and on logs.

The family CISIDÆ includes very small beetles, rarely exceeding 3 mm. in length, found under the bark of trees and in the dry and woody species of fungi. The body is cylindrical; the prothorax is prolonged over the head; the abdomen has five ventral segments, of which the first is longer than the others; and the tarsi are all four-jointed. There are nearly one hundred species in our fauna.

Family SCARABÆIDÆ

The Scarabæids or Lamellicorn Beetles

This very large family is represented in our fauna by nearly one thousand species, and includes beetles that exhibit a wide range of variation in size, form, and habits. They are mostly short, stout-bodied beetles, of which the well-known June-bugs or May-beetles represent the most familiar type. The most useful character for distinguishing these insects is the lamellate form of the club of the antennæ, the segments constituting it being greatly flattened, and capable of being brought close together. It is this character that suggests the name lamellicorn beetles. In the next family, the Trogidæ, which has recently been separated from this family, the antennæ are lamellate. The two families can be separated by the fact that in the Scarabæidæ the epimera of the mesothorax extend to the coxæ, while in the Trogidæ they do not.

According to their habits, the members of this family can be separated into two well-marked groups—the scavengers and the leaf-chafers.

THE LAMELLICORN SCAVENGERS

The lamellicorn scavengers in both the larval and adult states feed upon decaying vegetable matter. Nearly all the species live in dung, chiefly that of horses and cows; but a few species feed upon fungi. The following are the more common representatives of this division.

I. *The tumble-bugs.*—These are the most familiar of all dung-beetles, for their peculiar habits have attracted much attention from the earliest times. They are of rounded form, and the wing-covers are shortened so as to expose the tip of the abdomen. They are generally black, but some are colored with rich metallic hues. They vary greatly in size.

The name *tumble-bug* refers to the habit which many species exhibit of forming round balls of dung, which they roll long distances and then bury. They often work in pairs and it was formerly believed that such a pair was a male and a female working together to make provision for their progeny; but Fabre found by dissecting the beetles that the two members of a pair often proved to belong to the same sex; and concluded that the eager fellow-worker, under the deceitful pretense of lending a helping hand, nurses the scheme of purloining the ball at the first opportunity.

According to the observations of Fabre ('79 and '11), the balls made early in the year are devoured by the beetles, which bury themselves with them and feed upon them. Later other balls are made and buried, upon each of which an egg is laid. The larva hatching from this egg feeds upon the ball of dung, and when full-grown transforms within the cavity in which the ball was placed.

This strange habit of rolling these balls has occasioned much speculation as to its object, and has been the source of many superstitions, especially in ancient times. The only reasonable theory that we have met is that, as many predacious insects frequent the masses of dung from which the balls are obtained, in order to prey upon the larvæ which live there, the more intelligent tumble-bugs remove the food for their larvæ to a safe distance.

The most noted member of this group of genera is the sacred beetle of the Egyptians, *Ateuchus sācer*. This insect was held in high veneration by this ancient people. It was placed by them in the tombs with their dead; its picture was painted on sarcophagi, and its image was carved in stone and precious gems. These sculptured beetles can be found in almost any collection of Egyptian antiquities.

From the habits and structure of this scarabæid the Egyptians evolved a remarkable symbolism. The ball, which the beetles were supposed to roll from sunrise to sunset, represented the earth; the beetle itself personified the sun, because of the sharp projections on its head, which extend out like rays of light; while the thirty segments of its six tarsi represented the days of the month. All individuals of this species were thought to be males, and a race of males symbolized a race of warriors. This latter superstition was carried over to Rome and the Roman soldiers wore images of the sacred beetle set in rings.

Our common tumble-bugs are distributed among three genera: *Cănthon, Cōpris*, and *Phanæus*. In the genus *Canthon* the middle and posterior tibiæ are slender, and scarcely enlarged at the extremity. *Canthon lævis* is our most common species (Fig. 620). In *Copris* and *Phanæus* the middle and posterior tibiæ are dilated at the extremity. In *Phanæus* the fore tarsi are wanting, and the others are not furnished with claws; the species are brilliantly colored. *Phanæus cărnifex*, with its rough copper-colored thorax and green elytra, is one of our most beautiful beetles, and is our best-known species. It is about 16 mm. in length, and the head of the male is furnished with a prominent horn. In *Copris* all the tarsi are present and furnished with claws. *Copris carolīna* is a large, well-known species, which measures more than 25 mm. in length.

Fig. 620.

II. *The aphodian dung-beetles.*—These are small insects, our common species measuring from 4 mm. to 8 mm. in length. The body is oblong, convex, or cylindrical in form, and, except in one small genus, the clypeus is expanded so as to cover the mouth-parts entirely. These insects are very abundant in pastures in the dung of horses and cattle, and immense numbers of them are often seen flying through the air during warm autumn afternoons. More than one hundred and fifty North American species have been described; of these, one hundred belong to the genus *Aphōdius*. One of the more common species is *Aphodius fimetārius*, which is about 8 mm. in length, and is easily recognized by its red wing-covers.

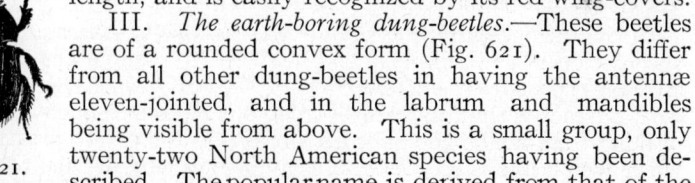

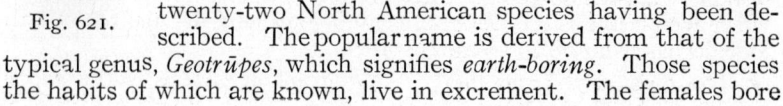

III. *The earth-boring dung-beetles.*—These beetles are of a rounded convex form (Fig. 621). They differ from all other dung-beetles in having the antennæ eleven-jointed, and in the labrum and mandibles being visible from above. This is a small group, only twenty-two North American species having been described. The popular name is derived from that of the typical genus, *Geotrŭpes*, which signifies *earth-boring*. Those species the habits of which are known, live in excrement. The females bore

Fig. 621.

holes into the earth either beneath the dung or near it; into these holes they convey a quantity of the dung; this is to serve as food for the larvæ, an egg being laid in each hole. This is an approach to the peculiar habits of the tumble-bugs.

THE LAMELLICORN LEAF-CHAFERS

The leaf-chafers are herbivorous insects which in the adult state usually feed upon the leaves of trees, but many of the species devour the pollen and petals of flowers. In the larval state some of these insects are found in rotten wood; others live in the ground, where they feed upon the roots of grass and other plants. These larvæ are thick, fleshy grubs, with well-developed legs (Fig. 622). The caudal segments of the abdomen are very large, and appear black on account of the large amount of dirt in the intestine. The body is strongly curved, so that the larvæ can crawl only with great difficulty; when in the ground they usually lie on their backs.

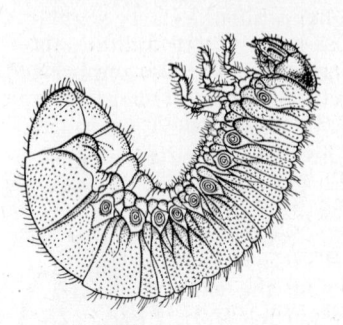

Fig. 622.

The following groups include the more important representatives of this division.

I. *The May-beetles or June-bugs.*—During the warm evenings of May and June we throw open our windows so that we may feel the refreshing coolness of the night air and the inspiration of the new summer. Suddenly, as we sit working or reading, our peace is disturbed by a buzzing object which whirls above us. Then comes a sharp thud and silence. A little later the scratching of six pairs of tiny claws tells us the whereabouts of the intruder. But so familiar are we with his kind that we need not look to know how he appears, the mahogany-brown blunderer, with yellowish wings sticking out untidily from under his polished wing-covers.

Although these insects are beetles, and attract our attention each year in May, they have received the infelicitous title of June-bugs. They are more properly termed May-beetles.

The May-beetles belong to the genus *Phyllŏphaga*, of which we have nearly one hundred species. The adults frequently do much injury by eating the foliage of trees. In the case of large trees this injury usually passes unnoticed; but small trees are often completely defoliated by them. When troublesome, they can be easily gathered by shaking them from trees upon sheets. Figure 623 represents a common species.

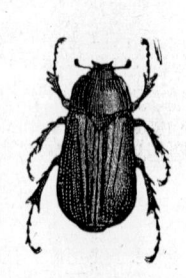

Fig. 623.

The larvæ of the different species of May-beetles are commonly classed together under the name "white-grubs." They are often great pests in meadows and in cultivated fields. We have known large strawberry plantations to be destroyed by them, and have seen large patches of ground in pastures from which the dead sod could be rolled as one would roll a carpet from a floor, the roots having been all destroyed and the ground just beneath the surface finely pulverized by these larvæ. No satisfactory method of fighting this pest has been discovered as yet. If swine be turned into fields infested by white-grubs they will root them up and feed upon them. We have destroyed great numbers of the beetles by the use of trap-lanterns, but many beneficial insects were destroyed at the same time.

II. *The rose-bugs.*—The common rose-bug, *Macrodăctylus subspinōsus*, is a well-known pest. It is a slender beetle, tapering before and behind, and measuring 9 mm. in length (Fig. 624). It is thickly clothed with fine, yellow, scale-like hairs, which give it a yellow color; the legs are long, slender, and of a pale red color. These beetles appear in early summer, and often do great injury to roses and other flowers, and to the foliage of various fruit-trees and shrubs. This is a very difficult pest to control. The best method now known is to use Paris green when safe to do so; in other cases the beetles should be collected by jarring them into a large funnel which is fitted into a can. The larvæ of rose-bugs feed on the roots of plants.

Fig. 624.

III. *The shining leaf-chafers.*—These insects resemble the May-beetles in form, but can be distinguished from them by the position of the hind pair of spiracles, which are visible on the sides below the edges of the wing-covers; and they differ from the other leaf-chafers in which the spiracles are in this position in that the tarsal claws are of unequal size, one claw of each pair being larger than the other. These beetles are usually polished, and many of them are of brilliant colors. To this group belong the most beautiful beetles known, many appearing as if made of burnished gold or silver, or other metal.

The goldsmith-beetle, *Cotălpa lanĭgera.*—This is one of our most beautiful species. It measures from 20 mm. to 26 mm. in length, and is a broad oval in shape. It is of a lemon-yellow color above, glittering like burnished gold on the top of the head and thorax; the under side of the body is copper-colored and thickly covered with whitish wool.

The spotted pelidnota, *Pelidnōta punctāta.*—This beetle is reddish brown above, with three black spots on each wing-cover and one on each side of the prothorax (Fig. 625). The scutellum, base of the head, and entire body beneath, are of a deep, bronzed-green color. The adult is commonly found feeding on the leaves of grape. The larva feeds upon decaying roots and stumps of various trees.

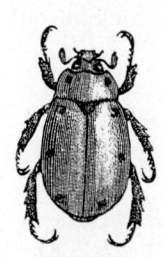

Fig. 625.

The Japanese beetle, *Popĭllia japŏnica.*—This is a very serious pest which feeds in the adult state on the foliage of many cultivated and wild plants, being practically omnivorous, and in the larval state feeds on the roots of grasses. It was first observed in this country in a limited area in Burlington County, New Jersey, in 1916, and has since spread over other counties of this state and into Pennsylvania. The adult insect is about the size of the Colorado potato-beetle, but slightly longer. The head and thorax are shining bronze-green in color, with the wing-covers tan or brownish,tinged with green on the edges. Along the sides of the abdomen are white spots, and two very distinct white spots at the tip of the abdomen below the wing-covers. The larva resembles the larvæ of May beetles.

This pest is regarded as of so great importance that a special laboratory, "The Japanese Beetle Laboratory," has been established for investigations regarding it at Riverton, N. J.

IV. *The rhinoceros-beetles.*—The name *rhinoceros-beetles* was suggested for this group by the fact that in many species the male bears a horn on the middle of the head. In addition to this horn there may be one or more horns on the thorax. These beetles are of medium or large size; in fact, the largest beetles known belong to this group. As with the flower-beetles, the claws of the tarsi are of equal size, but the fore coxæ are transverse, and not prominent.

One of the largest of our rhinoceros-beetles is *Dynăstes tĭtyrus*. This is of a greenish gray color, with scattered black spots on the wing-covers, or, if only recently transformed, of a uniform dark brown. The male (Fig. 626) bears a prominent horn on the top of the head, and a large one and two small ones on the prothorax. The female has only a tubercle on the head. This insect is found in the Southern States; the larva lives in rotten wood. In the Far West there is a closely allied species, *Dynăstes grăntii*, in which the large horn on the thorax is twice as long as in *D. tityrus*. In the West Indies there occurs a species, *Dynăstes hĕrcules*, which measures 150 mm. in length.

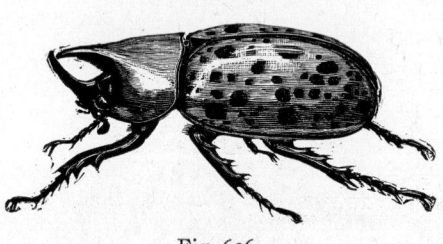

Fig. 626.

Several other genera occur in this country, in some of which the males have prominent horns; in others the horns are represented by tubercles, or are wanting. The following species represents the latter type.

The sugar-cane beetle, *Eucthēola rūgiceps*.—This beetle is a serious pest in the cane-fields of Louisiana, and it sometimes injures corn. Figure 627 represents the adult, and its method of attacking a plant.

V. *The flower-beetles.*—The flower-beetles are so called because many of them are often seen feeding upon pollen and flying from flower to flower. These beetles are somewhat flattened, or nearly level on the back; the claws of the tarsi are of equal size and the fore coxæ are conical and prominent. More than one hundred species occur in this country.

The hermit flower-beetle, *Osmodĕrma eremĭcola.*—This is one of the larger of our flower-beetles (Fig. 628). It is of a deep mahogany-brown color, nearly smooth, and highly polished. It is supposed that the larva lives on decaying wood in forest-trees.

The rough flower-beetle, *Osmodĕrma scābra*, is closely allied to the preceding. It is not quite as large, measuring about 25 mm. in length. It is purplish black, and the wing-covers are roughened with irregular, coarsely punctured striæ. It is nocturnal, concealing itself during the day in the crevices and hollows of trees. The larva lives in the decaying wood of apple and cherry, consuming the wood and inducing more rapid decay.

The genus *Euphōria* represents well the form of the more typical flower-beetles, which are distinguished by the margin of each wing-cover having a large, wavy indentation near its base, which renders the side pieces of the mesothorax visible from above. This indentation makes it unnecessary for these insects to raise or expand their wing-covers when flying, as most beetles do, as they are able to pass the wings out from the sides.

Fig. 627.—The sugar-cane beetle.

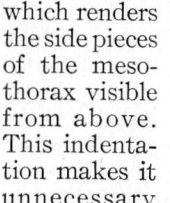

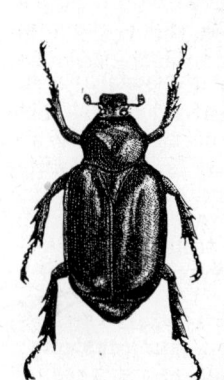

Fig. 628.

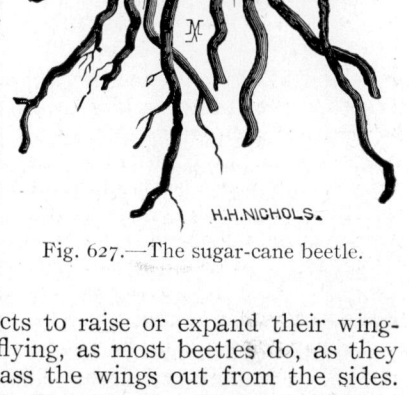

The bumble flower-beetle, *Euphōria ĭnda.*— The most common of our flower-beetles, at least in the North, is a yellowish brown one, with the wing-covers sprinkled all over with small, irregular, black spots (Fig. 629). It is one of the first insects to appear in the spring. It flies near the surface of the ground with a loud humming sound, like that of a bumble-bee, for which it is often mistaken. During the summer months it is not seen;

but a new brood appears about the middle of September. The adult is a general feeder occurring upon flowers, eating the pollen, upon corn-stalks and green corn in the milk, sucking the juices, and upon peaches, grapes, and apples. Occasionally the ravages are very serious.

The green June-beetle or fig-eater, *Cŏtinus nĭtida.*—This species extends over the Atlantic slope, and is very common in the South. It is a green, velvety insect, measuring from 16 mm. to 25 mm. in length. It is somewhat pointed in front, and usually has the sides of the thorax and elytra brownish yellow. These beetles often fly in great numbers at night, making a loud buzzing noise similar to that of the May-beetles. In fact, in the South the term *June-bug* is often applied to this insect. The larvæ feed upon the vegetable mold of rich soils; sometimes they injure growing vegetables by severing the roots and growing stalks; but the chief injury is due to the upheaval of the soil around the plants, which disturbs the roots; the larvæ are also often troublesome on lawns and golf greens by making little mounds of earth on the surface. Sometimes they leave the ground and crawl from one place to another. When they do so, they, strangely enough, crawl upon their backs, making no use of their short legs. On one occasion we saw them crawling over the pavements on the Capitol grounds at Washington in such numbers that bushels of them were swept up and carted away. The adults frequently attack fruit, especially figs, grapes, and peaches.

Fig. 629.

Family TROGIDÆ

The Skin-Beetles

This is a small family, which is represented in this country by twenty-five species. Until recently these insects were included in the preceding family; they can be distinguished from scarabæids by the fact that the epimera of the mesothorax do not extend to the coxæ as they do in the Scarabæidæ. The members of this family are oblong, convex species, in which the surface of the body and wing-covers is usually very rough, and covered with a crust of dirt, which is removed with great difficulty. They are small or of medium size; our most common species measure from 8 mm. to 12 mm. in length. The abdomen is covered by the elytra; the feet are hardly fitted for digging, but the femora of the front legs are greatly dilated.

These beetles feed upon dried, decomposing animal matter; many species are found about the refuse of tanneries, and upon the hoofs and hair of decaying animals.

Fig. 630.

Except a few species found in the Far West, all of our species belong to the genus *Trox* (Fig. 630).

Family LUCANIDÆ

The Stag-Beetles

The stag-beetles are so called on account of their large mandibles which in the males of some species are branched like the antlers of a stag. They and the members of the following family are distinguished by the form of the club of the antennæ, which is composed of flattened plates; but these plates are not capable of close apposition, as in the antennæ of the lamellicorn beetles. In the stag-beetles the mentum is not emarginate and the ligula is covered by the mentum or is at its apex.

The adult beetles are found in or beneath decaying logs and stumps. Some of them are attracted, at night, to lights. They are said to live on honeydew and the exudations of the leaves and bark of trees, for procuring which the brushes of their jaws and lips seem to be designed; but it seems probable that some species, at least, feed upon decomposing wood. They lay their eggs in crevices of the bark of trees, especially near the roots. The larvæ feed upon juices of wood in various stages of decay. They resemble the well-known larvæ of May-beetles.

The family is a small one; only thirty North American species are now known.

The common stag-beetle, *Lucānus dama*. —The most common of our stag-beetles is this species (Fig. 631). It flies by night with a loud buzzing sound, and is often attracted to lights in houses. The larva is a large, whitish grub resembling the larvæ of the lamellicorn beetles. It is found in the trunks and roots of old, partially decayed trees, especially apple, cherry, willow, and oak. The specimen figured here is a male; in the female the mandibles are shorter.

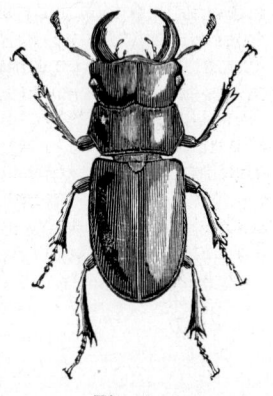

Fig. 631.

The giant stag-beetle, *Lucānus ĕlaphus*, is a large species found in the South. It measures from 35 mm. to 50 mm. in length, not including the mandibles, which in the case of the male are more than half as long as the body, and branched like the antlers of a stag.

The antelope-beetle, *Dŏrcus părallēlus*. —This beetle is somewhat smaller than the species of *Lucanus*, and differs in having the wing-covers marked with longitudinal striæ and the teeth on the outside of the fore tibiæ much smaller (Fig. 632).

Fig. 632.

Several species of stag-beetles that are much smaller than *Dorcus* are found in this country.

Family PASSALIDÆ

The members of this family resemble the stag-beetles in the form of the antennæ, but differ in that the mentum is deeply emarginate, with the ligula filling the emargination.

A single, widely distributed species is found in the United States; this is the horned passalus, *Păssalus cornūtus* (Fig. 633). It is a large, shining, black beetle, with a short horn, bent forwards, on the top of the head. This beetle and its larva are found in decaying wood. The larva appears to have only four legs, the hind legs being shortened and modified so as to form part of a stridulating organ. See Figure 101, page 89.

The beetles of this genus are common throughout the tropics of both hemispheres. According to the observations of Ohaus, which have been confirmed by Professor Wheeler ('23), these beetles are social. They form colonies, consisting of a male and female and their progeny, and make large, rough galleries in rather damp, rotten logs. The parent beetles triturate the rotten wood and apparently treat it with some digestive secretion which makes it a proper food for the larvæ, since their mouth-parts are too feebly developed to enable them to attack the wood directly. All members of the colony are kept together by stridulatory signals. The stridulatory organ of the adult consists of patches of minute denticles on the dorsal surface of the abdomen, which may be rubbed against similar structures on the lower surface of the wings.

Fig. 633.

Family CERAMBYCIDÆ

The Long-horned Beetles or Cerambycids

This is a very large family, there being more than eleven hundred described species in North America alone. As a rule the beetles are of medium or large size, and graceful in form; many species are beautiful in color. The body is oblong, often cylindrical. The antennæ are long, often

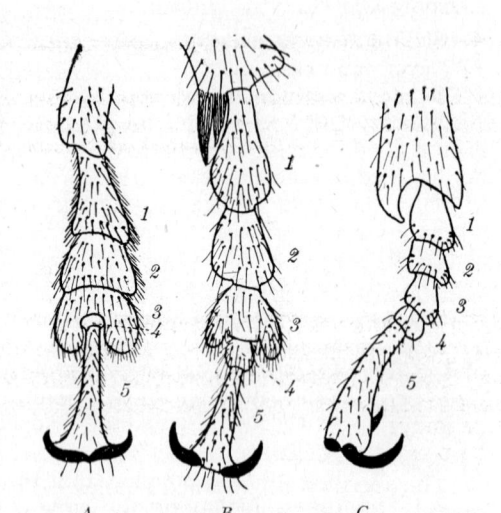

Fig. 634.—Tarsi of Phytophaga: *A*, typical; *B*, *Spondylis*; *C*, *Parandra*.

longer than the whole body; but except in one genus, *Prionus*, they are only eleven-jointed, as with most beetles. The legs are also long, and the tarsi are apparently four-jointed, the fourth segment being very small and hidden; the third segment of the tarsi is strongly bilobed (Fig. 634).

They are strong flyers and swift runners; but many of them have the habit of remaining motionless on the limbs of trees for long intervals, and when in this apparent trance they suffer themselves to be picked up. But, when once caught, many species make an indignant squeaking by rubbing the prothorax and mesothorax together.

The larvæ are borers, living within the solid parts of trees or shrubs, or beneath bark. They are white or yellowish grubs. The body is soft, and tapers slightly from head to tail (Fig. 635); the jaws are powerful, enabling these insects to bore into the hardest wood. The larval state usually lasts two or three years. The pupa state is passed within the burrow made by the larva; frequently a chamber is made by partitioning off a section of the burrow with a plug of chips; but sometimes the larva builds a ring of chips around itself just beneath the bark before changing to a pupa. The pupal state is comparatively short, lasting only a few days or weeks.

Fig. 635.

This family comprises three subfamilies, which are separated by LeConte and Horn as follows:

A. Sides of the prothorax with a sharp margin. p. 525............PRIONINÆ
AA. Prothorax not margined.
   B. Front tibiæ not grooved; palpi never acute at tip. p. 526. CERAMBYCINÆ
   BB. Front tibiæ obliquely grooved on the inner side; palpi with the last segment cylindrical and pointed. p. 528 ....LAMIINÆ

SUBFAMILY PRIONINÆ

The Prionids

The larger of the long-horned beetles constitute this subfamily. They are distinguished from other cerambycids by having the sides of the prothorax prolonged outwards into a thin margin, which is more or less toothed. The wing-covers are usually leathery in appearance, and of a brownish or black color. The following are our best-known species.

The aberrant long-horned beetles.—The beetles of the genus *Parandra* exhibit some striking differences from the more typical cerambycids, and were formerly placed in a separate family, the *Spondylidæ*; but they are now included in the Cerambycidæ. There are only four North American species of this genus. These live under bark of pine trees. The fourth segment of the tarsus, although much reduced in size, is distinctly visible; the first three segments are but slightly dilated, and the third is either bilobed or not (Fig. 634, C). The segments of the antennæ have deep impressions, in which are situated the organs of special sense (Fig. 636). The most common species is *Parăndra brŭnnea* (Fig. 637); this insect is of a mahogany-brown color, and measures from 9 mm. to 18 mm. in length.

The broad-necked prionus, *Priōnus laticŏllis.* —This is the largest of our common species; but the individuals vary from 22 mm. to 50 mm. in length. It is of a pitchy-black color, and of the form shown in Figure 638. The antennæ are twelve-jointed in both sexes. The larva is a large, fleshy grub, and infests the roots of grape, apple, poplar, and other trees.

Fig. 636.

Fig. 637.

The tile-horned prionus, *Priōnus imbricŏrnis*, is very similar to the preceding species but can be distinguished at a glance by the form of the antennæ. In the antennæ of the male the number of segments varies from eighteen to twenty, while in the female the number varies from sixteen to seventeen. The popular name refers to the fact that the segments of the antennæ of the male overlap one another like the tiles on a roof. The larva infests the roots of grape and pear, and also feeds upon the roots of herbaceous plants.

The straight-bodied prionid, *Derŏbrachus brŭnneus*, is also a common species. The body is long, narrow, and somewhat flattened; it measures from 25 mm. to 35 mm. in length, and is of a light brown color. The prothorax is short, and is armed on each side with three sharp spines. The sides of the wing-covers are very nearly parallel; this suggests the common name. The adult flies by night, and is often attracted to lights; the larva is supposed to infest pine.

Fig. 638.

Subfamily CERAMBYCINÆ

The Typical Cerambycids

In this subfamily the prothorax is rounded on the sides, the tibiæ of the fore legs are not grooved, and the palpi are never acute at the tip. There are nearly four hundred American species, representing more than one hundred genera. The few species mentioned below are those that the beginning student is most likely to meet.

The ribbed pine-borer, *Rhāgium lineātum.*—This is a gray beetle mottled with black, and has a narrow thorax, with a spine on each side (Fig. 639). It received its name because of the three ridges extending lengthwise on each wing-cover. Its larva bores in the wood of pine-trees. On one occasion the writer found many of them in a pine-tree eight inches in diameter, which they had bored through and through. When the larva is full-grown it makes a hole nearly through the thick bark of the tree, so that it may easily push its way out after its transformations; it then retreats a short distance and makes a little ring of chips around itself, between the bark and the wood, and changes to a pupa within this rude cocoon. The adult beetle remains in this pupal cell through the winter.

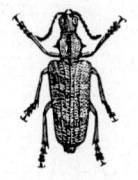

Fig. 639.

The cloaked knotty-horn, *Desmŏcerus palliātus.*—This beautiful insect is of a dark blue color, with greenish reflections. The basal part of the wing-covers is orange-yellow, giving the insect the appearance of having a yellow cape thrown over its shoulders (Fig. 640). The segments in the middle of the antennæ are thickened at the outer end, so that they look like a series of knots. The adult is quite common in June and July on elder, in the pith of which the larva bores.

Fig. 640.

The beautiful maple-borer, *Glycōbius speciōsus.*—This is a handsome insect, marked with black and yellow, as indicated in Figure 641. It lays its eggs in midsummer on the trunks of sugar-maples, in the wood of which the larvæ bore. If an infested tree be examined in the spring the presence of these borers can be detected by the dust that falls from the burrows. The larvæ can be destroyed at this time by the use of a knife and a stiff wire.

The locust-borer, *Cyllēne robīnæ.*—To the enthusiastic entomologist the goldenrod is a rich mine, yielding to the collector more treasures than any other flower. It gives up its gold-dust pollen to every insect-seeker; and because of this generous attitude to all comers it is truly emblematic of the country that has chosen it as its national flower.

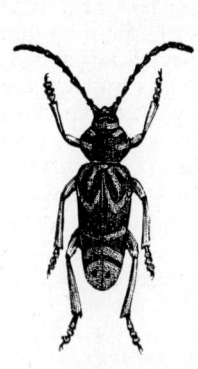

Fig. 641.

Fig. 642.

Among the insects that revel in this golden mine in the autumn is a black beetle with numerous transverse or wavy yellow bands (Fig. 642). This beetle is also found on locust-trees, where it lays its eggs. The larvæ bore under the bark and into the hard wood; they attain their growth in a little less than a year.

The locust-trees have been completely destroyed in some localities by the depredations of these larvae.

The painted hickory-borer, *Cyllēne cāryæ.*—This beetle resembles the preceding so closely that the same figure will represent either. But the hickory-borer not only infests a different kind of tree, but appears in the spring instead of the autumn. In this species the second segment of the hind tarsus is densely pubescent beneath, while it is glabrous in the locust-borer.

The oak-pruner, *Hypermăllus villōsus.*—The work of this insect is much more likely to attract attention than the insect itself. Frequently, in the autumn, the ground beneath oak-trees, and sometimes beneath apple-trees also, is strewn with small branches that have been neatly severed from the trees as if with a saw. These branches are sometimes nearly 25 mm. in diameter, and have been cut off by the larva of a beetle, which on account of this habit is called the oak-pruner. The beetle lays each of its eggs in a small twig. The larva eats out the inside of this twig, and works down into a larger branch, following the center of it towards the trunk of the tree. When full-grown the larva enlarges the burrow suddenly, so as nearly to sever the branch from the tree, leaving only the bark and a few fibers of wood. It then retreats up its burrow a short distance, and builds a plug of chips below it. The autumn winds break the branch from the tree. The larva remains in its burrow through the winter, and undergoes its transformations in the spring. No one has explained its object in severing the branch. The adult is a plain, brownish gray beetle. Whenever it becomes abundant its increase can be checked by gathering the fallen branches in the autumn and burning them before the beetles have escaped.

Subfamily LAMIINÆ

The Lamiids

As in the preceding subfamily, the prothorax is rounded with these beetles; but the lamiids are distinguished by having the fore tibiæ obliquely grooved on the inner side, and the last segment of the palpi cylindrical and pointed. The following are some of the more important species.

The sawyer, *Monŏchamus notātus.*—This beautiful brown and gray beetle is about 30 **mm.** long, with antennæ as long as the body in the case of the female and twice as long in the case of the male (Fig. 643). The larva bores in the sound wood of pine and of fir, making, when full-

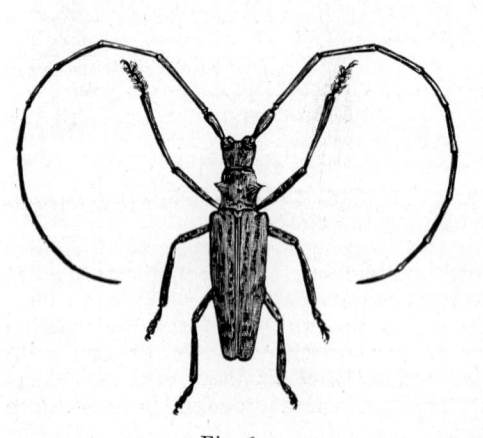

Fig. 643.

grown, a hole 12 mm. in diameter. The pupa state is passed within the burrow. It sometimes occurs in such numbers as to kill the infested trees.

The rounded-headed apple-tree borer, *Sapĕrda căndida.*—Excepting the codlin-moth, which infests the fruit, this is the worst enemy of the apple that we have. Its common name is used to distinguish it from the flat-headed apple-tree borer, already described, the larva of this species being nearly cylindrical in form (Fig. 635). The eggs are laid on the bark at the base of the tree late in June or July. The larva at first bores in the soft sap-wood, making a disk-shaped mine; after this it works in an upward direction in the harder wood, and at the close of its larval existence comes to the surface several inches above the place it entered. It requires nearly three years for this larva to attain its growth; it changes to a pupa, near the upper end of its burrow, about the middle of May, and emerges as a beetle in June. The beetle (Fig. 644) is of a pale brown color above, with two broad white stripes extending the whole length of the body. Although the larva is found chiefly in apple, it infests many other trees. The presence of the borers can be detected by the sawdust-like castings which the larvæ throw out at the entrances of their burrows.

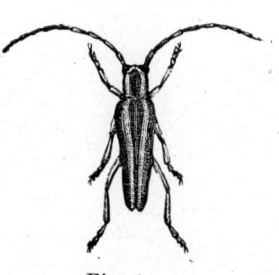

Fig. 644.

The two-spotted oberea, *Obĕrea bimaculāta*, is sometimes a serious pest, boring in the canes of blackberry and raspberry. The larva resembles that of the preceding species. The adult (Fig. 645) is about 12 mm. in length and of a deep black color, except the prothorax, which is yellow. There are usually two or three black spots on the pronotum, but frequently these are wanting.

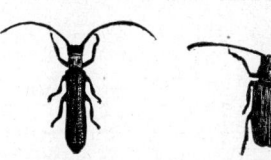

Fig. 645.    Fig. 646.

By cutting and burning all the canes after the crop has been picked, the borers in them can be destroyed.

The red milkweed-beetles, *Tetraŏpes.*—There are several species of bright red beetles that are common on milkweeds (*Asclepias*). These belong to the genus *Tetraopes*. Our most common species (Fig. 646) is *T. tetraophthălmus*. In this species there are four black spots on each wing-cover, and the antennæ are black and not ringed with a lighter color. The larva bores in the roots and the lower parts of the stems of milkweeds.

Family CHRYSOMELIDÆ

The Leaf-Beetles or Chrysomelids

The leaf-beetles are so called because they feed upon the leaves of plants both as larvæ and adults. They are usually short-bodied, and more or less oval in outline; the antennæ are usually of moderate length; and the front is not prolonged into a beak. The legs are usually short, and are furnished with tarsi of the same type as those of the preceding family (see Fig. 634, p. 524).

Although we are unable to cite any characteristic that will invariably distinguish these beetles from the preceding family, the student will rarely have any difficulty in making the distinction. The beetles of the genus *Donācia*, described below, are the only common ones that are liable to be misplaced. In other cases the more or less oval form of the body, and the comparatively short antennæ, and the leaf-feeding habits, will serve to distinguish the chrysomelids.

The leaf-beetles are nearly all comparatively small, the Colorado potato-beetle being one of our larger species.

The eggs are usually elongated and yellowish, and are laid upon the leaves or stems of the plants upon which the larvæ feed. Many of the larvæ live exposed on the leaves of plants; others that live in similar situations cover themselves with their excrement; some are leaf-miners; and a few, as the striped squash-beetle, bore in the roots or stems of plants.

This is a large family, of which nearly one thousand North American species are known. The following illustrations will serve to show the variations in form and habits.

The long-horned leaf-beetles, *Donācia*.—These are the common leaf-beetles that are liable to be mistaken for cerambycids. They are of elongated form, with slender antennæ (Fig. 647). They measure from 6 mm. to 12 mm. in length, and are of a metallic color —either greenish, bronze, or purplish. The lower side of the body is paler and is clothed with very fine hair which serves as a water-proof coat when the insect is submerged. The larvæ feed upon the roots or in the stems of aquatic plants; and the adults are found on the leaves of the same plants. We have many species, but they resemble one another so closely that it is difficult to separate them.

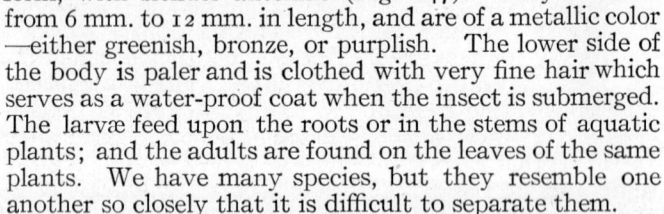

Fig. 647.

The three-lined lema, *Lēma trilineāta*.—This insect is common, feeding on the leaves of potato. The beetle is 6 mm. long, yellow, with three black stripes on the wing-covers. The eggs are usually laid in rows along the midrib on the lower side of the leaves. The larvæ feed on the leaves, and can be easily recognized by a habit they have of covering their backs with their own excrement. They transform in the ground in earthen cells. There are two broods each year; the second hibernates in the ground as pupæ.

The asparagus-beetle, *Criŏceris aspăragi*.—This is a small red, yellow, and black beetle, that gnaws holes into the heads of young asparagus, and lays oval, black eggs upon them. The larvæ, which are small, brown, slug-like grubs, also feed upon the young heads in the spring, and later in the season a second brood feed upon the full-grown plant. Figure 648 represents a head of asparagus bearing the eggs of this beetle, also a beetle and a larva enlarged. The beetle measures about 6 mm. in length. When this pest occurs, care should be taken to destroy all wild asparagus. This will force the beetles to lay their eggs upon the shoots that are cut for market. The larvæ hatching from such eggs will not have a chance to mature.

The grape root-worm, *Fĭdia lŏngipes.*—This insect is the most destructive enemy of the grape occurring east of the Rocky Mountains. The adult is a small, grayish brown beetle, measuring about 6 mm. in length. It feeds on the leaves in July, eating out characteristic chain-like holes. The eggs are laid beneath the loose bark of the vines. On hatching, the larvæ drop to the ground and burrow down to the roots, which they destroy, causing the death of the vine. Most of the larvæ do not transform till the following spring. The best means of fighting this pest is to poison the beetles while they are feeding on the leaves, and before they lay their eggs, by the use of a spray made by dissolving six pounds of arsenate of lead in 100 gallons of water.

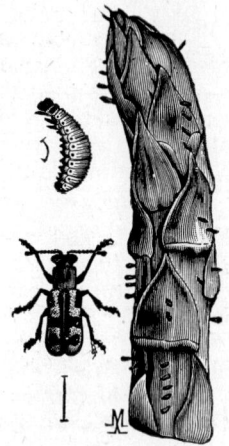

Fig. 648.

The Colorado potato-beetle, *Leptinotărsa decemlineāta.*—A good many insect tramps have come to us from Europe and from Australia, and appropriated whatever pleased them of our growing crops or stored grain. But two of our worst insect pests have swarmed out on us in hordes from their strongholds in the region of the Rocky Mountains. These are the Rocky Mountain locust and the Colorado potato-beetle (Fig. 649). The latter insect dwelt near the base of the Rocky Mountains, feeding upon the sand-burr (*Solanum rostratum*), until about the year 1859. At that time it began to be a pest in the potato-fields of the settlers in that region. Having acquired the habit of feeding upon the cultivated potato, it began its eastward march across the continent, spreading from potato patch to potato patch. At first the migration took place at about the rate of fifty miles a year, but later it was more rapid; and in 1874 the insect reached the Atlantic Coast.

Fig. 649.

The adult beetles hibernate in the ground; they emerge early in April or May, and lay their eggs on the young potato plants as soon as they appear; both larvæ and adult beetles feed on the foliage of the potato. The larvæ enter the ground to transform. This pest is usually controlled by the use of Paris green.

Labidomēra clivicŏllis.—This species is closely allied to the Colorado potato-beetle and resembles it in size and form. It is of a deep blue color, except the wing-covers, which are orange, with three dark-blue spots on each (Fig. 650). There is considerable variation in the size and shape of these spots; frequently the two near the base of the wing-covers are joined so as to make a continuous band extending across both wing-covers. The larva feeds on milkweed (*Asclepias*).

Fig. 650.

The diabroticas.—Several very important pests belong to the genus *Diabrŏtica*. In the East they are known as cucumber-beetles; but on the Pacific Coast, where they are more feared on account of their injuries to fruit and fruit-trees, they are commonly called the diabroticas. They are chiefly greenish yellow beetles, marked with black stripes or spots. The striped diabrotica, *D. vittata,* has two black stripes on each wing-cover. The adult feeds on the leaves of cucumber, squash, and melon; and the larva, which is a slender, worm-like creature, bores in the stems and roots of the same plants. The twelve-spotted diabrotica, *D. duodecimpunctāta*, and *Diabrŏtica sōror*, agree in having six black spots on each wing-cover. The former is very common in the East; the latter occurs on the Pacific Coast, and is the most destructive of all of the diabroticas. *Diabrŏtica longicŏrnis* is a green species, which feeds on the pollen and silk of corn and on the pollen of other plants. Its larva is known as the corn root-worm; it is very destructive to corn in the Mississippi Valley. Its injuries are greatest where corn is grown on the same land year after year; hence a rotation of crops should be practised where this pest is troublesome. The other species of *Diabrotica* mentioned above are difficult to combat, as the leaves of cucumber, melon, and squash are very apt to be injured by the use of arsenical poisons. The most practicable way of protecting these vines is to cover them while young with frames covered with netting. Where they infest fruit-trees they can be fought with Paris green; but this poison must be used with great care on such trees as prune and apricot. Squashes should not be grown in orchards, as is sometimes done in California.

The flea-beetles.—There is a group of leaf-beetles, of which we have many species, in which the hind legs are fitted for leaping, the thighs being very large. These are commonly called the flea-beetles.

The striped flea-beetle, *Phyllotrēta vittāta*, is exceedingly common on cabbage, turnip, radish, mustard, and allied plants. It is a small, black, shining beetle, with a broad, wavy, pale, dull yellow stripe upon each wing-cover (Fig. 651); it measures about 2.5 mm. in length. These beetles eat numerous little pits in the thicker leaves that they infest, and minute perforations in the thinner-leaved plants. The larva is a slender, white worm, about 8 mm. in length; it feeds on the roots of the plants infested by the adult. The adult beetles can be destroyed with kerosene emulsion.

Fig. 651.

The cucumber flea-beetle, *Ĕpitrix cucŭmeris*, is a common pest of melon and cucumber vines; it also attacks the leaves of potatoes, raspberry, turnip, cabbage, and other plants. This is a minute black species, measuring less than 2 mm. in length. The body is finely punctured and clothed with a whitish pubescence; there is a deep transverse furrow across the hind part of the prothorax; the antennæ are dull yellow, and the legs are of the same hue, except the posterior femora, which are brown. The adult beetles feed on the leaves of plants in the same manner as the preceding species; and the larvæ on the roots of the infested plants.

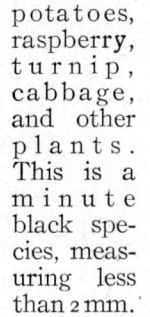

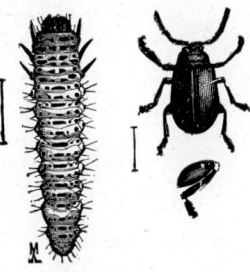

Fig. 653.

The grape flea-beetle, *Hăltica chalўbea*.—This is a larger species than the two preceding, measuring from 4 mm. to 5 mm. in length, and is of a dark, steel-blue color. It is a great pest in vineyards, eating into the buds of grape in early spring, and later gnawing holes in the leaves (Fig. 652). In May and June the brown, sluggish larvæ may also be found feeding upon the surface of the leaves. The full-grown larva is chestnut brown marked with black spots (Fig. 653). It drops to the ground and makes a cell in the earth in which it transforms. The most important injury caused by this pest is the destruction of buds in early spring, which causes a great loss of foliage and fruit. This pest is most easily controlled by spraying the vines with an arsenical poison between the middle of June and the middle of July, while the larvæ are feeding on the leaves.

Fig. 652.

The wedge-shaped leaf-beetles.—These insects are characterized by the peculiar form of the body, which is narrow in front and broad behind. In most of the species the body is much roughened by deep pits, and usually the pits on the wing-covers are in regular rows. These insects and the tortoise-beetles differ from other leaf-beetles in having the fore part of the head prominent, so that the mouth is confined to the under surface. Some of the larvæ feed externally upon the leaves and bear a parasol composed of their excrement;

other species are leaf-miners. *Baliōsus rubra* is a good representative of this group (Fig. 654). It varies in length from 3 mm. to 5 mm. It is of a reddish color, with the elevated portions of the elytra more or less spotted with black. The larva mines in the leaves of apple, forming a blotch-mine; the transformations are undergone within the mine. We have also found this species mining the leaves of basswood in great numbers.

Fig. 654.

The tortoise-beetles.—Among the more beautiful Coleoptera are certain bright golden, green, or iridescent beetles found on the leaves of sweet potato, morning-glory, nettle, and other plants. In these beetles the body is flattened below and convex above; the head is nearly or quite concealed beneath the prothorax; and the margins of the prothorax and elytra are broadly expanded, forming an approximately circular or oval outline, and suggesting a resemblance to the shell of a tortoise (Fig. 655). Not all of the species are iridescent; and in the case of those that are, the brightness of the colors is said to depend on the emotions of the insect. What a beautiful way to express one's feelings—to be able to glow like melted gold when one is happy! Unfortunately for the beauty of our collections, these bright colors disappear after the death of the insect.

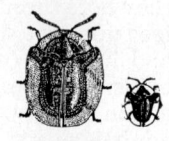

Fig. 655.

The larvæ of the tortoise-beetles are flattened, and have the margin of the body fringed with spines. At the caudal end of the body there is a forked appendage which serves a very strange purpose. This fork is bent forward over the back, and to it are attached the cast-off skins of the larva and its excrement; these constitute a parasol. When about to change to the pupa state these larvæ fasten the caudal end of the body to the under side of a leaf; the skin then splits open, and is forced back to this end of the body, where it remains.

The one-dotted or five-dotted tortoise, *Physonōta unipunctāta.*— The largest of our bright-colored tortoise-beetles is common in midsummer, feeding on the leaves of wild sunflower. It measures from 9 to 12 mm. in length, and is yellow, with the margins whitish. On the prothorax there are five black dots—two close together in front, and three more widely separated behind. Sometimes all but one of these dots are wanting. It was this form that was first described, hence the name *unipunctata*. We have found the larvæ abundant in July on the same plant with the adults.

The milkweed-tortoise, *Chelymŏrpha cassĭdea*, is a large, brick red species, which measures from 9 mm. to 12 mm. in length, and has the prothorax and wing-covers marked with many black spots. This species feeds on milkweed (*Asclepias*) and various other plants.

Family MYLABRIDÆ

The Pea-Weevil Family

These are small beetles, the larvæ of which live in the seeds of leguminous plants. The head of the adult is prolonged into a broad beak; and the wing-covers are rather short, so that the tip of the abdomen is always exposed (Fig. 656). This is a comparatively small family; ninety-three species are listed in our fauna, of which eighty-one belong to the genus *Mylabris*.

The pea-weevil, *Mylabris pisōrum*.—"Buggy peas" are well known in most sections of our country; but just how the "bugs" find their way into the peas is not so generally understood. The eggs of the pea-weevil are laid upon the pod while the peas are quite small; when the larvæ hatch they bore through the pod into the young peas. Here they feed upon the substance of the seed, which ripens, however, and in some cases will germinate when planted. The larva before transforming eats a circular hole on one side of the seed, leaving only a thin scale, which is easily pushed away by the mature beetle. The adult is about 5 mm. in length; it is dark brown, with a few white spots on the wing-covers, and one on the prothorax near the middle. Sometimes the beetles leave the peas during the autumn or winter; but as a rule they remain in the seed till spring, and are often planted with it. Seed peas should be placed in water, and the infested ones, which will float, should be picked out and destroyed. This species is not known to oviposit on dry peas.

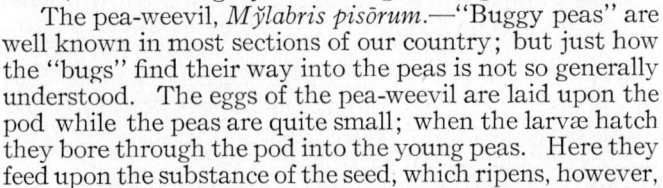

Fig. 656.

This and other grain-infesting insects can be destroyed by placing the grain in a closed receptacle with a small quantity of bisulphide of carbon.

The bean-weevil, *Mylabris obtĕctus*.—This species resembles the preceding quite closely; but it is a little smaller (Fig. 656), and lacks the white markings characteristic of *M. pisorum*. It infests beans, and often several individuals inhabit a single bean. The eggs are laid within the pod, being pushed through a slit which the female gnaws through the pod. This species will oviposit on dry beans, peas, and other grain, and will continue to breed for many generations in stored beans and peas.

SERIES VII.—THE RHYNCHOPHORA*

The six families included in this series constitute a well-marked division of the order, which has long been known as the Rhynchophora or snout-beetles. These names were suggested by the fact that in many of these insects the head is prolonged so as to form a snout or beak; but it should be remembered that, while these names are very appropriate for a large part of this series, in some members of it the head is not thus prolonged. This is especially true of the last two

---

*Rhynchŏphora: *rhynchos* (New Latin), snout; *phoros* (φόρος), bearing.

families, the bark-beetles and timber-beetles, in which the beak is either wanting or extremely short and broad.

The most distinctive features characterizing this series of families are the following: the suppression of the gula, the gular sutures being confluent (Fig. 657, gs); the absence of sutures between the prosternum and the episterna and epimera; the meeting of the epimera of the prothorax on the middle line behind the prosternum (Fig. 657, em); and the palpi being usually short and rigid.

A volume entitled "Rhynchophora or Weevils of North Eastern America" was published by W. S. Blatchley and C. W. Leng in 1916. This work includes descriptions of the then known species found in this region, with analytical keys, and many figures.

Fig. 657.

Family BRENTIDÆ

The Primitive Weevils

This family is confined chiefly to tropical regions; only six species are found in the United States, and but one of these in the North.

The northern brentid, *Eŭpsalis minūta*.—In the female the head is prolonged into a slender snout; but in the male the snout is broad and flat, and is armed with a pair of powerful jaws (Fig. 658). These are weapons of offence, for the males fight desperately for their mates; and too, the males are generally larger than the females. In these respects these insects resemble the stag-beetles, the males of which also fight for their mates.

The northern brentid is found beneath the bark of recently felled or dying oak, poplar, and beech trees, in the solid wood of which the larvæ bore; and is widely distributed over the United States and Canada.

Fig. 658.

Family PLATYSTOMIDÆ*

The Fungus Weevils

This family includes a small number of snout-beetles in which the beak is short and broad, and the labrum is present; the antennæ are not elbowed, and the terminal segments rarely form a compact club; the palpi are flexible; and the prothorax bears a transverse elevated ridge at or near its base.

The larvæ of many species infest woody fungi, others breed in the smut of corn and wheat, and still others bore in dead wood. The

---
*This family is the Anthribidæ of many authors.

larvæ of one cosmopolitan species, known as the coffee-bean weevil, *Aræŏcerus fasciculātus*, attack seeds of various plants.

Sixty-two species of this family are known to occur in America north of Mexico.

Family BELIDÆ

The New York Weevil

The family Belidæ is represented in our fauna by a single species, the New York weevil, *Ithy̆cerus noveboracĕnsis*. This is a large species, measuring from 12 mm. to 18 mm. in length. It is black, rather sparsely clothed with a mixture of ash-gray and pale brown prostrate hairs which give it a black-spotted appearance. The beak is short and broad. The mandibles are prominent, not very stout, and emarginate at the tip, with an inferior cusp. The antennæ are not elbowed; the first segment is longer than the second; and the terminal segments form a small, oval club.

This species breeds in the twigs and tender branches of oak, hickory, and possibly other forest trees. The adult beetles appear in early spring, and sometimes do much damage to fruit-trees by eating into buds, and gnawing the tender bark on new growth. They can be caught by jarring them on to sheets or by the use of a plum-curculio catcher.

Family CURCULIONIDÆ

The Curculios or Typical Snout-Beetles

The Curculionidæ is a very large family; it is represented in America north of Mexico by more than eighteen hundred species; to it belong four-fifths of all our Rhynchophora. This family includes the typical snout-beetles, the head being prolonged into a well-defined beak, which is usually long and curved downward.

The family Curculionidæ is divided into thirteen subfamilies; but several of these are very small. The seven subfamilies mentioned below will serve to illustrate the more important variations in structure and in habits, and they include the more important species from an economic standpoint.

The subfamily RHINOMACERINÆ, or pine-flower snout-beetles, includes a small number of snout-beetles in which the elytra have no fold on the lower surface near the outer edge, and in which the labrum is distinct. The head is prominent, not deflexed; the snout is as long as the prothorax, rather flat, narrowest about the middle, wider at base and tip; the elytra are rounded at the tip, and entirely cover the abdomen. These beetles infest the staminate flowers of coniferous trees, in which the eggs are laid. Six species are found in our fauna.

The subfamily RHYNCHITINÆ, or toothed-nose snout-beetles, includes snout-beetles in which the elytral fold is feeble, the labrum is wanting, and the mandibles are toothed on both the outer and the inner side. The mandibles can be spread widely, and when closed the outer tooth at the end of each projects forward so that two small, acute teeth seem to project from the mouth.

The most common member of the family is *Rhynchītes bīcolor* (Fig. 659). This is red above except the snout, and black below; the body, not including the snout, is about 6 mm. long, the snout is half that length. The adults are often abundant on wild roses, and less frequently on cultivated roses. The larvæ infest the hips of roses.

Fig. 659.

The subfamily ATTELABINÆ, or leaf-rolling weevils, is composed of beetles which have neither an elytral fold nor a labrum, and in which the mandibles are flat, pincer-shaped, and toothed on the inner side. The elytra do not entirely cover the abdomen, and each is separately rounded at the tip. Only five species are known from this country; all of these belong to the genus *Attĕlabus*. The females provide for their young in a very remarkable way. They make compact thimble-shaped rolls from the leaves of trees (Fig. 660), and lay a single egg in each. The larvæ feed on the inner parts of these rolls, and when full-grown enter the ground to transform. Sometimes these rolls are found hanging by a narrow piece to the leaf from which they were made, and sometimes they are found lying on the ground separated from the leaf.

Fig. 660.

The subfamily CYLADINÆ is represented in Florida, Louisiana, and Texas by a single species, the sweet-potato root-borer, *Cylas formicārius*. This beetle is somewhat ant-like in form; this fact suggested the specific name. It is about 6 mm. long; the color of the eltyra, head, and snout is bluish black, that of the prothorax reddish brown. Both larvæ and adults bore into the stems and tubers of the sweet potato, and sometimes do very serious damage. This species was formerly included in the Brentidæ.

The subfamily OTIORHYNCHINÆ, or scarred snout-beetles, is one of the larger of the subfamilies of the Curculionidæ; it is represented in our fauna by more than two hundred species. The most distinctive characteristic of these insects is the presence in the pupa state, and sometimes also in recently matured adults, of an appendage on each mandible, and in the adult state of a scar indicating the place from which the appendage has fallen. This scar is on the anterior face of the mandible, and frequently at the tip of a slight process. Many species of this family are beautifully ornamented with scales which resemble in a striking manner the scales on the wings of butterflies. Among the more important species are the following.

The imbricated snout-beetle, *Epicærus imbricātus*, is usually a dull, silvery white beetle with brown markings; but the species is quite variable in color. It is represented, somewhat enlarged, in Figure 661. It is omnivorous, gnawing holes in various garden vegetables, strawberry plants, and other fruits. The greater part of the insect is clothed with imbricated scales, which suggested the specific name.

Fig. 661.

Fuller's rose-beetle, *Pantŏmorus fŭlleri*.—This is an oval, black snout-beetle, lightly covered with dark brown scales, and about 6 mm. in length. It attacks roses and many other greenhouse plants. The adults feed on the foliage, flowers, and buds, the larvæ on the roots, of its food plants.

The strawberry crown-girdler, *Brachyrhīnus ovātus*.—This is a dark brown, almost black, snout-beetle, about 5 mm. in length, which often invades dwellings in search of shelter, in the Northern States and Canada. The larvæ feed on the roots of the strawberry, cutting them off near the crown. The adults feed on the foliage. In the adult, the hind wings are wanting and the elytra are grown together.

The black vine-weevil, *Brachyrhīnus sulcātus*.—This beetle is larger than the preceding species, measuring 9 mm. in length; it is black, with small patches of yellowish hairs on the elytra. The larvæ destroy the roots of strawberries; and both larvæ and adults infest various greenhouse plants.

The subfamily CURCULIONINÆ is represented in our fauna by more than one thousand species, among which are some very destructive pests. In this family there is on the lower side of each wing-cover a strong fold near the outer margin, which limits a deep groove in which the upper edge of the abdomen fits; the mandibles have no scar; the antennæ are usually elbowed, and have a ringed or solid club; the tarsi are usually dilated, with the third segment bilobed and spongy beneath; in a few cases the tarsi are narrow, but not spinose beneath.

The larvæ are soft, white, maggot-like grubs destitute of feet. They feed chiefly on fruits, seeds, and nuts, but all parts of plants are subject to their attacks.

In laying her eggs, the female first bores a hole with her snout, then drops an egg into this hole, and finally pushes the egg to the bottom of the hole with her snout. In many species the snout is highly developed for this purpose; sometimes it is twice as long as the remainder of the body. This is well shown in the acorn-weevils and the nut-weevils, which belong to the genus *Balăninus*. Figure 662 represents *Balăninus rĕctus* resting on an acorn; the specimen figured, when found, had her snout inserted in the acorn up to the antennæ. Of the closely allied species *Balaninus nasīcus* breeds in hickory-nuts, and *Balaninus proboscĭdeus* in chestnuts.

Fig. 662.

The following are some of the more important pests belonging to this subfamily:

The plum-curculio, *Conotrachēlus nĕnuphar*.—This is the insect that stings plums, often destroying a large portion of the fruit; the larva is also the well-known "worm" of "wormy" cherries. This species is the most destructive insect that infests plums, cherries, and other stone fruits; it also breeds in apple. Its presence in an orchard can be determined early in the season by a peculiar mark it makes when laying its eggs in the young fruit. The female beetle makes an incision, with her snout, through the skin of the fruit. In this incision she lays a single egg, which she pushes with her snout to the bottom of the cavity that she has prepared. She then makes a crescent-shaped incision in front of the one containing the egg. This last cut undermines the egg, leaving it in a little flap. The larvæ feed within the fruit. In the case of the plums the infested fruit falls to the ground; but not so with cherries. When full-grown the larvæ go into the ground to transform. This species infests nectarines, apricots, and peaches, as well as plums and cherries. This insect is fought in two ways: the beetles are jarred from the trees upon sheets in early spring, and destroyed before they have laid their eggs; they are also poisoned by spraying the trees with arsenate of lead, either alone or combined with a fungicide before the fruit is large enough for them to oviposit in it. The adult beetle feeds upon the foliage, and can thus be poisoned.

The apple-curculio, *Anthŏnomus quadrigĭbbus*, infests the fruit of apple, often in company with the plum-curculio. The specific name was suggested by the fact that there are two wart-like projections near the hind end of each wing-cover.

Fig. 663.

The strawberry-weevil, *Anthŏnomus signātus*, infests strawberry, blackberry, raspberry, and dewberry. The female beetle (Fig. 663) after laying an egg in the flower-bud causes it to fall by cutting the pedicel; the larva develops within the fallen bud.

The cotton-boll weevil, *Anthŏnomus grăndis*, is one of the most serious insect pests known in the United States. It infests only cotton. The egg is deposited in a young boll, which the larva destroys. The adults also feed upon the young bolls and upon the leaves, doing as much or more damage than that done by the larvæ. This species is a native of Central America. It spread through Mexico, and entered Texas about 1890. Since that time it has spread over a large part of the cotton-belt. Very extensive investigations of this pest have been made by the Federal Government and by several state governments; and much literature regarding it is available to those interested.

The subfamily CALENDRINÆ includes the bill-bugs and the grain-weevils; some of which are among our more common snout-beetles. The larvæ of the larger species feed upon the roots and bore in the stems of plants, especially grass and corn, while those of the smaller species infest grains and seeds.

Most of our larger species belong to the genus *Sphenophorus;* one of these is represented in Figure 664. These are of medium or rather large size, and are often marked in a very characteristic manner by longitudinal elevated bands of darker color; frequently, when collected, they are covered with a coat of clay. They are commonly known as the bill-bugs. One species, *Sphenŏphorus maidis*, is an important pest of corn in the South; it bores in the tap-root and lower part of the stalk. Most of the beetles hibernate in the corn-stubble, and can be destroyed by pulling out and burning the stubble.

Fig. 664.

Among the smaller members of this subfamily are two exceedingly important pests of stored grains; these are the granary-weevil, *Calăndra granāria*, and the rice-weevil, *Calăndra orȳzæ*. The rice-weevil is so called because it was first found in rice in India; but it infests various kinds of stored grain; and in the South it is fully as important a granary-pest as is the granary-weevil.

The two species are quite similar in appearance; but the granary-weevil is the larger, measuring from 3 mm. to 4 mm. in length; while the rice-weevil measures less than 3 mm. in length, and differs from the granary-weevil in having the elytra marked with four reddish spots. The thorax of the rice-weevil is closely pitted with round punctures; that of the granary-weevil, with sparse elongate punctures.

The adult female of both of these species gnaws a tiny hole in a kernel of grain and then deposits an egg in it. The larva feeds on the grain, becomes full-grown, and transforms within the kernel. The adult continues the injury begun by the larva, eating out the inside of the kernel.

The most effective method of destroying grain-weevils is by the use of carbon bisulphide. The grain is placed in a tight bin or other receptacle, and the carbon bisulphide is poured into a shallow tin pan placed on top of the grain, and then covered with blankets to keep in the fumes. Two or three pounds of carbon bisulphide should be used for each 1000 cubic feet of space. Care should be taken not to go near the bin with a lighted lantern or fire of any kind until after the blankets have been removed and the gas has been dissipated.

Family PLATYPODIDÆ

This is a small family, which is represented in our fauna by a single genus, *Plătypus*, of which only five species have been found in America north of Mexico; these are found chiefly in the South and the Far West.

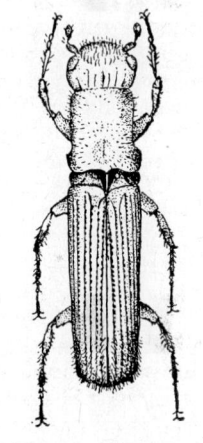

Formerly this group was classed as a subfamily of the Scolytidæ. It is distinguished from the Scolytidæ by the fact that the first segment of the anterior tarsi is longer than the second, third, and fourth together. The form of the body is cylindrical (Fig. 665); and the head is large, wider than the prothorax.

The species of this genus attack many kinds of conifers and deciduous trees. They bore deeply into the heart-wood, making "pin-holes" that often render lumber useless. The eggs are deposited in the galleries; and the larvæ feed on a fungus, which is cultivated by the beetles and is known as ambrosia. In this respect *Platypus* resembles several genera of the Scolytidæ, which also bore in solid wood and feed on ambrosia;

Fig. 665.—*Platypus wilsoni*, female. (After Swaine.)

all of these are known as ambrosia-beetles. The galleries of ambrosia-beetles are usually blackened by the fungus. See further account of the ambrosia-beetles in the discussion of the next family.

Family SCOLYTIDÆ

The Engraver-Beetles and the Ambrosia-Beetles

The members of the family Scolytidæ are mostly of cylindrical form (Fig. 666) and of small or moderate size; some species measure only 1 mm. in length, but others are much larger, attaining a length of 6 mm. or more. They are usually brown, sometimes black, and with many the hind end of the body is very blunt, as if cut off. The antennæ are elbowed or bent in the middle, and are clubbed at the tip; the tibiæ are usually serrate; and the first segment of the anterior tarsi is shorter than the second, third, and fourth together.

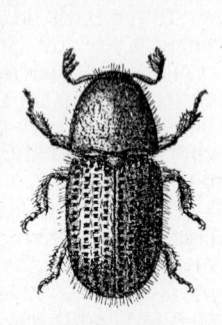

Fig. 666.—*Phthorophlœus liminaris.*

A few members of this family infest herbaceous plants; our most important one of these is the following.

The clover-root borer, *Hylăstinus obscūrus*.—This pest was introduced from Europe and has become the most serious enemy of clover, especially red clover and mammoth clover, in New York State and in other sections of the North. It bores in the roots of plants beginning their second year of growth and destroys them (Fig. 667). Where it is common it is practically impossible to keep fields in clover longer than the second summer after seeding. In these regions it is the common practice to seed with clover and timothy mixed; after the clover disappears the field becomes a timothy meadow. No practical method of control of this pest has been found.

Fig. 667.—Work of clover-root borer. (After Webster.)

Most scolytid beetles infest woody plants; among them are some of the most destructive enemies of forest-trees, and a few attack fruit-trees. As a rule they are more liable to attack sickly trees, but their injuries are not confined to these.

The scolytid beetles exhibit two radically different types of habits; and from this point of view they can be grouped into two groups: first, the engraver-beetles or bark-beetles; and second, the ambrosia-beetles or timber-beetles. These two groups, however, do not represent a natural division of the family based on structural characters. The peculiar habits of the ambrosia-beetles are believed to have arisen independently in different parts of the series of scolytid beetles, and in the family Platypodidæ as well.

The Engraver-Beetles or Bark-Beetles

If the bark be pulled from dead branches or trunks of trees, the inner layer and the sap-wood will be found, in many cases, to be ornamented with burrows of more or less regular form. The smoothly cut figures are the mines of engraver-beetles, which are also known as bark-beetles. Many kinds of these engravings can be found, each characteristic of a particular species of engraver beetles. A common pattern is shown in Figure 668.

Many figures and detailed descriptions of the burrows of engraver-beetles have been published by writers on forest-insects; among the more important papers on this subject published in America are those by Hopkins ('09) and Swaine ('18), in which can be found references to many other papers.

The different species of engraver-beetles vary so greatly in the details of their habits that it is difficult to make generalizations re-

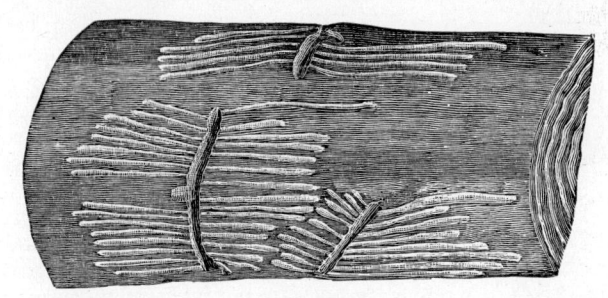

Fig. 668.

garding them in the space available here. In a common type, the adult beetle, after penetrating the bark, makes a tunnel in the inner layer of the bark or in the sap-wood or in both; this is known as the egg-tunnel, and may be either simple or branched. In the sides of the tunnel, most species make niches, the egg-niches, in which the eggs are laid. The larva when hatched feeds on the bark or sap-wood or both and thus makes a lateral tunnel. These lateral tunnels made by the larvæ often extend parallel in a more or less regular manner, as shown in Figure 668.

While most of the engraver-beetles infest forest-trees, the two following species are well-known pests of fruit-trees.

The fruit-tree bark-beetle, *Scŏlytus rugulōsus*.—This species infests apple, quince, plum, peach, and other stone-fruits. It is sometimes called the shot-hole borer by fruit-growers on account of the small entrance holes of its burrows. The adult beetle measures from 2 mm. to 2.5 mm. in length, and is dark brown or nearly black. It infests chiefly sickly trees.

The peach-tree bark-beetle, *Phthorophlæus limināris*.—This species resembles the preceding in size and habits, except that its injuries are confined chiefly to peach and cherry. It can be distinguished from the fruit-tree bark-beetle by the fact that the club of the antennæ is lamellate, an unusual feature in this family (Fig. 666).

The Ambrosia-Beetles or Timber-Beetles

Certain members of the family Scolytidæ differ in habits from the engraver-beetles or bark-beetles in a remarkable manner; these are those known as ambrosia-beetles or timber-beetles. They are termed

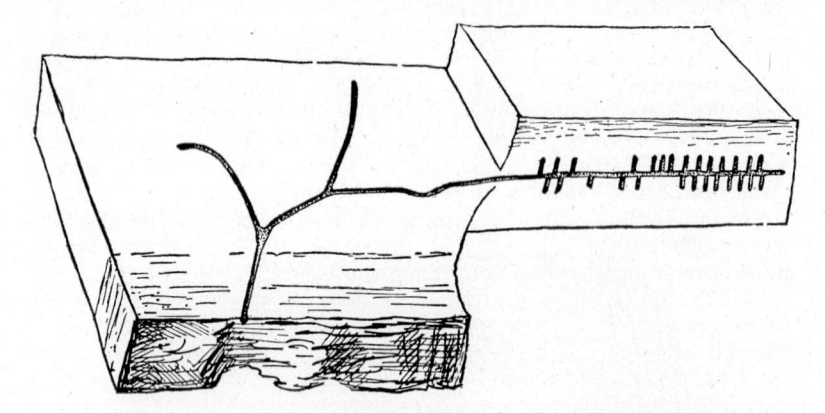

Fig. 669.—Gallery of *Monarthrum mali* in maple. (From Hubbard.)

ambrosia-beetles because they cultivate fungi, commonly called ambrosia, upon which they feed; and *timber-beetles*, because they burrow in the solid wood.

The galleries of the ambrosia-beetles can be distinguished from those of other wood-boring insects by the fact that in all of their ramifications they are of uniform size and free from wood-dust and other refuse, and their walls are stained black or brown by the fungus that is grown upon them.

The galleries of different species differ in form; but usually there is a main gallery, which extends deeply into the solid wood and is often branched; and extending from the sides of the main gallery there are short chambers, termed cradles, in each of which an egg is laid and a larva reared (Fig. 669). In some species, the female deposits her eggs loosely in the galleries, and the young and old live together in the same quarters.

The galleries are excavated by the adult beetles. In some species the gallery is started by a single female, in others the males assist the females in this work. The entrances through the bark to the galleries are similar to those made by the bark-beetles and like them are known as "shot-holes." Under favorable conditions colonies may continue their excavations during two or three generations.

The fungi upon which these beetles feed are carefully cultivated by them. So far as is known, each species of ambrosia-beetle cultivates only a single species of fungus, and only the most closely allied species have the same food-fungus. The fungus is started by the mother-beetle upon a carefully packed bed or layer of chips. It is probable that some conidia are brought for this purpose from the gallery in which the female was developed. The excrement of the larvæ is used in some and probably in all the species to form new beds for the propagation of the fungus.

In those species in which the larvæ are reared in separate cradles, "the mother-beetle is constantly in attendance upon her young during the period of their development, and guards them with jealous care. The mouth of each cradle is closed with a plug of the food-fungus, and as fast as this is consumed it is renewed with fresh material. The larvæ from time to time perforate this plug and clean out their cells, pushing out the pellets of excrement through the opening. This debris is promptly removed by the mother and the opening again sealed with ambrosia. The young transform to perfect beetles before leaving their cradles and emerging into the galleries."

A detailed account of the ambrosia-beetles of the United States was published by Hubbard ('97), from which I have drawn largely in the preparation of the account given here.

While the ambrosia-beetles are chiefly injurious to forest-trees, there are certain species that injure wine and beer casks; and one species, the pear-blight beetle, *Anisăndrus pyri*, sometimes infests the tips of pear and apple branches, causing an injury that is often mistaken for the bacterial disease known as pear-blight.

Nearly four hundred species of scolytid beetles, representing many genera, have been described from America north of Mexico.

CHAPTER XXIV

ORDER STREPSIPTERA*

The Stylopids or Twisted-winged Insects

The members of this order are small, endoparasitic insects, which prey on other insects. Only the males are winged; in this sex, the fore wings are reduced to club-shaped appendages; the hind wings are large compared with the size of the tiny body, fan-shaped, furnished with radiating wing-veins, and folded longitudinally when at rest. The adult female is larviform and legless. The mouth-parts are vestigial or wanting; the alimentation is probably by osmosis. Both sexes undergo a hypermetamorphosis.

The order Strepsiptera comprises insects that were formerly classed as a family of the Coleoptera, the Stylopidæ; for this reason, these insects have been known as the stylopids. Recently since the establishment of the order Strepsiptera, the name *the twisted-winged insects*, derived from the technical name of the order, has been proposed for them; but the old name is less cumbersome, and will probably continue to be used.

The stylopids are small insects which live parasitically within the bodies of other insects, chiefly bees, wasps, digger wasps, and certain Homoptera. Their small size and the fact that nearly their entire existence is passed within the bodies of their hosts result in their being rarely seen except by those who are searching for them. During the first stadium the young larvæ of both sexes are free, and the adult winged male leads a free existence for a brief period; but only the most skilled collectors are likely to observe these minute creatures during these periods, the only free stages of their existence.

The stylopids are most easily found by examining adult individuals of the species of insects that they infest, in which may be found adult females and male pupæ of the parasites. The presence of a stylopid is indicated by the projecting of the head end of the body from between two of the abdominal segments of the host (Fig. 670). Frequently a single host will contain several parasites. A female *Polistes* with eleven male stylopids has been recorded. If this projecting part of the parasite is a flat disk-like plate, it is the head end of a female; but if it is the rounded and tuberculate end of a cylindrical body, it is the head end of the puparium of a male. Adult male stylopids can be bred by keeping alive stylopized insects containing male pupæ.

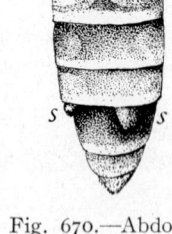

Fig. 670.—Abdomen of stylopized insect: *s, s*, stylopids.

*Strepsiptera: *strepsis* (στρέψις), a turning; *pteron* (πτερόν), a wing.

(546)

Figure 671 will serve to illustrate the appearance of an adult male stylopid. The more striking features are the flabellate antennæ; the large, stalked, compound eyes; the shortness of the prothorax and the mesothorax, and the great length of the metathorax; the reduction of the fore-wings to club-shaped appendages; and the large size of the hind wings.

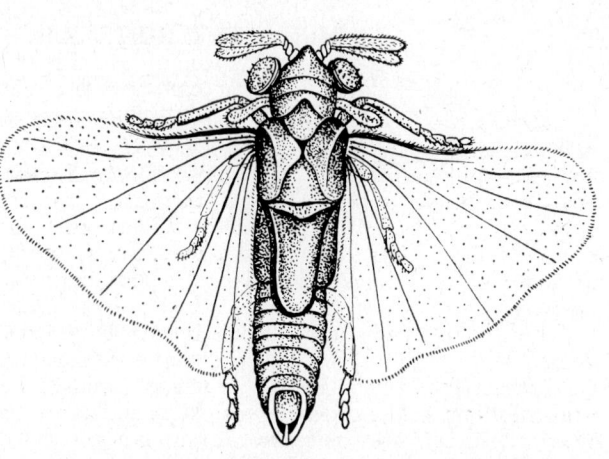

Fig. 671.—*Opthalmochlus duryi.* (After Pierce.)

The antennæ of adult males differ greatly in form in the different families of this order. The number of antennal segments varies from four to seven; the third segment is always furnished with a lateral prolongation, a flabellum, and one or more of the following segments may or may not be flabellate.

The compound eyes of adult males are large and more or less stalked. The facets are separated by densely ciliate walls.

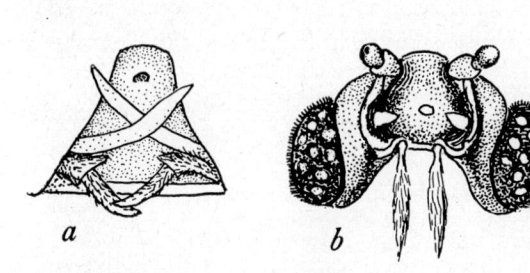

Fig. 672.—Mouth-parts of male stylopids: *a, Acroschismus bruesi.* (After Pierce.) *b, Pentozocera australensis.* (After Perkins.)

The mouth-parts are greatly reduced; those of two adult males are represented in Figure 672. The mouth opening is small. The labrum and labium are wanting as distinct parts. In *Acroschismus bruēsi* (Fig. 672, *a*) the mandibles are slender, curved, and scimitar-like; beneath the mandibles are the maxillæ; these are two-jointed; the second segment is believed to be the reduced palpus. In *Pentozŏcera austrălensis* (Fig. 672, *b*) the mandibles are greatly reduced, but the maxilllary palpi are quite large.

The three pairs of legs are similar in form. The tarsi are five-jointed in one family (Mengeidæ), and furnished with two claws;

in the other families they are two- to four-jointed and without claws.

The venation of the hind wings is degenerate. There is a variable number of radiating veins, which in the most generalized wings are eight in number. These are supposed, by Pierce ('09), to be the eight principal veins of the typical wing, the costa, subcosta, radius, media, cubitus, and the three anal veins, respectively (Fig. 673).

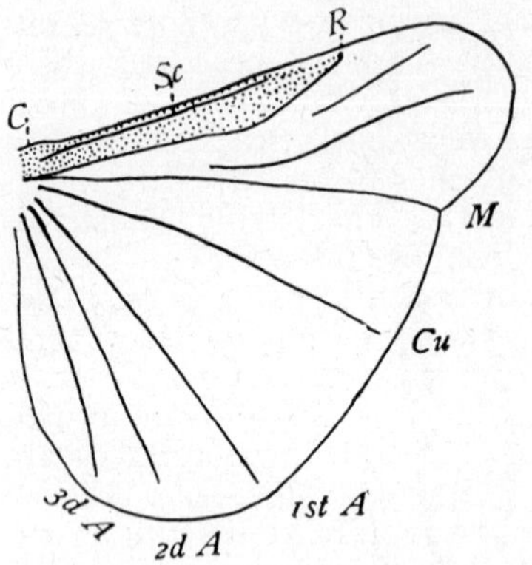

The abdomen is composed of ten segments.

The adult female is very degenerate in form. That part of the body which projects from the body of the host is the cephalothorax, the head and thorax being consolidated into a single disk-like region. The abdomen, which is within the body of the host, is a great sac filled with eggs. The body of the adult female is inclosed in the skin of the last larval instar, which is termed the puparium; but there is no pupal stage in this sex.

Fig. 673.—Wing of *Paraxenos eberi*. (From Pierce, after Saunders.)

The postembryonic development of the stylopids is very peculiar. In the adult female the eggs are free in the body cavity, where they hatch. The young larvæ are campodeiform and active. As they bear some resemblance to the triungulins of the parasitic blister-beetles, they are termed *triungulins* by some writers; but as they do not possess three tarsal claws, this term is inappropriate when applied to them. For this reason the first instar of a stylopid larva is termed a *triungulinid*.

The stylopids are very prolific; more than 2000 triungulinids produced by a single female have been counted. This fecundity is doubtless correlated with the uncertainty of any individual triungulinid being able to find its proper host.

The triungulinids escape from the body of the female through unpaired median genital apertures on the second to fifth abdominal segments. These apertures open into the space between the venter of the female and the puparium, which is termed the *brood chamber*. The triungulinids escape from this space through a slit in the cephalothorax of the puparium, between the head and the prothorax, and then crawl over the body of the host. This is the beginning of the most critical period in the life of the stylopids. For the continued existence of any individual of the brood it must find a larva or a nymph of the particular species that is its proper host. This is doubtless accomplished in different ways in the different species. Those that infest Homoptera and other insects that do not build nests must wander over the plants on which these insects live till they find a nymph of their host species. In the case of stylopids that infest social insects the problem is obviously not so difficult, especially if the triungulinids leave their host while it is in or near the nest. But those stylopids that infest solitary nest-building species are beset with more serious difficulties. It is believed that parasitized female bees and wasps are so weakened that they do not build nests; hence the triungulinids issuing from them, and from males as well, must attach themselves to other females of the same species in order to be carried to a nest where they can find their appropriate victims. This transfer is probably made in the flowers visited by these insects.

When a triungulinid finds a larva or a nymph of its host species it quickly bores into it, and begins its parasitic life. The most complete account of the metamorphosis of a stylopid yet published is that of *Xēnos vespārum* by Nassonow ('92). An abstract of this author's results is given by Pierce ('09, pp. 47–48); the more important features of them are the following.

The campodeiform triungulinid grows rapidly after entering the body of its host; at the first molt it loses its legs and becomes scarabæiform; later the body becomes cylindrical. From this point the development of the two sexes is different. In the case of the females, there are seven larval instars; in the fifth instar the head and thorax are fused, forming a cephalothorax; the seventh instar pushes its cephalothorax out between two of the abdominal segments of the host; the skin of this instar becomes the "puparium," in which the adult female is inclosed, and which she never leaves; the adult female is larviform; there is no pupal stage in this sex. In the case of males, the head and thorax of the fifth instar are fused, forming a cephalothorax; the seventh instar is inclosed in the skin of the sixth, and has strongly developed appendages; for this reason it may be termed a prepupa; during the seventh stadium the cephalothorax is exserted between two abdominal segments of the host; the true pupa is formed within the skin of the seventh instar; the adult male thrusts off the cap of the puparium and emerges as a winged individual.

The manner in which the female is fertilized, inclosed as she is in a puparium, has not been determined; it has been suggested that the seminal fluid is discharged into the space between the venter of the female and the puparium, the brood chamber. If this is true, the mobile spermatozoa probably pass from the brood chamber through the genital apertures into the abdominal cavity, where the eggs are massed free. The slit in the cephalothorax of the puparium, through which the triungulinids escape, may serve for the introduction of the seminal fluid into the brood chamber.

The order Strepsiptera is well represented in this country. Leng ('20) lists ninety-seven American species, and doubtless there are many undiscovered species here. The described American species represent five families and eighteen genera.

Students wishing to study the classification of these insects should consult the very complete monographs of the order by W. Dwight Pierce ('09, '11, and '18), and other papers listed in these works.

CHAPTER XXV

ORDER MECOPTERA*

The Scorpion-Flies and Their Allies

The winged members of this order have four wings; these are usually long, narrow, membranous, and furnished with a considerable number of cross-veins; the wings are wanting or vestigial in two genera. The head is prolonged into a deflexed beak, at the end of which chewing mouth-parts are situated. The metamorphosis is complete.

This is a small order composed of very remarkable insects. The most striking character common to all is the shape of the head, which is prolonged into a deflexed beak (Fig. 674). The dorsal wall of this beak is composed largely of the greatly elongated clypeus (Fig. 675, A, c); the central portion of the ventral wall is the greatly elongated submentum (Fig. 675, B, sm); and on each side of the submentum there

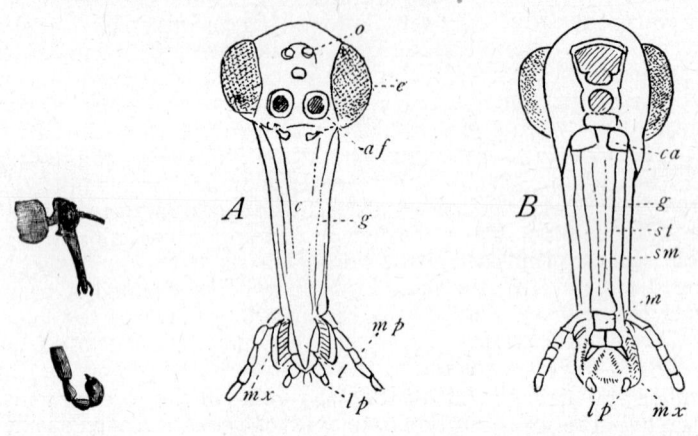

Fig. 674.—Head and tail of *Panorpa*.

Fig. 675.—Head of *Panorpa*: A, dorsal aspect; B, ventral aspect; af, antennal foramen; ca, cardo; e, eye; g, gena; l, labrum; lp, labial palpi; m, mentum; mp, maxillary palpi; mx, maxillæ; o, ocelli; sm, submentum; st, stipes. (After Miyake.)

is a greatly elongated stipes of the maxilla, at the distal end of which is borne the maxillary palpus. The mentum and labium are comparatively short; and from each side of the labium there extends a labial palpus. The mandibles are rather small and slender and are articulated to the apex of the beak, and can cross freely.

The antennæ are long, very slender, and many-jointed. The compound eyes are moderately large. There are usually three prominent ocelli, but these are wanting in *Merope* and in *Boreus*.

*Mecŏptera: mecos (μῆκος), length; pteron (πτερόν), a wing.

MECOPTERA 551

The prothorax is small; the mesothorax and metathorax are large. The legs are long and slender; the tarsi are five-jointed; in some genera there are two tarsal claws, in others only one.

The wings are membranous, and are usually long and narrow, but in two genera, *Notiothauma* and *Merope*, the representatives of which are very rare insects, the wings are comparatively broad. In the genus *Boreus* the wings are vestigial or wanting.

The type of the venation of the wings in this order is well shown by the wings of *Panorpa* (Fig. 676). In the species figured here, the

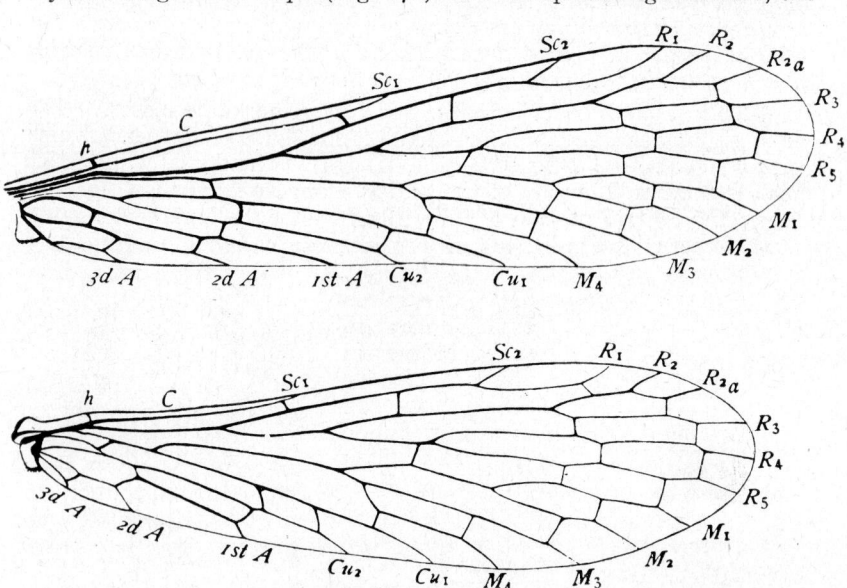

Fig. 676.—Wings of *Panorpa*.

number and arrangement of the wing-veins in the fore wings is that of the hypothetical primitive type, with the addition of a considerable number of cross-veins, and an accessory vein on vein R_2. The same is true of the hind wings except that each of the branches of cubitus anastomoses with the adjacent vein; that is, vein Cu_1 anastomoses with vein M, and vein Cu_2 with the first anal vein

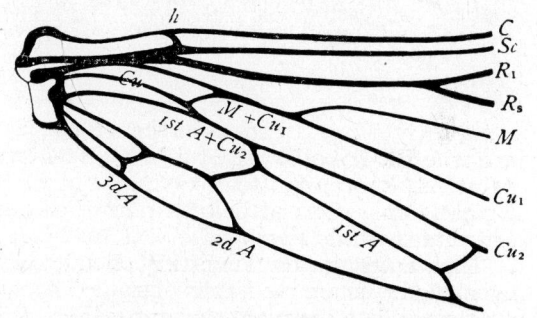

Fig. 677.—Base of hind wing of *Panorpa*.

(Fig. 677). For further details regarding the venation of the wings in this order, see "The Wings of Insects" (Comstock '18 a).

The metamorphosis is complete. The larvæ are caterpillar-like, with three pairs of thoracic legs and with or without abdominal prolegs. The pupæ are exarate, that is, the wings and legs are free, as in the Coleoptera and Hymenoptera.

This order is represented in our fauna by six genera; these can be separated by the following table:

A. With well-developed wings.
  B. Wings long and narrow; ocelli present.
    C. Tarsi with a single claw, and fitted for grasping............BĬTTACUS
    CC. Tarsi with two claws, and not fitted for grasping.
      D. Tarsal claws toothed...............................PANŎRPA
      DD. Tarsal claws simple..................... ....PANŎRPODES
  BB. Wings comparatively wide, with many cross-veins extending from the subcosta to the costa; ocelli wanting........................MERŌPE
AA. Wings wanting or imperfectly developed.
  B. Without ocelli; small insects, less than 6 mm., in length.......BŌREUS
  BB. Ocelli present; body about 20 mm., in length.......APTEROBĬTTACUS

Panŏrpa or the scorpion-flies.—The most common members of this order belong to the genus *Panorpa*, of which there are nearly twenty described North American species. Figure 678 represents a female of this genus. In our more common species the wings are yellowish, spotted with black. The males of this genus are remarkable for the peculiar form of the caudal part of the abdomen (Fig. 679). This at first sight reminds one of the corresponding part of a scorpion, and suggested the common name *scorpion-flies* for these insects. But in reality the two are very different; the last segment of the male *Panorpa*, instead of ending in a sting, like that of a scorpion, is greatly enlarged and bears a pair of clasping organs. The tarsal claws are toothed (Fig. 680, *a*).

Fig. 678.—*Panorpa*, female.

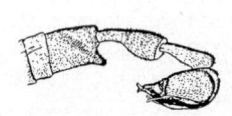

Fig. 679.—Abdomen of *Panorpa rufescens*.

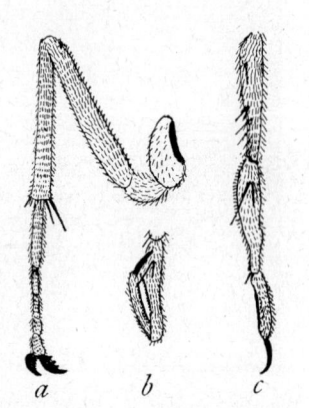

Fig. 680.—*a*, fore leg of *Panorpa*; *b*, last two segments of tarsus of *Bittacus*, apposed; *c*, last three segments of tarsus of *Bittacus*.

The adults are found resting on the surface of foliage of rank herbage growing on the banks of shaded streams and in damp woods where there is a luxuriant undergrowth of herbaceous plants. They feed on dead or injured insects and upon fruits; it appears that they rarely if ever capture living prey.

The females lay their eggs in crevices in the earth. The larvæ are caterpillar-like in form; they have three pairs of true legs and eight pairs of abdominal prolegs; and the body is armed with prominent spines (Fig. 681); the larvæ are carnivorous. The transformation takes place in a cell in the ground.

Panŏrpodes.—The members of this genus resemble *Panorpa* in general appearance, and as in that genus the abdomen of the male is furnished with a pair of clasping organs; but in *Panorpodes* the tarsal claws are simple. Only two species have been described from North America. These are not common; and but little is known regarding their habits.

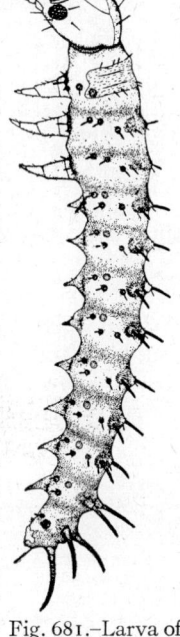

Merōpe.—This genus includes only a single known species, *Merope tuber*. This is an exceedingly rare insect. In this genus the wings are comparatively wide (Fig. 682); and there are many cross-veins extending from the subcosta to the costa. I have figured the venation of the wings in "The Wings of Insects" (Comstock '18 a). The ocelli are wanting. The abdomen of the male is terminated by a pair of long, stout forceps. This is probably a nocturnal insect as it is attracted to lights at night. Its life-history is unknown.

Borēus.—This genus includes small Mecoptera, our species measuring from 2.5 mm. to 5 mm. in length, which are often found on snow in winter. The wings of the female are vestigial or wanting; those of the male, imperfectly developed. The ocelli are wanting. The female has a long, protruding ovipositor, which in some species is nearly as long as the abdomen. The larva differs from that of *Panorpa* in lacking the abdominal prolegs. The pupa state is passed in an earthen cell in the ground. Four American species have been described, two from the East and two from the West.

Bĭttacus.—Insects belonging to this genus have long, narrow wings, long legs, and a slender abdomen. They resemble crane-flies very closely when on the wing, but can be distinguished by the presence of two pairs of wings. They are almost as common as

Fig. 682.—*Merope tuber*, slightly enlarged. (Photographed by J. G. Needham.)

Fig. 681.—Larva of *Panorpa rufescens*, first instar. (After Felt.)

Panorpa; and, like the scorpion-flies, are found among rank herbage growing on the banks of shaded streams and in damp woods where there is a luxuriant undergrowth of herbaceous plants. When at rest, they do not sit on the surface of foliage as does *Panorpa*, but hang suspended, by their front legs, from some support (Fig. 683).

The members of this genus capture and feed upon living insects. They are enabled to capture their prey by means of their curiously modified tarsi, the last two segments of which are armed with teeth, and the last segment can be folded back against the next to the last segment. In this way there is formed an efficient grasping organ (Fig. 680, *b, c*). It is an interesting fact that, while in other predacious insects the fore legs are the chief organs of prehension, in *Bittacus* the hind legs are used for this purpose fully as often as the others, especially when the *Bittacus* is hanging suspended by its fore legs and captures an insect that comes within reach of it.

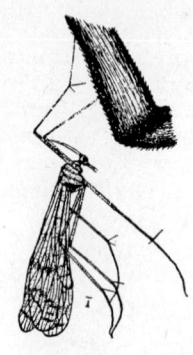

Fig. 683.—Natural position of *Bittacus*. (From Felt.)

Nine North American species of *Bittacus* have been described.

Apterobittacus.—This genus includes a single known species, *Apterobittacus apterus*, found in California. It resembles *Bittacus* except that the wings are completely wanting.

A review of the species of the Mecoptera of America north of Mexico was published by James S. Hine (Hine '01).

CHAPTER XXVI
ORDER TRICHOPTERA*
The Caddice-Flies

The members of this order have four wings; these are membranous and usually more or less densely clothed with long, silky hairs. In the more generalized members of the order, the venation of the wings corresponds closely to that of the hypothetical primitive type with but few or no accessory veins; in some of the more specialized members of the order, the venation of the wings is reduced. The mouth-parts of adults, except the palpi, are vestigial. The metamorphosis is complete.

The caddice-flies are moth-like insects, which are common in the vicinity of streams, ponds, and lakes, and are frequently attracted to lights at night (Fig. 684). The larvæ of these insects are the well-known caddice-worms; these live in the water, and most of them build cases about their bodies.

In the adult insect, the body-wall is soft, being membranous or at the most parchment-like, and is thickly clothed with hairs. The two pairs of wings are membranous and usually more or less clothed with long, silky hair. The fore wings are denser than the hind wings and are often slightly coriaceous; in a few forms the wings are naked. The hind wings are shorter than the fore wings; but they are usually broader; this is due to an expansion of the anal area of the hind wings. In a few species the hind wings are reduced so that they are smaller than the fore wings; in one species the female is apterous, and in another the wings of the female are vestigial. When not in use the wings are folded roof-like over the abdomen.

Fig. 684.—A caddice-fly.

The posterior lobe of the fore wings is specialized as a fibula, which is well developed in the more generalized forms, as *Rhyacophila*, but more or less reduced in the more specialized genera. The costal border of the hind wings is furnished with hamuli in some forms, as in the Leptoceridæ and some Hydropsychidæ.

In the more generalized forms the venation of the wings corresponds quite closely with the hypothetical primitive type; this is well shown by the wings of *Rhyacophila fuscula* (Fig. 685). The more important modifications of this type shown by the wings of *Rhyacophila* are the following: in the fore wing the tips of the second anal vein and two of the branches of the third anal vein coalesce; the subcosta bears an accessory vein; this, however, is unimportant; accessory veins borne by the subcosta exist in only a few genera of this order; the coalescence of veins Cu and 1st A at the base of the

---

*Trichŏptera: *trichos* (θρίξ, τρῐχός), the hair; *pteron* (πτερόν), a wing.

wing; and the formation of a serial vein consisting of the base of media, the posterior arculus (*pa*), and the distal part of vein Cu. In the hind wings, media has been reduced to a three-branched condition by the coalescence of veins M_3 and M_4.

In the more specialized members of this order the specialization of the preanal area of the wings is always by reduction. In the anal area of the hind wings the specialization is in some cases by addition, resulting in a broadly expanded anal area; in others it is by reduction.

The head is small; the antennæ are setaceous, and frequently several times as long as the body; the compound eyes are usually

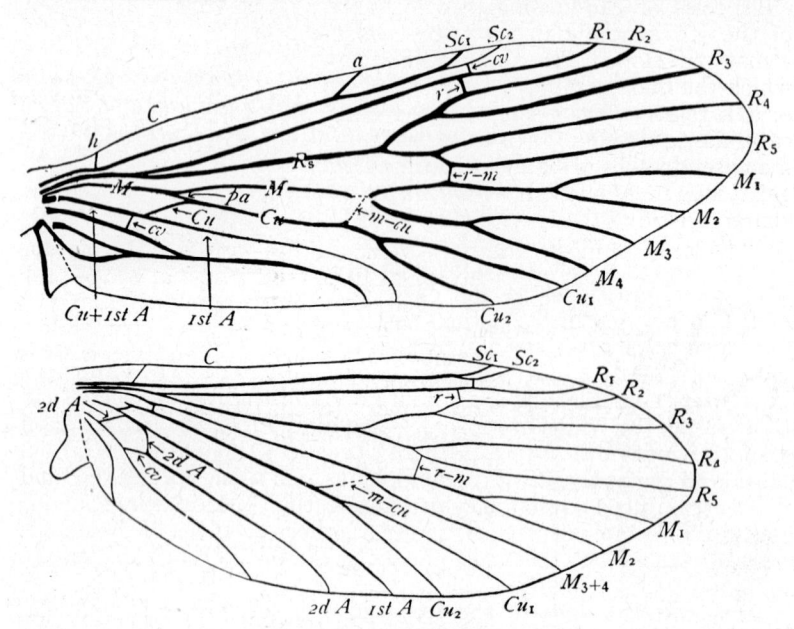

Fig. 685.—Wings of *Rhyacophila fuscula*.

small and with small facets; the ocelli are either present or absent; when present they are three in number; the mandibles are mere tubercles at the base of the labrum; the maxillæ are small, and ordinarily furnished with an obtuse maxillary lobe; the maxillary palpi are well-developed, and furnish characters which are much used in classification; the labium is usually well-developed, and bears three-jointed palpi.

The legs are long and usually slender; the coxæ are very large; the femora are long and slender, and generally without spines; the tibiæ are also long and slender; the tarsi are always five-jointed. The tibiæ and tarsi are often furnished with black or brown, sometimes yellow, spine-like setæ. In addition to the spine-like setæ, the tibiæ bear movable spurs either at the apex only, or also at some distance before the apex; these are larger than the spine-like setæ and are usually differently colored. The number of these spurs is much used in classification.

The eggs of caddice-flies are round or slightly oval in form. They are laid either in water or upon objects above water from which the larvæ when hatched can find their way into the water. Some species that lay their eggs in water descend below the surface in order to glue their eggs to some submerged support. So far as is known, all species of caddice-flies, except some of the Rhyacophilidæ, lay their eggs in a mass enveloped either in a cement, by which the mass is glued to some support, or in a gelatinous covering. In the latter case, the covering absorbs water and thus increases greatly in size. The form of the gelatinous mass and the arrangement of the eggs within it are often characteristic of the species (Fig. 686).

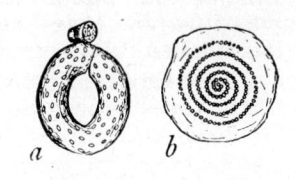

Fig. 686.—Two egg-masses of caddice-flies: *a, Phryganea interrupta; b, Triænodes sp.* (From Lloyd.)

The larvæ of most caddice-flies, the caddice-worms, are somewhat caterpillar-like (eruciform) in shape (Fig. 687); but some are more nearly campodeiform. Those that are eruciform build a portable case in which they live; most of the campodeiform larvæ do not build portable cases. In the eruciform larvæ the head is bent down, as in a caterpillar; in the campodeiform larvæ the head is horizontal, the mouth-parts projecting forward. Both types differ from the caterpillars in having only one pair of prolegs, the anal prolegs. These are each furnished with a chitinous hook. The mouth-parts are fitted for chewing. The thoracic legs are well developed. In the case-building forms, the first abdominal segment often bears three tubercles, one dorsal and one on each side; these are the "spacing-humps," and serve to keep a space between the insect and its case for the free circulation of water for respiration. In several families the larvæ possess abdominal tracheal gills; these are filamentous and are sometimes branched; they arise singly or in tufts. With the exception of a single European genus, *Enoicyla,* all caddice-worms are aquatic.

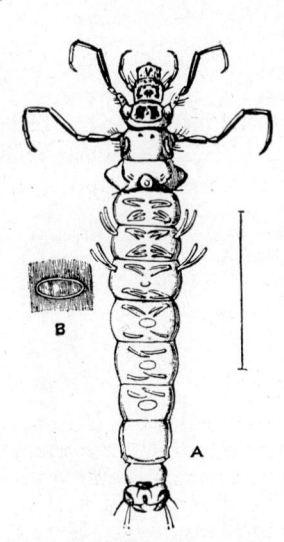

Fig. 687.—A caddice-worm, *Anabolia nervosa:* A, larva extracted from its case; B, one of the dorsal spaces of the abdominal segments more strongly magnified. (From Sharp.)

Most caddice-worms build portable cases in which they live and which they drag about wherever they go, projecting only the front end of the body and the legs from the case when they travel. The cases of different species differ greatly in form and in materials used in their construction; but silk is used in building all of them. This silk, like that of caterpillars, is secreted by modified salivary glands and is emitted through an opening in the labium; but in most cases it is not spun into a thread, but is poured forth in a glue-like sheet upon the objects to be cemented together; some species, however, build nets of silken strands.

Some caddice-worms build their cases entirely of silk; but most of the case-building species use other materials also; these may be grains of sand, small stones, bits of wood, moss, or pieces of leaves; and some species fasten shells of small mollusks to their cases. The materials used are glued together with silk; and the case is lined with silk, so as to form a suitable protection for the soft abdomen. Examples of different types of cases are figured later.

When the caddice-worms are full-grown they do not leave the water to transform, as do nearly all other aquatic larvæ, the pupæ being as truly aquatic as the larvæ. Some of the case-building species change the form and material of their cases at this time; and nearly all of them partly close their cases so as to keep out intruders and silt; but usually provision is made for the ingress of water for respiration. Some species merely cement a stone or grains of sand over each opening of the case; others build a silken lid with a slit in it; and still others build a silken grating in each end of the case. Frequently caddice-worms leave the open water in which the larval life has been spent and seek some more secluded place in which to transform, such as crevices in bark or among roots, or they may burrow into wood or into the soil.

The pupæ are of the exarate type, that is, the wings and legs are free (Fig. 688). Some pupæ have tracheal gills, others do not; this, however, is not correlated with the presence or absence of tracheal gills in the larva; tracheal gills may be present in either of these stages and absent in the other.

In the case of those caddice-flies that emerge from rapidly flowing water, as the net-building species, the wings expand instantly when the insect reaches the surface of the water and are then fitted for flight; it is evident that if much time were required for the wings to become fit for use, as is the case with most other insects, the wave succeeding that which swept the insect from the water would sweep it back again and destroy it.

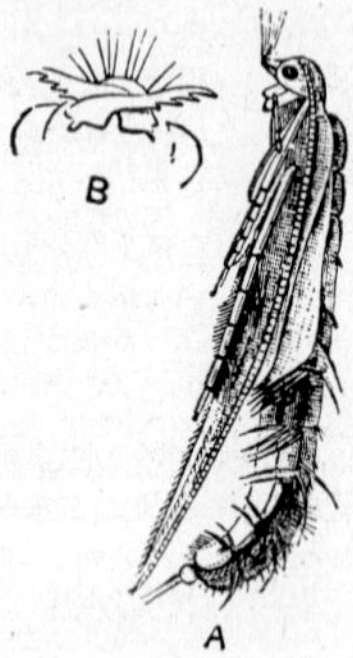

Fig. 688.—A, pupa of *Phryganea pilosa.* (After Pictet.) B, mandibles of pupa of *Molanna angustata.* (From Sharp.)

The Trichoptera can be regarded as beneficial insects, as the larvæ form an important element in the food of fishes, and especially of the brook trout. Sometimes in cities near rivers, the adults are annoying on account of the great numbers of them that are attracted to lights.

This order includes thirteen families, all of which are represented in North America. Nearly four hundred species have been described from this region. Among the more important works on the classification of these insects are McLachlan ('74–80), Ulmer ('07), and Ulmer ('09). This last-mentioned work is especially important for its accounts of the early stages of these insects.

The latest and most extended work on the life-histories of North American caddice-worms is that by Lloyd ('21). In this work there is a list of the more important papers on this subject, which, for this reason, need not be enumerated here. This monograph by Mr. Lloyd has been of great assistance to me in the preparation of the following account of the habits of representatives of the different families.

A monograph treating of all stages of North American Trichoptera has been prepared by Dr. Cornelius Betten and is to appear as a bulletin of the New York State Museum.

The following table of families is copied from Needham ('18).

TABLE OF FAMILIES OF THE TRICHOPTERA

For the Classification of Adults

A. Micro-caddice-flies; very small, moth-like, hairy, the fore wings bearing numerous erect clavate hairs; the marginal fringe of the wings longer than their greatest breadth; form of wings narrowly lanceolate; antennæ rather stout and not longer than the fore wings. p. 561............HYDROPTILIDÆ
AA. Larger caddice-flies, with broader wings; marginal fringes never as long as the wings are broad; antennæ usually longer than the fore wings.
  B. Maxillary palpi five-jointed.
    C. Last joint of the maxillary palpi simple, and not longer than the other joints.
      D. Ocelli present.
        E. Front tibiæ with two or three spurs, middle tibiæ with four spurs.
          F. The first two joints of the maxillary palpi short and thick, the third joint much longer and thinner. p. 560....RHYACOPHILIDÆ
          FF. The second joint of the maxillary palpi much longer than the first. Females. p. 564........................PHRYGANEIDÆ
        EE. Front tibiæ with a single spur, or with none; middle tibiæ with only two or three spurs. Females. p. 568.........LIMNOPHILIDÆ
      DD. Ocelli wanting.
        E. A closed cell in the principal fork of the median vein in the fore wings. p. 567.............................CALAMOCERATIDÆ
        EE. No closed cell in the median fork.
          F. A closed cell in the first fork of the radial sector.
            G. Both branches of the radial sector forked.
              H. Veins R_1 and R_2 confluent apically or connected by an apical cross-vein in the fore wing. Females. p. 567..ODONTOCERIDÆ
              HH. Veins R_1 and R_2 not connected apically. p. 569.................................SERICOSTOMATIDÆ
            GG. Only the anterior branch of the radial sector forked. p. 566. ....LEPTOCERIDÆ
          FF. No closed cell in the first fork of the radial sector. p. 566.. ....MOLANNIDÆ
  CC. Last joint of the maxillary palpi usually much longer than the others, twisted, and divided imperfectly into subsegments.
    D. Ocelli present. p. 563...........................PHILOPOTAMIDÆ
    DD. Ocelli wanting.
      E. Front tibiæ with three spurs. p. 563..........POLYCENTROPIDÆ
      EE. Spurs of front tibiæ fewer than three.

    F. Anterior branch of the radial sector in the fore wings forked. p. 562..HYDROPSYCHIDÆ
    FF. Anterior branch of the radial sector simple. p. 564...PSYCHOMYIDÆ
BB. Maxillary palpi with fewer than five joints.
  C. Maxillary palpi with four joints; ocelli present. Males. p. 564..PHRYGANEIDÆ
  CC. Maxillary palpi with two or three joints.
    D. Maxillary palpi filiform, with cylindric smooth joints; fore tibiæ with a single spur. Males. p. 568..................LIMNOPHILIDÆ
    DD. Maxillary palpi hairy or scaly, appressed against and often covering the face; fore tibiæ with two spurs. Males. p. 569..SERICOSTOMATIDÆ

TABLE OF TRICHOPTEROUS LARVÆ

The following table will aid in the classification of caddice-worms. It is based on a more detailed table of the family characters of trichopterous larvæ given by Lloyd ('21).

A. Anal prolegs not fused in median line to form an apparent tenth abdominal segment.
  B. Abdomen much wider than the thorax. p. 561.......HYDROPTILIDÆ
  BB. Abdomen not much wider than the thorax.
    C. Dorsal surface of the ninth abdominal segment with a chitinous shield. p. 560...RHYACOPHILIDÆ
    CC. Dorsal surface of the ninth abdominal segment without a chitinous shield.
      D. Tracheal gills present, branched. p. 562........HYDROPSYCHIDÆ
      DD. Tracheal gills absent.
        E. Labrum entirely membranous, white. p. 563..PHILOPOTAMIDÆ
        EE. Labrum entirely chitinized.
          F. Frons long, extending back to the caudal margin of the head. p. 563...POLYCENTROPIDÆ
          FF. Frons normal. p. 564.......................PSYCHOMYIDÆ
AA. Anal prolegs fused in median line so as to form an apparent tenth segment.
  B. Dorsal surface of the labrum with a row of twenty or more heavy bristles. p. 567...CALAMOCERATIDÆ
  BB. Dorsal surface of labrum normal.
    C. Labrum much longer than broad. p. 567............ODONTOCERIDÆ
    CC. Labrum broader than long.
      D. Metanotum with three pairs of plates. p. 568.......LIMNOPHILIDÆ
      DD. Metanotum soft.
        E. Mesonotum soft or with one pair of minute plates. p. 564...PHRYGANEIDÆ
        EE. Mesonotum chitinized.
          F. Femur of hind legs divided into two segments or apparently so. p. 566..LEPTOCERIDÆ
          FF. Femur of hind legs not divided.
            G. p. 569................................SERICOSTOMATIDÆ
            GG. p. 566...MOLANIDÆ

FAMILY RHYACOPHILIDÆ

The larvæ are campodeiform; they live in rapidly flowing streams with stony bottoms. The American species of this family represent two subfamilies. The members of one subfamily, the Rhyacophilinæ, do not build cases, but crawl about naked beneath stones seeking their food; they feed on small larvæ and filamentous algæ. The larva of our most common species, *Rhyacŏphila fŭscula*, when full-grown enters a crevice between two large stones and builds a wall of pebbles about itself; this wall is cemented in place with silk; and the chamber thus inclosed is much larger than the insect (Fig. 689); it then spins a parchment-like cocoon about its body, within which it transforms. The making of a cocoon is a family characteristic of the Rhyacophilidæ; only a few other caddice-worms spin cocoons.

Fig. 689.—A larva of *Rhyacophila* building its pupal chamber, exposed by lifting off the stone beneath which it was. (From Needham and Lloyd.)

The members of the subfamily Glossosmatinæ build cases out of sand or small stones. Our best-known species is *Glossosōma americāna*, the habits of which are described by Lloyd. Figure 690 represents a dorsal and a ventral view of the case. The larvæ live singly on the stones of the stream's bottom; but before pupating they congregate in dense colonies on the sides and bottoms of stones, with their cases placed edge to edge, sometimes one on top of another. At this time the floor of the case is cut away and the rim of the cup-like roof is glued to the supporting rock.

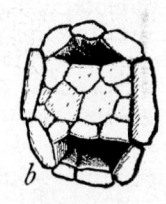

Fig. 690.—Case of *Glossosoma americana*: *a*, dorsal view; *b*, ventral view. (After Lloyd.)

Family HYDROPTILIDÆ

The Micro-Caddice-Flies

This family is composed of minute caddice-flies, which resemble tineid moths in appearance. The larvæ are found in both quiet water and rapid streams, and often occur in very great numbers. They build cases which differ in form in the different species, but are usually flat; some are elliptical, some flask-like, and others kidney-shaped: all are open at both ends; they are much larger than the larvæ. They are usually composed entirely of silk; but in some species grains of sand or minute bits of vegetable matter are used. "*Agraylea* decorates the parchment with filaments of Spirogyra, arranged concentrically over the sides in a single external layer." (Needham and Lloyd.) When moving about, the larva usually drags its case on one edge. There is one species, *Ithytrichia confusa*, which cements its case firmly to rocks in flowing water. These cases are common; they are parchment-like, elliptical, with a small opening at each end (Fig. 691, 2), and measure from 5 mm. to 6 mm. in length. They are incomplete, being cemented along the edges to the rock, with no floor below the larva. The larva is very remarkable in form (Fig. 691, 1). When feeding, it protrudes the narrower part of its body from its case and gathers food from the surface of the rock; the expanded abdominal segments are much wider than the openings in the case.

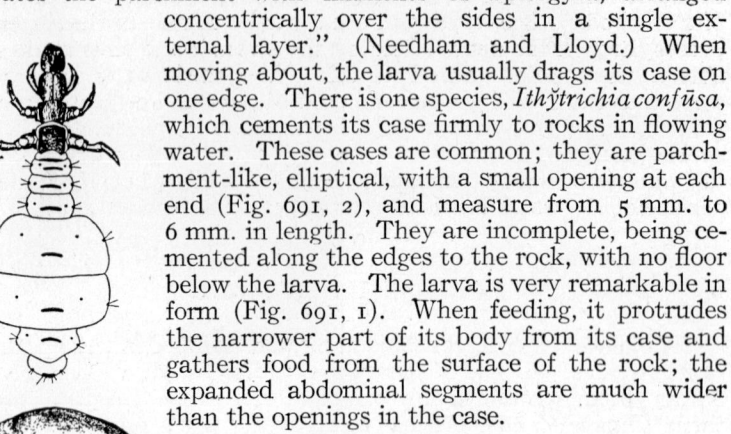

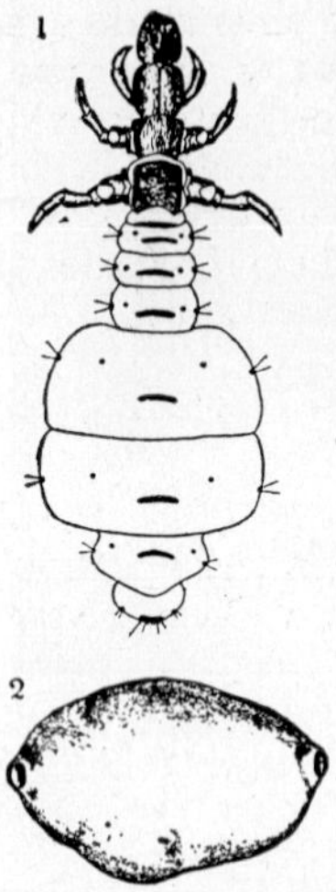

Family HYDROPSYCHIDÆ

Fig. 691.—*Ithytrichia confusa*: 1, larva; 2, case. (After Lloyd.)

The family Hydropsychidæ of the older authors has been divided into four families by Ulmer,—Hydropsychidæ, Philopotamidæ, Polycentropidæ, and Psychomyidæ. It is to this group of families that the net-spinning caddice-worms belong. The best-known of these are species of the genus *Hydropsyche*, the nets of which have been described by many writers.

The larvæ of *Hydropsyche* live only in rapid streams and on the wave-beaten shores of lakes. They are campodeiform, and do not build portable cases, but live in tubes composed of silk and debris, and fastened permanently in place; sometimes they establish themselves in old worm-holes in submerged wood. The most striking feature in their habits, however, is the fact that each one builds a net for the capture of its food. This net is built adjacent to the tube in which the larva lives; it is funnel-shaped and has at its downstream end an opening in which is built a strainer. This is a beautiful object, consisting of two sets of regularly spaced strands of silk extending across the opening at right angles to each

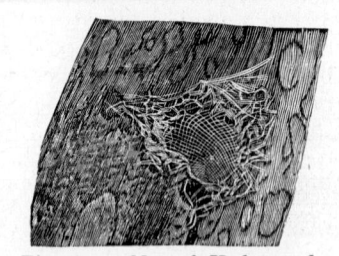

Fig. 692.—Net of *Hydropsyche*.

other (Fig. 692). These nets are often built in crevices between stones; but fully as often they are built up from a flat surface, as on the brink of a waterfall. In this case they are in the form of semi-elliptical cups, which are kept distended by the current. Much of the coating of dirt with which rocks in such places are clothed in summer is due to its being caught in these nets. Sometimes when the net is built up from a horizontal surface its sides are supported by bits of wood. Algæ, larvæ, and other small animals in the water that passes through the net are held by the strainer and thus made available to the caddice-worm for food. When the larva is full-grown it surrounds itself with a case composed of fine sand or gravel in which to transform; this case is firmly cemented in place, and, in some species at least, is closed at each end with a silken grating. The instantaneous flight of the newly emerged adult when it reaches the surface of the water has been referred to on an earlier page.

Family PHILOPOTAMIDÆ

The larvæ of members of this family are campodeiform and live in rapid streams. Several of them were studied by Miss Alice A. Noyes, but as yet an account of only one of them, *Chimărrha atĕrrima*, has been published (Noyes '14). This larva spins a delicate silken net resembling in shape the finger of a glove. The average size of the net of a growing larva is about 25 mm. long and 3 mm. wide. The nets are rarely found singly, but are generally placed five or six in a row (Fig. 693); sometimes they occur in great numbers, completely covering the stones with a thin, flocculent mass of dirty silk. There is a large opening at the end of the net facing the current, and a smaller opening at the hind end. The nets are fastened in place at the entrance; the rest of the sac floats freely, and is kept distended by the current. The net serves both as a hiding-place for the larva and as a sieve through which the flowing water is strained; the larva feeding on the organic particles that are entangled in it. The full-grown larva covers itself with an irregular dome of pebbles in which to transform, and spins about its body a delicate cocoon.

Fig. 693.—Nets of *Chimarrha aterrima*, natural size. (From Noyes.)

Family POLYCENTROPIDÆ

The larvæ are campodeiform; they usually live in flowing water, but some are found in standing water. They do not build portable cases, but make fixed silken tubes or nets. The nets of several European genera have been described; for an abstract of these accounts, see Noyes ('14). The nets of American species have been described by Clark ('91), Vorhies ('09), Noyes ('14), and Lloyd ('21).

"Several species of the genus *Polycĕntropus* live in still or slowly flowing water with sandy or muck bottoms. These larvæ spin subterranean tubes of silk which sometimes reach 10 centimeters in length. Often the tubes have one or more branches, and always they contain a bulbous swelling near the middle in which the larva probably rests, and in which pupation takes place. In natural position the tubes are beneath the ground, except about half an inch which projects upward into the water." (Lloyd.)

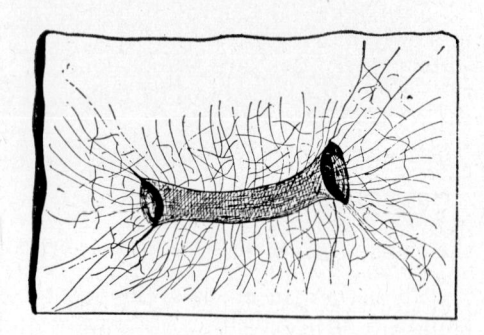

Fig. 694.—Dwelling of *Polycentropus sp.* (From Noyes.)

Two quite different larval tubes of members of this family are described and figured by Miss Noyes. One of these is represented in Figure 694. This tube is found on the under side of stones, and is fastened along its entire length. "It is 21 mm. long and 5.5 mm. wide, with an expanded opening at either end. Connected with each opening and along either side is a mass of tangled, silken threads, about 20 mm. square and loosely attached to the stone. This tangled mass may float partially over the tube and so obscure it."

"I have never observed the larvæ feeding, but do not doubt that Mayfly nymphs and chironomid larvæ become entangled in the meshes as they crawl about over the stones, for remains of these forms are abundant in the stomach contents." (Noyes.)

Family PSYCHOMYIDÆ

The larvæ are campodeiform. There is no account of the life-history of any American species published. The European species do not make portable cases; but the larvæ live on stones in long, loosely-spun galleries of silk and sand grains. They are found mostly in swift water, but also inhabit ponds and lakes.

Family PHRYGANEIDÆ

The larvæ are caterpillar-like, and usually live in standing water in which plants are growing, or in slowly moving streams of spring water. They make portable cases which are very regular in form. As these larvæ live in quiet water, they can be fed and reared in aquaria where their habits can be easily observed. The most extended account of the immature stages of these insects is that of Lloyd ('21), from which the following brief notes are compiled. This author discusses three species of *Neuronia* and three species of *Phryganea*.

Neurōnia.—The larvæ are found in slowly moving streams of spring water; rarely they are found along the edge of the large, warm streams where cool seepage enters. One species was found in a pond. The cases are cylindrical tubes of thin, rectangular bits of leaves arranged in a series of rings (Fig. 695, *a*). In the cases of old larvæ the rings are neatly fitted without overlapping; young larvæ sometimes leave the hind ends of the leaf-fragments protruding in long strips. Unlike other caddice-worms, these larvæ often abandon their cases and wander naked through the water. The form of the case indicates that they are not long retained; their uniform diameter proves that they are constructed more rapidly than the diameter of the larva increases. When the season for pupation draws near, the larvæ of *Neuronia* burrow into wood, or wedge themselves beneath bark, or in crevices, or, if the stream bottom be of clay, they may burrow into the soil. When entering the soil the larva stands on its head, with its case perpendicular to the bottom, and slowly enters, dragging its case with it.

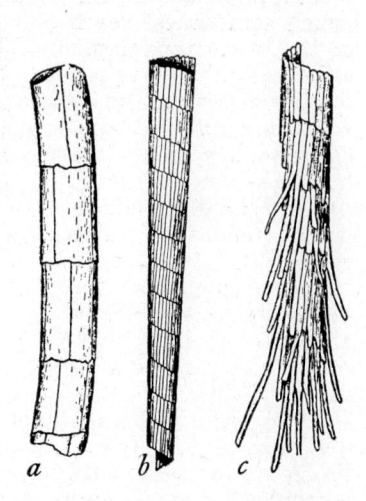

Fig. 695.—Cases of phryganeids: *a*, case of *Neuronia postica*; *b*, case of old larva of *Phryganea vestita*; *c*, case of young larva of *Phryganea vestita*. (After Lloyd.)

Phrygānea.—The larvæ live in ponds; they dwell, for the most part, among submerged plants above the bottom of the pond; hence they can be taken with a water net. They never abandon their cases as do the larvæ of *Neuronia*. The case is a straight tube composed of narrow strips of leaf arranged in spiral form around the circumference of the case (Fig. 695, *b*). Young larvæ often fail to cut the leaf-fragments used in the construction of the case into the rectangular form seen in the cases of old larvæ; but the bases of the untrimmed fragments are arranged in a spiral (Fig. 695, *c*). In preparing to pupate, the larvæ leave their abode among living plants and travel to some submerged log or chunk of wood and burrow into it until the last bit of the case is concealed. This operation sometimes requires several days of labor. When sufficient depth is reached, the larva spins a silken mesh across each end of the case.

The larva of a species of *Triænodes* of the family Leptoceridæ makes a case somewhat similar to that of *Phryganea;* this is described in the account of that family.

Family MOLANNIDÆ

The only members of this family the larvæ and cases of which have been described in this country belong to the genus *Molănna*. The larvæ are found on sandy bottoms of streams and of lakes. The cases have been figured by several writers, and are very characteristic in form (Fig. 696). The case is made of grains of sand, and, has on each side an extension, and at the head end a dorsal hood, which completely protects the larva when crawling or feeding.

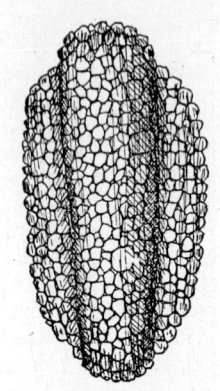

Fig. 696.—Case of *Molanna*. (After Lloyd.)

Family LEPTOCERIDÆ

The larvæ are caterpillar-like, and make portable cases. Most species live in standing water, as in lakes, ponds, and the bays of streams; but some are found in flowing water and on wave-beaten shores of lakes. The cases made by the different species differ greatly in form and in the materials used in their construction. Among the better-known species are the following.

Setōdes grăndis.—The larva of this species lives among aquatic vegetation in ponds and lakes. Its case is composed entirely of silk, and is translucent, so that the body of the larva can be seen through it. It is cylindrical, tapering, and slightly curved (Fig. 697, a). When ready to pupate, the larva fastens its case to the stem of a plant with a band of silk, and closes the anterior end of the case with a silken membrane, in which there is a central slit for the ingress of water.

Leptŏcerus ăncylus.—The larva is found on stones in the riffles of streams and on the stones of wave-beaten lake shores. It makes a case of grains of sand. The larvæ studied by Lloyd

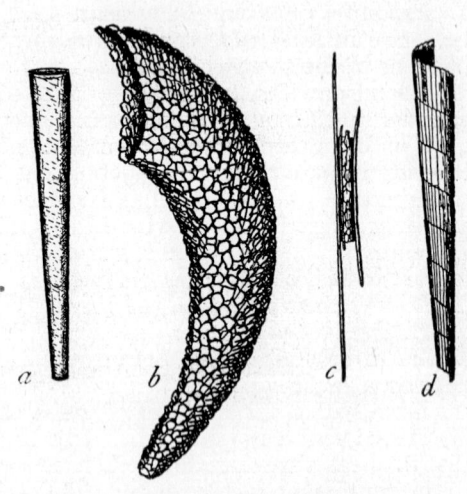

Fig. 697.—Cases of leptocerids: *a*, case of *Setodes grandis*; *b*, case of *Leptocerus ancylus*; *c*, case of *Mystacides sepulchralis*; *d*, case of *Triænodes*. (After Lloyd.)

at Ithaca, N. Y., made cases in the form of curved cornucopias (Fig. 697, *b*); those found in Wisconsin by Vorhies, who first described the species, make a case with decided lateral flanges and a hood that completely covers the head of the larva.

Mystăcides sepulchrālis.—The larva of this species and its case were described by Lloyd. It was found in ponds and in slow deep pools of creeks; it lives among the rubbish on the bottom. The case (Fig. 697, *c*) consists of a slightly tapering tube of sand or of minute fragments of bark, lined with silk; it measures about 12 mm. in length. On opposite sides are fastened pine needles, or grass stems, or slender sticks, which extend beyond both ends of the case. Before pupation a sheet of silk with a minute perforation in the center is spun across each end of the case.

Triænōdes.—The larvæ of species of this genus live in ponds and bays of creeks among branches of submerged plants. They are able to swim rapidly from place to place through the open water. The case (Fig. 697, *d*) is made of thread-like fragments of leaves arranged in a spiral. It resembles in form the case of *Phryganea* (Fig. 695, *b*), but is much smaller and more flexible, and the leaf-fragments are much narrower.

Family ODONTOCERIDÆ

The immature stages of only a single species belonging to this small family have been described in this country; the following notes regarding this species are from Lloyd ('21).

Psilotrēta frontālis.—The larvæ were found in upland streams and were confined to the riffles and the portions of the streams with stony bottoms. The case of the mature larva (Fig. 698) is a slightly curved cylinder made of sand; cases of immature larvæ differ only in being tapered toward the caudal end. The case of the pupa has a flat pebble set neatly within the aperture at each end. All spaces around these stones are tightly closed with heavy silk, leaving no apertures for the circulation of water; this is an unusual feature in the case-building Trichoptera.

During their early life the larvæ are free-moving, crawling separately over the bottom of the stream. But in the early spring, just before pupation, the larvæ develop a remarkable gregarious habit. Almost all of the larvæ within certain areas of the stream congregate on the sides of a few selected stones in such numbers that their cases are sometimes piled one on top of another to the depth of an inch or more, while other stones in the region are entirely uninhabited. The cases are always placed parallel to each other, with their cephalic ends directed toward the surface of the water.

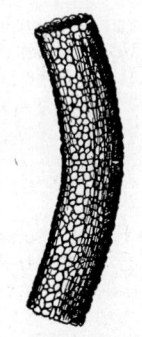

Fig. 698.—Case of *Psilotreta frontalis.* (After Lloyd.)

Family CALAMOCERATIDÆ

This is a small family of which only one American larva is known. The habits of this species have been described by Lloyd, from whose accounts I quote.

Ganonēma americāna.—The larvæ were found abundant in alder-bordered streams. The cases made by this species differ greatly from those of other described American caddice-worms. The case is made of a single piece of wood or bark or a twig; this is hollowed from end to end, and lined with silk. Although common, they are most inconspicuous among the debris on the bottom of the stream. Figure 699 represents a case with the silk tube cut away, except around the larva.

Family LIMNOPHILIDÆ

The larvæ of members of this family are caterpillar-like, and are found in a great variety of aquatic situations, but especially in ponds and slow-moving streams, even in those that become dry during the droughts of summer; a few, *Neophylax*, are found in rapids. Many of the larvæ that live in quiet water can be kept in aquaria.

The cases made by different members of this family differ greatly in form and in the materials used in their construction; in some species the case made by an old larva differs greatly from that made by it when young.

Fig. 699.–Case of *Ganonema americana*. (After Lloyd.)

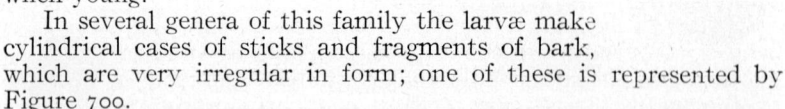

In several genera of this family the larvæ make cylindrical cases of sticks and fragments of bark, which are very irregular in form; one of these is represented by Figure 700.

To this family belong the larvæ that build cases of the "log-cabin type"; these are composed of sticks or of pieces of grass placed crosswise of the case (Fig. 701). A case closely resembling this in plan but differing in appearance is made of bits of moss.

Among the larvæ that change the form of their case when full-grown is *Limnŏphilus combinātus*, which is described by Lloyd. Dur-

Fig. 700.—Case of limnophilid larva.

Fig. 701.— Log-cabin type of case.

ing early life this larva frequents the grass and sedges that fringe the edges of streams, and makes a case of the cross-stick or log-cabin type. When the time for pupation draws near, it migrates away from the grassy area and makes a case differing entirely in appearance from the log-cabin type. Some individuals make a case composed of small chunks of bark (Fig. 702, *a*); others make cases composed almost entirely of shells of water snails (Fig. 702, *b*). Different combinations of these types are frequently found.

Some larvæ of this family make cases of leaves; these are either fastened so as to form a flat case, or arranged in three planes so as to form a tube, a cross-section of which is a triangle.

Larvæ of the genus *Neŏphylax* make cases of sand with large ballast stones at the sides; these are similar in form to those made by *Goera calcarata* of the next family, but are more slender, smaller, and made of lighter material.

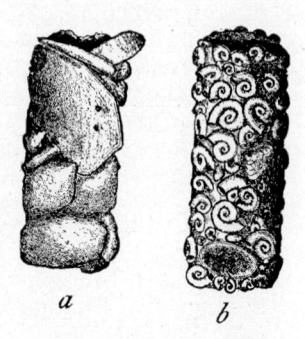

Fig. 702.—Case of *Limnophilus combinatus*. (After Lloyd.)

Family SERICOSTOMATIDÆ

The larvæ are caterpillar-like, and are found in streams and lakes. The cases made by members of the different genera differ greatly in form; the three following are our best-known examples.

Helicopsȳche boreālis.—The larvæ of this species are found in stony streams and along the rocky shores of lakes. They make a spiral case of grains of sand (Fig. 703). This case so closely resembles that of a snail in form that it has been described as the shell of a mollusk. When about to pupate, the larvæ fasten their cases to a submerged rock; at this time they display a gregarious instinct, large numbers of them congregating within a very small area. They are more easily found at this time than in their earlier stages when they are living free among the sand and gravel of the bottom of the stream.

Fig. 703.—Case of *Helicopsyche*. (From Lloyd.)

1 *Goera calcarāta.*—The arvæ of this species are found in the riffles of streams and on stones in wave-beaten areas of lake shores, where they crawl over the surface of bare, current-swept rocks. The larval case (Fig. 704) is a tube made of fine grains of sand on each side of which

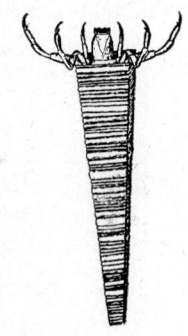

Fig. 704.—Case of *Goera calcarata*. (After Lloyd.)

Fig. 705.—Case of *Brachycentrus nigrisoma*. (From Lloyd.)

are fastened heavy ballast stones, usually two on each side.

Brachycĕntrus nigrisōma.—The larva of this species builds a case of the remarkable form shown in Figure 705. "It is constructed of minute twigs, root-fibers, and fragments of wood cut to the proper length to give even and straight edges, gradually diverging toward the anterior end. In cross-section the outer surface of the case is square; the interior is lined with a cylindrical tube of tough silk." "During the first six weeks of their lives the larvæ are active, crawling about in quiet eddies along the banks of streams in search of food. After this period they move to the center of the stream and live sedentary lives, with one edge of the larger end of their cases firmly cemented to submerged rocks or sticks. Always they inhabit positions on the exposed surface of their support and always they assume the position shown in Figure 705, protruding their heads slightly and extending their prothoracic legs straight forward. The mesothoracic legs are held upward while the metathoracic legs are extended to the sides. From this position they eagerly seize and quickly devour small larvæ or bits of vegetation that float within their grasp." (Lloyd.)

CHAPTER XXVII

ORDER LEPIDOPTERA*

The Moths, the Skippers, and the Butterflies

The winged members of this order have four wings; these are membranous, and covered with overlapping scales. The mouth-parts are formed for sucking. The metamorphosis is complete.

The members of this order, the moths, the skippers, and the butterflies, are well known to every observer of nature. Their most easily observed distinguishing characteristic is that which suggested the name of the order, the scaly covering of the wings and body. Every lad that lives in the country knows that the wings of moths and butterflies are covered with dust, which comes off upon one's fingers when these insects are handled. This dust when examined with a microscope is found to be composed of very minute scales of regular form; and when a wing is looked at in the same way, the scales are seen arranged with more or less regularity upon it (Fig. 706).

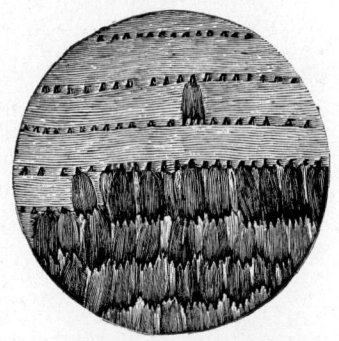

Fig. 706.—Part of a wing of a butterfly, greatly magnified.

The body, the legs, and other appendages are also covered with scales.

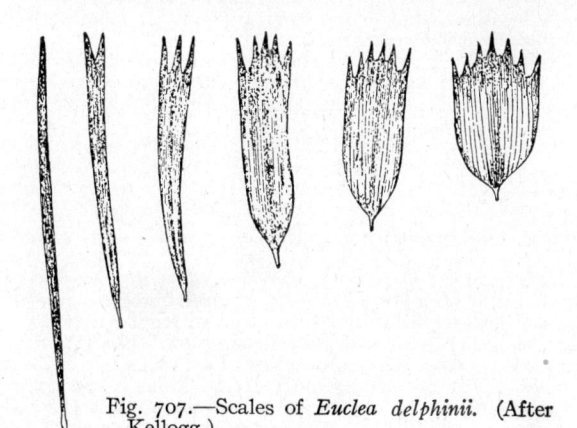

Fig. 707.—Scales of *Euclea delphinii*. (After Kellogg.)

*Lepidŏptera: *lepido* (λεπίs, λεπίδοs), scale; *pteron* (πτερόν), a wing.

It is well known that these scales are merely modified setæ. That is, they are setæ which, instead of growing long and slender as setæ usually do, grow very wide as compared with their length. Every gradation in form can be found, from that of the ordinary seta, which occurs most abundantly upon the body, to the short and broad scale, which is best seen upon the wings (Fig. 707). This fact was pointed out by Réaumur nearly two hundred years ago; and in recent times the morphological identity of setæ and scales has been established by studies of their development. Mayer ('96) gave a complete account of the development of scales and illustrated his paper by most excellent figures of all stages of this development.

The structure of scales is what would be expected from the fact that they are modified setæ, the scales, like setæ, being hollow; and the manner of their attachment to the cuticula of the body and its appendages is the same as that of the setæ, each scale being provided with a pedicel which fits into a cup-like socket in the cuticula.

A striking feature of the scales of Lepidoptera is the markings that exist on their exposed surface. These may consist merely of many very fine longitudinal ridges (Fig. 707); or they may be series of transverse ridges between the longitudinal ones (Fig. 708).

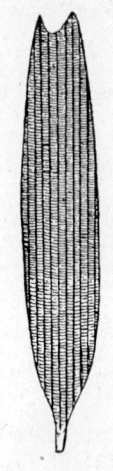

Fig. 708.—Scale of *Seryda constans*. (After Kellogg.)

A cross-section of certain scales indicates that the ridges are produced by foldings of the outer wall (*i. e.* the wall of the scale that is exposed when the scale is in place on the wing). Figure 709 represents cross-sections of a scale illustrating this condition. In some scales, however, the lumen of the scale has been filled to a considerable extent by chitin, and the origin of the ridges is not so obvious.

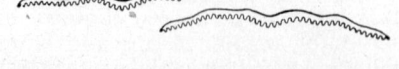

Fig. 709.—Cross-section of scales of *Parnassius smintheus*. (After Kellogg.)

The scales of the Lepidoptera were probably developed from that type of setæ known as clothing hairs, and were primarily merely protective in function. This is doubtless their chief, if not only, function on most parts of the body, where they form a very perfect armor.

The development of ridges on the surface of scales adds greatly to their stiffness, and thus increases their efficiency as a protective covering, as the corrugations in the sheets of iron used for covering the sides of buildings add to the stiffness of the metal.

Upon the wings a covering of rigid scales would serve not merely to protect the wings but would tend to stiffen them, and thus arose a secondary function of scales which has resulted in the perfecting of their arrangement upon the wings in the more specialized members of the order as already indicated.

There are great differences among the insects of this order regarding the regularity of the arrangement of the scales upon the wings. With some of the more generalized moths the scales are scattered irregularly over the surface of the wings. But if a wing of one of the more specialized butterflies be examined with a microscope, the scales will be found arranged in regular overlapping rows; the arrangement being as regular as that of the scales on a fish or of the shingles on a roof. Figure 706 represents a small portion of a wing of one of the more specialized butterflies, where the arrangement of the scales is most perfect. In the upper part of the figure the membrane is represented with the scales removed.

Even in those insects in which a very perfect arrangement of the scales upon the wings has been attained, great differences in the degree of perfection of this arrangement exist in the two wings of the same side and in the different parts of the same wing. The arrangement is most perfect in those wings and in those parts of each wing that are subjected to the greater strain during flight; and is more perfect in swift-flying species than in those of slow flight.

The taxonomic value of these differences in the arrangement of the scales of the wings of the Lepidoptera, and also of the different types of scales found in different divisions of the order, was investigated by Professor Kellogg ('94), to whose extended account the reader is referred for a discussion of this phase of the subject.

A secondary use of the scales of the Lepidoptera is that of ornamentation; for the beautiful colors and markings of these insects are due entirely to the scales, and are destroyed when the scales are removed.

The various colors of insects and of other animals are produced in quite different ways; and classifications of these colors have been proposed based on the methods of their production. The literature of this subject is too extensive to be referred to in detail here. A most enjoyable popular account is given by Professor Kellogg in his "American Insects" (Kellogg '08, pp. 583-614) and a detailed analysis of the methods of the production of color is given by Professor Tower in his "Colors and Color-Patterns of Coleoptera" (Tower '03).

Following the classification of Tower, the colors of the scales of the Lepidoptera may be either chemical, physical, or chemico-physical. The chemical colors are produced by pigments in the scales; the physical colors are produced either by reflection, refraction, or diffraction of light; and the chemico-physical colors are produced by either a reflecting, refracting, or diffracting structure overlying a layer of pigment. There are also what Tower calls combination colors due to a combination of the causes just mentioned.

As the production of colors by pigments is the most obvious method in nature, it is the one to which the colors of the Lepidoptera are commonly attributed. But it is now well known that a large proportion of the most beautiful colors of these insects are either physical or chemico-physical; this is true of the various metallic and iridescent colors so commonly found in butterflies and many moths.

Explanations of the methods of production of physical colors are given in textbooks on physics; it is, therefore, only necessary here to point out a feature in the structure of the scales of Lepidoptera that results in the production of these colors. This feature is the presence of the fine longitudinal striæ described above. When the striæ are very fine and close together they act in the same way as does a diffraction grating, producing the beautiful iridescent colors. Kellogg ('94) found that on certain scales from a species of *Morpho* the striæ were from .0007 mm. to .00072 mm. apart, or at the rate of about 35,000 to an inch.

The fact that certain colors are due to the way in which light is reflected from the scales can be shown by the following experiment. Place on the stage of a microscope the wing of a bright blue butterfly, and shade the specimen so that it is viewed only by transmitted light from the mirror of the microscope; when examined in this way the blue color will be absent. This is due to the fact that the light passing directly through the scales is not broken up, and only the colors produced by pigment are visible.

There is still another function of the scales of Lepidoptera; they may serve as the outlets of scent glands. As the scales that serve this purpose are found chiefly on the wings of males, they have received the special name of *androconia*, signifying *male dust*. See page 100.

In the suborder Jugatæ and in some of the more generalized families of the suborder Frenatæ, there are, in addition to the more obvious setæ and scales, many very small, hair-like structures, which differ from setæ in being directly continuous with the cuticula, and not connected with it by a joint (Fig. 710); these are termed the *fixed hairs* or *aculeæ*.

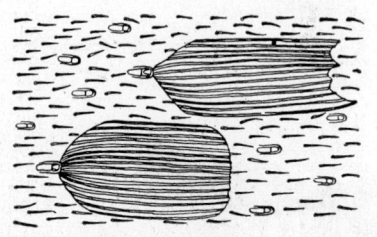

Fig. 710.—Part of a wing of an aculeate moth, with most of the scales removed so as to expose the aculeæ.

They are so small that they can be seen only by the aid of a microscope, and being covered by the scales they can be seen only in bleached and stained or denuded wings.

In the more generalized members of this order, the venation of the wings corresponds quite closely to the hypothetical primitive type. The most striking divergence from this type is the fact that vein M is only three-branched. This is probably due to a coalescence of veins M_4 and Cu_1. If this is true, the vein that is commonly designated as vein Cu_1 is really vein M_4 plus Cu_1; but for the sake of simplicity it seems best to designate it ordinarily as vein Cu_1. For a detailed discussion of this problem, see "The Wing's of Insects," pp. 334-337.

Although the wings of Lepidoptera, except in certain specialized forms, are broadly expanded, there are but few cross-veins, and normally no accessory veins.

In the more specialized members of this order the number of the wing-veins is reduced. This reduction is due in some cases to the atrophy of a vein or veins, as, for example, the loss of the main stem of vein M in many families; in other cases, it is due to the coalescence of adjacent veins, as, for example, the reduction of the number of branches of radius or of media which has taken place in many members of the order.

In many genera of this order the branches of radius of the fore wings anastomose so as to form one or more closed cells; these have been termed *accessory cells*.

There are several methods by which the fore and hind wings of Lepidoptera are held together in flight, in order to insure their synchronous action. In the suborder Jugatæ the posterior lobe of the fore wing functions either as a jugum (see p. 61) or as a fibula (see p. 62). In most moths the wings of each side are united by a frenulum (see p. 61). In some moths and in the skippers and butterflies, the humeral angle of the hind wing is greatly expanded and projects beneath the fore wing; this insures the synchronous action of the two wings and renders a frenulum unnecessary; in these forms, which doubtless descended from frenate ancestors, the frenulum has been lost.

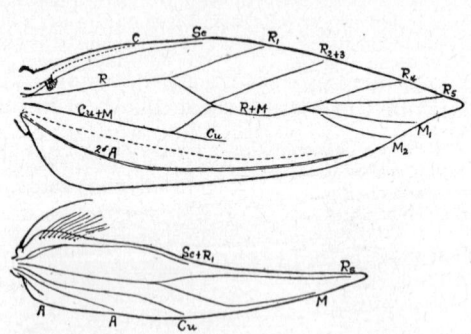

Fig. 711.—Wings of *Obrussa ochrefasciella*, male. (After Braun.)

The frenulum when well developed consists of a bunch of bristles situated at the base of the costa of the hind wings, on the costal sclerite. As a rule these bristles are separate in females, and consolidated into a single strong, spine-like organ in males.

In some of the more generalized Lepidoptera there is a series of slightly curved, spine-like setæ on the costa of the hind wing near the base, which aid in holding the wings together. These setæ lie beyond the costal sclerite, not on it as does the frenulum; they are termed by Braun ('19) the *costal spines*. The frenulum and costal spines are both present in some moths (Fig. 711).

In many moths, and especially in the Noctuidæ, the fore wings are marked by transverse lines or bands, and by spots that are so uniform in position in different species that they have been given names, which are used to designate them in the descriptions of those species in which they occur. Figure 712 is a diagram of a fore wing of a noctuid moth indicating the positions

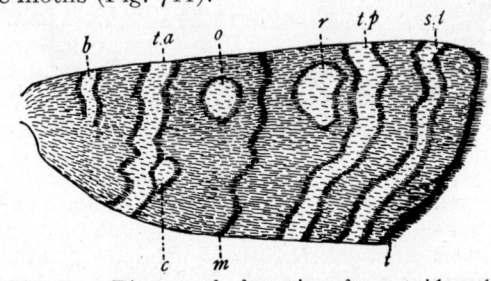

Fig. 712.—Diagram of a fore wing of a noctuid moth. The lettering is explained in the text. (After Crosby and Leonard.)

of the named lines or bands and spots. Six transverse lines or bands and three spots have been named, as follows:

The basal or subbasal band (Fig. 712, *b*).—This is a band extending halfway across the wing near its base.

The transverse anterior band (Fig. 712 *t. a*).—This is often designated as the *t. a. line*; in some English books it is termed the *first line*.

The median line (Fig. 712, *m*).

The transverse posterior band (Fig. 712, *t. p*).—This is often designated as the *t. p. line*; it is the *second line* of English authors.

The subterminal band (Fig. 712, *s. t*).

The terminal band (Fig. 712, *t*).

The orbicular or round spot (Fig. 712, *o*).—This is a round or oval spot situated in the discal cell.

The reniform spot (Fig. 712, *r*).—This is a somewhat kidney-shaped spot at the end of the discal cell.

The claviform spot (Fig. 712, *c*).—An elongate spot extending from the *t. a.* line toward the *t. p.* line in cell Cu.

The typical mouth-parts of adult Lepidoptera are fitted for sucking. In some families, the members of which do not take food during the adult stadium, the mouth-parts are vestigial; and in one family, the Micropterygidæ, which is doubtfully included in this order, the mouth-parts are of the mandibular type.

In those families in which the typical form of the mouth-parts is well shown, the only parts of these organs that are well developed are the maxillæ and the palpi, the other parts being either absent or

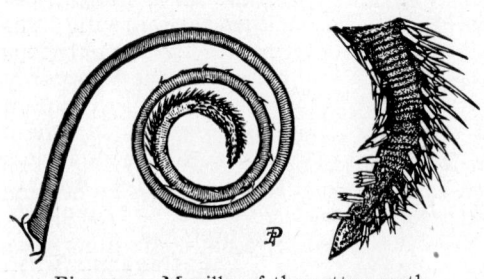

Fig. 713.—Maxillæ of the cotton-moth, and the tip of the same enlarged.

reduced to mere vestiges. When only one pair of palpi are developed they are the labial palpi; when maxillary palpi are present they can be distinguished by their attachment to the maxillæ.

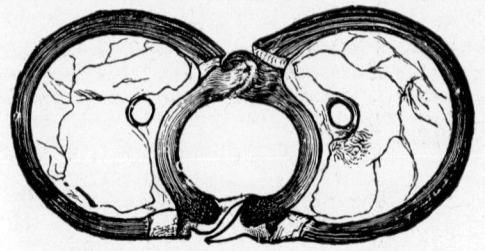

Fig. 714.—Cross-section of maxillæ.

If the head of a butterfly or of a moth in which the mouth-parts are not vestigial be examined, there will be found a long sucking-tube which when not in use is coiled on the lower side of the head between two forward-projecting appendages. This long sucking-tube is composed of the two maxillæ, greatly elongated, and fastened together side by side. In Figure 713 there is represented a side view of the maxillæ of a moth; and in Figure 714 a cross-section of these organs. Each maxilla is furnished with a groove, and the two maxillæ are so fastened together that the two grooves form a tube through which the liquid food is sucked. As a rule the maxillæ of insects of this order are merely fitted for extracting the nectar from flowers, but sometimes the tips of the maxillæ are armed with spines, as shown in Figure 713. This enables the insect to lacerate the tissue of ripe fruits and thus set the juice free, which is then sucked up. Many moths do not eat in the adult state; with these the maxillæ are wanting. The two forward-projecting organs between which the maxillæ are coiled when present are the labial palpi. In some moths the maxillary palpi are also developed.

The compound eyes are large and are composed of many small ommatidia. The ocelli, when present, are two in number; they are situated one on each side, above the compound eye and near its margin; the median ocellus is lacking throughout the order; and in the butterflies, the skippers, and some families of moths, all of the ocelli are wanting.

The antennæ are always conspicuous; they differ greatly in form in different divisions of the order, and, therefore, furnish characters that are much used in the classification of these insects. In some families the basal segment of the antennæ is greatly enlarged and forms what has been termed the *eye-cap*.

The prothorax is small, being reduced to a collar between the head and the wing-bearing segments. In many of the more specialized Lepidoptera the pronotum is produced on each side into a flat lobe which in some cases is even constricted at the base so as to become a stalked plate; these lobes are the *patagia*.

The legs are long and slender. In some families the front tibiæ bear on their inner aspect a mobile pad; this is termed the *epiphysis;* in some cases, at least, it is a combing organ used for cleaning the antennæ.

A special feature of the abdomen is the presence in the female of a bursa copulatrix; that of the female of the milk-weed butterfly is figured on page 160.

Close to the junction of the thorax and abdomen there is, in the majority of Lepidoptera, a pair of organs, which are known as the *tympana*. These are situated on each side near the first abdominal spiracle. Several types of these organs have been described by Forbes ('16) and by Eggers ('19), which are characteristic of certain families and groups of families.

The first type is that of the Geometridæ; it appears superficially as a hollow bulla located immediately below the spiracle, opening forward against the coxa of the hind leg. Some Pyralidæ have rudimentary tympana in the same position.

The second type is likewise wholly on the abdomen, but it is higher on the body, and its opening faces backward towards the second abdominal segment. It characterizes the Thyatiridæ and Drepanidæ.

The third type presents a variety of appearances. Its essential part is a membranous disk, the tympanum proper, on the metathorax just below the root of the wing. In the Dioptidæ, Notodontidæ, Agaristidæ, and a few noctuids and lithosians, the disk lies exposed or is merely sunk in a pit at the junction of the thorax and abdomen. In other moths having this type of tympana the disk is covered by a hood formed by the side of the first segment of the abdomen; in the Arctiidæ, Pericopidæ, Liparidæ, and the subfamily Herminiinæ of the Noctuidæ, this hood lies subdorsally, wholly above the spiracle; while in the majority of the Noctuidæ it is lower and incloses the spiracle, in some cases (Euteliinæ, etc.) being supplemented by a second hood formed by the alula of the hind wing.

The function of the tympana is probably auditory, as Eggers has described chordotonal organs in connection with them in several families.

In the Lepidoptera the metamorphosis is complete. The larvæ are known as caterpillars; they vary greatly in form and appearance, but are usually cylindrical, and provided with from ten to sixteen legs, —six thoracic legs and from four to ten abdominal legs. The thoracic legs have a hard external skeleton; and are jointed, tapering, and armed at the end with a little claw. The abdominal legs, which are shed with the last larval skin, are thick, fleshy, without joints, elastic or contractile, and armed at the extremity with numerous minute hooks (Fig. 715);

Fig. 715.—Larva of a hawk-moth.

they are termed prolegs. When all five pairs are present they are borne by the third, fourth, fifth, sixth, and tenth abdominal segments.

The hooks or crotchets with which the prolegs of caterpillars are armed vary in their arrangement in different families and thus afford useful characters for the classification of these larvæ. These hooks are usually arranged in a circle or in rows on the tip of the proleg. When they are in a single row or series, they are said to be *uniserial;* when in two concentric rows, *biserial;* when in several rows, *multiserial.* When the hooks of a row are uniform in length throughout or shorter towards the ends of the row, they are said to be *uniordinal;* when they are of two alternating lengths, *biordinal;* when of several lengths, *multiordinal.* The tip of a proleg on which the hooks or crochets are borne is termed the *planta.*

In most lepidopterous larvæ the clothing of setæ is comparatively inconspicuous; such larvæ are commonly termed *naked* in contradistinction to the hairy caterpillars. But in the so-called naked larvæ, each segment of the body, when not too highly specialized, is armed with a definite number of setæ which occupy definite positions. Each seta is borne on a small chitinous tubercle; the number of these *setiferous tubercles* and the positions they occupy differ in the different families, and, therefore, afford characters which are much used in the classification of Lepidoptera.

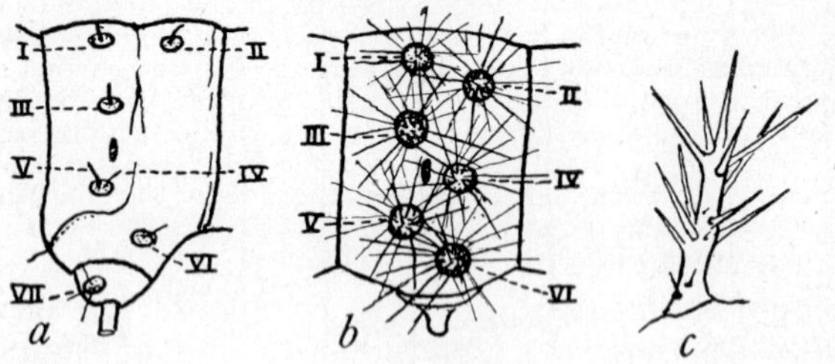

Fig. 716.—Types of setiferous tubercles. (*a* and *b* from Dyar.)

The small tubercle bearing a single seta (Fig. 716, *a*) is evidently the primitive form of setiferous tubercle; for it is the only form found in the more generalized families. In some of the more specialized families the tubercles are larger and many-haired (Fig. 716, *b*); this type of tubercles is termed a *verruca;* it is characteristic of the so-called hairy caterpillars, as, for example, the larvæ of most of the Arctiidæ. In the larvæ of the Saturnioidea and of certain butterflies, some of the tubercles are spinose projections of the body-wall (Fig. 716, *c*); such a projection is termed a *scolus.*

Some caterpillars are clothed with more or less numerous setæ which are scattered and which have no constant position; such setæ are termed *secondary setæ*, in contradistinction to those borne on setiferous tubercles which are of a definite number and occupy definite positions; these are termed *primary setæ.* Among the setiferous tubercles that are constant in position, there are a few that are not present in the first instar of generalized groups; although the setæ borne by these tubercles are regarded as primary setæ when contrasted with secondary setæ, they are distinguished from those found in the first instar as *subprimary setæ.*

In order to make use of the primary and subprimary setæ in classification, it is necessary that each of these setæ should be designated by a distinctive term. The terminology most generally used is that proposed by Dyar ('94), who was the first author to base a classification of lepidopterous larvæ on the variations in the arrangement of the setiferous tubercles.

The terminology of Dyar was based on a study of the tubercles of the abdominal segments. He recognized on each side of each abdominal segment, except the last two, eight tubercles, which he numbered with Roman numerals beginning with the one nearest the middle line of the back; the number VII was applied to a group of three tubercles on the outside of the proleg, or in a corresponding position in the legless abdominal segments. Subsequent studies, and especially those by Forbes ('10) and Fracker ('15), have revealed the presence of setiferous tubercles not numbered by Dyar. Figure 717,b, represents the arrangement of the tubercles of a middle abdominal segment of a noctuid larva as figured by Forbes. The tubercles are numbered according to the terminology of Dyar, with the addition of tubercles X, IIIa, and IX, not figured by Dyar.*

The arrangement of the setiferous tubercles on the thoracic segments of any caterpillar differs to a considerable extent from that on the abdominal segments of the same insect. In Figure 717, a represents the arrangement of the tubercles on the metathorax and b that of the tubercles of a middle abdominal segment of a noctuid larva as figured by Forbes ('10). This writer also figures and numbers the setæ on the head of a caterpillar.

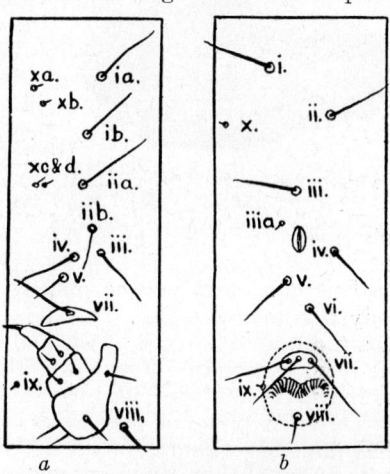

Fig. 717.—Arrangement of setiferous tubercles in a noctuid larva: a, tubercles of a metathorax; b, tubercles of a middle abdominal segment. (After Forbes.)

Fracker ('15) made an extended study of the classification of lepidopterous larvæ, which was based quite largely on the variations in the number and positions of the setiferous tubercles; and his paper is illustrated by a large number of setal maps. This writer proposes a new terminology for the setæ, using Greek letters instead of Roman numerals.

*In diagrams indicating the arrangement of setiferous tubercles, one side of a single segment is represented as if cut on the mid-dorsal and mid-ventral lines, and laid flat. The anterior edge is to the left, and the mid-dorsal line at the upper edge. In Figure 717 the positions of the spiracle and of the legs are also indicated.

Schierbeck ('16 and '17) proposes still another terminology for the setæ, applying a Latin name to each.

Most caterpillars, except, as a rule, the larvæ of butterflies, spin a cocoon. In some instances, as in the case of silk-worms, a great amount of silk is used in the construction of the cocoon; in others the cocoon is composed principally of the hairs of the larva, which are fastened together with a fine web of silk.

The pupæ of the Lepidoptera are typically of the obtected type; that is, the developing wings, legs, mandibles, maxillæ, and antennæ are glued to the surface of the body (Fig. 718); but in some of the more generalized forms these appendages are free. In the Microjugatæ, which are provisionally included in this order, these appendages are free, the pupæ resembling those of the Trichoptera; but in the Hepialidæ the appendages are glued to the surface of the body as in the specialized Frenatæ. In some of the more generalized Frenatæ, as the Nepticulidæ, and in the Heliozelidæ, the appendages are all free; between this condition and that of the truly obtected pupa of the more specialized Frenatæ, various intergrades exist.

Fig. 718.—Pupa of a moth.

The pupæ of this order vary also in the number of segments of the body that are movable. The eighth, ninth, and tenth abdominal segments are always fixed. All of the other segments are movable in the most generalized forms, and all are fixed in the most specialized forms; there are various intergrades between these two extremes.

Different pupæ of this order differ also in various other ways, thus affording characters that are of taxonomic importance. It is only recently that these characters have been used in an extended manner. A pioneer paper in this field is that of Miss Edna Mosher ('16).

More than nine thousand species of Lepidoptera are known to occur in America north of Mexico. These represent two suborders and seventy families.

In popular language the Lepidoptera includes two quite distinct groups of insects, the moths and the butterflies. Under the term *moths* are included all of the members of the first suborder, the Jugatæ, and the larger number of the families of the second suborder, the Frenatæ; under the term *butterflies* are included the remaining families of the suborder Frenatæ. These two groups are distinguished as follows.

The moths.—These are the insects that are commonly called *millers*. Most of the species fly by night and are frequently attracted to lights. When at rest the wings are either wrapped around the body, or spread horizontally, or folded roof-like on the abdomen; except in a few cases they are not held in a vertical position above the body. The antennæ of moths are of various forms; they are usually thread-like or feather-like; only in rare cases are they enlarged towards the tip. The moths have been termed the Heterocera* by many entomological writers, in contradistinction to Rhopalocera,* a term applied to the butterflies.

The butterflies.—All of our species of butterflies fly in the daytime; and, with few exceptions, they fold the wings together above the back in a vertical position when at rest. The antennæ are thread-like, and usually with a club at the tip. It was this feature that suggested the term Rhopalocera, which is applied to them.

The group butterflies as defined here includes the representatives of two quite distinct superfamilies, the Hesperioidea or skippers, and the Papilionoidea or true butterflies. The distinctive characters of these two superfamilies are discussed later.

The division of the Lepidoptera into moths and butterflies is an artificial one, the group moths including representatives of both of the two suborders into which the order is divided, as indicated above. In the natural classification, the primary division of the order is based on differences in the method of uniting the two wings of each side, and on differences in the venation of the hind wings. In one suborder, the Jugatæ, the posterior lobe of the fore wing is specialized so as to form an organ, a jugum or a fibula, which unites the fore and hind wings; and the venation of the hind wings is similar to that of the fore wings. In the other suborder, the Frenatæ, the two wings of each side are united by a frenulum in the more generalized forms and by a substitute for a frenulum in certain specialized forms; and the venation of the hind wings is quite different from that of the fore wings.

Hübner's Tentamen.—At some undetermined date, but previous to 1810 and probably in 1806, Jacob Hübner distributed a two-page work, giving a classification of the Lepidoptera. This work is commonly known as "Hübner's Tentamen," *tentamen* being the first word in its long Latin title. Entomologists differ regarding the standing of this work; some believe that it was merely privately printed, while others regard it as a published work and adopt the generic names that were used in it. This difference of opinion is the cause of serious confusion in the names of certain genera and families. It seems to the writer that the evidence supporting the view that the "Tentamen" was published is conclusive. See "Entomologists Record and Journal of Variation," Vol. 31 (1919), Supplement.

SYNOPSIS OF THE LEPIDOPTERA

The families comprising this order are grouped in various ways by different writers; none of these groupings can be regarded as final in the present state of our knowledge. The following provisional arrangement has been adopted for use in this book.

A. THE JUGATE LEPIDOPTERA.—Moths in which the two wings of each side are united by a jugum or by a fibula. p. 592............. SUBORDER JUGATÆ
  B. THE MICROJUGATÆ.
    C. The Mandibulate Jugates. p. 592...........Family MICROPTERYGIDÆ
    CC. The Haustellate Jugates. p. 593...........Family ERIOCRANIIDÆ
  BB. THE MACROJUGATÆ.
    The Swifts. p. 594.................................Family HEPIALIDÆ

---

*Heterŏcera: *hetero* (ἕτερος), other, different; *ceras* (κέρας), a horn.
*Rhopalŏcera: *rhopalon* (ῥόπαλον), a club; *ceras* (κέρας), a horn.

AA. The Frenate Lepidoptera.—Moths, skippers, and butterflies in which the two wings of each side are united by a frenulum or by its substitute, a large humeral area of the hind wings. p. 596................Suborder Frenatæ
  B. The Generalized Frenatæ.—Moths that are supposed to retain more nearly than other Frenatæ the form of the primitive Frenatæ, those that were the first to appear on earth.
    C. The Aculeate Frenatæ.—Moths in which the aculeæ are distributed over the general surface of the wings.
      The Incurvariids. p. 598.....................Family Incurvariidæ
      The Nepticulids. p. 600......................Family Nepticulidæ
    CC. The Non-aculeate Generalized Frenatae.—Moths in which the aculeæ are confined to small areas of the wings or are absent.
      The Carpenter Moths. p. 601....................Family Cossidæ
      The Smoky Moths, p. 604...................Family Pyromorphidæ
      The Dalcerids. p. 605.........................Family Dalceridæ
      The Flannel-moths. p. 606................Family Megalopygidæ
      The Slug-caterpillar-moths. p. 608................ Family Eucleidæ
      The Epipyropids. p. 610......................Family Epipyropidæ
  BB. The Specialized Frenatæ.—Moths, skippers, and butterflies that depart more widely than do the Generalized Frenatæ from the primitive type of Lepidoptera, being more highly modified for special conditions of existence. An indication of the specialized condition of these insects is the modified form of the wings. In nearly all the base of vein M has been lost and the branches of this vein joined to veins R and Cu.
    C. The Specialized Microfrenatæ.—Frenulum-bearing moths which are usually of small, often of minute, size. In many of these moths the anal area of the hind wings is not reduced, having three anal veins; in some others the hind wings are very narrow and a broad fringe acts as a substitute for the membrane of the anal area.
      The Acrolophids. p 611........................Family Acrolophidæ
      The Tineids. p. 611..............................Family Tineidæ
      The Bag-worm Moths. p. 613.....................Family Psychidæ
      The Tischeriids. p. 615........................Family Tischeriidæ
      The Lyonetiids. p. 616........................Family Lyonetiidæ
      The Opostegids. p. 617.......................Family Opostegidæ
      The Oinophilids. p. 617......................Family Oinophilidæ
      The Gracilariids. p. 617......................Family Gracilariidæ
      The Coleophorids. p. 620....................Family Coleophoridæ
      The Elachistids. p. 621........................Family Elachistidæ
      The Heliozelids. p. 622........................Family Heliozelidæ
      The Douglasiids. p. 623.......................Family Douglasiidæ
      The Œcophorids. p. 624....................Family Œcophoridæ
      The Ethmiids. p. 625..........................Family Ethmiidæ
      The Stenomids. p. 625........................Family Stenomidæ
      The Gelechiids. p. 625........................Family Gelechiidæ
      The Blastobasids. p. 628....................Family Blastobasidæ
      The Cosmopterygids. p. 629.............Family Cosmopterygidæ
      The Scythridids. p. 631.......................Family Scythrididæ
      The Yponomeutids. p. 631...................Family Yponomeutidæ
      The Plutellids. p. 632........................Family Plutellidæ
      The Glyphipterygids. p. 633..............Family Glyphipterygidæ
      The Heliodinids. p. 634.......................Family Heliodinidæ
      The Clear-winged Moths. p. 634................... Family Ægeriidæ
      Superfamily Tortricoidea
        The Olethreutids. p. 639...................Family Olethreutidæ
        The Typical Tortricids. p. 642..............Family Tortricidæ
        The Phaloniids. p. 643......................Family Phaloniidæ
        The Carposinids. p. 644..................Family Carposinidæ
    CC. The Pyralids and Their Allies
      Superfamily Pyralidoidea
        The Pyralids. p. 644.........................Family Pyralididæ
        The Plume-moths. p. 652.................Family Pterophoridæ

The Many-plume Moths. p. 653.............Family ORNEODIDÆ
The Window-winged Moths. p. 653...........Family THYRIDIDÆ
The Hyblæids. p. 655......................Family HYBLÆIDÆ

CCC. THE SPECIALIZED MACROFRENATÆ.—Specialized Frenatæ which are usually of medium or large size. This division includes certain moths and all skippers and butterflies. In these insects the anal area of the hind wings is reduced, containing only one or two anal veins.

D. *The Frenulum-conservers.*—Specialized Macrofrenatæ in which the two wings of each side are typically united by a frenulum; but in some highly specialized genera of some families (Sphingidæ, Geometridæ, and Drepanidæ) the supplanting of the frenulum by an expanded humeral angle of the hind wing is either far advanced or complete. This group of families includes only moths.

The hawk-moths or sphinxes. p. 655..........Family SPHINGIDÆ
Superfamily GEOMETROIDEA
    The Geometrids. p. 663.................Family GEOMETRIDÆ
    The Manidiids. p. 673..................Family MANIDIIDÆ
THE NOCTUIDS AND THEIR ALLIES
    The Dioptids. p. 673...........................Family DIOPTIDÆ
    The Prominents. p. 674..................Family NOTODONTIDÆ
    The Tussock-moths. p. 679..............Family LYMANTRIIDÆ
    The Noctuids. p. 683.......................Family NOCTUIDÆ
    The Foresters. p. 697.....................Family AGARISTIDÆ
    The Pericopids. p. 698...................Family PERICOPIDÆ
    The Arctiids. p. 699.......................Family ARCTIIDÆ
    The Euchromiids. p. 706................Family EUCHROMIIDÆ
The Eupterotids. p. 707....................Family EUPTEROTIDÆ
The Epiplemids. p. 708.....................Family EPIPLEMIDÆ
The Thyatirids. p. 709.....................Family THYATIRIDÆ
The Drepanids. p. 710.....................Family DREPANIDÆ

DD. *The Frenulum-losers.*—Specialized Macrofrenatæ, in which the frenulum has been supplanted by a greatly extended humeral area of the hind wings. In some of the more generalized forms a vestigial frenulum persists (Bombycidæ and Lacosomidæ). This division includes three groups of families: the Frenulum-losing moths, the skippers, and the butterflies. The grouping together of the families included in this division is merely provisional, as doubtless the loss of the frenulum has arisen independently several times.

E. *The Frenulum-losing Moths.*—In these moths the antennæ are usually pectinate; they are never enlarged into a club at the tip.

The Lacosomids. p. 712......................Family LACOSOMIDÆ
Superfamily SATURNIOIDEA
    The Royal-moths. p. 715................Family CITHERONIIDÆ
    The Giant Silk-worms. p. 719.............Family SATURNIIDÆ
The Silk-worms. p. 727....................Family BOMBYCIDÆ
The Lasiocampids. p. 728...............Family LASIOCAMPIDÆ

EE. *The Skippers.*—These are day-flying Lepidoptera which resemble butterflies in usually holding their wings erect when at rest, but are distinguished by the peculiar venation of the fore wings, vein R being five-branched, and all of the branches arising from the discal cell. The antennæ are enlarged into a club towards the tip.

Superfamily HESPERIOIDEA
    The Giant Skippers. p. 733..............Family MEGATHYMIDÆ
    The Common Skippers. p. 734............Family HESPERIIDÆ

EEE. *The Butterflies.*—Day-flying Lepidoptera that hold their wings erect when at rest, that have clubbed antennæ, and that differ from the skippers in the venation of the fore wings, some of the branches of vein R coalescing beyond the discal cell.

Superfamily PAPILIONOIDEA
    The Swallow-tails and the Parnassians. p. 740 Family PAPILIONIDÆ

The Pierids. p. 744.........................Family PIERIDÆ
The Four-footed Butterflies. p. 750.......Family NYMPHALIDÆ
The Metal-marks. p. 767....................Family RIODINIDÆ
The Gossamer-winged Butterflies. p. 768.....Family LYCÆNIDÆ

TABLES FOR DETERMINING THE FAMILIES OF LEPIDOPTERA

TABLE A

A. Wingless or with vestigial wings. This division includes only females. All males of Lepidoptera are winged.
  B. The larvæ case-bearers; the adult female either remains within the case to lay her eggs, or leaves the case and sits on the outside of it. p. 613. PSYCHIDÆ
  BB. The larvæ not case-bearers; the wingless adult not in a case.
    C. The adult female remains upon her cocoon to lay her eggs; the body of the adult is clothed with fine hairs. p. 679.............. LYMANTRIIDÆ
    CC. The adult female is active and lays her eggs remote from her cocoon; the body of the adult is closely scaled, or spined, or with bristling dark gray hair. p. 663...GEOMETRIDÆ
    CCC. In addition to the above there are some arctic species of the Noctuidæ and of the Arctiidæ in which the wings of the females are vestigial.
AA. With well-developed wings.
  B. Fore and hind wings similar in form and venation, the radius of the hind wings being, like that of the fore wings, five-branched (Suborder JUGATÆ).
    C. Minute moths resembling tineids in appearance.
      D. Adult moths with well-developed functional mandibles; subcosta of the fore wings forked near its middle. p. 592..........MICROPTERYGIDÆ
      DD. Mandibles of the adult vestigial; maxillæ formed for sucking; subcosta of fore wings forked near its apex. p. 593......ERIOCRANIIDÆ
    CC. Moths of medium or large size, without functional mouth-parts. p. 594 ....HEPIALIDÆ
  BB. Fore and hind wings differing in form and venation; the radial sector of the hind wings being unbranched, and vein R_1 of the hind wings usually coalesced with vein Sc (Suborder FRENATÆ).
    C. Antennæ of various forms, but rarely clubbed as in the skippers and butterflies; if the antennæ are clubbed the hind wings bear a frenulum.
      D. The fringe on the inner angle of the hind wings as long as, or longer than, the width of the wing; the hind wings often lanceolate, but never fissured. (Microfrenatæ.) Pass to Table B.
      DD. Hind wings much broader than their fringe, and not lanceolate.
        E. Wings fissured deeply.
          F. Each wing divided into six lobes. p. 653........ORNEODIDÆ
          FF. Wings never more than four-lobed; usually the fore wings. bilobed and the hind wings trilobed. p. 652...... PTEROPHORIDÆ
        EE. Wings not fissured or the front wings slightly fissured.
          F. Fore wings very narrow, the width at the middle less than one-fourth the length of the wing; a considerable part of the hind wings, and in many cases of the fore wings also, free from scales; inner margin of fore wings and costal margin of hind wings with a series of recurved and interlocking spines. p. 634.....ÆGERIIDÆ
          FF. Wings scaled throughout, or if clear with the fore wings triangular in outline; wings not interlocking at middle with series of recurved spines.
            G. With a double series of enlarged and divergent scales along vein Cu of the hind wings below; wings, body, and legs very long. (*Agdistis*.) p. 652...................PTEROPHORIDÆ
            GG. Without such scales on vein Cu of the hind wings.
              H. Hind wings with three anal veins. Care must be taken not to mistake a mere fold in the wing for a vein. When there is no thickening of the membrane of the wing along a fold, it is not counted as a vein.

I. Veins Sc + R₁ and Rs of the hind wings grown together for a greater or less distance between the apex of the discal cell and the apex of the wing, or in some cases separate but very closely parallel. p. 644...................... PYRALIDIDÆ
II. Veins Sc + R₁ and Rs of the hind wings widely separate beyond the apex of the discal cell.
  J. The fringe on the anal angle of the hind wings considerably longer than elsewhere (sometimes not obviously so in rubbed specimens); the spurs of the tibiæ more than twice as long as the width of the tibiæ. (Microfrenatæ.) Pass to Table B.*
  JJ. The fringe on the anal angle of the hind wings not longer than elsewhere or but slightly so; the spurs of the tibiæ about as long as the width of the tibiæ.
    K. Veins Sc + R₁ and Rs of the hind wings grown together to near the end of the discal cell (Fig. 734), or anastomosing beyond the middle of the cell (Fig. 730).
      L. Small moths, chiefly of a smoky black color, with thinly scaled wings. p. 604........ PYROMORPHIDÆ
      LL. Moths of medium size, and densely clothed with long, woolly hairs, which are light-colored or brown. p. 606........................... MEGALOPYGIDÆ
    KK. Veins Sc + R₁ and Rs of the hind wings separate or grown together for only a short distance.
      L. 1st and 2d anal veins of the fore wings united by a cross-vein.
        M. Accessory cell present (Hypoptinæ). p. 603. .... COSSIDÆ
        MM. Accessory cell absent. p. 613... PSYCHIDÆ
      LL. 1st and 2d anal veins not united by a cross-vein.
        M. Vein M₂ of the fore wings arising from the discal cell nearly midway between veins M₁ and M₃.
          N. Vein M₃ of both fore and hind wings coalesced with vein Cu₁ for a considerable distance beyond the end of the discal cell. p. 673.... DIOPTIDÆ
          NN. Veins M₃ and Cu₁ not coalesced beyond the end of the discal cell.
            O. Veins R₂ and R₃ coalesced at base, but separate from veins R₄ and R₅, which also coalesce at base. p. 712...... LACOSOMIDÆ
            OO. Veins R₂, R₃, R₄, and R₅ united at base. p. 727...................... BOMBYCIDÆ
        MM. Vein M₂ of the fore wings emerging from the discal cell nearer to cubitus than to radius, causing cubitus to appear four-branched.
          N. Fore wings with an accessory cell.
            O. Moths with heavy, spindle-shaped bodies, and narrow, strong wings. p. 601. COSSIDÆ
            OO. Moths in which the body is slender and the wings are ample.
              P. Wings ample (fore wings not half longer than wide); mouth-parts vestigial. p. 605. .... DALCERIDÆ
              PP. Wings more or less oblong, usually twice as long as wide; mouth-parts usually developed with scaled tongue. (Microfrenatæ.) Pass to Table B.
          NN. Fore wings without an accessory cell.

---

*A few of the Eucleidæ present these characters; but with these moths the wings are broad and the base of media extends through the middle of the discal cell.

O. With some of the branches of radius of the fore wings coalesced beyond the apex of the discal cell. p. 608...............EUCLEIDÆ
OO. With each of the five branches of radius of the fore wings arising from the discal cell. p. 655........................HYBLÆIDÆ

HH. Hind wings with less than three anal veins.
 I. Fore wings with two distinct anal veins or with the anal veins partly grown together so as to appear as a branched vein.
  J. Anal veins of fore wings partly grown together so as to appear as a branched vein. p. 613..........PSYCHIDÆ
  JJ. Fore wings with two distinct anal veins (*Harrisina*). p. 605.............................PYROMORPHIDÆ
 II. Fore wings with a single fully preserved anal vein. This is the second anal vein; the first anal vein is absent or represented merely by a fold; and the third anal vein is short, not reaching to the margin of the wing, or is wanting; usually when the third anal vein is present it is joined to the second anal vein, so that the latter appears to be forked towards the base.
  J. Frenulum present. In most cases, the humeral angle of the hind wings is not greatly expanded.
   K. The five branches of radius and the three branches of media of the fore wings all present, and each one arising separate from the discal cell. p. 653.......THYRIDIDÆ
   KK. With some of the branches of radius of the fore wings stalked, or else with some branches coalesced to the margin of the wing.
    L. The fringe on the anal angle of the hind wings considerably longer than elsewhere.
     M. Veins Sc and R of the hind wings seperate, but usually connected by a more or less distinct basal part of vein R_1. (Microfrenatæ.) Pass to Table B.
     MM. Veins Sc and R of the hind wings fused for a greater or less distance.
      N. Ocelli present. p. 683..........NOCTUIDÆ
      NN. Ocelli absent. p. 704........LITHOSIINÆ
    LL. The fringe on the anal angle of the hind wings not considerably longer than elsewhere.
     M. The basal part of vein R_1 of the hind wings, the part extending from radius to the subcosta, appearing like a cross-vein which is as stout as the other veins; veins Sc $+$ R_1 closely parallel to the end of the discal cell or beyond. p. 655. SPHINGIDÆ
     MM. The basal part of vein R_1 of the hind wings rarely appearing like a stout cross-vein; when it does appear like a cross-vein, veins Sc $+$ R_1 and R_s strongly divergent from the point of union of veins R_1 and Sc.
      N. Vein M_2 of the fore wings not more closely joined to cubitus than to radius, cubitus being apparently three-branched.
       O. The basal part of the subcosta of the hind wings extending from the base towards the apex of the wing in a regular curve.
        P. Vein M_2 of the hind wings arising nearer to cubitus than to radius; vein M_1 of the hind wings joined to radius before the apex of the discal cell. p. 709..........THYATIRIDÆ
        PP. Vein M_2 of the hind wings either wanting

or present, but when present arising either midway between radius and cubitus, or nearer to radius than to cubitus; vein M_1 of the hind wing joined to radius at or beyond the apex of the discal cell.
   Q. Tongue (maxillæ) wanting; fore wings with $R_2 + {}_3$ and $R_4 + {}_5$ stalked together, northern species with hyaline dots on fore wings. p. 707.....................EUPTEROTIDÆ
   QQ. Tongue present, often weak; fore wings fully scaled; usually with accessory cell, or with veins R_3 and R_4 stalked together. p. 674.....................NOTODONTIDÆ
OO. The basal part of the subcosta of the hind wings joined to radius for a short distance and then making a prominent bend towards the costal margin (Fig. 909). (See also OOO.)
   P. Veins R_3 and R_4 of the fore wings widely separated from each other, stalked respectively with R_2 and R_5. p. 712................
   ....LACOSOMIDÆ
   PP. Veins R_3 and R_4 long-stalked with each other, widely separated from R_5 which is stalked with M_1. p. 708......ÉPIPLEMIDÆ
OOO. The basal part of the subcosta of the hind wings making a prominent bend into the humeral area of the wing, and usually connected to the humeral angle by a strong crossvein (Fig. 817).
   P. Antennæ clubbed. p. 673....MANIDIIDÆ
   PP. Antennæ not clubbed. p. 663.............
   ....GEOMETRIDÆ
NN. Vein M_2 of the fore wings more closely joined to cubitus than to radius; cubitus being in most cases apparently four-branched.
O. Small moths, with the apex of the fore wings sickle-shaped. p. 710..............DREPANIDÆ
OO. Apex of the fore wings not sickle-shaped.
   P. Vein Sc of the hind wings apparently absent, being fused except at the extreme base with radius. Care should be taken not to mistake vein M_1 for radius (see Fig. 897). p. 706.....................EUCHROMIIDÆ
   PP. Veins Sc and R of the hind wings distinct and parallel to the point where vein R separates from the discal cell, and then approaching very close or fusing for a short distance. (See also PPP.)
      Q. Small moths with snow-white wings (*Eudeilinia*). p. 710............DREPANIDÆ
      QQ. Moths that are not white.
         R. Vein R_5 of the fore wings stalked with veins R_3 and R_4 (Chrysauginæ). p. 644.
         ....PYRALIDIDÆ
         RR. Vein R_5 free (*Meskea*). p. 653.....
         ....THYRIDIDÆ
   PPP. Veins Sc and R of the hind wings not as described under PP above.
      Q. Antennæ more or less thickened towards the tip. p. 697..............AGARISTIDÆ

QQ. Antennæ not clubbed.
R. Dorsal surface of the first abdominal segment with two prominent rounded bosses, the hoods of the tympana. These hoods are wholly above the spiracles, and separated by only about one-third of the width of the abdomen. Black moths with white or yellow bands or spots on the wings and often with metallic tints. Found only in the Far West or in the Gulf States. p. 698. ....PERICOPIDÆ
RR. Hoods of the tympana less conspicuous dorsally and more widely separated.
S. Veins Sc and R of the hind wings extending separate, or the two joined for a short distance near the base of the wing; ocelli present.
T. White or yellow species, with palpi not reaching the middle of the smooth-scaled front; vein Cu apparently four-branched in both fore and hind wings (*Haploa*). p. 700................ARCTIIDÆ
TT. Species with longer palpi, and vein Cu of the hind wings apparently three-branched, or species of a gray ground color. p. 683. ....NOCTUIDÆ
SS. Veins Sc and R of the hind wings fused or closely parallel near the middle of the discal cell, or connected by a short cross-vein (the free part of vein R_1); ocelli absent. (See also SSS.) p. 679. LYMANTRIIDÆ
SSS. Veins Sc and R of the hind wings united for one-fifth or more of the length of the discal cell.
T. Ocelli present (Arctiinæ). p. 700. ....ARCTIIDÆ
TT. Ocelli absent.
U. Fore wings with raised tufts of scales (Nolinæ). p. 705. ARCTIIDÆ
UU. Fore wings smoothly scaled.
V. Vein M_2 of the hind wings well developed and arising slightly nearer to vein M_3 than to vein M_1 (*Menopsimus*). p.683 ....NOCTUIDÆ
VV. Vein M_2 of the hind wings arising much nearer to vein M_3 than to vein M_1, or wanting (Lithosiinæ). p. 704. ....ARCTIIDÆ
JJ. Frenulum absent.
K. Vein Cu of both fore and hind wings apparently four-branched.
L. Small moths with slender bodies, and with the apex of the fore wings sickle-shaped; humeral veins absent. p. 710..........................DREPANIDÆ

LEPIDOPTERA

LL. Moths of various sizes, but with robust bodies, and with the apex of the fore wings not sickle-shaped; hind wings with humeral veins. p. 728.LASIOCAMPIDÆ
KK. Vein Cu of both fore and hind wings apparently three-branched.
L. Robust moths of medium or large size, with strong wings. p. 714......................SATURNIOIDEA
LL. Small moths with slender bodies and weak wings (*Dyspteris*). p. 667.................GEOMETRIDÆ
CC. Antennæ thread-like with a knob at the extremity; hind wings without a frenulum; ocelli wanting.
D. Radius of the fore wings five-branched, and with all the branches arising from the discal cell; club of antennæ usually terminated by a recurved hook. *The skippers*. p. 732....................HESPERIOIDEA
DD. With some of the branches of radius of the fore wings coalesced beyond the apex of the discal cell; club of antennæ not terminated by a recurved hook. *The Butterflies*. p. 739...............PAPILIONOIDEA

TABLE B

THE FAMILIES OF THE MICROFRENATÆ

Contributed by Dr. William T. M. Forbes

A. Basal segment of the antennæ enlarged and concave beneath, forming an eye-cap.
B. Fore wings with radius, media, and cubitus unbranched. p. 617.OPOSTEGIDÆ
BB. Fore wings with more complex venation.
C. Discal cell of fore wings very short and trapezoidal, or absent. p. 600. ....NEPTICULIDÆ.
CC. Discal cell more than half as long as the wing.
D. Discal cell oblique, its lower outer corner nearly touching the inner margin. (A few species only.) p. 628................BLASTOBASIDÆ
DD. Discal cell central in the wing.
E. Labial palpi minute and drooping, or absent. p. 616...LYONETIIDÆ
EE. Labial palpi moderate, upcurved. (*Phyllocnistis* in part, and one or two Florida genera.) p. 617......................GRACILARIIDÆ
AA. Basal segment of antennæ not forming an eye-cap.
B. Palpus with the first segment relatively very large, normally upcurved to the middle of the front; when the palpus is short the first segment is longer than the second. p. 611..............................ACROLOPHIDÆ
BB. First segment of palpus small.
C. Labial palpi bristled on the outer side of the second segment.
D. Aculeæ present over the general surface of the wings; female with piercing ovipositor; antennæ typically smooth and velvety-looking, with fine bristles, or narrow, closely appressed scales, sometimes very long. p. 598..INCURVARIIDÆ
DD. Aculeæ absent, or present only in a small area at the base of the discal cell; ovipositor membranous, retractile; antennæ typically rough, with an outer whorl of erect scales on each segment, rarely as in D. p. 611..TINEIDÆ
CC. Labial palpi scaled or loose-hairy only.
D. Maxillary palpi well developed and of the folded type.
E. Fore wings with all veins present and with vein R_5 running to the outer margin; hind wings narrow; vertex with a small, loose tuft only (*Acrolepia*). p. 632...............................PLUTELLIDÆ
EE. Fore wings with vein R_5 extending to the costa or absent.
F. Head smooth; hind wings narrow-lanceolate; fore wings down-curved at apex. p. 617......................OINOPHILIDÆ
FF. Vertex rough or rarely smooth in forms with ample hind wings; fore wings flat.
G. Aculeæ present, etc., as in D under C above. p. 598......
......................................INCURVARIIDÆ

GG. Aculeæ absent, etc., as in DD under C above. p. 611..TINEIDÆ
DD. Maxillary palpi porrect or vestigial.
E. Vertex and upper face at least with dense bristly hairs; third segment of labial palpi fusiform and equal to the second in length.
F. Aculeæ present, etc., as in D under C above. p. 598. INCURVARIIDÆ
FF. Aculeæ absent, etc., as in DD under C above. p. 611.TINEIDÆ
EE. Face at least smoothly and shortly scaled; third segment of labial palpus long and pointed, or very short in forms with roughest vestiture.
F. Hind wings ample, with well-marked anal angle, often wider than their fringe.
G. Hind wings with veins M_1 and M_2 both lost, only one vein being associated with the R-stem. p. 644.....CARPOSINIDÆ
GG. Hind wings with vein M_1 preserved, associated with the R-stem.
H. Vein Cu_2 of the fore wings arising from a point before the outer fourth of the discal cell; palpus more or less triangular, with a short, blunt, third segment, roughly scaled (short and nearly smooth in *Laspeyresia*, in which there is a strong fringe on base of vein Cu of the hind wings, save in *L. lautana*). p. 639, 642................OLETHREUTIDÆ and TORTRICIDÆ
HH. Vein Cu_2 of the fore wings arising from the outer fourth of the discal cell, save in a few Glyphipterygidæ, which have short, smooth-scaled palpi, or second segment tufted and third long and slender, and no fringe on vein Cu.
I. Vein 1st A of fore wings lost completely; hind wings with veins R_s and M_1 connate, approximate, or stalked.
J. Palpi with third segment long, slender, and tapering, often exceeding vertex, normally close-scaled, save in male *Anarsia* where veins R_4 and R_5 are stalked and both run to the costa. p. 625..................GELECHIIDÆ
JJ. Palpi with third segment short and blunt, roughly scaled; vein R_5 normally running to outer margin, and often free from vein R_4. p. 643............PHALONIIDÆ
II. Vein 1st A preserved, at least at the margin of the wing.
J. Hind wings with veins R_s and M_1 widely separate at origin, more or less parallel.
K. Palpi long, often exceeding vertex; tongue distinct.
L. Veins R_4 and R_5 stalked and both running to costa, or united.
M. Vein M_2 of the hind wings arising nearer to vein M_1 than to M_3. p. 625............ETHMIIDÆ
MM. Vein M_2 of the hind wings arising nearer to vein M_3 than to M_1. p. 624.........ŒCOPHORIDÆ
LL. Veins R_4 and R_5 long stalked; vein R_5 running to outer margin. (See also LLL.)
M. Ocelli very large and conspicuous (*Allononyma*.) p. 633.......................GLYPHIPTERYGIDÆ
MM. Ocelli small or absent.
N. Vein M_2 of the hind wings arising nearer to vein M_1 than to M_3. p. 625........ETHMIIDÆ
NN. Vein M_2 of the hind wings arising nearer to vein M_3 than to M_1. p. 624...ŒCOPHORIDÆ
LLL. Veins R_4 and R_5 separate, vein R_5 running to outer margin. p. 631..............YPONOMEUTIDÆ
KK. Palpi small, hardly exceeding the front, or obsolete; tongue obsolete; female with a brush-like tuft at end of abdomen (*Kearfottia, Solenobia*). p. 614....PSYCHIDÆ
JJ. Hind wings with veins R_s and M_1 coalesced or stalked.
K. Wings narrow; fore wings falcate; maxillary palpi well marked and porrect (*Cerostoma*, etc.). p. 631 ....PLUTELLIDÆ

KK. Wings broad, ample, not falcate; maxillary palpi of folded type, inconspicuous, invisible in *Setiostoma*. p. 625.................................STENOMIDÆ
FF. Hind wings with pointed apex and excavated below, rarely bifid. (See also FFF.) p. 625..................GELECHIIDÆ
FFF. Hind wings narrow-lanceolate and pointed or linear, and much narrower than their fringe.
G. Hind wings lanceolate, though sometimes very small, and with the principal vein running nearly through its center, widely separated from Sc.
H. Hind wings with a discal cell. p. 621........ELACHISTIDÆ
HH. Hind wings without a discal cell.
I. Vein R_s of hind wings separating from media near the middle of the length of the wing. p. 623.....DOUGLASIIDÆ
II. Vein R_s of hind wings separating from media near the apex of the wing. p. 634..................HELIOZELIDÆ
GG. Hind wings with vein R closely parallel with or fused to Sc near base. In the broad-winged Gracilariidæ, veins Sc and R are fused and the base of vein M is preserved, simulating the condition in G, but the combined base of Sc and R curves strongly into the lobed basal half of costa, and then approaches or fuses with M at middle of wing, unlike the relation of Sc and R in G.
H. Hind tarsi with strong spinules, usually near apices of segments, as well as tibiæ; posterior legs displayed when at rest. p. 634...............................HELIODINIDÆ
HH. Tarsi smooth-scaled, the spinules concealed in the scaling; the tibiæ often hairy, but rarely (*Acrocercops, Epermenia*) bristled.
I. Fore wings with only four veins running from the discal cell to the costa, and five or six to the inner margin. p. 631, 632........... YPONOMEUTIDÆ and PLUTELLIDÆ
II. Fore wings with five veins running to costa, or only four to inner margin.
J. Discal cell oblique in wing; vein Cu_2 very short, running directly across to inner margin.
K. Antennæ turned forward in repose; fore tibiæ slender, with a small epiphysis at the apex or none. p. 620. ....COLEOPHORIDÆ
KK. Antennæ turned back in repose; fore tibiæ with the epiphysis conspicuous, and often more than half as long as the tibia; the tibiæ rarely slender.
L. Hind wings with veins Sc and R **usually fused near** base; fore wings with a stigma, and with R_1 arising near the base of the discal cell and R_2 near the apex of the cell. p. 628..................BLASTOBASIDÆ
LL. Hind wings with veins Sc and R not fused; fore wings with the space between the origins of veins R_1 and R_2 only three or four times that between veins R_2 and R_3. p. 629..................COSMOPTERYGIDÆ
JJ. Discal cell not set obliquely in wing; vein Cu_2 normally long and parallel to the medial veins.
K. Male antennæ heavily ciliate; accessory cell of fore wings extending halfway to base of wing; head with a large, loose, but often obscure, semierectile tuft. p. 615. ....TISCHERIIDÆ
KK. Male antennæ rarely ciliate; accessory cell small or absent.
L. Palpi minute and drooping; vertex tufted; hind wings linear (*Bedellia*). p. 616..... LYONETIIDÆ
LL. Palpi moderate, with fusiform third segment; maxillary palpi often well developed and porrect. (See also LLL.) p. 617..........GBACILARIIDÆ

LLL. Palpi upturned, with acuminate third segment, often exceeding the vertex; maxillary palpi of folded type but very minute or obsolete.
M. Vein R_1 of the fore wings more than twice as long as vein R_2 and arising before the middle of the discal cell. p. 629............COSMOPTERYGIDÆ
MM. Vein R_1 of the fore wings but little longer than vein R_2, and arising beyond the middle of the discal cell. p. 631.....................SCYTHRIDIDÆ

Suborder JUGATÆ

This suborder includes those Lepidoptera in which the posterior lobe of the fore wing is specialized so as to form an organ which unites the fore and hind wings; and in which the venation of the hind wings is similar to that of the fore wings.

The Jugatæ includes the more generalized members of the order Lepidoptera now living, those which are believed to resemble most closely the primitive insects from which in ancient times the Lepidoptera were evolved. In fact the first two families here included in the Jugatæ may be of even more ancient origin, representing one or two lines of evolution distinct from the lepidopterous stem.

Several writers have called attention to indications of trichopterous affinities of the two families in question; and a study of the wing-venation of these insects led me to believe that they are more closely allied to the Trichoptera than to the Lepidoptera. For this reason, in "The Wings of Insects" I classed them with the Trichoptera.

Although these indications of trichopterous affinities are undoubted, it appears that the view now generally held is that, while they show a close community of descent of the Trichoptera and the Lepidoptera, they are not sufficient to warrant the removal of the families in question from the Lepidoptera. I, therefore, include them, provisionally, in this order in the following account. For a detailed discussion of this subject, see Braun ('19) and Crampton ('20 b).

The suborder Jugatæ, as now more commonly limited, includes several families, representatives of three of which have been found in America; these are the Micropterygidæ, the Eriocraniidæ, and the Hepialidæ.

The members of the first two of these families differ greatly in appearance from those of the third family, being very small moths which resemble the small tineids in size and appearance; our largest species has a wing expanse of from 12 to 14 mm. For this reason they may be known as the Microjugatæ. They have also been termed the Jugo-frenata, because, in addition to having the posterior lobe of the fore wing specialized so as to form an organ which serves in uniting the fore and hind wings, there is also a bunch of bristles borne by the hind wing near the humeral angle, which resembles a frenulum; these bristles, however, are not homologous with the frenulum, but are the costal spines described on page 575. On the other hand, the members of the third family are mostly large moths; many of them are very large; and the smaller species have a wing-expanse of 25 mm. The members of this family may be known as the Macrojugatæ.

Family MICROPTERYGIDÆ

The Mandibulate Jugates

The members of this family are small insects which resemble tineid moths in general appearance. As with other members of the

LEPIDOPTERA

suborder Jugatæ, the venation of the hind wings closely resembles that of the fore wings (Fig. 719). But these insects differ from all

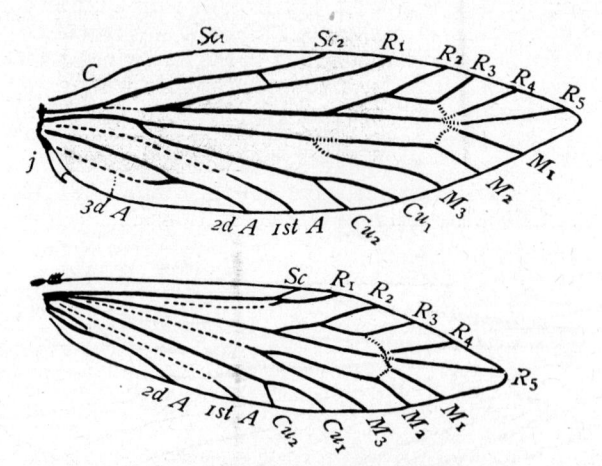

Fig. 719.—Wings of *Micropteryx*.

other Lepidoptera in having in the adult instar well-developed functional mandibles, and in that the females lack a bursa copulatrix. Chapman ('17) regards the presence of well-developed mandibles and the absence of a bursa copulatrix of sufficient importance to warrant the removal of these insects from the Lepidoptera and the establishment of a distinct order for them; for this order he proposed the name Zeugloptera.

Tillyard ('19) states that the wing-coupling apparatus in this family functions differently from that of the following family, in that in the Micropterygidæ the jugal lobe is bent under the fore wing and acts as a retinaculum for the bunch of costal spines, borne by the hind wings.

In this family, the subcosta of the fore wings is forked near its middle (Fig. 719); the abdomen of the adult female consists of ten distinct segments; and there is no ovipositor.

There is no published account of the transformations of our American species. The larvæ of certain exotic species have been described; they are very delicate, have long antennæ, and feed upon wet moss. The pupa state is passed in the ground; the pupa has large, crossed mandibles. The adults feed on pollen.

Two American species have been described; these are *Epimartȳria auricrinĕlla*, which is found in the East, and *Epimartȳria pardĕlla*, found in Oregon.

Family ERIOCRANIIDÆ

The Haustellate Jugates

The members of this family, like those of the preceding one, are small insects which resemble tineid moths in general appearance.

In this family the mandibles of the adult are vestigial; the maxillæ are formed for sucking, each maxilla forming half of a long sucking-tube, as in higher Lepidoptera. The females lack a bursa copulatrix but have a piercing ovipositor. An easily observed recognition character is the fact that the subcosta of the fore wings is forked near its apex (Fig. 720). The jugal lobe of the fore wing extends back above the base of the hind wing and is clasped over an elevated part of the hind wing, thus being of the type described as a fibula (see page 62).

Our best-known representative of this family is *Mnemŏnica auricyănea*. The structure and transformations of this species have

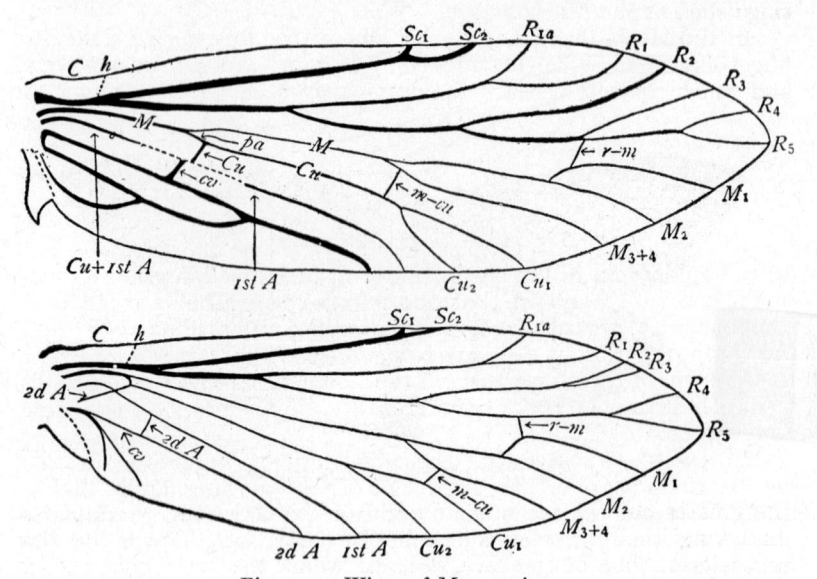

Fig. 720.—Wings of *Mnemonica*.

been described by Busck and Boving ('14). The adult has a wing-expanse of from 12 to 14 mm. The larva mines in the leaves of chestnut, oak, and chinquapin in early spring, making a large, bulgy blotch mine; it completes its growth within a week or ten days, and goes into the ground to transform, where it spins a tough cocoon; the change to pupa takes place in the following winter; the adult emerges in April. The pupa has long, arm-like toothed mandibles, with which it cuts the tough cocoon and with which it digs its way up to the surface of the ground. This species is found in the East.

Family HEPIALIDÆ

The Swifts or the Macrojugatæ

The members of this family are of medium or large size.

Figure 721 represents in natural size one of the larger of the American species, but many exotic species are larger than this one. Our smaller species have a wing-expanse of at least 25 mm. Our best-known species are brown or ashy gray in color, with the wings marked with silvery white spots.

It is said that these moths fly near the earth, and only in the evening after sunset, hiding under some low plant, or clinging to the stalk of an herb during the day. Some of them fly with extreme rapidity, with an irregular mazy flight, and have, therefore, been named *swifts* by collectors. So long as either or both of the two preceding families are retained in the suborder Jugatæ, the Hepialidæ may be distinguished as the Macrojugatæ.

In the Hepialidæ the posterior lobe of the fore wing is a slender, finger-like organ, which is stiffened by a branch of the third anal vein, and which projects beneath the costal margin of the hind wing. As

Fig. 721.—*Sthenopis purpurascens.*

the greater part of the inner margin of the fore wing overlaps the hind wing, the hind wing is held between the two. This is the type of posterior lobe of the fore wing to which the term *jugum* is applied. (Figs. 74 and 75.)

The larvæ are eruciform and furnished with sixteen legs; they feed upon wood or bark, and are found at the roots or within the stems of plants. They transform either in their burrows, or, in the case of those that feed outside of roots, within loose cocoons. The pupæ have transverse rows of teeth on the abdominal segments; these aid them in emerging from their burrows.

This family is represented in our fauna by two genera, *Hepialus* and *Sthenopis*.

Hepīalus.—This genus includes our smaller species, which range in wing-expanse from 25 to 55 mm. In *Hepialus* the apices of the fore wings are more rounded than in *Sthenopis*. Ten North American species have been described.

Sthenōpis.—This genus includes our larger species. In these the apices of the fore wings are more pointed than in *Hepialus*, and in some species are subfalcate. Four species have been found in our fauna; one of these, *Sthenōpis purpurăscens*, is represented in Figure 721.

The larva of *Sthenōpis argenteomaculātus* bores in the stems of the speckled or hoary alder (*Alnus incana*); that of *Sthenopis thule*, in willow.

Suborder FRENATÆ

The members of the Frenatæ are most easily recognized by the fact that the venation of the hind wings differs markedly from that of the fore wings, being much more reduced. In this suborder, vein R_1 of the hind wings coalesces with subcosta, the two appearing as a single vein, except that, in some cases, a short section of the base of R_1 is distinct, presenting the appearance of a cross-vein between

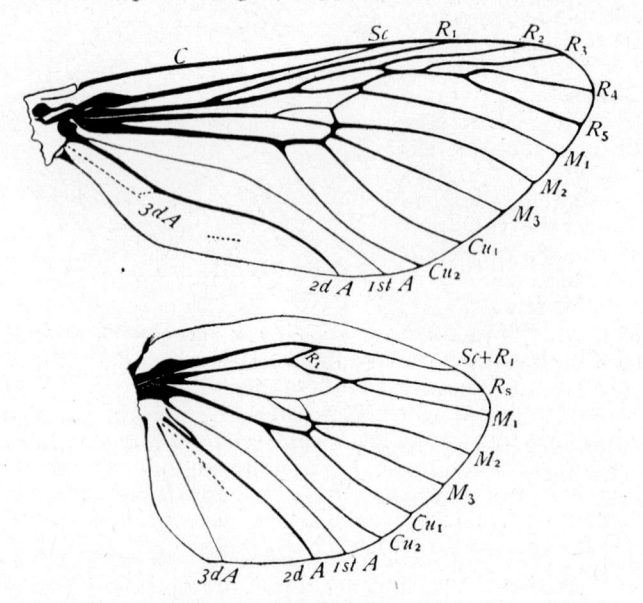

Fig. 722.—Wings of *Prionoxystus robiniæ*.

radius and subcosta (Fig. 722, R_1). After the separation of vein R_1, the radial sector continues unbranched to the margin of the wing (Fig. 722, R_s). Rarely, as in some members of the Gracilariidæ and of the Cosmopterygidæ, vein R_1 of the hind wings is free, not coalesced with vein Sc.

The essential characteristic of the Frenatæ is that they are descendants of those primitive Lepidoptera in which the two wings of each side were united by a frenulum. This fact should be clearly understood, for in many of the Frenatæ the frenulum has been lost. The loss of the frenulum in these cases is due to its having been supplanted by a substitute for it, by an enlarged humeral area of the hind wings, which causes the two wings of each side to overlap to a great extent. This overlapping of the two wings insures their synchronous action; and the frenulum, being no longer needed for this purpose, is lost. Illustrations of different stages in the reduction and loss of the frenulum are given in the discussions of family characters given later.

As a rule the frenulum of the female, when present, consists of several bristles, while that of the male consists of a single strong, spine-like organ. If one of the bristles of the compound frenulum of a female be examined, it will be found to be a typical seta, containing a single cavity. But if a frenulum of a male be examined, it will be found to contain several parallel cavities. Evidently the frenulum of the male is composed of several setæ, as is that of the female, but these setæ are grown together. This can be seen by examining a bleached wing that has been mounted in balsam; usually the cavities in the setæ contain air, which renders them visible.

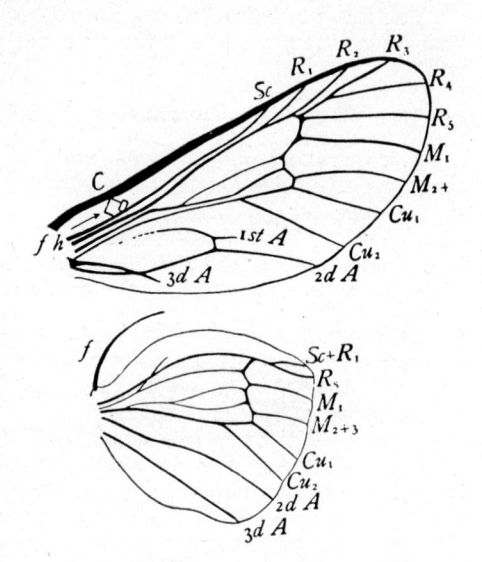

Fig. 723.—Wings of a moth: $f\,h$, frenulum-hook.

The frenulum-hook, which is present in the males of certain moths, is a membranous fold on the lower surface of the fore wing for receiving the end of the frenulum, and thus more securely fastening the two wings together (Fig. 723, $f\,h$). As a rule the frenulum-hook arises from the membrane of the wing near the base of cell C; but in some moths (*Castnia*) it seems to have been pulled back so that it arises from the subcostal vein.

THE GENERALIZED FRENATÆ

Under this heading are grouped those families of moths that are supposed to retain more nearly than any other Frenatæ the form of the primitive Frenatæ, those that were the first to appear on earth. In most of the families included here, the wings approach the typical form, except in the reduction of the number of branches of radius of the hind wings, which is true of all Frenatæ; usually the base of media of one or both pairs of wings is preserved throughout a considerable part, at least, of the discal cell; and the anal veins are well preserved, there being two or three in the fore wing and three in the hind wing. The frenulum is usually well preserved.

There are also included in this group of families those families in which the fixed hairs or aculeæ are retained over the general surface of the wings, even though in some cases, as in the Nepticulidæ, the venation of the wings may be greatly reduced. The presence of aculeæ distributed over the general surface of the wings is believed to indicate a generalized condition, as it is found elsewhere in the Lepidoptera only in the Jugatæ. As this condition is also found in the Trichoptera, it was probably inherited from the stem form from which the Lepidoptera and the Trichoptera were evolved. In the more specialized Lepidoptera the aculeæ are confined to small areas of the wing surface or have been lost.

Family INCURVARIIDÆ

This family and the following one differ from all other Frenatæ and agree with the Jugatæ in having retained aculeæ distributed over the general surface of the wing (Fig. 710). In this family the venation of the wings is but little reduced; the antennæ are without an eye-cap; and the females, so far as is known, are furnished with a piercing ovipositor. The moths are small or of moderate size. Many of the larvæ are miners when young, and later are case bearers.

The family Incurvariidæ includes three subfamilies, which are not very distinct but which, however, are treated as families by some writers.

SUBFAMILY ADELINÆ.—These tiny moths are characterized by the unusually long and fine antennæ of the males, which may be twice or more than twice as long as the wings. Some of the species are also conspicuous on account of their striking colors and markings.

The larvæ are elongate, cylindrical, with thoracic legs and five pairs of prolegs. They are at first miners; later they live in portable cases. They feed on the leaves of various herbs and shrubs; but none of our species is known to be of economic importance. Nearly all of our species belong to the genus *Adela*.

SUBFAMILY INCURVARIINÆ.—An interesting representative of this division of the family Incurvariidæ is the following well-known species.

The maple-leaf cutter, *Paraclemĕnsia acerifoliĕlla*.—The larva infests the leaves of maple, and occasionally is so abundant that it does serious injury. The larva is at first a leaf-miner, like other adelids; but later it is a case-bearer.

The leaves of an infested tree present a strange appearance (Fig. 724). They are perforated with numerous elliptical holes, and marked by many, more or less perfect, ring-like patches in which the green substance of the leaf has been destroyed but each of which incloses an uninjured spot. These injuries are produced as follows: The larva, after living for a time as a leaf-miner, cuts an oval piece out of a leaf, places it over its back, and fastens it down with silk around the edges. This serves as a house beneath which it lives. As it grows, this house becomes too small for it. It then cuts out a larger piece which it fastens to the outer edges of the smaller one, the larva being between the two. Then it crawls halfway out upon the leaf, and by a dexterous lifting of the rear end of its body turns the case over so that the larger piece is over its back. When it wishes to change its location it thrusts out its head and fore legs from the case and walks off, looking like a tiny turtle. When it wishes to eat, it fastens the case to the leaf and, thrusting its head out, eats the fleshy part of the leaf as far as it can reach. This explains the circular form of the patches, the round spot in the center indicating the position of the case. The insect passes the winter in the pupa state within its case, which falls to the ground with the infested leaf. The moth is of a brilliant steel-blue or bluish green color, without spots but with an orange-colored head; it appears in early summer.

SUBFAMILY PRODOXINÆ.—This subfamily includes the remarkable insects that are known as the yucca-moths and the closely allied bogus yucca-moths.

Fig. 724.— Leaf infested by the maple-leaf cutter.

The yucca-moths, *Tegetĭcula.*—Four species of this genus are now recognized; the best-known of these is *Tegeticula alba*. The life-history of this species was first described by Mr. C. V. Riley ('73), under the name *Prōnuba yuccasĕlla;* and in most of the accounts of this insect this name is used. The moth, however, was first described as *Tegeticula alba*. The most complete account of this and the allied species is that of Riley ('92).

This species infests *Yucca filamentosa*, a plant not fitted for self-pollination or for pollination by insects in the ordinary ways; in fact, it is pollinized only by moths of the genus *Tegeticula*, the larvæ of which feed on its seeds. This is one of the few cases in which a particular plant and a particular insect are so specialized that each is dependent upon the other for the perpetuation of the species. In the female moth, the maxillæ are each furnished with a long, curled, and spinose appendage, the *maxillary tentacle* (Fig. 725, *b*), fitted for the collection of pollen. After collecting a large load of pollen, often thrice as large as the head (Fig. 725,), the female moth places her eggs, by means of her long, extensile ovipositor, into an ovary, usually of another flower than that from which the pollen was collected. After oviposition, the moth runs up to the tip of the pistil and thrusts the pollen into the stigmatic opening. Thus is insured the development of seeds, upon which the larvæ hatched from the eggs placed in the ovary are to feed. As many more seeds are developed than are needed by the larvæ, the perpetuation of the yuccas is assured.

The full-grown larva leaves the yucca pod and makes its way to the ground, where it spins a dense cocoon several inches below the surface. The adult moth has a wing-expanse of about 25 mm. The front wings are silvery white above; the hind wings, semitransparent.

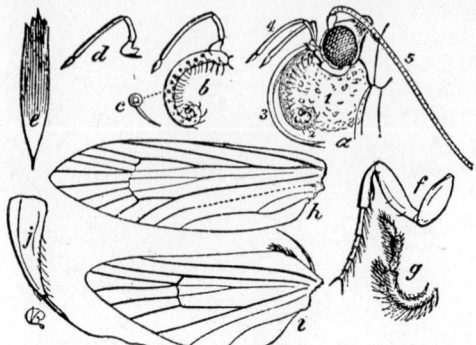

Fig. 725.—*Tegeticula alba:* a, side view of head and neck of female denuded; 1, load of pollen; 2, maxillary tentacle; 3, maxillæ; 4, maxillary palpi; 5, antennæ; b, maxillary tentacle and palpus; c, an enlarged spine; d, maxillary palpus of male; e, scale from front wing; f, front leg; g, labial palpus; h, i, venation of wings; j, last segment of abdomen of female, with ovipositor extruded. All enlarged. (From Riley.)

The bogus yuccamoths, *Prodŏxus*.—The moths of this genus are closely allied to the yucca-moths, but differ in the important particular that the females lack maxillary tentacles; they are consequently incapable of pollinating the yuccas as do the true yucca-moths. The larvæ of *Prodoxus* feed in the flower-stem or in the flesh of the fruit. But as, in *Yucca filamentosa* at least, the flowers drop and the flower-stem withers if the flowers are not pollinated, the bogus yucca-moths are dependent on the true yucca-moths for the conditions necessary for the development of their larvæ. The pupa state is passed in the burrow made by the larva. Eleven species of *Prodoxus* have been described.

Family NEPTICULIDÆ

In this family, as in the preceding one, fixed hairs or acuieæ are distributed over the general surface of the wings. In the Nepticulidæ the venation of the wings is much reduced; the basal segment of the antennæ is enlarged and concave beneath, so as to form an eye-cap; the female is without an ovipositor; the labial palpi are short; the maxillary palpi are long; and the maxillæ are vestigial.

This family includes the smallest of the Lepidoptera, some of the species having a wing-expanse of scarcely 3 mm.

Although this family presents characteristics which indicate that it should be placed among the generalized Lepidoptera, the venation of the wings is greatly reduced. This indicates that it represents a distinct line of development which in some respects has become more highly specialized than are the other families included in this division of the Lepidoptera.

The frenulum of the female consists merely of a group of small, functionless bristles; but in the male the frenulum is a strong, spine-like organ, which hooks into a well-developed frenulum hook (Fig. 726); in most cases the costal spines are well developed; this is shown in the accompanying figure; and the anal lobe of the fore wing is sometimes quite distinct.

With the exception of several gall-making species of *Ectœdĕmia*, the larvæ of all species of which the life-history is known are miners within the tissues of leaves (rarely in fruits) or in bark. They show a preference for trees and shrubs, but some mine in the leaves of herbaceous plants. The larva at first makes a very narrow linear mine. This mine may continue as a linear mine, gradually broadening throughout its course, or it may at some period abruptly enlarge into a blotch. When full-grown, the larva, with few exceptions, leaves the mine and, dropping to the ground, spins a dense, flattened cocoon amongst rubbish or on the loose surface soil. (Braun '17.)

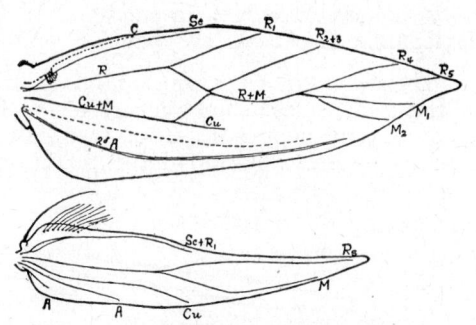

Fig. 726.—Wings of *Obrussa ochrefasciella*, male. (After Braun.)

More than seventy species have been described from our fauna, and doubtless many more are to be discovered. The Nepticulidæ of North America was monographed by Braun ('17).

Family COSSIDÆ

The Carpenter-Moths

This family includes moths with spindle-shaped bodies, and narrow, strong wings, some of the species resembling hawk-moths quite closely in this respect. The larvæ are borers; many of them live in the solid wood of the trunks of trees. The wood-boring habits of the larvæ suggest the popular name *carpenter-moths* for the insects of this family.

These moths fly by night and lay their eggs on the bark of trees, or within tunnels in trees from which adult carpenter-moths have emerged. The caterpillars are nearly naked, and, although furnished with pro-legs as well as true legs, are grub-like in form. The pupa state is passed within the burrow made by the larva. When ready to change to an adult, the pupa works its way partially out from its burrow. This is accomplished by means of backward-projecting saw-like teeth, there being one or two rows of these on each abdominal segment. After the moths have emerged, the empty pupa-skins can be found projecting from the deserted burrows.

The carpenter-moths are of medium or large size. The antennæ of the males are mostly bipectinate; those of the females are either very slightly bipectinate or ciliate. In a few species the antennæ are lamellate. The ocelli are wanting, and the maxillæ are vestigial.

The venation of the wings of our most common and most widely distributed species is shown in Figure 727. There are two well-preserved anal veins in the fore wing, and three in the hind wing. The base of media is preserved, and is forked within the discal cell. In the fore

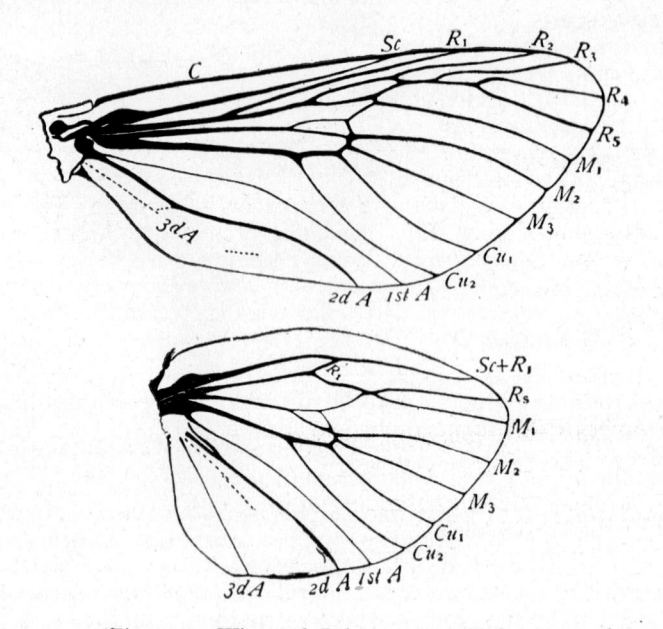

Fig. 727.—Wings of *Prionoxystus robiniæ*.

wing the veins R_3 and R_{4+5} anastomose, forming an accessory cell. The frenulum is vestigial in this genus; but in some other genera it is well developed.

Authors differ greatly regarding the appropriate position of this family in the series of families. Certain characteristics of the larvæ indicate that it belongs somewhere among the specialized Microfrenatæ; but I place it here at the beginning of the Non-aculeate Generalized Frenatæ on account of the generalized structure of the wings.

This family is represented in our fauna by thirty-four described species; it has been monographed by Barnes and McDunnough ('11). The family includes three subfamilies, which are separated as follows:

A. Anal veins of the fore wings united near the margin of the wing by a cross-vein...HYPOPTINÆ
AA. Anal veins of the fore wings not united near the margin of the wing by a cross-vein.
  B. Veins R_s and M_1 of the hind wings stalked or close together at the end of the discal cell; antennæ of male pectinate throughout..............COSSINÆ
  BB. Veins R_s and M_1 of the hind wings widely separate; antennæ of male pectinate on basal half only.............................ZEUZERINÆ

SUBFAMILY HYPOPTINÆ.—The members of this subfamily are distinguished by the presence of the anal cross-vein near the margin of the front wings. Nearly one-half our species belong to this subfamily. They have been described from Florida, Texas, Colorado, and westward to California. I have found no account of the early stages of any of them.

SUBFAMILY COSSINÆ.—This subfamily is represented in our fauna by six genera including fourteen species; but most of these are confined to the Far West and are known only in the adult state. Our best-known species are the following.

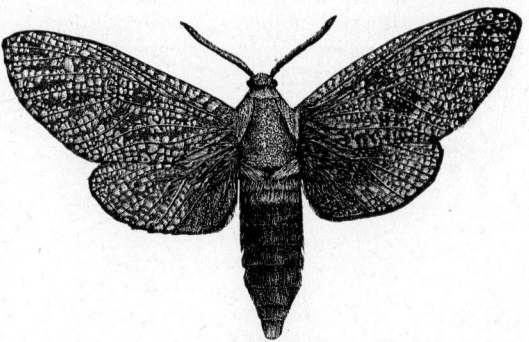

Fig. 728.—*Prionoxystus robiniæ*, female.

The locust-tree carpenter-moth, *Prionoxўstus robiniæ*.—Figure 728 represents the female, natural size. The male is but little more than half as large as the female. It is much darker than the female, from which it differs also in having a large yellow spot, which nearly covers the outer half of the hind wings. The moths fly in June and July; the larvæ bore in the trunks of locust, oak, poplar, willow, and other trees. It is supposed that the species requires three years to complete its transformations. It is found from the Atlantic Coast to California.

The lesser oak carpenter-worm, *Prionoxўstus macmŭrtrei*.—This is a slightly smaller species than the preceding. The larva bores in the trunks of oak in the East. The moth has thin, slightly transparent wings, which are crossed by numerous black lines. The male is much smaller than that of *P. robiniæ*, and lacks the yellow spot on the hind wings.

SUBFAMILY ZEUZERINÆ.—Excepting three little-known species of *Hamilcara*, found in Texas and Arizona, the following species is the only representative of this subfamily in our fauna.

The leopard-moth, *Zeuzēra pyrīna*.—This species is white, spotted with numerous small, black spots, which suggested its common name. The adult has a wing-expanse of from 40 to 60 mm. It is a European species, which was first observed in the vicinity of New York City in 1882; since that time it has spread to other parts of the East. The larva is a very injurious borer in many species of trees and shrubs. The young larvæ bore in the small twigs; later they migrate to larger limbs or to the trunk.

Family PYROMORPHIDÆ

The Smoky Moths

There are but few insects in our country pertaining to this family; only fifteen species are now recognized, but these represent six genera. These are small moths, which are chiefly of a smoky black color; some are marked with brighter colors; the wings are thinly scaled; and the maxillæ are well developed. The larvæ are clothed with tufted hair; they have five pairs of prolegs, which are provided with normal hooks.

Fig. 729.—*Acoloithus falsarius.*

Fig. 730.—Wings of *Acoloithus falsarius.*

A tiny representative of the family which seems to be not uncommon in the East is *Acoloithus falsārius*. This moth (Fig. 729) expands 16 mm. It is black, with the prothorax of an orange color. The venation of the wings (Fig. 730) is peculiar, in that subcosta and radius of the hind wings coalesce for only a short distance beyond the middle of the discal cell, and a stump of radius projects towards the base of the wing, from the point of union of the two veins. The larva feeds in early summer on the leaves of grape and of Virginia creeper. It is said that the pupa state lasts fourteen days and is passed within a parchment-like cocoon. The adults frequent flowers in the daytime.

Another well-known species is *Pyromŏrpha dimidiāta*. This is found in the Atlantic and Western States. The entire insect is smoky black, except the basal half of the fore wings in front of the second anal vein, and the basal half of the costa of the hind wings, which are yellow. The wings are thinly scaled and expand 25 mm. or a little more. The male is larger than the female and is more active. Figure 731 represents the venation of the wings. Some species of the genus *Pyromorpha* are remarkable in that none of the branches of radius of the fore wings coalesce beyond the discal cell.

Figure 732 represents the venation of the wings of *Pyromŏrpha martēni*, a species found in the Rocky Mountains.

The species of the genus *Harrisina* differ from the typical form of the family in that the anal area of the hind wings is greatly reduced, there being only two short, strongly curved, anal veins. As in other members of the family, there are two well-developed anal veins preserved in the fore wings. The following is the best-known species of this genus.

The grape-leaf skeletonizer, *Harrĭsina americāna* — The wings of this moth are long and narrow (Fig. 733); the abdomen is long, and widened towards the caudal end. It is greenish black in color, with the prothorax reddish orange. The larva feeds on the leaves of grape and of the Virginia creeper. An entire brood of these larvæ will feed side by side on a single leaf while young. This species rarely becomes of economic importance.

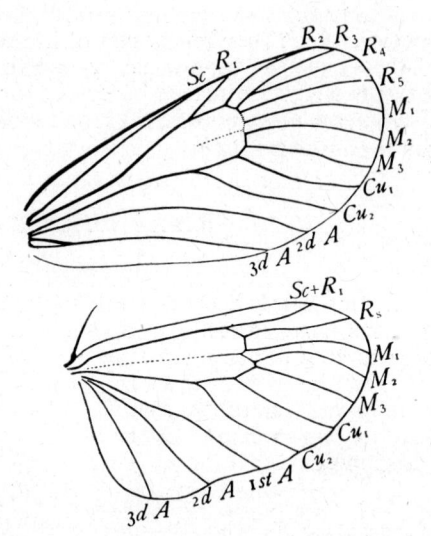

Fig. 731.—Wings of *Pyromorpha dimidiata*.

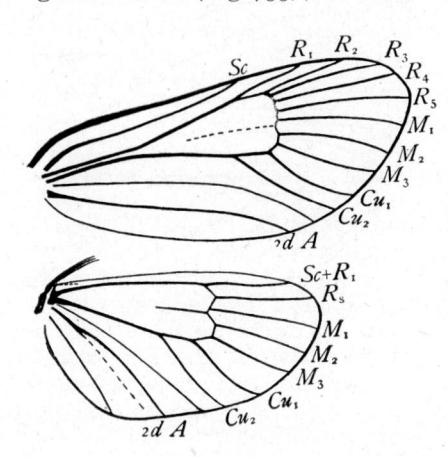

Fig. 732.—Wings of *Pyromorpha marteni*.

Fig. 733.—*Harrisina americana*.

Family DALCERIDÆ

In this family the body is small; the antennæ are short; and the wings are broad. In the fore wings there is a large accessory cell which is 1st R_3; and in the hind wings veins Sc and R are connected at a point.

The best-known species in our fauna is *Dalcĕrides ingĕnita*, found in Arizona. The expanse of the wings is about 25 mm. The wings are deep yellow, inclining to orange, without markings. The larva is unknown.

Another species, *Pincōnia coa*, which is not uncommon in Mexico has been reported from Arizona by Holland ('03).

Family MEGALOPYGIDÆ

The Flannel-Moths,

In this family the wings are heavily and loosely scaled, and mixed with the scales are long, curly hairs; these give the wings the appearance of bits of flannel. It is this that suggested the common name of these moths. The body is stout and clothed with long hairs.

The venation of the wings of our most common species, *Lagōa crispāta*, is represented in Figure 734. There are three anal veins in both fore and hind wings; but in the fore wings the second and third anal veins are partially grown together. The basal part of media is more or less distinctly preserved and divides the discal cell into two nearly equal parts. The subcosta and radius of the hind wings coalesce for nearly the entire length of the discal cell. In these moths the maxillæ are vestigial. The larvæ are remarkable for the possession of seven pairs of prolegs; these are borne by abdominal segments 2 to 7 and 10; but those of segments 2 and 7 are without hooks. The setiferous tubercles are verrucæ bearing large numbers of fine setæ; so that the body is densely hairy; and interspersed among the fine setæ are venomous setæ.

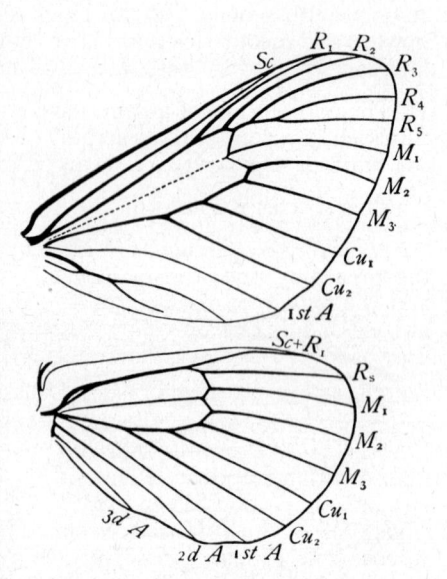

Fig. 734.—Wings of *Lagoa crispata*.

There are only ten North American species of this family; these represent four genera. Our most common species are the two following.

The crinkled flannel-moth, *Lagōa crispāta*.—This moth is cream-colored, with the fore wings marked with wavy lines of crinkled black and brownish hairs. The male is represented in Figure

735; the female is larger expanding, 40 mm. In the female the antennæ are very narrowly pectinate.

The larvæ feed on many trees and shrubs, including oak, elm, apple, and raspberry. They are short, thick, and fleshy, and are covered with a dense coat of long, silky, brown hairs, which project upward and meet to form a ridge or crest along the middle of the back; interspersed among these fine hairs are venomous setæ.

Fig. 735.—*Lagoa crispata*, male.

Fig. 736.—Old cocoon of *Megalopyge opercularis*.

The cocoons are of a firm, parchment-like texture, covered with a thin web of rather coarse threads. Mixed with the silk of the cocoon are hairs of the larva. The cocoon is provided with a hinged lid.

This species is found in the Atlantic States.

Megalopȳge operculāris.—This species is somewhat smaller than the preceding one; the male has a wing-expanse of about 25 mm., and the female of about 37 mm. The fore wings are umber brown at base, fading to pale yellow outwardly; they are marked with wavy lines of white and blackish hairs, and the fore margins are nearly black.

The larvæ are clothed with long, silky hairs, underneath which are venomous setæ. The cocoons are firmly attached to a twig of the infested tree, and are each furnished with a trapdoor. The old cocoons that one sees in collections present the appearance represented in Figure 736. But I found in Mississippi a cocoon, which I believe to be of this species, that is of the form shown in Figure 737. From this it appears that after the outer layer of the cocoon has been made, the larva constructs a hinged partition near one end of it, and adds no more silk to that part of the cocoon which is outside the partition. This part of the cocoon is quite delicate, and is destroyed when the moth emerges if not before.

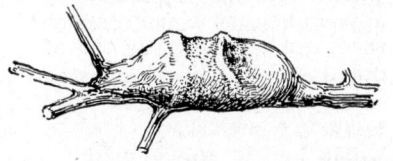

Fig. 737.—Complete cocoon of *Megalopyge*.

This species is found from North Carolina to Texas. The larva is a very general feeder; it is often found on oak.

Family EUCLEIDÆ*

The Slug-Caterpillar Moths

One often finds on the leaves of shrubs or trees, elliptical or oval larvæ that resemble slugs in the form of the body and in their gliding motion. As these are the larvæ of moths they have been termed *slug-caterpillars;* but they present very little similarity in form to other caterpillars. The resemblance to slugs is greatly increased by the fact that the lower surface of the body is closely applied to the object upon which the larva is creeping, the thoracic legs being small and the prolegs wanting. There is, however, on the ventral side of the abdomen a series of sucking-disks, which serve the purpose of prolegs. The head of the larva is small and retractile. In some species the body is naked; in others it is clothed with tufts of hairs; and in others there is an armature of branching spines. Several species bear venomous setæ.

Fig. 738.

The larvæ when full-grown spin very dense cocoons of brown silk; these are egg-shaped or nearly spherical, and are furnished at one end with a cap which can be pushed aside by the adult when it emerges (Fig. 738). The cocoons are usually spun between leaves.

The moths are of medium or small size; the body is stout, and the wings are heavily and loosely scaled. The maxillæ are vestigial. These moths vary greatly in appearance, and many of them are very prettily colored.

Considerable variation exists in the venation of the wings in this family (Fig. 739 and Fig. 740). The base of media may be preserved or wanting; in some species it is forked within the discal cell, in others not. There is also considerable variation in the coalescence of the branches of radius, but veins R_3 and R_4 coalesce to a greater extent than any other branches of this vein. There is no accessory cell. In the hind wings veins Sc

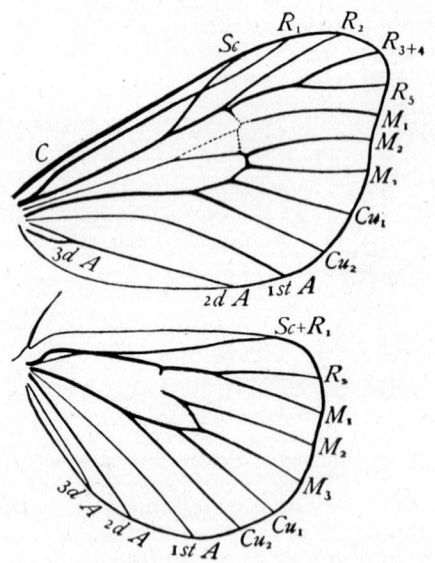

Fig. 739.—Wings of *Adoneta spinuloides*.

---

*This family is termed the Cochlidiidæ by some writers, and by others the Limacodidæ.

and R coalesce for a short distance at the point where vein R_1 joins vein Sc.

Only forty-three North American species of eucleids have been described; but these represent twenty genera. The larvæ are rarely abundant enough to be of economic importance; they are chiefly interesting on account of their remarkable forms. The following are some of the better-known species:

The saddle-back caterpillar, *Sibine stimulea*.—This larva can be recognized by Figure 741. Its most characteristic feature is a large green patch on the back, resembling a saddle-cloth, while the saddle is represented by an oval purplish brown spot. The moth is dark, velvety, reddish brown, with two white dots near the apex of the fore wings. The larva feeds on oaks and other forest trees. This is one of the species that are armed with venomous setæ.

Fig. 740.—Wings of *Packardia geminata*.

Fig. 741.—*Sibine stimulea*, larva.

The spiny oak-slug, *Euclea delphinii*.—This larva (Fig. 742) is one of the most common of our slug caterpillars and one of those that are armed with venomous setæ. It feeds on the leaves of oak, pear, willow, and other trees. The moth (Fig. 743) is cinnamon-brown, with a variable number of bright green spots on the fore wings.

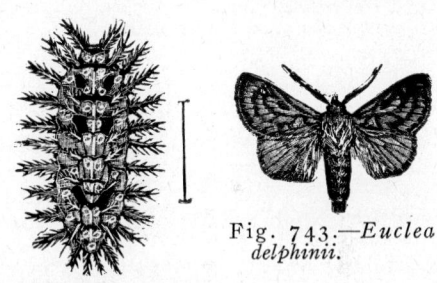

Fig. 742.—*Euclea delphinii*, larva.

Fig. 743.—*Euclea delphinii*.

The hag-moth, *Phobetron pithecium*.—The common name *hag-moth* is applied to the larva of this species on account of its remarkable appearance (Fig. 744). It bears nine pairs of fleshy appendages which are covered with brown hairs. In the full-grown larva the third, fifth, and seventh pairs of appendages are longest; these are twisted up and back, and suggest the disheveled locks of a hag. This larva feeds on various low shrubs and the lower branches of trees. At the time of spinning, the larva sheds the fleshy processes, and they remain on the outside of the cocoon.

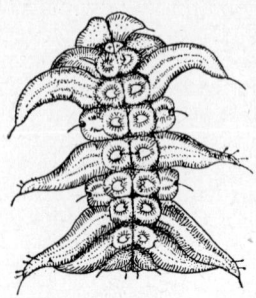

Fig. 744.—*Phobetron pithecium*, larva. (After Dyar.)

Fig. 745. — *Prolimacodes badia*, larva.

The skiff-caterpillar, *Prolimacōdes bādia*.—This remarkable larva (Fig. 745) is not uncommon on oak and other forest trees. It is pale apple-green, with a chestnut-brown patch on its back. The moth (Fig. 746) is light cinnamon-brown, with a tan-brown triangular spot on each fore wing.

Family EPIPYROPIDÆ

Fig. 746.—*Prolimacodes badia*.

This family is represented in our fauna by a single rare species which was found in New Mexico. Our species is *Epĭpyrops barberiāna*. Another species, *Epĭpyrops anŏmala*, has been described from China; and larvæ that are believed to belong to this genus have been found in Central America.

These insects are remarkable on account of the extraordinary habits of the larvæ, which are found firmly attached to living insects of the family Fulgoridæ. They are usually attached to the dorsal surface of the abdomen beneath the wings of their host. The body of the larva is covered with a cottony coat, causing it to resemble a *Coccus*. It is supposed that these larvæ feed on waxy matter excreted by the fulgorids.

For a detailed account of our species, see Dyar ('02).

THE SPECIALIZED MICROFRENATÆ

In the "Synopsis of the Lepidoptera" given on pages 581 to 584 I have grouped together under the heading "Specialized Microfrenatæ" twenty-six families of moths, which are more highly specialized than are the preceding families, and which as a rule are composed of small insects.

This group of families includes most of those families that were formerly classed together as the Microlepidoptera; but later studies have resulted in the removal from the old group Microlepidoptera of several families of small moths, hence this name is no longer distinctive. Among the families of small moths removed from the Microlepidoptera are the Micropterygidæ and the Eriocraniidæ, now placed in the suborder Jugatæ; the Incurvariidæ and the Nepticulidæ, placed at the beginning of the Frenatæ; and the group of families now known as the Pyralids, which are believed to be genetically quite distinct from the other families of small moths. On the other hand, in addition to the families here placed in this series some authors include the Cossidæ.

The families of the Microfrenatæ are grouped into superfamilies in various ways by different writers; but none of these groupings is sufficiently well established to be adopted here.

Family ACROLOPHIDÆ

These are large, stout, noctuid-like moths; some of the species have a wing-expanse of 30 mm. or more. The eyes are usually hairy, in which respect they differ from other "Micros." The antennæ are without an eye-cap. The labial palpi are large, and usually upcurved to the middle of the front; in the males of some species they are thrown back on the dorsum of the thorax, which they equal in length. The first segment is relatively very large; when the palpus is short it is longer than the second segment; the thorax is tufted. The venation of the wings is quite generalized; the base of media is more or less preserved, and all the branches of the branched veins are present; there are three anal veins in both fore and hind wings; in the fore wings the tip of the third anal vein coalesces with the second anal vein.

Forty-two species have been described from our fauna; these were formerly classed in several genera; but recent writers refer them all to the genus *Acrŏlophus*.

The burrowing web-worms, *A. arcanĕllus*, *A. mortipennĕllus*, and *A. popeanĕllus*.—The habits of these three species were described by Professor Forbes in his Twelfth Illinois Report (1905). The larvæ normally live in the ground feeding on the roots of grass. Each larva makes "a tubular web opening at the surface and leading down into a vertical cylindrical burrow about the diameter of a lead-pencil, and six inches to two feet, or even more in depth." The larva measures about 25 mm in length. Sometimes the larvæ injure young corn when planted on sod. They surround the base of each plant with a fine web mixed with earth and pellets, building this up in the lower blades, which they slowly eat away. As they get larger they eat the stripped plant to the ground. When disturbed they retreat into their web-lined burrows.

Family TINEIDÆ

The head is usually clothed with erect hair-like scales. The antennæ are shorter than the front wings. The maxillæ are usually small or vestigial. The maxillary palpi are usually large and folded.

612   AN INTRODUCTION TO ENTOMOLOGY

The labial palpi are short and clothed with but three or four bristles. In the typical genera the venation of the wings is quite generalized (Fig. 747), the base of media being preserved in both fore and hind wings and all of the veins characteristic of the Frenatæ being present; but in other genera the venation is somewhat reduced.

Many of the larvæ are case-bearers; many are scavengers or feed on fungi; some feed on fabrics, especially those that contain much wool; few if any feed on leaves.

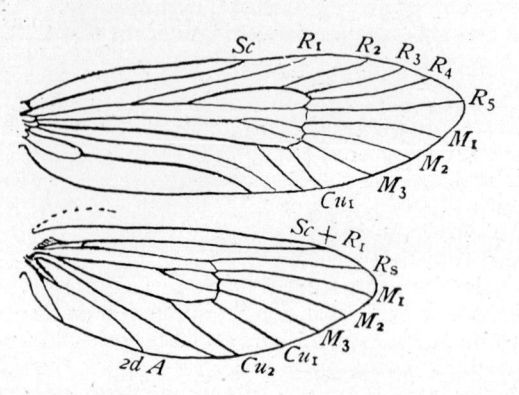

Fig. 747.—Wings of *Tinea parasitella*. (After Spuler.)

This is a large family. More than one hundred twenty-five North American species are already known; fifty of these belong to the genus *Tinea*. To this family belong the well-known clothes-moths.

The naked clothes-moth, *Tineola bisselliella*.—This is our most common clothes-moth. Although the larva spins some silk wherever it goes, it makes neither a case nor a gallery; it is, therefore, named the naked clothes-moth. But when the larva is full-grown it makes a cocoon, which is composed of fragments of its food-material fastened together with silk. The adult is a tiny moth with a wing-expanse of from 12 to 16 mm.; it is of a delicate straw-color, without dark spots on its wings.

The case-bearing clothes-moth, *Tinea pellionella*.—The larva of this species is a true case-bearer, making a case out of bits of its food-material fastened together with silk. The case is a nearly cylindrical tube open at both ends. The pupa state is passed within the case. The adult is a small, silky, brown moth, with three dark spots on each fore wing. It expands from 11 to 17 mm.

The tube-building clothes-moth or the tapestry-moth, *Trichŏphaga tapetiĕlla*.—The larva of this species makes a gallery composed of silk mixed with fragments of cloth. This gallery is long and winding and can be easily distinguished from the case of the preceding species. The pupa state is passed within the gallery. The moth differs greatly in appearance from the other two species, the fore wings being black from the base to near the middle, and white beyond. It expands from 12 to 24 mm.

Family PSYCHIDÆ

The Bag-Worm Moths

The bag-worm moths are so called on account of the silken sacs made by the larvæ, in which they live and in which they change to pupæ. In our more conspicuous and best-known species the sac is covered either with little twigs (Fig. 748) or, in the case of a species that feeds on cedar or arbor-vitæ, with bits of leaves of these plants. When the larva is full-grown it fastens its sac to a twig or other object and transforms within it.

In the adult state the two sexes differ greatly. The female is wingless, and in some genera the eyes, antennæ, mouth-parts, and legs are vestigial or wanting, the body being quite maggot-like. At the caudal end of the body there is a tuft of hair-like scales which are mixed with the eggs. In most species the female does not leave the sac before oviposition but deposits her eggs within it.

The male moths are winged; they are small or of moderate size. The wings are thinly scaled and in some species nearly naked; when clothed with scales they are usually of a smoky color without markings. The venation of the wings varies greatly within the family. Figure 749 represents the venation of our most common species.

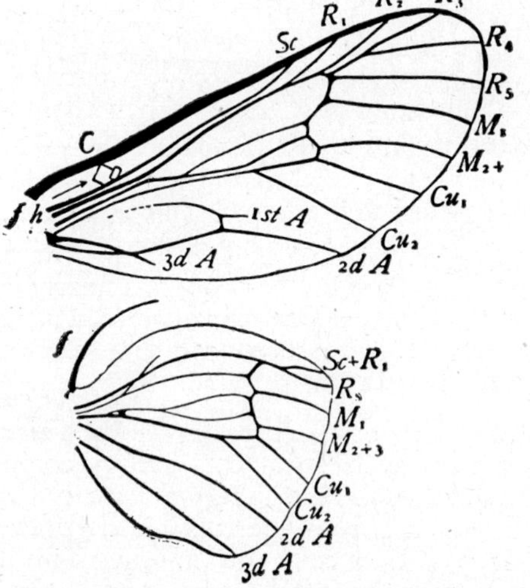

Fig. 748.—Bag of *Oiketicus abboti*.

Only about twenty species are known from our fauna, of which the following are most likely to be observed.

Abbot's bag-worm, *Oikĕticus abbŏti*.—This species occurs in the more southern part of our country. The larva makes a bag with sticks attached to it crosswise (Fig. 748). The adult male is sable brown, with a vitreous bar at the extremity of the discal cell of the fore wings; the narrow external edging of the wings is pale; the expanse of the wings is 33 mm.

Fig. 749.—Wings of *Thyridopteryx ephemeræformis*.

The evergreen bag-worm or the bag-worm, *Thyridŏpteryx ephemerǣfŏrmis*.—This species prefers red cedar and arbor-vitæ, and for this reason has been named the evergreen bag-worm; but it also feeds on many other kinds of trees, and as it is the species that is most likely to attract attention, and is sometimes a serious pest, it is often called the bag-worm. It is our best-known species, and its life-history will serve as an illustration of the habits of the members of the family Psychidæ.

The bag of this species is about the same size as that of Abbot's bag-worm (Fig. 748); but it differs in being covered with bits of leaves when it feeds on cedar or arbor-vitæ, or with twigs attached lengthwise when it feeds on other trees. When full-grown the larva fastens the bag to a twig with a band of silk, and then changes to a pupa. When the male is ready to emerge, the pupa works its way to the lower end of the bag and halfway out of the opening at the extremity. Then its skin bursts and the adult emerges. The male moth has a black, hairy body and nearly naked wings (Fig. 750). The adult female partly emerges from the pupa skin and pushes her way to the lower end of the bag, where she awaits the approach of the male. She is entirely destitute of wings and legs. The genitalia of the male can be greatly extended, making possible the pairing while the female is still in the bag. After pairing, the female works her way back into the pupa skin, where she deposits her eggs mixed with the hair-like scales from the end of her body. She then works her shrunken body out of the bag, drops to the ground, and perishes. The eggs remain in the pupa skin in the sac till the following spring.

Fig. 750.—*Thyridopteryx ephemerǣformis*.

Where this insect is a pest, two methods of control are practiced, first, the bags are collected and destroyed in the winter, while they still contain the eggs; second, when impracticable to collect the bags: on account of the height of the infested trees, a spray of arsenate of lead is used in the spring as soon as possible after the larvæ appear.

Eurycȳttarus confederāta.—This is a smaller species than the two preceding ones. Figure 751 represents the sac of a male with the empty pupa-skin projecting from the lower end, and Figure 752 the fully developed male.

Fig. 752.

Solenōbia walshĕlla.—This is a small tineid-like species; the male has a wing-expanse of about 13 mm. and the hind wings have a quite wide fringe. The fore wings are light gray speckled with brown. The bag of the larva is about 8 mm. long, made of silk, and covered with fine grains of sand or with particles of lichens and excrement of the larva. Chambers states that he has sometimes found small molluscan shells adhering to it. The larvæ are found on the trunks of trees and feed

Fig. 751.

upon lichens. Figure 753 represents the venation of the wings of a European species of this genus.

Family TISCHERIIDÆ

The vertex of the head is clothed with erect, broad, and short scales. The antennæ are long, with the first segment small. The maxillæ are longer than the head and thorax. The maxillary palpi are small or absent. The labial palpi are short, porrect, and without bristles on the outer side of the second segment. In the front wings (Fig. 754), the costal margin is strongly arched, the apex is prolonged into a sharp point, the discal cell is long and narrow, the accessory cell is very long, and the base of media is preserved. The hind wings are long and narrow and with greatly reduced venation. (Fig. 754). The hind tibiæ are very hairy.

Fig. 753.—Wings of *Solenobia*. (After Spuler.)

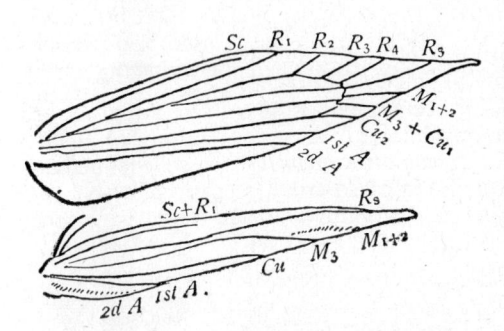

Fig. 754.—Wings of *Tischeria marginea*. (After Spuler.)

Nearly all of our species belong to the genus *Tischēria*. The larvæ lack thoracic legs; most of them make blotch mines in the leaves of oak; but the following one infests apple; and some other species infest blackberry and raspberry.

The trumpet-leaf miner of apple, *Tischēria malifoliĕlla*.—This species infests the leaves of apple over the Eastern half of the United States and Canada, and sometimes does serious injury. The larva makes a trumpet-shaped mine just beneath the epidermis on the upper side of the leaf; the first half of the mine is usually crossed by crescent-shaped stripes of white. There are two generations annually in the North, and several in the South. The larvæ pupate in their mines. The larvæ of the last generation line their mines with silk and pass the winter in them. They transform to pupæ in the spring and emerge as adults eight or ten days later. The adult moth expands about 6 mm.; it has shining dark brown front wings, tinged with purplish and dusted with pale yellowish scales.

To control this pest, plow the orchard after the leaves have fallen, or rake and burn the fallen leaves.

Family LYONETIIDÆ

Moths with the head smooth, at least on the front. The scape of the antennæ usually forms an eye-cap. The ocelli and the maxillary palpi are wanting. The labial palpi are usually very small. The wings are very narrow (Fig. 755); the hind wings are often linear, with the radial sector extending through the axis of the wing. The apices of the fore wings are usually warped up or down. The larvæ are leaf-miners or live in webs between leaves. The following species will serve as examples of this family.

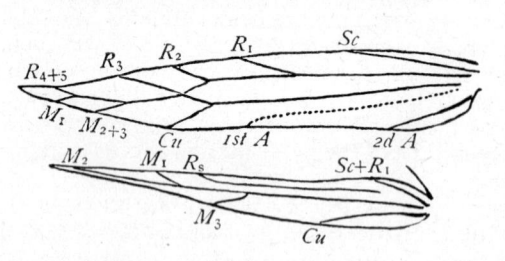

Fig. 755.—Wings of *Bedellia somnulentella*. (After Clemens.)

The morning-glory leaf-miner, *Bedĕllia somnulentĕlla*.—The young larva makes a serpentine mine with a central line of frass; later it leaves this mine and makes a blotch mine. The pupa is naked, and fixed by the caudal end to some cross-threads on the under side of the leaf. The adult is yellow and expands about 10 mm.

The apple bucculatrix, *Bucculātrix pomifoliĕlla*.—The larva of this species infests the leaves of apple, and when full-grown it makes a small white cocoon which is attached to the lower surface of a twig. These cocoons sometimes occur in great numbers, side by side, on the twigs of an infested tree (Fig. 756). They are easily recognized by their shape, being slender and ribbed lengthwise. It is these cocoons that usually first reveal the presence of the pest in an orchard. They are very conspicuous during the winter, when the leaves are off the trees. At this time each cocoon contains a pupa. The adult moth emerges in early spring. The eggs are laid on the lower surface of the leaves. Each larva when it hatches bores directly from the egg to the upper surface of the leaf, where it makes a brown serpentine mine. When these

Fig. 756.—Cocoons of *Bucculatrix pomifoliella*.

mines are abundant in a leaf, it turns yellow and dies. When the larva has made a mine from 12 to 18 mm. in length, which it does in from four to five days, it eats its way out through the upper surface. Then somewhere on the upper surface of the leaf it weaves a circular silken covering about 2.5 mm. in diameter. Stretched out on this network, the larva, which is now about 2.5 mm. long, makes a small hole in it near the edge, then, as one would turn a somersault, it puts its head into this hole and disappears beneath the silken covering, where it undergoes a change of skin. It remains in the molting cocoon usually less than 24 hours. After leaving this cocoon it feeds upon the leaves without making a mine; and in a few days makes a second molting cocoon which differs from the first only in being about 3 mm. in diameter. After leaving this it again feeds for a few days, and then migrates to a twig where it makes the long ribbed cocoon within which the pupa state is passed. The adult is a tiny, light brown moth, with the fore wings whitish, tinged with pale yellowish, freely dusted with brown; on the middle of the inner margin there is a dark brown oval patch.

The genus *Bucculatrix*, to which the above species belongs, is placed by some writers in a separate family, the *Bucculatrigidæ*.

The family OPOSTEGIDÆ has been established for the genus *Opostega*, of which only three species have been found in this country. These are moths with folded maxillary palpi, with the scape of the antennæ forming a large eye-cap, and with radius, media, and cubitus of the fore wings unbranched. The hind wings are linear. The combination of the eye-cap and the unbranched veins of the fore wings is a distinctive feature of this **family.**

The larvæ are very slender, cylindrical, without legs, and are bast-miners.

Family OINOPHILIDÆ

This family includes "strongly flattened moths, with flat coxæ closely appressed to the body, usually with smooth heads, rising to a rounded ridge between the antennæ, but often with a loose tuft on the vertex, and rather small maxillary palpi of the folded type. The labial palpi have a well-set-off, fusiform, terminal joint as in the Tineidæ, and are normally without bristles. The venation in the known genera is more or less reduced." (Forbes.)

Only one species representing this family has been found in our fauna. This is *Phæoses sabinella*, described by Forbes ('22), from Louisiana and Mississippi. It is a shining gray-brown (mouse gray) moth, with a wing-expanse of 9 mm.

The known larvæ of this family feed on decaying vegetable matter and fungi.

Family GRACILARIIDÆ

The vestiture of the head varies greatly; the vertex is clothed with prominent scales in some forms, in others it is smooth. The antennæ are long; the scape forms an eye-cap in some species and not in others. The fore wings are lanceolate, normal or with somewhat reduced venation (Fig. 757); usually without an accessory cell, but sometimes one is present in the genus *Parornix*. The hind wings are lanceolate or linear; in many members of the family they are expanded near the base, forming a more or less prominent hump in the costal margin, and in some species vein R_1 is free, not coalesced with vein Sc.

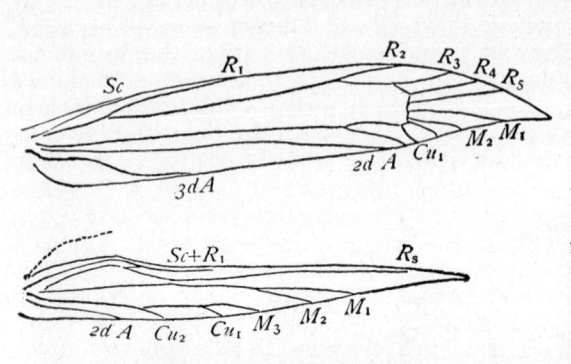

Fig. 757.—Wings of *Gracilaria*. (After Spuler.)

The adult moths when at rest elevate the front part of the body, the fore legs being held vertically so that the tips of the wings touch the surface on which the insect rests.

The larvæ are extraordinary; when young they are very much flattened and have thin, blade-like mandibles and vestigial maxillæ and labium; they merely slash open the cells of the leaf and suck up the cell-sap; later they usually have normal mouth-parts and eat the parenchyma. The young larvæ always make a flat blotch mine; later they make a blotch mine in which the epidermis of one side of the leaf is thrown into a fold by the growth of the leaf, *i. e.*, a *tentiform mine*, or they roll a leaf. The larvæ have only fourteen legs or none, never any on the sixth segment of the abdomen.

This is a large family; about two hundred North American species have been described, and doubtless many more are to be discovered.

About one-half of our described species belong to the genus which is commonly known as *Lithocollētis*, but which is termed *Phyllonoryc̆ter* by those who recognize the names in the "Tentamen" of Hübner. The following species will serve as an example of this genus.

The white-blotch oak-leaf miner, *Phyllonoryc̆ter hamadryadĕlla*.—This little miner infests the leaves of many different species of oaks, and is very common throughout the Atlantic States. The mine is a whitish blotch mine in the upper side of the leaf, and contains a single larva; but often a single leaf contains many of these mines (Fig. 758). The young larva is remarkable in resembling more the larva of a beetle than the ordinary type of lepidopterous larvæ (Fig. 758, *b*). It is nearly flat; the first thoracic segment is much larger than any of the others; the body tapers towards the hind end; and there are only the faintest rudiments of legs discernible. The larvæ molt seven times. At the seventh molt the form of the body undergoes a striking change.

LEPIDOPTERA 619

It now becomes cylindrical in form, there is a great change in the shape of the mouth-parts, and the fourteen feet are well developed. The full-grown cylindrical larva measures about 5 mm. in length. It spins a cocoon, which is simply a delicate, semi-transparent, circular sheet of white silk, stretched over a part of the floor of the mine. The pupa is dark brown in color, and bears a toothed crest upon its head (Fig. 758, *n*, *o*), which enables it doubtless to pierce or saw its way out from the cocoon. The moth is a delicate little creature, whose wings expand a little more than 6 mm. The fore wings

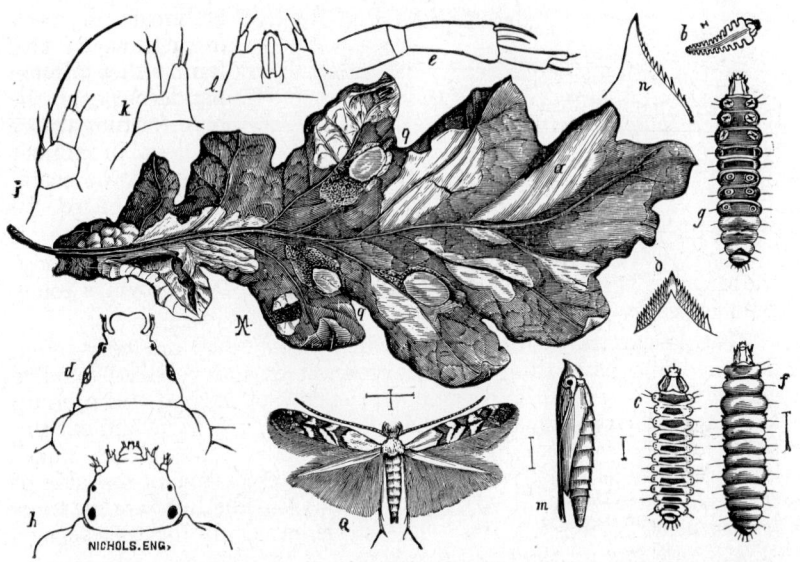

Fig. 758.—*Phyllonorycter hamadryadella*: *a*, mine; *b*, young larva; *c*, full-grown, flat-form larva; *d*, head of same, enlarged; *e*, antenna of same, enlarged; *f*, round-form larva from above; *g*, same from below; *h*, head of same, enlarged; *i*, antenna of same, enlarged; *k*, maxilla and palpus of same, enlarged; *l*, labium, labial palpi, and spinnerets of same; *m*, pupa; *n*, side view of pupal crest; *o*, front view of same; *q*, cocoon; Q, moth. (From the author's Report for 1879.)

are white, with three, broad, irregular, bronze bands across each, and each band is bordered with black on the inner side. The hind wings are silvery.

As this insect passes the winter as a larva within the dry leaves, the best way to check its ravages when it becomes a pest is to rake up and burn such leaves.

Another common oak-leaf miner in the East is *Phyllonorӳcter cincinnatiĕlla*. The larvæ form large blotch mines on the upper surface of leaves. In this species the larvæ are social, one mine often containing from several to a dozen larvæ. The loosened epidermis is brownish yellow, somewhat puckered, and often covers nearly the entire leaf. This species like most other gracilariids passes the winter as pupæ.

A common miner in the leaves of locust is *Parĕctopa robiniĕlla*. The larva makes on the upper surface of the leaf what has been termed a *digitate mine*, that is a blotch mine with a number of lateral galleries running out from it on each side.

Several members of this family make tentiform mines in the leaves of apple and other fruit trees; but these species are rarely of economic importance.

Family COLEOPHORIDÆ*

Moths with a smooth head, without ocelli, and without maxillary palpi. The labial palpi are of moderate size. The antennæ are held extended forward in repose. The wings are very narrow. The discal

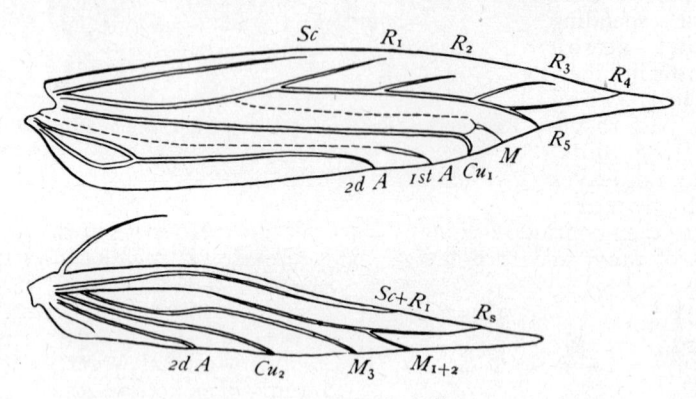

Fig. 759.—Wings of *Coleophora laricella*.

cell of the fore wing extends obliquely; vein Cu_1 and vein Cu_2 when present are very short (Fig. 759).

The larvæ are usually leaf-miners when young or feed within seeds; later, with few exceptions, they are case-bearers.

Nearly all of our species belong to the genus *Coleophora*, of which about ninety species have been found in this country. The two following species are those that have attracted most attention on account of their economic importance.

The pistol case-bearer, *Coleŏphora malivorĕlla*.—The larva of this species infests apple especially but is also found on quince, plum, and cherry. The larvæ hatch in mid-summer from eggs laid on the leaves and eat little holes in the leaves. They soon construct little pistol-

---

*The typical genus of this family is commonly known as *Coleophora*, the name used for it by Hübner in his "Tentamen." But those writers who do not recognize the "Tentamen" as a published work use the later name *Haploptĭlia* for the genus, and name the family the HAPLOPTILIIDÆ.

shaped cases composed of silk, the pubescence of leaves, and excrement. The larva projects itself out from the case far enough to get a foothold and eats irregular holes in the leaf, holding the case at a considerable angle with the leaf. About September first the larvæ migrate to the twigs where they fasten the cases to the bark (Fig. 760) and hibernate till April, spending about seven months in hibernation. They then pass to the swelling buds, expanding leaves and flowers, where they continue feeding.

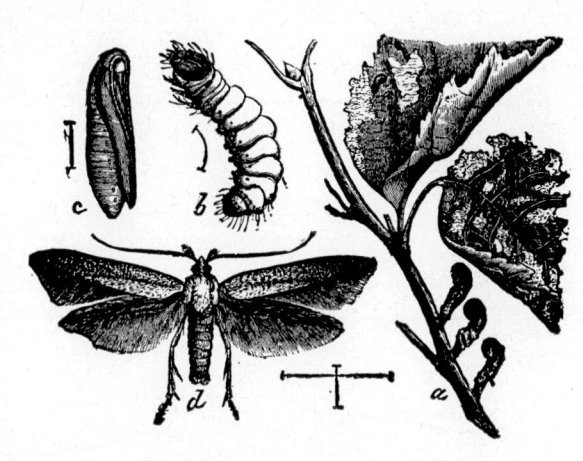

Fig. 760.—*Coleophora malivorella:* a, apple twig showing larval cases and work on leaves; b, larva; c, pupa; d, moth; b, c, d, enlarged. (After Riley.)

They become full-grown in the latter part of May, and then fasten their cases to the smaller branches. After the case is fastened to the branch the larva turns around in it, and changes to a pupa; consequently the moth emerges from the curved end of the case.

The cigar case-bearer, *Coleŏphora fletcherĕlla.*—This species, like the preceding one, is a pest of apple and other fruit trees, and resembles that species to a considerable extent in habits. In this species the young larvæ are miners in the leaves for two or three weeks before making their cases. The case (Fig. 761) is composed of fragments of leaves fastened together by silk.

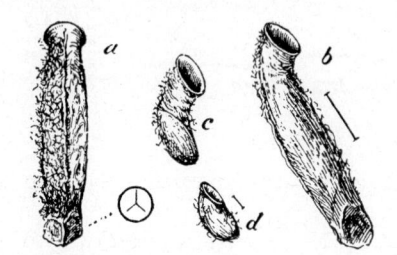

Fig. 761.—Cases of the cigar case-bearer. (After Hammar.)

Family ELACHISTIDÆ

The head is smooth. The scape of the antennæ does not form an eye-cap. The venation is but slightly reduced (Fig. 762). The hind wings are lanceolate, with a well formed discal cell.

The larvæ have sixteen legs. Most of the known species make blotch mines in grasses. And some at least when full-grown leave the mine and weave a slight web from which the pupa hangs suspended, like the pupa of a butterfly.

This is a small family; most of our species belong to the genus *Elachista*.

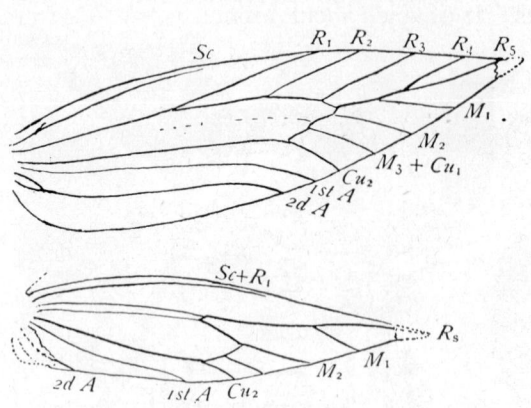

Fig. 762.—Wings of *Elachista quadrella*. (After Spuler.)

Family HELIOZELIDÆ

The antennæ are from one-half to two-thirds as long as the front wings; the scape is short and not enlarged so as to form an eye-cap. The wings (Fig. 763) are lanceolate; in the hind wings there is no discal cell, owing to the coalescence of the radial sector and media for nearly the entire length of the wing, vein R_s separating near the apex of the wing.

The habits of the larvæ are well-illustrated by the following species.

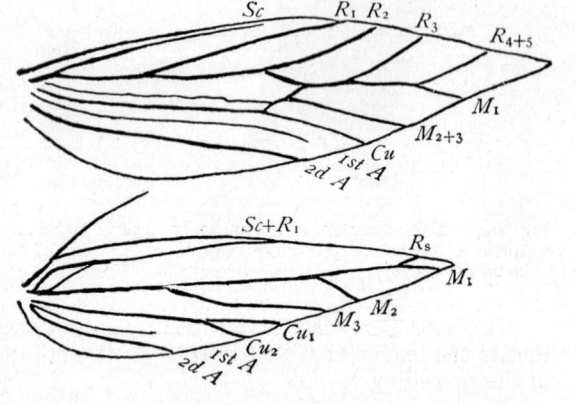

The resplendent shield-bearer, *Coptodĭsca splendorif-erĕlla*.—This species infests the leaves of apple, pear, quince, thorn-apple, and wild cherry. The larva is both a miner and a case-bearer. It at first makes a linear mine; but later this is enlarged into a blotch mine.

Fig. 763.—Wings of *Antispila pfeifferella*. (After Spuler.)

When full-grown the larva makes an oval case cut from the walls of its mine and lined with silk. It then seeks a safe place in which to fasten this case. This is usually on the trunk or on a branch of the infested tree (Fig. 764, *d*). There are two generations annually. The second generation pass the winter as larvæ within their cases. The adult (Fig. 764, g), is a brilliantly colored, golden-headed moth. The basal half of the front wings is leaden-gray with a resplendent luster and the remainder golden with silvery and dark brown streaks. It expands about 5 mm.

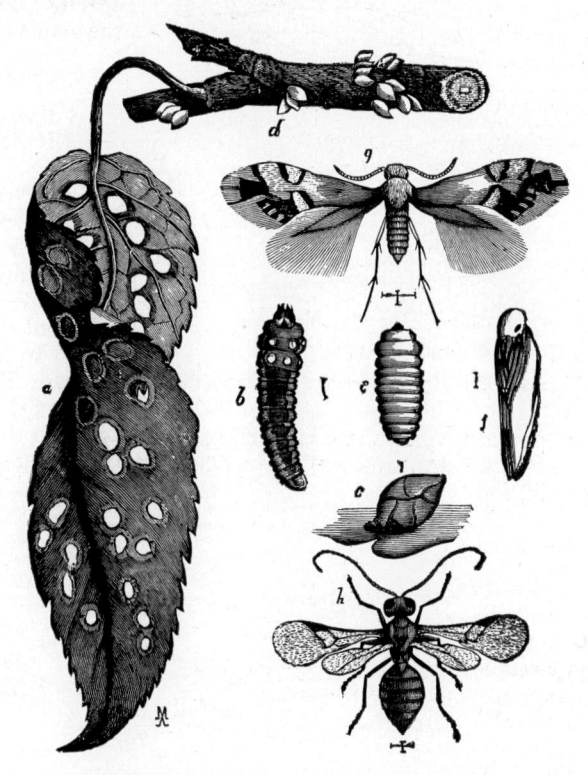

Fig. 764.—*Coptodisca splendoriferella:* *a*, leaf of apple showing work; *b*, summer larva; *c*, larva in case travelling; *d*, cases tied up for winter; *e*, hibernating larva; *f*, pupa; *g*, moth, *h*, parasite. (From the Author's report for 1879.)

The sour gum case-cutter, *Antispīla nyssæfoliĕlla.*—This species infests the leaves of *Nyssa sylvătica*. Its habits are similar to those of the preceding species.

Family DOUGLASIIDÆ

The scape of the antennæ is small and does not form an eye-cap. The first segment of the labial palpi is small. The ocelli are large. The hind wings are lanceolate and without a discal cell, owing to the coalescence of the radial sector and media. Vein R_s separates from media near the middle of the length of the wing (Fig. 765).

This family is represented in this country by a single species, *Tinăgma obscurofasciĕlla*, the larva of which is a leaf-miner in Rosaceæ.

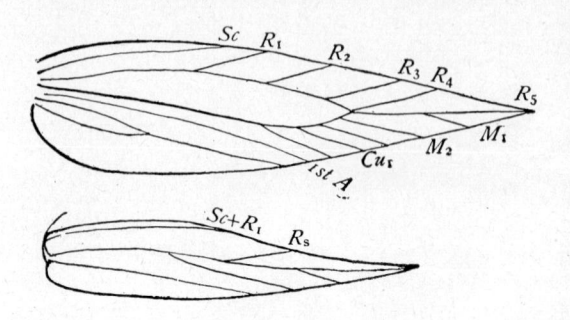

Fig. 765.—Wings of *Tinagma obscurofasciella*. (After Chambers.)

FAMILY ŒCOPHORIDÆ

The head is usually smooth, with appressed scales; sometimes with loose scales and spreading side tufts. The antennæ usually have a comb of bristles on the scape. The labial palpi are well-developed, generally curved upward; the terminal segment is acutely pointed. The maxillary palpi are vestigial. The wings are fairly broad, sometimes ample (Fig. 766). The venation is but little reduced. In the fore wings veins R_4 and R_5 are stalked or coalesce throughout; veins R_2 and Cu_2 arise well back from the end of the discal cell; and vein 1st A is preserved. In the hind wings veins R_s and M_1 are well separated and extend parallel. The posterior tibiæ are clothed with rough hairs above.

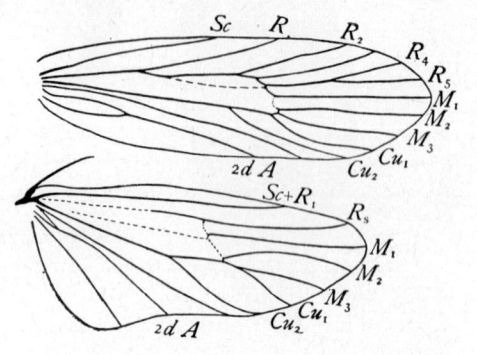

Fig. 766.—Wings of *Depressaria heracliana*.

The larvæ have sixteen legs; they are often prettily marked with dark tubercles on whitish or yellowish ground. The different species vary in their habits; the majority of them either live in webbed-together leaves or blossoms or feed in decayed wood; one species, *Ĕndrosis lacteĕlla*, is a stored-food pest in California and in Europe.

About one hundred species have been described from our fauna; many of them are common. A generic revision of the American species was published by Busck ('09 a). The following one is a well-known pest.

The parsnip webworm, *Depressāria heracliāna*.—The larvæ of this species web together and devour the unfolding blossom-heads of parsnip, celery, and wild carrot. After the larvæ have consumed the flowers and unripe seeds and become nearly full-grown, they burrow into the hollow stems and feed upon the soft lining of the interior. Here inside the hollow stem they change to pupæ. The moths appear in late July and early August, and soon go into hibernation in sheltered places.

Family ETHMIIDÆ

This family includes a small number of moths, which were formerly included in the family Œcophoridæ. The family Ethmiidæ was established by Busck ('09b), who states that the main structural character of the imago by which this family can be distinguished from the Œcophoridæ is the proximity of vein M_2 in the hind wings to vein M_1 instead of to vein M_3 as in the Œcophoridæ, it being radial not cubital. Fracker ('15) describes larval characters distinctive of the typical genus *Ĕthmia*.

The members of this family have broad wings. The fore wings are usually bright colored, with striking patterns, often black and white.

The larvæ, as a rule, are social, living in a light web. They feed chiefly on plants of the family Borraginaceæ.

Most of the species belong to the genus *Ĕthmia* of which about thirty are now known.

Family STENOMIDÆ

This family includes large moths as compared with most "micros." The wings are broad, especially the hind wings. In the fore wings all of the branches of the branched veins are typically present. In the hind wings vein M_1 is joined at its base to vein R_s.

The larvæ live in webs on leaves, especially of oak.

There are about twenty North American species, most of which belong to the genus *Stenoma*.

A common species in the Atlantic States is *Stenōma schlægeri*. This is one of our larger species, having a wing-expanse of 30 mm. The moth is of a dirty white color with the fore wings mottled with darker bands and spots, and with a conspicuous patch of brown scales near the base of the inner margin. When at rest on a leaf the insect folds its wings closely about its body, and resembles in a striking manner the excrement of a bird.

Family GELECHIIDÆ

The head is smooth or at most slightly ruffled. The labial palpi are long, curved, ascending, and usually with the terminal segment acutely pointed. The maxillary palpi are vestigial or wanting. The venation of the wings (Fig. 767) is more or less reduced; the stem of vein M is wanting; vein 1st A is wanting in the fore wings; and sometimes in the hind wings also. In the fore wings the second anal vein is forked at the base, *i. e.*, the tip of the third anal vein unites with it; and in some forms, veins R_4 and R_5 coalesce throughout their length, but they are usually stalked. The hind wings are usually more or less trapezoidal; and the outer margin is usually sinuate or emarginate below the apex.

The larvæ vary greatly in habits; some are leaf-miners; but more feed in rolled or spun together leaves or in stems or seed heads; and one is a serious pest of stored grain.

This is the largest family of the Microfrenatæ; more than four hundred species have been described from our fauna. A revision of the American moths of this family was published by Busck ('03).

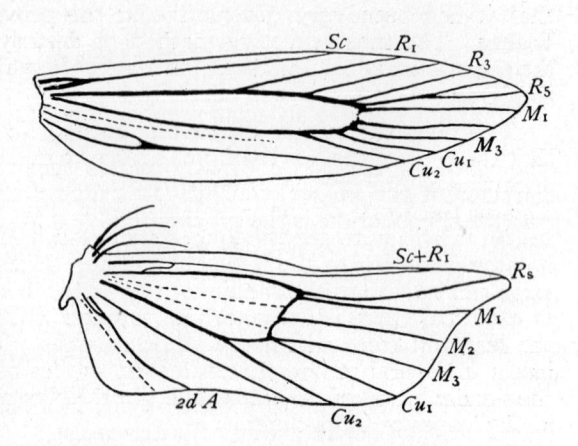

Fig. 767.—Wings of *Pectinophora gossypiella*. (After Busck.)

The Angoumois grain-moth, *Sitōtroga cereălĕlla*.—The larva of this moth feeds upon seeds, and especially upon stored grain. It occurs throughout our country; but it is especially destructive in the Southern States. In that part of the country it is extremely difficult to keep grain long on account of this pest and cer-

Fig. 768.—*Paralechia pinifoliella*: larva, pupa, adult, and leaves mined by the larva. (From the Author's Report for 1879.)

tain beetles that also feed on stored grain. The adult moth is of a very light grayish-brown color, more or less spotted with black; it expands about 12 mm. The common name is derived from the fact that it has been very destructive in the province of Angoumois, France. The most effective method of destroying this pest is by the use of carbon bisulphide in the manner in which it is used against the grain-weevils, already described.

The pine-leaf miner, *Paralēchia pinifoliĕlla.*—It often happens that the ends of the leaves of pine present a dead and brown appearance that is due to the interior of the leaf having been eaten out. This is the work of the pine-leaf miner (Fig. 768). At the right season it is easy to see the long, slender larva in its snug retreat by holding a leaf up to the light and looking through it; and later the pupa can be seen in the same way. Near the lower end of the tunnel in each leaf there is a round hole through which the larva entered the leaf and from which the adult emerges. We have found this insect in several of the stouter-leaved species of pine, but never in the slender leaves of the white pine. In the North it is most abundant in the leaves of pitch-pine.

The peach twig-borer, *Anărsia lineatĕlla.*—This pest is generally distributed throughout the United States and Canada, and sometimes it destroys a large part of the crop in some localities. The young larvæ hibernate in small cavities which they excavate in the bark of young twigs. In the spring the larvæ burrow into the tender shoots; the leaves of the buds unfold and then wither. There are several generations annually. The summer generations attack both twigs and fruit.

The solidago gall-moth, *Gnorimoschēma gallæsolidăginis.*—There are two kinds of conspicuous galls which are enlargements of the stems of golden-rod; one of these is a ball-like enlargement of the stem and is caused by the larva of a fly, *Eurosta solidaginis*, described in the next chapter; the other is spindle-shaped (Fig. 769) and is caused by the moth named above. The eggs are laid on the old plants in the fall and hatch in spring. The young larva crawls to a new shoot and boring down into it causes the growth of the gall. The larva becomes full-grown about the middle of July; then before changing to a pupa it eats a passage-way through the wall of the gall at its upper end, and closes the opening

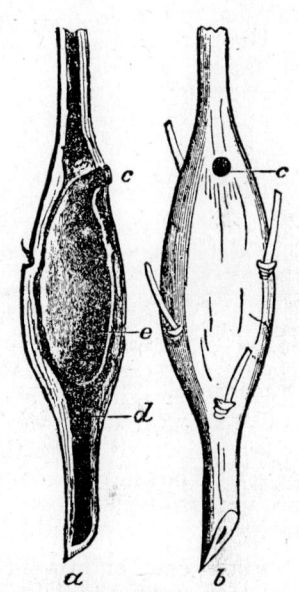

Fig. 769.—Gall of the solidago gall-moth. (After Riley.)

with a plug of silk, which is so formed that it can be pushed out by the adult moth when it is ready to emerge.

Some members of the family are leaf-rollers. Figure 770 represents a leaf rolled by a gelechiid larva, probably *Anacămpsis* **innocuella.** This species infests poplar.

The pink bollworm, *Pectinŏphora gossypiĕlla.*—This species is regarded as one of the most destructive cotton insects known and ranks among the half-dozen most important insect pests of the world. It often reduces the yield of lint fifty per cent. or more and materially lessens the amount of oil obtained from the seeds.

The adult is a small dark-brown moth, with a wing-expanse of from 15 to 20 mm. Figure 767 represents the shape and the venation of the wings. The larva eats the seeds and tunnels and soils the lint, causing the arrest of growth and the rotting or premature and imperfect opening of the boll (Busck).

A detailed account of this pest, illustrated by many figures was published by Busck ('17).

Family BLASTOBASIDÆ

The scape of the antennæ is armed with a fringe of strong bristles, or *pecten*. The labial palpi are slender and upturned or vestigial.

Fig. 770.— Leaf rolled by a gelechiid larva.

The discal cell of the fore wings (Fig. 771) is long compared with the lengths of the apical veins (R_2 to Cu_2); and these veins arise from the extreme end of the cell. As vein R_1 arises near the base of the wing it is unusually distant from vein R_2; to make up for the resulting weakening of the wing, the membrane is more or less thickened along the costa; this thickening is the so-called stigma. The hind wings are lanceolate, and rather narrower than the fore wings. Veins R_s and M_1 are well separated at the end of the discal cell. Veins M_2, M_3 and Cu_1 are close together or coincident.

About one hundred species have been described from our fauna; among them are the following.

The acorn-moth, *Valentĭnia glandulĕlla.*—The larva of this species lives as a scavenger in acorns that have been destroyed by acorn-weevils, **Balaninus.** The moth lays an egg in the destroyed acorn after the beetle has left it, and the larva hatching from this egg feeds upon the crumbs left by the former occupant. The larva passes the winter within the acorn. The moths emerge at various times throughout the summer.

The oak-coccid blastobasid, *Zenodŏchium coccivorĕlla*.—The larva is an internal parasite in the gall-like females of the coccid genus *Kermes*. I found it common at Cedar Keys, Fla.

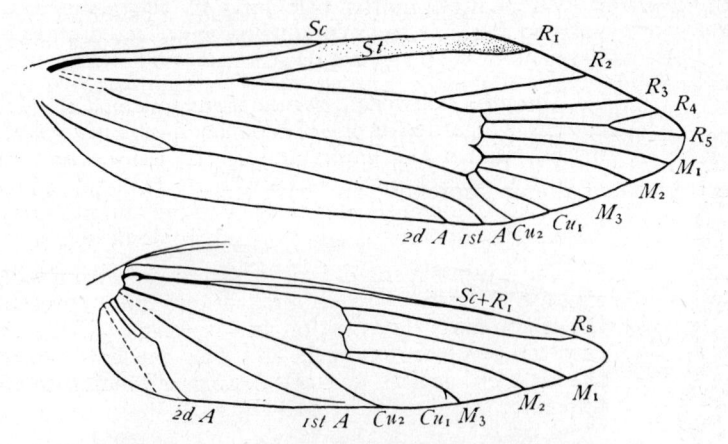

Fig. 771.—Wings of *Holcocera*. (After Forbes.)

Family COSMOPTERYGIDÆ

The moths grouped together in this family vary greatly in structure. The fore wings are lanceolate, sometimes caudate, *i. e.* with the apex greatly prolonged. Vein 1st A arises out of vein 2d A or is lost. The hind wings are lanceolate or linear. Vein R_1 is occasionally separate from vein Sc. Veins R_s and M_1 are close together.

The following species will serve as examples of members of this family.

The palmetto-leaf miner, *Homalĕdra sabalĕlla*.—This species occurs only in the South where the saw-palmetto grows; but it is of general interest as illustrating a peculiar type of larval habit. The larvæ can hardly be said to be leaf-miners; for they feed upon the upper surface of the leaf, destroying the skin as well as the fleshy part of the leaf. They are social, working together in small companies, and make a nest consisting of a delicate sheet of silk covering that part of the leaf upon which they are feeding; this sheet is covered with what appears like sawdust, but which is really a mass of the droppings of the larvæ (Fig. 772). The full-grown larva attains a length of 12 mm. The pupa state is passed within the nest made by the larvæ. The moth expands 15 mm. Its general color is a delicate silvery gray, with a tinge of lavender in some individuals.

The cat-tail moth, *Lymnæcia phragmitĕlla*.—The larva of this species feeds in the heads of cat-tail, *Typha*. It winters in the head, which presents a tattered and frayed appearance. The larvæ spin an abundance of silk, thereby tying the down or pappus together and

630   AN INTRODUCTION TO ENTOMOLOGY

keeping it from blowing away. The overwintering larvæ are half-grown. When full-grown some transform in the heads, but many go down and bore in the stems and transform there.

Cosmopteryx.—"The little moths belonging to the genus *Cosmopteryx* are probably familiar to anyone who has collected and observed insects in nature. Who has not occasionally on a warm midsummer day met with a slender little streak of gold and silver sitting in the sunshine on a leaf in a protected corner and twirling its long white-tipped antennæ in graceful motions If, when examined more closely, it is found to be a smooth shining little moth, brown with silvery lines on palpi and antennæ, and with a striking broad golden

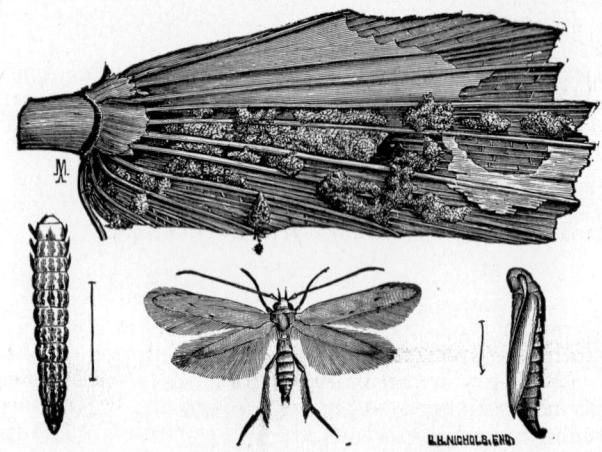

Fig. 772.—*Homaledra sabalella:* larva, pupa, adult, and part of injured leaf. (From the Author's Report for 1879.)

or orange fascia across the outer half of the wing, bordered on both sides by bright metallic scales, then you have a *Cosmŏpteryx.*"

"The larvæ are leaf-miners, and the mines are easily distinguished from most others by the scrupulous cleanliness with which the larva ejects all its frass through a hole, so that the mine remains clear and white. At maturity the larva changes its color from green to a vivid purple or wine-red, leaves the mine, and spins a matted flattened cocoon of silk." (Busck '06).

Among the better-known members of this family are the following:

Stagmatŏphora gleditschiæélla.—The larva burrows in the thorns of locust.

Mŏmpha eloisélla.—There are several species of *Mompha* that infest the fruit and pith of the evening primrose. The best-known of these is this one.

Psacăphora terminélla.—The larva is a miner in willow-herb, *Epilobium.*

Family SCYTHRIDIDÆ

This family includes a group of genera that are closely allied to the Yponomeutidæ and are included in that family by some writers. I do not find that any tangible characters of the adult insects separating the two families have been pointed out; but there appear to be differences in the setal characters of the larvæ (see Fracker '15).

The family is represented in our fauna by only two genera, *Scythris* and *Epermēnia*, including twenty-two species. None of these species has attracted attention on account of its economic importance.

The larvæ of *Scȳthris magnatĕlla* feed on *Epilobium*. They are solitary when small, folding over half of the leaf to the midrib in the central part of its length, attached with web. Later they form considerable web among the leaves. The pupa is formed in a delicate, flossy web. (Dyar).

The larva of *Epermĕnia pimpinĕlla* feeds by forming a puffy mine on *Pimpinella integerrima*. The pupa is inclosed in a frail, open-meshed cocoon on the under side of a leaf or in angles of leaf-stalks. (Murtfeldt.)

One of the more common representatives of this family is *Scȳthris eboracĕnsis*. The adult is a small black moth tinged with violet, with a wing-expanse of about 10 mm. It is found on flowers.

Family YPONOMEUTIDÆ*

In this family the ocelli are small or absent. In the more typical forms the wings are comparatively broad, with the venation but little reduced. In the fore wings all of the branches of the branched veins are usually separate, and vein R_5 extends to the outer margin. In the hind wings veins R_s and M_1 are well-separated. The first anal vein is distinct in both fore and hind wings.

Writers differ greatly as to the limits of this family; some include in it certain genera or groups of genera that by others are regarded as distinct families.

In its restricted sense the family Yponomeutidæ includes about fifty North American species; among these are the following.

The cedar tineid, *Argyrĕsthia thuiĕlla*.—This is a small narrow-winged moth, which expands about 8 mm. Its ground color is pearly-white, with the fore wings dotted and marked with brown, especially on the outer half of the wing. The larva feeds on the leaves of cedar, and when full-grown spins a small, conspicuous white cocoon attached to a leaf.

The apple fruit-miner, *Argyrĕsthia conjugĕlla*.—The larva of this species is a serious pest in the apple orchards of western Canada. It is pinkish white in color and about 9 mm. in length. It burrows in all directions through the fruit, causing it to decay. The winter is passed in the pupal state. The cocoons are made under the bark on the trunk of the tree or under leaves on the ground; they are white,

---

*An emended form of this family name, Hyponomeutidæ, is used by some writers.

and the outer layers have the threads arranged so as to form a beautiful openwork pattern. The adult moth has a wing-expanse of about 12 mm. It is figured by Slingerland and Crosby ('14).

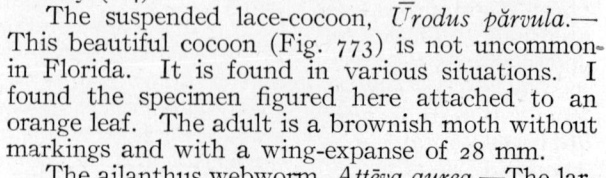

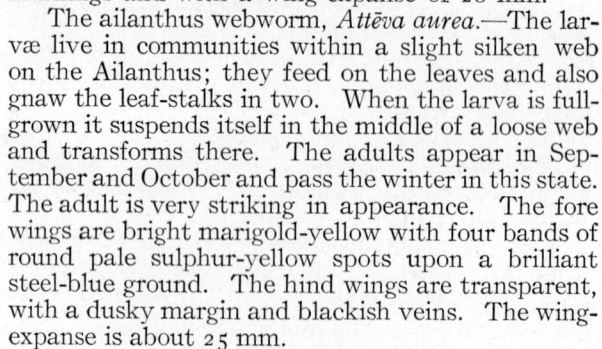

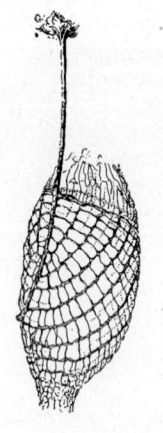

The suspended lace-cocoon, *Urodus părvula.*—This beautiful cocoon (Fig. 773) is not uncommon in Florida. It is found in various situations. I found the specimen figured here attached to an orange leaf. The adult is a brownish moth without markings and with a wing-expanse of 28 mm.

The ailanthus webworm, *Attēva aurea.*—The larvæ live in communities within a slight silken web on the Ailanthus; they feed on the leaves and also gnaw the leaf-stalks in two. When the larva is full-grown it suspends itself in the middle of a loose web and transforms there. The adults appear in September and October and pass the winter in this state. The adult is very striking in appearance. The fore wings are bright marigold-yellow with four bands of round pale sulphur-yellow spots upon a brilliant steel-blue ground. The hind wings are transparent, with a dusky margin and blackish veins. The wing-expanse is about 25 mm.

Fig. 773.—Cocoon of *Urodus parvula*

The ermine-moths, *Yponomeūta.*—There are several species of the typical genus of this family that have received the common name ermine-moths, because of the color of their fore wings, which are snowy white dotted with black. One of these, *Yponomeūta padĕlla,* is an introduced species which is an apple and cherry pest. The larvæ live in a common web, and in this they spin their cocoons.

The name ermine-moths is applied also, especially in England, to some of the Arctiidæ that are white spotted with black.

Family PLUTELLIDÆ

This family is closely allied to the Yponomeutidæ and is regarded by many writers as a subfamily of that family. These moths differ from the Yponomeutidæ in that they hold the antennæ extended forward in repose; in this respect they resemble the Coleophoridæ. The larvæ differ from those of the Yponomeutidæ and the Scythrididæ in that their prolegs are longer than wide.

About fifty North American species have been described; these represent nine genera. The most important species from an economic standpoint is the following one.

The diamond-back moth, *Plutĕlla maculipĕnnis.*—The larva of this species infests cabbage and other cruciferous plants, eating holes of variable size and irregular form in the leaves. It is sometimes also a pest in greenhouses, infesting stocks, wall-flowers, sweet alyssum, and candytuft. The larva when full-grown spins a lace-like cocoon attached to a leaf. The moth expands about 15 mm. The fore wings of the male are ash-colored, with a yellow stripe outlined by a wavy dark line extending along the inner margin. When the wings are closed, the united yellow stripes form a row of three diamond-shaped markings. These suggested the common name of the species. In the female the front wings are a nearly uniform gray. The hind wings in both sexes are a dull gray.

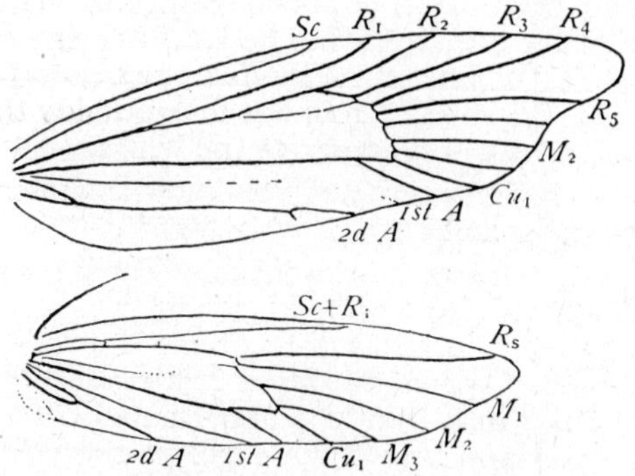

Family GLYPHIPTERYGIDÆ

The ocelli are usually large. The maxillæ are strong and clothed with scales. The maxillary palpi are vestigial or wanting. The labial palpi are upturned to the middle of the front or beyond, often beyond the vertex. In the fore wings veins R_4 and R_5 are usually separate and vein Cu_2 arises close to the angle of the discal cell. In the hind wings the second anal vein is strongly forked at the base.

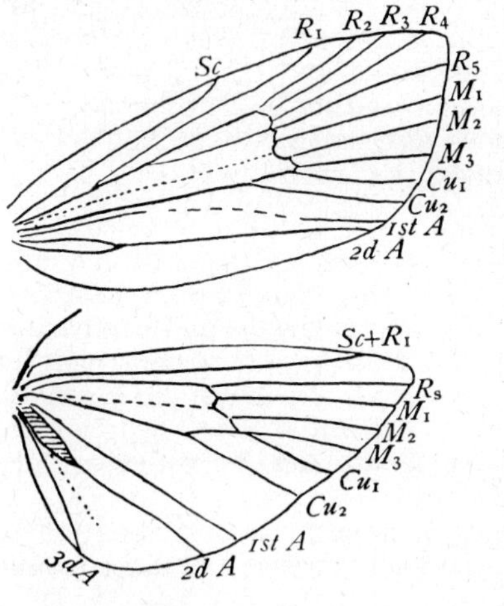

Fig. 774.—Wings of *Glyphipteryx thrasonella*. (After Spuler.)

Fig. 775.—Wings of *Simæthis fabriciana*. (After Spuler.)

Nearly forty North American species are now known. These represent two subfamilies.

Subfamily Glyphipteryginæ.—This subfamily is composed chiefly of the species of the genus *Glyphipteryx*. In this genus the wings are moderately broad (Fig. 774) and the fore wings have a lobe-like prolongation between veins R_4 and R_5. Ten species are now listed from our fauna.

Subfamily Choreutinæ.—In this subfamily the wings are broad and triangular (Fig. 775), and usually with narrow fringes. The moths bear a striking resemblance to tortricids. The larvæ live under webs on leaves or between leaves that are fastened together.

Family HELIODINIDÆ

The hind wings are narrow-lanceolate and pointed or linear and much narrower than their fringe. The maxillary palpi are minute and porrect. The labial palpi are very short and drooping. The maxillæ are strong. The tarsi are armed with more or less distinct whorls of bristles; the tibiæ are also often armed with stiff bristles. Usually when at rest the imago holds the posterior pair of legs elevated at the sides above the wings.

The larvæ are not well-known; those that have been described are of various habits.

Cyclŏplasis panicifoliĕlla.—The larva of this species mines in grass, *Panicum clandestinum*. Its mine is at first a long thread-like line; towards the latter part of the life of the larva it is enlarged into a blotch. When the larva has reached maturity, it cuts a perfectly circular disk from the upper cuticle of the leaf, folds it along its diameter and unites the edges of the circumference, so as to make a semicircle. When completed the larva, enclosed in its semicircular cocoon, lets itself fall to the ground, where it attaches the cocoon to some adjacent object. (Clemens).

Schreckensteinia erythriĕlla.—The larva feeds on sumac bobs. It is common; the body is uniform dark green, but the frass is scarlet. When full-grown it makes a lace-like cocoon on the outside of the bob.

Schreckensteinia festaliĕlla.—The larva of this species is an external feeder on *Rubus;* it is spiny, and when full-grown makes a lace-like cocoon.

Euclemĕnsia bassettĕlla.—The larva is an internal parasite in the gall-like females of the coccid genus *Kermes*. The adult is a beautiful greenish-black moth, which has its fore wings marked

Fig. 776.—*Euclemensia bassettella.*

with reddish orange (Fig. 776).

The genus *Euclemensia* is placed in this family provisionally.

Family ÆGERIIDÆ

The Clear-winged Moths

The clear-winged moths constitute a very remarkable family, many of them resembling bees or wasps in appearance more than they do ordinary moths, a resemblance due to their clear wings and in some cases to their bright colors (Fig. 777). There are a few moths in other families, as the

Fig. 777.

clear-winged sphinxes and certain euchromiids, that have a greater or less part of the wings devoid of scales; but they are exceptions. Here it is the rule that the greater part of one or both pairs of wings are free from scales; hence the common name, clear-winged moths. In a small number of members of this family the wings are scaled throughout.

These insects are of moderate size; as a rule they have spindle-shaped antennæ, which are terminated by a small silky tuft; sometimes the antennæ are pectinate; the margins of the wings and the veins of even the clear-winged species are clothed with scales; and at the end of the abdomen there is a fan-like tuft of scales.

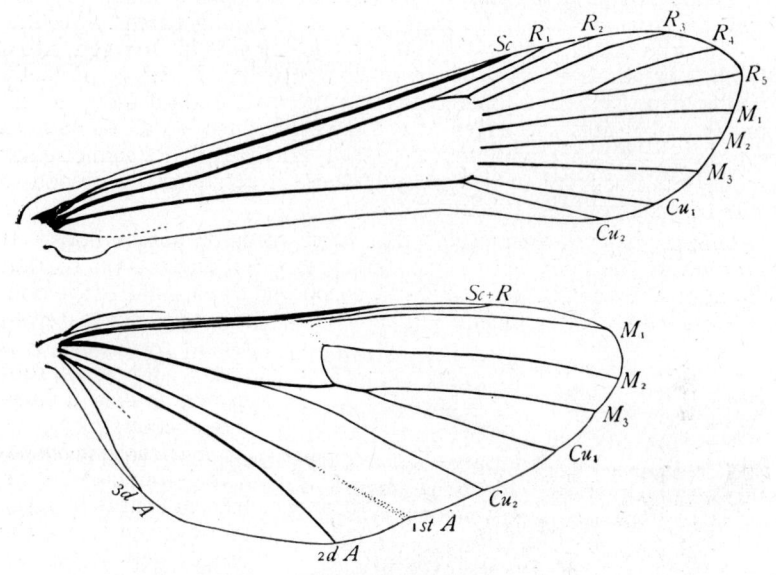

Fig. 778.—Wings of *Synanthedon exitiosa*.

The fore wings are remarkable for their extreme narrowness and the great reduction of the anal area (Fig. 778); while the hind wings have a widely expanded anal area. The number of anal veins in the hind wings varies greatly within the family, the number ranging from two to four; where there are four anal veins, it is probably the third anal vein that is forked.

Another remarkable feature of the wings of these insects is that in the female the bristles composing the frenulum are consolidated as in the male; this condition exists in the females of a few members of other families. The females of the Ægeriidæ possess a frenulum hook; but this is not so highly specialized as that of the male.

In addition to the presence of a highly specialized frenulum and a frenulum-hook, there is a unique provision for holding the fore and hind wings together. The inner margin of the fore wing is folded under; and the radius of the hind wing is armed with setæ, which hook into this fold.

The adults fly very swiftly and during the hotter part of the day. They frequent flowers thus increasing their resemblance to bees or to wasps. The larvæ are borers, living within the more solid parts of plants. Some species cause serious injury to cultivated plants. More than one hundred species have been found in America north of Mexico. Among the better known species are the following.

The blackberry crown-borer or the raspberry root-borer, *Bembēcia margināta*.—The larva of this species burrows in the roots and lower part of the canes of blackberries and raspberries, sometimes completely girdling the cane at the crown.

The peach-tree borer, *Synanthedon exitiōsa*.—This is the most important enemy of the peach-tree, except perhaps the San José scale in the North and the plum curculio in the South. In some parts of the country it is difficult to find a peach-tree that is not infested by it. The eggs are laid on the bark of the tree near the ground. The larvæ bore downward in the bark of the trunk just below the surface of the ground. Their burrows become filled by a gummy secretion of the tree. As this oozes out in large masses the presence of the borer is easily detected by it. The insect always passes the winter in the larval state. When full-grown the larva comes to the surface of the ground and makes a cocoon of borings fastened together with silk. The perfect insects appear from May till October; the date at which most of them appear varies in different sections of the country. There is a single generation each year. The adults differ greatly in appearance. The general color of both sexes is a glassy steel-blue. In the female (Fig. 779) the fore wings are covered with scales, and there is a bright orange-colored band on the abdomen. In the male both pairs of wings are nearly free from scales. The usual method of fighting this pest is to carefully watch the trees and remove the larvæ with a knife as soon as discovered. Recently the use of a toxic gas, paradichlorobenzene, has been found available on trees six years of age and older; and experiments are now being made to determine the practicability of its use on younger trees. See U. S. Dept. of Agr. Bull. 1169, and later bulletins when published.

Fig. 779.—*Synanthedon exitiosa*, female.

The Pacific peach-tree borer, *Synănthedon opalĕscens*.—On the Pacific Coast there is a peach-tree borer that is distinct from the above, and appears to be an even more serious pest. The larva is more difficult to remove from the tree, as it bores into the solid wood. The female of this species lacks the orange-colored band on the abdomen.

The lesser peach-tree borer, *Synănthedon pĭctipes*.—The larvæ of this species infest peach, plum, cherry, june-berry, beach-plum, and chestnut. They do not confine their attacks to the crown but more often occur on the trunk and larger branches. This borer rarely attacks perfectly sound, uninjured trees and is not a serious pest in orchards that receive good care. Both sexes of the adult resemble the male of the peach-tree borer, having both fore and hind wings transparent.

The imported currant borer, *Chamæsphēcia tipulifŏrmis.*—This is a small species, the adult having a wing-expanse of only about 18 mm. There are but few scales on either pair of wings except on the tip and discal vein of the fore wings and the outer margin of the hind wings. The eggs are laid on the twigs of currant. The larvæ penetrate the stem, and devour the pith; in this way they make a burrow in which they live and undergo their transformations. The perfect insects appear in June. Before this time the leaves of the infested plant turn yellow. If such plants be cut and burned in May the pest will be destroyed.

The squash-vine borer, *Melĭttia satyrinifŏrmis.*—The larva of this species (Fig. 780) does great damage by eating the interior of squash-vines; it also sometimes infests pumpkin-vines and those of cucumber and melon. It is most destructive to late squashes. When full-grown the larvæ leave the

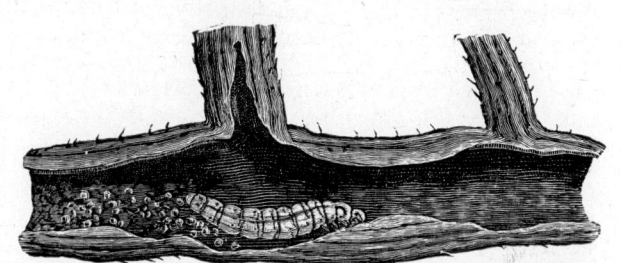

Fig. 780.—*Melittia satyriniformis*, larva in squash-vine

vines and enter the ground, where they make tough silken cocoons, a short distance below the surface, in which the winter is passed. The adults appear soon after their food-plants start growth. The fore wings of the adult are covered with scales and the hind legs are fringed with long, orange-colored scales. To check the ravages of this pest, the vines should be collected and destroyed as soon as the crop is harvested in order to destroy the larvæ that are still in them; the land should be harrowed in the fall to expose the cocoons and then plowed deeply the following spring in order to bury them so deeply that the moths can not emerge. If the vine is covered with earth two or three feet from its base it will produce a new root system which will sustain the plant in case the main stem is injured at the base. Where late squashes are grown early squashes can be used as trap plants. Borers can be removed from the vines with a knife; when this is done the vine should be cut lengthwise, and, after the larva is removed, the vine covered with earth; if this is done carefully the wound will soon heal.

The pine clear-wing moth, *Parharmŏnia pīni.*—Frequently there may be seen on the trunks of pine-trees large masses of resinous gum mingled with sawdust-like matter. These are the results of the work of the larvæ of this insect, which bore under the bark and into the superficial layers of the wood. The adult resembles the female of the peach-tree borer, but the abdomen is more extensively marked with orange beneath.

SUPERFAMILY

TORTRICOIDEA

The Tortricids

The tortricids are generally small moths; but as a rule they are larger than the members of most of the families of the Microfrenatæ. They have broad front wings, which usually end squarely. The costa of the front wings curves forward strongly near the base of the wing. When at rest the broad front wings fold above the body like a roof. The moths

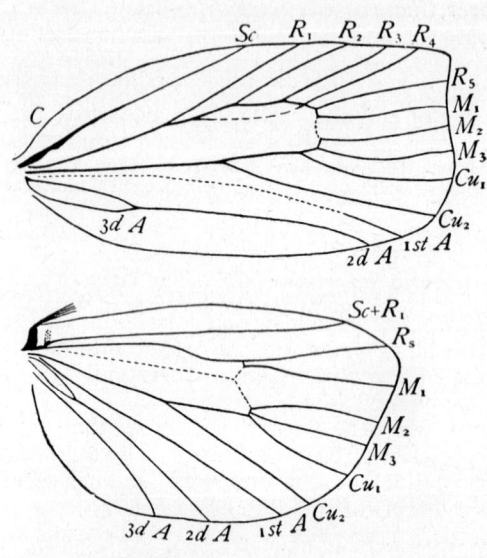

Fig. 781.—Wings of *Archips cerasivorana*.

are variegated in color, but are usually brown, gray, or golden rather than of brighter hues. As a rule the hind wings are of the color of the body and without markings. The venation of the wings of a common species is represented by Figure 781.

The larvæ vary greatly in habits. Many of them are leaf-rollers. It was this habit that suggested the name *Tortrix* for the typical genus, from which the names of one family and of the superfamily are derived. A large portion of the rolled leaves found upon shrubs and trees are homes of tortricid larvæ; but it should be remembered that the leaf-rolling habit is not confined to this family. While many are leaf-rollers probably a larger number are borers in stems, buds, or fruits.

About eight hundred North American species of the Tortricoidea have been decribed. This superfamily includes four families, which, can be separated by the following table.

A. Both veins M_1 and M_2 of the hind wings lost. p. 644..........CARPOSINIDÆ
AA. Vein M_1 of the hind wings present, vein M_2 either present or lost.
   B. With a fringe of long hairs on the basal part of vein Cu of the hind wings, on the upper side of the wing. Do not mistake a bunch of long hairs arising from the wing back of vein Cu for this fringe.

C. Fore wings with veins R_4 and R_5 stalked or united, veins M_2, M_3, and Cu_1 diverging or parallel. (A few species only). p. 642 ..... TORTRICIDÆ
CC. Fore wings with veins R_4 and R_5 separate, or with veins M_2, M_3, and Cu_1 converging strongly toward the margin of the wing. p. 639......
.... OLETHREUTIDÆ
BB. Without a fringe of long hairs on the basal part of vein Cu of the hind wings.
C. Fore wings with the distal part at least of vein 1stA preserved. Vein Cu_2 of the fore wings arising from a point before the outer fourth of the discal cell.
D. Veins M_1 and M_2 of the fore wings somewhat approximate at the margin of the wing (*Laspeyresia lautana*) p. 639..... OLETHREUTIDÆ
DD. Veins M_1 and M_2 of the fore wings divergent or parallel. p. 642
.... TORTRICIDÆ
CC. Both fore and hind wings with vein 1stA lost, vein Cu_2 of the fore wings arising from the outer fourth of the discal cell. p. 643 PHALONIIDÆ

Family OLETHREUTIDÆ*

As a rule the members of this family are easily distinguished from all other tortricids by the presence of a fringe of long hairs on the basal part of cubitus of the hind wings, on the upper side of the wing. This fringe is lacking in a few members of this family and is present in a few members of the next family.

This is the largest of the families of tortricids; more than four hundred North American species have been described. The following species are among those most likely to be observed, and will serve to illustrate the differences in habits of the different species.

The codlin-moth, *Carpocăpsa pomonělla*.—This is the best known and probably the most important insect enemy of the apple. The larva is the worm found feeding near the core of wormy apples. The adult (Fig. 782) is a beautiful little creature with finely mottled pale gray or rosy fore wings. There is a large brownish spot near the end of the fore wing, and upon this spot irregular, golden bands. The moth issues from the pupa state in late spring and lays its eggs singly on the surface of the fruit or on adjacent leaves. As soon as the larva hatches it burrows into the apple and eats its way to the core, usually causing the fruit to fall prematurely. When full grown the larva burrows out through the side of the fruit, and undergoes its transformations within a cocoon, under the rough bark of the tree, or in some other protected place. The number of generations annually varies in different parts of the country. As a rule there is in the North one full generation and usually a partial second; where the season is longer there are two or three generations. The larvæ winter in their cocoons transforming to pupæ during early spring.

Fig. 782.—*Carpocapsa pomonella.*

The method of combating this pest that is most commonly employed now is to spray the trees with a solution of arsenate of lead, four to six pounds of arsenate of lead in one hundred gallons of water, just after the petals fall and before the young apples are heavy

---
*This family is the Eucosmidæ of some writers, and the Epiblemidæ of others.

enough to droop. The falling spray lodges in the blossom end of the young apple, and many of the larvæ which attempt to enter at this point, the usual place of entrance, get a dose of poison with their first meal.

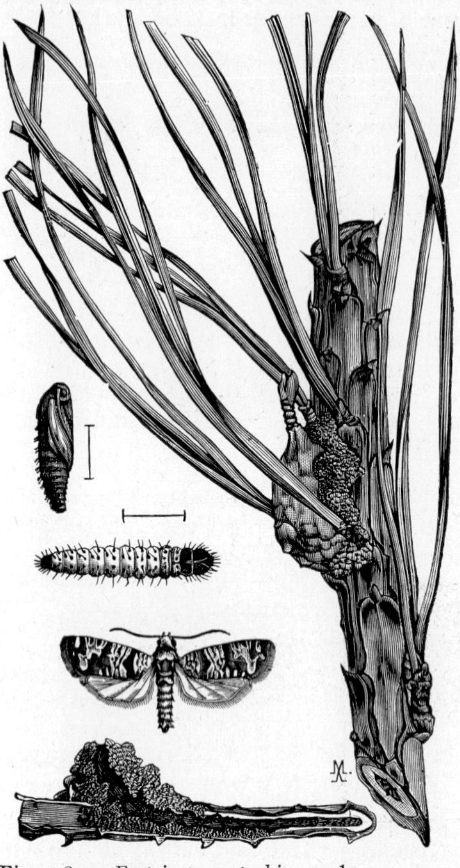

The pine-twig moths, *Evĕtria*.—The genus *Evetria* includes many species that infest the twigs and smaller branches of various species of pine. Some of our best known species were described under the generic name *Retinia* but now the older name *Evetria* is applied to them. The following species are well-known.

Evĕtria comstockiāna.—This species (Fig. 783) illustrates well the habits of the boring species. The larva infests the small branches of pitch-pine. It is a yellowish-brown caterpillar, which makes a burrow along the centre of the branch. Its presence may be detected by the resin that flows out of the wound in the twig and hardens into a lump. Two of these lumps are shown in the figure, one of them split lengthwise, and the other with a pupa-skin projecting from it. The larva, pupa, and adult are also figured. The moth is represented natural size; the darker shades are dark rust-color, and the lighter, light-gray. The insect winters as a larva; the adult emerges in May and June.

Fig. 783.—*Evetria comstockiana*: larva, pupa, adult, and work. (From the Author's Report for 1879.)

Evĕtria frustrāna.—This species infests the new growth of several species of pine, spinning a delicate web around the terminal bud, and mining both the twig and the bases of the leaves. The larva, pupa, and adult are represented somewhat enlarged in the figure. An infested twig is also shown (Fig. 784).

The grape-berry moth, *Polychrōsis viteāna.*—The most common cause of wormy grapes is the larva of this moth. The moth emerges in the spring from its cocoon on a fallen leaf where it has passed the winter in the pupa state. The first generation of larvæ make a slight web among the blossom buds into which they eat destroying many embryo grapes. When full-grown the larva passes to a leaf and makes a very peculiar cocoon. It cuts a semicircular incision in the leaf, bends over the flap thus made, fastens its free edge to the leaf, and lines the cavity thus enclosed with silk; here it transforms to a pupa. The moths of the second and later generations lay their eggs on the berries, and the larvæ bore into them and feed on the pulp and seeds. The most efficient method of control of this pest is by spraying with a solution of arsenate of lead, six pounds of the poison in one hundred gallons of water. The first application should be made shortly after the fruit sets, and one or two more at intervals of ten days.

The bud-moth, *Tmetŏcera ocellāna*.—The larva of this insect is a pest infesting apple-trees. It works in opening fruit-buds and leaf-buds, often eating into them, especially the terminal ones, so that all new growth is stopped. It also ties the young leaves at the end of a shoot together and lives within the cluster thus formed, adding other leaves when more food is needed. Sometimes so large a proportion of the fruit-buds are destroyed as to seriously reduce the amount of the crop. The pupa state is passed within the cluster of tied leaves or within a tube formed by rolling up one side of a leaf, and lasts about ten days. The moth expands about 15 mm.; it is of a dark ashen gray, with a large, irregular, whitish band on the fore wing.

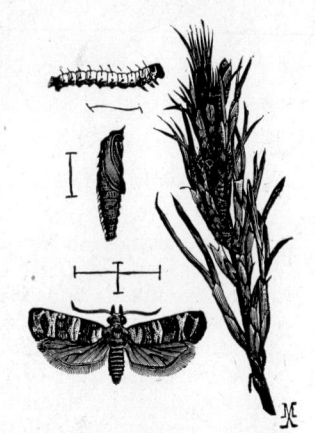

Fig. 784.—*Evetria frustrana*: larva, pupa, adult, and work. (From the Author's Report for 1879.)

The clover-seed caterpillar, *Laspeyrēsia interstinctāna*.—This is a common pest which feeds in the heads of clover, especially red clover, destroying many of the unopened buds and some of the tender green seeds, and spoiling the head as a whole. It sometimes greatly diminishes the crop of seed. There are three generations annually. The last generation passes the winter in the pupa state as a rule; but some larvæ hibernate under rubbish. The adult is a pretty brown moth, with a series of silvery marks along the costal margin of the fore wings, and two on the inner margin, which form a double crescent when the wings are closed on the back. This moth expands 10 mm. If the hay is cut early and stored the larvæ will be destroyed while still in the heads.

Family TORTRICIDÆ

The Typical Tortricids

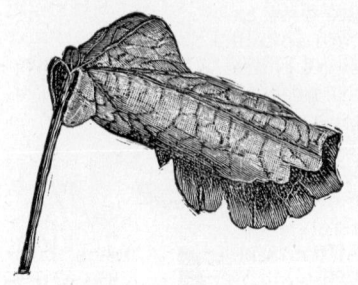

The tortricidæ differ, as a rule, from the preceding family in lacking a fringe of long hairs on the basal part of the cubitus of the hind wings. In the fore wings the distal part of the first anal vein is preserved, and vein Cu_2 arises from a point before the outer fourth of the discal cell.

Fig. 785.—Nest of *Archips rosana*.

Fig. 786.—*Archips rosana*.

In a recent list 165 North American species are enumerated; these represent 15 genera.

Several of our better-known members of this family belong to the genus *Archips*. This is the genus *Cacœcia* of those writers who do not recognize the names proposed by Hübner in his "Tentamen." These insects have been named the ugly-nest tortricids; ugly dwelling being the meaning of *Cacœcia*, and also descriptive of the nests of the larvæ of these insects. The four following species are common.

The rose ugly-nest tortricid, *Ărchips rosāna*.—The larva of this species feeds within the webbed-together leaves of rose and a number of other plants. Figure 785 represents the nest of a larva in a currant leaf; and Figure 786 the adult of this species. This moth expands about 20 mm. The fore wings are olive-brown, crossed by bands of darker color; the hind wings are dusky. This species differs from the two following in that each larva makes a nest for itself.

The cherry-tree ugly-nest tortricid, *Ărchips cerasivorāna*.—This species lives upon the choke-cherry and sometimes upon the cultivated cherry. The larvæ, which are yellow, active creatures, fasten together all the leaves and twigs of a branch and feed upon them (Fig. 787), an entire brood occupying a single nest. The larvæ change to pupæ within the nest; and the pupæ, when about to transform, work their way out and hang suspended from the outer portion of the nest, clinging to it only by hooks at the tail end of the body. Here they transform,

Fig. 787.—Nest of *Archips cerasivorana*.

leaving the empty pupa skins projecting from the nest, as shown in the figure. The moths vary in size, the wing expanse of those we have bred ranging from 20 mm. to nearly 30 mm. The wings are bright ochre-yellow; the front pair marked with irregular brownish spots and numerous transverse bands of leaden blue (Fig. 788, male; Fig. 789, female).

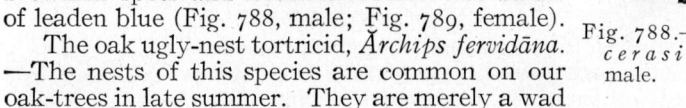

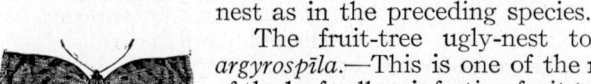

Fig. 788.—*Archips cerasivorana*, male.

The oak ugly-nest tortricid, *Archips fervidāna*. —The nests of this species are common on our oak-trees in late summer. They are merely a wad of leaves fastened together. Each nest contains several larvæ; later the empty pupa-skins may be found clinging to the outside of the nest as in the preceding species.

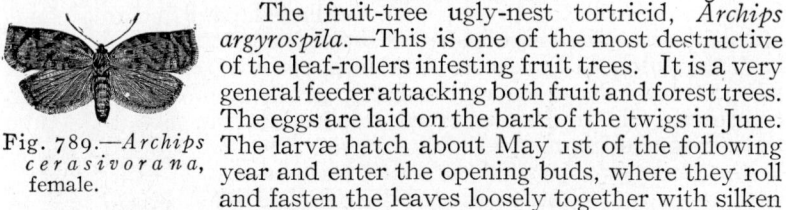

Fig. 789.—*Archips cerasivorana*, female.

The fruit-tree ugly-nest tortricid, *Archips argyrospīla*.—This is one of the most destructive of the leaf-rollers infesting fruit trees. It is a very general feeder attacking both fruit and forest trees. The eggs are laid on the bark of the twigs in June. The larvæ hatch about May 1st of the following year and enter the opening buds, where they roll and fasten the leaves loosely together with silken threads. After the fruits set, they are often included in the nests and ruined by the caterpillars eating large irregular holes in them.

The pine-leaf tube-builder, *Eulia pinatubāna*.—One of the most interesting of tortricid nests occurs commonly on white pine. Each nest consists of from six to fifteen leaves drawn together so as to form a tube, and is lined with silk. This tube serves as a protection to the larva, which comes out from it to feed upon the ends of the leaves of which the tube is composed; in this way the tube is shortened. I bred the moth from nests collected at Ithaca, N. Y.; and have found similar nests as far south as Florida. The moth expands 12 mm. Its head, thorax, and fore wings are of a dull rust-red color, with two oblique paler bands crossing the fore wings, one a little before the middle, the other a little beyond and parallel with it.

Family PHALONIIDÆ

In this family and in the following one the first anal vein is lost in both fore and hind wings and vein Cu_2 arises from the outer fourth of the discal cell. In this family vein M_1 of the hind wings is preserved, usually stalked with vein R_s. The palpi of the two sexes are alike.

More than one hundred North American species have been described, and constantly others are being found. Comparatively little is known about the habits of our species; but most of the European species whose habits are known are borers, chiefly in herbaceous plants.

The juniper web-worm, *Phalōnia rutilāna*.—This is an imported species which has attracted attention by its injuries to junipers, the leaves of which it fastens together with silk. In this way it makes a more or less perfect tube within which it lives, but from which it issues to feed. The moth expands about 12 mm. and has bright, glossy, orange fore wings, crossed by four reddish brown bands.

Family CARPOSINIDÆ

This family is distinguished from the preceding one by the fact that in the hind wings both vein M_1 and vein M_2 are completely lost, and the palpi of the male are short while those of the female are long. This is a small family, only five North American species are now listed, and very little is known regarding the habits of these.

The currant-fruit-worm, *Carpŏsina fernaldāna*.—In the unpublished notes of the late Professor M. V. Slingerland, I find an account of this insect. The larva feeds within the fruit of the currant, eating both the pulp and the seeds. The infested fruit soon drops. When full-grown the larva leaves the berry and goes into the ground to transform. The adult emerges in the following spring about the time the currants are turning red.

Superfamily PYRALIDOIDEA

The Pyralids and their Allies

This group of families includes a very large number of small or moderate-sized moths, of fragile structure, normally with firmly and finely scaled wings, and with the anal area of the hind wings broad. The first anal vein of the fore wings is almost always lost, and there is no accessory cell. In the hind wings, there are usually three anal veins; and veins Sc and R are separate along the discal cell, but grown together or closely parallel for a short distance beyond the cell. The maxillæ are scaled at their base; and the maxillary palpi when present are of the porrect type. The labial palpi often project beak-like.

The larvæ are characterized by the presence of only two setæ on the prespiracular wart of the prothorax, and by setæ IV and V of the abdomen being close together.

This superfamily includes the five following families, which can be separated by the characters given in Table A, page 584.

Family PYRALIDIDÆ

The Pўralids

The members of this family found in our fauna are mostly small moths, but a few are of moderate size; some tropical species, however, are quite large. So large a portion of the species are small that the family has been commonly classed with the preceding families as Microlepidoptera.

The members of the different subfamilies of this family differ so greatly in appearance that it is not possible to give a general description that will serve to distinguish it; a very large portion of the species have a special look, due to their thin and ample hind wings with large anal areas; it is necessary, however, to study structural characters to find evidences of a common bond.

The body is slender; the head is prominent; the ocelli are usually present; the antennæ are almost always simple, but frequently the antennæ of the male have a process on the scape or a notch and tuft on the clavola; and the palpi are usually moderate in size or long; but very often they project beak-like; for this reason the name snout-moths is often applied to this family.

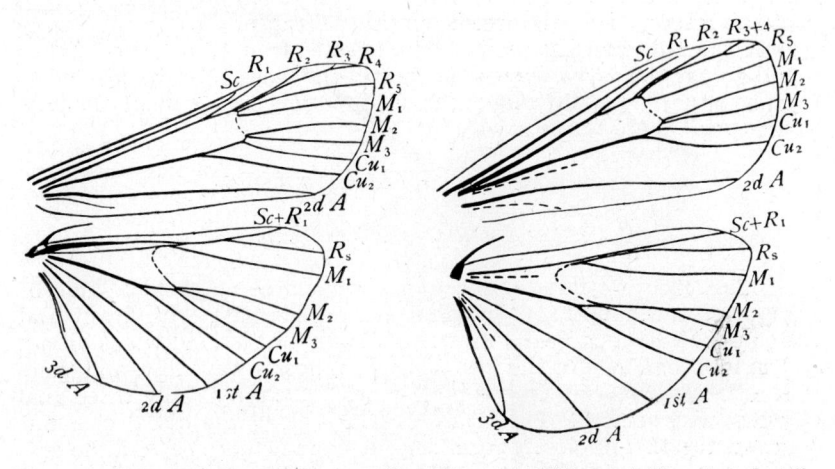

Fig. 790.—Wings of *Nomophila noctuella*. Fig. 791.—Wings of *Tlascala reductella*.

As a rule there are three anal veins in the hind wings and one in the fore wings. The discal cell is always well-formed, but there is no accessory cell. In most cases the pyralids can be recognized by the fact that the subcosta and radius of the hind wings are separate along the discal cell, but grown together for a short distance beyond the cell, after which they are again separate (Fig. 790). In some genera these two veins do not actually coalesce, but extend very near together for a short distance (Fig. 791). The two types, however, are essentially the same.

This is one of the larger families of the Lepidoptera; nearly one thousand species have been described from America north of Mexico alone. The family is divided into many subfamilies, representatives of fifteen of which are found in our fauna. The best known species, those that have attracted attention on account of their economic importance, belong to the subfamilies discussed below.

Subfamily PYRAUSTINÆ

The Pyraustids

This is one of the larger of the subfamilies of the Pyralididæ; about three hundred species have been described from America north of Mexico. This subfamily includes many small moths; but it contains also the majority of the larger species of pyralids; some of the species are very striking in appearance.

Fig. 792. — *Desmia funeralis*.

The members of this subfamily differ from other pyralids by the following combination of characters. There is no fringe of long hairs on the basal part of vein Cu of the hind wings; veins R_2 and R_5 of the fore wings arise from the discal cell distinct from vein R_4 (Fig. 790); and the maxillary palpi are never very large and triangular. Among our better known species are the following.

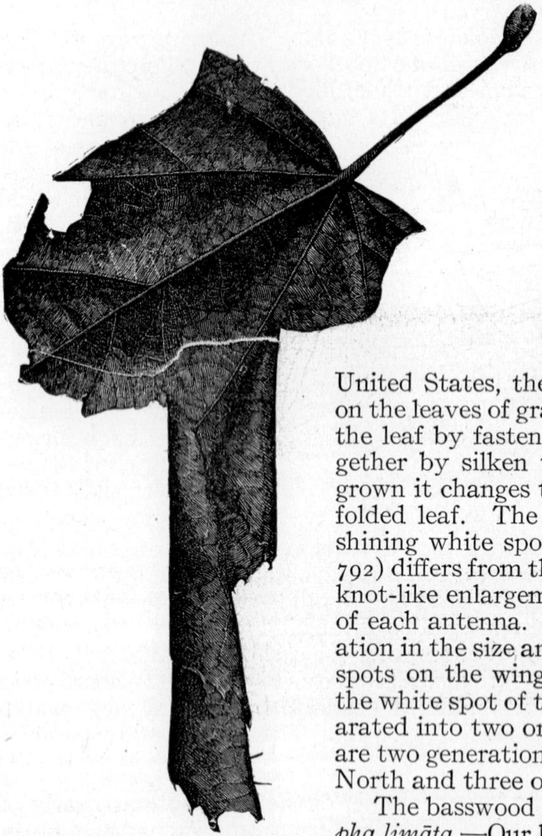

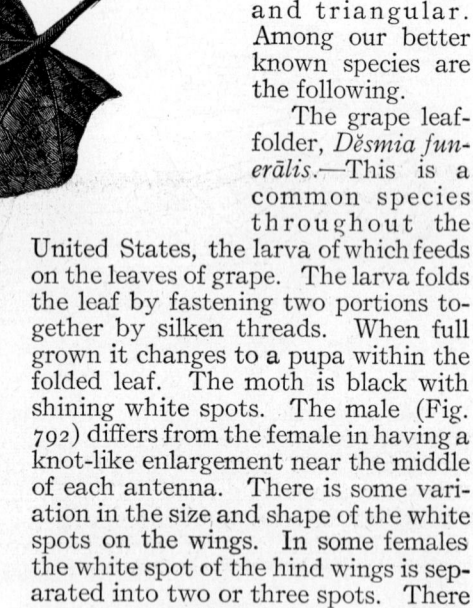

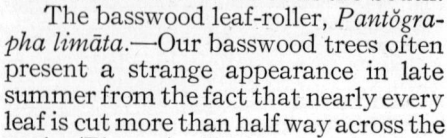

The grape leaf-folder, *Děsmia funerālis*.—This is a common species throughout the United States, the larva of which feeds on the leaves of grape. The larva folds the leaf by fastening two portions together by silken threads. When full grown it changes to a pupa within the folded leaf. The moth is black with shining white spots. The male (Fig. 792) differs from the female in having a knot-like enlargement near the middle of each antenna. There is some variation in the size and shape of the white spots on the wings. In some females the white spot of the hind wings is separated into two or three spots. There are two generations of this species in the North and three or more in the South.

The basswood leaf-roller, *Pantŏgrapha limāta*.—Our basswood trees often present a strange appearance in late summer from the fact that nearly every leaf is cut more than half way across the middle, and the end rolled into a tube (Fig. 793). Within this tube there

Fig. 793.—Nest of larva of *Pantographa limata*.

lives a bright green larva, with the head and thoracic shield black. When full grown the larva leaves this nest and makes a smaller and more simple nest, which is merely a fold of one edge of the leaf, or sometimes an incision is made in the leaf extending around about two-thirds of a circle and the free part bent over and fastened; in each case the nest is lined with silk, thus forming a delicate cocoon. Here the larvæ pass the winter in fallen leaves. At Ithaca, N. Y., Professor Slingerland found that the larvæ did not pupate till the following July, and that adults emerged in August. The adult moth expands about 33 mm.; it is straw-colored with many elaborate markings of olive with a purplish iridescence (Fig. 794).

Fig. 794.—*Pantographa limata.*

The melon-worm, *Diaphānia hyalināta.*—This beautiful moth (Fig. 795) is often a serious pest in our Southern States, where the larva is very destructive to melons and other allied plants. The young larvæ feed on the foliage; the older ones mine into the stems and fruit. The insect passes the winter as a pupa in loose silken cocoons in dead leaves or under rubbish. The moth is a superb creature, with glistening white wings bordered with black, and with a spreading brush of long scales at the end of the abdomen. This species appears to be injurious only in the Gulf States, but the moths have been taken as far north as Canada.

The most practicable method of protecting cantaloupes and cucumbers from this pest is by planting summer squashes among them as a trap crop at intervals of about two weeks so as to furnish an abundance of buds and blossoms during July and August. The earlier squash vines should be removed and destroyed before many worms have reached maturity on them; and after the crop is harvested the vines and waste fruits should be gathered and destroyed.

Fig. 795.—*Diaphania hyalinata:* larvæ, cocoon, and adults. (From the Author's Report for 1879.)

The pickle-worm, *Diaphānia nitidālis.*—This species is closely allied to the preceding one. The wings of the moth are yellowish brown with a purplish metallic reflection; a large irregular spot on the front wings and the basal two-thirds of the hind wings are semi-transparent yellow. The tip of the abdomen is ornamented with a brush of long scales, as in the preceding species. The range and habits of this species are quite similar to those of the melon-worm; and it should be fought in the same way.

The Webworms.—The larvæ of many pyralids have the habit of spinning a silken web beneath which they retreat when not feeding, and on this account have been termed webworms. Several species of these webworms belong to this family; among them are the following.

The cabbage webworm, *Hĕllula undālis.*—This species infests various cruciferæ in the Gulf and South Atlantic States. The larva is about 12 mm. in length, of a grayish yellow color, striped with five brownish-purple bands.

The garden webworm, *Loxŏstege similālis.*—This species is most injurious in the Southern States and in the Mississippi Valley. It infests various garden crops and corn and cotton. The larva varies in color from pale and greenish yellow to dark yellow and is marked with numerous black tubercles.

The European corn-borer, *Pyrausta nubilālis.*—This is a greatly feared pest which has recently appeared in this country. It is a borer in the stems of plants, in which it winters as a partly grown larva. Its favorite food appears to be corn and especially sweet corn; but it infests other cultivated plants, as dahlias and gladiolus, and many large-stemmed weeds. The full-grown larva measures about 20 mm. in length; the adult moth has a wing-expanse of from 25 to 30 mm. As this is written, efforts are being made by the National Government and by several State Governments to prevent the spread of this pest; and many circulars and bulletins are being published regarding it.

Subfamily NYMPHULINÆ

The Aquatic Pyralids

This subfamily is of especial interest as the larvæ are nearly all aquatic, differing in this respect from nearly all other Lepidoptera. The larvæ of most of the species live upon plants, like water lilies and pond weeds, that are not wholly submerged; but in some species the larva has a true aquatic respiration, being furnished with tracheal gills. In our best known species, these **tracheal** gills are numerous and form a fringe on each side of the body of long slender filaments, which are simple in some species and branched in others.

The larvæ vary greatly in habits; some species live free upon the plants they infest; in some species, each larva makes a case of two leaf fragments fastened together at the edges; most of the described larvæ live in quiet waters, lakes, ponds, or pools, but the larva **of**

Elŏphila fulicālis was found by Lloyd ('14) to live beneath sheets of silk spun over exposed surfaces of current-swept rocks, in a rapid stream.

In the case of several species whose life-history has been determined the pupal stadium is passed in a cocoon beneath the surface of the water.

An exception to the usual habits of larvæ of this subfamily is presented by *Eurrhўpara urticāta*, introduced from Europe to Nova Scotia; this species is not aquatic, but feeds on nettle.

Subfamily PYRALIDINÆ

The Typical Pyralids

This is a small subfamily; only twenty-four species are now enumerated in our lists, and these are mainly from the far Southwest. The best-known species are the two following:

The meal snout-moth, *Pўralis farinālis*.—The larva of this species feeds on meal, flour, stored grain, and old clover hay. It makes little tubes composed partly of silk and partly of the fragments of its food. It rarely occurs in sufficient numbers to do serious injury; and its ravages can be checked by a thorough cleaning of the infested places, or when practicable by the use of carbon bisulphide. The moth is commonly found near the food of the larva, but is often seen on ceilings of rooms sitting with its tail curved over its back. It expands about 25 mm.; the fore wings are light-brown, crossed by two curved white lines, and with a dark chocolate-brown spot on the base and tip of each.

The clover-hay worm, *Hypsopўgia costālis*.—The larva of this species sometimes abounds in old stacks of clover-hay, and especially near the bottom of such stacks. As the infested hay becomes covered with a silken web spun by the larva, and by its gunpowder-like excrement, much more is spoiled than is eaten by the insect. Such hay is useless and should be burned, in order to destroy the insects. The moth expands about 20 mm. It is a beautiful lilac color, with golden bands and fringes (Fig. 796).

Fig. 796.—*Hypsopygia costalis*.

Subfamily CRAMBINÆ

The Close-wings

Although this is not a large subfamily, there being only about one hundred and thirty species known in our fauna, the members of it are more often seen than any other pyralids. The larvæ of most of the species feed on grass; and the adults fly up before us whenever we walk through meadows or pastures. When at rest, the moths wrap their wings closely about the body; this has suggested the name close-wings for the insects of this family. When one of these moths alights on a stalk of grass it quickly places its body parallel with the stalk, which renders it less conspicuous (Fig. 797). Many of the species are silvery white or are marked with stripes of that color.

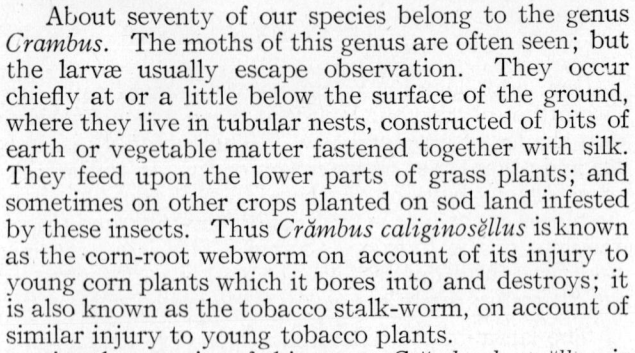

About seventy of our species belong to the genus *Crambus*. The moths of this genus are often seen; but the larvæ usually escape observation. They occur chiefly at or a little below the surface of the ground, where they live in tubular nests, constructed of bits of earth or vegetable matter fastened together with silk. They feed upon the lower parts of grass plants; and sometimes on other crops planted on sod land infested by these insects. Thus *Crămbus caliginosĕllus* is known as the corn-root webworm on account of its injury to young corn plants which it bores into and destroys; it is also known as the tobacco stalk-worm, on account of similar injury to young tobacco plants.

Another species of this genus, *Crămbus hortuĕllus*, is known as the cranberry girdler. This sometimes does considerable injury in cranberry bogs by destroying the bark of the prostrate stems of the vines.

Fig. 797.—*Crambus*.

To this subfamily belong the larger corn stalk-borer, *Diatræa zeacolĕlla*, which sometimes bores into the stalks of young corn in the Southern States, and the sugar-cane borer, *Diatræa saccharālis*, which bores into the stalks of sugar-cane.

Subfamily GALLERIINÆ

The Bee-moth Subfamily

This is a small subfamily, of which only seven species have been found in our fauna. The best known of these is the bee-moth, *Gallēria mellonĕlla*. The larva of this species is a well-known pest in apiaries. It feeds upon wax; and makes silk-lined galleries in the honey-comb, thus destroying it. When full grown the larva is about 25 mm. in length. It lies hidden in its gallery during the day, and feeds only at night, when the tired-out bees are sleeping the sleep of the just. When ready to pupate the caterpillar spins a tough cocoon against the side of the hive.

The moth has purplish-brown front wings, and brown or faded yellow hind wings. The fore wings of the male are deeply notched at the end, while those of the female (Fig. 798) are but slightly so. The female moth creeps into the hive at night to lay her eggs.

Fig. 798.—*Galleria mellonella*.

This pest is found most often in weak colonies of bees, which it frequently destroys. The best preventive of its injuries is to keep the colonies of bees strong. Of course the moths and larvæ should be destroyed whenever found.

Subfamily PHYCITINÆ

The Phycitids

Our most common members of this subfamily are small moths vita rather narrow but long fore wings, which are banded or mottled with shades of gray or brown. The subfamily is, however, a large one and other types of coloration occur. In this subfamily there is a fringe of long hairs on the basal part of vein Cu of the hind wings; the radius of the fore wings is only four-branched (Fig. 791); and the frenulum of the female is simple. This is a very large subfamily; more than three hundred species have been described from our fauna, and there are doubtless many undescribed species in this country.

The larvæ of the different species vary greatly in habits. Some live in flowers, some fold or roll leaves within which they live and feed; some are borers; others feed upon dried fruits, or flour and meal; and one, at least, is predacious, feeding on coccids. Usually the larva lives in a silken tube or case, lying concealed by day and feeding by night. The case made by certain of the leaf-eating species is very characteristic in form (Fig. 799), being strongly tapering and much curved; in this instance the case is composed largely of the excrement of the larva.

The following species are those that have attracted most attention on account of their economic importance.

Fig. 799.

The Indian-meal moth, *Plōdia interpunctĕlla*.—This is the best-known of the species that infest stored provisions. The larva is the small whitish worm, with a brownish-yellow head, that spins thin silken tubes through meal or among yeast-cakes, or in bags or boxes of dried fruits. The moth expands about 15 mm. The basal two-fifths of the fore wing is dull **olive** or cream colored; the outer part reddish brown, with irregular bands of blackish scales.

The Mediterranean flour-moth, *Ephĕstia kühniĕlla*, is an even more serious pest than the preceding species, which it resembles in habits. It has become very troublesome in recent years in flouring-mills. The moth expands about 25 mm. and is grayish in color.

When this pest or the Indian-meal moth infests a limited stock of flour, meal, or other cereal, the most economical way to combat it is to feed the infested product to stock, and then thoroughly clean the storage bin or pantry. In mills, where an entire building must be treated, fumigation with hydrocyanic acid gas is probably the best method of destroying the pest. This should be done under the direction of an expert.

Zimmermann's pine-pest, *Pinipĕstis zimmermănni*, is a common species, the larva of which is a borer. It infests the trunks of pine, causing large masses of gum to exude. The moths appear in midsummer.

The coccid-eating pyralid, *Lætĭlia coccidĭvora*, differs from the other members of this family in being predacious. It feeds on the eggs and young of various scale-insects (*Pulvinaria*, *Dactylopius*, and *Lecanium*). Figure 800 represents the different stages of this insect enlarged, and the moths natural size resting on eggsacs of *Pulvinaria*. Like other members of this family the larva spins a silken tube, within which it lives. On a thickly-infested branch these tubes may be found extending from the remains of one coccid to another.

To this subfamily belong also the gooseberry fruit worm, *Zophōdia grossulāriæ*, which feeds within the fruit of the gooseberry and currant, and the cranberry fruit-worm, *Minēola vaccĭnii*, which bores into cranberry fruit.

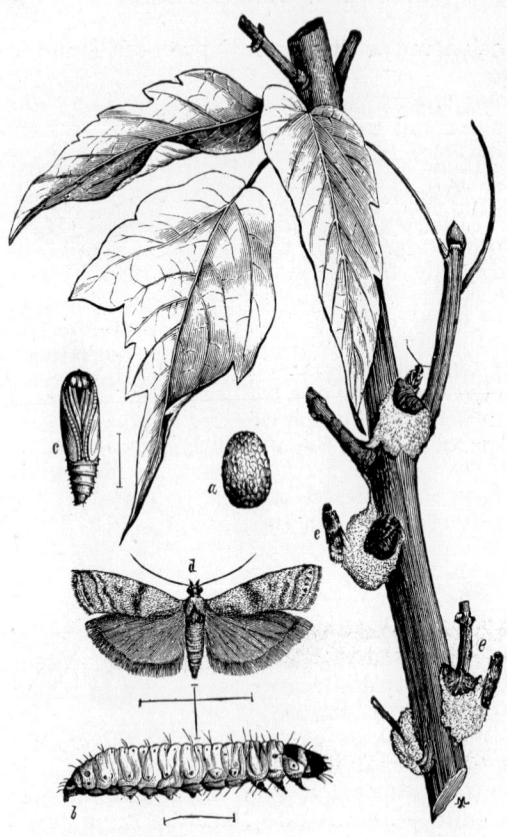

Fig. 800.—*Lætilia coccidivora*: *a*, egg; *b*, larva; *c*, pupa; *d*, adult; *e, e*, moths natural size, resting on egg sacs of *Pulvinaria*.

Family PTEROPHORIDÆ

The Plume-moths

The plume-moths are so called on account of the remarkable form of the wings in most species; the wings being split by longitudinal fissures into more or less plume-like divisions. In most species each forewing is separated into two parts, by a fissure extending about one-half the length of the wing; while each hind wing is divided into three parts by fissures extending farther towards the base of the wing. In a species found in California, *Agdistis adactyla*, the wings are not divided.

One hundred species belonging to this family have been found in America north of Mexico.

One of our most common species is the gartered plume, *Oxўptilus periscelidăctylus*. This is a small moth, expanding about 15 mm. It is of a yellowish brown color marked with dull whitish streaks and spots (Fig. 801). The larvæ hatch early in the spring and feed upon the newly-expanded leaves of the grape. They fasten together several of them, usually those at the end of a shoot, with fine white silk; between the leaves thus folded the caterpillars live either singly or two or three together. They become full-grown and change to pupæ early in June. The pupa is not enclosed in a cocoon, but is fastened to the lower side of a leaf by its tail by means of a few silken threads, in nearly the same way the as chrysalids of certain butterflies are suspended. The pupa state lasts about eight days.

Fig. 801.—*Oxyptilus periscelidactylus*.

Family ORNEODIDÆ

The Many-plume Moths

These insects resemble the plume-moths in having the wings fissured; but here the fissuring is carried to a much greater extent than in that family, each wing being divided into six plumes (Fig. 802).

As yet only a single species of this family has been found in North America. This is *Orneōdes hübneri*. It is an introduced species. European authors state that the larva feeds on the flowers of *Lonicera*, *Centaurea*, and *Scabiosa arvensis*. It transforms either in the flower-head or in the ground. This species has been mistaken for another European species, *Orneodes hexadactyla*, and is commonly known under this name.

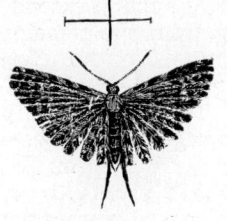

Fig. 802.—*Orneodes hübneri*.

Family THYRIDIDÆ

The Window-winged Moths

Excepting some subtropical species found in the Gulf States and California our members of this family can be easily recognized by the presence of curious white or yellowish translucent spots upon the wings; it is these spots that suggest the name window-winged moths for the family.

In this family the antennæ are either strictly filiform or slightly thickened in the middle; the ocelli are wanting; the palpi project horizontally, and are somewhat longer than the head; and the maxillæ are strongly developed. The venation of the wings of *Thyris* is represented by Figure 803. Here all of the branches of radius of the fore wings are present and each arises from the discal cell. This is a rather unusual condition, but it occurs in the next family, in certain genera of other families of moths, and in the skippers. In one of our thyridids, *Meskea*, which is found in the Gulf States, veins R_3 and R_4 are stalked.

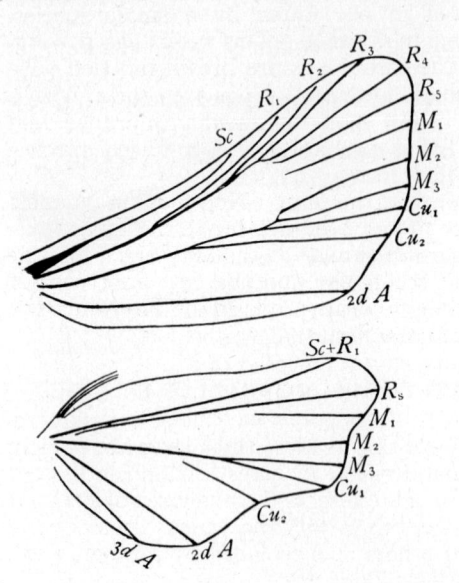

Fig. 803.—Wings of *Thyris maculata*.

This family is poorly represented in our fauna, only eleven species, representing six genera have been described from the United States.

The spotted thyris, *Thyris maculāta*.—This is the most common representative of this family in the Eastern and Middle States and it occurs also in the West. This species (Fig. 804) is brownish black, sprinkled with rust-yellow dots; the outer margin of the wings, especially of the hind wings, is deeply scalloped, with the edges of the indentations white. There is on each wing a translucent white spot, that of the hind wing is larger, kidney-shaped, and almost divided in two.

Fig. 804.—*Thyris maculata*.

Fig. 805.—*Thyris lugubris*.

The mournful thyris, *Thyris lugūbris*.—This is a larger species found in the Southern States and as far north as New York. It can be recognized by Figure 805. It is brownish black, marked with yellow, and with the translucent spots yellowish. The larva is said to infest grape.

Dysōdia oculatāna.—This is a yellow and brown species, with a single translucent spot in each wing; those of the hind wings are crescentic. The larvæ infest various flowers and seeds, and beans.

Family HYBLÆIDÆ

This family is represented in our fauna by a single species, *Hyblæa pūera*, which is found in Florida. This moth has a wing-expanse of about 35 mm. The fore wings are brown mottled with indistinct spots of a darker shade; the hind wings are brown, with a median band of three bright yellow spots margined with orange, and a similar terminal spot. This is probably an introduced species. In India the larva is a leaf-roller on teak.

This species has been placed in the family Noctuidæ in our lists of Lepidoptera; but it is much more closely related to the Thyrididæ. The venation of the wings is quite similar to that of *Thyris;* but the maxillary palpi are large and triangular and the first anal vein of the hind wings is present although weak; while in the Thyrididæ the maxillary palpi are minute and the first anal vein is lost.

THE SPECIALIZED MACROFRENATÆ

In the families included under this heading the insects are usually of medium or large size. This division includes certain moths and all skippers and butterflies. In these insects the anal area of the hind wings is reduced, containing only one or two anal veins. In some the frenulum is well-preserved, in others it is replaced by a broadly expanded anal area of the hind wing.

Family SPHINGIDÆ

The Hawk-moths or Sphinxes

Hawk-moths are easily recognized by the form of the body, wings and antennæ. The body is very stout and spindle-shaped; the wings are long, narrow and very strong; the antennæ are more or less thickened in the middle or towards the tip, which is frequently curved back in the form of a hook; rarely the antennæ are pectinated. The sucking-tube (maxillæ) is usually very long, being in some instances twice as long as the body; but in one subfamily it is short and membranous. When not in use it is closely coiled like a watch-spring beneath the head. None of the species has ocelli.

The venation of the wings (Fig. 806) is quite characteristic; the most distinctive feature is the prominence of the basal part of vein R_1 of the hind wing, the part that extends from the stem of radius to the subcosta. This free part of vein R_1 has the appearance of a cross-vein and is as stout as the other veins. In the comparatively few cases in other families where the free part of the vein R_1 has the appearance of a cross-vein it is rarely as strong as the other veins. In the hawk-moths the frenulum is usually well-preserved, but in a few it is wanting or vestigial. In many genera veins R_2 and R_3 of the fore wings coalesce throughout their length, which results in the radius being only four-branched.

Some of the hawk-moths are small or of medium size; but most of them are large. They have the most powerful wings of all Lepidoptera in our fauna. As a rule they fly in the twilight, and have the habit of remaining poised over a flower while extracting the nectar, holding themselves in this position by a rapid motion of the wings. This attitude and the whir of the vibrating wings give them a strong resemblance to humming-birds, hence they are sometimes called humming-bird moths; but they are more often called hawk-moths, on account of their long, narrow wings and strong flight.

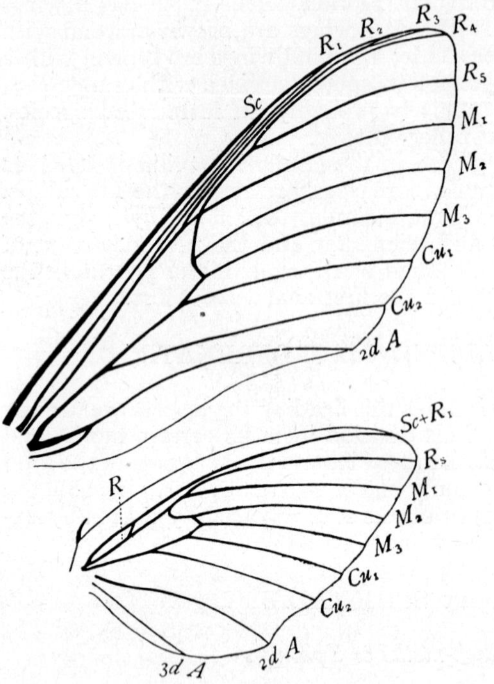

Fig. 806.—Wings of *Protoparce quinquemaculata*.

Of all the beautifully arrayed Lepidoptera some of the hawk-moths are the most truly elegant. There is a high-bred tailor-made air about their clear-cut wings, their closely fitted scales and their quiet but exquisite colors. The harmony of the combined hues of olive and tan, ochre and brown, black and yellow, and greys of every conceivable shade, with touches here and there of rose color, is a perpetual joy to the artistic eye. They seldom have vivid colors except touches of yellow or pink on the abdomen or hind wings, as if their fastidious tastes allowed petticoats only of brilliant colors always to be worn beneath quiet-colored overdresses.

The larvæ of the Sphingidæ feed upon leaves of various plants and trees and are often large and quite remarkable in appearance (Fig. 807). The body is cylindrical and naked and usually has a horn on the eighth abdominal segment. Sometimes instead of the horn there is a shiny tubercle or knob. We cannot even guess the use of this horn, unless it is ornamental, for it is never provided with a sting. These caterpillars when resting rear the front end of the body up in the air, curl the head down in the most majestic manner, and remain thus rigid and motionless for hours. When in this attitude they are supposed to resemble the Egyptian Sphinx, and so the typical genus was named *Sphinx* and the family the Sphingidæ.

Most species pass the pupa state in the ground in simple cells made in the earth; some, however, transform on the surface of the ground in imperfect cocoons composed of leaves fastened together with silk.

One hundred species of hawk-moths occur in this country. The following are some of the more common ones.

The modest sphinx, *Pachȳsphinx modĕsta.*—It was probably the quiet olive tints in which the moth is chiefly clothed that suggested the name *modesta* for it, but it is one of the most beautiful of our hawk-moths. The body and basal third of the fore wings are pale olive; the outer third of the fore wings is a darker shade of the same color; while the middle third is still darker (Fig. 808). The hind wings are dull carmine red in the middle or, in the eastern race, a deeper crimson; there is a bluish-gray patch with a curved black streak over it near the anal angle. The larva feeds on poplar and cotton-wood. When full-grown it is 75 mm. long, of a pale-green

Fig. 807.—*Sphinx chersis*, larva.

Fig. 808.—*Pachysphinx modesta*.

color, and coarsely granulated, the granules studded with fine white points, giving the skin a frosted appearance; these are wanting in the western **race.**

The twin-spotted sphinx, *Smerĭnthus geminātus.*—This exquisitely-colored moth expands about 60 mm. The thorax is gray with a velvety dark brown spot in the middle. The fore wings are gray, with a faint rosy tint in some specimens, and tipped and banded with brown as shown in Figure 809. The hind wings are deep carmine at the middle, and are bordered with pale tan or gray. Near the anal angle there is a large black spot in which there is a pair of blue spots, which suggested the name *geminatus*. The larva feeds upon the leaves of apple, plum, elm, ash, and willow.

Fig. 809.—*Smerinthus geminatus*.

Harris's sphinx, *Lăpara bombycoides*.—This sphinx has interested us chiefly on account of the habits and markings of its larva (Fig. 810). It feeds upon the foliage of pine, and is colored with alternating green and white longitudinal stripes; the dorsal stripe is green spotted with red. It has a way of hanging head downward in a pine tassel that conceals it entirely from the sight of all but very sharp eyes, its stripes giving it a close resemblance to a bunch of pine leaves. The moth expands about 50 mm.; it is gray with the fore wings marked by several series of small brown spots.

The pen-marked sphinx, *Sphinx chersis*.—This moth is of an almost evenly distributed ashy-gray color. This sombre color is relieved somewhat by a black band on each side of the abdomen, marked with four or five white transverse bars; by two dark brown, smoky bands which cross the hind wings; and by a series of black dashes on the fore wings, one in each cell between the apex of the wing and the anal vein. These dashes appear as if drawn casually with a pen. The larva (Fig. 807) is not uncommon upon ash and lilac; it is greenish or bluish white above, and darker below; there are seven oblique yellow bands on the sides of the body, each edged above with dark green. When disturbed it assumes the threatening attitude shown in the figure.

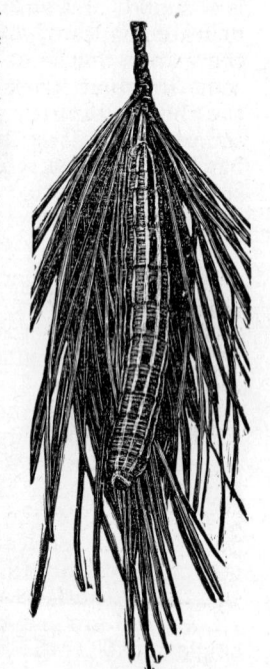

Fig. 810.—*Lapara bombycoides*, larva.

The tomato-worm, *Protopărce quinquemaculāta*.—This larva is the best known of all our sphinxes, as it may be found feeding on the leaves of tomato, tobacco, or potato wherever these plants are grown in our country. It resembles in its general appearance the larva of *Sphinx chersis* (Fig. 807); but is stouter and has a series of pale longitudinal stripes low down on each side, in addition to the oblique stripes; and its favorite attitude is with the fore end of the body slightly raised. It is usually green, but individuals are often found that are brown, or even black. There appear at frequent intervals in the newspapers accounts of people being injured by a poison excreted by the caudal horn of this larva; but there is absolutely no foundation whatever for such stories. The pupa (Fig. 811) is often ploughed up in gardens, and attracts attention on account of its currious tongue-case a part of which is free resembling the handle of a pitcher. The moth is a superb creature, expanding four or five inches. It is of many delicate shades of ash-gray, marked with black or very dark gray; there are a few short black dashes on the fore part of the thorax, and some irregular black spots edged with white on the posterior part; the abdomen is gray with a black middle line, and five yellow, almost square spots along each side. Each of these spots is bordered with black, and has a white spot above and below, on the edge of the segment. The hind wings are crossed by four blackish lines, of which the two intermediate are zigzag.

Fig. 811.—*Protoparce quinquemaculata*, pupa.

The most practicable method of control of this pest in a small garden or in a larger field where the larvæ are not numerous is by hand-picking; when they are numerous they can be destroyed by spraying with arsenate of lead; use two or three pounds of the paste dissolved in fifty gallons of water. Paris green is liable to burn the foliage of tomato.

The tobacco-worm, *Protopărce sexta*.—This species closely resembles the preceding and the two are often mistaken the one for the other. The larvæ have similar habits, feeding on the same plants; but in this species the larva lacks the series of longitudinal stripes characteristic of the tomato worm. The moths are easily distinguished; this species is brownish gray instead of ashy gray; at the end of the discal cell of the fore wings there is a distinct white spot; and the two dark bands crossing the middle of the hind wings are not zigzag, and are less distinctly separate; often they are united into a single broad band.

The hog-caterpillar of the vine, *Ampelæca myron*.—There is a group of hawk-moths the larvæ of which have the head and first two thoracic segments small, while the two following segments are greatly swollen. These larvæ from a fancied resemblance to fat swine have been termed hog-caterpillars; and the present species, which is common on grape, has been named the hog-caterpillar of the vine. It is a comparatively small species, the full-grown larva being but little more than 50 mm. long. There is a row of seven spots varying in color from red to pale lilac, each set in a patch of pale yellow, along the middle of the back. A white stripe with dark green margins extends along the side from the head to the caudal horn, and below this are seven oblique stripes. This larva is often infested by braconid parasites; and it is a common occurrence to find one of them with the cocoons of the parasites attached to it (Fig. 812). The pupa state is passed on the surface of the ground within a rude cocoon made by fastening leaves together with loose silken threads. The adult expands about 55 mm. The fore wings are olive gray, with a curved, olive-green, oblique band crossing the basal third, a discal point of the same color, and beyond this a large triangular spot with its apex on the costa and its base on the inner margin.

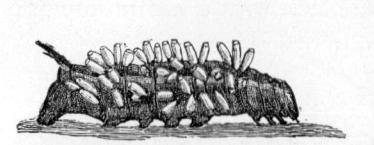

Fig. 812.—*Ampelœca myron*, larva with cocoons of parasites.

The pandorus Sphinx, *Pholus pandōrus*.—This magnificent moth expands from 100 to 112 mm. The ground color of its wings is pale olive, verging in some places into gray; the markings consist of patches and stripes of dark, rich velvety olive, sometimes almost black (Fig. 813). Near the inner margins of both pairs of wings the lighter color shades out into pale yellow, which is tinged in places with delicate rose-color. These markings show a harmony of contrasting shades

Fig. 813.—*Pholus pandorus.*

rarely equalled elsewhere by nature or art. The larva is one of the hog-caterpillars. It feeds upon the leaves of Virginia-creeper. When young it is pinkish in color, and has a long caudal horn; as it matures it changes to a reddish brown, and the horn shortens and curls up like a dog's tail and finally disappears, leaving an eye-like tubercle. The caterpillar has on each side five or six cream-colored oval spots, enveloping the spiracles.

The white-lined sphinx, *Celērio lineāta*.—This moth can be easily recognized by Figure 814. Its body and fore wings are olive-brown; there are three parallel white stripes along each side of the thorax;

the outer one of these extends forward over the eyes to the base of the palpi; on the fore wings there is a buff stripe extending from near the base of the inner margin to the apex, and veins R_5 to 2d A are lined with white; the hind wings are black with a central reddish

Fig. 814.—*Celerio lineata*.

band. The larva is extremely variable in color and markings. It feeds on many plants, among which are apple, grape, plum, and currant.

The thysbe clear-wing, *Hæmorrhāgia thysbe*.—There is a group of hawk-moths that have the middle portion of the wings transparent, resembling in this respect the Ægeriidæ and certain of the Euchromiidæ; but they are easily recognized as hawk-moths by the form of the body, wings, and antennæ. One of the more common of these is the thysbe clear-wing (Fig. 815). The scaled portions of the wings are of a dark reddish brown; but this species is most easily distinguished from all our other species by a line of scales dividing the discal cell lengthwise and representing the position of the base of vein M. The larva of this species feeds on the different species of *Viburnum*, the snow-berry, and hawthorn.

Fig. 815.—*Hæmorrhagia thysbe*.

The bumblebee hawk-moth, *Hæmorrhāgia diffinis*.—This clear-wing appears to be about as common as the preceding, and resembles it somewhat. It lacks, however, the line of scales in the discal cell, and the body is more nearly yellow in southern specimens. This color probably suggested the name bumblebee hawk-moth, given to this insect nearly one hundred years ago by Smith and Abbott. The larva feeds on the bush honeysuckle (*Diervilla*) and the snow-berry (*Symphoricarpus*).

Superfamily GEOMETROIDEA

The Geometrids or the Measuring-worms

The superfamily Geometroidea is composed of those moths the larvæ of which are known as measuring-worms, span-worms, or loopers. These larvæ are very familiar objects, attracting attention by their peculiar manner of locomotion. They progress by a series of looping movements. They first cling to the supporting twig or leaf by their thoracic legs; then arch up the back while they bring forward the hinder part of the body and seize the support, at a point near the thoracic legs, by the prolegs at the caudal end of the body; then, letting loose the thoracic legs (Fig. 816), they stretch the body forward, thus making a step; this process is then repeated.

Fig. 816.—A measuring-worm.

It was this peculiar manner of locomotion that suggested the name of the typical genus, *Geometra*, from the Greek word meaning a land-measurer.

Correlated with this mode of walking there has been a loss in nearly all members of the family of the first three pairs of prolegs.

Frequently measuring-worms when resting cling by their caudal prolegs and hold the body out straight, stiff, and motionless, appearing like a twig; this is doubtless a protective resemblance.

The geometrid larvæ are leaf-feeders, and some species occur in such large numbers as to be serious pests.

The pupæ are slender, and some species are green or mottled in this state. The pupa state is passed in a very flimsy cocoon or in a cell in the ground.

Fig. 817.—Wings of *Caripeta angustiorata*.

The moths are of medium size, sometimes small, but only rarely very large. Usually the body is slender, and the wings broad and delicate in appearance. This appearance is due both to the thinness of the membrane and to the fineness of the scales with which the wings are clothed. These moths occur on the borders of woods and in forests, rarely in meadows and pastures. Their flight is neither strong nor long sustained. Many species when at rest hold the wings horizontally and scarcely overlapping; but other species assume other positions.

In the geometrids the frenulum is usually well-preserved, but in a few it is wanting or vestigial. A striking feature of the wing-venation is the fact that the basal part of the subcosta of the hind wings makes a prominent bend into the humeral area of the wing and is usually connected to the humeral angle by a strong cross-vein (Fig. 817).

A monograph of the geometrid moths found in the United States was published by Packard ('76).

The superfamily Geometroidea includes two families; but one of these, the Manidiidæ is represented in our fauna by a single rare species.

Family GEOMETRIDÆ

In this family the antennæ are not clubbed, as in the next family. The other distinctive features of the Geometridæ are those given above in the characterization of the superfamily Geometroidea.

There occur in our fauna representatives of six subfamilies of the Geometridæ; these can be separated by the following table:—

A. Eyes small and oval. p. 664.................................BREPHINÆ
AA. Eyes round and usually large.
  B. Vein M_2 of the hind wings vestigial, being represented merely by a fold in the wing or by a non-tubular thickening (Fig. 817). p. 670..GEOMETRINÆ
  BB. Vein M_2 of the hind wings well-preserved.
    C. Vein M_2 of the hind wings arising much nearer to vein M_1 than to vein M_3 (Fig. 820). Wings usually green. p. 665..............HEMITHEINÆ
    CC. Vein M_2 of the hind wings arising nearly midway between veins M_1 and M_3 or nearer to vein M_3 than to vein M_1. Wings rarely green.
      D. Veins $Sc + R_1$ and Rs of the hind wings extending distinctly separate from each other, except that they are connected by the free part of vein R_1 near the middle of the discal cell (Fig. 821). p. 666.... LARENTIINÆ
      DD. Veins $Sc + R_1$ and Rs of the hind wings approximated or coalesced for a greater or less distance.
        E. Veins $Sc + R_1$ and Rs of the hind wings closely approximated but not coalesced along the second fourth (more or less) of the discal cell. With transverse rows of spines on abdominal segments. (*Palæacrita*). p. 670...................................GEOMETRINÆ
        EE. Veins $Sc + R_1$ and Rs of the hind wings coalesced for a greater or less distance. Abdomen without transverse rows of spines.
          F. Veins $Sc + R_1$ and Rs of the hind wings coalesced for a short distance near the beginning of the second fourth of the discal cell, thence rapidly diverging (Fig. 821). p. 666........ACIDALIINÆ
          FF. Veins $Sc + R_1$ and Rs of the hind wings coalesced to or beyond the middle of the discal cell (Fig. 823), or with a short fusion near the end of the discal cell.
            G. Fore wings with one or two accessory cells. p. 666.. LARENTIINÆ
            GG. Fore wings without an accessory cell. p. 664..ŒNOCHROMINÆ

Subfamily BREPHINÆ

The members of this subfamily are most easily distinguished from other geometrids by the fact that their eyes are small and oval. It is represented in our fauna by only five species, of which the following one is the best-known.

Brēphos ĭnfans.—This interesting species has been found only in the northeastern part of our country; its range is from Labrador to New York. It is a blackish-brown moth with the fore wings marked with pinkish white and the hind wings with reddish orange (Fig. 818). The specimen figured is a male. In the female the black border on the outer margin of the hind wings is narrower, and the subterminal, light band on the fore wings is more distinctly marked. In the larva the prolegs are all present; but the first three pairs are stunted. The full-grown larva measures 30 mm. in length. The color of the larva on the dorsum varies from apple-green to blue-green according to age. The food-plant is white birch.

Fig. 818.—*Brephos infans.*

Subfamily ŒNOCHROMINÆ

This subfamily is represented in our fauna by only three species of which only the following one is well known.

The fall canker-worm, *Alsŏphila pometāria*.—The canker-worms are well-known pests, which are often very destructive to the foliage of fruit-trees and shade-trees. Although they attack many kinds of trees, the apple and the elm are their favorite food-plants.

There are two species of canker-worms which resemble each other to such an extent that they were long confounded; but they differ structurally, being members of different subfamilies; and they differ also in habits. The two species agree in being loopers or measuring-worms in the larval state, in the possession of ample wings by the adult male, and in the adult female being wingless. They are easily distinguished however, in all stages, the eggs, larvæ, and adults differing markedly.

The fall canker-worm is so called because the greater number of the moths mature in the autumn and emerge from the ground at this season; but a considerable number come out of the ground in the winter during warm weather, and in the spring. As the females are wingless they are forced to climb up the trunks of trees in order to lay their eggs in a place from which the larvæ can easily find their food. The eggs appear as if cut off at the top, and have a central puncture and a brown circle near the border of the disk. They are laid side by side in regular rows and compact batches, and are generally exposed. They hatch in the spring at the time the leaves appear; and the larvæ mature in about three weeks. In this species there is a pair of vestigial prolegs on the fifth abdominal segment. The pupa state is passed beneath the ground in a perfect cocoon of fine densely spun silk. The adult male is represented by Figure 819. There is a distinct whitish spot near the apex of the fore wings. The moths of both sexes lack the abdominal spines characteristic of the spring canker-worm. In the fall canker-worm veins $Sc + R_1$ and R_s of the hind wings coalesce for a considerable distance along the second fourth of the discal cell; and veins R_s and M_1 of the hind wings separate at the apex of the discal cell.

Fig. 819.—*Alsophila pometaria*.

Control of canker-worms.—The two species of canker-worms are sufficiently alike in habits to warrant our combating them by similar methods. The fact that in each the female is wingless and is thus forced to climb up the trunks of trees in order to place her eggs in a suitable place has suggested the method of defence that has been most generally used in the past. This is to place something about the trunks of the trees which will make it impossible for the wingless female to ascend them. Some viscid substance, as tar, printers' ink, or melted rubber, either painted on the trunk of the tree or upon a paper band which is tacked closely about the tree, is the means usually adopted. Many other devices have been recommended. In the use of this method of prevention, operations should be begun in the autumn, even when it is the spring canker-worm that is to be combated; for in this species some of the moths emerge in the fall or during the winter.

Although the method just described is still the most available one when tall shade-trees are to be protected, it is now rarely used in orchards. Here the spraying of the trees with Paris-green or arsenate of lead soon after the leaves appear is found more practicable. This method has also the advantage of enabling the fruit-grower to reach other important pests, as the codlin-moth, at the same time.

Subfamily HEMITHEINÆ

The Green Geometrids

As a rule the members of this subfamily are bright green in color; and as we have but few other geometrids of this color, the subfamily may be well termed the green geometrids. The distinctive structure that characterizes this subfamly is the fact that vein M_2 of the hind wings arises much nearer to vein M_1 than to vein M_3 (Fig. 820).

This is a comparatively small subfamily, including 17 genera and 64 species. The following species will serve as an example.

The raspberry geometer, Synchlōra ærāta.—The larva of this species feeds on the fruit and foliage of raspberry, but chiefly on the fruit. It, like some other members of this subfamily, has the curious habit of covering its body by attaching to it bits of vegetable matter, so that it is masked beneath a tiny heap of rubbish. The wings of the adult are of a delicate pale green color crossed by two lines of a lighter shade; the face is green; and the abdomen is not marked with pink and white ocellate spots, as is the case in certain allied species.

Subfamily ACIDALIINÆ

The members of this subfamily are most easily recognized by the venation of the hind wings (Fig. 821). In these veins $Sc+R_1$ and R_s coalesce for a short distance near the beginning of the second fourth of the discal cell and then diverge rapidly. The greater number of our common species are of medium size, with whitish wings crossed by from two to four indistinct lines, and with the head black in front; some are pure white, and others are brown marked with reddish lines. Eighty-six species are now listed from this country.

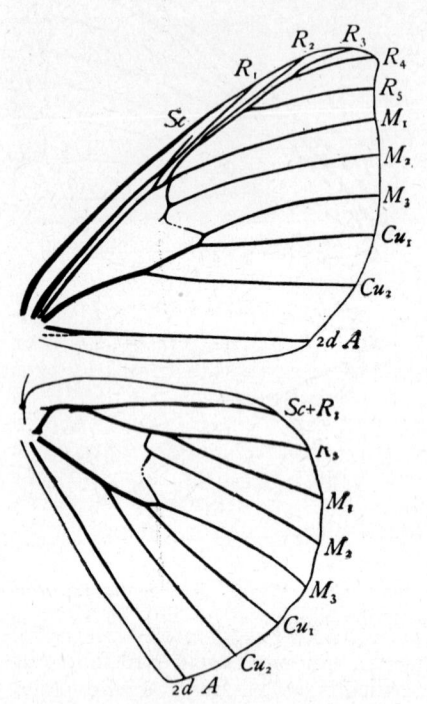

Fig. 820.—Wings of *Dichorda iridaria*.

The chickweed geometer, *Hæmătopis gratāria*.—This little moth (Fig. 822) is very common in our meadows and gardens during the summer and autumn months. Its wings are reddish yellow, with the fringes and two transverse bands pink. It is found from Maine to Texas. The larva feeds on the common chickweed, *Alsine media*.

The sweet-fern geometer, *Cosȳmbia lumenāria*.—This moth is grayish white, with three rows of black dots extending across the wings, one marginal, one submarginal, and one near the base of the wings; near the center of each wing there is a small reddish ring. The moth has a wing expanse of from 20 to 25 mm. The larva is common on sweet-fern, *Comptonia;* it also feeds on birch.

Subfamily LARENTIINÆ

In this subfamily the branches of radius of the fore-wings anastomose so as to form one or two accessory cells, this anastomosis involving vein R_1 (Figs. 823 and 824). In the hind wings in most of the genera veins $Sc+R_1$ and R_s coalesce to or beyond the middle of the discal cell or with a short fusion near the end of the discal cell (Fig. 823); but in certain genera, where the costal area of the hind wings is greatly expanded, these veins extend distinctly separate from each other, except that they are connected by the free part of vein R_1 near the middle of the discal cell (Fig. 824). In a few genera belonging to other subfamilies veins $Sc+R_1$ and R_s of the hind wings coalesce to the middle of the discal cell, but these genera lack the accessory cell in the fore wings characteristic of this subfamily.

This subfamily ranks second in size among the geometrid subfamilies, including 365 North American species; these represent 58 genera. Many of the species are very common; among them are the following.

Fig. 821.—Wings of *Acidalia enucleata*.

Fig. 822.—*Hæmatopis grataria*.

The white-striped black, *Trichodēzia albovittāta*.—This beautiful little moth, which occurs from the Atlantic to the Pacific, is the most easily recognized member of the family. It expands about 22 mm. and is of a uniform black color, with a single, very broad white band extending across the fore wing from the middle of the costa to the inner angle, where it is usually forked. The fringe of the wings is white at the apical and inner angles of both pairs; sometimes the white is lacking on the inner angle of the hind wings. The early stages are unknown.

The bad-wing, *Dўspteris abortivāria*.—It is easy to recognize this moth (Fig. 825) by the peculiar shape of its wings,

Fig. 823.—Wings of *Eudule mendica*.

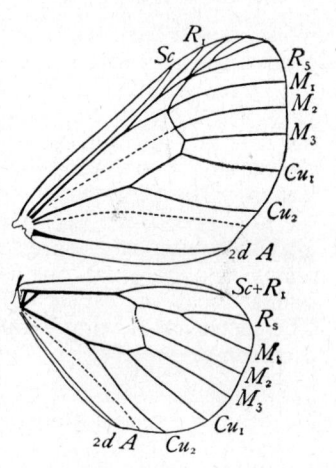

the hind wings being greatly reduced in size. It is of a beautiful pea-green color, with two white bands on the fore wings and one on the hind wings. In color it resembles members of the Hemitheinæ; but the structure of its wings (Fig. 824) shows that it belongs in the Larentiinæ. The larva feeds on the leaves of grape, which it rolls.

The scallop-shell moth, *Calocălpe undulāta*.—This is a pretty moth, with its yellow wings crossed by so many fine, zigzag, dark brown lines that it is hard to tell which of the two is the ground-color (Fig. 826). It lays its eggs in a cluster on a leaf near the tip of a twig of cherry, usually wild cherry. The larvæ make a snug nest by fastening together the leaves at the end of the twig; and within this nest (Fig. 827) they live, adding new leaves to the outside as more food is needed. The leaves die and become brown, and thus render the nest conspicuous. There are two generations in the year. The larvæ of the fall brood are black above, with four white or green stripes, and flesh-colored below; the larvæ of the summer brood are black only on the sides. When full-grown they descend to the ground to transform, and pass the winter in the pupa state.

Fig. 824.—Wings of *Dyspteris abortivaria*.

The diverse-line moth, *Lȳgris diversilineāta*.—This moth has pale ochre-yellow wings, with a brownish shade near the outer margin, and crossed by many diverging brown lines (Fig. 828). It varies from 37 to 50 mm. in expanse. We have often found this moth on the side of our room, resting on the wall, head downward, and with its abdomen hanging down over its head in a curious manner. The larva feeds on the leaves of grape. There

Fig. 825.—*Dyspteris abortivaria*.

Fig. 826.—*Calocalpe undulata*.

are two broods; the first brood infests the vines during June; the second, in the autumn and early spring, wintering as larvæ.

The spear-marked black, *Rheumăptera hastāta*.—This is a black-and-white species, which is found from the Atlantic to the Pacific. It is much larger than the white-striped black described above, expanding 35 mm. It is black, striped and spotted with white; It varies greatly as to the number and extent of the white markings. The most constant mark is a broad white band crossing the middle of the fore wings, and often continued across the hind wings. Near the middle of its course on the fore wing this band makes a sharp angle pointing outward; and just beyond the apex of this angle there is usually a white spot. This spot and angular band together form a mark shaped something like the head of a spear. In some individuals the white predominates, other individuals are almost entirely black, excepting the spear mark. In the East, there is more white on the fore wings than on the hind wings; this form is the variety *gothicāta;* in some parts of the West and in Europe there is more white on the hind wings than on the fore wings. According to European authorities the larva is brown or blackish brown, with a darker line along the middle of the back, and a row of horse-shoe-shaped spots on the sides. It feeds on birch and sweet gale. It is gregarious, a colony of larvæ spinning together the leaves of the food-plant, and thus forming a nest within which they live and feed.

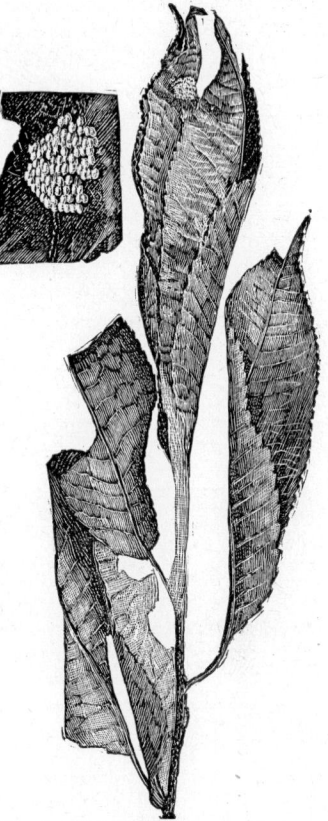

Fig. 827.—Eggs and nest of *Calocalpe undulata*.

Fig. 828.—*Lygris diversilineata.*

Fig. 829.—*Eudule mendica.*

The beggar, *Eudūle mendīca.*— One of the most delicate-winged moths that we have in the northern Atlantic States is this species (Fig. 829). Although the wings are yellowish white in color they are almost transparent. On the fore wings there are two transverse rows of pale gray spots, and a single spot near the outer margin between veins M_3 and Cu_1. (This spot was indistinct in the specimen figured.) The moth is common in midsummer. The larva feeds on violet.

Subfamily GEOMETRINÆ

Nearly all of the members of this subfamly can be easily recognized as such by the fact that vein M_2 of the hind wings is wanting, being represented merely by a fold.

This is by far the largest of the subfamilies of the Geometridæ; it includes more than 500 North American species; these represent 124 genera. The following are some of the more common species.

The currant span-worm, *Cymatŏphora ribeāria.*—There are several species of insects that are popularly known as currant-worms. The most common of these are larvæ of saw-flies, which can be easily recognized by the large number of prolegs with which the abdomen

Fig. 830.—*Cymatophora ribearia.*

Fig. 831.—*Cingilia catenaria.*

is furnished. In addition to the saw-flies there is a yellow looper spotted with black, which often appears in such great numbers on currant and gooseberry bushes as to suddenly strip them of their foliage.

This larva has been named the currant or gooseberry span-worm. When full-grown it measures about 25 mm. in length, and is of a bright yellow color, with white lines on the sides and with numerous black spots and round dots. It has only four prolegs. There is only one generation a year; the larva matures in May or June; the pupa state lasts about a fortnight; the moth flies during the summer months and oviposits on the twigs of the plants; and the eggs remain unhatched till the following spring. The moth (Fig. 830) is pale yellow, with the wings marked by irregular dusky spots, which sometimes form one or two indefinite bands across them.

The chain-dotted geometer, *Cingĭlia catenāria.*—This moth has snow-white wings marked with zigzag lines and with dots of black as shown in Figure 831. The head is ochreous-yellow in front; and the thorax is yellowish at the base of the patagia. The moth flies during

September and October. The larva feeds on various shrubs and trees. The pupa state is passed in a slight but well-formed web of yellow threads, which is formed between twigs or leaves, and through which the pupa can be seen.

The evergreen nepytia, *Nepÿtia semiclusāria.*—This beautiful moth (Fig. 832) is common in the vicinity of pines, spruce, fir, and hemlock during August and September. It varies from a smoky-ash color to almost snow-white; the wings are marked with black. The larva feeds on the leaves of Conifers. It is reddish yellow above, with lateral yellow bands below, while on each side are two pairs of black hair-lines. There are black spots above on the segments. When full-grown it is a little more than 25 mm. long and spins a loose cocoon among the leaves. The chrysalid is green with white stripes and is very pretty.

Fig. 832.—*Nepytia semiclusaria.*

The spring canker-worm, *Paleăcrita vernāta.*—The eggs are ovoid in shape, and are secreted in irregular masses, usually under loose scales of bark or between the leaflets of the expanding buds. The larvæ hatch about the time the leaves expand, and become full-grown in from three to four weeks. They vary greatly in color, and are marked on the back with eight narrow, pale, longitudinal lines which are barely discernible; the two lateral lines of each side are much farther apart than the others; and there are no prolegs on the fifth abdominal segment as in the fall canker-worm. The pupa state is passed below the surface of the ground in a simple earthen cell, which is lined with very few silken threads. The adult moths usually emerge early in the spring before the leaves expand; but they sometimes appear late in the fall, or on warm days during the winter when the ground is thawed. In both sexes the adult of this species is distinguished by the presence of two transverse rows of stiff reddish spines, pointing backwards, on each of the first seven abdominal segments.

Regarding the control of canker-worms see page 665.

The lime-tree winter-moth, *Erănnis tiliāria.*—This species (Fig. 833) resembles the canker-worms in many particulars. The larva is a looper which infests both fruit and forest trees; and in the adult state the male has well-developed wings, while the female is wingless.

The eggs are oval, of a pale yellow color, and covered with a network of raised lines. They are thrust by the female under loose bark and in crevices on the trunk and large limbs. They hatch in May, and the larvæ attain their full growth in the latter part of June. The larva is yellow, marked with ten crinkled black lines along the top of the back; the head is rust-colored, and the venter yellowish white. There is a second form of the larva which is brown above with slate color towards the sides. When full-grown the larva measures about 30 mm. in length. The pupa state is passed in the ground. The moths issue in October or November; and then the wingless females ascend the trees to oviposit as do the females of the canker-worms. The female is represented in the lower left-hand part of the figure.

Fig. 833.—*Erannis tiliaria*. (From The Author's Report for 1879.)

She is grayish in color, with two black spots on the back of each segment except the last, which has only one. The male has pale yellow and brown or buff fore wings, with a central spot and a band beyond the middle, while the hind wings are much lighter. This insect can be combated by the same methods as are used against canker worms.

The notched-wing geometer, *Ĕnnomos magnārius*.—This is one of the larger of our geometrids. The larva is a common looper upon maple, chestnut, and birch trees, and measures about 58 mm. in length when full-grown. It spins a rather dense, spindle-shaped cocoon within a cluster of leaves. The moth (Fig. 834) is ochre-yellow with reddish tinge. The wings are shaded towards the outer margin with brown, and are thickly spotted with small brown dots.

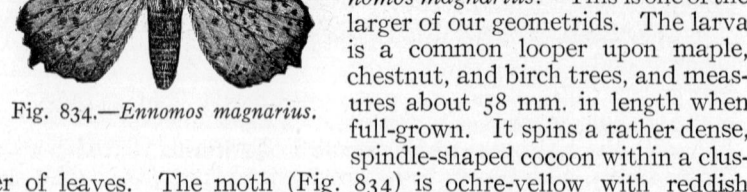

Fig. 834.—*Ennomos magnarius*.

The pepper-and-salt currant-moth, *Amphidasis cognataria*.—
This moth (Fig. 835) differs remarkably in appearance from most
geometrids, the body being stouter, and the wings appearing heavier.

Fig. 835.—*Amphidasis cognataria*.

Fig. 836.—*Phryganidia californica*.

It can be easily recognized by its evenly distributed pepper-and-salt
markings. The larva feeds on various plants, but is found most often
on currant.

Family MANIDIIDÆ

This family is represented in our fauna by a single, recently discovered species, *Anurăpteryx crenulāta*, found in Arizona. In the genus *Anurapteryx* the antennæ are gradually enlarged toward the tip forming a **well-marked club**; the maxillæ are well-developed; the eyes are hairy and overhung by long cilia; and the frenulum is well-developed.

Our species was described by Barnes and Lindsey in "Entomological News," vol. 30, p. 245.

Family DIOPTIDÆ

The Dioptids

The only member of this family that is well known in this country is

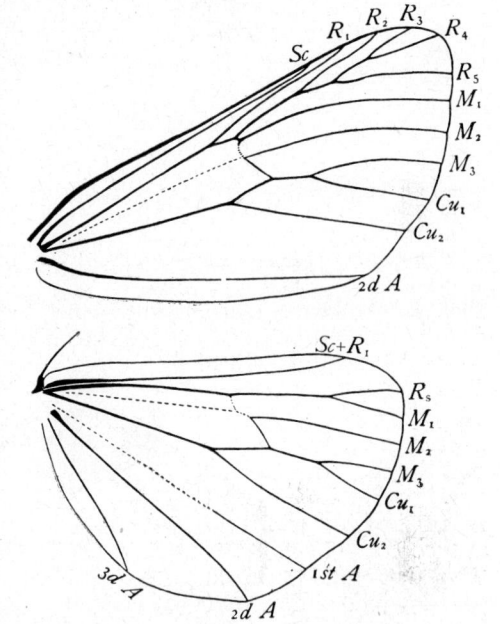

Fig. 837.—Wings of *Phryganidia*.

Phryganidia califŏrnica, which occurs in California. This is a pale-brown insect, with nearly transparent wings (Fig. 836). The veins of the wings are dark, which renders them prominent. In the males there is a yellowish spot just beyond the discal cell. The venation of the wings (Fig. 837) is very different from that of any other insect that occurs in this country.

The larvæ feed upon the leaves of live-oaks, and sometimes occur so abundantly as to almost strip the trees of their foliage. They are said to feed singly, and appear to make little if any use of the anal feet as a means of locomotion, generally carrying the last segment of the body elevated in the air.

FAMILY NOTODONTIDÆ

The Prominents

This family includes moths of moderate size, only a few of the larger ones expanding more than 50 mm. The body is rather stout and densely clothed with hair, and the legs, especially the femora, are clothed with long hairs. The wings are strong, and not very broad, the anal angle of the hind wings rarely reaching the end of the abdomen. In their general appearance many of these moths bear a strong resemblance to noctuids; but they can be easily distinguished from the Noctuidæ by the position of vein M_2 of the fore wings, which does not arise nearer to vein Cu than to vein R, as it does in that family; and the fact that in this family veins $Sc+R_1$ and R_s of the hind wings do not coalesce (Fig. 838). The first anal vein is wanting in both fore and hind wings; and in some species an accessory cell is present in the fore wings.

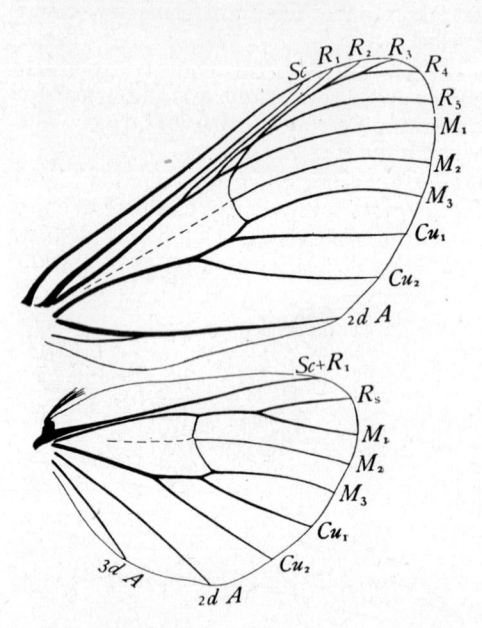

Fig. 838.—Wings of *Hyperæschra stragula*.

Fig. 839.—*Pheosia rimosa*.

In some species the front wings have a prominence or backward projecting lobe on the inner margin, which suggested the common name of *prominents* for these insects (Fig. 839). The name is more generally appropriate, however, for the larvæ, as a much larger proportion of them than of the adults bear striking prominences.

The larvæ feed upon the leaves of shrubs and trees. Our most common species live exposed; but some species live in folded leaves. They are either naked or clothed with hairs. Many species have only well-developed prolegs, the anal pair being rudimentary, or transformed into elongated spikes. Some species are hump-backed; and spines or fleshy tubercles are often present. The transformations occur in slight cocoons or in the ground.

The family Notodontidæ is represented in this country by about one hundred species. A monograph of the family was published by Packard ('95) in which there are many colored figures of larvæ. The following are some of the more common species.

The handmaid moths, *Datana*.—Among the more common representatives of the Notodontidæ are certain brown moths that have the fore wings crossed with bars of a different shade (Fig. 840) and that bear on the fore part of the thorax a conspicuous patch of darker color. In most of our species the fore wings are also marked with a dot near the center of the discal cell and a bar on the discal vein. These moths belong to the genus *Datāna*. The common name, handmaid, is a translation of the specific name of our most common species, *D. minĭstra*. But as this species is now generally known as the yellow-necked apple-tree worm, and as all of our species are dressed in sober attire as becomes modest servants, we have applied the term handmaid moths to the entire genus.

Fig. 840.—*Datana*.

The larvæ of the handmaid moths are easily recognized by their peculiar habits. They are common on various fruit and forest trees, but especially on apple, oak, and hickory.

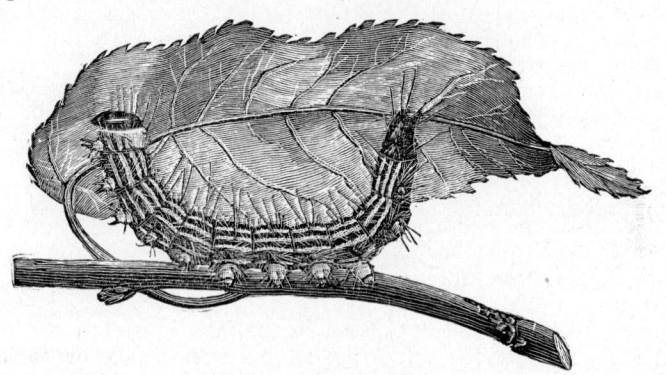

Fig. 841.—*Datana*, larva.

They feed in colonies; and have the habit of assuming the curious attitude shown in Figure 841. The body is black or reddish, marked with lines or stripes of yellow or white. Owing to the gregarious habits of these larvæ they can be easily collected from the trees they infest.

All the species that we have studied agree in being single-brooded, the moths appearing in midsummer; the eggs are laid in a cluster on a leaf; the larvæ are conspicuous in August and September. In some of the species the larvæ have the curious habit of leaving the branch upon which they are feeding when the time to molt arrives, the whole colony gathering in a large mass on the trunk of the tree, where the molt takes place. The pupa state is passed in the ground, in a very light cocoon or in none at all, and lasts about nine months in the species that we have bred.

The white-tipped moth, *Symmerista albifrons*.—This beautiful moth, which is quite common, can be easily recognized by the accompanying figure (Fig. 842, a); the white patch, which extends along the costa of the fore wing for half the length from the tip, being very characteristic. The larva (Fig. 842, b) is quite common in the autumn on leaves of oak. It is known as the red-humped oak-caterpillar; it is smooth and shining, with no hairs; along each side of the back there is a yellow stripe, and between these, on the back, fine black lines on a pale lilac ground; on each side below the yellow stripes there are three black lines, the lowest one just above the spiracles. The head is orange-red; and there is an orange-red hump on the eighth abdominal segment.

Fig. 842a.—*Symmerista albifrons*.

The two-lined prominent, *Heterocampa bilineata*.—The larva of this species (Fig. 843) is much more apt to be observed than the adult. It is common in the latter part of the summer and in early autumn, feeding on the leaves of elm, beech, and basswood. It measures when full-grown about 37 mm. in length. Its ground-color is usually green, but sometimes claret-red. There is a pale yellow stripe along the middle of the back, and on each side a stripe of the same color. The course of these side stripes is very characteristic; passing back from the head, they converge on the prothorax; on the mesothorax and metathorax they are separated from the dorsal line only by a narrow band of red or purple; on the first abdominal segment they diverge to the lateral margin of

Fig. 842b.—*Symmerista albifrons*, larva.

Fig. 843.—*Heterocampa bilineata*, larva.

the back, but converge again on the seventh and eighth abdominal segments. This yellow subdorsal line is bordered without by a milk-white stripe; and extending from this stripe over the side of the body there is a whitish shade which fades out below. The moth is ash colored, with the fore wings crossed by two wavy lines between which the wing is darker; between the outer wavy line and the outer margin of the wing there is a faint band.

Antlered larvæ.—Among the remarkable forms exhibited by notodontian larvæ are those of the freshly-hatched larvæ of the species of *Heterocampa*. Figure 844 represents the first instar of *Heterocampa vāria*, which has on the first thoracic segment a pair of large antler-like horns, and other horns on several of the abdominal segments. In the second instar all of these horns are wanting except small vestiges of the first pair. This species feeds on oak.

The freshly hatched larva of *Heterocampa guttivitta* is also antlered. The horns borne by the prothorax are four-branched and there are eight-pairs of horns on the abdomen. As in the preceding species all of these horns are wanting in the second instar except vestiges of the first pair. This species feeds especially on maple, but has been found on other trees.

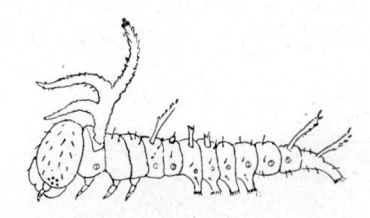

Fig. 844.—*Heterocampa varia*, larva. (After Packard.)

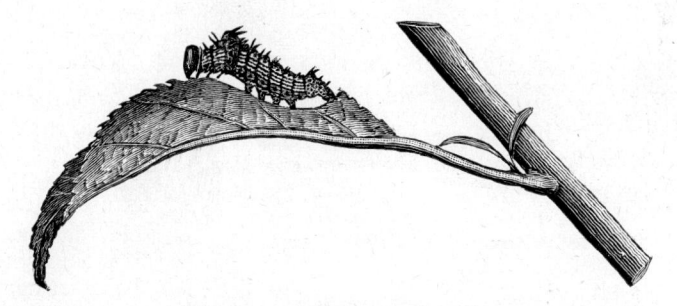

Fig. 845.—*Schizura concinna*, larva.

The red-humped apple-worm, *Schizūra concĭnna.*—The larva of this species (Fig. 845) is common on apple and allied plants. The head is coral-red, and there is a hump of the same color on the back of the first abdominal segment; the body is striped with slender black, yellow, and white lines, and has two rows of black spines along the back, and other shorter ones upon the sides. When not eating, the larvæ remain close together, sometimes completely covering the branch upon which they rest. This species passes the winter in the pupa state. The adults appear in June and July.

The Mocha-stone moths, *Melălopha*.—To the genus *Melalopha* belong several species of brownish-gray moths, whose fore wings are crossed by irregular whitish lines. It was these peculiar markings, resembling somewhat those of a moss-agate, that suggested the popular name given above. The larvæ feed on poplar and willow, and conceal themselves within nests made by fastening leaves together. Our most common species is the following.

Melălopha inclūsa.—The adult (Fig. 846) is a brownish-gray moth with the fore wings crossed by three irregular whitish lines. The basal line is broken near the middle of the wing; and the intermediate one forms an inverted Y, the main stem of which joins the third line near the inner margin of the wing, making with it a prominent V. These lines are bordered without by rust-red; there is a chocolate-colored spot near the apex of the fore wings, and an irregular row of blackish dots near the outer margin. The hairs of the thorax form a prominent crest, the fore side of which is a rich dark brown. The hind wings are crossed by a wavy band, which is light without and dark within.

The eggs are nearly spherical and smooth; they are deposited in a cluster a single layer deep on a leaf (Fig. 847). When the larvæ hatch they make a nest either by fastening several leaves together or, as is the case when they infest poplar, by folding the two halves of a single leaf together; frequently in the latter case the tip of the leaf is folded in as shown in the figure. Within this nest the entire colony lives, feeding on the parenchyma, and causing the leaf to turn brown. Later other leaves are added to this nest or additional nests are made among adjoining leaves. All of these infested leaves are securely fastened to the twig by bands of silk. When the larvæ become large they leave their nests at night to feed upon other leaves. These they entirely consume excepting the petioles, midribs, and larger veins. We have seen on poplar a nest composed of only three leaves which contained one hundred and twenty-five half-grown larvæ; all of the leaves, about thirty in number, arising from the end of the branch bearing this nest had been consumed.

Fig. 846.—*Melalopha inclusa*.

Fig. 847.—Eggs, larva, and nest of *Melalopha inclusa*.

The full-grown larva measures 35 mm. in length. It is striped with pale yellow and brownish black, and bears a pair of black tubercles close together on the first abdominal segment, and a similar pair on the eighth abdominal segment. The cocoon is an irregular thin web; it is made under leaves or other rubbish on the ground. The insect remains in the pupa state during the winter, and emerges as a moth in the latter part of June or later. In the South this species infests willow as well as poplar, and is double-brooded.

Fig. 848.—*Schizura ipomeæ*, larva.

Among the more grotesque larvæ belonging to this family are those of the genus *Schizura*, of which we have several species. Figure 848 represents the larva of *Schizura ipomeæ*. At the left in the figure is shown a front view of the longest tubercle. This species feeds on oak, maple, and many other plants. In the Gulf States it feeds on *Ipomea coccinea*, which fact suggested its specific name.

Family LYMANTRIIDÆ

The family Liparidæ of some writers

The Tussock-moths

The larvæ of these moths are among the most beautiful of our caterpillars, being clothed with brightly-colored tufts of hairs; and it is to this characteristic clothing of the larvæ that the popular name *tussock-moths* refers.

The adult moths are much plainer in appearance than the larvæ; and in the genera *Hemerocampa* and *Notolophus*, to which our most common species belong, the females are practically wingless, the wings being at most short pads, of no use as organs of flight.

The tussock-moths are of medium size, with the antennæ of both sexes when winged pectinated, those of the males very broadly so; the wingless females have serrate or narrowly pectinate antennæ. The ocelli are wanting. The legs are clothed with woolly hairs; when the insect is at rest the fore legs are usually stretched forward, and are very conspicuous on account of these long hairs. The venation of the wings is quite similar to that of the Noctuidæ, but in the Lymantriidæ the point at which veins $Sc + R_1$ and R_s of the hind wings anastomose is farther from the base of the wing (Fig. 849). In some genera these two veins are separate being connected only by the free part of vein R_1. The tussock-moths are chiefly nocturnal; but the males of some of them fly in the daytime.

The larvæ of our native species are very characteristic in appearance. The body is hairy; there are several conspicuous tufts of hairs on the dorsal aspect of the abdomen, and at each end of the body there are long pencils of hairs; on the sixth and seventh abdominal segments there is on the middle of the back of each an eversible gland supposed to be a scent-organ similar to the osmateria in the larvæ of *Papilio*, and it is stated that a fine spray of liquid is sometimes thrown from them.

Fig. 850.—*Hemerocampa leucostigma.*

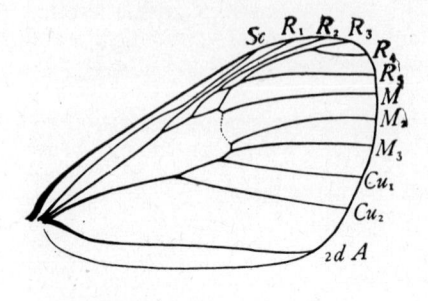

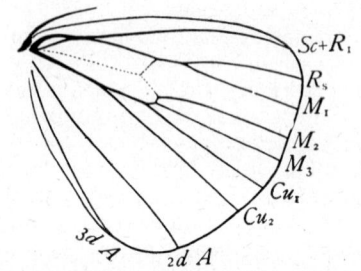

Fig. 849.—Wings of *Hemerocampa leucostigma.*

Excepting a few rare forms our native species pertain to the genera *Hemerocampa*, *Notolophus*, and *Olene*. In the first two of these genera the males are winged and the females are nearly wingless. In *Olene* both sexes are winged. Our best known of the native members of this family are the following.

The white-marked tussock-moth, *Hemerocămpa leucostĭgma*.—This is our most common representative of the family. It frequently occurs in such great numbers that it seriously injures the foliage of shade-trees and orchards. The male (Fig. 850) is of an ashy gray color; the fore wings are crossed by undulated bands of darker shade and bear a conspicu-

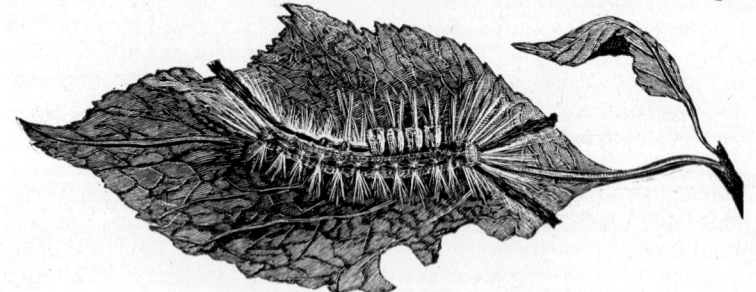

Fig. 851.—*Hemerocampa leucostigma*, larva.

ous white spot near the anal angle. The female is white and resembles a hairy grub more than a moth. She emerges from her cocoon and after pairing lays her eggs upon it, covering them with a frothy mass. The larva (Fig. 851) is one of the most beautiful of our caterpillars.

The head and the glands on the sixth and seventh abdominal segments are bright vermilion red. There is a velvety black dorsal band, bordered with yellow subdorsal stripes; and there is another yellow band on each side just below the spiracles. The prothorax bears on each side a pencil of long black hairs with plume-like tips; a similar brush is borne on the back of the eighth abdominal segment, and the first four abdominal segments bear dense brush-like tufts of cream-colored or white hairs.

When this insect becomes a pest the larvæ can be destroyed by spraying the infested trees with Paris-green water; or the egg-bearing cocoons can be collected during the winter and destroyed. These cocoons are attached to the trunks of the trees and to neighboring objects, or to twigs. In the latter case they are usually partially enclosed in a leaf. Cocoons not bearing eggs should not be destroyed, as many of them contain parasites. Owing to the wingless condition of the female this pest spreads slowly.

The well-marked tussock-moth, *Hemerocampa plagiata*.—The male, like that of the preceding species, is of an ashy gray color; but the markings of the fore wings are much more distinct. The female is light brown. She lays her eggs in a mass on her cocoon, covering them with hair from her body. The larva closely resembles the preceding species in the form and arrangement of its tufts of hair, but differs markedly in color, being almost entirely light yellow. There is a dusky dorsal stripe and a velvety black spot behind each of the tufts of the first four abdominal segments. The head and the glands on the sixth and seventh abdominal segments are, like the body, light yellow.

The California tussock-moth, *Hemerocămpa vetūsta*.—The two species of *Hemerocampa* described above are found only in the East; this species is found in California, where it is common on live oak and yellow lupin trees, and has injuriously infested apple and cherry orchards. The larvæ have black heads, crimson hair-bearing warts and prolegs, and the four tussocks or brush-like tufts of hairs on the back are often dark gray with brownish crests. In general the life-history of this species is similar to that of the two eastern species.

The old tussock-moth, *Notŏlophus antīqua*.—The male is of a rust-brown color; the fore wings are crossed by two deeper brown bands and have a conspicuous white spot near the anal angle. The body of the grub-like female is black, clothed with yellowish white hairs; she lays her eggs on her cocoon, but, unlike the three preceding species, does not cover them with anything. The larva differs from either of the preceding in having an extra pair of pencils of plume-like hairs arising from the sides of the second abdominal segment; the head is jet-black; the glands on the sixth and seventh abdominal segments are vermilion-red or sometimes bright orange; and the tubercles on the sides of the back of the second and third thoracic and the sixth and seventh abdominal segments are orange-red or yellow margined with pale yellow.

The gypsy moth, *Porthētria dĭspar*.—This is a European species which was introduced into Massachusetts in 1866 by a French naturalist who was conducting experiments with silk-worms. Some of the insects escaped from him into a neighboring woodland and became established there; but they did not attract particular attention till about twenty years later. It was then realized that this species is a serious pest. Since then millions of dollars have been expended by the State of Massachusetts and the Federal Government in an unsuccessful effort to exterminate it. It has spread over a large part of New England, and isolated colonies have been found in New York. The larva has a wide range of food-plants, feeding on the foliage of most forest and fruit trees. The male moth is yellowish brown; the female white (Fig. 852). In each the fore wings are crossed by many dark lines and bear a black lunule on the discal vein. The specimen figured is unusually small. The eggs are laid in a mass on any convenient object and are covered with hair from the abdomen of the female. The larva differs greatly in appearance from those of the preceding genera, lacking the peculiar pencils and tufts of hair; but the characteristic glands of the sixth and seventh abdominal segments are present and are red. The body is dark brown or black, finely reticulated with pale yellow, and with narrow yellow dorsal and subdorsal lines. On the dorsal aspect of each segment there is a pair of prominent, rounded tubercles bearing spiny black hairs. The first five pairs of these tubercles are bluish, the others dark crimson-red. There are also two rows of tubercles on each side of the body which bear longer hairs.

Fig. 852.—*Porthetria dispar*.

The brown-tail moth, *Euprŏctis chrysorrhœa*.—The brown-tail moth is another European pest, which was introduced into Massachusetts at some unknown date. It first attracted attention by its ravages in 1897, and since then has spread over a considerable part of New England and has extended into New Brunswick and Nova Scotia. The wings of the female moth are white; and the tip of the abdomen bears a tuft of yellowish brown hairs, hence the popular name of the insect. The female expands about 37 mm. The male is a little smaller than the female; and the brownish tuft at the end of the abdomen is not so conspicuous as in the female. The larva feeds on the foliage of fruit-trees and of almost all kinds of shade-trees except conifers. The eggs are laid in an elongate mass on the underside of a leaf, during July. The egg-mass is covered with brownish hairs from the tip of the abdomen of the female. The eggs hatch in two or three weeks. The larvæ hatching from an egg-mass feed together on adjoining leaves at the tip of a branch. These they web together with silk, making a nest within which they pass the winter in a partially grown condition. In early spring the larvæ leave their winter quarters and feed on the expanding foliage. They become full grown in five or six weeks; and then spin thin cocoons of white silk in curled leaves, crevices in bark of trees, or under any convenient shelter. About three weeks later the moths emerge.

The full-grown larva of the brown-tail moth measures about 37 mm. in length. It is nearly black in ground color, clothed with tufts of brownish barbed hairs, and has a row of nearly white tufts on each side of the body. In the center of the sixth and seventh abdominal segments are small, red, retractile tubercles. The barbed hairs borne by the subdorsal and lateral tubercles are venomous and produce an inflamation of the skin of man much like that caused by poison ivy. As the cast skins of the larvæ are blown about by the wind, people are frequently badly poisoned where this pest is common.

To control this pest the nests in which the larvæ hibernate should be collected during the winter and burned.

Family NOCTUIDÆ

The Noctuids or the Owlet-Moths

If only our fauna be considered, this is the largest of all of the families of the Lepidoptera; more than 2500 species of noctuids are now know to exist in America north of Mexico. The great majority of the moths that fly into our houses at night, attracted by lights, are members of this family. The nocturnal habits of these insects, and the fact that often when they are in obscurity their eyes shine brightly suggested the name of the typical genus, *Noctua*, from the Latin for owl, as well as the popular name owlet-moths, by which they are known. Similar popular names have been given them in several other languages.

Although there exist within the limits of the family great differences in size, form, and coloring, most of the species are dull-colored moths of medium size.

In the typical noctuids, the body is large in proportion to the size of the wings; the front wings are strong, somewhat narrow, and elongated, the outer margin being shorter than the inner margin; and when at rest, the wings are folded upon the abdomen, giving the insect a triangular outline. The antennæ are thread-like, or fringed with hairs, or **brush-like, often** pectinate in the males. Two ocelli are almost always present. The labial palpi are well developed, and in some species quite prominent. The maxillæ are quite long and stout in most species. The thorax is heavy and stout. In the majority of the species the scales or the dorsal surface of the thorax are turned up more or less, forming tufts. The abdomen is conical and extends beyond the anal angle of the hind wings when these are spread. The venation of the wings of a member of this family is represented by Figure 853. Vein M_2 of the fore wings arises much nearer to vein M_3 than to vein M_1; there is usually an accessory cell; the first anal vein is wanting, and the third anal vein may be present with its tip joined to the second anal vein near its base. On the hind wings veins $Sc + R_1$ and vein R_s coalesce for a short distance near the base of the wing; vein M_2 may be either well preserved or much weaker than the other veins, or in a few cases lost; and there is considerable variation in the point of origin of this vein.

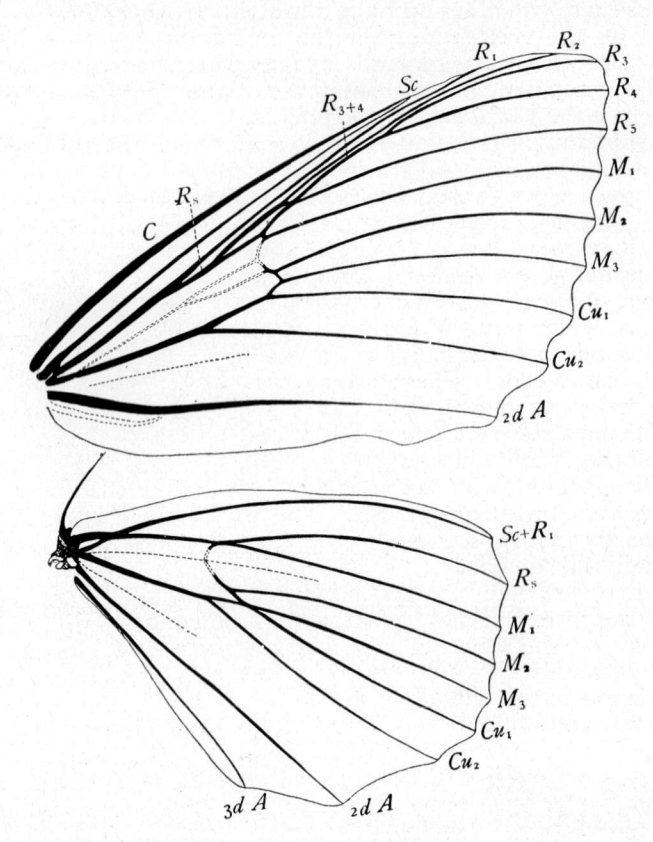

Fig. 853.—Wings of *Catocala fraxini*.

The majority of the larvæ are naked, of dull colors, and provided with five pairs of prolegs. As a rule they feed on the leaves of plants, but some are borers and some gnaw into fruits. Among them are some of the more important insects injurious to agriculture.

The family Noctuidæ has been divided into many subfamilies. In the following pages the more important of those represented in our fauna are briefly discussed, in order to show, as well as possible in a limited space, the variations in form included in this family, and to indicate the position of our more important species

There is a group of moths, the *deltoids*, which are placed at the foot of this family on account of their apparent relationship to the geometrids and to the pyralids. These moths are usually of dull colors and of medium size. The name deltoids was suggested by the triangular outline of the wings when at rest, which is well represented by the Greek letter delta. When in this position the wings slope much less than with other noctuids, the attitude being more like that assumed by the geometrids; but the hind wings are more nearly covered than with the geometrids. Many of the deltoids have very long palpi, resembling in their size those of the pyralids. The deltoids include the two following subfamilies.

The subfamily HYPENINÆ.—A representative of this subfamily is the following species.

The green clover-worm, *Plathypēna scābra.*—This is a common deltoid. The usual food-plant of the larva is clover, but it occasionally defoliates peas, beans, and lima beans. It is a slender green worm measuring when full-grown 16 mm. in length and only about 2.5 mm. in width in its widest part; it has a narrow subdorsal whitish line and a lateral one of the same color. When ready to transform it webs together several leaves and passes the pupa state in the nest thus made. The adult (Fig. 854) is a blackish brown moth, with an irregular grayish shade on the outer half of the fore wings, and with very broad hind wings. The palpi, which are not well shown in the figure, are long, wide, and flattened; they project horizontally like a snout.

Fig. 854.—*Plathypena scabra.*

The hop-vine deltoid, *Hypēna hūmuli.*—This species is closely allied to the preceding and has often been confounded with it. The larva feeds on the leaves of hop, and is sometimes a serious pest.

The subfamily HERMINIINÆ.—The following species will serve as an example of this subfamily.

Epizeuxis lubricālis.—This is one of the most abundant of our deltoids. In this species (Fig. 855) the fore wings are chocolate-brown, crossed with yellowish lines; the hind wings are much lighter. The palpi are long; but they are curved over the head, so that they appear short when seen from above, as represented in the figure. The larva feeds on dead leaves.

Fig. 855.—*Epizeuxis lubricalis.*

The subfamily EREBINÆ.—More than 120 species belonging to this subfamily are now listed from our fauna. The three following will serve as examples.

The black witch, *Ĕrebus odōra.*—This is the most magnificent in size of all of the noctuids found in this country (Fig. 856). There is much variation in the depth of coloring. The individual figured is a female; in the male the fore wings are more pointed at the apex and the median band is indistinct. It is a native of the West Indies; but it is believed that it breeds in the extreme southern portion of the United States. Isolated individuals are found in the North in

Fig. 856.—*Erebus odora*.

late summer or autumn. These are found as far north as Canada and west to Colorado, and even in California. These have doubtless flown north from their southern breeding places, possibly from Cuba or Mexico.

The larva feeds on certain tropical leguminous trees, *Cassia fistula*, *Pithecolobium*, and *Saman*.

Fig. 857.—*Scoliopteryx libatrix*.

The scalloped owlet, *Scoliŏpteryx libātrix*.—This moth is easily recognized by the shape of its wings, the outer margins of which are deeply cut and scalloped (Fig. 857). The color of the fore wings is soft brownish gray, slightly powdered with rust-red, and frosted with white along the costa. There is an irregular patch of rust-red reaching from the base to the middle of the wing, a single, white, transverse line before the middle, and a double one beyond the middle. The larva feeds on willow. This species is found in all parts of the United States and in Europe.

The cotton-worm, *Alabāma argillācea*.—Excepting perhaps the cotton-boll weevil, this is the most important insect pest in the cotton-growing states. The adult insect (Fig. 858) is a brownish moth with its fore wings crossed with wavy lines of darker color and marked with a bluish discal spot and two white dots as shown in the figure. This moth is found in the Northern States and even in

Canada in the latter part of the summer and in the autumn. But this occurrence in the North is due to migrations from the South, as the insect can not survive the winter north of the Gulf States. The larva feeds on the foliage of cotton; and as there are five or six generations in a year, the multiplication of individuals is very rapid, and the injury to the cotton great.

Fig. 858.—*Alabama argillacea.*

Fig. 859.—*Autographa falcifera.*

The best-known way of combating this pest is by the use of Paris green. Dusting machines drawn by horses are in common use.

The subfamily PLUSIINÆ includes nearly seventy North American species. In a large number of these the fore wings are marked with metallic-colored scales. The most common form of this marking is a silvery spot, shaped something like a comma, near the center of the wing (Fig. 859). In some of the species the metallic markings cover a large portion of the fore wings, in others they are wanting.

Most larvæ have only three pairs of prolegs, the first two pairs being wanting; due to this fact they walk with a looping motion (Fig. 860) resembling somewhat that of the geometrids.

The two following species have attracted attention by their injuries to cultivated plants; the celery looper, *Autŏgrapha falcīfera*, and the cabbage looper, *Autŏgrapha brăssicæ*.

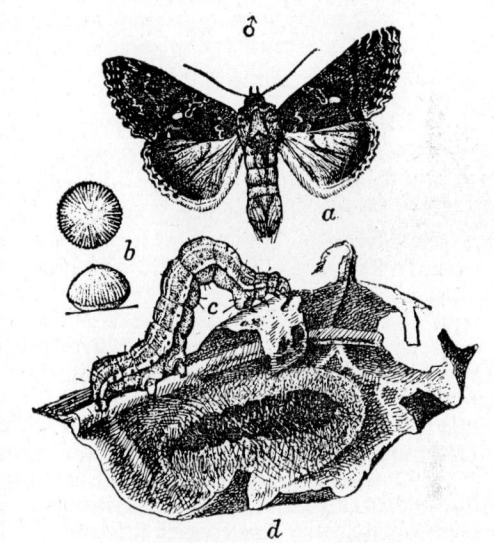

The subfamily CATOCALINÆ.—To this subfamily belong the "*underwings*" and their allies. Here belong nearly two hundred North American species. The following are some of those most likely to attract attention.

Fig. 860.—The cabbage-looper, *Autographa brassicæ:* a, male moth; b, egg; c, full-grown larva; d, pupa in cocoon. (After Howard and Chittenden.)

The underwings or catocalas, *Catŏcala*.—The most striking in appearance of the noctuids, if we except the black witch and one or two allied species, are the moths belonging to the genus *Catocala*. These moths are of large size, often expanding 75 mm. or more. The fore wings are usually brown or gray, marked with wavy or zigzag lines. The ground-color of the hind wings is black; but in many species these wings are conspicuously banded with red, yellow, or white. This peculiarity has suggested the name underwings by which these insects are commonly known in England. The genus is a very large one; more than 100 species are now known from this country; and many of these are extremely variable, so that nearly twice that number of named forms are now recognized. The ilia underwing, *Catŏcala ĭlia*, will serve as an example (Fig. 861). The larvæ of the underwings feed on the leaves of various forest-trees. Many species infest oak and hickory. By careful search both the adults and larvæ can be found resting on the trunks of these trees; but it needs sharp eyes to do it, as the colors of these insects are usually protective, the bright-colored hind wings of the moths being covered by the fore wings when the insect is at rest.

Fig. 861.—*Catocala ilia*.

The clover-looping-owlets, *Cænŭrgia*.—Among the more common noctuids that occur in our meadows and pastures, and that fly up before us as we walk through them, are two species belonging to the genus *Cænurgia*. These may be called the clover-looping owlets; for the larvæ feed on clover, and, as they have only three pairs of prolegs they walk in a looping manner. One of these species is *Cænŭrgia erĕchtea*. This moth (Fig. 862) has dark or light drab-gray fore wings, which are marked by two large dark bands, as shown in the figure. These bands are always separate, distinct, and well defined towards the inner margin in the male; in the female the markings are much less distinct, the bands usually invisible.

Fig. 862.—*Cænurgia erechtea*

The other common species of this genus is *Cænŭrgia crassiŭscula*. In this species the fore wings have either a distinct violaceous brown or a red or buffy shade, with the two large dark bands very variable, often shading into the ground-color on the outer edge or coalescing near the inner margin; all the markings are equally distinct in both sexes.

Parallelia bistriaris.—This moth (Fig. 863) is brownish in color, and has the fore wings crossed by two parallel lines. The larva feeds on the leaves of maple.

Zale lunata.—This is a brownish moth with marbled wings. It varies greatly in its markings. Figure 864 represents the **female**

Fig. 863.—*Parallelia bistriaris.*

Fig. 864.—*Zale lunata.*

which was once called *edusa*, and which does not show well the lunate mark on the hind wings that probably suggested the name of the species. The larva feeds on the leaves of rose, willow, maple, plum, and other plants.

The subfamily ERASTRIINÆ.—In this subfamily the moths are of small or moderate size; and some of them bear a strong resemblance to tortricids. Many of the species are marked with bright colors, and especially with white. The two following species will serve to illustrate this group.

Chămyris cerĭntha.—This moth (Fig. 865) is white, with the fore wings marked with shades of olive, brown, and blue. The hind wings have a narrow border of dark scales, within which there may be a cloudy shade as shown in the figure, or this shade may be wanting. The larva feeds on the leaves of apple.

Fig. 865.—*Chamyris cerintha.*

Tarachĭdia candefăcta.—This species (Fig. 866) is also largely white, with the fore wings marked with shades of olive, brown, and yellow. The amount of yellow varies greatly in different individuals. The larva feeds on the leaves of *Ambrōsia artemisiæfōlia*.

Fig. 866.—*Tarachidia candefacta.*

The subfamily APATELINÆ.—This is a large subfamily, including more than 600 North American species. The various species grouped together here exhibit great differences in appearance. Among those that are most likely to attract attention are the following.

The typical genus, *Apatēla*, includes nearly 100 North American species. This genus is named *Acronȳcta* by those authors who do not recognize the names proposed by Hübner in his "Tentamen," of which *Apatela* is one. The fore wings of these moths are generally light gray with dark spots, and in many species have a dagger-like mark near the anal angle. On this account they have received the name of *daggers*. The larvæ exhibit much diversity in appearance; those of some species are hairy like the larvæ of arctiids, while others are nearly naked.

Fig. 867.—*Apatela morula*.

The ochre dagger, *Apatēla mŏrula*.—This moth (Fig. 867) is pale gray with a yellowish tinge. Besides the black line forming part of the dagger near the anal angle of the fore wing, there is a similar black line near the base of the wing, and a third near the outer margin between veins M_1 and M_2. The larva feeds on elm and basswood. When full-grown it is mottled brown and greenish like bark; it is clothed with but few scattered hairs, and has a hump on the first, fourth, and eighth abdominal segments.

The American dagger, *Apatēla americāna*.—This is a gray moth resembling in its general appearance the preceding, but with the black lines on the fore wings much less distinct. Its larva, however, is very different (Fig. 868). This larva looks like an arctiid, being densely clothed with yellow hairs. But these hairs are scattered over the surface of the body instead of growing from tubercles as with the larvæ of arctiids. Along the sides of the body and at each end are a few scattered hairs that are longer than the general clothing, and there are two pairs of long black pencils borne by first and third abdominal segments, and a single pencil on the eighth abdominal segment. When at rest the larva remains curled sidewise on a leaf, as shown in the figure. It feeds on maple, elm, and other forest trees.

The witch-hazel dagger, *Apatēla hamamēlis*.—In the latter part of summer and in autumn what is believed to be the larva of this species is common on the leaves of witch-hazel, oak, and other forest trees. It differs greatly in appearance from the preceding species, being nearly naked (Fig. 869). When at rest it usually lies curled as shown in the figure.

Fig. 868.—*Apatela americana*, larva.

It varies in color from light yellow to reddish brown. Its most char-

LEPIDOPTERA

acteristic feature is a double row of milk-white spots along the middle of the back.

The copper hindwing, *Amphipyra pyramidoides.*—The fore wings of this moth (Fig. 870) are dark brown, shaded with paler brown, and with dots and wavy lines of a glassy gray or dull whitish hue. The hind wings, except the costal third, are reddish with more or less of a coppery luster. This suggests the popular name. The larva feeds on the leaves of grape and of Virginia-creeper.

The many-dotted apple-worm, *Balsa malāna.* In June, and again in August or September, there is sometimes found on apple-trees, in considerable numbers, a rather thick, cylindrical, light-green worm, an inch or more in length, with fine, white, longitudinal lines and numerous whitish dots. These are the larvæ of the little moth represented by Figure 871. The fore wings of this moth are ash-gray, marked by irregular, blackish lines. The larvæ feed on the leaves of many other trees beside apple. The moth has been found throughout the eastern half of our country.

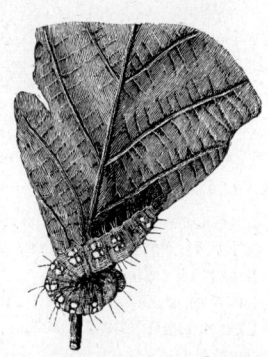

Fig. 869.—*Apatela hamamelis*, larva.

The hop-plant borer, *Gortȳna immānis.*—This is a well-known pest in the hop-growing regions. The moths deposit their eggs on the tips of the hop-vines just as they begin to climb. The young larva burrows into the vine just below the tip and spends the early part of its life in the vine at this point, causing the injury called by growers "mufflehead." Later the larva burrows to the base of the vine, where it feeds upon the stems. In this stage it is known as the hop-grub. The pupa stage is passed in the ground near the infested roots. The moths emerge in the autumn or in the following spring. To check the ravages of this pest the muffleheads should be picked off and destroyed while the larvæ are still in them.

Fig. 870.—*Amphipyra pyramidoides.*

The divers, *Bellūra.*—The genus *Bellura* contains three North American species, *B. melanopyga*, *B. diffūsa*, and *B. gortynoides*. The first two of these species were bred by the writer from the leaf-stalks of the yellow pond-lily, the habits of the third species are as yet undescribed. The larvæ of the first two species are able to descend into water and remain there for a long time; for this reason the common name the divers is proposed for them.

Fig. 871.—*Balsa malana.*

The black-tailed diver, *Bellūra melanopȳga.*— Only the female of this species has been described. In this sex the thick tuft of hair at the caudal end of the body is black or blackish. The larva of this species was first observed in Florida (Comstock '81). A detailed account of its habits was later published by Welch ('14), who studied it at Douglas Lake, Michigan. It is at first a leaf-miner in the leaves of the yellow pond-lily, later it is a borer in the leaf-stalks. Its habits are similar to those of the following species.

The brown-tailed diver, *Bellura diffūsa*.—Shortly after the discovery of the preceding species in Florida, the writer studied larvæ with similar habits at Ithaca, N. Y. From these larvæ were bred moths which proved to be *Bellura diffusa*. In this species the anal tuft of the female is dark brown. In the male there is a series of dark tufts on the basal abdominal segments.

The young larvæ of this species were not observed; doubtless they are leaf-miners like those of the preceding species. The older larvæ live in the leaf-stalks of the pond-lily, a single larva in a leaf-stalk. The larva bores a hole from the upper side of the leaf into the petiole, which it tunnels in some instances to the depth of two feet or more below the surface of the water. This necessitates its remaining below the surface of the water while feeding. The writer has seen one of these larvæ remain under water voluntarily for the space of a half-hour. The tracheæ of these larvæ are unusually large, and we believe that they serve as reservoirs of air for the use of the insect while under water. The form of the hind end of the larva has also been modified, so as to fit it for the peculiar life of the insect. The last segment appears as if the dorsal half had been cut away; and in the dorsal part of the hind end of the next to the last segment, which, on account of the peculiar shape of the last segment, is free, there open a pair of spiracles much larger than those on the other segments. When not feeding the larva rests at the upper end of its burrow, with the segment bearing these large spiracles projecting from the water.

The white-tailed Bellura, *Bellūra gortynoides*.—In this species the anal tuft of the adult female is white. The habits of the larva have not been described.

The cat-tail noctuids, *Arzama* and *Archanara*.—Two or more species of noctuids infest the cat-tail plant, *Typha*, in this country. The larvæ of both are at first leaf-miners, later they bore in the stalks. Our most common species is *Arzāma oblīqua*. According to the observations of Claassen ('21) the full-grown larva overwinters in its burrow in the cat-tail plant and transforms in the spring. But the late Professor D. S. Kellicott, who made a special study of this species, informed me in a letter written in 1882, that the larva leaves the cat-tail plant in the fall and conceals itself under bark, in old wood and even in the ground until spring when it pupates, and emerges as a moth in May. It is evident, therefore, that individuals of this species differ as to the location in which they pass the winter.

Figure 872 represents either a variety of this species or a closely allied species. It was determined for me by Grote in 1882 as *Arzama obliqua*. I collected larvæ of this form in winter from under bark of fence-posts near water.

The grape-vine epimenis, *Psychomŏrpha epimēnis.*—This is a velvety-black species with a large white patch on the outer third of the front wings and a brick-red patch on the hind wings (Fig. 873). The larva resembles somewhat that of *Alypia* figured on a later page, but it is bluish and has only four light and four dark stripes on each segment. It feeds upon the terminal shoots of grape and Virginia-creeper in spring, drawing the leaves together by a weak silken thread and destroying them. When ready to transform, which is usually towards the end of May, it either enters the ground or bores into soft wood to form a cell. Within this it remains until the following spring.

Fig. 872.—*Arazama obliqua.*

The beautiful wood-nymph, *Euthisanōtia grāta.*—This moth (Fig. 874) well deserves the popular name that has been applied to it. Its front wings are creamy white, with a glassy surface; a wide brownish purple stripe extends along the costal margin, reaching from the base to a little beyond the middle of the wing, and on the outer margin is a band of the same hue, which has a wavy white line running through it, and is margined internally with a narrow olive-green band. On the inner margin is a yellowish olive-green cloud. The hind wings are pale ochre-yellow, with a brown band on the outer margin. The wing expanse is about 40 mm. The moth appears during the latter part of June or early in July. The larva of this species is pale bluish, crossed by bands of orange and many fine black lines. It also bears a resemblance to that of *Alypia*, but may be distinguished by having only six transverse black lines on each segment. It has the same food-plants as the species described above. It transforms in a cell in the ground or in soft wood.

Fig. 873.—*Psychomŏrpha epimenis.*

Fig. 874.—*Euthisanotia grata.*

The pearl wood-nymph, *Euthisanōtia ūnio.*—This moth closely resembles the species just described, but is smaller, expanding a little less than 37 mm. The outer border of the front wings is paler and mottled; and the band on the hind wings extends from the anal angle to the apex. The larva resembles that of *E. grata;* it feeds upon the leaves of *Epilobium coloratum*, and perhaps on grape also.

The subfamily CUCULLIINÆ.—This subfamily is of considerable size, 264 North American species are now listed. Among them are the following.

The hooded owlets, *Cucŭllia*.—We have several common grayish moths, in which the fore wings are marked with numerous irregular dashes of dark color, and in which the thorax is furnished with a prominent tuft of scales. These moths belong to the genus *Cucŭllia*. Figure 875 represents *Cucŭllia spĕyeri*.—These insects evidently have the power of moving this tuft of scales; for sometimes it projects forward over the head as shown in the figure, while in other specimens of the same species it may be directed backward; in this case it is much less conspicuous. The larvæ of the hooded owlets feed upon the flowers of goldenrod and other Compositæ.

Fig. 875.—*Cucullia speyeri.*

The subfamily HADENINÆ.—About 370 North American species are included in this subfamily; among them are the following.

The army-worm, *Cĭrphis unipŭncta*.—The army-worm is so called because it frequently appears in great numbers, and, after destroying the vegetation in the field where the eggs are laid, marches like an army to other fields. This insect occurs throughout the United States east of the Rocky Mountains and is present every year; but it attracts attention only when it appears in great numbers. The larva is from 40 to 50 mm. long when full-grown, and is striped with black, yellow, and green. The adult is of a dull brown color, marked in the center of each fore wing with a distinct white spot (Fig. 876). In seasons of serious outbreak of this pest it usually appears first in limited areas, in meadows or pastures. If it is discovered before it has spread from these places it can be confined by surrounding the field with a ditch, or it may be destroyed by spraying the grass with Paris green water. Ordinarily, however, the worms are not observed until after they have begun to march and are wide spread. In such cases it is customary to protect fields of grain in their path by surrounding them with ditches with vertical sides; it is well to dig holes like postholes at intervals of a few rods in the bottom of such ditches. The worms falling into the ditch are unable to get out, and crawl along on the bottom and fall into these deeper holes. We have seen these insects collected by the bushel in this way.

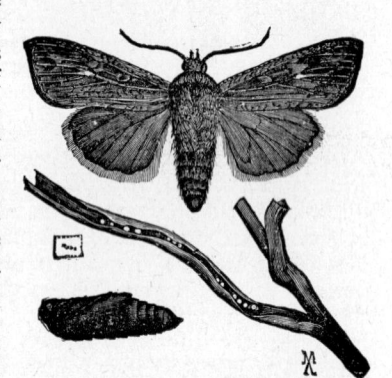

Fig. 876.—*Cirphis unipuncta.*

The zebra-caterpillar, *Ceramīca picta*.—Cabbage and other, garden vegetables are often subject to the attacks of a naked caterpillar, which is of a light yellow color, with three broad, longitudinal black stripes, one on each side and the top of the back. The stripes on the sides are broken by numerous pure white lines (Fig. 877). When full-grown the larva enters the ground where it makes a slight silken cocoon in which to transform. There are two generations a year. The adult (Fig. 878) has dark chestnut-brown fore wings and pale yellowish hind wings.

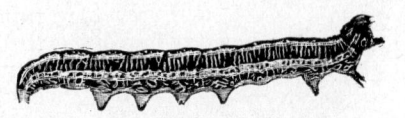

Fig. 877.—*Ceramica picta*, larva.

Certain members of this subfamily have attracted attention on account of their ravages as cutworms. Several of these belong to the genus *Pōlia*, the *Mamĕstra* of some authors, which includes more than 100 North American species. The majority of our described cutworms pertain to the next subfamily.

The subfamily AGROTINÆ.—This is one of the larger of the subfamilies of noctuids, including more than 550 described North American species. Here belong the larger number of those noctuids that are known as cutworms; but other members of this subfamily exhibit quite different habits.

Fig. 878.—*Ceramica picta*.

The corn ear-worm or the cotton boll-worm, *Heliōthis obsolēta.*—This is a widely distributed pest, the larva of which infests many different plants. It is often found feeding on the tips of ears of growing corn, especially of sugar-corn; in fact it is the worst insect pest of sugar-corn. And it is also one of the more important of the pests of cotton, ranking next to the boll-weevil and the cotton-worm; the larva bores into the pods or bolls of the cotton, destroying them. It frequently infests tomatoes, eating both the ripe and the green fruit. Occasionally it is found within the pods of peas and of beans, eating the immature seeds. It also bores into the buds, seed-pods, and flower-stalks of tobacco. The full-grown larva measures from 30 to 40 mm. in length. It varies greatly in color and markings. The pupa state is passed in the ground. The number of generations annually varies according to latitude; there is probably only one in Canada, but in the Gulf States there are from four to six. Like the larva, the moth is extremely variable in color and markings.

The evening primrose moth, *Rhodŏphora flŏrida.*—This is a very beautiful moth with most interesting habits. It is quite common, flying at night about evening primroses, both wild and cultivated, and hiding during the day in the partially closed flowers. It expands about 30 mm. The fore wings are bright pink or rosy red from the base to the subterminal line, beyond which they are pale yellow, like the flowers of the evening primrose. The hind wings are white. The fading petals of the primrose turn pinkish, and the pink color of the closed fore wings renders the moth invisible when in old flowers, while the yellow tips of the fore wings protruding from a flower still fresh and yellow, forms an equally perfect protection from observation. This moth in its passage from flower to flower transports pollen and is the special means of insuring the cross-fertilization of the evening primrose. It attaches its eggs to the stalks of the flower buds or near them. The larvæ feed on the petals of the flowers and bore into the buds and seed-vessels. They are bright green, covered with numerous, elevated, white granules; when full-grown they measure 30 mm. in length. Their color is protective. There is a single generation each year.

Cut worms.—Few pests are more annoying than the rascally little harvesters that nightly, in the spring, cut off our corn and other plants before they are fairly started. There are many species of these cut-worms, but they are all the larvæ of owlet-moths. In general their habits are as follows: The moths lay their eggs during midsummer. The larvæ soon hatch, and feed upon the roots and tender shoots of herbaceous plants. At this time, as the larvæ are small and their food is abundant, they are rarely observed. On the approach of cold weather they bury themselves in the ground and here pass the winter. In the spring they renew their attacks on vegetation; but now, as they are larger and in cultivated fields the plants are smaller, their ravages quickly attract attention. It would not be so bad if they merely destroyed what they eat; but they have the unfortunate habit of cutting off the young plants at the surface of the ground, and thus destroy much more than they consume. They do their work at night, remaining concealed in the ground during the daytime. When full-grown they form oval chambers in the ground in which they pass the pupa state. The moths appear during the months of June, July, and August.

There are some exceptions to these generalizations: some species of cut-worms ascend trees during the night and destroy the young buds; many pass through two generations in the course of a year; and a few pass the winter in the pupa state.

Cut-worms can be destroyed by poisoned baits of fresh clover or other green vegetation, or with poisoned dough made of bran. Much can be done by making holes in the ground with a sharpened stick, as a broom-handle. The holes should be vertical, a foot deep, and with smooth sides. On the approach of day the cut-worms will crawl into such holes to hide and will be unable to crawl out again. One of our cut-worms, which is known as the spotted cut-worm, is the larva of the black-c owlet, *Agrotis c-nigrum.* This moth (Fig. 879) is one of the most common species attracted to lights. It occurs throughout our country and in Europe.

Fig. 879.—*Agrotis c-nigrum.*

Family AGARISTIDÆ

The Foresters

The validity of this family is in doubt. Some of the best-known genera that were formerly included in it have been transferred to the Noctuidæ; and it is an open question whether or not the remaining genera should be similarly transferred.

The character that is used to distinguish these moths from the Noctuidæ is that the antennæ are more or less thickened towards the tip, while in the Noctuidæ the shaft of the antennæ tapers regularly. The venation of the wings (Fig. 880) is very similar to that of some noctuids.

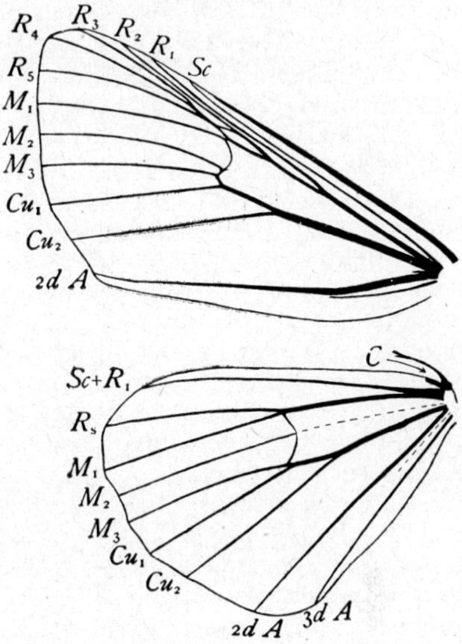

The larvæ are but slightly clothed and live exposed on the leaves of plants. They are distinguished from those of the Noctuidæ only in color, nearly all of the species being transversely striped. Our more common species feed chiefly on grape and Virginia-creeper, which they sometimes injure to a serious extent. In such cases they can be destroyed by the use of arsenical poisons, even in vineyards in the East, as the application would have to be made early in the season and the summer rains would wash the poison from the vines. The pupa state is passed either in an earthen cell or in a very slight cocoon.

Fig. 880.—Wings of *Copidryas gloveri*.

The family as now restricted is one of limited extent, only sixteen North-American species are known. The larger number of these occur in the Far West or in the Gulf States. The following are the best-known species.

The eight-spotted forester, *Alўpia octomaculāta*.—This species is of a deep velvety-black color. The front wings have two large sulphur-yellow spots; and the hind wings, two white spots. The tegulæ are sulphur-yellow. In markings both sexes of this species closely resemble the male of the following species represented in Figure 882.

The larva (Fig. 881) feeds upon the leaves of grape and Virginia-creeper, and sometimes occurs in such large numbers as to do serious injury. The ground-color of the larva is white, with eight black stripes on each segment, and a broader orange band, bounded by the two middle stripes; the orange bands are marked by black, conical, elevated spots. There are usually two broods each year, the moths appearing on the wing in May and August, the caterpillars in June and July, and in September. The pupa state is passed in an earthen cell in the ground.

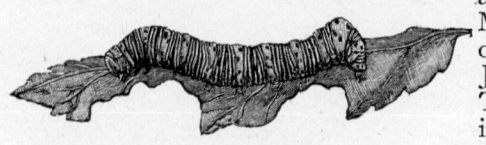

Fig. 881.—*Alypia octomaculata*, larva.

This species is found in the Atlantic States from Massachusetts to Texas.

Langton's forester, *Alỹpia langtōnii*.—This species resembles the preceding in general appearance, but the females can be readily distinguished by the hind wings bearing only a single spot, which is yellow. The males are dimorphic; in one form the males resemble the females in having a single spot on the hind wings, in the other form there are two spots (Fig. 882). This species is found in northern New York, the mountains of New Hampshire, Canada to the Pacific Coast, and the mountains of California. The larva feeds on fireweed, *Epilobium angustifolium*.

Fig. 882.—*Alypia langtonii*.

Family PERICOPIDÆ

The Pericopids

These beautiful insects occur within the limits of our country only in the far West and in the Gulf States. They resemble the wood-nymph moths in their strongly contrasting colors; but can be distinguished from them by the position of the origin of vein M_2 of the hind wings, which appears to be a branch of cubitus (Fig. 883).

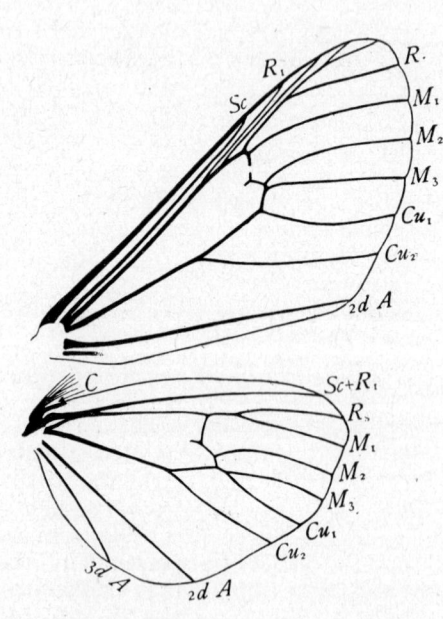

This family is represented in our fauna by only four species; but these represent three genera. Our most common species is *Gnophæla*

Fig. 883.—Wings of *Gnophæla latipennis*.

latipĕnnis, which is found in the Rocky Mountains and in the Pacific States, in the foot-hills of the Sierra Nevadas. The wings of this species are black spotted with yellow. There is some variation in the number and size of the spots on the wings. Figure 884 represents a specimen taken in Colorado. This is the variety known as *vermiculāta*.

The larva feeds on *Mertensia;* when full-grown it measures about 30 mm. in length. The body is black with sulphur-yellow interrupted bands and steel-blue tubercles. There are three pairs of the blue tubercles on each side of each segment; each tubercle bears some short whitish hairs.

Fig. 884.—*Gnophæla latipennis.*

Family ARCTIIDÆ
The Tiger-Moths and Footman-Moths or Arctiids

The Arctiidæ includes stout-bodied moths, with moderately broad wings, which in the majority of cases are conspicuously striped or spotted, suggesting the popular name tiger-moths; some of the species, however, are unspotted. A large proportion of the species are exceedingly beautiful; this renders the family a favorite one with collectors. As a rule, when at rest, the wings are folded, roof-like upon the body. Most of the moths fly at night, and are attracted to lights.

The ocelli are present in the first subfamily, absent in the other two. The

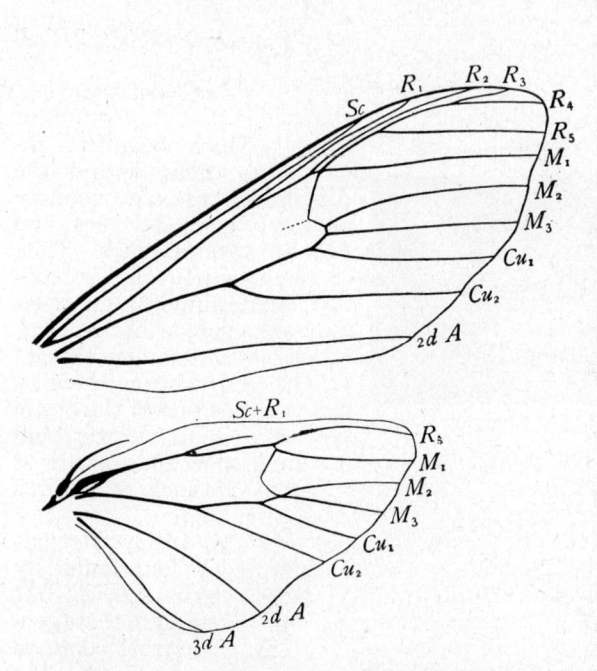

Fig. 885.—Wings of *Halisidota sp.*

palpi are short, usually but little developed. The maxillæ are present, but they are often weak. The most important features in the venation of the wings are the following; first, the union of veins M_2 and M_3 of the fore wings with cubitus, making it apparently four-branched, in a few lithosiids these branches of media are wanting; and second, the coalescence of the subcosta and radius of the hind wings for a considerable distance (Fig. 885). The extent of the union of these two veins varies greatly in the different genera; it is for at least a fifth, usually a half of the length of the discal cell, but not beyond the end of the cell.

The larvæ of the tiger-moths, except that of *Utetheisa*, are clothed with dense clusters of hairs. In fact a large proportion of our common hairy caterpillars are members of this family. In some species, certain of the clusters of hairs are much larger than the others, resembling in this respect the clothing of the tussock-moths. Most larvæ of the arctiids feed upon herbaceous plants, and many species seem to have but little choice of food-plant; but certain common species feed upon leaves of forest-trees.

The family Arctiidæ is divided into three subfamilies, each of which is regarded as a distinct family by some writers. These subfamilies can be separated as follows:

A. Ocelli present. p. 700..................................ARCTIINÆ
AA. Ocelli absent.
  B. Fore wings with raised scale-tufts. p. 705...................NOLINÆ
  BB. Fore-wings smoothly scaled. p. 704....................LITHOSIINÆ

Subfamily ARCTIINÆ
The Tiger-Moths

The presence of ocelli distinguishes the members of this subfamily from those of the other two. It is the largest of the three subfamilies, including about 125 North American species. The following are some of the more common representatives.

The genus *Haploa*.—Among the more beautiful of the tiger-moths is a genus the species of which are snow-white or light yellow with the fore wings banded with brown. In most species the hind wings are unspotted and are snow-white, but in some the hind wings are yellow.

Fig. 886.—*Haploa contigua*.

These moths constitute the genus *Haploa*. A species common in the Atlantic States and represented by Figure 886 is *Hăploà contĭgua*. The insects of this genus vary greatly in their markings.

The Bella-moth, *Utetheisa bĕlla*.—This is a whitish moth with lemon-yellow or orange-colored fore wings, crossed by six transverse white bands, each

Fig. 887.—*Utetheisa bella*.

containing a series of black dots (Fig. 887); the hind wings are pink, with a black outer margin, which is bordered within by a narrow white line. The species occurs in the Atlantic States and west to Texas.

The harlequin milkweed caterpillar, *Euchætias ĕgle*.—This larva is the most common caterpillar found on milkweed. It is clothed with tufts of orange, black, and white; those at each end of the body are longer than the others and are arranged radiately (Fig. 888). When full grown the larva makes a felt-like cocoon composed largely of its hairs. The adult has mouse-gray wings; the abdomen is yellow, with a row of black spots along the middle of the back.

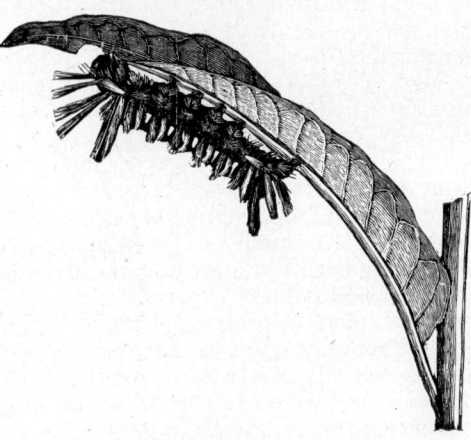

Fig. 888.—*Euchætias egle*, larva.

The genus *Apantēsis*.—A very large number of species of tiger-moths belong to the genus *Apantesis*. These are perhaps the most striking in appearance of all members of the family. The fore wings are velvety black marked with yellowish or pink bands; in some species the lighter color predominates, so that the fore wings appear to be yellow or pink, spotted with black. The hind wings are red, pink, or yellow, and are margined or spotted with black. The thorax is usually marked with three black stripes, of which the lateral ones are borne by the patagia and tegulæ. There is also a black line or a row of black spots along the middle of the back of the abdomen, and a similar row of spots on each side. Our most common species of this genus is *virgo* (Fig. 889). The larva of this species feeds on pigweed and other uncultivated plants, and winters in the larval state.

Fig. 889.—*Apantesis virgo*.

The salt-marsh caterpillar, *Estigmēne acræa*.—The popular name of this insect was given to it by Harris, nearly a century ago, and was suggested by the fact that the salt-marsh meadows near Boston, where is now the Back Bay quarter of the city, were overrun and laid waste in his time by swarms of the larvæ. But the name is misleading, as the species is widely distributed throughout the

United States, and infests a great variety of grasses and garden crops. The moth (Fig. 890) is white, marked with yellow and black. There are many black dots on the wings, a row of black spots on the back of the abdomen, another row on the venter, and two rows on each side. The sexes differ greatly in the ground-color of the wings; in the female, this is white throughout; in the male, only the upper surface of the fore wings is white, the lower surface of the fore wings and the hind wings above and below being yellow. The number and size of the black spots on the wings vary greatly. There are usually more submarginal spots on the hind wings than represented in our figure.

Fig. 890.—*Estigmene acræa.*

The fall webworms, *Hyphăntria cūnea* and *Hyphăntria tĕxtor.*— A very common sight in autumn in the North and in midsummer in the South is large ugly webs enclosing branches of fruit or forest trees. These webs are especially common on apple and on ash; but the insects that make them infest more than one hundred kinds of trees. These webs differ from those made by the apple-tree tent-caterpillar in being much lighter in texture and in being extended over all of the leaves fed upon by the colony; and they are also made later in the year. Each web is the residence of a colony of larvæ which have hatched from a cluster of eggs laid on a leaf by the parent moth.

It is a disputed point whether there are one or two species of fall webworms. In the North the adults are all snow-white in color and there is only a single generation annually. This form is the *Hyphantria textor* of those who believe that there are two species.

In the South, some of the moths have the fore wings thickly studded with dark brown points, some are pure white, and every gradation exists between these two types. Of this southern form there are two generations annually. This form is known as *Hyphantria cunea;* which name should be applied to both the northern and southern forms if they prove to be specifically identical, *cunea* being the older specific name.

Both forms winter in the pupa state.

The Isabella tiger-moth, *Isia isabĕlla.*—"Hurrying along like a caterpillar in the fall" is a common saying among country people in New England, and probably had its origin in observations made upon the larva of the Isabella tiger-moth. This is the evenly clipped, furry caterpillar reddish brown in the middle and black at either end, which is seen so commonly in the autumn and early spring (Fig. 891). The extent of the black color varies in different individuals; rarely, especially on the West Coast the body is all brown. In the spring after

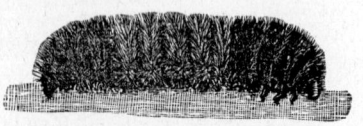

Fig. 891.—*Isia isabella,* larva.

feeding for a time the larva makes a blackish-brown cocoon composed largely of its hair. The adult is of a dull grayish tawny-yellow, with a few black dots on the wings, and frequently with the hinder pair tinged with orange-red. On the middle of the back of the abdomen there is a row of about six black dots, and on each side of the body a similar row of dots.

The yellow-bear, *Diacrisia virginica*.—The larva of this species is one of the most common hairy caterpillars found feeding on herbaceous plants. It was named by Harris the yellow-bear on account of the long yellow hairs with which the body is clothed. These hairs are uneven in length, some scattered ones being twice as long as the greater number of hairs. The long hairs are more numerous near the caudal end than elsewhere, but are nowhere gathered into pencils as with the tussock-caterpillars. This larva varies greatly in color. The body is most often of a pale yellow or straw color, with a black, more or less interrupted, longitudinal line along each side, and a more or less distinct transverse line of the same color between each of the segments. Sometimes the hairs are foxy red or light brown, and the body brownish or even dark brown. The head and the ends of the feet and forelegs are yellowish, and the venter is dusky. The larva feeds on almost any plant. The cocoon is light, and is composed almost entirely of the hairs of the caterpillar. This insect passes the winter in the pupa state; and it is probable that there are usually two or more broods each year; but these are not well marked. The moth (Fig. 892) is snowy white, with the wings marked by a few black dots; these vary in number, but there are rarely more than three on either wing. There is a row of

Fig. 892.—*Diacrisia virginica*.

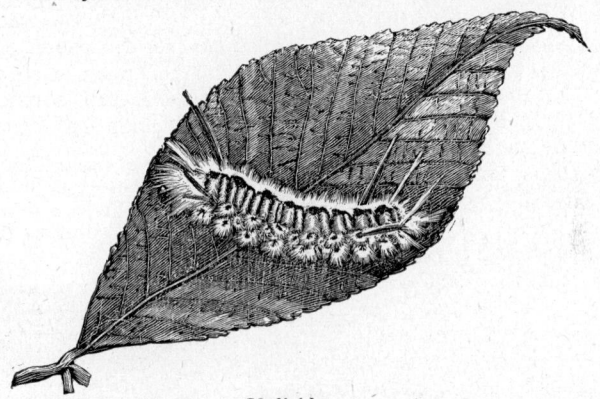

Fig. 893.—*Halisidota caryæ*, larva.

black spots on the back of the abdomen, and another on each side, and between these a longitudinal deep yellow stripe.

The hickory tiger-moth, **Halisidota** *căryæ*.—One of the most abundant of caterpillars in the Atlantic States and westward during the months of August and September is one clothed with dense tufts of finely barbed white hairs. (Fig. 893); there is a ridge or crest of black hairs on the middle of the back of the abdominal segments, a few long white hairs projecting over the head from the thorax, and others projecting back from the last segment; there are also two pairs of pencils of black hairs, one on the first and one on the seventh abdominal segment, and a similar pair of pencils of white hairs on the eighth abdominal segment. This larva feeds on hickory, butternut, and other forest-trees. Its grayish cocoons, composed almost entirely of the hair of the larva, are often found under stones, fences, and other similar places. The fore wings of the adult (Fig. 894) are dark brown spotted with white.

Fig. 894.—*Halisidota caryæ.*

Subfamily LITHOSIINÆ

The Footman-Moths

The Lithosiinæ include small moths with rather slender bodies, filiform antennæ, and usually narrow front wings and broad hind wings. As a rule they are closely scaled insects of sombre colors, a fact that has won for them the title of footman-moths; but in case of some of the species their livery is very gay. Some species fly by day, while others are attracted to lights at night.

The Lithosiinæ differ from the preceding subfamily and agree with the following one in lacking ocelli. They differ from the following subfamily in having the fore wings smoothly scaled. The venation of the wings differs greatly in the different genera. In some genera veins M_2 and M_3 of the fore wings are wanting.

The larvæ are cylindrical and covered with short, stiff hairs. The majority of the species whose transformations are known feed upon lichens. They transform in very delicate cocoons or have naked pupæ.

This subfamily includes about fifty North American species, of which the following are some of the more common ones.

The striped footman, *Hypoprĕpia miniāta*.—This beautiful moth is of a deep scarlet color, with three broad lead colored stripes on the front wings. Two of the stripes extend the entire length of the wings; while the third is between these and extends from the end of the discal cell to the outer margin (Fig. 895). The outer half of the hind wings is also slate-colored. Vein M_2 of the fore wings is

Fig. 895.—*Hypoprepia miniata.*

present; but Vein M_2 of the hind wings is wanting. The larva feeds upon lichens, and may be found under loose stones or on the trunks of trees. It is dusky, and thinly covered with stiff, sharp, and barbed black bristles, which grow singly from small warts. The cocoon is thin and silky.

The painted footman, *Hypoprēpia fucōsa*.—This species is very similar to the preceding one and has been confounded with it. With the painted footman the ground-color of the fore wings is partly yellow and partly pink.

The clothed-in-white footman, *Clemĕnsia albāta*.—The specific name of this insect is somewhat misleading; for although the general color of the moth is white, there are so many ashen and gray scales, and dark spots, that the general effect is gray. On the front wings the more prominent black spots are six or seven on the costa, one on the discal vein, and a row of small ones on the outer margin. The hind wings are white, but finely dusted with gray scales. With this species Vein M_2 is present in both fore and hind wings.

The banded footman, *Illice unifasciāta*.—This little beauty (Fig. 896) occurs in the Atlantic States from New York to Texas. The fore wings are lead-colored, and crossed by a yellow band, which extends also along the inner margin to the base of the wings. The hind wings are pink except the apex, which is lead-colored. There is much variation in the width of the yellow band.

Fig. 896.—*Illice unifasciata.*

There are several closely allied species which are difficult to distinguish from this one.

The pale footman, *Crambĭdia păllida*.—This moth is of a uniform drab color with the abdomen and the inner part of the hind wings paler; it expands 22 mm. The moths of the genus *Crambidia* can be recognized by the fact that veins M_2 and M_3 of the fore wings are both wanting, leaving cubitus only two-branched.

The two-colored footman, *Tigrioides bīcolor*.—This is larger than the preceding species, expanding from 25 to 37 mm. It is slate-colored, with the palpi, the prothorax, the costa of the fore wings and the tip of the abdomen yellow. Vein M_2 of the fore wings is wanting, leaving cubitus apparently three-branched.

Subfamily NOLINÆ

The Nolinæ are small arctiids in which the ocelli are wanting and in which there are tufts of raised scales on the fore wings. It is a small subfamily including only fifteen North American species. Our most common species is the following.

Cĕlama triquetrāna.—This is a gray moth with a wing-expanse of 17 to 20 mm. On the fore wings there is a short black or dark brown stripe at the base of the costa, and beyond this two spots of the same color, the outer one is near the middle of the length of the costa. The larva infests the foliage of apple, but not in sufficient numbers to be a pest.

Family EUCHROMIIDÆ

The family Syntomidæ of various lists.

These moths are most easily distinguished from the allied families that are represented in this country by the structure of the hind wings (Fig. 897); in these the subcosta is apparently absent except at the base of the wing, where it is separate from radius for a short distance. Occasionally forms are found in which the tip of subcosta is separated from radius. In some of the more specialized forms, the hind wings are greatly reduced in size and the venation is reduced.

Among the better-known representatives of this family are a small number of bluish-black or brown moths which have more or less vermilion or yellow on the head, prothorax, and patagia. These moths are of medium size, expanding from 30 to 50 mm. The dull color of the wings is usually relieved by the bright color of the head and patagia; and by a layer of blue scales covering the thorax and abdomen; but in some species these are wanting. The larvæ feed on grasses. Some of them strongly

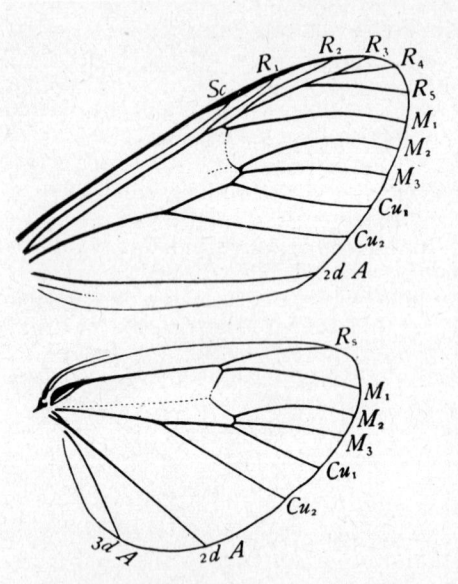

Fig. 897.—Wings of *Ctenucha virginica*.

Fig. 898.—*Ctenucha virginica*.

Fig. 899.—*Scepsis fulvicollis*.

resemble those of the Arctiidæ in appearance as well as in habits, being thickly clothed with hair; they also spin cocoons similar to those of arctiids. Our common forms of this group represent two genera, *Ctenucha* and *Scepsis*. In the East we have only a single species of each of these genera, *Ctenūcha virgĭnica*, which is represented by Figure 898, and *Scĕpsis fulvicŏllis*, represented by Figure 899. The larvæ of both of these species feed on grasses.

Closely allied to these is another species, which is common in the East, *Lycomŏrpha phōlus*. This is black with the basal half of the fore wings and the basal third of the hind wings yellow (Fig. 900). A variety of this species occurs in California and other parts of the West in which the lighter parts of the wings are pinkish instead of yellow. These moths occur in stony places, where the larvæ feed on lichens growing on rocks.

Fig. 900.—*Lycomorpha pholus*.

In the extreme southern part of this country and in the regions south of that, there occur highly specialized members of this family, in which the hind wings are greatly reduced in size, and the veins of the hind wings coalesce to a remarkable degree. In some of these forms the discal portion of the wings bears but few if any scales. *Cosmosōma myrodōra* from Florida (Fig. 901) will serve as an example of these. In this species the body and legs are bright red, with the head end of abdomen, and a dorsal band blue-black; the veins and borders of the wings are also black.

Fig. 901.—*Cosmosoma myrodora*.

Family EUPTEROTIDÆ

This family is represented in North America by a single genus, *Apatelōdes*, of which only three species occur in our fauna. These moths bear a striking resemblance to the Notodontidæ, but differ in lacking maxillæ. The moths usually have hyaline dots on the fore wings near the apex. The venation of the wings is very similar to that of the Notodontidæ. Vein Cu of the hind wings is apparently three-branched, and in our species the frenulum is normal. The egg is flat, wafer-like, unlike that of the Notodontidæ which is spherical with the micropyle at the top. The larvæ are cylindrical and are covered with numerous secondary setæ, some short, others much longer; there are no fleshy protuberances or verrucæ present. The mediodorsal setæ are usually grouped into a distinct tuft on each segment, sometimes forming long pencils. The pupa state is passed in the ground. Our two best-known species are the following.

Apatelōdes torrefăcta.—The moth is soft velvety ashen. The fore wings are falcate and are crossed by four wavy, brown lines; there is a hyaline dot near the apex, margined externally with reddish brown; there is a double reddish brown spot near the base of the inner margin. The wings expand from 45 to 50 mm.

The larva is a yellowish or whitish, long-haired caterpillar, about 50 mm. long. There are conspicuous pencils of dark hairs on the dorsimeson of the last two thoracic segments and the eighth abdominal segment. It occurs in midsummer on various shrubs and trees.

Apatelōdes angĕlica.—The moth (Fig. 902) is of a pale soft steel-gray, with the outer margins of the wings toothed. The fore wings are crossed by two bands of a darker shade. The hyaline spot near the apex of the wings is usually doubled; and there is no brown spot near the base of the inner margin as in *A. torrefacta*. The wings expand from 47 to 50 mm.

Fig. 902.—*Apatelodes angelica*. (From Packard.)

The larva (Fig. 902) feeds on ash and on lilac. It is grayish brown: the setæ of the dorsimeson are comparatively short, but are grouped in a small tuft on each body segment; no pencils are present.

Family EPIPLEMIDÆ

This family includes moths in which the body is slender and the wings ample. In their general appearance, these moths resemble geometrids; but can be distinguished from them by the venation of the hind wings in which veins Sc+R_1 and R_s separate near the base of the wing and are strongly divergent, resembling in this respect the Lacosomidæ. In the fore wings, vein Cu is apparently three-branched and veins R_5 and M_1 are stalked and are well separated from vein R_4. The frenulum is present in our species. The moths rest with the fore wings spread and the hind wings separated from them and partly rolled about the body.

Only five North American species of this family have been described, but these represent four genera. Two of these genera, *Philagraula* and *Schidax,* are each represented by a single species found in Florida.

Callizzia amorāta.—This is the best-known of our species; it is found both in Canada and the United States. The moth expands about 20 mm. It is pearly-ash colored. Both pairs of wings are crossed near the middle of their length by two wavy dark lines, which are connected by a bar of the same color near the inner margin of the fore wings. On the fore wings there is a third similar line near the outer margin. The larva feeds on the leaves of *Lonicera dioica.* The pupa state is passed at the surface of the ground.

Calledăpteryx dryoptĕrata.—This species is found in the Atlantic States. The moth is pale ochreous in color, sometimes of a pale wood-brown. Both pairs of wings are crossed by two transverse dark lines. The wing expanse is 20 mm. The larva feeds on *Viburnum nudum.* The pupa state is passed between the leaves.

Family THYATIRIDÆ

The Thyatirids

The family Thyatiridæ includes moths of medium size with elongated wings. The front wings are usually slightly widened at the anal angle (Fig. 903), and in our more common species are conspicuously marked with wavy or zigzag lines. The antennæ are filiform and more or less velvety or pubescent in the male, and the maxillæ are well developed. The moths fly by day, and when at rest fold their wings roof-like upon the abdomen.

The venation of the wings is illustrated by Figure 894. The important features to be noted are the following: In the front wing vein M_2 arises midway between veins M_1 and M_3. In the hind wing vein Sc+R_1 and vein R_s are closely parallel for a space beyond the end of the discal cell and vein M_1 is jointed to vein R_s by a comparatively long cross-vein (Fig. 904, *c.v.*), so that the two appear to separate before the end of the discal cell. In the males the tip of the frenulum is knobbed.

Fig. 903.—*Habrosyne scripta.*

The larvæ are naked, and live upon the leaves of shrubs and trees. They often conceal themselves in a case, made by loosely fastening together leaves, or by folding a single leaf.

Only twelve species are known in our fauna; these represent five genera.

One of the more common species is *Habrŏsyne scrĭpta*. This has fawn-colored front wings, conspicuously marked with light bands and zigzag lines (Fig. 903). According to Thaxter, it lays its eggs late in July, in chains of five or six, on the leaves of raspberry, upon which the larvæ feed. The mature larva is rich yellow-brown, often almost black, with a distinct dorsal black line. The lateral portions are more yellow with blackish mottlings. When at rest the larva either elevates the cephalic and caudal ends of the body, like the notodontids, so that the head rests upon the caudal segments, or conceals itself in a case formed by curling down the edge of a leaf. It makes a very slight cocoon late in August.

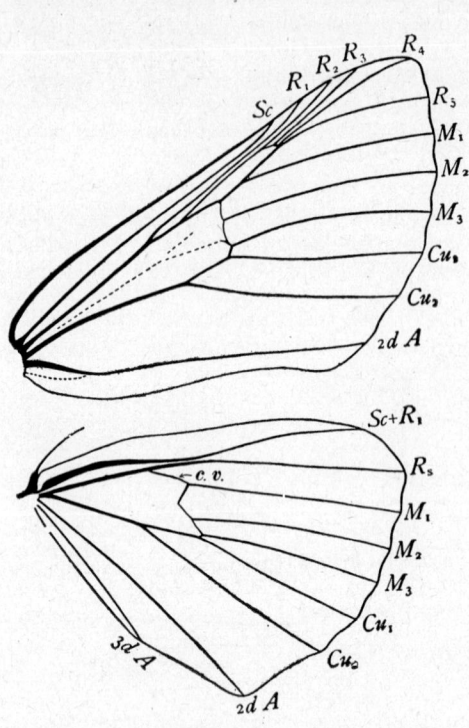

Fig. 904.—Wings of *Habrosyne scripta*.

Another common species is *Pseudothyatīra cymatophoroides*. This species is slightly larger than the preceding one, expanding nearly 50 mm. The front wings are silky gray tinted with rose. They are marked with a black spot at the base, a double or triple line, forming a black band at the end of the basal third of the wing, two black spots on the outer half of the costa, a black spot at the anal angle, and a row of black points on the outer margin. There is a variety, *expŭltrix*, which lacks the black band and the four black spots. The larva of this species has been found on red oak; it is of a rich yellow-brown, mottled with fine dark lines, and lives in a case made by fastening leaves together; some specimens have several cream-white spots. It makes a slight cocoon late in September; the adult emerges in June.

Family DREPANIDÆ
The Drepanids

The typical members of this family are small, slender-bodied moths, which can be easily recognized by the sickle-shaped apex of the front wings (Fig. 905). An approach to this form of wing is represented by some saturnians and by certain geometrids; but the former are larger, stout-bodied moths, and both differ in wing-venation, cubitus of the fore wings appearing only three-branched with them, whereas it appears to be four-branched with the drepanids.

In addition to the more typical members of this family, which are known as the *hook-tip moths*, there occurs in our fauna a single species, *Eudeilinia herminiāta*, in which the fore wings are not falcate (Fig. 906).

Fig. 905.—*Drepana arcuata*.

In this family veins $Sc+R_1$ and vein R_s of the hind wings are closely parallel or coalesced for a space beyond the end of the discal cell, resembling in this respect the Thyatiridæ. But the Thyatiridæ

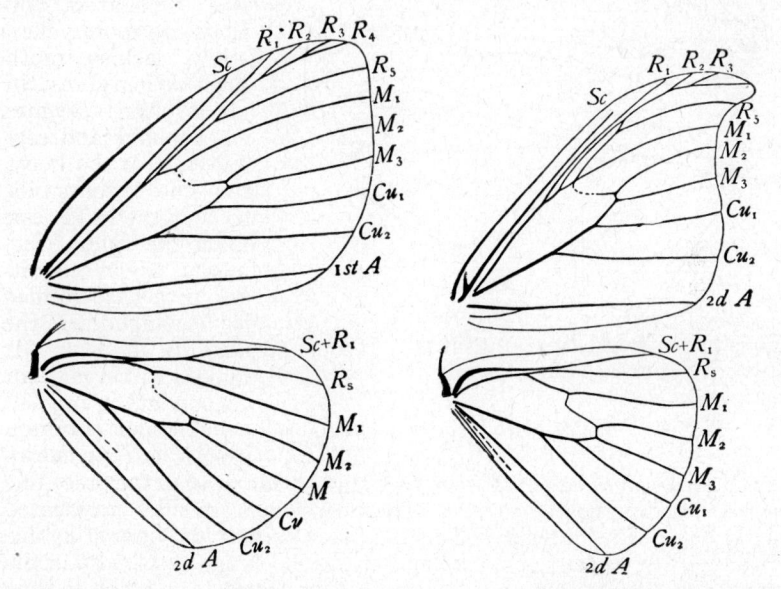

Fig. 906.—Wings of *Eudeilinia herminiata*. Fig. 907.—Wings of *Drepana arcuata*.

are true frenulum-conservers, while the Drepanidæ exhibit a very anomalous condition as regards the preservation or loss of the frenulum.

While the form of the humeral angle of the hind wings in the Drepanidæ is that characteristic of the frenulum-losers, some of these moths retain the frenulum and in others it is lost (Fig. 907 and 908). When the frenulum is present it is borne at the end of a long costal sclerite.

712   AN INTRODUCTION TO ENTOMOLOGY

The larvæ are remarkable in having the anal prolegs vestigial, and the caudal segment prolonged into a more or less lizard-like tail. They live upon the foliage of shrubs and trees, and transform in a web between leaves, or in a case in a rolled leaf.

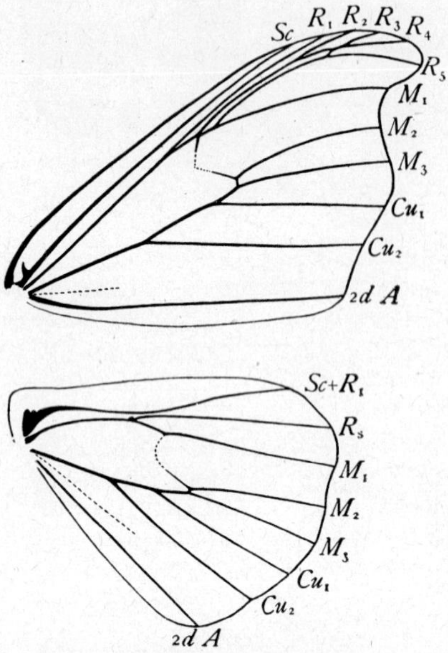

Only six species belonging to this family occur in our fauna. These represent three genera; the venation of the wings of a species of each of these genera is figured here.

Our most common hook-tip moth is *Drepăna arcuāta*. The typical form of this species is of a dirty white color marked with dark brownish lines and bands as shown in Figure 905. A summer form of this species differs in being of a light ochre-yellow color and in the course of the wavy lines on the front wings; this was described as a distinct species under the specific name *genĭcula*.

Fig. 908.—Wings of *Oreta rosea*.

These two forms are found in the Atlantic States. A third form of this species occurs in California; this was described under the specific name *siculifer*.

Our single representative of this family that is not a hook-tip moth is *Eudeilĭnia herminiāta*. This is a small moth with delicate snow-white wings, which expands from 18 to 25 mm. The venation of the wings is shown in Figure 906. The larva lives on cornel; the caudal prolongation of the body is very short. This species is found in the Atlantic States.

Family LACOSOMIDÆ

This family is of special interest on account of the structure of the wings of its members. While these moths clearly belong to the series of frenulum-losing moths, having the humeral angle of the hind wings greatly expanded so that a frenulum is not needed to insure the synchronous action of the fore and hind wings, they retain a vestige of a frenulum (Fig. 909). This vestige, however, is very small and is probably no longer of any use. It was the presence of this vestige that first suggested to the writer that those families of the

Lepidoptera which he termed Frenulum-losers were descended from frenulum-bearing ancestors (Comstock '93).

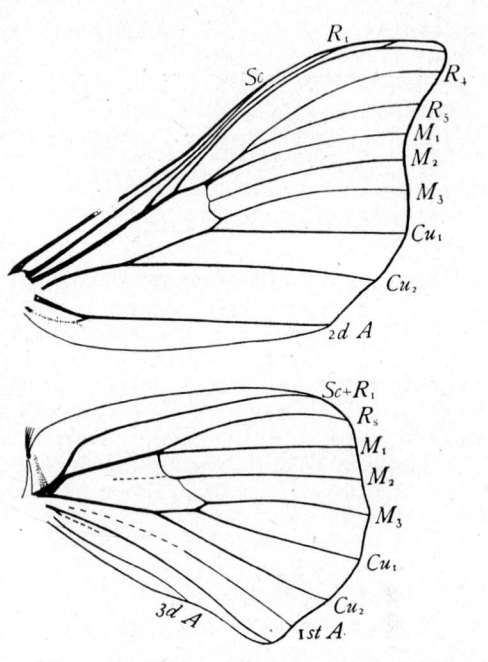

The Lacosomidæ seem to be the sole survivors of a very distinct line of descent. In many respects they appear to be closely allied to the Bombycidæ and to the Saturnioidea; but they differ markedly both in the structure and in the habits of the larvæ; and, too, the wings of the adult, although at first sight resembling those of the silk-worm, are really quite different. In the coalescence of the branches of radius of the fore wings veins R_3 and R_4 remain widely separate, while in the Bombycidæ and in the Saturnioidea these are the first branches to coalesce.

This is a small New World family. Members of it are distributed over a large part of the Western Hemisphere; but so far as is now known only three species occur in the United States. Two of our species are found in the East; the third one, *Lacosōma arizŏnicum*, was described from Arizona.

Fig. 909.—Wings of *Cicinnus melsheimeri*.

Melsheimer's sac-bearer, *Cicĭnnus melsheimeri*.—The larva feeds on the leaves of various species of oak. The habits of the young larvæ have not been described. The older larvæ make cases of leaves in which they live and which they carry about (Fig. 910). The adult moth (Fig. 911) is of a reddish gray color, finely sprinkled all over with minute black dots; there is a small black spot at the end of the discal cell of the fore wings; and both pairs of wings are crossed by a narrow blackish band.

Fig. 910.—Case of larva of *Cicinnus*.

Fig. 911.—*Cicinnus melsheimeri*.

Lacosŏma chiridōta.—Although this is the rarer of our two eastern species its complete life-history has been published by Dyar ('oo). He found the larvæ common on scrub oaks on Long Island. The eggs are laid on the edge of the leaf or on one of its points. The first three instars live under a net of silken threads on the upper surface of a leaf. At the end of the third stadium the larva begins to make a case; but the larva does not leave its net and construct a complete case until during the fifth stadium. At the end of the sixth stadium "the larva spins up one end of the case and hibernates. Pupation in the spring. A single brood in the year." The moth is somewhat smaller than the preceding species, and darker yellowish brown in color; the outer margins of the fore wings are more scalloped.

Superfamily SATURNIOIDEA

The Saturnians

The superfamily Saturnioidea includes the largest of our native moths; in fact nearly all of our very large moths belong to it, but it also includes a considerable number of species of moderate size.

These moths are most easily distinguished from other moths by the structure of their wings. Here, as with the skippers and the butterflies, the frenulum is lost and its place is taken by a greatly expanded humeral angle of the hind wing (Fig. 912), which, projecting under the fore wing, insures the acting together of the two in flight without the aid of a frenulum. This losing of the frenulum is also characteristic of the Lasiocampidæ and of some members of the Drepanidæ; but the saturnians differ from these moths in that vein M_2 arises midway between radius and cubitus or is more closely united to radius than to cubitus, leaving the latter apparently three-branched while in the Lasiocampidæ and in the Drepanidæ cubitus appears to be four-branched. In the Lacosomidæ and in the Bombycidæ the humeral angle of the hing wings is greatly expanded, but in each of these families a vestige of a frenulum is retained.

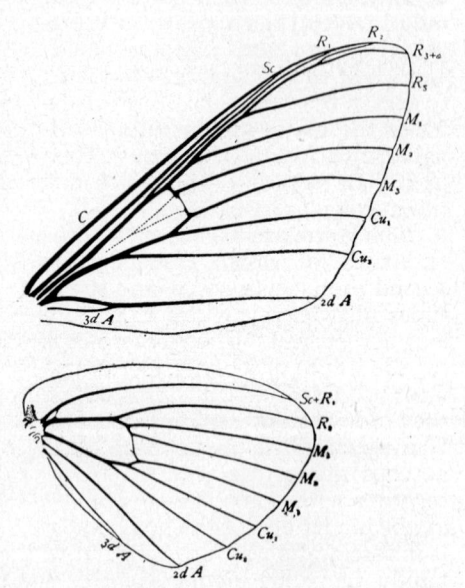

Fig. 912.—Wings of *Citheronia regalis*.

In the Saturnioidea the branches of radius of the fore wings are crowded closely together and at least one of them is lost. In all of our species the antennæ are naked or bear very few scattered scales.

This superfamily includes two families, the North American forms of which can be separated as follows.

A. Vein M_1 of the fore wings coalesced with radius to a point beyond the apex of the discal cell; vein M_1 of the hind wings joined to radius by the cross-vein *r-m* (Fig. 912), Antennæ of males pectinated but little more than half way to the apex. p. 715...CITHERONIIDÆ
AA. Vein M_1 of both fore and hind wings joined to radius by the cross-vein *r-m* (Fig. 919) or rarely (*Coloradia*) coalesced at its base with radius in both fore and hind wings. Antennæ of males pectinated to the apex. p. 719...
....SATURNIIDÆ

Family CITHERONIIDÆ

The Royal-Moths

The royal-moths are stout-bodied and hairy, with sunken heads and strong wings. The species are of medium or large size, some of them being nearly as large as the largest of our moths. There are two anal veins in the hind wings; vein M_1 of the fore wings separates from radius beyond the apex of the discal cell (Fig. 912 and 913); veins M_1 and M_2 of the hind wings are joined to radius by vein *r-m*. The antennæ of the males are broadly pectinated, but for only little more than half their length. The palpi and the maxillæ are very small.

The larvæ are armed with horns or spines, of which those on the second thoracic segment, and sometimes also those on the third, are long and curved. These caterpillars eat the leaves of forest-trees, and go into the ground to transform, which they do without making cocoons. The rings of the pupa bear little notched ridges, the teeth of which, together with some strong prickles at the hinder end of the body, assist it in forcing its way upwards out of the earth. A monograph of this family including many colored figures of moths and larvæ was published by Packard ('05).

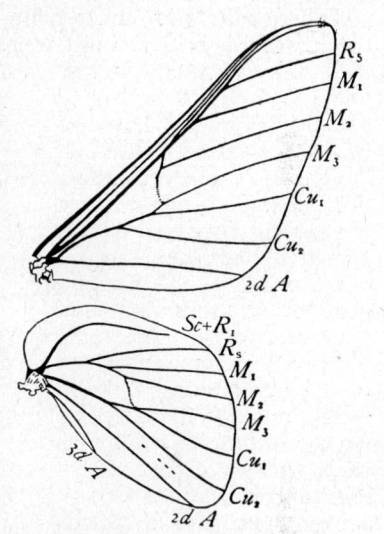

Fig. 913.—Wings of *Anisota virginiensis*.

This is a small family; it is not represented in Europe, and less than twenty species are known to occur in this country. The more common ones are the following.

Fig. 914.—*Citheronia regalis.*

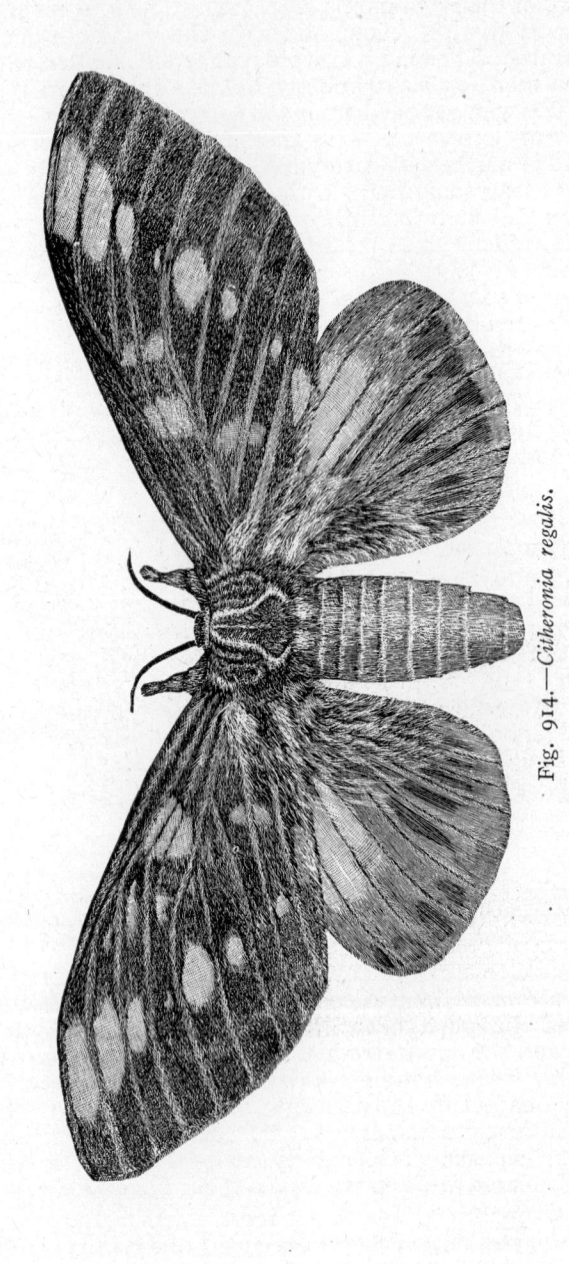

The regal-moth, *Citherōnia regālis*.—This is the largest and most magnificent of the royal-moths (Fig. 914). The fore wings are olive-colored, spotted with yellow, and with the veins heavily bordered with red scales. The hind wings are orange-red, spotted with yellow, and with a more or less distinctly marked olive band outside the middle. The wings expand from 100 to 150 mm.

When fully grown the larva measures from 100 to 125 mm. in length. It is our largest caterpillar, and can be readily recognized by the very long spiny horns with which it is armed. Those of the mesothorax and metathorax are much longer than the others. Of these there are four on each segment; the intermediate ones measure about 15 mm. in length This larva feeds on various trees and shrubs. It is known in some regions as the *hickory horned devil*.

The imperial-moth, *Basilōna imperiālis*.—This moth rivals the preceding species in size, expanding from 100 to 137 mm. It is sulphur-yellow, banded and speckled with purplish brown. The full grown larva (Fig. 915) measures from 75 to 100 mm. in length. It is thinly clothed wth long hairs, and bears prominent spiny horns on the second and third thoracic segments. In the early larval stages

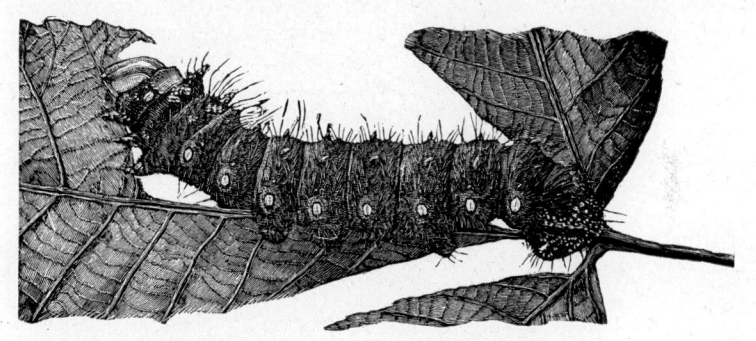

Fig. 915.—*Basilona imperialis*, larva.

these thoracic horns are very long and spiny, resembling those of the larva of the regal-moth. The larva feeds on hickory, pine, oak, butternut, and other forest-trees.

The two-colored royal-moth, *Adelocĕphala bīcolor*.—In this species the upper side of the fore wings and the under side of the hind wings are yellowish brown, speckled with black. The under side of the fore wings and the upper side of the hind wings are to a considerable extent pink. There is usually a dark discal spot on the fore wings, upon which, especially in the males, there may be two white dots. This species is more common in the Southern States than in the North. The expanse of wings in the male is 50 mm.; in the female, 60 mm. The larva feeds on the leaves of the honey-locust and of the Kentucky coffee-tree.

Anisota.—The genus *Anisōta* contains four species of moths found in the Northeastern United States. These moths are dark yellow, purplish red, or brownish in color, and agree in having the fore wings marked with a white discal dot. The larvæ feed on the leaves of oak; they are more or less striped and are armed with spines. These insects hibernate as pupæ.

In determining these moths, the student should remember that the two sexes of the same species may differ more in appearance than do individuals of different species but of the same sex. The sexes can be distinguished, as already indicated, by the antennæ. The three species can be separated as follows.

The rosy-striped oak-worm, *Anisōta virginiĕnsis*.—The wings of the female are purplish red, blended with ochre-yellow; they are very thinly scaled, and consequently almost transparent; and are not speckled with small dark spots (Fig. 916). The wings of the male are purplish brown, with a large transparent space on the middle (Fig. 917). The larva is of an obscure gray or greenish color, with dull brownish yellow or rosy stripes, and with its skin rough with small white warts. There is a row of short spines on each segment, and two long spines on the mesothorax.

Fig. 916.—*Anisota virginiensis*, female.

The orange-striped oak-worm, *Anisōta senatōria*.—The wings of the female are more thickly scaled than in the preceding species and are sprinkled with numerous blackish dots; in other respects the two are quite similar in coloring. The male differs from that of *A. virginiĕnsis* in lacking the large transparent space on the middle of the wings. The larva is black, with four orange-yellow stripes on the back and two along each side; its spines are similar to those of the preceding species.

The spiny oak-worm, *Anisōta stĭgma*.—The female closely resembles that of *A. senatoria;* and as both species are variable it is sometimes difficult to determine to which a given specimen belongs. In *A. stigma* the wings are rather darker and have a greater number of blackish spots, and the hind wings are furnished with a middle band which is heavier and more distinct than in *A. senatoria*. The male differs from that of the other two species in quite closely resembling the female in coloring, and in having the wings speckled. The larva differs from the other species of *Anisota* in having long spines on the dorsal aspect of the third thoracic and each abdominal segment in addition to the much longer spines on the mesothorax. It is of a bright tawny or orange color, with a dusky stripe along its back and dusky bands along its sides.

Fig. 917.—*Anisota virginiensis*, male.

The rosy Anisota, *Anisota rubicŭnda*.—The wings of this moth (Fig. 918) are pale yellow, banded with rose-color. The distribution of the color varies greatly in different specimens. In some the pink of the fore wings predominates, the yellow being reduced to a broad discal band, while in one variety the ground-color is yellowish white and the pink is reduced to a shade at the the base and a narrow stripe outside the middle. The hind wings may be entirely yellow, or may have a pink band outside the middle. The expanse of wings in the male is 35 to 43 mm. in the female 50 mm. or more.

Fig. 918.—*Anisota rubicunda*.

The larva of this species is known as the green-striped maple-worm, and is sometimes a serious pest on soft-maple shade-trees. It measures when full grown about 37 mm. It is pale yellowish green, striped above with eight very light, yellowish green lines, alternating with seven of a darker green, inclining to black. There are two prominent horns on the second thoracic segment, and two rows of spines on each side of the body, one above and one below the spiracles. And on the eighth and ninth abominal segments there are four prominent dorsal spines. The species is one- or two-brooded, and winters in the pupa state.

Family SATURNIIDÆ

The Giant Silk-Worms

The large size of members of this family and the ease with which cocoons of some of the species can be collected render them well known to every beginner in the study of entomology. They are stout-bodied, hairy moths with more or less sunken heads and strong wide wings. The palpi are small, and the maxillæ but little developed, often vestigial. The sexes of these moths can be ditinguished by the fact that the antennæ of the males are more broadly pectinated than are those of the females.

The family includes our largest lepidopterous insects and all of our species are above medium size. They can be distinguished from the Citheroniidæ, some of which rival them in size, by the form of the antennæ of the males, which are pectinated to the apex; and in all of our genera, except *Colorādia*, which is found in the Rocky Mountains, vein M_1 of both fore and hind wings is joined to radius by the cross-vein *r-m* (Fig. 919).

The wings are often furnished with transparent, window-like spots. The frenulum is completely lost. The humeral angle of the hind wings is largely developed, and is usually strengthened by a deep

720 AN INTRODUCTION TO ENTOMOLOGY

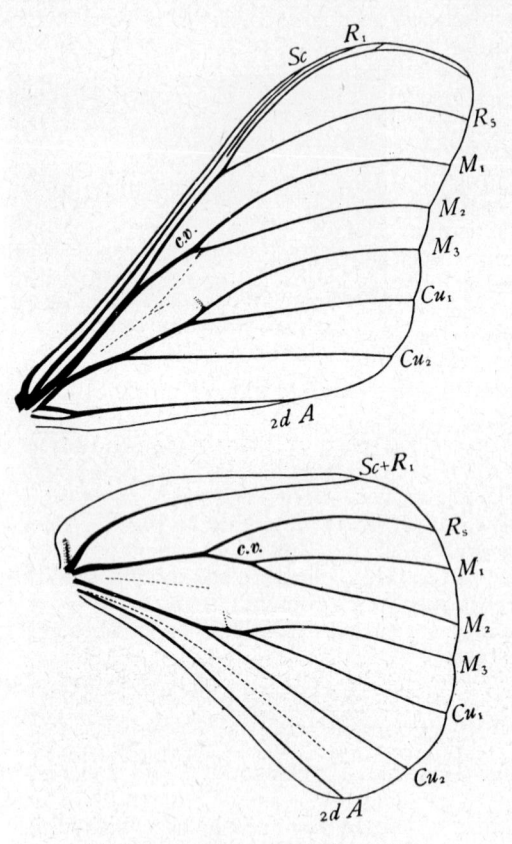

Fig. 919.—Wings of *Samia cecropia*

furrow, the bottom of which is sometimes thickened so as to appear like a humeral vein (Fig. 919).

The larvæ of most of our species live exposed on the leaves of trees and shrubs; but some of them, as the New Mexico range-caterpillar, feed on grass. They are more or less armed with tubercles and spines and are very conspicuous on account of their large size. Most of them transform within silken cocoons, which are usually very dense, and in some cases have been utilized by man. These cocoons are often attached to trees and shrubs, and are sometimes inclosed in a leaf. They can be easily collected during the winter months, and the adults bred from them. The larvæ of some members of the family, as *Hemileuca māia*, enter the ground to transform.

The family Saturniidæ as now recognized includes what were formerly regarded as two distinct families, the Hemileucidæ and the Saturniidæ. Our latest list includes only 34 species, of which the following are the better known.

The Maia-moth, *Hemileuca maia*.—The genus *Hemileuca* is represented in our fauna by eleven species, but only one of these is found in the East. In this species (Fig. 920) the wings are thinly

Fig. 920.—*Hemileuca maia*.

scaled, sometimes semi-transparent; they are black with a common white band near their middle; and the discal veins are usually white and broadly bordered with black. There are great variations in the width of the white band on the wings. The larva feeds on the leaves of oak; it is brownish black, with a lateral yellow stripe; and is armed on each segment with large, branching, venomous spines. The larva almost always enters the ground to transform.

The New Mexico range-caterpillar, *Hemileuca oliviæ*.—Of the ten western species of *Hemileuca* this is doubtless of the greatest economic importance. It is a grass-feeding species, which has been very destructive in certain sections of the cattle-range in northeastern New Mexico. It was estimated that in 1909 the total infested area was at least 15,000 square miles, and that there were an average of 10 caterpillars to the square rod over this region. For a full account see U. S. Dept. Agr. Bull. No. 85, Part V.

Fig. 921.—*Pseudohazis hera*.

Pseudohāzis.—In the West there occur two species of *Pseudohazis*. These are *P. hera* in which the ground-color of the wings is white (Fig. 921), and *P. eglanterīna*, in which the ground-color is buff or salmon. Both species are spotted and striped with black as shown in the figure.

Colorādia pandōra.—This is a brownish gray species found in the Rocky Mountains. The wings are only moderately broad, and each is marked with a small black spot at the end of the discal cell. The hind wings are semi-transparent. The expanse of the wings is from 75 to

100 mm. This genus is easily recognized by the fact that vein M_1 of both fore and hind wings is coalesced at its base with radius.

The larvæ live in the tops of pines and are abundant in alternate years; they are dried and eaten by Indians.

The Io-moth, *Autŏmeris io.*—This is a common species in the East. The female is represented by Figure 922. In this sex the ground-color of the fore wings is purplish red. The male differs greatly in appearance from the female, being somewhat smaller and of a deeper yellow color, but it can be easily recognized by its general resemblance to the female in other respects.

Fig. 922.—*Automeris io.*

The larva is one that the student should learn to recognize in order that he may avoid handling it; for it is armed with spines the prick of which is venomous (Fig. 923). The same is true of the larva of the Maia-moth, but that is much less common. The larva of the Io-moth is green, with a broad brown or reddish stripe, edged below with white, on each side of the abdomen. The spines are tipped with black. This larva feeds on various trees and shrubs.

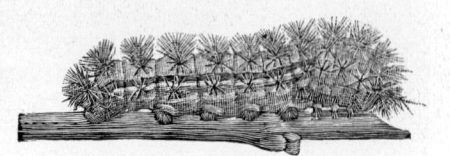

Fig. 923.—*Automeris io*, larva.

The polyphemus-moth, *Tēlea polyphēmus.*—This is a yellowish or brownish moth with a window-like spot in each wing. There is a gray band on the costal margin of the fore wings; and near the outer margin of both pairs of wings there is a dusky band, edged without with pink; the fore wings are crossed by a broken dusky or reddish line near the base, edged within with white or pink. The transparent spot on each wing is divided by the discal vein, and encircled by yellow and black rings. On the hind wings the black surrounding the transparent spot is much widened, especially toward the base of the wing, and is sprinkled with blue scales. The wings expand from 125 to 150 mm.

The larva (Fig. 924) feeds on oak, butternut, basswood, elm, maple, apple, plum and other trees. When full grown it measures 75 mm. or more in length. It is of a light green color with an oblique yellow line on each side of each abdominal segment except the first and last; the last segment is bordered by a purplish-brown V-shaped mark. The tubercles on the body are small, of an orange color with metallic reflections. The cocoon (Fig. 925) is dense and usually enclosed in a leaf; it can be utilized for the manufacture of silk. When

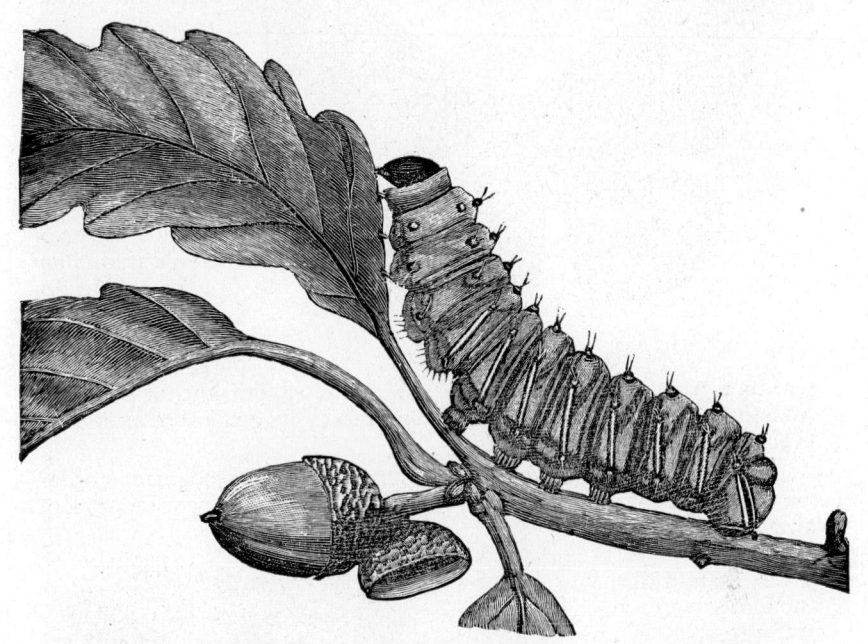

Fig. 924.—*Telea polyphemus*, larva.

the adult is ready to emerge, it secretes a fluid which softens the cocoon at one end, and breaking the threads by means of a pair of stout spines, one on each side of the thorax at the base of the fore wings, it makes its exit through a large round hole.

The Luna-moth, *Tropæa luna*.—This magnificent moth (Fig. 926) is a great favorite with amateur collectors. Its wings are of a delicate light green color, with a purple-brown band on the costa of the fore wings; there is an eye-like spot with a transparent center on the discal vein of each wing; and the anal angle of the hind wings is greatly prolonged. The larva feeds on the leaves of walnut, hickory, and other forest-trees. It measures when full

Fig. 925.—Cocoon of *Telea polyphemus*.

724 AN INTRODUCTION TO ENTOMOLOGY

Fig. 926.—*Tropæa luna*.

Fig. 927.—*Callosamia promethea*, female.

grown about 75 mm. in length. It is pale bluish green with a pearl-colored head. It has a pale yellow stripe along each side of the body, and a transverse yellow line on the back between each two abdominal segments. The cocoon resembles that of the preceding species in form, but is very thin, containing but little silk.

The Promethea-moth, *Callosāmia promēthea*.—This is the most common of the giant silk-worms. The wings of the female (Fig. 927) are light reddish brown; the transverse line crossing the middle of the wings is whitish, bordered within with black; the outer margin of the wings is clay-colored, and each wing bears an angular discal spot. The discal spots vary in size and distinctness in different specimens. The male differs so greatly from the female that it is liable to be mistaken for a distinct species. It is blackish, with the transverse lines very faint, and with the discal spots wanting or very faintly indicated. The fore wings also differ markedly in shape from those of the female, the apex being much more distinctly sickle shaped. The males fly by day. The larva when full grown measures 50 mm. or more in length. It is of a clear pale bluish-green color; the legs and anal shield are yellowish; and the body is armed with longitudinal rows of tubercles. The tubercles are black, polished, wart-like elevations, excepting two each on the second and third thoracic segments, which are larger and rich coral-red, and one similar in size to these but of a yellow color on the eighth abdominal segment. This larva feeds on the leaves of a large proportion of our common fruit and forest trees; but we have found it more frequently on wild cherry, lilac, tulip-tree, and ash than on others. The cocoons can be easily collected during the winter from these trees. This is the best way to obtain fresh specimens of the moths, which will emerge from the cocoons in the spring or early summer. The cocoon (Fig. 928) is interesting in structure. It is greatly elongated and is enclosed in a leaf, the petiole of which is securely fastened to the branch by a band of silk extending from the cocoon; thus the leaf and enclosed cocoon hang upon the tree throughout the

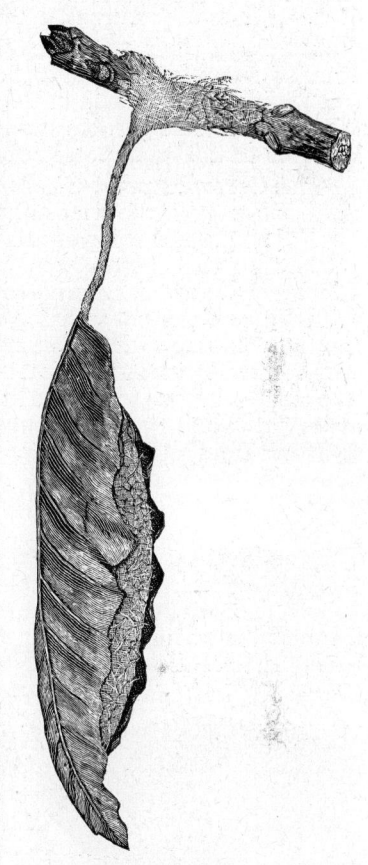

Fig. 928.—*Callosamia promethea*, cocoon.

winter. At the upper end of the cocoon there is a conical valve-like arrangement which allows the adult to emerge without the necessity of making a hole through the cocoon. This structure is characteristic of the cocoons of the moths of this and the following genus. See Figure 211, page 189.

The angulifera-moth, *Callosāmia angulĭfera.*—This is a somewhat rare insect which closely resembles the Promethea-moth. Specimens of it are usually a little larger than those of *C. promethea*, and the transverse line and discal spots are more angular. The most important differences, however, are presented by the male, which quite closely resembles the female of the Promethea-moth in color and markings, and thus differs decidedly from the male of that species. The male of this species is nocturnal, differing in this respect from *C. promethea*.

The larva feeds on the leaves of the tulip-tree and of Magnolia. It make its cocoon within a leaf or it crawls down the trunk of the tree and spins its cocoon in the grass or fastens it to some object on the ground. The cocoon usually has no stem and when made in a leaf falls to the ground in it when the leaf falls.

The Cecropia-moth, *Sāmia cecrōpia.*—This is the largest of our giant silk-worms, the wings of the adult expanding from 125 to 160 mm. The ground color of the wings is a grizzled dusky brown, especially on the central area. The wings are crossed beyond the middle by a white band, which is broadly margined without with red, and there is a red spot near the apex of the forewing just outside of a zigzag line. Each wing bears near its center a crescent-shaped white spot bordered with red. The outer margin of the wings is clay-colored. The larva is known to feed on at least fifty species of plants, including apple, plum, and the more common forest trees. When full grown it measures from 75 mm. to 100 mm. in length and is dull bluish green in color. The body is armed with six rows of tubercles, extending nearly its entire length, and there is an additional short row on each side of the ventral aspect of the first five segments following the head. The tubercles on the second and third thoracic segments are larger than the others, and are coral-red. The other dorsal tubercles are yellow, excepting those of the first thoracic and last abdominal segments, which with the lateral tubercles are blue; all are armed with black bristles. The pupa is represented by Figure 929 and the cocoon by Figure 930.

Fig. 929.—*Samia cecropia*, pupa.

The Cecropia-moth occurs from the Atlantic Coast to the Rocky Mountains. In the North Atlantic States there is another species which resembles it in general appearance but is much smaller, expanding from 75 to 100 mm. and is much less common; this is *Sāmia colŭmbia*. In the Far West the place of the Cecropia-moth is taken by two very closely allied species. In these the ground-color of the wings is usually reddish or dusky brown. *Sāmia glŏveri* is found in the Rocky

Mountain region and in Arizona; and *Samia rubra* in the Pacific States. In *Samia rubra* the crescent-shaped white spot near the center of the hind wings is more elongate and pointed than in the other species.

The Ailanthus-worm, *Philosāmia Walkeri*.—This is an Asiatic species which has been introduced into this country. It has become

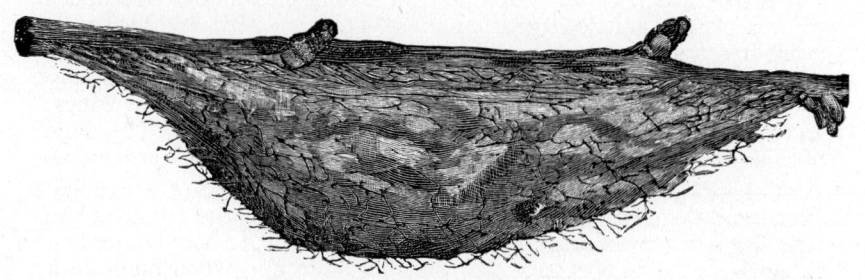

Fig. 930.—*Samia cecropia*, cocoon.

a pest in the vicinity of New York City, where it infests the Ailanthus shade trees. The moth differs from all our native species of this family in having rows of tufts of white hairs on the abdomen. Its cocoon resembles that of the Promethea-moth. The specific identity of this species is in doubt.

Family BOMBYCIDÆ

The Silk-Worm

The family Bombycidæ is not represented in our fauna; but a single species, the silk-worm, is frequently bred in this country, and is usually present in collections of Lepidoptera.

The silk-worm, *Bombyx mori*.—The moth (Fig. 931) is of a cream-color with two or three more or less distinct brownish lines across the fore wings and sometimes a faint double bar at the end of the discal cell. The head is small; the antennæ are pectinated broadly in both sexes; and the ocelli, palpi, and maxillæ are wanting. A striking feature of the venation of the wings (Fig. 932) is the obvious presence of the base of vein R_1 in the hind wings.

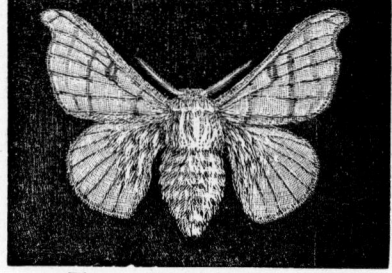

Fig. 931.—*Bombyx mori*.

The usual food of the silk-worms is the leaf of the mulberry. Our native species, however, are not suitable. The species that are most used are the white mulberry

(*Morus alba*), of which there are several varieties, and the black mulberry (*Morus nigra*); the former is the better. The leaves of osage orange (*Maclura aurantiaca*) have also been used as silk-worm food to a considerable extent. In case silk-worms hatch in the spring before either mulberry or osage-orange leaves can be obtained, they may be quite successfully fed, for a few days, upon lettuce-leaves.

The newly-hatched larva is black or dark-gray, and is covered with long stiff hairs, which spring from pale-colored tubercles. The hairs and tubercles are not noticeable after the first molt, and the worm becomes lighter and lighter, until in the last larval period it is of a cream-white color. There is a prominent tubercle on the back of the eighth abdominal segment, resembling those borne by certain larvæ of the Sphingidæ.

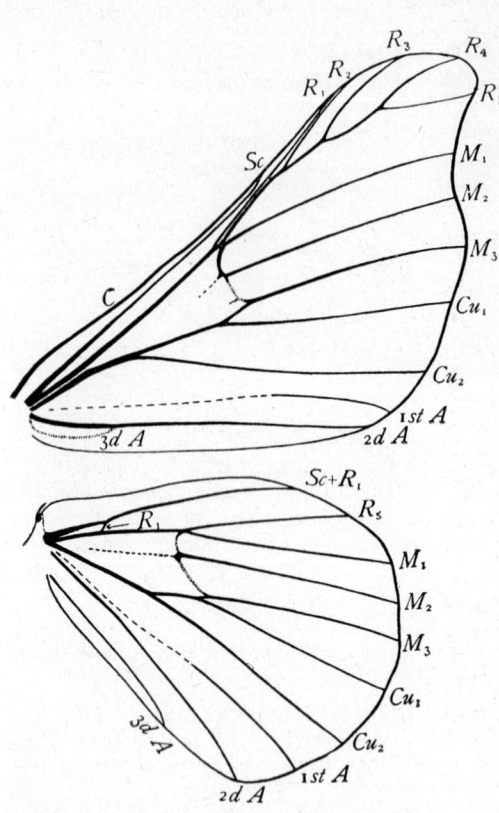

Fig. 932.—Wings of *Bombyx mori*.

There are many special treatises on this insect, some of which should be consulted by any one intending to raise silk-worms.

Family LASIOCAMPIDÆ

The Lasiocampids

The best-known representatives of this family are the tent-caterpillars and the lappet-caterpillars. The adults are stout-bodied, hairy moths of medium size. The antennæ are pectinated in both sexes, and from one-fourth to one-half as long as the front wings; the teeth of the antennæ of the male are usually much longer than those of the female. The ocelli and the maxillæ are wanting; and the palpi are usually short and woolly. But the most distinct characteristic is found in the wings. The frenulum is wanting, there being instead, as in the Saturnioidea, a largely expanded humeral angle of the hind wings. But these moths differ from the Saturnioidea in having cubitus apparently four-branched and in having the humeral angle of the hind wings strengthened by the development of some extra veins, the humeral veins (Fig. 933 h. v.).

The larvæ of the Lasiocampidæ feed upon the foliage of trees, and are frequently very destructive.

The family is a small one, less than thirty North American species are known; but these represent eleven genera. Our more common species represent three genera: *Malacosoma*, which includes the tent-caterpillars, and *Tolype* and *Epicnaptera*, which include the lappet-caterpillars.

There are several species of tent-caterpillars in this country. Most of them belong to the Pacific coast; but two are common in the East. Of these the most common one is the apple-tree tent-caterpillar, *Malacosōma americāna*.—This is the insect that builds large webs in apple and wild cherry trees in early spring. Figure 934 represents its transformations. The moth is dull reddish brown, with two transverse whitish or pale yellowish lines on the fore wing. The figure represents a male; the female is somewhat larger. These moths appear early in the summer. The eggs are soon laid, each female laying all her eggs in a single ring-like cluster about a twig; and here they remain unhatched for about nine months. This cluster is covered with a substance which protects it during the winter. The eggs hatch in early spring, at the time or just before the leaves appear. The larvæ that hatch early feed upon the unopened buds till the leaves expand. The larvæ are social, the entire brood that hatch from a cluster of eggs keeping together and building a tent in which they live when not feeding. The figure represents a specimen in our collection. In this case the tent was begun near the cluster of eggs. But usually the larvæ soon after hatching migrate down the branch towards the trunk of the tree until a fork of considerable size is reached before they begin their tent. This is necessary, as the completed tent often measures nearly two feet in length.

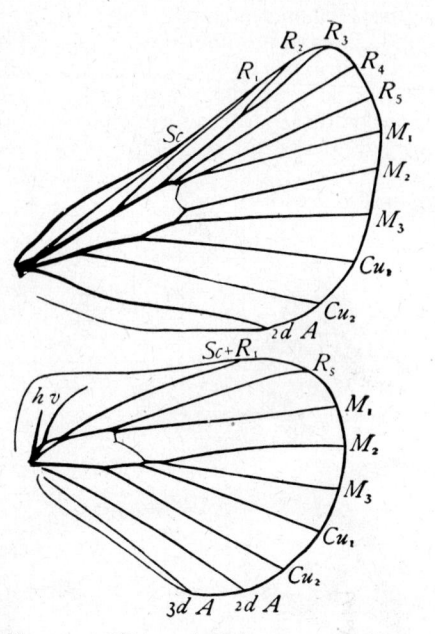

Fig. 933.—Wings of *Malacosma americana*.

The larvæ leave the nest daily in order to feed; and spin a silken thread wherever they go. The larvæ become full grown early in June; one of them is represented on a partially-eaten leaf in the figure. When ready to transform they leave the trees and make their cocoons in some sheltered place. These cocoons are quite peculiar in appearance, having a yellowish white powder mixed with the silk. The pupa state lasts about three weeks. The easiest way to fight this pest is to destroy the webs containing the larvæ as soon as they appear in the spring. This should be done early in the morning, or late in the afternoon, or on a cold day, when the larvæ are not scattered over the tree feeding.

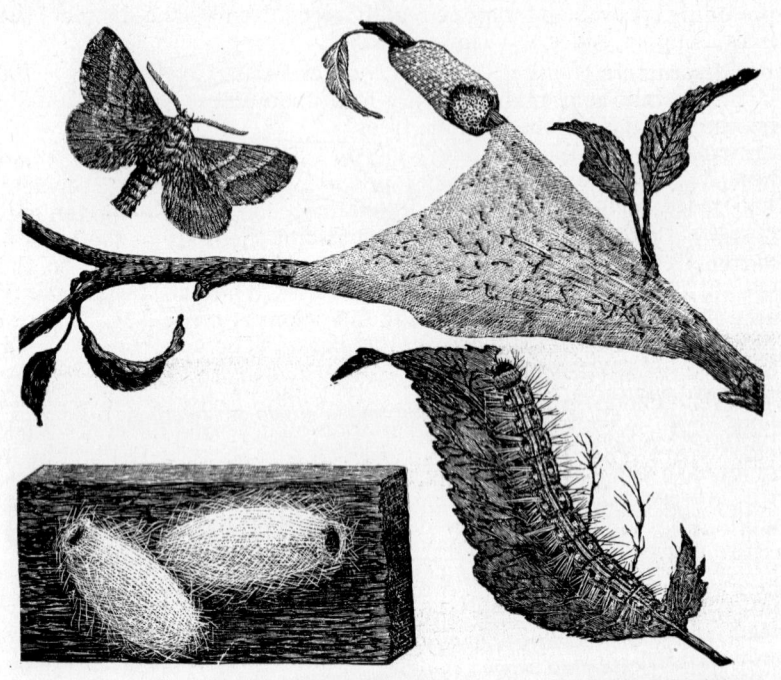

Fig. 934.—*Malacosma americana*, eggs, tent, larva, cocoons, and adult.

Another species of the genus *Malacosōma* found in the East is the so-called forest tent-caterpillar, *Malacosōma disstria*. The range of this species extends throughout the United States and Canada. It differs from the preceding species in that the larvæ do not construct a true tent. It feeds on the leaves of many forest and fruit trees, but maple is its favorite food-plant. In other respects its life history is quite similar to that of the apple-tree tent-caterpillar. The moth differs from *M. americana* in having the oblique lines on the wings dark instead of light; the larva differs in having a row of spots along the back instead of a continuous narrow line; the egg-masses differ in ending squarely instead of being rounded at each end; and the cocoon is more fragile, with less powder, and distinctly double.

The Great Basin tent-caterpillar, *Malacosōma frăgilis*.—This species is found throughout the northern portions of the Great Basin, extending from the Rocky Mountains to the Cascades and Sierra Nevadas, and has been found in California. It feeds on *Ceanothus* and many other wild shrubs.

The California tent-caterpillar, *Malacosōma califŏrnica*, feeds normally on oak but also attacks fruit trees. The caterpillars are orange-colored and about 25 mm. long.

Malacosōma constrĭcta.—The larva is somewhat larger than the preceding species, and may be readily recognized by the distinct blue lines along the sides. It feeds on oaks.

Malacosōma pluviālis.—This is another Pacific Coast species. The larvæ are buff-colored and usually feed upon alder, but occasionally become quite injurious to apple trees.

The lappet-caterpillars.—The larvæ of the species of *Tŏlype* and of *Epicnăptera* are remarkable for having on each side of each segment a little lappet or flat lobe; from these many long hairs are given out, forming a fringe to the body. When at rest the body of the larva is flattened, and the fringes on the sides are closely applied to the surface of the limb on which the insect is. Thus all appearance of an abrupt elevation is obliterated; the colors of these larvæ are also protective, resembling those of the bark. The following are our better-known species.

The Velleda lappet, *Tŏlype vĕlleda*.—The body of the moth is milk-white, with a large blackish spot on the middle of its back (Fig. 935). That part of this spot which is on the thorax is composed of erect scales, the caudal part of recumbent hairs. The wings are dusky gray, crossed by white lines as shown in the figure. The figure represents the male; the female is much larger. The moths are found in August and September. The larva feeds upon the leaves of apple, poplar, and syringa. Its body is bluish gray, with many faint longitudinal lines; and across the back of the last thoracic segment there is a narrow velvety-black band. The larva reaches maturity during July. The cocoon is brownish gray, and is usually attached to one of the branches of the tree on which the larva has fed.

Fig. 935.—*Tolype velleda*.

The larch lappet, *Tŏlype lăricis*.—This is a smaller species, the females being about the size of the male of the preceding species, and the males expanding only about 30 mm. The wings of the females are marked much like those of *T. velleda*, except that the basal two-thirds of the front wings are much lighter, and the dark band on the outer third is narrower and much darker than the other dark bands.

The males are bluish black, with the markings indistinct. The larva feeds upon the larch. When mature it is of a dull brown color and less than 40 mm. in length. When extended, the front of the first thoracic segment is pale green, and the incision between the second and third is shining black. The larva matures during July. The cocoon is ash-gray, flattened and moulded to the limb to which it is attached, and partially surrounding it. The moths appear in August or September. The winter is passed in the egg state.

The American lappet, *Epicnăptera americāna*—This species is found from the Atlantic to the Pacific. It is somewhat variable, and the different varieties were formerly regarded as distinct species. The moth (Fig. 936) is reddish brown, with the inner angle of the front wings and the costal margin of the hind wings deeply notched. Beyond the middle of each wing there is a pale band edged with zigzag, dark brown lines. The larva lives upon apple, cherry, oak, birch, maple and ash. When full grown it measures 60 mm. in length and 12 mm. in breadth. The upper side is slate-gray, mottled with black, with two transverse scarlet bands, one on the second and one on the third thoracic segments. There is a black spot on each end and in the middle of each of these bands. The larva is found during July and August. It is said that the cocoons are attached to limbs like those of *Tolype;* but the larvæ of this species that we have bred made their cocoons between leaves, or in the folds of the muslin bag enclosing the limb upon which they were feeding. The species passes the winter in the pupa state; and the moth appears in June, when it lays its eggs upon the leaves of the trees it infests.

Fig. 936.—*Epicnaptera americana.*

Superfamily HESPERIOIDEA

The Skippers

The skippers are so-called on account of their peculiar mode of flight. They fly in the daytime and dart suddenly from place to place. When at rest most species hold the wings erect in a vertical position like butterflies; in many the fore wings are thus held while the hind wings are extended horizontally; and a few extend both pairs of wings horizontally. The head is wide; the antennæ are widely separated; they are thread-like, and enlarged toward the tip; and in most cases the extreme tip is pointed and recurved, forming a hook. The abdomen is usually stout, resembling that of a moth rather than that of a butterfly. The skippers are most easily distinguished by the peculiar venation of the wings, vein R of the fore wings being five-branched, and all of the branches arise from the discal cell (Fig. 937). In some butterflies all of the branches of vein R appear to arise from the discal cell; but this is because two of the branches coalesce to the margin of the wing. In such butterflies vein R appears to be only four-branched.

The North American skippers represent two families. In Australia there is a skipper-like insect, *Euschēmon rafflēsiæ*, which has a distinct frenulum. If this belongs to the Hesperioidea, it represents a third family, the Euschemonidæ. Our two families can be separated as follows.

A. Head of moderate size; club of antenna large, neither drawn out at the tip nor recurved. Large skippers, with wing expanse of 40 mm. or more. p. 733........ .... MEGATHYMIDÆ
AA. Head very large; club of antenna usually drawn out at the tip, and with a distinct recurved apical crook. In a few forms the crook of the antennæ is wanting; such forms can be distinguished from the Megathymidæ by their smaller size, the wing expanse being less than 30 mm. p. 734... .... HESPERIIDÆ

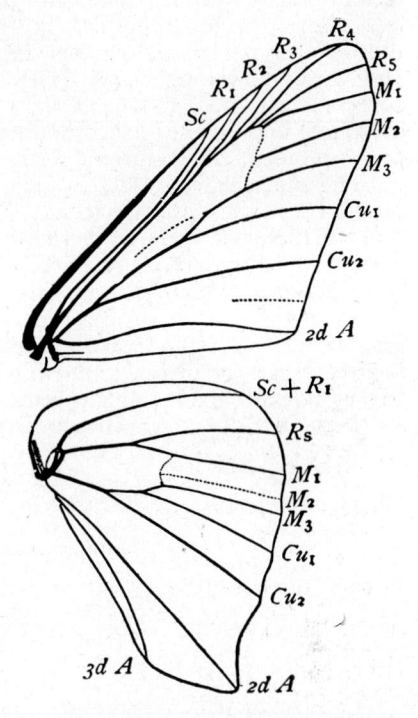

Fig. 937.—Wings of *Epargyreus tityrus*.

FAMILY MEGATHYMIDÆ

The Giant Skippers

This family includes a small number of large skippers, which are found in the South and far West. In the adult insect the head is of moderate size, the width, including the eyes, being much less than that of the metathorax. The club of the antennæ is large; and, although the tip is turned slightly to one side, it is neither drawn out to a point nor recurved. The body is very robust, even more so than in the Hesperiidæ. These insects fly in the daytime and with a rapid darting, flight. When at rest they fold their wings in a vertical position.

So far as is known the larvæ in the later stages of their growth are borers in the stems and roots of various species of *Yucca* and *Agave* and the young larvæ spin silken tubes between the young and tender shoots of these plants.

A monograph of the family was published by Barnes and Mc-Dunnough ('12). It is represented in the United States by a single genus, *Megathymus*, of which eight species have been found in the United States.

Megathȳmus strĕckeri (Fig. 938) will serve as an example of the giant skippers. The specimen figured is a female of the variety

Fig. 938.—*Megathymus streckeri.*

known as *texana*.

A much better known species is the yucca-borer, *Megathymus yuccæ*. The female of this species differs from that of the preceding in having much darker wings, all of the spots being smaller, and in having only one or two white spots on the lower surface of the hind wings. The male lacks the erect hairs on the hind wings. The larva bores in the stem and root of the Yucca or Spanish bayonet. It differs greatly in appearance from the larvæ of the Hesperiidæ, having a small head. This species is widely distributed through the southern part of our country.

Family HESPERIIDÆ

The Common Skippers

This family includes all of our skippers except the very small number that belong to the preceding family, the giant skippers. The two families can be separated by the table given above.

The larvæ of the common skippers present a very characteristic appearance, having large heads and strongly constricted necks (Fig. 939). They usually live concealed in a folded leaf or in a nest made of several leaves fastened together. The pupæ are rounded, not angular, resembling those of moths more than those of butterflies. The pupa state is passed in a slight cocoon, which is generally composed of leaves fastened together with silk, and thinly lined with the same substance.

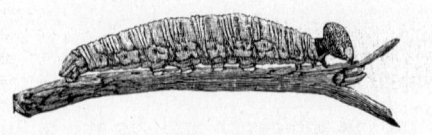

Fig. 939.—*Epargyreus tityrus*, larva.

A monograph of the North American species was published by Lindsey ('21); and the species of the Eastern United States are described and figured in natural colors by Comstock and Comstock ('04).

The family Hesperiidæ includes four subfamilies; but only three of them are represented in this country, the fourth being confined to the Old World. Our forms can be separated as follows.

A. Club of antennæ large; the entire club reflexed. p. 735.....Pyrrhopyginæ
AA. Club of antennæ variable, but never with the entire club reflexed.
   B. Vein M_2 of the fore wings arising nearer to vein M_1 than to vein M_3. p. 735.
....Hesperiinæ

BB. Vein M_2 of the fore wings arising midway between veins M_1 and M_3 or nearer to vein M_3 than to vein M_1.
   C. Vein M_2 of the fore wings arising nearly midway between veins M_1 and M_3.
     D. Discal cell of fore wings more than two-thirds as long as the costa. Mid tibiæ without spines. Males usually with costal fold in fore wings p. 735...HESPERIINÆ
     DD. Discal cell of fore wings less than two-thirds as long as the costa. Mid tibiæ spined. Males usually with a discal patch, the brand, on the **fore wings.** p. 737.......................................PAMPHILINÆ
   CC. Vein M_2 of the fore wings arising much nearer to vein M_3 than to vein M_1 p. 737..PAMPHILINÆ

Subfamily PYRRHOPYGINÆ

The distinguishing feature of this subfamily is that the antennal club is large and the entire club is recurved. In the other members of the Hesperiidæ if the antennal club is recurved it is only the terminal part of the club that is bent back. This subfamily includes a large number of South and Central American species; but only one has been found north of Mexico. This is *Apўrrothrix arăxes* variety *arizōnæ*. This has been found only in Arizona. It is a large skipper, having a wing-expanse of 50 mm. In general appearance it resembles our common silver-spotted skipper, *Epargўreus tĭtyrus*, except that the ground color is a darker brown and the spots on the fore wings and the lighter parts of the fringes are snow white.

Subfamily HESPERIINÆ

Skippers with a Costal Fold and their Allies

This subfamily includes the larger of the common skippers, as well as some that are of moderate size. Most of the species are dark brown, marked with white or translucent, angular spots. The antennæ usually have a long club, which is bent at a considerable distance from the tip (Fig. 940) and vein M_2 of the fore wings retains its primitive position nearly midway between veins M_1 and M_3 or is nearer to vein M_1 than to M_3 at base (Fig. 937). But the most distinctive feature of the subfamily is exhibited by the males alone, and is lacking in some species. It consists of a fold in the fore wing near the costal margin, which forms a long slit-like pocket, containing a sort of silky down. This is a scent-organ. When this pocket is tightly closed it is difficult to see it. It is known as the *costal fold*.

Fig. 940.—*Thanaos martialis.*

More than eighty species belonging to this subfamily have been found in America north of Mexico. The following are some of the more common of these.

The silver-spotted skipper, *Epargȳreus tĭtyrus*.—This is one of the larger of our common skippers, having a wing-expanse of nearly or quite 50 mm. It is dark chocolate brown, with a row of yellowish spots extending across the fore wing and with a large silvery-white spot on the lower side of the hind wing (Fig. 941). It is found in nearly the whole United States and in southern Canada. The larva (Fig. 939) feeds upon various papilionaceous plants. It is common on locust. It makes a nest, within which it remains concealed, by fastening together, with silk, the leaflets of a compound leaf (Fig. 942). This is one of the very few skippers that winter in the

Fig. 941.—*Epargyreus tityrus;* under surface at left. (From Scudder.)

Fig. 942.—Nest of larva of *Epargyreus tityrus*.

pupa state; most species winter as larvæ, either partly grown or in their cocoons.

The bean leaf-roller, *Goniūrus prōteus*.—This skipper by the shape of its wings reminds one of a swallow-tail butterfly, the hind wings being furnished with long tails. It expands about 43 mm. and the greatest length of the hind wings is about 30 mm. The wings are very dark chocolate-brown. The front wings contain several silvery-white spots; and the body and base of the wings bear metallic-green hairs. The larvæ feed upon both Leguminosæ and Cruciferæ. In the South it is sometimes a pest in gardens, cutting and rolling the leaves of beans, turnips, and cabbage, and feeding within the rolls thus formed. It is found on the Atlantic border from New York southward into Mexico.

There are two common skippers which are nearly as large as the two described above, but which have neither the yellow band of the first nor the long tails of the second; neither do they have the brown spots characteristic of the following genus. These two skippers belong to the genus *Thorybes*. The wings are of an even dark brown; the fore wings are flecked with small or very small irregular white spots, and the hind wings are crossed beneath by two rather narrow, parallel, inconspicuous darker bands. These skippers are distinguished as follows.

The northern cloudy-wing, *Thŏrybes pȳlades*.—In this species the white spots on the fore wing are usually mere points, although their number and size vary. The species is found in nearly all parts of the United States. The larva commonly feeds on clover.

The southern cloudy-wing, *Thŏrybes daunus*.—In this species the white spots are larger than in the preceding, almost forming a continuous band. It differs also in that the males do not have a costal fold. It is widely distributed over the Eastern United States, except the more northern portions.

To the genus *Thanaos* belong a large number of species which on account of their dark colors have been named *dusky-wings*. These species resemble each other so closely in markings that it is very difficult to separate them without longer descriptions than can be given here. The one following will serve as an example.

Martial's dusky-wing, *Thănaos martiālis*.—The wings are grayish brown with many dark brown spots evenly distributed and with several minute white ones on the outer half of the fore wings (Fig. 940). This skipper is found throughout the greater part of the United States east of the Rocky Mountains, and in Canada.

Among the smaller members of this subfamily are the skippers of the genus *Pholisora*. The most widely distributed species of this genus is the sooty-wing, *Pholisōra catŭllus*. The expanse of the wings is a little more than 25 mm. The wings are nearly black, marked with minute white spots, which vary in size and number. This species is found throughout the United States and southern Canada.

The genus *Hesperia* includes a considerable number of small skippers, which are easily recognized by their checkered markings of white upon a dark brown ground. Small white spots on the wings are common in this subfamily, but in this genus the white spots are unusually large, so large in some cases that they occupy the greater part of the wing. One of the more common species is the variegated tessellate, *Hespēria tessellāta*. This is distributed from the Atlantic to the Pacific, and is the only one common in the Eastern United States. In this species more than one-half of the outer two-thirds of both fore and hind wings is white.

Subfamily PAMPHILINÆ

Skippers with a Brand and their Allies

This subfamily includes the greater number of our smaller skippers. Some of the species, however, surpass in size many of the Hesperiinæ. To the Pamphilinæ belong all of our common tawny skippers, as well as some black or dark brown species. The antennæ usually have a stout club, with a short, recurved tip; sometimes this tip is wanting.

In the majority of our species the males can be recognized at a glance by a conspicuous patch crossing the disk of the fore wings, which usually appears to the naked eye like a scorched, oblique streak, and which on this account is termed the *brand* (Fig. 943). The brand is a complicated organ, composed of tubular scales, the androconia, that are the outlets of scent-glands, and of other scales of various shapes; in some species the brand is wanting. In this subfamily vein M_2 of the fore wings arises much nearer to vein M_3 than to M_1, the base of the vein usually curving noticeably toward vein M_3 (Fig. 944).

Fig. 943.—*Atryone conspicua.*

This subfamily is an exceedingly difficult one to study. One hundred and twenty-five species have been described from America north of Mexico; and in many cases the differences between allied species are not well marked. The following two are named merely as examples. The first one is easily recognized.

The least skipper, *Ancyloxipha nūmitor*.—This skipper is the smallest of our common species, and is also remarkable for lacking the recurved hook at the tip of the antennæ. The wings are tawny, broadly margined with dark brown. In the females the fore wings are almost entirely brown. The larger individuals expand about 25 mm. The larva feeds upon grass in damp places.

The black-dash, *Atrўtone conspĭcua*.—The male of this species is represented by Figure 943. It is blackish brown, with considerable yellow on the basal half of the fore wings. The brand is velvety black. This species is distributed from Massachusetts to Nebraska.

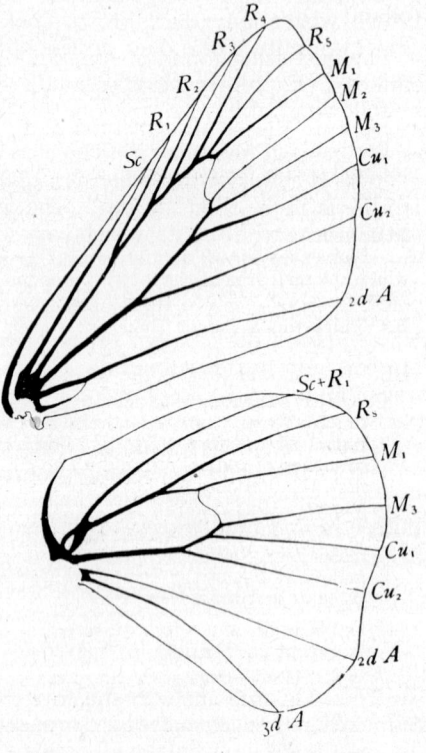

Fig. 944.—Wings of *Pamphila sassacus*.

Superfamily PAPILIONOIDEA

The Butterflies

The butterflies differ from moths in that they have clubbed antennæ, fly only in the daytime (except some species in the tropics), hold the wings erect above the back when at rest, and have no frenulum. Some moths present one or more of these characteristics, but no moth presents all of them. Butterflies can be distinguished from skippers by the venation of the front wings, as indicated above in the characterization of the Hesperioidea.

Among the many works treating of American butterflies the two following are especially useful for the classification of our species, each of these works is illustrated by many full-page plates representing the insects in their natural colors: "How to Know the Butterflies, A Manual of the Butterflies of the Eastern United States" by J. H. and A. B. Comstock ('04), and "The Butterfly Book, A Popular Guide to a Knowledge of the Butterflies of North America" by W. J. Holland (1931).

The butterflies found in America north of Mexico represent five families. Our representatives of these families can be separated by the following table.

A. Butterflies in which the cubitus is apparently four-branched; and in which the anal area of the hind wings is more reduced than the anal area of the fore wings, as in the fore wings there are always two anal veins, and usually all three are at least partly preserved, while in the hind wings there is only a single anal vein. p. 740..PAPILIONIDÆ
AA. Butterflies in which the cubitus is apparently three-branched; and in which the anal area of the fore wings is more reduced than the anal area of the hind wings, the former having a single anal vein and the latter two.
  B. Palpi much longer than the thorax (Subfamily Libytheinæ). p. 766 ....NYMPHALIDÆ
  BB. Palpi not as long as the thorax.
    C. With only four well-developed legs in both sexes, the front legs being much shorter than the others, and folded on the breast like a tippet; radius of the fore wings five-branched. To determine the number of branches of radius, count the two cubital and the three medial branches first, the branches left between vein M_1 and the Subcosta belong to radius. p. 750..NYMPHALIDÆ
    CC. With six well-developed legs in the females and with the fore legs of the males more or less reduced, only slightly reduced except in the metalmarks; radius of the fore wings (except in some orange-tips, p. 747) only three or four branched.
      D. Vein M_1 of the fore wings arising at or near the apex of the discal cell (except in *Feniseca*, p. 772).
        E. Hind wings with the costa thickened out to the humeral angle (Fig. 978) and with a humeral vein. p. 767............RIODINIDÆ
        EE. Costa of hind wings not thickened at base; humeral vein absent. p. 768..LYCÆNIDÆ
      DD. Vein M_1 of the fore wings united with a branch of radius for a considerable distance beyond the apex of the discal cell (Fig. 952). p. 744... ....PIERIDÆ

Family PAPILIONIDÆ

The Swallow-tails and the Parnassians

This family includes the swallow-tail butterflies, which are common throughout our country, and the parnassians, which are found only on high mountains or far north. These insects are distinguished from all other butterflies found in our fauna by the fact that vein M_2 of the fore wings appears to be a branch of cubitus, making this vein appear to be four-branched, and by the fact that the anal area of the hind wings is more reduced than the anal area of the fore wings, the former containing only a single anal vein, the latter two in the parnassians and three in the swallow-tails.

This family includes two well-marked subfamilies, which are distinguished as follows.

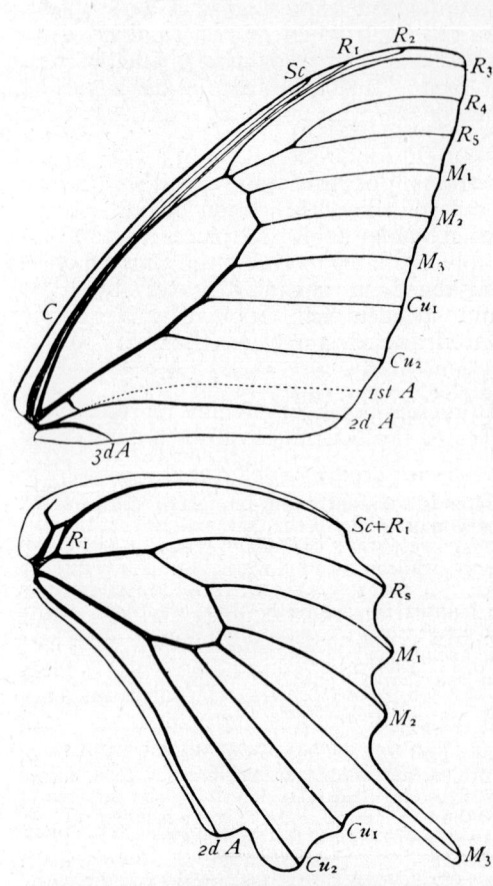

Fig. 945.—Wings of *Papilio polyxenes*.

A. The outer margins of the hind wings usually with one or more tail-like prolongations; ground color of wings black or yellow; the base of the first anal vein preserved as a spur-like branch of vein Cu (Fig. 945); radius of the fore wings five-branched. p. 740 ....PAPILIONINÆ

AA. The outer margin of the hind wings rounded, without a tail-like prolongation; ground color of wings white; radius of fore wings four-branched; the first anal vein wanting (Fig. 951). p. 744.................PARNASSIINÆ

Subfamily PAPILIONINÆ

The Swallow-tails

These magnificent butterflies are easily recognized by their large size and usually by their tail-like prolongations of the hind wings.

The ground color of the wings is black, which is usually marked with yellow, and often with metallic blue or green. The larvæ of our swallow-tails are never furnished with spines, but are either naked or clothed with a few fine hairs. In a single species that is widely distributed in the United States, *Laërtias philēnor* (Fig. 946) and in the genus *Ithobalus*, which is represented in our fauna only in the extreme South, the body of the larva bears fleshy filaments. An osmeterium is always present; this is bright colored, forked, and is thrust out from the upper part of the prothorax when the caterpillar is disturbed. It diffuses a disagreeable odor. See page 101.

The chrysalids are thickened in the middle and taper considerably at each end; they are more or less angulated, and always have two prominent projections at the anterior end. They are suspended by the tail and by a loose girth around the middle (Fig. 947).

There are about twenty species of swallow-tails in America north of Mexico. The following well-known species will serve as illustrations.

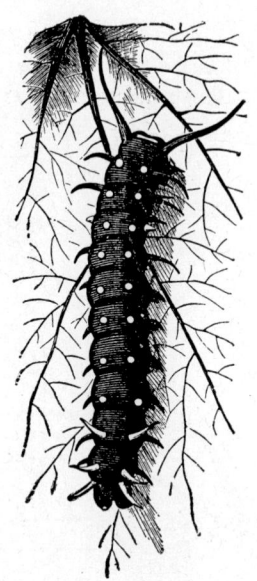

Fig. 946.—*Lærtias philēnor*, larva. (From Riley.)

The black swallow-tail, *Papĭlio polўxenes*.—The larva of this swallow-tail (Fig. 948) is well-known to most country children. It is the green worm, ringed with black and spotted with yellow, that eats the leaves of caraway in the back yards of country houses. It feeds also on parsnips and other umbelliferous plants. The first instar of this larva is black, banded about the middle and caudal end with white. There are two generations annually in the North and at least three in the South.

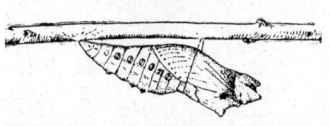

Fig. 947.—Chrysalis of *Papilio*.

In the adult the wings are black, crossed with two rows of yellow spots, and with marginal lunules of the same color. The two rows of spots are much more distinct in the male than in the female, the inner row on the hind wing forming a continuous band crossed with black lines on the veins. Between the two rows of spots on the hind wings there are many blue scales; these are more abundant in the female. Near the anal angle of the hind wing there is an orange spot with a black center. On the lower surface of the wings the yellow markings become mostly orange and are heavier.

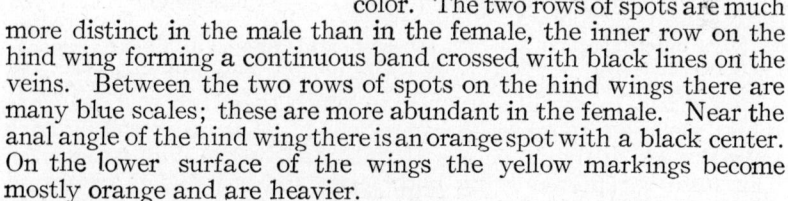

This species is found throughout the United States and in the southern parts of Canada. In California the black swallow-tail is replaced as a celery and parsley pest by a related species, *Papilio zolicaon.*

The tiger swallowtail, *Papilio glaucus.*— The larva of this butterfly (Fig. 949) is even more striking in appearance than that of the preceding species. When full grown it is dark green, and bears on each side of the third thoracic segment a large greenish-yellow spot, edged with black, and enclosing small black spots like a figure 10. This caterpillar has the curious habit of weaving upon a leaf a carpet of silk, upon which it rests when not feeding; when nearly full grown, instead of spinning a simple carpet as before it stretches a web across the hollow of a leaf and thus makes a spring bed upon which it sleeps (Fig. 949).

The larva of this species feeds on birch, poplar, ash, wild cherry, fruit-trees, and many other trees and shrubs.

Fig. 948.—*Papilio polyxenes*, larva.

Fig. 949.—*Papilio glaucus*, larva upon its bed.

In the adult state two distinct forms of this insect occur. These differ so greatly in appearance that they were long considered distinct species. They may be distinguished as follows.

(1) The turnus form, *Papilio glaucus turnus.*—The wings are bright straw-yellow above, and pale, faded straw-yellow beneath, with a very broad black outer margin, in which there is a row of yellow spots. On the fore wings there are four black bars, extending back from the costa; the inner one of these crosses the hind wings also. This form is represented by both sexes, and is found in nearly all parts of the United States and Canada.

(2) The glaucus form, *Papilio glaucus glaucus.*—In this form the disk of the wings is entirely black, but the black bands of the turnus

form are faintly indicated, especially on the lower surface, by a darker shade. The marginal row of yellow spots is present, and also the orange spots and blue scales of the hind wings. This form is represented only by the female sex, and occurs only in the more southern part of the range of the species, i. e., from Delaware to Montana and southward. It was the first of the two forms to be described, hence the species bears the name *glaucus*.

The zebra swallow-tail, *Iphiclides marcellus*.—This butterfly (Fig. 950) differs from all other swallow-tails found in the eastern half of the United States in having the wings crossed by several bands of greenish white. This is one of the most interesting of our butterflies, as it occurs under three distinct forms, two of which were considered for a long time distinct species. Without taking into account the more minute differences these forms can be separated as follows.

(1) The early-spring form, *Iphiclides marcellus marcellus*.—This is the form figured here. It expands from 65 to 70 mm.; and the tails of the hind wings are about 15 mm. in length and tipped with white.

Fig. 950.—*Iphiclides marcellus*.

(2) The late-spring form, *Iphiclīdes marcĕllus telamŏnides*.—This form is a little larger than the early spring form and has tails nearly one-third longer; these tails are bordered with white on each side of the distal half or two-thirds of their length.

(3) The summer form, *Iphiclīdes marcĕllus lecŏntei*.—The summer form is still larger expanding from 80 to 87 mm., and has tails nearly two thirds longer than the early spring form.

The life history of this species has been carefully worked out by Mr. W. H. Edwards. He has shown that there are several generations each year, and that the winter is passed in the chrysalis state. But the early-spring form and the late-spring form are not successive broods; these are both composed of individuals that have wintered as chrysalids, those that emerge early developing into *marcellus marcellus*, and those that emerge later developing into *marcellus telamonides*. All of the butterflies produced from eggs of the same season, and there are several successive broods, are of the summer form, *marcellus lecontei*.

The larva feeds upon papaw (*Asimina*). This insect is found throughout the eastern half of the United States except in the extreme north.

This species was formerly supposed to be the *Papilio ajax* of Linnæus, and the specific name *ajax* has been commonly applied to it in this country.

Subfamily PARNASSIINÆ

The Parnassians

The parnassians are butterflies of medium size in which the ground color of the wings is white shaded with black, and marked with round red or yellow spots margined with black.

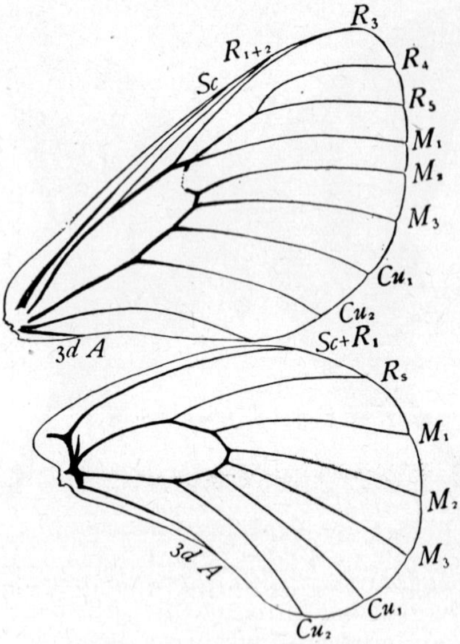

In structure the parnassians are closely allied to the swallow-tails; but in their general appearance they show little resemblance to them, differing in the ground color of the wings, and in lacking the tail-like prolongation of the hind wings in all of our species.

In the venation of the wings (Fig. 951) they differ from the swallow-tails in that radius of the fore wings is only four-branched and the first anal vein is wanting. They agree with the swallow-tails and differ from all other butterflies in that the cubitus of the fore wings is apparently four-branched.

Fig. 951.—Wings of *Parnassius*.

The larvæ possess osmeteria similar to those of the larvæ of swallow-tails. When about to pupate the larva either draws a leaf or leaves about its body by a few threads or it merely hides under some object on the ground. The pupa is cylindrical and rounded, not angulate like those of swallow-tails.

Only four species have been found in North America; they all belong to the genus *Parnăssius*. Of the four species, two are Alaskan; the others occur in the mountains of the Pacific States, in Wyoming, and in the Rocky Mountains. Of each of the two latter there are several named varieties.

Family PIERIDÆ

The Piĕrids

These butterflies are usually of medium size, but some of them are small; they are nearly always white, yellow, or orange, and are usually marked with black. They are the most abundant of all our butterflies, being common everywhere in fields and roads. Some species are so abundant as to be serious pests, the larvæ feeding on cultivated plants.

The characteristic features of the venation of the wings are the following (Fig. 952): Vein M_2 of the fore wings is more closely connected with radius than with cubitus, the latter appearing to be three-branched; vein M_1 of the fore wings coalesces with radius for a considerable distance in

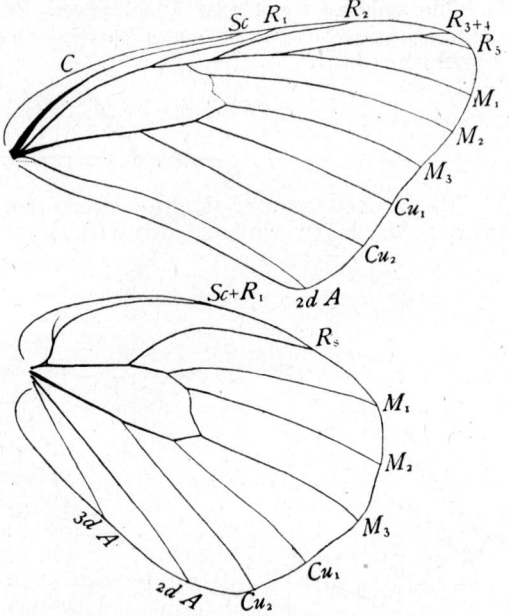

Fig. 952.—Wings of *Pieris protodice*.

all of our species; and only three or four of the branches remain distinct except in some orange-tips.

In this family the fore legs are well developed in both sexes, there being no tendency to their reduction in size, as in the three following families.

The larvæ are usually slender green worms clothed with short, fine hairs; the well-known cabbage-worms are typical illustrations (Fig. 953). The chrysalids are supported by the tail and by a girth around the middle. They may be distinguished at a glance by the presence of a single pointed projection in front (Fig. 953).

Our genera of this family can be separated into three groups, which seem hardly distinct enough to be ranked as subfamilies. These are the whites, the yellows and the orange tips.

I. THE WHITES

The more common representatives of this group are the well-known cabbage-butterflies. They are white butterflies more or less marked with black. Occasionally the white is tinged with yellow; and sometimes yellow varieties of our white species occur. About a dozen North American species of this group are known.

The cabbage-butterfly, *Pieris rāpæ*.—The wings of this butterfly are dull white above, occasionally tinged with yellowish, especially

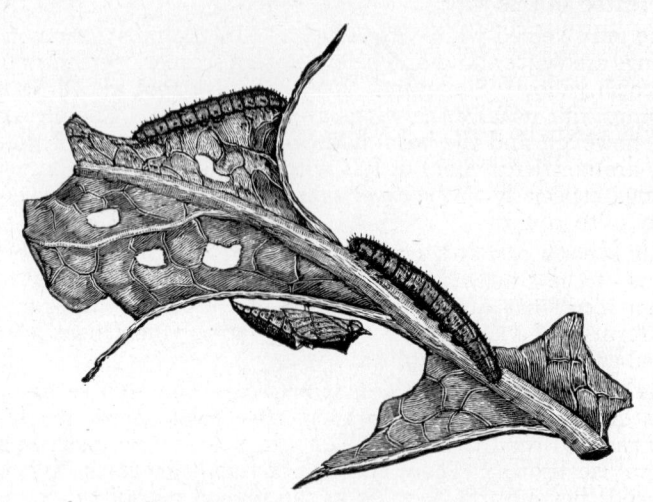

Fig. 953.—*Pieris rapæ*, larva and pupa.

in the female; below, the apex of the fore wings, and the entire surface of the hind wings are pale lemon yellow. In the female there are two spots on the outer part of the fore wing besides the black tip, in the male only one (Fig. 954).

There is considerable variation in the intensity of the black markings, and in the extent of the yellow tinge of the wings.

The larva of this species (Fig. 953) feeds principally on cabbage, but it also attacks many other cruciferous plants. Its color is the green of the cabbage-leaf, with a narrow, greenish, lemon-yellow dorsal band, and a narrow, interrupted stigmatal band of the same color. The body is clothed with very fine short hairs.

Fig. 954.—*Pieris rapæ*.

Pīeris rāpæ is without doubt the most injurious to agriculture of all our species of butterflies. It is an introduced species, but has spread over the greater part of this country. As it is three-brooded in the North and more in the South, it is present nearly the entire season, so that it needs to be fought constantly. The larvæ can be easily killed by spraying the plants with Paris green, one pound in fifty gallons of water. Chemical analysis of sprayed plants has shown that there is practically no danger from eating cabbages that have been treated in this way.

The gray-veined white, *Pīeris nāpi.*—In the most common form the wings are white above and below, with a scarcely perceptible tinge of greenish yellow. Sometimes there is a dark spot in cell M_3 of the fore wings, but usually the wings are immaculate. The base of the wings, however, and the basal half of the costal margin of the front wings, are powdered more of less with dark scales, and the veins of the wings, especially on the lower side are grayish. The wing-expanse is from 42 to 50 mm.

This species occurs throughout Canada and the more northern portions of the United States. It appears in many different forms; eleven named forms are now recognized in the United States, and still other forms are known in Europe. This polymorphism is partly seasonal and partly geographical.

The checkered white, *Pīeris protŏdice.*—The two sexes of this species differ greatly in appearance, the female being much more darkly marked than the male. The wings are white, marked above with grayish brown. There is a bar of this color at the end of the discal cell; beyond this there is in the male a row of three more or less distinct spots, and in the female an almost continuous band of spots. Besides these there is in the female a row of triangular spots on the outer margin of both fore and hind wings, and on the hind wings a submarginal zigzag bar.

The larva of this species is colored with alternating stripes of bright golden yellow and dark greenish purple, upon which are numerous black spots. It feeds upon cabbage and other cruciferous plants, and occurs in nearly the whole of the United States. Both this and the preceding species seem to become greatly lessened in numbers by the increase of the imported *Pieris rapæ*.

II. THE ORANGE-TIPS

These, like the butterflies comprising the preceding group, are white, marked with black. Their most characteristic feature is the presence on the lower surface of the hind wings of a greenish network, or a marbled green mottling. This usually shows through the wing so as to appear as a dark shade when the wings are seen from above (Fig. 955). Many species have a conspicuous orange spot on the apical portion of the front wings. This has suggested the common name *orange-tips* for the group. But it should be remembered that some species lack this mark, and that in some others

Fig. 955.—*Euchloe ausonides.*

it is confined to the males. Nearly all of our species are confined to the Far West. The two following occur in the East.

The falcate orange-tip, *Euchloe genūtia*.—In this species the apex of the fore wings is hooked, reminding one of the hook-tip moths. In the males there is a large apical patch. This butterfly is found throughout the southeastern part of the United States, not including Florida. It occurs as far north as New Haven, Conn. It is nowhere abundant. The larva feeds on rock-cress, bitter cress, shepherd's purse, *Sisymbrium*, and other Cruciferæ that are slender in form.

The olympia orange-tip, *Sȳnchloe olȳmpia*.—In this species the orange patch is wanting in both sexes. There is a conspicuous black bar at the end of the discal cell of the fore wings, and the apical portion of these wings is gray including a large irregular white band.

The larva is striped lengthwise with pale slate color and bright yellow; the feet, legs, and head are grayish green. It feeds on hedge-mustard and other Cruciferæ.

III. THE YELLOWS

The yellows are easily recognized by their bright yellow colors, although in some species whitish forms occur. They abound almost everywhere in open fields, and are common about wet places in roads. To this group belong the larger number of our pierids.

The roadside butterfly or the clouded sulphur, *Eurymus philŏdice*. —The wings above are rather pale greenish yellow, with the outer borders blackish brown. Figure 956 represents the male; in the female the border on the fore wings is broader, and contains a sub-marginal row of yellow spots. The discal dot of the fore wings is black, that of the hind wings is orange. The under surface is sulphur-yellow.

This species is dimorphic. The second form is represented only by the female sex, and differs in having the ground-color of the wings white instead of yellow.

Fig. 956.—*Eurymus philodice*.

This butterfly often occurs in large numbers in muddy places in country roads, for this reason it may be known as the roadside butterfly. It is also known as the clouded sulphur. Its range extends from the mouth of the St. Lawrence to South Carolina and westward to the Rocky Mountains. Its larva feeds upon clover and other Leguminosæ.

The orange sulphur, *Eurymus eurȳtheme*.—This species closely resembles the preceding one in size, shape and markings. The typical form differs from *E. philodice* in being of an orange color above instead of yellow. This butterfly is found chiefly in the Mississippi Valley and west to the Pacific Ocean; it is also found in the Southwestern

States, and occurs occasionally north to Maine. It is one of the most polymorphic of all butterflies; the forms differ so much in appearance that four or five of them have been described as distinct species. The larva feeds on clover and allied plants, and is sometimes a pest in alfalfa fields.

The dog's head, *Zerēne cæsōnia*.—The wings are lemon-yellow above bordered on the outer margin with black. On the hind wings the border is narrow, but on the fore wings it is broad. The outline of the yellow of the fore wings suggests a head of a dog or of a duck, a prominent black spot on the discal vein serving as the eye. This is an abundant species in the Southeastern and Southwestern States, extending from the Atlantic to the Pacific. The larva feeds on clover.

The sleepy yellow, *Eurēma nicĭppe*.—The wings above are bright orange, marked with blackish brown as follows: on the fore wings a narrow bar at the apex of the discal cell, the apical portion of the wings, and the outer margin; on the hind wings, the outer margin. In the female the outer marginal band is interrupted at the anal angle of each wing, and on the hind wings it may be reduced to an apical patch. The expanse of the wings is from 40 to 47 mm.

The common name, sleepy yellow, was suggested by the fact that the black spot near the middle of each fore wing is reduced to a narrow transverse line, which looks like an eye almost closed. This species occurs from southern New England to Florida and west to Lower California. The larva feeds on several species of *Cassia*.

The little sulphur, *Eurēma eutĕrpe*.—Although this species is larger than the following one it is considerable below the average size of our yellows, the larger individuals expanding less than 37 mm. The wings are canary-yellow above, with the apex of the fore wings and the outer margin of both fore and hind wings blackish brown. The border of the hind wings is narrow and sometimes wanting. There is a red-brown splash on the apex of the hind wings below.

The distribution of this species is similar to that of the preceding one. The larva feeds on *Cassia*.

The dainty sulphur, *Nathālis īole*.—This little butterfly can be distinguished from all others described here by its small size, as it expands only from less than 25 mm. to 30 mm. It is of a pale canary-yellow color, with dark brown markings. There is a large apical patch on the fore wings, and a broad band parallel with the inner margin; on the hind wings there is a stripe on the basal two-thirds of the costa, and spots on the ends of the veins; these are more or less connected on the margin of the wing, especially in the female.

This species also is found from Southern New England to Florida and west to Lower California. It, too, feeds on *Cassia*.

The cloudless sulphur, *Catopsĭlia eubūle*.—This large butterfly differs greatly in appearance from those described above. It expands 62 mm. The wings above are of uniform bright canary-yellow. In the male they are without spots, except frequently an inconspicuous brown dot at the tip of each vein, and a lilac-brown edging of the costal border. In the female there is a discal dot on the fore wings and a marginal row of brown spots at the ends of the veins.

This is a southern species which occasionally extends as far north on the coast as New York City, and in the Mississippi Valley as far as Southern Wisconsin. The larva feeds on *Cassia*.

Family NYMPHALIDÆ

The Four-footed Butterflies

The family Nymphalidæ, includes chiefly butterflies of medium or large size, but a few of the species are small. With a single exception, *Hypatus*, these butterflies differ from all others in our fauna in having the fore legs very greatly reduced in size *in both sexes*. So great is the reduction that these legs cannot be used for walking, but are folded on the breast like a tippet. Dr. W. T. M. Forbes has observed members of this family use their reduced fore legs for cleaning their antennæ, and in case of *Basilarchia* to make a sound.

More or less reduction in the size of the fore legs occurs in the Lycænidæ and Riodinidæ, but there it occurs only in the males, and to a much less degree than in this family. The Nymphalidæ differ from these two families in retaining all of the branches of radius of the fore wings, this vein being five-branched, except in the genus *Anæa*.

This is the largest of the families of butterflies. It not only surpasses the other families in number of species, but it contains a greater number and variety of striking forms, and also a larger proportion of the species of butterflies familiar to every observer of insects. There may be in any locality one or two species of yellows or of whites more abundant, but the larger number of species commonly observed are four-footed butterflies.

Five subfamilies of the Nymphalidæ are represented in our fauna. These can be separated by the following table. Each of these subfamilies is regarded as a distinct family by some writers.

A. Palpi much longer than the thorax. p. 766................LIBYTHEINÆ
AA. Palpi not as long as the thorax.
  B. Vein 3d A of the fore wings preserved; antennæ apparently naked, p. 765.
....DANAINÆ
  BB. Vein 3d A of the fore wings wanting; antennæ abundantly clothed with scales, at least above.
    C. Discal cell of the hind wings closed by a well-preserved vein.
      D. With some of the veins of the fore wings greatly swollen at the base (Fig. 972). p. 761..SATYRINÆ
      DD. With none of the veins of the fore wings unusually swollen at base. p. 764...HELICONINÆ
    CC. Discal cell of the hind wings either open or closed by a mere vestige of a vein. p. 750..NYMPHALINÆ

Subfamily NYMPHALINÆ

The Nymphs

The nymphs can be distinguished from the other four-footed butterflies as follows: the palpi are not longer than the thorax as in the long-beaks; the veins of the fore wings are not greatly swollen at the base as in the meadow-browns, except in the genera *Mestra* and *Eunica*, which are found only in the extreme South; the discal cell of the hind wings is not closed as in the heliconians; and the antennæ do not appear to be naked as in the milkweed butterflies.

The larvæ are nearly or quite cylindrical, and are clothed to a greater or less extent with hairs and sometimes with branching spines.

The chrysalids are usually angular, and often bear large projecting prominences; sometimes they are rounded. They always hang head downwards, supported only by the tail which is fastened to a button of silk.

Our genera of nymphs represent six quite distinct groups, as follows.

I. THE FRITILLARIES

The fritillaries are butterflies varying from a little below to somewhat above medium size. The color of the wings is fulvous, bordered and checkered with black, but not so heavily bordered as in the next subgroups. The lower surface of the wings is often marked with curving rows of silvery spots. The common name fritillary is from the Latin *fritillus*, a dice box, and was suggested by the spotted coloration of these butterflies.

In the larvæ there is an even number of rows of spines on the abdomen, due to the fact that there are none on the middle of the back. The larvæ feed upon the leaves of violets.

There are many species of fritillaries, about fifty occur in America north of Mexico, and it is difficult to separate the closely allied species.

The great spangled fritillary, *Argynnis cўbele*.—This species (Fig. 957) will serve to illustrate the appearance of the larger members of this group, those belonging to the genus *Argynnis*. In this genus vein R_2 of the fore wings arises before the apex of the discal cell.

There are a number of common fritillaries which resemble the preceding in color and markings but which are much smaller, the wings expanding considerable less than 50 mm. These belong to the genus *Brenthis*. In this genus vein R_2 of the fore wings arises beyond the apex of the discal cell.

Fig. 957.—*Argynnis cybele.*

The variegated fritillary, *Euptoieta claudia*.—This butterfly agrees with the smaller fritillaries (*Brenthis*) in the origin of vein R_2 of the fore wing beyond the apex of the discal cell, but differs from them in the shape of the fore wing, the apex of which is much more produced (Fig. 958) and the outer margin, except at the apex, concave; it is also considerably larger.

This species occurs throughout the United States east of the Rocky Mountains; but is very rare in the northern half of this region. The larva feeds on the passion-flowers.

Fig. 958.—*Euptoieta claudia.*

II. THE CRESCENT-SPOTS

This group includes some of the smaller members of the Nymphalidæ. The color of the wings is sometimes black, with red and yellow spots; but it is usually fulvous, with the fore wings broadly margined, especially at the apex, with black, and crossed by many irregular lines of black.

In the larva there is an odd number of rows of spines on the abdomen, due to the presence of spines on the middle of the back of some of the abdominal segments.

Sixty-three species of crescent-spots have been described from America north of Mexico; but nearly all of these are restricted to the Far West.

The Baltimore, *Euphy̆dryas pha̅eton.*—The wings are black above, with an outer marginal row of dark reddish-orange spots, and two parallel rows of very pale yellow spots; on the fore wings a third row is more or less represented. The wings expand 50 mm. or more.

The larvæ feed on a species of snakehead, *Chelone glabra;* they are gregarious in the fall and build a common nest in which they pass the winter; but separate after hibernation. They are very striking in appearance. The head and first two thoracic segments are shining black and the last three abdominal segments are black with two orange bands around each. All the other segments have a ground color of orange with various narrow transverse lines of black. This species occurs in Ontario and the northern half of the United States east of the Rocky Mountains. It is very local, the butterflies remaining near the bogs or moist meadows where the food-plant of the larva is found.

The butterflies of the genus *Phyciōdes* and the allied genera abound throughout our country. They are of small size, and of a fulvous color, heavily marked with black. Each species varies considerably in markings, and

Fig. 959.—*Phyciodes tharos.*

Fig. 960.—1, *Lycæna argiolus;* 2, *Polygonia faunus;* 3, *Polygonia comma.* 4, *Incisalia niphon;* 5, *Euvanessa antiopa;* 6, *Mitoura damon;* 7, *Lycæna argiolus;* 8, *Polygonia interrogationis.*

different species resemble each other closely, making this a difficult group for the beginning student. Figure 959 represents a common species.

The larvæ feed on asters and other Compositæ.

III. THE ANGLE-WINGS

To this group belong many of our best-known butterflies; there are twenty-five species in our fauna. With these the outer margin of the fore wings is usually decidedly angular or notched as if a part had been cut away. A large proportion of the species hibernate in the adult state, and some of them are the first butterflies to appear in the spring. Some of the hibernating species, however, remain in concealment till quite late in the season.

The red admiral, *Vanĕssa atalănta*.—The wings are purplish black above. On the fore wing there is a bright orange-colored band beginning near the middle of the costa, and extending nearly to the inner angle; between this and the apex of the wing are several white spots as shown in Figure 961; on the hind wing there is an orange band on the outer margin inclosing a row of black spots.

Fig. 961.—*Vanessa atalanta.*

The larva feeds chiefly on elm, nettle, and hop. When first hatched it folds together a half-opened leaf at the summit of the plant; when larger it makes its nest of a lower expanded leaf. There are two broods; both butterflies and chrysalids hibernate. This butterfly occurs over nearly the whole of the European and North American continents.

The painted beauty, *Vanĕssa virginiĕnsis*.—Figure 962 represents the upper side of this butterfly. The darker parts of the wings are very dark brownish black, the lighter parts a golden orange, sometimes with a pinkish tinge. In the apical portion of the fore wings there are several white spots as shown in the figure; the largest of these, the proximal one, is salmon or flesh-colored in the female. A characteristic feature of this species is the presence of *two* submarginal eye-like spots on the lower side

Fig. 962.—*Vanessa virginiensis*

of the hind wings. The larva feeds on everlasting (*Antennaria*) and allied plants. This species occurs in Ontario and nearly the whole of the United States, also in South America and the Canary Islands.

The painted beauty has been commonly known in this country as *Vanessa huntera;* but *Vanessa virginiensis* is the older name.

The cosmoplite, *Vanĕssa cărdui.*—The Butterfly resembles the preceding very closely in color and markings. There is however, a smaller proportion of orange markings; and on the lower surface of the hind wings there is a submarginal row of *four* or *five* eye-like spots.

The larva feeds upon Compositæ, especially thistles. This species is very remarkable for its wide distribution. Mr. Scudder states "with the exception of the arctic regions and South America it is distributed over the entire extent of every continent."

The American tortoise-shell, *Ăglais milbĕrti.*—The wings above are brownish black, with a broad orange-fulvous band between the middle and the outer margin. There are two fulvous spots in the discal cell of the front wings (Fig. 963). The larvæ feed on nettle (*Urtica*) and are gregarious in habits. This species occurs in the northern portions of the United States and in Canada.

Fig. 963.—*Aglais milberti.*

The mourning-cloak, *Euvanĕssa antīopa.*—The wings above are purplish brown, with a broad yellow border on the outer margin sprinkled with brown, and a submarginal row of blue spots. The upper surface is represented by Figure 964, the lower by Figure 960, 5. The larvæ live on willow, elm, poplar and *Celtis;* they are gregarious, and often strip large branches of their leaves. The species is usually two-brooded. "This butterfly is apparently distributed over the entire breadth of the Northern Hemisphere below the Arctic Circle as far as

Fig. 964.—*Euvanessa antiopa.*

the thirtieth parallel of latitude." (Scudder.)

The Compton tortoise, *Eugōnia j-ălbum*.—This butterfly (Fig. 965) resembles in its general appearance those of the genus *Polygonia*, but it is sharply distinguished from them by the inner margin of the fore wings being nearly straight, by the heavier markings of the fore wings, and by the presence of a whitish spot on both fore and hind wings, near the apex, and between two larger black patches. On the lower surface of the hind wings there is a small L-shaped silvery bar. This species occurs throughout Canada and the northern portion of the United States east of the Rocky Mountains. It is double-brooded.

Fig. 965.—*Eugonia j-album*.

Polygōnia.—The butterflies of this genus resemble the preceding species in having a metallic spot on the lower surface of the hind wings, but differ in having the inner margin of the fore wings roundly notched beyond the middle.

Ten species occur in this country. These differ principally in the coloring and markings of the under surface of the hind wings. The following are some of the more common ones.

The green comma, *Polygōnia faunus*.—The silvery mark of the hind wings is usually in the form of a C or a G, the ends being more or less expanded (Fig. 960, 2) but sometimes it is reduced to the form of an L. The lower surface of the wings is more greatly variegated than in any other species of this genus; and there is a larger amount of green on this surface than in any other of the eastern species, there being two nearly complete rows of green spots on the outer third of each wing.

The larva feeds upon black birch, willow, currant, and wild gooseberry. This is a Canadian species; but it is also found in the Mountains of New England and of New York, and in the northern portions of the Western States, extending as far south as Iowa. It is single-brooded.

The hop-merchant, *Polygōnia cŏmma*.—As in the preceding species, the silvery mark of the hind wings is in the form of a C or a G (Fig. 960, 3) but the general color of the lower surface of the hind wings is very different, being marbled with light and dark brown; and the green spots so characteristic of *faunus* are represented here by a few lilaceous scales on a submarginal row of black spots.

Two forms of this species occur. In one, *P. comma dryas*, the hind wings above are suffused with black on the outer half, so that the submarginal row of fulvous spots is obsured, and on the lower side the wings are more yellowish than in the other form. The latter is the typical form of *P. comma comma*.

The larva feeds upon hop, elm, nettle, and false-nettle. It is often abundant in hop-yards, and the chrysalids are commonly known as hop-merchants, from a saying that the golden or silvery color of the metallic spots on the back of the chrysalis indicates whether the price of hops is to be high or low. This species is found in Canada and the northern part of the eastern half of the United States; its range extends south to North Carolina, Tennessee, Arkansas and Indian Territory. It is double-brooded in the North, and at least three-brooded in the South.

The gray comma, *Polygōnia prŏgne*.—In its general appearance this butterfly closely resembles *P. comma*, but it can be readily distinguished by the form of the silvery mark, which is L-shaped and tapers towards the ends. It is much grayer below with a finer striate pattern in the male.

The larva feeds on currant, wild gooseberry, and rarely elm. This species occurs in Canada and in the northern portion of the United States except in the extreme West.

The violet tip, *Polygōnia interrogatiōnis*.—This butterfly (Fig. 960, 8.) is somewhat larger than the preceding species of *Polygonia* and differs in the form of the silvery mark, which consists of a dot and a crescent resembling a semicolon. It received its scientific name from the Greek note of interrogation, which is identical with our semicolon. On the upper side, the outer margins of the wings and the tails of the hind wings are tinged with violet, this fact suggested its common name.

This species is dimorphic; and the two forms differ so constantly and in such marked manner that they were described as distinct species. In *P. interrogationis interrogationis* the upper surface of the hind wings is not much darker than that of the fore wings, and there is a submarginal row of fulvous spots in the broad ferruginous brown border. In *P. interrogationis umbrōsa* the outer two-thirds of the upper surface of the hind wings is blackish, and the submarginal fulvous spots are obliterated, except sometimes faint traces near the costal margin.

This species is found in Canada and throughout the United States east of the Rocky Mountains.

IV. THE SOVEREIGNS

The members of this group differ from other Nymphalidæ in that the first three veins of the hind wings separate at the same point (Fig. 966); in the other nymphs the humeral vein arises beyond this point. The club of the antennæ is very long, and increases in size so gradually that it is difficult to determine where it begins. In its thickest part it is hardly more than twice as broad as the stalk. The palpi are slender, and the wings are rounded.

The larvæ present a very grotesque appearance, being very irregular in form, and strongly mottled or spotted in color.

The following are our best-known species.

The banded purple, *Basilarchia arthemis*.—The upper surface of the wings is velvety chocolate-black, marked with a conspicuous white bow (Fig. 967).

This is a Canadian species which extends a short distance into the northern part of the United States; the larva feeds on birch, willow, poplar, and many other plants.

The red spotted purple, *Basilărchia astỹanax*.—The upper surface of the wings is velvety indigo-black, tinged with blue or green. There are three rows of blue or green spots on the outer third of the hind wings; the spots of the inner row vary greatly in width in different individuals. On the lower surface there is a reddish orange spot in the discal cell of the fore wings, and one on the discal vein; on the hind wings there are two orange spots similarly situated, a third at the base of cell R_1 and a row of seven spots just within a double row of submarginal blue or green spots.

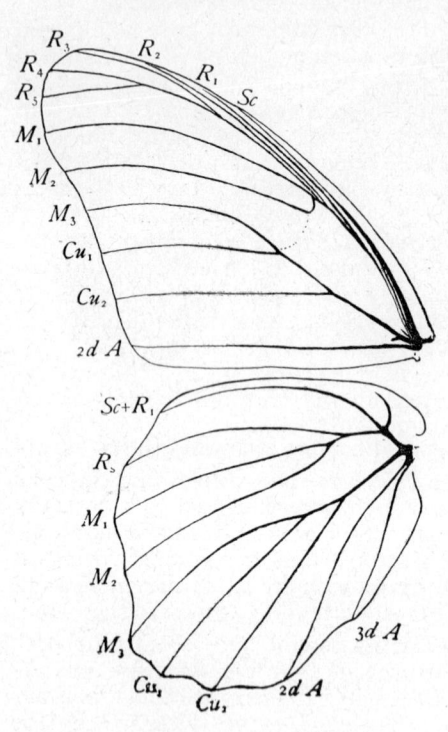

Fig. 966.—Wings of *Basilarchia astyanax*.

This species occurs throughout nearly the whole of the Eastern United States south of the 43rd parallel of latitude. The larva feeds on many plants; among them are plum, apple, pear, and gooseberry.

The hybrid purple, *Basilarchia prosĕrpina*.—There is a form of *Basilarchia* which was described as a distinct species under the name *proserpina*, but which is believed to be a hybrid between *B. astyanax*

Fig. 967.—*Basilarchia arthemis*.

and *B. arthemis*. See Field ('10). This butterfly has the coloring of *B. astyanax*, with a portion of the white bow of *B. arthemis*. It occurs in a narrow belt of country extending from southern Wisconsin and northern Illinois eastward to the Atlantic coast of New England. This is the region which forms the southern limit of the range of *B. arthemis* and the northern limit of the range of *B. astyanax*, the place where the two species meet. The hybrids vary greatly in the extent of the white band and the red spots.

Fig. 968.—*Basilarchia archippus*.

The viceroy, *Basilărchia archĭppus*.—The wings vary in color from a dull yellow orange tinged slightly with brown to a dark cinnamon color; they are bordered with black, and all the veins are edged with the same color (Fig. 968). The fringe of the wings is spotted with white, and the black border on the outer margin contains a row of white spots.

This species is remarkable for its resemblance to the monarch *Danaus archippus* (Fig. 974). But aside from the structural characters separating the two subfamilies which these butterflies represent, the viceroy can be easily distinguished from the species it mimics by its smaller size, and by the presence of a transverse black band on the hind wings. As *Danaus archipppus* has been termed the monarch, this species is aptly called the viceroy.

The larva (Fig. 969, *a*) when full-grown is about 30 mm. in length. The body is humped and naked, with many tubercles. In color it is dark brownish yellow or olive green, with a pale buff or

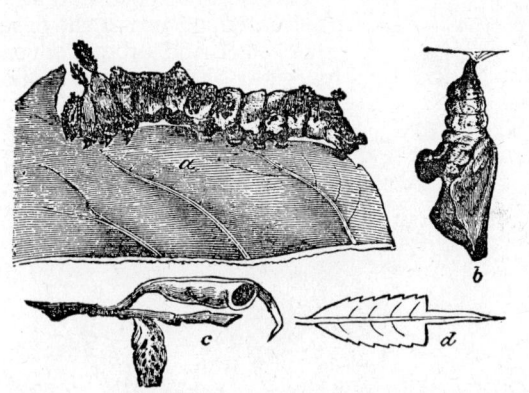

Fig. 969.—*Basilarchia archippus*: *a*, larva, *b*, pupa, *c*, nest; *d*, partly eaten leaf before rolled to form nest. (From Riley.)

whitish saddle on the middle segment of the abdomen. The tubercles on the second thoracic segment are club-shaped and spiny.

The larva of the viceroy feeds upon willow, poplar, balm of gilead, aspen, and cottonwood. The species is two- or three-brooded and hibernates as a partially grown larva in a nest made of a rolled leaf. (Fig. 969, c). This nest is lined with silk, and the leaf is fastened to the twig with silk so that it cannot fall during the winter. So far as is known all of the species of the sovereigns hibernate as larvæ in nests of this kind. It is worthy of note that only the autumn brood of caterpillars make these nests; so that the nest-building instinct appears only in alternate generations, or even less frequently when the species is more than two-brooded. *B. archippus* is found over nearly the whole of the United States as far west as the Sierra Nevada Mountains, and has been found sparingly even to the Pacific coast near our northern boundaries.

The vice-reine, *Basilărchia archĭppus floridĕnsis*.—This is a variety of *Basilarchia archippus* that is much darker than the typical form; the ground color of the wings resembling that of the queen, *Danaus berenice*. As it is found in the same region as the queen it is supposed to mimic that species, hence the popular name suggested above.

V. THE EMPERORS

This group is poorly represented in our fauna; our best-known species are the two following, which occur in the South.

The tawny emperor, *Chlorĭppe clȳton*.—In this and the following species the apex of the front wings and the anal angle of the hind wings are considerably produced in the males, but more rounded in the females. The male is represented in Figure 970 and the dotted line at the left indicates the contour of the wings of the female. This excellent figure is from the sixth Missouri report by C. V. Riley, where a detailed account of the life-history of the species is given. The wings of this butterfly are more or less obscure tawny, marked with blackish brown, and with pale spots. There is a submarginal row of six eyelike spots on the hind wings.

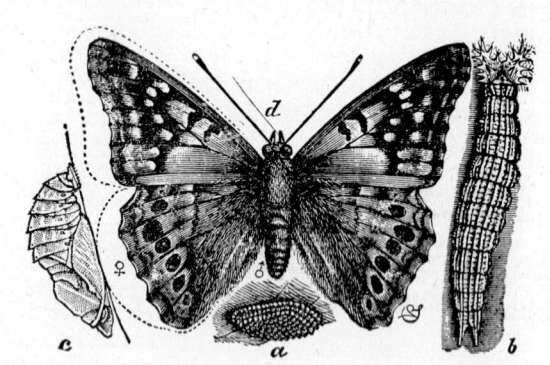

Fig. 970.—*Chlorippe clyton: a*, eggs; *b*, larva; *c*, pupa; *d*, upper surface of male butterfly; the dotted line at left indicates the contour of the wings of the female. (From Riley.)

LEPIDOPTERA

The species is dimorphic; the dimorphism affects both sexes and is independent, so far as is known, of season, as there is only one brood each year. It is the typical form *Chlorippe clyton clyton* that is figured here. The second form, *Chlorippe clyton prosĕrpina*, differs in having the hind wings darker and the submarginal row of eyelike spots wanting.

The larva (Fig. 970, *b*) feeds on hackberry.

The gray emperor, *Chlorĭppe cĕltis*.—In this species the wings are russety brown marked with blackish brown. In addition to the submarginal row of six eye-like spots on the hind wings, there is one in cell Cu_1 of the fore wings.

The larva of this species also feeds on hackberry.

VI. THE ANÆAS

The butterflies of the genus *Anæa* are quite distinct from any of the preceding divisions of the Nymphalinæ, although they have been classed with the emperors. There are three species found in the United States, *A. pŏrtia* from Florida, *A. morrisōnii* from Arizona, and the following one.

The goat-weed butterfly, *Anæa ăndria*.—The female of this species can be easily recognized by Figure 971. The male is smaller, with wings of a rich dark orange, margined with brown, and without the light-colored band characteristic of the female. This species is found in the Mississippi Valley from Illinois to Texas.

Fig. 971.—*Anæa andria*.

Subfamily SATYRINÆ

The Meadow-browns

This subfamily includes chiefly brown butterflies whose markings consist almost entirely of eyelike spots. Some western species, however, are bright-colored. In our species some of the veins of the fore wings are greatly swollen at the base (Fig. 972). This character is not quite distinctive; for in some species of the Nymphalinæ that are found in southern Florida and in Texas near the Mexican border some of the veins of the fore wings are swollen at the base.

The larvæ are cylindrical, tapering more or less towards each end. The caudal segment is bifurcated, a character that distinguishes them from all other American butterfly larvæ excepting those of the emperors, *Chlorippe.*

The pupæ are rounded; in some cases the transformation takes place beneath rubbish on the ground without any preparation of cell or suspension of the body.

Nearly sixty species belonging to this subfamily have been described from America north of Mexico.

The eyed brown, *Satyrōdes cănthus.*—The upper surface of the wings is soft mouse-brown on the basal half and paler beyond, considerably so in the female; each wing bears a row of four or five small black eye-like spots (Fig. 973). This species is found in Ontario, and throughout the eastern half of the United States in wet places. The larva feeds on swamp grasses; its head and caudal segment are each adorned with a pair of red cone-shaped tubercles.

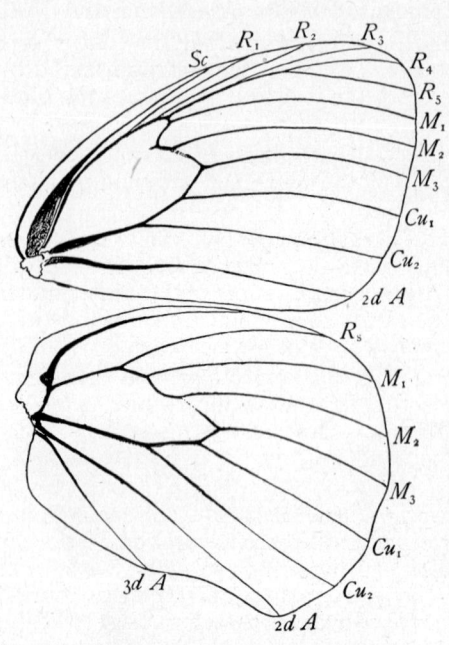

Fig. 972.—Wings of *Cercyonis alope.*

The grayling, *Cercyonis ălope.*—This species is found from the Atlantic to the Pacific; it occurs under several forms, some of which have been described as distinct species. The most common forms found East of the Rocky Mountains are the first two described below and intergrades between these. The expanse of the wings is from 50 to 62 mm. The larva feeds on grass.

(1) The blue-eyed grayling, *Cercyonis alope alope.*—The upper surface of the wings is dark brown; on the outer half of the fore wings there is a distinct yellow band, which extends from vein R_5 to the anal vein; in this band there are two dark spots with a white or bluish center. The hind wings usually bear a small spot in cell Cu_1, which is narrowly rimmed with yellow and has a minute white pupil. The lower surface of the hind wings is either with or without eye-like spots, usually with six of them.

Fig. 973.—*Satyrodes canthus.*

This is a Southern form, which extends into the southern portions of New England, New York, Michigan, Wisconsin, Iowa, and Nebraska; and into the northern portions of Illinois, Indiana, and Ohio.

(2) The dull-eyed grayling, *Cercyonis alope nĕphele*.—In this form the yellow band of the fore wings is either absent or represented by a faint pallid cloud. In other respects it closely resembles the blue-eyed grayling.

This is a Northern form; the southern limits of its range overlap the northern limits of the range of the blue-eyed grayling as given above.

(3) The hybrid graylings.—In that narrow belt where the ranges of the two forms of **Cercyonis** *alope* described above overlap, all variations between the two types occur. In most of these intergrades the eye-spots of the upper side of the fore wings are surrounded by yellowish rings, or each of them is on a yellowish patch.

(4) The sea-coast grayling, *Cercyonis alope maritima*.—In a narrow belt along the Atlantic coast there occurs a form which is smaller than those described above, and of a dark color; this form is easily recognized by the color of the band bearing the eye-spots, it being reddish yellow.

The White Mountain butterfly, *Œnēis semĭdea*.—The genus *Œneis* is composed of cold-loving arctic species whose natural habitat is the Far North; but some members of this genus are found within the limits of the United States. Their presence here and their distribution are extremely interesting. The best-known of these forms is the White Mountain butterfly.

This butterfly is found only on the higher parts (above 5,000 feet) of the White Mountains in New Hampshire, and on the highest peaks of the Rocky Mountains of Colorado, above 12,000 feet.

These two widely separated colonies of this butterfly are believed to be the remnants of an arctic fauna which was forced southward during the Ice Age. At the close of this period, as the arctic animals followed the retreating ice northward, the tops of these mountains became colonized by the cold-loving forms. Here they found a congenial resting place, while the main body of their congeners, which occupied the intervening region, was driven northward by the increasing heat of the lower land. Here they remain, clinging to these islands of cold projecting above the fatal sea of warmth that fills the valleys below.

The White Mountain butterfly is a delicate-winged species. The upper surface of the wings is grayish brown, without spots, except sometimes a minute one in cell M_1 of the fore wings; the fringe of the wings is brownish white interrupted with blackish brown at the ends of the veins. On the hind wings the marbling of the lower surface shows through somewhat. On the lower surface, the tip of the fore wings and the greater part of the hind wings are beautifully marbled with blackish brown and grayish white. The expanse of the wings is 43 mm.

The larva feeds on *Carex*. The species is either single-brooded or requires two years for the development of a brood.

A closely allied species, *Œnēis katăhdin*, is found on Mount Katahdin, Maine. This is called the Katahdin butterfly.

Subfamily HELICONIINÆ
The Heliconians

This subfamily consists chiefly of tropical butterflies. They are of medium or rather large size, and are easily recognized by their narrow and elongated fore wings, which are usually more than twice as long as broad. Most of the species are striking in appearance, being black banded with yellow or crimson, and sometimes with blue. The discal cell of the hindwings is closed by a well-preserved vein. The following species is the only one found in our fauna that unquestionably belongs to this subfamily.

The zebra, *Helicōnius charitōnius*.—This is a black butterfly with its wings banded with lemon yellow. There are three bands on the fore wings; on the hind wings there is a broad band parallel with the front wings when they are spread, a submarginal row of about fifteen spots, and a row of dots on the outer margin near the anal angle. The wings expand from 62 mm. to 100 mm. The larva feeds upon the passion-flower. This species is found in the hotter portions of the Gulf States.

There are two other species found in the United States that are placed in this subfamily by some writers; but in each of these the discal cell of the hind wings is open as in the Nymphalinæ. These are the following.

The julia butterfly, *Colænis jūlia*.—This butterfly resembles the true heliconians in having very long and narrow fore wings, these being more than twice as long as broad. The upper side of the wings is dark reddish orange, with the margins of the wings black, and with a more or less distinct black band cutting off the outer third of the fore wings. On the lower surface the wings are pale rusty-red, mottled with a darker shade. The wings expand about 85 mm. This is a very common species in the tropics of America and is found in the extreme southern part of the United States.

The gulf fritillary, *Diōne vanillæ*.—In this species the front wings are about twice as long as broad, but the markings of the wings resemble those of a fritillary more than those of an heliconian. The wings are reddish fulvous above; the veins of the front wings are black on the outer two-thirds of the wing; the black expands into spots at the ends of veins M_3 to anal; there are two white spots in the discal cell and one at the apex of it, each of these spots is surrounded with black; cells M_3, Cu_1, and Cu_2 each contains a round black spot. The outer margin of the hind wings has a broad black border, which contains a fulvous spot in each cell. The wings expand from 55 mm. to 75 mm.

The larva feeds on the passion-flower. In addition to the six rows of thorny spines, which characterize the caterpillars of many other fritillaries, this one has on the head a pair of backward bending spines branched like the others.

This species occurs from New Jersey and Pennsylvania southward also in Arizona and California.

Subfamily DANAINÆ

The Milkweed Butterflies

These butterflies are of large size, with rounded and somewhat elongate wings, the apical portion of the fore wings being much produced. The discal cells of the wings are closed; the third anal vein of the fore wings is preserved; and the antennæ are apparently without scales. Only a very few species of this family occur in our fauna. The two following are the best-known; the others are found only in the extreme South or in California.

The monarch, *Dānaus archippus*.—The upper surface of the wings is light ruddy brown, with the borders and veins black, and with two rows of white spots on the costal and outer borders as shown in Figure

Fig. 974.—*Danaus archippus*.

974. The figure represents a female; in the male the veins of the wings are more narrowly margined with black, and there is a black pouch next to vein Cu_2 of the hind wings, containing scent-scales or androconia.

The larva feeds upon different species of milk-weed, *Asclepias*. When full grown it is lemon or greenish yellow, broadly banded with shining black. It is remarkable for bearing a pair of long fleshy filaments on the second thoracic segment, and a similar pair on the seventh abdominal segment (Fig. 975). The chrysalis is a beautiful object; it is bright green dotted with golden spots, and about 25 mm. in length (Fig. 976).

766   AN INTRODUCTION TO ENTOMOLOGY

This species occurs throughout the greater part of the United States, and is distributed far beyond our borders. It is believed, however, that the species dies out each year in a large part of the Northern States, and that those butterflies which appear first in this

Fig. 975.—*Danaus archippus*, larva. (From Riley.)

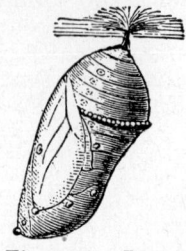

Fig. 976.—*Danaus archippus*, chrysalis. (From Riley.)

region, in June or July, have flown hither from the South, where they hibernate in the adult state. In the extreme South they fly all winter. Great swarms, including many thousands of individuals of this species, are sometimes seen, late in the year; and these swarms appear to be migrating southward.

The queen, *Dānaus berenīce*.—This species is found in the Southern States. The upper surface of the wings is reddish chocolate-brown with the costal margin of the front wings and the outer margins of both pairs bordered with black. There are two partial rows of white dots near the costal and outer margins of the front wings; and there is a larger white spot in each of the cells R_5 to Cu. The male possesses a black pouch containing androconia next to vein Cu of the hind wings as in the preceding species. The wings expand from 60 to 88 mm.

There is a well-marked variety, *Dānaus berenīce strigōsa*, in which on the upper surface of the hind wings the veins are narrowly edged with grayish white.

The larva of this species feeds on milkweed. This larva bears three pairs of long, brown, whiplash filaments; these are on the second thoracic and the second and eighth abdominal segments.

Subfamily LIBYTHEINÆ

The Long-beaks

The long-beaks can be easily recognized by their excessively long, beak-like palpi, which are from one-fourth to one-half as long as the body and project straight forward (Fig. 977). The outer margin of the fore wings is deeply notched; the males have only four well-developed legs, while the females have six.

Fig. 977.—*Hypatus bachmanni*.

Only two species are listed from the United States, one from Texas and one from the East; and these may be merely varieties of one species.

The snout butterfly, *Hўpatus bachmănni.*—The wings are blackish brown above, marked with orange and white spots. This species occurs throughout the Eastern United States, excepting the northern part of New England and the southern part of Florida. The larva feeds on hackberry, and in the West where hackberry does not occur, it feeds on wolfberry.

Family RIODINIDÆ

The Metal-marks

The metal-marks are small butterflies, which bear some resemblance to the gossamer winged butterflies. They are distinguished from the gossamer-winged butterflies by the presence of a humeral vein in the hind wings, and from them and all other butterflies by the fact that the costa of the hind wings is thickened out to the humeral angle (Fig. 978). The fore legs are reduced and brushlike in the males, normal in females.

Only twelve species have been found in our fauna, and nearly all of these are from the Far West or Southwest. The two following species occur in the East.

The small metal-mark, *Calephēlis virginiënsis.*—The upper surface of the wings is rust-colored, and is crossed by four or five more or less sinuous blackish lines on the basal two-thirds, and on the outer third by two lines of shining scales, that look like cut steel, and an intermediate row of black spots. The under surface is of a brighter rust color and marked as above. The expanse of the wings is 20 mm. This species occurs in the Southern States.

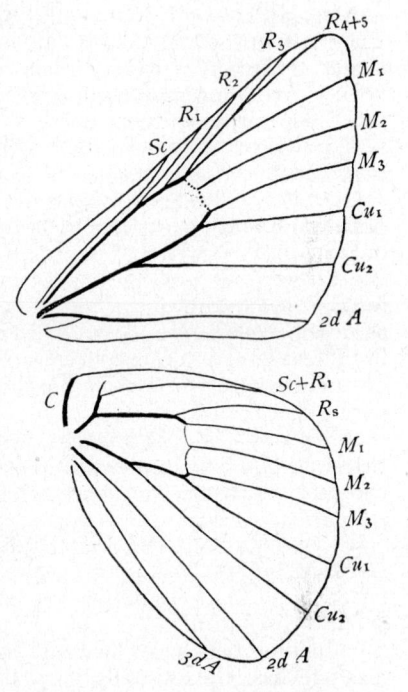

Fig. 978.—Wings of *Emesis zela.*

The large metal-mark, *Calephēlis boreālis.*—The upper surface of the wings is dull brownish yellow, crossed by obscure transverse stripes; on the outer half of the wings are two lead-colored lines, with a row of black dots between them. The under surface is of a rather dark and a pale orange; paler and duller next the base, marked with transverse black lines and dots, and transverse series of steel-colored spots. The wings expand from 25 to 30 mm.

This is a rare butterfly; it has been taken in New York, New Jersey, West Virginia, Michigan, and Illinois.

Family LYCÆNIDÆ

The Gossamer-winged Butterflies

The family Lycænidæ includes butterflies which are of small size and delicate structure. In size they resemble the smaller Hesperiidæ; but they can be distinguished at a glance from the skippers, as they present an entirely different appearance. The body is slender, the wings delicate and often brightly colored, and the club of the antenna straight. The antennæ are nearly always ringed with white; each is situated very closely to the edge of an eye, often flattening it; they are not in pits; and a conspicuous rim of white scales encircles the eyes.

An easily-observed combination of characters by which the members of this family can be distinguished is the absence of one or two of the branches of radius of the fore wings, this vein being only three- or four-branched, and the origin of vein M_1 of the fore wings at or near the apex of the discal cell (Fig. 979). In all other butterflies occurring in our fauna in which radius is only three- or four-branched (except *Parnassius*), vein M_1 of the fore wings coalesces with radius for a considerable distance beyond the apex of the discal cell.

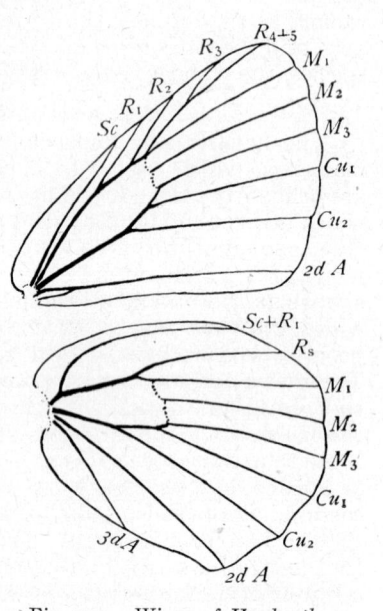

Fig. 979.—Wings of *Heodes thoe*.

An exception to the characters of the Lycænidæ is presented by *Feniseca*, as indicated in the table of families, p. 739.

A characteristic of this family is that while in the female the front legs are like the other legs, in the male they are shorter, without tarsal claws, and with the tarsi more or less aborted.

The caterpillars of the Lycænidæ present a very unusual form, being more or less slug-like, reminding one of the larvæ of the Eucleidæ. The body is short and broad; the legs and prolegs are short and small, allowing the body to be closely pressed to the object upon which the insect is moving—in fact some of the species glide rather than creep; and the head is small, and can be retracted more or less within the prothorax. The body is armed with no conspicuous appendages; but some of the species are remarkable for having osmeteria which can be pushed out from the seventh and eighth abdominal segments, and through which honeydew is excreted for the use of ants. Certain other species are remarkable in being carnivorous; one American species feeds exclusively upon plant-lice.

The chrysalids are short, broad, ovate, and without angulations. They are attached by the caudal extremity, and by a loop passing over the body near its middle. The ventral aspect of the body is straight and often closely pressed to the object to which the chrysalis is attached.

The family Lycænidæ is represented in our fauna by three well-marked groups of genera, which are hardly distinct enough to be ranked as subfamilies; these are known as the hair-streaks, the coppers, and the blues respectively. In addition to these there is a single species, the wanderer, the relationship of which is uncertain.

I. THE HAIR-STREAKS

The hair-streaks are usually dark brown, with delicate striped markings, which suggested their common name; but some species are brilliantly marked with metallic blue, green, or purple. The hind wings are commonly furnished with delicate tail-like prolongations (Fig. 980), and the eyes are hairy. The fore wings of the male often bear a small dull oval spot near the middle of the costal part of the wings, the discal stigma, which is filled with the peculiar scent-scales known as androconia. The males are also distinguished by having a tuft of hair-like scales, the beard, on the front; this is wanting or very thin in the females. More than sixty species occur in America north of Mexico; of these nearly twenty occur in the eastern half of the United States.

The banded hair-streak, *Thĕcla călanus*.—In the northeastern United States the most common of the hair-streaks is this species (Fig. 980). The upper surface of the wings is dark brown or blackish brown. The under surface is blackish slate-brown nearly as dark as the upper surface, and marked as shown in the figure.

Fig. 980.—*Thecla calanus*.

The larva feeds on oak and hickory. Excepting the southern portion of the Gulf States, the species is found throughout our territory east of the Rocky Mountains, and in the southern part of Canada.

The olive hair-streak, *Mitoura dāmon*.—The upper surface of the wings is dark brown, with the disk more or less deeply suffused with brassy yellow in the male or tawny in the female; the hind wing has two tails, one much longer than the other, both black tipped with white. The lower surface of the hind wings is deep green; both fore and hind wings are marked with white bars bordered with brown. (Fig. 960, 6).

Southern individuals have much longer tails than the one shown in the figure; and there is a variety, *patersōnia*, in which the upper surface of the wings is all dark brown.

The larva feeds on red cedar and smilax. This species occurs from Massachusetts to Florida and westward to Dakota and Texas.

The banded elfin, *Incisālia nīphon*.—In the butterflies of the genus *Incisalia* the fringe of the outer margin of the hind wings is slightly prolonged at the end of each vein, giving the wings a scalloped outline; they also lack tail-like prolongations of the hind wings.

There are several species occurring on both sides of the continent. One of these, the banded elfin, is represented in Figure 960, 4. In this species there is a distinct white or whitish edging near the base of the under side of the hind wing which limits a darker band that occupies the outer two-thirds of the basal half of the wing.

This species occurs in the Eastern and Middle States. The larva feeds on pine.

The hair-streaks described above are of moderate size and modest colors. The two following will serve to illustrate a somewhat different type.

The great purple hair-streak, *Ătlides halēsus*.—This is the largest of our eastern hair-streaks, the larger individuals expanding 50 mm. In the male the greater part of the upper surface of the wings is bright blue; the discal stigma, the outer fourth of the fore wings, the apex of the inner margin of the hind wings, and the tails are black. In the female the outer half of the wings is black.

This species occurs in the southern half of the United States and southward. It has been found as far north as Illinois. The larva is said to feed on oak.

The white-m hair-streak, *Thecla m-ălbum*.—This is a smaller species, expanding about 37 mm. The upper surface of the disk of the wings is a rich, glossy dark blue, with green reflections; a broad outer border and costal margin are black. The hind wing has two tails, and a bright dark orange spot preceded by white at the anal angle. The under surface is brownish gray, and on this surface both wings are crossed by a common, narrow white stripe which forms a large W or reversed M on the hind wings.

This species occurs in the southern half of the United States. The larva feeds on oak and on milk-vetch.

II. THE COPPERS

The coppers, as a rule, are easily distinguished from other gossamer-winged butterflies by their orange-red and brown colors, each with a coppery tinge, and conspicuous black markings. They are the stoutest of the Lycænidæ. Among the exceptions to the more common coloring of these insects are the following: In the male of *Heodes epixanthe*, a small species which frequents cranberry bogs, the wings have a purple tinge; and in *Heodes heteronea*, a species found from California to Colorado, the male is blue.

Eighteen species of the group are now listed in our fauna; the two following will serve as examples:

The American copper, *Heōdes hypophlæas.*—This is the most common of our coppers in the Northeastern States and in Canada. Its range extends also along the boundary between the United States and Canada to the Pacific Ocean, and southward into California; and in the east along the Alleghany Mountains south to Georgia. The fore wings are orange-red above, spotted with black, and with a blackish brown outer border; the hind wings are coppery brown, with a broad orange-red band on the outer margin; this band is indented by four black spots.

The larva feeds on the common sorrel (*Rumex acetosella*).

The bronze copper, *Heodes thoe.*—This is larger than the preceding species, the wings expanding 37 mm. or more. In the male the wings are coppery brown above, spotted with black, and with a broad orange-red band on the outer margin of the hind wings. The female differs in having the fore wings orange-red above, with prominent black spots.

This species occurs in the Middle and Western States from the Connecticut Valley to Nebraska. The larva feeds on curled dock (*Rumex crispus*).

III. THE BLUES

The blues can be distinguished from the other gossamer-winged butterflies by the slender form of the body, and the blue color of the upper surface of the wings of the males at least; in many species the upper surface of the wings of the female is much darker than that of the male. Thirty-eight North American species have been described; but most of these occur only in the Far West. This is a rather difficult group to study owing to the fact that in several cases a single species exists under two or more distinct forms, and also that the two sexes of the same species may differ greatly. It often happens that two individuals of the same sex but of different species resemble each other more closely in the coloring of the upper surface than do the two sexes of either of the species.

The spring azure, *Lycæna argiolus.*—In this species the hind wings are without tails, the eyes are hairy, and the lower surface of the wings is pale ash-gray. This combination of characters will distinguish it from all other blues occurring in the Eastern United States. But the species is not confined to this region, as it occurs in nearly all parts of the United States, in a large part of Canada, and most of the Old World.

This butterfly exhibits polymorphism to the greatest degree of any known species. In this country alone there are thirteen or more named forms. Some of these are geographical races; some are seasonal forms; and some are distinct forms that exist at the same time and place as the more typical form. In the Old World many other forms of this species have been described. Two forms are represented in Figure 960, 1 and 7.

The larva feeds on the buds and flowers of various plants, especially those of *Cornus*, *Cimicifuga*, and *Actinomeris*. They are frequently attended by ants for the sake of the honey-dew which they excrete from osmeteria which they push out from the seventh and eighth abdominal segments.

The tailed blue, *Everes comўntas*.—The butterflies of the genus *Everes* can be distinguished from our other blues by the presence of a small tail-like prolongation of the hind wing. This is borne at the end of vein Cu₂. Our common species (*E. comyntas*) is distributed over nearly all parts of North America. The male is dark purplish violet above, bordered with brown; the female is dark brown, sometimes flecked with bluish scales. In the Eastern United States this is the only species of the genus.

The larva feeds upon clover and other leguminous plants.

IV. GENUS FENISECA

The wanderer, *Feniseca tarquinius*.—This is the only known member of the genus *Feniseca*, the affinities of which have not been determined. It does not seem to belong to either of the three groups of genera mentioned above. A distinctive feature of this genus is the fact that vein M₁ of the fore wings coalesces with a branch of radius for a considerable distance beyond the apex of the discal cell; in this respect it differs from all other members of the Lycænidæ found in our fauna.

Fig. 981.—*Feniseca tarquinius*.

The upper surface of the wings of this butterfly (Fig. 981) is dark brown, with a large irregular, orange-yellow patch on the disk of the fore wing, and one of the same color next the anal angle of the hind wing.

This species is of unusual interest, as the larva is carnivorous in its habits. It feeds on plant-lice; and, so far as observed, it feeds only on the woolly aphids. It is found more often in colonies of the alder blight (*Schizoneura tessellata*) than in those of the allied species. It is found from Maine to Northern Florida and westward to Kansas. It is a very local insect, being found only in the neighborhood of water where alder grows.

I do not know why the name the wanderer was applied to this butterfly, it may have been on account of its local appearance in widely separated places, or because in habits the larva deviates far from the more usual habits of caterpillars. The name is also appropriate as its nearest relatives are found in Africa and in Asia.

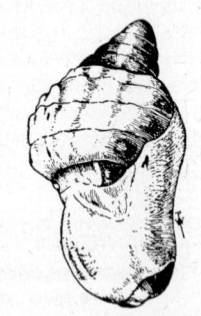

Fig. 982.—Chrysalis of *Feniseca*. Enlarged.

The chrysalis of *Feniseca* presents a remarkable appearance (Fig. 982); the anterior half when viewed from above bears a curious resemblance to a monkey's face; and it differs from all other lycænid pupæ in our fauna in having on each side a row of small rounded tubercles.

CHAPTER XXVIII

ORDER DIPTERA*

The winged members of this order have only two wings; these are borne by the mesothorax. The second pair of wings is represented by a pair of knobbed, thread-like organs, the halteres, these are present in nearly all flies, even when the mesothoracic wings are wanting. The mouth-parts are formed for sucking. The metamorphosis is complete.

To the order Diptera belong all insects that are properly termed flies, and only these. The word fly forms a part of many compound names of insects of other orders, as butterfly, May-fly, and chalcis-fly; but when used alone, it is correctly applied only to dipterous insects. To some flies other common names have been applied, as mosquito, gnat, and midge.

The presence of a single pair of wings and of a pair of halteres is sufficient to distinguish members of this order from those of all other orders, except in the case of a few wingless forms.

This is a large order both in number of species and of individuals. Aldrich ('05) gives a list of about eight-thousand North American species, distributed in more than a thousand genera.

Different species differ greatly in habits. Some are very annoying to man. Familiar examples are the mosquitoes, which attack his person, the flesh-flies, which infest his food, the botflies and the gadflies that torment his cattle, and the gallgnats that destroy his crops. Some species are extremely noxious, being disease carriers, as for example the mosquitoes that transmit malaria and yellow fever. Other species are beneficial. Those belonging to the Syrphidæ and the Tachinidæ destroy many noxious insects; and very many species, while in the larval state, feed upon decaying animal and vegetable matter, thus acting as scavengers.

There are certain structural features of flies that are used in the classification of these insects and to which special terms have been applied. The more important of these terms are defined below; others are defined later in the discussion of chætotaxy.

The head and its appendages.—The head is very mobile, being connected to the thorax by a slender neck. It is variable in shape and in its relative size.

The *compound eyes* are usually large, sometimes occupying a large part of the surface of the head. When the eyes are contiguous on the upper side of the head they are termed *holoptic;* when they are separated more or less broadly they are termed *dichoptic.* In some flies each compound eye is divided into two parts, one of which is a day-eye and the other a night-eye. See page 144.

The *ocelli* are usually three in number.

---

*Diptera: *dis* (δις), two; *pteron* (πτερόν), a wing.

774   *AN INTRODUCTION TO ENTOMOLOGY*

The *antennæ* vary greatly in form in the different families. In

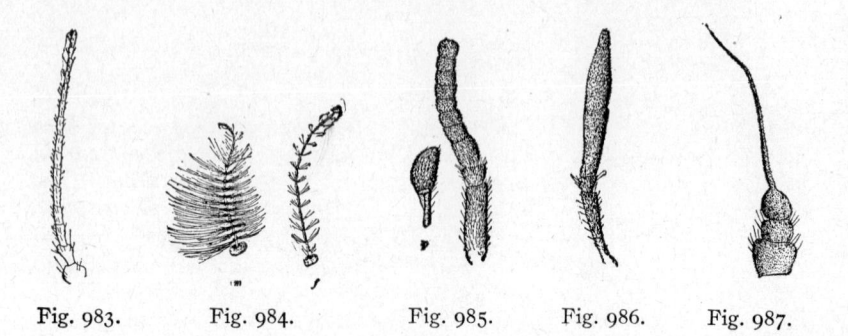

Fig. 983.   Fig. 984.   Fig. 985.   Fig. 986.   Fig. 987.

the more generalized families the antennæ consist of many segments, which, except the basal two, are, similar in form (Fig. 983). Frequently such antennæ bear whorls of long hairs (Fig. 984). In the more specialized families there is a reduction in the number of segments of the antennæ. This is brought about either by a more or less complete consolidation of the segments beyond the second into a single segment (Figs. 985 and 986), or by a dwindling of the terminal segments, so that they form merely a slender style (Fig. 987) or bristle (Fig. 988). Such a bristle is termed the *arista*. In most cases where a style or arista exists it is borne by the third segment, and this segment is then usually greatly enlarged. When the enlargement of this segment has taken place evenly the style or arista is terminal; but frequently one part of the third segment is expanded so that it projects beyond the insertion of the arista (Fig. 989); then the arista is said to be dorsal.

Fig. 988.

The *mouth-parts* of flies are formed for sucking, and sometimes also for piercing. Their structure differs greatly in different families; and in some cases it is exceedingly difficult to determine the homologies of the different parts. In the more typical forms the mouth-parts consist of six bristle-like or lance-like organs enclosed in a sheath and a pair of jointed palpi. There are differences of opinion as to the homologies of these parts, but according to the most generally accepted view they are as indicated below. The mouth-parts of a mosquito will serve as an example of a comparatively generalized type.

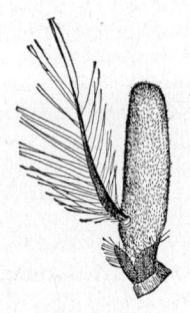

Fig. 989.

Figure 990 represents a side view of the head of *Anopheles* with the bristle-like organs removed from the sheath. The parts are as follows:

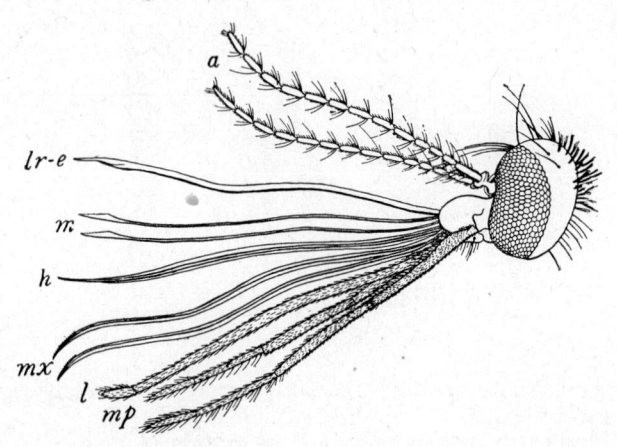

Fig. 990.—Head of *Anopheles*. (After Nuttall and Shipley.) The lettering is explained in the text.

the antennæ (*a*); the labrum or labrum-epipharynx (*lr-e*); the hypopharynx (*h*); the two mandibles (*m*); the two maxillæ (*mx*); the labium (*l*); and the maxillary palpi (*mp*). The labium is the sheath in which the six bristle-like organs are normally enclosed; the maxillary palpi are not enclosed in the sheath. At the tip of the labium there is on each side a lobe-like appendage; these are termed the *labella*. The labella are believed by some writers to be the labial palpi; but it seems more probable that they are the paraglossæ. The labella of certain flies are quite large; in the house-fly, for example, they are expanded into broad plates, which are fitted for rasping.

The *frontal lunule* (Fig. 991, *f. l*) is a small crescent-shaped sclerite immediately above the antennæ, which is characteristic of the second suborder, the Cyclorrapha. In most members of this suborder there is a suture

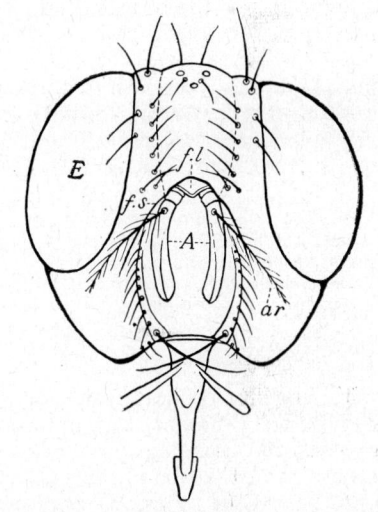

Fig. 991.—Head of a fly: A, antennæ; *ar*, arista, E, eye; *f. l.*, frontal lunule; *f.s.*, frontal suture.

separating the frontal lunule from that part of the head above it; this is termed the *frontal suture*. Frequently the frontal suture extends down on each side to near the mouth (Fig. 991, *f. s*).

The *ptilinum* is a large bladder-like organ which exists in those flies that have a frontal suture. The ptilinum is pushed out through this suture when the adult is about to emerge from the puparium. In this way the head end of the puparium is forced off, making a large opening through which the adult escapes; afterwards the ptilinum is withdrawn into the head. If a specimen is captured soon after its emergence from the puparium, there may be seen instead of the frontal suture the bladder-like ptilinum projecting from the head, immediately above the antennæ.

The thorax and its appendages.— The thoracic region of the body consists chiefly of the mesothorax, both the prothorax and the metathorax being greatly reduced in size. The thorax of a crane-fly (Fig. 992), will serve to illustrate the structure of this part in the more generalized members of this order, and will also serve as a type with which to compare the thorax of the more specialized forms.

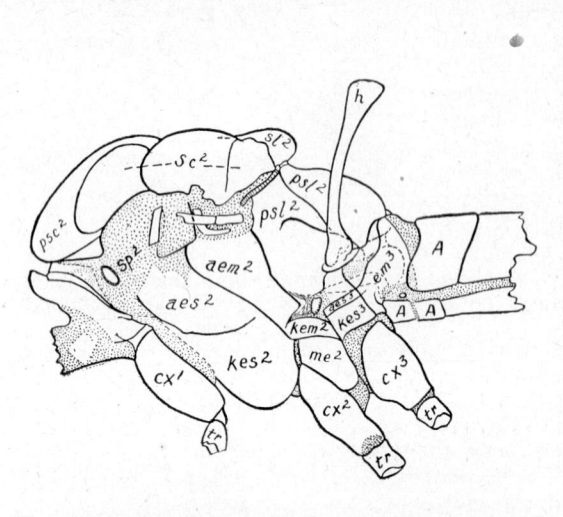

Fig. 992.—Lateral aspect of thorax of *Pachyrhina ferruginea*. (After Young). The thoracic segments to which the sclerites belong are indicated by the numbers 1, 2, and 3. A, A, A, first abdominal segment; *aem*, anepimerum, the upper part of the epimerum; *aes*, anepisternum, the upper part of the episternum; *cx*, coxa; *em*, epimerum of the metathorax; *h*, halter; *kem*, katepimerum, the upper part of the epimerum; *kes*, katepisternum, the lower part of the episternum; *me*, meron; *psc*, prescutum; *psl*, postscutellum; *sc*, scutum; *sl*, scutellum; *sp*, spiracle; *tr*, trachantin.

There are differences of opinion among writers on this order regarding the homologies of certain thoracic sclerities. The most extended investigation of this subject is that of Young ('21) who studied and figured the thorax of representatives of fifty-five of the fifty-nine families found in our fauna. I have adopted this writer's conclusions regarding the homologies of the sclerites in question.

The most distinctive feature of the wings of the Diptera is the fact that only the first pair are developed as organs of flight; the second pair being greatly reduced in size. The second pair of wings are known as the *haltēres*, they are thread-like, enlarged at the end, and bear organs of special sense, the function of which has not been definitely determined. They are present in nearly all members of the order, even when the front wings are wanting; they can be easily seen in a crane-fly (Fig. 993).

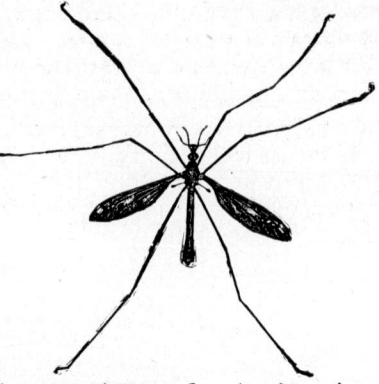

The *fore wings* are thin, membranous, and usually either naked or clothed with microscopic setæ; but with mosquitoes the wings bear a fringe of scale-like setæ on the margin and usually also on each of the wing-veins, and in the moth-like flies (Psychodidæ) and in some others the clothing of setæ is very conspicuous.

Fig. 993.—A crane-fly, showing wings and halteres.

In the more generalized members of this order the venation of the mesothoracic wings corresponds quite closely to the hypothetical primitive type. Neither accessory nor intercalary veins are ever developed, and only the principal cross-veins are present. The most

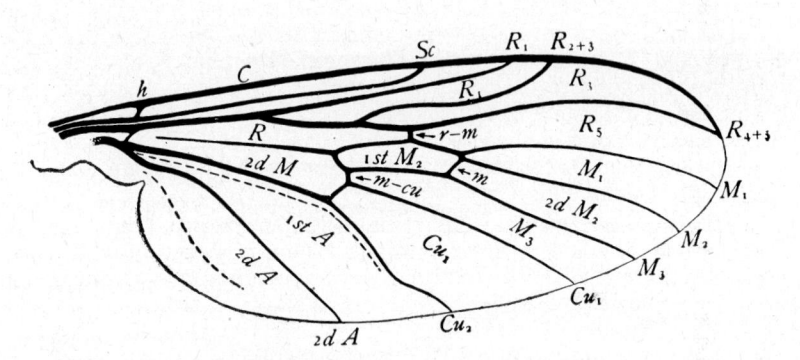

Fig. 994.—Wing of *Anisopus*.

striking divergence from this type is the fact that vein M is only three-branched. The wing of *Anisopus* (Fig. 994) is a good example of a generalized dipterous wing except that the branches of radius have been reduced. In the more specialized forms the typical arrangement of the veins has been greatly modified by the approaching and coalescing of the tips of adjacent veins, as shown later.

In many families there is a notch in the inner margin of the wing near its base (Fig. 995, *a. e*); this is the *axillary excision;* that part of the wing lying between the axillary excision when it exists, and the axillary membrane is the *posterior lobe* (Fig. 995, *l*). In certain families the axillary membrane, the **membrane** of the wing base, is expanded so as to form a lobe or lobes which fold beneath the base of the wing when the wings are closed; this part of the wing is the *alula* or *alulet*. The alulæ are termed the *squamæ* by some writers and the *calypteres* by others. The alulæ are well developed in the common

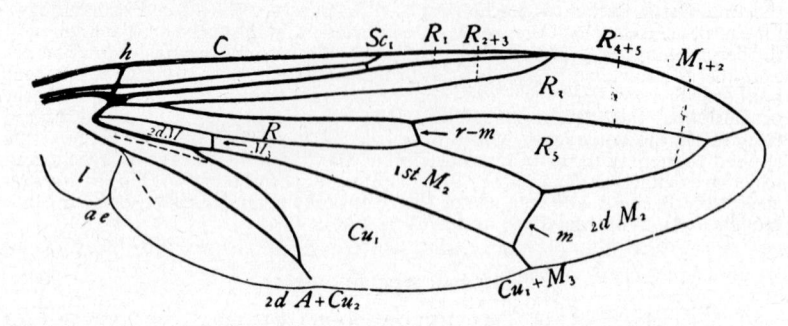

Fig. 995.—Wing of *Conops*.

House-fly. Each alula, in those species where the alulæ are well developed, consists of two lobes which fold over each other when the wings are closed. These two lobes are designated as the *upper squama* or *squamula alaris* and the *lower squama* or *squamula thoracalis* respectively. The alulæ are called the tegulæ by many writers on Diptera; but the term tegula was first used in insect anatomy for the cuplike scale which covers the base of the wing in certain insects, as most Hymenoptera, and should be restricted to that use. The terms alula and alulet are also often misapplied, being used to designate the posterior lobe of the wing.

The *legs* vary greatly in length and in stoutness. The coxæ are usually long, and in most of the fungus-gnats (Mycetophilidæ) they are very long. When pulvilli are developed they are membranous pads, one beneath each tarsal claw. A third appendage, the *empodium*, often exists between the two pulvilli of each tarsus. The empodia may be bristle-like or tapering (Fig. 996) or membranous, resembling the pulvilli in form (Fig. 997) in the last case they are described as pulvilliform.

Fig. 996.

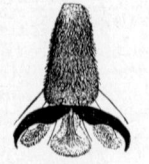

Fig. 997.

In descriptions of flies the number of tibial spurs borne by the different pairs of legs is often indicated by a brief formula, as, for example; "Tibial spurs 1:2:2" indicates that the fore tibiæ bear each one spur; the middle tibiæ, two; and the hind tibæ two.

CHÆTOTAXY

OR THE ARRANGEMENT OF THE CHARACTERISTIC BRISTLES OF DIPTERA

In certain families of the Diptera some of the setæ with which the body is clothed are stout bristles, termed *macrochætæ*. In the classification of these families much use is made of the number, position, and arrangement of these bristles. This has made necessary the establishment of a set of terms by which the different bristles or sets of bristles can be designated. Such a terminology was proposed by Osten-Sacken in 1881, and is still in use with additions.

In the choice of terms Osten-Sacken and later writers have used those that indicate the places of insertion of the bristles. But owing to the fact that the homologies of the sclerites of the head and thorax had not been definitely determined at the time Osten-Sacken wrote he proposed a "purely conventional terminology" for the areas upon which the bristles are inserted; and in this he has been followed to the present time. The result is that some of the terms are misleading; as for example, the so-called frontal bristles are not inserted on what is really the front but on the vertex. But the use of these terms is so firmly established that it is not probable that they will be changed. In the following account I have endeavored to indicate the homologies of the parts named in those cases where the terms applied to them differ from those used in accounts of other orders of insects, and which are defined in Chapter II. In defining the special terms used by writers on chætotaxy I have made free use of the definitions given by Osten-Sacken ('81), Hough ('98), Williston, ('08) and Walton ('09).

THE PARTS OF THE HEAD

The homologies of the areas of the head were determined by Peterson ('16) who studied and figured the heads of representatives of nearly all of the families of Diptera found in our fauna, and who gives a diagram representing a hypothetical type of the head-capsule of Diptera (Fig. 998). The conclusions of Peterson are based on comparisons of heads of the more generalized Diptera with those of the more generalized members of other orders of insects; for descriptions of the latter see above, pages 37–40 and 96–97.

The more important landmarks for determining the homologies of the areas of the cephalic aspect of the head, the region in which the greatest confusion exists, are the stem of the epicranial suture (Fig. 998, *s. e. s*); the arms of the epicranial suture (Fig. 998, *a. e. s*); and the positions of the invaginations of the dorsal arms of the tentorium (Fig. 998, *i. d.*); and of the anterior arms of the tentorium (Fig. 998, *i. a.*)

In Figure 999 is given a diagram illustrating the terms applied by writers on chætotaxy to the areas of the head. These terms are defined below.

The antennal fossa, fovea, or *groove.*—Depressed areas of the fronto-clypeus in which the antennæ rest (Fig. 999, *a. f.*)

The Bucca.—That part of the wall of the head on each side that is ventrad of the transverse impression, and ventrad of the eye, extending ventrad to the edge of the mouth opening, cephalad to the vibrissal ridge and continuing caudad on the gena to the caudal margin of the head (Fig. 999, *b.*)

The cheeks.—This term is used differently by different systematists; by some is applied to the space on each side of the head that is between the lower border of the eye and the oral margin, differing from the bucca only in that it does not extend over the caudal aspect of the head; by others it is applied to this space and the so-called gena of writers on chætotaxy; and by others to the so-called gena alone.

The cheek-grooves.—A more or less distinct depression on each side below the eye.

The clypeus.—See fronto-clypeus.

The epistoma.—The oral margin and an indefinite space immediately contiguous thereto.

The face.—That part of the cephalic aspect of the head lying below an imaginary horizontal line passing through the base of the antennæ. (Fig. 999, *Fa.*).

The facial depression.—See antennal fossa.
The facialia or facial ridges.—See vibrissal ridges.

The front.—(*a*) The true front (Fig. 998, *fr.*) is the first of the unpaired sclerites between the arms of the epicranial suture. See above pages 37 and 38. This term has been generally applied to this sclerite since it was proposed by Kirby (Kirby and Spence) nearly a century ago. (*b*) The so-called front of writers on chætotaxy is that part of the vertex that extends from the base of the antennæ to the upper margin of the head (Fig. 999, *Fr.*).

The frontalia.—See frontal vitta.
The frontal orbits.—See genovertical plates.
The frontal triangle.—In holoptic flies, those in which the eyes are contiguous on the upper side of the head, the triangle between the eyes and the antennæ, the apex of which is above, is termed the frontal triangle. Sometimes this term is applied to a triangle indicated by color or a depression in the corresponding position in flies with dichoptic eyes.

The frontal lunule.—See page 775.
The frontal suture.—See page 776.
The frontal vitta. The median portion of the so-called front, extending from the base of the antennæ to the ocelli (Fig. 999, f. v.).

The fronto-clypeus.—This term is applied to the combined front and clypeus when the suture between them is obsolete, as is usually the case in Diptera. It is the part bounded above by the arms of the epicranial suture (Fig. 998, *a. e. s*) and below by the clypeo-labial suture (Fig. 998, *c. l. s.*).

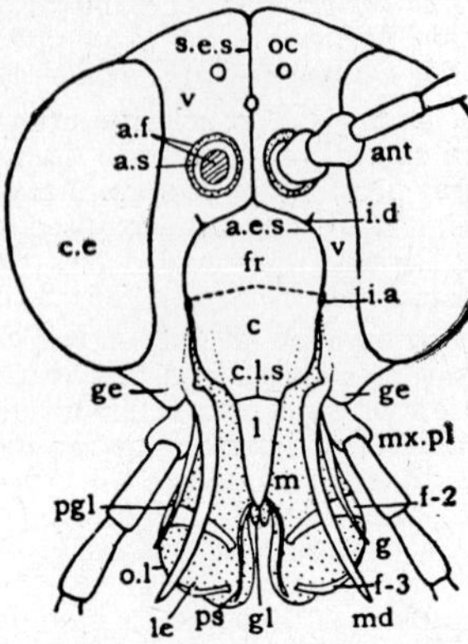

Fig. 998.—Hypothetical type of head-capsule of Diptera; *a. e. s*, arms of epicranial suture; *a, f*, antennal fossa; *ant*, antenna; *a. s*, antennal sclerite; *c*, clypeus; *c. e*, compound eye; *c. l. s.* clypeo-labial suture; *f2*, furca; *f3*, furca; *g*, galea, *ge*, gena; *gl*, glossa; *i. a.* and *i. d*, invaginations of the tentorium; *l*, labrum; *le*, labella; *m*, membrane; *md*, mandible; *mx. pl*, maxillary palpus; *oc*, ocelli; *o. l*, oral lobe; *pgl*, paraglossa; *s. e. s*, stem of epicranial suture; *v*, vertex. (From Peterson.)

The genæ.—(*a*) The true genæ (Fig. 48, G, page 39). The term genæ was introduced into entomology nearly a century ago by Kirby (Kirby and Spence) and was applied to the lateral portions of the epicranium, that part on each side of the head lying beneath and behind the eye; and has been generally used by writers on insect morphology in this sense. (*b*) The so-called genæ of writers on chætotaxy are portions of the cephalic aspect of the head, that part on each side which is dorsad of the transverse impression, laterad of the arm of the frontal suture and mesad of the eye (Fig. 999, *g*). This region is "the sides of the face" of older descriptions and the parafacialia of some later writers.

The genovertical plates.—The so-called front of writers on chætotaxy (see above) is usually distinctly divided into three parts, a median, the *frontal vitta* or *frontalia*, and two lateral, the *genovertical plates* or *parafrontalia*. (Fig. 999, *g. p.*).

The interfrontalia.—Specialized stripes on the middle of the so-called front, formed from the enlarged ocellar triangle.

The occiput.—The term occiput is applied by writers on the classification of the Diptera to the caudal aspect of the head, this includes the genæ and post-genæ described on page 39.

The ocellar plate or *ocellar triangle.*—A triangle indicated by grooves or depressions on which the ocelli are situated.

The orbits.—That part of the epicranium on each side immediately contiguous to the compound eye. The orbit is sometimes indicated by structural characters, at other times it is indefinite.

The parafacials.—The so-called genæ of writers on chætotaxy.
The parafrontals.—The genovertical plates.
The peristome.—The region around the mouth.

The postgenæ.—See page 39.
The ptilinum.—See page 776.
The transverse impression.—See cheek-groves.

The tormæ.—A sclerite situated between the clypeus and the labrum in the more specialized Diptera. It is composed of two sclerites which belong to the lateral portions of the epipharynx and are internal in the more generalized insects; but they become exposed and united on the middle line in the more specialized Diptera. In such cases the tormæ are sometimes incorrectly termed the clypeus. See Peterson ('16 p. 19).

The vertex.—(a) The term vertex, as defined by Kirby (Kirby and Spence, 1815-1826), is the dorsal portion of the epicranium; or, more specifically, that portion which is next the front and between the compound eyes (Fig. 998, *v. v. v.*) (b) This term is often applied merely to the top of the head.

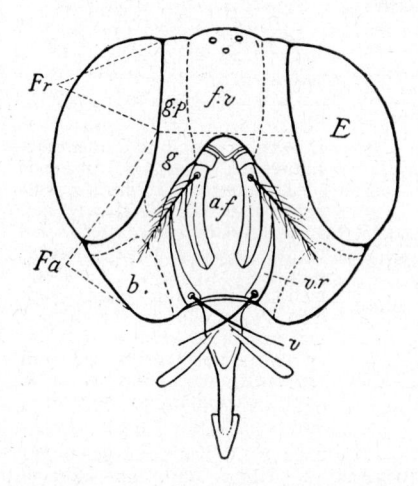

Fig. 999.—Diagram illustrating the terms applied by writers on chætotaxy to the areas of the head: *a. f.* antennal fossa; *b*, bucca; *E*, eye; *Fa*, face; *Fr*, the so-called front; *f.v*, frontal vitta; *g*, the so-called gena; *g.p*, genovertical plate; *v*, vibrissa; *v. r*, vibrissal ridge.

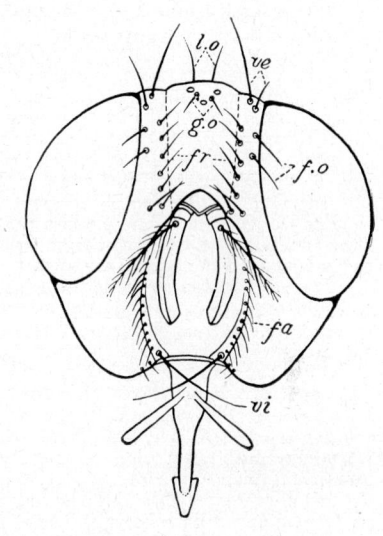

Fig. 1000.—Cephalic bristles: *fa*, facial, *fr*, frontal; *f. o.*, fronto-orbital; *g. o*, greater ocellar; *l. o*, lesser ocellar; *ve*, vertical; *vi*, vibrissæ.

The vertical triangle.—See the ocellar plate.

The vibrissal angles.—Two prominences at the lower ends of the vibrissal ridges upon which are borne the vibrissæ.

The vibrissal ridges.—Two ridges, one on each side, inside the arms of the frontal suture, constituting the lateral boundaries of the antennal fossa, and bearing the vibrissæ (Fig. 999, *v. r*). These are also termed the *facialia* or *facial ridges*.

THE CEPHALIC BRISTLES

The ascending frontal bristles.—See frontal bristles.

The beard.—A clothing of hair borne by the lower part of the so-called occiput and on the buccæ.

The cilia of the posterior orbit.—A row of bristles along the posterior orbit of the eye.

The cruciate bristles.—A pair of bristles on the lower part of the frontal vitta, directed inward and forward.

The facial bristles.—A series of bristles on either side borne by the vibrissal ridge, above the vibrissa (Fig. 1000, *fa.*).

The facio-orbital bristles.—Bristles borne on that portion of the face on each side next to the orbit, the so-called gena.

The frontal bristles.—A row of bristles on each side on the boundary line between the frontal vitta and the genovertical plate. (Fig. 1000, *fr.*) the uppermost of these, from one to four in number, are termed the *ascending frontal bristles;* the lower ones, which are often directed across the frontal vitta, are termed the *transfrontral bristles.*

The fronto-orbital bristles.—A bristle or bristles on the genovertical plate, immediately below the vertical bristles. (Fig. 1000, *fo.*). So named because they are on that part of the so-called front next to the orbit.

The lateral facial bristles.—One or two bristles sometimes present on the sides of the face below, towards the eye.

The lower fronto-orbital bristles.—These are situated on the lower part of the genovertical plates near the eyes and are not quite in line with the fronto-orbitals. They are not of frequent occurrence.

The ocellar bristles.—(*a*) The *greater ocellars or* the *ocellar pair*, a pair of bristles on the ocellar triangle just back of the median ocellus (Fig. 1000, *g. o.*). (*b*) The *lesser ocellar bristles*, from three to twelve pairs of bristles, usually inserted in two parallel lines, sometimes in four, which begin very close to the insertion of the greater ocellar bristles and extend backward a variable distance (Fig. 1000, *l. o.*).

The occipito-central bristles.—A pair of bristles on the upper part of the occiput just below and almost in line with the inner vertical pair.

The occipito-lateral bristles.—A pair of bristles borne, one on each side, a little back of the outer vertical bristles.

The orbital bristles.—See fronto-orbital bristles.

The postorbital bristles.—See cilia of the posterior orbit.

The postvertical bristles.—The hinder pair of the lesser ocellar bristles.

The preocellar bristles.—A pair of small bristles sometimes found below the median ocellus.

The transfrontal bristles.—See frontal bristles.

The vertical bristles.—Two pairs of bristles, an inner and outer pair, inserted more or less behind the upper and inner corners of the eyes (Fig. 1000, *ve.*).

The vibrissæ.—A pair of stout bristles, one on each side of the face, near or a little above the oral margin (Fig. 1000, *vi*). These are the longest or strongest of the bristles borne on the vibrissal ridges.

THE THORACIC SUTURES

The transverse suture.—The suture between the prescutum and the scutum of the mesothorax.

The notopleural or dorsopleural suture.—The suture on each side separating the mesonotum from the pleurum of the mesothorax.

The mesopleural suture.—The suture on each side separating the episternum and the epimerum of the mesothorax.

The sternopleural suture.—The suture on each side separating the mesopleurum and the sternopleurum.

THE PLEURAL DIVISIONS

The propleura.—The pleura of the prothorax (Fig. 1001, *pr.*).

The notopleura.—A sclerite on each side at the end of the transverse suture in the presutural depression. (Fig. 1001, *np*.).

The mesopleura.—The upper part of the episterna (anepisterna) of the mesothorax (Fig. 1001 *mes*.).

The sternopleura.—The lower part of the episterna (katepisterna) of the mesothorax (Fig. 1001, *st*.).

The pteropleura.—The upper part of the epimera (anepimera) of the mesothorax, (Fig. 1001, *pt*.).

The hypopleura.—The lower part of the epimera (katepimera) of the mesothorax (Fig. 1001, *hy*.).

The metapleura.—The pleura of the metathorax.

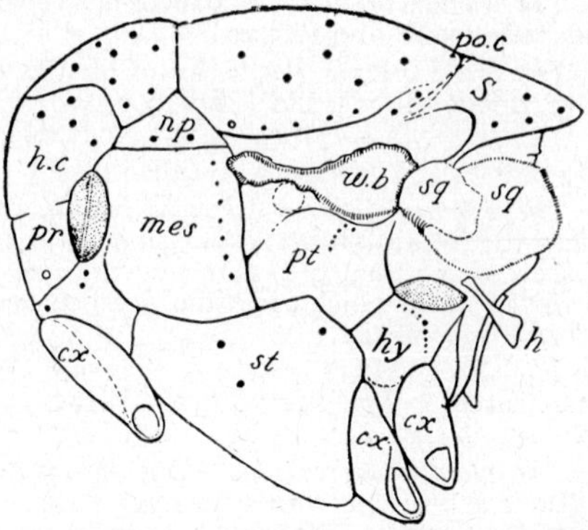

OTHER TERMS FOR PARTS OF THE THORAX

The alar frenum.—A little ligament dividing the supraalar cavity into an anterior and a posterior part.

The humeral callus.—Each of the anterior lateral angles of the prescutum of the mesothorax, usually a more or less rounded tubercle. (Fig. 1001, *h. c*.).

The prealar callus.—A not very prominent projection, situated before the root of the wing, on each side of the mesonotum, just back of the outer end of the transverse suture.

The postalar callus.—A more or less distinct rounded swelling on each side, situated between the root of the wing and the scutellum. (Fig. 1001, *po. c*.).

The presutural depression.—A depression, usually triangular in shape, at the outer end of the transverse suture, near the notopleural suture (Fig. 1001, *np*).

The supraalar groove or cavity.—A groove on the mesothorax immediately above the root of the wing.

The scutellar bridge.—A small ridge on either side of the scutellum connecting it with the scutum, crossing the intervening suture.

THE THORACIC BRISTLES

The acrostichal bristles.—Two rows of bristles, one on each side of the median line of the mesonotum, the two rows nearest to the median line (Fig. 1002, *a*). Those in front of the transverse suture are termed the *anterior acrostichals* or *preacrostichals;* those behind this suture, the *posterior acrostichals* or *postacrostichals.*

The anterior acrostichals.—See the acrostichal bristles.
The discal scutellar bristles.—See the scutellar bristles.
The dorsocentral bristles.—A row of bristles on each side next to and parallel with the acrostichals bristles (Fig. 1002, *dc*). Those before the transverse suture are termed the *anterior*, those behind, the *posterior*, or *postsutural dorsocentrals.*

The humeral bristles.—One or more bristles inserted on the humeral callus. (Fig. 1002, *h m*).

Fig. 1001.—Diagram of the thorax of a fly illustrating the terms applied by writers on chætotaxy to the areas of the thorax. The positions of the more important bristles are indicated by dots: *cx, cx, cx*, coxæ; *h. c*, humeral callus; *h*, halter; *hy*, hypopleura; *mes*, mesopleura; *np*, notopleura or presutural depression; *po. c*, postalar callus; *pr*, propleura; *pt*, pteropleura; *s*, scutellum; *sq, sq*, squamæ or calypteres; *st*, sternopleura; *w. b.* wing-base. (After Riley and Johannsen.)

The hypopleural row.—A row of bristles extending in a more or less vertical direction on the hypopleura, usually directly above the hind coxæ. They are sometimes grouped into a tuft. (Fig. 1001).

The inner dorsocentral bristles.—These are the acrostichal bristles.

The intraalar bristles.—A row of two of three bristles between the supraalar group and the dorsocentral bristles (Fig. 1002, *i. a*).

The intrahumeral bristles.—These are the same as the presutural bristles.

The marginal scutellar bristles.—See the scutellar bristles.

The mesopleural row.—A row of bristles inserted on the mesopleurum, near its dorso-caudal angle, or along its caudal margin. (Fig. 1001).

The metapleural bristles.—A fan like row on the metapleurum conspicuous in some families but not found in the Calypteratæ.

The notopleural bristles.—Usually two bristles, inserted immediately above the dorsopleural suture, between the humeral callus and the root of the wing, on the notopleura. (Fig. 1002, *n p l*).

The postacrostichals.—See the acrostichal bristles.

The postalar bristles.—Bristles on the postalar callus back of the supraalar bristles. (Fig. 1002, *pa*).

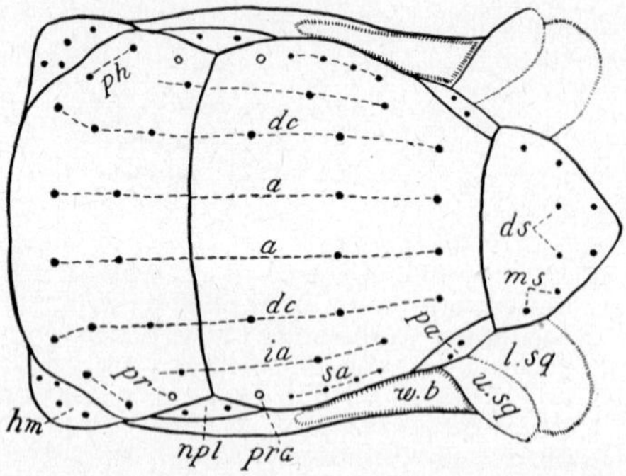

The posthumeral bristles.—One or more bristles situated on the prescutum near the inner margin of the humeral callus. (Fig. 1002, *ph*.)

Fig. 1002.—Thoracic bristles: *a*, acrostichal; *dc*, dorsocentral; *ds*. dorsoscutellar; *hm*, humeral; *ia*, intraalar; *ms*, marginal scutellar; *npl*, notopleural; *pa*, postalar; *ph*, posthumeral; *pr*. presutural; *sa*, supraalar; *l. sq.*, lower squama or calypter; *u. sq.* upper squama or calypter; *w. b*, wing-base. (After Riley and Johannsen.)

The preacrostichals.—See acrostichal bristles.

The prealar bristle.—A bristle found in the Anthomyiidæ inserted just back of the transverse suture in line with the supraalar bristles. (Fig. 1002, *pra*).

The prescutellar row.—A row of bristles in front of the scutellum consisting of the hindermost dorsocentral and the acrostichal bristles.

The presutural bristles.—One or more bristles situated immediately in front of the transverse suture, above the presutural depression. (Fig. 1002, *pr*.).

The propleural bristles.—A bristle or bristles inserted on the lower part of the pleurum of the prothorax, immediately above the front coxa.

The prothoracic bristle.—The same as the propleural bristle.

The pteropleural bristles.—Bristles inserted on the pteropleura.

The scutellar bristles.—(*a*). The *dorso-scutellar bristles.*, usually a single pair of bristles, borne on the dorsal portion of the scutellum, one on each side of the median line, slightly behind its middle. (Fig. 1002, *ds*). (*b*) *The marginal scutellar bristles*, usually a distinct row of large bristles borne on the margin of the scutellum (Fig. 1002, *ms*).

The sternopleural bristles.—One or several bristles on each sternopleurum below the sternopleural suture and near it.

The supraalar bristles.—Usually one to four bristles above the root of the wing, between the notopleural and the postalar bristles. (Fig. 1002, *sa*).

The trichostichal bristles.—The same as the metapleural bristles.

DIPTERA

THE BRISTLES OF THE LEGS

The extensor row.—A row of bristles on the upper surface of femur.

The flexor row.—One or more rows of bristles placed along the lower surface of the femur.

The preapical bristle.—A large bristle found on the extensor side on the distal third of the tibia in some families of the Acalyptratæ; it is quite distinct from the tibial spurs. This term is sometimes used for a bristle on the femur.

The tibial spurs.—One or more bristly spurs placed at the distal end of the tibia.

THE ABDOMINAL BRISTLES

The discal bristles.—Usually one or more pairs of bristles inserted near the middle of the dorsal wall of the abdominal segments before the hind margin.

The lateral bristles.—One or more bristles situated on or near the lateral margins of the abdominal segments, above.

The marginal bristles.—Bristles inserted on the posterior margin of the abdominal segments, above.

Flies undergo a complete metamorphosis. The larvæ are commonly called *maggots*. These are usually cylindrical and are footless. In the more generalized families the larvæ possess a distinct head; but in the more specialized Diptera there is an anomalous retarding of the development of the head; with these the rudiments of the head are invaginated within the body of the larva and the head does not become exposed until the pupal stage is reached. The development of the head in these insects is described in Chapter IV. The pupæ are usually either naked or enclosed in the last or the next to the last larval skin. A few are enclosed in cocoons. When the pupa state is passed within the last larval skin the body of the pupa separates from the larval skin more or less completely; but the larval skin is not broken till the adult fly is ready to emerge. In this case the larval skin, which serves as a cocoon, is termed a *puparium*. In some families the puparium retains the form of the larva; in others the body of the larva shortens, assuming a more or less barrel-shaped form before the change to a pupa takes place.

SYNOPSIS OF THE DIPTERA

Suborder ORTHORRHAPHA. The Straight-seamed Flies. p. 794.
  Series I.—NEMOCERA. The Long-horned Orthorrhapha. p. 795.
    Subseries A.—The TRUE NEMOCERA.
      The Crane-flies. p. 795. Superfamily TIPULOIDEA
        The Primitive Crane-flies. p. 796. Family TANYDERIDÆ
        The Phantom Crane-fly Family. p. 796. Family PTYCHOPTERIDÆ
        The So-called False Crane-flies. p. 797. Family ANISOPIDÆ
        The Typical Crane-flies. p. 798. Family TIPULIDÆ
      The Dixa midges. p. 800. Family DIXIDÆ
      The Moth-like Flies. p. 801. Family PSYCHODIDÆ
      The Midges. p. 802. Family CHIRONOMIDÆ
      The Mosquitoes. p. 804. Family CULICIDÆ
      The Fungus-gnats. p. 810. Family MYCETOPHILIDÆ
      The Gall gnats. p. 813. Family CECIDOMYIIDÆ
    Subseries B.—The ANOMALOUS NEMOCERA.
      The March-flies. p. 820. Family BIBIONIDÆ
      The Scatopsids. p. 821. Family SCATOPSIDÆ

The Black-flies. p. 821. Family SIMULIIDÆ
The Net-winged Midges. p. 824. Family BLEPHAROCERIDÆ
The Solitary-midge. p. 828. Family THAUMALEIDÆ
Series II.—BRACHYCERA. The Short-horned Orthorrhapha. p. 828.
Subseries A.—THE ANOMALOUS BRACHYCERA.
The Horse-flies. p. 829. Family TABANIDÆ
The Soldier-flies. p. 830. Family STRATIOMYIIDÆ
The Xylomyiids. p. 832. Family XYLOMYIIDÆ
The Xylophagids. p. 833. Family XYLOPHAGIDÆ
The Cœnomyiids. p. 834. Family CŒNOMYIIDÆ
Subseries B.—THE TRUE BRACHYCERA.
The Snipe-flies. p. 834. Family RHAGIONIDÆ
The Tangle-veined Flies. p. 836. Family NEMESTRINIDÆ
The Small-headed Flies. p. 837. Family ACROCERIDÆ
The Bee-flies. p. 838. Family BOMBYLIIDÆ
The Stiletto-flies. p. 839. Family THEREVIDÆ
The Window-flies. p. 839. Family SCENOPINIDÆ
The Robber-flies. p. 840. Family ASILIDÆ
The Mydas-flies. p. 842. Family MYDAIDÆ
The Apiocerids. p. 842. Family APIOCERIDÆ
The Long-legged Flies. p. 843. Family DOLICHOPODIDÆ
The Dance-flies. p. 845. Family EMPIDIDÆ
The Spear-winged Flies. p. 846. Family LONCHOPTERIDÆ
Suborder CYCLORRHAPHA. The Circular- seamed Flies.
Series I.—ASCHIZA. Cyclorrhapha without a frontal suture.
The Humpbacked Flies. p. 847. Family PHORIDÆ
The Flat-footed Flies. p. 848. Family PLATYPEZIDÆ
The Big-eyed Flies. p. 849. Family PIPUNCULIDÆ
The Syrphus-flies. p. 850. Family SYRPHIDÆ
Series II.—SCHIZOPHORA. Cyclorrhapha with a frontal suture.
Section I.—MYODARIA. The Muscids.
Subsection I.—ACALYPTRATÆ. The Acalyptrate Muscids.
The Thick-headed Flies. p. 853. Family CONOPIDÆ
The Dung-flies. p. 854. Family CORDYLURIDÆ
The Clusiids. p. 854. Family CLUSIIDÆ
The Helomyzids. p. 854. Family HELOMYZIDÆ
The Borborids. p. 855. Family BORBORIDÆ
The Phycodromids. p. 855. Family PHYCODROMIDÆ
The Sciomyzids. p. 855. Family SCIOMYZIDÆ
The Sapromyzids. p. 856. Family SAPROMYZIDÆ
The Lonchæids. p. 856. Family LONCHÆIDÆ
The Ortalids. p. 856. Family ORTALIDÆ
The Trypetids. p. 858. Family TRYPETIDÆ
The Tanypezids. p. 858. Family TANYPEZIDÆ
The Micropezids. p. 858. Family MICROPEZIDÆ
The Sepsids. p. 858. Family SEPSIDÆ
The Piophilids. p. 858. Family PIOPHILIDÆ
The Psilids. p. 859. Family PSILIDÆ
The Diopsids. p. 859. Family DIOPSIDÆ
The Canaceids. p. 859. Family CANACEIDÆ
The Ephydrids. p. 859. Family EPHYDRIDÆ
The Chloropids. p. 860. Family CHLOROPIDÆ
The Asteiids. p. 860. Family ASTEIIDÆ
The Drosophilids. p. 860. Family DROSOPHILIDÆ
The Geomyzids. p. 861. Family GEOMYZIDÆ
The Agromyzids. p 861. Family AGROMYZIDÆ
The Milichiids. p. 862. Family MILICHIIDÆ
The Ochthiphilids. p. 862. Family OCHTHIPHILIDÆ
Subsection II. CALYPTRATÆ. The Calyptrate Muscids.
Superfamily ANTHOMYIOIDEA
The Anthomyiids. p. 863. Family ANTHOMYIIDÆ

Superfamily MUSCOIDEA.
　The Bot-flies of Horses. p. 864. Family GASTROPHILIDÆ
　The Œstrids. p. 866. Family ŒSTRIDÆ
　The Phasiids. p. 868. Family PHASIIDÆ
　The Megaprosopids. p. 869. Family MEGAPROSOPIDÆ
　The Blow-fly Family p. 869. Family CALLIPHORIDÆ
　The Sarcophagids. p. 870. Family SARCOPHAGIDÆ
　The Tachina-flies. p. 871. FAMILY TACHINIDÆ
　The Typical Muscids. p. 872. Family MUSCIDÆ
Section II. PUPIPARA.
　The Louse-flies. p. 874. Family HIPPOBOSCIDÆ
　The Bat-ticks in part. p. 875. Family STREBLIDÆ
　The Bat-ticks in part. p. 875. Family NYCTERIBIIDÆ
　The Bee-lice. p. 876. Family BRAULIDÆ

TABLES FOR DETERMINING THE FAMILIES OF THE DIPTERA

Table A.—DIPTERA WITH WELL DEVELOPED WINGS.

A. Flies in which the abdomen is distinctly segmented, and the two legs of each thoracic segment are not widely separated. Habits various, but the adults do not live parasitically upon either birds or mammals.
　B. Antennæ consisting of more than three segments. (Note that a style or arista borne by the third segment is not counted as a segment.)
　　C. Antennæ consisting of more than five distinct segments, the segments beyond the second not consolidated; cell 1st A of the wings but slightly narrowed at the margin of the wing, if at all; palpi usually elongate, and composed of from three to five segments.
　　　D. Small moth-like flies, with the body and wings densely clothed with hairs and scales. Wings with from nine to eleven longitudinal veins but with no cross-veins except sometimes near the base of the wings. (Fig. 1014). p. 801...................................PSYCHODIDÆ
　　　DD. Flies that do not resemble moths in appearance.
　　　　E. Mesonotum with a distinct V-shaped transverse suture.
　　　　　F. The radial sector four-branched. p. 796..........TANYDERIDÆ
　　　　　FF. The radial sector with less than four branches.
　　　　　　G. With only one anal vein. p. 796............PTYCHOPTERIDÆ
　　　　　　GG. With two anal veins. p. 798..................TIPULIDÆ
　　　　EE. Mesonotum without a distinct V-shaped suture.
　　　　　F. Media three-branched (*Anisopus* and *Trichocera*). p. 797..
　　　　　..ANISOPIDÆ
　　　　　FF. Media simple, two-branched, or wanting; cell M_2 is not divided by a cross-vein.
　　　　　　G. Wings with a network of fine lines near the outer and inner margins in addition to the veins (Fig. 1048). p. 824.......
　　　　　　..BLEPHAROCERIDÆ
　　　　　　GG. Wings without a network of fine lines.
　　　　　　　H. The margin of the wings and each of the wing-veins fringed with scales (Fig. 1019). p. 804...............CULICIDÆ
　　　　　　　HH. The wing-veins with or without a fringe of hairs, but without a fringe of flat scales.
　　　　　　　　I. Anal veins entirely wanting; vein M wanting or at most represented by a single unbranched fold. p. 813.....
　　　　　　　　..CECIDOMYIIDÆ
　　　　　　　　II. Anal veins usually present or represented by folds; vein M present or at least represented by a fold which is usually branched.
　　　　　　　　　J. Ocelli present.
　　　　　　　　　　K. Antennæ shorter than the **tho**rax; coxæ not unusually long.

    L. Cross-vein m-cu present. p. 820...Bibionidæ
    LL. Cross-vein m-cu wanting. p. 821.Scatopsidæ.
   KK. Antennæ usually longer than the thorax; legs slender and usually with greatly elongate coxæ.
    L. Vein m-cu present, the first branch of the radial sector, vein R_{2+3}, arising slightly proximad of vein m-cu. (*Mycetobia*) p. 797.....Anisopidæ
    LL. Vein m-cu present or absent, when present the forking of the radial sector is distad of it.
     M. Eyes rounded or oval. p. 810.Mycetophilidæ
     MM. Each eye with a narrow expansion, the two expansions extending behind the antennæ and in front of the ocelli, and meeting on the middle line of the head or nearly so.
      N. With tibial spurs; larvæ with well-developed head (Sciarinæ) p. 812.....Mycetophilidæ
      NN. Without tibial spurs, larvæ with a poorly developed head (Lestremiinæ). p. 816......
      ....Cecidomyiidæ
 JJ. Ocelli absent.
  K. Antennæ short, not clothed with long hairs, and with most of the segments wider than long; wings very broad. p. 821.........................Simuliidæ
  KK. Antennæ either bushy, being densely clothed with long hairs, or slender with narrow segments; wings narrow or moderately broad.
   L. Wing-veins well-developed on all parts of the wing
    M. Vein R_1 ending at or near the end of the second third of the costal margin. p. 828. Thaumaleidæ
    MM. Vein R_1 ending on the outer margin of the wing. p. 800..........................Dixidæ
   LL. Wing veins much stouter near the costal margin of the wing than elsewhere. p. 802.Chironomidæ

CC. Antennæ either consisting of four or five distinct segments or consisting of five or more segments, with those beyond the second more or less closely consolidated so as to appear as a single segment consisting of several sub-segments. (Figs. 985, 986); cell 1st A closed by the coalescence of the tips of veins Cu_2 and 2nd A, or greatly narrowed at the margin of the wing; palpi rarely elongate, and composed of from one to three segments.
 D. Antennæ consisting of four or five distinct segments, empodia wanting or bristle-like.
  E. The first branch of media terminating at or before the apex of the wing. p. 842.......................................Mydaidæ
  EE. The first branch of media terminating on the outer border of the wing.
   F. The vertex of the head sunken, the eyes bulging and never contiguous. p. 840...................................Asilidæ.
   FF. The vertex of the head plane or convex, the eyes not bulging; eyes of males often contiguous. p. 838...........Bombyliidæ.
 DD. Antennæ consisting of five or more segments but with those beyond the second more or less closely consolidated; empodia resembling pulvilli in form.
  E. The alulets large. p. 829.......................Tabanidæ
  EE. The alulets small or vestigial.
   F. The branches of radius crowded together near the costal margin; tibiæ without spurs. p. 830................Stratiomyiidæ
   FF. The branches of radius not crowded together near the costal margin; at least some of the tibiæ with spurs.
    G. Cell M_3 closed. p. 832....... ....Xylomyiidæ
    GG. Cell M_3 open.

> H. Scutellum without spinous protuberances. p. 833.....
> ....Xylophagidæ
> HH. Scutellum with two spinous protuberances. p. 834...
> ....Cœnomyiidæ

BB. Antennæ consisting of not more than three segments; the third segment either with or without a style or an arista, but not divided into subsegments.

> C. Antennæ consisting apparently of a single globular segment bearing a long arista; wings with some stout veins near the costal margin and other weaker veins extending across the wing unconnected by cross-veins (Fig. 1099). p. 847..............   ....Phoridæ
> CC. Flies that do not present the type of wing-venation represented by Figure 1099. p 847.
>> D. Cells M and 1st M_2 not separated. (See Figures 1094 and 1095, p. 844 for examples of this type of wing-venation).
>>> E. Vein R with a knot-shaped swelling at the point of separation of veins $R_2 + {}_3$ and $R_4 + {}_5$; the cross-vein r-m at or near this swelling when present; frontal suture absent. p. 843.......Dolichopodidæ
>>> EE. Vein R with or without a swelling at the point of separation of veins $R_2 + {}_3$ and $R_4 + {}_5$; the cross-vein r-m more remote from the base of the wing; the frontal suture present. (Muscoidea). Pass to Table B.
>> DD. Cells M (or 2d M) and 1st M_2 separate.
>>> E. Radial sector three-branched.
>>>> F. Venation intricate due to an unusual anastomosing of the veins (Fig. 1077) p. 836....................................Nemestrinidæ
>>>> FF. Venation not of the type represented by Figure 1077.
>>>>> G. Vertex of the head distinctly hollowed out between the eyes; eyes never contiguous. p. 840...................Asilidæ
>>>>> GG. Vertex of the head not hollowed out between the eyes; eyes often contiguous in males.
>>>>>> H. Alulets very large. p. 837..............Acroceridæ
>>>>>> HH. Alulets small or rudimentary.
>>>>>>> I. Cell M_3 present.
>>>>>>>> J. Vein R_5 ending before the apex of the wing. p. 842..
>>>>>>>> ....Apioceridæ
>>>>>>>> JJ. Vein R_5 not ending before the apex of the wing.
>>>>>>>>> K. Empodia pulvilliform. p. 834.....Rhagionidæ
>>>>>>>>> KK. Empodia wanting. p. 839......Therevidæ
>>>>>>> II. Cell M_3 obliterated by the coalescence of veins M_3 and Cu_1.
>>>>>>>> J. Third segment of the antennæ without a style or an arista; vein M_1 ending at or before the apex of the wing. p. 839..................................Scenopinidæ
>>>>>>>> JJ. Third segment of the antennæ usually with a style or an arista; vein M_1 ending beyond the apex of the wing.
>>>>>>>>> K. Vein Cu_2 extending free to the margin of the wing or coalesced with vein 2d A for a short distance at the margin of the wing. p. 838...........Bombyliidæ
>>>>>>>>> KK. Vein Cu_2 joining vein 2d A far from the margin of the wing, often extending towards the base of the wing p. 845................................Empididæ
>>> EE. Radial sector with not more than two branches.
>>>> F. Wings lanceolate and with no cross-veins except at the base. (Fig. 1097) p. 846......................Lonchopteridæ
>>>> FF. Wings not of the type represented by Figure 1097.
>>>>> G. Empodia pulvilliform. p. 837..............Acroceridæ
>>>>> GG. Empodia not pulvilliform.
>>>>>> H. Vein Cu_2 not coalesced with vein 2d A to such an extent as to cause the free part to appear like a cross-vein.
>>>>>>> I. Antennæ with a terminal style or arista.
>>>>>>>> J. Antennæ with a terminal arista. p. 848. Platypezidæ
>>>>>>>> JJ. Antennæ with a terminal style.

    K. Front with grooves or a depression beneath the
     antennæ. p. 853........................Conopidæ
    KK. Front convex beneath the antennæ. p. 850....
    ....Syrphidæ
  II. Antennæ with a *dorsal* arista.
   J. Head extremely large, and with nearly the entire surface
    occupied by the eyes. (Fig. 1103). p. 849. Pipunculidæ
   JJ. Head not of the type represented by Figure 1103.
    K. Wings with a vein-like thickening, the spurious vein,
     between veins R and M. p. 850.......Syrphidæ
    KK. Wings without a spurious vein.
     L. Front with grooves or a depression beneath the
      antennæ. p. 853......................Conopidæ
     LL. Front convex beneath the antennæ. p. 850..
     ....Syrphidæ
  HH. Vein Cu_2 appearing as a cross-vein or curved back towards
   the base of the wing (Figs. 1096 and 1115).
   I. Antennæ with a terminal style or arista. p. 845. Empididæ
   II. Antennæ with a dorsal arista.
    J. Proboscis vestigial; mouth opening small; palpi wanting
     (Bot-flies) Pass to Table B.
    JJ. Proboscis not vestigial; palpi present in most cases.
     K. Frontal suture present (Myodaria). Pass to Table B.
     KK. Frontal suture absent. p. 845.......Empididæ
AA. Flies in which the abdomen is indistinctly segmented, and the two legs of
 each thoracic segment are widely separated by the broad sternum. The *adults*
 live parasitically upon birds, or mammals.
 B. Head sunk in an emargination of the thorax; eyes round or oval; palpi
  forming a sheath for the proboscis, not projecting in front of the head. p. 874.
  ....Hippoboscidæ
 BB. Head with a fleshy movable neck; eyes wanting or vestigial; palpi projecting leaf-like in front of the head. p. 875.................Streblidæ

TABLE B.—THE FAMILIES OF THE MYODARIA

A. The alulæ or calypteres small or rudimentary; the subcostal vein often indistinct or vestigial, but sometimes well-preserved; vein R_1 shortened and
 often very short; thorax without a complete transverse suture; postalar callus
 usually absent. (Subsection 1—Acalyptratæ).
 B. Subcosta distinctly separated from vein R_1 and ending in the costa noticeably before R_1, the latter ending near or beyond the middle of the wing in
  most cases.
  C. Oral vibrissæ present and distinctly differentiated from the hairs of the
   peristome.
   D. A distinct costal break or scar proximad of the tip of R_1 near the
    apex of Sc.
    E. The postvertical bristles are divergent or parallel or wanting.
     F. The frontals are convergent (lacking in *Hydromyza*) and stand
      nearer the median line than the fronto-orbitals. Abdominal
      spiracles in most cases in the chitin. p. 854........Cordyluridæ
     FF. The frontals absent, or if present and convergent stand in line or
      laterad of the line of the fronto-orbitals; transverse suture in-
      terrupted in the middle; anal vein does not reach the wing-margin;
      cross veins in most cases approximated; abdominal spiracles in the
      conjunctivæ. p. 854.................................Clusiidæ
    EE. Postvertical bristles convergent, costa of wing in nearly all cases
     with a row of spines projecting beyond the ciliation. p. 854......
     ....Helomyzidæ
   DD. Costa without a sign of a break; palpi vestigial in most species.
    E. Palpi vestigial; "front" never bristly near the antennæ; anal vein
     not produced to the wing margin. p. 858................Sepsidæ

EE. Palpi well-developed; "front" and face bristly; anal vein produced to the wing-margin. Seashore flies. p. 855... PHYCODROMIDÆ
CC. Oral vibrissæ not differentiated from the peristomal hairs.
D. Legs very long and stilt-like; tibiæ usually without preapical bristle; cell R_5 constricted at the wing-margin.
E. Proboscis greatly elongate and folding near its middle; arista terminal; ovipositor very long (*Stylogaster*) p. 853... CONOPIDÆ*
EE. Proboscis short; arista dorsal; ovipositor not lengthened.
F. Buccæ and posterior orbits narrow. p. 858.... TANYPEZIDÆ
FF. Buccæ broad. p. 858...................MICROPEZIDÆ
DD. Legs not long and stilt-like.
E. Preapical tibial bristles present; "ovipositor" membranous and retractile.
F. Postvertical bristles well-developed and converging.
G. Anal vein produced to the wing margin; last tarsal segment enlarged and flat. p. 855................... PHYCODROMIDÆ
GG. Anal vein not produced to the wing margin; last tarsal segment normal. p. 856.................... SAPROMYZIDÆ
FF. Postvertical bristles parallel or diverging, rarely lacking. p. 855. ....SCIOMYZIDÆ
EE. Preapical tibial bristles absent, if present in exceptional cases, the ovipositor horny and not wholly retractible, or the cell 1st A of the wing drawn out into a lobe, or vein R_1 bristly above.
F. Ovipositor membranous, retractile; vein R_1 bare above; cell 1st A without acute process or lobe.
G. Palpi well-developed; postvertical bristles converging or wanting. p. 862.......................... OCHTHIPHILIDÆ
GG. Palpi vestigial; postvertical bristles divergent; the "front" not bristly except at the vertex. p. 858............. SEPSIDÆ
FF. Ovipositor horny, not wholly retractible; postvertical bristles not converging; palpi present.
G. The costal break or scar is located just proximad of the tip of vein Sc; cell M and the cell 1st A small, the latter with rounded apex; only one fronto-orbital bristle. p. 856..... LONCHÆIDÆ
GG. Costa either unbroken or if broken there is also a trace of a break basally, indicated by a dark or light scar or constriction; cell M and cell 1st A large, the latter in many cases with a sharp angle or prolonged into a lobe; *or* the costa with a strongly marked stigma due to an abrupt double curve in vein R_1 near its tip and the second tergite of the abdomen with at least one long bristle on each side. p. 856...................... ORTALIDÆ
BB. Subcosta absent, or vestigial, or running very close to vein R_1 and ending with it in the costa before the middle of the wing, or evanescent at the tip.
C. Head produced on each side into a lateral process bearing the eye. p. 859. ....DIOPSIDÆ
CC. The eyes not stalked.
D. First segment of hind tarsi swollen and in most cases shorter than the second segment; oral vibrissæ present. p. 855........... BORBORIDÆ
DD. First segment of hind tarsi normal.
E. Subcostal vein evanescent at the tip, where it turns sharply forward at some distance before the tip of vein R_1; wings nearly always pictured; cell 1st A angular or drawn out into an acute lobe; no preapical tibial bristle. p. 856.......................... TRYPETIDÆ
EE. The subcosta runs very close to R_1 or is fused with it either wholly or in part but only in a few cases suddenly interrupted at the tip, in which case the wing with a fold extending across from the costal fracture to the tip of cell M.
F. Cell 1st A wanting.

---

*See Table A for distinctive characters of those Conopidæ in which the free part of vein Cu_2 does not appear like a cross-vine.

G. Subcosta vestigial or only basally indicated as a fold; costa fractured but once; cell M and anal vein wanting; ocellar triangle large and conspicuous; head bristles but feebly developed. p. 860..CHLOROPIDÆ

GG. Subcosta developed basally at least; ocellar triangle in most cases not conspicuous; head bristles well-developed.

H. Costa twice fractured, basally and near the tip of R_1; arista never feathered below. p. 859................EPHYDRIDÆ

HH. Basal fracture of the costa indistinct; cell R_5 very long, the bounding veins converging. p. 860.........ASTEIIDÆ

FF. Cell 1st A present though in some cases quite small.

G. Basal cells M and 1st A large; wings with a fold extending across from the costal fracture to the tip of cell M; frontal triangle conspicuous, p. 859......................PSILIDÆ

GG. Basal cells M and 1st A small; wings without a fold.

H. Arista plumose, in rare cases pectinate; wing with two costal fractures; vibrissæ present. p. 860..........DROSOPHILIDÆ

HH. Arista bare or pubescent; if in a few cases plumose then costa of wing with but one fracture which is situated distinctly before the tip of R_1.

I. Tormæ large and distinctly projecting; vein Sc distinctly isolated at its extremity; ocellar triangle large nearly attaining the base of the antennæ; costal fracture close to the tip of R_1. p. 859.........................CANACEIDÆ

II. Tormæ small and not projecting, or differing in other characters.

J. Anterior part of the "front" not bristly; postvertical bristles not converging; cells M and 1st M_2 not confluent; Sc distinct to the tip; arista bare. p. 858...PIOPHILIDÆ

JJ. Not such flies.

K. Costa broken twice; proboscis in most cases geniculate; postvertical bristles converging, rarely parallel or wanting; anal vein in most cases vestigial or wanting. p. 862..............................MILICHIIDÆ

KK. Costa broken but once or proboscis not geniculate.

L. Postvertical bristles divergent, in exceptional cases wanting; basal segment of the arista minute, shorter than broad; the so-called genæ narrower than the buccæ, except in *Phytomyza* in which cell 1st M_2 is open distally. p. 861.................AGROMYZIDÆ

LL. Postvertical bristles converging, or if wanting Cell R_5 long and narrowed in the margin of the wing; or cells M and 1st M_2 confluent, or proboscis geniculate, or arista plumose; basal segment of arista longer than wide; the so-called genæ as broad as or broader than the buccæ. p. 861....GEOMYZIDÆ

AA. At least the lower lobe of the alulæ or calypteres well-developed; the subcostal vein distinct in its whole course; vein R_1 never very short; thorax with a complete transverse suture; and the postalar callus present. (Subsection II.—Calyptratæ).

B. Proboscis usually much reduced or vestigial, not functional; mouth-opening small. (Bot-flies).

C. Costa ends at or slightly beyond the tip of vein $R_4 + _5$; vein $M_1 + _2$ extends in a nearly straight line toward the outer margin of the wing p. 864...GASTROPHILIDÆ

CC. Costa extends to the tip of vein $M_1 + _2$; vein $M_1 + _2$ with a bend so that cell R_5 is much narrowed or closed at the margin of the wing. p. 866. ....ŒSTRIDÆ

BB. Mouth-opening normal; mouth-parts functional.

C. Hypopleural bristles absent. Cell R$_5$ very slightly or not at all narrowed at the margin of the wing. (Anthomyioidea). p. 863.... ANTHOMYIIDÆ
CC. Either the hypopleural or the pteropleural bristles or both present. Cell R$_5$ narrowed or closed. (Muscoidea).
  D. Both hypopleural and pteropleural bristles present.
    E. Clypeus more or less produced below the vibrissal angles. Like the bridge of a nose. Abdomen not armed with stout bristles. The conjunctivæ of the ventral sclerites of the abdomen present. p. 868. ....PHASIIDÆ
    EE. Clypeus flattened, at most slightly produced. Abdomen bearing some stout bristles. The conjunctivæ of the ventral sclerites of the abdomen not visible.
      F. Clypeus receding and short; the cheeks very broad; vibrissæ located near the middle of the face; antennæ short. p. 869.... ....MEGAPROSOPIDÆ
      FF. Clypeus long and never conspicuously receding; the oral margin more or less prominent; vibrissal angles near the oral margin; antennæ usually long.
        G. Second ventral sclerite of the abdomen lying with its edges either upon or in contact with the ventral edges of the corresponding dorsal sclerite.
          H. Hindermost posthumeral bristle almost always lower (more ventrad) in position than the presutural bristle; body color very frequently metallic green or blue, or yellow; arista plumose. p. 869......................................CALLIPHORIDÆ
          HH. Hindermost posthumeral bristle on a level with or higher (more dorsad) than the presutural bristle; arista bare, pubescent, or plumose only on the basal two-thirds; body coloring gray or silvery, tessellated or changeable pollinose.
            I. Fifth ventral abdominal sclerite of the male either wanting or with the caudal margin straight, presutural intraalar bristle rarely present. p. 870...........SARCOPHAGIDÆ
            II. Fifth ventral abdominal sclerite of the male cleft to beyond the middle. p. 871........................TACHINIDÆ
        GG. Second ventral abdominal sclerite, as well as the others, more or less covered, sometimes wholly, by the edges of the dorsal sclerites. p. 871...............................TACHINIDÆ
DD. Either the hypopleural or the pteropleural bristles present; basal bristles of the abdomen reduced; arista plumose to the tip. p. 872.... ....MUSCIDÆ

TABLE C.—DIPTERA IN WHICH THE WINGS ARE WANTING OR VESTIGIAL

A. Mesonotum with a complete V-shaped transverse suture (*Chionea*) p. 798. ....Family TIPULIDÆ
AA. Mesonotum without a V-shaped transverse suture.
  B. Flies in which the abdomen is distinctly segmented and the two legs of each thoracic segment are not widely separated. The adults do not live parasitically upon either birds, mammals or the honey-bee.
    C. Nematocerous flies; antennæ more or less thread-like and consisting of six or more segments.
      D. Wings short, strap-like, thickened, and without distinct venation (*Eretmoptera*). p. 802...............................CHIRONOMIDÆ
      DD. Wings and halteres wholly wanting (*Pnyxa*, female) p. 810...... ....MYCETOPHILIDÆ
    CC. Brachycerous flies.
      D. Antennæ consisting apparently of a single segment, which bears a long, three-jointed arista. p. 847........................PHORIDÆ

DD. Antennæ three-jointed, third joint with an arista.
    E. Hind metatarsi shorter than the second segment and more or less thickened. p. 855.................................Borboridæ
    EE. Hind metatarsi longer than the second segment and slender. p. 859..Ephydridæ
BB. Flies in which the abdomen is indistinctly segmented (except *Braula*), and the two legs of each thoracic segment are widely separated by the broad sternum. The adults are parasites.
  C. Flies parasitic upon birds or mammals.
    D. Head folded back on the dorsum of the thorax. p. 875...Nycteribiidæ
    DD. Head not folded back on the thorax.
      EE. Head sunk in an emargination of the thorax; eyes round or oval; palpi forming a sheath for the proboscis, not projecting in front of the head. p. 874............................Hippoboscidæ
      EE. Head with a fleshy movable neck; eyes wanting or vestigial; palpi projecting leaf-like in front of the head. p. 875. Streblidæ
  CC. Flies parasitic upon the honey-bee. p. 876...........Braulidæ

MEIGEN'S FIRST PAPER ON DIPTERA

In the year 1800, J. A. Meigen published a paper on the classification of the Diptera, in which many generic names were proposed. This was followed by a second paper published in 1803, in which nearly all of the generic names used in his first paper were discarded and new names proposed. The first paper was evidently not widely distributed for it was practically unknown for more than one hundred years. Attention was called to it by Mr. Fr. Hendel in 1908, and since then an effort has been made to substitute the generic names proposed by Meigen in 1800 for those used by him in 1803. If this were done, not only would these generic names be changed but the well-known names of many families based on these generic names would need to be changed also. Fortunately this revolution in nomenclature is not necessary, even according to the law of priority; for the names published by Meigen in 1800 were not adequately defined and *no type species were indicated.*

Suborder ORTHORRHAPHA.*

The Straight-seamed Flies

This suborder includes those flies in which the pupa escapes from the larval skin through a T-shaped opening, which is formed by a lengthwise split on the back near the head and a crosswise split at the front end of this (Fig. 1003), or rarely through a crosswise split between the seventh and eighth abdominal segments. The adults do not have a frontal lunule.

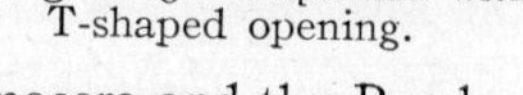

Fig. 1003.—Puparium with T-shaped opening.

The families included in this suborder are commonly grouped in two series: the Nemocera and the Brachycera.

---

*Orthŏrrhapha; *orthos* (ὀρός), straight; *rhaphe* (ῥαφή), a seam.

Series I.—NEMOCERA*

The Long-horned Orthorrhapha

This series of families is termed the Nemocera from the fact that in the more typical forms the antennæ are elongate and slender; but in some families placed at the end of the series, the Anomalous Nemocera, the antennæ are shorter and less thread-like than in the more typical forms. The antennæ are composed of from six to thirty-nine segments, usually from eight to sixteen. The palpi are pendulous and consist of from one to five segments, usually of four. Except in a few genera, cell 1st A is not narrowed towards the margin of the wing. In those cases where the radial sector is three-branched, it is veins R_4 and R_5 that have coalesced; in the Brachycera veins R_2 and R_3 are the first to coalesce.

Subseries A.—The True Nemocera

In this subseries the antennæ are usually long and frequently bear whorls of long hairs, especially in the males. The legs are long and slender, and the abdomen is usually long and slender.

Superfamily TIPULOIDEA

The Crane-flies

The crane-flies are mosquito-like in form; but they are usually very much larger than mosquitoes. The body is long and slender, the wings narrow, and the legs very long (Fig. 1004). This family includes the larger members of that series of families in which the antennæ are thread-like, the Nemocera; but it also includes some species that are not larger than certain mosquitoes.

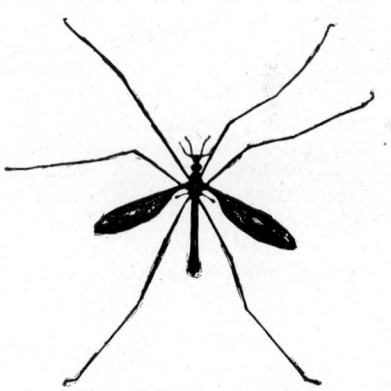

Fig. 1004.—A crane-fly.

Most crane-flies differ from all other Nemocera in that the transverse suture of the mesonotum is V-shaped; but one small family the Anisopidæ, lack the V-shaped suture.

This superfamily includes the four following families; these can be separated by the characters indicated in the table of families page 787.

*Nemŏcera: *nema* (νῆμα), thread; *ceras* (κέρας), horn.

Family TANYDERIDÆ

The Primitive Crane-flies

This family is of especial interest as it includes the most generalized of living crane-flies. It is a small family, only ten species representing three genera being known. Of these a single genus, *Protoplasa*, represented by three species, has been found in North America. *Protŏplasa vĭpio* and *Protŏplasa vanduzēei* are found in the West and *Protŏplasa fĭtchii*, in the East.

The life-history of no member of this family is known. Alexander ('20) described what is probably the larva of *Protoplasa fitchii*. It was found in a much decayed maple log in Fairfax County, Virginia.

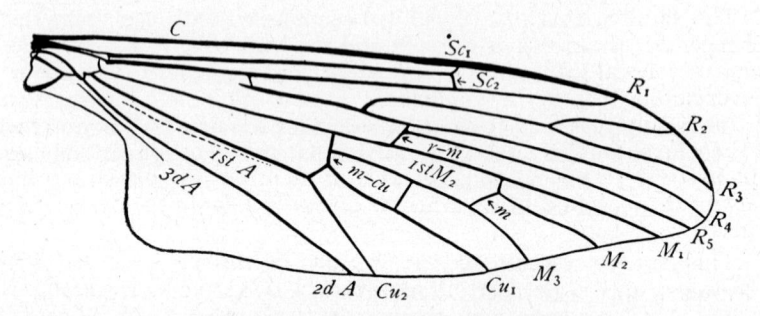

Fig. 1005.—Wing of *Protoplasa fitchii*.

The venation of a wing of *Protoplasa fitchii* is represented by figure 1005. The generalized condition of this wing is shown by the following features; both branches of vein Sc are preserved; the forking of the other branched veins is nearer the base of the wing than in the typical crane-flies; and all of the branches of vein R are distinct.

Family PTYCHOPTERIDÆ

The Phantom Crane-fly Family

This is a small family of which only six or seven species have been found in our fauna; and of these only three are found in the East. These flies differ from the typical crane-flies in having only one anal vein preserved, and the transverse suture of the mesonotum is rather poorly defined.

The larvæ are found in decaying vegetable matter rich in organic mud, usually in swamps, swales, or wet meadows, but sometimes in shaded woods. They feed on decaying vegetable matter, diatoms, and the organic mud in which they live.

The phantom crane-fly, *Bittacomŏrpha clăvipes*.—This remarkable insect is the member of this family that is most likely to attract attention. Its long legs are banded with black and white and the metatarsi are conspicuously enlarged and swollen. In its progress through the air the legs are held outspread like the spokes of a wheel with the metatarsi hanging vertically. It uses its wings but little in flight but is borne along by currents in the air. The black and white banding of its legs makes it a very conspicuous object as it drifts phantom-like through the air.

Family ANISOPIDÆ*

The So-called False Crane-Flies

The family Anisopidæ has not been classed with the crane-flies till recently; the presence of ocilli and the lack of a V-shaped transverse mesonotal suture in this family having been regarded as characters excluding it from the Tipuloidea. On the other hand the members of this family resemble crane-flies in certain features of the venation of the wings; for this reason they have been known as false crane-flies. But a study of the larvæ and pupæ of members of this family has shown that it should be regarded as one of the families of the Tipuloidea.

This family is represented in our fauna by three genera, *Anisopus Trichocera*, and *Mycetobia;* of these the last two have been commonly classed in other families; but the immature stages of the three genera are very similar.

Anisōpus.—The adults are mosquito-like insects with spotted wings, which often enter houses, where they are found on windows. I

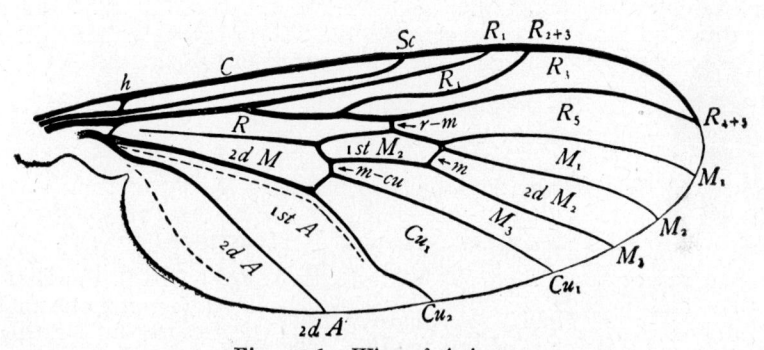

Fig. 1006.—Wing of *Anisopus*.

have also observed them in considerable numbers just at nightfall, feeding on sugar which had been placed on trees to attract moths.

*This family has been known as the Rhyphidæ.

They feed on over-ripe fruit, the exuding sap of trees, and upon the nectar of flowers. Figure 1006 represents the venation of the wings and Figure 1007 the form of the antennæ. Only four species of this genus are recorded from the United States.

The larvæ are found in decaying vegetable matter, in manure, in sewage, and in similar material.

Trichŏcera.—The members of this genus often attract attention by appearing in swarms in the autumn and early spring, and sometimes on warm, sunny days in winter. The swarms vary greatly in size, sometimes one includes thousands of individuals. They are usually from five to twenty-five feet above ground; and all members of a swarm face the wind.

Fig. 1007.

These flies are often found during the winter months in cellars, resting on the windows. Nine species have been described from our fauna. In this genus the radial sector is three-branched and there are two distinct anal veins.

The larvæ are found in decaying vegetable matter, beneath dead or decaying leaves, and in fungi. They have also been found in stored roots and tubers, especially potatoes.

Mycetōbia.—A single species of these genus, *M. divergens* is found in North America. This is a small fly measuring from three to four millimeters in length and resembling superficially a fungus-gnat more than a crane-fly. For this reason it has been commonly classed in the Mycetophilidæ. In this genus cell 1st M_2 is lacking, the radial sector and media are each two-branched, and there is only one distinct anal vein.

The larvæ are common in wounds on trees from which sap is exuding, and in decaying wood.

Family TIPULIDÆ

The Typical Crane-Flies

To this family belong the far greater number of the crane-flies, the other three families of the Tipuloidea including but few species. The typical crane-flies differ from the Anisopidæ in having a V-shaped transverse mesonotal suture (Fig. 1008), from the Tanyderidæ in that the radial sector has less than four branches, and from the Ptychopteridæ in having two distinct anal veins.

Fig. 1008.—Thorax of a crane-fly showing the V-shaped suture.

Figure 1009 represents the venation of a wing of a member of this family. The most striking feature of this venation is the fact that the forking of the branched veins is near the distal end of the wing. This gives the wing a very distinctive appearance.

Crane-flies are seen most often in damp localities, especially where there is a rank growth of vegetation; but sometimes they occur in great numbers flying over meadows and pastures. In most cases their power of flight does not seem to be well developed for they fly slowly, and only a short distance at a time. Some species, however, sustain themselves in the air for long periods. This is especially true of some of the smaller species; which often collect in swarms at twi-

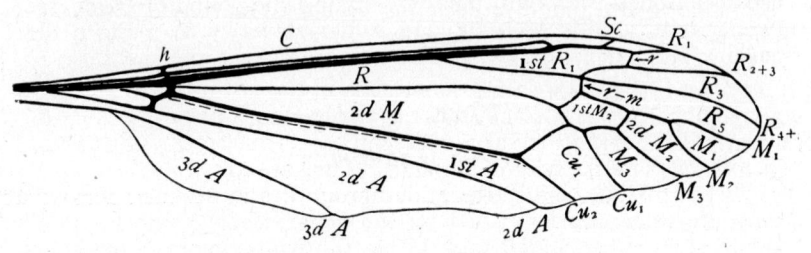

Fig. 1009.—Wing of *Tipula abdominalus*.

light, forming a small cloud, and dancing up and down like some of the midges. Their ability to walk is also poor, for they use their long legs awkwardly, as if they were in the way. Little is known regarding the feeding habits of the adult crane-flies; but some species have been observed to feed on the nectar of flowers. Many species are attracted to lights.

The larvæ of crane-flies vary greatly in habits both as to the situations in which they live and as to the nature of their food. Some are aquatic; *Antocha* lives in silken cases on rocks in swiftly flowing streams; and members of several other genera live on submerged plants. Some live in or beneath damp cushions of moss. Many live in mud or sand along the margins of streams, in swamps, or in shaded woods, while others are strictly terrestrial, burrowing in the soil of meadows and pastures.

The larvæ of most species are scavengers feeding on decaying vegetable matter, but some feed on living vegetable tissue, and still others are carnivorous. For a detailed account of the life-histories and the structural characteristics of the early stages of the different groups of crane-flies see Alexander ('20).

The Tipulidæ is a large family; nearly 3000 species are known and about 500 species have been described from North America alone. Among those that are of especial interest are the following:

The snow-flies, *Chiōnea*.—To the genus *Chiōnea* belong several species of crane-flies in which the wings are vestigial, being reduced to mere knobs, much smaller than the halteres. These flies are most often seen in winter crawling about on the snow; but they are occasionally found in the spring and fall in leaf-mold.

The meadow-maggots or leather-jackets.—The larvæ of some species of crane-flies, most of which belong to the genus *Tipula*, often do considerable damage in meadows, pastures, and grain fields by devouring the roots of the plants. The full-grown larvæ are about 25 mm. long and of a dirty-grayish color. As the body-wall is of a tough leathery texture these larvæ are commonly known as leather-jackets. Serious outbreaks of these pests have occurred at various times in Ohio, Indiana, Illinois, and California. In the case of the species infesting ranges, pastures, and grain and alfalfa fields in California it was found that the larvæ usually come out upon the surface of the ground during the night and could be destroyed by the use of poisoned-bran bait, made by mixing one pound of Paris green, twenty-five pounds of bran, and sufficient water to make a flaky mash. The bait is applied with a broadcast grain seeder.

Family DIXIDÆ

The Dixa-Midges

These midges closely resemble mosquitoes in size and form; but they are easily distinguished by the venation of their wings, (Fig. 1010).

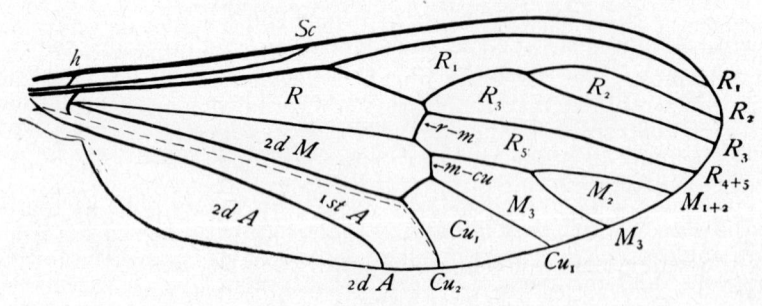

Fig. 1010.—Wing of *Dixa*.

The wing-veins are not furnished with scales, and are distinct over the entire surface of the wing; the costa is prolonged into an ambient vein, the subcosta is well developed, but is short, ending in the margin of the wing near its middle, and before the first fork of the radius; the radius is four-branched, the vein R_1 extends parallel to the margin of the wing to a point on the outer end of the wing; the media is two-branched; and the medial cross-vein is wanting. The antennæ (Fig. 1011) are sixteen-jointed, and differ but slightly in the two sexes; the legs are long and slender; and the caudal end of the abdomen of the male is enlarged.

Fig. 1011.

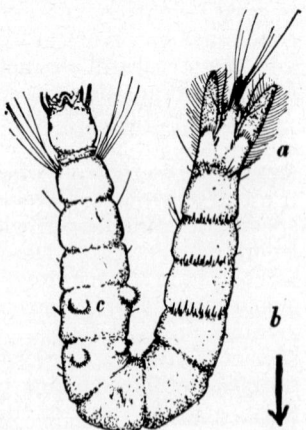

Fig. 1012.—Larva of *Dixa*, (After Needham and Lloyd.)

The family includes only a single genus, *Dixa*, of which eight species have been found in North America.

The adult midges occur in the vicinity of streams and in swampy places.

The larvæ are aquatic, living in ponds or slowly running water; they resemble somewhat those of *Anopheles* but the body is almost always bent so that the head and tail come close together. They progress by alternate thrusts of the two ends of the body the bent portion traveling foremost (Fig. 1012). The first and second abdonimal segments each bear a pair of pseudopods on the ventral surface. These larvæ feed on algæ.

Family PSYCHODIDÆ
The Moth-like Flies

There may be found frequently upon windows and on the lower surface of the foliage of trees small flies which have the body and wings densely clothed with hair and which resemble tiny moths in appearance. The wings are broad, and when at rest slope at the sides in a roof-like manner or are held horizontally in such a way as to give the insect a triangular outline (Fig. 1013).

The moth-like appearance of these insects is sufficient to distinguish them from all other flies. The venation of the wings, (Fig. 1014) is also very peculiar. All of the longitudinal veins separate near the base of the wing except veins R_2 and R_3 and veins M_1 and M_2. In some forms veins R_4 and R_5 are distinct, as shown in the figure, in others they coalesce completely, so that radius is only four-branched. Cross-veins are wanting in most cases.

Fig. 1013.—A moth-like fly.

The antennæ are long and slender, and are clothed with whorls of

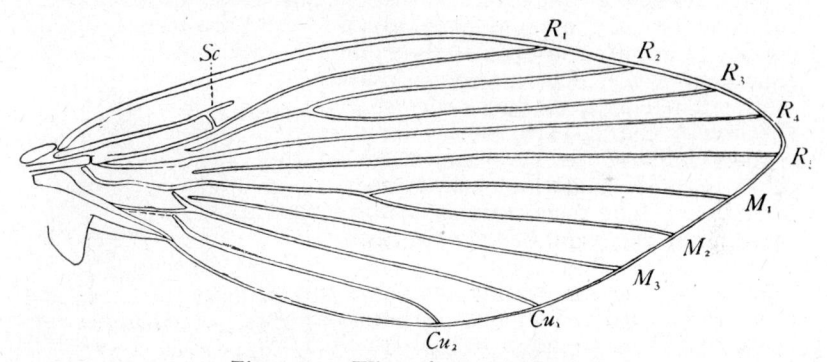

Fig. 1014.—Wing of a moth-like fly.

hairs (Fig. 1015). Those of the male are longer; and in the species figured the two basal segments are clothed with scales like those of the Lepidoptera. Scales of this form occur also on the wings, palpi, and legs of certain species.

The moth-like flies are often very minute and rarely exceed 4 mm. in length. Most of the species, so far as is known, feed on nectar or other fluid matter other than blood; but the species of the genus *Phlebŏtomus* are blood-suckers, feeding upon the blood of various reptiles, amphibians and mammals, including man; and it has been found that some exotic species transmit certain diseases of man, as the European pappatici fever, or three day fever, and the Peruvian veruga (Riley and Johannsen '15). A single species of this genus, *Phlebotomus vexātor*, has been found in the United States; this is a minute species, measuring 1.5 mm. in length. It was taken on Plummer's Island, Maryland.

The larvæ of members of this family are found in various situations; in decaying vegetable matter, in sewage, in cow dung, in exuding sap on tree-trunks, and in streams.

About thirty species have been described from the United States.

Family CHIRONOMIDÆ

The Midges

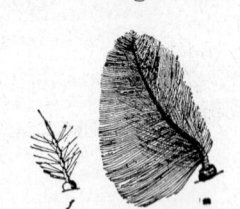

Fig. 1015.—Antennæ of *Psychoda: m*, antenna of male and the second segment of the same more enlarged; *f*, antenna of female and the tip enlarged.

The members of this family are more or less mosquito-like in form, but are usually more delicate than mosquitoes. The abdomen is usually long and slender; the wings narrow; the legs long and delicate; and the antennæ, especially in the males, strongly plumose (Fig. 1016). In fact many of these insects are commonly mistaken for mosquitoes; but only a few of them can bite, the greater number being harmless.

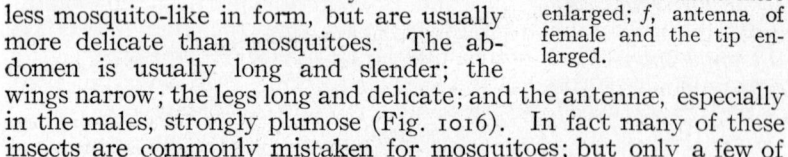

The midges are most easily distinguished from mosquitoes by the structure of the wings (Fig. 1017). These are furnished with fewer and usually less distinct veins; and the veins, although sometimes hairy, are not fringed with scale-like hairs. There is a marked contrast between the stouter and darker colored veins near the costal border of the wing and those on the other parts of the wing, which seem to be fading out. The costal vein is not prolonged into an ambient vein, beyond the apex of the wing.

Fig. 1016.—Antennæ of *Chironomus*. *f*, female; *m*, male.

In several genera of this family the wings are either absent or vestigial; of these a single species has been found in our fauna. This is *Eretmŏptera browni*, a species described by Professor Kellogg from tide-pools on the Pacific Coast. In this species the wings are short, strap-like, thickened, and without distinct venation.

The name midge has been used in an indefinite way, some writers applying it to any minute fly. It is much better, however, to restrict it to members of this family except where it has become firmly established as a part of a specific name. The wheat-midge and the clover-seed midge are examples of names of this kind; it would not be wise to attempt to change these names, although the insects they represent belong to the gall-gnat family, and hence are not true midges.

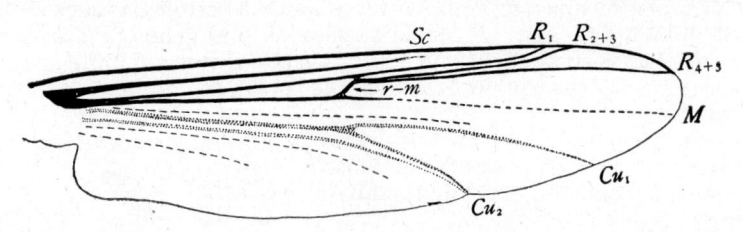

Fig. 1017.—Wing of *Chironomus*.

Midges often appear in large swarms, dancing in the air, especially towards the close of day. Professor Williston states that, over meadows in the Rocky Mountains, he has seen them rise at nightfall in most incredible numbers, producing a buzzing or humming noise like that of a distant waterfall, and audible for a considerable distance.

Most larvæ of midges are aquatic; but some live either in manure, in decaying vegetable matter, under bark, or in the ground. Some of the pupæ are free and active, others are quiescent; some of the latter remain partly enclosed in the split larval skin. The larvæ and pupæ of the aquatic species are of much importance as fish-food.

Many of the aquatic larvæ live in tubes which they build of bits of dead leaves and particles of sand fastened together with viscid threads. These tubes are frequently seen upon the surface of dead leaves, stones, and sticks; and they are often made in the mud of the bottom of a pool, in which case they open at the surface of the mud. Many of the species are blood-red in color, and hence are frequently known as *blood-worms*.

The aquatic larvæ feed on algæ, decaying vegetable matter, diatoms, and small crustacea; the terrestrial species, on manure or decaying vegetable matter. There are a few cases reported of the larvæ of midges infesting living plants.

To the genus *Culicoides* belong the small midges commonly known as sandflies or punkies. Certain minute species are sometimes very abundant, and extremely annoying on account of their bites. They are exceedingly troublesome in the Adirondack Mountains, in the White Mountains, and along mountain streams generally; they are also abundant in some places at the seashore.

More than 200 species of the Chironomidæ have been described from our fauna. The family was monographed by Professor Johannsen ('05 and '08).

Family CULICIDÆ

The Mosquitoes

The form of mosquitoes is so well known that it would be unnecessary to characterize the Culicidæ were it not that there are certain mosquito-like insects that are liable to be mistaken for members of this family.

The mosquitoes are small flies, with the abdomen long and slender, the wings narrow, the antennæ plumose in the males, (Fig. 1018), and usually with a long, slender, but firm proboscis. The thorax lacks the transverse V-shaped suture characteristic of the crane-flies; and vein M of the wings is only two-branched. But the most distinctive feature of mosquitoes is the fringe of scale-like setæ on the margin of the wings and also in most cases on each of the wing-veins (Fig. 1019).

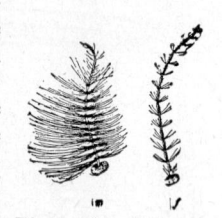

Fig. 1018.—Antennæ of mosquitoes, m, male; f, female.

The eyes are large, occupying a large part of the surface of the head. The ocelli are wanting. The antennæ are composed of fifteen segments, of which the first segment, the scape, is concealed by the large globular pedicel (Fig. 173. p. 153) and has been over-looked by many describers of mosquitoes. The pedicel contains the Johnston's organ described on pages 152 to 154. The form of the mouth-parts differs in the two subfamilies; those of *Anopheles* are represented by Figure

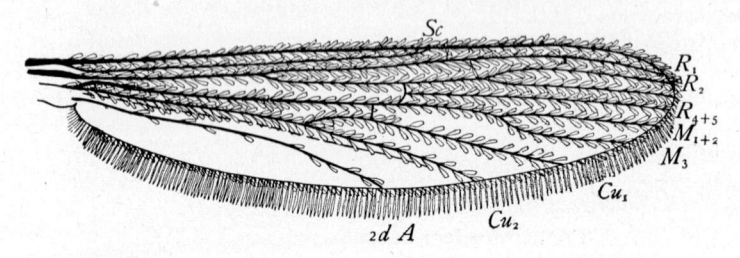

Fig. 1019.—Wing of a mosquito.

990 on page 775; in the Corethrinæ they are short and not adapted for piercing.

The larvæ of mosquitoes are all aquatic. They are well known and are commonly called "wigglers," a name suggested by their wriggling motion as they swim through the water. They vary in details of structure but the larva of *Culex* will serve to illustrate the general form of the body (Fig. 1020). The head and thorax are large and the abdomen is slender. The next to the last abdominal segment, the eighth, bears a breathing-tube; and when the larva is at rest it hangs head downward in the water, with the opening of this tube at the surface. (Fig. 1021). At the end of this tube there is a rosette of plate-like lobes (Fig. 1022, *a*) which resting on the surface

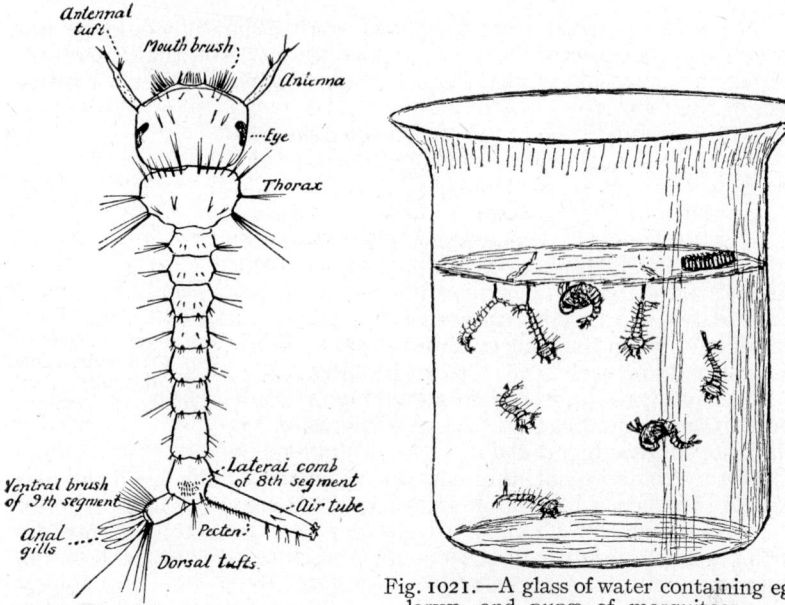

Fig. 1020.—Larva of *Culex* showing details of external structure. (From Riley and Johannsen.)

Fig. 1021.—A glass of water containing eggs, larvæ, and pupæ of mosquitoes.

film, keeps the larva in position. At the end of the last abdominal segment there are one or two pairs of tracheal gills. About the mouth, on the antennæ, and on the caudal segments of the abdomen are tufts of setæ that afford characters much used in the classification of mosquito larvæ. These various tufts have received special names as indicated in Figure 1020.

The food of mosquito larvæ varies with the different species, with most of them it consists of organic matter in suspension in the water, or floating upon the surface, or settled or growing upon the bottom. Some mosquito larvæ are cannibalistic; those of the Corethinæ are all predacious and seize their prey with the antennæ.

So far as is known, there are four larval instars in all species of mosquitoes, with the fourth molt the larva becomes a pupa.

Fig. 1022.—*a*, end of breathing tube of larva; *b*, breathing tube of pupa.

The pupæ of mosquitoes like the larvæ are aquatic, but they differ greatly in form from the larvæ. (Fig. 1023). The head and thorax are greatly enlarged and are not distinctly separated, while the abdomen is slender and flexible. With the change to the pupa state a remarkable change takes place in the respiratory system. There are now two breathing tubes, and these are borne on the thorax. One of these is represented greatly enlarged by Figure 1022, *b*. At the tail-end of the body there is a pair of leaf-like appendages, with which the insect swims, for the pupæ of mosquitoes, and also of certain midges, differ from the pupæ of most other insects in being active; but the pupæ take no food. The duration of the pupal stage is brief, usually not more than two or three days; then the skin splits down the back, and the winged mosquito carefully works itself out and cautiously balances itself on the cast skin, using it as a raft, until its wings are hardened so that it can fly away.

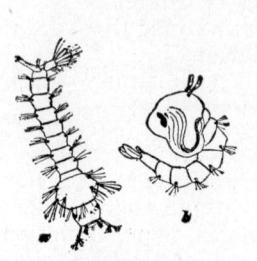

Fig. 1023.—Mosquitoes. *a*, larva; *b*, pupa.

All adult mosquitoes are commonly regarded as blood-sucking insects and are feared on that account; but there are many species that never suck blood at all, and of the blood-sucking species many attack by choice birds and mammals other than man. It is only the females that suck blood; the mouth-parts of males are not fitted for piercing the skin of animals. The males feed on nectar, the juices of ripe fruits, and other sweet substances; this is also true to a certain extent of females.

The different species of mosquitoes differ greatly in their manner of oviposition. Those most often observed about water-barrels, *Culex*, lay their long, slender eggs side by side in a boat-shaped mass, on the surface of the water (Fig. 1021); species of *Anopheles* deposit their eggs separately upon the surface of the water; and many *Aëdes* lay their eggs on the ground after the pools in which they were developed have dried out. In this case the eggs remain unhatched until later rains or melting snow refill the pools. The eggs of some mosquitoes hatch very soon after they are laid; but with the majority of species the winter is passed in the egg state; and in the case of certain species it is believed that the eggs may remain on dry ground several years awaiting rain and then hatch.

The family *Culicidæ* is divided into two subfamilies, the Corethrinæ and the Culicinæ.

Subfamily CORETHRINÆ

This is a small group of mosquitoes including but few species. It is distinguished from the Culicinæ by the comparative shortness of the proboscis, which is not much longer than the head and is not fitted for sucking blood.

The larvæ of members of this subfamily are transparent; they are predacious and capture their prey with their antennæ; they feed on infusoria, small crustacea, and small larvæ, including those of mosquitoes; they are free-swimming and are found most abundantly beyond the line of shore vegetation. The pupæ are also transparent at first but become darker colored just before transforming. The females of *Corēthra plumicŏrnis*, as observed by Professor Needham, deposit their eggs on the surface of the water, laying them down flatwise, in a spiral held together by scanty gelatine.

A monograph of this subfamily was published by Johannsen ('03).

Subfamily CULICINÆ

To this subfamily belong by far the greater number of mosquitoes. With these the proboscis is longer than the head and thorax taken together; this character is sufficient to distinguish them from the Corethrinæ.

The Culicinæ have received much attention in recent years. Since the discovery that certain species are carriers of diseases of man many investigators have studied mosquitoes, and thousands of papers have been published regarding them. Fortunately the more important results of these investigations have been summarized by several writers and published in easily available books. The most important contribution to this subject is the monograph by Howard, Dyar, and Knab ('12-'17). This is a large work in four volumes. A more available and more recent monograph is that of Dyar published in 1922.

It has been demonstrated that malaria, yellow fever, filariasis, and dengue are each caused by a parasitic organism, which has a complex life-cycle, part of which is passed in man and part in certain mosquitoes; and that it is only by being bitten by an infected mosquito that one contracts any of these diseases. There are also diseases of other mammals and of birds that are transmitted in a similar way.

In each case the parasitic organism is restricted in host relations, infecting in turn only certain species of Vertebrates and certain species of mosquitoes.

Representatives of eleven genera of the Culicinæ have been found in the United States. Some of these occur only in the extreme South and others are either rare or rarely attack man. The species that are our most serious pests are included in the genera *Culex*, *Anopheles*, and *Aëdes*.

Fig. 1024.—Normal position of *Culex* and *Anopheles* on a wall; *Culex* above, *Anopheles* below. (From Riley and Johannsen.)

Cūlex.—To this genus belong our common house mosquitoes that have unspotted wings and short palpi in the females and which when at rest on a vertical wall hold the body parallel with the wall or with the tip of the abdomen inclined toward it (Fig. 1024). These are very annoying pests; but although many of the species of this genus transmit blood diseases of birds and animals they do not play an important role in human diseases.

The eggs are laid in boat-shaped masses or "rafts" on the surface of the water of ponds of a permanent nature and on water of artificial containers, as water-barrels. A larva of *Culex* is described and figured above (Fig. 1020).

Anŏpheles.—To this genus belong those mosquitoes that have been found to be the carriers of malaria. Nine species of *Anopheles* have been found in the United States, of which four are known to be carriers of this disease. In this genus the palpi of both sexes are nearly or quite as long as the proboscis and the wings are frequently spotted. When at rest on a vertical wall the body is usually held at an angle with the vertical (Fig. 1024). Some species often enter houses. They hibernate in the adult state and can be found during the winter in cellars.

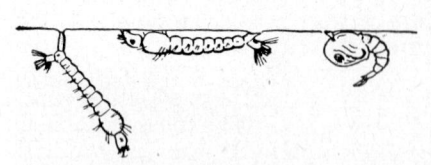

Fig. 1025.—Normal position of the larvæ of *Culex* and *Anopheles* when at rest. *Culex*, left, *Anopheles*, middle; *Culex* pupa, right hand figure.

The eggs are laid singly in small numbers upon the surface of water. The larva when at rest floats in a horizontal position beneath the surface film (Fig. 1025). There is no respiratory tube but instead a flattened area on the eighth abdominal segment into which the two spiracles open.

Malaria is a well-known and widely distributed disease. It is most common in the vicinity of swamps, and is more virulent in the South than in the cooler parts of the country. It is caused by unicellular parasites in the blood which feed upon the red blood corpuscles. These parasites belong to the genus *Plasmodium* of the Protozoa. Three species of these malarial parasites are now recognized, each of which causes a distinct type of malaria.

The life-cycle of the malarial parasites is an exceedingly complicated one. Our knowledge of it is the result of extended investigations by several workers. The history of these investigations is a most interesting one; but space can not be taken to narrate it here. Accounts of these investigations, with details of the results obtained are given by Howard, Dyar, and Knab ('12) and Riley and Johannsen ('15). The following summary of the life-cycle is condensed from the accounts by these writers.

The malarial infection is introduced into the blood circulation of man by the bite of a mosquito that has previously bitten a person having this disease. It is only certain species of mosquitoes of the genus *Anopheles* that transmit this disease to man. The infecting organism, which in that stage of its development is known as a *sporozoite*, penetrates a red blood corpuscle and becomes an amoeboid *schizont*. This lives at the expense of the blood corpuscle and as it develops there are deposited in its body scattered black or reddish black particles. These are generally called melanin granules, but are much better referred to as *hæmozoin*, as they are not related to melanin. The hæmozoin is the most conspicuous part of the parasite, a feature of advantage in diagnosing from unstained preprations. As the schizont matures, its nucleus breaks up into a number of daughter nuclei, each with a rounded mass of protoplasm about it, and finally the corpuscles are broken down and these rounded bodies are liberated in the plasma as spores which are known as *merozoits*. These spores infect new corpuscles, where they again go through the stages of schizonts and merozoits, and thus the asexual cycle is continued. The malarial paroxysm is coincident with sporulation.

Parallel with the asexual cycle sexual elements or *gametes* are produced by schizonts. These sexual elements, however, can not copulate within the human organism on account of the unfavorable temperature. To enable them to carry out this function, and to develop further they must be transferred to the alimentary canal of an *Anopheles*, which is done when one of these mosquitoes sucks the blood in which they are. Here the union of the male and female gametes takes place and there results a stage known as the *migratory oökinete*. The oökinete penetrates the wall of the midintestine of the mosquito and there transforms to the *oöcyst*. In the process of growth of the oöcyst further stages occur, first by its division the *sporoblasts*, and from these, by further division, the *sporozoits*, when the oöcyst is mature it bursts, liberating the sporozoits which thus pass into the general body cavity of the host. The sporozoits now find their way into the salivary glands of the host and there remain until the mosquito, in biting, forces them along with the saliva, through its proboscis into a human being. Then the asexual cycle begins in the blood of a new host.

Aëdes.—This is a very large genus of world-wide distribution Dyar ('22) describes 73 species that have been found in the United States. The species vary greatly in habits; but with most of them the larvæ develop from over-wintering eggs in early spring pools. Some species, however, breed in water-barrels, and other artificial containers; one of these is the carrier of yellow fever.

The yellow-fever mosquito, *Aëdes ægÿpti*, is distributed throughout all tropical regions of the world and is often carried by commerce into temperate regions. But as it is destroyed by frost it can not become established where frosts occur. Hence outbreaks of yellow fever in the North are checked naturally as winter approaches, and with our present knowledge of the methods of control of this disease it is not probable that it will be permitted to become epidemic again in the United States.

When yellow fever appears the patients should be kept in mosquito-proof rooms, so that they may not serve as centers of distribution of the disease; and the breeding places of mosquitoes should be drained or screened, or oiled.

The yellow fever mosquito breeds in cisterns, water-barrels, flower-vases, and in the various water receptacles about houses. The life-cycle under favorable conditions is completed in from twelve to fifteen days. This is essentially a domesticated species. It is rarely found far from the habitations of man.

The fact that yellow fever is transmitted by this mosquito has been definitely established; but it is not certain that the causative organism is known, although some investigators claim to have found it.

The yellow fever mosquito was first described **by Linnæus in 1762** under the name *Culex ægypti*. But the Linnæan genus *Culex* has been divided and this species pertains to the genus *Aëdes* established by Meigen in 1818; hence its correct name is *Aëdes ægypti*. Unfortunately a score of other names have been applied to it; those most commonly found are *Aëdes calōpus, Stegomÿia fasciāta,* and *Stegomÿia calopus.*

Other mosquito-borne diseases of man.—In tropical countries there are, in addition to malaria and yellow fever, two other diseases of man that have been found to be transmitted by mosquitoes; these are *dengue* and *filariasis*. The causative organism of dengue has not been discovered; but it is believed to be a protozoan of ultra-microscopic size. Filariasis is due to the presence in the blood, the lymphatics, the mesentery, and subcutaneous connective tissue of nemotode worms belonging to the family Filariidæ, and which pass part of their life-cycle in the bodies of mosquitoes. One of these parasites, *Filaria bancrofti*, is the cause of the extraordinary deformities of different parts of the human body known as *elephantiasis*.

Mansōnia.—In this genus "the larvæ are peculiar in having the air tube adapted for piercing the vascular roots of certain aquatic plants, from which they get their supply of air. The eggs are deposited in rafts in swamps where suitable plants grow, and the young larvæ descend to the roots, never coming to the surface again." (Dyar, '22). *Mansōnia pertūrbans* is widely distributed in the United States and Canada. Its larva lives attached to the roots of a species of *Carex* growing in marshes or the edges of ponds. The winter is passed as half-grown larva.

Wyeomy̆ia smĭthii.—This species is remarkable on account of its habits. "The larvæ live in the water in the leaves of pitcher plants (*Sarracenia purpurea*), passing the winter frozen up in the ice cores. The eggs are laid on the still, dry, newly opened leaves and hatch when water collects in them" (Dyar '22). The adult can be distinguished from all other Culicinæ found north of Southern Florida by the presence of a tuft of setæ on the metonotum.

Family MYCETOPHILIDÆ

The Fungus-Gnats

These flies are of medium or small size and more or less mosquito-like in form. They are most easily recognized by the length of the coxæ (Fig. 1026). The ocelli are

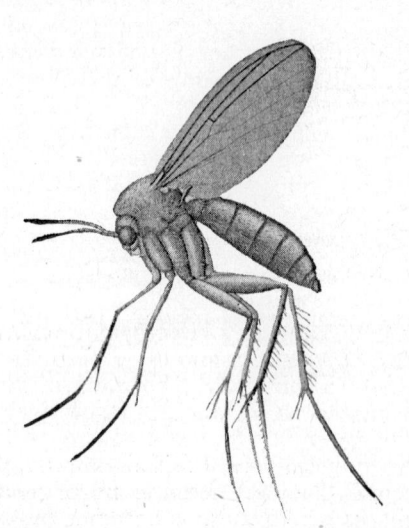

Fig. 1026.—*Mycetophila punctata.* (After Johannsen.)

Fig. 1027.

present; the antennæ, as a rule, lack whorls of hairs (Fig. 1027) and all the tibiæ are furnished with spurs.

At first sight considerable variation seems to exist in the venation of the wings as shown in the three wings represented in Figure 1028; but in reality the variations are comparatively slight. The costa extends along the margin of the wing to the end of the radial sector. Radius preserves three branches in the more generalized forms (Fig. 1028, a); in some genera veins R_1 and R_{2+3} coalesce from the apex of the wing backward for a greater or less distance so that the basal part of vein R_{2+3} appears like a cross-vein (Fig. 1028, b); in some genera

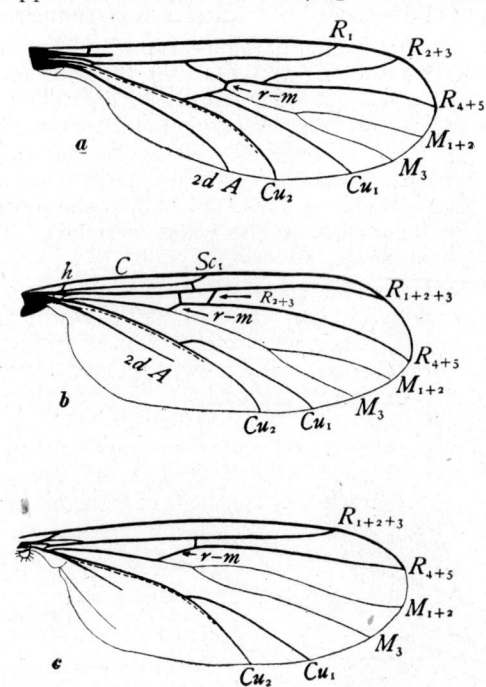

Fig. 1028.—Wings of fungus-gnats. (The drawings are after Winnertz; the lettering is original.)

radius is only two-branched, this condition may have been brought about by the complete coalescence of veins R_1 and R_{2+3} or by the coalescence of veins R_{2+3} and R_{4+5}, whichever may be the case the two branches are commonly designated as R_1 and R_s respectively (Fig. 1028, c).

The fungus gnats are exceedingly numerous both in number of individuals and in number of species. They are often found in great numbers on fungi and in damp places where there is decaying vegetable matter. They are active and leap as well as fly.

A monograph of the known species of the world, not including the Sciarinæ, was published by Professor Johannsen ('09a); and in a later series of papers he published a synopsis of the species of North America, including the Sciarinæ (Johannsen '09-'12).

The larvæ of most species live upon and destroy mushrooms, usually the wild plants, but sometimes they are pests in mushroom cellars; other species are found in decaying wood; and certain species of the subfamily Sciarinæ are sometimes pests of cultivated plants, destroying seed corn, seed potatoes, and the roots of other plants.

In this family the larva is more or less cylindrical, smooth, soft, whitish in color, and with a small strongly chitinized head, which is usually brown or black, and is provided with mandibles and maxillæ. There are usually eight pairs of spiracles.

The pupa is not enclosed in the skin of the larva; but in some genera the transformations are undergone in a delicate cocoon.

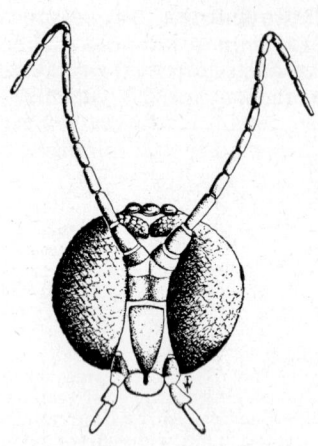

Fig. 1029.—Eyes of *Sciara*.

The subfamily *Sciarinæ*.— The family Mycetophilidæ is divided into nine subfamilies, eight of which are represented in our fauna. One of these subfamilies, the Sciarinæ, merits special mention in this place. The members of this subfamily differ from the more typical fungus-gnats as follows; the coxæ are not so greatly elongated; the eyes differ in shape, there being a narrow ex-

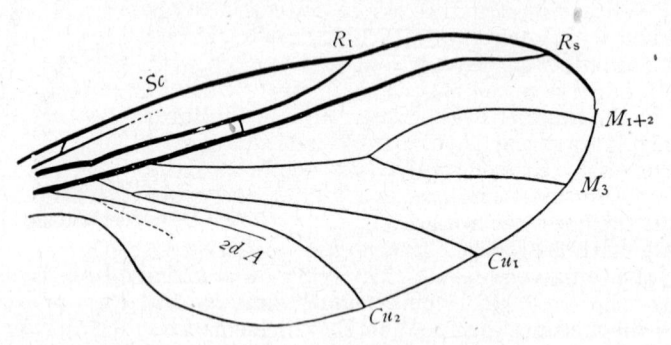

Fig. 1030.—Wing of *Sciara*. (After Enderlein.)

pansion of each eye extending above the base of the antennæ and meeting or nearly meeting the expansion of the other eye, (Fig. 1029), while in other fungus-gnats the eyes are either round, oval, or kidney-shaped, but not markedly narrowed above; and the cross-vein r-m is in the same right line with the second section of the radial sector (Fig. 1030).

The larvæ of some species of the genus *Sciara* often attract attention on account of a strange habit they have of sticking together in dense patches. Such assemblages of larvæ are frequently found under the bark of trees. But what is more remarkable is the fact that when the larvæ are about to change to pupæ an assemblage of this kind will march over the surface of the ground, presenting the appearance of a serpent-like animal. Such a congregation is commonly spoken of as a Sciara-army-worm. Examples have been described that were four or five inches wide and ten or twelve feet long, and in which the larvæ were piled up from four to six deep. The larvæ crawl over each other so that the column advances about an inch a minute.

THE FAMILY SCIARIDÆ OF ENDERLEIN

The establishment of a family to be known as the Sciaridæ was proposed by Enderlein ('11 & '12 a). The proposed family includes the subfamily Sciarinæ of the Mycetophilidæ and the subfamily Lestremiinæ of the Cecidomyiidæ and is characterized by the form of the eyes; the two subfamilies agreeing in having the type of eyes described above (Fig. 1029) and differing in this respect from other fungus-gnats and gall-gnats.

This proposed grouping of these two subfamilies has not been generally accepted. While they agree in the shape of their eyes, they differ in the presence of tibial spurs in the Sciarinæ and the absence of these spurs in the Lestremiinæ; and they differ markedly in the form of their larvæ. The larvæ of the Lestremiinæ have, like those of other Cecidomyiidæ, an undeveloped head, indistinct mouth-parts, and a well-developed sternal spatula; the larvæ of the Sciarinæ have, on the contrary like other Mycetophilidæ, a well-developed head which is strongly chitinized, and strong, toothed mandibles, and do not have a sternal spatula.

Family CECIDOMYIIDÆ*
The Gall-Gnats

The gall-gnats are minute flies which are extremely delicate in structure. The body and wings are clothed with long hairs, which are easily rubbed off. The antennæ are usually long and clothed with whorls of hairs (Fig. 1031); but they vary greatly in length, in the number of their segments, in the form of the segments of the flagellum, and in the nature of their clothing. Except in the first subfamily, the ocelli are wanting. The legs are slender and quite long, but the coxæ are not greatly elongated and the tibiæ are without spurs. Except in the first subfamily, the wing-veins are greatly reduced in number (Fig. 1032), the anal veins being entirely wanting, and the media wanting or merely represented by a slight, unbranched fold; in the first subfamily, the Lestremiinæ, vein M is well preserved.

Fig. 1031.—Antennæ of gall-gnats: *m*, male; *f*, female, enlarged more than that of the male.

A striking feature of this family is the presence in most species (*i. e.* in all except the first two small subfamilies) of what have been generally known as arched filaments

---

*This family is named the Itonididæ by some writers, those who recognize the names published by Meigen in 1800; see page 794 for a discussion of these names.

on the antennæ (Fig. 1033). These filaments occur in series, and there may be one, two or three of these series on each segment of the flagellum. As each of these series has the appearance of a looped

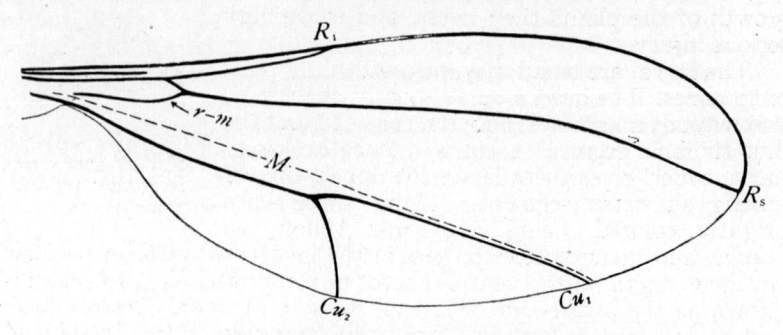

Fig. 1032.—Wing of a gall-gnat.

thread extending around the segment of the antenna they are termed *circumfili* by Dr. Felt, who has described and figured many forms of them in his series of papers on this family.

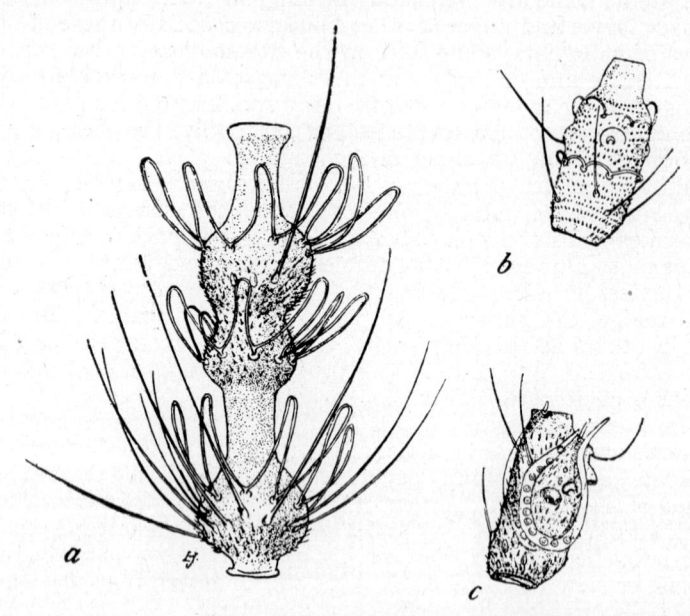

Fig. 1033.—Antennal segments with circumfili: *a*, fifth segment of antenna of *Karschomyia viburni*, male; *b*, fifth antennal segment of a *Rhopalomyia*, female; *c*, sixth antennal segment of *Winnertzia calciequina*, female. (From Felt.)

To this family belong the smallest of the midge-like flies. On account of their minute size the adult flies are not apt to attract the attention of the young student. But the larvæ of many species cause the growth of galls on plants, some of which are sure to be found by any close observer. Other species arrest the growth of the plants they infest, and cause very serious injury.

The larvæ are small maggots, with nine pairs of spiracles. The head is small, poorly developed, and without mandibles; between the head and the first thoracic segment there is a large neck-segment, which gives these larvæ the appearance of having an extra segment. Many species are brightly colored, being red, pink, yellow, or orange, and many species possess in the last larval instar a peculiar chitinous organ on the ventral aspect of the prothorax; this organ is known as the *breast-bone* or *sternal spatula*, or *anchor process* (Fig. 1034). It varies in form in different species; different views are held regarding its function, none of which seems well established.

Fig. 1034.—Head end of larva showing the breast-bone.

The larval mouth-parts are fitted only for taking liquid food; but the nature of this food differs greatly in different members of the family. Some species are parasitic in the bodies of aphids and other Homoptera; some are predacious feeding on either aphids, coccids, mites, or larvæ and pupæ of other Diptera, especially those of other species of gall-gnats; some feed on the excrement of other larvæ or that of cattle and of birds; but most species are vegetable feeders. Among those that feed on plants many species produce galls. The larvæ of several genera of the second subfamily, the Heteropezinæ give birth to living young, as described later.

Different modes of pupation have been observed among the gall-gnats; in some the pupa is naked; in others the change to the pupa state takes place within a puparium, but this puparium differs from that of most Diptera in being formed by the next to the last larval skin, the last larval molt taking place within it; in some species as the wheat-midge, the puparium consists only of the peunltimate larval skin, in others as the Hessian fly, it is lined with a delicate silken layer (Marchal '97); and in still others the pupa is enclosed in a delicate cocoon instead of in a puparium.

The literature regarding this family is very extensive; hundreds of articles have been written about those species that are of economic importance, and very many papers have been published on the classification of these insects; A monograph of the species of the world was published by Kieffer in 1813 and a review of the American species is being published by Dr. Felt in his annual reports as State Entomologist of New York. See also Felt ('18) for descriptions and figures of the galls produced by members of this family.

The family Cecidomyiidæ is separated into three subfamilies, which can be separated by the following table:

A. Ocelli present; vein M, preserved either simple or forked. p. 816. LESTREMIINÆ
AA. Ocelli wanting; vein M wanting or represented merely by a fold.
  B. Antennæ without either circumfili or horseshoelike appendages; the first segment of the tarsi usually longer than the second p. 816. HETEROPEZINÆ

BB. Antennæ with circumfili or (*Winnertzia*) with horseshoelike appendages; the first segment of the tarsi shorter than the second p. 817. CECIDOMYIINÆ

Subfamily LESTREMIINÆ

The members of this subfamily differ from other gall-gnats in having ocelli, in the shape of their eyes, these resembling those of *Sciara* (Fig. 1029), and in the less reduced venation of their wings, vein M being preserved; in some genera this vein is forked (Fig. 1035) in others it is unbranched.

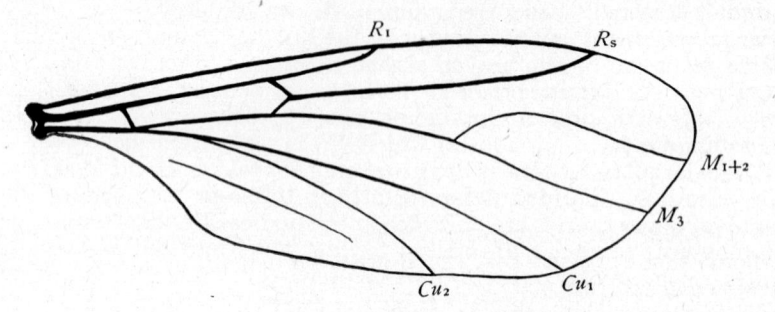

Fig. 1035.—Wing of *Lestremia*. (After Kieffer.)

Most of the known larvæ of this subfamily live in decaying vegetable matter, especially in rotten wood under bark.

Subfamily HETEROPEZINÆ

This subfamily includes comparatively few species, none of which is known to be of economic importance. The known larvæ live in the decaying bark of trees. Some of them are remarkable for the fact that they give birth to living young.

This type of reproduction is termed pædogenesis (See page 192). It was first discovered by Nicholas Wagner in 1862 and has been investigated by several other Europeans. It has also been studied in this country by Dr. Felt ('11) who gives an extended account of it as observed by him in *Miăstor americāna*, and by Professor Hegner ('12 and '14) who gives the history of the germ cells in the pædogenetic larva of this species.

The larva of *Miastor americāna* possesses two ovaries, one on either side of the body in the tenth or eleventh segments. Each ovary consists of typically thirty-two oöcytes, each of which is accompanied by a group of nurse-cells, and with them is surrounded by a folicular epithelium. The nurse-cells furnish nutrition to the growing oöcytes, gradually becoming reduced as the oöcytes increase in size. Finally the oöcyte with accompanying nurse-cells, still surrounded by the follicular epithelium, becomes separated from the rest of the ovary and is forced by the movements of the larva into some other part of its body. Here it continues its growth and development at the expense of the tissues of the mother-larva. Not all of the oöcytes complete their development, since usually only from five to seventeen young are produced by a single mother-larva (Hegner).

When the tissues of the mother-larva are consumed, the young larvæ break forth from the skin of their parent and continue their growth. These larvæ may in turn produce another generation of larvæ in the same manner. It is believed that this asexual, pædogenetic reproduction may continue through many generations covering a period of two or three years. Finally a generation of larvæ are produced which do not reproduce in this manner, but which when full grown transform first to pupæ and then to adults, which reproduce sexually. According to the observations of Kieffer ('13) the adult females of those species in which pædogenesis occurs produce each only four or five eggs, while other gall-gnats produce a large number of eggs

Subfamily CECIDOMYIINÆ

This subfamily includes the larger number of the gall-gnats; to it belong those species that have attracted attention on account of their economic importance, and others that are well known on account of the conspicuous galls produced by them. Much has been published regarding some of these species; but unfortunately they are discussed under different generic names by different writers. For this reason the common names will be found more useful when one is attempting to learn what has been published regarding these species.

The pine-cone willow-gall. —One of the most common and conspicuous of the galls made by gall-gnats is the pine-cone willow-gall (Fig 1036). This often occurs in great abundance on the tips of twigs of the heart-leaved willow, *Salix cordata*. The gnat that causes the growth of this gall is *Rhabdŏphaga strobiloides*. The larva remains in the heart of the gall throughout the summer and winter, changing to a pupa early in the spring. The adult emerges soon afterward, and lays its eggs in the newly-started buds of the willow.

Fig. 1036.—The pine-cone willow-gall.

The pine-cone willow-gall guest, Cecidomyia albovittāta.—This species breeds in large numbers between the leaves composing the pine-cone willow-gall. The larvæ of this gnat do not seem to interfere in any way with the development of their host, there being abundant food in the gall both for the owner of the gall and for its numerous guests.

The clover-leaf midge, *Dasyneura trifōlii.*—The leaflets of white clover are sometimes infested by white or orange-colored maggots which fold the two halves of the leaflet together. From one to twenty of these larvæ may be found in a single leaflet. When full grown the larvæ make cocoons, and undergo their transformations within the folded leaflet. In Figure 1037 an infested leaf containing cocoons is represented natural size, also a larva and an adult gnat, greatly enlarged.

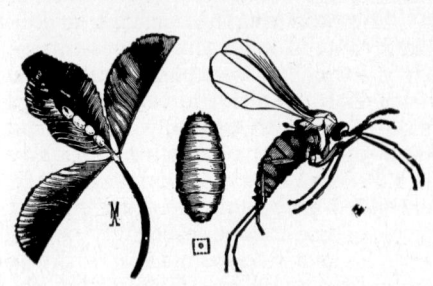

Fig. 1037.—The clover-leaf midge.

The clover-flower midge, *Dasyneura leguminĭcola.*—This is a very serious pest. The larvæ live in the heads of clover and destroy the immature seed. Different kinds of clover are infested by this pest; but red clover is its chief food plant; and in some parts of this country it has seemed impossible to raise clover-seed on account of this insect.

The larva of the clover-flower midge passes the winter on or slightly below the surface of the ground, usually but not always, in a cocoon; it changes to a pupa early in the spring, and emerges an adult in late April or early May. The eggs are laid in the small green clover heads, many eggs in a single head, as each larva infests a single floret. The larvæ mature in about four weeks, and then drop to the ground to transform. Two or three weeks later a second generation of midges appear and lay their eggs.

The most efficient method of combating this pest is to make the first cutting of clover early, before the first generation of midges have matured, that is, *early* in June; the drying of the clover heads will result in the destruction of the larvæ in them; and thus the second crop of clover will not be infested. Care should be taken to cut at the same time any clover that may be growing wild in fence-corners by roadsides, or elsewhere.

The Hessian-fly, *Phytŏphaga destrŭctor.*—This is the most serious pest infesting wheat in this country. The larvæ live at the base of a leaf between it and the main stalk, where they draw their nourishment from the plant. There are two or three broods of this insect in the course of the year. The larvæ of the fall brood infest the young wheat-plants near the surface of the ground. When full-grown each changes to a pupa within a brown puparium, which resembles a flax-seed. Here they remain throughout the winter. In the spring the adult gnats emerge and lay their eggs in the sheaths of leaves some distance above the ground. The infested plants are so weakened by the larvæ that they produce but little if any seed, and often bend or even break off at the weakened spot.

There is no method by which a crop of wheat can be saved from the ravages of this pest after it is infested by it; the only means of control are those that prevent infestation. Where practicable all infested stubble should be plowed under immediately after harvest and the soil rolled or lightly harrowed, thus preventing the emergence of the fall brood of gnats; but this can not be done where clover or grasses are sown with the wheat. The most available means of control is to sow wheat moderately late, that is, after the fall generation of gnats has disappeared, but early enough to secure the maximum yield of grain. This safe date varies with the latitude, longitude, and elevation above sea level. This date has been carefully determined for the different parts of the country and can be ascertained for any locality by application to a State or Federal Entomologist. Dr. A. D. Hopkins, who has made a very extended investigation of this subject states that the first general coloring of the foliage, especially on the hickories, dogwood, birch, ash, etc., is, as a rule, coincident with the safest and best time to begin sowing wheat on any farm within the range of winter wheat culture.

The wheat-midge, *Thecodiplōsis mosellāna.*—This gnat is also a very serious enemy of wheat. It deposits its eggs in the opening flowers of wheat. The larvæ feed on the pollen and the milky juice of the immature seeds, causing them to shrivel up and become comparatively worthless. When full-grown the larvæ drop to the ground, where the transformations are undergone near the surface. The adults appear in May or June. No effective method of control of this pest has been devised.

Until recently our common wheat-midge was supposed to be the same as the European species the specific name of which is *tritici*, and which has been placed successively in the following genera; *Tipula, Cecidomyia, Diplosis,* and *Contarinia*. But it has been found to be another European species the *Thecodiplosis mosellana*.

The resin-gnat, *Retinodiplōsis resinīcola.*—This species infests the branches of various species of pine. I have found it throughout the Atlantic region from New York to Florida. The larvæ live together in considerable numbers within a lump of resin.

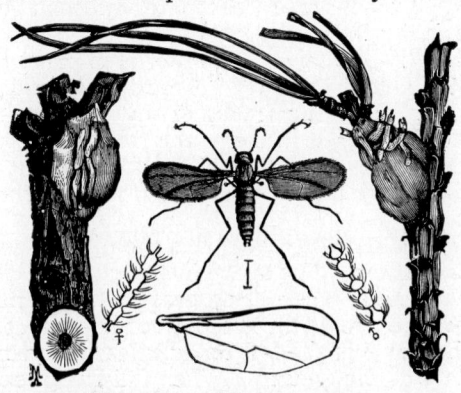

Fig. 1038.—The resin-gnat.

They derive their nourishment from the abraded bark of the twig; and the resin exuding from the wound completely surrounds and protects them. The transformations are undergone within the lump of resin. After the gnats emerge the empty pupa-skins project from the lump of resin as shown at the right in Figure 1038. In this figure the gnat, a single wing, and a part of the antenna of each sex are represented, all greatly enlarged.

The pear-midge, *Contarinia pyrivora*.—The female of this species deposits her eggs by means of her long ovipositor, in the interior of the unopened blossoms of pear. The young fruit is destroyed by the larvæ. There is a single annual generation. The winter is passed in the ground, usually as pupæ but sometimes as larvæ. This is an introduced European species, which has not yet become a serious pest in this country.

The chrysanthemum gall-midge, *Diarthronomyia hypogœa*.—This species causes the growth of galls on the leaf, stem, and flower-head of the chrysanthemum plant, and is sometimes a serious pest in greenhouses. A detailed account of it is given in Bulletin No. 833 of the U. S. Department of Agriculture.

Subseries B.—The Anomolous Nemocera

In this subseries the antennæ are composed of many segments, but are shorter than the head and thorax, and are without whorls of long hairs. The segments of the flagellum of the antennæ are short and broad and are closely pressed together. The abdomen is comparatively stout, and the legs are shorter and stouter than in the True Nemocera.

Family BIBIONIDÆ

The March-Flies

In this family the abdomen is often comparatively robust, and the legs shorter and stouter than in most of the families with thread-like antennæ (Fig. 1039). The antennæ are rarely longer than the head and thorax, and composed of short, broad, and closely-pressed-together segments (Fig. 1040). These insects resemble the fungus-gnats in having ocelli; but they differ from them in the shortness of the antennæ, in the fact that the coxæ are not greatly elongate, and that tibial spurs of any magnitude are confined to the front tibiæ. The venation of the wings of the typical genus is represented by Figure 1041. The cross-vein m-cu is present, and vein Cu forks at a considerable distance from the base of the wing.

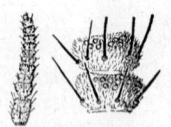

Fig. 1039.—*Bibio*.

Fig. 1040.—Antenna of *Bibio*.

The adult flies are generally black and red, sometimes yellow. They are most common in early spring; which has suggested the name March-flies; but some occur later in the season, and even in the autumn.

The larvæ vary in habits; some species feed on decaying matter, while others attack the roots of growing plants, especially of grass.

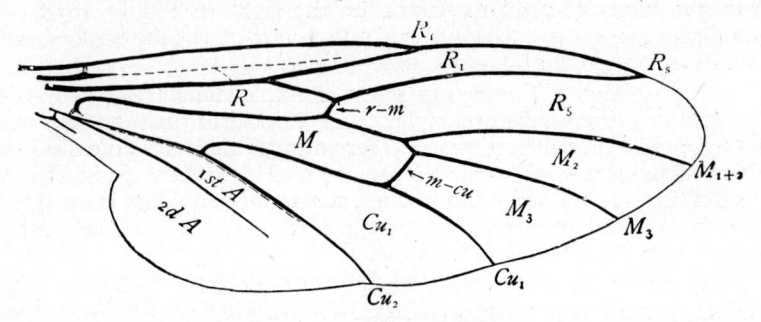

Fig. 1041.—Wing of *Bibio*.

They have ten pairs of spiracles, which is a rather large number, although there are other insects with as many. The pupæ are usually free.

For descriptions of our species of bibionid flies see McAtee ('21).

Family SCATOPSIDÆ

This family includes minute black flies; our known species measure in length from less than one millimeter to three millimeters. Formerly these flies were included in the Bibionidæ; but they differ markedly in the venation of their wings from members of that family. In the Scatopsidæ vein Cu forks at or very near the base of the wing (Fig. 1042) and the cross-vein m-cu is wanting. In some species there is a vestige of an anal vein but usually there is none.

This is a small family, only about a score of species are known from our fauna; these are described by Melander ('16).

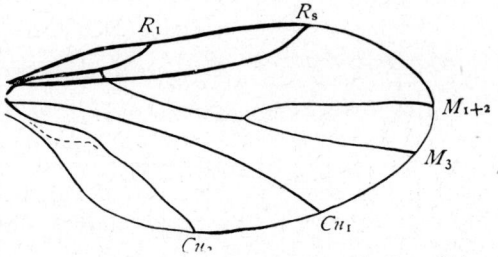

Fig. 1042.—Wing of *Reichertella collaris*. (After Melander.)

Most of the known larvæ live in excrement. One species, *Cobŏldia formicārium*, is believed to be myrmecophilous, for the adult was taken as it crawled from a populous nest of the carpenter ant. In this species the wings are vestigial.

Family SIMULIIDAE

The Black-Flies

The common name, black-flies given to the members of this family is not distinctive, for there are many species of other families that are of this color; but like many other names that are descriptive in form, it has come to have a specific meaning distinct from its original one. It is like the word blackberry; some blackberries are white, and not all berries that are black are blackberries.

In this family the body is short and stout; the thorax is much arched, giving the fly a humpbacked appearance (Fig. 1043); and the legs are comparatively short. The antennæ are scarcely longer than the head and are eleven jointed; the segments of the flagellum are short and closely pressed together (Fig. 1044), they are clothed with fine hairs, but do not bear whorls of long hairs. The ocelli are absent. In the male the eyes are very large and contiguous, and divided; the upper half of each has the facets very much larger than the lower, from which they are distinctly divided by a horizontal line. The upper half of each is doubtless a night-eye, while the lower half is a day-eye. In the female, the eye facets are of almost uniform size; and the two eyes are widely separated. The proboscis is not elongated, the small labella are horny, and the palpi are

Fig. 1043.—*Simulium*.

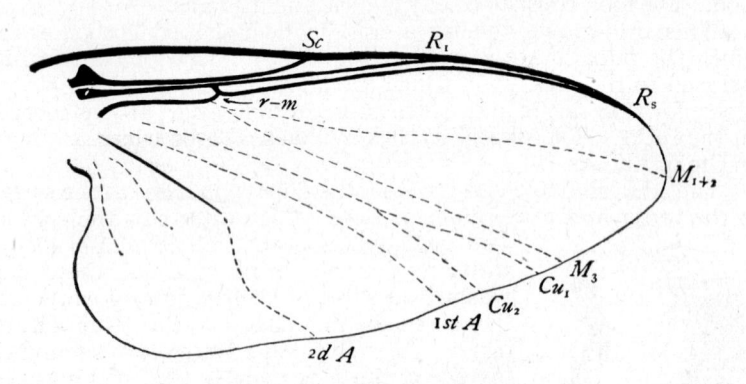

Fig. 1044.

Fig. 1045.—Wing of *Simulium*.

four-jointed. The wings are broad, iridescent, and not clothed with hairs. The veins near the costal border are stout; those on the other parts of the wing are vestigial (Fig. 1045), and are usually represented merely by folds.*

The females of many species suck blood and are well-known pests. Unlike mosquitoes and midges, the black-flies like heat and strong light. They are often seen in large numbers disporting themselves in the brightest sunshine.

---

*The forked fold that I believe to be a vestige of vein Cu is not so regarded by some writers, who refer to it as the fold between media and cubitus and label two folds of the anal area as cubitus. I can see no reason for this conclusion; in no other flies that I have studied is there a fold between media and cubitus.

The larvæ are aquatic; and usually live in swiftly-flowing streams, clinging to the surface of rocks in rapids or on the brinks of falls. They sometimes occur in such large numbers as to form a moss-like coating over the rocks. There is a disk-like sucker fringed with little hooks at the caudal end of the body by means of which the larva clings to the rocks; and just back of the head there is a fleshy proleg which ends in a similar sucker fringed with hooks (Fig. 1046). By means of these two organs the larva is able to walk with a looping gait similar to that of a measuring-worm. It has also the power of spinning silk from its mouth, which it uses in locomotion. The hooks on the caudal sucker and at the end of the proleg are well adapted to clinging to a thread or to a film of silk spun upon the rock to which the larva is clinging. Respiration is accomplished by means of blood-gills, which appear on the dorsal side of the last abdominal segment, but are evaginations of the ventral wall of the rectum, and lie, when retracted completely within the rectal cavity (Headlee '06). The head bears two large fan-shaped organs, which aid in procuring food. The food consists chiefly of algæ and diatoms.

Fig. 1046.—Head of larva

When full-grown the larva spins a boot-shaped cocoon within which the pupal state is passed (Fig. 1047). This cocoon is firmly fastened to the rock upon which the larva has lived or to other cocoons, for they occur in dense masses, forming a carpet-like covering on the rocks. The pupa breathes by tracheal gills which are borne on the prothorax.

The adult fly, on emerging from the pupa-skin, rises to the surface of the water and takes flight at once. Soon after this the eggs are laid. I have often watched *Simūlium pictipes* hovering over the brink of a fall where there was a thin sheet of swiftly-flowing water, and have seen the flies dart into the water and out again. At such times I have always found the surface of the rock more or less thickly coated with eggs, and have no doubt that an egg is fastened to the rock each time a fly darts into the water. Malloch ('14) states that the eggs are deposited in many cases on blades of grass, twigs, or leaves of trees which are dipping in running water.

Fig. 1047.—Cocoon and larva.

Until recently all members of this family were included in the genus *Simulium;* consequently in nearly all of the published accounts of these insects the various species are placed in this genus. But later writers have divided the old genus *Simulium* into several genera, Enderlein ('21) now recognizes fifteen genera of which seven are represented in North America.

Monographic papers on the North American species of this family have been published by Coquillett ('98), Johannsen ('03) and Malloch ('14).

This is a comparatively small family. Malloch ('14) in his "Catalogue of North American and Central American Simuliidæ" lists 38 species. The species that have attracted most attention in the United States are the following.

The Adirondack black-fly *Prosimūlium hĭrtipes*.—This is a widely distributed species but it has attracted most attention in the mountains of the Northeastern States, where fishermen find it to be a scourge in May and June. In this species the radial sector is distinctly forked.

The southern buffalo-gnat, *Cnĕphia pecuārum*.—This is the "Buffalo-gnat" of the Mississippi Valley, which in the past has been a terrible pest of mules and other domestic animals, sometimes causing their death· but it seems to be much less common now than in former years. In this species the radial sector is very indistinctly forked at the apex. The popular name of this insect refers to a fancied resemblance in the shape of the insect when viewed from one side to that of a buffalo.

The turkey-gnat, *Simūlium meridionāle*.—This species closely resembles the preceding in habits, infesting all kinds of domestic animals, especially in the Mississippi Valley. As it appears at the time that turkeys are setting and causes great injury to this fowl, it is commonly known as the turkey-gnat. In this species the radial sector is not forked.

The white-stockinged black-fly, *Simūlium venūstum*.—This species is widely distributed and is one of the more common species of the genus. It can be distinguished from the other species mentioned here by the fact that the tibiæ are silvery white above. In the Adirondacks it appears later than *Prosimulium hirtipes* and is not so serious a pest. Professor Needham writes: "Guides have a saying, that, when the black-flies put on their white stockings in June, the trouble is about over. This species has the white stockings".

The innoxious black-fly, *Simūlium pĭctipes*.—This black-fly is very widely distributed and at Ithaca it is our most common species. Although it may abound where during many summers I have taken my classes for study of aquatic insects, I have never known it to bite.

Family BLEPHAROCERIDÆ

The Net-winged Midges

The net-winged midges are extremely remarkable insects; for in certain respects the structure of the adults is very peculiar, and the larvæ appear much more like crustaceans than like insects.

The adults are mosquito-like in form; but they differ from all other insects in having the wings marked by a net-work of fine lines which extend in various directions and are not influenced at all by the veins of the wing (Fig. 1048). They are, however, quite constant in their position in the species that I have studied.

When a wing is examined with a microscope, the fine lines are seen to be slender thickenings extending along the courses of slight folds in the wing. The significance of these folds is evident when a net-winged midge is observed in the act of issuing from its pupa-skin

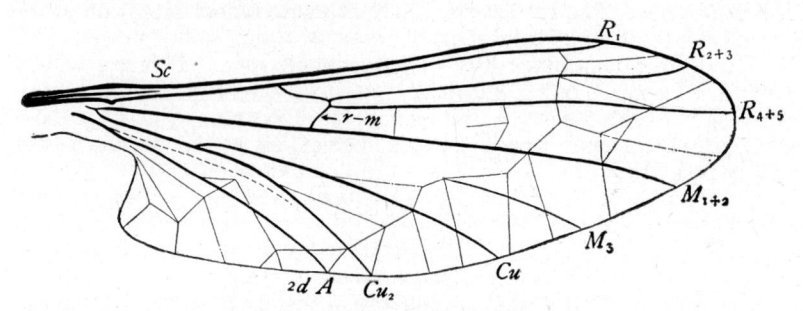

Fig. 1048.—Wing of *Blepharocera tenuipes*.

When the wing is first pulled out of the wing-sheath of the pupa, that part of it which is crossed by the fine lines is plaited somewhat like a fan and folded over the other portion. By this means the wing, which is fully developed before the adult emerges, is packed within the wing-sheath of the pupa, which is much shorter and narrower than the wing. When the wing is finally unfolded, it does not become perfectly flat, but slight, alternating elevations and depressions remain, showing the positions of the former folds, a permanent record of the unique history of the wings of these insects.

Ordinarily the wings of insects, while still in the wing-sheaths of the pupa, are neither longer nor wider than the wing-sheaths, but expand after the adult emerges from the pupa skin. Usually it takes considerable time for the wings to expand and become fit for flight and during this interval the insect is in an almost helpless condition. In certain caddice-flies that emerge from swiftly-flowing water, the time required for the expansion of the wings has been reduced to the minimum. In the net-winged midges, which also emerge from swiftly-flowing water the difficulty is met by the wings reaching their full development before the adult leaves the pupa-skin. It is only necessary when the adult emerges from the water that it should unfold its wings to be ready for flight.

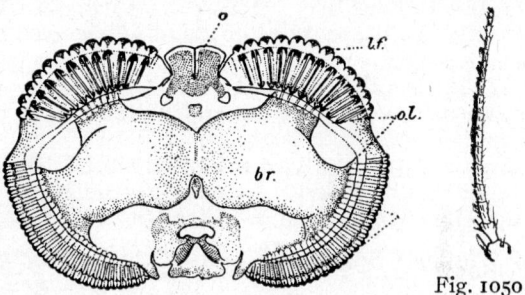

Fig. 1049.—Section of head through the eyes of *Blepharocera tenuipes*: *o*, ocelli; *br*, brain; *l. f*, large facets; *s. f*, small facets, *o. l*, optic lobes. (From Kellogg.)

Fig. 1050

The members of this family have three ocelli. The compound

eyes are usually divided in both sexes; the upper part of each eye being composed of large facets, characteristic of night-eyes, and the lower part, of smaller facets, characteristic of day-eyes, in a few species the eyes are divided only in the males. Figure 1049 represents a section of the head of the *Blepharŏcera tenūipes* through the eyes. The antennæ are thread-like, but are not furnished with whorls of long hairs (Fig. 1050). The mouth parts are elongate; the females have slender flattened elongate mandibles (Fig. 1051) the males are without mandibles. The legs are very long. On the dorsum of the mesothorax there is on each side, beginning just in front of the base of the wing, a well-marked suture like that of the crane-flies; but the two do not meet so as to form a continuous V-shaped suture as in the Tipulidæ.

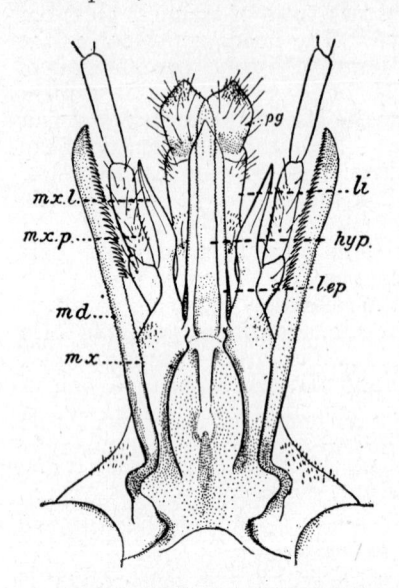

Fig. 1051.—Mouth-parts of female of *Bibiocephala doanei: l. ep*, labrum-epipharynx; *md*, mandibles; *mx. l*, maxillary lobe; *mx. p*, maxillary palpus; *hyp.* hypopharynx; *li*, labium; *pg*, paraglossa. (From Kellogg.)

In some species at least there are two kinds of females, which differ somewhat in the shape of the head. These two forms also differ in habits, one being blood-sucking, the other feeding upon nectar. The adults may be found resting on the foliage of shrubs or trees on the margins of mountain-brooks, or dancing in the spray of waterfalls.

The immature forms of these insects are even more wonderful than are the adults. The larvæ live in water, in swiftly-flowing streams, where the water flows swiftest. I have observed the transformations of *Blepharocera tenuipes*, which is abundant in some of the ravines near Ithaca, N. Y.

The larvæ of this species are readily seen on account of their black color, and are apt to attract attention on account of their strange form. (Fig. 1052, a). At first sight the body appears to consist of only seven segments, but careful examination reveals the presence of smaller segments alternating with these. Each of the larger segments except the last bears a

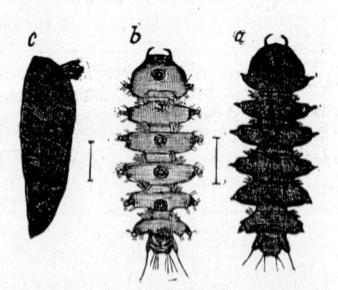

Fig. 1052.—*Blepharocera: a*, larva, dorsal view; *b*, larva, ventral view; *c*, puparium, side view.

pair of conical, leg-like appendages. On the ventral side of the body (Fig. 1052, b) each of the seven larger segments except the last bears a sucker, the cavity of which extends far into the body, and each of these segments except the first bears two tufts of tracheal gills; but those of the last segment are united. The head, which forms the front end of the first of the seven larger divisions, bears a pair of slender antennæ, each of these consists of a very short basal segment and two long segments; at the tip of the last of these there is a pair of minute appendages and a bristle. The suture between the head and the remaining parts of the first division is best seen on the ventral side of the body. On the dorsal side a suture may be seen dividing the last division into two segments.

The pupa-stage is passed in the same place as the larval. Like the larvæ the pupæ are very conspicuous on account of their black color, and are apt to occur like the larvæ closely clustered together. The pupa is not enclosed in the larval skin, and differs greatly in form from the larva. On the dorsal side the skin is hard, forming a convex scale over the body (Fig. 1052, c); and the thorax bears a pair of breathing organs, each composed of four flattened leaves, two of them delicate tracheal gills, and the other two protecting chitinized plates; on the ventral side the skin is very delicate, soft, and transparent; so that the developing legs and wings may be easily seen when the insect is removed from the rock. The pupæ cling to the rock by means of six suckers, three on each side near the edge of the lower surface of the abdomen; and so firmly do they cling that it is difficult to remove specimens without breaking them.

I have watched the midges emerge from their pupa-skins and escape from the water. The pupæ occurred in groups so as to form black patches on the rocks. Each one was resting with its head down stream. Each midge on emerging forced its way out through a transverse rent between the thorax and abdomen. It then worked its body out slowly and in spite of the swift current held it vertical. The water covering the patch of pupæ varied from 6 mm. to 25 mm. in depth. In the shallower parts the adult had no trouble in working its way to the surface still clinging to the pupa-skin by its very long hind legs.

While still anchored by its legs the midge rests on the surface of the water for one or two seconds and unfolds its wings; then freeing its legs it takes flight. The adults emerging from the deeper water were swept away by the current before they had a chance to take wing. The time required for a midge to work its way out of the pupa-skin and take flight varied from three to five minutes.

The larvæ of the net-winged midges live only in swift-flowing streams; they are found, therefore, only in mountainous or at least hilly regions. It is believed that they feed chiefly on algæ and diatoms. It does not seem probable that these delicate midges can deposit their eggs on the rocks in the swift-running water where the larvæ live, as do the females of *Simulium*. It is more likely that the eggs are deposited on the wet rocks at the margins of the stream and that the larvæ migrate to the center of the stream.

This family is a small one; but it is world-wide in distribution. Representatives of it have been found in both North and South America, in Europe, and in Australia and New Zealand. A monograph of the North American species was published by Kellogg ('03) and one of those of the world by the same author ('07), a table of the North American species is given by Kellogg in Williston ('08).

Family THAUMALEIDÆ

The Solitary-Midge

Only a single species of this family is known to occur in North America; this is *Thaumālea americāna*. It is a small fly measuring about 8 mm. in length, and is found on the banks of streams.

The antennæ are short, about as long as the head, and nearly of the same structure in the two sexes; the segments of the antennæ except those at the base are slender and are clothed with a few short hairs. The ocelli are wanting. The eyes are large and meet in front in both sexes. The venation of the wings is illustrated by Fig. 1053 vein R_1 ends at or near the end of the second third of the costal margin; the radial sector is two-branched; media is simple; and there are no anal veins.

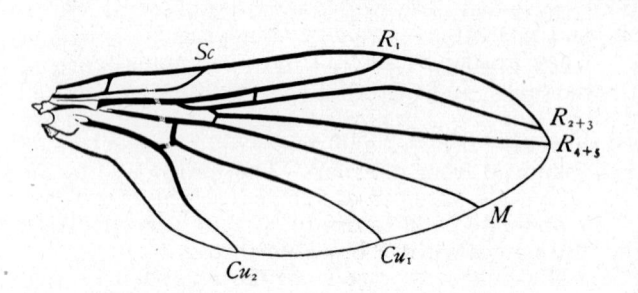

Fig. 1053.—Wing of *Thaumalea americana*.

The larvæ live in streams and resemble those of Chironomidæ.

This family has been commonly known as the Orphnephilidæ; but it has been shown by Bezzi ('13) that the typical genus was first described under the name *Thaumalea*.

Series II—BRACHYCERA*

The Short-horned Orthorrhapha

In most of the families included in this division of the Orthorrhapha the antennæ are short and three-jointed, the flagellum being reduced to a single segment, with or without a style or arista; but in the first subseries, the Anomalous Brachycera, the flagellum is more or less distinctly segmented. In all of the Brachycera the palpi are porrect and one- or two-jointed; and the first anal cell is either closed or narrowed towards the margin of the wing.

*Brachўcera: *brachy*, short; *ceras* (κερας), a horn.

Subseries A—The Anomalous Brachycera

In the families constituting this subseries the antennæ consist of five or more segments; but those beyond the second, the flagellum, are usually more or less consolidated. In some cases the antennæ do not differ markedly in form from those of certain Anomalous Nemocera; but the Brachycera are sharply distinguished from the Nemocera by the palpi being porrect and only one or two-jointed and by the fact that the anal cell is either closed or narrowed towards the margin of the wing. In this subseries the head and thorax are not furnished with strong bristles and the empodia are pulvilli-form.

Family TABANIDÆ

The Horse-Flies

The horse-flies are well-known pests of stock, and are often extremely annoying to man. They appear in summer, are common in woods, and are most abundant in the hottest weather.

In this family the flagellum of the antennæ is composed of from four to eight, more or less closely consolidated segments and is never furnished with a distinct style or arista (Figs. 1054, 1055). The wing-veins (Fig. 1056) are evenly distributed over the wing, as the branches of vein R are not crowded together as in the following family; the costal vein is continued as an ambient vein which extends

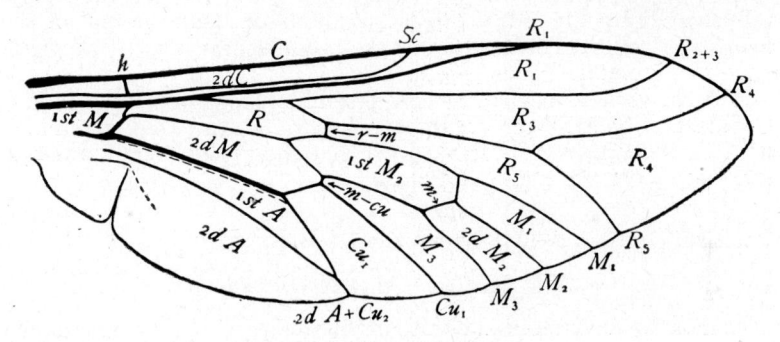

Fig. 1054.—Antenna of *Tabanus*.

Fig. 1055.—Antenna of *Chrysops*.

Fig. 1056.—Wing of *Tabanus*.

completely around the wing; the alulets are large, in other Anomalous Brachycera they are small or vestigial.

The flight of these flies is very powerful, they are able to outstrip the swiftest horse. The males feed on the nectar of flowers and on sweet sap. The mouth-parts of the female are fitted for piercing the, skin and sucking the blood of men and quadrupeds; the females, however, also feed on sweets of plants when they cannot obtain blood.

The larger species, as well as some of moderate size, belong to the genus *Tabānus* of which nearly two hundred species are listed from North America. One of the most common of these is the mourning horse-fly, *Tabānus atrātus*. This insect is of uniform black color throughout, except that the body may have a bluish tinge (Fig. 1057). The species of this genus attack cattle and other farm animals almost exclusively.

Fig. 1057.—*Tabanus atratus*.

Fig. 1058.—*Chrysops niger*.

To the genus *Chrȳsops* belong the smaller and more common horse-flies with banded wings (Fig. 1058). The species of this genus attack man as well as other animals. To this genus belong the well-known deer-flies familiar to fishermen and hunters. Sixty-three North American species of this genus are listed by Aldrich.

The eggs are deposited in large masses on plants or on exposed stones in the bed of a stream.

The larvæ are aquatic or semi-aquatic. As far as known, they are predacious, feeding on various small animals, some upon snails, others upon the larvæ of insects. In most cases they have a single pair of spiracles, which is situated at the hind end of the body; some have a pair of spiracles at each end of the body. Figure 1059 represents a larva of *Tabanus*.

The pupa is not enclosed in the skin of the larva.

Hine ('03) redescribes all Ohio Species and gives a table of the North American genera.

Family STRATIOMYIIDÆ

The Soldier-Flies

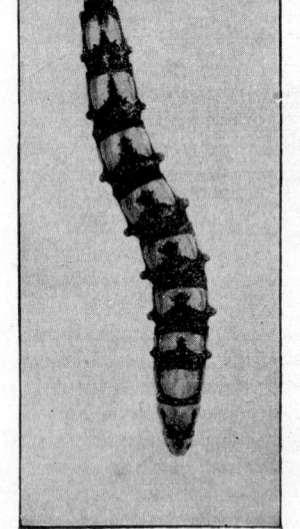

Fig. 1059.—Larva of *Tabanus*. (Photo. by M. V. Slingerland.)

The soldier-flies are so called on account of the bright-colored stripes with which some of the species are marked.

In the more typical members of this family the abdomen is broad and greatly flattened (Fig. 1060) and the wings when at rest lie parallel upon each other over the abdomen. But in some genera the abdomen is narrow and considerably elongate.

The antennæ vary greatly in form, in some genera the flagellum is long and consists of several quite distinct segments (Fig. 1061) in others it is short with but few indistinctly separated segments and with an arista (Fig. 1062) as in the True Brachycera.

The most distinctive characteristic is the peculiar venation of the wings (Fig. 1063). The branches of vein R are crowded together near the costal border of the wing; and the first cell M_2 is unusually short and broad; the branches of vein M and vein Cu_1 are comparatively weak, and the tibiæ are without spurs.

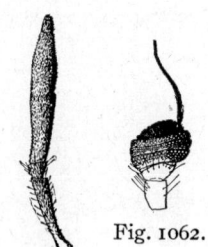

Fig. 1060.—*Stratiomyia*.

Fig. 1061.

Fig. 1062.

These flies are found on flowers and leaves, especially in the vicinity of water, and in bogs and marshes, but some species are found far from water.

The larvæ occur in various situations; some are aquatic and feed upon algæ, decaying vegetable matter, and small Crustacea; some live in privies, in cow-dung, and in other decaying

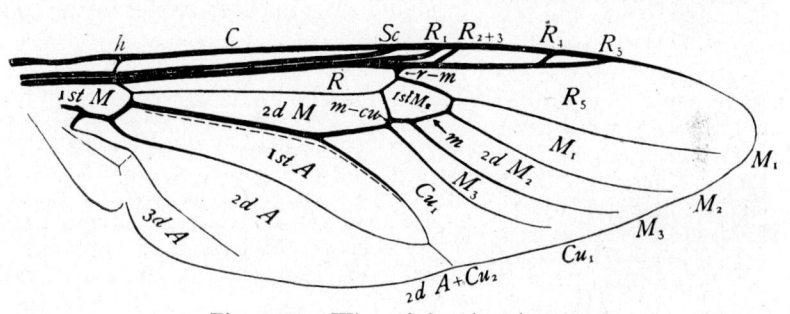

Fig. 1063.—Wing of *Stratiomyia*.

matter; some are found under bark of trees that has become slightly loosened and feed upon sap and upon insect larvæ; and some have been found in nests of Hymenoptera and in those of rodents, where they act as scavengers.

The larvæ are spindle-form or elliptical and flattened, and with the surface of the body finely shagreened. The posterior pair of spiracles is situated in a cleft or chamber at the end or near the end of the body. In the aquatic forms the apical respiratory chamber is furnished along its margins with long plumose hairs. When the larva is at rest hanging from the surface of the water these hairs are spread radiatingly upon the surface film; they thus form a means of support and prevent the water from entering the respiratory chamber.

Fig. 1064.—Puparium of *Odontomyia*.

The aquatic species leave the water to

transform. The pupæ of the Stratiomyiidæ are enclosed within the last larval skin, differing in this respect from other Brachycera (Fig. 1064).

This is a large family; more than three hundred species, representing more than forty genera, have been described from North America

Family XYLOMYIIDÆ

I group together here, provisionally, two genera, as representing

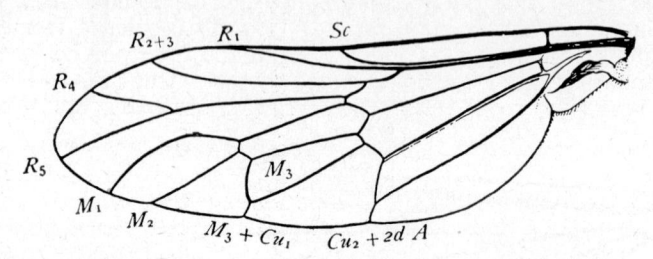

Fig. 1065.—Wing of *Xylomyia*. (After Verrall.)

a separate family, that have been placed by some writers in the Stratiomyiidæ and by others in the Rhagionidæ. These genera are *Xylomyia* and *Rhachicerus*. Both differ from the Rhagionidæ, as restricted here, in the form of the antennæ, they clearly belong with the Anomalous Brachycera. They differ from the Stratiomyiidæ in that the branches of radius are not crowded together near the costal border of the wing and in the possession of tibial spurs. They agree with each other and differ from all other Anomalous Brachycera found in our fauna in that cell M_3 is closed (Fig. 1065).

Xylōmia.—This genus includes rather elongate flies, somewhat Ichnenmon-like in appearance, which are mainly of black coloration with more or less yellow markings. The flagellum of the antennæ consists of eight closely consolidated segments, the last of which usually bears a tiny style (Fig. 1066). Six species have been described from our fauna.

The larvæ of *Xylomȳia păllipes* have been found under the bark of fallen trees and are predacious.

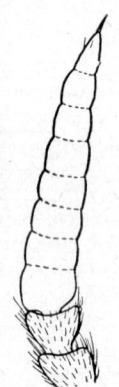

Fig. 1066.—Antenna of *Xylomyia*. (After Verrall.)

Rhachicerus.—The members of this genus resemble *Xylomyia* in the form of the body but differ markedly in the structure of the antennæ. In *Rhachicerus* the flagellum of the antennæ consists of from twenty to thirty-five segments. The segments of the flagellum are more or less cup-shaped; and in some species

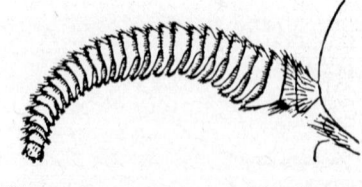

Fig. 1067.—Antenna of *Rhachicerus*. (From Williston.)

one edge of each segment is prolonged so that the antennæ are pectinate (Fig. 1067). Four species of this genus have been found in America north of Mexico.

The larva of *Rhachicerus nitidus* has been found in a decayed log.

Family XYLOPHAGIDÆ

This family, like the preceding one, includes slender flies, which are Ichneumon-like in appearance. It is distinguished from other Anomalous Brachycera as follows: from the Stratomyiidæ in that the branches of radius are not crowded together near the costal border of the wing (Fig. 1068) and in the possession of tibial spurs; from the

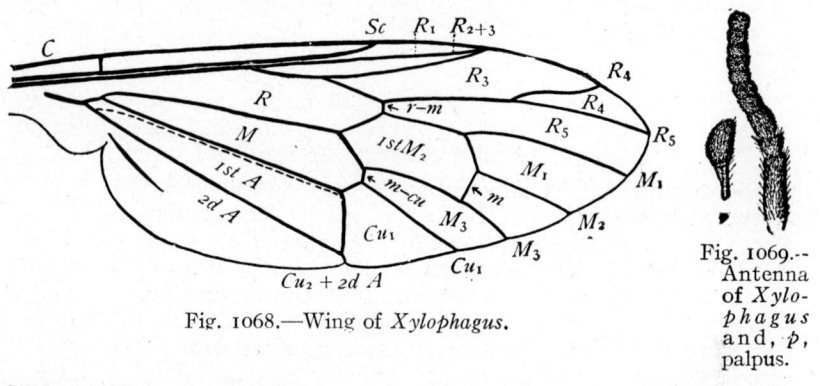

Fig. 1068.—Wing of *Xylophagus*.

Fig. 1069.—Antenna of *Xylophagus* and, *p*, palpus.

Xylomyiidæ in that cell M_3 is open; and from the Cœnomyiidæ in the absence of spinous protuberances on the scutellum. The flagellum of the antennæ consists of several closely consolidated segments; the antenna of a member of the typical genus is represented in Figure 1069.

This family is represented in our fauna by the following genera:

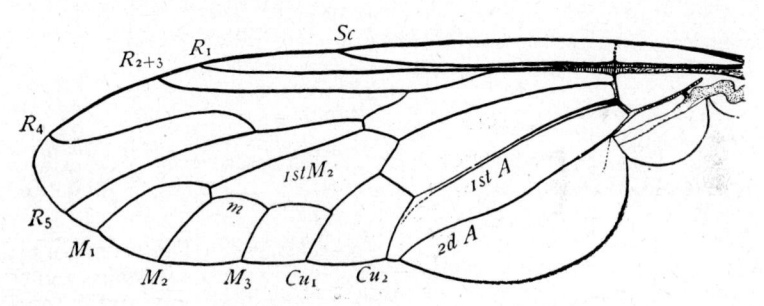

Fig. 1070.—Wing of *Cœnomyia ferruginea*. (After Verrall.)

Xylophagus, Glutops, Arthoceras, Arthropeas, and *Misgomyia*. These genera include sixteen species described from our fauna.

834   AN INTRODUCTION TO ENTOMOLOGY

The larvæ that are known live in earth or under the bark of rotten trees and feed upon the larvæ of other insects.

Family CŒNOMYIIDÆ

The members of this family differ from the Stratiomyiidæ in that the branches of radius are not crowded together near the costal border of the wing (Fig. 1070) and in the possession of tibial spurs; and they differ from all other Anomalous Brachycera in that the scutellum is armed with two spinous protuberances. The eyes are pubescent; the flagellum of the antennæ consists of eight closely consolidated segments (Fig. 1071); veins M_3 and Cu_1 anastomose for a considerable distance; and the tips of veins Cu_2 and 2nd A are narrowly separated or, rarely, united for a short distance.

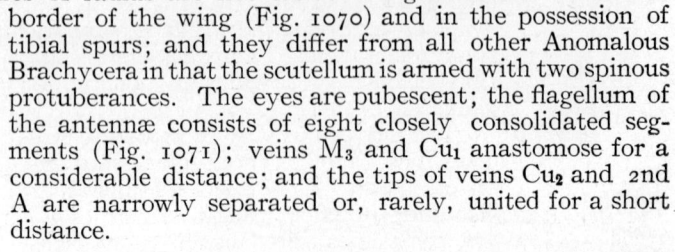

This family is represented in our fauna by a single species, *Cænomyia ferruginea*. This is a large thick-bodied fly, often measuring 25 mm. in length; but it varies greatly in size. It is of a pale yellowish brown color.

Fig. 1071.—Antenna of *Cænomyia*.

The larvæ are usually found in the ground, but they are sometimes found in decaying wood; they are predacious, feeding upon insect larvæ.

Subseries B—the true Brachycera

In the families constituting this subseries the antennæ are usually three-jointed, but in some cases they are four- or five-jointed; the third segment is not ringed, but usually bears a style or an arista. A similar type of antenna is possessed by the Cyclorrhapha, which were formerly on this account included in the Brachycera; but this term is now restricted to the Short-horned Orthorrapha.

In the first three families the head and thorax are not furnished with strong bristles and the empodia are pulvilliform.

Family RHAGIONIDÆ*

The Snipe-Flies

These trim-appearing flies have rather long legs, a cone-shaped abdomen tapering towards the hind end (Fig. 1072) and sometimes a downward-projecting proboscis, which with the form of the body and legs has suggested the name snipe-flies.

The body is naked or hairy, but it is not clothed with strong bristles. Frequently the hairy covering, though short, is very dense and is of strongly-contrasting

Fig. 1072.—*Chrysopila thoracica*.

*This family has been commonly known as the Leptidæ, but Rhagionidæ is the older name and is now coming into use.

colors. Three ocelli are present. The antennæ are only three-jointed and the third segment bears a style or an arista (Fig. 1073).
The proboscis is usually short, only a few members of the family having it long like the bill of a snipe. The wings are broad, and when at rest are held half open. The empodia are pulvilliform (Fig. 1074).

The venation of the wings is comparatively generalized. Figure 1075 represents the venation of a wing of the typical genus.

The flies are usually of moderate size. They may be found about low bushes and on tall grass. They are sometimes sluggish and, therefore, easily caught. They are sometimes predacious upon other insects, and the species of *Symphoromyia* suck blood as do horseflies.

Fig. 1073.—Antenna of *Chrysopila*.

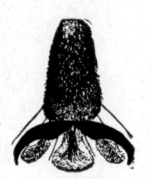

Fig. 1074.

The females of the genus *Athērix* have the remarkable habit of clustering in large numbers on branches or rocks overhanging water, where they deposit their eggs in common, and dying as they do so, add their bodies to the common mass, which may contain thousands

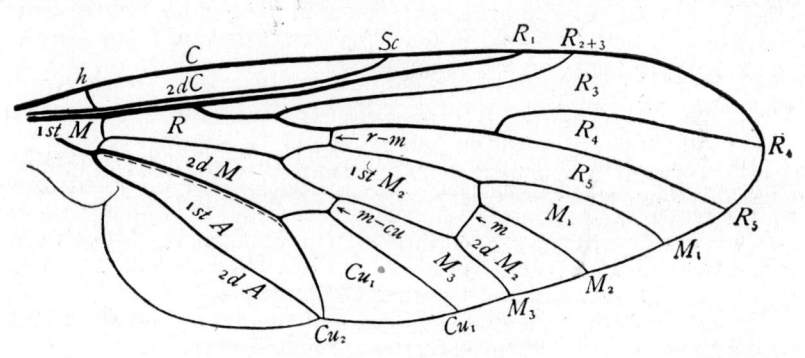

Fig. 1075.—Wing of *Rhagio*.

of individuals. Figure 1076, copied from Sharp ('99) represents a European species, *Atherix ibis*, natural size, and a mass of dead flies, much reduced. It is said that the larvæ feed upon the bodies of the dead mothers until the mass is loosened and falls into the water, where the larvæ complete their growth. Other writers state that the larvæ drop into the water when hatched.

Large masses of these flies have been observed in various parts of this country; and formerly, in the Far West, they were collected by the Indians and used for food after being cooked. It is said that as many as a hundred bushels of flies could be collected in a single day. For an account of the methods of collecting the flies and of preparing them for food practiced by the Indians see Aldrich ('12).

The larvæ of this family are found in various situations; some as those of *Atherix* live in water, but a larger number live in earth, in decaying wood, or in sand. The larvæ of *Vermileo* resemble ant-lions in habits, digging pitfalls in sand for trapping their prey. Only one species, *Vermileo comstŏcki*, has been found in America; this lives in the mountains of California; other species are well-known in Europe. I propose the common name *ant-tigers* for the larvæ of this genus.

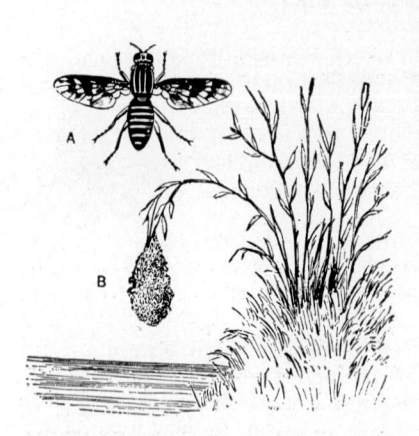

Fig. 1076.—*Antherix ibis:* A. The fly, natural size; B, mass of dead flies overhanging water, much reduced. (From Sharp.)

Family NEMESTRINIDÆ

The Tangle-veined Flies

The members of this family are of medium size; some of them resemble horse-flies, and others bee-flies. They can be recognized by the peculiar venation of the wings, there being an unusual amount of anastomosing of the veins (Fig. 1077), which gives the wings a very characteristic appearance.

The antennæ are small and short; the third segment is simple and furnished with a slender, jointed, terminal style. The proboscis is usually long, sometimes very long and fitted for sucking nectar

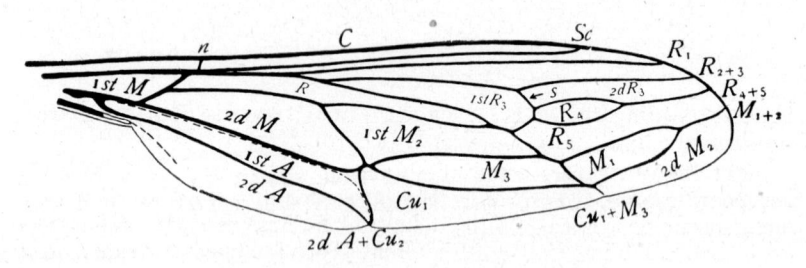

Fig. 1077.—Wing of *Parasymmictus clausa*.

from flowers. The head and thorax are not armed with strong bristles, and the empodia are pulvilliform.

But little is known regarding the habits of the larvæ; one species has been found to be an internal parasite of coleopterous larvæ.

Only six species have been found in America north of Mexico.

Family ACROCERIDÆ*

The Small-headed Flies

These flies are easily recognized by the unusually small head, the large humpbacked thorax, the inflated abdomen, and the very large alulets (Fig. 1078). The body is devoid of bristles and the empodia are pulvilliform.

The head is composed almost entirely of eyes, and in some genera is minute. The eyes are contiguous in both sexes or nearly so. The antennæ are three-jointed, and are furnished with a style or an arista in some genera, in others not. Sometimes the antennæ are apparently two-jointed, the first segment being sunken in the head. The venation of the wings varies greatly in the different genera. The accompanying figure (Fig. 1079) represents a single genus rather than the family.

Fig. 1078.—*Pterodontia misella.*

The flies are generally slow and feeble in their movements. In some species that feed upon flowers the proboscis is very long, some-

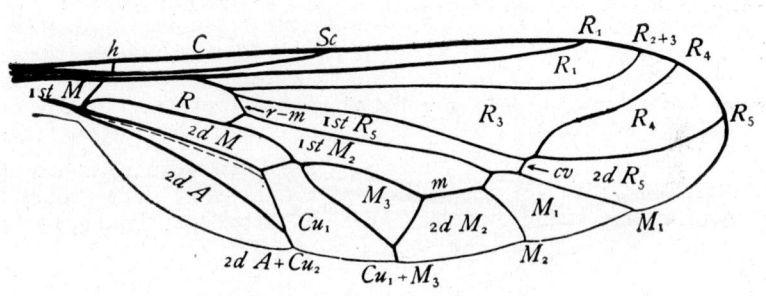

Fig. 1079.—Wing of *Eulonchus.*

times exceeding the body in length. Other species take no nourishment in the adult state, and have no proboscis.

The larvæ of only a few members of this family have been observed; these are parasitic in the egg-sacs or in the bodies of spiders. The life-history of an American species, *Pterodŏntia flăvipes* has been described by King ('16). The females of this species deposit their eggs on the bark of trees; they produce a large number of eggs; in one of the cases observed these numbered 3,977. This production of many eggs is doubtless an adaptation made necessary by the fact that many of the young larvæ will fail to find spiders to attack and will consequently perish. A larva that succeeds in finding a spider bores into its body and there lives till fully grown; it leaves the body of its host to transform.

---

*This is the family Cyrtidæ of some authors; but Acroceridæ is the older name.

A monograph of the North American species of this family was published by Cole ('19).

Family BOMBYLIIDÆ

The Bee-Flies

These flies are mostly of medium size, some are small, others are rather large. In some the body is short and broad and densely clothed with long, delicate hair (Fig. 1080). Other species resemble the horse-flies somewhat in appearance, especially in the dark color or markings of the wings; but these can be distinguished from the horse-flies by the form of the antennæ and the venation of the wings.

Fig. 1080.—*Bombylius.*

The antennæ are usually short; they are three-jointed with or without a style; in some genera the style is so large that it may be considered a fourth segment. The ocelli are present. The proboscis is sometimes very long and slender, and sometimes short and furnished with fleshy lips at the extremity.

The radial sector is three-branched; cell R_3 is sometimes divided by a sectorial cross-vein (Fig. 1081, *s*); cell M_3 is obliterated by the coalescence of veins M_3 and Cu_1; in a few genera cell M_1 is also obliterated by the coalescence of veins M_1 and M_2; cell 1st A is narrowly

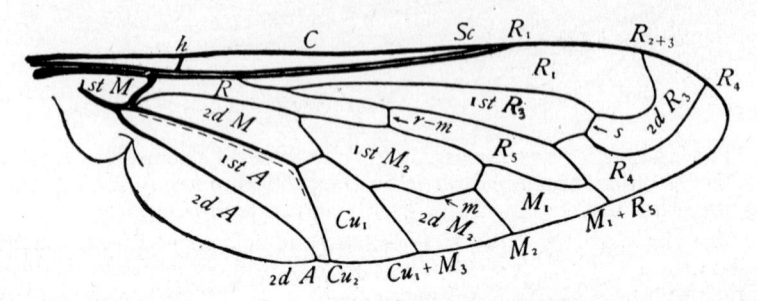

Fig. 1081.—Wing of *Pantarbes capito.*

open, or is closed at or near the border of the wing. The alulets are small or of moderate size. The empodia are pulvilliform.

The adult flies feed on pollen and nectar, and are found hovering over blossoms, or resting on sunny paths, sticks or stones; they rarely alight on leaves.

The larvæ are parasitic, infesting hymenopterous and lepidopterous larvæ and pupæ and the egg-sacs of Orthoptera. The pupæ are free.

The family is a large one; more than four hundred and fifty North American species, representing forty-one genera are listed by Aldrich ('05).

Family THEREVIDÆ

The Stiletto-Flies

With the flies of this family the head is transverse, being nearly as wide as the thorax; and the abdomen is long and tapering, suggesting the name stiletto-flies. These flies are small or of medium size; they are hairy or bristly. The antennæ are three-jointed; the third segment is simple, and usually bears a terminal style; but this is sometimes wanting. Three ocelli are present. The legs are slender and bristly; the empodia are wanting.

The radial sector is three-branched, (Fig. 1082) and the last branch, vein R_5, terminates beyond the apex of the wing; the three branches of media are separate; the cross-vein m-cu is present; and the first anal cell is usually closed.

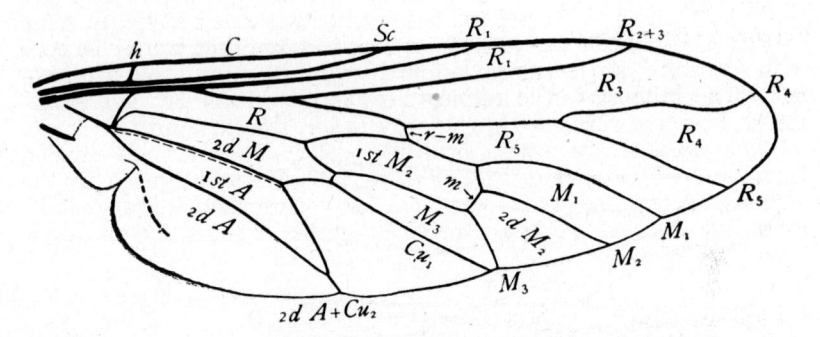

Fig. 1082.—Wing of *Thereva*.

The adult flies are predacious; they conceal themselves among the leaves of low bushes or settle on the ground in sandy spots, waiting for other insects, chiefly Diptera, upon which they prey.

The larvæ are long and slender, and the body is apparently composed of nineteen segments. They are found in earth, fungi, and decaying wood. They feed on decaying animal and vegetable matter and are said to be predacious also. The pupæ are free.

The family is a comparatively small one, including but few genera and species.

Family SCENOPINIDÆ

The Window-Flies

The window-flies are so-called because the best-known species are found almost exclusively on windows; but the conclusion that these are the most common flies found on windows should not be drawn from this name; for such is not the case.

These flies are of medium size, our most common species measuring 6 mm. in length. They are usually black, and are not clothed with bristles. The thorax is prominent, and the abdomen is flattened and somewhat bent down, so that the body when viewed from the side presents a hump-backed appearance (Fig. 1083). When at rest, the wings lie parallel, one over the other, on the abdomen; when in this position they are very inconspicuous. There are three ocelli. The antennæ are three-jointed; the first and second segments are short, the third is long and bears neither a style nor an arista (Fig. 1084).

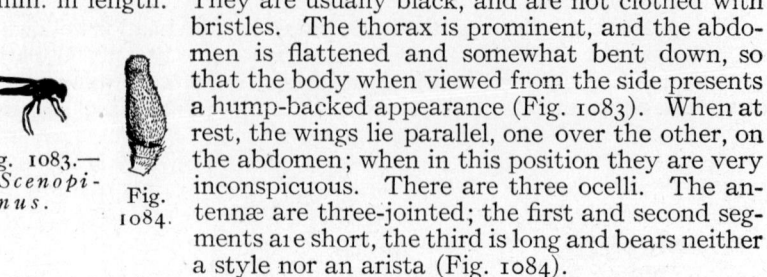

Fig. 1083.—*Scenopinus*.

Fig. 1084.

The venation of the wings is represented by Figure 1085. The radial sector is three-branched; cells M_1 and M_3 are both obliterated by the coalescence of the veins that bound them; the first anal cell is closed at a considerable distance before the margin; and cell R is much longer than cell 2d M.

The larvæ, which are sometimes found in dwellings under carpets or in furniture, are very slender, and are remarkable for the apparently large number of the segments of the body, each of the abdom-

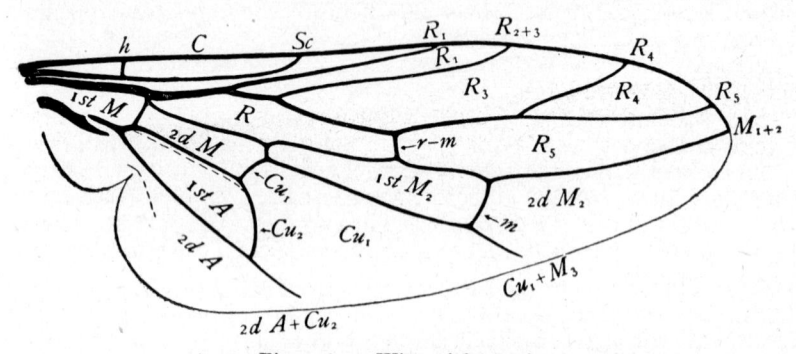

Fig. 1085.—Wing of *Scenopinus*.

inal segments except the last being divided by a strong constriction. They are also found in decaying wood, and are supposed to be carnivorous.

The family is a very small one. The most common species is *Scenŏpinus fenestrālis*.

Family ASILIDÆ

The Robber-Flies

These are mostly large flies, and some of them are very large. The body is usually elongate, with a very long, slender abdomen (Fig. 1086); but some species are quite stout, resembling bumblebees in form. This resemblance is often increased by a dense clothing of black and yellow hairs.

In this and the following family the vertex of the head is hollowed out between the eyes (Fig. 1087). In this family the proboscis is pointed and does not bear fleshy lips at the tip. The antennæ project forward in a prominent manner. They are three-jointed, and with or

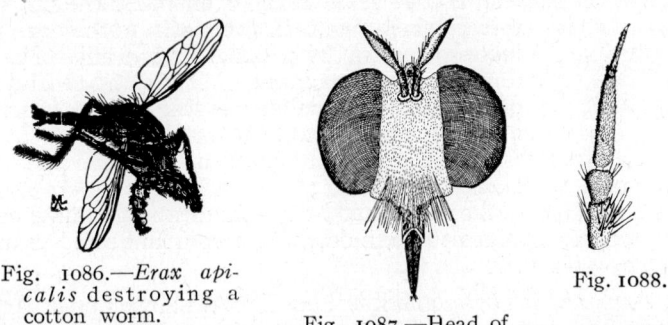

Fig. 1086.—*Erax apicalis* destroying a cotton worm.

Fig. 1087.—Head of a robber-fly.

Fig. 1088.

without a terminal style. The style when present sometimes appears like one or two additional segments (Fig. 1088).

Vein M_1 (Fig. 1089) does not terminate at or before the apex of the wing as in the following family. Cell M_3 is present, but is usually closed by the coalescence of the tips of veins M_3 and Cu_1. The tips of veins Cu_2 and 2d A may or may not coalesce for a short distance.

The robber-flies are extremely predacious. They not only destroy other flies, but powerful insects, as bumblebees, tiger-beetles, and dragon-flies, fall prey to them; they will also feed upon larvæ. They are common in open fields and are as apt to alight on the ground as on elevated objects.

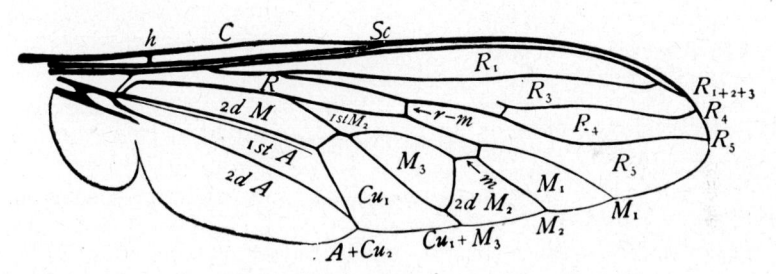

Fig. 1089.—Wing of *Erax*.

The larvæ live chiefly in the ground or in decaying wood, where they prey upon the larvæ of beetles; some, however, are supposed to feed upon the roots of plants. The pupæ are free.

More than five hundred North American species of this family, representing seventy-five genera, have been described.

Family MYDAIDÆ

The Mydas-Flies

The Mydas-flies rival the robber-flies in size, and quite closely resemble them in appearance. As in that family, the vertex of the head is hollowed out between the eyes; but these flies can be distinguished by the form of the proboscis, which bears a pair of fleshy lobes at the tip, by the form of the antennæ, which are four-jointed, long and more or less clubbed at the tip (Fig. 1090), and by the peculiar venation of the wings (Fig. 1091), vein M_1 terminating at or before the apex of the wing, and the branches of vein R coalescing near the apex of the wing in an unusual way.

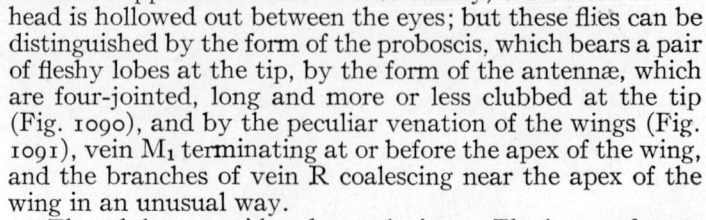

The adults are said to be predacious. The larvæ of some species, at least, live in decaying wood, and some are known to prey upon the larvæ of beetles.

Fig. 1090.

The family is a small one, there being only about one hundred known species, of these nearly fifty have been found in

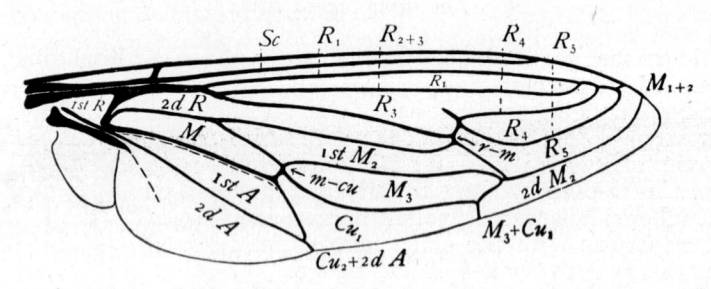

Fig. 1091.—Wing of *Mydas*.

North America, but most of these occur south of the United States. The family includes the largest known Diptera.

Family APIOCERIDÆ

The Apiocerids

This family includes only a small number of species, which are rare and occur in the Far West. They are rather large and elongate, and are found upon flowers.

The head is not hollowed out between the eyes; the ocelli are present; the antennæ are three-jointed, with or without a short simple style; the proboscis is not adapted for piercing. The radial sector is usually three-branched, but sometimes it is only two-branched; all of the branches of vein R end before the apex of the wing (Fig. 1092);

cell M_3 is present but closed by the coalescence of veins M_3 and Cu_1

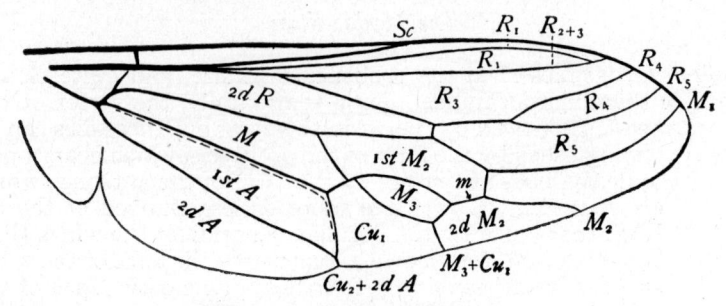

Fig. 1092.—Wing of *Apiocera*.

at the margin of the wing; and the medial cross-vein is present. The empodia are wanting. The larvæ are unknown.

Family DOLICHOPODIDÆ

The Long-legged Flies

These flies are of small or medium size and usually bright metallic green or blue in color. The legs are much longer than is usual in the families belonging to the series of short-horned flies (Fig. 1093). This suggested the name *Dolĭchopus*, which means long-footed, for the typical genus; and from this the family name is derived. It should be remembered however, that these flies are long-legged in comparison with the allied families, and not in comparison with crane-flies and midges.

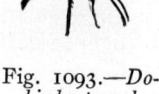

Fig. 1093.—*Dolichopus lobatus*.

The members of this family are easily distinguished as such by the peculiar venation of the wings, the most characteristic features of which are the following: cells M and 1st M_2 are rarely if ever separated by a complete vein; the basal part of vein M_3 being either atrophied Fig. 1094) or represented by a short spur (Fig. 1095); veins R_{2+3} and R_{4+5} separate near the base of the wing, and there is usually at the point of separation a more or less knot-shaped swelling; the cross-vein r-m, when present, is near this swelling, so that cell R is very short.

The members of this family have three ocelli; the antennæ are three-jointed; the second segment of the antennæ is sometimes vestigial and the third segment bears an arista; the palpi are one-jointed; and the empodia are not pulvilliform.

The adults are predacious and hunt for smaller flies and other soft-bodied insects. They are usually found in damp places, covered with rank vegetation. Some species occur chiefly on the leaves of aquatic plants, and about dams and waterfalls; and some are able to run over the surface of water. Others occur in dry places.

In the genus *Melandĕria*, of which two species have been found on the Pacific Coast, the outer lobe of the labella is mandible-like,

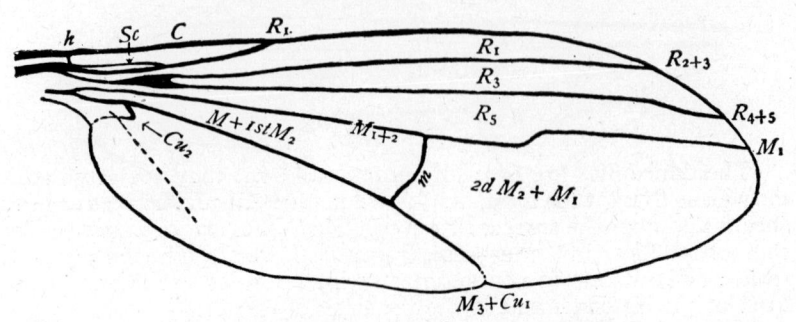

Fig. 1094.—Wing of *Dolichopus coquilletti*.

and doubtless functions as a mandible. For descriptions of these remarkable flies see Aldrich ('22).

The males of some species have the fore tarsi exceedingly elongated and slender, with the last segment in the shape of a comparatively large, oval, black disk, as shown in Figure 1093; in others the front tarsi are plain but the middle ones are elongated, thickened, and very black; and in still others the first two segments of the antennæ are ornamented with coarse black hair and the arista covered with a black pubescence. In several cases the males have been observed to display these ornaments before the female when courting. Detailed accounts of these observations are given by Professor Aldrich in the monograph of *Dolichopus* by Van Duzee, Cole, and Aldrich ('21).

The larvæ live in a variety of situations, some in earth or decomposing vegetable matter, some in the burrows of wood-boring larvæ

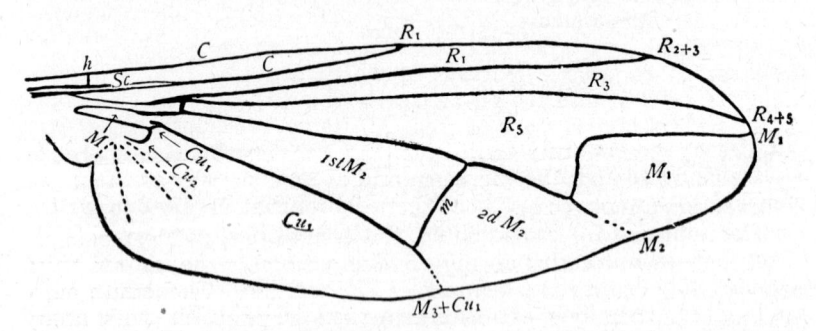

Fig. 1095.—Wing of *Psilopodius sipho*.

and also under bark; some in the stems of plants; and some are aquatic. But little is known regarding the habits of the larvæ; it is said that some species feed on decaying vegetation, while others are believed to be predacious.

This is a very large family; and representatives of it are common everywhere in the United States and Canada.

Family EMPIDIDÆ

The Dance-Flies

The dance-flies are of medium or small size; they are often seen in swarms flying with an up and down movement under trees or near shrubs and over the surface of water. These flies are predacious, like the robber-flies, but they also frequent flowers. The family is a rather difficult one to characterize owing to great variations in the form of the antennæ and in the venation of the wings.

The branches of vein Cu coalesce with the adjacent veins (vein Cu_1 with vein M_3 and vein Cu_2 with vein 2d A) from the margin of the wing towards the base for a considerable distance (Fig. 1096). In most genera this coalescence is carried so far that the free parts of the branches of vein Cu appear like cross-veins. The only other

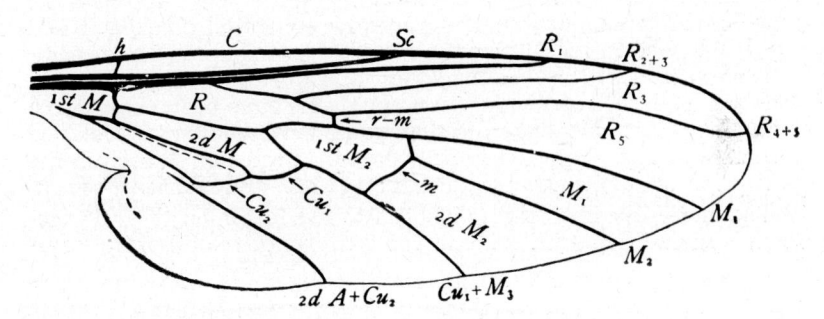

Fig. 1096.—Wing of *Rhamphomyia*.

families of the suborder Orthorrhapha in which this occurs are the Dolichopodidæ and the Lonchopteridæ, and the venation of the wings in each of these is very different from that of the Empididæ.

The antennæ are three-jointed, the first and second segments are often very small, and then appear like a single segment, the third segment may or may not bear a style or an arista. The mouth-parts are in many cases long, and extend at right angles to the body or are bent back upon the breast.

The larvæ live in various situations, some in the ground or in decaying wood, and some species are aquatic; they are believed to be either predacious or scavengers. The pupæ are free.

This family is a large one. It was monographed by Coquillett ('96) and by Melander ('02).

Family LONCHOPTERIDÆ
The Spear-winged Flies

These are minute flies, which measure from 2 mm. to 4 mm. in length, and are usually brownish or yellowish but never green nor metallic in color. When at rest the wings are folded flat, one over the other, on the abdomen. The apex of the wing is pointed, and the wing as a whole is shaped somewhat like the head of a spear (Fig. 1097). This suggested the family name.

The venation of the wings is very characteristic, and is sufficient to distinguish these flies from all others. The cross-veins r-m and m-cu are oblique, and near the base of the wing. Vein Cu_2 is very short, and extends towards the base of the wing. In the females

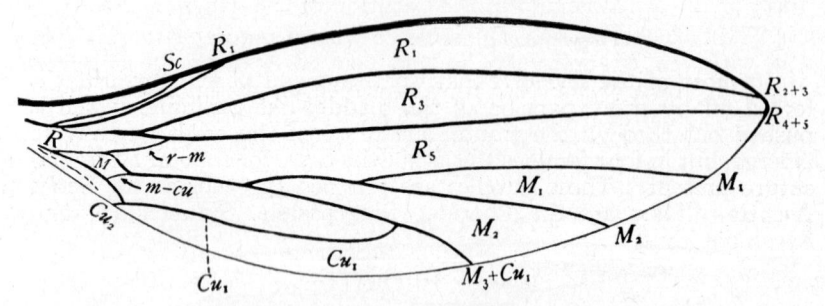

Fig. 1097.—Wing of *Lonchoptera*, female.

vein Cu_1 coalesces with vein M_3 as shown in the figure, but in the males the tip of vein Cu_1 is free. The posterior lobe is wanting.

Three ocelli are present. The antennæ are three-jointed; the third segment is globular, and bears a long arista.

These flies are common from spring till autumn, in damp grassy places. They frequent the shores of shady brooks, where the atmosphere is moist. The males are very rare in this country. Professor Aldrich examined over 2,000 specimens and found only two males among them.

"The larvæ live under leaves and decomposed vegetable matter. The larva transforms into a sort of semipupa within the last larval skin, and later into a true pupa" (Williston '08).

The family includes a single genus, *Lonchŏptera*, of which only three North American species are known.

Suborder CYCLORRHAPHA*

The Circular-seamed Flies

To this suborder belong those families of flies in which the pupa is always enclosed in a puparium from which the adult escapes through

---

*Cyclŏrrhapha: *cyclos*. (κύκλος), a circle; *rhaphe*, (ῥαφή), a seam.

a round opening made by pushing off the head-end of it. (Fig. 1098), the cap thus pushed off is often split lengthwise, as shown in the figure. The adult flies possess a frontal lunule and except in the first four families a frontal suture, through which the ptilinum is pushed out, when the adult is about to emerge from the puparium.

The antennæ are three-jointed, with a terminal or dorsal arista, rarely with a terminal style; in the Pupipara the antennæ are apparently only one- or two-jointed and sometimes lack the arista. The radius is not more than three-branched; cells M_1 and M_3 are wanting. The empodia are never pulvilliform.

Fig. 1098. Puparium of a muscid.

SERIES I—ASCHIZA*

Cyclorrhapha without a frontal suture

In most of the Cyclorrhapha the head end of the puparium is forced off by the expansion of the bladder-like ptilinum, which is pushed out through the frontal suture when the adult is ready to emerge; but in four families this is not the case, there being no frontal suture present. These families are grouped together as the Series Aschiza. They are the Phoridæ, Platypezidæ, Pipunculidæ, and Syrphidæ.

Family PHORIDÆ

The Hump-backed Flies

These are minute, dark-colored, usually black flies, which can be easily recognized by their humpbacked form, their peculiar antennæ, and the peculiar venation of the wings. Certain species are often found running about rapidly on windows, others on fallen leaves. Sometimes they are seen in swarms dancing up and down in the air. Many species measure less than 2 mm. in length, and some less than 1 mm.

The head is small; the thorax large and humped; and the abdomen rather short. The antennæ are three-jointed; but the first segment is exceedingly small, and the second is enclosed in the third, so that they appear as single-jointed. The third segment bears an arista, composed of two short basal segments and a long, usually more or less plumose third segment. The legs are large and strong and well adapted to jumping. The femora, especially of the hind legs are often very stout and flattened. The wings (Fig. 1099) are large, and are furnished with a series of strong veins near the costal border, which extend but a short distance beyond the middle of the wing. From these strong veins from three to five weak ones extend across the wing.

In the females of some species that live in the nests of ants and termites the wings are absent or very much reduced in size.

*Aschiza: *a* (ά), without; *schizo* (σχίζω), a cleft.

The larvæ of the different species differ greatly in habits, some feed on decaying vegetable matter, dead insects, snails, etc.; some are common in mushrooms, and are sometimes a pest in mushroom cellars; some are internal parasites of other insects, as bees, wasps, ants, saw-flies, etc.; several species are known to live in the nests of ants, some as parasites and others as commensals. One of these *Metŏpina*

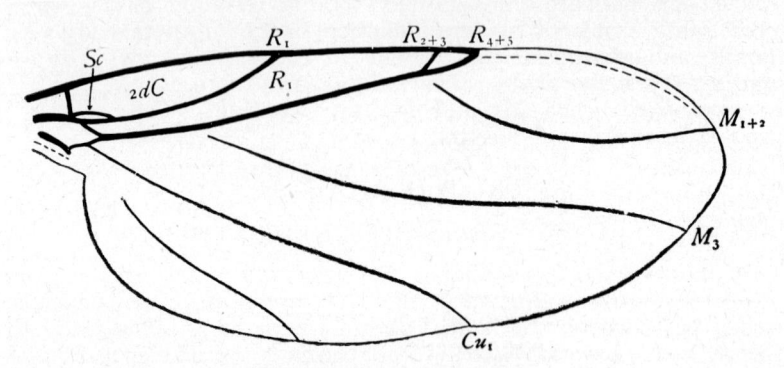

Fig. 1099.—Wing of *Phora*.

pachycŏndyle lives curled about the neck of its host ant-larva, partaking of the food given the latter by the attendant worker ants; and one has been bred from an egg-sac of a spider.

A monograph of the North American species of this family was published by Brues ('03) and one by Malloch ('13). About 150 species have been described from this region.

The Phoridæ are classed among the Brachycera by some writers and among the Cyclorrhapha by others. Morris ('22) states that when the adult fly of a species studied by him, *Hypŏcera incrassāta*, emerges from the puparium a circular cap, consisting of the skin of the cephalic region and thoracic segments of the larva, is split off by means of a fissure passing round the body between the third thoracic and first abdominal segments. This confirms the opinion that this family should be placed in the Cyclorrhapha.

Family PLATYPEZIDÆ

The Flat-footed Flies

These flies resemble the house-fly somewhat in appearance but are very much smaller. They hover in the air in shady places and alight frequently on the leaves of low plants, where they run about in circles with great rapidity.

The head is hemispherical or spherical, and as broad as or broader than the thorax. The antennæ are three-jointed with a terminal arista. The legs are short and stout, and the tarsi of the hinder pair are often very broad and flat (Fig. 1100); but they vary greatly in form in different genera. The wings are rather large, and when at rest lie parallel upon the abdomen; the axillary excision is prominent, but the posterior lobe of the wing is small (Fig. 1101); the alulets are minute.

The radial sector is two-branched; veins M_1 and M_2 either coalesce throughout or separate near the margin of the wing; the medial cross vein is present in some members of the family and absent in others; cells 2d R, M, and 1st A are short.

This family is represented in North America by about twenty-five species, and these are usually rare. The larvæ live in mushrooms, the puparia are not very different from the larvæ in form.

Fig. 1100.—Leg of *Platypeza*, *a*, forked hairs of leg greatly enlarged.

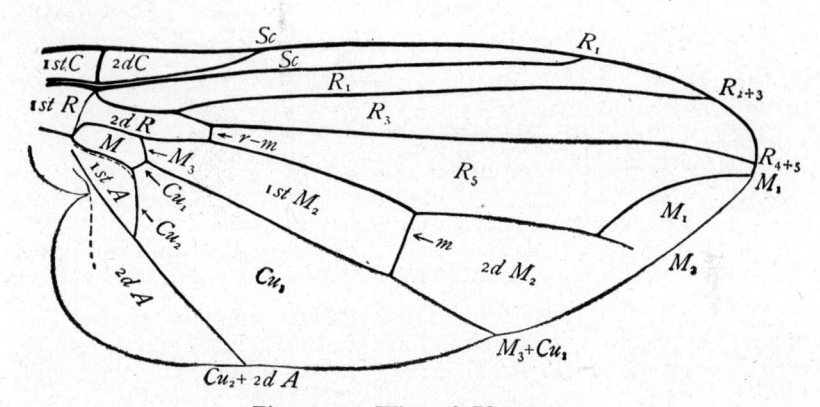

Fig. 1101.—Wing of *Platypeza*.

Family PIPUNCULIDÆ

The Big-eyed Flies

The members of this family are small flies with very large heads composed almost entirely of eyes (Fig. 1102). The head is nearly spherical and broader than the thorax. The antennæ are small, short, three-jointed, with a dorsal arista. The ocelli are present. The abdomen is somewhat elongate with the sides nearly parallel. The body is thinly clothed with hair or nearly naked. The wings are much longer than the abdomen, and when at rest they lie parallel to each other upon it. The venation of the wings (Fig. 1103) closely resembles that of the Conopidæ. The radial sector is two-branched. Veins R_{4+5} and M_{1+2} approach each other at their tips. Vein M_3 coalesces with vein Cu_1 for nearly its

Fig. 1102.—*Pipunculus*.

entire length. Veins Cu_2 and 2d A coalesce at their tips, except in *Chalarus*. In this genus vein M_{1+2} is atrophied and the medial crossvein absent. Cells R and M are long.

The flies hover in shady places. They are sometimes found on flowers, and may be swept from low plants; our most common species

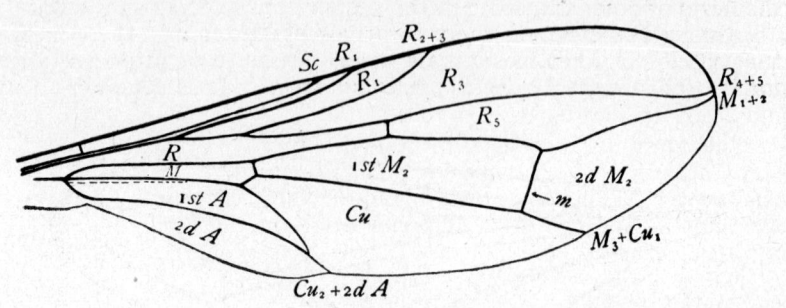

Fig. 1103.—Wing of *Pipunculus*.

measure about 3 mm. in length, not including the wings. The larvæ so far as known are parasitic upon bugs.

This small family is represented in North America by about thirty species, nearly all of which belong to the genus *Pipŭnculus*.

Family SYRPHIDÆ

The Syrphus-Flies

The family Syrphidæ includes many of our common flies; but the different species vary so much in form that no general description of their appearance can be given. Many of them mimic hymenopterous insects, thus some species resemble bumblebees, others the honeybee, and still others wasps; while some present but little resemblance to any of these.

The most distinctive characteristic of the family is the presence of a thickening of the membrane of the wing, which appears like a longitudinal vein between veins R and M. This is termed the *spurious vein*, and is lacking in only a few members of the family; it is represented in Figure 1104 by a band of stippling. Vein R_{4+5} is never forked; the tips of vein R_{4+5} and M_{1+2} coalesce; and the first anal cell is closed.

The antennæ are three-jointed; the third segment usually bears a dorsal arista, but sometimes it is furnished with a thickened style. The face is not furnished with longitudinal furrows to receive the antennæ as in the Muscidæ. The frontal lunule is present, but the frontal suture is wanting.

The adults frequent flowers and feed upon nectar and pollen. Some fly with a loud humming sound like that of a bee, others hover motionless except as to their wings for a time, and then dart away suddenly for a short distance, and then resume their hovering.

The larvæ vary greatly in form and habits. Some prey upon plant lice, and are often found in the midst of colonies of these insects; some feed in the stems of plants and in bulbs; others feed on decaying vegetable matter, and live in rotten wood, in mud and in water; and others live in ordure or in decomposing animal remains. Some are found in the nests of ants; and some in the nests of bumblebees and wasps.

Among the common representatives of this family there is one that so closely resembles a male honey-bee as to be often mistaken for it. This is the Drone-fly, *Eristalis tēnax*. It is common about

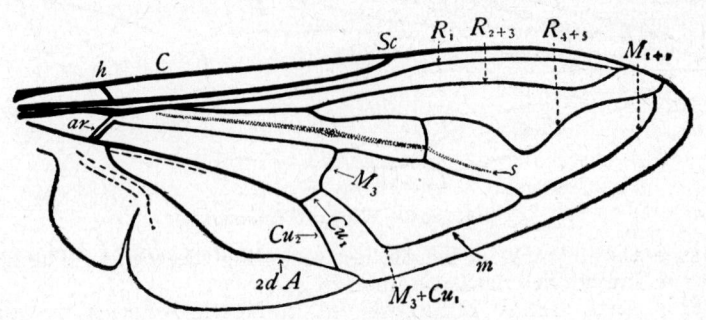

Fig. 1104.—Wing of *Eristalis*.

flowers. The larva lives in foul water, where it feeds on decaying vegetable matter; it is of the form known as "rat-tailed," which is described below.

The larvæ of the genus *Volucĕlla* live as scavengers in the nests of bumblebees and of wasps (*Vespa*). Some of the species in the adult state very closely resemble bumblebees.

The larvæ of the genus *Mĭcrodon* are hemispherical, slug-like creatures (Fig. 1105), which resemble mollusks more than ordinary maggots; they are common in ants' nests.

The larvæ of several species that live in water as well as some that live in rotten wood are known as rat-tailed maggots on account of the long, tail-like, appendage, with which the hind end of the body is furnished. This is a tube, like that of a diver, which enables the insect to obtain air when its body is submerged beneath several inches of water or decaying matter. This tube being telescopic can be lengthened or shortened as the insect may need it; and at its tip there is a rosette of hairs, which, floating on the surface of the water, keeps the tip from being submerged. The larva has on the ventral side of its body several pairs of tubercles armed with spines, which serve as prolegs.

Fig. 1105.—*Microdon*, adult and larva.

Fig. 1106.—*Syrphus*.

Among the more common members of this family are the yellow-banded species belonging to the genus *Syrphus* (Fig. 1106). The larvæ of these live in colonies of aphids, and do much good by destroying these pests.

This family is a very large one; more than 300 species representing about 60 genera have been described from America north of Mexico. Williston ('86) published an extended monograph of our species.

SERIES II—SCHIZOPHORA*

Cyclorrhapha with a frontal suture

This series of families includes all of the Cyclorrhapha except the four preceding families in which the frontal suture is wanting. In this series, the Schizophora, there is a frontal suture through which the ptilinum is extruded in order to force off the head end of the puparium when the adult is ready to emerge.

This series is divided into two quite distinct sections, the Myodaria and the Pupipara.

SECTION I—MYODARIA

The Muscids

This section of the order Diptera is a very large group, including many families and probably more than half of all the living species of this order. As many of the species are very common, it usually happens that a large portion of the flies in a collection belong to the families included in this section.

Excepting the first family, the Conopidæ, the families included in the Myodaria differ from all of the preceding families in that vein Cu_2 coalesces with vein 2d A to such an extent that it appears like a cross-vein or is curved back towards the base of the wing; and they differ from the following families, the Pupipara, in having the abdomen distinctly segmented and the two legs of each thoracic segment not widely separated.

The antennæ are three-jointed; the third segment bears an arista which is almost always dorsal in position. The radial sector is not more than two-branched; cells M_1 and M_3 are wanting; and the two branches of cubitus coalesce with the adjacent veins (Vein Cu_1 with M_3 and Cu_2 with 2d A) for nearly their entire length, except in the Conopidæ.

The families included in this section are grouped in two subsections, the Acalyptratæ and the Calyptratæ.

---

*Schizŏphora: *schizo* (σχίζω), a cleft; *phoros* (φορός), bearing.

Subsection I—THE ACALYPTRATÆ*

The Acalýptrate Muscids

In the families included in this subsection of the Myodaria the alulæ or calypteres are always small or rudimentary; the subcostal vein is often indistinct or vestigial, but well preserved in some forms; vein R_1 is shortened and is often very short; the thorax is without a complete transverse suture; the posterior callus is usually absent; and the abdominal spiracles, with some exceptions, are in the conjunctivæ. The flies are usually small or very small, they are never **large.**

The subsection Acalyptratæ includes many families, twenty-three of which are represented in our fauna. Some of these families include well-known species that have attracted attention on account of their economic importance or for other reasons; but most of the families have been studied comparatively little in this country.

Family CONOPIDÆ

The Thick-headed Flies

With the members of this family the head is large, being broader than the thorax. The body is more or less elongate; it may be naked or thinly clothed with fine hair, but it is rarely bristly.

The ocelli may be either present or absent. The antennæ are prominent and project forward; they are three-jointed; and the third

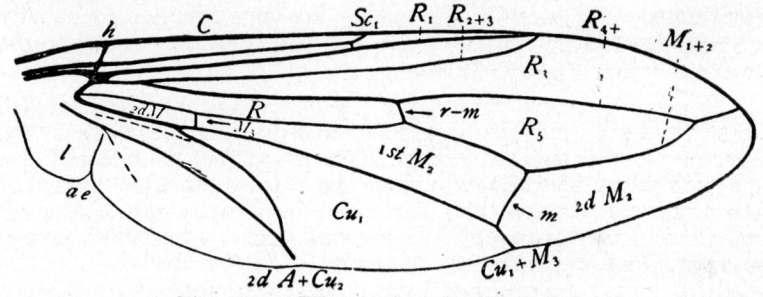

Fig. 1107.—Wing of *Physocephala affinis*.

segment bears either a dorsal arista or a terminal style. The radial sector is only two-branched (Fig. 1107); veins R_{4+5} and M_{1+2} end near together or coalesce at their tips. The medial cross-vein is present. Vein M_3 coalesces with vein Cu_1 for nearly its entire length. Veins Cu_2 and 2d A coalesce at their tips, and sometimes for nearly the entire length of vein Cu_2.

---

*Acalyptrātæ: *a* (ά); without; *calypter* (καλυπτήρ), a sheath.

The adult flies are found on flowers. In some genera the abdomen is long, with a slender, wasp-like pedicel (Fig. 1108). In others the abdomen is of the more usual form. The larvæ are parasitic, chiefly upon bumblebees and wasps, but some species infest locusts. The eggs are deposited by the female, in some cases at least, directly upon the bodies of the bees or wasps during flight. The newly hatched larvæ burrow within the abdominal cavity of their host.

Fig. 1108.—*Conops.*

Nearly one hundred North American species have been described.

Family CORDYLURIDÆ

The Dung-Flies

The members of this family are often of considerable size for Acalyptratæ, they are never very small. The subcostal vein is distinctly separated from vein R_1 and ends in the costa; vein R_1 is nearly half as long as the wing, cell M is not minute; the frontal vitta is usually well differentiated from the orbits; and the vibrissæ are present.

Although members of several families of flies frequent excrement certain species of this family and of the Borboridæ are so commonly observed about dung and refuse that they have received the common names of dung-flies. Among these are those of the genus *Scatŏphaga;* these are rather slender flies, which have the body clothed with yellowish hair, and which are often abundant especially about fresh cow-dung. Other members of this family are found in meadows and in moist places; some feed on other insects which they capture.

The larvæ of some species have been bred from excrement; some live in the stems of plants; and some are said to be parasitic in caterpillars.

This family is named the Scatophagidæ by some writers, and by some it is classed with the Calyptratæ.

The family CLUSIIDÆ is a small family of rather small flies; only a few representatives of which are found in our fauna. In this family the subcostal vein is distinctly separated from vein R_1 and ends in the costa; vein R_1 is less than one-third as long as the wing; the so-called front is broad and bristly to or nearly to the base of the antennæ; the frontal vitta is not differentiated from the orbits; the ocellar bristles are usually present; the postvertical bristles are divergent; and the vibrissæ are present. Larvæ of this family have been found in decaying wood and under the bark of trees.

This family is named the Heteroneuridæ by some writers.

Family HELOMYZIDÆ

The members of the Helomyzidæ can be recognized by the following combination of characters; the wings are armed with a row of spines along the costa (Fig. 1109); the subcostal vein is distinct; the oral vibrissæ are present; the postvertical bristles, which are located on the back of the head somewhat behind the ocelli, are convergent; and the tibiæ are armed with spurs and with preapical bristles.

The flies are found in shady and damp places; many of them have been found in caves; and some species on windows.

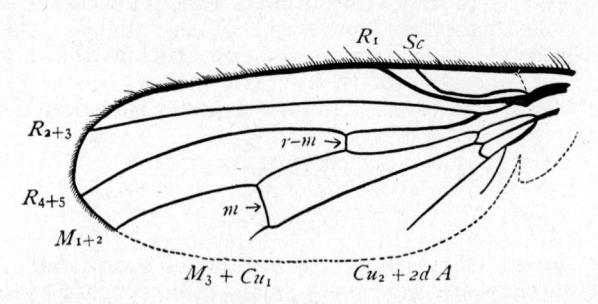

Fig. 1109.—Wing of *Leria*. (After Williston.)

"The larvæ of *Lēria* have been bred from bat and rabbit dung; those of *Helomȳza* from truffles, decaying wood, etc." (Williston)

The family was monographed by Aldrich and Darlington ('08); they recognized ten genera and about two score species found in the United States.

The family BORBORIDÆ is composed of rather small or very small black, brown or obscurely yellowish flies, having a quick short flight. Some of the species are very common, occurring in great numbers about excrement or near water. These "dung-flies" differ from those of the Cordyluridæ in that the subcostal vein of the wings is wanting or indistinct and in having the hind metatarsi dilated and usually shorter than the following segment.

The larvæ of some of the species, at least, live in excrement.

The family PHYCODROMIDÆ includes only a few species of flies that are found among sea-weeds along the sea-shore. Two species are listed from Alaska and one from California. This family is named the Cœlopidæ by Hendel ('22).

Family SCIOMYZIDÆ

The members of this family are usually brown or brownish yellow in color; and in many species the wings are spotted or infumated. They are usually found in moist situations, as along the banks of streams. The larvæ are aquatic.

These flies lack vibrissæ. The face in profile forms a sharp often very acute angle with that of the oral margin. The postvertical bristles are divergent when present.

This family is represented in this country by more than twenty genera and by nearly one hundred described species. It has been monographed by Cresson ('20) and also by Melander ('20a) under the name Tetanoceridæ.

Family SAPROMYZIDÆ

The head is as broad as or a little broader than the thorax; the legs are of moderate length; the hind tibiæ bear a preapical bristle; the abdomen is short; the ovipositor is neither flat nor drawn out; the subcostal vein is distinctly separated from vein R_1 and ends in the costa; vein 2d A does not reach the margin of the wing; the oral vibrissæ are absent; the second antennal segment bears a dorsal bristle; the postvertical bristles are convergent; and one or more sternopleural and a mesopleural bristle are present.

This family is composed of small flies, which are seldom more than 7 mm. in length. Nearly one hundred species have been described from America north of Mexico. A synopsis of the family was published by Melander ('13 a).

The larvæ of *Sapromyza* live in decaying vegetable matter.

This family is named the Lauxanidæ by some writers.

The family LONCHÆIDÆ includes the genera *Lonchæa* and *Pallŏptera*, which are included in the Sapromyzidæ by some writers. These genera differ from the Sapromyzidæ in that the hind tibiæ are without a preapical bristle; and the ovipositor is flattened and more or less projecting. A synopsis of our species is included in Melander ('13a).

Lonchæa polita has been reared from a decaying fungus (*Polyporus*) and from human excrement.

Family ORTALIDÆ

This family like the Trypetidæ is a large one and contains many common species which have the wings beautifully marked with dark spots or bands. The members of this family differ from the Trypetidæ in that the subcostal vein extends to the margin of the wing in the usual way and in that the lower fronto-orbital bristles are wanting.

Comparatively few species of the Ortalidæ have been bred and these differ greatly in habits. The larvæ of some have been found under bark of dead trees, others in excrement, some are parasitic on lepidopterous larvæ, and several infest growing plants. Among the latter are *Chætŏpsis ænea* and *Tritŏxa flexa* which sometimes infest onions; but the most important onion maggot is *Hylemȳia antiqua* of the family Anthomyiidæ.

The family includes six subfamilies each of which is given family rank by Hendel ('22).

Family TRYPETIDÆ

This is a very large family including many common species with pictured wings, in which it resembles the preceding family, the

Ortalidæ. These two families can be separated by differences in the subcostal vein; in the Trypetidæ the distal part of the subcostal vein is abruptly turned forward and usually becomes very weak. In this family the vibrissæ are wanting; the fronto-orbital bristles are numerous and extend down to the antennæ; cells M and 1st M_2 are separated by a cross-vein; the legs are moderately long; the tibiæ lack preapical bristles; and the ovipositor is flattened and more or less projecting.

The larvæ of the species that have been bred infest living plants. Some are leaf-miners, some live in the stems of plants, some make galls, and some are pests that infest fruit. Among the better-known species are the following.

The apple-maggot, *Rhagolētis pomonělla*.—The adult is blackish with the head and legs yellowish; the abdomen is crossed by three or four white bands (Fig. 1110) and the wings are crossed by four dark confluent bands. The female punctures the skin of the apple with her ovipositor and lays her eggs in the pulp. The larvæ bores tunnels in all directions through the fruit. Early maturing varieties of apples are

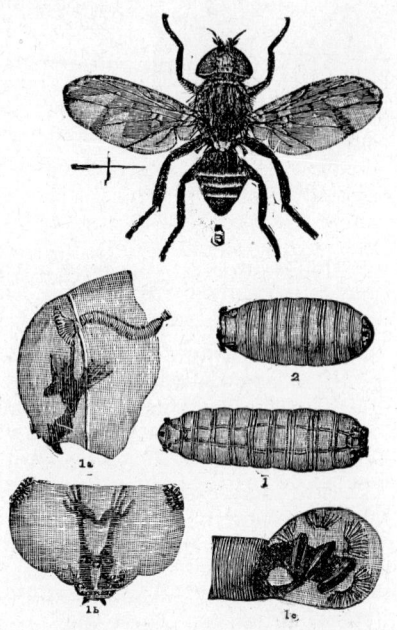

Fig. 1110.—The apple-maggot: *1*, larva; *2*, puparium; *3*, adult; *1a*. head of larva from side showing the oral hooks and spiracle; *1b*, head of larva from below; *1c*, caudal spiracle of larva.

especially attacked. When full-grown the larva goes into the ground to transform where it hibernates in a brownish puparium. This is a serious pest in the Eastern States and in Canada. This is a native species, originally feeding in the fruit of wild thorn. It has been found that most of the flies can be destroyed before they lay their eggs by applying a spray of arsenate of lead, four pounds in one hundred gallons of water, during the first week of July. The flies lap up drops of moisture from the fruit and foliage and are thus poisoned. The fallen apples should be collected or hogs allowed to run in the orchard.

The cherry fruit-flies, *Rhagolētis cingulāta* and *Rhagolētis fausta*. These two species, which are closely allied to the apple-maggot, infest cherries, but not so commonly as does the plum curculio. The cherry fruit-flies can be destroyed by the use of the arsenate of lead spray early in June.

The currant fruit-fly, *Epōchra canadēnsis*.—The larva of this species is a small white maggot which feeds within currants and gooseberries. The infested fruit colors prematurely and usually falls to the ground. No practicable method of controlling this pest has been suggested.

The round goldenrod gall.—One of the most familiar of abnormal growths on plants is a ball-like enlargement of the stem of goldenrod (Fig. 1111). This is caused by a maggot, which lives within it, and which develops into a pretty fly with banded wings; this is *Eurŏsta solidăginis*. The larva hibernates in the gall; the adult emerges in May. The gall of this species is easily distinguished from that of the solidago gall-moth, described in the preceding chapter, by its rounded form.

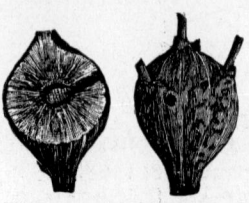

Fig. 1111.—The round goldenrod gall.

The family MICROPEZIDÆ is represented in America north of Mexico by a few species; most of the species of this family occur in South America. In our representatives of the family, the subcostal vein is distinctly separated from vein R_1; cell R_5 is closed or narrowed at the margin; the head is subspherical; the buccæ are broad; the face is retreating; the vibrissæ are wanting; and the proboscis is short.

The family TANYPEZIDÆ includes the genus *Tanypēza*, which differs from the preceding family in that the buccæ and the posterior orbits are narrow. The described species are chiefly from South and Central America. These flies are rare in the United States.

The family SEPSIDÆ includes only a few species. These are small slender flies, which are principally scavengers, feeding and breeding in decaying organic matter. They are not rare, and can be obtained by sweeping grass in meadows and pastures, especially where there are droppings of horses and cattle. "Species of *Sepsis* are particularly abundant and can be found on fresh dung, where they run about with vibrating wings pirouetting in a unique and pretty dance."

The family has been monographed by Melander and Spuler ('17).

Family PIOPHILIDÆ

This family includes only a few species of small flies, rarely exceeding 5 mm. in length. They are usually glistening black or slightly bluish metallic in luster. They are found about either decaying organic matter, preserved meats, or cheese. The best known species is the following.

The cheese-maggot, *Piŏphila cāsei*.—This fly lays its eggs on cheese, ham, and bacon. The larvæ live in these substances and are often serious pests. They are commonly known as "skippers" on account of the remarkable leaps they can make. This is accomplished by first bringing the head and tail ends together and then suddenly straightening the body; in this way they can leap several inches.

The members of this family were formerly included in the Sepsidæ. The family Piophilidæ was monographed by Melander and Spuler ('17).

Family PSILIDÆ

The flies of this family are of moderate size and slender. In many of the species the antennæ are very long and decumbent, but in others they are of moderate length. The vibrissæ are wanting; the costa is interrupted near the end of vein R_1; and cells 1st A and M are complete and relatively large. Five genera including thirty-three species have been described from our fauna. For a synopsis of the family see Melander ('20 b). The following is our best-known species.

The carrot rust-fly, *Psila rosæ.*—The larva of this species infests carrots, celery, parsnips, and parsley. In the case of carrots and parsnips the larvæ perforate the roots in all directions; their burrows are of a rusty color, hence the common name of the insect. When celery is attacked the fibrous roots are eaten off and destroyed.

The family DIOPSIDÆ is represented in North America only by the following species.

The stem-eyed fly, *Sphyracĕphala brevicŏrnis.*—This is a very singular fly, which is found on the leaves of skunk-cabbage and the foliage of other plants in shady glens. On each side of the head there is a lateral process, upon which the eye is situated. The life-history of the species is unknown.

The family CANACEIDÆ is represented in our fauna by a single described species, *Cănace snodgrăssi*, recorded from New Jersey.

Family EPHYDRIDÆ

These are small or very small, black or dark-colored flies, that live in wet places. The subcostal vein is coalescent for the greater part of the length with vein R_1, being distinct only at its proximal end; cells M and 1st M_2 are not separated by a cross-vein; the hind metatarsi are not thickened and are longer than the following segment; the vibrissæ are wanting; and in some species the mouth cavity is very large.

Most of the species live about fresh water; but to this family belong the "Brine-flies" the larvæ of which live in salt or strongly alkaline waters. These are common in pools about salt-works; and in the far West and in Mexico these larvæ occur in the alkaline lakes in countless numbers, and are washed ashore in such quantities that bushels of them can be collected. They are gathered by the Indians, who dry them and use them for food, which they call koo-tsabe, accented on the first syllable. The best-known "brine-flies" belong to the genus *Ĕphydra*.

Still more remarkable are the habits of the larva of the petroleum-fly, *Psilōpa petrōlei*, which lives, feeds, and swims about in the pools of crude petroleum, which are numerous in the various oil-fields of

California. For an account of the structure and habits of this larva see Crawford ('12).

Family CHLOROPIDÆ

This family includes a considerable number of species that are common in meadows and other places where there is rank growing grass, in such situations they can be collected in large numbers by a sweep-net.

The members of this family are small bare species; with moderately short or very short wings; the subcostal vein is vestigial cells M and 1st M_2 are not separated by a cross-vein; the antennæ are usually short and with the third joint rounded; the vibrissæ are rarely present; and the postvertical bristles are convergent.

The larvæ of the different species differ in their habits; many species infest the stems of wheat, oats, rye, clover, and grasses; some live in burrows or cavities in plants made by other insects; a few feed upon the egg-shells and cast-off skins of insects; some live in excrement; and species of *Gaurax* develop in the egg-sacs of spiders. Among the more important members of this family is the following species.

The European frit-fly, *Ŏscinis frit*.—This is a minute black species, measuring from 1.1 to 2 mm. in length. It was first described by Linnæus in Sweden, where it was a very serious pest of barley, the larvæ feeding upon the immature kernels. The light and worthless kernels resulting from this the Swedes called "frits", hence the common name of the species. There are several generations annually. The larvæ of the late fall generation winter as stem miners in winter grain; and spring grain is attacked in the same way by the spring generation. In this country the commonest form of injury is to the stems of wheat close to the ground. The larva of this species can be easily distinguished from the larva of the hessian fly by the fact that it works in the center of the stem and crawls actively when removed. For a detailed account of this species see Aldrich ('20).

This family is named the Oscinidæ by some writers.

The family ASTEIIDÆ includes a few genera, mostly exotic, that were formerly classed in the Drosophilidæ by some writers and in the Chloropidæ (Oscinidæ) by others; it can be separated from these families by the characters given in Table B. The best-known representative of the family in our fauna is *Sigaloëssa flavĕola*, which is widely distributed in the Atlantic states.

Family DROSOPHILIDÆ

The Pomace-flies and their Allies.

There are certain small yellowish flies from three to four millimeters in length which are very common about the refuse of cider-mills, decaying fruit, and fermenting vats of grape pomace; these are the pomace-flies (Fig. 1112); their larvæ live in the decaying fruit.

The pomace-flies and their allies constitute the family Drosophilidæ. In this family the costa is microscopically broken twice, once just beyond the humeral cross-vein and again just before the end of vein R_1; the subcostal vein is vestigial; the arista are almost invariably plumose; the vibrissæ are present; the postvertical bristles are convergent; and the foremost fronto-orbital bristles are proclinate.

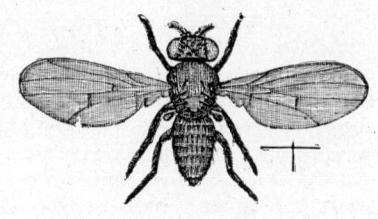

Fig. 1112.—A pomace-fly.

The larvæ of most species of this family, so far as is known, live in decaying fruit or in fungi; a few are leaf-miners; and some exotic species have been found feeding on other insects, *Aleurodes* and *Clastoptera*.

One of the pomace-flies, *Drosŏphila melanogăster*, which is easily bred and which has a short life-cycle, is widely used in laboratories in the study of heredity. This species has been commonly known as *Drosŏphila ampelŏphila*; but *melanogaster* is the older specific name.

A monograph of this family was published by Sturtevant ('21).

The family GEOMYZIDÆ is a group of small flies of which nearly fifty species have been described from our fauna. In these flies the postvertical bristles are convergent when present; the clypeus is large, the foremost fronto-orbital bristles are directed backward; and the fringe of the calypteres is not dense.

The larvæ of the few species of which the habits are known live in the stems of plants or mine in leaves.

The family was monographed by Melander ('13 b).

Family AGROMYZIDÆ

Fig. 1113.—Mine of *Phytomyza aquilegia*.

This family includes small or minute flies in which the costa is broken only at the end of the subcostal vein; the oral vibrissæ are present, the arista of the antennæ is closely short-pubescent, the post-vertical bristles are divergent, and the lower fronto-orbital bristles are convergent.

The genus *Cryptochætum* differs from the typical members of this family in having the costa twice broken and in that the antennæ lack the arista. One or two species of this genus have been introduced into California from Australia, as they are parasites of the cottony-cushion scale.

About one hundred species of the Agromyzidæ as restricted here have been described from our fauna. See monograph of the family by Melander ('13 b).

So far as is known the larvæ of most members of this family feed on living plants by forming burrows or mines in various parts of them, but principally in the leaves. A common species, *Phytomyza aquilēgiæ*, makes serpentine mines in the leaves of wild columbine, *Aquilegia canadensis*. (Fig. 1113).

The family MILICHIIDÆ is a small group of flies that is often classed with the Agromyzidæ. The members of this family differ from the agromyzids in that the costa is broken twice, once beyond the humeral cross-vein at which place there is usually a stronger costal bristle, and again just before the end of vein R_1, and the postvertical bristles are convergent. See monograph of this family by Melander ('13 b).

The family OCHTHIPHILIDÆ is also a small group of flies which is often classed with the Agromyzidæ. In this family the oral vibrissæ are wanting or not differentiated; the postvertical bristles convergent, the subcostal vein ends in the costa; and the clypeus is small. See monograph of the family by Melander ('13 b).

Only thirteen species, representing three genera, are recognized by Melander from our fauna.

Larvæ of species of *Leucōpis* prey upon aphids and upon coccids. This family is named the Chamæyidæ by Hendel ('22).

SUBSECTION II.—CALYPTRATÆ

The Calyptrate Muscids

To this subsection belong our most familiar representatives of the muscid flies, of which the house-fly and the flesh-flies are good illustrations. In the families included here the alulæ or clypteres are well developed or of moderate size, not rudimentary; the subcostal vein is always distinct in its whole course; vein R_1 is never very short; there is a complete transverse suture on the thorax; the postalar callus is present and separated by a distinct suture from the dorsum of the thorax; and the abdominal spiracles, with but few exceptions, are in the chitin. The flies are usually of moderate or considerable size, never very small.

The subsection Calyptratæ includes a series of families that are exceedingly difficult to differentiate. In fact no two of the authorities on this group of flies agree either as to the number of families that should be recognized or as to the limits of certain families that are generally recognized. This is believed to be due to the fact that this group of flies is of recent origin and the different types included in it have not become segregated by the dropping out of intermediate forms. It is the dominant group of flies, including an immense number of species.

Family ANTHOMYIIDÆ

The Anthomyiids

The anthomyiids are very common flies of which about five hundred species have been described from North America. They are somewhat similar to the house-fly in appearance but structurally distinct.

In this family cell R_5 of the wings is very slightly or not at all narrowed, vein M_{1+2} extending in a nearly straight line to the margin of the wing (Fig. 1114) and not bent in its outer part towards the tip of the vein R_{4+5} as in the house-fly. The hypopleural and often the pteropleural bristles are absent; and the proboscis is never adapted

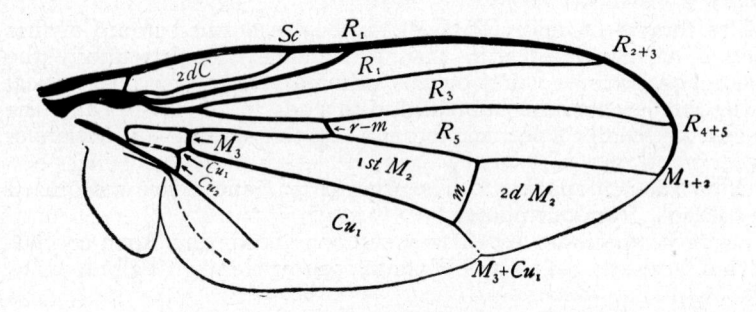

Fig. 1114.—Wing of *Lispa*.

for bloodsucking. The adult flies are found on leaves and flowers, and are also often found on windows in our dwellings.

The larval habits are variable. Most species live in decaying vegetable matter; many live in excrement, and doubtless are conveyers of typhoid fever, like the house-fly or typhoid-fly. Several species have been found to be the cause of internal myiasis, having been taken into the alimentary canal with vegetables and continuing to live there. A few species are parasitic within living insects. And some attack growing plants. Among the latter are certain well-known pests of garden crops. The more important of these are the following.

The cabbage-root maggot, *Hylemÿia brăssicæ*.—This insect in its larval state feeds on the roots of cabbage, radish, turnip, and cauliflower; it also attacks the roots of various weeds belonging to the same family of plants. There are two or more generations of this pest each year. The first generation infests the young plants; the eggs of the second generation are laid late in June or in July; later generations, if they occur, do but little harm.

The most practicable methods of control of this pest are to protect the seed-beds with a covering of cheesecloth in order to exclude the flies; to protect the plants when they are set out by fitting around the stem of each next to the ground a tarred paper card, these cards can be obtained from seedsmen and dealers in garden supplies; and by the use of a solution of corrosive sublimate crystals, one ounce dissolved in ten gallons of water. Two applications of one-half cupful to a plant are made, one, three or four days after the plants are set, and another eight or ten days later. The solution is poured on the stem and at the base of the plant. Great care should be taken to keep the supply of this poison where children or animals can not get it.

The onion maggot, *Hylemȳia antīqua*.—The larva of this species is often exceedingly destructive to onions, destroying young plants in the spring, and when the plants are older, burrowing into the bulb and causing decay. This species is difficult to control. As the flies require from ten days to two weeks after emergence in which to mature their eggs, many of them can be destroyed before they are ready to oviposit by a poisoned bait spray composed of one-fifth ounce sodium arsenite, one pint molasses, and one gallon water. There are two or three generations annually of this pest.

The raspberry-cane maggot, *Hylemȳia rubĭvora*.—The larva of this species burrows in the new canes of black and red raspberries and blackberries and kills them. The eggs are laid on the young shoots in the spring. The larva bores into the pith of the shoot, and tunnels downward; when about half way to the ground it girdles the wood beneath the bark. The part of the shoot above the girdle soon wilts, shrinks in size and droops over. The larva continues its burrow downward in the pith to the surface of the ground, transforms to a pupa without leaving its burrow in late June or early July; but the adult does not emerge till the following April. To check the ravages of this pest, cut off and burn the wilting canes as soon as observed.

The beet or spinach leaf-miner, *Pegomȳia hyoscȳami*.—This leaf miner infests the leaves of beets, sugar-beets, spinach, orach, mangels, and chard. The mine at first is thread-like but is soon enlarged to form a blotch. Several larvæ usually occupy the same leaf and their mines usually coalesce. There are several generations each year, and the winter is passed in the pupal state under fallen leaves in the soil. Where practicable the infested leaves should be picked and burned. By plowing the field deeply as soon as the crop is removed the overwintering pupæ can be buried.

The kelp-flies, *Fucĕllia*.—The larvæ of these flies live in brown sea-weeds, cast up by waves along ocean beaches. The adults can be found all summer long on the masses of these weeds often in immense numbers. The North American species, of which thirteen are known, were monographed by Aldrich ('18).

Family GASTROPHILIDÆ

The Bot-flies of Horses

This family includes the well-known pests the larvæ of which infest the alimentary canal of horses and which are commonly known

DIPTERA

as bots. These insects constitute the genus *Gastrophilus*, three species of which are now well established in the United States and Canada. In the adult flies the oral opening is small and the proboscis vestigial. The members of this genus can be distinguished from those of the following family by the venation of the wings; the most striking feature of which is that vein M_{1+2} extends in a nearly straight line towards the margin of the wing (Fig. 1115).

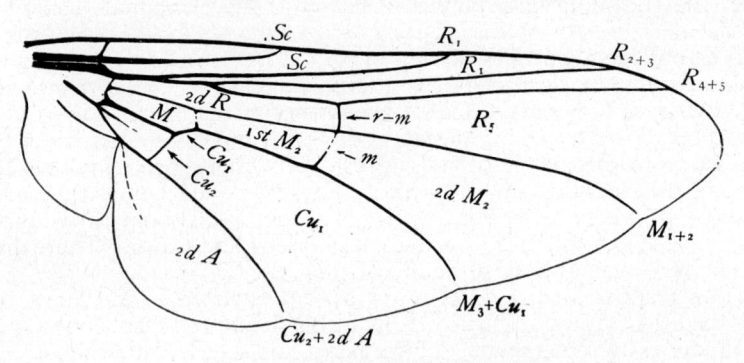

Fig. 1115.—Wing of *Gastrophilus*.

The genus *Gastrophilus* has been commonly included in the family Œstridæ but it is now believed to represent a distinct line of development. The three species that are established in this country are the following.

The common bot-fly or the stomach bot, *Gastrŏphilus intestinālis*. —The adult fly closely resembles the honey-bee in form except that the female (Fig. 1116) has the end of the abdomen elongate and bent forward under the body. The wings are transparent with dark spots, those near the center form an irregular, transverse band. This fly is most often seen flying about horses, which have an instinctive fear of it. The eggs are laid on different parts of the host, but preferably on the long hairs investing the inside of the forelegs. The eggs rarely hatch when left untouched; but the horse by scratching the forelegs with the teeth removes the small cap of the egg-shell and inadvertently takes the larva into its mouth. The larvæ thus taken into the mouth are carried with the food or water to the stomach. When the larvæ reach the stomach they fasten themselves to the inner coat of it, and remain there until full-grown. Then they pass from the animal with the dung, and crawl into some protected place, where they transform within a puparium. The adult fly measures about 18 mm. in length. This species is found throughout the United States and Canada where horses are present. It has been commonly known as *Gastrophilus equi*, but *intestinalis* is the older specific name.

Fig. 1116.

The chin-fly or the throat-bot, *Gastrŏphilus nasālis*.—This species is smaller than the common bot-fly and the wings are not marked with dark spots as in that species. The female usually deposits its eggs upon hairs under the jaws, and for this reason is commonly known as the chin-fly; but sometimes the eggs are laid upon the flanks or forelegs of the host. The manner in which the larvæ reach the mouth of the horse has not been definitely determined; but having reached the mouth they are carried down the alimentary canal. In some cases the larvæ attach themselves to the pharynx, and as this is the only species that is known to do this, the larva is known as the throat-bot. But this species is sometimes found in the stomach, and it attaches by preference in the duodenum. When the larvæ are matured they pass out from the horse and burrow in the manure or soil to transform.

This species is widely distributed in the United States and Canada.

The red-tailed bot-fly or the nose-fly, *Gastrŏphilus hæmorrhoidālis*.—The adult fly is easily distinguished by the bright orange-red tip of the abdomen. The wings are unspotted as in the chin-fly but differ from those of both of the preceding species in that the cross-vein m-cu is much farther from the base of the wing than is cross-vein r-m. The female oviposits on the lips of the horse; the flight of the fly about the nose of the horse when attempting to oviposit on its lips, suggested the common name, the nose-fly. The larvæ of this species attach themselves during their early stages within the stomach; but later loosen themselves and reattach in the rectum, from which they gradually move to the anus, where they remain for a short time before dropping to the ground to transform.

This species is found in the North-Central and Northern Rocky Mountain States and in the western provinces of Canada.

Family ŒSTRIDÆ

Bot-Flies (except Gastrophilus) and Warble-Flies

This family includes flies that are large or of medium size; most of the species resemble bees in appearance; some, the honey-bee, others, bumblebees. The mouth-opening is small, and the mouth-parts are usually vestigial. The venation of the wings differs from that of the preceding family in that vein M_{1+2} is bent so that cell R_5 is much narrowed or closed at the margin of the wing.

The larvæ are parasitic upon mammals; some develop in tumors under the skin and others, in the pharyngeal and nasal cavities of their hosts. As a rule each species infests a single species of mammal; and closely allied œstrids are parasitic, in a similar manner, upon closely allied mammals. In addition to the species that infest our domestic animals, other species infest rabbits, squirrels, deer, and reindeer. One that lives beneath the skin of the neck of rabbits is very common in the South.

The sheep bot-fly, *Œstrus ovis*.—This species is viviparous; the female fly deposits larvæ, which have hatched within her body, in the nostrils of sheep. The larvæ pass up into the frontal sinuses and into the horns when they are present. Here they feed upon the mucus. They are very injurious to sheep, causing vertigo or the disease known as "staggers." When full-grown they pass out through the nostrils and undergo their transformations beneath the surface of the ground.

The ox-warble-flies, *Hypodĕrma bovis* and *Hypodĕrma lineātum*.— If during the later winter months the backs of cattle be examined by rubbing the hand over them, there will be found present in many cases small lumps or swellings in each of which there is an opening through the skin; these swellings are known as warbles, and each contains a maggot, which when full grown measures nearly or quite 25 mm. in length.

The maggots that produce these warbles are the larvæ of flies, which for this reason are known as warble-flies. Two species of warble-flies both of which were introduced from Europe, infest cattle in this country, and are very serious pests.

The warble-flies when attempting to oviposit annoy cattle, which have an instinctive fear of them and run about in an effort to escape them; this leads to decreased milk yield. The larvæ as parasites injuriously affect the health of the cattle. And the holes in the skin through which the larvæ escape from the warbles very seriously reduce the value of the hide when made into leather. A careful estimate made by the Department of Agriculture of Canada showed that the annual loss in value of hides in Canada due to warbles is between 25 and 30 per cent. of the total value of the hides.

Our two species of warble-flies have much in common. The adults measure from 12 to 14 mm. in length and are bumblebee-like in appearance. They attach their eggs to hairs of cattle, usually on the hind legs, more rarely on the flanks. The newly hatched larva crawls down to the hair follicle where it penetrates the skin. Later the second instar of the larva is found in the wall of the œsophagus. The exact course of the migration from the hind legs to the œsophagus has not been determined; but it is believed that the larvæ travel in the loose connective tissues under the skin to the region of the throat and into the œsophagus where the muscles bifurcate. This part of their migration occupies about four months. They remain in the œsophagus about three months, and then migrate to the lumbar region. This part of the migrations of the larvæ is better-known than the earlier part. Hadwen ('19) states as follows: "The last larvæ to leave the œsophagus are at the paunch end. They pass out under the pleura and go to the neural canal, either up the crura of the diaphragm, or up the posterior border of the ribs entering the canal by the posterior foreamen. The larva evidently makes use of the canal as an easy means of access to the lumbar region, the part of of the animal which is best suited for passing its last stages within the host. The larvæ follow connective tissue exclusively, no larvæ have been discovered in muscular tissue."

The larvæ complete their growth in the lumbar region, causing as they increase in size the formation of the warbles. The hole through the skin into the cavity of the warble serves as a breathing hole for the larva and as a means of escape when it is full-grown. The mature larvæ leave the warbles and drop to the ground to transform. This takes place during the first half of the year. The average pupal period is about one month. The adults live only a short time as they are unable to feed.

The above account refers to both of our species of warble-flies; the following details chiefly compiled from Hadwen ('19) will serve to distinguish the two.

The bomb-fly, *Hypodĕrma bōvis*.—The adult fly measures 14 mm. in length; there is yellow hair on the anterior part of the thorax; the alulæ are bordered with reddish brown; and the tail end of the abdomen is orange-yellow. As a rule the flies lay their eggs while the cattle are running; the eggs are laid singly at the roots of the hairs; the flies are clumsy insects and strike at the animals blunderingly. The presence of one of these flies in a herd of cattle causes them to scatter and stampede just as a crowd of people would do if a bomb were thrown in their midst. For this reason Mr. R. C. Shannon ('22) has suggested "bomb-fly" as a common name for this species. The name "European warble-fly" that is often applied to it is not distinctive, as both of our species are of European origin. The eggs are laid mostly on the outside of the hind quarters and on the legs above the fetlocks; they are laid during June and July. In the larva the segment in front of the spiracular segment is unarmed.

The heel-fly, *Hypodĕrma lineātum*.—The adult fly is 12.7 mm. in length; the anterior part of the thorax is black and shining; the alulæ are uniformly white; and the tail end of the abdomen is reddish-orange. The eggs are laid mostly when the animals are recumbent and on all parts which the fly can reach when it is resting on the ground. Even when the animals are standing the fly is able to lay eggs on those hairs which are close to the ground, namely on the heels. The frequency with which this species lays its eggs in this place has caused it to be known as the heel-fly. This species irritates the cattle much less than does *Hypoderma bovis*, and consequently is able to lay several eggs on a single hair. The eggs are usually laid during May. In the larva the segment in front of the spiracular segment is spinose.

Family PHASIIDÆ

The Phasiids

This is a comparatively small family, which is composed of certain genera that were formerly included in the Tachinidæ but which are now regarded as representatives of a distinct branch of the Muscoidea.

In this family the clypeus is more or less produced below the vibrissal angles, like the bridge of a nose. The conjunctivæ of the ventral sclerites of the abdomen are present, and frequently well-developed, surrounding the sclerites. Vein M_{1+2} is bent so that cell R_5 is narrowed or closed at the margin of the wing; in some genera vein M_{1+2} joins R_{4+5} at a considerable distance before the margin of the wing. The abdomen is not armed with macrochætæ.

There are only a few records regarding the habits of members of this family. Some species are parasitic on adult Coleoptera, and other on nymphs and adults of Hemiptera.

Family MEGAPROSOPIDÆ

This is a small group of flies the rank of which is in question; some authorities regard it as a distinct family and others, as a subfamily of the Tachinidæ. With these flies the clypeus is receding and short; the cheeks are very broad; the vibrissæ are located near the middle of the face; and the antennæ are short.

This family is represented in our fauna by two genera, *Megaprosōpus* and *Microphthălma*. *Microphthalma disjuncta* has been bred from the larva of *Phyllŏphaga arcuāta*.

Family CALLIPHORIDÆ

The Blow-fly Family

Certain members of this family are very familiar objects and are commonly known as blow-flies, bluebottle-flies, or greenbottle-flies. With these, and with most other members of the family as well, the body, especially the abdomen, is metallic blue or green in color. This fact has suggested the common names bluebottle-flies and greenbottle-flies, that have been applied to certain species. These names, however, merely indicate in each case the more usual color of the species; for in all of the metallic colored species of this family the color varies; it may be either violet, green, blue, or copper color.

In this family the arista of the antennæ is plumose; both the hypopleural and the pteropleural bristles are present; the hindermost posthumeral bristle is almost always more ventrad in position than the presutural bristle; and the second ventral sclerite of the abdomen lies with its edges upon or in contact with the ventral edges of the corresponding dorsal sclerites.

The larvæ of the different species vary in habits; some have been bred from cow-dung; some feed on fresh or decaying meat and on the bodies of dead animals; one frequently infests wounds on animals, and two are blood-sucking parasites of nestling birds. The following are our best-known species.

The blow-flies *Callĭphora*.—The blow-flies normally live out-of-doors, but they often enter houses in search of material upon which to deposit their eggs. They then attract attention by their large size, much larger than that of the house-fly, and by the buzzing noise that they make. They lay their eggs upon meat, cheese, and other provisions. The eggs soon hatch and the larvæ develop rapidly. There are several species of this genus in our fauna; all of them have reddish palpi; bluish black, opaque thorax; metallic blue or green, more or less whitish pollinose abdomen; two or three posthumeral bristles; and black legs. There are two common species *Callĭphora erythrocĕphala*, in which the bucca is reddish brown and the beard black; and *Callĭphora vomitōria*, in which the bucca is black and the beard reddish.

The large bluebottle-fly, *Cynomȳia cadaverīna*.—This is a common species which resembles the blow-flies in size and in habits; it differs from them in that the abdomen is without silvery pruinosity, and there is only one posthumeral bristle on each side.

The greenbottle-fly, *Lucĭlia cæsar*.—This also is a common species, which resembles the blow-flies in habits; but it is smaller and its cheeks are bare. The abdomen is sometimes bluish but more often greenish.

The screw-worm fly, *Chrysomȳia macellāria*.—This is a bright metallic-green fly, with three black stripes on the thorax, and a yellow face. It measures from 8 to 10 mm. in length. It lays its eggs on decaying animal matter and also in wounds, sores, and the nostrils and ears of men and cattle. The larvæ living in these situations often cause serious sickness, and sometimes even death. This is a widely distributed species; but it has attracted most attention in the Southwestern States where it is a serious pest of stock.

The cluster-fly, *Pollēnia rudis*.—The cluster-fly is so-called because of its habit of entering houses in the autumn and hiding away in protected nooks in large groups or clusters. It is a dark colored, slow-moving species, slightly larger than the house-fly. The thorax is thickly beset with soft woolly hair in addition to the bristles; the abdomen is brown with white pollinose spots.

The calliphorid parasites of nestling birds, *Protocallĭphora*.—Two species at least are found in our fauna, *P. avium* and *P. splendida*. Both of these have been found to be external blood-sucking parasites of nestling birds, often causing the death of the nestlings. Plath ('19) found that of 63 nests of five species of birds studied by him 39 were infested.

Family SARCOPHAGIDÆ

The Sarcophagids

This family has been commonly known as the flesh-flies because some of them lay their eggs in bodies of dead animals, resembling in habits the blow-fly, which belongs to the family Calliphoridæ; but a wider knowledge of the habits of various members of this family shows that this name is misleading.

The Sarcophagidæ as limited by Aldrich ('16) in his monograph of the North American species includes all of the Muscoidea that agree in having the following characteristics: The coloration is gray or silvery, tessellated or changeable pollinose; vein M_{1+2} has an almost angular bend and ends considerably before the apex of the wing; the sides of the face are hairy; and the arista of the antennæ is plumose above and below for nearly half its length or a little more. None of the species has discal machrochætæ on the abdominal segments, hairy eyes, long proboscis, rudimentary palpi, or more than a single pair of discal scutellar bristles.

So far as is known all species of this family are larviparous. The different species show a wide range in larval habits; but by far the greater number of the species that have been bred are parasitic in other arthropods. They have been bred from various insects, from scorpions, and from the egg-sacs of spiders. Several species have been bred from dead fish; and a considerable number from the excrement of mammals. Five or six species live only in the tubular cups of pitcher-plants (*Sarracenia*), feeding on the dead insects found there. It has been found that *Sarcŏphaga hæmorrhoidālis* is sometimes the source of intestinal myiasis in man, and several cases of cutaneous myiasis caused by larvæ of *Wohlfăhrtia vigil* have been described by Walker ('22 a).

For the determination of species of this family one should consult the monograph by Aldrich ('16).

Family TACHINIDÆ

The Tachina-Flies

The tachina-flies are often found about flowers and rank vegetation. They are usually short, stout and bristly (Fig. 1117). They differ from the following family, the Muscidæ, in that with the tachina-flies both the hypopleural and the pteropleural bristles are present; and they differ from the two preceding families, the Calliphoridæ and the Sarcophagidæ, in that in this family the second ventral abdominal sclerite, as well as the others, is more or less covered, sometimes wholly, by the edges of the dorsal sclerites.

This is a very large family, more than fourteen hundred species are listed from North America alone; and from the standpoint of the agriculturist it is the most beneficial family of the Diptera.

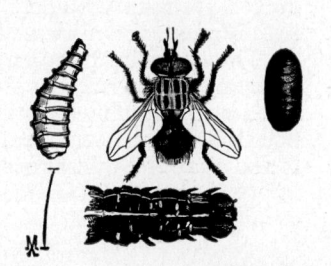

Fig. 1117.—A tachina-fly. Larva, adult, puparium, and eggs upon the fore part of an army-worm.

This family includes two subfamilies; the Dexiinæ, and the Tachininæ, each of which is regarded as a separate family by some writers.

The larvæ are parasitic, chiefly within caterpillars, but they have been bred from members of several other orders of insects. An extended list of tachinid parasites and their hosts is given by Coquillett

('97). The manner in which the larva finds its way into the body of its host differs greatly in different species of tachinids. Many observations on this have been made at the Gipsy Moth Laboratory and reported by Townsend ('08 b). In many species the female fastens her eggs to the skin of the caterpillar (Fig. 1117); when the larvæ hatch they bore their way into their host and live there till they are full-grown. In some of the viviparous species the female punctures the skin of the caterpillar with the sheath of her ovipositor and deposits the larva within the body of the host. Some species deposit their eggs on the leaves of the food-plant of their host; these eggs are swallowed when the leaves are eaten. But most remarkable of all is the method practiced by *Eupeletēria magnicŏrnis;* this is a viviparous species which infests the larva of the brown-tail moth. It attaches its larvæ to the surface of stems and leaves by a thin membranous case, which is cup-shaped and surrounds the anal end of the larva. Attached to the stem or leaf by this base, the maggot is able to reach out in all directions as far as its length will permit. As the maggot is deposited on the silken thread with which the caterpillar marks its trail as it leaves its nest, it is in a position where it can attach itself to the caterpillar when it is on its way back to the nest.

Family MUSCIDÆ

The Typical Muscids

To this family belong the house-fly and many other well-known members of the Muscoidea. In this family either the hypopleural or the pteropleural bristles are present, the basal bristles of the abdomen are reduced, and the arista of the antennæ are plumose to the tip. Among the more important species are the following.

The house-fly or typhoid-fly, *Musca domĕstica.*—This is the most familiar representative of the order Diptera, as it abounds in our dwellings. The flies lay their eggs preferably on horse-manure, but will oviposit on other decaying vegetable matter, when horse-manure is not available. A single female may deposit from 120 to 160 eggs at one laying, and they have been observed to make as many as four layings. The larvæ become full-grown in from five to seven days; the pupa state lasts from five to seven days; and in about fourteen days after the flies emerge they are ready to oviposit. Hence there may be at least a generation a month during the warm season; and from a few overwintering flies an immense number may be developed. The house-fly is not only an exceedingly annoying pest in our dwellings, but as it will breed in human excrement, especially where there are open closets, it is doubtless often a carrier of the germs of typhoid fever, dysentery, and other enteric diseases. For these reasons Dr. Howard has suggested that this species be known as the *typhoid-fly* and the "Swat-the-fly" crusades have been urged. Various means of protection from this pest, as window-screens and different kinds of traps, are well-known; but as Howard has so well put it, "the truest and simplest way of attacking the fly problem is to prevent them from breeding, by the treatment or abolition of all places in which they can breed." Garbage cans should be kept tightly closed and emptied at least once a week. Manure should be stored in tight receptacles or treated with borax, one-half pound of borax to eight bushels of manure. The borax should be applied immediately after the removal of the manure from the barn. Apply the borax particularly around the edges of the pile with a flour sifter or any fine sieve, and sprinkle two or three gallons of water over the borax treated manure. It is estimated that the cost of the borax will be about one cent per horse per day, (Cook, Hutchison, and Scales '14).

The stable-fly, *Stomŏxys călcitrans.*—This species resembles the house-fly in appearance, but it is a trifle larger and has its mouth-parts fitted for piercing and for sucking blood. It annoys cattle greatly; and before storms and in the autumn it enters our dwellings and attacks us. The popular belief that the house-fly bites more viciously just before a rain is due to invasions of this species at such times. The mouth-parts of the true house-fly are not fitted for piercing. The stable-fly is especially common in barns. It breeds in vegetable refuse, manure and excrement.

The horn-fly, *Hæmatōbia ĭrritans.*—This is an exceedingly annoying pest of horned cattle. It resembles the house-fly in appearance, but is less than half as large. These flies cluster in great numbers around the base of the horns; they also settle upon the back. The larvæ live in fresh cow-manure. The flies can be killed by spraying the cattle with kerosene emulsion or with crude petroleum.

The tsetse-fly, *Glossīna morsĭtans.*—This species, which is closely allied to the stable-fly, is widely distributed in Africa and is the carrier of the blood parasite that causes the disease of cattle known as nagana and the sleeping sickness of man.

SECTION II—PUPIPARA

Under this head are classed several families of flies that are parasitic in the adult state. In most cases the adults live like lice on the bodies of birds or of mammals; but two species are parasites of the honey-bee. The name Pupipara was suggested by the fact that in the best-known forms the larvæ attain their full growth within the body of the female fly, where they are nourished by the product of glands specialized for this purpose. It was formerly believed that the young are born as pupæ; but it is now known that the change to the pupa state does not take place until after the larva is born. It is also known that this remarkable manner of development is not restricted to this group of families as it is characteristic also of the tsetse-fly. But the name Pupipara has been so generally used for this group of families that it seems best to retain it.

In the Pupipara the eyes are never large and in some forms they are either vestigial or wanting; the ocelli are present in some genera in others they are wanting. The antennæ are apparently only one- or

874   AN INTRODUCTION TO ENTOMOLOGY

two-jointed, and in some genera lack the arista; the mouth-parts are short and not at all retractile; the wings are well developed in some forms, in others they are vestigial or wanting; the abdomen is indistinctly segmented in most cases and leathery in appearance.

Family HIPPOBOSCIDÆ

The Louse-Flies

The louse-flies are very abnormal flies that, in the adult state, live like lice, parasitically, upon the bodies of birds and mammals. Some species are winged, others are wingless, and still others are winged for a time and then lose their wings.

The body is depressed; the head is closely attached to the thorax which is notched to receive it. The antennæ are apparently one-jointed, with a terminal arista or style; they are situated in depres-

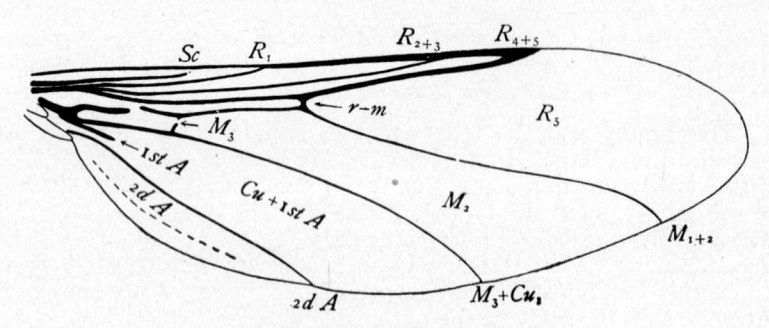

Fig. 1118.—Wing of *Lynchia*.

sions near the mouth. The legs are broadly separated by the sternum; they are comparatively short and stout; the tarsal claws are strong and are often furnished with teeth. The winged forms vary greatly in the venation of the wings. The veins near the costal border are usually strong while the others are weak. Figure 1118 represents the venation of *Lynchia*.

The sheep-tick, *Melŏphagus ovīnus*.—This well-known pest of sheep is the most common member of the Hippoboscidæ found in this country. It is wingless and its halteres are vestigial (Fig. 1119). It is about 6 mm. in length, of a reddish or gray-brown color, and with the entire body covered with long bristly hairs. This pest is often very injurious, especially to lambs after shearing time, as it tends to migrate from the old sheep to the lambs at this period.

Fig. 1119.

The life-history of this species illustrates well that type of development which suggested the name Pupipara for the Hippoboscidæ and allied flies. The structure of the female genital tract is described by Pratt ('99). A striking feature of it is the presence of two pairs of much branched glands, the "milk-glands", which secrete a fluid for the nourishment of the larva. The larvæ become full-grown in the uterus of the female and are born one at a time at intervals of several weeks. In about twelve hours after the larva is born the puparium is completed; and the adult emerges in from nineteen to twenty-four days later.

To control this pest the sheep should be dipped twice after shearing, in some good "dip" of which several kinds are on the market (See Farmers' Bull. 798, U. S. Dept. Agr.)

Among the more common representatives of this family, besides the sheep-tick, found in this country are the following.

Lynchia americāna.—This is a yellowish winged species rather common on owls and other raptorial birds, and on the partridge or ruffed grouse.

Lipŏptena deprĕssa.—This species is found on deer. The young adults are winged and probably fly about in search of their host; but later after becoming established on a deer they shed their wings.

Family STREBLIDÆ

The Bat-Ticks in part

This family and the following one include small flies that are parasitic upon bats. In this family the head is of moderate size, with a freely movable neck, but is not bent back upon the dorsum of the thorax, as in the following family. The eyes are vestigial or wanting; the ocelli are wanting; the palpi are broad and project leaf-like in front of the head; the wings are sometimes wanting or vestigial.

In this family, as in the Hippoboscidæ, the larva becomes fully grown within the body of the parent female.

For figures and descriptions of some of our species of this family see Ferris ('16).

The genus *Ascodipteron*, species of which are found in the Australian region and in other parts of the Eastern Hemisphere, is of great interest, because the females become endoparasites. The adults of both sexes are winged at first. Later the female, probably after copulation, cuts a hole through the skin of a bat and after shedding her wings and legs nearly completely imbeds herself in her host. The insect then increases greatly in size and becomes a flask-shaped creature, both head and thorax becoming invaginated so that they are not visible. The caudal end of the body projects from the cavity in which the insect lies. The larvæ, which become full-grown one at a time, are ejected from the uterus and fall to the ground, where they pupate. For a more detailed account see Muir ('12).

Family NYCTERIBIIDÆ

The Bat-Ticks in part

This family includes small, spider-like, wingless flies, which are parasitic upon bats. The head is narrow, and when at rest is folded back in a groove on the dorsum of the thorax. The eyes and ocelli are vestigial; the antennæ are short and only two-jointed; the legs are long, and the tarsal claws of ordinary form; although these insects are wingless, the halteres are present, but sometimes vestigial.

The reproduction of these flies is of the pupiparous type. When the mature larva is born it is fastened in some cases to the host, (Muir, '12) and in others the female leaves the host for a short time and fastens the larva to the perch to which the bats cling (Scott, '17). In each case the larva is pressed against the supporting object to which it adheres firmly.

Only a few species have been found in this country, for figures and descriptions of these see Ferris ('16 and '24).

Family BRAULIDÆ

The Bee-Lice

This family includes only a single genus, *Braula*, of which there is only one well-known species, *Braula cæca*. This is a minute insect, 1.5 mm. in length, which is parasitic upon the honey-bee (Fig. 1120). It is found clinging to the thorax of queens and drones. It is wingless and also lacks halteres; the head is large; the ocelli are wanting; the eyes are vestigial; the legs are comparatively short; and the last segment of the tarsi is furnished with a pair of comb-like appendages.

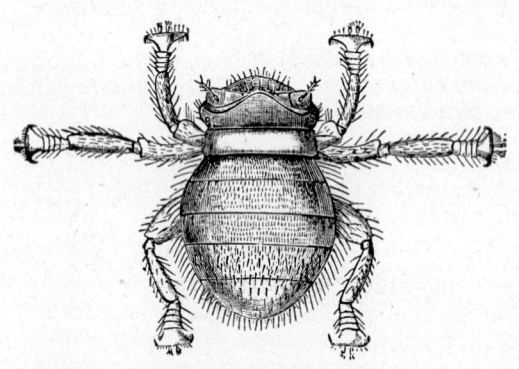

Fig. 1120.—*Braula cæcca*. (From Starp after Meinert.)

The bee-louse was described by Réaumur nearly two hundred years ago and remained the only known species of this family till 1914, when another species, *Braula kohli*, was described from the Belgian Kongo; this species is parasitic on an African honey-bee, *Apis mellifica* var. *adamsoni*.

The affinities of this family are in doubt. Until recently *Braula* has been supposed to be similar in its **mode** of development to the sheep tick, and for this reason the family classed with the Pupipara. But it is now known that *Braula* lays eggs; and the developmental stages have been found in tunnels under the capping of sealed honey (See "Monthly Letter" of the U. S. Bureau of Entomology, Number 113, September 1923).

It is maintained by both Börner and Bezzi that *Braula* should be classed near the Phoridæ. But Muggenburg ('92) has shown that there is a ptilinum in this genus. It, therefore, does not belong to the series Aschiza.

CHAPTER XXIX
ORDER SIPHONAPTERA*
The Fleas

The members of this order are small, wingless insects, in which the body is laterally compressed, so that the transverse diameter is small, the vertical one great. The mouth-parts are formed for piercing and sucking. The metamorphosis is complete.

The name of this order refers to the form of the mouth-parts and to the wingless condition of these insects.

These tiny tormentors are best known to us in the adult state; for it is only the adults that annoy us and our household pets. The larvæ and pupæ are rarely observed except by students who search for them.

The body of the adult is oval and greatly compressed, which allows the insect to glide through the narrow spaces between the hairs of its host. The integument is smooth, quite hard, and armed with bristles, which are arranged with great regularity (Fig. 1121) and thus afford good characters for distinguishing the different species. The smoothness and firmness of the body make it easy for the insect to escape when caught between the fingers of man or the teeth of lower animals. When once out of the clutch of an enemy it quickly leaps away.

Fig. 1121.—The dog-flea and its larva.

The head is broadly joined to the thorax. There are no compound eyes; but on each side of the head there is usually an unfaceted eye; these, however, are sometimes wanting. Each antenna lies in a groove somewhat behind and above the eye (Fig. 1122). The antennæ are three-jointed; the third joint, the flagellum, often called the club, may be unsegmented, segmented on the posterior border only, or completely segmented into several, usually nine, more or less separate pseudo-segments (Fig. 1123A). There is usually an internal thickening of the body-wall extending over the vertex from one antennal groove to the other (Fig. 1122, *f*), this is known as the *falx* or sickle-shaped process. That part of the dorsal wall of the head in front of the antennal groove and this thickening is termed, by writers on the Siphonaptera, the *frons;* the part behind them, the *occiput;* and the lateral aspects of the head, below and behind the eyes, the *genæ* or cheeks.

A remarkable feature of the head of the Siphonaptera is the fact that in the more generalized forms it is divided into two distinct

---

*Siphonăptera: *siphon* (σίφων) a tube; *apteros* (ἄπτερος) without wings.

parts with a somewhat flexible articulation between them. (Fig. 1124); the anterior part bears the mouth-parts and the eyes, while the antennæ are joined to the posterior part. The falx (Fig. 1122, *f*), which is present in many of the more specialized fleas is evidently a vestige of the articulation between the two parts of the head; even this vestige is wanting in many fleas.

The mouth-parts are formed for piercing and sucking. When seen without dissection these parts are apparent: the maxillæ, which are triangular plates (Fig. 1122, *mx*); the maxillary palpi, which are long and four-jointed (Fig. 1122, *mx. p*); and the proboscis, (Fig. 1122, *p*).

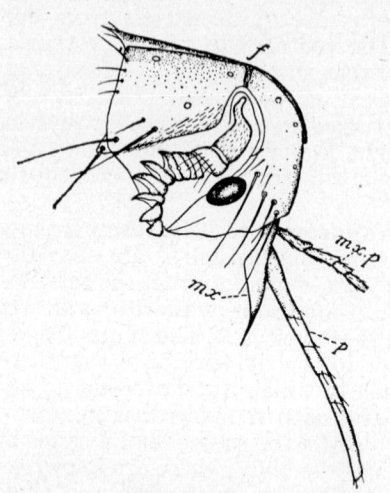

Fig. 1122.—Head of a flea, *Ceratophyllus multispinosus*: *f*, falx; *mx*, maxilla; *mx.p*, maxillary palpi; *p*, proboscis. (After Baker.)

The proboscis consists of an elongated labrum-epipharynx, two long and slender mandibular blades, and a sheath formed by the labium and the labial palpi. The space between the labrum-epipharynx, in the lower side of which there is a groove,

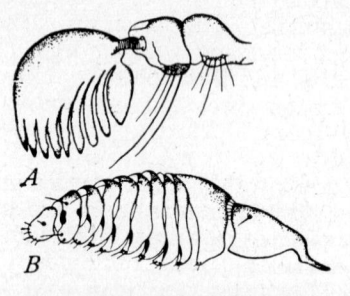

Fig. 1123.—Antennæ of fleas: A, *Ctenocephalus felis*; B, *Ceratophyllus fasciatus*. (After Patton and Cragg.).

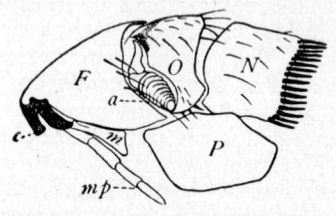

Fig. 1124.—Head and prothorax of *Ischnopsyllus*: *F*, frons; *O*, occiput; *N*, pronotum; *P*, propleurum; *a*, antenna; *c*, ctenidia; *m*, maxilla; *mp*, maxillary palpus.

and the mandibular blades, which are closely applied to the labrum-epipharynx, serves as a food-canal, through which the blood taken from the host is sucked into the alimentary canal. There is also a second canal formed by the apposition of two grooves, one on the inner side of each of the mandibles; through this canal the salivary secretion is forced into the wound. The piercing organ is the mandibles; the distal part of each mandible is beset with recurved teeth; and the proximal end of the blade is connected with a chitinous lever, which in turn articulates with the head-capsule; by the action of muscles attached to this lever, the blade of the mandible can be forced in and out. The number of segments of the labial palpi varies from two to seventeen.

The three segments of the thorax are quite distinct from one another although closely joined. The ventral part of the prothorax extends forward under the head, so that the first pair of legs appear to depend from the head. There is usually one, but there may be two or three, transverse rows of bristles on each thoracic segment. There are no vestiges of wings. The legs are long and strong and fitted for leaping; the hinder pair are largest and the middle pair next in size.

The abdomen is composed of ten segments; the first seven of these are comparatively simple in structure; the last three are specially modified for sexual purposes. The variations in form of the sclerites of these segments and of the genital appendages afford characters much used in the classification of these insects. The ninth tergite bears what is evidently a sensory plate; on the surface of this there is a number of clear areas from each of which there projects a long slender and extremely fine hair. The function of this organ is unknown.

A conspicuous feature of many fleas is the presence of a series of short, stout spines on various parts of the body; these are known as combs or *ctenidia*. The presence or absence of ctenidia and their location when present are important distinctive characters. In the dog-flea (Fig. 1121) there are ctenidia on the genæ and on the posterior border of the prothorax.

Large bristles placed at the dorsal angle of the seventh abdominal segment are termed the *antipygidial bristles*.

The eggs of fleas are scattered about the floors of dwellings and in the sleeping-places of infested animals. The larvæ are slender, worm-like creatures, with a distinct head and without legs. (Fig. 1121). They have biting mouth-parts, and feed upon the decaying particles of animal and vegetable matter always to be found in the dirt in which they live. When full-grown the larva spins a cocoon within which the pupa state is passed.

Fleas are parasitic only in the adult state. Some species infest birds, but by far the larger number prey upon mammals, and most mammals are subject to the attacks of these parasites. Although the different species of fleas infest different hosts they are not so restricted in their host relations as are many parasites, and may pass from their normal host to another species; for example, both the dog-flea and the cat-flea frequently attack man.

Formerly fleas were regarded as merely annoying pests of man and his pets; but it has been found that fleas are the carriers of bubonic plague; this fact has greatly increased the interest in these insects. A result of this increased interest is that extended studies of the Siphonaptera have been made recently and are being made now.*

*It is now known that the bubonic plague, which has caused the death of many millions of people, is a specific infectious disease caused by one of the bacteria, *Bacillus pestis;* and that it is primarily a disease of rodents, especially of rats. It has been shown by experiments that this disease is transmitted from one ro-

880   AN INTRODUCTION TO ENTOMOLOGY

To rid a dog or cat of fleas Persian insect powder should be carefully rubbed into its hair, or powdered naphthalene or moth balls used in the same way. The animal should be treated on paper spread on the floor, and the stupified fleas that come to the surface of the hair or drop out collected and burned. The bedding in kennels should be of some substance that can be replaced frequently, as shavings or straw, and when replaced the old bedding should be burned, and the floors wet with kerosene emulsion or some other insecticide that will destroy the eggs and larvæ. The animals should be kept from beneath dwellings where fleas may breed rapidly, and where it is difficult to reach the breeding places.

In regions where fleas abound much relief can be obtained by the use of rugs on the floors of dwellings instead of carpets. The frequent shaking of the rugs and cleaning of the floors will prevent the breeding of these pests within the house. As a single flea will inflict many bites, it often happens that a house will seem to be overrun by them when only a few are present. In such cases a careful search for and the capture of the offenders will soon remedy the evil.

People that suffer from the attacks of these pests can also gain much relief by dusting the upper part of their stockings each morning with Persian insect powder, and by sprinkling a small quantity of this powder between the sheets of their beds at night.

The destruction of rats and the fleas that they harbor in a region where bubonic plague exists is the most important means of preventing the spread of this disease. This has been done very efficiently by the United States Marine Hospital Service in those cases where the plague has been introduced into this country.

Except where bubonic plague is present the bites of fleas are not likely to cause serious results, although they may be very annoying. The irritation caused by them can be relieved by the use of some cooling application as menthol, camphor, or carbolated vaseline. Scratching the bites should be avoided as that aggravates the inflammation.

This order includes about 500 described species, and additions to the list are constantly being made. A revision of the American species was published by Baker ('04 and '05). Later Oudemans ('10) published a new classification of the order in which he gave a table of the families and genera of the world. The more striking features of the classification by Oudemans are the division of the order into two suborders and the proposed establishment of several new families, some of which have not been adopted by other authors. In the following account reference is made only to North American forms.

The division of the order into two suborders is based on the structure of the head. In the first suborder, the *Fracticipita*, are included those fleas in which the head is jointed; the second suborder, the *Integricipita*, includes those fleas in which the head is not jointed.

---

dent to another by fleas; and from rodents to monkeys in the same manner. It is concluded, therefore, that the disease is carried to man also by fleas. See textbooks of medical entomology.

Suborder FRACTICIPITA

The Broken-headed Fleas

In the members of this suborder the head is divided into two distinct parts with a somewhat flexible articulation between them. On the dorsal wall of the head the anterior part, the frons, overlaps the posterior part, the occiput (Fig. 1124).

Representatives of two subfamilies of this suborder occur in our fauna; these families can be separated as follows.

A. Ctenidia of the head consisting only of two broad teeth on each side in front of the maxillary palpi, none between the palpi and the antennal groove; apex of maxillæ truncate, p. 881 .... ISCHNOPSYLLIDÆ

AA. Genal ctenidia, when present, placed between the maxillary palpi and the antennal grooves or extending from the antennal groove to the anterior margin of the head; apex of maxillæ pointed. p. 881 .... LEPTOPSYLLIDÆ

Family ISCHNOPSYLLIDÆ

This family includes those Fracticipita in which the ctenidia of the head consist only of two broad teeth on each side close to the lower anterior angle of the head, in front of the maxillary palpi, (Fig. 1124), and in which the apex of the maxillæ is truncate. It is represented in our fauna by only a few described species; these infest bats and are occasionally found on mice of the genus *Peromyscus*.

Family LEPTOPSYLLIDÆ

To this family belong most of our members of the suborder Fracticipita. In these the genal ctenidia, when present, are placed between the maxillary palpi and the antennal grooves or extend from the antennal groove to the anterior margin of the head; the apex of the maxillæ is pointed.

Many Amerian species have been described; the greater number of these infest rodents, a few species are found on birds, and some of those infesting rats will attack man.

Suborder INTEGRICIPITA

The Unbroken-headed Fleas

In the members of this suborder the head is not divided into two distinct parts by an articulation, although a vestige of the segmentation, the falx, may be present (Fig. 1122, *f*). The frons does not overlap the occiput on the dorsal wall of the head.

Representatives of three families of this suborder occur in our fauna, these families can be separated as follows.

A. Abdomen with small, sharply pointed spines on apices of tergites; inner surface of anterior portion of hind coxæ without small spines. although a patch of bristles is sometimes present. p. 882 .... CERATOPHYLLIDÆ

AA. Abdomen without apical spines on the tergites; inner surface of anterior portion of hind coxæ with one row of small spines, rarely with more than one row.
  B. Thorax longer than the head, not shorter than the first abdominal tergite, p. 882..PULICIDÆ
  BB. Thorax shorter than the head, also shorter than the first abdominal tergite. p. 882...ECHIDNOPHAGIDÆ

Family CERATOPHYLLIDÆ

This family includes those Integricipita in which there are small, sharply pointed spines on the apices of some of the abdominal tergites. It is represented in this country by many species; these infest various rodents, mink, birds, and man.

Family PULICIDÆ

This family includes those Integricipita without apical spines on the abdominal tergites in which the thorax is longer than either the head or the first abdominal tergite. It includes some of our most common and best-known species; among these are the following.

The cat-flea, *Ctenocĕphalus felis*.—This is the species that is most often found in our dwellings in the East and in the South, and the one that most often attacks man in these regions. Both the genæ of the head and the pronotum are armed with ctenidia; the genal ctenidia extend from the antennal groove to the anterior margin of the head. The first spine of the genal ctenidia is about as long as the second. This species infests dogs as well as cats.

The dog-flea, *Ctenocĕphalus canis*.—This species is closely allied to the cat-flea, but in this species the first spine of the genal ctenidia is only about one-half as long as the second spine. This flea infests dog, cat, and man.

The human flea, *Pulex irrītans*.—On the Pacific Coast this is the species that is most often found in houses attacking man. It is easily distinguished from the two preceding species by the fact that its head and thorax lack ctenidia. Man is its natural host; but it will infest various other animals temporarily.

The Indian rat-flea, *Xenopsȳlla chēopis*.—Among the various species that are supposed to transmit the bubonic plague, this cosmopolitan species is regarded as one of the more important. It resembles the human flea in lacking ctenidia, but can be distinguished from that species by the fact that the mesosternite is broad, with a rod-like internal thickening extending from the insertion of the coxa upward.

Hoplopsȳllus anŏmalus.—This is the plague carrier, of the California ground squirrels, and it also infests rats. In this species there is a ctenidium on the pronotum, but none on the head.

Family ECHIDNOPHAGIDÆ

In this family the thoracic segments are short, the three segments together being shorter than the first abdominal segment in the dorsal line; the genal margin of the head is produced into a triangular process at the ventral oral angle; and ctenidia are lacking. The two following species are the best-known members of this family.

The sticktight flea, *Echidnŏphaga gallinācea*.—This is a very serious pest of poultry especially in the southern and southwestern portions of the United States. It is a small, dark brown species, which is often found in dense masses attached to its host; heads of chickens are often covered with dark patches of these fleas. This species is known as the sticktight flea because it seldom hops about, biting here and there, as do most fleas, but settles down on its host, deeply inserting its mouth-parts, and remains for days or weeks. While this species is chiefly a pest of poultry, it is often found in dense masses on the ears of dogs and cats.

The chigoe (chig'ō) or jigger, *Tŭnga pĕnetrans*.—This is a small flea found in the West Indies, Mexico, Central and South America, and in tropical Africa. It has been reported from Florida but it is not known to be established in the United States. The males and the unimpregnated females live in dry, sandy soil; they are only about one millimeter in length, and behave in the ordinary manner of fleas, feeding on the blood of man and many other animals, domestic and wild, and even birds. When impregnated, the female burrows into the skin of her host. Soon after this the abdomen becomes distended with eggs and acquires the size of a small pea. This species often causes serious injury to man by burrowing beneath the skin of the foot, causing the formation of a sore, which may become infected with bacteria, and cause the loss of a toe or a leg.

In the southern United States the names chigoe and jigger are improperly applied to the harvest-mites, which are the immature six-legged forms of various mites that attach themselves like ticks to the skin and become gorged with blood.

CHAPTER XXX
ORDER HYMENOPTERA*
Bees, Wasps, Ants, and others

The winged members of this order have four wings; these are membranous and have the wing-venation more or less reduced. The hind wings are smaller than the fore wings. The mouth-parts are formed for chewing or for both chewing and sucking. The abdomen in the females is usually furnished with a sting, piercer, or saw. The metamorphosis is complete.

The Hymenoptera is a very large order, including a vast number of species. The bees, wasps, and ants are among the better-known

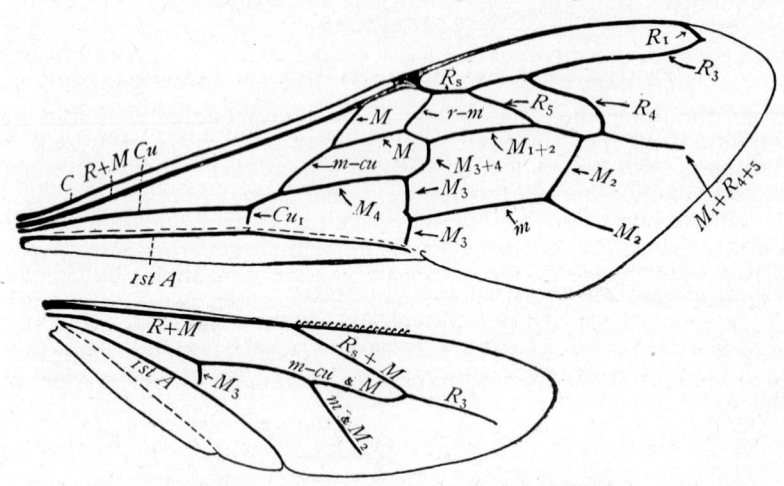

Fig. 1125.—Wings of *Apis* showing hamuli.

members of it; but in addition to these it includes a large number of less familiar forms. Many of these are minute parasites of other insects; others cause the growth of galls on plants; and still others, in their larval state, feed on the foliage of plants or are borers in the stems of bushy or herbaceous plants or in the limbs and trunks of trees.

The members of this order are chiefly of small or moderate size, and many of them abound wherever flowers bloom. From very early times some of them have been favorites with students of the habits of animals, for among them we find wonderful developments of instinctive powers. Many volumes have been written regarding their ways, and much remains to be discovered, even concerning our most common species.

---

*Hymenŏptera: *Hymen* (ὑμήν), membrane; pteron (πτερόν), wing.

The membranous nature of the wings, which suggested the name of the order, is not a distinctive characteristic, for it is possessed by the wings of many other insects.

The two pairs of wings are similar in texture. The wings of each side are held together by a row of hooks, the *hămuli*, on the front mar-

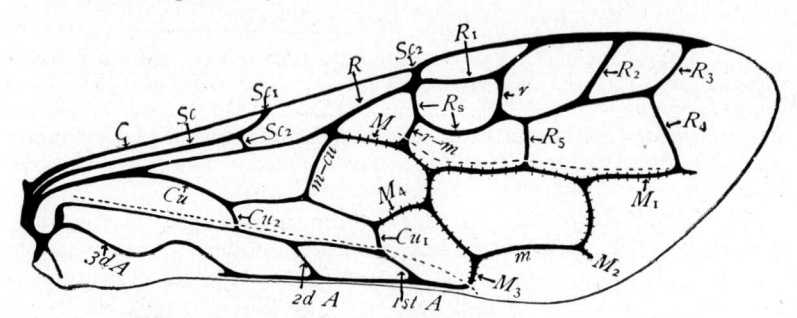

Fig. 1126.—The veins of a typical hymenopterous wing.

gin of the hind wing (Fig. 1125); these hooks fasten to a fold in the hind margin of the front wing, so that the two wings present a continuous surface. The hind wings are smaller than the fore wings and have a more reduced venation. Some forms are apterous.

This is one of the orders in which in the specialization of the wings the wing-venation is reduced. In the more generalized members of the order this reduction of the wing-venation is slight, but in the more specialized forms it is extreme. Even in the more generalized forms, where nearly all of the veins are preserved, the courses of the branches of the forked veins have been greatly modified. This has been brought about by the coalescence of veins from the margin of

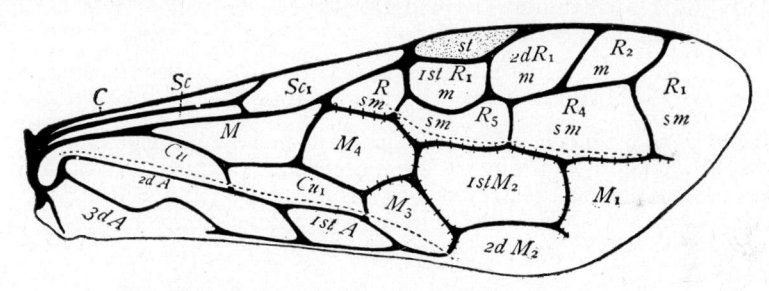

Fig. 1127.—The cells of a typical hymenopterous wing.

the wing inward. To understand this one should study a series of wings of Diptera in which all stages of the modification of the venation in this way are illustrated, for in the Hymenoptera only the later stages are shown. The series of figures illustrating the coalescence of veins Cu_2 and 2d A in the Diptera will aid in understanding what has happened in the Hymenoptera.

Figures 1126 and 1127 represent what may be regarded as a

typical hymenopterous wing; in the former the veins are lettered, in the latter, the cells. These are figures of a fore wing of *Pamphilius* (Fig. 1135) except that vein R_2, which is lacking in this genus, is added. This vein is well preserved in *Macroxyela* (Fig. 1134); but in *Macroxyela* vein Cu_2 is lost; the position of the last forking of the cubitus is indicated, however, by a bend in this vein. In these figures of the typical hymenopterous wing the lines indicating the course of the free part of media, after it separates from radius, are crossed by short lines.

The cells marked *m, m, m,* in Figure 1127 are termed the *marginal cells;* and those marked *sm, sm, sm, sm,* the *submarginal cells;* the three cells, M_4, 1st M_2, and M_3 are termed the *discal cells.*

The working out of the various ways in which the wing-venation has been reduced in the more specialized families is an exceedingly difficult problem, one that is beyond the scope of this book. A general discussion of it has been published by the writer (Comstock '18); a special paper on the venation of the Chalastogastra has been published by Professor A. D. MacGillivray ('06); and a very detailed account of the modifications of the wing-venation in the Clistogastra has been prepared by Professor J. C. Bradley and will probably soon be published.

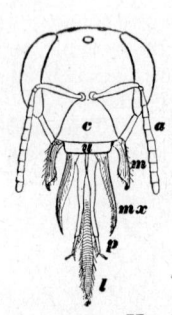

The mouth-parts are formed for chewing in all Hymenoptera, and in the more specialized members of the order they are fitted for both chewing and for sucking or lapping liquid food. In the saw-flies, for example, the mouth-parts resemble quite closely the orthopterous type, while in the bees they differ markedly from this type; and intermediate forms exhibit intermediate degrees of modification of the mouth-parts.

In the long-tongued bees the labrum and mandibles retain the form characteristic of chewing insects and the mandibles function as organs for crushing or cutting; but the labium and maxillæ are elongated, the maxillæ form a sheath to the labium, the three organs thus constituting a suctorial apparatus (Fig. 1128). In this figure the maxillæ are represented separated from the labium.

Fig. 1128.—Head of a honey-bee: *a,* antenna; *c,* clypeus; *u,* labrum; *m,* mandible; *mx,* maxilla; *p,* labial palpus; *l,* labium.

The legs of the Hymenoptera present characters that are much used in the classification of these insects. Among the more striking of these are the following; the trochanter may consist of two segments (Fig. 67, B) or of only one; the metatarsus of the hind legs is greatly enlarged in bees (Fig. 67, C); and in several families the fore legs are fitted with an organ which is used in cleaning the antennæ, the *antenna cleaner* or *strigilis.* This consists of a curved, comb-like, movable spur on the distal end of the fore tibia (Fig. 1129) and opposite this, on the base of the metatarsus, a concavity fringed with hairs. In cleaning an antenna it is drawn through the space between these two parts of the strigilis.

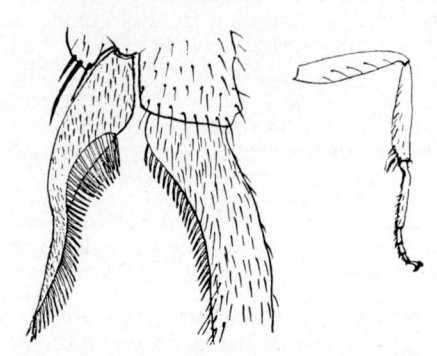

In addition to the terms defined above the following are used in descriptions of Hymenoptera.

The malar space.—The area on each side of the head included between the proximal end of the mandible and the ventral end of the compound eye.

Fig. 1129.—Leg of an ant, and strigilis enlarged. (From A. B. Comstock, Handbook of Nature Study.)

The propōdeum.—The first abdominal segment when it forms a part of the alitrunk or wing-bearing region of the body. See characterization of the suborder Clistogastra. This is often called the *epinotum* by writers on ants.

The parăpsides or scăpulæ.—In many Hymenoptera the prescutum of the mesothorax is prolonged backward to a greater or less extent; in some it extends a considerable distance toward the scutellum but does not reach it (Fig. 1130, B); in others it reaches the scutellum dividing the scutum into two parts (Fig. 1130, A); these separated halves of the scutum are commonly called the *parapsides* or *scapulæ*. (Fig. 1130, *par*)

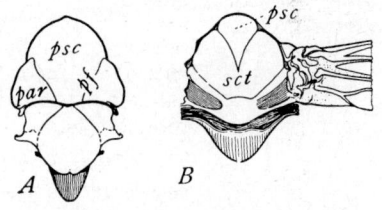

The parăpsidal furrows or notauli.—The sutures separating the prescutum from the parapsides. (Fig. 1130, *p.f*).

Fig. 1130.—A. Mesonotum of *Eurytoma*. B. Mesonotum of *Cimbex*. (After Snodgrass.) *psc*, prescutum, *sct*, scutum, *par*, parapsides; *p.f*, parapsidal furrows.

The posterior lobes of the pronotum.—A distinctly differentiated rounded lobe, on each side covering the spiracle, which forms the lateral extension of the pronotum of Sphecoidea.

The prepĕctus.—An area along the cephalic margin of the episternum of the mesothorax which in some Hymenoptera is separated by a suture-like furrow.

The epicnĕmium.—This is the same part as the prepectus.

The cĕnchri.—A pair of membranous lobes or areas on the metanotum of all Chalastogastra.

The gaster.—The swollen portion of the abdomen behind the pedicel in the suborder Clistogastra.

The pygidial area.—In many of the aculeate or stinging Hymenopetra there is an area on the pygidium which is bounded on each side by a carina, the two carinæ meeting posteriorly on the middle line of the segment; this area is known as the *pygidial area*.

The anal lobe.—The posterior lobe of the wings, which is defined on page 61 (Fig. 71, *l*) is also called the anal lobe. In the suborder Clistogastra the presence or absence of an anal lobe in the hind wings is an important taxonomic character.

The preaxillary excision.—In the hind wings of some Hymenoptera there is in addition to the axillary excision, defined on page 61, another notch, the *preaxillary excision*. In the hind wings of the Hymenoptera the axillary excision, when present, is at the apex of

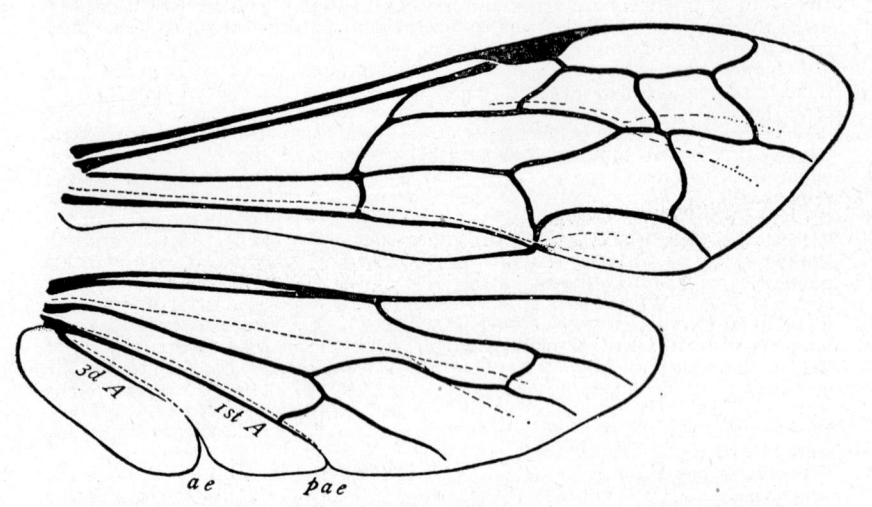

Fig. 1131.—Wings of Elis: *ae*, axillary excision; *pae*, preaxillary excision.

the second anal furrow, which lies between the second and the third anal veins (Fig. 1131, ae); the notch may be present in forms in which both the furrow and the veins are lacking. The preaxillary excision is situated at the apex of the first anal fold, which is just cephalad of the first anal vein (Fig. 1131, pae).

The preanal lobe.—That portion of the anal area of the hind wings that lies between the axillary excision and the preaxillary excision constitutes the preanal lobe.

In the Hymenoptera the metamorphosis is complete. The larvæ of the Chalastogastra are caterpillar-like in form and are furnished with thoracic legs and usually with abdominal prolegs; but in some, mostly borers or internal feeders, the prolegs are wanting. In all

Clistogastra the larvæ are maggot-like in form and have no legs. The pupæ are of the exarate type, that is, the legs and wings are free, as in the Coleoptera. With many species the larva, before changing to a pupa, spins a cocoon about its body. With some this cocoon is composed of comparatively loose silk, and resembles somewhat the cocoon of a moth. In others the cocoon is of a dense parchment-like texture, and in still others it resembles a very delicate foil.

Parthenogenesis.—The production of young by females that have not mated is known to occur in members of several families of this order. In some species the young thus produced are all males; in others they are all females; and in still others both males and females are developed from unfertilized eggs. Among the well-known examples of parthenogenetic reproduction are the following. Sometimes a queen honey-bee produces eggs before she has mated; from such eggs only males are developed. The eggs produced by fertile worker bees and fertile worker ants, neither of which mate, develop only into males. In certain gall-flies there is an alternation of a generation consisting of males and females and a generation consisting only of females, which reproduce parthenogenetically; the young of the latter are males and females. In some species of the Tenthredinidæ the reproduction is believed to be entirely parthenogenetic, males of these species being unknown.

Polyembryony.—In several genera of minute parasitic Hymenoptera the number of young produced is not dependent upon the number of eggs laid, for with these insects many embryos are developed from a single egg. This type of development is termed polyembryony; and has been investigated by several workers. A recent paper on this subject is that of Dr. R. W. Leiby ('22), in which there is a list of the earlier papers. Dr. Leiby traced the development of *Copidosoma gelechiæ*, a parasite of the solidago gall-moth, the insect that makes the spindle-shaped gall on golden-rod (Fig. 769). The parasite oviposits in the eggs of the moth which it finds on leaves or stems of goldenrod in early fall. By the arrival of cold weather the developing egg or parasite body is found as a polynuclear mass within the completely formed host embryo, which has developed synchronously. When the host larva hatches in the spring the parasite body is found lodged in the fat-body of the host. As the host larva grows the polygermal mass becomes a mass of embryos, which are later set free into the body-cavity of the host larva as parasitic larvæ. These larvæ feed upon the body content of the host devouring the blood, muscles, fatty tissue, and in fact everything except the chitinous parts. An average of 163 adult-parasites is developed from a single egg. The details of this development are described at length by Dr. Leiby and are illustrated by many figures.

In a later paper Dr. Leiby and C. C. Hill ('23) described the development of *Platygaster heimales*, a parasite of the Hessian fly. From some of the eggs of this parasite a single larva is developed; but from others two larvæ are produced. This species is of great interest as illustrating the beginning of polyembryony.

Aquatic Hymenoptera.—It has long been known that the adults of certain parasitic Hymenoptera descend beneath the surface of water in order to oviposit. One of these is a parasite of caddice-worms, the others whose hosts are known lay their eggs in the eggs of various aquatic insects. Most of the observations on these insects have been made in Europe but recently Professors Matheson and Crosby ('12) have described the habits of three minute species which have been reared at Ithaca, N. Y.; they also give a list of the known aquatic Hymenoptera.

The classification of the Hymenoptera.—The classification of the Hymenoptera, *i. e.*, the sequence of the families and the groupings of these families into superfamilies, adopted in this chapter is that recently worked out by Messrs. J. C. Bradley, S. A. Rohwer, and J. Bequaert.

Authorities are not in agreement as to the proper application of certain generic family and other group names in the order Hymenoptera. The matter is a technical one, and is before the International Commission on Zoological Nomenclature for a definite decision. Until such a decision is rendered, we prefer to retain the long established usage of these names, as indicated below, where is also shown the equivalent name used by some recent authors.

Argidæ instead of Cryptidæ for a family of sawflies.
Cimbicidæ instead of Crabronidæ for a family of sawflies.
Proctotrupidæ instead of Serphidæ for a family of parasitic wasps.
Proctotrupoidea instead of Serphoidea for a superfamily of parasitic wasps.
Ceraphronidæ instead of Calliceratidæ for a family of parasitic wasps.
Toryminæ instead of Callimominæ for a subfamily of chalcid-flies.
Lasius instead of Acanthomyops or Donisthorpea for a genus of ants.
Pompilidæ instead of Psammocharidæ for the family of spider-wasps.
Bethylidæ instead of Psilidæ for a family of parasitic wasps.
Prosopidæ instead of Hylæidæ for a family of bees.
Prosopis instead of Hylæus for a genus of bees.
Bombidæ instead of Bremidæ for the family of bumblebees.
Bombus instead of Bremus for a genus of bees.
Anthophora instead of Lasius or Podalirius for a genus of bees.
Ammophila instead of Sphex for a genus of thread-waisted **wasps.**

SYNOPSIS OF THE HYMENOPTERA

Suborder CHALASTOGASTRA. The sawflies and horn-tails. p. 891.
    The Xyelid sawflies. p. 896.............................Family XYELIDÆ
    The Web-spinning and the leaf-rolling Sawflies. p. 897..Family PAMPHILIIDÆ
    The Horn-tails. p. 898..................................Family SIRICIDÆ
    The Xiphydriid sawflies. p. 899.......................Family XIPHYDRIIDÆ
    The Stem Sawflies. p. 900.............................Family CEPHIDÆ
    The Cimbicid Sawflies. p. 902.........................Family CIMBICIDÆ
    The Typical Sawflies. p. 902.......................Family TENTHREDINIDÆ
    The Argid Sawflies. p. 904............................Family ARGIDÆ
    The Oryssids. p. 905..................................Family ORYSSIDÆ
Suborder CLISTOGASTRA or APOCRITA.
  Superfamily ICHNEUMONOIDEA
    The Stephanids. p. 919...............................Family STEPHANIDÆ
    The Braconids. p. 919................................Family BRACONIDÆ
    The Ichneumon-flies. p. 922.........................Family ICHNEUMONIDÆ
    The Trigonalids. p. 929..............................Family TRIGONALIDÆ
    The Aulacids. p. 929..................................Family AULACIDÆ
    The Gasteruptiids. p. 930.........................Family GASTERUPTIIDÆ
  Superfamily PROCTOTRUPOIDEA.
    The Roproniids. p. 931...............................Family ROPRONIIDÆ
    The Helorids. p. 931..................................Family HELORIDÆ
    The Vanhorniids. p. 931............................Family VANHORNIIDÆ
    The Proctotrupids. p. 931........................Family PROCTOTRUPIDÆ
    The Belytids. p. 932..................................Family BELYTIDÆ
    The Ceraphronids. p. 932.........................Family CERAPHRONIDÆ
    The Pelecinids. p. 932................................Family PELECINIDÆ
    The Scelionids. p. 933...............................Family SCELIONIDÆ
    The Platygasterids. p. 933.......................Family PLATYGASTERIDÆ
  Superfamily CYNIPOIDEA. The Cynipids.
    The Gall-flies or Gall-wasps and their Allies. p. 934.....Family CYNIPIDÆ
  Superfamily CHALCIDOIDEA.
    The Chalcid-flies. p. 941............................Family CHALCIDIDÆ
  Superfamily EVANIOIDEA.
    The Ensign-flies. p. 949.............................Family EVANIIDÆ
  Superfamily VESPOIDEA. The Vespoid-wasps.
    The Spider-wasps. p. 950..........................Family POMPILIDÆ
    The Embolemids. p. 951..........................Family EMBOLEMIDÆ
    The Cleptids. p. 951.................................Family CLEPTIDÆ

The Cuckoo-wasps. p. 951........................Family Chrysididæ
The Anthoboscids. p. 952......................Family Anthoboscidæ
The Sapygids. p. 952..............................Family Sapygidæ
The Thynnids. p. 952.............................Family Thynnidæ
The Tiphiids. p. 953..............................Family Tiphiidæ
The Velvet-ants. p. 953..........................Family Mutillidæ
The Scoliids. p. 954..............................Family Scoliidæ
The Ants. p. 954..................................Family Formicidæ
The Bethylids. p. 965............................Family Bethylidæ
The Rhopalosomids. p. 965..................Family Rhopalosomidæ
The Typical Wasps or Diploptera. p. 965..............Family Vespidæ
Superfamily Sphecoidea. The Sphecoid-wasp and the Bees.
   I. The Sphecoid-Wasps.
The Ampulicids. p. 978..........................Family Ampulicidæ
The Dryinids. p. 978.............................Family Dryinidæ
The Typical Sphecoid-wasps. p. 979.............Family Sphecidæ
   II. The Bees. p. 989
The Bifid-tongued Bees. p. 993.................Family Prosopidæ
The Andrenids. p. 995..........................Family Andrenidæ
The Leaf-cutter Bees and their Allies. p. 999.......Family Megachilidæ
The Bumblebees. p. 1001......................Family Bombidæ
The Honey-bees. p. 1005......................Family Apidæ

KEY TO THE SUBORDERS OF HYMENOPTERA

A. Base of abdomen not slender but broadly joined to the thorax by a more or less immovable joint; subanal vein of the fore wing present (except in the genus Acordulecera); wings always present. p. 894. Suborder Chalastogastra
AA. Base of the abdomen constricted to a slender pedicel and joined to the thorax (alitrunk) by a narrow movable joint; subanal vein of the fore wings absent; wings often absent. p. 907................Suborder Clistogastra

KEY TO THE FAMILIES OF CHALASTOGASTRA

By Dr. H. K. Townes

A. Front tibia with two apical spurs; body usually rather short and broad.
  B. Third antennal segment very long, about as long as or longer than all of the following segments together; fore wing usually with vein R_2. p. 896. Xyelidæ
  BB. Third segment of antenna not remarkably long or the antenna consisting of three segments only; fore wing without vein R_2.
    C. Subcostal vein present and distinct; antenna with thirteen or more segments, filiform. p. 897...............................Pamphiliidæ
    CC. Subcostal vein absent or present as a trace; antenna either with less than thirteen segments or pectinate or serrate.
      D. Antenna ending in a knob; abdomen with the lateral margin sharp. p. 902...Cimbicidæ
      DD. Antenna not ending in a knob, filiform, somewhat enlarged toward the tip, or of other form; abdomen with the lateral margin rounded.
        E. Antenna with three, six, or thirteen or more segments. p. 904. Argidæ
        EE. Antenna with seven to twelve segments. p. 902. Tenthredinidæ
AA. Front tibia with one apical spur; body usually more elongate.
  B. Antenna attached under a ridge just above the mouth, apparently arising from the mouth. p. 905......................................Oryssidæ
  BB. Antenna attached normally, near the middle of the face.
    C. Abdomen terminating in a hard spike-like or triangular point (in females, above the ovipositor). p. 898....................................Siricidæ
    CC. Abdomen not terminating in a hard point.
      D. Pronotum rectangular, saddle-shaped, with lateral and dorsal surfaces; abdomen compressed. p. 900.............................Cephidæ
      DD. Pronotum a narrow collar extending around the front of the thorax, presenting a lateral and cephalic but no dorsal surfaces; abdomen cylindrical. p. 899.......................................Xiphydriidæ

KEY TO THE FAMILIES OF THE COMMONER CLISTOGASTRA*
By Dr. H. K. Townes

A. Wings present.
  B. Either the antenna with more than thirteen segments or the hind wing without closed cells (= cells completely surrounded by veins); legs usually with two trochanters (parasitic Hymenoptera).
    C. Antenna with seventeen or more segments, rarely as few as fifteen; costal cell of front wing obliterated by the fusion or close approximation of the costal and R + M veins; hind wing nearly always with closed cells.
      D. Fore wing with vein M_2 so that cells M_1 and 1st M_2 are separate. p. 922 ....ICHNEUMONIDÆ
      DD. Fore wing lacking vein M_2, cells M_1 and 1st M_2 confluent. p. 919 ....BRACONIDÆ
    CC. Antenna with sixteen or fewer segments; costal cell of fore wing present though usually very narrow, or rarely completely absent.
      D. Abdomen attached high on the thorax (alitrunk) so that there is quite a distance between its base and the bases of the hind coxæ.
        E. First abdominal segment cylindrical, set off from the rest of the abdomen which is small and oval; anal lobe present. p. 949. EVANIIDÆ
        EE. First abdominal segment gradually enlarged toward the apex, not set off from the rest of the abdomen; anal lobe absent.
          F. Hind tibia strongly swollen toward the tip; fore wing can be folded lengthwise. p. 930......................GASTERUPTIIDÆ
          FF. Hind tibia not swollen toward the tip; fore wing can not be folded. p. 929.....................................AULACIDÆ
      DD. Abdomen attached low on the thorax so that its base is next to those of the hind coxæ.
        E. Hind margin of hind wing with a deep notch which sets off an "anal lobe."
          F. Body metallic greenish or bluish; abdomen with three or rarely four exposed tergites and concave beneath. p. 951..CHRYSIDIDÆ
          FF. Body not greenish or bluish; abdomen with more than four exposed tergites and convex beneath.
            G. Antenna with ten segments; front tarsus of female usually chelate. p. 978...................................DRYINIDÆ
            GG. Antenna with twelve or thirteen segments; front tarsus of female simple. p. 965.........................BETHYLIDÆ
        EE. Hind margin of hind wing without a notch.
          F. Hind corner of pronotum not reaching the tegula, separated from it by a more or less distinct sclerite, the prepectus; middle tibia with one apical spur or rarely none; fore wing with only a single distinct vein which is usually forked at the tip; antenna elbowed. p. 941...CHALCIDIDÆ
          FF. Hind corner of pronotum reaching the tegula, the prepectus absent; middle tibia usually with two apical spurs; antenna usually not elbowed.
            G. Abdomen more or less compressed, polished, and most of it covered by a single tergite; costal vein absent; cell 2d $R_1 + R_2$ large. p. 934.....................................CYNIPIDÆ
            GG. Abdomen cylindrical or depressed; costal vein present or cell 2d $R_1 + R_2$ small or absent. (Proctotrupoidea)
              H. First segment of hind tarsus one fourth as long as the second; second to fourth abdominal segments of female each as long as the head and thorax together; abdomen of male shorter and clavate; large species. p. 932............PELECINIDÆ
              HH. First segment of hind tarsus longer than the second; second to fourth segments of abdomen not exceedingly long; mostly very small species. (other families of Proctotrupoidea) See other key at II on page 909.

---

*A complete key to the Clistogastra including the uncommon and aberrant forms will be found on page 908.

BB. Antenna with thirteen or fewer segments; hind wing with closed cells; legs each with a single trochanter (aculeate or stinging Hymenoptera).
  C. Hind corner of pronotum not in the form of a rounded lobe and always extending back to touch the tegula (Fig. 1195, A).
    D. An erect scale or one or two knot-like swellings on the stalk between the main body of the abdomen and the thorax (alitrunk); front wing without vein M_2 . . . (ants). p. 954.....................FORMICIDÆ
    DD. No scale or node between the abdomen and the thorax; front wing with vein M_2; antenna usually with twelve segments in the female and thirteen in the male.
      E. Cell M_4 of fore wing longer than cell $Cu + Cu_1$ (Fig. 1182); wings can usually be folded lengthwise. p. 965..................VESPIDÆ
      EE. Cell M_4 of fore wing shorter than cell $Cu + Cu_1$; wings can not be folded.
        F. Mesopleurum divided by a transverse groove into an upper and a lower half; legs unusually long, the hind femur reaching approximately to the tip of the abdomen. p. 950...........POMPILIDÆ
        FF. Mesopleurum not divided by a transverse groove; legs shorter.
          G. Hind coxæ widely separated by the broad plate-like metasternum which overlaps their bases; apical part of wing membrane with fine parallel ridges; eye margin strongly notched opposite the antenna. p. 954..........................SCOLIIDÆ
          GG. Hind coxæ not widely separated and not overlapped at the base.
            H. Antenna with twelve segments (females). p. 953. TIPHIIDÆ
            HH. Antenna with thirteen segments (males).
              I. Abdomen without any or with two visible terminal spines. p. 953...MUTILLIDÆ
              II. Abdomen with a single upcurved terminal spine. p. 953. ....TIPHIIDÆ
  CC. Hind corner of pronotum in the form of a rounded lobe and not touching the tegula (Fig. 1195, B). If the thorax is so hairy that the pronotum is invisible, the specimen is a bee and belongs here.
    D. First segment of hind tarsus not dilated; hairs not plumose; abdomen often petiolate. p. 979...............................SPHECIDÆ
    DD. First segment of hind tarsus enlarged and flattened; some of the hairs plumose, especially those on top of the thorax . . . (bees).
      E. Hind tibia without apical spurs . . . (honey-bee). p. 1005. APIDÆ
      EE. Hind tibia with apical spurs.
        F. Cell below the stigma of fore wing with a hair-like line extending downward from the base of the stigma; anal lobe absent; hind tibia of females and workers with a corbicula except in *Psithyrus* . . . (bumblebees). p. 1001.............................BOMBIDÆ
        FF. Cell below the stigma of fore wing without a hair-like line *or* the hind wing with an anal lobe; hind tibia without a corbicula.
          G. Tongue short, its apex divided (bifid). p. 993....PROSOPIDÆ
          GG. Tongue long or short but its apex never divided and frequently pointed.
            H. Front wing with three cells beneath the stigma and the cell beyond the stigma. p. 995....................ANDRENIDÆ
            HH. Front wing with two cells beneath the stigma and the cell beyond the stigma.
              I. Labrum not large and free, usually entirely concealed by the clypeus, if visible then strongly inflexed; most females with a ventral abdominal pollen-collecting brush; pygidial area absent. p. 999......................MEGACHILIDÆ
              II. Labrum large, free, and uncovered; female without a ventral abdominal pollen-collecting brush; pygidial area usually present. p. 995....................ANDRENIDÆ
AA. Wings absent or reduced so that they are useless for flight and the normal venation is disturbed.
  B. An erect scale or one or two knot-like swellings on the stalk between the main body of the abdomen and the thorax (alitrunk); antenna elbowed . . . (ants). p. 954..FORMICIDÆ

BB. No scale or nodes between the abdomen and thorax.
  C. Pronotum not movable, fused with the rest of the thorax; very hairy ant-like insects, usually red and black. p. 953............MUTILLIDÆ
  CC. Pronotum movable, jointed to the rest of the thorax; less hairy insects.
    D. Antenna elbowed; hind corner of pronotum not reaching the tegula, separated from it by a special sclerite, the prepectus. p. 941. CHALCIDIDÆ
    DD. Antenna not elbowed; hind corner of pronotum reaching the tegula.
      E. Second and third abdominal sternites membranous, in dried specimens with a longitudinal fold; ovipositor always exposed beyond the tip of the abdomen. p. 922......................ICHNEUMONIDÆ
      EE. Second and third abdominal sternites sclerotized, without a longitudinal fold; ovipositor usually retracted into the abdomen.
        F. Front tarsus chelate; antenna with ten segments. p. 978. DRYINIDÆ
        FF. Front tarsus normal.
          G. Abdomen compressed, covered mostly by a single tergite. p. 934 ....CYNIPIDÆ
          GG. Abdomen depressed or cylindrical.
            H. Head elongate with the antenna inserted close to the anterior end. p. 965..BETHYLIDÆ
            HH. Head of normal shape, not elongate; antenna with twelve segments. p. 953.............................TIPHIIDÆ

Suborder CHALASTOGASTRA or SYMPHYTA*

The Sawflies and Horn-tails

This suborder includes the more generalized members of the Hymenoptera, those in which the form of the body is less modified and the venation of the wings less reduced than is the case with other members of the order.

The basal segments of the abdomen are similar in form and the abdomen is broadly joined to the thorax as in the more generalized orders of insects. The first abdominal segment is not closely anchylosed to the thorax, forming a propodeum, as is the case in the Clistogastra, and its tergum is usually longitudinally divided on its middle line.

The adult forms of the Chalastogastra have never developed the highly specialized habits and instincts exhibited by many members of the suborder Clistogastra, especially by the wasps, ants and bees. There are no parasitic forms in this suborder except in the single genus *Oryssus*, p. 907.

There are no wingless forms of the Chalastogastra. In the more generalized members of the suborder nearly all of the wing-veins are preserved, although the courses of the branches of the forked veins have been greatly modified, as indicated on an earlier page, and as is shown in the figures of wings given later.

The ovipositor of the females is well developed and complicated in structure. It is fitted for making incisions in the leaves or stems of plants and is more or less saw-like in form. It is this fact that

---

*The name Chalastogastra is the one most commonly applied to this suborder and for that reason is used in this work; but some authors use Symphyta, which is really the older name. The etymology of these names is as follows:—
    Chalastogăstra; *chalastos* (χαλαστός), loose; *gastros* (γαστρός), the belly.
    Sÿmphyta: *sym* (σύν), with; *phyton* (φυτόν), plant.

suggested the common name sawflies which is applied to members of this order.

The ovipositor and its sheath consists of three pairs of appendages or gonapophyses; one pair arising from the sternum of the eighth abdominal segment and two pairs from the sternum of the ninth abdominal segment. The outer pair of the ninth abdominal segment constitute the sheath of the ovipositor, so called because when the ovipositor is not in use it is enclosed between the two members of this pair of gonapophyses. The ovipositor is a double organ, consisting of two similar blades situated side by side. Each blade consists of two gonapophyses, an upper or posterior one, known as the support or lance, and a lower or anterior one, the so-called saw or lancet. The supports are the inner gonapophyses of the ninth abdominal segment, and the saws are the gonapophyses of the eighth abdominal segment.

Although each of the saws is closely joined to its support it can be moved backward and forward along it. Figure 1132 represents one of the blades of the ovipositor of *Cimbex americana*.

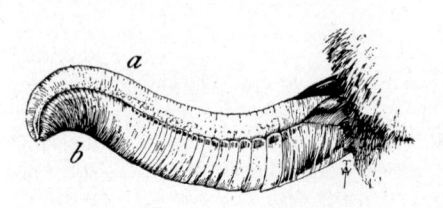

Fig. 1132.—Blade of ovipositor of *Cimbex americana*; *a*, support; *b*, lancet.

The ovipositor of this sawfly is fitted for cutting slits in leaves in which the eggs are deposited. In those members of this suborder that deposit their eggs in the stems of plants or the trunks of trees, as the Siricidæ, the ovipositor is slender and long. After a slit has been cut or a hole drilled in the trunk of a tree, as the case may be, an egg is forced down between the blades of the ovipositor to the nidus prepared for it.

The larvæ of the Chalastogastra are all plant-feeders. With the exception of those that are leaf-miners they are caterpillar-like in form. Prolegs are present in the Xyelidæ, Cimbicidæ, Tenthredinidæ and Argidæ; but these are not provided with hooks as are the prolegs of caterpillars.

A striking feature of the larvæ of this suborder is the possession of a pair of ocelli, one on each side, which in their position and in their structure agree with the ocelli of adult insects, that is, they are primary ocelli. This characteristic distinguishes these larvæ from the larvæ of Lepidoptera, which have only adaptive ocelli, usually several on each side (see page 136).

A classification of the larvæ of this suborder was published by Yuasa ('22).

Family XYELIDÆ
The Xyelid Sawflies

The members of this family can be recognized by the form of the antennæ and the venation of the wings. The basal segments of the flagellum are consolidated, thus forming what appears to be a very long third segment of the antenna and the remaining segments of the flagellum are small (Fig. 1133). Except in *Neoxyēla albĕrta*, a species recently described from Banff, Alberta, the members of this family differ from all other Hymenoptera in that the free part of vein R_2 of the fore wings is present (Fig. 1134).

The posterior margin of the pronotum is straight or nearly so. The mesonotum is short and never extends much beyond the anterior margins of the tegulæ. The anterior tibiæ are armed with two apical spurs. In some species the ovipositor is very long, in others it is of moderate length.

The described larvæ feed on the foliage of hickory, butternut, pecan, elm, and the staminate flowers of pine. In the larvæ each of the ten abdominal segments bears a pair of prolegs, although in some species those of the first and ninth segments are smaller than the others.

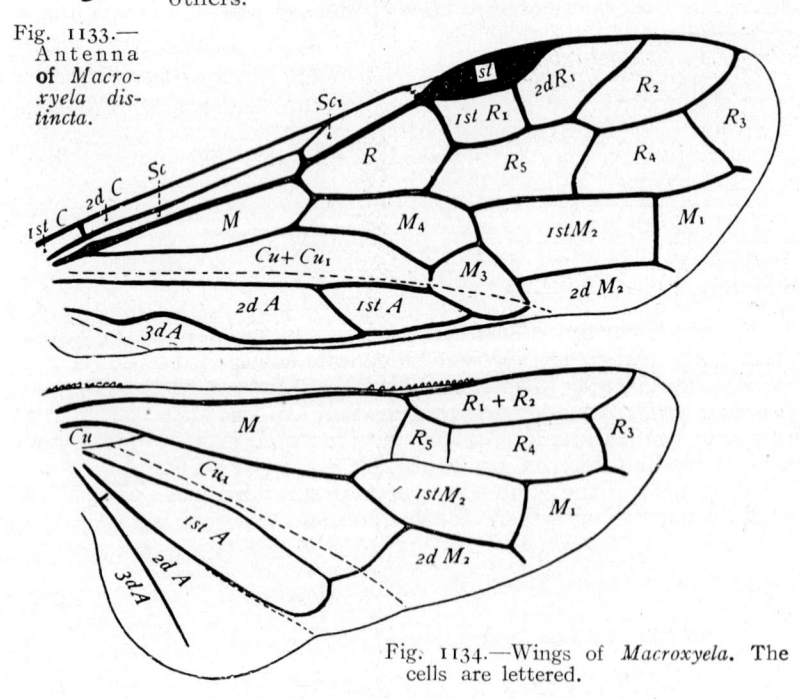

Fig. 1133.—Antenna of *Macroxyela distincta*.

Fig. 1134.—Wings of *Macroxyela*. The cells are lettered.

Family PAMPHILIIDÆ

The Web-spinning and the Leaf-rolling Sawflies

The common names given above were suggested by the fact that the larvæ of some species build nests by tying the leaves of their food plants together with a web of silk, and others build nests by rolling the edge of a leaf and live inside the tube so formed. The larvæ of some species are gregarious. The larvæ of members of this family have long, seven-jointed antennæ, well-developed thoracic legs, but lack abdominal prolegs.

In the fore wings of the adult (Fig. 1135) vein Sc is preserved as a distinct vein; the free part of vein R_2 is wanting; and vein Cu_2 is usually preserved, at least as a short spur. In the hind wings vein Sc is more or less distinctly preserved.

The body of the adult is robust. The posterior margin of the pronotum is straight or nearly so. The mesonotum is short and never extends much beyond the anterior margins of the tegulæ. The anterior tibiæ are armed with two apical spurs. The ovipositor of the female is short.

More than fifty species have been described from America north of Mexico; but the larvæ of only a few of these are known; among these are the following.

The plum web-spinning sawfly, *Neurŏtoma inconspĭcua*.—The larvæ of this species feed on the foliage of plum and cherry; they are gregarious and form unsightly nests by spinning webs over the leaves; frequently these webs cover an entire tree. The injury is done in early summer. When full-grown the larvæ find their way to the ground, where they pass the remainder of the summer and winter in earthen cells, they transform to pupæ in the spring, and the adults emerge in May or June. This pest is controlled by spraying or dusting the infested trees, with lead arsenate.

The peach sawfly, *Pamphĭlius pĕrsicus*.—This pest of the peach is one of the leaf-rolling species. The adults emerge from the ground late in May or early in June and lay their eggs on the leaves; the eggs soon hatch; each larva cuts a slit in a leaf and then rolls over a portion of the leaf, making a case within which it stays during the daytime, feeding chiefly at night. There is a single generation a year. The larva passes the winter in the ground. The same method of control is used as with the preceding species.

898　AN INTRODUCTION TO ENTOMOLOGY

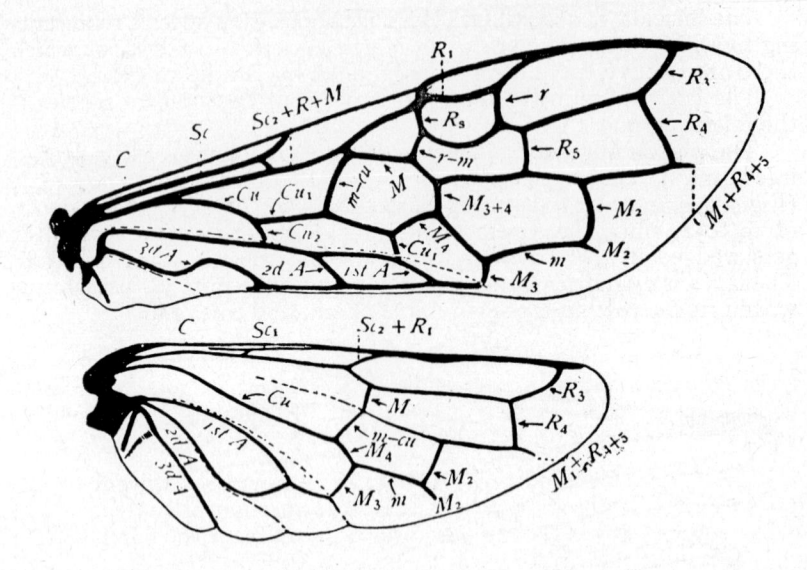

Fig. 1135.—Wings of *Pamphilus*. The veins are lettered.

Family SIRICIDÆ

The Horn-tails

The common name horn-tails is applied to members of this family because the last abdominal segment bears a more or less horn-like prolongation. This is short and triangular in the males, and is a prolongation of the last ventral segment; in the females it is long and often spear-shaped, and is a prolongation of the last dorsal segment.

The body is cylindrical (Fig. 1136); the head large and widened behind the eyes; the pronotum is right-angled, so that it presents both a strictly dorsal and a cephalic aspect, the latter concave; vein Sc_1 of the front wings is absent (Fig. 1137); the propodeum is divided longitudinally; the anterior tibiæ each with only one apical spur; the sheath of the ovipositor is very long and exserted beyond the end of the abdomen; the ovipositor is fitted for boring.

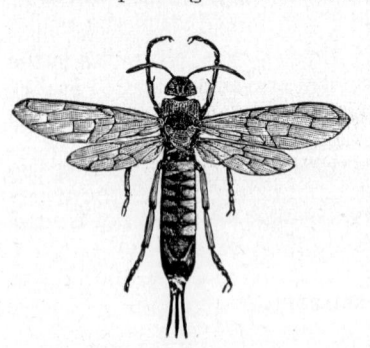

Fig. 1136.—*Tremex columba*.

The Siricidæ is a small family; only about fifty species representing five genera are known. The North American species, of which there are twenty, have been monographed by Bradley ('13).

The larvæ bore in the trunks of trees; our best-known species is the following one.

The pigeon horn-tail, *Trēmex colŭmba*.—The larva of this species infests maple, elm, apple, pear, beech, oak, and sycamore. The female (Fig. 1136) in order to oviposit pierces the wood of a tree to the depth of 10 to 12 mm; the eggs are laid singly; sometimes her ovipositor gets wedged in the wood and holds her a prisoner until she dies. The larva is cylindrical and attains a length of 40 mm. It transforms within its burrow, in a cocoon made of silk and fine chips.

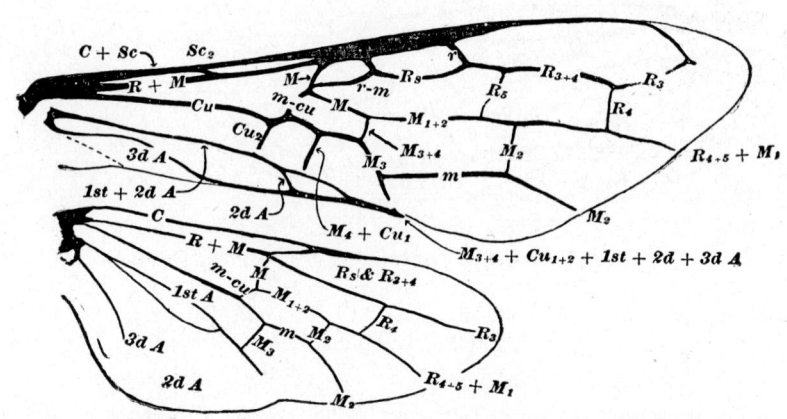

Fig. 1137.—Wings of *Sirex juvencus*. (From Bradley.)

The adults of this species vary in color and marking; based on these variations, three fairly distinct races have been recognized. which to a considerable extent are geographical, although their ranges overlap. In the typical form, race *columba*, the abdomen is black, with ochre-yellow bands and spots along the sides; this is the common form in Quebec, Ontario, and the northeastern United States. In the race *aureus* the ground color of the abdomen is yellow and the markings black; this is the common form in the Rocky Mountains and is found on the Pacific Coast. In the race *sericeus* the entire body is fulvous, the legs beyond the femora yellow, and the wings dark reddish brown; this race is found in the southeastern United States and as far north as Pennsylvania and West to Utah.

Family XIPHYDRIIDÆ

The Xiphydriid Sawflies

This family is composed of a small number of species which are closely allied to the Siricidæ but which differ from them in several important particulars. As with the horn-tails the body is cylindrical but the last abominal segment is not terminated by a triangular or lanceolate process. The back of the head is separated from the pronotum by an elongate neck; the pronotum is very short medially and not angulate laterally; vein Sc_1 is present in the front wings as a transverse vein (Fig. 1138) and the sheath of the ovipositor is seldom longer than the last tergite.

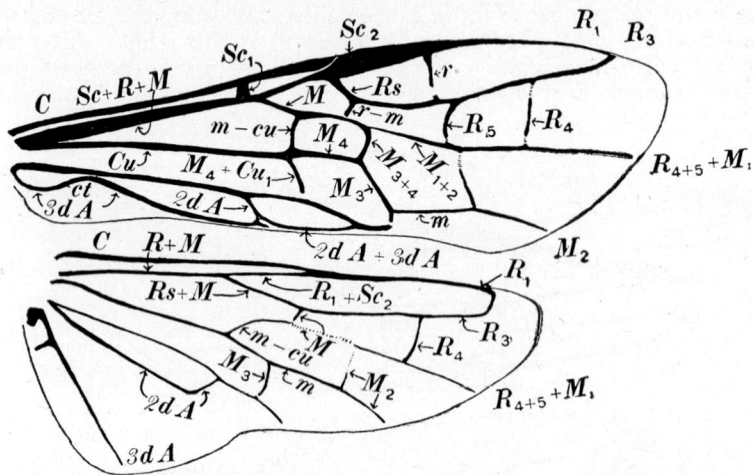

Fig. 1138.—Wings of *Xiphydria maculata*. (From MacGillivray.)

The members of this family are of moderate size. Less than a dozen species have been described from North America.

The known larvæ bore in dead and decaying wood of deciduous trees.

Family CEPHIDÆ

The Stem Sawflies

The stem sawflies are so-called because the larvæ bore into the stems of plants or in the tender shoots of trees and shrubs. The adults are slender, elongate insects of moderate size. The pronotum is more or less quadrate and longer than is usual in the Hymenoptera. The front wings are without a distinct cell between the costa and vein Sc+R+M, and with cross-vein *m-cu* joined to vein M at or near its separation from vein R (Fig. 1139). The anterior tibiæ are armed with one terminal spur.

This family is of moderate size; less than a score of species have been found in our fauna; but these represent nine genera. Some of the species are of economic importance. Several species bore in the stems of grains and grasses, the following species illustrate the habits of these.

The wheat-sawfly-borer, *Cēphus pygmæus*.—The larvæ of this species bore in the stems of wheat, a single larva in a stem, dwarfing and stunting the growth of the plant. As the grain becomes ripe the larva works its way toward the ground; and at the time of harvest the greater number of them have penetrated the root. Here, in the lowest part of the cavity of the straw, they make preparations for passing the winter, and even for their escape from the straw, as adults, the following year. This is done by cutting the straw circularly on the inside, nearly severing it a short distance from the ground, so that a strong wind will cause it to break off at this point. After the circular cut has been made, the larva fills the cavity of the straw just below it for a short distance with a plug of borings. Between this

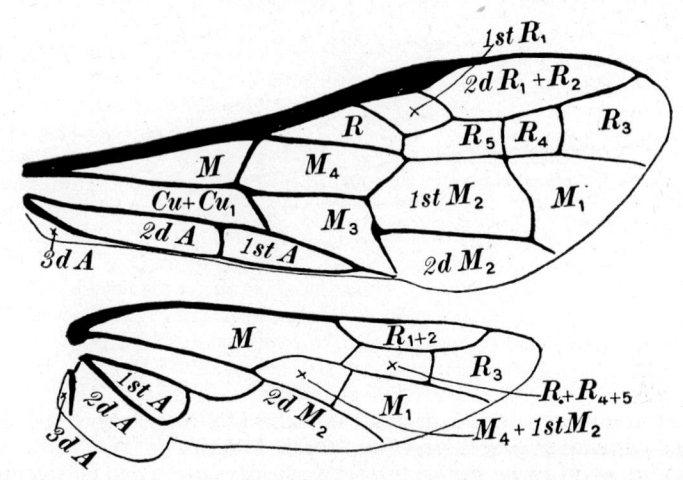

Fig. 1139.—Wings of *Cephus pygmæus*. The cells are lettered. (From Mac-Gillivray.)

plug and the lower end of the cavity, the wall of the cavity is lined with silk forming a cocoon within which the larva passes the winter and changes to a pupa in March or April. The adult insects emerge early in May.

The currant-stem girdler, *Jānus ĭnteger*.—The larva of this species bores in the upper portion of the canes of currant. Its presence is indicated by the wilting and drooping, in late spring, of the new growth at the tip of the infested cane. This is due to the fact that the parent sawfly after depositing her egg in the cane moves up a short distance above where the egg is deposited and with her ovipositor girdles the cane, sometimes nearly severing it. This killing of the tip, and thus checking the growth of the cane, seems to be necessary for the development of the egg and larva. The larva bores in the pith of the cane. In the fall it eats a hole through the woody wall of the cane to the outer bark, thus making provision for the escape of the adult, and then spins a cocoon in which it hibernates. The change to the pupa state takes place in April and the adult emerges in May. The obvious method of control of this pest is to remove and burn the infested portion of the canes while the larvæ are in them.

Family CIMBICIDÆ

The Cimbicid Sawflies

This is a small family, which is represented in our fauna by a few genera and a limited number of species. In this family the body is stout and often very large; there are distinct pleural sclerites in the abdomen (Fig. 1140) and the antennæ are clavate. The anterior tibiæ and metatarsi bear ribbon-shaped or spatulate hairs; the pulvilli are large, broadly sessile on the last tarsal segment, and are not retractile. The sheath of the ovipositor extends but little if at all beyond the end of the abdomen.

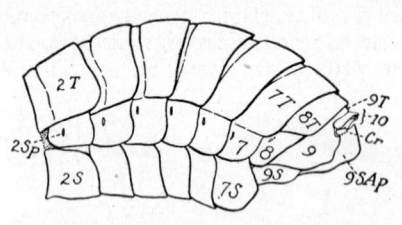

Fig. 1140.—*Cimbex americana*. Abdomen except first segment: *7, 8, 9*, pleurites; *T, T, T*, tergites; *S, S, S*, sternites; *cr*, cercus; *sp*, spiracle. (After Snodgrass.)

The body of the larvæ is cylindrical, stout, and covered with a waxy bloom when living; the thoracic legs are well-developed and five-jointed; and the abdomen bears eight pairs of prolegs. The larvæ live free upon foliage upon which they feed.

The American sawfly, *Cimbex americāna*.—This is the largest of our common sawflies. The female is about 18 mm. in length and has a black head and thorax, a steel-blue or purplish abdomen, with four yellowish spots on each side, smoky brown wings, and black legs, while her feet and short, knobbed antennæ are pale yellow. The male is longer and slenderer and differs somewhat in color. Several varieties of this species, differing in color, have been described. The eggs are laid in June in crescent-shaped slits made in leaves. The food plants are elm, birch, linden, and willow. The larva is greenish yellow, with black spiracles and a black stripe down its back. When disturbed it spurts forth a fluid from glands just above the spiracles. It clings to the upper surface of a leaf and feeds on the edge of the leaf. When not feeding it rests on one side with the body curled up in a spiral form. There is but one generation each year. When the larva is full-grown it burrows in the ground, makes an oval, brownish cocoon, and there spends the winter, not changing to a pupa until spring. The adults appear in May or June.

Family TENTHREDINIDÆ

The Typical Sawflies

This is a very large family, including more than seven-eighths of all of the members of the suborder Chalastogastra. To this family belong all of the Chalastogastra in which the radial cross-vein of the fore wings is opposite cell R_4, or cell R_3, or is wanting, except the small family Cimbicidæ. The typical sawflies differ from the Cimbicidæ in lacking pleural sclerites in the abdomen and in that the antennæ are not clavate. The anterior tibiæ and tarsi do not bear ribbon-shaped or spatulate hairs, as in the Cimbicidæ; and the pulvilli are inserted on the end of the last tarsal segment and are retractile, like the finger of a glove. The sheath of the ovipositor extends but little if at all beyond the end of the abdomen.

The larvæ are caterpillar-like; the thoracic legs are always present and are usually well developed, but are vestigial in some species. Prolegs are usually present; these are borne on abdominal segments 2-7 and 10 or 2-8 and 10, rarely the prolegs are vestigial. The larvæ of the different species differ greatly in size, varying from 10-40 mm. in length.

The larvæ of the majority of the species live free on the foliage of plants, upon which they feed (Fig. 1141). The plants infested by the different species include trees, both deciduous and conifers, shrubs, herbs, grasses, and ferns. The larvæ of some species are leafminers; others fold the edges of leaves; some make galls on leaves, especially of willow and poplar; others make galls in the stems of these plants; and one species, *Caulocămpa acericaulis*, bores in the petioles of maple leaves. Among the species that have attracted attention on account of their economic importance are the following.

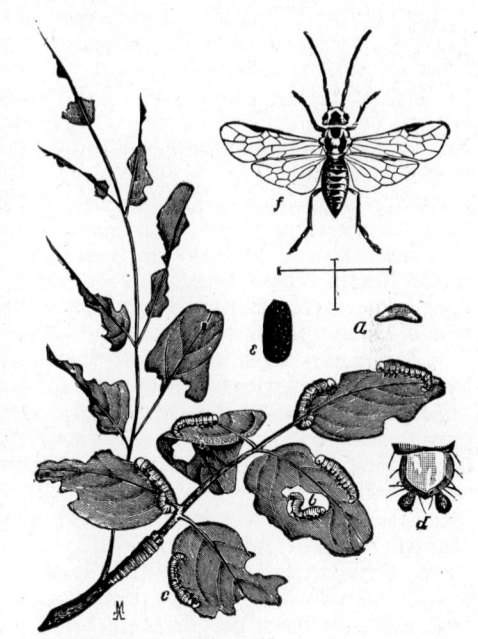

Fig. 1141.—The locust saw-fly, *Pteronidea trilineata*: *a*, egg; *b*, young larva; *c*, full-grown larva; *d*, anal segment of full-grown larva; *e*, cocoon; *f*, adult.

The imported currantworm, *Pteronĭdea ribēsi*.— This is the commonest and best-known of the garden pests. The adult sawflies appear early in the spring and the females lay their eggs in rows along the principal veins on the underside of the leaves of currants and gooseberries. The eggs are glued to the leaf-veins and not inserted in slits, as is usually the case with sawflies, and they increase considerably in size before hatching. They hatch in a week or ten days; and the larvæ begin at once to feed upon the leaves. Often by the time the larvæ become full-grown the infested bush is completely stripped of its foliage. The larvæ are at first whitish, as they increase in size the color changes to green; after the first molt the body becomes covered with many black spots and the head is black; at the last molt they lose their black spots and assume a uniform green color tinged with yellow at the ends. When full-grown the larvæ descend to the ground and spin their cocoons, either just below the surface of the ground or beneath rubbish; sometimes the cocoons are attached to the stems or leaves some distance from the ground. A second generation of the sawflies appears late in June or early in July; and sometimes a third generation is developed; this makes it necessary to fight this pest throughout the spring and summer. The larvæ can be easily destroyed in the spring by spraying the bushes with Paris green or with arsenate of lead; later when the fruit is near maturity fresh hellebore should be used at the rate of four ounces in two or three gallons of water, or as a dry application, one pound in five pounds of flour or air-slacked lime.

The pear-slug, *Calĭroa cĕrasi.*—This is a well-known pest of pear, cherry, and plum. It causes the leaves of the infested tree to turn brown. When such leaves are examined it is found that the injury is due to small, slimy, slug-like larvæ, which have eaten off the upper surface of the leaves, leaving the skeleton of veins and the lower epidermis to turn brown, wither and fall; sometimes trees are entirely defoliated in this way by midsummer. When full-grown the larvæ descend and burrow into the ground a short distance, where each constructs an earthen cell in which it transforms. A second generation of the sawflies appear and lay their eggs about three weeks later. The larvæ can be destroyed by the use of arsenate of lead spray or by dusting the leaves with freshly slaked lime.

The rose-slug, *Clădius isŏmerus.*—Often in the summer our rose-gardens look as if fire had swept over them, so scorched and brown are the leaves. The cause of this apparent conflagration is a transparent jelly-like slug, greenish above and yellowish below, which eats the upper surface of the leaves, leaving patches of the lower surface and the veins. These slugs usually feed by night and remain hidden on the lower surface of the leaves by day. When ready to pupate they crawl down or drop to the ground and burrow beneath the surface; here each makes a little cell and then transforms. The adult fly is shining black with smoky wings and with the fore and middle legs grayish or dirty-white. The female is 5 to 6 mm. in length. There are two broods a year, one in June and one in August. The last brood passes the winter in the ground. This pest can be destroyed with a solution of whale-oil soap, or with kerosene emulsion.

Family ARGIDÆ
The Argid Saw-flies

This family has been recently separated from the Tenthredinidæ, from which it is distinguished by the absence of a post-tergite. This is a distinct apical plate borne by the scutellum, which is present in all of the Tenthredinidæ and absent in this family.

The family Argidæ consists chiefly of tropical insects, but a few representatives of the family are found in this country. Among these are two species of *Sterictiphora*, the larvæ of which occasionally infest sweet potatoes to an injurious extent.

Family ORYSSIDAE

The Oryssids

In former editions of this book the oryssids were given the rank of a suborder, Idiogastra. The adults resemble those of the Chalastogastra in the shape of the abdomen but the form and habits of the larvæ are those characteristic of the Clistogastra. However, there does not seem to be sufficient justification for giving the family the rank of a suborder and it has therefore been referred back to the suborder Chalastogastra in which it was originally placed. In sequence of relationships the family should probably precede the Cimbicidæ and Tenthredinidæ and stand in closer relation to the Cephidæ. The Oryssidæ is a small family of rare insects, only a single species of larva being known in this country.

In the shape of the body (Fig. 1142, A) the members of this family strongly resemble the Siricidæ. They are easily distinguished from all of the other Chalastogastra by the anomalous position of the antennæ, which are inserted far below the eyes, immediately above the mandibles, under a transverse ridge (Fig. 1142, B); by the more reduced venation of the wings; and by the remarkable form of the ovipositor.

In the fore wings (Fig. 1143) the transverse part of vein M_2 is wanting; and in the hind wings R_4 is wanting; therefore, there are no closed submarginal cells in the hind wings. This combination of characters distinguishes the Oryssidæ from all of the Chalastogastra. In the Oryssidæ the first anal cell of the fore wings is preserved, in which respect the members of this family differ from all of the Clistogastra.

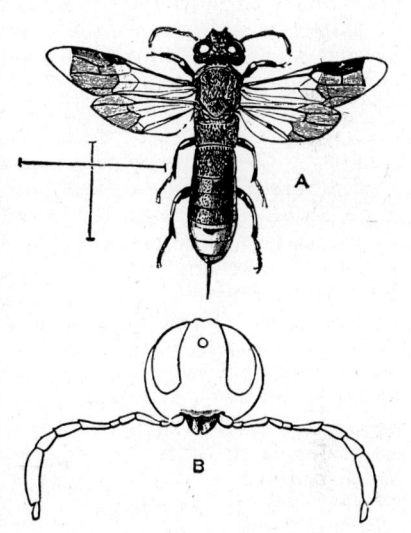

Fig. 1142.—*Oryssus sayi;* A, female; B, head seen from in front. (From Sharp.)

In the form of the ovipositor and in its position when at rest the

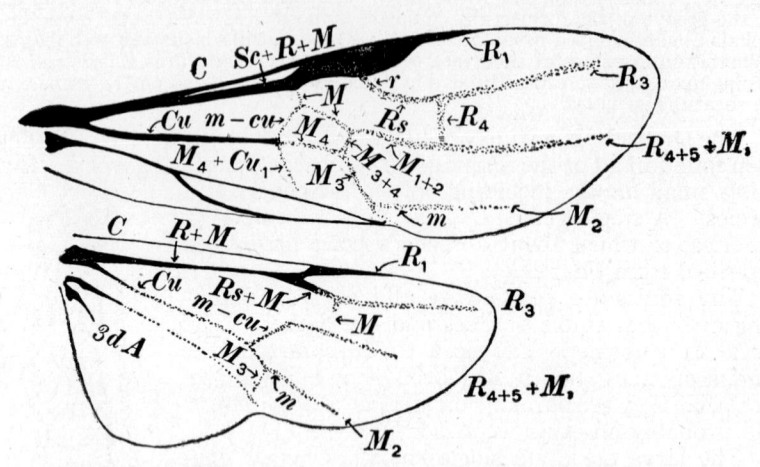

Fig. 1143.—Wings of *Oryssus abietinus*. (From MacGillivray.)

Oryssidæ differ from all other Hymenoptera. The following account of this organ is that of Rohwer and Cushman ('17).

Lying below and on each side of the eighth tergite in the female is a large heavily chitinized plate, the two together forming ventrally a channel for the reception of the ovipositor, and each bearing at its tip a small triangular appendage. These plates apparently represent the fused ninth and tenth tergites which are longitudinally divided dorsally (Fig. 1144, 9T and 10T), and the appendages are apparently the cerci (Fig. 1144, c); the eighth sternite is internal and lies above and somewhat behind the ninth, and is represented by two triangular plates, from the upper angle of which originate the lancets (first gonopophyses) (Fig. 1145, *lt*);

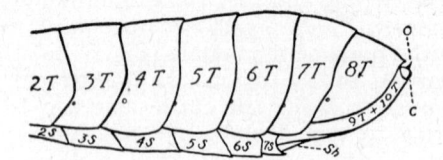

Fig. 1144.—Lateral aspect of abdomen of the female of *Oryssus*. (After Rohwer and Cushman.)

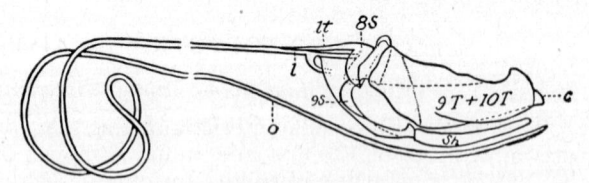

Fig. 1145.—Details of the ovipositor of Oryssus: 9 T, 10 T, ninth and tenth tergites; 8 S, eighth sternite; 9 S, ninth sternite; *lt*, lancet; *l*, lance; *sh*, sheath; *o*, ovipositor; *c*, cerci. (After Rohwer and Cushman.)

the ninth sternite is also internal, lying below and in front of the eighth and represented by two more or less triangular plates which extend posteroventral; the lance (second gonopophyses) originates from the inner ends of these plates and becomes fused a short distance cephalad of its origin (Fig. 1145, *l*);

the two parts of the sheath (third gonopophyses) arise from the apices (Fig. 1145, Sh). Shortly cephalad of the origin of the lance and lancets the latter enter the groove of the former, the complete ovipositor as thus formed extending cephalad in an inverted position enclosed within a membranous sac, probably invaginated intersegmental skin, into the mesothorax, where it is coiled, and returning upon itself continues caudad in its normal position and enters the base of the sheath (Fig. 1145).

The Oryssidæ is a widely distributed family, members of it having been found in all of the major geographical regions of the world. But it is a small family, including only a few genera and species. A single genus, *Oryssus*, is found in North America, of which about a dozen species have been described from this region.

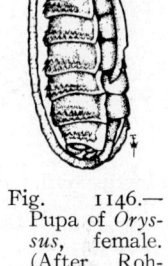

The adults are very active and are found running over the trunks of trees and on timber. The larvæ were formerly supposed to be borers in the trunks of trees; but it has been shown by Burke ('17) that they are parasitic on the larvæ of *Buprestis* and probably on other wood-boring larvæ.

The larva of only a single species, *Oryssus occidĕntalis*, is known. This is white, subcylindrical, about one-third as thick as long, and legless; but the positions of the legs are indicated by chitinized disks. The mouth-parts are very simple, the labrum, labium, and maxillæ being merely fleshy lobes, but the mandibles are heavily chitinized; the antennæ are tubercle-like and set at the summits of rounded elevations.

In the pupa of the female (Fig. 1146) the terminal portion of the ovipositor is external and extends over the back the entire length of the body. Referring to this, Rohwer and Cushman ('17) state as follows:

Fig. 1146.—Pupa of *Oryssus*, female. (After Rohwer and Cushman.)

The reason for the formation in the pupa of the long external ovipositor is inexplicable, and its reduction to the form existing in the adult is equally inexplicable. This is rendered all the more difficult to understand by the fact that in the prepupa the ovipositor is coiled as it is in the adult, while in the pupa it forms a simple loop in the thorax.

Suborder CLISTOGASTRA or APOCRITA*

Parasitic Hymenoptera, ants, wasps and bees

This most striking characteristic of this suborder is the fact that what appears to be the first abdominal segment, but which is really the second, is greatly constricted forming a slender petiole or waist between the larger portion of the abdomen and the alitrunk or wing-bearing region of the body (Fig. 1147).

---

*The name Clistogastra is the one most commonly applied to this suborder and for that reason is used in this work; but some authors use Apocrita, which is really the older name. The etymology of these names is as follows: Clistogăstra: *clistos* (χλειστός), closed; *gaster* (γαστήρ), the belly. Apŏcrita: *apocritos* (ἀπόκριτος), separated; *apo* (ἀπό), from, *crino* (κρίνω), to separate.

In this suborder the intermediate region of the body is not merely the thorax but includes also the first abdominal segment, only the tergum of which is preserved in the adult. This is known as the *median segment*, or the *propodeum* and can be identified by its spiracles, the third pair of this region of the body. It should be remembered that the thorax bears only two pairs of spiracles (see page 115). From the above it follows that what appears to be the first abdominal segment in the Clistogastra, and which is usually so-called, is really the second.

Fig. 1147.

In the Clistogastra the ovipositor and its sheath are composed of the same morphological elements as are those of the sawflies described on an earlier page; but these parts differ greatly in form in different members of this suborder. In some the ovipositor is a boring instrument by means of which deep holes are made into trees and eggs placed in these holes; in others it is used for thrusting the eggs into the bodies of other insects; and in still others it is modified so as to form a sting with which poison glands are connected.

AN INCLUSIVE TABLE OF FAMILIES OF THE CLISTOGASTRA FOR ADVANCED STUDENTS

By Dr. J. C. Bradley

A. With well-developed wings.
  B. Hind wings without an anal lobe.*
    C. No erect scale or node between the gaster and the propodeum.
      D. The costal cell of the fore wings eliminated by the coalescence of the costal and subcostal veins, except in the case of two or three rare genera. The venter is membranous and has in dried specimens a longitudinal fold.
        E. The transverse part of vein M_2 of the fore wings wanting, causing the union of cells M_1 and 1st M_2 (Fig. 1148).
          F. The abdomen not very long and slender and strongly compressed. p. 919.....................................BRACONIDÆ
          FF. The abdomen very long, slender, and strongly compressed. (The genus *Hymenopharsalia*) p. 922............ICHNEUMONIDÆ
        EE. Cells M_1 and 1st M_2 separated by the transverse part of vein M_2 (Fig. 1152). p. 922..............................ICHNEUMONIDÆ
      DD. The costal cell of the fore wings present. The venter chitinized.
        E. Abdomen borne on the dorsal surface of the propodeum far above the hind coxæ.
          F. The transverse part of vein M_2 present in the front wings, which have at least two closed submarginal cells. p. 929......AULACIDÆ
          FF. The transverse part of vein M_2 wanting in the front wings, which do not have two closed submarginal cells. p. 930.........
          ....GASTERUPTIIDÆ
        EE. Abdomen borne between the hind coxæ, or on the end of the propodeum slightly above them.
          F. The transverse part of vein M present and situated close to the stigma.

---

*In the hind wings of the insects belonging under this category neither the second anal furrow nor the axillary excision is present, but there is sometimes present a weak preaxillary excision, more rarely (some genera of Braconidæ) a pronounced notch, but never forming a deep slit. See also footnote on page 911.

G. Antennæ in both sexes of more than fifteen segments; trochanters clearly two-segmented.
  H. Two or three closed submarginal cells in the forewings. p. 929 .... TRIGONALIDÆ
  HH. Only one closed submarginal cell in the forewings. p. 919 .... STEPHANIDÆ
GG. Antennæ of thirteen segments in the male and twelve in the female; trochanters one-segmented.
  H. Pronotum without posterior lobes (see Fig. 1195) its lateral extensions reaching the tegulæ.
    I. Cell M_4 of the forewings shorter than cell $Cu + Cu_1$ or absent (Mutillinæ). p. 953 .... MUTILLIDÆ
    II. Cell M_4 of the forewings present and longer than cell $Cu + Cu_1$ (Fig. 1182). (Vespinæ.) p. 965 .... VESPIDÆ
  HH. Pronotum with posterior lobes terminating at a distance from the tegulæ (Fig. 1195). (This distance is short in the Ampulicinæ.)
    I. Abdomen of male with only six exposed segments, the fourth and following scarcely exposed; that of the female with compressed apex. Prothorax elongate, usually with a median longitudinal groove. Nude insects, often brilliantly metallic. (Ampulicinæ.) p. 978 .... AMPULICIDÆ
    II. Abdomen of male with seven exposed segments; that of the female not compressed at apex. Prothorax short without median longitudinal groove. Usually hairy insects, some of the hairs plumose. (A few genera in the various families of bees. Pass to EE on page 914.)
FF. The transverse part of vein M situated about two-thirds of the way from the wing base to the end of the costal cell $(C + Sc_1)$ or wanting.
  G. Wings not longitudinally plaited in repose. Ovipositor not carried along the mid-dorsal line.
    H. The pronotum laterally reaching the tegulæ. No prepectus present.
      I. Hind metatarsi one-fourth the length of the following segment. Large insects; the abdomen of the female filiform and several times the length of the head and thorax together; that of the male shorter and clavate. p. 932 .... PELECINIDÆ
      II. Hind metatarsi at least as long as the following segment.
        J. Mandibles in a reversed position, their apices directed outwardly, away from the mouth-opening. p. 931 .... VANHORNIIDÆ
        JJ. Mandibles in a normal position.
          K. Cells $Cu + Cu_1$ and M_3 of the forewings fully enclosed and separated from each other by perfect veins.
            L. Cell M_4 of the forewings triangular; antennæ composed of sixteen segments. p. 931 .... HELORIDÆ
            LL. Cell M_4 of the forewings irregularly six-sided; antennæ composed of fourteen segments. p. 931 .... ROPRONIIDÆ
          KK. Cells $Cu + Cu_1$ and M_3 partly enclosed by brown lines, or altogether wanting. Claws not pectinate.
            L. Abdomen sharply margined by a carina along the sides; antennæ arising near the clypeus.
              M. Antennæ of never more than ten segments, rarely with only eight or nine. Front wings without vein C or "stigmal" vein $(Sc_2 + R_1)$, often without any venation. p. 933 .... PLATYGASTERIDÆ
              MM. Antennæ usually of twelve segments, more rarely of eleven segments, or if of seven or eight segments the club is unsegmented. p. 933 .... SCELIONIDÆ

LL. Abdomen immargined laterally (acute in Telenominæ but without a carina).
M. Forewings with a distinct stigma.
N. A closed, usually very short, marginal cell (2d $R_1 + R_2$) present. Antennæ of thirteen segments. Abdomen with a short cylindrical petiole, the second segment much longer and larger than the others. p. 931..Proctotrupidæ
NN. No closed marginal cell (2d $R_1 + R_2$) but the basal part of the marginal vein (vein r) often present in the forewings.
O. Antennæ of eleven segments. (Megaspilinæ.) p. 932..............Ceraphronidæ
OO. Antennæ of twelve or more segments (a few genera of Diapriinæ). p. 932...Belytidæ
MM. Forewings without a distinct stigma.
N. Costal cell (C + Sc_1) either closed along the margin, or if open very narrow; the marginal cell (2d $R_1 + R_2$) if present narrowly triangular, its proximal margin a straight line: Abdomen not compressed nor dorsally keeled. Hypopygium of the female not divided, but closely applied to the pygidium, the ovipositor issuing from between them at the tip of the abdomen.
O. Hind wings with a closed median cell (M). Forewings almost invariably with a closed marginal cell (2d $R_1 + R_2$). Antennæ of fourteen or fifteen segments. (Belytinæ.) p. 932 ....Belytidæ
OO. Hind wings without any closed cells; forewings without a closed marginal cell (2d $R_1 + R_2$).
P. Abdomen margined laterally, the margin acute but not sharply carinate. Antennæ arising from near the clypeus and composed of ten to twelve segments. Scutellum not divided into three lobes. (Telenominæ.) p. 933......................Scelionidæ
PP. Abdomen not at all margined laterally. Scutellum divided by two oblique curved impressed lines into three lobes. (Calliceratinæ.) p. 932..........Ceraphronidæ
NN. Costal cell (C + Sc_1) open along the costal margin and abnormally wide. The marginal cell (2d $R_1 + R_2$) present; often closed; often open along the costal margin, sometimes at tip; it is always four-sided, acute at apex and at its base. Abdomen more or less strongly compressed and with a mid-dorsal keel, if rarely but little compressed and without a keel, it is more or less swollen dorsally. Hypopygium of the female divided, the ovipositor issuing from the cleft thus formed, anterior to the apex of the abdomen. p. 934....................Cynipidæ
HH. The pronotum not reaching the tegulæ, separated therefrom by a chitinized sclerite, the prepectus. Antennæ elbowed, never more than thirteen-segmented. Fore wings with a short, narrow costal cell open along its anterior margin (Fig. 1166); its apex remote from the stigmal vein (Fig. 1166, d); a more or less elongate marginal vein (Fig. 1166, b); postmarginal and a stigmal spur (Fig. 1166, d). An occasional

trace of the transverse part of vein M is the only additional vein present, there being never any closed cells. p. 941......
.... CHALCIDIDÆ

GG. Wings longitudinally plaited in repose, ovipositor carried along the mid-dorsal line. Pronotum reaching the tegulæ, the prepectus not being distinctly set off. (*Leucospis.*) p. 941.....
.... CHALCIDIDÆ

CC. An erect scale or one or two nodes between the propodeum and the gaster. p. 954.. FORMICIDÆ

BB. Hind wings with an anal lobe.* If there are any closed cells in the hind wings the antennæ are thirteen-segmented in the male, twelve- in the female, except in a few instances where the number is reduced by fusion, but then the apical segments always form a club, or are abruptly recurved or otherwise strikingly modified (except that in some species of *Crabro* both sexes have twelve-segmented, otherwise normal antennæ).

C. Hind wings without closed cells. Number of antennal segments variable, but never thirteen in the male and twelve in the female, nor are the apical segments in the male either formed into a distinct club, or strongly reflexed or otherwise peculiarly modified. Vein dissolution extensive.

D. Abdomen attached to the dorsal surface of the propodeum. p. 949...
.... EVANIIDÆ

DD. Abdomen attached normally, at the apex of the propodeum between or slightly above the hind coxæ.

E. Antennæ composed of ten segments, or if of thirteen in the female (*Ampulicomorpha*) then the pronotum is elongate and has a median longitudinal sulcus.

F. Antennæ inserted on a frontal prominence distant from the clypeus; mouth ventral. Fore tarsi simple. p. 951...... EMBOLEMIDÆ

FF. Antennæ inserted close to the clypeus. Fore tarsi of the female usually chelate. p. 978............................. DRYINIDÆ

EE. Antennæ usually composed of thirteen segments, more rarely of twelve or eleven segments, or multiarticulate.

F. Abdomen with six exposed segments or less. Forewings always with cells M and Cu + Cu_1 closed. Ovipositor an extensile jointed tube.

G. Venter convex; abdomen with six exposed segments. p. 951..
.... CLEPTIDÆ

GG. Venter strongly concave; abdomen with at most five usually three or four exposed segments. Brilliantly metallic. p. 951....
.... CHRYSIDIDÆ

FF. Abdomen with eight exposed segments, the petiolar segment very short and scarcely perceptible. Ovipositor a true sting. p. 965.
.... BETHYLIDÆ

CC. Hind wings with at least a closed median cell (cell M). Males with thirteen, females with twelve antennal segments, except in rare instances, where they have been reduced slightly below that number in the males, in which case they usually either end in a jointed club or the last segments are recurved or hooked or otherwise modified. Venation usually well preserved.

---

*If there is a very deep or slit-like incision on the margin of the hind wing, the insect is certain to come under this heading. There are some genera of Sphecidæ (of the tribes Larrini, Astatini and the subfamily Sphecinæ) in which the axillary excision or both axillary and preaxillary excisions are reduced to small and inconspicuous notches, close to one another, and in some cases the axillary excision is altogether lacking. But in all such cases the second anal furrow is distinct, its apex close to that of the first anal furrow, and it delimits a large anal area behind it, which therefore lacks only the notch itself in order to become a distinct anal lobe. In all insects falling under grouping B this furrow is wanting, due to the reduction or entire absence of the area which in a more primitive condition exists behind it and forms the anal or posterior lobe of the wing.

D. The pronotum extending laterally directly to the tegulæ, where its lateral prolongations do not terminate in the form of a rounded "posterior lobe" covering the spiracles.

E. Cell M_4 of the forewings longer than cell $Cu + Cu_1$. Lateral prolongations of pronotum forming a posterior angle which lies above the tegulæ. Wings usually longitudinally plaited. p. 965.......VESPIDÆ

EE. Cell M_4 of the forewings shorter than cell $Cu + Cu_1$ or absent. Lateral prolongations of pronotum bluntly rounded, not lying dorsad of the tegulæ. Wings never longitudinally plaited.

F. Mesopleura divided by a transverse suture into an upper and lower plate. First and second abdominal sternites imbricate. Coxæ very large and long; the legs long and usually distinctly spinose. p. 950................................POMPILIDÆ

FF. Mesopleura not divided by a transverse suture. Coxæ and legs not unusually long.

G. First abdominal segment united by a ball and socket joint to the second, and itself forming an almost completely separated "scale" or "node." Hypopygium of male unciform. Females winged; a worker caste present. (Some more primitive genera of ants.) p. 954.......................................FORMICIDÆ

GG. First abdominal sternite attached to the second by a suturiform articulation or more or less imbricate, the first segment not forming a "scale" or "node" between the propodeum and the gaster.*

H. Mesosternum not forming with the metasternum a continuous plate overlying the bases of the hind and middle coxæ Axillary excision of the hind wings in normal position, apex of male abdomen without three retractile spurs between the last tergite and its sternite.

I. Vein $M_4 + Cu_1$ of the forewings opposite vein *m-cu* or nearly so. Second and third tarsal segments of the female not dilated.

J. Mesosternum with two laminæ which overlie the bases of the middle coxæ.

K. Little or no constriction between the first and second abdominal sternites, which are almost or somewhat imbricate. A trace of a vein (base of R_s) often divides the first submarginal cell ($R + 1st\ R_1$). Hypopygium of male not unciform. Apex of the marginal cell ($2d\ R_1 + R_2$) distant from the wing apex by not more than the length of the cell. Both sexes winged. p. 952.......
....ANTHOBOSCIDÆ

KK. First and second abdominal sternites separated by a strong and distinct suturiform articulation and either the hypopygium of the male is unciform, or the females are wingless and carried about by the males in mating.

L. Hypopygium of the male not unciform; but sometimes it is tridentate at apex with the middle tooth long and spiniform. Females apterous and carried about by the males in mating. First submarginal cell usually divided by a weak vein (base of the radial sector). p. 952....................THYNNIDÆ

LL. Hypopygium of male unciform, known American females winged. First submarginal cell not divided (base of R_s wanting).

M. Diurnal insects with normal eyes and ocelli. Females winged. (Tiphiinæ.) p. 953. TIPHIIDÆ

---

*Some genera of Mutillidæ in which the first or first and second abdominal segments are more or less nodose may be recognized as falling in this category by the unciform hypopygium of the males, the apterous females, and the absence of a neuter caste.

MM. Nocturnal insects with enlarged eyes and ocelli. The marginal cell (2d $R_1 + R_2$) removed from the apex of the wing by several times its length. Females unknown but presumably apterous. (Brachycistinæ.) p. 953....................MUTILLIDÆ
JJ. Mesosternum simple or with two minute erect teeth between the bases of the coxæ.
K. Marginal cell (2d $R_1 + R_2$) removed by less than its length from the wing apex; the fourth submarginal cell (R_3) not traversed by an adventitious vein. Abdomen never petiolate.
L. No constrictions between the abdominal segments (except between the first and the second), all tergites and all but the first and second sternites loosely imbricated plates. The last tergite of the male a small simple lamina. Edges of the hypopygium of the female turned upwards and meeting on the mid-dorsal line, often fused, enclosing the sting in a cone. Both sexes winged. Vein m and M_2 of the hind wings wanting. Mesosternum unarmed, upper surface of hind coxæ simple. p. 952.................SAPYGIDÆ
LL. Strong constrictions between each of the abdominal segments, the tergites and sternites all heavily chitinized and not loosely imbricate; the last tergite modified and hood-like. Female apterous. Upper surface of the hind coxæ, at least in the male, with a lamella. (Methocinæ and Myrmosinæ.) p. 953.
....TIPHIIDÆ
KK. Marginal cell (2d $R_1 + R_2$) removed by two or more times its length from the apex of the wing; cell R_3 when present usually traversed by a longitudinal adventitious vein. Often nocturnal insects with large eyes and ocelli. Females apterous. (Several subfamilies.) p. 953.
....MUTILLIDÆ
II. Vein $M_4 + Cu_1$ of the forewings more than two-thirds its length apicad of vein m-cu. Second and third tarsal segments of the female dilated, deeply excised, and filled with membrane between the lobes. Nocturnal insects with very large eyes and ocelli. Petiole long and slender. Hypopygium unarmed. p. 965....................*RHOPALOSOMIDÆ
HH. Mesosternum and metasternum together forming a continuous plate overlying the bases of the hind and middle coxæ, separated from each other by a transverse suture. Axillary excision of the hind wings almost opposite the apex of cell $M_3 + Cu_1 + Cu$. Abdomen of male with three spines, retractile between the last sternite and tergite. Tongue elongate. Wings with membrane striolate. p. 954.....SCOLIIDÆ
DD. Pronotum terminating behind laterally in the form of two clearly differentiated rounded "posterior lobes," covering the spiracles. These lobes reach the tegulæ in North American forms only in the extremely rare genus *Dolichurus*.
E. Posterior metatarsus not dilated. No plumose hairs. Females without pollen-collecting apparatus, but often with a comb on the anterior tarsi. Maxillæ rarely elongate; if so, either the ocelli are distorted, or the abdomen has a petiole composed only of the first sternite.
F. Abdomen of the male with only three or four exposed tergites. Last sternite of the female enclosing the sting, its edges fused in the mid-dorsal line. (Dolichurinæ.) p. 978.............AMPULICIDÆ
FF. Abdomen of the male with seven exposed tergites. Sting not enclosed by the hypopygium. p. 979...................SPHECIDÆ

EE. Posterior metatarsi elongate and dilated. Some of the hairs, especially of the thorax, plumose. A pollen-collecting brush or a corbicula present in the majority of females. Maxillæ usually with either the stipes or the lacinia elongate; the latter often very long and covering the tongue; the ocelli never distorted; the abdomen rarely petiolate, and never with a petiole composed only of the first sternite.
  F. Hind tibiæ with apical spurs. If the marginal cell ($2d\ R_1 + R_2$) is long and slender, reaching nearly to the wing apex, the anal lobe is short and fully separated. Cell M_4 usually as large as cell 1st M_2.
    G. Females without a corbicula. First submarginal cell ($R + $ 1st R_1) rarely divided (by the base of R_s) in which case there is a large anal lobe present. In case the marginal cell ($2d\ R_1 + R_2$) is longer than the three submarginals, taken together, there is usually a well-marked anal lobe in the hind wings.
      H. Wings with two submarginal cells (very rarely less).
        I. Tongue short and the basal segments of the labial palpi not sheath-like; or the labrum is large and free and uncovered. Females without a ventral pollen-collecting brush; often with a pygidial area.
          J. Tongue short, its apex bifid; labial palpi normal. Female only rarely with a pygidial area. Mesepisternal suture present. Labrum hidden (Prosopinæ). p. 993
....PROSOPIDÆ
          JJ. Tongue long or short, but its apex acute; the labial palpi normal or their basal segments sheath-like. Mesepisternal suture wanting. (Many.) p. 995..ANDRENIDÆ
        II. Tongue elongate, the basal segments of the labial palpi sheath-like. Labrum not large and free, usually entirely covered by the clypeus, or if somewhat visible, then strongly inflexed. Females without a pygidial area; those of the non-parasitic species with a ventral pollen-collecting brush. p. 999................................MEGACHILIDÆ
      HH. Wings with three submarginal cells. Females without a ventral pollen-collecting brush; often with a pygidial area.
        I. Tongue short, its apex bifid. Labial palpi normal. Females rarely with a pygidial area. Mesepisternal furrow present. Labrum hidden. (Colletinæ.) p. 993..PROSOPIDÆ
        II. Tongue long or short, but its apex acute. The labial palpi normal, or the basal segments sheath-like. Mesepisternal suture rarely present. (Most.) p. 995.ANDRENIDÆ
    GG. Females and workers with corbiculæ (except *Psithyrus*). First submarginal cell divided by a transverse, hair-like chitinized streak (base of R_s), rarely indistinct. Marginal cell ($2d\ R_1 + R_2$) rather long and pointed or appendiculate, usually as long as the three submarginal cells taken together, and extending far beyond the apex of the third (R_4). Malar space large and distinct. Hind wings stalked, the anal lobe absent. Tongue very long; the two basal segments of the labial palpi and the laciniæ elongate, and forming a sheath. Social insects with normally a worker caste (except in *Psithyrus*). p. 1001...........BOMBIDÆ
  FF. Hind tibiæ without apical spurs. Social insects with a worker caste. Workers with corbiculæ; females without functionally developed ones. Marginal cell ($2d\ R_1 + R_2$) long and slender reaching nearly to the wing apex. Anal lobe of the hind wing long and scarcely separated. Cell 1st M_2 much larger than cell M_4. Eyes hairy. p. 1005..................................APIDÆ
AA. Apterous or subapterous (the wings so reduced as to interfere with the normal venation).
  B. The ventral segments membranous, more or less concave with a longitudinal fold in dried specimens.

C. Second and third abdominal tergites connate.*
 D. Abdomen sessile, or if petiolate, the petiole not curved, or expanded at apex. (A few genera.) p. 919..........................BRACONIDÆ
 DD. Abdomen with an elongate petiole, which is strongly decurved and expanded at apex. (*Thaumatotypidea* in Stilpnini.) p. 922 ICHNEUMONIDÆ
CC. Second and third abdominal tergites not connate.
 D. Abdomen petiolate, the first segment elbowed and enlarged posteriorly (Cryptinæ in part). p. 922........................ICHNEUMONIDÆ
 DD. Abdomen sessile (Aphidiinæ in part). p. 919..........BRACONIDÆ
BB. The ventral segments chitinized, convex and without a longitudinal fold.
 C. Pronotum separated from the tegulæ laterally by the interposition of a chitinized sclerite, the prepectus. Antennæ elbowed. Hypopygium of the female divided, the ovipositor issuing from anterior to the tip of the abdomen. (A few genera.) p. 941.......................CHALCIDIDÆ
 CC. Pronotum reaching the tegulæ laterally or the latter altogether absent; no prepectus present.
  D. First abdominal segment not forming a "scale" or "node" between the propodeum and the gaster. Thorax sometimes with sutures largely obliterated.
   E. Fore tarsi chelate. (Females of Anteoninæ.) p. 978....DRYINIDÆ
   EE. Fore tarsi normal.
    F. Hypopygium of the female divided, the ovipositor issuing from before the tip of the abdomen, which is more or less strongly compressed and with a mid-dorsal keel. Second abdominal tergite the longest. Petiole very short, cylindrical, scarcely visible. Wingless or subapterous forms are usually agamic females. (Mostly Cynipinæ.) p. 934................................CYNIPIDÆ
    FF. Hypopygium of the female not divided, but closely applied to the pygidium, the ovipositor or sting issuing from between the two, at the tip of the abdomen. Abdomen rarely or never strongly compressed, without a mid-dorsal ridge.
     G. Mesopleura not divided by a transverse suture. Coxæ and legs not unusually long.
      H. Abdomen margined laterally, the sides sometimes only acute, but usually with a blade-like carina.
       I. Antennæ composed of ten segments, rarely less, but in that case without an unsegmented club. p. 933. PLATYGASTERIDÆ
       II. Antennæ composed of eleven or twelve segments, or if of seven or eight they have an unsegmented club. p. 933 ....SCELIONIDÆ
      HH. Abdomen neither acute nor margined laterally.
       I. Males.
        J. Antennæ of ten segments (*Myrmecomorphus*). p. 951... ....EMBOLEMIDÆ
        JJ. Antennæ of thirteen segments. p. 953.....MUTILLIDÆ
        JJJ. Antennæ of fifteen segments. (Belytinæ.) p. 932.... ....BELYTIDÆ
       II. Females.
        J. Antennæ of ten or eleven segments.
         K. Antennæ of eleven segments. (Six genera.) p. 932 ....CERAPHRONIDÆ
         KK. Antennæ of ten segments.
          L. Antennæ inserted close to the mouth. (Several genera.) p. 932..................CERAPHRONIDÆ
          LL. Antennæ inserted on a frontal prominence in the middle of the face (*Myrmecomorphus*). p. 951...... ....EMBOLEMIDÆ
        JJ. Antennæ of twelve segments. (See also JJJ.)

---

*If the suture is entirely obliterated, the presence of two pairs of spiracles will indicate that the apparent second segment consists really of the second and third.

K. Prothorax, mesothorax, and propodeum more or less intimately fused into a chitinous box; usually with little or no trace of sutures, more rarely with the sutures visible but connate; still more rarely the prothorax is fully separated.
L. First tergite not reaching the base of the petiole, but forming an apical gibbous cap thereto. Prothorax separated. (Apterogyninæ and Chyphotinæ.) p. 953..............................MUTILLIDÆ
LL. First tergite reaching the base of the petiolar segment.
M. Upper surface of posterior coxæ with an erect inner lamella at base. Pronotum separated from mesothorax by a movable suture. (Myrmosinæ.) p. 953...............................TIPHIIDÆ
MM. Posterior coxæ simple. Pronotum and propodeum fused to the thorax, the sutures lacking. Ocelli wanting. p. 953..............MUTILLIDÆ
KK. Prothorax, mesothorax, and propodeum fully separated.
L. Hind edge of mesosternum with two horizontal lamellæ between the middle coxæ, which they partly overlap. p. 952................THYNNIDÆ
LL. Mesosternum without such lamellæ overlapping the bases of the coxæ.
M. Mesosternum with two minute, erect teeth between the middle coxæ. Mesothorax elongate. Moderately large ant-like insects. (Methocinæ.) p. 953...............................TIPHIIDÆ
MM. Mesosternum unarmed. Mesothorax rarely elongate.
N. Head oblong, porrect; antennæ inserted close to the mouth, never geniculate. Femora, especially of the fore-legs, usually strongly thickened in the middle. Scutellum without pits. p. 965............................BETHYLIDÆ
NN. Head usually globular or transverse. Antennæ inserted in the middle of the face. Femora not medially strongly thickened. Scutellum usually foveolate.
O. Antennæ geniculate. Scutellum with deep anterior foveæ; or if these are absent it is not distinctly separated from the mesonotum. (Sixteen genera.) p. 932..........BELYTIDÆ
OO. Antennæ not geniculate. Scutellum distinct, without foveæ. (New genus in Myrmosinæ.) p. 953................TIPHIIDÆ
JJJ. Antennæ of from thirteen to fifteen segments.
K. Antennæ inserted close to the mouth. Head oblong, porrect. Anterior femora spindle-shaped. p. 965......
....BETHYLIDÆ
KK. Antennæ inserted on the middle of the face. Head globular or transverse.
L. Abdomen produced behind into a pointed tube. Spur of the anterior tibiæ without a ventral blade-like lamella. (Three genera.) p. 931...PROCTOTRUPIDÆ
LL. Abdomen not produced behind into a tube. Spur of the anterior tibiæ with two prongs and no ventral lamina. Anterior femora clavate. p. 932. BELYTIDÆ

GG. Mesopleura divided by a transverse suture into an upper and lower plate. Coxæ very large and long; the legs long and usually distinctly spinose. p. 950........................POMPILIDÆ
DD. First abdominal segment forming a "scale" or "node" between the propodeum and the gaster. Thorax with its sutures distinct, but the mesothorax usually much reduced in size. p. 954.........FORMICIDÆ

SUPERFAMILY ICHNEUMONOIDEA

The Larger Parasitic Hymenoptera

Often after a great outbreak of insects people are astonished that the hordes of pests have disappeared so quickly and completely. Few of them realize that the subjugation of the pests was probably due either to the weather or to the activities of parasitic insects. In this superfamily are included some of the parasites most often responsible for the control of insect outbreaks. They perform also the less noticeable but more important role of keeping most species of herbivorous insects below the level of economic importance.

Entomologists have made many attempts to supplement the activities of native parasites with introduced species. Usually this has been a matter of going to the native home of an introduced pest for the purpose of bringing in the insects found to attack it there. Some introduced parasites have found conditions in their new home favorable for rapid and continued reproduction and have effected spectacular control. Pests of major importance have been reduced to insignificant numbers in a few years. Most attempts to introduce beneficial insects have met with failure or with very moderate success due to the inability of the parasite to establish itself or to build up sufficient numbers. A single success in this field of work, however, more than compensates for many failures.

In the parasitic superfamily Ichneumonoidea are included the Stephanidæ, Braconidæ, Ichneumonidæ, Trigonalidæ, Aulacidæ, and Gasteruptiidæ. In these families are most of the larger species of parasitic Hymenoptera. They have rather complete venation in the fore wing, no anal lobe in the hind wing, and an indefinitely large number of antennal segments, usually seventeen to sixty or more. The multi-segmented antenna is the only distinctive character of the superfamily and this is subject to the exception that the Aulacidæ and Gasteruptiidæ have only fourteen antennal segments. This obvious lack of cohesion makes one doubt the naturalness of the group.

Correlated with its parasitic habits the larva of ichneumonoids is degenerate, legless, and maggot-like. Except for the mandibles the head appendages are more or less vestigial. The body has thirteen segments and in the mature larva nine pairs of spiracles. Ichneumonoid larvæ differ from those of parasitic Diptera in lacking the mouth hooks and pharyngeal skeleton so characteristic of the latter. They differ from chalcidoid larvæ in having the labium and maxillæ present or represented by sensory areas. These organs are not detectable in the chalcidoids.

The ichneumonoids are all parasites upon other arthropods.

Larvæ and pupæ of holometabolous insects are the usual hosts. Some braconids parasitize plant lice and a very few attack adult Coleoptera. A few genera of ichneumonids are parasites in the eggs, cocoons or on the bodies of spiders.

In a typical life history of an ichneumonoid, the adult female locates the host larva by searching in a likely spot, identifies it with a few gentle taps on her antenna, and then quickly deposits an egg inside the larva by inserting her sting-like ovipositor. The larva writhes and thrashes about but is usually not able to drive away the parasite before the deed is done. The parasite's egg soon hatches and the tiny maggot-like larva that issues from it lies in the body cavity of its host. It grows gradually by feeding on the blood and fat body, avoiding the vital organs. The host larva develops in a normal manner but perhaps is unusually lethargic. When the host is nearly ready for pupation the parasite begins more rapid feeding and growth. Soon it begins to gorge itself with the vital organs of its host which rapidly sickens and dies, leaving a limp and shrivelled carcass of which the parasite makes short work. The cocoon is spun by the mature parasite larva either in the remains of its host or a short distance away. Throughout its feeding life the parasite has not excreted any visible wastes. Just before pupation it discharges the accumulated wastes in its intestine in a mass called the meconium. This dries to a hard pellet in the posterior end of the cocoon. Pupation takes place and a few days or weeks later the adult parasite gnaws a hole in one end of its cocoon and escapes fully hardened and colored. This is usually at a time when a new generation of the host is ready for parasitism.

Some species are obligatory secondary parasites and will feed only on some other parasite. Others are either primary or secondary parasites according to whether or not other parasites are in the hosts in which they develop.

Instead of living in the inside of their hosts, the larvæ of some species are external feeders. In these cases the parasite egg is placed on the host or near it and the larva feeds through cuts in the host's skin. External parasites are liable to be dislodged and therefore are seldom found on exposed and active hosts. The usual victims are larvæ or pupæ in tunnels, nests, or cocoons. The externally parasitic species usually permanently paralyze the host by a sting from the ovipositor before depositing the egg. This immobilization of the host gives further protection to an externally feeding larva.

An ichneumonoid larva feeding externally upon a paralyzed host in its burrow has habits which differ but little from those of tiphiid wasps whose larvæ feed upon scarabæid beetle grubs permanently paralyzed in their burrows by a sting from the parent wasp. It is a short step from this type of life history to that so common among the wasps usually called predators. In these the female provisions a nest for her offspring with one or several paralyzed insects. The difference is in the amount of care given to her offspring's food by the adult wasp rather than in the habits of the larva itself. It is thus difficult to distinguish between the parasitic and the predacious habit among the Hymenoptera. Recently some authors have been using the term *parasitoid* for the economy of the parasitic Hymenoptera (see Wheeler (23), page 26), for although the host is kept alive and usually developing while it is being consumed the parasite eventually devours its prey completely as does a true predator. With the infinite variation in types of parasitism and the frequent intergradation with other modes of life it seems useless to try to restrict the term "parasite" too closely.

Adult ichneumonoids feed upon the honey-dew secreted by Homoptera and at flowers such as Umbelliferæ whose nectar can be reached by the insect's usually short mouth parts. For water, of which they need a good supply, they lap up dew or rain drops from leaves. The females frequently feed upon the host juices that exude from punctures made by the ovipositor. Some species will attack a larva or a pupa just for this purpose, puncturing it again and again to lap up the juices, but never depositing an egg. Since only the larva is parasitic, adult ichneumonoids show none of the degenerative specializations for a parasitic habit that would otherwise be expected.

Polyembryony, or the development of several larvæ from a single egg, is known to occur among the ichneumonoids only in the braconid genus *Macrocĕntrus*. Parthenogenesis is possible in most species. Unmated females usually produce only males. In a few cases females are produced. In these species the male is usually either extremely rare or entirely unknown.

Family STEPHANIDÆ

The Stephanids

The stephanids are among the strangest of the Hymenoptera. The spherical head is on a long neck and has a crown of teeth. The very slender antennæ arise just above the mouth. The hind femur is enlarged and has a series of teeth beneath. These insects are rare outside of the tropics and even there are not common. Their life history is unknown although they are supposed to be parasitic on the larvæ of wood-boring beetles or in the nests of solitary bees or wasps.

Family BRACONIDÆ*

The Braconids

The Braconidæ have a close superficial resemblance to the Ichneumonidæ. They differ from this family in always lacking vein M_2 so that cells M_1 and 1st M_2 are confluent (Fig. 1148) and, except in the subfamilies Aphidiinæ and Paxylommatinæ, in having the second

---

*Ashmead ('00) has given a key to the genera of Braconidæ, but unfortunately this is difficult to use and contains many errors. Some recent revisions of North American species and genera are as follows: Aphidiinæ, *Bull. Md. Agr. Exp. Sta.* 1911, 152 : 147–200; Opiinæ, *Proc. U. S. Nat. Mus.* 1915, 49 : 63–95; Bracon, *Proc. U. S. Nat. Mus.* 1917, 52 : 305–343; Apanteles, *Proc. U. S. Nat. Mus.* 1920, 58 : 483–576; Neoneurinæ and Microgasterinæ, *Proc. U. S. Nat. Mus.* 1922, 61 : 1–76; Meteorus, *Proc. U. S. Nat. Mus.* 1923, 63 : 1–44; Microbracon, *Proc. U. S. Nat. Mus.* 1925, vol. 67, art. 8, 85 pages; Braconinæ, *Proc. U. S. Nat. Mus.* 1927, vol. 69, art. 16, 73 pages; Macrocentrus, *Proc. U. S. Nat. Mus.* 1932, vol. 80, art. 23, 55 pages; and Euphorinæ, *Misc. Pub. U. S. Dept. Agr.* 1936, no. 241, 36 pages.

and third abdominal segments inflexibly joined together. There is usually also a vein crossing the large cell situated below the stigma. This is wanting in the Ichneumonidæ. The braconids as a rule are smaller than the ichneumonids. Many of the North American braconids large enough to be confused with ichneumonids have black wings and a red and black body, a color combination rare in the latter group.

In this family are included probably over two thousand North American species. The great majority are parasitic upon caterpillars. The rest attack mainly the larvæ of weevils, leaf-beetles, borer beetles, and cyclorrhaphous Diptera. The subfamily Aphidiinæ attacks plant lice, the subfamily Ichneutinæ sawfly larvæ, and the rare subfamilies Paxylommatinæ and Neoneurinæ are ant parasites. A few Alysiinæ are secondary parasites.

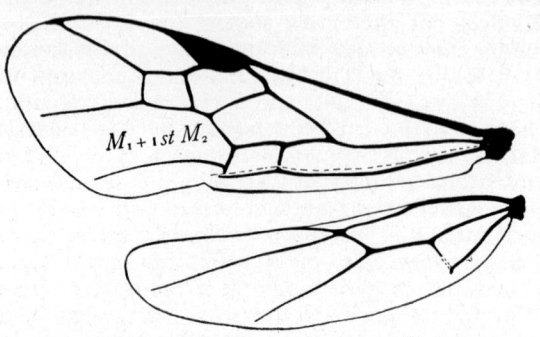

Fig. 1148.—Wings of a braconid.

Because of the excellent characters offered by their venation, the braconids are easier to classify than the ichneumonids, yet because of their small size they have escaped the attention of the average entomologist so that many facts about them are still to be learned. About eighteen subfamilies are now recognized, of which the Alysiinæ, Aphidiinæ, and Paxylommatinæ are sometimes treated as separate families. Some of the larger and more interesting groups are discussed below.

The subfamilies Vipiinæ (= Braconinæ), Spathiinæ, Hecabolinæ, Doryctinæ, Hormiinæ, and Aleiodinæ are all closely related and perhaps should be made a single subfamily. In these the clypeus is arched upwards and the labrum is very concave so that there is a circular opening above the mandibles, hence the name Cyclostomi which is applied to this group. Members of the Cyclostomi usually parasitize caterpillars or beetle larvæ living in tunnels or nests. The female permanently paralyzes the host larva by stinging it with her ovipositor, then deposits one or several eggs on its body. She may stop to lap up the juices exuding from the ovipositor wound. The eggs hatch into larvæ which feed through holes cut in the skin of the host. When mature, they spin cocoons near the host remains.

The subfamily Cheloninæ has the abdomen in the shape of an oval shield which looks like an inverted bath tub or a turtle shell. In some genera there are evident sutures separating the three tergites which make up the dorsal shield. In this subfamily the parasite inserts its egg into the egg of the host. The parasitized host larva, except for being somewhat stunted, develops normally to maturity, when the nearly grown parasite larva emerges from its body and finishes it off in one big meal. *Ascogăster quadridentātus* was introduced into this country from Europe to control the codlin moth. Its life history is typical of the subfamily. In midsummer about thirty-five days are required for a generation. Although its activities are helpful, this species does not effectively control the codlin moth.

The Microgasterinæ includes species with hairy eyes and with the abdomen so short that it is surpassed by the hind femur. Except in the genus *Apănteles* there is a tiny cell near the center of the fore wing. Most of our species have eighteen antennal segments. *Microgăster*, *Microplītis*, and *Apănteles* are the genera of importance. All are parasitic on caterpillars. The genus *Apănteles* contains nearly two hundred North American species and is probably the most beneficial among the braconids. Each species attacks a particular type of caterpillar—leaf miners, or cutworms, or larvæ of Sphingidæ, etc. Many are gregarious parasites, often a hundred or more developing from a single host. The female may insert ten to thirty eggs at a single thrust of her ovipositor and may make several thrusts into a single caterpillar. The parasite larvæ all mature at exactly the same time and bore to the outside simultaneously. Often they spin their yellow or white cocoons at the point on the caterpillar from which they emerged, when they look superficially more like eggs than cocoons (Fig. 1149). Other species may spin a mass of cocoons bound together by more or less loose silk near the body of the caterpillar (Fig. 1150). The parasitized caterpillar lingers on for a few days after the parasites have left it and finally dies. The species commonly attacking sphingid larvæ is *Apănteles congregātus*. *Apănteles glomerātus* attacks larvæ of the cabbage butterfly. Through the efforts of Dr. C. V. Riley this European species was introduced and it is now widely distributed in North America. Other species of *Apănteles* have been introduced to control the gypsy moth and the European corn borer.

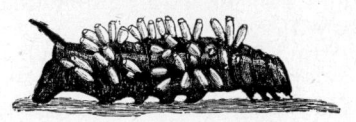

Fig. 1149.—Caterpillar with cocoons of a braconid.

The Braconinæ (= *Agathidinæ*) includes species with the cell next to the stigma extending less than half-way beyond the end of the stigma to the wing tip. There is a very small cell in the center of the wing below the stigma. These are the braconids most frequently collected on flowers. Many of the species have black wings and the abdomen or entire body red. Some species have elongate mouth parts for reaching nectar and the head prolonged below into a sort of beak. *Eărinus limitāris* is a black species about nine millimeters

Fig. 1150.

long, common in the early spring. Most Braconinæ parasitize caterpillars.

The Macrocentrinæ includes the genus *Macrocĕntrus* with about forty species in North America. They are yellowish brown or more or less black in color. They are slender with long legs and with an ovipositor as long as the body. Caterpillars living in nests or in tunnels are the usual hosts. *Macrocĕntrus ancylivōrus* is a native species first discovered parasitizing the strawberry leaf-roller (*Ancylis comptāna*) in New Jersey. It has been found the most effective control of the oriental fruit moth, a notorious pest of peaches, and has now been colonized in peach orchards all over the eastern United States and in some foreign countries. In some localities it parasitizes more than eighty per cent of the fruit moth larvæ. *Macrocĕntrus gifuĕnsis* is a polyembryonic species from the Old World introduced a few years ago to help control the European corn borer. It is one of the most effective parasites of this pest.

The Alysiinæ includes species with the teeth of the mandibles pointing outward instead of inward toward the mouth. The only other Hymenoptera with this type of mandible are the Vanhorniidæ and the ichneumonid genus *Lysiŏgnatha*, very rare insects. The Alysiinæ are parasites of fly maggots, and the paddle-like mandibles are used by some for digging in filth in search of hosts.

The Aphidiinæ includes delicate little species with a movable joint between the second and third abdominal segments and with reduced wing venation. All are parasites of plant lice. The female aphidiine approaches an aphid and identifies it with a few taps of her antennæ. She then stands high and ducks her abdomen down between her legs and forward in front of her head to plunge her ovipositor into the plant louse. In a few days the developing parasite larva has eaten out the inside of the aphid. It cuts a hole in the bottom of the aphid's empty skin and through this glues the carcass to the substratum. Then it makes its cocoon in the aphid's abdomen. The adult braconid emerges by cutting a circular lid (Fig. 1151). In the genus *Prāon*, the cocoon is spun beneath instead of inside the host and the dead aphid is used as a roof for the cocoon. Plant lice with braconid cocoons inside have an inflated appearance and brownish color that makes them conspicuous among a colony of living aphids. These parasites are of prime importance in reducing the numbers of aphids and were it not for them, the lady beetles, the lace-wing flies, and the syrphids, plant lice would be serious pests.

Fig. 1151.

Family ICHNEUMONIDÆ*

The Ichneumon Flies

The Ichneumonidæ includes most of the larger parasitic Hymenoptera. They are similar to the Braconidæ and differ from most

---

*The following papers should be consulted for identification of species:

other Hymenoptera in having the narrow cell just behind the costal vein obliterated by the fusion or close approximation of the costal vein and the vein behind it and usually by having some of the abdominal sterna membranous. They differ from braconids in having cells M_1 and 1st M_2 of the fore wing separated by a vein (Fig. 1152) and the second and third abdominal tergites movable. There are a very few exceptions to one or the other of these characters. Except for the frequent absence of the areolet, a small cell near the middle of the wing (Fig. 1152), ichneumonids have a very constant and distinctive venation in the fore wing. Many species are wasp-like in appearance but may at once be recognized as ichneumonids by the long many-segmented antenna.

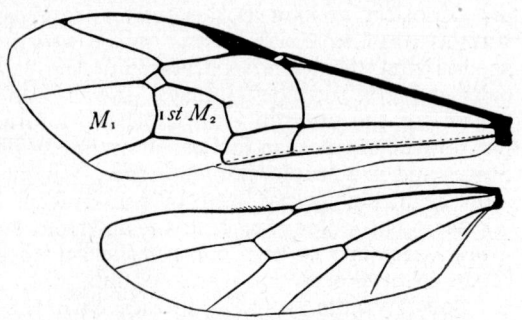

Fig. 1152.—Wings of an Ichneumon fly.

Except perhaps for the Chalcididæ this is the largest family of the Hymenoptera. Probably about six thousand species occur in the United States. All are parasitic and many are important enemies of destructive insects. The taxonomy of the family is difficult because the subfamilies and tribes are not easily characterized and must be learned by experience. Five subfamilies are usually recognized, but these are admittedly unnatural in some degree and any definition of them is subject to exceptions. The student, however, can place most of the common species by means of the key below. Some of the more important tribes of ichneumon flies are discussed under the different subfamilies, but space will not permit all of them to be mentioned.

Tribes of Pimplinæ, *Proc. U. S. Nat. Mus.* 1921, vol. 60, art. 4, pages 1–7; Labenini, Rhyssini, Xoridini, Odontomerini, and Phytodietus, *Proc. U. S. Nat. Mus.* 1920, 57 : 405–474; Accenitini, *Proc. U. S. Nat. Mus.* 1920, 57 : 502–523; Lycorini, Polysphinctini, and Theronia, *Proc. U. S. Nat. Mus.* 1920, 58 : 7–48; Pimplini (except Theronia), *Proc. U. S. Nat. Mus.* 1920, 58 : 327–362; Odontomerus, *Proc. U. S. Nat. Mus.* 1930, vol. 77, art. 3, 15 pages; Cremastini, *Proc. U. S. Nat. Mus.* 1917, 53 : 503–551 and *Proc. U. S. Nat. Mus.* 1920, 58 : 268–288; Cryptini with a very small areolet, *Proc. U. S. Nat. Mus.* 1929, vol. 74, art. 16, 58 pages; Exetastes, *Proc. U. S. Nat. Mus.* 1937, 84 : 243–312; Netelia, *Lloydia* 1938 (1939), 1 : 168–231; Trogini, *Trans. Amer. Ent. Soc.* 1939, 65 : 307–346; Ichneumoninæ, *Trans. Amer. Ent. Soc.* 1877, 6 : 129–212; Diplazonini, *Trans. Amer. Ent. Soc.* 1895, 22 : 17–30; Tryphoninæ (except Diplazonini), *Trans. Amer. Ent. Soc.* 1897, 24 : 193–348; Glypta and Lissonota (= "Lampronota"), *Trans. Amer. Ent. Soc.* 1870, 3 : 151–166. The last four papers cited are out of date and somewhat unreliable, but are the best available. Ashmead ('00) has given a key to all the genera, but the student will find this unreliable.

KEY TO THE SUBFAMILIES OF ICHNEUMONIDÆ

By Dr. H. K. Townes

A. Spiracles of first abdominal segment situated definitely behind the middle of the segment; sternite of first abdominal segment immovably fused with the tergite or if rarely free, then the abdomen strongly compressed.
  B. Abdomen more or less compressed; areolet (= a small cell near the center of the wing) more or less triangular or absent. p. 927..........OPHIONINÆ
  BB. Abdomen depressed or cylindrical; areolet more or less pentagonal, quadrangular, or occasionally absent.
    C. Sternaulus (a horizontal groove in lower part of mesopleurum) sharp and usually more than half as long as the mesopleurum; ovipositor usually surpassing tip of abdomen. p. 927..........................CRYPTINÆ
    CC. Sternaulus absent or poorly defined and short; ovipositor not or but slightly surpassing tip of abdomen. p. 928................ICHNEUMONINÆ
AA. Spiracles of first abdominal segment situated near or in front of the middle of the segment; sternite of first abdominal segment usually more or less free from the tergite.
  B. Face and clypeus forming an even undivided surface; male claspers drawn out into a pair of long spines; ovipositor sheaths slightly shorter than the first abdominal segment, rather broad, flat, and somewhat polished; areolet large and rhomboidal. p. 927..............(tribe *Mesochorini*) OPHIONINÆ
  BB. Face more or less separated from clypeus by a groove, or insect otherwise not entirely agreeing with the above.
    C. Ovipositor surpassing tip of abdomen; abdomen often with definite elevations and depressions above and usually elongate. p. 924..PIMPLINÆ
    CC. Ovipositor not or but slightly surpassing tip of abdomen; abdomen without definite elevations and depressions above and not elongate unless also petiolate. p. 926..................................TRYPHONINÆ

SUBFAMILY PIMPLINÆ

Most species of Pimplinæ are black with ferruginous or banded legs. They are parasites mainly of caterpillars and wood-boring beetle larvæ. Their long ovipositors are adapted to reaching hosts in tunnels in wood or in weed stems.

Megarhỹssa (= *Thalĕssa*) is our largest ichneumon fly (Fig. 1153). It is a parasite of the wood-boring larva of the pigeon horn-tail, *Trēmex colŭmba*. When a female finds a tree infested by this borer she selects a place which she judges to be opposite a *Trēmex* burrow and making a derrick out of her body proceeds with great skill and precision to drill a hole in the tree. At the beginning of the process the excess length of the ovipositor is coiled into a sack formed by the very elastic membrane between the sixth and seventh abdominal segments. When most of the ovipositor is deep in the wood the ovipositor sheaths which do not enter with it form a loop over her back (Fig. 1153). After the *Trēmex* burrow is reached, a long egg flows down the tube in the slender ovipositor and is left on or near the host. The larva which hatches from the egg feeds upon the *Trēmex* larva by sucking its blood and eventually destroys it entirely. When full grown it pupates in the *Trēmex* burrow. The adult gnaws a hole through the bark to emerge. If the adult is a female, there are males waiting just outside and she is mated just after emergence or even before she can extricate herself from the hole. Females of the four North American species are easy to distinguish. *Megarhÿssa atrāta* has a black body and yellow head. The northern species *M. nŏrtoni* has yellow circular spots on the sides of the abdomen. *M. lunātor* and *M. greenei* have brownish abdomens with angulate yellow stripes on the sides (Fig. 1153). *M. lunātor* has an ovipositor twice as long as its body while that of *M. greenei* is about one and a half times as long as its body.

The tribe Polysphinctini comprises small species usually without an areolet in the fore wing and with a rather short ovipositor. They are parasites of spiders. The female temporarily paralyzes a spider by stinging it and then glues an egg to its abdomen. The spider recovers and resumes its normal activities. The maggot-like parasite larva hatches and keeps its tail end attached to the egg shell to maintain its position on the spider's abdomen. It feeds through the skin of the spider and finally kills it and then spins a cocoon and soon emerges as an adult.

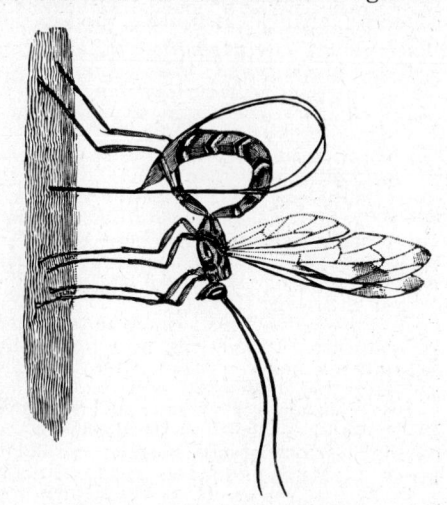

Fig. 1153.—*Megarhyssa lunator*.

The tribe Pimplini includes robust species with stout ovipositors about half as long as the abdomens. In the genus *Therōnia* are species that are almost entirely fulvous in color. The other three genera, *Pĭmpla*, *Apĕchthis*, and *Itoplĕctis* are black and can usually be distinguished from other Pimplinæ by the hind tibia which is white at the middle and has the apex and base (including the extreme base) black. Members of this tribe are parasitic upon pupæ of Lepidoptera. Each species parasitizes a great variety of hosts differing considerably both in size and relationships. Due to the varying sizes of the hosts the adult parasites have a remarkable variation in size. Usually males develop in the smaller hosts and females in the larger so that males average smaller than females. This is a frequent phenomenon among parasitic Hymenoptera but no satisfactory explanation for it has been advanced. *Itoplĕctis conquĭsitor* is a common parasite on tent caterpillars and *Therōnia atalăntæ* is a common hyperparasite of *Itoplĕctis conquĭsitor*. Sometimes *Itoplĕctis* is itself a secondary parasite. Species of *Pĭmpla* and *Apĕchthis* give off a strong pungent odor when captured.

The tribe Lissonotini includes species with a tiny notch near the tip of the ovipositor and a single bulla or weak place in vein M_2. The propodeum usually has a single semicircular carina. The larvæ are internal rather than external parasites, a biological characteristic found in the present subfamily only in this tribe and in the Pimplini. Caterpillars boring in twigs or weeds are the usual hosts of those with long ovipositors. The rest attack cutworms and similar caterpillars. *Glўpta rufiscutellāris* is a common parasite of the oriental fruit moth. In some localities it gives a moderately effective control of this pest. The genus *Ceratogăstra* contains wasp-like species with conical pointed abdomens. They are commonly found on goldenrod flowers.

Subfamily TRYPHONINÆ

The Tryphoninæ includes most of the ichneumon flies that attack sawflies, and the group is most abundant in our northern forests. In the tropics, where sawflies are scarce, few species occur. Some Tryphoninæ are slender, pale brown in color, and have large eyes. They fly at night and are often confused with the superficially similar Ophioninæ but may be at once distinguished from these by their possession of an areolet and by the characters given in the key to subfamilies.

The tribe Tryphonini contains short stocky species with the propodeum divided into many areas by cross and longitudinal carinæ. Most of the species are black with the abdomen and legs largely reddish. The life history is quite characteristic of this and related groups. The egg is large and oval with a slender stalk. The female inserts the stalk of the egg into the skin of a sawfly larva near its head where the larva cannot turn to reach it with its mandibles. The stalk stays embedded in the larva's skin, anchoring the egg in place. After the sawfly spins its cocoon the egg hatches into a small larva which feeds on the host until it is consumed. The parasite pupates to emerge as an adult the following spring to parasitize the next generation of sawflies. A female may often be seen with an egg ready for deposition attached by its stalk to her ovipositor, and in the genus *Polyblăstus* a dozen or more eggs may be carried on the ovipositor at once. The older eggs, those near the tip of the ovipositor, contain fully developed larvæ. It is said that if a host is not found soon enough the oldest eggs are discarded.

The tribes Mesoleiini and Mesoleptideini contain species usually more slender than the Tryphonini and with the propodeum not divided into numerous areas by carinæ. These tribes also parasitize sawflies, but the egg does not have a stalk and is inserted into the host larva.

The tribe Metopiini contains species that are usually black, with a very protuberant face, and short stout legs. Most of the species are small, though the genus *Metōpius* contains fairly large individuals. The Metopiini are parasites in lepidopterous pupæ. The genus *Metacælus* contains two cosmopolitan parasites of clothes moths.

The tribe Diplazonini contains small stocky bright-colored species with the upper tooth of the mandible broad and chisel-shaped. All are internal parasites of aphid-eating syrphid larvæ. The adult emerges from the host puparium. *Diplăzon lætatōrius* is a species that is common all over the world. It may be recognized by its black-, white-, and red-banded hind tibia.

Subfamily OPHIONINÆ

This subfamily may usually be recognized by its compressed abdomen, though a few species have the abdomen indistinctly compressed. All are internal parasites, usually solitary. The majority parasitize caterpillars.

The tribe Mesochorini is characterized in the key to subfamilies. The species are mostly small. All are secondary parasites. The female of *Mesochōrus* finds a host that is already parasitized and probes with her ovipositor until the parasite larva is reached, whereupon she oviposits within the body of the parasite. The braconid genus *Apănteles* is frequently attacked by *Mesochōrus*.

The tribe Campoplegini includes a great number of species. The face is evenly convex, not separated from the clypeus by a groove, and is covered by short, thick silvery-white hair. The head and thorax are usually entirely black and the abdomen more or less reddish. Many species are important enemies of injurious caterpillars. A colony of *Anisōta* larvæ is sometimes found with many dead individuals stuck to the leaves and twigs. They are swollen at the middle and shrivelled at both ends. These have been parasitized by *Hyposōter fugĭtivus* whose cocoon makes a swelling in the middle of the dead caterpillar. Many Campoplegini spin exposed cocoons. These are closely woven, oval in shape, and often whitish encircled with irregular dark bands.

The tribe Ophionini includes species with an unusual venation in the fore wing. The large cell just above cell 1st M_2 (Fig. 1154) extends further distad than does cell 1st. M_2. Most are night-flying species about eighteen to twenty millimeters long, yellowish brown in color, and with large eyes. These belong to the genera *Ōphion* and *Enicospĭlus* (Fig. 1154). They parasitize cutworms and similar caterpillars. One very large species of *Enicospĭlus* infests the caterpillars of the polyphemus moth and its relatives. The caterpillar lives until it spins a cocoon, but does not change to a pupa. The ichneumonid larva, when full grown, spins a dense brownish cocoon within the cocoon of the caterpillar. *Thyrĕodon atricolor* is a large coal-black species with bright orange antennæ. It is often seen flying along the edges of woods in search of its hosts, sphingid larvæ. This species and other large ichneumonids will sting with the ovipositor if not handled carefully.

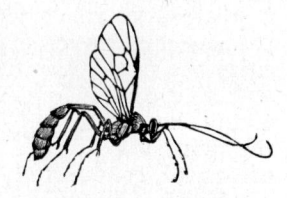

Fig. 1154.—*Ophion*.

The tribe Therionini includes relatively large species with very slender abdomens and the hind tarsi more or less swollen. They are parasites of various caterpillars.

Subfamily CRYPTINÆ

The Cryptinæ can usually be recognized by the exserted ovipositor somewhat shorter than the uncompressed abdomen and by the fact that the first abdominal segment is narrow at the base, decurved and broadened at the apex and with the spiracles much nearer the apex than the base. The species are external parasites of pupæ and prepupæ in cocoons. A few parasitize wood-boring beetle larvæ or are internal parasites of dipterous larvæ.

The tribe Hemitelini includes small species with the outer vein of the areolet lacking so that the areolet is incompletely formed. *Hemitēles* parasitizes a variety of pupæ enclosed in cocoons and often attacks cocoons of other ichneumonids and of braconids, thus becoming a secondary parasite. *Gēlis* is very similar to *Hemitēles* in appearance and habits except that the females are always wingless. The males may be either with or without wings, even in the same species. The wingless forms have a superficial resemblance to ants. Besides attacking various small insect cocoons, *Gēlis* often parasitizes the egg cocoons of spiders.

The tribe Phygadeuonini has the areolet complete and the propodeum with both longitudinal and transverse carinæ. It includes a large number of rather small species parasitic on Lepidoptera and Diptera.

The tribe Cryptini differs from the two preceding tribes by having two transverse and no longitudinal carinæ on the propodeum. The larger species of the subfamily belong here. *Agrothereutes extremātis* infests the cocoons of the cecropia moth. The odor of fresh silk draws the female parasite to the newly spun cocoon of the host. About thirty parasites develop in each cocoon. *Agrothereutes nŭncius* parasitizes the promethea moth, and a third species of the genus attacks the polyphemus.

Subfamily ICHNEUMONINÆ

Members of this subfamily resemble the Cryptinæ in general appearance but lack a definite sternaulus, have a short ovipositor, and are more stocky in build. They are all parasitic upon lepidopterous pupæ. The female oviposits into either the host larva or the pupa, but the adult emerges from the pupa in both cases.

In the tribe Phæogenini the propodeal spiracles are round. Other members of the subfamily have these spiracles long and oval. The Phæoginini are small species parasitic upon Microlepidoptera.

The tribe Ichneumonini comprises many species about thirteen millimeters long and somewhat wasp-like in appearance. It may be distinguished from the preceding tribe by the long oval propodeal spiracles and from the following tribe by the fact that the scutellum is usually flat or convex, and not subconically elevated. In this tribe the females are usually of a more stocky build than the males and often so differently colored that it is difficult to decide which males belong with the various females. The sexes also differ in habits, the males being entirely free-living and the females of many species spending much of their time searching for hosts in grass tufts and under dead leaves and debris. Females may be found overwintering in rotten logs, under bark, and in tufts of grass. Some species enter hibernating quarters in August, long before the first signs of cold weather. Each kind occurs in a particular type of shelter and is not to be found elsewhere. Species hibernating in logs often congregate in groups of three to twenty in old beetle burrows and other cavities. It is strange that *Ichneumon ŭltimus* and *I. mĕndax*, two of the commonest species collected in hibernation, have never in the author's knowledge been found during the growing season, and their males are apparently unknown. The genus *Hoplismenus*, with a conical scutellum, is parasitic on nymphalid butterflies. The other species, most of which have been included in the genera *Ichneumon* and *Amblўteles*, parasitize largely Noctuidæ and Geometridæ.

The tribe Trogini includes large species about twenty millimeters long with subconical scutella. *Trōgus vulpīnus* is a common red species with black wings. The female oviposits into caterpillars of swallowtail butterflies, and the adult parasite emerges through a hole cut in the side of the chrysalid. Most of the other Trogini are parasitic on hawk moths.

Family TRIGONALIDÆ

The Trigonalids

The family Trigonalidæ includes a small number of rare species. The adults look very much like sawflies. They differ from other ichneumonoids in having a distinct costal cell in the fore wing and more than fourteen antennal segments. Female trigonalids lay their minute flattened eggs, about ten thousand in number, on the underside of leaves just back of the margin. In some species they are inserted into the leaf tissue. The eggs hatch when eaten with the leaves by caterpillars or sawfly larvæ, provided that the shell is broken in the process. The young larva enters the body cavity of the caterpillar and develops either as a primary parasite or as a secondary parasite on some other parasite of its host. Some species are parasitic in vespid nests. It is not known how these reach their hosts. One species found parasitic in vespid nests has been observed to insert its eggs into the margins of leaves. For a classification of the family see Schulz, *Genera Insectorum* 1907, fasc. 61, 24 pp., 3 pls.

Family AULACIDÆ

The Aulacids

The aulacids have the abdomen attached high on the propodeum far above the bases of the hind coxæ as in the Gasteruptiidæ and the Evaniidæ. They may be distinguished from these families by the venation (Fig. 1155), the linear hind tibia, and the lack of a separate petiolar segment to the abdomen. The ovipositor is somewhat longer than the abdomen. There are three Nearctic genera: *Pristaulacus* with three or more teeth on the tarsal claws; *Odontaulacus* with one or two rather blunt teeth on the claws; and *Aulacus* (= *Pammegĭschia*) with the claws apparently simple but each with a tooth at the extreme base. *Aulacus* parasitizes the larvæ of *Xiphўdria*, a wood-boring sawfly. The other two genera attack larvæ of wood-boring

Coleoptera. All may be collected on the trunks of dead or dying trees but are seldom common.

Female aulacids have the inner side of each hind coxa notched to form a channel for the guidance of the ovipositor when the coxæ are brought together over it. It is interesting to note that the braconid genus *Capitōnius* and the ichneumonid genera *Labēna* and *Apechonēura* have the hind coxæ notched for the same purpose. These genera further resemble the Aulacidæ in having the abdomen attached high on the propodeum and in being parasitic on wood-boring Coleoptera. Apparently the habit of supporting the ovipositor between the hind coxæ when it is in use has some causal connection with the high attachment of the abdomen.

Fig. 1155.—Wings of an aulacid.

Revisions of this family were included in their papers on Evaniidæ by Bradley ('08) and Kieffer ('12).

Family GASTERUPTIIDÆ

The Gasteruptiids

The family Gasteruptiidæ, often spelled incorrectly Gasteruptionidæ, resembles the Aulacidæ and the Evaniidæ in having the abdomen attached high on the propodeum. It differs from these families in having the hind tibia bulbously swollen toward the tip and in having venation like Figure 1157. The front wing can be folded lengthwise.

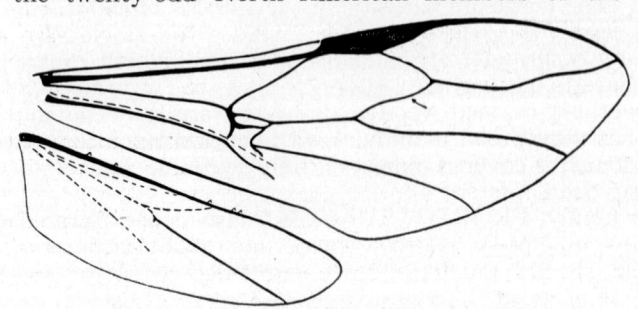

Fig. 1156.

Figure 1156 represents a species of *Gasterŭption*, in which genus are included nearly all of the twenty-odd North American members of the family.

Fig. 1157.—Wings of *Gasteruption incertus*.

Except that some have long ovipositors and some short, the various species look much alike. They are parasitic in the nests of solitary bees and wasps. The adults are common on umbelliferous flowers and around logs in which their hosts are nesting.

A monograph of this family was included by Kieffer ('12) in his paper on Evaniidæ.

Superfamily PROCTOTRUPOIDEA

The Proctotrupoids

The Proctotrupoidea includes mostly very small species superficially resembling chalcids or braconids. Most of them are black with the legs and antenna often brownish or reddish. The first three families listed have a fairly complete venation in the front wing. In the other families it is more or less reduced. The hind wing has never more than one closed cell and lacks an anal lobe. The ovipositor issues from between dorsal and ventral plates at the tip of the abdomen. In most other parasitic hymenoptera it issues from a subterminal cleft on the ventral side of the abdomen. These insects may be distinguished from chalcids by the fact that the pronotum extends back to the tegula. They differ from the braconids and ichneumonids in having never more than fifteen antennal segments.

A few proctotrupoids are inquilines. The rest are internal parasites of other insects or their eggs.

The North American Proctotrupoidea were monographed by Ashmead ('93). Note that the first three subfamilies of Ashmead's classification, those in which the hind wings possess an "anal lobe," are now included in the Vespoidea and Sphecoidea.

The family ROPRONIIDÆ includes the single genus *Roprōnia* with a few rare species in North America. The adults are about eight millimeters long and have a subcircular compressed abdomen on a petiole. They may be collected among rank herbage in moist woods. The immature stages are unknown.

The family HELORIDÆ includes the genus *Helōrus* with the single species *H. paradŏxus* in North America. It is a black species, about four millimeters long with fifteen-segmented antenna and a petiolate abdomen. The adults may be collected on umbelliferous flowers. The larvæ are parasitic in the cocoons of *Chrysōpa*.

The family VANHORNIIDÆ includes the single rare species *Vanhŏrnia eucnemidārum* occurring in eastern North America from Massachusetts to Virginia. It is a parasite of *Isorhĭpis rufĭcŏrnis*, a beetle working in dead wood that is fairly sound. The adult *Vanhŏrnia* has very broad mandibles with the teeth pointing outwards. The abdomen is covered almost entirely by single dorsal and ventral sclerites. See Crawford ('09).

The family PROCTOTRUPIDÆ,* also called Serphidæ, contains several closely related genera and about two score North American species ranging from about three to six millimeters in

---

*For a monograph of the family see Kieffer ('14) and for a recent revision of our species see Brues, *Jour. New York Ent. Soc.* 1919, 27 : 1–19.

length. They may be recognized by the large stigma in the fore wing just beyond which is a short but well-defined cell smaller than the stigma. The abdomen is circular in cross-section and tapers to a conical point. In the male there is a pair of pointed claspers and in the female a heavy tubular ovipositor sheath. The species of *Proctotrūpes* have a red abdomen. The proctotrupids whose habits are known are all solitary or gregarious internal parasites of beetle larvæ.

The family PELECINIDÆ is represented by a single species, *Pelecīnus polyturātor* (Fig. 1158). The females are often confused with ichneumon flies but may be easily recognized by the long cylindrical abdomen. The abdomen of the male is much shorter, with about the same length and shape as that of a *Sphĕx* wasp. This sex is very rare in the United States, but is common in some parts of South America. The species is a solitary internal parasite on the larvæ of June-beetles. The adult females are common in August and September. This family and the Proctotrupidæ have many structural resemblances. They are doubtless closely related.

Fig. 1158.

The family BELYTIDÆ* differs from the following proctotrupoid families in that the antennæ arise from an elevated portion of the face at some distance above the clypeus. It differs from the preceding families in that the fore wing lacks a stigma. There are two subfamilies: the Belytinæ with a closed cell in the hind wing, and the Diapriinæ without a closed cell in the hind wing.

The subfamily Belytinæ contains several common genera whose species resemble Figure 1159. Some species are known to parasitize larvæ of Mycetophilidæ and other Diptera breeding in fungi.

The subfamily Diapriinæ contains, among others, the remarkable genus *Gălesus*. This has an oblong head with a turret of teeth on top. The face slopes sharply backwards to the mandibles which are long and point backwards. The wings can be folded lengthwise. *Gălesus* and other Diapriinæ are parasites in dipterous puparia.

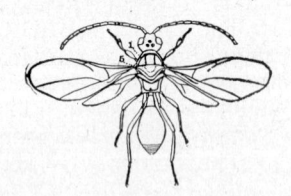

Fig. 1159. Proctotrupid

In the family CERAPHRONIDÆ, often called Calliceratidæ, the nine to eleven-segmented antennæ are inserted next to the clypeus, but unlike the Scelionidæ and Platygasteridæ the lateral margin of

---
*For a monograph of the family see Kieffer ('16).

the abdomen is rounded. In the subfamily Megaspilinæ the fore wing has a large stigma. This is absent in the other subfamily, Ceraphroninæ. The ceraphronids attack a variety of hosts, including Cecidomyiidæ and other Diptera. Certain species of *Lygŏcerus* are secondary parasites of braconids which attack aphids.

The family SCELIONIDÆ have the usually twelve-segmented antennæ attached near the clypeus. The abdomen is more or less depressed and has an acute, usually carinate, lateral margin. Except possibly for the Platygasteridæ, this is the largest family of the Proctotrupoidea. The species are all parasitic in insect or spider eggs. *Bǣus* parasitizes the eggs of spiders, *Trissolcus* and *Hadronōtus* parasitize eggs of Hemiptera, and *Telenōmus* attacks eggs of Lepidoptera and Hemiptera. *Scĕlio* attacks the eggs of locusts. A species of *Scĕlio* has been observed to penetrate the loose soil above a locust's egg pod and with her very extensible ovipositor reach every egg in the pod for oviposition. Sometimes she first chews a hole in the pod and backs into it for oviposition.

A few scelionids exhibit the interesting phenomenon called phoresy—they attach to another insect for transportation. *Phanūrus benefĭciens* is a common parasite of the eggs of pyralidid moths in the Oriental Region. The adult *Phanūrus* attaches to a female moth, and when the latter oviposits the parasite leaves the body of the moth to parasitize the freshly laid eggs. The European *Riĕlia manticīda* has comparable habits. The female takes up a position on the body of an adult *Măntis*, usually a female. The wings are then discarded and the scelionid lives as a true external parasite. The usual position is under the wings of the host, but other more or less protected areas are inhabited also. When the *Măntis* lays her egg mass the *Riĕlia* enters the frothy, not yet hardened, covering of the egg pod to parasitize the eggs. After oviposition the parasite returns to the *Măntis*.

Scelionid first stage larvæ are like those of certain other hymenopterous egg parasites in being of a bizarre yet undifferentiated form which is often termed cyclopoid from its general resemblance to *Cȳclops*, the water flea. The body is sack-like and unsegmented. At the anterior end is a pair of long mandibles, and posteriorly is a long ventrally curved tail that is often forked. Near the middle of the body are bands or tufts of long spines. The long mandibles and whip-like tail serve for locomotion and for disorganizing the contents of the egg. The second stage larva is of more normal form.

The family PLATYGASTERIDÆ includes many minute black species. The usually ten-segmented antennæ are attached next to the clypeus, the abdomen is depressed with a sharp lateral margin, and the wings are veinless (Platygasterinæ) or with a single short vein (Inostemminæ). Most of the species parasitize the larvæ of Cecidomyiidæ. The egg is inserted into that of the host, but development of the parasite is retarded until the host egg has hatched. Sometimes young cecidomyiid larvæ are attacked. There are a few records of platygasterids being reared from mealy bugs, white flies, and other insects. Some species of *Platygáster* are polyembryonic. For a revision of the Platygasterinæ see Fouts, *Proc. U. S. Nat. Mus.* 1924, vol. 64, art. 15, 145 pages.

Superfamily CYNIPOIDEA

The Cynipids

Most members of this superfamily are small insects and many of them are minute; for this reason they are not commonly observed; but the galls produced by some species, especially those that are found on oaks and roses, are very familiar objects. Not all cynipids, however, produce galls; some are parasites and others are inquilines, living in galls produced by other species.

The antennæ of the cynipids are not elbowed and only rarely composed of more than sixteen segments; the pronotum is produced on each side so as to reach the tegula or is separated from it only by a membranous area; the wings lack a stigma and have at most five closed cells; the wings are rarely wanting; the abdomen is strongly compressed. In most genera the tergites of the basal abdominal segments differ greatly in length; in some the second is half as long as the entire abdomen (Fig. 1160).

Fig. 1160.—*Amphibolips.*

For papers on the classification of the Cynipoidea see Ashmead ('03) and Dalla Torre and Kieffer ('02 and '10). Dalla Torre and Kieffer recognize only a single family of cynipids, the Cynipidæ; this they divide into ten subfamilies, seven of which are represented in our fauna.

These are the Ibaliinæ, Anacharitinæ, Aspicerinæ, Figitinæ, Eucoilinæ, Charipinæ, and Cynipinæ. The first six of these subfamilies include comparatively few species. These are parasitic, infesting chiefly dipterous larvæ and aphids. The last subfamily, the Cynipinæ merits a more detailed discussion.

Subfamily CYNIPINÆ

This subfamily is composed of the gall-flies or gall-wasps as they are termed by some writers. Most of the species cause the growth of galls on plants, but some species, the guest gall-flies, are inquilines living in galls produced by other species.

Although these insects are known as the gall-flies it should be remembered that galls are produced by many insects that do not belong to this subfamily. Galls made by plant-lice, flies, and moths have been described in the preceding pages, and galls are also produced by beetles and certain other insects; but the great majority of these strange growths are made either by gall-midges, mites, plant-lice, or true gall-flies (Cynipinæ).

The galls made by mites and plant-lice have open mouths, from which the young of the original dweller escape. But in the case of the gall-flies the gall is closed, and a hole must be made by the insect in order to emerge. Moreover, there is no reproduction of insects within the galls of gall-flies, as there is within the galls of mites and plant-lice.

It is a remarkable fact that each species of gall-insects infests a special part of one or more particular species of plants, and the gall produced by each species of insect is of a definite form. Hence when an entomologist who has studied these insects sees a familiar gall, he knows at once what species of insect produced it.

Naturalists have speculated much as to the way galls are made to grow. It has been supposed that at the time the egg is laid there is deposited in the tissue of the plant with it a drop of poison, which causes the abnormal growth. By this theory the differences between the galls of different insects was explained by supposing that the fluid produced by each species of insect had peculiar properties. There are certain kinds of galls which may be produced in this way. Thus it is said that the wound made by a certain saw-fly in the leaves of willow causes an abundant formation of plant-cells, and the gall thus formed attains its full growth at the end of a few days, and before the larva has escaped from the egg. But with the gall-flies the gall does not begin to grow until the larva is hatched; but as soon as the larva begins to feed, the abnormal growth of the plant commences. In this case, therefore, if the gall is produced by a poison, this poison must be excreted by the larva.

Galls produced by the different species of cynipids differ greatly in form and are found on all parts of plants. A most useful manual for the identification of galls is the "Key to American Insect Galls" by Dr. Felt ('18) in which, are figures of the galls of many of our gall-making insects, including those of all orders.

There are two terms that are frequently used in the descriptions of galls; these are *monothalamous*, indicating that the gall contains a single larval cell, and *polythalamous* indicating that the gall is compound, containing more than one larval cell.

Certain insect-galls have been found valuable for various purposes; they have been used in medicine, in the manufacture of ink, for tanning, and for dyeing. A summary of the literature dealing with the uses of insect-galls was published by **Fagan** ('18).

There exists in many species of gall-flies an alternation of generations; that is, the individuals of one generation do not resemble their parents, but are like their grandparents. In many cases the two succeeding generations of a species differ so greatly that they have been considered as distinct species until by careful studies of the life-cycle one has been found to be the offspring of the other.

In those species where an alternation of generations exists, one generation consists only of agamic females while the other consists of both males and females, which reproduce sexually. In some cases the galls produced by the two generations are quite similar; but in others they are very different and are found on different parts of the host plant. For an example of this see the account of the hedgehog gall-fly, *Andricus erinacei*, given below.

The guest gall-flies or inquilines.—Some species of this subfamily do not form galls but lay their eggs in the galls made by other species. The larvæ of these inquilines feed upon the galls produced by their hosts and in some cases do not discommode the owners of the galls in the least. But some guest gall-flies are parasites as well as guests. For example, Triggerson in his study of the hedgehog gall-fly ('14) found that the larva of *Synĕrgus erinācei*, a guest in the hedgehog gall, mines from cavity to cavity of the gall and feeds on the occupant of each in turn.

Among the more conspicuous of our cynipid galls are the following.

The oak hedgehog gall, Andricus erinācei.—A common gall on the leaves of white oak is one known as the oak hedgehog gall. This gall is rounded or oblong, with the surface finely netted with fissures, and more or less densely covered with spines (Fig. 1161, *a*). It varies in length from 10 mm. to 15 mm., and occurs on both sides of the leaves. The point of attachment is generally on the midrib, though it is often found on the lateral veins. When young the gall is yellowish green, but in autumn it becomes yellowish brown. This gall is polythalamous, containing from two to eight larval cells (Fig. 1161, *b*).

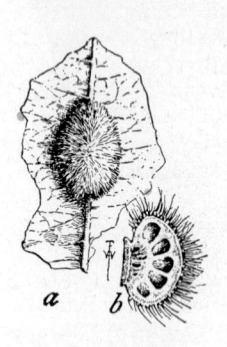

Fig. 1161.—The oak hedgehog gall: *a*, gall on leaf; *b*, section of gall.

Within the hedgehog galls is developed one generation, the agamic one, of *Andricus erinacei*. The alternating generation, the sexual one, is developed in very different galls made on the terminal growing points of buds and bud-scales. These are small, thin-walled, elongate, egg-shaped galls, from 2 to 3 mm. in length, and are monothalamous.

The life-cycle of this species was very carefully worked out by Triggerson ('14), from whose account the data given here are compiled. The two generations are distinguished by the use of trinominal names, as follows.

Andricus erinacei erinacei.—This is the agamic form, which is developed in the oak hedgehog galls. In this form the wings are vestigial not twice the length of the scutellum. The adults emerge in November and deposit their eggs in the leaf- and flower-buds.

Andricus erinacei bicolens.—This is the sexual form, which is developed in the galls formed in the buds. The larvæ hatch in May, from eggs laid by the agamic form the previous autumn and produce the galls known as the soft oak-bud galls. The adults, male and female, emerge early in June, and the females lay their eggs in the leaf-veins; from these eggs hatch the larvæ that cause the growth of the oak hedgehog galls. In the sexual form the wings are well-developed. The name *bicolens* was proposed for this form by Kinsey ('20).

The oak-apples.—There are various kinds of galls found on the leaves and stems of oaks that are commonly known as oak-apples, a name suggested by the spherical form and large size of some of them. Several of these are quite similar in external appearance but are markedly different in internal structure. In all there is a firm outer wall and a small, central larval cell (Fig. 1162). The part of the gall between the larval cell and the outer wall differs in structure in the galls of different species of gall-flies; in some it is filled with a spongy mass of tissue, in others the larval cell is held in place by a small number of filaments that radiate from it to the outer wall.

Fig. 1162.—An oak-apple.

The large oak-apple, *Amphibolips confluens*.—This is the largest of our common oak-apples, measuring from 18 to 50 mm. in diameter. It occurs on several species of oak and is usually attached to a vein or the petiole of a leaf. The space between the larval cell and the outer wall of the gall is filled with a spongy mass of tissue, in which in some of the galls there are many radiating fibers, as shown in the figure above, but in other galls these fibers are indistinct, the space being filled with an amorphous mass of tissue.

In spite of the fact that these galls are common and conspicuous the life-cycle of the species that produces them has not been fully worked out. What is known of it is based chiefly on the observations of Walsh (1864), little has been added during the long interval since this publication.

The oviposition has not been observed. The galls appear on the leaves early in the spring; from some of them there emerges in June a generation of gall-flies consisting of both males and females; and from other galls there emerges in the autumn a generation of gall-flies which consists only of females.

The large empty oak-apple, *Amphĭbolips inānis*.—There are two oak-apples which are very similar in structure, and which may be termed the empty oak-apples. In these the space between the central larval cell and the outer shell contains only a few, very slender, silky filaments, which hold the larval cell in place (Fig. 1163). The larger of these two galls, measures from 25 to 35 mm. in diameter, and is found on the leaves of the scarlet oak and red oak. Externally this gall resembles that of the preceding species but is easily distinguished by its internal structure. The adult gall-flies emerge in June and early in July; they are male and female; an agamic form of this species is not known.

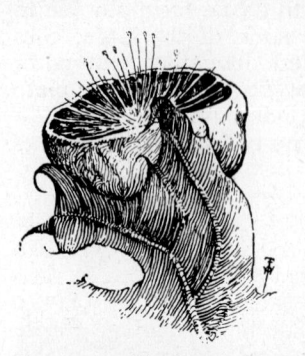

Fig. 1163.—The large empty oak-apple.

The smaller empty oak-apple, *Diplŏlepis centrĭcola*.—This gall is found on the lower side of leaves of the post-oak and measures from 15 to 20 mm. in diameter. It is sometimes tinged with pink and covered with a white bloom. The adult gall-flies emerge in October and are all females. A sexual generation of this species is not known.

The oak-bullet-gall, *Disholcăspis glŏbulus*.—One of the most common galls on white oak, chestnut oak, and scrub chestnut oak is a bullet-shaped gall which is attached to the small twigs of these trees. This gall measures from 8 to 16 mm. in diameter and occurs singly or in clusters of two, three, or more. Internally it is of a compact, rather hard, corky texture, and contains a free, oval, larval cell, resembling an egg.

Fig. 1164.—The mossy rose-gall.

The adult gall-flies emerge during October and November; these are all females; a sexual generation of this species has not been identified.

The giant oak-gall, *Ăndricus califŏrnicus*.—This is the most common oak-gall of the Pacific Coast. It is very abundant on the twigs and branches of the California white oaks, and during the winter, when the trees are bare, it is a very conspicuous object, on account of its abundance and large size. It varies in shape from globose to reniform and also varies greatly in size; some of the larger ones measure more than 100 mm. in their greatest diameter. The outer surface of the gall is white and usually smooth; the interior is more or less filled with a compact soft material, and contains from one to a dozen larval cells. Several varieties of the gall-fly that produces this gall are described by Kinsey ('22).

The mossy rose-gall, *Rhodītes rōsæ*.—This is a very common polythalamous gall, which is formed on the stems of rose bushes, especially of the sweetbrier. The gall consists of a large mass of moss-like filaments surrounding a cluster of hard kernels (Fig. 1164). In each of these kernels a gall-fly is developed. The galls appear early in the summer, but the adults do not emerge till the following spring. These are male and female; there is no alternation of generations in this species.

Superfamily CHALCIDOIDEA

The Chalcid-flies

This superfamily is the most highly evolved of the Hymenoptera. It includes the family Chalcididæ with a very large number of genera and species. A few species are as large as honey-bees, but the vast majority are minute. Most are black or metallic greenish in color, but some are brown or yellow. The most distinctive feature of the group is the presence of a separate sclerite, the prepectus, in front of the mesopleurum. This sclerite is interposed between the pronotum and the tegula so that the two are not in contact as in most other Hymenoptera. In a few chalcids the prepectus is very small or entirely absent. Chalcids have the antenna elbowed and never with more than thirteen segments. The ovipositor usually does not extend beyond the tip of the abdomen but may be quite long. Each wing has only a single vein. In some genera a few other indistinct veins can be traced.

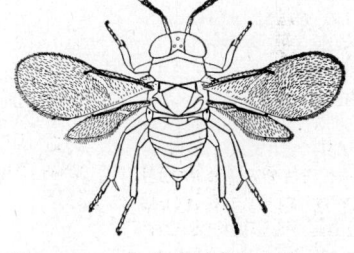

Fig. 1165.—A chalcid-fly, *Aphycus eruptor*.

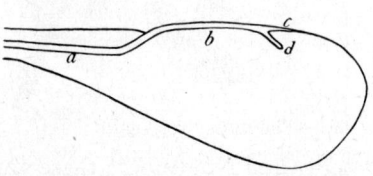

Fig. 1166.—Fore wing of a chalcid-fly.

The vein that persists in the wings of chalcid-flies consists probably of the coalesced subcosta, radius, media, the stigma, the radial crossvein, and the base of vein R_3. Writers on the Chalcidoidea make use of a special set of terms in describing the different parts of this compound vein. These are the *submarginal* vein or *subcostal* vein (Fig. 1166, *a*), the *marginal* vein (Fig. 1166, *b*), the *postmarginal* vein (Fig. 1166, *c*), and the *stigmal* vein (Fig. 1166, *d*).

The chalcid-flies constitute an exceedingly important group of insects from an economic standpoint; and nearly all of them are beneficial, being parasites that do much to keep in check noxious insects. A few species, however, are phytophagous; among these are those of the genus *Isosoma* that infest the stalks of growing grain, and species of several genera that infest the seeds of various plants. While these are noxious, the fig-insects, forming the subfamily Agaoninæ, although phytophagous are very beneficial to man.

Insects in all stages of their development suffer from the attacks of chalcid-flies, eggs, larvæ, pupæ, and even adults in a few cases being attacked by them. The larvæ of chalcid-flies usually feed within the body of their host, but some species are external parasites of other larvæ.

While the development of most species of chalcid-flies is a normal one, in certain species remarkable modifications of the usual course exists. The phenomenon of polyembryony which has been observed in several species is discussed on an earlier page. Another modification of the usual course of development is the existence of a hypermetamorphosis that occurs in some genera.

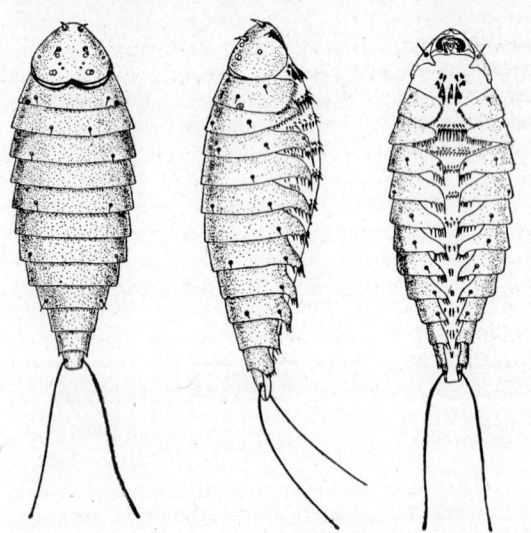

Fig. 1167.—Planidium of *Perilampus*. From left to right, dorsal, lateral, and ventral views. (After Ford.)

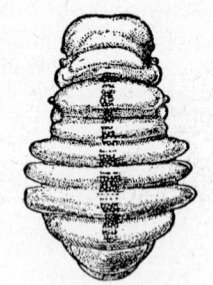

Fig. 1168.—*Perilampus hyalinus*. Mature larva, greatly enlarged. (From Smith.)

In these cases the larvæ when they leave the egg differ greatly in form from ordinary chalcid-larvæ (Fig. 1167). and are active, moving about in search of their prey. This active instar was termed by Professor Wheeler, who first described it, the *planidium*, from the Greek meaning diminutive wanderer. The planidea of the species of several genera have been described. Two of these, *Orasēma viridis*, described by Wheeler ('07a) and *Psilogăster fasciivēntris* described by Brues ('19) are parasites of ants, and *Perilămpus hyalīnus* described by H. S. Smith ('12) is a secondary parasite of the tachinid and ichneumonid parasites of the fall webworm. The planidium figured above was found by Miss Norma Ford within the bodies of dissected specimens of one of the meadow-grasshoppers, *Conocephalus fasciatus;* it is probably a secondary parasite of some parasite of the grasshopper (Ford '22).

The development of *Perilampus hyalinus* will serve as an illustration of the life-cycle of these remarkable parasites. The egg has not been observed, but it seems probable that it is deposited upon the food plant of the fall webworm in the vicinity of a colony of this insect. The planidia, which measure less than 0.3 mm. in length and are therefore almost invisible to the unaided eye, were found first on the exterior of the caterpillars; later within their bodies, having bored through the cuticula of the caterpillar by means of their well-developed mandibles; still later the planidia were endoparasitic within the larvæ of parasites of the caterpillars. After feeding for a time the planidium molts; the second instar is ovate in shape, with the head bent underneath. After another short period of feeding the larva molts a second time and becomes greatly changed in form (Fig. 1168). Finally after the primary parasite has left the caterpillar and pupated the larva of *Perilampus* becomes an ectoparasite. It then soon completes its growth and pupates.

Family CHALCIDIDÆ

Many authors regard as distinct families the groups treated here as subfamilies. Since there is much difference of opinion regarding the extent of the various groups and since the definition of many of them is difficult, it seems unwise to call them families. The classification adopted is that of Ashmead ('04), except that the Leucospidinæ are separated from the Chalcidinæ and that Ashmead's families are reduced to subfamilies. Some authors recognize additional groups, among them the following: Eupelmidæ, Tanaostigmatidæ, and Signiphoridæ here included in the Encyrtinæ; Leptofœnidæ (= Pelecinellidæ) here included in the Cleonyminæ; Tridymidæ included in the Miscogasterinæ; Spalangidæ included in the Pteromalinæ; and Aphelinidæ, Elachertidæ, Entedontidæ, and Tetrastichidæ here included in the Eulophinæ. A thorough revision of the genera and higher groups is greatly needed. Such a work would do much to help the study of this difficult yet most interesting and important family.

The key to the subfamilies will serve for most species, though the diversity of form in many groups and the present state of our knowledge makes the construction of a reasonably perfect key quite difficult.

KEY TO THE SUBFAMILIES OF CHALCIDIDÆ

By Dr. H. K. Townes

A. Tarsi with three segments; wing hairs usually arranged in lines; very small species. p. 949...................................Trichogrammatinæ
AA. Tarsi with four or five segments.
  B. Hind wing linear; tarsus with four or five segments; minute species. p. 948
....Mymarinæ
  BB. Hind wing not linear, wider near the middle.
    C. Tarsi with four segments, or sometimes five-segmented; axilla of fore wing produced on to mesonotum, its anterior margin on or in advance of

an imaginary line drawn from tegula to tegula; anterior tibial spur small and weak
- D. Hind coxa extremely large and disc-like; hind tarsus very long. p. 948 .... ELASMINÆ
- DD. Hind coxa of approximately normal size, more or less cylindrical; hind tarsus of normal length. p. 947 .... EULOPHINÆ
- CC. Tarsi with five segments; axilla normal, not produced on mesonotum forward of the tegula; anterior tibial spur larger and strong.
  - D. Hind femur much enlarged and with a row of teeth beneath.
    - E. Hind tibia ending in a long spine beyond the insertion of the tarsus and with a single weak apical spur.
      - F. Prepectus a large triangular sclerite; thorax metallic green (genus *Podagrion*). p. 943 .... ORMYRINÆ
      - FF. Prepectus a small almost completely hidden sclerite; thorax never metallic green. p. 944 .... CHALCIDINÆ
    - EE. Hind tibia rather squarely truncate and with two apical spurs.
      - F. Wings folded longitudinally in repose; ovipositor recurved over dorsum of abdomen. p. 947 .... LEUCOSPIDINÆ
      - FF. Wings never folded; ovipositor not recurved. p. 946 CLEONYMINÆ
  - DD. Hind femur of more normal size and with not more than two teeth beneath.
    - E. Mesopleurum large and evenly convex, without a groove for the femur; apical spur of middle tibia long and stout. p. 946 ENCYRTINÆ
    - EE. Mesopleurum with a groove for the reception of the femur; apical spur of middle tibia not stout.
      - F. Pronotum above with only an anterior face, not visible from the dorsal aspect; mandible sickle-shaped. p. 944 .... EUCHARINÆ
      - FF. Pronotum above with a dorsal face so that it is easily visible from above.
        - G. Thorax very robust and short; abdomen small, oval, polished, and entirely or largely enclosed by the second or the fused second and third tergites. p. 944 .... PERILAMPINÆ
        - GG. Thorax and abdomen not as above, of more normal proportions.
          - H. Head of female long, oblong, and with a deep longitudinal groove above; front and hind femora and tibiæ stout, the middle ones slender; male wingless with three- to nine-segmented antenna. p. 943 .... AGAONTINÆ
          - HH. Not as above.
            - I. Hind tibia with one apical spur; head often large, in front often wide and flat with a reticulate surface and striæ converging to the mouth. p. 947 .... PTEROMALINÆ
            - II. Hind tibia with two apical spurs, one of which may be quite small; head usually not as above.
              - J. Hind coxa usually large and with more or less of a ridge above; abdomen usually compressed and the ovipositor usually long. p. 943 .... ORMYRINÆ
              - JJ. Hind coxa not enlarged, rather cylindrical in cross section; ovipositor not or slightly surpassing the tip of the abdomen.
                - K. Pronotum large, quadrate and as wide as the mesonotum; color black or brown, never metallic; abdomen usually compressed. p. 945 .... EURYTOMINÆ
                - KK. Pronotum short transverse, conical or conically produced anteriorly and narrower than the mesonotum; color often metallic greenish; abdomen usually depressed.
                  - L. Mesepisternum large; front or hind femur often enlarged. p. 946 .... CLEONYMINÆ
                  - LL. Mesepisternum not unusually large; neither front nor hind femur enlarged; species resembling the Pteromalinæ. p. 947 .... MISCOGASTERINÆ

Subfamily ORMYRINÆ

The subfamily Ormyrinæ has gone also under the names Toryminæ and Callimominæ. Most species are more or less metallic green with compressed abdomens and long ovipositors. They are parasitic largely on cynipid and cecidomyiid gall insects. The commonest genus is *Callimōme* (= *Torymus*). Although most of its species are parasites of gall insects, *Callimōme drupārum* infests the seeds of apple. For a revision of the genus see Huber, *Proc. U. S. Nat. Mus.* 1927, vol. 70, art. 14, 144 pages. *Monodontomērus* is parasitic upon a great variety of hosts. *Megastĭgmus* includes brownish species with a large black stigma in the fore wing. The larvæ develop in seeds. Some are common in rose hips, and others destroy the seeds of coniferous forest trees. See Crosby ('09). *Podăgrion* is a parasite in the eggs of Mantidæ. The genus *Ormyrus* is parasitic on gall insects. It has the abdomen tapering to a conical tip and peculiar bands of large punctures on the tergites. The ovipositor is not exserted.

Subfamily AGAONTINÆ

The Fig Insects

This subfamily is composed of those remarkable insects that live within figs and fertilize them. It is represented in the United States by a single species, *Blastŏphaga psēnes*, that was introduced into California in order to make possible the production of the Smyrna fig in that state.

The fruit of the fig tree consists of a hollow receptacle on the lining of which the flowers are borne. At the apex of the fig there is a more or less distinct opening leading into the interior. It is through this opening that the female fig insect leaves the fig in which she was developed and enters a young fig in order to oviposit.

The eggs are laid at the base of a modified form of pistillate flower, known as gall-flower, that is found in wild figs, and the larvæ produce little galls in which they develop. The female fig insect when leaving the fig in which she was developed becomes covered with pollen which is carried into the young fig which she enters to oviposit, and thus the flowers in this fig are fertilized.

The male fig insect is wingless. It crawls about over the galls in the fig in which it was developed, and when it finds a gall containing a female it gnaws a hole in it and then thrusting the tip of its abdomen through the puncture fertilizes the female.

It is only in the wild figs that the gall-flowers are developed. For this reason only the wild figs are suitable for the development of the fig insects. But the female insect will enter the cultivated figs seeking a place to oviposit and will thus fertilize them.

Although the numerous varieties of common cultivated figs do not require the stimulus of pollination and the resulting fertilization of the ovary to make the fruit set, in the case of the Smyrna fig, which is the most desirable variety grown, without this stimulation the young figs soon turn yellow and drop. It is the oily kernel of the fertile seed that gives the Smyrna figs their superior quality.

The fertilization of the edible figs is termed caprification. It is brought about by placing in the fig trees fruit of the wild figs containing the fig insects. In order to produce the Smyrna figs it is necessary to grow also the wild figs, or caprifigs as they are termed.

There are many species of fig insects living in the wild figs of tropical and semi-tropical countries.

Subfamily PERILAMPINÆ

The members of this subfamily are short species with large heads, robust thoraces, and short oval abdomens. Some are metallic green and thus resemble cuckoo wasps. Others are black. Most of our species belong to the genus *Perilămpus*. *Perilămpus* lays its eggs on leaves partially inserted into incisions made by the ovipositor. The first stage larva is a tiny active animal with chitinous armor and series of spines beneath (Fig. 1167). The name *planĭdium* has been given to this type of larva. The planidium attaches to a passing caterpillar and bores into its body. There it locates and penetrates the larva of a tachinid, ichneumonid, or other parasite of the caterpillar. After the primary parasite of the caterpillar has completed its growth and pupated, the *Perilămpus* larva emerges from its body and finishes development as an external parasite of the parasite pupa. The other larval instars lose the specializations of the planidium stage and look more like normal parasite larvæ (Fig. 1168). Most species of *Perilămpus* are secondary parasites, but *P. chrysōpæ* is a primary parasite of lacewing-flies.

Subfamily EUCHARINÆ

The subfamily Eucharinæ includes a few uncommon or local species parasitic on ants. In the tropics species and individuals are more numerous. Some genera have a bizarre pair of spines extending backwards from the scutellum. The eggs are laid in the leaves or buds of plants. The first stage larva is a planidium resembling that of the Perilampinæ. Probably the planidia attach to worker ants and are thus carried to the ant larvæ and pupæ which they parasitize.

Subfamily CHALCIDINÆ

This subfamily is easily recognized by its very swollen hind femur, though certain members of other groups have this character also (see the key to subfamilies). The peculiar hind legs appear to be suited for grasping prey. Females occasionally grasp host larvæ with them for oviposition. The Chalcidinæ includes most of the large species of the family. Some of our species of *Spilochălcis* are as large as yellow jacket wasps and on account of their similar black and yellow color are easily confused with them while on the wing. Common genera are *Spilochălcis*, with the abdomen on a noticeable stalk and with an apical spur on the middle tibia; and *Chălcis* with an oval unstalked abdomen. *Spilochălcis mariæ* is a common parasite in the cocoons of the large silk moths. Some species of *Chălcis* are parasites of lepidopterous pupæ and others of fly maggots. Some oviposit into tachinid maggots while they are still inside the bodies of their caterpillar hosts. The tachinid larvæ complete their development and pupate before succumbing to the *Chălcis* hyperparasites.

Subfamily EURYTOMINÆ

This subfamily is remarkable on account of the diversity of its habits. Primary and secondary parasites of a variety of insects, egg parasites, and phytophagous species are included.

Members of the genus *Harmolīta* (= *Isosōma*) infest the stems of growing grasses, either forming gall-like swellings or living in the center of the stems. The two following species are of economic importance.

The wheat joint-worm, *Harmolīta trĭtici*, is a well-known pest that infests the stalks of growing wheat and certain grasses. It causes a woody growth which fills up the cavity of the stalk, and sometimes also causes a joint to swell and the stalk to bend and lop down. The presence of this insect is often indicated by pieces of hardened straw coming from the threshing machine with the grain. There is but a single generation of this species in a year. The insect remains in the straw and stubble during the winter, the adults emerging in the spring. The methods of control of this pest are rotation of crops, burning or deep ploughing under of stubble when practicable, or harvesting of stubble in spring with a horserake and burning it before the adults emerge.

The wheat straw-worm, *Harmolīta grăndis*, is often a serious pest of wheat west of the Mississippi River. In the East it is less injurious than the wheat joint-worm. There is a summer generation which consists only of winged females and a winter and spring generation which consists of both males and females. These are smaller than the summer form and are frequently wingless. The adults of the winter and spring generations emerge in April, and the females deposit their eggs in the young wheat plants; the larvæ eat out and totally destroy the forming heads of wheat. The adults of the second generation deposit their eggs about the time the wheat is heading, just above the youngest and most succulent joints which are not so covered by the enfolding leaf-sheaths as to be inaccessible to them. The larvæ pupate by October, and the winter is passed in the straw or stubble.

Evoxysōma vītis infests the seeds of wild grapes and occasionally attacks cultivated varieties. *Bruchŏphagus funēbris* infests the seeds of red and crimson clovers and alfalfa. The genus *Decătoma* attacks gall-making insects. *Rīleya* parasitized Cecidomyiidæ. *Macrorīleya* parasitizes the eggs of tree crickets. *Axima* lives as a parasite in the nests of *Ceratīna dupla*, the little carpenter bee. *Eurȳtoma* is a large genus parasitic on many types of insects. A few species infest seeds. Recent revisions of United States species are: Harmolita, *Proc. U. S. Nat. Mus.* 1919, 55:433-471; Decatoma, *Proc. U. S. Nat. Mus.* 1932, vol. 79, art. 28, 95 pages. For an account of the seed-infesting species see Crosby ('09).

Subfamily ENCYRTINÆ

This is a large and diverse group parasitic on many kinds of insects. It can be recognized by the broad mesopleurum without the usual groove for the femur and by the large heavy spur on the middle tibia. This spur is used in jumping. Many parasitize Coccidæ, Aphidæ, and Aleyrodidæ. At least four highly successful cases of biological control are due to the introduction of these insects to parasitize scale insects or mealy bugs.

Polyembryony is a common phenomenon in the Encyrtinæ and reaches its highest development in this group. The eggs of the polyembryonic species are inserted into those of Lepidoptera. The parasite larvæ mature in the host caterpillar just before it pupates. *Copidosōma truncatĕllum* is a common parasite of *Autŏgrapha brăssicæ*, the cabbage looper. The female lays one or two eggs into the egg of *Autŏgrapha*. Usually the second egg deposited is without a sperm and is therefore unfertilized. Each egg develops inside the growing host caterpillar into hundreds of parasite larvæ. The fertilized egg produces female larvæ and the unfertilized egg produces male larvæ. Beside these sexual larvæ which mature into adults, there are many sexless larvæ in which the reproductive and respiratory systems are not developed. These die without maturing. The host larva is usually much larger than an unparasitized caterpillar and somewhat more sluggish. Just before pupation it is killed by its parasites which by that time practically fill its body. The parasite larvæ pupate in the skin of their host and emerge as adults through holes cut to the outside. Twelve hundred to three thousand may develop in a host. A single egg usually gives rise to about one thousand adults.

Some Encyrtinæ have the body rather long and narrow, the disc of the mesonotum more or less concave, and notauli present. These are often put in a separate subfamily, the Eupelminæ. The species are mostly egg parasites or secondary parasites. They are good jumpers, and their peculiar ability to turn both the head and abdomen back over the thorax seems to be used to help them leap much as the click beetles use the movable prothorax for this purpose. Many have short wings with a joint at the middle so that they can be folded upwards when the abdomen is raised.

Subfamily CLEONYMINÆ

This is a relatively small heterogeneous group that is probably polyphyletic. The species that have been reared are mostly from Coleoptera.

Subfamily LEUCOSPIDINÆ

This subfamily includes the single genus *Leucŏspis* in the Nearctic Region. Our only widespread species is *L. ăfflinis*, a large black and yellow chalcid frequently found on flowers. It has been bred from nests of a leaf-cutter bee, *Megachīle*. For a revision of the species see Weld: *Proc. U. S. Nat. Mus.* 1922, vol. 61, art. 6, 43 pages.

Subfamily MISCOGASTERINÆ

This group is related to the Pteromalinæ. It differs in having two instead of one spur on the hind tibia. Species have been bred from gall insects, dipterous larvæ, and other hosts.

Subfamily PTEROMALINÆ

This is one of the larger subfamilies of the Chalcididæ. Many of the species are of economic importance either as beneficial primary parasites of common pests or as harmful secondary parasites.

Pterŏmalus puparum is a common parasite of the cabbage butterfly. Oviposition is into the pupa. It is said that if a female *Pterŏmalus* finds a caterpillar of this species preparing to pupate, she will wait patiently until the pupa is formed and will then oviposit into it. Many *Pterŏmalus* larvæ develop in a single host. In a few weeks they become adults and emerge from a hole chewed in the pupal shell.

Dĭbrachys boucheānus is a widespread species attacking many kinds of insects. The principal requirement for a host seems to be that it be enclosed in a cocoon. Since braconid and ichneumonid cocoons are the ones most commonly attacked, the species is usually a secondary parasite and therefore harmful. It is often reared from *Apanteles* cocoons. After a host cocoon is located and identified with the antennæ, the female inserts her ovipositor and stings the enclosed larva or pupa once to several times until it is paralyzed. If the host is lying against the wall of the cocoon the female laps the juices that flow through the ovipositor puncture. If the host is away from the cocoon wall a feeding tube is constructed. To do this the female parasite pushes the tip of her ovipositor just through the cocoon wall. A mucilaginous substance then exudes from the tip of the ovipositor. Additional quantities of this substance are occasionally added as the ovipositor is slowly advanced toward the pupa. Eventually there is a mucilaginous tube enclosing the ovipositor from the cocoon wall to the paralyzed host. The ovipositor is withdrawn and the parasite then sucks the juices of the host through the tube. When she has finished feeding she seals the tube with a drop of the same material and then lays several eggs on the host. The habit of making a feeding tube is a common one among the Pteromalinæ.

Subfamily EULOPHINÆ

Species of this subfamily are very numerous and mostly quite small in size. They are diverse in habits and in host selection.

Euplēctrus platyhypēnæ is a common parasite of army worms. It is unusual in being an external parasite of an exposed host. The eggs are deposited on the caterpillar's skin in groups of about twenty to thirty. The larvæ form a compact mass with the heads attached to feeding holes in the skin of the host. The caterpillar does not die until they become mature and detach from their feeding holes. The parasites pupate under the dead host.

Melittōbia is common in the nests of solitary bees and wasps. The female gains access to the larval cell, stings the enclosed larva or pupa to paralyze it and then deposits her eggs on it. The parasite larvæ develop externally. Several generations develop on the same host before it is completely consumed. The males are short-winged and have the antenna modified into a pincher-like structure for holding the female's antenna during courtship. They are very pugnacious and often kill each other in battle. The females are either short- or long-winged.

Tetrăstichus is a large important genus parasitic on many kinds of insects. *T. aspăragi* oviposits in the eggs of the asparagus beetle. The beetle egg hatches and the larva lives to maturity but is killed in its pupal cell by its parasites. About a half dozen parasite larvæ develop in each host.

A large and important group of Eulophinæ, sometimes put in the separate subfamily Aphelininæ, are parasites of scale insects. A few attack aphids and aleyrodids. Several species of this group have been introduced into various countries to control certain species of scale insects and have proved remarkably efficient. In some species of *Coccŏphagus* and related genera the male and female larvæ are strikingly different in morphology and in habits. The male and female eggs also are different. The female larva develops as a primary parasite of a scale insect while the male larva is a secondary parasite in the scale, either upon a female larva of its own species or on some other chalcid. For a revision of *Coccŏphagus* see Compere: *Proc. U. S. Nat. Mus.* 1931, vol. 78, art. 7, 132 pages.

Subfamily ELASMINÆ

Elăsmus is the only North American genus of the Elasminæ. It contains small black or brown species somewhat triangular in cross section. Most species are solitary or gregarious external parasites of caterpillars.

Subfamily MYMARINÆ

The Fairy-flies

The Mymarinæ includes exceedingly minute species parasitic in the eggs of other insects. They may be recognized by the linear hind wings pedunculate at the base (Fig. 1169). Species of the genus *Allăptus* are some of the smallest insects known. One has a body length of only 0.21 millimeters. The entire development of the Mymarinæ is in the host egg. There are two larval instars of which the first is cyclopoid, resembling that of the Scelionidæ. As many as fifty parasites have been reared from a single egg, although one is the usual number.

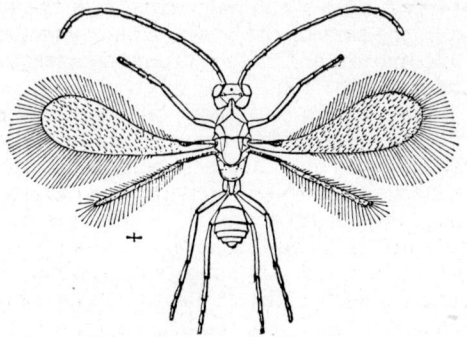

Fig. 1169.—*Cosmocoma elegans.*

Subfamily TRICHOGRAMMATINÆ

This subfamily comprises minute egg parasites with three-segmented tarsi. *Trichogrămma evanĕscens* (= *minūtum*) is an ubiquitous species parasitizing almost any insect egg that can be pierced with its ovipositor. It is an important parasite of many economic pests and has frequently been reared in insectaries for liberation in fields and orchards to control various insects. The success of these attempts has not been encouraging. Several other forms of *Trichogrămma* resemble *T. evanĕscens* very closely, differing somewhat in color and in length of the life cycle when reared at a constant temperature. Since they will not interbreed with *T. evanĕscens* they are now regarded as distinct species. *Hydrophỹlax aquivōlans* is an aquatic species parasitic probably upon damsel fly eggs. It has narrow wings and uses them as oars for "flying" through the water. Some species of Trichogrammatinæ have two larval instars, the first of cyclopoid form as in the Mymarinæ and Scelionidæ. Others have a single instar, a bag of cells with a mouth. This larva gets the egg contents on the inside of its body and then pupates.

Superfamily EVANIOIDEA

Family EVANIIDÆ

The Ensign-flies

The family Evaniidæ is so distinctly separated from all other families of the Hymenoptera that it is regarded as constituting a separate superfamily. We have in this case a superfamily represented by a single family.

The family Evaniidæ differs from all of the preceding families in the presence of a well-marked anal lobe in the hind wings (Fig. 1170) and it differs from all of the following families in that the petiole of the abdomen is attached to the dorsal surface of the propodeum (Fig. 1171) instead of at the hind end of it, as it is in the following families. The abdomen is short and carried aloft—like a flag; this fact suggested the common name ensign-flies for these insects.

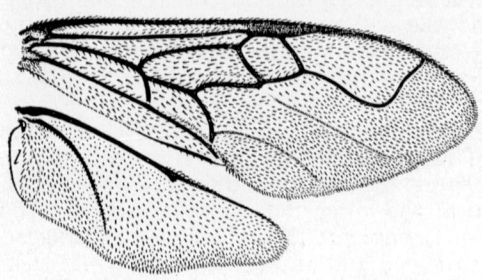

Fig. 1170.—Wings of *Evania appendigaster*; *l*, anal lobe.

Fig. 1171.—*Evania appendigaster*

The venation of the wings is greatly reduced, and in the hind wings there are no closed cells; the trochanters are two-segmented; and the ovipositor is not at all or but little exserted.

All of the species are parasitic in the eggs of cockroaches.

This family was monographed by Bradley ('08) and Kieffer ('12) It is represented in our fauna by two genera, *Evănia* and *Hўptia*.

Superfamily VESPOIDEA

The Vespoid-Wasps

This superfamily is one of three superfamilies, the Evanioidea, the Vespoidea, and the Sphecoidea, in which the hind wings are typically furnished with an anal lobe; but in some of the more specialized members of the Vespoidea and of the Sphecoidea the anal lobe has been lost. This is the case in certain genera of the Formicidæ, Mutillidæ, and Vespidæ; these exceptional forms can be placed by the table of families on pages 906 to 915.

The members of Vespoidea differ from the Evanioidea in that the petiole of the abdomen is attached to the hind end of the propodeum; and they differ from the Sphecoidea in that the lateral extensions of the pronotum, which reach the tegulæ (except in the Cleptidæ and Chrysididæ), are not in the form of well-differentiated rounded lobes, as is the case in the Sphecoidea. See Figure 1195 on a later page.

The Vespoidea is represented in our fauna by fourteen families; these can be separated by the table of families referred to above.

Family POMPILIDÆ

The Spider-Wasps

The members of this family are commonly called spider-wasps, because they provision their nests with spiders; this habit, however, is not distinctive as certain other wasps use spiders for this purpose.

The members of this family are slender in form, with long spiny legs, (Fig. 1172). The pronotum extends back on each side to the tegula; and the abdomen is sessile. Many of the species are of medium size, but some are very large; in fact, the largest of our Hymenoptera belong to this family.

Fig. 1172.— A spider-wasp.

Most of the Pompilidæ make their nests in the ground. The wasp first finds a spider and stings it until it is paralyzed, and then digs a burrow, which is enlarged at the lower end, forming a cell for the reception of the spider; the spider is then dragged down into the cell and an egg attached to it; then the passage leading to the cell is filled with earth. Detailed accounts of the actions of these spider-wasps when making and provisioning their nests are given by Peckham and Peckham ('98) and by Rau and Rau ('18).

Among the giants of this family are the well-known tarantula-hawks of the genus *Pĕpsis* of the Southwest, which store their burrows with tarantulas. Many a hard-fought battle do these spider-wasps have with these enormous spiders, and sometimes they are conquered and ignominiously eaten.

Not all members of this family are digger-wasps, for some are mason-wasps. The species of the genus *Pseudagēnia* make thimble-shaped cells, of mud, attached to the lower surface of stones, in chinks of walls, under bark and in various other situations; and at least one species of *Cerŏpales* is said to be parasitic in the nests of *Pseudagēnia*.

More than one hundred species belonging to this family have been described from our fauna. A classification of the family was published by Banks ('11).

The family EMBOLEMIDÆ includes only a few rare species the habits of which are unknown.

Family CLEPTIDÆ

Cleptes

This family includes only the genus *Cleptes*, which was formerly included in the following family. But the genus *Cleptes* differs from the Chrysididæ in that there are six exposed segments in the abdomen and the venter is convex. It is believed that these wasps are parasitic in the cocoons of saw-flies; as one of them infests the currant-worm in Europe. Several native species are found in the Far West.

Family CHRYSIDIDÆ

The Cuckoo-Wasps

The cuckoo-wasps are wonderfully beautiful creatures, being usually a brilliant metallic green in color. The species are of moderate size, the largest being only about 12 mm. in length. They can be distinguished from other Hymenoptera by the form of the abdomen, in which there are at most five and usually only three or four exposed segments (Fig. 1173), and which is strongly concave below, so that it can be readily turned under the thorax and closely applied to it. In this way a cuckoo-wasp rolls itself into a ball when attacked leaving only its wings exposed.

In this family and in the preceding one the antennæ are 13-segmented in both sexes; the pronotum does not reach the tegulæ; there are no closed cells in the hind wings; and the ovipositor is an extensile jointed tube.

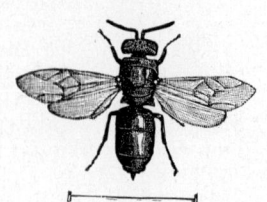

Fig. 1173.—*Chrysis nitidula*.

The cuckoo-wasps are so-called because they are parasitic in the nests of solitary wasps and solitary bees. A cuckoo-wasp seeks until it finds a wasp or bee, building its nest, and when the owner of the nest is off collecting provisions steals in and lays its egg, which the unconscious owner walls in with her own egg. Sometimes the cuckoo-wasp larva eats the rightful occupant of the nest, and sometimes starves it by eating up the food provided for it. The bees and wasps know this foe very well, and tender it so warm a reception that the brilliant-coated little rascal has reason enough to double itself up so the righteous sting of its assailant can find no hole in its armor. There is one instance on record where an outraged wasp, unable to sting one of the cuckoo-flies to death, gnawed off her wings and pitched her out on the ground. But the undaunted invader waited until the wasp departed for provisions, and then crawled up the post and laid her egg in the nest before she died.

A monograph of the North American species was published by Aaron ('85), one of the species of the world by Mocsary ('89), and another by Bischoff ('12).

The family ANTHOBOSCIDÆ is represented in North America by a single species, *Sierolomŏrpha ambĭgua*. This is shining black, with an oval abdomen, the first segment of which is constricted off from the rest. It measures 4.5 to 6 mm. in length.

Family SAPYGIDÆ

The Sapygids

This is a small family including only three North American genera, and but little more than twenty species. These insects are of moderate size, with short legs, and are usually black, spotted or banded with yellow, rarely entirely black. So far as their habits are known. they are inquilines in the nests of solitary wasps and solitary bees,

The family THYNNIDÆ is represented in our fauna by a single rare species of the genus *Glyptomĕtopa*, found in California.

Family TIPHIIDÆ

The Tiphiids

The tiphiids are quite closely related to the following family, the velvet-ants. In fact, some of the genera now included in the Tiphiidæ have been classed in the Mutillidæ. The characters separating these two families are indicated in the table of families of the Hymenoptera given above. The family Tiphiidæ includes three subfamilies.

The subfamily TIPHIINÆ is represented in our fauna by five genera; these are *Pterombus, Epomidiopteron, Elis (Myzine), Tiphia,* and *Paratiphia*. One of the most common species is *Tiphia inornāta* (Fig. 1174). This is a shining black species; the male measures from 7 to 11 mm. in length and has an upward projecting spine near the tip of the abdomen. The female measures from 12 to 14 mm. in length. The accompanying figure represents a female; in this sex the antennæ are curled much more than is indicated in the figure. This species is parasitic upon white-grubs, the larvæ of the May-beetles. For descriptions of species of *Tiphia* see Malloch ('18).

Fig. 1174.—*Tiphia inornata.*

The subfamily METHOCINÆ includes forms in which the two sexes are very dissimilar; the males are winged, the females are wingless and resemble ants. **Our only species is *Methoca stygia*.**

The subfamily MYRMOSINÆ includes two genera, *Myrmōsa* and *Myrmŏsula*. The males are winged and the females are wingless. Our most common species is *Myrmosa unĭcolor*. The subfamily has been monographed by Bradley ('17).

Family MUTILLIDÆ

The Velvet-ants

These handsome insects resemble ants in the general form of the body, but lack the scale-like knot of the pedicel of the abdomen characteristic of the true ants, although there is sometimes a constriction between the first and second abdominal segments (Fig. 1175). The body is often densely clothed with hair, which gives the insects the appearance of being clothed in velvet; and as the body is usually ringed or spotted with two or more strongly contrasting colors, they are very conspicuous. But in many species the body is naked. The colors most commonly worn by the velvet-ants are black and scarlet. The males are winged and frequent flowers. The females are wingless, but they run very rapidly and they sting severely. In the western states there are many straw yellow species, which are nocturnal.

Fig. 1175.

These insects are abundant in the warmer portions of our country, and several species occur in the North. A large species, *Dasymutĭlla occidĕntalis*, which measures from 16 to 30 mm. or more in length, is known in the South as the "cow-killer ant" because of the popular superstition that its sting is very dangerous to live stock.

Andre ('03) states that the mutillids of which the habits have been observed are parasites of nest-building Hymenoptera in the cells of which they deposit their eggs. The larvæ attack those of the owners of the nest without touching the provisions which the cell may contain.

In this country a species has been reared from the cells of the solitary bee, *Nomia patteni*, and one from the nests of the mud-dauber, *Chalybion cæruleum*. In Europe several species are parasitic in the nests of bumblebees; and in Africa several species have been found to be parasitic on the tsetse fly.

The Mutillidæ of North America have been monographed by Fox ('99) and those of the Eastern United States by Bradley ('16).

Family SCOLIIDÆ

The Scoliids

The scoliids are quite closely related to the preceding family but differ in their general appearance, resembling wasps rather than ants. They are parasitic on white grubs, the larvæ of Scarabæidæ. In their habits they do not exhibit as much intelligence as do most digger wasps; for they do not build nests and transport prey to them for their carnivorous larvæ. Instead of this they dig in the ground where the white grubs are, and finding one they sting it in order to paralyze it, work out a crude cell about it, and attach an egg to a ventral abdominal segment of the grub. The larva of the scoliid consumes the grub and then spins a cocoon and completes its development in this place.

The members of this family are very striking in appearance, being of large size and with the abdomen marked with conspicuous spots. Two genera are represented in our fauna, *Scōlia* and *Campsomēris* (*Elis*). In *Scolia* the transverse part of vein M_2 of the fore wings is wanting, in *Campsomeris* it is present.

Family FORMICIDÆ

The Ants

The great number of ants and their wide distribution render them the most familiar of all insects except perhaps the house-fly. As has been said by Professor Wheeler, an indefatigable investigator of these insects, "Ants are to be found everywhere, from the arctic regions to the tropics, from timberline on the loftiest mountains to the shifting sands of dunes and seashores, and from the dampest forests to the driest deserts. Not only do they outnumber in individuals all other terrestrial animals, but their colonies even in very circumscribed localities often defy enumeration." The present time has been termed the "age of insects" and of all insects the Formicidæ is the dominant family.

The habits of ants have attracted the attention of students of animal behavior from very early times and many volumes have been written on this subject. Among those most often quoted and to be found in most public libraries are those of Gould (1747), P. Huber (1810), and Lubbock (1894). The most comprehensive contributions have been made by Forel and Emory, each of whom has published more than one hundred papers in various European journals, and by Wheeler in this country. Among the other American writers who have made important contributions to our knowledge of the ways of ants are Buckley, Miss Fielde, Leidy, Lincecum, McCook, Pricer, Mrs. Treat, and Turner. But the most important work on this subject is Professor Wheeler's "Ants, their Structure, Development and Behavior" ('10). In the following pages there is space for only the more important generalizations that can be made regarding this family.

Ants are easily recognized by the well-known form of the body. The most distinctive feature is the form of the pedicel of the abdomen; this consists of either one or two segments, and these segments are either nodiform or bear an erect or inclined scale (Fig. 1176).

Fig. 1176.

When the pedicel of the abdomen consists of a single segment it is known as the *petiole;* when it consists of two segments the first segment is termed the *petiole* and the second segment the *postpetiole*. The swollen portion of the abdomen behind the pedicel is known as the *gaster*.

Another striking characteristic of ants is that in the antennæ of females and workers and of the males of some species the basal joint, the scape, is long and the antennæ are abruptly elbowed at the extremity of this joint.

The ants are all social insects, there being no solitary species. Each colony consists of three castes, the males, the female or queen, and the workers. As with the social bees and the social wasps, and unlike the termites, the workers are all modified females. With most ants the males and the queens are winged and the workers wingless; the wings of queens, however, are deciduous. In certain genera that live as parasites in the nests of other ants the worker caste is wanting, and in some species the females are wingless.

With many ants the polymorphism is not restricted to the presence of three uniform castes for one or more of the castes may be represented by more than one form. Of the males there may be either an unusually large form, or dwarfs, or ergatoid males, that is, males that resemble workers in having no wings and in the structure of the antennæ. The queens exhibit a similar series of forms; those of unusually large stature; dwarfs which are sometimes smaller than the largest workers; and ergatoid queens, which are a worker-like form, with ocelli, large eyes, and a thorax more or less like that of the normal queens, but without wings. The workers are even more polymorphic than the sexual forms. In many species the workers are of two distinct sizes, the worker majors and the worker minors. In colonies that are founded by an isolated female the first brood of workers is of the worker minor form. With many species a worker form exists in which the head and the mandibles are very large, the soldier caste. And with the honey-ants in some of the workers, the repletes, the gaster is a large spherical sac, being distended by the crop which is used as a reservoir for storing honey-dew to be used later by the colony.

Although all ants are social, great differences exist among them as to the size of their colonies. In the more primitive species the fully developed colony consists of only a few dozen individuals with comparatively feeble caste development; while in the more highly specialized forms a colony may consist of hundreds of thousands of individuals and exhibit an elaborate polymorphism.

The different species of ants differ also in their nesting habits. By far the greater number of species excavate their nests in the ground. Certain species are often seen burrowing in paths or other open places; but many more are to be found under small flat stones or other objects lying on the ground. Some species, especially those in which the colonies become large, build large mounds of the excavated material. These mounds are very familiar objects in many parts of our country.

A striking difference between the nests of ants and those of wasps and bees is that the ants do not construct premanent cells for their brood. The eggs, larvæ and pupæ are stored in chambers of the nest and are moved from one to another in order to take advantage of the changes in temperature and moisture. Thus the brood may be brought near the surface of the nest during the warmer portion of the day and removed to deeper chambers at nightfall.

While most species of ants nest in the soil, there are many that build their nests in wood, in timbers, in the trunks of decaying trees, in or under bark, or in hollow stems. Others, especially certain tropical species, build in cavities of living plants; and still others, as *Cremastogaster*, build carton nests.

Large swarms of winged ants are often seen. These are composed of recently matured males and females that have emerged at the same time from many different nests, probably from all of the nests of the particular species involved that exist in the immediate region, and in which young queens and males have been developed. The object of these flights is mating, and they render probable the pairing of males and females from different nests, thus preventing too close interbreeding. The factors that determine the occurrence of the nuptial flights from all the nests of a species in one locality at the same time are not understood. In the case of those species in which the female is wingless the mating must take place either in the nest or on the ground outside.

After the pairing of the sexes the males soon die and each female proceeds to found a new colony if she is not captured by workers and taken into a colony already established or finds her own way into one. Except among the parasitic ants the method of founding a colony is as follows: The female breaks off her wings; then seeks out a small cavity under a stone or under bark or makes one in the ground. She closes the entrance to this cavity and remains isolated without food for weeks or months while the eggs in her ovaries are developing. During this period there is a histolysis of the large wing-muscles the products of which are used as food. When the eggs are mature they are laid and the larvæ that hatch from them are fed by the female, or queen as she is termed, with her saliva till they are ready to pupate. As the young queen takes no food during this period, that fed the larvæ must be derived from the fat stored in her body and the dissolved wing-muscles. The adults that are developed from this first brood or larvæ are workers, but owing to the limited amount of food that they have received they are abnormally small; that is, of the form known as worker minors. These open the chamber in which they were developed and go forth to collect food for themselves and for the queen, and they take charge of the second brood of larvæ, which being supplied with abundant food develop into larger workers. The nest is now enlarged by the addition of new chambers and the growth of the colony continues. A few years later numerous males and females are developed, which at the proper time leave the nest for their nuptial flight.

The method of founding colonies described above is the usual one. But in some species the females have lost the power of establishing a colony unaided and must be adopted by workers of her own species or by workers of an alien species. The adoption of a queen by workers of an alien species explains the existence of some of the mixed colonies which are sometimes observed. The practice of slave-making described later, is the explanation of others. In certain highly parasitic species the worker class is wanting and the queens must become established in the nest of an alien species.

The worker ants are so-called because upon this caste devolve all the labors of the colony after they appear on the scene in the foundation chamber. As a rule workers are sterile; but sometimes, as with bees, and wasps, fertile workers occur. It is believed that only males are developed from eggs laid by workers.

The feeding habits of ants differ greatly in different members of the family. Some of the more primitive forms are strictly carnivorous, feeding on insects and other small animals that they can destroy; while others add vegetable substances to their diet. Many feed on sweet fluids, as sap exuding from wounded stems, the nectar excreted by extrafloral nectar glands, and honey-dew produced by aphids, membracids, the larvæ of certain butterflies, and other insects, and the leaf-cutting ants cultivate fungi upon which they feed.

Ants also lick their larvæ in order to feed on the exudates excreted by them. This exchange of nourishment between the workers and their wards, which is known as trophallaxis, is discussed in the chapter on Isoptera (pp. 279-280).

The study of the habits of ants in the field is often supplemented by observations on colonies kept in artificial nests. Several types of such nests are in use; for descriptions of them see Wheeler ('10).

The family Formicidæ includes seven subfamilies, all of which are represented in the United States; but two of these subfamilies, the Cerapachyinæ and the Pseudomyrminæ, are confined to tropical and subtropical regions and their range extends only into the southern part of our territory, where they are represented by only a very small number of species. The workers of the other subfamilies can be separated by the following table, which is based on one published by Professor Wheeler in "The Hymenoptera or Wasp-like Insects of Connecticut" (Vierick '16). This work includes also tables of the genera and subgenera of ants found in America north of Mexico. For keys to the subfamilies, genera, and subgenera of the world see Wheeler ('22).

KEY TO THE SUBFAMILIES OF ANTS

(Includes only the workers)

A. Anal orifice round, terminal, surrounded by a fringe of hairs; abdominal pedicel consisting of a single segment; no constriction between the first and second segments of the gaster; pupæ rarely naked, most frequently in a cocoon. p. 963..FORMICINÆ
AA. Anal orifice ventral, in the shape of a slit; pedicel of the abdomen consisting of one or two segments.
  B. Pedicel of the abdomen consisting of a single segment; no constriction between the first and second segments of the gaster; sting vestigial; pupæ naked. p. 962..DOLICHODERINÆ
  BB. Pedicel of the abdomen consisting of one or two segments, when only of one, a distinct constriction between the first and second segments of the gaster, sting developed, sometimes very small but capable, nevertheless, of being exserted from the abdomen.
    C. Pedicel of the abdomen consisting of a single segment; gaster with a distinct constriction between its first and second segments; frontal carinæ separated or close together, when close together, dilated to form oblique or horizontal laminæ partly covering the insertions of the antennæ; pupæ always enclosed in cocoons. p. 959......CERAPACHYINÆ and PONERINÆ
    CC. Abdominal pedicel consisting of two segments; pupæ naked.
      D. Frontal carinæ very close together, almost vertical, not at all covering insertions of antennæ; eyes always very small or absent; tropical and subtropical. p. 958..DORYLINÆ
      DD. Frontal carinæ of a different conformation and covering the antennal insertions; eyes rarely vestigial or absent; cosmopolitan. p. 959, 960 ..............................PSEUDOMYRMINÆ and MYRMICINÆ

SUBFAMILY DORYLINÆ

The Legionary or Visiting Ants

The members of this subfamily are largely confined to Equatorial Africa and tropical America. The colonies are nomadic, wandering from place to place in search of prey, and forming only temporary nests. Some of the species travel in vast armies and often overrun houses in the tropics, clear out the vermin with which they may be infested, and compel the human inhabitants to leave for a time.

The subfamily is represented in our fauna by a single genus, *Ĕciton*, species of which occur from North Carolina and Colorado southward. Our species, however, do not form large armies, though they hunt in files like the tropical species, and the colonies of some of the species may consist of thousands of individuals. Some of the species are fond of kidnapping the brood of other ants. The females are wingless and much larger than the workers. The workers are polymorphic.

The subfamily CERAPACHYINÆ is represented in our fauna by two genera, *Cerăpachys* and *Acanthŏstichus*, species of which occur in Texas. These genera were formerly included in the subfamily Ponerinæ.

Subfamily PONERINÆ

The Ponerine Ants

In the ants of this subfamily the pedicel of the abdomen consists of a single segment and there is a distinct constriction between the first and second segments of the gaster (Fig. 1177). The constriction between the first and second segments of the gaster distinguishes these ants from those other ants in which the pedicel of the abdomen consists of a single segment.

Fig. 1177.—A ponerid.

Our representatives of this subfamily are rare or of local occurrence in the North, where they form small colonies, often of a few dozen individuals. They make their nests in the soil or in old logs. As a rule the queens are but little larger and the males but little smaller than the workers, and there is only a single form of worker in a species. The pupa stage is passed within a cocoon. These ants are carnivorous, feeding on other insects and do not collect honey-dew. In the South *Odontomachus* is common, forming large colonies of active ants of large size, under old logs.

The subfamily PSEUDOMYRMINÆ includes four genera only one of which, *Pseudomўrma*, is represented in North America. This is a neotropical genus, species of which are found from Florida to Texas, and in southern California. They are very slender ants and make their nests in hollow twigs or other cavities of plants. The larvæ are of a remarkable form; the body is long, the head large and ventrally placed, and the thoracic and first abdominal segments are furnished with peculiar exudatory papillæ, which form a cluster about the mouth. These ants were formerly included in the subfamily Myrmicinæ.

Subfamily MYRMICINÆ

The Myrmicine Ants

In this subfamily the pedicel of the abdomen consists of two segments (Fig. 1178) and the frontal carinæ cover the antennal insertions. This is a large subfamily; more than half of the species of ants found in America north of Mexico belong to it. The following species will serve to illustrate the remarkable differences in habits of its different members.

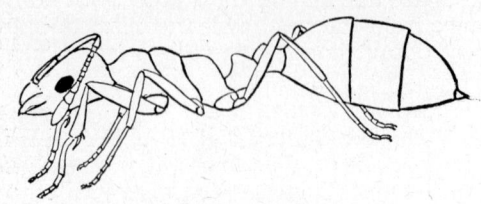

Fig. 1178.—A myrmicid.

The little yellow house-ant, Monomōrium pharaōnis.—This is the species commonly known as the "little red ant" although it is light yellow in color. It is the most troublesome of all ants that invade our dwellings. When these ants build their nests within the walls or beneath the foundations of a house it is almost impossible to dislodge them. By trapping and destroying the workers their numbers can be lessened somewhat; but so long as the queens are undisturbed in their nests the supply of workers will continue. Sometimes the nests can be reached by pouring carbon bisulphid into the crevices from which the workers come.

The thief ant, Solenŏpsis mŏlestus.—This is a species with minute yellow workers and much larger brown females and blackish males, which is common in open grassy places, where it may have independent nests under stones. But they often make burrows in the walls of nests of other and much larger ants, from which they emerge to prey upon the larvæ and pupæ of the larger ants, which are unable to follow them into their tenuous burrows.

The harvesting ants.—Several genera of myrmicine ants feed on seeds, and as they collect these seeds and store them in their nests they are known as harvesting ants. It was to these ants that Solomon referred. They have also been known as agricultural ants; for it was formerly believed that they sow around their nests seeds of the plants from which they collect the grain that they use. But this has been disproved. Most of our harvesting ants are confined to the warm and arid regions of the Southwest, where insect prey is scarce and the ants are compelled to feed on seeds. A single species, *Pheidōle pilĭfera*, which is a southern species, occurs along the coastal plain as far north as Massachusetts.

The shed-builder ant, Cremastogăster lineolāta.—In the tropics ants belonging to several genera build carton nests attached to branches of trees. One of these genera is *Cremastogăster* of which we have a common species, *C. lineolāta*, in the Northern States and Canada. This is a small ant, the workers measuring from 3 to 4.5 mm. in length.

It is usually yellowish brown, with a black abdomen; but it varies greatly in color. Its favorite nesting-place is under stones or underneath and within the decayed matter of old logs and stumps. Out of this material the ants sometimes make a paper-like pulp with which they build a nest attached to the side of a log, or even to the branches of a shrub at some distance from the ground. While such nests are uncommon these ants often build small sheds at some distance from the nest, over the herds of aphids or coccids from which they obtain honey-dew (Fig. 1179). In these cases the aphids or

Fig. 1179.—A "cow-shed" built by ants. (From A. B. Comstock, Handbook of Nature Study.)

coccids are huddled together on a branch, from which they are deriving their nourishment, and are completely covered by the "cow-shed" built by the ants.

An inquiline or guest ant, *Leptothōrax emersōni*.—This ant, the habits of which are described in detail by Wheeler, lives only associated with another species, *Myrmīca brevinōdis*. The Myrmica "builds its nest in the soil of bogs, in clumps of moss (*Polytrichum*) or under logs and stones, and the Leptothorax excavates small cavities near the surface and communicating by means of short, tenuous galleries with those of its host. The broods of both species are brought up separately. The Leptothorax, though consorting freely with the Myrmica workers in their galleries, resents any intrusion of these ants into its own chambers. The inquilines do not leave the nest to forage but obtain all their food in a very interesting manner, from their hosts. Both in the natural and artificial nests the Leptothorax are seen to mount the backs of the Myrmicas and to lick or shampoo their surfaces in a kind of feverish excitement. This shampooing has a two-fold object: to obtain the oleaginous salivary secretion with which the Myrmicas cover their bodies when they clean one another, and to induce these ants to regurgitate the liquid food stored in their crops."

The fungus-growing ants.—Among the many remarkable examples of insect behavior none is more marvellous than the habits of the fungus-growing ants, although analogous habits are exhibited by certain termites and by the ambrosia beetles, discussed in earlier chapters of this book.

The fungus-growing ants constitute one of the tribes, the Attii, of the Myrmicinæ, of which about 100 species, subspecies, and varieties have been described. They are confined almost exclusively to tropical and subtropical America; but one species is found as far north as New Jersey. Many accounts have been published regarding these insects, which have been commonly known as the leaf-cutting ants or the parasol ants. These names were suggested by the fact that these ants cut pieces from the leaves of trees and carry them, like parasols, into their nests.

The use that the ants make of the leaves that they carry into their nests was long a mystery. But it is now known that the leaves are used as a culture medium upon which they cultivate a fungus, which they eat and feed to their larvæ, and which is their only food.

Professor Wheeler ('07 b) has published a monograph of the fungus-growing ants of North America in which is given a résumé of the writings of previous students of the Attii; and a chapter is devoted to these insects in his volume on "Ants" ('10).

Subfamily DOLICHODERINÆ

In this subfamily, as in the following one, the pedicel of the abdomen consists of a single segment and there is no constriction between the first and second segments of the gaster; but these ants can be distinguished from the Formicinæ by the fact that the anal orifice is in the form of a slit. These ants, also, often possess in addition to the poison glands, anal glands which excrete a foul smelling, sticky fluid, which is used as a means of defense in their combats with other ants.

Only about a dozen species have been described from our fauna and most of these are southern. Certain tropical species build carton nests attached to trees and some of our species make carton nests under stones. The members of this subfamily are especially fond of honey-dew and attend aphids and coccids to secure it. Some of the species, "establish their nests on or near the nests of larger ants and either feed on the refuse food or waylay the workers when they return home and compel them to give up their booty" (Wheeler). The most important species of the subfamily in our fauna is the following.

The Argentine ant, Iridomyrmex humilis.—This is an introduced species, which has become an exceedingly serious pest in the Gulf States and in Southern California. Its injuries are of two kinds:

first, as a household pest, entering and overrunning dwellings; and second, as an orchard pest. Its injuries in orchards are due to the fact that it protects aphids and coccids in order to secure the honey-dew that they excrete. The ants drive away the insect enemies of the aphids and coccids, which as a result multiply to an abnormal extent. It has been found that this ant can be exterminated in houses and orchards by the use of an arsenical poisoned syrup. Detailed directions for the preparation and use of this syrup are given in bulletins published by the U. S. Department of Agriculture.

Subfamily FORMICINÆ

The Typical Ants

This subfamily is characterized by the form of the anal orifice, which is round, terminal, and surrounded by a fringe of hairs. The pedicel of the abdomen consists of a single segment and there is no constriction between the first and second segments of the gaster (Fig. 1180). The following are some of our more common species.

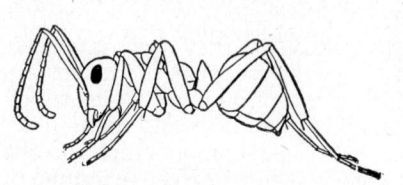

Fig. 1180.—A formicid ant.

The carpenter ant, *Campo-nōtus herculeanus pennsylvānicus*.—This is one of the largest of our common ants. It is the large black species that builds its nests in the timbers of buildings, in logs and in the trunks of trees. Frequently it builds in the dead interior of a living tree, excavating a complicated series of chambers.

The mound-building ant, *Fŏrmica exsectoides*.—This species is the builder of our largest ant-hills; these are often one meter in height and two meters across, and sometimes they are much larger than this. New colonies are often formed by fission, a portion of the colony emigrating and founding a new colony with one or more queens. In this way many colonies are often established in a limited area. The head and thorax of this ant are rust-red, while the legs and abdomen are blackish brown.

The blood-red slave-maker, *Fŏrmica sanguĭnea*.—More than a century ago Pierre Huber called attention to the fact that this species which is common in both Europe and America, keeps in its nests the workers of other species of *Formica*, which aid in performing the labors of the colony. The relations of the two species thus associated have been commonly regarded as that of slaveholders and slaves. The slaveholders obtain their slaves by making periodical forays on the colonies of the common black *Formica fusca*, and of other species of *Formica*; and bringing to their own nest the worker larvæ and pupæ. Some of these are eaten, but others are reared, and these knowing no other home take their place as active members of the colony.

In the blood-red slave-maker the gaster is black or brown and there is a notch in the margin of the clypeus. The nests of this species are low obscure mounds of earth or are excavated under stones or logs or around stumps. Many subspecies and varieties of this species are recognized, some of which do not keep slaves.

The shining amazon, *Polyĕrgus lūcidus*.—The species of the genus *Polyergus* were named amazons by Pierre Huber on account of their warring habits. Species of this genus occur in this country as well as in Europe. The shining amazon is a beautiful, brilliant red species widely distributed in the Eastern and Middle states. The species of this genus are slave-makers that have become absolutely dependent on their slaves. They cannot build their own nests or feed themselves or care for their young, but have only retained the power of fighting to get more slaves. Their mandibles are sickle-shaped and fitted only as weapons of offence. Like *Formica sanguinea*, these ants make periodical forays on the colonies of other species of *Formica* and carry home the worker larvæ and pupæ. The workers developed from these perform all of the labors of the colony except that of the making of forays on the colonies of other ants, in which they take no part. The young queens of *Polyergus*, being unable to work, establish new colonies of their species by securing adoption in some small weak colony of another species of Formica after killing its queen by piercing her head.

The corn-field ant, *Lāsius niger americānus*.—To the genus *Lasius* belong several common species of small brown ants that make small mounds in various situations. These ants are fond of honey-dew and not only care for the aphids from which they obtain it but collect the eggs of the aphids and store them in their nests through the winter, and in the spring place the recently hatched plant-lice on the stems and roots of the plants on which they feed. A well-known species of this genus is the corn-field ant, the habits of which are discussed in the account of the corn-root aphis, p. 419.

The honey-ants, *Myrmecocȳstus*.—The ants of this genus are found in the arid regions of the Southwest, from the city of Mexico to Southern California and to Denver, Colorado. They have received the name of honey-ants from the remarkable fact that with them some of the workers function as honey-pots or reservoirs for storing the honey-dew collected by other workers, from nectar excreting galls on trees and from aphids and coccids. The individuals in which the honey-dew is stored are known as *repletes*. The workers that collect the honey-dew swallow it and carry it in their crop to the nest. There they regurgitate it and feed it to a replete, which in turn swallows it and retains it in its crop. The crop of the replete becomes so greatly distended that the gaster becomes a translucent sphere, as large as a pea, on the surface of which the sclerites appear as isolated patches separated by the tense, pelucid, yellowish, intersegmental membrane (Fig. 1181). The repletes are unable to go

Fig. 1181.

about but remain quiet clinging to the roof of a chamber of the nest. When the season for obtaining honey-dew is passed, these living cells disgorge their supply through their mouths, for the use of the colony.

There are several species and subspecies of *Myrmecocystus* in some of which the replete form has not been found.

Family BETHYLIDÆ

The Bethylids

This is a large family of parasitic wasps, including many genera and species. The family is widely distributed, representatives of it being found in all parts of the world. Our species are of small or moderate size. Those whose habits are known prey upon either coleopterous or lepidopterous larvæ, and before pupating most of them spin cocoons.

In this family there are eight exposed segments in the abdomen, the petiolar segment is very short and scarcely perceptible, and the ovipositor is a true sting. The majority of genera comprise species that are winged in both sexes; but in a few genera the males alone are winged. Some of the wingless females are ant-like in appearance.

Among those species that prey upon important insect pests are *Neoscleroderma tarsalis*, which is parasitic on the beetle, *Silvanus surinamensis*, and an undescribed species of *Goniozus*, which is parasitic on the codlin-moth in Kansas. A detailed account of the life-history of *Lælius trogodermatis*, which is an external parasite of dermestid larvæ, is given by Howard ('01). This family was monographed by Kieffer ('14).

The family RHOPALOSOMIDÆ although widely distributed is a very small family. It is represented in our fauna by *Rhopalosoma poëyi*, the larva of which was found by Hood ('13) to be an external parasite of a bush-cricket, *Orocharis saltator*, in Maryland. The adult is nocturnal; it has very large eyes and ocelli, and the petiole of the abdomen is long and slender.

Family VESPIDÆ

The Typical Wasps or Diploptera

The family Vespidæ includes our most familiar wasps, the hornets, and the yellow-jackets, and their near allies. All members of this family are winged and nearly all of them when at rest fold their wings lengthwise like a fan; for this reason they are often termed the Diploptera or the diplopterous wasps. In the habit of folding their wings when at rest, the typical wasps differ from all other Hymenoptera except the Gasteruptiidæ, the chalcid genus *Leucospis* and the genus *Galesus* of the family Belytidæ. In this family the lateral extensions of the pronotum are angular extensions behind and above the tegulæ; cell M_4 of the fore wings is longer than cell $Cu+Cu_2$ (Fig. 1182); and there are closed cells in the hind wings.

The typical wasps found in America north of Mexico represent seven subfamilies of the family Vespidæ. These can be separated

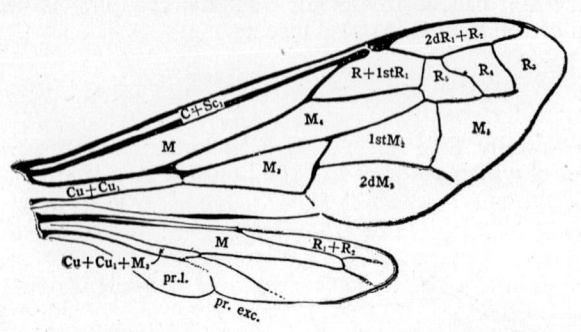

Fig. 1182.—Wings of *Vespa diabolica: pr. l,* preanal lobe; *pr. exc.,* preaxillary excision. (From Bradley.)

by the following table, for which I am indebted to Professor J. C. Bradley.

A. Fore wings with two submarginal cells; antennæ clavate. p. 967..Masarinæ
AA. Fore wings with three submarginal cells.
  B. Vein $M_4 + Cu_1$ of the fore wings elongate (Fig. 1183); cell M_3 four-sided; wings not plaited. p. 967..Euparagiinæ
  BB. Vein $M_4 + Cu_1$ of the fore wings exceedingly short (Fig. 1189); cell M_3 a scalene triangle (Fig. 1189); wings longitudinally plaited when at rest.
    C. Hind wings with an anal lobe (Fig. 1189).
      D. Tarsal claws bifid; middle tibiæ with one apical spur; solitary wasps without a worker caste.
        E. Mandibles short, obliquely truncate and toothed at the apex, folding above each other beneath the clypeus or very slightly crossing; head quadrate; abdomen petiolate, the apex of the petiole globose and strongly constricted before the second segment. p. 968. Zethinæ
        EE. Mandibles elongate, crossing each other or placed parallel in a long sharp beak; if the abdomen is petiolate the head is transverse. p. 969...Eumeninæ
      DD. Tarsal claws simple; middle tibiæ with two apical spurs; social wasps building open or closed paper nests.
        E. Extensory muscle of the abdomen fixed on the thorax in an oval slit between the apical scales of the propodeum; the slit always broadly rounded at its upper angles. p. 973..........Epiponinæ
        EE. Extensory muscle of the abdomen fixed on the thorax in a narrow and much compressed slit between the apical scales of the propodeum. p. 974..Polistinæ
  CC. Hind wings without an anal lobe, somewhat stalked (Fig. 1182); abdomen conical; social wasps with a worker caste, building closed paper nests; tarsal claws simple; middle tibiæ with two apical spurs. p. 975 ....Vespinæ

If we take into account only the habits of these insects the subfamilies of the typical wasps can be separated into two groups, the solitary Diploptera, those in which a single female makes a nest for her young, and the social Diploptera or social wasps, in which many individuals work together to make a nest. This grouping of the subfamilies, however, is not regarded as a natural division of the family Vespidæ, as each of the two groups is believed to be polyphyletic, and too, F. X. Williams ('19) has shown that in the genus *Stenogaster*, found in the Oriental and Australian regions, some species are solitary and others are social; but this grouping is useful in a discussion of the habits of these insects

THE SOLITARY DIPLOPTERA

The subfamily EUPARAGIINÆ includes the genus *Euparagia*, two species of which are found in the Southwest. These wasps differ from other Vespidæ in that cell M_3 of the fore wings is four-sided (Fig. 1183). Very little is known regarding the habits of these insects. For figures and descriptions of the species see Bradley ('22).

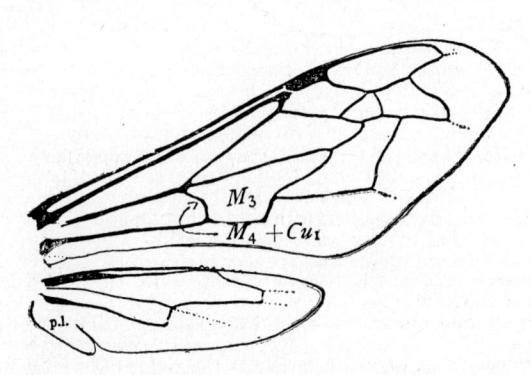

Fig. 1183.—Wings of *Euparagia scutellaris*. (After Bradley.)

The subfamily MASARINÆ is a very widely distributed group but it is represented in our fauna only by the genus *Pseudomăsaris*, of which thirteen species have been described; these are found in the Far West and Southwest. In these wasps there are only two submarginal cells in the fore wings and the antennæ are clavate. The North American species were monographed by Bradley ('22).

There are but few accounts of the nest-building habits of masarid wasps. Giraud ('71) describes the habits of *Cerămius lusitănicus*, a species found in France. This is a mining wasp, which digs a burrow in the ground, leading to a cell, in which the larva lives. The larva is fed by the mother, who brings to it from time to time a supply of a paste, described as being somewhat like dried honey. When the larva is full-grown it lines its cell with a layer of silk, otherwise the pupa is naked.

Ferton ('10) describes the habits of *Celonītes abbreviātus*, another species found in France. This is a mason wasp, which makes earthen cells attached to the sides of rocks or to stems of plants. A figure of one of these nests is given by Sharp ('99) page 89. In this figure the cells are represented as opening downward. This species also provisions its nest with a paste made of pollen and honey; and the full-grown larva lines its cell with silk.

The most remarkable feature in the habits of these two species is that they provision their nests with a paste made of pollen and honey. In this respect they differ from other solitary wasps in the same way that the solitary bees differ from other sphecoid wasps. But this difference in habits is not true of all masarid wasps as is shown by the habits of the following species, which provisions its nests with larvæ.

The only published account of the nest-building habits of an American masarid is that by Dr. Anstruther Davidson ('13) who described the nest of *Pseudomăsaris vespōīdes*. This is our largest and most handsome species; it measures from 15 to 22 mm. in length, is black marked with yellow, and is widely distributed in the Far West; but is not common. Dr. Davidson's account of the nest follows.

Their nests, a combination of cells as shown in the accompanying illustration, are built after the manner of the common mud dauber wasp and when completed are plastered over with a further layer of clay. They are usually attached to a twig in a low bush, the one in the illustration being found on a Audibertia shrub. When the cell is completed the opening is closed by a stopper of clay which is, however, always depressed below the rim of the cell so that the top shows as a series of miniature cups. The clay used is that common to the neighborhood, but in the process of building it is mixed with some secretion that makes the whole of such stony hardness, that it seems impossible any insect could possibly cut its way through it. Perhaps the cup shaped depression on top may be a device to conserve the rain necessary to soften the stopper and render the exit of the wasp possible. That rain or excessive moisture is necessary before the insect can successfully emerge is suggested by the results attained in indoor hatching. In those nests kept indoors in dry receptacles while the wasp usually attains the mature state, it only exceptionally cuts its way out. Kept under these conditions the larvæ do not always mature in the following spring as the following record makes evident. Of a cluster of cells gathered in June, 1902; in April, 1903, I opened two of them to find one had pupated while the other was still in the larval state. It remained in this state till March, 1905, when it died. The other cells were then opened, one contained a live larva, the other four or five contained perfect insects all dead, apparently unable to emerge. The capability of insects to survive for more than one season in the larval stage is probably an evolutionary acquirement, and a necessity to those insects living on a food supply that is wholly dependent on climatic conditions. As the writer has shown elsewhere in recording a similar experience with *Anthidium consimile*, this is a very necessary acquirement in a country where, as sometimes happens, no rain at all may fall, and no food supply would in those seasons be available. The cells are stored with small larvæ of what species I am unable to determine.

As in the nests of the European species, the cells are lined with a layer of silk. Figure 1184, *a* represents a nest given me by Professor Doane of Stanford University; this is an incomplete nest in which the cells have not been plastered over with an additional layer of clay; Fig. 1184, *b* is a diagram of a longitudinal section of a single cell showing the cup at the upper end; Figure 1184, *c* represents a completed nest; this was given me by Dr. Davidson.

The subfamily ZETHINÆ is represented in our fauna by a single genus *Zethus*. of which there are two species, *Zēthus spĭnipes* and *Zethus slossōnæ*, common in the southeastern part of the United States.

Ashmead ('94) states that our *Zēthus spĭnipes* builds globular cells of clay or sand and mud mixed, which are attached by a small pedicel to some shrub or tree. I find no other account of the habits of this species, and the other species of *Zethus* the habits of which have been described build nests of a very different type. Saussure ('75) states that *Zēthus romandīnus*, found in Cayenne, "constructs with woody fibres and gummy materials several rounded cells, with thick walls toward the bottom and irregularly united, recalling a little those of *Bombus*." Ducke ('14) describes and figures the nest of a Brazilian species, *Zēthus lobulātus*. This nest consists of a long mass of cells, suspended from a twig and is composed of fragments of leaves cement-

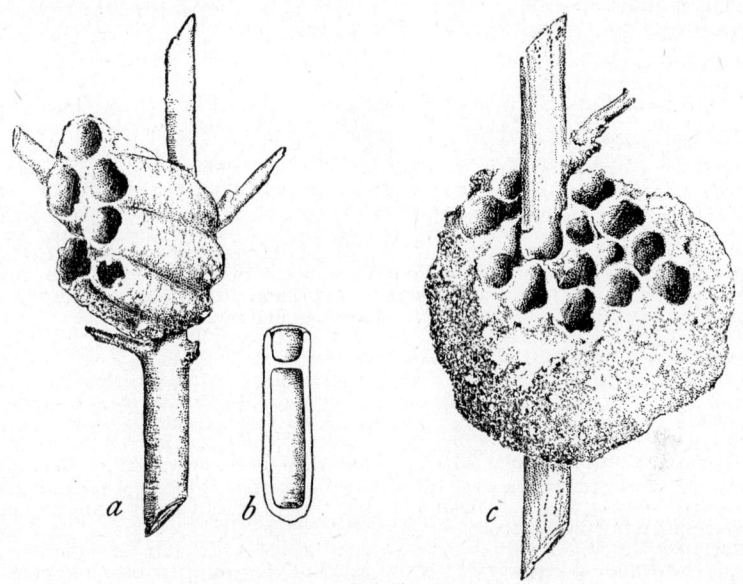

Fig. 1184.—Nests of *Pesudomasaris vespoides*: *a*, incomplete nest; *b*, diagram of a cell showing the cup at the upper end; *c*, completed nest.

ed together by a resinous substance. Williams ('19) describes the nest of *Zēthus cyanŏpterus*, which he observed in the Philippines; this nest is made of bits of leaves which "are chewed along one or more of their edges which makes them adhere the more firmly to the nest." This author says nothing about the use of a gummy material for cementing together the fragments of leaves.

Subfamily EUMENINÆ

The Eumenids

This subfamily includes by far the greater number of our species of the solitary Diploptera; and is represented in our fauna by about eight genera. The distinguishing features of these wasps are indicated in the table of subfamilies given above.

The different species of eumenids differ greatly in habits; many are miners digging burrows in the earth leading to cells in which provisions are placed for their young; some make burrows in wood, which they divide into cells by partitions of mud; some build their nests in the stems of pithy plants or make use of any suitable cavity that they find; and others are mason or potter-wasps, making cells of earth, which are built in holes, or on the surface of the ground, or attached to twigs.

Although the adult eumenids do not confine themselves to a carnivorous diet but often visit flowers to obtain nectar, they all provision their nests with insects, which they have paralyzed with their sting; usually only a single species of caterpillar is used for this purpose by each wasp.

A remarkable feature that has been observed in the nesting habit of many eumenids and perhaps is true of all, is that after the cell is prepared the egg is suspended by a slender thread from the ceiling or side of the cell. In some cases, at least, this is done before the provisioning of the nest is begun. An African species, *Odynērus tropicālis*, the habits of which are described by Roubaud ('16) does not provision its cell with prey amassed in advance, but feeds its larva from day to day with small, entire, paralyzed caterpillars, and does not close the cell until the larva has completed its growth.

The following examples will serve as illustrations of the habits of members of the subfamily. Among the more detailed accounts of the activities of some of our species are those of Peckham and Peckham ('05), Hartman ('05), Isley ('13) and Rau and Rau ('18). For a general account of the habits of these insects see Roubaud ('16).

Odynerus.—The greater number of our species of eumenids belong to the genus *Odynērus*. In this genus the abdomen is sessile. The shape of the body and frequently the coloration resemble those of the social wasps known as yellow-jackets, although usually the body is more slender and smaller. The common species are quite neighborly; and, owing to this resemblance to the yellow-jackets they inspire us with a fear that is out of all proportion to their will or ability to inflict pain.

Many species of *Odynerus* are miners. Their burrows are to be found both in level ground and in the sides of cliffs. Branching from these burrows are short passages, each leading to a cell, from the ceiling of which an egg is suspended by a slender thread; and in which food is stored for the larva. In the species that have been studied, this food consists of small, paralyzed caterpillars. Some of the mining species while digging the burrow build a turret over the entrance of it, made of pellets of mud removed from the burrow; one of these turrets is figured by the Raus (Fig. 1185). The material of which the turret is composed is used to fill up the burrow after the cells are finished.

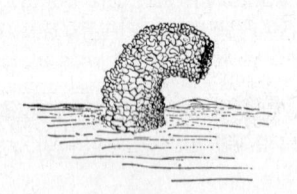

Fig. 1185.—Turret over the burrow of *Odynerus geminus*. (After Rau and Rau.)

In digging the burrow and in tearing down the turret the earth is softened with water, which the wasp brings in her mouth from some pool or stream.

Not all species of *Odynerus* mine in the ground; many burrow in the stems of pithy plants, making a series of cells separated by partitions of mud; other species will avail themselves of any convenient cavity in which to make their nest, frequently utilizing the deserted nests of the sphecoid-wasps known as mud-daubers. In this case a

Fig. 1186.—*Eumenes fraternus* and its nests.

single cell of a mud-dauber is divided by a transverse partition, making two cells for the smaller *Odynerus*. One year these wasps plastered up many of the keyholes in our house, including those in bureaus.

Some species of *Odynerus* are masons constructing nests entirely of mud. One of our species, *Odynērus birenimaculātus*, makes a nest about the size of a hen's egg. This is composed of hard clay, fastened to a twig of a bush, and contains many cells.

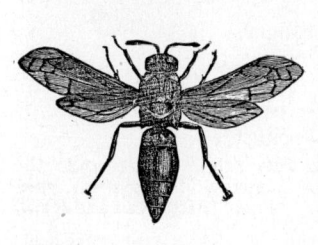

Fig. 1187.—*Monobia quadridens*.

The jug-builders, *Eumenes*.—The wasps of the typical genus of this subfamily are potter-wasps which build nests that appear like miniature water-jugs. The nests of our common species, *Eŭmenes fratĕrnus*, are often found attached to twigs (Fig. 1186). In this genus the abdomen of the adult is petiolate. These wasps provision their nests with caterpillars and frequently with cankerworms.

Fabre, who studied the habits of a European species of *Eumenes* observed what goes on within the nest by making a window in the side of it. The egg is suspended from the ceiling of the nest by a slender thread;

when the larva hatches it at first makes use of the egg-shell as its habitation and stretches down to feed on the caterpillar below it; if disturbed it retreats up its support. Later when the larva has increased in size and strength it descends to the mass of food.

Monōbia quădridens.—This species (Fig. 1187) is common in most of the states east of the Mississippi. It is larger than the jug-builders, and the abdomen of the adult is sessile. Figure 1188 represents a nest of this species, now in the Cornell University collection, which was made in a board in the side of a barn. The partitions are made of mud. Each cell contained a pupa when the nest was opened, hence it was not evident what the food of the larvæ had been; but several observers state that this species stores its nests with large cutworms; and it is doubted that this species is a carpenter-wasp. It seems probable that the nest figured here was made in a deserted burrow of the large carpenter-bee, *Xylocopa virginica*. It differs from a nest of this bee only in that the partitions are made of mud.

THE SOCIAL WASPS

Since the social Diploptera are the only wasps that are social they are commonly referred to as the social wasps instead of the more technical name.

As with the ants the colonies of social wasps consist of three castes, the female or queen, the workers, and during the later part of the season, the males. The workers are females in which the reproductive organs are imperfectly developed. In the genus *Belonogaster* a worker caste is believed to be lacking. In *Polistes* it is very difficult to make a distinction between females and workers, for they can apparently all become fertile.

In the temperate regions the colonies exist for only one season; the males and the workers die in the autumn; the females hibernate and each starts a new colony in the spring. At first the female performs the functions of both worker and queen, starting the building of the nest and laying the eggs. In the early part of the season only workers are developed; after they appear they carry on the labors of the colony, expanding the nest and procuring the food for the larvæ; the only function of the

Fig. 1188.—Nest of *Monobia quadridens*.

queen then is to produce the eggs. In the later part of the season males and females are developed.

The social wasps are predacious, and they feed their larvæ upon insects which they have malaxated. The wasps are also fond of the sweets of flowers, the juices of fruits, and of honey-dew. They also feed upon a liquid which the wasp larva emits from its mouth. This exchange of nourishment between the larvæ and the adults, termed trophallaxis, is discussed in the Chapter on Isoptera, page 279.

In the temperate regions the multiplication of colonies is brought about by the production of many males and females in the nest in the later part of the season; these pair, the females hibernate, and each female founds a new colony in the spring. But in the Tropics many of the Epiponinæ form large perennial colonies, which from time to time give off swarms, in a way quite similar to the well-known swarming of the honey-bee.

Representatives of three of the subfamilies of social wasps are found in America north of Mexico. The distinguishing characters of these are indicated in the table of subfamilies of the Vespidæ given above.

Subfamily EPIPONINÆ

This is a large group of wasps including a great variety of forms, which exhibit great differences in the architecture of their nests.

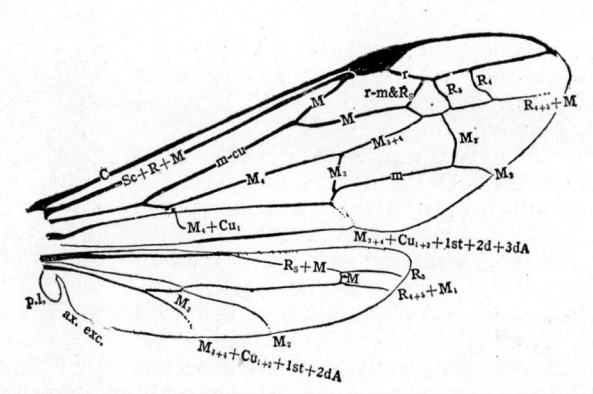

Fig. 1189.—Wings of *Mischocyttarus labiatus*: *pl*, posterior or anal lobe; *ax. exc.*, axillary excision. (From Bradley.)

The species are mostly confined to tropical America; but three are found in the southern portions of the United States. Figure 1189 represents the venation of the wings of *Mischocyttarus labiatus*.

Brachygăstra lecheguāna.—This species is found within the limits of our territory only along the Mexican border. Its nest resembles in form externally those of hornets (*Vespa*) but the combs are attached to the envelope. This wasp is especially interesting from the fact that it frequently stores honey in the combs of its nest; but the honey is probably not an exclusive or essential constituent of the larval food.

Mischocȳttarus.—This tropical genus is represented in our fauna by two species, one, *Mischocyttarus cubĕnsis,* found in Florida and in Southern Georgia, one, *Mischocyttarus flavitărsis,* found in the Southwest, Colorado, and California.

These wasps make small, few-celled paper nests, without an envelope like those of *Polistes;* but the wasps are easily distinguished from *Polistes* by the form of the first segment of the abdomen which is slender and elongate, forming a pedicel. The nests of *M. cubensis* are found on palmetto leaves.

Subfamily POLISTINÆ

Polistes

The wasps of the genus *Polistes* and their nests are very familiar objects. The nests consist each of a single comb suspended by a peduncle, and the comb is not enclosed in an envelope (Fig. 1190). These nests are often built under the eaves of buildings, in garrets, and in sheds and barns; they are also often made under flat stones in fields, and sometimes attached to bushes. The combs of our species of *Polistes* are horizontal; but the nests of *Polistes linĕatus,* which I

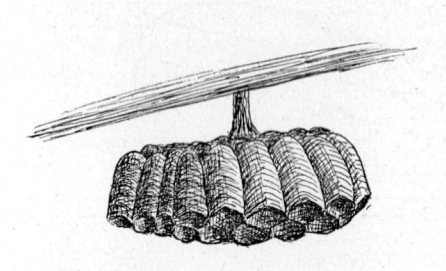

Fig. 1190.—Nest of *Polistes.*

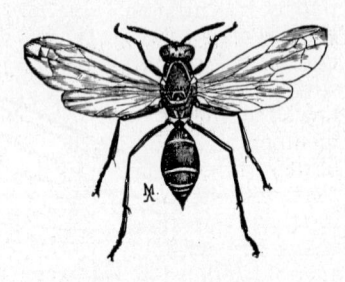

Fig. 1191.—*Polistes.*

found hanging from the ceiling of a cave in Cuba, are long, narrow, vertical combs, from one to two inches in width and from twelve to eighteen inches in length.

The nests are made of a grayish paper-like material, composed of fibers of weather-worn wood, which the wasps collect from the sides of unpainted buildings, fences, and other places, and convert into a paste by the action of the jaws and the addition of some fluid, probably an oral secretion. The nests of *Polistes* are usually comparatively small; but some have been found in Texas that measured more than a foot in diameter.

In this genus the abdomen is long and spindle-shaped (Fig. 1191).

HYMENOPTERA

Several of the species are known to store small quantities of honey in their combs.

These wasps are often infested by stylopids.

Subfamily VESPINÆ

The Hornets and the Yellow-Jackets

This subfamily includes those wasps that are commonly known as hornets and the yellow-jackets. With these insects the body is

Fig. 1192.—*Vespa*.

comparatively short and rather stout (Fig. 1192); the abdomen is attached to the thorax by a very short pedicel; and the color is black, spotted and banded with yellow or yellowish white. The members of this subfamily differ from other vespoid wasps in that the hind wings are without an anal lobe (Fig. 1182).

These wasps make their nests of paper, which in some cases is composed of fibers of weather-worn wood, like that of *Polistes* described above, in other cases of fragments of more or less decayed wood. These nests consist of a series of horizontal combs suspended one below another and all enclosed in a paper envelope (Fig. 1193).

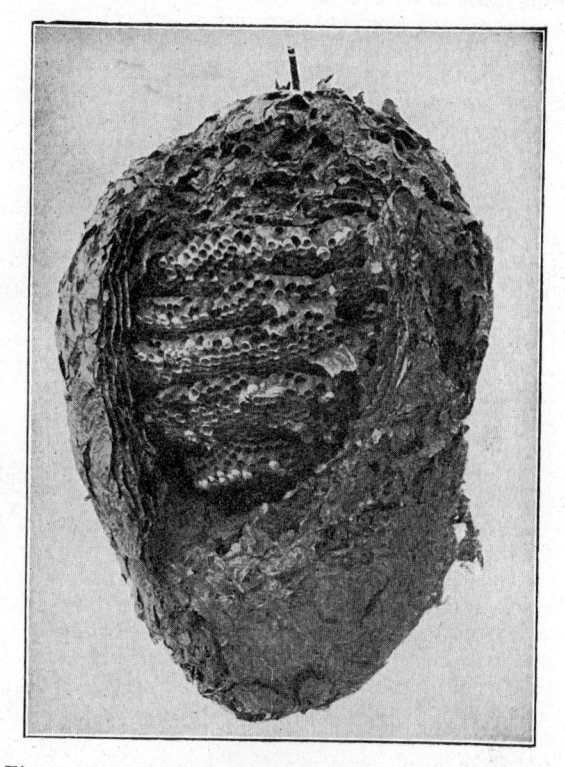

Fig. 1193.—Nest of *Vespa*, with side removed. (From A. B. Comstock, Handbook of Nature Study.)

When the wasps wish to enlarge their nest they remove the inner layers of the envelope, and add to the sides of the combs, build additional combs below, and put on new layers on the outside of the envelope. By these additions the nest may become of large size by the end of the season.

Very small empty nests consisting of a single comb with but few cells and enclosed in an envelope of only one or two layers of paper are often found (Fig. 1194). Such a nest is evidence of a tragedy. A queen wasp, in the spring, had started to found a colony. It was necessary for her to go back and forth in the fields collecting material for her nest and food for her larvæ; and before a brood of workers were developed to relieve her of this dangerous occupation she became the prey of some bird and the development of the colony was wrecked.

Two quite different types of nests are made by different species of these wasps, and these are made in quite different situations. One kind is built above ground; these are attached to bushes or trees, or beneath the eaves of buildings; they are made of a grayish paper composed of fibers of weather-worn but not decayed wood. This paper is comparatively strong, so that the envelope of the nest is composed of sheets of paper of considerable size, a single sheet often completely enveloping the nest.

Fig. 1194.—Early stage of nest of *Vespa*.

The other kind of nest is built in a hole in the ground, which is enlarged by the wasps as they need more room for the expansion of the nest. The paper of which these nests are made is brownish in color and is made out of partially decayed wood; it is very fragile and would not be suitable, therefore, for use in nests built in exposed places. Even though the nest is built in a protected place, the use of this fragile material necessitates a different style of architecture. The enveloping layers of the nest, instead of being composed of sheets of considerable size, are made up of small, overlapping, shell-like portions, each firmly joined by its edges to the underlying parts.

If a completed hornet's or yellow-jacket's nest be examined it will be found that some of the later-built combs consist wholly or in part of cells that are larger than those in the first-made combs; the smaller cells are those in which workers were developed; the larger ones those in which the sexual forms were reared.

It has been found that at least two species of this subfamily are social parasites. In these species the worker caste has been lost, there being only males and females. The female enters the nest of another species of *Vespa* and lays her eggs, and her larvæ are reared by the rightful owners of the nest. The species that are known to be parasites are *Věspa ărctica*, which infests the nests of *Věspa diabŏlica*, and **Vespa austriaca**, which infests the nests of *Věspa rūfa* in Europe. **Vespa austriaca** has been found in this country but *Vespa rufa* is not known to occur here. The American host of *Vespa austriaca* has not been definitely ascertained, but is thought to be *Věspa consōbrina*. See Wheeler and Taylor ('21).

The members of the subfamily Vespinæ found within the limits of our territory are commonly included in a single genus, *Vespa;* but some writers place all of our species except *Vespa crabro* in a separate genus *Vespula;* Only a few of our species can be mentioned here.

The giant hornet, *Věspa crābro.*—This is our largest species, measuring from 18 to 22 mm. in length. It is brown and yellow in color and is found around New York City, on Long Island, and in Connecticut. It builds its nests in hollow trees and within buildings suspended from the roof.

The white-faced hornet, *Věspa maculāta.*—This is the common, large black and white hornet. It is widely distributed in the United States and Canada. The nest, which is sometimes very large is usually attached to the limb of a tree.

The yellow-jackets.—This common name is applied to several small, black and yellow species of *Vespa*, which are so closely related that it is difficult to distinguish them. Most of the species build their nests in the ground; these are the brownish paper nests described above. Sometimes the nest is built in a stump or under some object lying on the ground. On one occasion I found a fine large nest under the base-board of one of my bee-hives, and into which I inadvertantly thrust my toes, with sad results, while examining the hive. The nest is now in the Cornell collection.

Superfamily SPHECOIDEA

The Sphecoid-wasps and the Bees

The superfamily Sphecoidea resembles the two preceding superfamilies in the presence of an anal lobe in the hind wings, except in

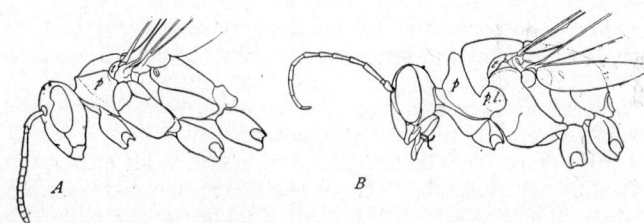

Fig. 1195.—A, Head and thorax of a vespoid-wasp, *p*, pronotum; *t*, tegula. B, Head and thorax of a sphecoid-wasp, *p*, pronotum, *p. l.*, posterior lobe of the pronotum; *t*, tegula.

some specialized genera of the Ampulicidæ and some bees; it differs from the Evanioidea in that the petiole of the abdomen is attached to the hind end of the propodeum; and it differs from the Vespoidea in that there is on each side a lateral extension of the pronotum in the form of a distinctly differentiated rounded lobe, which covers the spiracle (Fig. 1195, B); these lobes are known as the posterior lobes of the pronotum; they do not reach the tegulæ except in some Dryinidæ and Ampulicidæ. These exceptional forms can be placed in their families by the table on pages 906 to 914.

The families constituting the superfamily Sphecoidea, which can be separated by the table referred to above, represent two quite distinct groups of families, known respectively as the sphecoid-wasps and the bees. These two groups of families are distinguished as follows.

A. First segment of the posterior tarsi cylindrical and naked; or with but little hair; hairs clothing the thorax simple; nests provisioned with animal food. p. 978..The Sphecoid-Wasps
AA. First segment of the posterior tarsi elongate and dilated; some of the hairs, especially of the thorax, plumose; nests provisioned with honey and pollen. p.989 ....The Bees

THE SPHECOID-WASPS

The group known as the Sphecoid-wasps includes three families, the Ampulicidæ, the Dryinidæ, and the Sphecidæ. These three families and certain families of the Vespoidea were formerly classed together as the Fossores or digger-wasps; which names were suggested by the fact that most of the species belonging to these families make nests for their young by digging burrows in the ground or in wood. But this group is no longer regarded as a natural one notwithstanding the striking similarity in habits exhibited by its members.

The family AMPULICIDÆ is represented in our fauna by only two genera, *Rhinŏpsis* and *Dolichūrus*, the species of which are very rare. So far as is known the members of this family prey on cockroaches, with which they store their nests.

Family DRYINIDÆ

The Dryinids

This family is composed of small parasitic wasps; it is widely distributed over the world and is represented in our fauna by many genera.

The fore wings have a lanceolate or ovate stigma; the hind wings are without closed cells; the antennæ consist of ten segments, the anterior tarsi of the female are usually chelate; and either the pronotum has a longitudinal sulcus or the antennæ are borne close to the clypeus. The females of the genus *Gonătopus* are wingless, ant-like, and are without a scutellum.

These parasites confine their attacks to the homopterous insects belonging principally to the families Fulgoridæ, Membracidæ, and Cicadellidæ.

The female dryinid seizes her victim with her raptorial fore legs; one pair of pincers usually grips the neck of the prey, the other pair grips the abdomen towards the apex or the hind legs. The wasp then inserts her egg into the body of the bug. A few days later the immature larva of the parasite appears outside the body of its host enclosed in a sac composed of molted skins. Here it remains, with its head in the opening in the body-wall of its host, until it has completed its growth. It then leaves its host and spins a silken cocoon, which in some cases is furnished with an outer covering formed of the larval sac or of round patches of epidermis stripped off from the leaf surface.

A detailed account of the habits of these remarkable insects is given by Perkins ('05) and the family Dryinidæ was monographed by Kieffer ('07), and Kieffer ('14) in a paper on the Bethylidæ.

Family SPHECIDÆ

The Typical Sphecoid Wasps

In this family the hind wings have an anal lobe and some closed cells; the abdomen of the male has seven exposed tergites; the sting of the female is not enclosed by the hypopygium; the posterior metatarsi are not dilated as in the bees and there are no plumose hairs. All members of the Sphecidæ are winged.

To this family belong all of our common nest-building sphecoid wasps. These differ from the bees in that they provision their nests with animal food, insects or spiders, which they have paralyzed by stinging them. Different members of the family differ greatly in their nesting habits; some are mason-wasps, building cells of earth; many burrow in the ground; and others burrow in the stalks of pithy plants or make use of cavities that they find.

Most members of the Sphecidæ, after preparing their nest, rapidly accumulate an amount of prey sufficient to enable the young to develop to maturity, lay an egg with it, and then close the cell before the egg has hatched. This method is termed *mass provisioning*. But certain members of the family, *Bembex* and some others, feed their young from day to day as long as they remain in the larval state. This method is termed *progressive provisioning*. As each larva requires constant attention for a considerable time only a few young can be reared by a single female in this way.

Many of these wasps after stinging their prey and before placing it in their nest malaxate (*i. e.* chew) its neck or some other part of the body and lap up the exuding juices.

The family Sphecidæ is divided into six subfamilies, some of which include two or more quite distinct groups of genera or tribes. As many of these tribes are given subfamily rank by some writers each of those represented in our fauna is defined below. These tribes can be separated by the following key, which has been kindly prepared for me by Professor J. Chester Bradley.

A KEY TO THE TRIBES OF SPHECIDÆ OCCURRING IN THE UNITED STATES

A. Postscutellum with squamae which project backward, and base of propodeum with a median spine. p. 989..................................Oxybelini
AA. Postscutellum and propodeum simple.
  B. Only 1 submarginal cell. Inner margins of eyes emarginate or strongly converging toward the clypeus (except in *Anacrabro*, which may be recognized by its flat venter).
    C. Inner margins of eyes entire. p. 989......................Crabronini
    CC. Inner margins of eyes deeply emarginate. p. 982.....Trypoxylonini
  BB. Usually 3 submarginal cells, sometimes two, rarely only one, in which case the inner margins of the eyes are neither emarginate nor strongly convergent toward the clypeus.
    C. Anal lobe large, reaching to opposite or beyond the apex of the cell $M_3 + Cu_1 + Cu$.
      D. The marginal cell not appendiculate. Abdomen with a cylindrical petiole. Middle tibiæ with two apical spurs. p. 983.........Sphecini
      DD. The marginal cell appendiculate. Abdomen not petiolate.
        E. Ocelli normal. Middle tibiæ with two apical spurs. p. 981...Astatini
        EE. Ocelli distorted. Middle tibiæ with one apical spur. p. 981...Larrini
    CC. Anal lobe small, not reaching to opposite the apex of the cell $M_3 + Cu_1 + Cu$.
      D. Ocelli normal, circular and convex.
        E. Antennæ inserted low on the face, near the base of the eyes.
          F. No epicnemium.
            G. First abdominal segment sessile or with a cylindrical petiole which is composed only of the sternite. Middle tibiæ with a single apical spur.
              H. An appendiculate cell present, or the mandibles with an external notch, usually both. p. 982..............Dinetini
              HH. No appendiculate cell. Mandibles without an external notch. p. 985...........................Pemphredonini
            GG. First abdominal segment, petioliform, the petiole consisting however of both sternite and tergite, nodose at apex. Middle tibiæ with two apical spurs. p. 986.....*Mellinus* in Nyssonini
          FF. An epicnemium present. Middle tibiæ with two apical spurs. p. 986..Nyssonini
        EE. Antennæ inserted near the middle of the face above the bases of the eyes.
          F. Abdomen without a cylindrical petiole which is composed only of the sternite.
            G. Transverse part of vein M distant from the stigma by less, usually much less, than twice the distance between the apex of the cell $2d\ R_1 + R_2$ and the apex of the wing. Labrum rarely exserted.
              H. Hind femora without a transversely expanded reniform apical plate.
                I. Upper margin of clypeus extending across in a straight line or arch, without a median lobe. Cell R_5 usually small and triangular, or if not, the middle tibiæ have two apical spurs. p. 986..Nyssonini
                II. Upper margin of clypeus with a median lobe extending upward toward the antennæ. Cell R_5 four-sided. Middle tibiæ with one apical spur. p. 986..............Philanthini
              HH. Hind femora broadened at apex, there forming a transverse reniform plate. p. 986.........................Cercerini

GG. Trànsverse part of vein M distant from the stigma by two or more times the distance between the apex of cell 2d $R_1 + R_2$ and the apex of the wing. Labrum exserted. p. 987..STIZINI
FF. Abdomen with a cylindrical petiole which is composed only of the sternite. p. 985.................................PSENINI
DD. Ocelli distorted. p. 988............................BEMBICINI

Subfamily LARRINÆ

Tribe ASTATINI

The members of this tribe are rather small, seldom more than 12 mm. in length. They are usually black or black and red. As in the following tribe the anal lobe of the hind wings is large, the marginal cell of the fore wings is appendiculate, and the abdomen is not petiolate; but these wasps differ from the Larrini in that the ocelli are normal and the middle tibiæ are armed with two apical spurs.

The habits of *Astata unicolor* and of *Astata bicolor* are described by the Peckhams ('98); these species burrow in the ground and provision their nests with bugs.

Tribe LARRINI

Most members of the Larrini are of moderate size, but our species range from 3 mm. to about 23 mm. in length. They are usually rather stoutly built insects (Fig. 1196). The anal lobe of the hind wings is long and scarcely separated from the preanal lobe (Fig. 1197); the marginal cell of the fore wings is appendiculate; the ocelli are distorted; and the middle tibiæ are armed with one apical spur.

Fig. 1196.—*Tachysphex terminatus.*

Nearly all of the species burrow in sandy places and provision their nests with orthopterous insects or with bugs; but

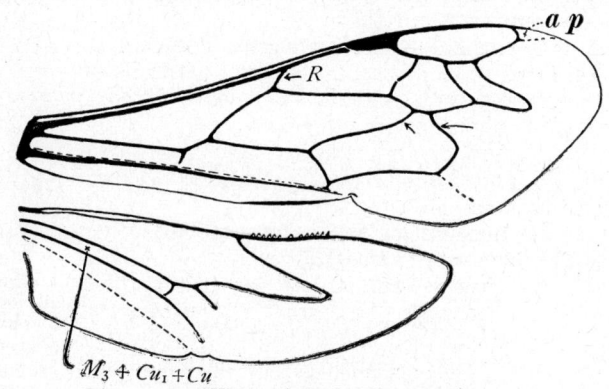

Fig. 1197.—Wings of *Tachysphex terminatus.*

Williams ('13) states that a few of the smaller species make their nests in brambles.

Some members of the Larrini (*Tăchysphex*) dig short burrows in the ground at the bottom of which the prey is placed with an egg, and then the burrow is closed with loose sand, there being no well-formed cell; several of these burrows are often made and stored in a single day. Other members of the Larrini make deeper burrows which contain from a few to many cells.

For a monograph of this tribe see Fox ('93).

Tribe DINETINI

In this tribe the anal lobe of the hind wings is small, not reaching to opposite the apex of cell M_3+Cu_1+Cu; the ocelli are normal, circular and convex.

The tribe is composed chiefly of small and little known insects except the genus *Lyroda*. A common species of this genus, *Lyrōda sŭbita*, practises progressive provisioning, feeding its young from day to day with crickets of the genus *Nemobius;* its nest is made in the ground. The small wasps of the genus **Miscophus** prey on spiders.

For a monograph of this tribe see Fox ('93).

Subfamily TRYPOXYLONINÆ

This subfamily includes a single tribe, the Trypoxylonini. In these wasps the inner margin of the eyes is deeply incised, the ocelli are normal, the marginal cell of the fore wings is not appendiculate, and there is only one submarginal cell. The body is black, slender, and of medium size.

I have found, in New York, the nests of *Trypŏxylon frĭgidum* very common in branches of sumac (Fig. 1198), more common than those of any other insect except the little carpenter-bee, *Ceratina*. The cells of the nests of *Trypoxylon* are separated by partitions of mud and are stored with spiders. The larva of *Trypoxylon frigidum* when full-grown makes a very slender cocoon, with the upper end rounded and sometimes slightly swollen, and the lower end blunt and of denser texture than the remainder of the cocoon.

An extended account of the habits of two other species of *Trypoxylon*, *T. albopilōsum* and *T. rubrocĭnctum*, is given by the Peckhams ('89), who studied these species in Wisconsin. Of special interest are the observations made by these writers on the cooperation of the males and females during the nest-building period. They state as follows.

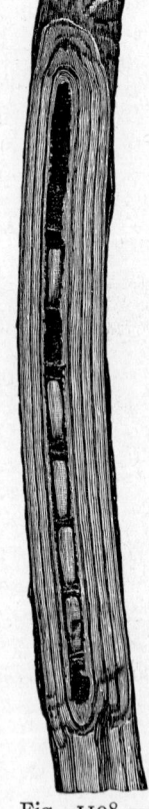

Fig. 1198.—Nest of *Trypoxylon frigidum*.

"With both species when the preliminary work of clearing the nest and erecting the inner partition has been performed by the female, the male takes up his station inside the cell, facing outward, his little head just filling the opening. Here he stands on guard for the greater part of the time until the nest is provisioned and sealed up, occasionally varying the monotony of his task by a short flight." "We have frequently seen him drive away the brilliant green *Chrysis* fly which is always waiting about for a chance to enter an unguarded nest." "In one instance, with *rubrocinctum* where the work of storing the nest had been delayed by rainy weather, we saw the male assisting by taking the spiders from the female as she brought them and packing them into the nest, leaving her free to hunt for more."

Some species of *Trypoxylon* are mud-daubers. *Trypoxylon albitarsis*, a shiny black species with white tarsi, builds large nests of mud, which consist of several parallel tubes, often three inches or more in length, placed side by side. These nests are known as pipe-organ nests. Each tube is divided by transverse partitions into several cells, which are provisioned with spiders. The tubes when completed are not covered with an extra layer of mud as is commonly the case in the nests of other mud-daubers. When an adult is ready to emerge from the cell in which it was developed, it mkes a hole through the exposed side of the tube.

For a monograph of *Trypoxylon* see Fox ('93).

Subfamily SPHECINÆ

The Thread-waisted Wasps

These insects are termed the thread-waisted wasps on account of the great length of the petiole of the abdomen (Fig. 1199). With these wasps the marginal cell of the fore wings (2d R_1+R_2) is not appendiculate; the anal lobe of the hind wings is large, extending to the apex of cell $M_3+Cu_1+Cu_2$ or beyond (Fig. 1200); and the middle tibiæ bear two apical spurs.

Fig. 1199.—*Sceliphron cementarium*.

These are the most commonly observed of all of our sphecoid wasps, as certain species build their mud nests in the attics of our houses; and, too, the

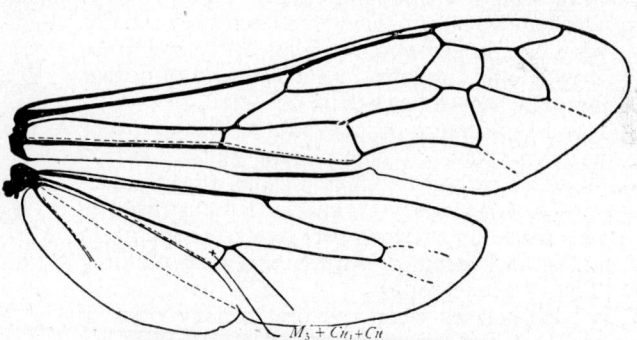

Fig. 1200.—Wings of *Sceliphron cementarium*.

peculiar shape of the body makes them very conspicuous. Most of the species burrow in the ground and store their nests either with caterpillars or with Orthoptera. But those best known to us are the mud-daubers.

The mud-daubers make nests of mud attached to the lower surface of flat stones or to the ceilings or walls of buildings. These nests usually consist of several tubes about twenty-five millimeters in length placed side by side (Fig. 1201) and are provisioned with spiders. The mud-daubers may be seen in damp places collecting mud for their nests, or exploring buildings in search of a place to build. They have a curious habit of jerking their wings frequently in a nervous manner. There are in this country two widely distributed and common species of mud-daubers; these are the blue mud-dauber, *Chalȳbion cærulium*, which is steel blue with blue wings, and the yellow mud-dauber, *Scĕliphron cementārium*, which is black or brown with yellow spots and legs. The latter of these species has been commonly described under the generic name *Pelopæus*.

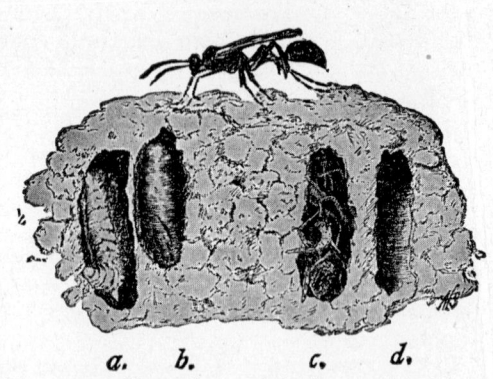

Fig. 1201.—Nest of a mud-dauber removed from a wall exposing the cells: *a*, larva full-grown; *b*, cocoon, *c*, young larva feeding on its spider-meat; *d*, an empty cell. (From A. B. Comstock, Handbook of Nature Study.)

The tool-using wasps, *Ammŏphila*.—Among the members of the Sphecinæ, that burrow in the ground and store their nest with caterpillars are certain species of the genus *Ammŏphila*. These are of especial interest on account of the habit, first observed by the Peckhams, of pounding down the earth with which they close their burrow by taking a stone or some other object in their mandibles and using it as a hammer.

The genus *Chlōrion*, formerly known as *Sphex*, includes species which are among the most common of flower visitors in the warmer parts of our country, and are among the largest and most handsome, and therefore most often observed of our wasps. In the West the common, very large, all metallic green *Chlōrion cyāneum* is a very striking insect; and in the East *Chlorion ichneumōneum*, which is brownish red with the end of the abdomen black is the most noticeable.

Subfamily PSENINÆ

This subfamily includes two tribes, the Psenini and the Pemphredonini; each of these has been regarded as a separate subfamily or family.

Tribe PSENINI

The Psenini are small sphecoid-wasps in which the base of the abdomen is slender, forming a petiole much like that of the Sphecinæ, but differing in being flattened and usually furrowed above (Fig. 1202), and these wasps are much smaller than the true thread-waisted wasps. The antennæ are inserted at the level of the middle of the eyes; and there are three complete submarginal cells in the fore wings. These wasps make their burrows either in sand or in the pith of brambles, and provision them with aphids or other small Homoptera.

Fig. 1202.— A psenid.

Tribe PEMPHREDONINI

In this tribe the antennæ are inserted low on the face, near the level of the base of the eyes; and there are at most two submarginal cells present, sometimes a trace of vein $r+m$ and R_s

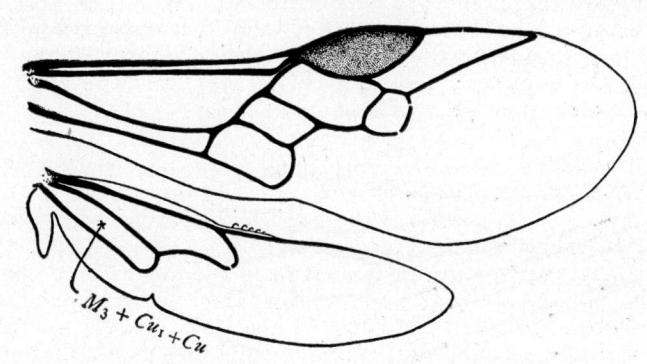

Fig. 1203.—Wings of *Stigmus podagricus*. (From Bradley).

incompletely indicates a third (Fig. 1203). The abdomen is sometimes sessile.

The pemphredonids usually burrow in the pith of dry branches and provision their burrows with plant-lice. A very common species in the East is *Stigmus fratĕrnus*. This insect measures 5 mm. in length, and makes tortuous burrows in the pith of sumac (Fig. 1204). Other common members of the tribe are larger. Some species of *Xylocēlia* (*Diodŏntus*) have been found to burrow in the ground.

This tribe was monographed by Fox ('92).

Fig. 1204.— Nest of *Stigmus fraternus*.

Subfamily BEMBICINÆ

Tribe PHILANTHINI

In this tribe the upper margin of the clypeus is suddenly expanded into a broad median lobe, usually rhomboidal, extending upward toward the antennæ (Fig. 1205).

These wasps burrow in the ground; some species provision their nests with ants others with bees.

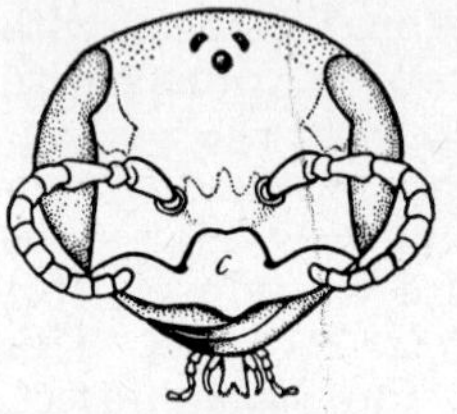

Tribe CERCERINI

Fig. 1205.—Face of a philanthid; *c*, clypeus.

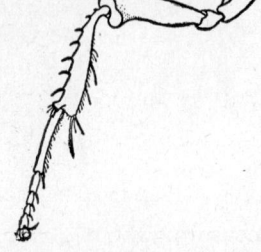

Fig. 1206.—Hind leg of *Cerceris clypeata*.

In the Cercerini the hind femora are broadened at the apex, there forming a transverse reniform plate (Fig. 1206); in both male and female there is a distinct pygidial area; the transverse part of vein M is distant from the stigma by less than twice the distance between the apex of the cell $2dR_1+R_3$ and the apex of the wing (Fig. 1207); and the anal lobe of the hind wings is large.

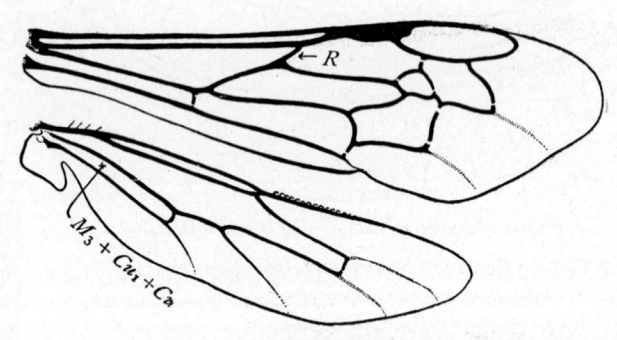

Fig. 1207.—Wings of *Cerceris*.

The species of *Cerceris* usually burrow in the ground and provision their nests with beetles; but some provision their nests with bees (*Halictus*)

Tribe NYSSONINI

In the Nyssonini the ocelli are never distorted and the upper margin of the clypeus extends directly across in a straight line or an arch without a median lobe. In most genera the middle tibiæ have a single apical spur and the second submarginal cell (R_5) is small and triangular; but in some the middle tibiæ have two apical spurs and cell R_5 is four-sided.

These wasps burrow in the ground; some species store their nests with Homoptera, others with Orthoptera.

Our best-known representative of this tribe is the Cicada-Killer, *Sphēcius speciōsus*. This is a formidable insect, measuring about 30 mm. in length (Fig. 1208). It is black, sometimes of a rusty color, and has the abdomen banded with yellow. It digs deep burrows in the earth and provisions each with a Cicada. Figure 1209 represents its wings.

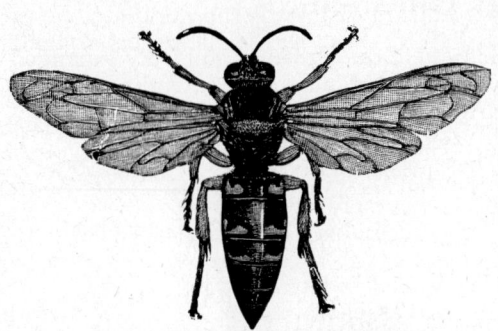

Fig. 1208.—*Sphecius speciosus*.

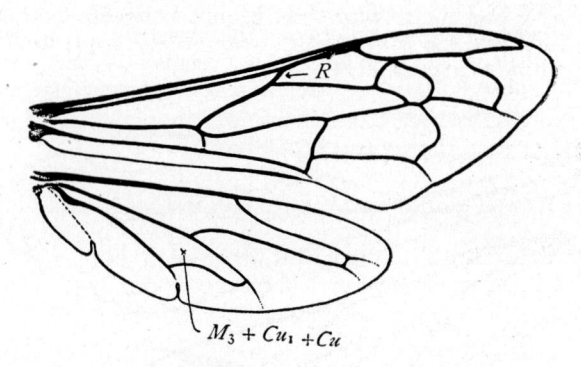

Fig. 1209.—Wings of *Sphecius speciosus*.

Tribe STIZINI

In the Stizini the transverse part of vein M is distant from the stigma by two or more times the distance between the apex of cell $2dR_1+R_2$ and the apex of the wing; the labrum is usually transverse, rarely long and pointed; the ocelli are normal; and the middle tibiæ are armed with two apical spurs.

The species of the genus *Stizus* are gregarious, many individuals building their burrows near together. The common and conspicuous western *Stizus unicinctus* is believed to lay eggs in nests of *Chlorion (Priononyx) atratrum*.

Tribe BEMBICINI

With these wasps the ocelli are distorted, the middle tibiæ are armed with only one evident apical spur; and the labrum is elongate, pointed, rostriform (Fig. 1210).

Our best-known representatives of this tribe belong to the genus *Bembix;* these are stout-bodied wasps, usually black with greenish or greenish-yellow bands. They burrow in the sand and provision their nests with flies. Some species at least practise progressive provisioning. After excavating its burrow and making a cell, the wasp captures a fly and stings it to death, then places it on the floor of the cell and attaches an egg to it. After the larva has hatched, the mother collects flies from day to day, feeding the larva till it is ready to change to a pupa, closing the nest behind her each time she leaves it.

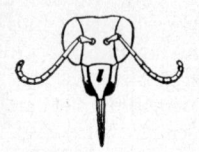

Fig. 1210.—Face of *Bembex: l,* labrum.

A common and well-known member of this tribe in the South is *Sticta carolina,* which is called the "horse guard." This is a large species which hunts about horses in order to capture flies.

Microbembex monodonta is one of the most abundant of wasps along the seashore everywhere on the Atlantic, Gulf, and Pacific coasts. This species is black with greenish-white markings; the pleuræ and mesoscutum are black and the wings are slightly dusky.

The North American Bembicini were monographed by Parker ('17).

Subfamily CRABRONINÆ

This subfamily differs from all other Sphecidæ except the Trypoxylonini in having only one submarginal cell (Fig. 1211), and it

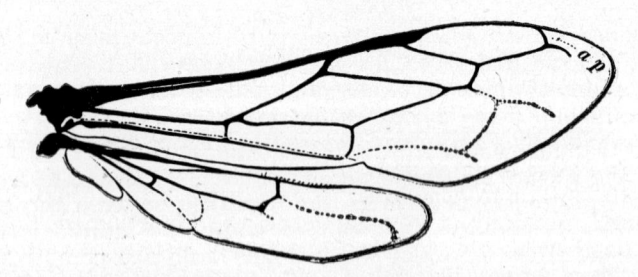

Fig. 1211.—Wings of *Crabro singularis; ap,* appendiculate cell.

differs from the Trypoxylonini in that the inner margin of the eyes is not emarginate. It includes two tribes, each of which is classed as a subfamily by some writers.

Tribe CRABRONINI

In the Crabronini the postscutellum and the base of the propodeum are unarmed; the eyes are usually much widened below, their inner margins strongly converging towards the clypeus; and the longitudinal free part of vein M of the fore wings is complete. The head is generally large and square when viewed from above, and sometimes broader than the thorax (Fig. 1212).

The different members of this tribe vary greatly in their nesting habits. Some mine in the pith of such plants as sumac and elder; some bore in more solid wood; some dig burrows in the ground; and others make use of any suitable hole they can find, often the deserted burrow of some other insect. These insects usually provision their nests with flies.

Fig. 1212.—*Crabro.*

The North American Crabronini was monographed by Fox ('95).

Tribe OXYBELINI

The Oxybelini are easily distinguished from all other sphecoid wasps by the two squamæ projecting back from the metanotum and by a median spine borne by the base of the propodeum (Fig. 1213). The inner margin of the eyes is convex, not converging toward the clypeus; and the longitudinal free part of vein M is lost or present as a trace.

These wasps nest in sand and provision their nests with flies.

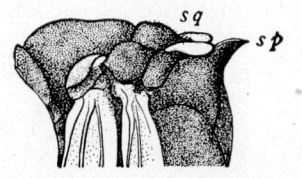

Fig. 1213.—Metanotum and propodeum of *Oxybelus:* *sq.*, squama; *sp.*, spine.

THE BEES

Superfamily Apoidea of Authors

The bees constitute a very large group of insects, including besides the well-known honey-bee and the bumblebees thousands of other species, many of which can be observed visiting flowers on any pleasant summer day. Friese ('23) states that 12,000 species of bees have been described, of which 2,500 are from North America and estimates that there are 20,000 living species in the world.

The bees differ from all other Hymenoptera, except some members of a small subfamily of vespoid wasps, the Masarinæ, in that they provision their nest with pollen and honey instead of with animal food, as do other nest-building Hymenoptera. The honey is obtained from flowers in the form of nectar, which is swallowed and transported to the nest in the crop. While in the crop the nectar undergoes a chemical change, which is probably due to a mixture with it of a ferment derived from the salivary glands, and becomes what is known as honey.

The distinctive characteristics of bees that have been recognized are chiefly those that are correlated with the habit of collecting pollen and nectar for provisioning their nests. These consist in specializations of the form and arrangement of some of the hairs, fitting them for collecting and carrying pollen; in the dilation of the metatarsus of the hind legs, which forms a part of the pollen-collecting apparatus; and in varying degrees of specialization of the maxillæ and labium to form a proboscis fitted for extracting nectar from flowers.

These characteristics are easily recognized in the higher bees, but in the most generalized bees (*Prosopis*) they are feebly developed, and too as male bees do not collect and carry pollen to nests they do not possess organs for this purpose; this is also true of both sexes of the parasitic bees, the females of which have acquired the habit

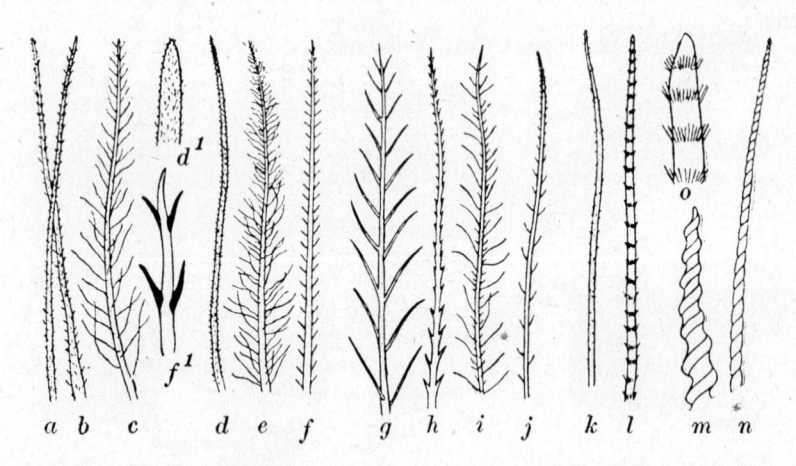

Fig. 1214.—Hairs of various bees: *a-f*, of bumblebees; *g-j*, of *Melissodes* sp.; *k-n*, of *Megachile* sp. (After John B. Smith.)

of laying their eggs in the nests of other bees, and consequently, have become degenerate so far as their pollen-collecting apparatus is concerned.

A characteristic of bees found in only a few other Hymenoptera is the presence, especially on the thorax, of plumose hairs. Many forms of these hairs exist; some of them are represented in Figure 1214. In this figure there is also represented (Fig. 1214, n) another type of hair which is spirally grooved; this type is found in the pollen brush of leaf-cutter bees, Megachilidæ. It has been suggested that the plumose hairs serve to hold the grains of pollen that become entangled among them when a bee visits a flower; but they occur in males and in parasitic bees neither of which gathers pollen; they are lacking, however, in some parasitic bees.

Female bees, excepting those of the genus *Prosopis* and of the parasitic bees, are furnished with pollen-brushes or *scopæ*, for collecting and transporting pollen. In most bees these consist of brushes of hairs borne by the hind legs, but in the Megachilidæ the brush is on the ventral side of the abdomen.

In some bees the pollen brushes are restricted to the tibia and the metatarsus of the hind legs, in others they are borne on these two segments and on the femur, trochanter and coxa as well (Fig. 1215). With the queens and workers of the nest-building bumblebees and with the workers of the honey-bee the pollen carrying apparatus is

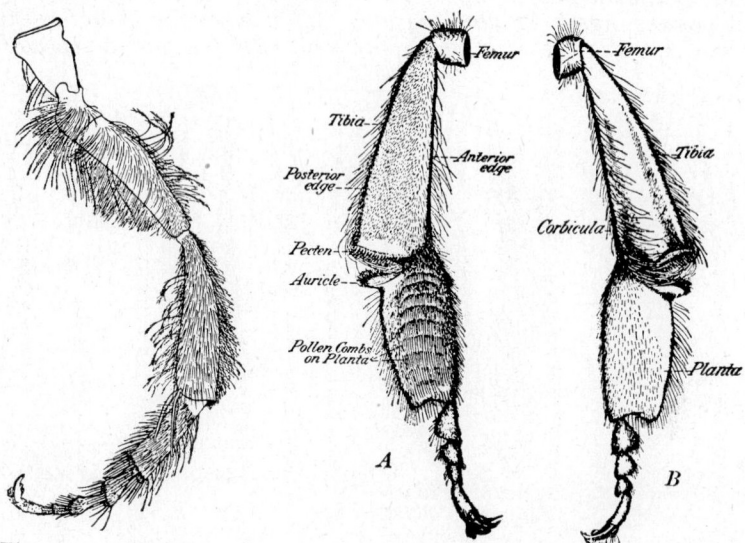

Fig. 1215.—Hind leg of female of *Colletes*. (From Brane.)

Fig. 1216.—A. Inner surface of the left hind leg of a worker honey-bee; B. Outer surface of the same. (After D. B. Casteel.)

very highly specialized (Fig. 1216). On the outer surface of the tibia of the hind legs there is a smooth area which is margined on each side by a fringe of long curved hairs; this structure is known as the pollen basket or *corbicula;* and on the inner surface of the metatarsus, termed *planta* by some writers, there is a brush of stiff hairs by means of which the bee gathers the pollen from its body. In the honey-bee the hairs composing this brush are arranged in transverse rows and are termed the pollen combs.

The mouth-parts differ greatly in form in the different groups of bees; this is especially true of the maxillæ and labium, which together constitute the proboscis, used for extracting nectar from flowers. The mandibles are fitted for chewing and do not vary so much in form.

In the most generalized bees, the Prosopidæ, the proboscis is comparatively short and the labium is either notched at the tip (*Prosopis*, Fig. 1217) or is quite deeply bifid (*Colletes*, Fig. 1218). In all other bees the labium is pointed at the tip. Among the bees with a pointed labium the proboscis varies greatly in length; in some (*Sphecodes*, Fig. 1219) it is comparatively short, while in the more specialized forms, as in *Apis* (Fig. 1220) it is greatly elongate.

Fig. 1217.—Proboscis of *Prosopis*. (After Saunders.)

Fig. 1218.—Proboscis of *Colletes*. (After Saunders.)

Fig. 1219.—Proboscis of *Sphecodes*. (After Saunders.)

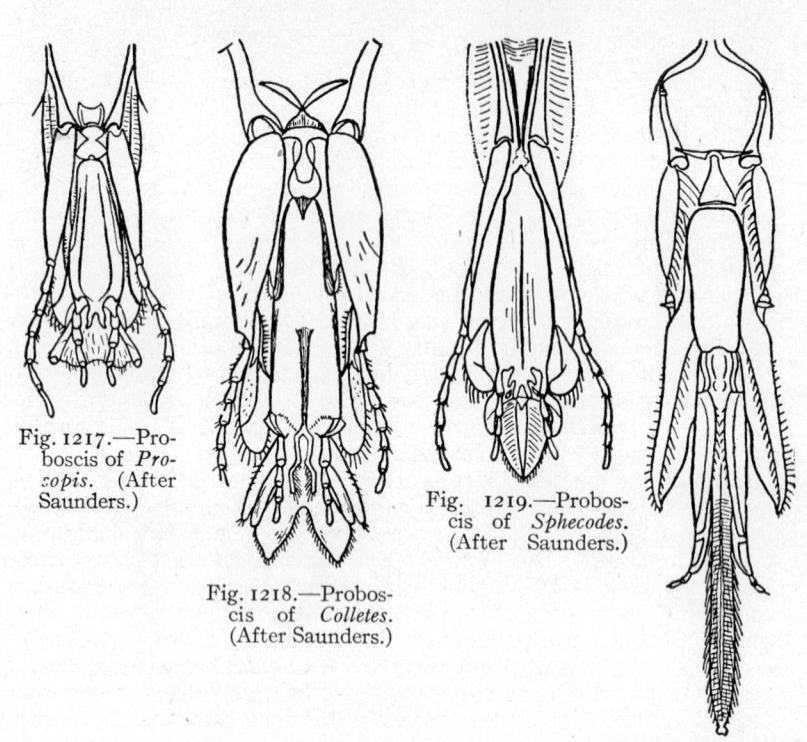

Fig. 1220.—Labium of the honey-bee. (After Saunders.)

The two sexes of bees differ in the number of abdominal tergites exposed to view; in the male there are seven, in the female, only six.

The different species of bees exhibit great differences in habits; some are solitary; each female providing a nest for her young; some are parasitic, the females laying their eggs in the nests of other bees and the larvæ feeding on the provisions stored by their hosts; and some are social, living in colonies consisting of many individuals.

The social bees are the honey-bees, the bumblebees, and the stingless honey-bees of the Tropics. In all of these, as with the social wasps and the ants, there is in addition to the males and the egg-laying females a worker caste; with all other bees there are only two forms, the males and the females.

The parasitic bees do not constitute a natural division of the group of bees, as was formerly supposed, instead of that it is evident that members of several of the families of bees have acquired the parasitic habit. The bees of the genus *Psithyrus*, which are parasitic in the nests of bumblebees, are closely allied to the bumblebees and should be placed with them in the family Bombidæ; the parasitic genera *Stelis* and *Cœlioxys* are evidently members of the leaf-cutter-bee family, the Megachilidæ; and there are many parasitic genera belonging to the family Andrenidæ.

The nests of solitary bees, like those of the digger-wasps, are of many forms. The mining-bees dig tunnels in the ground; the mason-bees build nests of mortar-like material; the carpenter-bees make tunnels in the stems of pithy plants or bore in solid wood; and some bees make nests of comminuted vegetable matter. The distinctive characteristic of the nests of bees is the fact that they are always provisioned with honey and pollen. In many cases closely allied species of bees differ in their nesting habits; for example, different species of the genus *Osmia* build very different kinds of nests.

Although many entomologists have studied the bees intensively, no classification of them has been proposed that is generally accepted. Some writers regard them as constituting a single family, the Apidæ; other writers recognize several families and restrict the term Apidæ to the honey-bee family; but these writers differ among themselves as to the number of families that should be recognized.

When several families of bees are recognized they are commonly grouped together as the superfamily Apoidea; but the writers whose classifications I have adopted believe that the bees are a group of sphecoid wasps that have acquired the habit of provisioning their nests with honey and pollen, and should not, merely for this reason, be placed in a separate superfamily. An analogous case is that of the subfamily Masarinæ some members of which differ from other Vespidæ in nest-provisioning habits in the same way that bees differ from other sphecoid wasps.

Family PROSOPIDÆ

The Bifid-tongued Bees

The members of this family differ from all other bees in having the tip of the labium either shallowly emarginate at the apex or deeply bifid. In all of them the labium is comparatively short. This family has been commonly known as the wasp-like bees. It includes two quite distinct subfamilies.

Subfamily PROSOPINÆ

This subfamily is represented in our fauna by a single genus, *Prosopis*. The members of this genus, of which there are many species, are small black bees, with pale, usually yellow, marks. They are the least specialized of the bees. The body is almost bare, but an examination with a microscope will reveal the presence on the thorax of a few of the plumose hairs characteristic of bees; the labium is short and broad and shallowly emarginate at the apex (Fig. 1217) and the hind legs of the females are not furnished with pollen brushes.

The numerous species of this genus build nests in the stems of pithy plants, or in burrows in the ground, or in crevices in walls. I have found them in dead branches of sumac. In some cases, at least, the burrow used was an old burrow made by some other pith-mining bee or wasp. After the burrow is made or selected, the walls of it are coated with a glistening substance, probably silk, which is sometimes dense enough to form a distinct membrane. Then a cell is formed at the bottom of the burrow of the same material; and at the bottom of the cell a denser circular disk is spun, which makes a quite firm partition, the edges of which extend slightly up the sides of the cell.

The cell is provisioned with a semi-liquid paste consisting largely of honey but containing also some pollen. It is said that when collecting provisions for its nest the bee swallows both pollen and nectar, brushing the pollen to the mouth by aid of the front legs.

Usually several cells are made, one above another, in the burrow; although the walls of the cells are quite delicate, the cells are firmly separated by the dense silken partition at the bottom of each.

Subfamily COLLETINÆ

In this subfamily the labium is short, and deeply emarginate at the apex; (Fig. 1218). The body, especially the head and thorax, is more or less densely clothed with hair; and in the female the hind legs are furnished with pollen-brushes. Our most common representatives belong to the genus *Colletes*.

Colletes.—In most species of this genus the abdomen is marked with pubescent white bands. All of the species, the habits of which have been described, burrow in soil, either that which is level or in banks, or sometimes in the interstices of walls. In favorable situations, some of the species are gregarious, many individuals digging their tunnels in a limited area. Sharp ('99) in writing of *Colletes* states: "They have a manner of nesting peculiar to themselves; they dig cylindrical burrows in the earth, line them with a sort of slime, that dries to a substance like gold-beater's skin, and then by partitions arrange the burrow as six to ten separate cells, each of which is filled with food that is more liquid than usual in bees."

Professor J. B. Smith ('01) in his account of *Colletes compácta* states that this species digs a burrow which extends from 18 to 28 inches down; from this, lateral branches from two to six inches in length are made, at the end of each of which a cell is formed. The bee begins making cells from the bottom of the burrow and works up, never making more than four and rarely more than two cell-bearing laterals from one upright. How many such burrows an individual female may make was not determined.

Family ANDRENIDÆ

The Andrenids

The family includes those solitary nest-building bees and their parasitic allies in which the tongue is either short or long but is pointed at the apex, and in which the pollen-brushes of the nest-building females are borne by the hind legs. To this family belong a large portion of the species and genera of our bees. Space can be taken here to discuss only a few of these.

Halictus.—Among the more common of our mining bees are those of the genus *Halictus*. This is a large genus including very many species, among which are the smallest of our bees. The nests of some

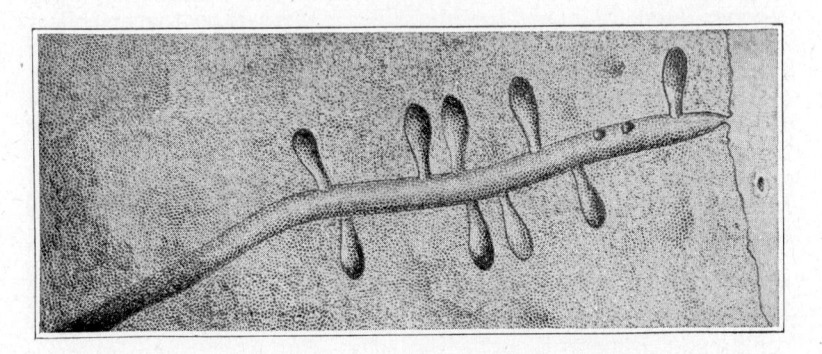

Fig. 1221.—Diagram of part of a nest of *Halictus*.

species are excavated in level ground; other species dig tunnels in the vertical sides of banks. These bees are often gregarious, hundreds of nests being built near together in the side of a bank.

If these nests be studied in midsummer, each will be found to consist of a burrow extending into the bank (Fig. 1221) and, along the sides of this main burrow or corridor, smaller short burrows each leading to a cell, the sides of which are lined with a thin coating of firm clay. In each of these cells that is closed will be found either a mass of pollen and nectar with an egg upon it or a larva feeding on the food stored for it.

The most striking feature of these nests is the fact that several bees use the corridor as a passage way to the cells they are building and provisioning. But this corridor is not a public one; it is constricted at its outer end and is guarded by a sentinel whose head nearly fills the opening. When a bee comes that has a right to enter the sentinel backs into the wider part of the corridor and allows it to pass and immediately thereafter resumes its guarding position with its head closing the opening of the corridor.

The explanation of this association of several bees, in a single nest was worked out by Fabre. He found that in the spring each female *Halictus* that has survived the winter makes a nest and rears a brood. Then the old bee and the young ones together clean out the nest, enlarge it, and use it as a carefully guarded apartment house, each bee having her own group of cells.

Halictus (Augochlōra).—A detailed account of the habits of one species of this subgenus, *A. humerālis*, was given by Professor J. B. Smith ('01). This species is a mining bee which digs very deep burrows. Certain other species of this genus have very different nesting habits. These burrow in decomposing sap-wood beneath the bark of trees and make their cells of bits of decayed wood agglutinated together.

Anthŏphora.—The genus *Anthophora* is widely distributed and includes many species, more than eighty have been described from North America alone; but the habits of only a few of these have been described.

The nests of those American species the habits of which are well known are usually built in steeply inclined or perpendicular banks of earth, preferably in those of compact clay; they are also excavated in the clumps of clay held between the roots of stumps in stump-fences. In the West a favorite nesting place of these bees is in the walls of sun-dried bricks of the adobe houses. Like *Halictus* and *Andrena*, the bees of this genus are gregarious, hundreds of individuals building their nests close together in the same bank of earth.

A striking feature of these nests is the presence of a cylindrical tube of clay extending outward and downward from the entrance of the tunnel (Fig. 1222). This tube is rough on the outside but smooth within. It is composed of small pellets of earth compacted together. These pellets when brought out from the tunnel are wet and easily molded into the desired form, but soon become dry and firm. The wetness of the pellets of clay brought out from the tunnel in a hard dry bank is explained by the fact that these bees when nest-building go to some place where water can be had and after lapping up a supply of it fly to their nest. This water is obviously used for softening the hard clay (Frison '22).

The tunnel extends into the bank a variable distance and leads to a cluster of oval cells. The layer of earth forming the wall of a cell is made firm by some cementing substance; this is shown by the fact that when a lump of earth containing nests is broken apart the cells retain their form and may be readily separated from the earth surrounding them. Nininger ('20) in his notes on the life-history of *Anthophora stanfordiāna* state: "At the bottom of a tunnel five to seven inches deep, the bee excavated an oval chamber about three-fourths inch in diameter by one inch deep, and then built up within this a nest-cell to fit, made of pellets of clay and worked smooth on the inner side, after which it was coated with a thin layer of water-proofing which seemed to be a salivary secretion."

The water-proofing of the wall of the cell is an essential feature for without it the semi-fluid mass of pollen and nectar with which the cell is provisioned would be partially absorbed by the wall of the cell.

Fig. 1222.—Section of a bank with nests of *Anthophora*. (Photographed by Miss P. B. Fletcher.)

The larvæ remain in their cells throughout the winter, and transform to pupæ in the spring. The duration of the pupa state is short, the adult bees appearing early in the summer. The parasitic beetles *Hornia* are often found in the nests of *Anthophora*.

Andrēna.—Among the larger of our common mining bees are certain species of the genus *Andrena*; some of these nearly or quite equal in size the workers of the honey-bee. They build their nests in road sides and in fields that support a scanty vegetation. They sink a vertical shaft with broad cells branching from it. These bees, though strictly solitary, each female building her own nest, frequently build their nests near together, forming large villages. I once received from a correspondent a description of a collection of nests of this kind which was fifteen feet in diameter, and in the destruction of which about two thousand bees were killed; what a terrible slaughter of innocent creatures!

The small carpenter-bee, *Cerătina dūpla*.—The nests of this bee are built in dead twigs or sumac and in the hollows of brambles and other plants. They are more common than those of any other of our solitary bees that build in these situations. This is a dainty little bee, about 6 mm. in length, and of a metallic blue color. She always selects a twig with a soft pith which she excavates with her mandibles, and so makes a long tunnel. Then she gathers pollen and nectar and puts it in the bottom of the nest, lays an egg on it, and then makes a partition out of pith-chips, which serves as a roof to this cell and a floor to the one above it. This process she repeats until the tunnel is nearly full (Fig. 1123), then she rests in the space above the last cell, and waits for her young to grow up. The lower one hatches first; and after it has attained its growth, it tears down the partition above it, and then waits patiently for the one above to do the same. Finally after the last one in the top cell has matured, the mother leads forth her full-fledged family in a flight in the sunshine. After the last of the brood has emerged from its cell, the substance of which the partitions were made, and which has been forced to the bottom of the nest by the young bees when making their escape, is cleaned out by the family, the old bee and the young ones all working together. Then the nest is used again by one of the bees. ·I have collected hundreds of these nests and by opening different nests at different seasons, have gained an idea of what goes on in a single nest. There are two broods each year. The mature bees of the fall brood winter in the nests.

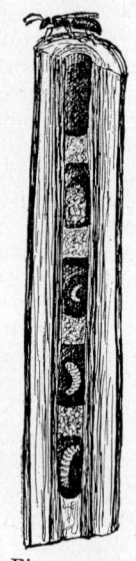

Fig. 1223.— Nest of *Ceratina dupla*.

The large carpenter-bee, *Xylŏcopa virgĭnica*.—This is a large insect, measuring from 22 to 24 mm. in length and resembling a bumblebee in size, and somewhat in appearance. But it can be easily distinguished from a bumblebee, as the female has a dense brush of hairs on the hind leg, instead of a basket for carrying pollen. This bee builds its nest in solid wood, and sometimes excavates a tunnel a foot in length, which it divides into several cells. The partitions between the cells are made of chips of wood, securely cemented together, and arranged in a closely-wound spiral. This arrangement of chips is easily seen when the lower side of a partition is examined; but the upper side of a partition which forms the floor of the cell above it is made concave and very smooth, so that the arrangement of the chips is not visible. The nest of *Monobia quadridens* described on an earlier page (Fig. 1188) was probably made in a deserted tunnel of *Xylocopa*. *Monobia*, however, makes the partitions of its nest of mud.

This species is distributed generally throughout the United States and is the only species of *Xylocopa* found in the Northeastern part of this country. Eight other species have been described from the South and the West. A monograph of the species of *Xylocopa* of the United States was published by Ackerman ('16).

Family MEGACHILIDÆ
The Leaf-cutter Bees and their Allies

To this family belong those bees in which the pollen brush of the female is borne on the ventral side of the abdomen and the parasitic bees that are allied to them. In this family the tongue is long and there are only two submarginal cells of approximately equal size in the fore wings. Among the better-known representatives of the family are the following.

The leaf-cutter bees, *Megachile*.—The bees of the genus *Megachile* have a curious habit of making cells for their young out of neatly-cut pieces of leaves. These cells are packed away in such secure places that one does not often find them; but it is a very easy thing to find

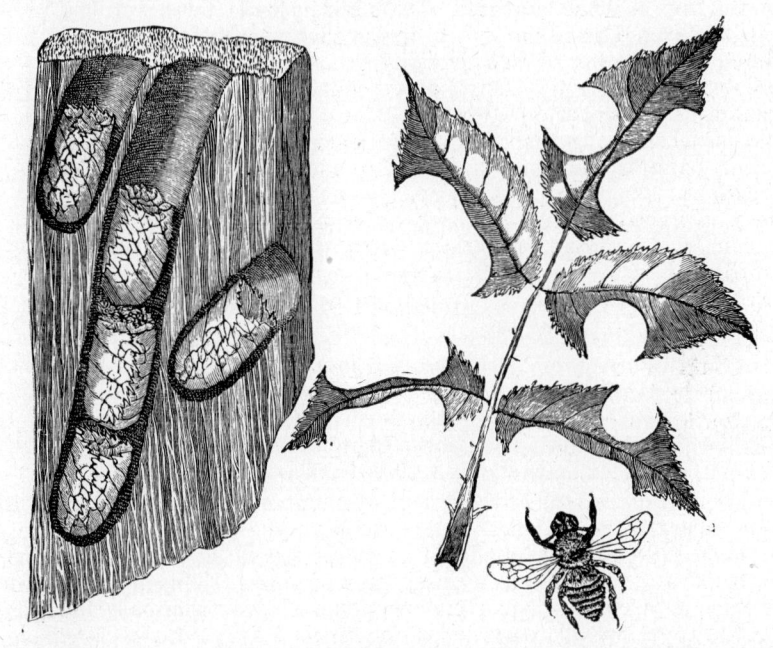

Fig. 1224.—A leaf-cutter bee, *Megachila latimanus*, its nest, and rose-leaves cut by the bee.

fragments of leaves from which the pieces have been cut by bees. The leaves of various plants are used for this purpose, but rose-leaves are used more frequently than any other kind. In Figure 1224 there are represented one of these bees, its nest, and a spray of rose-leaves from which pieces have been cut by the bee.

The nests are made in various situations. The specimen figured was taken from a piece of hemlock timber in which many of these bees had bored tunnels to receive their cells. I have also found nests of these bees in a tunnel in the ground under a stone, between shingles on a roof, in the cavity of a large branch of sumac, in the cavity of a lead pipe, and in Florida in the tubular leaves of a pitcher-plant.

When a suitable tunnel has been made or found the bee proceeds to build a thimble-shaped tube at the bottom of it. For this purpose it cuts from leaves oblong pieces, each of which forms a part of a side and the bottom of the thimble-shaped tube. Two such pieces had been cut from the lower leaf on the left side of the spray figured here. When the thimble-shaped tube is completed, the bee partially fills it with a paste of pollen and honey, and then places an egg upon the supply of food. She then cuts several circular pieces of leaves, the diameter of which is a little greater than the diameter of the tube, and forces them into the open end of it, thus making a tightly fitting plug; three of these circular pieces have been cut from the spray figured. Usually several cells of this kind are placed end to end in a burrow; and sometimes many bees will build their nests near together in the same piece of wood.

Alcidāmea prodŭcta.—Among the more common members of the Megachilidæ is this species which builds its nests in branches of sumac and other pithy plants. I have collected many of the nests during winter, from which the bees emerged the following spring. A distinctive feature of the nest is the fact that the partitions between the cells are composed of comminuted plant fibers. The larva when full-grown spins a silken cocoon, which fills the cell. The adult is a black bee, 7 mm. long, with white marginal bands on the abdominal segments; these are often interrupted on the middle line.

Trachūsa lateralis.—This is a parasitic species which is very common in the nests of *Alcidamea producta*. This bee is somewhat smaller than its host and has on each side of the three basal abdominal segments a rather small ovate yellowish-white spot. The cocoons of this species are denser than those of *Alcidamea producta*, do not fill the cell, and bear at the apex a tiny nipple.

Other illustrations of the habits of members of this family can be referred to here only briefly.

The petals of various flowers and especially those of *Pelargonium* often have pieces cut from them shaped like those cut from the leaves of rose by the leaf-cutter bees. This is probably the work of some species of *Osmia*, as certain European species of this genus are known to build in their burrows thimble-shaped tubes resembling those of the leaf-cutter bees, except that they are composed of pieces of petals.

Other species of *Ŏsmia* make their cells of comminuted vegetable fibers. These are placed in various situations. There are in our collection several old cells of mud-wasps in each of which are several cells of this kind, from which were bred a small species of *Osmia*.

Some species of *Osmia* make use of empty shells of snails, *Helix*, in which to build their cells, and some European species are known to cover the snail shell thus used with a mound of fragments of grass or of pine needles.

Many bees make their cells of a cement-like substance made of a mixture of earth and some fluid which is believed to be secreted by the salivary glands. These bees are commonly known as the mason-bees.

The firm-cement-like nature of the nests of some of the mason-bees leads to the belief that the earth of which they are made is mixed with some other fluid than water. They are much firmer than are the tubes built by *Anthophora*, which are made of a mixture of earth and water.

A remarkable accumulation of the nests of an Old World species of mason-bee, known as the wall-bee, *Chalicŏdoma murāria*, was ob-

Fig. 1225.—Nests of the wall-bee on the Temple of Dendera.

served by the writer on the walls of the Temple of Dendera in Egypt. This temple, which was buried by drifting sands long ago has been excavated by modern archæologists; but the inscriptions on the walls of the temple are being rapidly buried again beneath a layer of the cement-like nests of the wall-bee (Fig. 1225).

Family BOMBIDÆ

The Bumblebees

The family Bombidæ includes the well-known nest-building bumblebees and certain parasitic bumblebees, *Psĭthyrus*, that infest the nests of the nest-building species. The members of this family are large bees or of medium size, they are robust with oblong bodies and a rather dense covering of hair. They are common, and are conspicuous on account of their noisy flight and striking coloration, which is usually yellow and black. They are called bumblebees on account of the sound they make in flight; in England they are commonly known as humblebees.

The distinctive characters of this family are given in the table of the Clistogastra on page 912. Most writers recognize only two genera in the Bombidæ, *Bombus* and *Psithyrus;* but some have separated certain species from *Bombus* and placed them in a separate genus, *Bombias*. As there is considerable doubt regarding the validity of this genus it will not be discussed here.

The nest-building bumblebees, *Bŏmbus*.—The members of this genus are social insects, each species consisting as in other social insects of three castes, the queens, the workers, and the males. In this genus the queens as well as the workers possess pollen-baskets or corbiculæ on the hind legs; as the queen when founding a colony must collect pollen.

With the bumblebees the queens are larger than either the workers or the males and, in temperate regions, are the only ones that live through the winter; as in these regions the colonies, like those of our northern species of social wasps, break up in the autumn and all of the bees, except the young queens perish. These crawl away into some protected place and pass the winter. In the spring each queen that has survived the winter founds a new colony, performing, until a brood of workers has been developed, both the duties of queen and of worker. In South America, where according to von Ihering, bumblebee colonies are perennial, new nests are formed by swarming as among the social wasps of the same region.

In selecting a place for her nest the queen usually chooses a deserted mouse-nest, within which she builds her nest; sometimes an old bird's nest is used for this purpose. In certain European species the queen, sometimes at least, constructs her nest entirely without making use of a nest of another animal. This she does by making use of moss or soft dead grass, which she combs together with her mandibles and legs, for this reason these species are often known as "carder-bees."

Many observers have studied the founding and development of colonies of bumblebees; among these is Sladen ('12) who has made very detailed studies of the species found in England. The following condensed summary is based on the statements of this author.

Having found a suitable nest the queen spends a good deal of time in it, the heat of her body gradually making its interior perfectly dry. She then gathers the finest and softest material she can find into a heap and in the center of this makes a cavity with an entrance at the side just large enough for her to pass in and out. In the center of the floor of this cavity she forms a lump of paste made of pollen moistened with honey. Upon the top of this lump she builds with her jaws a circular wall of wax, and in the little cell so formed she lays her first batch of eggs, and seals it over with wax. The queen now sits on her eggs day and night to keep them warm, only leaving them to collect food when necessary. In order to maintain animation and heat through the night and in bad weather when food cannot be obtained, it is necessary for her to lay in a store of honey. She therefore sets to work to construct a large waxen pot to hold the honey. This pot is built in the entrance passage of the nest (Fig. 1226).

The eggs hatch four days after they are laid. The larvæ devour the paste which forms their bed and also fresh food furnished by the queen. To feed the larvæ the queen makes a small hole with her mandibles in the skin of wax that covers them. While the larvæ remain small they are fed collectively, but when they grow large they are fed individually. As the larvæ grow the queen adds wax to their covering, so that they remain hidden. When the larvæ are full-grown, each one spins around itself an oval cocoon, which is thin and

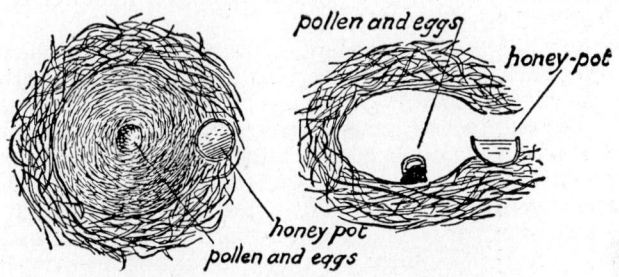

Fig. 1226.—Honey-pot. (From Sladen.)

papery but very tough. The queen now clears away most of the brown wax covering, revealing the cocoons, which are pale yellow. These first cocoons number from seven to sixteen, according to the species and the prolificness of the queen. These cocoons are incubated by the queen, who spends much time sitting on them, with her abdomen stretched to about double its usual length so that it will cover as many cocoons as possible.

The bees that are developed during the early part of the summer are all workers; these relieve the queen of all duties except laying the eggs. They feed the larvæ, construct honey-pots and special receptacles for pollen or store these substances in cocoons from which workers have emerged. The appearance of a nest in mid-summer is represented by Figure 1227. Later in the summer males and queens are developed; and in the autumn the colony breaks up.

The bumblebees play a very important role in the fertilization of certain flowers, as those of red clover, in which the tubular corolla is so long that the nectar can not be reached by bees with shorter tongues.

A monograph of the Bombidæ of the New World was published by Franklin ('12–'13).

The parasitic bumblebees, *Psĭthyrus*.—To this genus belong those parasitic bees that infest the nests of bumblebees. They closely resemble bumblebees in appearance and in structure, except that, as in other parasitic bees, the females do not possess organs for collecting and carrying pollen. Although the females of *Psithyrus* are easily distinguished from those of *Bombus* by the absence of the pollen-baskets or corbiculæ in the former, the males of the two genera are very similar. In *Psithyrus* there is no worker caste.

The conclusions of different observers as to the extent of the parasitism of *Psithyrus* differ widely. Sladen ('12) from his studies of English species regards them as the deadliest enemies of the

Fig. 1227.—Nest in mid-summer. (From Sladen.)

bumblebees whose nests they infest. He found "That it is the practice of the *Psithyrus* female to enter the nest of the *Bombus*, to sting the queen to death, and then get the poor workers to rear her young instead of their own brothers and sisters." This conclusion is not in accord, however, with those of other European writers; and the American species of *Psithryus* whose habits have been studied, rarely, if ever, kill the host queen. For a detailed account of the relations of these parasites and their hosts and for references to the literature of this subject see Plath ('22). For descriptions of the New World species see Franklin ('12–'13).

Family APIDÆ

The Honey-bees

The family Apidæ, as restricted here, includes only a single genus *Apis*, of which only four species are known, and one of these is doubtfully distinct. In this country a single introduced species, the honey-bee, *Āpis mellīfica*, is found. This species has been widely distributed over the world by man. The other species are restricted to the Indomalyan region; these are *A. dorsāta*, *A. flōrea*, and *A. indica*. The last named species is probably a variety of *Apis mellifica*. The colonies of *A. dorsata* and *A. florea* build a single pendent comb from the lower sides of a branch, and are not available for cultivation. *A. mellifica* and *A. indica* nest in cavities, as hollow tree-trunks and caverns, and will make use of hives prepared for them.

This family consists of social bees in which the hind tibiæ are without apical spurs; the workers are furnished with pollen-baskets or corbiculæ on the hind legs, but the queens are without functionally developed ones. Unlike the queen of the nest-building bumblebees the queen of the honey-bee is unable to found a colony or even to exist apart from workers of her own species.

The honey-bee was introduced into America more than three centuries ago, and escaping swarms have stocked our forests with it; for when free, swarms almost invariably build their nests in hollow trees. These nests include a variable number of vertical combs, which have cells on both sides, instead of a single series as is the case in the combs of our native social wasps. The cells of which the comb is composed are used both for storing the food of the colony and for rearing the brood.

The three castes of bees of which a colony is composed are easily distinguished. The workers are the well-known form that we see collecting pollen and nectar from flowers and entering and leaving the hive in large numbers. They constitute the greater part of the colony; an average strong colony will include from 35,000 to 50,000 workers. They are females in which the reproductive organs are imperfectly developed; they do not ordinarily lay eggs, and when they do the eggs develop only into males. The workers do not pair with males, consequently their eggs are unfertilized, and unfertilized eggs of the honey-bee produce only males. The workers are so-called because they perform all the labors of the colony. Young workers attend to the inside work of the hive; they take care of the young brood, and for this reason are termed nurse-bees, they build combs, and protect the entrance of the hive against robbers. The older workers go into the field to collect pollen, nectar and propolis.

The drones are larger than the workers, and are reared in larger cells. If honeycombs be examined, some sheets will be seen to be composed of larger cells than those of the more common type. It is in cells of this kind that the eggs are laid which are to develop into males. In shape the drones are broader and blunter than the workers.

They are few in number and are only present in the hive during the early summer. After the swarming season is over, these gentlemen of leisure are driven out of the hive by the workers or are killed by them.

The queen is larger than a worker, and has a long pointed body. She is developed in a cell which differs greatly from the ordinary hexagonal cell of honey-comb. This cell is large, cylindrical, and extends vertically. In Figure 1228 the beginnings of two queen cells are represented on the lower edge of the comb, and a completed cell extends over the face of the comb near the left side. From the lower end of this cell hangs a lid, which was cut away by the workers to allow the queen to emerge.

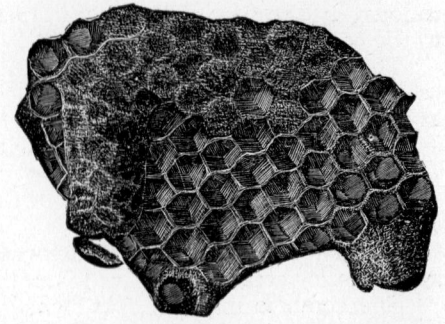

Fig. 1228.—Comb of honey-bee with queen-cells.

The queen larva is fed with a substance called royal jelly. This is a substance which resembles blanc-mange in color and consistency. It is excreted from the mouth by the nurse-bees, and is very nutritious food. The origin of this food, whether it is a secretion from special glands of the nurse-bees, or is regurgitated from their stomachs is not at present known. During the first three days of the larval stage of worker bees they are also fed with royal jelly after which they are fed with honey and bee-bread.

It has been demonstrated that in the egg state there is no difference between a worker and a queen. When the workers wish to develop a queen they tear down the partitions between three adjacent cells containing eggs that under ordinary conditions would develop into workers. Then they destroy two of the eggs, and build a queen-cell over the third. When the egg hatches they feed the larva with royal jelly, and it develops into a queen.

In early summer several queen-cells are provided in each colony. As soon as a queen is developed from one of these the old queen attempts to destroy her. But the young queen is guarded by the workers, and then the old queen with a goodly portion of her subjects swarm out, and they go to start a new colony.

The swarming of the honey-bee is essential to the continued existence of the species; for in social insects it is as necessary for the colonies to be multiplied as it is that there should be a reproduction of individuals. Otherwise, as the colonies were destroyed the species would become extinct. With the social wasps and with the bumble-bees the old queen and the young ones remain together peacefully in the nest; but at the close of the season the nest is abandoned by all as an unfit place for passing the winter, and in the following spring each young queen founds a new colony. Thus there is a tendency towards a great multiplication of colonies. But with the honey-bee the habit of storing food for the winter, and the nature of the habitations render it possible for the colonies to exist indefinitely. And thus if the old and young queens remained together peacefully there would be no multiplication of colonies, and the species would surely die out in time. We see, therefore that what appears to be merely jealousy on the part of the queen honey-bee is an instinct necessary to the continuance of the species.

The sting of a queen-bee is no ignoble weapon, but it is rarely used except against a rival queen. When several young queens mature at the same time there is a pitched battle for supremacy, and the last left living on the field becomes the head of the colony. One morning we found the lifeless bodies of fifteen young queens cast forth from a single hive—a monument to the powers of the surviving Amazon in triumphant possession within.

The materials used by bees are wax and propolis, which serve as materials for construction; and honey and bee-bread used for food.

The comb is made of wax, which is an excretion of the bees. When a colony needs wax, many of the workers gorge themselves with honey and then hang quietly in a curtain-like mass, the upper bees clinging to the roof of the hive, and the lower ones to the bees above them. After about twenty-four hours there appear on the lower surface of the abdomen of each bee little plates of wax that are forced out from openings between the ventral abdominal segments called wax-pockets. Other workers attend to this curtain and collect the wax as fast as it appears, and use it at once in constructing comb.

Propolis is a cement used for cementing up crevices, and is made of a resin which the bees collect from the buds of various trees, but especially of the poplar.

Honey is made from the nectar of flowers and is taken into the crop of the bee, and there changed into honey, and then regurgitated into the cells of the comb.

Bee-bread is made from the pollen of flowers, which the bees bring in on the plates fringed with hairs on the hind legs, the corbiculæ.

Very many books have been written regarding the habits of the honey-bee; some of these are to be found in most public libraries. There are also many manuals for the use of those who wish to keep bees; among these is a small one for beginners by Mrs. A. B. Comstock ('20) and a cyclopedia by A. I. and E. R. Root ('17). The U. S. Department of Agriculture has published many bullletins on this subject; one of a general nature is "Farmers Bulletin 447."

BIBLIOGRAPHY

The following list includes only the titles of the books and papers to which references have been made in the preceding pages.

AARON, S. F. ('85). "The North American Chrysididæ." Trans. Am. Ent. Soc. Vol. 12, pp. 209–248.

ACKERMAN, A. J. ('16). "The carpenter-bees of the United States of the genus Xylocopa." Jour. N. Y. Ent. Soc. Vol. 24, pp. 196–232.

ADELUNG, N. VON ('92). "Beiträge zur Kenntnis des tibialen Gehörapparates der Locustiden." Zeit. wiss. Zool. Vol. 54, pp. 316–385.

ALDRICH, J. M. ('05). "A catalogue of North American Diptera." Smithsonian Misc. Coll. Part of Vol. 46.

ALDRICH, J. M. ('12). "Flies of the leptid genus *Atherix* used as food by California Indians (Dipt.)." Ent. News. Vol. 23, pp. 159-163.

ALDRICH, J. M. ('16). "Sarcophaga and allies in North America." The Thomas Say Foundation of the Ent. Soc. of Am.

ALDRICH, J. M. ('18). "The kelp-flies of North America (genus Fucellia, family Anthomyidæ)." Proc. Cal. Acad. Sci. Vol. 8, pp. 157–179.

ALDRICH, J. M. ('20). "European frit fly in North America." Jour. Agr. Research. Vol. 18, pp. 451–473.

ALDRICH, J. M. ('22). "A new genus of two-winged fly with mandible-like labella." Proc. Ent. Soc. Wash. Vol. 24, pp. 145–148.

ALDRICH, J. M., AND DARLINGTON, P. S. ('08). "The dipterous family Helomyzidæ." Trans. Am. Ent. Soc. Vol. 34, pp. 67–100.

ALEXANDER, C. P. ('19). "The crane-flies of New York. Part I. Distribution and taxonomy of the adult flies." Cornell Univ. Agr. Exp. Sta. Memoir 25.

ALEXANDER, C. P. ('20). "The crane-flies of New York. Part II. Biology and phylogeny." Cornell Univ. Agr. Exp. Sta. Memoir 38.

ANDRÉ, E. ('03). "Hymenoptera Fam. Mutillidæ." Genera Insect., Fasc. 11.

ANTHONY, MAUDE H. ('02). "The metamorphosis of Sisyra." The American Naturalist. Vol. 36, pp. 615–331.

ASHMEAD, W. H. ('93). "A monograph of the North American Proctotrypidæ." Bull. U. S. Nat. Mus. No. 45.

ASHMEAD, W. H. ('94). "The habits of the aculeate Hymenoptera." Psyche. Vol. 7, pp. 19, 39–46, 59–66, 75–79.

ASHMEAD, W. H. ('00). "Classification of the ichneumon-flies, or the superfamily Ichneumonoidea." Proc. U. S. Nat. Mus. Vol. 23, pp. 1–220.

ASHMEAD, W. H. ('03). "Classification of the gall-wasps and parasitic cynipoids, or the superfamily Cynipoidea." Psyche. Vol. 10, pp. 7, 59, 140, 210.

ASHMEAD, W. H. ('04). "Classification of the chalcid flies, or the superfamily Chalcidoidea." Memoires of the Carnegie Museum, Vol. I, No. 4. Carnegie Institute, Pittsburgh, Pa.

AUDOUIN, J. V. (1824). "Recherches anatomiques sur le thorax des animaux articulés." Ann. Sci. Nat. Tome I, pp. 97–135, 416–432.

BAKER, A. C. ('20). "Generic classification of the hemipterous family Aphididæ." U. S. Dept. Agr. Bull. 826.

BAKER, C. F. ('04). "A revision of American Siphonaptera, or fleas, together with a complete list and bibliography of the group." Proc. U. S. Nat. Mus. Vol. 27, pp. 365–469.

BAKER, C. F. ('05). "The classification of the American Siphonaptera." Proc. U. S. Nat. Mus. Vol. 29, pp. 121–170.

BANKS, NATHAN ('05). "A revision of the nearctic Hemerobiidæ." Trans. Am. Ent. Soc. Vol. 32, pp. 21–51.

BANKS, NATHAN ('07). "A revision of the nearctic Coniopterygidæ." Proc. Ent. Soc. Wash. Vol. 8, pp. 77–86.

BANKS, NATHAN ('11). "Psammocharidæ: Classification and descriptions." Jour. N. Y. Ent. Soc. Vol. 19, pp. 219–237.
BANKS, N., AND SNYDER, T. E. ('20). "A revision of the nearctic Termites by Nathan Banks with notes on biology and geographic distribution by Thomas E. Snyder." U. S. Nat. Mus. Bull. 108.
BARBER, H. S. ('13a). "Observations on the life history of *Micromalthus debilis* Lec." Proc. Ent. Soc. Wash. Vol. 15, pp. 31–38.
BARBER, H. S. ('13b). "The remarkable life-history of a new family (Micromalthidæ) of beetles." Proc. Biol. Soc. of Wash. Vol. 27, pp. 185–190.
BARNES, W., AND McDUNNOUGH, J. H. ('11). "Revision of the Cossidæ of North America." Contrib. Nat. Hist. Lepidoptera N. Am. Vol. 1, No. 1. Decatur, Ill. The Review Press.
BARNES, W., AND McDUNNOUGH, J. H. ('12). "Revision of the Megathymidæ." Contrib. Nat. Hist. Lep. N. Am. Vol. 1, No. 3. Decatur, Ill. The Review Press.
BARNES, W., AND McDUNNOUGH, J. ('17). "Check list of the Lepidoptera of boreal America." Decatur, Ill. Herald Press.
BELLESME, J. ('78). "Note au sujet d'un travail adressé a l'Acad. par M. Perrez sur la bourdonnement des insectes." Compt. Rend. Acad. d. Science. Vol. 87, p. 535.
BERLESE, ANTONIO ('96). "Le Cocciniglie italiane viventi sugli agrumi. Part III. I Diaspiti." Firenze.
BERLESE, ANTONIO ('09a). "Gli Insetti." Vol. 1 (1909) 4to, pp. x + 1004. 1292 text figures and 10 plates. Società Editrice Libraria, Milano.
BERLESE, A. ('09b). "Monografia dei Myrientomata." Redia, Firenze.
BEZZI, M. ('13). "Taumaleidi (Orfnefilidi) italani." Portici. Boll. Lab. Zool. Gen. Agr. Vol. 7, pp. 227–266.
BISCHOFF, H. ('13). "Hymenoptera Fam. Chrysididæ." Genera Insect., Fasc. 151.
BLANC, M. L. ('90). "La tête du *Bombyx mori* à l'état larvaire." Extrait du volume des travaux du laboratoire d'études de la soie. Lyons, 1889–90.
BLATCHLEY, W. S. ('10). "An illustrated catalogue of the Coleoptera or beetles (exclusive of the Rhynchophora) known to occur in Indiana." The Nature Publishing Co., Indianapolis.
BLATCHLEY, W. S. ('20). "Orthoptera of northeastern America." The Nature Publishing Co., Indianapolis.
BLATCHLEY, W. S., AND LENG, C. W. ('16). "Rhynchophora or weevils of north eastern America." The Nature Publishing Co., Indianapolis.
BOISE, M. P. ('90). "Note on *Braula cæca*." Bull. Soc. Ent. France, 1890. p. cc.
BÖRNER, C. ('04). "Zur Systematik der Hexapoden." Zool. Anz. Vol. 27, pp. 511–533.
BÖRNER, C. ('08a). "Eine monographische Studie über die Chermiden." Arb. Kais. Biol. Anstalt. Bd. VI, Heft 2, 1908, pp. 224–245.
BÖRNER, C. ('08b). "Ueber Chermesiden." Zool. Anz. Vol. 33, pp. 612–616.
BOUVIER, E. L. ('05–07). "Monographie des Onychophores." Ann. Sci. Nat. 9e série, tome 2, p. 1–383, Pl. I–XIII; tome 5, p. 61–318.
BRADLEY, J. C. ('08). "The Evaniidæ, ensign-flies, an archaic family of Hymenoptera." Trans. Amer. Ent. Soc. Vol. 27, pp. 101–194, with 11 plates.
BRADLEY, J. C. ('13). "The Siricidæ of **North** America." Jour. Entom. and Zool. Vol. 5, pp. 1–30, with 5 plates.
BRADLEY, J. C. ('16). "Contribution toward a monograph of the Mutillidæ and their allies of America north of Mexico." Trans. Am. Ent. Soc. Vol. 42, pp. 309–336.
BRADLEY, J. C. ('17). Contributions toward a monograph of the Mutillidæ and their allies of America north of Mexico. IV. A review of the Myrmosidæ." Trans. Am. Ent. Soc. Vol. 43, pp. 247–290.

BRADLEY, J. C. ('22). "The taxonomy of the masarid wasps, including a monograph on the North American species." Univ. Cal. Publ., Tech. Bull. Vol. I, pp. 369-464. Berkeley, Cal.
BRAUER, F. ('69). "Beschreibung der Verwandlungsgeschichte der *Mantispa styriaca*, und Betrachtung über die sogenannte Hypermetamorphose Fabre's." Verh. d. Zool. Bot. Ges. Wien. Vol. 19.
BRAUER, F. ('76). "Die Neuropteren Europas." Wien.
BRAUER, FRIEDRICH ('85). "Systematisch-zoologische Studien." Sitzb. der Kais. Akad. Wissensch. 1885, pp. 237-413.
BRAUN, A. F. ('17). "Nepticulidæ of North America." Trans. Am. Ent. Soc. Vol. 43, pp. 155-209.
BRAUN, A. F. ('19). "Wing structure of Lepidoptera and the phylogenetic and taxonomic value of certain persistent trichopterous characters." Ann. Ent. Soc. Am. Vol. 12, pp. 349-366.
BRINDLEY, H. H. ('98). "On certain characters of reproduced appendages in Arthropoda, particularly in Blatta." Proc. Zool. Soc. Lond. 1898, pp. 924-958.
BRUES, C. T. ('03). "A monograph of the North American Phoridæ." Trans. Am. Ent. Soc. Vol. 29, pp. 331-404, with 5 plates.
BRUES, C. T. ('19). "A new chalcid-fly parasitic on the Australian bull-dog ant." Ann. Ent. Soc. Am. Vol. 12, pp. 13-21.
BRUES, C. T., AND MELANDER, A. L. ('15). "Key to the families of North American insects." Published by the authors.
BUGNION, E., AND POPOFF ('11). "Les pièces buccales des Hémiptères." Archives de Zoologie Expérimentale et Générale. Series 5, Vol. 7 (1911), pp. 643-675.
BURGES, E. ('80). "Contribution to the anatomy of the milkweed butterfly." Anniv. Mem. Bost. Soc. Nat. Hist.
BURKE, H. E. ('17). "Oryssus is parasitic." Proc. Ent. Soc. Wash. Vol. 19, pp. 87-88.
BURR, MALCOLM ('11). "Dermaptera." Genera Insect. Fasc. 122.
BUSCK, AUGUST ('03). "A revision of the American moths of the family Gelechidæ, with descriptions of new species." Proc. U. S. Nat. Mus. Vol. 25, pp. 767-938.
BUSCK, AUGUST ('06). "A review of the American moths of the genus Cosmopteryx." Proc. U. S. Nat. Mus. Vol. 30, pp. 707-713.
BUSCK, AUGUST ('09a). "A generic revision of American moths of the family Œcophoridæ, with descriptions of new species." Proc. U. S. Nat. Mus. Vol. 35, pp. 187-207.
BUSCK, AUGUST ('09b). "Notes on Microlepidoptera with descriptions of new North American species. Proc. Ent. Soc. Wash. Vol. 11, pp. 91-103.
BUSCK, AUGUST ('14). "On the classification of the Microlepidoptera." Proc. Ent. Soc. Wash. Vol. 16 (1914), pp. 47-54.
BUSCK, AUGUST ('17). "The pink bollworm, *Pectinophora gossypiella*." Jour. Agr. Research. Vol. 9 (1917), pp. 343-370.
BUSCK, A., AND BÖVING, A. ('14). "On *Mnemonica auricyanea* Walsingham." Proc. Ent. Soc. Wash. Vol. 16, pp. 151-163, with 8 plates.
CARLET, G. ('77). "Memoire sur l'appareil musical de la cigale." Ann. Sci. Nat. Zool. 6e serie, tome. 5.
CARPENTER, G. H. ('03). "On the relationships between the classes of the Arthropoda." Proc. Royal Irish Acad. Vol. 24, pp. 320-360.
CARPENTER, GEO. H. ('06). "Notes on the segmentation and phylogeny of the Arthropoda, with an account of the Maxillæ in *Polyxenus lagurus*." Quart. Jour. Micr. Sci. Vol. 49, pp. 469-491.
CARRIÈRE, J., AND BURGER, O. ('97). "Die Entwickelungsgeschichte der Mauerbiene (*Chalicodoma muraria*, im Ei." Nova Acta, Kais. Acad. Leop.-Carol in. Deutsch. Akad· d. Naturf. Vol. 69.
CAUDELL, A. N. ('03). "The Phasmidæ or walking-sticks of the United States." Proc. U. S. Nat. Museum. Vol. 26, pp. 863-885, with plates.
CAUDELL, A. N. ('04). "An orthopterous leafroller." Proc. Ent. Soc. Wash. Vol. 6, No. 1, pp. 46-49.

CAUDELL, A. N. ('07). "The Decticinæ (A group of Orthoptera) of North America." Proc. U. S. Nat. Mus. Vol. 32, pp. 285–410.

CAUDELL, A. N. ('16). "The genera of the tettiginid insects of the subfamily Rhaphidophorinæ found in America north of Mexico." Proc. U. S. Nat. Mus. Vol. 49, pp. 655–690.

CAUDELL, A. N. ('18). "*Zorotypus hubbardi*, a new species of the order Zoraptera from the United States." Canad. Entom. Vol. 50, pp. 375–381.

CAUDELL, A. N. ('20). "Zoraptera not an apterous order." Proc. Ent. Soc. Wash. Vol. 22, pp. 84–97.

CHAPMAN, T. A. ('17). "Micropteryx entitled to ordinal rank; Order Zeugloptera." Trans. Ent. Soc. London, 1916, pp. 310–314 (April '17).

CHESHIRE, FRANK R. ('86). "Bees and bee-keeping." London, L. Upcott Gill.

CHOLODKOVSKY, N. A. ('15). A paper on Chermes injurious to conifers, published in Russian. Petrograd, 1915. An abstract in English is given in the Review of Applied Entomology, Vol. 3 (1915), pp. 592–599.

CHILD, C. M. ('94). "Beiträge zur Kenntniss der antennalen Sinnesorgane der Insekten." Zeit. wiss. Zool. Vol. 58, pp. 475–525–528, mit 2 Taf.

CLAASSEN, P. W. ('21). "Typha insects: their ecological relationships." Cornell Univ. Agr. Exp. Sta. Memoir 47.

CLARKE, CORA H. ('91). "Caddis-worms of Stony Brook." Psyche. Vol. 6, pp. 153–158.

COLE, F. R. ('19). "The dipterous family Cyrtidæ in North America." Trans. Am. Ent. Soc. Vol. 45, pp. 1–79.

COMSTOCK, MRS. A. B. ('05). "How to keep bees. A handbook for the use of beginners." Doubleday, Page & Co, Garden City, N. Y.

COMSTOCK, J. H. ('81a). "An aquatic noctuid larva." Papilio. Vol. I, pp. 147–149.

COMSTOCK, J. H. ('81b). "Report of the Entomologist of the U. S. Dept. Agr. for 1880." Includes report on scale-insects.

COMSTOCK, J. H. ('82). "Report on insects for the year 1881." Ann. Rept. U. S. Dept. Agr. for the year 1881.

COMSTOCK, J. H. ('83). "Report of the Department of Entomology." Includes second report on scale insects. Second Rept. Dept. Ent., Cornell Univ. Exp. Sta.

COMSTOCK, J. H. ('93). "Evolution and taxonomy." The Wilder Quarter-Century Book, pp. 37–114, Pl. I–III.

COMSTOCK, J. H. ('01). "The wings of the Sesiidæ." (In Monograph of the Sesiidæ by Wm. Beutenmüller. Memoirs Amer. Mus. Nat. Hist. Vol. 1, p. 220.

COMSTOCK, J. H. ('12). "The spider book." Doubleday, Page & Co., Garden City, N. Y.

COMSTOCK, J. H. ('18a). "The wings of insects." The Comstock Publishing Company, Ithaca, N. Y.

COMSTOCK, J. H. ('18b). "Nymphs, naiads, and larvæ." Ann. Ent. Soc. Am. Vol. 2, pp. 222–224.

COMSTOCK, J. H. AND A. B. ('04). "How to know the butterflies." With forty-five full-page plates in colors. The Comstock Publishing Company, Ithaca, N. Y.

COMSTOCK, J. H., AND KOCHI, C. ('02). "The skeleton of the head of insects" The American Naturalist. Vol. 36 (1902), pp. 13–45, with 29 text figures.

COMSTOCK, J. H., AND NEEDHAM, JAMES G. ('98–'99). "The wings of insects." A series of articles on the structure and development of the wings of insects, with special reference to the taxonomic value of the characters presented by the wings. Reprinted from The American Naturalist, with the addition of a table of contents. 124 pages, 90 figures. Ithaca, N. Y., 1899. The articles appeared originally in The American Naturalist, Vol. XXXII (1898), pp. 43, 81, 231, 237, 240, 243, 249, 253, 256, 335, 413, 420, 423, 561, 769, 774, 903; Vol. XXXIII (1899), pp. 118, 573, 845, 851, 853, 858.

COMSTOCK, J. H., AND SLINGERLAND, M. V. ('91). "Wireworms." Cornell Univ. Agr. Exp. Sta. Bull. 33.

COOK, F. C., HUTCHINSON, R. H., AND SCALES, F. M. ('14). "Experiments in the destruction of fly larvæ in horse manure." U. S. Dept. Agr. Bul. 118, pp. 1–26.

COQUILLETT, D. W. ('96). "Revision of the North American Empidæ—A family of two-winged insects." Proc. U. S. Nat. Mus. Vol. 18, pp. 387–440.

COQUILLETT, D. W. ('97). "Revision of the Tachinidæ of America north of Mexico." U. S. Dept. Agr., Div. Ent. Tech. Ser. No. 7.

COQUILLETT, D. W. ('98). "The buffalo-gnats, or black-flies, of the United States." U. S. Dept. Agr., Div. Ent. Bull. No. 10, New Series, pp. 66–69.

CRAMPTON, G. C. ('09). "A contribution to the comparative morphology of the thoracic sclerites of insects." Proc. Acad. Nat. Sci. Phila., 1909, pp. 3–54.

CRAMPTON, G. C. ('14). "Notes on the thoracic sclerites of winged insects." Ent. News. Vol. 25 (1914), pp. 15–25.

CRAMPTON, G. C. ('15). "The thoracic sclerites and the systematic position of *Grylloblatta campodeiformis* Walker, a remarkable annectent, 'orthopteroid' insect." Ent. News. Vol. 26, pp. 337–350.

CRAMPTON, G. C. ('17). "The nature of the veracervix or neck region in insects." Ann. Ent. Soc. Am. Vol. 10, pp. 187–197.

CRAMPTON, G. C. ('18). "A phylogenetic study of the terminal abdominal structures and genitalia of male Apterygota, ephemerids, Odonata, Plecoptera, Neuroptera, Orthoptera, and their allies." Bull. Brooklyn Ent. Soc. Vol. 13 (1918), pp. 49–68, pl. 2–7.

CRAMPTON, G. C. ('20a). "Some anatomical details of the remarkable winged zorapteron, *Zorotypus hubardi* Caudell, with notes on its relationships." Proc. Ent. Soc. Wash. Vol. 22, pp. 98–106.

CRAMPTON, G. C. ('20b). "A comparison of the external anatomy of the lower Lepidoptera and Trichoptera from the standpoint of phylogeny." Psyche. Vol. 27, pp. 23–45.

CRAMPTON, G. C. ('21). "The origin and homologies of the so-called 'superlinguæ' or 'paraglossæ' (paragnaths) of insects and related arthropods." Psyche. Vol. 28, pp. 84–92.

CRAWFORD, D. L. ('12). "The petroleum fly in California, *Psilopa petrolei*." Pomona Col. Jour. Ent. Vol. 4, pp. 687–697.

CRAWFORD, DAVID L. ('14). "A monograph of the jumping plant-lice or Psyllidæ of the New World." U. S. Nat. Mus. Bull. 85, pp. IX+186, with 30 plates.

CRAWFORD, J. C. ('09). "A new family of parasitic Hymenoptera." Proc. Ent. Soc. Wash. Vol. 2, pp. 63–64.

CRESSON, E. T. ('20). "A revision of the nearctic Sciomyzidæ." Trans. Am. Ent. Soc. Vol. 46, pp. 27–89.

CROSBY, C. R. ('09). "On certain seed-infesting chalcis-flies." Cornell Univ. Agr. Exp. Sta. Bull. 265.

CROSBY, C. R. ('11). "The apple redbugs." Cornell Univ. Agr. Exp. Sta. Bull. 291.

CROSBY, C. R., AND LEONARD, M. D. ('18). "Manual of vegetable-garden insects." pp. XV+391, many figures. The Macmillan Co., New York.

CROSBY AND SLINGERLAND ('14). "Manual of fruit insects." The Macmillan Co, New York.

DALLA TORRE, K. VON ('08). "Anoplura." Genera Insect., Fasc. 81.

DALLA TORRE, W. VON, AND KIEFFER, J. J. ('02). "Hymenoptera Fam. Cynipidæ." Genera Insect., Fasc. 9 and 10.

DALLA TORRE, K. W. VON, AND KIEFFER, J. J. ('10). "Cynipidæ." Das Tierreich. 24 Lieferung, pp. I–XXXV+1–891.

DAVIDSON, ANSTRUTHER ('13). "Masaria Vespoides." Bull. Southern Cal. Acad. Sci. Vol. 12, p. 17.

DAVIS, K. C. ('03). "Sialididæ of North and South America." N. Y. State Mus. Bull. 68, pp. 442–487.

DEWITZ, H. ('84). "Ueber die Fortbewegung der Thiere an senkrechten, glatten Flachen vermittelst eines Secretes." Pflüger's Archiv. f. d. ges. Phys. Vol. 33.
DOYÈRE, M. (1840). "Mémoire sur les Tardigrades." Ann. des Sci. Nat. (2). Vol. 14, pp. 269–361.
DUCKE, A. ('14). "Ueber Phylogenie und Klassifikation der sozial Vespiden." Zool. Jahrb. Vol. 36, pp. 303–330.
DUFOUR, LEON (1824). "Recherches anatomiques sur les Carabiques et sur plusieurs autres insectes Coléoptères." Ann. Sci. Nat. Vol. 2 (1824), pp. 462–498.
DUNCAN, C. D. ('24). "Spiracles as sound producing organs." The Pan-Pacific Entomologist. Vol. I, pp. 42–43.
DYAR, H. G. ('90). "The number of molts of lepidopterous larvæ." Psyche. Vol. 5, pp. 420–422.
DYAR, H. G. ('94). "A classification of lepidopterous larvæ." Ann. N. Y. Acad. Sci. Vol. 8, pp. 194–232.
DYAR, H. G. ('00). "Notes on the larval-cases of Lacosomidæ (Perophoridæ) and life-history of *Lacosoma chiridota* Grt." Jour. N. Y. Ent. Soc. Vol. 8, pp. 177–180.
DYAR, H. G. ('02). "A lepidopterous larva on a leaf-hopper." Proc. Ent. Soc. Wash. Vol. 5, pp. 43–45.
DYAR, H. G. ('22). "The mosquitoes of the United States." Proc. U. S. Nat. Mus. Vol. 62, Art. 1, pp. 1–119. (No. 2447.)
EATON, A. E. ('83–'85). "A revisional monograph of recent Ephemeridæ." London Trans. Linn. Soc. Vol. 3.
EGGERS, FRIEDRICH ('19). "Das thoracal bitympanale Organ einer Gru.ppe der Lepidoptera Heterocera." Zool. Jahrb. Vol. 41, Abt. f. Anat., pp. 273–376.
ENDERLEIN, G. ('09). "Klassifikation der Plecopteren, sowie Diagnosen neuer Gattungen und Arten." Zool. Anz. Vol. 34, pp. 385–419.
ENDERLEIN, G. ('11). "Die phyletischen Beziehungen der Lycoriiden (Sciariden) zu den Fungivoriden (Mycetophiliden) und Itoniididen (Cecidomyiiden) und ihre systematische Gliederung." Archiv. für Naturg. 1911. I. 3. Suppl., pp. 116–208.
ENDERLEIN, G. ('12a). "Embiidinen Monographisch Bearbeitet." Coll. Zool. Selys Longchamp Tasc. III.
ENDERLEIN, G. ('12b). "Zur Kenntnis der Zygophthalmen. Ueber die Gruppierung der Sciariden und Scatopsiden." Zool. Anz. Vol. 40, pp. 261–282.
ENDERLEIN, G. ('21). "Die systematische Gliederung der Simuliiden." Zool. Anz. Vol. 53, pp. 43–46.
EXNER, S. ('91). "Die Physiologie der Facettirten Augen von Krebsen und Insecten." 1891, 8vo., pp. I–VIII, 1–206, 7 plates. Franz Deuticke, Leipzig und Wien, 1891.
FABRE, J. H. (1876–1904). "Souvenirs entomologiques; études sur l'instinct et les moeurs des insectes." 10 vol. Paris, Librairie Ch. Delagrave, 1879–1904.
FABRE, J. H. ('11). "The life and love of the insect." Translated by Alexander Teixeira de Mattos. London: Adam and Charles Black.
FAGEN, M. M. ('18). "The uses of insect galls." Amer. Nat. Vol. 52, pp. 155–176.
FELT, E. P. "Studies in Cecidomyiidæ "and "A study of gall midges." A series of papers published in the annual reports of the State Entomogist of New York.
FELT, E. P. ('96). "The scorpion-flies." 10th Report N. Y. State Ent. for 1894.
FELT, E. P. ('05–'06). "Insects affecting park and woodland trees." N. Y. State Mus. Memoir 8, 2 vols.
FELT, E. P. ('11). "*Miastor americana* Felt, An account of pedogenesis." 26th Report of the State Ent. of N. Y. N. Y. State Mus. Bull. 147.
FELT, E. P. ('18). "Key to American galls." N. Y. State Mus. Bull. 200.
FERRIS, G. F. ('16). "Some ectoparasites of bats (Dipt)." Ent. News. Vol. 27, pp. 433–438.

Ferris, G. F. ('19). "Some records of Polyctenidæ (Hemiptera)." Jour. N. Y. Ent. Soc. Vol. 27, pp. 261–263, 1 plate.

Ferris, G. F. ('24). "The New World Nycteribiidæ (Diptera Pupipara). Ent. News. Vol. 35, pp. 191–199.

Ferton, Ch. ('19). "Notes détachées sur l'instinct des Hyménoptères mellifères et ravisseurs." Ann. Soc. Ent. France, Vol. 79, pp. 145–178.

Field, W. L. W. ('10). "The offspring of a captured female of *Basilarchia proserpina.*" Psyche. Vol. 17, pp. 87–89.

Folsom, J. W. ('99). "The anatomy and physiology of the mouth-parts of the collembolan *Orchesella cincta.*" Bull. Mus. Comp. Zool. Vol. 35, No. 2.

Folsom, J. W. ('00). "The development of the mouth-parts of *Anurida maritima.*" Bull. Mus. Comp. Zool. Vol. 36, pp. 87–157.

Folsom, J. W. ('13). "Entomology," pp. vii+402, 304 text figures, and 4 plates P. Blakiston's Sons & Co., Philadelphia.

Folsom, J. W. ('16). "North American collembolous insects of the subfamilies Achorutinæ, Neanurinæ, and Podurinæ." Proc. U. S. Nat. Mus. Vol. 50, pp. 477–525.

Forbes, W. T. M. ('10). "A structural study of some caterpillars." Ann. Ent. Soc. Am. Vol. 3, pp. 94–143.

Forbes, W. T. M. ('14). "A structural study of the caterpillars: III, The somatic muscles." Ann. Ent. Soc. Am. Vol. VII, pp. 109–124, with 9 plates.

Forbes, W. T. M. ('16). "On the tympanum of certain Lepidoptera." Psyche. Vol. 23, pp. 183–192.

Forbes, W. T. M. ('22a). "Five strange Lepidoptera (Oinophilidæ, Noctuidæ, Gelechiidæ)." Ent. News. Vol. 33, pp. 97–104.

Forbes, W. T. M. ('22b). "The wing-venation of the Coleoptera." Ann. Ent. Soc. Am. Vol. 15, pp. 328–345, with 7 plates.

Ford, Norma ('22). "An undescribed planidium of Perilampus." Canad. Ent. Sept. 1922.

Fox, W. J. ('92). "The North American Pemphredonidæ." Trans. Am. Ent. Soc. Vol. 19, pp. 307–326.

Fox, W. J. ('93). The North American Larridæ. Proc. Acad. Nat. Sci. Phila., 1893, pp. 467–551.

Fox, W. J. ('95). "The Cerabraninæ of boreal America." Trans. Am. Ent. Soc. Vol. 22, pp. 129–226.

Fox, W. J. ('99). "The North American Mutillidæ." Trans. Am. Ent. Soc. Vol. 25, pp. 219–300.

Fracker, S. B. ('12). "A systematic outline of the Reduvidæ of North America." Proc. Iowa Acad. Sci. Vol. 19 (1912), pp. 217–247.

Fracker, S. B. ('15). "The classification of lepidopterous larvæ." Illinois Biological Monographs, Vol. 2, No. 1. Univ. Ill., Urbana, Ill.

Franklin, H. J. ('12–'13). "The Bombidæ of the New World." Trans. Amer. Ent. Soc. Vol. 38, pp. 177–486; Vol. 39, pp. 73–200. 22 plates.

Freiling, H. H. ('09). "Duftorgane der weiblichen Schmetterlinge nebst Beiträgen zur Kenntnis der Sinnesorgane auf dem Schmetterlingsflügel und der Duftpinsel der Männchen von Danais und Euploea." Zeit. wiss. Zool. Vol. 92, pp. 210–290, 6 plates.

Friese, H. ('23). "Die europäischen Bienen (Apidæ). Das Leben und Wirken unserer Blumenwespen," pp. 1–456, with 33 colored plates. Berlin und Leipzig.

Frison, T. H. ('22). "Notes on the life-history, parasites and inquiline associates of *Anthophora abrupta* Say, with some comparisons with the habits of certain other Anthophorinæ (Hymenoptera). Trans. Am. Ent. Soc. Vol. 48, pp. 137–156.

Fuller, Claude ('12). "White ants in Natal." Agr. Jour. Union of S. Africa. Sept. and Oct. 1912.

Fulton, B. B. ('15). "The tree crickets of New York: life history and bionomics." N. Y. Agr. Exp. Sta. Tech. Bull. 42.

Funkhouser, W. D. ('17). "Biology of the Membracidæ of the Cayuga Lake Basin." Cornell Univ. Agr. Exp. Sta. Memoir 11.

LINDSEY, A. W. ('21). "The Hesperioidea of America north of Mexico." Univ. of Iowa Studies, Vol. 9, No. 4. Published by the University, Iowa City, Iowa.

LINNÆUS (CARL VON LINNÉ). "Systema naturæ, sine regna tria naturæ systematice proposita per classes, ordines, genera et species, etc." First edition, 1735; twelfth edition, 1768. To be obtained of dealers in second-hand books.

LLOYD, J. T. ('14). "Lepidopterous larvæ from rapid streams." Jour. N. Y. Ent. Soc. Vol. 22, pp. 145–152.

LLOYD, J. T. ('15). "Notes on *Ithytrichia confusa* Morton." Canad. Ent. Vol. 47, pp. 117–121.

LLOYD, J. T. ('21). "The biology of North American caddice-fly larvæ." Cincinnati, Ohio. Bull. 21 of the Lloyd Library.

LOZINSKI, PAUL ('08). "Beitrag zur Anatomie und Histologie der Mundwerkzeuge der Myrmeleonlarven." Zool. Anz. Vol. 33, pp. 473–484, with 9 figures.

LUBBOCK, J. ('73). "Monograph of the Collembola and Thysanura." London Roy. Soc., 1873.

LUBBOCK, SIR JOHN ('94). "Ants, bees and wasps." Revised Ed. Internat. Sci. Ser., Appleton & Co., New York.

LUGGER, OTTO ('98). "The Orthoptera of Minnesota." Third Ann. Report of the Entomologist of the State Exp. Station of the Univ. of Minnesota, for the year 1897. This is a reprint of Bull. No. 55 (Dec. 1897) of the same station.

LYONET, P. (1762). "Traité anatomique de la chenille qui ronge le bois de saule." Amsterdam, 1762.

McATEE, W. L. ('08). "Notes on an orthopteron leaf-roller." Ent. News. Vol. 19, pp. 488–491.

McATEE, W. L. ('21). "Notes on nearctic bibionid flies." Proc. U. S. Nat. Mus. Vol. 60, Art. II.

McCLENDON, J. F. ('02). "The life history of *Ulula hyalina*." The Amer. Nat. Vol. 36, pp. 421–429.

McCLENDON, J. F. ('06). "Notes on the true Neuroptera." Ent. News, 1906, pp. 116–121.

MACGILLIVRAY, A. D. ('06). "A study of the wings of the Tenthredinoidea, a superfamily of Hymenoptera." Proc. U. S. Nat. Mus. Vol. 39, pp. 569–654, with 24 plates.

MACGILLIVRAY, ALEX. D. ('21). "The Coccidæ." Scarab Co., Urbana, Ill., 1921.

McINDOO, N. E. ('14). "The olfactory sense of the honey bee." Jour. Exp. Zool. Vol. 16, pp. 265–346.

McLACHLAN, ROBERT ('74–'80). "A monograph revision and synopsis of the Trichoptera of the European fauna." London, John VanVoorst, 1874–1880.

MALLOCH, J. R. ('13). "The insects of the dipterous family Phoridæ in the United States National Museum." Proc. U. S. Nat. Mus. Vol. 43, pp. 411–529.

MALLOCH, J. R. ('14). "American black flies or buffalo gnats." U. S. Dept. Agr., Bur. Ent. Bull. No. 26, Tech. Ser.

MALLOCH, J. R. ('17). "A preliminary classification of Diptera, exclusive of Pupipara, based upon larval and pupal characters, with keys to imagines in certain families. Part I." Bull. Ill. State Lab. Nat. Hist. Vol. 12, Art. III.

MALLOCH, J. R. ('18). "The North American species of the genus *Tiphia* (Hymenoptera Aculeata) in the collection of the Illinois State Natural History Survey." Ill. Nat. Hist. Survey Bulletin. Vol. 13, pp. 1–24.

MARCHAL, P. ('97). "Les Cécidomyies des céréales et leurs parasites." Ann. Soc. Ent. France. Vol. 66, pp. 1–105.

MARCHAL, P. ('13). "Contribution a l'étude de la biologie des Chermes." Ann. Sci. Nat. 9th Series (Zoology), Vol. 18 (1913), pp. 153–385.

MARLATT, C. L. ('95). "The hemipterous mouth." Proc. Ent. Soc. Wash. Vol. 3, pp. 241–249.

MARLATT, C. L. ('98). "The principal enemies of the grape." U. S. Dept. Agr. Farmers' Bull. No. 70.
MARLATT, C. L. ('07). "The periodical cicada." U. S. Dept. Agr., Bur. Ent. Bull. 71, 1907.
MATHESON, ROBERT ('12). "The Haliplidæ of North America, north of Mexico." Jour. N. Y. Ent. Soc. Vol. 20, pp. 156-193.
MATHESON, ROBERT ('14). "Life-history notes on two Coleoptera (Parnidæ)." Canad. Ent. Vol. 46, pp. 185-189.
MATHESON, R., AND CROSBY, C. R. ('12). "Aquatic Hymenoptera in America." Ann. Ent. Soc. Am. Vol. 5, pp. 65-71.
MAYER, A. G. ('96). "The development of the wing scales and their pigment in butterflies and moths." Bull. Mus. Comp. Zool. Vol. 29, pp. 209-236.
MAYER, A. M. ('74). "Experiments on the supposed auditory apparatus of the mosquito." The American Naturalist. Vol. 8, p. 577.
MEEK, W. J. ('03). "On the mouth-parts of the Hemiptera." Kansas Univ. Sci. Bull. Vol. 2, pp. 257-277, 5 plates.
MEIJERE, J. C. H. DE ('01). "Ueber das letzte Glied der Beine bei den Arthropoden." Zool. Jahrb. Anat. Vol. 14 (1901).
MELANDER, A. L. ('02a). "Two new Embiidæ." Biol. Bull. Vol. 3, pp. 16-26.
MELANDER, A. L. ('02b). "Notes on the structure and development of *Embia texana*." Biol. Bull. Vol. 4, pp. 99-118.
MELANDER, A. L. ('02c). "A monograph of the North American Empididæ." Trans. Am. Ent. Soc. Vol. 28, pp. 195-367.
MELANDER, A. L. ('03). "Notes on North American Mutillidæ, with descriptions of new species." Trans. Am. Ent. Soc. Vol. 29, pp. 291-330.
MELANDER, A. L. ('13a). "A synopsis of the Sapromyzidæ." Psyche. Vol. 20, pp. 57-82.
MELANDER, A. L. ('13b). "A synopsis of the dipterous groups Agromyzinæ, Milichiinæ, Ochthiphilinæ, and Geomyzinæ." Jour. N. Y. Ent. Soc. Vol. 21, pp. 219-273, 283-300.
MELANDER, A. L. ('16). "The dipterous family Scatopsidæ." State Coll, Wash. Agr. Exp. Sta. Bull. 130.
MELANDER, A. L. ('20a). "Review of the nearctic Tetanoceridæ." Ann. Ent. Soc. Am. Vol. 13, pp. 305-332.
MELANDER, A. L. ('20b). "Synopsis of the dipterous family Psilidæ." Psyche. Vol. 27, pp. 91-101.
MELANDER, A. L., AND SPULER, A. ('17). "The dipterous families Sepsidæ and Piophilidæ." State Coll. Wash. Agr. Exp. Sta. Bull. 143.
MERCER, W. F. ('00). "The development of the wings in Lepidoptera." Jour. N. Y. Ent. Soc. Vol. 8, pp. 1-20.
MIALL, L. C. ('95). "The natural history of aquatic insects," pp. IX+395, with 116 figures. Macmillan & Co., London and New York.
MIALL, L. C., AND DENNY, A. ('86). "The structure and life-history of the cockroach." London, 1886.
MOCSARY, A. ('89). "Monographia Chrysidarum Orbis Terrarum Universi." Budapestini Typis Societatis Franklinianæ. 4to, pp. 1-643.
MORGAN, ANNA HAVEN ('12). "Homologies in the wing-veins of May-flies." Ann. Ent. Soc. Am. Vol. 5, pp. 89-106.
MORGAN, ANNA H. ('13). "A contribution to the biology of May-flies." Ann. Ent. Soc. Am. Vol. 6, pp. 371-413.
MORGAN, T. H., AND SHULL, A. F. ('10). "The life cycle of *Hormaphis hamamelidis*." Ann. Ent. Soc. Am. Vol. 3, pp. 144-146.
MORRILL, A. W. ('03). "The greenhouse aleyrodes (*A. vaporariorum* Westw.) and the strawberry aleyrodes (*A. packardi* Morrill)." Hatch Exp. Sta. Mass. Tech. Bull. No. 1.
MORRILL, A. W., AND BACK, E. A. ('11). "White flies injurious to citrus in Florida." U. S. Dept. Agr., Bur. Ent. Bull. No. 92.
MORRIS, H. M. ('22). "On the larva and pupa of a parasitic phorid fly—*Hypocera incrassata*." Parasitology. Vol. 14, pp. 70-74.
MORSE, A. P. ('20). "Manual of the Orthoptera of New England." Proc. Bost. Soc. Nat. Hist. Vol. 35, No. 6.

Mosher, Edna ('16). "A classification of Lepidoptera based on characters of the pupa." Bull. Ill. State Lab. Vol. 12, pp. 12–159.
Moulton, Dudley ('11). "Synopsis, catalogue, and bibliography of North American Thysanoptera." U. S. Dept. Agr., Bur. Ent. Tech. Ser., Bull. No. 21, 1911.
Müggenburg, F. H. ('92). "Der Rüssel der Diptera pupipara." Arch. f. Naturgesch. Vol. 58, pp. 287–392.
Muir, F. ('12). "Two new species of Ascodipteron." Bull. Mus. Comp. Zool. Vol. 54, pp. 351–366.
Muir, F., and Kershaw, J. C. ('12). "The development of the mouth-parts in the Homoptera, with observations on the embryo of Siphanta." Psyche. Vol. 19, pp. 77–89.
Müller, Johannes (1826). "Zur vergleichenden Physiologie des Gesichtssinnes des Menscher und der Thiere." Leipzig, 1826.
Muttkowski, R. A. ('10). "Catalogue of the Odonata of North America." Bull. Pub. Mus. Milwaukee. Vol. 1, Article 1. Milwaukee, Wis., 1910.
Needham, J. G. ('97). "The digestive epithelium of dragon-fly nymphs." Zool. Bull. Vol. 1, pp. 103–113.
Needham, J. G. ('01). "Aquatic insects in the Adirondacks." N. Y. State Mus. Bul. 47 (1901), pp. 384–560, 573–596.
Needham, J. G. ('03). "A genealogic study of dragon-fly wing-venation." Proc. U. S. Nat. Mus. Vol. 26, pp. 703–764, plates 31–54.
Needham, J. G. ('05). "May-flies and Midges of New York." N. Y. State Mus. Bull. 86.
Needham, J. G. ('18). "Aquatic insects," in "Freshwater biology" by Ward and Whipple, pp. 876–946. John Wiley & Sons, New York, 1918.
Needham, J. G., and Lloyd, J. T. ('16). "The life of inland waters," pp. 1–438, with 244 figures. The Comstock Publishing Co., Ithaca, N. Y., 1916.
Newcomer, E. J. ('18). "Some stoneflies injurious to vegetation." Jour. Agr. Research. Vol. 13, pp. 37–41.
Newport, G. (1839). The article "Insecta." Todd's Cycl. of Anat. and Physiol. London, 1839.
Nininger, H. H. ('20). "Notes on the life-history of *Anthophora stanfordiana*." Psyche. Vol. 27, pp. 135–137.
Noyes, Miss A. A. ('14). "The biology of the net-spinning Trichoptera of Cascadilla Creek." Ann. Ent. Soc. Am. Vol. 7, pp. 251–276.
Nuttall, G. H. F., and Shipley, A. E. ('01). "Studies in relation to malaria. II. The structure and biology of Anopheles." Jour. Hygiene. Vol. 1, pp. 451–483.
Oestlund, O. W. ('18). "Contribution to knowledge of the tribes and higher groups of the family Aphididæ (Homoptera)." Seventeenth Rept. State Ent. of Minn., pp. 46–72.
Osborn, Herbert ('96). "Insects affecting domestic animals." U. S. Dept. Agr., Div. Ent. Bull. No. 5, new series, 1896.
Osten-Sacken ('81). "An essay of comparative chætotaxy, or the arrangement of characteristic bristles of Diptera." Mittheil. Münchener Ent. Ver., 1881, pp. 121–140.
Oudemans, A. C. ('10). "Neue Ansichten über die Morphologie des Flohkopfes, sowie über die Ontogenie, Phylogene und Systematik der Flöhe." Novit. Zool. Vol. 16, pp. 133–158.
Oudemans, J. T. ('87). "Beiträge zur Kenntniss der Thysanura und Collembola." Original in Dutch (*Acad. Proefsch*. Amsterdam, 1887).
Oudemans, J. T. ('88). "Beiträge zur Kenntniss der Thysanura und Collembola." *Bijdragen tot de Dierkunde*. Amsterdam.
Packard, A. S. ('76). "A monograph of the geometrid moths or Phalænidæ of the United States." Rept. U. S. Geol. Survey of the Territories. F. V. Hayden, Vol. 10.
Packard, A. S. ('80). "On the anatomy and embryology of Limulus." Anniversary Memoires of the Boston Soc. of Nat. Hist., 1880.

PACKARD, A. S. ('95). "Monograph of the bombycine moths of America north of Mexico, including their transformations and origin of the larval markings." Mem. Nat. Acad. Sci. Vol. 7.

PACKARD, A. S. ('98). "A text-book of entomology," pp. xvii+729, with 654 text figures. The Macmillan Co., London and New York.

PACKARD, A. S. ('05). "Monograph of the bombycine moths of North America. Part II. Family Ceratocampidæ, subfamily Ceratocampinæ." Mem. Nat. Acad. Sci. Vol. 9.

PANKRATH, O. ('90). "Das Auge der Raupen und Phryganidenlarven." Zeit. wiss. Zool. Vol. 49, pp. 690-708.

PARKER, J. B. ('17). "A revision of the bembicine wasps of America north of Mexico. Proc. U. S. Nat. Mus. Vol. 52, pp. 1-155.

PATCH, EDITH M. ('09). "Homologies of the wing veins of the Aphididæ, Psyllidæ, Aleurodidæ, and Coccidæ." Ann. Ent. Soc. Am. Vol. 2, pp. 101-129, with 6 plates.

PATCH, EDITH M. ('10). "Gall aphids of the elm." Maine Agr. Exp. Sta. Bull. No. 181.

PATTON, W. S., AND CRAGG, F. W. ('13). "A textbook of medical entomology." Christian Literature Soc. for India, London, Madras, and Calcutta.

PECKHAM, G. W. AND E. G. ('98). "On the instincts and habits of the solitary wasps." Wis. Geol. and Nat. Hist. Survey. Bull. No. 2.

PECKHAM, G. W. AND E. G. ('05). "Wasps social and solitary." Houghton, Mifflin & Co., Boston and New York.

PEREZ, J. ('78). "Sur les causes de bourdonnement chez les insectes." Comptes Rend. Acad. des Sci. Vol. 87, p. 378.

PERGANDE, THEO. ('01). "The life-history of two species of plant-lice, inhabiting both the witch-hazel and birch." U. S. Dept. Agr., Div. Ent. Tech. Ser., Bull. No. 9.

PERGANDE, THEO. ('04). "North American Phylloxerinæ affecting Hicoria (Carya) and other trees." Proc. Davenport Acad. Sci. Vol. 9, pp. 185-273, with 21 plates.

PERKINS, R. C. L. ('05). "Leaf-hoppers and their natural enemies (Pt. I. Dryinidæ)." Exp. Sta., Hawaiian Sugar Planters' Association. Bull. No. 1, Part 1.

PETERSON, ALVAH ('15). "Morphological studies on the head and mouth-parts of the Thysanoptera." Ann. Ent. Soc. Am. Vol. 8, pp. 20-66.

PETERSON, ALVAH ('16). "The head-capsule and mouth-parts of Diptera." Illinois Biol. Monogr. Vol. 3, No. 2. Univ. of Ill., Urbana, Ill.

PIERCE, W. DWIGHT ('09). "A monographic revision of the twisted winged insects comprising the order Strepsiptera Kirby." U. S. Nat. Mus. Bull. No. 66.

PIERCE, W. DWIGHT ('11). "Strepsiptera." Genera Insect., Fasc. 121.

PIERCE, W. DWIGHT ('18). "The comparative morphology of the order Strepsiptera together with records and descriptions of insects." Proc. U. S. Nat. Mus. Vol. 54, pp. 391-501.

PLATH, O. E. ('19). "A muscid larva of the San Francisco Bay region which sucks the blood of nestling birds." Univ. Cal. Pub. Zool. Vol. 19, pp. 191-200.

PLATH, O. E. ('22). "Notes on Psithyrus, with records of two new American hosts." Biol. Bull. Vol. 43, pp. 23-44.

POCOCK, R. I. ('10). Article "*Centipede*" in the eleventh edition of the Encyclopædia Britannica.

POCOCK, R. I. ('11). Article "*Millipedes*" in the eleventh edition of the Encyclopædia Britannica.

PRATT, H. S. ('99). "The anatomy of the female genital tract of the Pupipara as observed in *Melophagus ovinus*." Zeit. wiss. Zool. Vol. 66, pp. 16-42.

QUAINTANCE, A. L., AND BAKER, A. C. ('13). "Classification of Aleyrodidæ." U. S. Dept. Agr. Tech. Ser. Bull. No. 27, pp. 1-114.

QUAINTANCE, A. L., AND BAKER, A. C. ('17). "A contribution to our knowledge of the white flies of the subfamily Aleyrodinæ (Aleyrodidæ)." Proc. U. S. Nat. Mus. Vol. 51, pp. 335-445.

RATH, O. VOM ('96). "Zur Kenntnis der Hautsinnesorgane und des sensiblen Nervensystems der Arthropoden." Zeit. wiss. Zool. Vol. 61, pp. 499-539.

RAU, PHIL., AND RAU, NELLIE ('18). "Wasp studies afield." Princeton Univ. Press.

REDIKORZEW, W. ('00). "Untersuchungen über den Bau der Ocellen der Insekten." Inaugural-Dissertation, Universität Heidelberg. Leipzig, Wilhelm Engelmann, 1900.

REES, J. VAN ('88). "Beiträge zur Kenntnis der inneren Metamorphose von *Musca vomitoria*." Zool. Jahrb. Abt. Anat. Bd. 3 (1888).

REHN, A. G., AND HEBARD, M. ('12). "A revision of the genera and species of the group Mogoplistii (Orthoptera; Gryllidæ) found in North America north of the Isthmus of Panama." Proc. Acad. N. S. Phil., June, 1912.

RICHARDSON, HARRIET ('05). "A monograph of the isopods of North America." U. S. Nat. Mus. Bull. No. 54 (1905).

RILEY, C. V. ('69). "First annual report on the noxious, beneficial and other insects of the State of Missouri." Jefferson City, Mo., 1869.

RILEY, C. V. ('73). "On a new genus in the lepidopterous family Tineidæ, with remarks on the fertilization of yucca." Fifth Rept. Ins. Mo., pp. 150-160.

RILEY, C. V. ('77). "On the larval characters and habits of the blister-beetles belonging to the genera *Macrobasis* Lec. and *Epicauta* Fabr.; with remarks on other species of the family Melvidæ." Trans. Acad. Sci. St. Louis. Vol. 3, pp. 544-565.

RILEY, C. V. ('79). "Philosophy of the pupation of butterflies and particularly of the Nymphalidæ." Proc. Am. Ass. Adv. Sci. Vol. 28, 1879.

RILEY, C. V. ('92). "The yucca moth and yucca pollination." Missouri Botanical Garden. Third Ann. Rept., St. Louis, Mo., 1892.

RILEY, WM. A., AND JOHANNSEN, O. A. ('15). "Handbook of medical entomology," pp. IX+348, with 174 figures. The Comstock Publishing Co., Ithaca, N. Y., 1915.

ROGER, O. ('75). "Das Flügelgeäder der Kafer." Erlangen, 1875, 90 p.

ROHWER, S. A., AND CUSHMAN, R. A. ('17). "Idiogastra, a new suborder of Hymenoptera, with notes on the immature stages of Oryssus." Proc. Ent. Soc. Wash. Vol. 19, pp. 89-98.

ROLLESTON, GEORGE ('70). "Forms of animal life." Oxford, Clarendon Press.

ROOT, A. I. AND E. R. ('17). "The A B C and X Y Z of bee culture." The A. I. Root Co., Medina, Ohio.

ROUBAUD, E. ('16). "Recherches biologiques sur les guêpes solitaires et social d'Afrique." Ann. Sci. Nat. Zool. 10e série, Vol. 1, pp. 1-157.

SANDERSON, E. D. ('12). "Insect pests of farm, garden and orchard," pp. XII+684, with 513 figures. John Wiley & Sons, New York, 1912.

SAUNDERS, E. ('91). "On the tongues of the British Hymenoptera." Proc. Linn. Soc. London. Vol. 23, pp. 410-432, with 8 plates.

SAUSSURE, HENRI DE ('75). "Synopsis of American wasps." Washington, Smithsonian Institution.

SAVIGNY, J. C. (1816). "Mémoires sur les animaux sans vertebres." Paris, 1816.

SCHENK, OTTO ('02). "Die antennal Hautsinnesorgane einiger Lepidopteren und Hymenopteren mit besonderer Berücksichtigung der sexual Unterschiede." Zool. Jahrb. Anat. u. Ontogene. Vol. 17, 1902.

SCHIERBEEK, A. ('16). "On the setal pattern of caterpillars." Koninkljke Akademie van Wetenschappen te Amsterdam. Proceedings. Vol. 19, No. 1.

SCHIERBEEK, A. ('17). "On the setal pattern of caterpillars and pupæ." Onderzoekingen verricht in het Zoölogisch Laboratorium der Ryksuniversiteit Gronigen. Vi. E. J. Brill, Leiden, 1917.

SCHIODTE, J. C. ('61-'83). "De metamorphosi eleutheratorum." Kjobenhaven, 1861-1883.

SCHMIEDEKNECHT, OTTO ('09). "Fam. Chalcididæ." Genera Insect. Fasc.. 97.
SCHMIEDER, R. G. ('22). "The tracheation of the wings of early larval instars of Odonata Anisoptera, with special reference to the development of the radius." Ent. News. Vol. 33, Nos. 9-10.
SCHNEIDER, ANTON ('85). "Die Entwicklung der Geschlechtsorgane bei den Insekten." Zool. Beiträge, Breslau. Vol. 1.
SCHWABE, JOSEF ('06). "Beiträge zur Morphologie und Histoligie der tympanalen Sinnesapparate der Orthopteren." Zoologica. Heft 50. Stuttgart, 1906.
SCOTT, HUGH ('17). "Notes on Nycteribiidæ, with descriptions of two new genera." Parasitology. Vol. 9, pp. 593-610.
SCUDDER, S. H. ('93). "'The songs of our grasshoppers and crickets." Twenty-third Annual Report of the Entomological Society of Ontario, 1892, pp. 62-78.
SCUDDER, S. H. ('97). "Guide to the genera and classification of the North American Orthoptera." Edward M. Wheeler, Cambridge, Mass.
SCUDDER, S. H. ('00). "Catalogue of the described Orthoptera of the United States and Canada." Proc. Davenport Acad. Nat. Sci. Vol. 8, pp. 1-101.
SEATON, FRANCES ('03). "The compound eyes of Machilis." Amer. Naturalist. Vol. 37, pp. 319-329.
SEILER, W. ('05). "Ueber die Ocellen der Ephemeriden." Zool. Jahrb. Bd. 22, pp. 1-40.
SEURAT, L. G. ('99). "Contributions à l'étude des Hyménoptères entomophages." Ann. Sci. Nat. Zool. Vol. 10 (1899), pp. 1-159.
SHANNON, R. C. ('22). "The bot-flies of domestic animals." The Cornell Veterinarian, July 1922.
SHARP, DAVID ('95). "Insects." Part I. The Cambridge Natural History. Vol. 5, pp. 83-584, figures 47-371. Macmillan & Co., London and New York, 1895.
SHARP, DAVID ('99). "Insects." Part II. The Cambridge Natural History. Vol. 6, pp. 626, with 293 text figures. Macmillan & Co., London and New York, 1899.
SHIPLEY, A. E. ('98). "An attempt to revise the family Linguatulidæ." Archives de Parasitoligie. Vol. 1.
SIEBOLD, C. T. VON ('44). "Ueber das Stimm und Gehörorgane der Orthopteren." Arch. Naturg. Vol. 10, pp. 52-81, 1 Taf.
SILVESTRI, F. ('96). "I Diplododi." Ann. Mus. Genova, (2). Vol. 16 (1896).
SILVESTRI, F. ('05). "Contribuzione alla conoscenza della metamorfosi e dei costumi della *Lebia scapularis*, con descrizione dell' apparato sericiparo della larva." Estratto dal "Redia." Vol. 2 (1904).
SILVESTRI, F. ('07). "Descrizione di un nuovo genere di insetti apterigoti." Boll. Lab. Zool. Portici, 1907.
SILVESTRI, F. ('13). "Descrizione di un nuovo ordine di insetti." Boll. Zool. Gen. e. Agr. Portici. Vol. 7, pp. 193-209.
SLADEN, F. W. L. ('12). "The humble-bee, its life-history and how to domesticate it, with descriptions of all the British species of Bombus and Psithyrus." Macmillan & Co., London.
SLINGERLAND, M. V. ('94). "The cabbage root maggot, with notes on the onion maggot and allied insects." Cornell Univ. Agr. Exp. Sta. Bull. No. 78.
SLINGERLAND, M. V., AND CROSBY, C. R. ('14). "Manual of fruit insects." pp. xvi+503, with 396 figures. The Macmillan Co., New York.
SMITH, H. S. ('12). "Technical results from the gipsy moth parasite laboratory." U. S. Dept. Agr., Bur. Ent. Tech. Series, Bull. No. 19, Part IV.
SMITH, J. B. ('92). "The structure of the hemipterous mouth." Science. Vol. 19, No. 478.
SMITH, J. B. ('01). "Notes on some digger bees." Jour. N. Y. Ent. Soc. Vol. 9, pp. 29-40, 52-72.
SMITH, L. W. ('17). "Studies of North American Plecoptera." Trans. Am. Ent. Soc. Vol. 43, pp. 433-489.

SMITH, R. C. ('21). "A study of the biology of the Chrysopidæ." Ann. Ent. Soc. Am. Vol. 14, pp. 27-35.
SMITH, R. C. ('22). "The biology of the Chrysopidæ." Cornell Univ. Agr. Exp. Sta. Memoir 58.
SNODGRASS, R. E. ('05). "A revision of the mouth-parts of the Corrodentia and the Mallophaga." Trans. Am. Ent. Soc. Vol. 31, pp. 297-307.
SNODGRASS, R. E. ('09). "The thorax of insects and the articulation of the wings." Proc. U. S. Nat. Mus. Vol. 36, pp. 511-595.
SNODGRASS, R. E. ('10a). "The thorax of the Hymenoptera." Proc. U. S. Nat. Mus. Vol. 39, pp. 37-91, with 16 plates.
SNODGRASS, R. E. ('10b). "The anatomy of the honey bee." U. S. Dept. Agr., Bur. Ent. Tech. Series, Bull. No. 18.
SNYDER, T. E. ('15). "Biology of the termites of the eastern United States, with preventive and remedial measures." U. S. Dept. Agr., Bur. Ent. Bull. 94, Part II, 1915.
STILES, C. W. ('91). "Bau u. Entwicklungsgeschichte v. *Pentastomum proboscideum* Bud. und *Pentastomum subcylindicum* Dies." Zeit. wiss. Zool. Bd. 52, pp. 85-157.
STRAUS-DURKHEIM, H. E. (1828). "Considérations générales sur l'anatomie comparée des animaux articulés, auxquelles on a joint l'anatomie descriptive du Hanneton vulgaire." Paris, 1828.
STURTEVANT, A. H. ('21). "The North American species of the Drosophila." Carnegie Institution of Wash. Publication No. 301.
SWAINE, J. M. ('18). "Canadian bark-beetles, Part II." Dom. Canada. Dept. Agr., Ent. Branch. Bull. 14.
THOMPSON, C. B. ('17). "Origin of castes of the common termite *Leucotermes flavipes* Kol." Jour. Morph. Vol. 30, No. 1, Dec.
THOMPSON, C. B. ('19). "The development of the castes of nine genera and thirteen species of termites." Biol. Bull. Vol. 36, pp. 379-398.
THOMPSON, C. B., AND SNYDER, T. E. ('19). "The question of the phylogenetic origin of termite castes." Biol. Bull. Vol. 36, pp. 115-130.
TILLYARD, R. J. ('16). "Studies in Australian Neuroptera. No. 3. The wing-venation of the Chrysopidæ." Proc. Linn. Soc. New South Wales. Vol. 41, pp. 221-248.
TILLYARD, R. J. ('19). "On the morphology and systematic position of the family Micropterygidæ (sens. lat.)." Proc. Linn. Soc. New South Wales. Vol. 44 (1919), pp. 95-136.
TILLYARD, R. J. ('21). "A new classification of the order Perlaria." Canad Ent. Vol. 53, pp. 35-43.
TILLYARD, R. J. ('22). "New researches upon the problem of the wing-venation of Odonata." Ent. News. Vol. 33, pp. 1-7, 45-51.
TOWER, D. G. ('14). "The mechanism of the mouth-parts of the squash bug, *Anasa tristis* De Geer." Psyche. Vol. 21, pp. 99-108.
TOWER, W. L. ('03). "The development of the colors and color patterns of Coleoptera, with observations upon the development of color in other orders of insects." Decennial Publ. Univ. Chi. Vol. 10, pp. 33-70, 3 plates.
TOWER, W. L. ('06). "Observations on the changes in the hypodermis and cuticula of Coleoptera during ecdysis." Biol. Bull. Vol. 10, pp. 176-192.
TOWNSEND, A. B. ('04). "The histology of the light organs of *Photinus marginellus*." Am. Nat. Vol. 38, pp. 127-148.
TOWNSEND, C. H. ('08a). "The taxonomy of the muscoidean flies, including descriptions of new genera and species." Smithsonian Misc. Coll. Vol. 51, No. 1803.
TOWNSEND, C. H. ('08b). "A record of results from rearings and dissections of Tachinidæ." U. S. Dept. Agr., Bur. Ent. Tech. Ser., Bull. No. 12, Part VI.
TRIGGERSON, C. J. ('14). "A study of *Dryophanta erinacei* (Mayr) and its gall." Ann. Ent. Soc. Am. Vol. 7, pp. 1-34.
ULMER, GEORGE ('07). "Trichoptera." Genera Insect., Fasc. 60.
ULMER, G. ('09). "Trichoptera." Heft 5/6 of "Die Süsswasserfauna Deutschlands" herausgegeben von A. Braur. Jena. Verlag von Gustav Fischer.

Van Duzee, E. P. ('17). "Catalogue of the Hemiptera of America north of Mexico excepting the Aphididæ, Coccidæ, and Aleurodidæ." Univ. Cal. Pub., Tech. Bull. Ent. Vol. 2, pp. I–XIV, 1–902.

Van Duzee, M. C., Cole, F. R., and Aldrich, J. M. ('21). "The dipterous genus *Dolichopus* Latreille in North America." U. S. Nat. Mus. Bull. 116.

Van Rees.—See Rees, J. van.

Verhoeff, K. W. ('03). "Beiträge zur vergleichenden Morphologie des Thorax der Insekten mit Beruchsichtigun der Chilopoden." Nova Acta, Kais. Leop.-Carolin. Deut. Akad. der Naturf. Vol. 81 (1903), pp. 63–109, pls. 7–13.

Verson, E. ('04). "Evoluzione postembrionale degli arti cefalici e toracali nel filugello." Atti del Reale Istituto Veneto di scienze lettere ed arti. Tomo 63 (1903–1904), pp. 49–87, with 3 plates.

Verson and Bisson ('91). "Cellule glandulari ipostigmatiche nel *Bombyx mori*." Bull. Soc. Ital. Vol. 23, pp. 3–20.

Viereck, H. L. ('16). "The Hymenoptera, or wasp-like insects of Connecticut." By H. L. Viereck, with the collaboration of A. D. MacGillivray, C. T. Brues, W. M. Wheeler, and S. A. Rohwer. State Geol. and Nat. Hist. Survey of Connecticut. Bull. 22.

Vogel, Richard ('11). "Ueber die Innervierung der Schmetterlingsflugel und über den Bau und Verbreitung der Sinnesorgan auf denselben." Zeit. wiss. Zool. Vol. 98, pp. 68–134.

Vorhies, C. T. ('09). "Studies on the Trichoptera of Wisconsin." Trans. Wisconsin Acad. Sci., Arts and Letters. Vol. 16, Part I, No. 6.

Wagner, Nicholas ('62). "Spontane Fortpflanzung bei Insectenlarven." Denkschrift d. kais. Kansan'schen Univers.

Wagner, Nic. ('63). "Beiträge zur Lehre von Fortpflanzung der Insectenlarven." Zeit. wiss. Zool. Vol. 13, pp. 514–527, 2 Taf.

Wagner, Nic. ('65). "Über den viviparen Gallmuckenlarven." Zeit. wiss. Zool. Vol. 15, pp. 106–117, 1 Taf.

Walker, E. M. ('14). "A new species of Orthoptera, forming a new genus and family." Canad. Ent. Vol. 46, p. 93.

Walker, E. M. ('19). "The terminal abdominal structures of orthopteroid insects: a phylogenetic study." Ann. Ent. Soc. Am. Vol. 12, pp. 267–316, with 9 plates.

Walker, E. M. ('22a). "Some cases of cutaneous myiasis, with notes on the larvæ of *Wohlfahrtia vigil*." Jour. Parasitology. Vol. 9, pp. 1–5.

Walker, E. M. ('22b). "The terminal structures of orthopteroid insects: a phylogenetic study." Ann. Ent. Soc. Am. Vol. 15, pp. 1–76, with 11 plates.

Walton, W. R. ('09). "An illustrated glossary of chætotaxy and anatomical terms used in describing Diptera." Ent. News. Vol. 20, pp. 307–319, with 4 plates.

Weele, H. W. van der ('08). "Ascalaphiden monographisch Bearbeit." Coll. Zool. du Baron Edm. Longchamps. Fasc. 8, Bruxelles, 1908.

Weele, H. W. van der ('10). "Megaloptera monographic revision." Coll. Zool du. Baron Edm. Longchamps. Fasc. 5, Bruxelles, 1910.

Weismann, A. ('64). "Die nachembryonale Entwicklung nach Beobachtungen an *Musca vomitoria* und *Sarcophaga carnaria*." Zeit. wiss. Zool. Bd. 14 (1864).

Welch, Paul S. ('14). "Habits of the larva of *Bellura melanopyga* Grote (Lepidoptera)." Biol. Bull. Vol. 27, pp. 97–114.

West, Tuffen ('61). "The foot of the fly; its structure and action: elucidated by comparison with the feet of other insects," etc. Part I. Trans. Linn. Soc. Lond. Vol. 23 (1862), pp. 393–421, 3 plates.

Wheeler, W. M. ('00). "The habits of *Myrmecophila nebrascensis* Brunner." Psyche. Vol. 9, pp. 111–115.

Wheeler, W. M. ('07a). "The polymorphism of ants." Bull. Am. Mus. Nat. Hist. Vol. 23, Art. I.

Wheeler, W. M. ('07b). "The fungus-growing ants of North America." Bull. Am. Mus. Nat. Hist. Vol. 23, pp. 669–807.

WHEELER, W. M. ('10). "Ants, their structure, development and behavior." The Columbia University Press, New York.
WHEELER, W. M. ('18). "A study of some ant larvæ, with a consideration of the origin and meaning of the social habit among insects." Proc. Amer. Phil. Soc. Vol. 57, pp. 293-343.
WHEELER, W. M. ('22). "Keys to the genera and subgenera of ants." Bull. Am. Mus. Nat. Hist. Vol. 45, pp. 631-710.
WHEELER, W. M. ('22-'23). "Social life among the insects." Lowell Lectures. Published in *The Scientific Monthly*, Vols. 24-26.
WHEELER, W. M. ('23). "Social life among the insects." Harcourt, Brace & Co., New York.
WHEELER, W. M., AND TAYLOR, L. H. ('21). "*Vespa arctica* Rohwer, a parasite of *Vespa diabolica* De Saussure." Psyche. Vol. 28, pp. 135-144.
WILLEM, VICTOR ('00). "Recherches sur les Collemboles et les Thysanoures." Mém. cour. Mém. sav. étr. Acad. Roy. Belgique. Vol. 58, pp. 1-144.
WILLIAMS, F. X. ('13). "The Larridæ of Kansas." The Kansas Univ. Sci. Bull. Vol. 8, No. 4.
WILLIAMS, F. X. ('19). "Philippine wasp studies." Exp. Sta. Hawaiian Sugar Planters' Assn. Bull. 14, pp. 19-186.
WILLISTON, S. W. ('86). "Synopsis of the North American Syrphidæ." U. S. Nat. Mus. Bull. No. 31.
WILLISTON, S. W. ('96 and '08). "Manual of the families and genera of North American Diptera." Second Edition. New Haven, James T. Hathaway. Third Edition, 1908.
YOUNG, B. P. ('21). "Attachment of the abdomen to the thorax in Diptera." Cornell Univ. Agr. Exp. Sta. Memoir 44.
YUASA, HACHIRO ('22). "A classification of the larvæ of the Tenthredinoidea." Ill. Biol. Monographs. Vol. 7, No. 4. Univ. Ill., Urbana, Ill.
ZIMMERMANN, O. ('80). "Ueber eine eigenthümaliche Bildung des Rückengefässes bei einigen Ephemeridenlarven." Zeit. wiss. Zool. Vol. 34, pp. 404-406.

References to go with *Chitin*, p. 30

For definitions of the words Chitin, Chitinization, Chitinize, see the large English Dictionaries: A New English Dictionary on Historical Principles, Vol. II, Oxford, 1893, Webster's New International Dictionary of the English Language, 1924, Funk and Wagnall's New Standard Dictionary of the English Language, 1928.
NEWPORT, G. Article, Insecta, in Todd's Cyclopaedia of Anatomy and Physiology, Vol. II, p. 801 (1836-39).
ROLLESTON, GEORGE. "Forms of Animal Life." Pp. civ and cxxxiii. 1870.
ODIER, AUGUSTE. "Mémoir sur la composition chemique des parties cornée des insectes." Mém. de la Soc. d'Hist. Natrl. Paris t. i pp. 29-42 (1823).
    Translation of the above with additional remarks by J. G. Children, F. R. S. The Zoological Journal, Vol. I, pp. 101-115 (1825).
FERRIS and CHAMBERLIN. "On the use of the word 'chitinized'." Entomological News, Vol. XXXIX, pp. 212-215 (1928).
    The authors discuss the inconsistency of the use of this word to indicate the hardening of the integument of insects when the term was introduced by Odier to indicate the soft, colorless cuticula after the hardening substances and pigment had been removed by boiling in caustic potash solution.
SNODGRASS, R. E. "Some further errors of body-wall nomenclature in entomology." Entomological News, Vol. XL, pp. 150-154 (1929).
    He emphasizes the objections of Ferris and Chamberlin for the use of chitinize, etc., for hardening the cuticula, and makes some suggestions for improvements.
CAMPBELL, F. L. Detection and estimation of insect chitin. Annals of the Entomological Soc. of America. Vol. xxii, (1929) pp. 401-426 (60 references to chitin).

INDEX

Figures in boldface type refer to pages bearing illustrations.

Aaron, S. F., 952
Abedus, **367**
Acalles, 88
Acalyptratæ, 786, 853
Acanthaclisis, 304
Acanthocephala femorata, **390**
Acanthomyops, 890
Acanthostichus, 959
Accessory cells, 574
Accessory circulatory organs, 122
Accessory glands, 162
Accessory veins, 68
Acerentomidæ, 218
Acerentomon, 218
Acerentomon doderoi, **25**
Acerentulus, 218
Acetabula, 52
Achorutes armatus, 228; *A. nivicola*, **228**; *A. maritima*, 229; *A. socialis*, 228
Acidalia enucleata, wings of, **667**
Acidaliinæ, 663, 666
Acilius, 483
Ackerman, A. J., 998
Acoenitini, 923
Acoloithus falsarius, wings of, **604**
Acone eyes, 141
Acorn-moth, 628
Acrida turrita, **134**
Acrididæ, 252
Acroceridæ, 786, 789, 837
Acrolophidæ, 582, 589, 611
Acrolophus, 611; *A. arcanellus*, 611; *A. mortipennellus*, 611; *A. popeanellus*, 611
Acronycta, 689
Acroschismus bruesi, **547**
Acrydiinæ, 253, 259
Acrydium arenosum obscurum, **260**; *A. granulatum*, **260**
Aculeæ, **573**
Abdomen, 75; appendages of the, 76; segments of the, 75
Adalia bipunctata, 512
Adaptive ocelli, 135, 136
Adela, 598
Adelges, 429; *A. abietis*, **432**; *A. abietis*, gall of, **432**; A. green-winged, **432**; A. pine-bark, 432; A. pine-leaf, 431; *A. pinicorticis*, 432; *A. pinifoliæ*, 431; gall of, **432**; A. wings of, **414**, 428
Adelginiæ, 429
Adelinæ, 598
Adelocephala bicolor, 717
Adelung, N. von., 150

Adephaga, 468, 469, 476
Adipose tissue, 123
Adirondack black-fly, 824
Adoneta spinuloides, wings of, **608**
Adventitious veins, 70
Aëdes, 806, 809; *A. ægypti*, 809; *A. calopus*, 809
Ægeriidæ, 582, 584, 634
Æolothripidæ, 344
Æolothrips fasciatus, 344; *A. nasturtii*, **342**
Æolus dorsalis, **500**
Æschna cyanea, hind-intestine and part of tracheal system of naiad of, **319**
Æschnidæ, 320
Agamic forms of aphids, 415, 416
Agaontinæ, 942, 943
Agaristidæ, 583, 587, 697
Agathidinæ, 921
Agdistis adactyla, 653
Aglais milberti, **755**
Agrilus ruficollis, 503
Agrion, 324
Agrionidæ, 324
Agrionids, true, 324
Agromyzidæ, 786, 792, 861
Agrothereutes extrematis, 928; *A. nuncius*, 928
Agrotinæ, 695
Agrotis c-nigrum, **696**
Ailanthus webworm, 632
Ailanthus-worm, 727
Air-sacs, 118
Akers, Elizabeth, 78
Alabama argillacea, 686, **687**
Alar frenum, 783
Alaus oculatus, **501**
Alcidamea producta, 1000
Alder blight, 421
Alder-flies, 285
Aldrich, J. M., 830, 844, 860, 864, 870
Aleiodinæ, 920
Aleurochiton forbesii, 437, **440**
Aleyrodes, 439
Aleyrodidæ, 177, 400, 437
Aleyrodid, an, **437**
Aleurodothrips fasciapennis, 346
Alimentary canal, 107
Alitrunk, 49
Allaptus, 948
Alleculidæ, 474, 512
Alsophila pometaria, 664, **665**
Alternation of generations, 935
Alveolus, 32

1029

Alula or alulet, 60, 778
Alypia langtonii, **698**; *A. octomaculata*, 697; larva, **698**
Alysiinæ, 920, **922**
Ambient vein, 74
Amblycera, 337
Amblycheila, 477, 478; *A. cylindriformis*, 478
Amblycorypha, 237; *A. oblongifolia*, **236**, 237; *A. rotundifolia*, 237; *A. uhleri*, 237
Amblyteles, 929
Ambrosia-beetles, 542, 544
Ambrysus, 367
Ametabola, 174
Ametabolous development, 174
Ammophila, 890, 984
Ampelœca myron, 659, **660**
Amphibolips confluens, 937; *A. inanis*, 938
Amphicerus bicaudatus, 515
Amphidasis cognataria, **673**
Amphipneustic, 115
Amphipyra pyramidoides, **691**
Amphizoidæ, 470, 481
Ampulicidæ, 891, 909, 913, 977, 978
Anabolia nervosa, **557**
Anacampsis innocuella, 628
Anacharitinæ, 934
Anacrabro, 980
Anajapyx, A. vesiculosus, 224
Anæa, 761; *A. andria*, **761**; *A. morrisonii*, 761; *A. portia*, 761
Anal angle, 60
Anal area, 75; the veins of the, 65
Anal furrow, 73
Anal lobe, **888**
Anal lobes, 445
Anal plates, 445
Anal ring, 445
Anal ring setæ, 445
Anal setæ, 445
Anamorphosis, 218
Anaphothrips striatus, 345
Anarsia lineatella, 627
Anasa tristis, **389**
Anastomosis, 326
Anastomosis of veins, 70
Anax, 317
Anaxipha, 242; *A. exigua*, 243, **244**
Anchor process, 815
Ancylis comptana, 922
Ancyloxipha numitor, 738
André, E., 954
Andrena, 996, 997
Andrenidæ, 891, 893, 914, 995
Andricus erinacei, 936; *A. californicus*, 938
Androconia, **100**, 573
Anepimerum, 51
Anepisternum, 51

Angles of wings, 60
Angle-wings, the, 754
Angoumois grain-moth, 626
Angulifera-moth, 726
Anisandrus pyri, 545
Anisolabis annulipes, 462; *A. maritima*, 462
Anisomorpha buprestoides, 262
Anisopidæ, 785, 787, 788, 797
Anisoptera, 316
Anisopus, 797; wing of, 797
Aniso a, 717, 927; A. rosy, **719**; *A. rubicunda*, 719; *A. senatoria*, 718; *A. stigma*, 718; *A. virginiensis*, **718**
Anobiidæ, 471, 514
Anobium, 79
Anopheles, 808; head of, **775**
Anoplura, 211, 214, 347
Anosia plexippus, head of, 109
Ant, see Ants
Antecoxal piece, 54
Antecubital cross-veins, 319
Antelope-beetle, 523
Antenna cleaner, **886**
Antennæ, 40, **41**; the development of, 199
Antennal fossa, fovea, or groove, 779
Antennal sclerites, 39
Antenodal cross-veins, 319
Anteoninæ, 915
Anterior arculus, 72
Anthicidæ, 475, 498
Anthidium consimile, 968
Anthoboscidæ, 891, 912, 952
Anthocharis genutia = Euchlœ genutia, 748
Anthocoridæ, 356, 358, 377
Anthomyiidæ, 786, 793, 863
Anthomyioidea, 786
Anthonomus grandis, 540; *A. signatus*, **540**; *A. quadrigibbus*, 540
Anthony, Maude H., 113, 282
Anthophora, 890, 996; nests of, **997**; *A. stanfordiana*, 996
Anthrenus museorum, 507; *A. scrophulariœ*, 506; *A. verbasci*, 507
Antispila pfeifferella, wings of, 622
Antlered larvæ, **677**
Ant-lion, head and mouthparts of, larva of, **282**
Ant-lions, **303**
Antocha, 799
Ants, 954; Amazon, 964; carpenter, 963; corn-field, 964; fungus-growing, 962; harvesting, 960; honey, **964**; inquiline or guest, 961; mound-building, 963; slave-maker, 963; thief, 960; typical, 963
Anuraphis maidi-radicis, 419; *A. roseus*, 418
Anurapteryx, 673; *A. crenulata*, 673

Anurida, 47; *A. maritima*, post-antennal organ of, **226**, 229
Anus, 113
Aorta, 122
Apanteles, 921, 927, 947; *A. congregatus*, 921; *A. glomeratus*, 921
Apantesis, 701; *A. virgo*, **701**
Apatela, 689; *A. americana*, **690**; *A. hamamelis*, 690; larva, **691**; *A. morula*, **690**
Apatelinæ, 689
Apatelodes, **707**; *A. angelica*, **708**; *A. torrefacta*, 707
Apechoneura, 930
Apechthis, 925
Apex of the wing, 60
Aphelinidæ, 941
Aphelininæ, 948
Aphididæ, 177, 400, 415, 417; different types of individuals in, 416
Aphidiinæ, 919, 920, **922**
Aphidinæ, bark-feeding, 418; leaf-feeding, 418; root-feeding, 419
Aphidoidea, 400, 412
Aphids, 412; giant hickory-, 418; strawberry-root, 419; typical, 415; woolly, 420
Aphis, corn-root, 419; *A. forbesi*, 419; *A. gossypi*, **413**; *A. melon*, **413**; *A. pomi*, 418; *A.* apple-leaf, 418; rosy apple-, 418; *A.* woolly apple, 421
Aphis-lions, 299
Aphodian dung-beetle, 517
Aphodius, 517; *A. fimetarius*, 517
Aphrophora quadranotata, 403
Aphycus eruptor, 939
Apidæ, 108, 891, 893, 914, **1005**
Apis, 992, 1005; *A. dorsata*, 1005; *A. florea*, 1005; *A. indica*, 1005; *A. mellifica*, 1005; wings of, **884**
Apiocera, wing of, **843**
Apioceridæ, 786, 789, 842
Apocrita, 907
Apodemes, 95, 98
Apoidea, 989, 993
Apophyses, 31
Appendages, the development of, 194
Apple fruit-miner, 631
Apple-maggot, 857
Apple-worm, many-dotted, 691
Apposed image, 143
Aptera, 207
Apterobittacus, 554; *A. apterus*, 554
Apterogyninæ, 916
Apterygogenea, 206
Apterygota, 206, 211, 217
Apyrrothrix araxes, 735
Aquatic Hymenoptera, 889
Arachnida, 9
Aradidæ, 357, 358, 359, 388
Aradus acutus, **389**

Aræocerus fasciculatus, 537
Archanara, 692
Archips argyrospila, 643; *A. cerasivorana*, **642**; *A. fervidana*, 643; *A. rosana*, **642**
Arctiidæ, 583, 588, 699
Arctiinæ, 700
Arculus, **72**
Areolet, 928
Argentine Ant, 962
Argidæ, 891, 895, **904**, 905
Argynnis cybele, **751**
Argyresthia conjugella, 631; *A. thuiella*, 631
Arilus cristatus, **381**
Arixenia esau, 463; *A. jacobsoni*, 463
Arixeniidæ, 463
Armored Scales, 454
Army-worm, **694**
Arolium, 58
Arthoceras, 833
Arthromacra ænea, 514
Arthropeas, 833
Arthropleona, 228
Athropoda, 1
Articular membrane of the setæ, 32
Articular sclerites of the legs, 53; of the wings, 54, **55**
Arzama, 692; *A. obliqua*, 692, **693**
Ascalaphidæ, 284, 305
Aschiza, 786, 847
Ascodipteron, 875
Ascogaster quadridentatus, 921
Ashmead, W. H., 921, 922, 929, 968
Asilidæ, 786, 788, 840
Asparagus-beetle, **530**
Aspicerinæ, 934
Aspidiotus perniciosus, 458
Astatini, 980, 981
Astatus bicolor, 981; *A. unicolor*, 981
Asteiidæ, 786, 792, 860
Asterochiton packardi, 440; *A. vaporariorum*, **439**
Ateuchus, 88; *A. sacer*, 516
Atherix, 835; *A. ibis*, 835, **836**
Atlanicus, 240; *A. davisi*, **240**; *A. testaceus*, **240**
Atropidæ, 333
Atropos divinatoria, 80; *A. pulsatoria*, 334
Atryone conspicua, **738**
Attelabinæ, 538
Attelabus, **538**
Atteva aurea, 632
Attii, 962
Auditory pegs, 147
Audouin, J. V., 49
Augochlora, 996
Aulacaspis rosæ, 458
Aulacid, wing of, **930**
Aulacidæ, 890, 892, 908, 917, **929**

Aulacus, 929
Autographa brassicæ, **687**; *A. falcifera*, **687**, 946
Automeris io, **722**
Axillaries, 54
Axillary cord, 60; A. excision, 61, 778; A. furrow, 74; A. membrane, 60
Axima, 945

Bacillus pestis, 879
Back-swimmers, 362
Bad-wing, 667
Bæus, 933
Bag-worm Moths, 613; Abbot's, **613**; evergreen, 614
Baker, A. C., 419
Baker, C. F., 880
Balaninus, **539**; *B. nasicus*, 539; *B. proboscideus*, 539; *B. rectus*, 539
Baliosus rubra, **534**
Balsa malana, **691**
Baltimore, the, 752
Banded elfin, 770
Banded purple, 758
Banks, Nathan, 280, 296, 303, 307, 951
Barber, H. S., 494
Bark-beetles, 543; fruit-tree, 544; peach-tree, 544
Barnacle scale, **454**
Barnes and Lindsey, 673
Barnes and McDunnough, 602
Basal anal area, 320
Basal anal cell, 327
Basal or subbasal band, 575
Basement membrane, 31, 109, 118
Basilarchia, 750; *B. archippus*, **759**; *B. archippus floridensis*, 760; *B. arthemis*, 758; *B. astyanax*, **758**; *B. proserpina*, 758
Basilona imperialis, **717**
Basswood leaf-roller, 646
Bat-ticks, 875
Bean leaf-roller, 736
Bean-weevil, 535
Bear animalcules, **12**
Beard, 781
Beaver-parasite, 486
Bedbug, 103, 355, 378
Bedbug, big, 382
Bedbug-hunter, masked, 381
Bedellia somnulentella, **616**; wings of, 616
Bee-bread, 1006, 1007
Beech-tree blight, 422
Bee-flies, 838
Bee-lice, **876**
Bee-moth, 650
Bees, 952, 977, 978, 989; bifid tongued, 993; hairs of various, 990; leaf-cutter, 999; three castes of, 1005
Beet or Spinach leaf-miner, 864
Beetles, 464

Beetle, ventral aspect of a, **465**
Beggar, 669
Belidæ, 476, 537
Bella-moth, **700**
Bellesme, J., 92
Bellura, 691; *B. diffusa*, 691, 692; *B. gortynoides*, 691, 692; *B. melanopyga*, 691; B. white tailed, 692
Belonogaster, 972
Belostoma, 367; *B. fluminea*, **367**
Belostomatidæ, 356, 357, 365
Belytidæ, 890, 910, 915, 916, 932
Belytinæ, 910, 915, 932
Bembecia marginata, 636
Bembicini, 981, 986, 988
Bembex, 979, **988**
Benacus, 366; *B. griseus*, 367
Bequaert, J., 889
Berlese, A., 25, 106, 113, 128, 132, 133, 134, 151, 155, 398
Berothidæ, 284, 298
Berotha insolita, **298**
Bethylidæ, 891, 892, 894, 911, 916, 965, 979
Betten, Dr. Cornelius, 559
Bezzi, M., 828, 876
Bibionidæ, 785, 788, 820
Bibio, 820; wing of, 821
Bibiocephala doanei, mouth parts of female, **826**
Big-eyed Flies, 849
Bill-bugs, 540
Bird-lice, 335
Bischoff, H., 952
Bittacomorpha clavipes, 797
Bittacus, 553, **554**
Blackberry crown-borer, 636
Black-dash, 738
Black-flies, 821; innoxius, 824; white-stockinged, 824
Black scale, 453
Black witch, **685**
Blastobasidæ, 582, 589, 591, 628
Blastophaga, 59; *B. psenes*, 943
Blasturus, 312
Blatchley, W. S., 236, 502
Blatchley and Leng, 475, 536
Blatta orientalis, **266**
Blattella germanica, **265**, 266
Blattidæ, 234, 263
Blattoidea, 263
Blepharocera, 144; *B. tenuipes*, 826; wing, 825; section of head, **825**; larva, **826**
Blepharoceridæ, 786, 787, 824
Blissus leucopterus, **387**
Blister-Beetles, 495
Blood, 122
Blood-gills, 114, 120
Blow-fly family, 869
Blues, the, 771

Blue, tailed, 772
Bluebottle-fly, large, 870
Body-segments, 34
Body-wall, 29, 34
Boletotherus cornutus, 513
Bollworm, pink, 628
Bombardier-beetles, 479
Bomb-fly, 868
Bombias, 1002
Bombidæ, 890, 891, 893, 914, 993, 1001, 1003
Bombus, 890, 1002, 1004; nest of, 1004
Bombycidæ, 583, 585, 727
Bombyliidæ, 786, 788, 789, 838
Bombylius, 838
Bombyx mori, 128, 727
Book-lice, 331, 333
Book-louse, mouth-parts of, 331
Boophilus annulatus, 2
Borboridæ, 786, 791, 794, 855
Boreus, 550, 551, 553
Boriomyia, 297
Borner, C., 430, 432, 876
Bot-flies, 864; common, 865; of Horses, 864; red-tailed, 866; sheep, 867; stomach, 865
Bothropolys multidentatus, 21
Bostrichidæ, 471, 515
Brachelytra, 467, 468; families of the, 470
Brachinus, 479, 480
Brachyacantha, 512
Brachycentrus nigrisoma, 569, 570
Brachycera, 786, 828; Anomalous, 786, 829; True, 786
Brachygastra lecheguana, 973
Brachypauropodidæ, 19
Brachyrhinus ovatus, 539; *B. sulcatus*, 539
Brachystola magna, 257
Bracon, 919
Braconidæ, 890, 892, 908, 915, 917, 919
Braconinæ, 920, 921
Bradley, Prof. J. C., 886, 889, 908, 950, 953, 954, 966, 979
Brand, 738
Brathinidæ, 471, 486
Brathinus, 486
Brauer, F., 206
Braula, 876; *B. cæca*, 876; *B. kohli*, 876
Braulidæ, 787, 794, 876
Braun, A. F., 575, 592, 601
Breast bone, 815
Bremidæ, 890
Bremus, 890
Brenthis, 752
Brentidæ, 476, 536
Brephinæ, 663, 664
Brephos infans, 664
Brindley, H. H., 172

Brine-flies, 859
Bristle-tails, 219
Bristles, antipygidial, 879; the abdominal of Diptera, 785; the cephalic of Diptera, 781; of the legs of Diptera, 785; the thoracic of Diptera, 783, 784
Brown-tail moth, 682
Bruchophagus funebris, 945
Brues, C. T., 931
Brychius, 481
Bubonic plague, 879
Bucca, 779
Bucculæ, 353
Bucculatrix, apple, 616; *B. pomifoliella*, 616; cocoons of, 616
Bucculatrigidæ, 617
Buckley, 955
Bud-moth, 641
Buenoa, 364
Buffalo-gnat, Southern, 824
Bugnion and Popoff, 398
Bugs, ambush-, 382; assassin-, 380; burrower-, 391, 392; chinch-, 386, 387; creeping water-, 367; flat-, 388; flower-, 377; four-lined leaf-, 375; giant water-, 365; harlequin cabbage-, 391; insidious flower-, 378; lace-, 384; leaf-, 375; many-combed, 379; negro, 391, 392; shield-backed, 392; shore, 369; squash-, 389; stilt-, 388; stink-, 390; tarnished plant-, 376; thread-legged, 382; toad-shaped, 368; true, 350; unique-headed, 383
Bullæ, 74
Bumblebees, 1001, 1002; parasitic, 1004
Buprestidæ, 472, 502
Buprestid, Virginian, 503
Buprestis, 907
Burges, E., 109, 160
Burke, H. E., 907
Bursa copulatrix, 159
Busck, A., 625, 628
Busck and Boving, 594
Butterflies, 571, 581, 583, 739
Buzzing of flies and bees, 91
Byrrhidæ, 471, 508
Byturidæ, 472, 510
Byturus, 510; *B. unicolor*, 510

Cabbage-butterfly, 746; development of the wings of, 196
Cabbage looper, 687
Cabbage-root maggot, 863
Cacœcia, 642
Caddice-Flies, 555; Micro-, 561
Caddice-worms, 557
Cadelle, 508
Cæcum, 113
Cænis, 309, 312

Cænomyia ferruginea, wing of, 833
Cænurgia, 688; *C. crassiuscula*, 688; *C. erechtea*, **688**
Calamoceratidæ, 559, 560, 567
Calandra granaria, 541; *C. oryzæ*, 541
Calendrinæ, 540
Calephelis borealis, 767; *C. virginiensis*, 767
Caliroa cerasi, 904
Calledapteryx dryopterata, 709
Callibælis, 312
Calliceratidæ 890, 932
Callimome druparum, 943
Callimominæ, 890, 943
Calliphora, 869; *C. erythrocephala*, 870; *C. vomitoria*, 870
Calliphoridæ, 787, 793, 869
Callosamia angulifera, 726; *C. promethea*, **724, 725**; cocoon of, **189**
Callizzia amorata, 709
Calocalpe undulata, **668, 669**; eggs and nest of, 669
Caloptenus italicus, 149
Calopteron reticulatum, 491
Calopteryx, 324
Calosoma calidum, 479, **480**; *C. scrutator*, **479**; *C. sycophanta*, 479
Calypteratæ, 786, 862
Calypteres, 778
Camel-crickets, 241
Campodea, 157, 161; *C. staphylinus*, **224**
Campodeidæ, 224
Campodeiform, 184
Camponotus herculeanus pennsylvanicus, 963
Campoplegini, 927
Campsomeris, 954
Camptonotus carolinensis, **240**
Campylenchia latipes, 405
Canaceidæ, 786, 792, 859
Canace snodgrassi, 859
Candle-fly, Chinese, 409
Canker-worm, spring, 671; fall, 664; control of, 665
Cantharidæ, 473, 492
Cantharis vesicatoria, larva of, **117**
Canthon lævis, **517**
Capitate, 41
Capitonius, 930
Capnia, 179, 330; *C. pygmæa*, 326; *C.* wing of, 330
Capniidæ, 330
Caprification, 944
Caprifigs, 944
Capsidæ, 375
Carabidæ, 470, 478
Carabus auratus, alimentary canal of, **110**
Cardo, 44
Caripeta angustiorata, wings of, **662**
Carlet, G., 89, 90

Carolina locust, 82
Carpenter, G. H., 17
Carpenter-bee, large, 972, 998; small, 997
Carpenter-moth, locust-tree, **603**
Carpenter-moths, 601
Carpenter-worm, lesser oak, 603
Carpet-beetle, 506
Carpocapsa pomonella, **639**
Carposina fernaldana, 644
Carposinidæ, 582, 590, 638, 644
Carrière and Burger, 103
Carrion-Beetles, 487
Carrot rust-fly, 859
Case-bearer, cigar, 621
Cat-flea, 882
Catocala, 688; *C. fraxini*, wings of, **684**; *C. ilia*, **688**
Catocalinæ, 687
Catopsilia eubule, 749
Cat-tail moth, 629; noctuids, 692
Caudell, A. N., 240, 241, 272
Caulocampa acericaulis, 903
Cave-crickets, 241
Cebrionidæ, 473, 499
Cecidomyia albovittata, 817
Cecidomyiidæ, 785, 787, 788, 813
Cecidomyiinæ, 816, 817
Cecropia-moth, 726
Cedar-beetles, 499
Cedar tineid, 631
Celama triquetrana, 705
Celerio lineata, 660, **661**
Celery looper, 687
Cells of the wing, terminology of the, 72
Celonites abbreviatus, 967
Cenchri, 887
Centipedes, 20
Cephaloidæ, 474, 494
Cephidæ, 890, 891, 900
Cephus pygmæus, 901; wings of, **901**
Cerambycidæ, 475, 524
Cerambycids, typical, 526
Cerambycinæ, 526
Ceramica picta, **694**; larva, **695**
Ceramius lusitanicus, 967
Cerapachyinæ, 958, 959
Cerapachys, 959
Ceraphronidæ, 890, 910, 915, **932**
Ceraphroninæ, 933
Cerastipsocus venosus, **331**
Ceratina dupla, 946, 997; nest of, **998**
Ceratocombus, 374
Ceratogastra, 926
Ceratophillidæ, 882
Ceratophyllus fasciatus, 878; *C. multispinosus*, 878
Ceratopogon, 136
Ceratubæ, 446
Cercerini, 980, 986
Cerceris clypeata, **986**; *C.* wings of, **986**

Cerci, 24, **77**, 232
Cercopidæ, 400, 402
Cercyonis alope, **762**; *C. alope nephele*, 763; *alope maritima*, 763
Ceresa bubalus, **404**; *C. diceros*, 405
Ceropales, 951
Cerophytidæ, 472, 499
Cerophytum, 499
Ceroplastes, 453; *C. cirripediformis*, **454**
Cerores, **446**
Cervical sclerites, 40
Ceuthophilus, 232, **241**; *C. lapidicola*, **233**; *C. maculatus*, **241**; *C. uhleri*, 241
Chætopsis ænea, 856
Chain-dotted geometer, **670**
Chalarus, 850
Chalastogastra, 890, 891, 894, 895
Chalcid-flies, **939**
Chalcid-fly, wing of, **939**
Chalcididæ, 892, 894, 911, 915, 941
Chalcidinæ, 941, 942, 944
Chalcidoidea, 890, 939, 940
Chalcis, 945
Chalicodoma muraria, 1001; nests of, 1001
Chalcophora virginica, **503**
Chalybion cærulium, 954, 984
Chamæsphecia tipuliformis, 637
Chamæyidæ, 862
Chamyris cerintha, **689**
Chapman, T. A., 593
Charipinæ, 934
Chauliodes, 288; *C. pectinicornis*, 288; *C. rastricornis*, 288
Chauliognathus, 492; *C. marginatus*, **492**; *C. pennsylvanicus*, **492**
Cheeks, 779
Cheek-grooves, 779
Cheese-maggot, 858
Chelonariidæ, 471, 506
Chelonarium lecontei, 506
Cheloninæ, 920
Chelophores, 11
Chelymorpha cassidea, 534
Chemical sense organs, 130, 132
Chermes, 429; *C. abieticolens*, 432; *C. viridis*, 432
Chermidæ, 400, 410
Cherry-fruit-flies, 857
Cherry-tree ugly-nest tortricid, **642, 643**
Cheshire, F. R., 102, 103
Chiasognathus, 88
Chickweed geometer, 666
Chigoe, 883
Child, C. M., 153, 154
Chilopoda, 20
Chimarrha aterrima, **563**
Chinch-bug, 386
Chin-fly, 866
China wax, 102, 441

Chionaspis pinifoliæ, **456**, 458
Chionaspis furfura, **441**, 457
Chionea, 799
Chiromeles torquatus, 463
Chironomidæ, 785, 788, 793, 802
Chironomus, 120, 147, 148; wing of, **803**
Chirotonetes albomanicatus, wing of, 309, 310
Chitin, 30
Chitinized tendons, 95
Chloealtis conspersa, **259**
Chlorion, 984; *C. cyaneum*, 984; *C. ichneumoneum*, 984; *C. atratum*, 987
Chlorippe celtis, 761; *C. clyton*, **760**
Chloropidæ, 786, 792, 860
Chordotonal ligament, 147
Chordotonal organs, 145, **146**, 147, **148**; of the Acridiidæ, 148, **149**; of the Locustidæ and of Gryllidæ, 149
Choreutinæ, 633
Chortophaga viridifasciata, 257
Choruses, 93
Cholodkovsky, N. A., 432
Chrysalid, 186
Chrysalis, 186
Chrysanthemum gall-midge, 820
Chrysididæ, 891, 892, 911, 950, 951
Chrysis nitidula, 952
Chrysobothris femorata, **503**
Chrysomelidæ, 475, **530**
Chrysomphalus tenebricosus, pygidium of, **448**
Chrysomyia macellaria, 870
Chrysopa, 170, 171, 300; *C. nigricornis*, wings of, **300, 301**, 302
Chrysopidæ, 109, 284, 299
Chrysops, 830; *C. niger*, **830**
Chylestomach, 111
Chyphotinæ, 916
Cicada, head of, **397, 398**; the musical organs of a, 89, **90**; tracheation of wings of, **395**, 396
Cicada-killer, 987
Cicada, periodical, 402
Cicada plebia, 89
Cicadas, 177, 401
Cicadidæ, 400, 401
Cicadellidæ, 400, 406, 978
Cicindela, 476, **477**, 478; maxilla of, **45**
Cicindelidæ, 469, 476
Cicinnus melsheimeri, **713**
Cigarette-beetle, 515
Cimbex americana, **902**; blade of ovipositor of, **895**
Cimbicidæ, 890, 891, 895, 902
Cimex, 378; *C. lectularius*, 378, **379**; *C. pilosellus*, 378
Cimicidæ, 356, 358, 378
Cingilia catenaria, **670**
Circular-seamed Flies, 846
Circulation of the blood, 122

Circulatory system, 121
Circumfili, 814
Cirphis unipuncta, **694**
Cisidæ, 474, 515
Citheronia regalis, **716**, 717; wings of, **714**
Citheroniidæ, 583, 715
Claassen, P. W., 327, 692
Cladius isomerus, 904
Clambidæ, 470, 488
Clastoptera, 403; *C. cbiusa,* 404; *C. proteus,* 403
Clavate, 41
Clavicornia, 467, 469
Claviform spot, 575
Clavigeridæ, 470, 490
Clavola, 41
Clavus, 350
Clear-winged moths, **634**
Clear-wing, thysbe, **661**
Clemensia albata, 705
Cleonyminæ, 941, 942, 946
Cleptes, 951
Cleptidæ, 890, 911, 950, 951
Cleridæ, 473, 493
Click-beetles, **499**
Climacia dictyona, 292; cocoon and cocoon-cover of, **292**
Clisiocampa americana, 170
Clistogastra, 890, 891, 894, 907, 908
Cloaked knotty-horn, **527**
Cloëon, head of, **144**; *C. dipterum,* 312
Close-wings, 649
Closing apparatus of the tracheæ, **116**
Clothes-moths, 612; case-bearing, 612; naked, 612; tube-building, 612
Clothilla pulsatoria, 80
Clothing hairs, 33
Clouded sulphur, 748
Cloudy-wing, northern, 737; southern, 737
Clover-flower midge, 818
Clover-hay worm, **649**
Clover-leaf midge, 818
Clover-looping-owlets, 688
Clover-root borer, **542**
Clover-seed caterpillar, 641
Clover-seed chalcid, 945
Clover-worm, green, 685
Clusiidæ, 786, 790, 854
Cluster-fly, 870
Clypeus, 38, 779
Cnaphalodes strobilobius, 430
Cnephia pecuarum, 824
Coarctate pupæ, 191
Coboldia formicarium, 821
Coccidæ, 177, 400, 401, 440
Coccid-eating pyralid, **652**
Coccids, the Cochineal, 450; the ensign, 450; the giant, 449; metamorphosis of, 448; mouth-parts of a, **443**; the pseudogall, 454
Coccinæ, 450
Coccinella novemnotata, 512
Coccinellidæ, 473, 511
Cochineal, 441
Cochlidiidæ, 608
Cockroach, head of a, **38**; head and neck of, **39**; internal anatomy of, **107**; labium of a, **46**; tentorium of a, **96**; the base of a leg of a, **53**
Cockroach, American, 266, **267**; common wood-, 266, **267**; oriental, 266; wing of nymph of, **265**
Cockroaches, 263, 950
Coccophagus, 948
Coccus cacti, 441, 450
Cocoon, 188; modes of escape from the, 188
Codlin-moth, **639**
Cœlioxys, 993
Cœnagrionidæ, 324
Cænomyia ferruginea, 834
Cœnomyiidæ, 786, 789, 834
Coffee-bean weevil, 537
Colænis julia, 764
Cole, F. R., 838, 844
Coleophora, 620; *C. fletcherella,* **621**; *malivorella,* **620**, 621
Coleophoridæ, 582, 591, 620
Coleoptera, 109, 211, 213, 214, 215, 218, 464; synopsis of the, 467
Colleterial glands, 160
Colletes, 991, 992; *C. compacta,* 994
Colletinæ, 914, 994
Collembola, 214, 217, 225
Collophore, 76, 227
Collops quadrimaculatus, **493**
Colonici, 434
Colopha eragrostidis, 423; *C. ulmicola,* 422, **423**
Coloradia, 719; *C. pandora,* 721
Colorado potato-beetle, **531**
Colors of butterflies and moths, 573
Colydiidæ, 474, 510
Colymbetes, 483
Comma, gray, 757
Comma, green, 756
Commissure, 125
Complete metamorphosis, 180
Compound eyes, 134, 139; absence of, 135; dioptrics, 141
Compton tortoise, **756**
Comstock, Mrs. A. B., 1007
Comstockaspis perniciosa, 458
Concave veins, 73
Coniopterygidæ, 307
Conjunctiva, 34
Connectives, 123
Connexivum, 355
Conocephalinæ, 235, 238

Conocephalus, 86, 87, **233**, **238**; *C. fasciatus,* **233**, 941
Conopidæ, 786, 790, 791, 853
Conops, **854**; wing of, 60
Conorhinus, 382
Conotrachelus nenuphar, 539
Contarinia pyrivora, 820
Convex veins, 73
Cook, F. C., 873
Copidosoma gelechiæ, 889; *C. truncatellum,* 946
Copiphorinæ, 235, 239
Copper, American, 771; bronze, 771
Copper hindwing, 691
Coppers, 770
Copris, 517; *C. carolina,* 517
Coptodisca splendoriferella, 622, **623**
Coquillett, D. W., 823, 845, 871
Corbicula, 991
Cordulegaster, 320
Cordyluridæ, 786, 790, 854
Coreidæ, 357, 359, 389
Corethra, 121; *C. culiciformis,* **154**; *C. plumicornis,* 807
Corethrinæ, 806
Corisa, 360
Corium, 350
Corixa, 360; *C.,* eggs of, 362; *C. mercenaria,* 362
Corixidæ, 356, 357, 360
Corn-borer, European, 648
Cornea, 138, 139
Corneagen, 138
Corneal hypodermis, 138, 139
Corn ear-worm, 695
Corneas of the compound eyes, **36**; of the ocelli, 37
Corn stalk-borer, larger, 650
Corrodentia, 212, 213, 214, 217, 331
Corrugations of the wings, 73
Corydalinæ, 286
Corydalus, 62, 111, 119, 125, 126, 136; head of, **39**; head of a larva of, **38**, 137; *C.,* larva of, **288**; *C. cornutus,* 287, **288**; wing of pupa of, **287**
Corylophidæ, 471, 488
Corynetidæ, 473, 493
Corythucha arcuata, **384**
Cosmocoma elegans, **949**
Cosmoplite, 755
Cosmopterygidæ, 582, 591, 592, 629
Cosmopteryx, 630
Cosmosoma myrodora, **707**
Cossidæ, 582, 585, 601
Cossinæ, 603
Cossus ligniperda, 104, 105
Costa, 64
Costal fold, 735
Costal margin, 60
Costal spines, 575
Cosymbia lumenaria, 666

Cotalpa lanigera, 519
Cotinus nitida, 522
Cotton-boll weevil, 540
Cotton-boll worm, 695
Cotton-moth, maxillæ of the, **575**
Cotton-stainer, 385
Cotton-worm, 686
Cottony maple-scale, **453**
Cow-killer ant, 954
"Cow-shed" built by ants, **961**
Coxa, 56
Coxal cavities, 52
Coxites, 222
Crabro, 989; *C. singularis,* wings of, **988**
Crabronidæ, 890
Crabroninæ, 988
Crabronini, 980, 989
Crambidia pallida, 705
Crambinæ, 649
Crambus, **650**; *C. caliginosellus,* 650; *C. hortuellus,* 650
Crampton, G. C., 40, 43, 49, 233, 272, 592
Cranberry fruit-worm, 652
Crane-flies, **795**; phantom, 797; primitive, 796; so-called false, 797; typical, 798
Crawford, D. L., 860, 931
Cray-fishes, 6
Cremaster, 187
Cremastini, 923
Cremastogaster lineolata, 960
Creophilus maxillosus, 489
Crescent-spots, 752
Cresson, E. T., 856
Cricetomys, 463
Cricket, head of a, **37**, **40**, 136; part of the tentorium of a, **96**
Crickets, 242; ant-loving, 249; field-, 247; house-, 248; larger brown bush-, 244; larger field-, 248; mole-, 250; pigmy mole-, 251; smaller field-, 248; sword-bearing, 243; tree-, 245; wingless bush-, 250
Crioceris asparagi, **530**
Crista acustica, 152
Crochets, 578
Crop, 110
Crosby, C. R., 377, 459, 889, 943, 946
Crosby and Leonard, 575
Cross-veins, 64, **71**
Crotch, 88
Croton-bug, **265**, 266
Crumena, 444
Crura cerebri, 123
Crustacea, 6
Cryptidæ, 890
Cryptinæ, 924, 927
Cryptini, 923, 928
Cryptochætum, 861
Cryptophagidæ, 472, 474, **510**

Cryptoptilum trigonipalpum, **250**
Crystalline cone-cells, 140
Ctenidia, 879
Ctenocephalus canis, 882; *C. felis*, 878, 882
Ctenucha virginica, **706**
Cuban termite, nest of, **279**
Cubital area, 320
Cubito-anal fold, 73
Cubitus, 64
Cuckoo-wasps, 951, 952
Cucujidæ, 472, 474, 509
Cucujo, 165
Cucujus clavipes, **509**
Cucullia, 694; *C. speyeri*, **694**
Cuculliinæ, 693
Cucumber flea-beetle, 533
Culex, **153**, 806, 807; larva of, **805**; *C. ægypti*, 809
Culicidæ, 785, 787, 804
Culicinæ, 807
Culicoides, 803
Cupesidæ, 471, 494
Curculio, apple-, 540
Curculionidæ, 476, 537
Curculioninæ, 539
Curculios, 537
Curicta, 364
Currant borer, imported, 637
Currant fruit-fly, 858
Currant fruit-worm, 644
Currant moth, pepper-and-salt, **673**
Currant span-worm, **670**
Currant stem-girdler, 901
Currant-worm, imported, 903
Cushman, R. A., 906, 907
Cuticula, 30
Cuticular nodules, 31
Cut worms, 696
Cybister, 483
Cycloplasis panicifoliella, 634
Cyclops, **6**
Cyclorrhapha, 786, 846; with a frontal suture, 852; without a frontal suture, 847
Cyclostomi, 920
Cydnidæ, 357, 359, 391
Cydninæ, 392
Cyladinæ, 538
Cylas formicarius, 538
Cylisticus convexus, **7**
Cyllene caryæ, 528; *C. robinæ*, **527**
Cymatophora ribearia, **670**
Cynipidæ, 890, 892, 894, 915, 934
Cynipinæ, 915, 934
Cynipoidea, 890, 934
Cynomyia cadaverina, 870
Cypridopsis, **6**
Cyrtidæ, 837
Cyrtomenus mirabilis, **392**
Cyrtophyllus concavus, 93

Cyrtoxipha columbiana, 244

Dactylopius, **28**
Dactyls, 251
Dagger, American, 690; ochre, 690; witch-hazel, 690
Dalceridæ, 582, 585, 605
Dalcerides ingenita, 606
Dalla Torre, 347, 934
Damsel-flies, 314, **315**, 321; naiad of, 323; stalked-winged, 324; tracheal gill of a, **120**, 323
Danainæ, 750, 765
Danaus archippus, 759, **765**; chrysalis, **766**; larva, **766**; *D. berenice*, 766; *D. berenice strigosa*, 766
Dance-flies, 845
Daphnia, **6**
Darkling Beetles, 513
Darwin, Charles, 88, 181
Dascillidæ, 472, 505
Dascillus cervinus, 505
Dasymutilla occidentalis, 953
Dasyneura leguminicola, 818; *D. trifolii*, **818**
Datana, **28, 675**; *D. ministra*, 675; larva, **675**
Davidson, Dr. Anstruther, 968
Day-eyes, 142
Death-watch family, 514
Death-watch, 80, 515
Decatoma, 945
Decticinæ, 235, 239
Decticus verrucivorus, 150, **151, 152**
Definitive accessory veins, 69
Deltoids, 685
De Meijere, 58
Dendroleon, 304
Dengue, 810
Densariæ, 447
Dentes, 228
Depressaria heracliana, 624; wings of, **624**
Dermaptera, 212, 214, 460
Dermestes lardarius, **506**
Dermestidæ, 472, 506
Dermis, 31
Derobrachus brunneus, 526
Derodontidæ, 473, 510
Desmia funeralis, **646**
Desmocerus palliatus, **464, 527**
Deutocerebrum, 124
Development without metamorphosis, 174
Dewitz, H., 101
Dexiinæ, 871
Diabrotica, 532; *D. duodecimpunctata*, 532; *D. longicornis*, 532; *D. soror*, 532; *D. vitata*, 532
Diacrisia virginica, **703**
Dialeurodes citri, 440

Diamond-back moth, 632
Diaphania hyalinata, **647**; *D. nitidalis*, 648
Diapheromera, 169; *D. femorata*, **261**
Diapriinæ, 910, 932
Diarthronomyia hypogæa, 820
Diaspidinæ, 454
Diatræa saccharalis, 650; *D. zeacolella*, 650
Dibrachys boucheanus, 947
Dicælus, 480
Dicerca divaricata, **503**
Dichorda iridaria, wings of, **666**
Digitate mine, 620
Digitules, 442
Digitus, 45
Dilar americanus, 297
Dilaridæ, 284, 297
Dinetini, 980, 982
Dineutus, 484
Diodontus, 985
Dione vanillæ, 764
Diopsidæ, 786, 791, 859
Dioptidæ, 583, 585, 673
Diplazoninæ, 923
Diplazon lætatorius, 926
Diplolepis centricola, 938
Diplopoda, 15
Diploptera, 966; The Solitary, 967
Dipsocoridæ, 356, 358, 374
Diptera, 212, 214, 215, 217, 773
Diptera, synopsis of the, 785
Discal cell, 74, 886
Discal vein, 74
Disholcaspis globulus, 938
Dissosteira carolina, 82, 233, **234**, **258**
Distal retinula cells, 140
Divers, 691
Diver, black-tailed, 691; brown-tailed, 692
Diverse-line moth, 668
Divided eyes, 144
Diving-beetles, the predacious, **482**
Dixa, larva of, 800; wing of, 800
Dixidæ, 785, 788, 800
Doane, Prof. R. W., 968
Dog-flea, 882, **877**
Dog's head, 749
Dolichoderinæ, 958, 962
Dolichopodidæ, 786, 789, 843
Dolichopus coquilletti, wing of, **844**; *D. lobatus*, 843
Dolichurus, 978
Donacia, 530
Donisthorpea, 890
Dorcus parallelus, 523
Dorsal diaphragm, 121, 162
Doru aculeatum, **463**
Doryctinæ, 920
Dorylinæ, 958
Douglasiidæ, 582, 591, 623

Doyère, M., 12, 13
Dræculacephala, 407; *D. reticulata*, 407
Dragon-flies, 314, 316
Dragon-fly, exuviæ of naiad of, **319**
Drepana arcuata, **711**, 712
Drepanidæ, 583, 587, 588, 710
Drone-fly, 851
Drones, 1005
Drosophila ampelophila, 861; *D. melanogaster*, 861; egg of, 168
Drosophilidæ, 786, 792, 860
Dryinidæ, 891, 892, 911, 915, 978, 979
Drug-store beetle, 515
Dryopidæ, 471, 504
Ducke, A., 969
Dufour, L., 110
Dung-beetles, earth-boring, **517**
Dung-flies, 854, 855
Dusky-wing, Martial's, 737
Dyar, H. G., 33, 173, 579, 610, 807, 810
Dynastes grantii, 520; *D. hercules*, 26, 520; *D. tityrus*, **520**
Dysdercus suturellus, 385, **386**
Dysodia oculatana, 654
Dyspteris abortivaria, **667**; wings of, 668
Dytiscidæ, 470, 482
Dytiscus, **483**

Earinus limitaris, 921
Earwig, European, 463; hind wing of an, **461**; handsome, 462; little, **462**; ring-legged, 462; seaside, 462; spine-tailed, 463
Earwigs, **460**
Eaton, A. E., 311
Ecdysis, 171
Echidnophaga gallinacea, 883
Echidnophagidæ, 882
Eciton, 959
Ectoderm, 29
Ectognatha, 222
Ectotrophi, 220
Egg, 166; -burster, 171; -calyx, 159; -follicles, 158; -tooth, 171
Egg-masses of caddice-flies, **557**
Eggers, F., 577
Ejaculatory duct, 162
Elachertidæ, 941
Elachista quadrella, wings of, **622**
Elachistidæ, 582, 591, 621
Elasminæ, 942, 948
Elasmus, 948
Elateridæ, 472, 499
Eleodes, **513**
Elephantiasis, 810
Elis (*Myzine*), 953
Ellipes, 252; *E. minuta*, 252
Elm-gall colopha, cockscomb, 422
Elm-gall tetraneura, cockscomb, 424
Elmidæ, 471, 504

Elophila fulicalis, 649
Elytra, 59, 212, 464
Embia major, 340; *E. sabulosa*, **338**; *E. texana*, **339**
Embiidina, 209, 211, 213, 215, 338
Embolemidæ, 890, 911, 912, 915, 951
Embolium, 351
Emesa brevipennis, **382**
Emory, 955
Emperor, tawny, **760**; gray, 761
Empididæ, 786, 789, 790
Empoasca fabæ, 407
Empodium, 58, **778**
Empoa rosæ, 407
Enchenopa binotata, **405**
Enchroma gigantea, **465**
Encoptolophus sordidus, **257**
Encyrtinæ, 941, 942, 946
Enderlein, G., 338, 813, 823
Endomychidæ, 473, 474, 511
Endo-skeleton, 95
Endothorax, 97
Endrosis lacteella, 624
Eneopterinæ, 243, 244
Engraver-beetles, 542, 543
Enicocephalidæ, 356, 358, 383
Enicocephalus formicina, 383
Enicospilus, 927
Ennomos magnarius, **672**
Ensign-flies, 949
Entedontidæ, 941
Entognatha, 224
Entomobryidæ, 229
Eosentomidæ, 218
Eosentomon, 218
Epargyreus tityrus, **734**, **736**
Epermenia pimpinella, 631
Ephemera, 312; *E. simulans*, 312; *E. varia*, 178, **311**
Ephemerella, 312
Ephemerida, 180; ocelli of, 139, 209, 211, 212, 213, 308
Ephemeridæ, 312
Ephestia kuhniella, 651
Ephydra, 859
Ephydridæ, 786, 792, 794, 859
Epiblemidæ, 639
Epicærus imbricatus, **538**
Epicauta, 497; *E. cinerea*, 497; *E. pennsylvanica*, **497**; *E. vittata*, **496**
Epicnaptera americana, **732**
Epicnemium, 887
Epidermis, 31
Epicranial suture, 37
Epicranium, 38
Epilachna borealis, 512; *E. corrupta*, **512**
Epimartyria auricrinella, 593; *E. pardella*, 593
Epimerum, 51
Epinotum, 887

Epipharynx, 46
Epiphysis, 576
Epiplemidæ, 583, 587, 708
Epipleuræ, 74
Epiponinæ, 966, 973
Epipyropidæ, 582, 610
Epipyrops barberiana, 610; *E. anomala*, 610
Episternum, 51
Epistoma, 779
Epithelium, 109, 118; of mid-intestine, 112
Epitrix cucumeris, 533
Epizeuxis lubricalis, **685**
Epochra canadensis, 858
Epomidiopteron, 953
Erannis tiliaria, 671, **672**
Erastriinæ, 689
Erax apicalis, **841**
Erax, wing of, **841**
Erebinæ, 685
Erebus agrippina, 26
Erebus odora, 685, **686**
Eretmoptera browni, 802
Ericerus pe-la, 441
Eriococcinæ, 450
Eriococcus araucariæ, **444**
Eriocraniidæ, 581, 584, 593
Eriosoma americana, wings of, **414**
Eriosoma lanigera, 421
Eriosomatinæ, 420; gall-making, 422
Eristalus, wing of, **851**; *E. tenax*, 851
Ermine-moths, 632
Erotylidæ, 472, 474, 509
Eruciform, 184
Erythroneura comes, 406, **407**
Erythrothrips arizonæ, fore wing of, **344**
Estigmene acræa, 701, **702**
Ethmia, 625
Ethmiidæ, 582, 590, 625
Euchætias egle, larva, **701**
Eucharinæ, 942, 944
Euchloe genutia = *Anthocharis genutia*, 748
Euchromiidæ, 583, 587, 706
Eucinetidæ, 472, 505
Euclea delphinii, **609**; scales of, **571**
Eucleidæ, 582, 586, 608
Euclemensia bassettella, **634**
Eucnemidæ, 472, 502
Eucoilinæ, 934
Eucone eyes, 141
Eucosmidæ, 639
Euctheola rugiceps, **520**
Eudeilinia herminiata, 711, 712
Eudule mendica, **669**; wings of, **667**
Euglenidæ, 475, 499
Eugonia j-album, **756**
Eulia pinatubana, 643
Eulonchus, wing of, **837**
Eulophinæ, 941, 942, 947, 948

Eumenes, 971; *E. fraternus*, **971**
Eumeninæ, 966, 969
Eunica, 751
Euparagia, **967**
Euparagiinæ, 966, 967
Eupelereria magnicornis, 872
Eupelmidæ, 941
Eupelminæ, 946
Euphoria, 521; *E. inda*, **521**
Euphorinæ, 919
Euphydryas phaeton, 752
Euplectrus platyhypenæ, 948
Euproctis chrysorrhœa, 682
Eupsalis minuta, **536**
Eupterotidæ, 583, 587, 707
Euptoieta claudia, 751, **752**
Eurema euterpe, 749; *E. nicippe*, 749
Eurosta solidaginis, 627, 858
Eurrhypara urticata, 649
Eurycyttarus confederata, **614**
Eurygaster alternatus, **393**
Eurymus eurytheme, 748; *E. philodice*, **748**
Euryophthalmus succinctus, 386; hemelytron of, **385**
Eurypauropidæ, 20
Eurypauropus ornatus, 19; *E. spinosus*, **19, 20**
Eurystethidæ, 474, 498
Eurystethus subopacus, 498
Eurytoma, 946
Eurytomids, seed-infesting, 946
Eurytominæ, 942, 945
Euscelis exitiosus, **406**
Euschemon rafflesiæ, 733
Eusternum, 52
Euthisanotia grata, **693**; *E. unio*, 693
Euthrips citri, 345; *E. fuscus*, 345; *E. pyri*, 345; *E. tritici*, 345
Euvanessa antiopa, **753, 755**
Evania, 950; *E. appendigaster*, **950**; wings of, **950**
Evaniidæ, 890, 892, 911, 949
Evanioidea, 890, 949, 950, 978
Evening primrose moth, 695
Everes comyntas, 772
Evergreen nepytia, 671
Evetria, 640; *E. comstockiana*, **640**; *E. frustrana*, **640**
Evoxysoma vitis, 945
Exarate pupæ, 190
Exetastes, 923
Exner, S., 141, 143
Exuviæ, 171
Eye-cap, 576
Eyed brown, **762**
Eyes of insects, two types of, 134; with double function, 143

Fabre, J. H., 516, 954
Face, 779

Facial depression, 780
Facialia or facial ridges, 780
Fairy flies, 948
Falcicula hebardi, 244
Fall webworms, 702
Falx, 877
Felt, E. P., 814, 815, 816
Femur, 57
Fenestra, 274
Feniseca, 772; *F. tarquinius*, 422, **772**
Ferris, G. F., 875, 876
Ferton, Ch., 967
Fibula, 62; of Corydalus, **63**
Fidia longpipes, 531
Fielde, Miss A., 955
Fiery hunter, 479
Fig-eater, 522
Fig-insects, 943
Figitinæ, 934
Filaria bancrofti, 810
Filariasis, 810
Filiform, 41
Fiorinia fioriniæ, 455, **456**
Fire-brat, 223
Firefly Family, **491**
Fish-flies, 286
Fish-moth, 223
Fixed hairs, 31, **573**
Flannel-moths, 606
Flannel-moth, crinkled, 606
Flat-headed apple-tree borer, **503**
Flat-footed Flies, 848
Flask-like sense-organ, 131
Flea-beetles, 532
Fleas, 877; antennæ of, 878; head of, 878; broken-headed, 881; unbroken-headed, 881
Fletcher, Miss P. B., 997
Flower-beetles, 520
Flower-beetle, hermit, **521**; rough, 521
Fluvicola, 504
Follicular epithelium, functions of the, 159
Folsom, J. W., **43, 47**, 226
Fontanel, 274
Fontanelle, 274
Footman-moths, 699, 704
Footman, banded, **705**; clothed-in-white, 705; painted, 705; pale, 705; footman, striped, 704; footman, two-colored, 705
Forbes, S. A., 419, 611
Forbes, Wm. T. M., 465, 577, 579, 589, 617, 750
Ford, Norma, 941
Fore-intestine, 108, 109
Forel, 955
Forester, eight-spotted, 697; Langton's, **698**
Foresters, 697
Forficula auricularia, **463**

Forked fungus-beetle, **513**
Formica exsectoides, 963; *F. sanguinea*, 963, 964; *F. fusca*, 963
Formicidæ, 108, 891, 893, 911, 917, 954
Formicinæ, 958, 963
Fossores, 978
Four-footed Butterflies, 750
Fox, W. J., 954, 982, 983, 985, 989
Fracker, S. B., 579, 625
Fracticipita, 880, 881
Franklin, H. J., 1003, 1004
Frenatæ, 582, 596; Aculeate, 582; Generalized, 582, 597; Non-Aculeate Generalized, 582; Specialized, 582
Frenulum, 61
Frenulum-conservers, 583; Frenulum-losers, 583
Frenulum hood, 61
Frenulum-hook, **597**
Friese, H., 989
Frison, T. H., 996
Frit-fly, European, 860
Fritillaries, 751
Fritillary, great spangled, 751; gulf, 764; variegated, 751
Frog-hoppers, 402
Frons, 877
Front, 37, 780; so-called, 780
Frontalia, 780
Frontal lunule, 775, 780; orbits, 780; suture, 776, 780; triangle, 780; vitta, 780
Fronto-clypeus, 780
Froth-glands of spittle insects, 102
Fruit-tree ugly-nest tortricid, 643
Fucellia, 864
Fulgoria candelaria, 409
Fulgoridæ, 400, 408, 978
Fungus-gnats, **810**; wings of, **811**
Fungus weevils, 536
Funicle, 41
Furcæ, 98
Furcæ maxillares, 332
Furcula, **228**
Furrows of the wing, 73

Gage, S. H., 444
Gahan, J., 88
Galerita janus, 480
Galesus, 932, 965
Galgulidæ, 368
Galgulus, 368
Gall-aphid, vagabond, 424
Galleria mellonella, **650**
Galleriinæ, 650
Gall-flies, 936
Gall-gnats, 813; wing of, **814**
Gallicolæ, 434
Gametes, 809
Gamogenetic eggs, 416

Ganonema americana, **568**
Garden-flea, 229
Gartered plume, **653**
Gaster, 888, 955
Gasteruptiidæ, 892, 908, 917, 930, 965
Gasteruption incertus, wings of, **930**
Gastric cæca, 112
Gastrophilidea, 787, 792, 864
Gastrophilus, 865; *G. equi*, 865; *G. hæmorrhoidalis*, 866; *G. intestinalis*, **865**; *G. nasalis*, 866; wing of, 865
Gaurax, 860
Gelastocoridæ, 356, 357, 368
Gelastocoris oculatus, **368**
Gelechiidæ, 582, 590, 625
Gelis, 928
Genacerores, 447
Genæ, 39, 780, 877; so-called, 780
Geniculate, 41
Genital appendages, the development of the, 201
Genital claspers, 76
Genitalia, 76
Genovertical plates, 780
Geometridæ, 583, 584, 587, 589, 663
Geometrids, 662; green, 665
Geometrinæ, 663, 670
Geometroidea, 583, 662
Geomyzidæ, 786, 792, 861
Geophilus flavidus, **21**
Georyssidæ, 474, 505
Georyssus, 505
Geotrupes, 517
Germarium, 158
Gerridæ, 356, 357, 370
Gerris conformis, **371**
Giraud, J., 967
Glands, 98; connected with setæ, **99**
Glandular hairs, 33
Glenurus, 304
Glischrochius (*Ips*) *fasciatus*, 508
Glossina morsitans, 194, 873
Glossosoma americana, Case of, 561
Glover's scale, 457
Glow-worms, 165
Glutops, 833
Glycobius speciosus, **527**
Glyphipterygidæ, 582, 590, 633
Glyphipteryginæ, 633
Glyphipteryx, 633; *G. thrasonella*, wings of, **633**
Glypta rufiscutellaris, 926
Glyptocombus saltator, **374**
Glyptometopa, 952
Gnathochilarium, **16**
Gnophæla latipennis, **698**, 699
Gnorimoschema gallæsolidaginis, **627**
Goat-weed, butterfly, **761**
Goera calcarata, 569; case of, **569**
Goldenrod-gall, round, **858**
Golden-eyed-flies, 300

Gomphines, 318
Gomphus descriptus, wings of, **318**; wings of naiads of, 317
Gonapophyses, 76, 232, 895
Gonatopus, 978
Gonin, J., 199
Goniodes stylifer, **336**
Goniozus, 965
Goniurus proteus, 736
Gooseberry fruit worm, 652
Gortyna immanis, 691
Gossamer-winged butterflies, 768
Gould, W. R., 955
Graber, V., 146, 149, 150
Gracilaria, wings of, **618**
Gracilariidæ, 582, 589, 591, 617
Gradual metamorphosis, 175
Graham, S. A., 466
Grain-weevils, 540
Granary-weevil, 541
Grape-berry moth, 640
Grape flea-beetle, **533**
Grape-leaf skeletonizer, 605
Grape leaf-folder, 646
Grape root-worm, 531
Grape-seed chalcid, 945
Grape-vine epimenis, **693**
Grasshopper, lubber, 257; western, 254
Grasshoppers, cone-headed, 239; leaf-rolling, 240; long-horned, 234; meadow, 238; shield-backed, 239; shorthorned, 252
Grassi, B., 157, 172
Grayling, 762; blue-eyed, 762; dull-eyed, 763; hybrid, 763; sea-coast, 763
Greenbottle-fly, 870
Grimm, O., 192
Ground-beetles, 478
Gryllacrinæ, 235, 240
Gryllidæ, 234, 242
Gryllinæ, 243
Grylloblatta campodeiformis, **268**
Grylloblattidæ, 215, 268
Gryllotalpa borealis, 251; chirp of, **93**; *G. hexadactyla*, **251**
Gryllotalpinæ, 243, 250
Gryllus, 83, **248**; ventral aspect of the meso- and metathorax of, **98**; *G. assimilis*, 248; *G. assimilis luctuosus*, 248; *G. assimilis pennsylvanicus*, 248; *G. domesticus*, 248, **249**
Guenther, K., 132
Guest gall-flies, 936
Guilbeau, B. H., 102, 403
Gula, **39**
Gynandromorph, 156
Gypona, 408
Gyponinæ, 407
Gypsy moth, **682**
Gyretes, 484
Gyrinidæ, 470, 484

Gyrinus, 484
Gyropidæ, 337

Habrosyne scripta, **709, 710**
Hadeninæ, 694
Hadronotus, 933
Hadwen, S., 867
Hæmatobia irritans, 873
Hæmatopinidæ, 349
Hæmatopinus asini, **349**; *H. eurysternus*, **349**; *H. suis*, 349
Hæmatopis grataria, 666, **667**
Hæmatosiphon inodorus, 378
Hæmorrhagia diffinis, 661; H. thysbe, **661**
Hæmozoin, 808
Hagen, H. A., 113, 171
Hag-moth, 609
Hair-streaks, 769
Hair-streak, banded, 769; olive, 769; white-m, 770; purple, 770
Halictus, 986, 995; nest of, **995**; *H. humeralis*, 996
Haliplidæ, 470, 481
Haliplus, 481, 482
Halysidota, *sp.*, wings of, **699**
Halobates, 372; *H. micans*, 372; *H. sericeus*, 372
Haloptiliidæ, 620
Halteres, 59
Haltica chalybea, **533**
Halysidota caryæ, **704**; larva, **703**
Hamamelistes spinosus, **426, 427**
Ham-beetle, red-legged, 493
Hamilcara, 603
Hammar, A. G., 125, 126
Hamuli, 61, 885
Handlirsch, A., 260, 262, 263, 284, 289
Handmaid moths, 675
Hansen, H. J., 23, 24, 43
Hapithus agitator, 245
Haploa, **700**; *H. contigua*, **700**
Haploptilia, 620
Harlequin milk-weed caterpillar, **701**
Harmolita, 945; *H. grandis*, 945; *H. tritici*, 945
Harpalus caliginosus, **480**; head of, **468**; labium of, **45, 52**; prothorax of, **468**
Harrisina americana, **605**
Hartman, C. G., 970
Harvestmen, **9**
Hatching of young insects, 171
Hatching spines, 171
Hautsinnesorgane, 130
Hawk-moth, bumblebee, **661**
Hawk-moths, 655; larva of a, **577**
Head, 36
Headlee, T. J., 823
Head measurements of larvæ, 173

Hearing, organs of, 145
Heart, 121
Hebridæ, 356, 358, 359, 372
Hebrus, 373
Hecabolinæ, 920
Heel-fly, 868
Hegner, R. W., 816, 817
Helichus lithophilus, **504**
Heliconiinæ, 750, 764
Heliconius charitonius, 764
Helicopsyche borealis, **569**
Heliocharis, base of wing of, **322**
Heliodinidæ, 582, 591, 634
Heliothis obsoleta, 695
Heliothrips fasciatus, 345; *H. hæmorrhoidalis*, 345
Heliozelidæ, 582, 591, 622
Helix, 1000
Hellula undalis, 648
Helodidæ, 472, 505
Helomyzidæ, 786, 790, 854
Heloridæ, 890, 909, 931
Helorus paradoxus, 931
Hemelytra, 59, 350
Hemerobiidæ, 284, 294
Hemerobius, 301; larva of, 297; *H. humuli*, wings of, **295**
Hemerocampa, 679, 680; *H. leucostigma*, **680**; *H. plagiata*, 681; *H. vetusta*, 681
Hemileuca maia, **720**; *H. oliviæ*, 721
Hemimeridæ, 269, 463
Hemimerus bouvieri, 463; *H. talpoides*, 463; *H. deceptus*, 463; *H. vicinus*, 463; *H. advectus*, 463; *H. vosseleri*, 463; *H. sessor*, 463; *H. hanseni*, 269, 463
Hemimetabola, 179
Hemimetabolous, development, 178
Hemiptera, 212, 213, 214, 215, 216, 350
Hemispherical scale, 453
Hemiteles, 928
Hemitelini, 928
Hemitheinæ, 663, 665
Hendel, Fr., 794, 856, 862
Henicocephalus, culicis, 384
Henneguy, L., 117, 124
Heodes epixanthe, 770; *H. heteronea*, 770; *H. hypophlæas*, 771; *H. thoe*, 771; wings of, **768**
Hepialid, wings of a, **62**
Hepialidæ, 581, 584, 594, 595
Hepialus, 595
Hermininæ, 685
Herrick, G. W., 379
Hesperia, 737; *H. tessellata*, 737
Hesperiidæ, 583, 734
Hesperiinæ, 735
Hesperioidea, 589, 732
Hesperoctenes longiceps, **379**
Hesperophylum heidemanni, 377
Hesse, R., 136, 137, 139

Hess, W. N., 146, 147, 148
Hessian-fly, 818
Hetærina, 324; base of wing of, 323
Heterocampa bilineata, 676; larva, **676**; *H. guttivitta*, 677; *H. varia*, **677**
Heterocera, 581
Heteroceridæ, 474, 505
Heterocerus, 505
Heterocordylus malinus, 376
Heterogamy, 177, 415
Heteropezinæ, 815, 816
Heteroptera, 209, 350
Hewitt, C. G., 202
Heymons, R., 174, 354, 398, 399
Hexagenia, 312
Hexapoda, 26
Hexapoda, sub-classes and orders of, 211; table of orders, 212
Hickory-borer, painted, 528
Hickory horned devil, 717
Hicks, B., 155
Hill, C. C., 889
Hilton, W. A., 128, 129, 132, 133
Hind-intestine, 108, 112
Hinds, W. E., 341, 342
Hine, J. S., 554, 830
Hippiscus apiculatus, 258, **259**
Hippoboscidæ, 108, 787, 790, 874
Hispopria foveicollis, 88
Histeridæ, 470, 490
Histoblast, **195**, 205
Histogenesis, 204
Histolysis, 204
Hochreuter, R., 155
Hoeck, P. P. C., 11
Hofer, B., 127
Hog-caterpillar of the vine, 659
Holcocera, wings of, **629**
Holland, W. J., 739
Holmgren, E., 99, 280
Holometabola, 180
Holometabolous development, 180
Holorusia rubiginosa, **197**
Holosphyrum boreale, 250
Homaledra sabalella, 629, **630**
Homochronous heredity, 181
Homologizing of the sclerites, **35**
Homoptera, 209, 211, 212, 215, 394
Honey, 1007
Honey-bee, 158, **992**, 1005; African, 876
Honey-bees, stingless, 992
Honey-pot, 1003
Hood, J. D., 965
Hooded owlets, 694
Hook-tip moths, 711
Hoplismenus, 929
Hoplopsyllus anomalus, 882
Hop-merchant, 756
Hop-plant borer, 691
Hop-vine deltoid, 685
Hormaphidinæ, 424

Hormaphis hamamelidis, **425**
Hormiinæ, 920
Hornet, giant, 977; white-faced, 977
Hornets, 975
Hornia, 997
Horn-fly, 873
Horn-tails, 890, 894, 898
Horse-flies, 829
Horse-guard, 988
Horseshoe-crabs, 8
House-fly, 872; larva of the, 202
Howard, L. O., 807, 872, 965
Huber, P., 955
Hübner, J., 620
Hübner's Tentamen, 581
Human flea, 882
Humeral, angle, 60; callus, 783; crossvein, 71; suture, 274; veins, 74
Hump-backed Flies, 847
Hungerford, H. B., 361, 364, 365
Hutchison, R. H., 873
Huxley, T. H., 40
Hyblæa puera, 655
Hyblæidæ, 583, 586, 655
Hybrid purple, 758
Hydrometra, 373; *H. australis*, 373; *H. martini*, **373**; *H. wileyi*, 373
Hydrometridæ, 356, 357, 373
Hydrophilidæ, 471, 473, 485
Hydrophilus, 486; egg sac of, 170; embryo of, **76**; maxilla of, **44**; *H. obtusatus*, 486
Hydrophylax aquivolans, 949
Hydropsyche, 562; net of, **562**
Hydropsychidæ, 560, 562
Hydroptilidæ, 559, 560, 561
Hydrous, 486; *H. triangularis*, 486
Hylæidæ, 890
Hylæus, 890
Hylastinus obscurus, 542
Hylemyia antiqua, 864; *H. brassicæ*, 863; *H. rubivora*, 864
Hymenopharsalia, 908
Hymenoptera, 108, 210, 211, 213, 214, 884
Hymenopterous wing, typical, **885**
Hypatus bachmanni, **766**, 767
Hypeninæ, 685
Hypena humuli, 685
Hyperæschra stragula, wings of, **674**
Hypermallus villosus, 528
Hypermetamorphosis, 191
Hyphantria cunea, 702; *H. textor*, 702
Hypocera incrassata, 848
Hypoderma bovis, 867; *H. lineatum*, 867, 868
Hypodermal structures, 95; glands, 98
Hypodermis, 29
Hyponomeutidæ, 631
Hypopharynx, 47, 400
Hypopleura, 783

Hypoprepia fucosa, 705; *H. miniata*, 704
Hypoptinæ, 603
Hypopygium, 75
Hyporhagus, 514
Hyposoter fugitivus, 927
Hypothetical tracheation of a wing of the primitive nymph, **63**
Hypothetical type of the primitive wing-venation, 62
Hypsopygia costalis, **649**
Hyptia, 950
Hyslop, J. A., 501

Ibaliinæ, 934
Icerya, 449; *I. purchasi*, **449**, 512
Ichneumon-flies, 922
Ichneumonidæ, 890, 892, 894, 908, 915, 917, 922
Ichneumoninæ, 923, 924, 928
Ichneumonoidea, 917
Ichneutinæ, 920
Ichneumon ultimus, 929; *I. mendax*, 929
Illice unifasciata, **705**
Imaginal disc, 195, 205
Imago, 191
Imperforate intestines, 108
Imperial-moth, 717
Incisalia niphon, 770, **753**
Incisurae, 447
Incomplete metamorphosis, 178
Incurvariidæ, 582, 589, 590, 598
Incurvariinæ, 598
Indian-meal moth, 651
Inner margin of wing, 60
Inocellia, 289
Inostemminæ, 933
Inquilines, 936, 952
Insects, 26
Instars, 172
Integricipita, 880, 881
Intercalary veins, 69
Interfrontalia, 780
Intermediate organ, 152
Internal anatomy, 94
Internal organs, the transformations of the, 204
Internal skeleton, 95; sources of the, 95
Intersegmental plates, 40
Intima, 109, 117
Invaginations of the body-wall, 95
Io-moth, **722**
Iphiclides marcellus, **743**
Iridomyrmex humilis, 962
Isabella tiger-moth, **702**
Ischnocera, 337
Ischnopsyllidæ, 881
Ischnopsyllus, **878**
Isia isabella, **702**
Isley, D., 970
Isogenus sp., wings of, **329**

Isometopidæ, 356, 359, 374
Isometopus pulchellus, **375**
Isoptera, 209, 211, 213
Isorhipis ruficornis, 931
Isosoma, 940
Ithobalus, 741
Ithycerus noveboracensis, 537
Ithytrichia confusa, **562**
Itonididæ, 813
Itoplectis conquisitor, 925

Jalysus perclavatus, 388; *J. spinosus*, **388**
Janet, C., 87
Janus integer, 901
Japanese beetle, 519; Laboratory, 520
Japygidæ, 224
Japyx, 161, 220, 224; *J. solifugus*, 220; ovary of, **222**, 461
Jigger, 883
Johannsen, Oskar A., 803, 807, 811, 823
Johnston, Christopher, 152
Johnston's organ, 152
Joint-worms, grass and grain, 945; wheat, 945
Jones, P. R., 342
Judeich and Nitsche, 116
Jugatæ, 581, 584, 592
Jugates, haustellate, 593; mandibulate, **592**
Jug-builders, **971**
Jugular sclerites, 40
Jugum, 61; of a hepialid, **63**
Julia butterfly, 764
Julus, **16**
June-beetle, green, 522
June-bug, 515, **518**, 522
Juniper web-worm, 643

Karschomyia viburni, **814**
Katepimerum, 51
Katepisternum, 51
Katydid, chirp of the, **93**; angular winged, 237; northern bush-, 237; round-winged, 237; Uhler's, 237; the false, 236
Kellicott, D. S., 692
Kellogg, Vernon L., 100, 197, 199, 200, 336, 573, 828
Kelp-flies, 864
Kenyon, F. C., 18, 19
Kermes, 454, **455**, 634; *K. ilicis*, 441
Kermesiinae, 454
Kieffer, J. J., 815, 817, 931, 932, 934, 950, 965, 979
King-crabs, 8
Kinsey, A. C., 939
Kirby and Spence, 97
Kirkaldy, G. W., 362
Knab, F., 807

Korschelt and Heider, 203
Kowalevsky, A., 202
Krecker, F. H., 443

Labena, 930
Labenini, 923
Labia minor, **460**, **462**
Labial palpi, 46
Labidomera clivicollis, **531**
Labium or second maxillae, 45
Labrum, 38, 43
Lac-dye, 441
Lace-bud, hawthorn, 384
Lace-cocoon, suspended, **632**
Lace-like cocoon, 188
Lacewing-flies, 299
Lacinia, 45
Lac-insect, 440
Lacosoma arizonicum, 713; *L. chiridota*, 714
Lacosomidæ, 583, 585, 587, 712
Lady-bugs, **511**, 512; bean, **512**; nine-spotted, 512
Lælius trogodermatis, 965
Lærtias philenor, larva, **741**
Lætilia coccidivora, **652**
Lagoa crispata, **606**, **607**
Lagriidæ, 474, 514
Lake-flies, 312
Lamellate, 41
Lamellicorn beetles, 515; leaf-chafers, 518; scavengers, 516
Lamellicornia, 468, 469; the families of the, 475
Laminæ, 528
Lampronota, 923
Lampyridæ, 473, 491
Landois, H., 91
Languria, 510; *L. mozardi*, **510**
Lantern-fly of Brazil, great, 408
Lapara bombycoides, **658**
Lappet-caterpillars, 729, 731; american, **732**; larch, 731; velleda, **731**
Larder-beetle, **506**
Larentiinæ, 663, 666
Large-intestine, 113
Larrinæ, 981
Larrini, 980, 981
Larvæ, adaptive characteristics of, 181; the different types of, 183; the term defined, 180
Lasioderma serricorne, 515
Lasiocampidæ, 583, 589, 728
Lasius, 451, 964; *L. americanus*, 419; *L. niger americanus*, 964
Laspeyresia interstinctana, 641
Lateral conjunctivæ, 35
Laternaria phosphorea, **408**
Lathridiidæ, 473, 511
Latzel, R., 19, 21, 23, 24
Larentiinæ, 663, 666

Leach, W. E., 174, 347
Leaf-beetles, 530
Leaf-beetles, long-horned, **530**
Leaf-chafers, shining, 519
Leaf-hoppers, 406; apple, 407; destructive, 406; grape-vine, 406
Leaf-insects, 260
Leaf-miner, morning-glory, 616
Leather-jackets, 709
Lebia grandis, **480**
Lecaniinæ, 451
Lecanium, **445**; *L. hesperidum*, **451**, 452
Legionary or Visiting ants, 958
Legs, **56**; the development of, 197
Leiby, R. W., 889
Leidy, 955
Lema trilineata, 530
Leng, C. W., 467
Lentigen layer, 138
Leon, N., 354
Leopard-moth, 603
Lepidoptera, 213, 214, 215, 216, 571; frenate, 582; metamorphosis of, 577
Lepidosaphes, **445**; *L. gloverii*, **457**; *L. pinnæformis*, 456, **457**; L. ulmi, 457
Lepisma saccharina, 48, **78**, **223**
Lepismatidæ, 223
Leptidæ, 834
Leptinidæ, 470, 487
Leptinillus aplodontiæ, 487; *L. validus*, 487
Leptinotarsa decemlineata, **531**
Leptinus testaceus, 487
Leptoceridæ, 559, 560, 566
Leptocerus ancylus, 566; case of, **566**
Leptocoris trivittatus, hemelytron of, 389
Leptofœnidæ, 941
Leptophlebia, 312
Leptopsyllidæ, 881
Leptothorax emersoni, 961
Leptysma marginicollis, **257**
Lepyronia quadrangularis, **403**
Leria, wing of, **855**
Lestes rectangularis, wing of, **321**, 322
Lestremia, wing of, **816**
Lestremiinæ, 813, 815, 816
Lethocerus, 366; *L. americanus*, **366**; head of, **352**; last segment of beak of, 354
Leucocytes, 122
Leucopis, 862
Leucospidinæ, 942, 947
Leucospis affinis, 947
Libellula luctuosa, **316**
Libellulidæ, 318, 321
Libytheinæ, 750, 766
Lice, true, 347; jumping plant-, 410
Liénard, V., 125
Ligament of the ovary, 159; of the testes, 162

Light-organs, 164
Limacodidæ, 608
Lime-tree winter-moth, 671, **672**
Limnobates, 373
Limnophilidæ, 559, 560, 568
Limnophilus combinatus, 568; case of, 569
Limnoporus, 371; *L. rufoscutillatus*, 371
Limulus polyphemus, 8
Lincecum, 955
Lingula, 438
Lingua, 47
Linguatula, **14**
Linguatulids, 14
Linnæus, 206, 252
Linognathus piliferus, **349**; *L. vituli*, **349**
Liotheidæ, 337
Liparidæ, 679
Lipoptena depressa, 875
Lispa, wing of, **863**
Lissonota, 923
Lithocolletis, 618
Lithosiinæ, 586, 704
Lloyd, J. T., 559, 564, 567, 570, 649
Lloydia, 923
Locust borer, **527**
Locust, Boll's, 258; Carolina, 258; clouded, 257; coral-winged, 258; red-legged, 256; Rocky Mountain, 254; seventeen-year, 402
Locusta viridissima, 128
Locustidæ, 234, 252
Locustinæ, 253, 254
Locusts, 252; band-winged, 257; northern green-striped, 257; pigmy, 259; slant-faced, 259; spur-throated, 254
Lomamyia, 298
Lonchæa, 856; *L. polita*, 856
Lonchæidæ, 786, 791, 856
Lonchoptera, 846; wing of, **846**
Lonchopteridæ, 786, 789, 846
Long-beaks, 766
Long-horned beetles, 524
Longistigma caryæ, 418
Longitudinal veins, 64
Long-legged flies, 843
Louse, body-, 348; crab-, 349; dog-, 349; head-, 348; hog-, **349**; horse-, **349**; long-nosed ox-, **349**; short-nosed ox-, **349**
Louse-flies, 874
Loxostege similalis, 648
Lubbock, J., 18, 48, 106, 955
Lucanidæ, 475, 523
Lucanus dama, **523**
Lucanus elaphus, 523
Lucilia cæsar, 870
Luna-moth, 723
Lycæna argiolus, **753**, 771
Lycænidæ, 584, 739, 768
Lycidæ, 491

Lycomorpha pholus, **707**
Lycorini, 923
Lyctidæ, 471, 515
Lygæidæ, 357, 359, 386
Lygidea mendax, 377
Lygris diversilineata, 668, **669**
Lygus pratensis, 376
Lymantriidæ, 583, 584, 588, 679
Lymexylidæ, 473, 493
Lymexylon navale, 493
Lymnæcia phragmitella, 629
Lynchia, wing of, **874**; *L. americana*, 875
Lyonet, P., 104, 105, 106
Lyonetiidæ, 582, 589, 591, 616
Lyreman, 401
Lyroda subita, 982
Lysiognatha, 922

McAtee, W. L., 241
McClendon, J. F., 300
McCook, H. C., 955
MacGillivray, A. D., 446, 458, 886
Machilidæ, 222
Machilis, 220; mandibles of, **220**; ovary of, **222**; *M. alternata*, 174; *M.* ommatidium of, **139**; leg of, **57**; ventral aspect of, **77**; tracheæ of, 116, **117**
McIndoo, N. E., 155
Macrobasis, **497**; *M. unicolor*, **497**
Macrocentrinae, 922
Macrocentrus, 919; *Macrocentrus ancylivorus*, 922; *M. gifuensis*, 922
Macrocephalus, 383
Macrochætæ, 779
Macrodactylus subspinosus, **519**
Macrofrenatæ, Specialized, 583, 655
Macrojugatæ, 594
Macrorileva, 945
Macrovelia harrisii, 370
Macroxyela, 886; wings of, **896**
Macroxyela distincta, 896
Maia-moth, 720
Malacosoma americana, 729, **730**; *M. californica*, 731; *M. constricta*, 731; *M. disstria*, 730; *M. fragilis*, 731; *M. pluvialis*, 731
Malar space, 887
Malarial Infection, 808
Malloch, J. R., 823, 824, 848, 953
Mallophaga, 211, 214, 335
Malpighian vessels, 113; as silk-glands, 113
Mammal-nest beetles, 487
Mandibles, 43
Mandibular sclerites, 353, 398
Manidiidæ, 583, 587, 673
Mansonia, 810; *M. perturbans*, 810
Mantidæ, 234, 262
Mantis religiosa, 263

Mantispa, **290**; hypermetamorphosis of, **290**; *M. styriaca*, 290
Mantispidæ, 289
Mantoidea, 262
Manubrium, 228
Many-plume moths, 653
Maple-borer, beautiful, **527**
Maple-leaf cutter, **598**
Marchal, P., 431, 815
March-flies, **820**
Marey, 81
Marginal accessory veins, 69
Marginal cells, **886**
Margins of wings, 59, 60
Marlatt, C. L., 398
Masarinæ, 966, 967, 989, 993
Mass provisioning, 979
Matheson, R., 481, 505
Maxillæ, 43; cross-section of, **576**
Maxillary, palpus, 44; pleurites, 40; sclerites, 353, 398; tentacle, 599
Maxillulæ, 16, 43
May-beetle, heart of a, **121**; leg of a, **106**
May-beetles, 515, **518**
Mayer, A. G., 572
Mayer, A. M., 154
May-flies, **308**, 312
May-fly, metamorphosis of, **311**; wings of a, **70**
Meadow-browns, 761
Meadow-maggots, 799
Meal snout-mouth, 649
Meal-worm, **513**
Mealy-bugs, 448, 450
Measuring-worms, **662**
Mechanical sense-organs, 130
Mecoptera, 210, 211, 213, 214, 215, 550
Media, 64
Medial cross-vein, 71
Median, caudal filament, 78; furrow, 74; line, 575; plates, 55; segment, 49, 908; sutures, 35
Medio-cubital cross-vein, 71
Mediterranean flour-moth, 651
Meek, W. J., 398, 399
Megachile, 947, 999; *M. latimanus*, nest of, **999**
Megachilidæ, 891, 893, 914, 990, 999
Megalodachne, 509; *M. heros*, 510; *M. fasciata*, 510
Megalomus mæstus, wings of, **296**
Megaloptera, 284
Megalopyge opercularis, **607**; cocoon of, 189
Megalopygidæ, 582, 585, 606
Megamelus notula, antenna of, **409**
Megaphasma dentricus, 262
Megaprosopidæ, 787, 793, 869
Megaprosopus, 869
Megarhyssa lunator, **925**; *M. atrata*, 925; *M. nortoni*, 925; *M. greenei*, 925

INDEX 1049

Megaspilinæ, 909, 933
Megastigmus, 943
Megathymidæ, 583, 733
Megathymus, 733; *M. streckeri*, 733, **734**; *M. yuccæ*, 734
Meigen, J. A., 794, 813
Meinert, 347
Melalopha, 678; *M. inclusa*, **678**
Melander, A. L., 339, 340, 845, 856, 861, 862
Melandryidæ, 474, 475, 514
Melanin granules, 808
Melanoplus, 160; ental surface of the pleurites of the meso- and metathorax of, **96**; head of, **97**; tentorium of, **97**
Melanoplus bivittatus, **255**, 256; *M. differentialis*, **256**; *M. femur-rubrum*, **254**, 256; *M. spretus*, **254**
Melissodes, 990
Melittia satyriniformis, **637**
Mellinus, 980
Mellitobia, 948
Meloe, 497; *M. angusticollis*, **498**
Meloidæ, 109, 475, 495
Melolontha vulgaris, larva of, 185
Melon-worm, **647**
Melophagus ovinus, 194, 874
Melsheimer's sac-bearer, **713**
Melyridæ, 473, 493
Membracidæ, 400, 404, 978
Membrane, 350
Mentum, 46
Mercer, W. F., 196
Merian, Maria Sibylla, 408
Merope, 550, 551, 553; *M. tuber*, **553**
Merozoits, 809
Merragata, 373
Mesenteron, 108, 111
Mesochorini, 924, 927
Mesochorus, 927
Mesogenacerores, 447
Mesoleiini, 926
Mesoleptideini, 926
Mesonotum, 50
Mesophragma, 97
Mesopleura, 783
Mesothorax, 48
Mesovelia douglasensis, 372; *M. mulsanti*, 372
Mesoveliidæ, 356, 358, 372
Mestra, 751
Metacœlus, 926
Metallic wood-borers, 502
Metal-mark, large, 767; small, 767
Metameres, 34
Metamorphosis of insects, 166
Metanotum, 50
Metaphragma, 97
Metapleura, 783
Metapneustic, 115

Metathorax, 48
Meteorus, 919
Methoca stygia, 953
Methocinæ, 913, 916, 953
Metopiini, 926
Metopius, 926
Metrobates, 371; *M. hesperius*, 371
Miastor americana, 816
Microbembex monodonta, 988
Microbracon, 919
Microcentrum, 237; *M. retinerve*, 237; *M. rhombifolium*, 236, 237
Microdon, **851**
Microfrenatæ, Specialized, 582, 610; families of, 589
Microgaster, 921
Microgasterinæ, 921
Microlepidoptera, 610
Microplitis, 921
Micromalthidæ, 473, 494
Micromalthus debilis, 494
Micromus, 297
Micropezidæ, 786, 791, 858
Microphthalma, 869; *M. disjuncta*, 869
Micropterygidæ, 575, 581, 584, 592
Microvelia, 370
Micropteryx, wings of, **593**
Micropyle, 167
Midges, 802
Mid-intestine, 108, 111
Migrants, 435
Milichiidæ, 786, 792, 862
Milk-glands, 875
Milkweed-beetles, red, **529**
Milkweed Butterflies, 765
Milkweed butterfly, reproductive organs of the, **160**; transformations of the, 187
Millers, 580
Millipedes, 15
Milne-Edwards, 47
Mindarinæ, 419
Mindarus, 419; *M. abietinus*, **420**
Mineola vaccinii, 652
Miridæ, 356, 358, 359, 375
Mischocyttarus, 974; *M. cubensis*, 974; *M. flavitarisis*, 974; *M. labiatus*, wings of, **973**
Miscogasterinæ, 941, 942, 947
Miscophus, 982
Misgomyia, 833
Mites, 9
Mitoura damon, **753**, 769
Mnemonica auricyanea, 594; M. wings of, **594**
Mocha-stone moths, 678
Mocsary, A., 952
Mogoplistinæ, 243, 250
Molanna angustata, 558
Molanna, case of, **566**

Molannidæ, 559, 560, 566
Molting fluid, 172; glands, 99
Molting of Insects, 171
Mompha eloisella, 630
Monarch, the, 765
Monarthrum mali, Gallery of, **544**
Moniliform, 41
Monobia quadridens, **971, 972**, 998
Monochamus notatus, **528**
Monodontomerus, 943
Monommidæ, 474, 514
Monothalamous, 935
Mononyx, 368; *M. fuscipes*, 368
Monophlebinæ, 449
Monotomidæ, 473, 509
Mordellidæ, 475, **494**
Mordwilkoja vagabunda, 424
Morgan, Miss A. H., 70, 311
Morgan, T. H., 426
Morrill, A. W., 438
Morris, H. M., 848
Mosaic vision, theory of, 141, **142**
Mosher, Miss Edna, 580
Mosquitoes, **804**; antennæ of, **153**
Moth, pupa of a, **580**
Moth-like Flies, **801**
Moth-like fly, wing of, **80**
Moths, 571
Mourning-cloak, **755**
Mouth-parts, **42**; the development of, 200
Mucrones, 228
Mud-daubers, 984
Mufflehead, 691
Muggenburg, F. H., 876
Muir and Kershaw, 398
Muller, Fritz, 181
Muller, J., 141
Muller's organ, 149
Murgantia histrionica, 391
Murmidiidæ, 474, 511
Musca domestica, 872
Muscidæ, 787, 793, 872; development of the head in the, 202
Muscids, 852; typical, 872
Muscles, 104
Muscoidea, 787
Museum pests, 507
Music of flight, 80
Musical notation of the songs of insects, 92
Musical organs of insects, 78
Mutillidæ, 891, 894, 909, 913, 915, 916, 950, 953, 954
Muttkowski, R. A., 320, 323
Mycetæidæ, 474, 511
Mycetobia, 798
Mycetophagidæ, 472, 474, 510
Mycetophilidæ, 785, 788, 793, 810, 813
Mydaidæ, 786, 788, 842
Mydas, wing of, **842**

Myiasis in man, 871
Myiomma cixiiformis, 375
Mylabridæ, 475, 535
Mylabris obtectus, 535; *M. pisorum*, **535**
Mymarinæ, 941, 948, 949
Myodaria, 786, 852
Myriapoda, 15
Myrientomata, 24
Myrmecia, wings of, **74**
Myrmecocystus, 964, 965
Myrmecomorphus, 915
Myrmecophila pergandei, **249**
Myrmecophilinæ, 243, 249
Myrmeleon immaculatus, 304; *Myrmeleon*, wings of, **304**
Myrmeleonidæ, 284, 303
Myrmica brevinodis, 961; stridulating organ of, 87
Myrmicinæ, 958, 960
Myrmosa, 953; *M. unicolor*, 953
Myrmosinæ, 913, 916, 953
Myrmosula, 953
Mystacides sepulchralis, **567**; case of, **566**
Mytilaspis citricola, 456; *M. pomorum*, 457

Nabidæ, 356, 358, 380
Nabis, 380; *N. ferus*, 380; *N. subcoleoptratus*, **380**
Nagana, 873
Naiad, the term defined, 179
Nassonow, N., 549
Nasuti, **277**
Nathalis iole, 749
Naucoridæ, 356, 357, 367
Naupliiform, 185
Necrobia rufipes, 493
Necrophorus, **487**
Needham, J. G., 112, 178, 312, 313, 317, 807
Needham and Lloyd, 561
Neelidæ, 229
Neelus, 229
Neides muticus, 388
Neididæ, 357, 359, 388
Nemestrinidæ, 786, 789, 836
Nemobius, 84, 242, 248, 982; *N. fasciatus*, **249**; *N. palustris*, **249**
Nemocera, 785, 795; N. Anomalous, 785, 820; N. the true, 785, 795
Nemognatha, 498
Nemoura sp., wings of, **329**
Nemouridæ, 330
Neoconocephalus, 239; *N. ensiger*, **239**
Neohermes, 288; *N. californicus*, 288
Neoneurinæ, 920
Neophylax, 569
Neoscleroderma tarsalis, 965
Neoteinia, 194
Neoxabea, 245; *N. bipunctata*, **245**
Neoxyela alberta, 896

Nepa apiculata, **364**, 365
Nepidæ, 356, 357, 364
Nepticulidæ, 582, 589, 600
Nepytia semiclusaria, **671**
Nerthra, 368; *N. stygiea*, 368
Nerves, 123
Nervous system, 123
Nestling birds, parasites of, 870
Netelia, 923
Net-winged midges, 824
Neuronia, 565; lateral aspect of the mesothorax, **57**; *N. postica*, 565
Neuropore, 130
Neuroptera, 207, 209, 211, 213, 215, 281; N. mantis-like, 289; N. mealy-winged, 307
Neurotoma inconspicua, 897
Newcomer, E. J., 326
Newport, G., 106
New York weevil, 537
Nidi, **112**
Niggers, 511
Night-eyes, 143
Nigronia, 288; *N. fasciatus*, 288; *N. serricornis*, 288
Nininger, H. H., 996
Nitidulidæ, 470, 472, 473, 508
Noctuid moth, Diagram of a fore wing of a, **575**
Noctuidæ, 583, 586, 588, 683
Noctuids, 583, 683
Nodal furrow, 74
Nolinæ, 705
Nomia patteni, 954
Nomophila noctuella, wings of, **645**
Nosodendridæ, 471, 508
Nosodendron, 508
Notauli, 887
Notched-wing geometer, **672**
Notiothauma, 551
Notodontidæ, 583, 587, 674
Notolophus, 679, 680; *N. antiqua*, 681
Notonecta, 363
Notonecta undulata, **362**
Notonectidæ, 356, 357, 362
Notopleura, 782
Notopleural suture, 782
Notostigma, 22
Notoxus, 498
Notum, 49
Noyes, Miss Alice A., 563, 564
Nurse-bees, 1006
Nurse-cells, 158
Nycteribiidæ, 787, 794, 875
Nymph, the term defined, 176
Nymphalidæ, 584, 739, 750
Nymphalinæ, 750
Nymphon hispidum, **11**
Nymphs, 750
Nymphulinæ, 648
Nyssonini, 980, 986

Oak-apples, **937**; large, 937; large empty, **938**; smaller empty, 938
Oak-bullet gall, 938
Oak-coccid blastobasid, 629
Oak-gall, giant, 938
Oak hedgehog gall, 936
Oak-leaf miner, white-blotch, **618**
Oak-pruner, 528
Oak-slug, spiny, 609
Oak ugly-nest tortricid, 643
Oak-worm, orange-striped, **718**; rosy-striped, **718**; spiny, 718
Oberea bimaculata, **529**
Oblique vein, 319
Obrussa ochrefasciella, Wings of, **574**, 601
Obtected pupæ, 191
Occiput, 39, 780, 877
Ocellar triangle, 780; plate, 780
Ocelli, 134, 135
Ochteridæ, 356, 357, 368
Ochterus, 368; *O. americanus*, 369
Ochthiphilidæ, 786, 791, 862
Ocular sclerites, 39
Odonata, 180, 212, 213, 314
Odontaulacus, 929
Odontoceridæ, 560 567
Odontomachus, 959
Odontomerini, 923
Odontomerus, 923
Odontomyia, puparium of, 831
Odynerus, **970**, 971; *O. geminus*, 970; *O. tropicalis*, **970**; *O. birenimaculatus*, 971
Œcanthinæ, 243
Œcanthus, **84**, **85**, **86**, 242, 245; table of species, **246**; *Œ. argentinus*, 246; *Œ. californicus*, 246; *Œ. nigricornis*, **247**; *Œ. niveus*, 93
Œciacus vicarius, 378
Œcophoridæ, 582, 590, 624
Œdemeridæ, 474, 494
Œdipodinæ, 253, 257
Œneis katahdin, 764; *Œ. semidea*, 763
Œnochrominæ, 663, 664
Œnocytes, 163
Œsophageal, sympathetic nervous system, **125**, **127**; valve, 111
Œsophagus, 110
Œstridæ, 787, 792, 866
Œstrus ovis, 867
Ohr-Wurm, 460
Oiketicus abboti, **613**
Oinophilidæ, 582, 589, 617
Olene, 680
Olethreutidæ, 582, 590, 639
Olfactory pores, 131, 154; pore of McIndoo, **155**
Oligotoma saundersi, fore wing of, **339**
Ommatidium, **135**; structure of, 139
Omophronidæ, 470, 481

Omus, 478
Oncometopia, 407; *O. undata*, **407**
Onion maggot, 864
Oniscoida, 7
Ontholestes, cingulatus, 489
Onychii, 58
Onychophora, 4
Oocyst, 809
Ookinete, migratory, 809
Ootheca, 170; of a cockroach, **264**
Operculum, 438
Ophion, 927
Ophioninæ, 924, 927
Opostega, 617
Opostegidæ, 582, 589, 617
Opthalmochlus duryi, **547**
Oral hooks, 201
Orange-tips, The, 747; falcate, 748; olympia, **748**
Orasema viridis, 941
Orbicular or round spot, 575
Orbits, 780
Orchelimum, 238; *O. vulgare*, **238**
Orchesella, 229
Oreta rosea, wings of, **712**
Organs of sight, 130
Ormenis, 410; *O. septentrionalis*, **410**
Ormyrinæ, 942, 943
Ormyrus, 943
Orneodes hexadactyla, 653; *O. hubneri*, **653**
Orneodidæ, 583, 584, 653
Orocharis saltator, **245**, 965
Orphnephilidæ, 828
Ortalidæ, 786, 791, 856
Orthezia, **102**, 448, 450
Ortheziinæ, 450
Orthoptera, 212, 214, 215, 230
Orthopteroid insects of uncertain kinship, 267
Orthorrhapha, 785, 794; short-horned, 828
Oryssidæ, 891, **905**, 907
Oryssus, 894, 906; ovipositor of, **906**; *O. abietinus*, wing of, **906**; *O. occidentalis*, 907; *O. sayi*, **905**
Osborn, H., 337, 349
Oscinidæ, 860
Oscinis frit, 860
Osmeteria, **101**
Osmia, 993, 1000
Osmoderma eremicola, **521**; *O. scabra*, 521
Osmylidæ, 109
Osmylus hyalinatus, wings of, **68, 69**
Ostia of the heart, 121
Ostomidæ, 472, 508
Othniidæ, 474, 498
Othnius, 498
Otiocerus, 409; *O. coquebertii*, **409**
Otiorhynchinæ, 538

Oudemans, J. T., 117, 880
Outer margin, 60
Ovarian tubes, 157, **158**
Ovaries, 156
Oviduct, 156, 159
Ovigerous legs, 11
Ovipara, 416
Oviparous, 191
Ovipositor, 76
Owlet-Moths, 683
Ox-warble-flies, 867
Oxybelini, 980, 989
Oxybelus, **989**
Oxyptilus periscelidactylus, **653**
Oyster-shell scale, 457

Pachysphinx modesta, **657**
Pachypsylla celtidis-mamma, **411**
Packard, A. S., 149, 189
Packardia geminata, Wings of, **609**
Pæcilocapsus lineatus, 375, **376**
Pædogenesis, 192, 816
Pædogenitic, larvæ, 192; pupæ, 192
Painted beauty, **754**
Palæ, 361
Palæodictyoptera, 210, 211
Palæostracha, 8
Paleacrita vernata, 671
Palloptera, 856
Palmetto-leaf miner, 629, **630**
Palpicornia, 467, 469
Palpifer, 44
Palpiger, 46
Palpognaths, 21
Pammegischia, 929
Pamphila sassacus, wings of, **738**
Pamphiliidæ, 890, 891, 897
Pamphilinæ, 737
Pamphilius, persicus, 897; wings of, 67, **898**
Pankrath, 137
Panorpa, **552**; Head of, **550**; wings of, **551**
Panorpodes, 553
Pantarbes capito, wing of, **838**
Pantographa limata, **646, 647**
Pantomorus fulleri, 539
Papilio glaucus, 742; larva, **742**; *glaucus glaucus*, 742; *glaucus turnus*, 742; *polyxenes*, 741; larva, **742**; wings of, 740; *P. thoas*, 173; larva of, **101**; *P. zolicaon*, 742
Papilionidæ, 583, 739, 740
Papilionoidea, 589, 739
Papilioninæ, 740
Papirius, 229
Paraclemensia acerifoliella, 598, 599
Parafacials, 780
Parafrontals, 780
Paraglossæ, 43
Paragnatha, 43; of Machilis, **221**

Paralechia pinifoliella, **626**, 627
Parallelia bistriaris, **689**
Parandra, **469**, **524**, 525; *P. brunnea*, **526**
Paraphyses, 448
Paraprocts, 232
Parapsidal furrows, **887**
Parapsides, 51, **887**
Paraptera, 51
Parasites, Respiration of, 120
Parasitoid, 919
Parasymmiclus clausa, wing of, **836**
Paratenodera sinensis, 263
Paratiphia, 953
Parcoblatta pennsylvanica, 266
Parectopa robiniella, 620
Parharmonia pini, 637
Parker, J. B., 988
Parnassiinæ, 744
Parnassius, **744**; *P. smintheus*, cross-section of scales of, **572**
Parsnip webworm, 624
Parthenogenesis, 889, 919
Paraxenos eberi, Wing of, **548**
Parornix, 618
Passalidæ, 475, 524
Passalus cornutus, **524**; stridulating organ of a larva of, **89**
Patagia, 50, 576
Patch, Dr. Edith, 422, 432
Paurometabola, 176
Paurometabolous development, 175
Pauropodidæ, 20
Pauropoda, 18
Pauropus huxleyi, **18**
Paxylommatinæ, 919, 920
Peach sawfly, 897
Peach-tree borer, 636; lesser, 636; Pacific, 636
Peach twig-borer, 627
Pear-blight beetle, 545
Pear-midge, 820
Pear-slug, 904
Pea-weevil, **535**; family, 535
Peckham and Peckham, 957, 970, 981, 982
Pecten, 628
Pectinæ, 448
Pectinate, 41
Pectinophora gossypiella, 628; wings of, **626**
Pedicel, 41
Pedilidæ, 475, 498
Pegomyia hyoscyami, 864
Pelecinidæ, 890, 892, 909, 932
Pelecinellidæ, 941
Pelicinus polyturator, **932**
Pediculidæ, 348
Pediculus, capitis, 348; *P. corporis*, 348
Pe-la, 441
Pelidnota punctata, **519**

Pelobius, 120
Pelocoris, 367; *P. femoratus*, **367**
Pelogonus, 368
Pelopæus, 984
Peltodytes, 481, 482
Pemphigus acerfolii, **422**
Pemphredonini, 980, 984, 985
Penis, 162
Pentastomida, 14
Pentatomidæ, 103, 357, 359, 390
Penthe, 514; *P. obliquata*, **514**; *P. pimelia*, 514; prothorax of, **53**
Penthima americana, 408
Pentozocera australensis, **547**
Pepsis, 951
Perceoreille, 460
Pérez, J., 92
Pericardial, cells, 164; diaphragm, 163
Pericopidæ, 583, 588, 698
Perilampinae, 942, 944
Perilampus hyalnius, 940, 941; *P. chrysopæ*, 944
Peripatoides novæ-zealandicæ, **4**
Peripatus, 1, 4
Peripheral sensory nervous system, 128, **129**
Periplaneta americana, 266; *P. orientalis*, **107**, 127
Peripneustic, 115
Peripodal, cavity, 197; membrane, 197
Peristome, 780
Peritoneal membrane, 109
Peritremes, 52
Peritrophic membrane, **111**, 112
Perkins, R. C. L., 979
Perlidæ, 328
Peterson, A., 341
Petiole, 955
Petroleum-fly, 859
Pettit, R. H., 365
Phæogenini, 928
Phæoses sabinella, 617
Phagocyte, 164, 204
Phagocytic organs, 164
Phagocytosis, 164, 204
Phalacridæ, 472, 511
Phalonia rutilana, 643
Phaloniidæ, 582, 590, 639, 643
Phanæus, 517; *P. carnifex*, 517
Phaneropterinæ, 235, 236
Phanurus beneficiens, 933
Pharynx, 109
Phasgonuridæ, 234
Phasiidæ, 787, 793, 868
Phasma, 121
Phasmidæ, 234, 260
Phasmoidea, 260
Pheidole pilifera, 960
Phengodidæ, 473, 492
Pheosia rimosa, **674**

Philagraula, 709
Philanthini, 980, **986**
Philopotamidæ, 559, 560, 563
Philopteridæ, 337
Philosamia walkeri, 727
Phlebotomus, 802; *P. vexator*, 802
Phlœothripidæ, 346
Phobetron pithecium, 609, **610**; larva, 610
Pholisora, 737; *P. catullus*, 737
Pholus pandorus, **660**
Phonapate, 88
Phora, wing of, **848**
Phoresy, 933
Phoridæ, 786, 789, 793, 847
Photinus marginellus, 165
Photurus pennsylvanicus, 165
Phragmas, 97
Phryganea, 565; *P. pilosa*, pupa of, **558**; *P. vestita*, 565
Phryganeidæ, 559, 560, 564
Phryganeids, Cases of, 565
Phryganidia californica, **673**
Phthirius pubis, 349
Phthorophlæus liminaris, **542**, 544
Phyciodes, 752; *P. tharos*, **752**
Phycitinæ, 651
Phycodromidæ, 786, 791, 855
Phygadeuonini, 928
Phyllium, 261; *P. scythe*, **262**
Phyllonorycter, 618; *P. cincinnatiella*, 619; *P. hamadryadella*, **618**, 619
Phyllophaga, 518
Phylloscyrtus pulchellus, 244
Phyllotreta vittata, **532**
Phylloxera, 433; gall-inhabiting form, **436**; gall of, **434**; grape, 433; root-inhabiting form, 435, **436**; *P. vastatrix*, 433; wings of, **429**
Phylloxeridæ, 400, 428
Phylloxerinæ, 433
Phymata, 383; *P. erosa*, **383**
Phymatidæ, 356, 358, 382
Physocephala affinis, wing of, **853**
Physonota unipunctata, 534
Phytodietus, 923
Phytomyza aquilegiæ, 861; mine of, **861**
Phytophaga, 468, 469; families of the, 475
Phytophaga destructor, 818
Pickle-worm, 648
Pièces jugulaires, 40
Pierce, W. Dwight, 548, 549
Pieridæ, 584, 739, 744
Pieris napi, 747; *P. protodice*, 747; wings of, **745**; *rapæ*, **746**
Piesma cinerea, 385
Piesminæ, 385
Pigeon horn-tail, 899
Pigment cells, accessory, 138, 140; iris, 140

Piliferous tubercles of larvæ, 35
Pill-beetles, 508
Pimpinella integerrima, 631
Pimpla, 925
Pimplinæ, 923, 924
Pimplini, 925
Pinacate-bugs, **513**
Pinconia coa, 606
Pine clear-wing moth, 637
Pine-cone willow gall, **817**; guest, 817
Pine-leaf, miner, 627; scale, 458; tube-builder, 643
Pine-pest, Zimmermann's, 652
Pine-twig moths, 640
Pinipestis zimmermanni, 652
Piophila casei, 858
Piophilidæ, 786, 792, 858
Pipunculidæ, 786, 790, 849
Pipunculus, **849**; wing of, **850**
Pistol case-bearer, 620
Planidium, 941; of *perilampus*, **940**
Planta, 578, 991
Plant-lice, **412**; jumping, 410
Plasma, 122
Plasmodium, 808
Plastoceridæ, 473, 499
Plates, 448
Plathemis lydia, **314**
Platygaster heimales, 889
Platygasteridæ, 890, 909, 915, 933
Platygasterinae, 933
Plathypena scabra, **685**
Platypeza, wing of, **849**
Platypezidæ, 786, 789, 848
Platypodidæ, 476, 541
Platypsyllidæ, 470, 486
Platypus, wilsoni, **541**
Platysomidæ, 476, 536
Plea, 364
Plecoptera, 136, 209, 211, 213, 325
Pleura, 34
Pleurites, 35
Pleurostigma, 21
Plodia interpunctella, 651
Plum-curculio, 539
Plume-moths, 652
Plum web-spinning sawfly, 897
Plusiinæ, 687
Plutella maculipennis, 632
Plutellidæ, 582, 589, 591, 632
Pocock, R. I., 17, 21
Podagrion, 943
Podalirius, 890
Podical plates, 232
Podisus, 391; *P. maculiventris*, **391**
Podura aquatica, 229
Poduridæ, 115, 228
Polistes, 546, 972; nest of, **974**; *P. lineatus*, 974
Polistinæ, 966, 974
Pollen brushes, 991

Pollenia rudis, 870
Polyblastus, 926
Polycentropidæ, 559, 560, 563
Polycentropus, **564**
Polychrosis viteana, 640
Polyctenidæ, 356, 379, 359
Polyembryony, 168, 889, 946
Polyergus, 964; *P. lucidus*, 964
Polyformia, 467, 469
Polygonia, 756
Polygonia comma, **753**, 756; *P. comma comma*, 757; *P. comma dryas*, 756; *P. faunus*, **753**, 756; *P. interrogationis*, **753**, 757; *progne*, 757
Polymitarcys, 312
Polyphaga, 467, 468, 485
Polyphemus-moth, 722
Polysphinctini, 923, 925
Polystæchotes punctatus, **299**; *P. vittatus*, 299
Polystœchotidæ, 284, 298
Polythalamous, 935
Polyxenus, 16, **17**
Pomace-flies, 860, **861**
Pomocerus aquatica, ommatidium of, 225
Pompilidæ, 890, 893, 912, 916, 950, 951
Ponerinæ, 958, 959
Pontia rapæ, 195
Popillia japonica, 519
Poplar-leaf gall aphid, **424**
Popoff, 398
Pore-plate, 131
Porocephalus, *P. annulatus*, 14; *P. proboscideus*, 14, **14**
Porthetria, dispar, **682**
Postalar callus, 783
Postantennal organ, 227
Postcubital cross-veins, 319
Postembryonic molts, number of, 172
Posterior arculus, 72
Posterior lobe, 778; of the wing, 61; of the pronotum, **887**
Postgenacerores, 447
Postgenæ, 39, 781
Postnodal cross-veins, 319
Postnotum, 50
Postpetiole, 955
Postphragma, 98
Postscutellum, 50
Poststernellum, 52
Powder-post beetles, 515
Prætarsus, 58
Praon, 922
Pratt, H. S., 874
Praying mantes, 262; eggs of, **170**
Prealar callus, 783
Preanal area, 75
Preanal lobe, 888
Preaxillary excision, **888**

Preepisternum, 51
Pregnacerores, 447
Prepectus, 887
Prephragma, 98
Prepupa, 185
Prescutum, 50
Presternum, 52
Presultural depression, 783
Pricer, 955
Primary ocelli, 135; structure of, **137**, **138**
Primitive weevils, 536
Primordial germ-cells, 158
Prionid, straight-bodied, 526
Prioninæ, 525
Priononyx, 987
Prionoxystus macmurtrei, 603; *P. robiniæ*, **603**; wings of, **70**, **596**, **602**
Prionus, broad-necked, 526; *P. imbricornis*, 526; *P. laticollis*, 526
Pristaulacus, 929
Prociphilus imbricator, 422; *P. tessellatus*, 421, 422
Proctodæum, 108
Proctotrupes, 932
Proctotrupidæ, 108, 890, 910, 916, 931
Proctotrupoidea, 890, 931
Prodoxinæ, 599
Prodoxus, 600
Progrediens type, 431
Progressive provisioning, 979
Projapygidae, 224
Projapyx, 224
Prolabia pulchella, 463; *P. burgessi*, 462
Prolegs of larvæ, 78; the development of, 182
Prolimacodes badia, **610**
Promethea-moth, **725**
Prominent, two-lined, 676
Prominents, 674
Pronotum, 50
Pronuba yuccasella, 599
Prophragma, 97
Propleura, 782
Propneustic, 115
Propodeum, 49, 887, 908
Propolis, 1007
Propupa, 343
Propygidium, 75
Prosimulium hirtipes, 824
Prosopidæ 890, 891. 893, 914, 992, 993
Prosopinæ, 914, 993
Prosopis, 890, 990, 991, 992, 993
Protapteron, 218
Prothorax, 48
Protocalliphora, 870; *P. avium*, 870; *P. splendida*, 870
Protocerebrum, 47, 124
Protoparce quinquemaculata, 658; pupa, **659**; wings of, **656**; *P. sexta*, 659

Protoplasa, 796; *P. fitchii*, **796**; *P. vanduzeei*, 796; *P. vipio*, 796
Protosialis americana, 286
Protura, 218
Proventriculus, 110, **111**
Psacaphora terminella, 630
Psammocharidæ, 890
Psectra, 294; *P. diptera*, 294
Pselaphidæ, 470, 489
Pseninæ, 984
Psenini, 981, 984, 985
Psephenidæ, 471, 503
Psephenus, 503; *P. lecontei*, **504**
Pseudagenia, 951
Pseudococcus citri, **451**; *P. (Dactylopius) destructor*, eyes of, **443**; *P. longispinosus*, **451**; wing of, **442**
Pseudocone eyes, 141
Pseudo-cubitus-one, 301
Pseudo-halteres, 59
Pseudohazis, 721; *P. eglanterina*, 721; *P. hera*, **721**
Pseudomasaris, 967, 968, **969**; *P. vespoides*, 968; nests of, 969
Pseudo-media, 301
Pseudomyrma, 959
Pseudomyrminæ, 958, 959
Pseudophyllinæ, 235, 238
Pseudothyatira cymatophoroides, 710
Pseudova, 191, 416
Psila rosæ, 859
Psilidæ, 786, 792, 859, 890
Psilogaster fasciiventris, 941
Psilopa petrolei, 859
Psilopodius sipho, wing of, **844**
Psilotreta frontalis, 567; case of, **567**
Psithyrus, 993, 1001, 1002, 1004
Psocid, wings of, **332**
Psocidæ, 333
Psocids, 331
Psychidæ, 582, 584, 585, 586, 613
Psychodidæ, 785, 787, 801
Psychomorpha epimenis, **693**
Psychomyidæ, 560, 564
Psyllia, floccosa, **410**; pear-tree, 411; *P. pyricola*, **411**
Pterodontia flavipes, 837; *P. misella*, 837
Pteromalinæ, 941, 942, 947
Pteromalus puparum, 947
Pterombus, 953
Pteronarcella, 328; *P. badia*, wings of, **328**
Pteronarcidæ, 328
Pteronarcys, 120, 328; head of, **136**; *P. dorsata*, 325
Pteronidea ribesi, 903; *P. trilineata*, **903**
Pterophoridæ, 582, 584, 652
Pterophylla camellifolia, **237**, 238
Pteropleura, 783
Pterostigma, 74, 327

Pterygogenea, 206
Pterygota, 211, 230
Ptilinum, 190, 776, 781
Ptinidæ, 471, 514
Ptinus fur, 514
Ptychopteridæ, 785, 787, 796
Pulex irritans, 882
Pulicidæ, 882
Pulsations of the heart, 122
Pulvilii, 58
Pulvinaria, 453; *P. acericola*, 453; *P. innumerabilis*, 170; *P. vitis*, **453**
Punkies, 803
Pupa, 186; of a beetle, 466
Pupæ, active, 187; the different types of, 190
Puparium, 190, 815
Pupipara, 193, 873, 876
Pycnogonida, 10
Pygidial area, 888
Pygidium, 75, 445; diagram of a, **446**, 447
Pyralididæ, 582, 585, 587, 644
Pyralidinæ, 649
Pyralidoidea, 582, 644
Pyralids, 582, 644; aquatic, 648; typical, 649
Pyralis farinalis, 649
Pyrausta nubilalis, 648
Pyraustinæ, 646
Pyrochroidæ, 475, **498**
Pyromorpha, 604; *P. dimidiata*, 604; wings of, **605**; *P. marteni*, wings of, **605**
Pyromorphidæ, 582, 585, 586, 604
Pyrrhocoridæ, 357, 359, 385
Pyrrhopyginæ, 735
Pythidæ, 474, 498

Quadrangle, 322
Queen, the, 766, 1006

Radial cross-vein, 71
Radicicolæ, 434
Radio-medial cross-vein, **71**
Radius, 64
Ramphocorixa acuminata, 362
Ranatra fusca, **365**
Range-caterpillar, New Mexico, 721
Raphidia, 289
Raphidiidæ, 289
Raphidioidea, 289
Raspberry fruit-worm, 510; geometer, 665; root-borer, 636
Raspberry-cane maggot, 864
Rasping organs, 87
Rat-flea, Indian, 882
Rath, O. vom, 132
Rat-tailed maggots, 851
Rau and Rau, 951, 970
Reaumur, R. A. F. de, 572, 876
Rectum, 113

Red admiral, **754**
Red-bug, 385; apple-, 376; false apple-, 377; hop-, 377
Red-humped apple-worm, 677
Redikorzew, W., 137
Red-necked agrilus, **503**
Red spotted purple, 758
Reduviidæ, 356, 358, 380
Reduvius personatus, **381**
Regal-moth, **716**, 717
Regions of the body, 36
Reichertella collaris, wing of, **821**
Reighardia, 14
Reniform spot, 575
Repletes, 964
Reproduction of lost limbs, 173
Reproductive organs, 156; of the female, **157**; of the male, 160, **161**
Resin-gnat, **819**
Respiratory organs, the closed or apneustic types of, 119; the open or holopneustic types of, 114
Respiratory system, 113
Resplendent shield-bearer, 622, **623**
Reticulitermes, 279; *R. (Leucotermes) flavipes*, 278
Retina, 138
Retinodiplosis resinicola, **819**
Retinula, 138, 140
Rhabdom, 137
Rhabdomere, 137
Rhabdophaga strobiloides, 817
Rhachicerus, 832; *R. nitidus*, 833
Rhagio, wing of, **835**
Rhagionidæ, 786, 789, 834
Rhagium lineatum, **526**
Rhagoletis cingulata, 857; *R. fausta*, 857; *R. pomonella*, **857**
Rhagovelia, **370**; *R. obesa*, 370
Rhamphomyia, wing of, **845**
Rhaphidophorinæ, 235, 241
Rheumaptera hastata, 669
Rheumatobates, 372
Rhinoceros-beetles, **520**
Rhinomacerinæ, 537
Rhinopsis, 978
Rhipiceridæ, 472, 499
Rhipiphoridæ, 475, **494**
Rhizophagidæ, 472, 474, 508
Rhizophagus, 508
Rhodites rosæ, **939**
Rhodophora florida, 695
Rhopalocera, 581
Rhopalomyia, 814
Rhopalosoma poeyi, 965
Rhopalosomidæ, 891, 913, 965
Rhyacophila, 555; *R. fuscula*, 555, 561; wings of, **556**
Rhyacophilidæ, 559, 560
Rhynchites bicolor, **538**
Rhynchitinæ, 537

Rhynchophora, 468, 469, 535, 537; families of the, 475
Rhynchophorus, head and prothorax of, **469**
Rhysodidæ, 470, 508
Ribbed pine-borer, **526**
Rice-weevil, 541
Rielia manticida, 933
Rileva, 945
Riley, C. V., 171, 177, 187, 496, 599, 921
Riley and Johannsen, 378
Ring-joints, 41
Riodinidæ, 584, 739, 767
Ripersia, 451
Ripipteryx, 252
Roadside butterfly, **748**
Robber-flies, 840
Rodolia cardinalis, 511
Rohwer, S. A., 906, 907
Rolleston, 107
Root, A. I., and E. R., 1007
Root-cage, **501**
Ropronia, 931
Roproniidæ, 890, 909, 931
Rose-beetle, Fuller's, 539
Rose-bugs, **519**
Rose-gall, mossy, 938, 939
Rose-slug, 904
Rose ugly-nest tortricid, 642
Roubaud, E., 970
Round-headed apple-tree borer, **529**
Rove-beetles, 488
Royal jelly, 1006
Royal-moth, two-colored, 717
Royal-moths, 715
Ruptor ovi, 171

Sabine stimulea, 609
Sacred beetle of the Egyptians, 516
Saddle-back caterpillar, 609
Saissetia hemisphærica, **453**; *S. oleæ*, **452**, 453
Saldidæ, 356, 358, 369
Salivary glands, 103, 104
Saltatorial orthoptera, 177
Salt-marsh caterpillar, 701
Samia cecropia, 726; pupa, **726**; cocoon, **727**; wings of, **720**; *S. columbia*, 726; *S. gloveri*, 726; *S. rubra*, 727
Sand-crickets, 242
Sandflies, 803
San José scale, 458
Saperda candida, **529**
Sapromyza, 856
Sapromyzidæ, 786, 791, 856
Sapygidæ, 913, 952
Sarcophaga hæmorrhoidalis, 871
Sarcophagidæ, 787, 793, 870
Saturniidæ, 583, 719
Saturnoidea, 583, 589, 714

Satyrinæ, 750, 761
Satyrodes canthus, **762**
Saussure, H. de, 969
Saw-flies, argid, 904; cimbicid, 902; leaf-rolling, 897; stem, 900; typical, 902; web-spinning, 897; xiphydriid, 899
Sawfly, American, **902**; locust, **903**
Sawyer, **528**
Scale-insects, 440; control of, 459
Scales of butterflies and moths, **571, 572, 573**
Scalloped owlet, 686
Scallop-shell moth, **668, 669**
Scape, 40
Scaphidiidæ, 470, 490
Scapulæ, **887**
Scarabæidæ, 475, 515, 954
Scarabeiform, 184
Scatophaga, 854
Scatopsidæ, 785, 788, 821
Scelio, 933
Scelionidæ, 890, 909, 910, 915, 933, 949
Sceliphron cementarium, **983**, 984; wings of, 983
Scenopinidæ, 786, 789, 839
Scenopinus, **840**; wing of, 840; *S. fenestralis*, 840
Scent-glands of females, 100
Scepsis fulvicollis, **706, 707**
Schierbeek, A., 580
Schidax, 709
Schiodte, J. C., 88, 185
Schistocerca americana, **256**
Schizoneura americana, 421; *S. pinicola*, 420; *S. rileyi*, 421; *S. ulmi*, 421
Schizont, 808
Schizophora, 786, 852
Schizopteridæ, 356, 358, 373
Schizura, 679; *S. concinna*, larva, **677**; *S. ipomeæ*, larva, **679**
Schmiedknecht, O., 929
Schneider, A., 192
Schreckensteinia erythriella, 634; *S. festaliella*, 634
Schwabe, J., 150, 151
Sciara, eyes of, **812**; wing of, **812**, 813
Sciara armyworm, 813
Sciaridæ, 813
Sciarinæ, 812, 813
Sciomyzidæ, 786, 791, 855
Sclerites, 35
Scolia, 954
Scoliidæ, 891, 893, 913, 954
Scoliopteryx libatrix, **686**
Scolopale, 146
Scolops, **409**
Scolus, 578
Scolytidæ, 476, 542
Scolytus rugulosus, 544

Scopæ, 991
Scorpion, 9
Scorpion-flies, 550, 552
Scorpions, lateral ocelli of, 137
Screw-worm fly, 870
Scudder, S. H., 92, 235
Scudderia, 237; *S. mexicana*, 237; *S. septentrionalis*, 237
Scurfy scale, 457
Scutellar bridge, 783
Scutelleridæ, 357, 359, 392
Scutellum, 50
Scutigera forceps, 22
Scutigerella, 24
Scutum, 50
Scydmænidæ, 470, 488
Scythrididæ, 582, 592, 631
Scythris eboracensis, 631; *S. magnatella*, 631
Searcher, the, **479**
Seaton, Frances, 139
Second antecoxal piece, 54
Secondary sexual characters, 157
Sectorial cross-vein, 71
Segmentation of the appendages, 34; of the body, 34
Segments of the head, 47, 48
Seiler, W., 139
Semidalis aleurodiformis, **306**
Seminal vesicle, 162
Sense-cones, 131
Sense-domes, 154, **155**
Sense-hairs, 33
Sense-organs, classification of the, 129; cuticular part of the, 130; of unknown functions, 154
Sensillum, ampullaceum, 131; basiconicum, 131; chœticum, 131; cœloconicum, 131; placodeum, 131; trichodeum, 130, 132
Sepsidæ, 786, 790, 791, 858, 859
Sepsis, 858
Serial veins, 67
Sericostomatidæ, 559, 560, 569
Serphidæ, 890
Serrate, 41
Seryda constans, scale of, **572**
Setæ, **32**; classification of, 33; primary, 578; subprimary, 578; secondary, 578; taxonomic value of, 33
Setiferous sense-organs, 130
Setiferous tubercles, arrangement of, **579**; types of, **578**
Setodes grandis, 566; case of, **566**
Sexuales, 436
Sexuparæ, 435
Sharp, David, 87, 88, 89, 144, 194, 301, 505, 506, 967, 994
Sheep-tick, **874**
Shellac, 441
Shull, A. F., 426

INDEX 1059

Sialidæ, 284
Sialis infumata, **285**, 286; larva of, **286**
Sibine stimulea, 609
Siebold, C. T. von, 92, 145
Siebold's organ, 152
Sierolomorpha ambigua, 952
Sigalæssa flaveola, 860
Signiphoridæ, 941
Sight, organs of, 134
Silk-glands, cephalic, 103
Silk-Worm, **114, 727**; sense hairs of the, **133**
Silk-Worms, giant, 719
Silpha, **488**
Silphidæ, 470, 471, 487
Silvanus surinamensis, 509, 965
Silverfish, 223
Silvestri, F., 16, 25, 113
Simæthis fabriciana, wings of, **633**
Simuliidæ, 786, 788, 821
Simulium, 120; head of larva of, **200**; larva of, 111; wing of, **822**
Simulium meridionale, 824; *S. pictipes,* 824; *S. venustum,* 824
Siphlurus alternatus, caudal end of abdomen of, **308**
Siphonaptera, 211, 214, 877
Siphunculata, 347
Sirex juvencus, wings of, **899**
Siricidæ, 890, 891, 898
Sistens, 431
Sisyra flavicornis, **292**; *S. umbrata,* **291**; larva of, **292**; silk-organs of, 282, 283
Sisyridæ, 284, 291
Sitodrepa panicea, 515
Sitotroga cerealella, 626
Skiff-caterpillar, **610**
Skimmers, 321
Skin-beetles, **522**
Skipper, least, 738; silver-spotted, **736**
Skippers, 571, 583, 732; common, 734; giant, 733
Skippers with a costal fold, 735
Skippers with a brand, 737
Sladen, F. W. L., 1002, 1004
Sleeping sickness of man, 873
Slingerland, M. V., 417, 459, 644, 647
Slingerland and Crosby, 459
Slug-caterpillar moths, 608
Small-headed flies, 837
Small-intestine, 113
Smell, organs of, 132
Smerinthus geminatus, 657, 658
Sminthuridæ, 229
Sminthurus, 115, 229; *S. hortensis,* 229
Smith, J. B., 398, 994, 996
Smith, R. C., 300, 302
Smoky moths, 604
Snake-flies, 289
Snipe-flies, 834

Snodgrass, R. E., 49, 50, 55, 57, 98, 205
Snout-beetles, 535; imbricated, **538**; pine-flower, 537; scarred, 538; toothed-nose, 537; typical, 537
Snout-butterfly, **767**
Snow-flea, **228**
Snow-flies, 799
Snyder, T. E., 280
Soldier-beetles, 492
Soldier-flies, 830
Solenobia, wings of, **615**; *S. walshella,* 614
Solenopsis molestus, 960
Solidago gall-moth, **627**
Solitary-midge, 828
Solpugida, 9
Somites, 34
Soothsayers, 262
Sooty-wing, 737
Sovereigns, the, 757
Sow-bugs, 7
Spalangidæ, 941
Spathiinæ, 920
Spatula, sternal, 813
Spear-marked black, **669**
Spear-winged flies, 846
Spermatazoa, 160
Spermatheca, 159
Spermathecal gland, 160
Spermatophores, 162
Sphæriidæ, 471, 490
Sphærites glabratus, 490
Sphæritidæ, 470, 490
Spharagemon bolli, **258**
Sphecidæ, 891, 893, 913, 979
Sphecinæ, 983
Sphecini, 980
Sphecius speciosus, **987**; wings of, **987**
Sphecodes, 992
Sphecoidea, 891, 931, 950, 977
Sphecoid-wasp, head and thorax of, **977**
Sphecoid-wasps, 977, 978, 979; typical, 962
Sphenophorus, **540**; *S. maidis,* 540
Sphex, 890, 984
Sphindidæ, 474, 515
Sphingidæ, 583, 586, 655
Sphinx, 657; *chersis,* **657**, 658; Harris's, 658; modest, 657; *pandorus,* 660; pen-marked, 658; twin-spotted, 657; white-lined, 660
Sphinxes, 655
Sphyracephala brevicornis, 859
Spiders, 9
Spider-wasps, 950
Spilochalcis mariæ, 945
Spines, 32, 445
Spiracles, 52, 113, 114; structure of, **116**
Spiracular musical organs, 91
Spirostreptus, **16**
Spittle-insects, 402

Spondylidæ, 525
Spondylis, **469**, **524**
Spongilla-flies, 291; labia of, **293**
Spongophorus ballista, **404**; *S. querini*, **404**
Sporoblasts, 809
Sporozoite, 808, 809
Spotted pelidnota, **519**
Spring azure, 771
Spring of the Collembola, 76
Spring-tails, 225
Spuler, A., 859
Spurious vein, 70
Spurs, 32
Squama, 778
Squamæ, 60
Squash-bug, egg-mass of the, 170
Squash-vine borer, 637
Stable-fly, 873
Stadia, 172
Stag-beetles, **523**
Stagmatophora gleditschiæella, 630
Stagmomantis carolina, **263**
Staphylinidæ, 470, 488
Staphylinus maculosus, 489; *S. vulpinus*, 489
Stegomyia calopus, 809; *S. fasciata*, 809
Stelis, 993
Stem-eyed fly, 859
Stem-mother, 415
Stenelimis bicarinatus, **505**
Stenobothrus, **82**
Stenogaster, 967
Stenoma, **625**; *S. schlægeri*, 625
Stenomidæ, 582, 591, 625
Stenopelmatinæ, 235, 242
Stenopelmatus, **242**; ventral aspect of the metathorax, **98**
Stephanidæ, 890, 909, 917, 919
Sterictiphora, 905
Sternal spatula, **815**
Sternellum, 52
Sternites, 35
Sternopleura, 783
Sternum, 34, 52
Sthenopis, 595; *S. argenteomaculatus*, 596; *S. pupurascens*, **595**, 596; *S. thule*, 596
Sticktight flea, 883
Sticta carolina, 988
Stigma, 74
Stigmata, 113
Stigmus fraternus, 985; *S. podagricus*, wings of, 985
Stiletto-flies, 839
Stink-bugs, source of odor, 355
Stink-flies, 300
Stink-glands, 102, 462
Stipes, 44
Stizini, 981, 987
Stizus unicinctus, 987

Stomach, 111
Stomodæum, 108
Stomoxus, calcitrans, 873
Stone-flies, 325
Stone-fly, naiad of, **327**
Stratiomyia, 831; wing of, **831**
Stratiomyiidæ, 786, 788, 830
Straus Durckheim, 40, 106, 121
Strawberry crown-girdler, 539
Streblidæ, 787, 790, 794, 875
Strepsiptera, 59, 194, 211, 212, 546
Stridulating organs, 81; of corixidæ, 362; of the locustidae, 82; of the gryllidæ and the tettigoniidæ, 83; of *Ranatra*, 365
Strigilis, **886**
Striped flea-beetle, **532**
Sturtevant, A. H., 861
Styli, 56, 76, 222
Stylopidæ, **109**, 546
Stylopids, **546**, 975; mouth-parts of male, **547**
Stylus, 442
Subcosta, 64
Subcostal fold, 73
Subgalea, 44
Subimago, 312
Submarginal cells, **886**
Submentum, 46
Subnodus, 319
Subœsophageal commissure 125; ganglion, **123**, **124**
Subquadrangle, 323
Subterminal band, 575
Sugar-cane, beetle, **520**; borer, 650
Sulphur, cloudless, 749; dainty, 749; little, 749; orange, 748
Sumac bobs, 634
Superimposed image, 143
Superlinguæ, 43, 226
Supertriangle, 319
Supplements, 70
Surpa-alar groove or cavity, 783
Supra-anal plate, **231**
Supra-tympanal or subgenual organ, 151
Suspensoria of the viscera, **162**; thread-like, 163
Sutures, 35; the thoracic of Diptera, 782, **783**
Swallow-tail, black, 741; larva, **742**; tiger, larva, **742**; zebra, **743**
Sweet-fern geometer, 666
Sweet-potato root-borer, 538
Swifts, 594, 595
Symmerista albifrons, larva, **676**
Symphasis varia, 291
Sympherobiidæ, 284, 293
Sympherobius, 294; *S. amiculus*, wings of, **294**
Symphoromyia, 835

INDEX

Symphyla, 23
Symphypleona, 229
Symphyta, 894
Synanthedon exitiosa, **636**; wings of, **635**; *S. opalescens*, 636; *S. pictipes*, 636
Synchloe olympia, 748
Synchlora ærata, 665
Synergus erinacei, 936
Syntomidæ, 706
Syrphidæ, 786, 790, 850
Syrphus, **851**
Syrphus-flies, 850
Systelloderus biceps, **383**

Tabanidæ, 786, 788, 829
Tabanus, 830; *atratus*, 830; wing of, **66**, **829**
Tachardia lacca, 440
Tachina-flies, **871**
Tachinidæ, 787, 793, 871
Tachininæ, 871
Tachysphex terminatus, **981**; wings of, **981**
Tæniopteryx, 326; *T. pacifica*, 326
Tafalisca lurida, 245
Tanaostigmatidæ, 941
Tangle-veined flies, 836
Tanyderidæ, 785, 787, 796
Tanypeza, 858
Tanypezidæ, 786, 791, 858
Tapestry-moth, 612
Tapetum, 144
Tarachidia candefacta, **689**
Tarantula hawks, 951
Tardigrada, **12**
Tarsal claws, 58
Tarsus, 57
Taste and smell, organs of, 132
Tegeticula, 599; *T. alba*, 599, **600**
Tegmina, 59, 230
Tegula, 54
Telamona, **405**
Telea polyphemus, 722; larva, **723**
Telenomus, 933
Telson, 75
Telson-tails, 211, 218
Tenaculum, 228
Tenagogonus, 371; *T. gillettei*, 371
Tenebrio molitor, 466, **513**
Tenebrionidæ, 474, 513
Tenebroides mauritanicus, 508
Tenent hairs, 58, 100, **101**
Tentamen, 581, 620
Tent-caterpillars, 729; apple-tree, 729, **730**; California, 731; forest, 730; Great Basin, 731
Tenthredinidæ, 890, 891, 895, 902
Tentorium, 96
Terebrantia, 344
Tergites, 35

Tergum, 34
Termatophylidæ, 356, 359, 377
Termes gilvus, **276**
Terminal band, 575
Terminal filament, 158
Termite, queen, **276**
Termites, 158, 194, 273
Termitoxinia, 156
Termopsis angusticollis, wings of, **274**
Testes, 160
Testicular follicle, structure of a, 161
Tetanoceridæ, 856
Tetracha, **478**
Tetraneura colophoides, 424; *T. graminis*, 424
Tetraopes, **529**; *T. tetraophthalmus*, 529
Tetrastichidæ, 941
Tetrastichus asparagi, 948
Tettigoniidæ, 234
Thalessa, 924
Thanaos, 737; *T. martialis*, **735**, 737
Thaumalea americana, 828; wing of, **828**
Thaumaleidæ, 786, 788, 828
Thaumatotypidea, 915
Thecabius populicaulis, **424**
Thecla calanus, **769**; *T. m-album*, 770
Thecodiplosis mosellana, 819
Thereva, wing of, **839**
Therevidæ, 786, 789, 839
Therionini, 927
Thermobia domestica, 223
Theronia atalantæ, 923, 925
Thick-headed flies, 853
Thompson, C. B., 278
Thorax, 48; diagram of, **50, 51**
Thorybes daunus, 737; *T. pylades*, 737
Thread-waisted wasps, 983
Three-lined lema, 530
Thripidæ, 344
Thrips, **341**; banded, 344; bean, 345; grass, 345; greenhouse, 345; immature forms of citrus-, **343**; onion, 345; orange, 345; pear, 345; strawberry, 345; *tabaci*, 345; tobacco, 345
Throat-bot, 866
Throscidæ, 472, 502
Thyatiridæ, 583, 586, 709
Thynnidæ, 891, 912, 916, 952
Thyreocorinæ, 392
Thyreocoris ater, **392**; *T. pulicarius*, 392
Thyreodon atricolor, 927
Thyrididæ, 583, 586, 587, 653
Thyridopteryx ephemeræformis, **614**; wings of, **61**, **613**
Thyris lugubris, **654**; *T. maculata*, **654**; wings of, **654**; T. mournful, 654; T. spotted, 654
Thysanoptera, 178, 209, 211, 212, 215, 341
Thysanura, 211, 219

Tibia, 57
Tibicen linnei, **401**
Tibicina septendecim, 402
Tiger-beetles, 476
Tiger-moths, 699, 700; tiger-moth, hickory, 704
Tigrioides bicolor, 705
Tillyard, R. J., 300, 593
Timber-beetles, 544
Tinagma obscurofasciella, 624; wings of, **624**
Tinea, 612; *T. parasitella*, wings of, **612**; *T. pellionella*, 612
Tineidæ, 582, 589, 590, 611
Tineola biselliella, 612
Tingidæ, 357, 384
Tinginæ, 384
Tiphia, 953; *T. inornata*, **953**
Tiphiidæ, 891, 893, 894, 912, 913, 916, 953
Tiphiinæ, 953
Tipula, 799; *T. abdominalis*, larva of, **2**; wing of, **799**
Tipulidæ, 785, 787, 793, 798
Tipuloidea, 785, 795
Tischeria, 615; *T. malifoliella*, 615; *T. marginea*, wings of, **615**
Tischeriidæ, 582, 591, 615
Tlascala reductella, wings of, **645**
Tmetocera ocellana, 641
Tobacco-worm, 659
Tolype laricis, 731; *T. velleda*, **731**
Tomato-worm, 658
Tomocerus plumbens, 225
Tool-using wasps, 984
Tormæ, 781
Tortoise-beetles, **534**; -scales, 451; -shell, American, **755**
Tortricidæ, 582, 590, 639, 642
Tortricids, 638; typical, 642
Tortricoidea, 582, 638
Tortrix, 638
Toryminæ, 943
Torymus, 943
Touch, organs of, 131
Tower, W. L., 99, 172, 354, 573
Townes, H. K., 891, 892, 941
Townsend, Miss A. B., 165
Townsend, C. H., 872
Toxicognaths, 21
Tracheæ, 113, 116; the structure of the, **117**
Tracheal gills, **119**; the development of, 182
Tracheation of wing of imago of *Calosoma*, **466**
Tracheation of wing of imago of *Dytiscus verticalis*, **466**
Tracheoles, 113, 118
Trachusa lateralis, 1000
Tramea, 317

Transverse, anterior band, 575; conjunctivæ, 34; cord, 326; impression, 781; posterior band, 575
Treat, Mrs., 955
Tree-hoppers, 404; buffalo, 404; two-horned, 405; two-marked, 405
Tremex columba, 169, **898**, 899, 924; *T. aureus*, 899; *T. sericeus*, 899
Trepobates, 372; *T. pictus*, 372
Triænodes, 565, 567; case of, **566**
Triangle, 319
Triatoma sanguisuga, 382
Trichocera, 798
Trichodectes equi, 336; *T. latus*, **336**; *T. scalaris*, **336**; *T. spherocephalus*, **336**
Trichodectidæ, 337
Trichodes nuttalli, **493**
Trichodezia albovittata, 667
Trichogens, 30
Trichogramma evanescens, 949; *T. minutum*, 949
Trichogrammatinæ, 941, 949
Trichophaga tapetiella, 612
Trichopore, 32, 130
Trichoptera, 211, 213, 555
Trichopterous Larvæ, table of, 559
Trichopterygidæ, 471, 490
Trichostibus parvula (Urodus parvula), 188
Tridactylinæ, 243, 251
Tridactylus, 252; *T. apicalis*, **252**
Tridymidæ, 941
Trigonalidæ, 890, 909, 917, 929
Trigonidiinæ, 243
Triphelps insidiosus, 378
Trissolcus, 933
Tritocerebrum, 47, 124
Tritoxa flexa, 856
Triungulin, 495
Triungulinid, 548
Triungulins, 548
Trochanter, 57
Trochantin, 53; of the mandible, 40
Troctes divinatorius, 331, 333
Trogidæ, 475, 522
Trogini, 923, 929
Trogus vulpinus, 929
Tropœa luna, 723, **724**
Trophallaxis, 280, 958, 973
Tropidacris latreillei, 26
Tropisternus, 486; *T. californicus*, 486; *T. glabra*, 486
Trox, **522**
Trumpet-leaf miner of apple, 615
Truxalinæ, 253, 259
Trypetidæ, 786, 791, 856
Tryphoninæ, 923, 924, 926
Tryphonini, 926
Trypoxylon, 982; *T. albitarsis*, 983; T albopilosum, 982; *T. frigidum*, 982; *T. rubrocinctum*, 982, 983; nest of, 982

Trypoxyloninæ, 982
Trypoxylonini, 980, 982, 988
Tsetse-fly, 873
Tubulifera, 345
Tumble-bugs, 516
Tunga penetrans, 883
Turkey-gnat, 824
Turner, 955
Tussock-moth, 679; California, 681; old, 681; well-marked, 681
Twisted-winged insects, 546
Two-spotted oberea, **529**
Tylus, 353
Tympana, **145**, 577
Typhoid-fly, 872

Udamoselis, wings of, **437**
Ululodes hyalina, **305**; larva of, **306**; wings of, **305**
Underwings, 687
Ungues, 58
Unguiculus, 227
Unguis, 227
Urodus parvula (*Trichostibus parvula*), 632
Utetheisa, 700; *U. bella*, **700**

Vagina, 159
Valentinia glandulella, 628
Valvulæ, 232
Van der Weele, 284, 285, 306
Van Duzee, E. P., 362, 844
Van Dyke, E. C., 498
Vanessa atalanta, **754**; *V. cardui*, 755; *V. huntera*, 755; *V. virginiensis*, **754**
Vanhornia eucnemidarum, 931
Vanhorniidæ, 890, 909, 922, 931
Van Rees, 202
Vas deferens, 156, 162
Vasiform orifice, 438
Vedalia cardinalis, 512
Velia, 370
Veliidæ, 356, 357, 369
Velvet-ants, **953**
Venomous setæ and spines, 100
Ventral diaphragm, 163; heart, 163; sacs, **222**; sympathetic nervous system, **127**; tube, 227
Ventriculus, 111
Verhoeff, 49
Vermiform, 185
Verruca, 578
Verson, 114, 199
Vertex, 39, 781
Vertical triangle, 781
Vespa, **975**; *V. arctica*, **977**; *V. austriaca*, 977; *V. consobrina*, 977; *V. crabro*, 977; *V. diabolica*, 977; wings of, **966**; *V. maculata*, 977; nest of, **975**, **976**; *V. rufa*, 977

Vespidæ, 108, 891, 893, 909, 912, 950, 965, 993
Vespinæ, 966, 975
Vespoidea, 931, 950, 978
Vespoid-wasp, head and thorax of, **977**
Vespoid-wasps, 950
Vespula, 977
Viallanes, 47
Vibrissæ, 782
Vibrissal angles, 781; ridges, 781
Vice-reine, 760
Viceroy, the, 759
Viereck, H. L., 958
Violet tip, 757
Vipiinæ, 920
Visiting ants, 958
Visual cell, structure of a, **137**
Vitellarium, 158
Vitreous layer, 138
Viviparity, 192, 193
Viviparous insects, 191; adult agamic females, 192
Vogel, R., 155
Volucella, 851

Wagner, Nicholas, 192, 816
Walker, E. M., 233, 252, 268
Walking-sticks, 260
Wall-bee, 1001; nests of, **1001**
Wanderer, the, **772**
Warble-flies, 866
Wasps, social, 972; typical, 965
Water-beetles, the crawling, 481
Water-boatmen, 360, **361**
Water-bugs, giant, 365
Water-measurers, 373
Water-penny, **504**
Water-scavenger beetles, 485
Water-scorpions, 364
Water-striders, 370; broad-shouldered, 369
Water-tigers, **483**
Wax, 1007
Wax-glands, **102**; outlets of, 445
Webworms, 648; burrowing, 611; cabbage, 648; garden, 648
Wedge-shaped leaf-beetles, 533
Weevil, black vine-, 539; strawberry-, **540**
Weevils, leaf-rolling, 538
Weisman, A., 202, 203
Wheat joint-worm, 945
Wheat-midge, 819
Wheat-sawfly-borer, 901
Wheat straw-worm, 945
Wheeler, W. M., 280, 524, **940**, 941, 954, 955, 958, 961, 962, 977
Whirligig-beetles, 484
White-ants, 273
White, checkered, 747
White flies, 437

White fly, citrus, 440; greenhouse, 439; maple, 440; strawberry, 440
White, gray-veined, 747
White-grubs, **518**
White marked tussock-moth, **680**
White Mountain butterfly, 763
White-striped black, 667
Whites, the, 745
White-tipped moth, **676**
Williams, F. X., 967, 969, 981
Williston, S. W., 852
Window-flies, 839
Window-winged moths, 653
Wings, 58; the development of, 182, 195; specialization of, 212
Wings of the heart, **121**, 162
Wing-veins, reduction of the number of, 65; the chief branches of the, 64; the increase of the number of, 68; the principal, 64
Winnertzia calciequina, **814**
Wire-worms, **500**
Witch-hazel cone-gall, **425**
Witch-hazel-gall, the spiny, **426**
Wohlfahrtia vigil, 871
Wollaston, 88
Wood-nymph, beautiful, **693**; pearl, 693
Workers, 1005
Wyeomyia smithii, 810

Xenopsylla cheopis, 882
Xenos vesparum, 549
Xestobium rufovillosum, **515**
Xiphidium, 238
Xiphosura, 8
Xiphydria maculata, wings of, 900, 929

Xiphydriidæ, 890, 891, 899
Xoridini, 923
Xyelidæ, 890, 891, 895, 896
Xylocelia (Diodontus), 985
Xylocopa virginica, 972, 998
Xylomyia, **832**; *X. pallipes*, 832
Xylomyiidæ, 786, 788, 832
Xylophagidæ, 786, 789, 833
Xylophagus, 833; wing of, **833**

Yellow, sleepy, 749
Yellow-bear, **703**
Yellow-fever mosquito, 809
Yellow-jackets, 975, 977
Yellows, the, 748
Yponomeuta, 632; *Y. padella*, 632
Yponomeutidæ, 582, 590, 591, 631
Yucca-borer, 734
Yucca-moths, 599; bogus, 600

Zale lunata, **689**
Zebra, the, 764
Zebra-caterpillar, 694, **695**
Zenodochium coccivorella, 629
Zerene cæsonia, 749
Zethinæ, 966, 968
Zethus, 968; *Z. cyanopterus*, 969; *Z. lobulatus*, 969; *Z. romandinus*, 969; *Z. slossonæ*, 968; *Z. spinipes*, 968
Zeuzera pyrina, 603
Zeuzerinæ, 603
Zophodia grossulariæ, 652
Zoraptera, 215, 270
Zorotypidæ, 270
Zorotypus, 270; *Z. hubbardi*, **271**; *Z. snyderi*, **271**, 272
Zygoptera, 321